MAMMALOGY

Jones and Bartlett Titles in Biological Science

MAMMALOGY

FIFTH EDITION

TERRY A. VAUGHAN
Northern Arizona University
Professor Emeritus

JAMES M. RYAN
Hobart and William Smith Colleges

NICHOLAS J. CZAPLEWSKI
University of Oklahoma

JONES AND BARTLETT PUBLISHERS
Sudbury, Massachusetts
BOSTON   TORONTO   LONDON   SINGAPORE

World Headquarters

Jones and Bartlett Publishers
40 Tall Pine Drive
Sudbury, MA 01776
978-443-5000
info@jbpub.com
www.jbpub.com

Jones and Bartlett Publishers
Canada
6339 Ormindale Way
Mississauga, Ontario L5V 1J2
Canada

Jones and Bartlett Publishers
International
Barb House, Barb Mews
London W6 7PA
United Kingdom

Jones and Bartlett's books and products are available through most bookstores and online booksellers. To contact Jones and Bartlett Publishers directly, call 800-832-0034, fax 978-443-8000, or visit our website, www.jbpub.com.

Substantial discounts on bulk quantities of Jones and Bartlett's publications are available to corporations, professional associations, and other qualified organizations. For details and specific discount information, contact the special sales department at Jones and Bartlett via the above contact information or send an email to specialsales@jbpub.com.

Production Credits
Publisher, Higher Education: Cathleen Sether
Acquisitions Editor: Molly Steinbach
Senior Editorial Assistant: Jessica S. Acox
Editorial Assistant: Caroline Perry
Production Director: Amy Rose
Production Editor: Dan Stone
Senior Marketing Manager: Andrea DeFronzo
V.P., Manufacturing and Inventory Control: Therese Connell
Composition: Auburn Associates, Inc.
Senior Photo Researcher: Christine Myaskovsky
Photo and Permissions Associate: Emily Howard
Cover Design: Kristin E. Parker
Cover Image: Courtesy of Terry Vaughan; background © Ragnarock/ShutterStock, Inc.
Printing and Binding: Courier Stoughton
Cover Printing: Courier Stoughton

Library of Congress Cataloging-in-Publication Data
Vaughan, Terry A.
    Mammalogy / Terry A. Vaughan, James M. Ryan, Nicholas J. Czaplewski.—
5th ed.
        p. cm.
    Includes bibliographical references and index.
    ISBN 978-0-7637-6299-5
    1. Mammals. I. Ryan, James M. (James Michael), 1957– II. Czaplewski,
Nicholas J. III. Title.
    QL703.V38 2009
    599—dc22
                                                                    2010001259
    6048

Printed in the United States of America
14 13 12 11 10   10 9 8 7 6 5 4 3 2

Brief Contents

Chapters 27 and 28 are available online only, at
http://www.jbpub.com/catalog/9780763762995/

Contents

22 Echolocation 467

PART IV

ECOLOGY, BEHAVIOR, AND CONSERVATION 493

23 Ecology 495

24 Behavior 547

Chapters 27 and 28 are available online only, at http://www.jbpub.com/catalog/9780763762995/.

Preface

During the decade since the publication of the fourth edition of *Mammalogy*, there has been an explosive increase in the volume of literature concerning mammals. The fifth edition of this text represents a significant revision, because although the basic framework of the book remains intact, all of the chapters have been extensively rewritten and updated with new information. We have not attempted to present an exhaustive review of the world literature on mammals, but we have tried to include many of the latest and most interesting discoveries. Because no two instructors cover the same material in exactly the same way, we treat the biology of mammals broadly enough to make the book appropriate for the variety of mammalogy courses currently being taught.

The first part of this book (Chapters 1 through 4) defines mammals and summarizes their origins. Chapter 2 details the structural features that characterize mammals. Chapter 3 summarizes the origin of mammals, and Chapter 4 introduces classification. In the second part of the book (Chapters 5 through 19), the orders and families of mammals are discussed in detail. These taxonomic chapters include information on the fossil history, current distribution, morphological characteristics, and basic behavior and ecology of each family of mammals. The third part (Chapters 20 through 26) treats special topics such as mammalian echolocation, physiology, behavior, ecology, and zoogeography. From this coverage of the subject, students can gain a general understanding and an appreciation of the form and function of mammals. Two additional chapters (27 and 28), both new to this edition, cover mammalian domestication and mammalian disease and zoonoses. These two chapters are available for download from http://www.jbpub.com/catalog/9780763762995/.

In the interest of reflecting the extraordinary recent activity in mammalian biology, one of the most important changes in the fifth edition is the more thorough use of phylogenetically based mammalian relationships. The widespread application of phylogenetic systematics in recent years has led to a new general agreement about the higher-level evolutionary relationships among living mammals. In turn, this has led to a reorganization of mammalian orders and a new classification of the Mammalia. Accordingly, we have included cladograms to illustrate the hypothesized relationships among mammalian lineages. Because phylogenetic systematics, and particularly molecular systematics, are relatively recent advances, there are still many conflicting views and often many competing phylogenetic hypotheses. We point out those relationships that remain controversial and cite sources from both sides of the debate to facilitate class discussions. As a consequence of adopting a more cladistic approach, we have tried to place studies of ecology, behavior, reproduction, and physiology in an evolutionary context that better reflects the interdisciplinary nature of the field of mammalogy.

New to the Fifth Edition

We have completely revised and updated our discussion of the fossil history of mammals, reorganized the ecology and behavior chapters with many new topics, completely updated the chapter on mammalian zoogeography, and included two entirely new online chapters: "Mammalian Domestication" and "Mammalian Disease and Zoonoses." We have also added many new photographs and drawings to illustrate key points made in the text, while retaining the original anatomical drawings by senior author Terry Vaughan. These drawings, which we regard as essential to the ordinal chapters, should help students to understand the descriptions of structure and function. We have made liberal use of these drawings, most of which illustrate skulls, teeth, or feet. The profiles of skulls are usually of the right side, and most occlusal views of teeth show the right upper or the left lower tooth row. When other teeth are shown, the legend so indicates. The chapter-opening illustrations were also provided by Terry Vaughan.

In an effort to provide additional pedagogical tools and resources for students, we have placed key terms in bold font the first time they appear in text (these terms are listed at the end of each chapter and defined in the Glossary), included chapter summaries, and provided a list of useful books and journal articles at the end of each chapter. For the ordinal chapters, we also provide a list of characteristics; these traits are not necessarily synapomorphies for the group, but should help students appreciate the morphological specializations of the group's members.

Finally, as in the previous edition, we have included a detailed glossary and an updated bibliography. The literature citations have been brought up to date, with many citations from the period of 2000 to 2009. To help students access the primary literature, we have, where possible, selected literature that is synthetic, accessible (in widely available journals and online), and current. The advent of Web-based literature searches makes finding these resources much easier than in the past; the bibliography uses a minimal punctuation format in recognition of this fact. If this text is useful, holds the interest of students, and is respected by members of the far-flung community of mammalogists, it is primarily due to the work of the researchers on whose studies we have relied. The organization, choice of topics, and selection of illustrative material are ours, but the book should be credited largely to those whose names appear in the Bibliography.

Ancillaries

An *Instructor's Media CD-ROM*, compatible with Windows® and Macintosh® operating systems, is available to instructors using *Mammalogy, Fifth Edition*. The media CD includes the following traditional ancillaries:

- The *PowerPoint® Image Bank* provides the illustrations, photographs, and tables (to which Jones and Bartlett holds the copyright or has permission to reproduce digitally) inserted into PowerPoint slides. The images can be easily incorporated into existing lecture slides.
- The *PowerPoint Lecture Outline Slides* presentation package provides lecture outline notes, prepared by the authors of the main text, for each chapter of *Mammalogy*. Instructors can customize these outlines utilizing art from the PowerPoint Image Bank.

Two **new chapters** on Mammalian Domestication and Mammalian Disease and Zoonoses are available online for instructors and students. The chapters are available for download on the Jones and Bartlett Publishers catalog page, http://www.jbpub.com/catalog/9780763762995/.

Acknowledgments

Help on this and earlier editions was generously provided by the following friends and colleagues. We are enormously grateful for their contributions.

M. Adera
J. S. Altenbach
R. Angliss
J. D. Archibald
J. Arroyo-Cabrales
R. J. Asher
M. Augee
A. D. Barnosky
J. A. Baskin
G. Bateman
R. Baxter
G. D. Bear
C. J. Bell
W. Bogdanowicz
F. Bonaccorso
D. Bos
E. Bostelman
P. Boveng
R. Bowker
R. T. Bowyer
C. Brain
J. Braun
A. Brouwer
R. Cadenillas Ordinola
A. A. Carlini
K. Catania
R. Cifelli
W. Clemens
E. Colbert
L. Consiglieri
J. A. Cook
L. P. Costa
M. Cozzuol
D. Croft
L. Dávalos
M. Dawson
M. M. Díaz
W. Downs

J. Eisenberg
E. Eizirik
J. Estes
M. Brock Fenton
T. Fleming
J. J. Flynn
C. Francis
P. Freeman
F. J. Goin
G. Goslow
C. L. Gordon, Jr.
M. Griffin
T. Griffiths
E. Hall
S. Hand
E. Harwell
J. R. Henschel
I. A. Henschel
J. Hermanson
P. Hill
H. Hoekstra
C. Hood
D. Hosking
T. Huels
E. W. Jameson, Jr.
J. U. M. Jarvis
F. Jenkins, Jr.
C. Johnson
R. F. Kay
R. W. Kays
A. Kellner
G. L. Kirkland
C. Koford
K. Koopman
G. Kooyman
T. Kunz
T. Lacher
J. A. Lackey

W. F. Laurance
T. A. Lawlor
W. Lawton
W. Lidicker, Jr.
J. Lillegraven
U. M. Lindhe Norberg
R. Lindsay
E. L. Lindsey
S. L. Lindstedt
R. MacMillen
R. MacPhee
R. H. Madden
H. Mantilla-Meluk
M. A. Mares
L. Marshall
R. A. Martin
C. W. May
J. McDonald
J. I. Mead
R. Medellín
S. Mizroch
G. S. Morgan
P. Myers
K. Norris
R. Ojeda
T. J. O'Shea
U. Pardiñas
B. D. Patterson
W. D. Peachey
L. Radinsky
G. Rathbun
N. Reeve
O. J. Reichman
K. Reiss
A. D. Rincón
W. L. Robinette
D. Rubinstein
M. Sánchez-Villagra

G. Schaller
D. A. Schlitter
G. D. Schnell
M. Seely
N. Siepel-Hyatt
M. Silva
N. B. Simmons
K. S. Smith
J. Stanton
H. Stanton
P. Stapp
J. States
A. Tejedor
M. F. Tejedor
B. Thomson
R. Timm
M. Tuttle
N. Vandemey
H. M. Van Deusen
J. Varnum
P. Vogel
J. Waggoner
J. Wahlert
L. T. Wasserthal
C. Wemmer
R. S. White, Jr.
T. Whitham
J. Wible
K. Wilkins
E. Willerslev
D. E. Wilson
M. Wolsan
C. Yahnke
S. Zeveloff
C. Ziegler

We would also like to thank the following experts who reviewed portions of the fifth edition manuscript:

David Armstrong, University of Colorado, Boulder
Troy Best, Auburn University
Jack Cranford, Virginia Tech University
Brock Fenton, University of Western Ontario
Patricia Freeman, University of Nebraska
Timothy Gaudin, University of Tennessee, Chattanooga
Chris Goguen, Pennsylvania State University
Dawn Goley, Humboldt State University
Clay Green, Texas State University, San Marcos
John Harder, Ohio State University
Margaret Lewis, The Richard Stockton College of New Jersey
Michael Mares, University of Oklahoma

Daniela Monk, Washington State University
Thomas Nupp, Arkansas Tech University
David Roon, University of Idaho
Robert Rose, Old Dominion University
Benjamin Sacks, California State University, Sacramento
Michele Skopec, Weber State University
Jerry Svendsen, University of Ohio
Monte Thies, Sam Houston State University
Kaci Thompson, University of Maryland
Francis Villablanca, California Polytechnic State University
Don White, University of Arkansas, Monticello
Kenneth Wilkins, Baylor University

We also gratefully acknowledge the help of the staff of Jones and Bartlett Publishers. Special thanks are extended to Shoshanna Goldberg, Molly Steinbach, Caroline Perry, Dan Stone, and Christine Myaskovsky, each of whom contributed important advice, critical comments, and forbearance.

We offer our sincerest gratitude to our wives, Rosemary Vaughan, Gillian Ryan, and Cheryl Czaplewski, who helped with the preparation of this fifth edition in many ways. Finally, we want to acknowledge the patience of our daughters Kaylee Ryan and Jessica Czaplewski during the writing of this book.

Terry A. Vaughan
James M. Ryan
Nicholas J. Czaplewski

Introduction to the Mammalia

PART

I

Introduction

These are exciting times for people interested in the study of mammals. Since the first edition of *Mammal Species of the World* was published in 1982, nearly 760 new mammal species have been described. Of those, over 400 mammal species were discovered since 1993 (Ceballos & Ehrlich 2009). These discoveries were the result of recent explorations in remote parts of the world, taxonomic revisions of problematic groups, and evaluation of DNA evidence for **cryptic species**. Indeed, the cumulative number of new mammal species is still on the rise. Recently discovered fossils continue to change the landscape of mammalian evolution. In addition, new tools from the field of molecular biology are helping to address questions in animal behavior and mammalian phylogeny. The study of mammalian relationships is being revolutionized by molecular data that suggest new hypotheses that remain to be tested (Murphy et al. 2001; Springer et al. 2007). Advances in radiotelemetry and the use of satellites to track far-ranging mammals increase our understanding of the life histories of many secretive species. The use of radioisotopes to measure field metabolic rates is changing the way mammalogists think about mammalian physiology. New computer models that simulate changes in populations allow mammalogists to generate testable predictions that would be impossible to formulate without such software. And now as in the past, careful, patient observations and experiments in the field are contributing to our understanding of mammals.

Although our scientific knowledge is increasing dramatically, numerous important questions remain to be answered by future generations of mammalogists. The advances described above have yielded new and exciting results. For example, the discovery that elephants can communicate using infrasound, or that kangaroo rats use foot-drumming as a form of seismic communication, has opened new areas of research on these well-studied mammals. Indeed, a combination of technological advances and innovative thinking on the part of many mammalogists has contributed to the development of entirely new areas of research. The fields of functional genomics, landscape ecology, and conservation medicine are three examples. Modern mammalogy is a dynamic and exciting field in need of curious minds (**Box 1-1**).

As a discipline, mammalogy occupies the efforts of a diverse group of scientists. Vertebrate zoologists study such aspects as the structure, **taxonomy**, distribution, and life histories of mammals, physiologists consider mammalian hibernation and water metabolism, physicists and engineers study mammalian echolocation and locomotion, vertebrate paleontologists and molecular biologists outline the patterns of mammalian evolution, and ecologists and behaviorists consider where and how mammals live and adapt to changing environments. In addition, perceptive observers without formal zoological training contribute a wealth of information.

Mammals are worthy of study for many reasons. Practical aspects of mammalogy attract some. By studying various kinds of laboratory mammals, we gain practical knowledge about mammalian histology and about the effects of diseases and drugs. Work on domesticated breeds of mammals improves meat production, and research on game species shows how sustained yields of these animals may be achieved through appropriate management techniques. To most students and researchers, however, practical applications are not foremost. Because we human primates are mammals, we are fascinated by our relatives. Mammals are beautiful and fascinating creatures that show physiological, structural, and behavioral adaptations to an amazing array of lifestyles. Thus, living mammals in their natural settings are the focal point of interest. The adaptations themselves, how they evolve, how they enable mammals efficiently to exploit demanding environmental conditions, and the interaction between mammals and their environment are all fascinating lines of inquiry. The most productive studies are the result of the intense interest of researchers in a biological relationship rather than their preoccupation with solving a practical problem. In this book, we deal primarily with the impressive literature on mammals that results from such basic research. Far from being impractical, the perspective gained from such work must guide our decisions affecting the recovery and survival of threatened species of mammals and, indeed, of entire ecosystems.

Basic research during the last half century has expanded our knowledge of mammalian biology tremendously. Echolocation (animal sonar) has been intensively studied in bats and marine mammals. The remarkable ability of some mammals to live in conditions of extreme aridity with no drinking water has been partially explained. Experiments on the circulatory and metabolic adaptations associated with temperature regulation and metabolic economy have been tested. Adaptations to deep diving in marine mammals have been examined. Hibernation and migration and the mechanisms that influence them have

BOX 1-1   New Methods for Studying Mammals

In recent years mammalogists have increasingly adopted new methods in their fieldwork and research. Some methods draw on varied technological improvements in video, molecular chemistry, DNA sequencing, and geochronology. In particular, several new (and some old low-technology) noninvasive methods save time, money, and equipment, relative to radiocollaring and radiotracking methods.

Numerous noninvasive or minimally intrusive techniques have been very widely adopted globally in the study of wild mammals (Evans & Yablokov 2004; Long et al. 2008). One of the most popular has been **camera-trapping**, placing automatic cameras along animal trails or grids to capture still photos or stop-action videos of mammals unmolested by the presence of humans (**Fig. 1-1**). The photos are analogous to museum specimens that provide vouchers and are especially valuable for nocturnal, rare, and wide-ranging species such as carnivores (Kays & Slauson 2008; Rowcliffe et al. 2008). Some cameras capable of taking short bursts of sequential photos can even provide videos useful for studying behavior (Kays et al. 2009a, 2009b). More high-technology methods use thermal-imaging cameras and night-vision devices. In a reversal of who is taking the photos, "crittercams," animal-borne video systems, harness the animals with minicameras to provide data potentially pertinent to biological studies (Moll et al. 2007). An ancient method, following and identifying mammal tracks and signs, has seen a great resurgence in the last few years as a noninvasive research technique (Elbroch 2003; Liebenberg 1990;

CyberTracker software: www.cybertracker.org/). These kinds of tracking data will soon be compiled in a new web repository called Movebank (www.movebank.org) for data sharing. "Signs" such as mammal scats and hair samples, collected at scent posts, can provide samples of bile acids, DNA, or stable isotopes, that will lead to individual recognition of such secretive animals as wolverines. These methods are especially popular with mammalogists who receive help from enthusiastic students and the public (**citizen scientists;** Silvertown 2009).

Advances in computing, molecular extraction, amplification, analysis, automation, and DNA sequencing methods, combined with improvements in data sharing and transfer (e.g., GenBank, Morphobank) and phylogenetic analysis software, have revolutionized and sped up our ways of reconstructing mammal phylogeny. Even recovery of fragmentary DNA from fossils and subfossils (e.g., Australian Centre for Ancient DNA: http://www.adelaide.edu.au/acad/) together with improved geochronometry (dating of fossils and rocks) and biochronology (placement of fossils in geological time) has increased our understanding of extinct mammal relationships and timing of divergence of lineages as well as phenomena such as mammal domestication. **Stable isotopes** found in various tissues of mammals have been used recently in the study of mammal migrations and diets (Herrera et al. 2001; Hobson & Wassenaar 2008). Using the Global Positioning System satellites to track mammals carrying transmitters facilitated research on long-distance movements of marine or land mammals.

FIGURE 1-1   Photograph of a coyote from a camera trap in New York.

been studied. There have been important contributions to our knowledge of mammalian population cycles and the factors that may control them. Studies of functional morphology have increased our understanding of mammalian terrestrial, aquatic, and aerial locomotion. Probably no field has been slower to develop than that of conservation biology. This field, however, has advanced rapidly, perhaps in response to the belated realization that time is growing short for the study of particular species in their natural environments. Studies of behavior and population genetics have contributed tremendously to our appreciation of how finely tuned mammals are to their environments. There is no doubt that mammals play vital roles in shaping regional and global ecosystems.

Mammalogists were quick to recognize that pristine ecosystems, including many of their favorite study sites, were in peril. Global **biodiversity**, the result of millions of years of evolution, is declining because of the cumulative impact of more than a quarter million new people added to the planet each day. Worldwide extinction of plant and animal species is accelerating at an alarming rate, primarily because of the fragmentation or outright loss of habitat (Cardilio et al. 2006; Davies et al. 2008). Many large or highly specialized mammals are now threatened with extinction. Since the last edition of this book was published in 2000, however, several new mammalian genera and species (including primates, rodents, and ungulates) have been discovered living in remote or poorly explored parts of the world. Ironically, these discoveries were made by mammalogists attempting to describe and conserve the remaining biodiversity before it disappears.

Classification

In any careful study, one of the vital early steps is the organization and naming of objects. As stated by Simpson (1945) with reference to animals, "It is impossible to examine their relationships to each other and their places among the vast, incredibly complex phenomena of the universe, in short to treat them scientifically, without putting them into some sort of formal arrangement." The arrangement of organisms is the substance of taxonomy, but modern taxonomists, perhaps better termed systematists, are less interested in identifying and classifying animals than in studying their evolution. These systematists bring information from such fields as genetics, ecology, behavior, and paleontology to bear on the subjects of their research. They attempt to base their classifications on the most reliable evidence of evolutionary relationships. Excellent discussions of the importance of **systematics** to our knowledge of animal evolution are given by Mayr (1963), McKenna

and Bell (1997), Nei and Kumar (2000), Hall (2007), and Simpson (1945).

Because of difficulties that arise when a single kind of animal or plant is given different common names by people in different areas or by many common names by people in one area, scientists more than 200 years ago adopted a system of naming organisms that would be recognized by biologists throughout the world. Each known kind of organism has been given a binomial (two-part) scientific name. The first, the generic name, may be applied to a number of related kinds, but the second name refers to a specific kind, a **species**. As an example, the blacktail jackrabbit of the western United States is *Lepus californicus*. To the genus *Lepus* belongs a number of similar, but distinct, long-legged species of hares, such as *L. othos* of Alaska, *L. europaeus* of Europe, and *L. capensis* of Africa. Because considerable geographic variation frequently occurs within a species, a third name is often added to designate **subspecies**. Thus, the large-eared and pale-colored subspecies of *L. californicus* that lives in the deserts of the western United States is *L. c. deserticola*; the smaller-eared and dark-colored subspecies from coastal California is *L. c. californicus*.

The species is the basic unit of classification. A once widely accepted definition of species was given by Mayr (1942): "Species are groups of actually or potentially interbreeding natural populations, which are reproductively isolated from other such groups." Each species is generally separated from all other species by a "reproductive gap," but within each species, there is the possibility for gene exchange. According to Dobzhansky (1950), all members of a species "share a common gene pool."

The hierarchy of classification, based on the starting point of the species, has been developed to express degrees of **phylogenetic relationship** among species and groups of species. The taxonomic scheme includes a series of categories, each category more inclusive than the one below it. Using our example of the hares, many long-legged species are included in *Lepus*. This genus and other genera containing rabbit-like mammals form the Leporidae; this family and the Ochotonidae (the pikas) share certain structural features not possessed by the other mammals and belong to the order Lagomorpha. This order and all other mammalian orders form the class Mammalia, members of which differ from all other animals in the possession of hair, mammary glands, and many other features. Mammals, birds, reptiles, amphibians, and fish all possess a bony or cartilaginous endoskeleton, and these groups (in addition to some others) form the phylum Chordata. All of the phyla of animals (Porifera, Cnidaria, Platyhelminthes, and so forth) are united in the animal kingdom. The classification of our jackrabbit can be outlined as follows:

Kingdom Animalia
  Phylum Chordata
    Class Mammalia
      Order Lagomorpha
        Family Leporidae
          Genus *Lepus*
            Species *Lepus californicus*
              Subspecies *Lepus californicus deserticola*

Further subdivision of this classification scheme may result from the recognition of additional intermediate categories, such as subclass, superorder, or subfamily. Most ordinal names end in –a, as in Carnivora; all family names end in –idae, and all subfamily names end in –inae. In this book, contractions of the names of orders, families, or subfamilies are often used as adjectives for the sake of convenience: leporid will refer to Leporidae, leporine to Leporinae, lagomorph to Lagomorpha, and so on.

Some similarities between different kinds of animals are due to **parallelism** or to **convergence**, two forms of the more general concept of **homoplasy**. Parallelism occurs when two closely related kinds of animals pursued similar modes of life and evolved similar structural adaptations. The similar specializations of the skull and dentition (elongate snouts and reduced number of teeth) that occur in a number of genera of nectar-feeding Neotropical bats are examples of parallelism. Convergence involves the development of similar adaptations to similar (or occasionally nearly identical) styles of life by species in different orders. The golden moles of Africa and the "marsupial moles" of Australia are convergently evolved. These animals belong to different mammalian infraclasses (Eutheria and Metatheria, respectively; see Table 4-1), and their lineages have been separate for more than 70 million years. Their habits are much the same, however, and structurally they resemble each other in many ways.

Chapter 4 provides an outline of the classification of mammals used in this book. It is based largely on the classification by Wilson and Reeder in 2005. We wish to stress that no universal agreement has been reached on the classification of mammals. Our knowledge of many groups of mammals is incomplete, and future study may demonstrate that some of the families listed here can be discarded because they contain animals best included in another family. Other species and families are yet to be described. The present classification, then, is not used by all mammalogists, and it is by no means immutable.

Phylogeny Reconstruction

Since the 18th century, taxonomists have based their systems of classification on overall similarity among organisms. In the 1960s, Willi Hennig proposed a more objective method to make the classification reflect the actual evolutionary history of the group. This system came to be known as phylogenetic systematics or **cladistics** (Wiley 1981). In biology, therefore, the subdiscipline of systematics primarily deals with the classification and **phylogeny** of organisms. Classification is simply a way of ordering species into hierarchical groups and giving names to them so that they can be recognized and discussed. In phylogenetics, researchers develop hypotheses to reconstruct the evolutionary history or relatedness among species. In recent decades classifiers of organisms have incorporated phylogenetic information into classification schemes so that the named groups comprise evolutionarily related species. This is much easier said than done. There are also increasing efforts to use and integrate data from a variety of sources, including molecular, paleontological, morphological, biogeographic, ecological, and behavioral studies into phylogenetic hypotheses. Molecular data in particular are revolutionizing the way we view relationships and classify mammals, and even changing our concept of species (Baker & Bradley 2006; Riddle et al. 2008).

Methods used to infer phylogenetic relationships are explained in Kitching et al. (1998), Nei and Kumar (2000), and Swofford (1998). Detailed descriptions of the theory and practice of classifying mammals and reconstructing their evolutionary relationships are beyond the scope of this book, but we give a few brief notes, including some of the technical jargon. Establishing the pattern of relationships among mammals or any other group of organisms is often referred to as phylogeny reconstruction. The reconstruction of a phylogeny usually involves cladistics as a technique of analysis (**Box 1-2**). The goal of cladistic analysis is to produce a hypothesis of phylogenetic relationships of a group of organisms. The hypothesis is presented as a **cladogram** (such as that in **Fig. 1-2**), a branching, tree-like diagram in which the ends of the branches represent species or **taxa** and the branching points or **nodes** indicate the point at which species separated from one another to follow their own evolutionary pathways. A cladogram may show the relationships of species of mammals within a group to one another, or it may show the relationships of a group of mammals to other living organisms. The data used to produce a cladogram of mammals are called **characters** and usually consist of morphological features or molecules such as DNA or proteins.

The most important feature of cladistics is its reliance on shared derived characters (called **synapomorphies**) to establish relationships instead of overall similarity. In building a cladogram or phylogeny, the characters can have various **character states** (e.g., in the character "number

Taxa	Sequence data	Trait X	Trait Y
H	AAA GCT ACT	yes	no
G	AAA GCT ACT	yes	no
F	AAA GGT ACT	yes	no
E	CAA GGT ACT	yes	no
D	CAA GGT ACT	yes	no
C	CAA GGT ACG	yes	yes
B	CAA GGT ACG	yes	yes
A	CAA GGT ACG	yes	yes

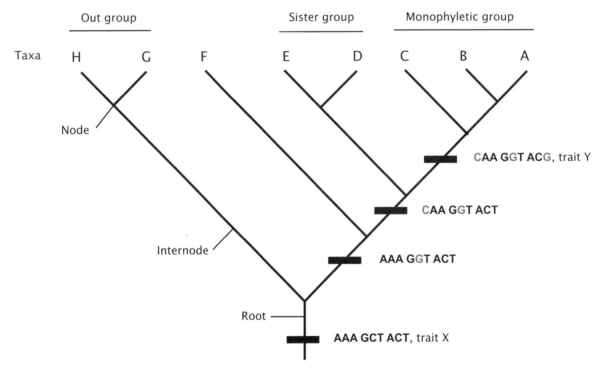

FIGURE 1-2 A phylogeny of eight taxa A through H based on DNA sequence data and two morphological traits (X and Y). The character matrix for the sequence data and the morphological traits is given above an example of one phylogenetic hypothesis. In the cladogram shown, Taxa A, B, and C form a monophyletic group. Taxa A through E also form a monophyletic group; however, a grouping of taxa E, D, C, and B would be paraphyletic. A node represents a speciation event where two lineages diverge from each other. An outgroup is a group of taxa that serves as a reference for determining the relationships of other taxa. A sister group is a group of taxa that is the most closely related to the ingroup.

of toes on hind feet," the possible character states might be "five toes" or "four toes" or "two toes"). The various character states must be analyzed, and their **polarity** (direction of evolutionary transformation) must be determined. In this example, the trend is toward reduction from five digits, the primitive number of toes in mammals and other terrestrial vertebrates, to fewer than five digits. The best way to determine character polarity is by including in the analysis an **outgroup**, consisting of a lineage that is closely related but outside of the **ingroup** (the group being studied). The underlying assumption is that if an outgroup itself is not too **derived**, it will tend to share **ancestral** (**plesiomorphic**) but not derived (**apomorphic**) features with the ingroup, thereby indicating which are ancestral character states.

Primitive or ancestral features shared by the outgroup and ingroup are called **symplesiomorphies** and provide no information about phylogenetic relationships. Because all mammals have hair, for example, this shared primitive characteristic would be of no use in understanding the evolutionary relationships of rodents (or any other mammalian group). Strict application of cladistics requires that only shared derived features (synapomorphies) be used to construct a hypothesis of relationships. The presence of a small pair of peg-like upper incisors immediately behind the first pair of incisors, therefore, is a synapomorphy of lagomorphs that clearly distinguishes them from rodents. Synapomorphies indicate a close relationship unless a character state arises by convergent evolution in an unrelated species. Ideally, synapomorphic character

BOX 1-2   Using Molecules to Construct Phylogenies

Phylogenetic systematics is a field that seeks to reconstruct the evolutionary history of groups of organisms. Because all life on Earth is related, like a family tree, there can only be one true history of descent from the first living organism that appeared some 3.8 billion years ago. Obviously, that true history was not observed and recorded, but it did leave behind tangible clues that biologists can use to reconstruct the evolutionary history of particular groups of organisms. Biologists have two major lines of evidence: morphology and molecules.

Fossils and living species have long been studied to yield morphological characteristics useful for constructing phylogenies (evolutionary trees). The use of molecules (DNA, RNA, proteins) is relatively new. By comparing homologous molecules from different organisms, it is possible to establish their evolutionary relationships. The basic premise behind molecular phylogenies is that all life uses the same DNA code, many of the genes are homologous across diverse taxa, and the DNA code is modified over time by mutations. Thus, species that diverged from each other relatively recently should have few such mutations, whereas those that diverged long ago will have accumulated many changes.

Molecular systematists begin by extracting DNA or proteins from several taxa of interest. They determine the sequence of DNA or amino acids for a specific region of the gene(s) or protein(s), and using a variety of sophisticated tools, they align those sequences for all the taxa (Fig. 1-2). The changes that have accumulated in those sequences are now apparent, and the goal is to find the evolutionary tree that best explains the sequences changes (using computer-assisted phylogenetic analysis software with appropriate assumptions). For example, in the hypothetical phylogeny illustrated here, the common ancestor of taxa A, B, C, D, E, F, G, and H, all share the ancestral DNA sequence (AAA GCT ACT) and morphological trait X (symplesiomorphies). The clade including taxa A through F accumulated one mutation in their DNA for this gene (C becomes G at position 5), and this character is a synapomorphy for the clade A–F. Another mutation occurs in position 1 (A becomes C) after lineage F diverges from the clade A–E. Finally, a third mutation (T becomes G in position 9) occurs in the common ancestor of taxa A, B, and C, and these taxa also evolve a new morphological trait Y. If we assumed that taxon C was more closely related to taxa E and D, then it would have required trait Y and the third mutation in the DNA sequence to evolve twice in two separate lineages—a highly unlikely event. Combining molecular and morphological evidence is useful where one type of evidence alone leads to ambiguous results. Many of the phylogenies in this text are based on combinations of molecular and morphological data.

It is important to remember that cladistic analysis generates an evolutionary tree, which represents a hypothesis of relationships among the organisms supported by the data analyzed; but, new data could emerge in support of a different hypothesis. In general, the more homologous characters used, the more robust the hypothesis.

---

states represent **homologous** not **analogous** features, but convergent evolution of characters is very difficult to discern and frequently causes problems for systematists. Old World jerboas and New World kangaroo rats independently evolved elongate hindlimbs and bipedal hopping, a particularly good example of convergent evolution. Barring any convergent characters in other species or groups under study, a group that exclusively shares derived characters is said to be **monophyletic** and to form a **clade**. A monophyletic clade is a phylogenetic lineage that arose from a single ancestor and includes only the ancestor and all of its descendants. The clade or branch of a cladogram nearest to a monophyletic clade is termed the **sister group** (Fig. 1-2).

A cladogram indicates only the relative timing of separation of species or lineages, but often it is desirable to know the numerical timing of divergence of lineages. For this, the cladogram must be linked with the geologic time scale. When the morphological data and resultant cladogram are combined with independent **geochronological** data from the rocks in which the fossils were buried, it is possible to estimate the time of appearance of fossil species and produce an evolutionary tree that is calibrated to the geologic time scale.

Many molecular geneticists use DNA and other molecules to construct cladograms or a **molecular phylogeny** of a group of mammals (Box 1-2). By using the so-called **molecular clock** (**Box 1-3**), they can also link a time scale to the phylogeny. Instead of measuring the time of first appearance of a recognizable member of a certain mammalian group in the fossil record, molecular systematists are able to estimate the time of divergence or lineage splitting of two species or lineages of mammals before the lineages may have become well differentiated morphologically. Often the molecular and morphological data agree, giving greater confidence in the accuracy of the phylogenetic hypothesis. The phylogenetic information, in turn, can be linked to biogeographic data and aspects of geological history, such as the past positions of the continents, islands, or tectonic plates and the timing of their collision or separation. When morphological and molecular phylogenies and various aspects of geological history are combined, the result can be an eclectic, robust model for the evolutionary history of a mammalian group.

Virtually all systematists agree that named taxa should share a common evolutionary origin. Naming taxa based on cladistic hypotheses is an entirely different problem from determining phylogenetic relationships, and one that is intractable and controversial. Some **cladists** prefer to name only monophyletic groups (at their nodes in a cladogram), but in practice, this can result in a proliferation of names and taxa of unequivalent ranks that fall between

BOX 1-3   The Molecular Clock—Does It Run on Time?

The molecular clock is a concept that assumes that the time that two species diverged from one another can be calculated by measuring the number of molecular differences between the species' DNA sequences or proteins. Instead of measuring time, the molecular clock measures the number of silent mutations (those that do not result in amino acid sequence changes) that accumulate in the gene sequences of different species over time (**Fig. 1-3**). Evolutionary biologists use this information to estimate the date when two species diverged on the geologic timeline.

In 1962, Zuckerkandl and Pauling recognized that the number of amino acid differences in the protein hemoglobin was random and occurred at a constant rate. This led to the proposal that the number of mutations in a given stretch of DNA could be used as a measure of time since two or more species diverged. Like all clocks, however, the "molecular clock" has to be calibrated. This is accomplished by using known dates from the fossil record for the taxa being studied. Thus, if the mutation rate for a specific nonfunctional gene is 3 mutations per million years (r) and there are 21 mutations in the DNA sequences between two species (d), the estimated time (t) those two species began to diverge is calculated as $t = d/2r$, or approximately 3.5 million years ago.

Later research showed that different genes evolve at different rates. Functionally important genes are more often conserved and mutate more slowly than genes with less vital functions. Consequently, more recent speciation events are dated using genes with higher mutation rates, and slow evolving genes are used to date more ancient divergence events.

The molecular clock hypothesis has certain limitations. As Ayala (1999, 2000) pointed out, the molecular clock is un-

likely to give accurate divergence dates if (1) the species being compared have different generation times; (2) population sizes of the species differ by orders of magnitude (e.g., comparing small isolated mole-rat populations with those of the nearly cosmopolitan *Rattus* sp.); (3) the species differ markedly in metabolism, evolutionary history, or life history traits; (4) the intensity of natural selection changes dramatically over time; or (5) the function of the protein or gene being used is altered over time.

Instead of abandoning the molecular clock entirely, the assumptions of the molecular clock have been relaxed, allowing the mutation rate to vary across the branches of the evolutionary tree. Nevertheless, it is important to remember that even relaxed-clock methods need to be calibrated using the absolute age of some evolutionary divergence event from the fossil record.

Carefully calculated and calibrated, the molecular clock can provide important date estimates, especially for fossil-poor taxa, and it remains an important tool in evolutionary biology (**Fig. 1-4**; Kumar 2005). In addition, as Ho (2008) pointed out, "With the rapid accumulation of new genetic data, particularly as a result of the many genomic sequencing projects that are currently underway, it seems that the molecular clock will continue to shed light on the tempo and time scale of evolution for years to come." With respect to mammals, Baker & Bradley (2006) suggested that changing from a morphological species concept to a genetic species concept could reveal as many as 2,000 additional "genetic species" of mammals hidden among the 5,400 known by their morphology.

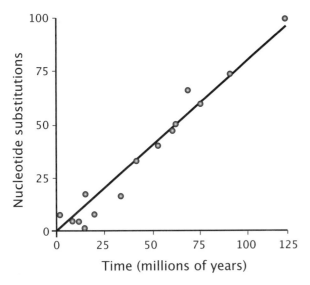

FIGURE 1-3   A graph of the combined number of nucleotide substitutions from 7 proteins and the species divergence time for 15 pairs of mammalian species. (Modified from Fitch WM. 1976. Molecular evolutionary clocks, pgs. 160-178, in *Molecular Evolution* (FJ Ayala, ed.) Sinauer Associates, Sunderland, MA.)

the standard categories of Class, Order, Family, and so on. To reconcile this problem, many systematists informally name nodes but do not provide them with a rank or category. Others name **crown groups**, which consist of the hypothetical ancestors of all the living members of a group and all its descendants, extinct and extant. For Mammalia, for example, this would include the common ancestor of Monotremata, Metatheria, and Eutheria and all of the common ancestor's descendants (Rowe 1988; Rowe & Gauthier 1992).

In addition to these works cited, students are referred to Wilson and Reeder (2005) and McKenna and Bell (1997). As an entry into the vast literature about the phylogeny of mammals, see Rose and Archibald (2005), Szalay et al. (1993a, 1993b), Honeycutt and Adkins (1993), Novacek (1992a), and also "The Tree of Life" web site that deals with mammals and their extinct relatives (http://tolweb.org/Mammalia).

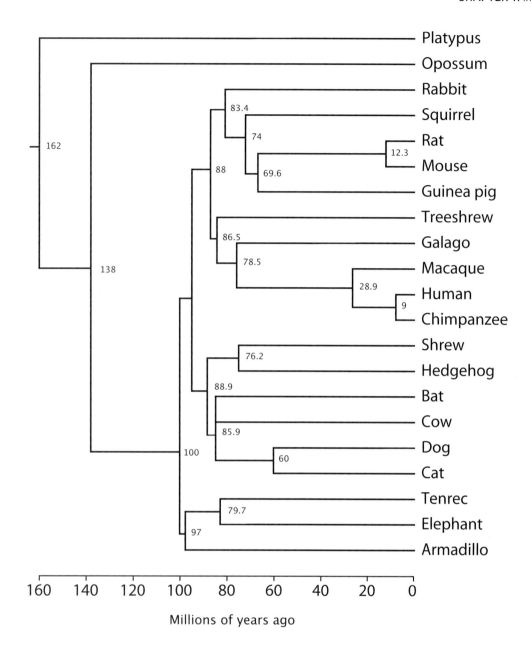

FIGURE 1-4   A molecular phylogeny of mammals showing estimated divergence times (millions of years ago) based on analysis of 3,012 genes (2,844,615 nucleotides) from a total of 22 species. (Adapted from Hallström, BM and A Janke, *BMC Evolutionary Biology* 8 (2008): 1-13.)

SUMMARY

These are exciting times for people interested in the study of mammals. Over 400 mammal species were discovered since 1993 as a result of recent explorations in remote parts of the world, taxonomic revisions of problematic groups, and the evaluation of DNA evidence for cryptic species. Indeed, the cumulative number of new mammal species is still increasing. Recently discovered fossils also continue to reshape the landscape of mammalian evolution. In addition, new tools from the field of molecular biology are helping to address questions in mammalian behavior, phylogeny, and other fields. Although our scientific knowledge is increasing dramatically, numerous important questions remain to be answered by the next generation of mammalogists. Among the most important are questions that address the causes and repercussions of mammalian extinctions. Many mammals are now threatened with extinction primarily by human activities, even as we struggle to catalog and conserve species. In recent years, mammalogists have increasingly studied wild mammals by using noninvasive techniques such as camera- and video-trapping, thermal imaging, satellite tracking, and identification of tracks and signs. Laboratory analyses used in modern studies include the recovery of bile acids, DNA, stable isotopes, or other evidence from mammal scats and hair samples and extracting and sequencing fragments of ancient DNA from fossils and subfossils. Contributions during fieldwork by enthusiastic students and the public as citizen scientists are important also.

The classification of mammals, must be based on evolutionary relationships. Each known kind of organism is given a binomial scientific name. The first, or generic name, may be applied to a number of related kinds, but the second name refers to a specific kind, a species. Each species shares a common gene pool and is generally separated from all other species by a "reproductive gap." A classification that expresses the true evolution of mammals uses a series of categories, each category more inclusive than the one below it in the classification hierarchy. The present classification, proposed by Wilson and Reeder in 2005, will continue to change as the study of mammals advances.

Researchers develop hypotheses to reconstruct the evolutionary history or relatedness among species using phylogenetic information. The hypothesis is presented as a phylogeny or cladogram, a branching, tree-like diagram in which the ends of the branches represent species or higher level taxa, and the branching points or nodes indicate the point at which species separated from one another to follow their own evolutionary pathways. The data used to produce a phylogeny are shared characters and usually consist of morphological features and molecular ones, such as DNA sequences. Systematists strive to use only shared derived features (synapomorphies) to construct a hypothesis of relationships. When a cladogram is combined with independent geochronological data from fossil-bearing rocks, it can indicate the timing of evolutionary events. Molecular geneticists use DNA and other molecules to construct cladograms or molecular phylogenies, and use the molecular clock to link a time scale to the phylogeny.

The fifth edition of this text represents a significant change from the previous edition, published in 2000. Although the basic framework remains intact, we needed to incorporate more than a decade of new research. All of the chapters were extensively rewritten and updated. In the first part of this book (Chapters 1 through 4), we define mammals and summarize their origins. In the second part (Chapters 5 to 19), we present the orders and families of mammals in detail. In the third part (Chapters 20 to 26), we treat special topics such as mammalian echolocation, physiology, behavior, ecology, and zoogeography. From this coverage, students can gain a general understanding and an appreciation of the form and function of mammals. Two additional new chapters, Chapter 27, Domesticated Mammals and Chapter 28, Mammalian Disease and Zoonoses, are available on the web site that accompanies the text.

KEY TERMS

Analogous	Crown group	Phylogenetic relationship
Ancestral	Cryptic species	Phylogeny
Apomorphy	Derived	Plesiomorphy
Biodiversity	Geochronological	Polarity
Camera trap	Homologous	Sister group
Character states	Homoplasy	Species
Characters	Ingroup	Stable isotope
Citizen science	Molecular clock	Subspecies
Clade	Molecular phylogeny	Symplesiomorphy
Cladist	Monophyletic	Synapomorphy
Cladistics	Nodes	Systematics
Cladogram	Outgroup	Taxa
Convergence	Parallelism	Taxonomy

RECOMMENDED READINGS

Baker, RJ, & RD Bradley. 2006. Speciation in mammals and the genetic species concept. *Journal of Mammalogy* 87(4):643–662.

Hall, BG. 2007. *Phylogenetic Trees Made Easy: A How-to Manual*, 3rd ed. Sinauer Associates, Sunderland, MA.

Kumar, S. 2005. Molecular clocks: four decades of evolution. *Nature Reviews Genetics*, 6, 654–662.

McKenna, MC & SK Bell. 1997. *Classification of Mammals Above the Species Level*. Columbia University Press, NY.

Murphy, MJ, et al. 2001. Molecular phylogenetics and the origins of placental mammals. *Nature, 409,* 614–618.

Nei, M & S Kumar. 2000. *Molecular Evolution and Phylogenetics.* Oxford University Press, NY.

Wilson, DE & DM Reeder. 2005. *Mammal Species of the World, A Taxonomic and Geographic Reference,* 3rd ed., 2 vols. Johns Hopkins University Press, Baltimore, MD.

Mammalian Characteristics

Mammals owe their spectacular success to many features. Perhaps the most important and diagnostic mammalian characteristics are those that enhance intelligence and sensory ability, promote endothermy, or increase the efficiency of reproduction or of securing and processing food. The senses of sight and smell are highly developed, and the sense of hearing has undergone greater specialization in mammals than in any other vertebrates. Efficient gathering and utilization of a wide variety of foods are aided by specializations of the dentition and the digestive system. Endothermy has allowed mammals to remain active under a wide array of environmental conditions. Specializations of the postcranial anatomy, particularly the limbs and feet, have enabled mammals to make effective use of endothermically driven physical activity. In some species, extended periods of parental care have increased the length of time during which the young can learn demanding foraging patterns and complex social behavior.

The basic structural plan of the mammalian body was inherited more than 200 million years ago from the non-mammalian synapsids (formerly known as mammal-like reptiles) of the order Therapsida (see Chapter 3). Members of this ancient order followed an evolutionary path that diverged strongly from the sauropsid ("reptilian") path from which arose the spectacular and successful Mesozoic dinosaurs, birds, and a wide diversity of reptiles. The key to the persistence of the therapsids through the Triassic period (see Table 3-1) was perhaps their ability to move and to respond to their environment more quickly than their archosaurian contemporaries. These same abilities probably enabled the descendants of the therapsids, the mammals, to survive through the Jurassic and Cretaceous periods, when the dinosaurs dominated the terrestrial scene. Also of major importance to early mammals was their highly specialized dentition, which probably allowed them to use certain foods more efficiently than could turtles, most crocodilians, or reptiles (for modern herpetological taxonomy see www.cnah.org/taxonomy.asp).

An important morphological trend in the therapsid–mammalian line was toward skeletal simplification. In general, the therapsid skeleton evolved greater efficiency by reducing the number of parts while retaining the effective performance of a particular function. In the skull and lower jaw, which in primitive reptiles and other sauropsids consisted of many bones, a number of bones were lost, reduced in size, or put to other uses. The limbs and limb girdles also were simplified to some extent and reduced in massiveness. The skeleton of egg-laying mammals (order Monotremata) roughly resembles or retains features similar to those of advanced therapsids, but the limbs of some therapsids were less laterally splayed than today's specialized monotremes.

When mammals first appeared in the Triassic period, they represented no radical structural departure from the therapsid plan but had attained a level of development (involving a dentary/squamosal jaw articulation) that is interpreted by most vertebrate paleontologists as a key indication that the animals had crossed the nonmammalian–mammalian boundary (therefore, they represent an evolutionary grade, not a clade). Many of the mammalian characters discussed in this chapter resulted from evolutionary trends that are clearly characteristic of therapsids. Unfortunately, the fossil record cannot directly indicate when various important features of the soft anatomy became established, and only indirect evidence can be used to judge whether advanced therapsids had such features as mammae, hair, or a four-chambered heart.

Endothermy probably began to develop in the therapsid ancestors of mammals (Bennett & Ruben 1986; Tracy et al. 1986; Turner & Tracy 1986). We know that today the metabolic "machinery" of mammals differs markedly from that of turtles, crocodilians, and squamates (formerly collectively known as reptiles, now considered separate classes, including as Class Reptilia only the snakes and lizards;

www.cnah.org/taxonomy.asp). Mammals have a threefold to sixfold greater capacity for energy production than do reptiles and a standard metabolic rate some 8 to 10 times higher (Else & Hulbert 1981). Relative to total body size, the internal organs of mammals are larger than those of reptiles. Mitochondrial membrane surface areas for the heart, kidney, and brain are far greater in mammals, as are mitochondrial enzyme activity and thyroid activity. A much greater absorptive surface area in the lungs enables a higher oxygen uptake in mammals. Similarly, an increased absorptive surface area in the digestive system enables a higher nutrient uptake in mammals than that in reptiles (Karasov & Diamond 1994). These differences are clearly related to the mammalian capacity for high energy production (**Table 2-1**) and were probably part of a suite of anatomical and physiological features that allowed mammals (perhaps even the earliest ones) to be nocturnal.

The characteristics proposed as being diagnostic of mammals and their closest relatives (extinct synapsids sometimes called mammaliaforms, see Chapter 3) are discussed by Gauthier et al. (1988), Rowe (1988, 1996), Wible (1991), and other authors. A list of these characteristics (called "characters") of the hard and soft anatomy were repeated and discussed by McKenna and Bell (1997). Some of these are included in **Table 2-2**. Only one molecular character is included in this definition of Mammalia (Table 2-2, number 13).

Soft Anatomy

Skin Glands

The skin of mammals contains several kinds of glands not found in other vertebrates; the most important of these are the **mammary glands** (the key feature after which Mammalia is named; Table 2-2, number 24). These glands in females provide nourishment for the young during their postnatal period of rapid growth. Mammary glands consist of a complex system of ducts that reach the surface of the skin usually through a prominence called a nipple or teat (**Fig. 2-1**). During late pregnancy, the epithelium of the ducts is stimulated by endocrine secretions (estrogens and progesterone) and divides rapidly to produce secretory alveoli, which produce milk after birth. Secretions of the anterior lobe of the pituitary (prolactin and growth hormone) stimulate the production of milk. Nursing and emptying of milk from the mammary glands result in nervous stimulation of the anterior lobe of the pituitary and continued production of prolactin and milk (**Box 2-1**).

The composition of milk differs from species to species. Cow's milk contains about 85% water; the dry weight includes approximately 20% protein, 20% fat, and 60% sugars (largely lactose), as well as vitamins and salts in roughly the proportion in which they are found in blood. The milk of mammals with young that grow unusually rapidly contains high levels of protein and fat. Seals, for example, produce

TABLE 2-1   Some Contrasts Between Adult Ectothermic Reptiles (Lizards) and Endothermic Small Mammals

Ectotherms (Lizards)	Endotherms (Small Mammals)
Lower size limit, 1g	Lower size limit approximately 2g
Body shape usually elongate	Body shape more spherical
50% to 90% of energy from anaerobic metabolism	High level of aerobic energy production
High activity for only brief periods	High activity over long periods
Low blood pressure (30–50 mm Hg)	High blood pressure (80–200 mm Hg)
Low hematocrit	High hematocrit
Large-diameter, widely spaced capillaries	Small-diameter, closely spaced capillaries
Blood oxygen capacity 25% to 50% that of endotherms	
Capillary length (frogs) 155 mm/mm^3 of tissue	Capillary length (mouse) 3,500 mm/mm^3
Daily energy requirement of nontorpid lizard 3% to 4% that of mammal	
Can use highly ephemeral food source (e.g., *Heloderma* eat bird eggs; *Coleonyx* can store enough energy in 4 days to last 9 months)	Mammals need more long-term food sources
Lizards allocate 90% of energy budget to new biomass	Mammals allocate 90% of energy budget to thermoregulation
Minor consumers but often (e.g., in deserts) major producers of biomass	Major consumers but often low producers of biomass

TABLE 2-2   Diagnostic Characters of Mammalia

1. Accessory jaw bones shifted away from the dentary-squamosal joint in adults to become associated with the cranium alone. These include the middle ear bones, the stapes (columella auris of nonmammals), incus (quadrate), and malleus (articular), as well as other bones formerly associated with the jaw.

2. Stapes very small relative to skull size.

3. Atlas intercentrum and neural arches fused to form single, ring-shaped osseous structure.

4. Epiphyses on the long bones and girdles.

5. Heart completely divided into four chambers with a thick, compact myocardium (muscular wall).

6. Heart with atrioventricular node ("pacemaker") and Purkinje fibers.

7. Single aortic trunk.

8. Pulmonary artery with three semilunar valves.

9. Erythrocytes lack nuclei at maturity.

10. Endothermy.

11. Central nervous system covered by three meninges.

12. Cerebellum folded.

13. Nerve filaments with three polypeptides.

14. Brain with divided optic lobes.

15. Strong representation of the facial nerve field in the motor cortex of the brain.

16. Superficial musculature expanded onto the face and differentiated into muscle groups associated with the eye, ear, and snout.

17. Muscular diaphragm encloses pleural (lung) cavities, and consequent development of diaphragmatic breathing.

18. Complex lung structure with division of the lungs into lobes, bronchioles, and alveoli.

19. Epiglottis.

20. Skin with erector muscles and dermal papillae.

21. Hair.

22. Sebaceous glands.

23. Sweat glands.

24. Mammary glands.

25. Loop of Henle in the kidney.

Selected from a much longer list in McKenna and Bell (1997) and sources therein. (By this definition, Mammalia does not include the extinct near-mammals, the Mammaliaformes: Morganucodonta, Docodonta, and Haramyoidea. Our use in this book (see Chapter 3) considers those groups to be Mammals. The first character forms the arbitrary dividing feature separating Mammalia from nonmammalian synapsids.

milk with roughly 12 times as much fat and 5 times as much protein as is in cow's milk.

The period of association between the mother and her young during lactation and suckling is one that encourages close social bonds. For many mammals, especially those with complex foraging and social behavior patterns, this is an extremely important time of rapid learning by the young that prepares them for an adult life independent from their mothers. Recently, Francis et al. (1994) discovered that in some Malaysian and New Guinean populations of the Dyak fruit bat *Dyacopterus spadiceus*, the males also appear to lactate in a process similar to galactorrhea in humans (Racey et al. 2009).

In most mammals, the young suck milk from the projecting nipples. In monotremes, however, nipples are lacking, and the young suck milk from tufts of hair on the

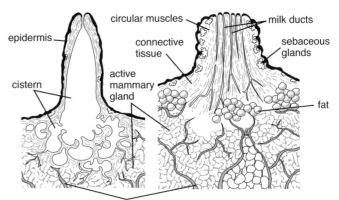

FIGURE 2-1   A section through the nipple and associated tissues of a primate (right) and the nipple, or teat, of an artiodactyl (left). (Adapted from Hildebrand, M. *Analysis of Vertebrate Structure.* John Wiley & Sons, 1974.)

BOX 2-1   Evolution of Lactation and Suckling

One of the key innovations in mammalian evolution is lactation. All mammals feed their newborn young with milk, a liquid secretion from specialized skin glands that is a rich, nutritious, and precisely appropriate food for a growing infant. Because soft tissues such as skin and glandular tissues rarely fossilize, there is very little direct evidence that can be derived from the fossil record informing us about the evolution of milk production. As a result, there is more speculation than hard data on the topic. New molecular evidence helps somewhat. Oftedal (2002) provided an interesting, in-depth scenario for the gradual evolution of lactation along with the gradual evolution of other mammalian characters through their Paleozoic and Mesozoic ancestry, partially summarized here.

Because all existing mammals nourish their young with milk, lactation probably evolved in an early common ancestor of monotremes (egg-laying mammals), marsupials, and placental mammals. Moreover, similarities in mammary glands and milk composition (as well as considerable differences) also point to a single evolutionary origination of lactation, possibly even in mammals' ancestors, the synapsids. Primitive mammals were small and probably agile, suggesting they already had a relatively high metabolism and were endothermic (warm blooded) in the basalmost Mesozoic members of the lineage. In addition, early synapsids have long bones that show lines of arrested growth, or seasonal growth lines, whereas later synapsids have uniformly vascularized bone indicating sustained growth and probably reflecting a warm, constant body temperature. A diet of milk would have permitted endothermic babies to grow quickly. Interestingly, birds, which are also endothermic, must also heavily feed their fast-growing young. A few kinds of birds even produce a milk-like substance to feed their hatchlings, but they have otherwise followed a very different evolutionary trajectory than mammals with respect to parental care.

When the platypus genome was sequenced recently, Brawand et al. (2008) found that the platypus, an egg-laying mammal, retained ancestral genes for egg yolk proteins, but these same genes were lost during evolution in the placental and marsupial mammals. These authors further showed that genes called caseins related to milk production (which function somewhat like egg yolk gene proteins in nourishing a developing embryo) probably evolved in a common mammalian ancestor during the Mesozoic era. They interpret the evidence to show that the development of lactation and placentation in marsupials and placentals allowed for the evolutionary loss of the egg and yolk genes of monotremes and other basal mammals. Oftedal (2002) suggests that milk with casein proteins, which provide calcium and phosphate, had already evolved by the Triassic because fossils of juvenile synapsids show extensive bone development before the young would have been capable of independent feeding.

Along with many other evolutionary transformations in the skull, synapsids developed a secondary palate of bone and soft tissue (Sidor 2003). The palate separates airflow in the nasal chamber from that in the oral chamber. This not only allows

the animals to breathe while chewing food, but in newborns, it allows breathing while they suckle. The soft tissues (cheeks) also facilitate both the suckling and chewing processes.

Early therians (the mammalian lineage to which placentals and marsupials but not monotremes belong) share an anatomical feature of the skeleton, a pair of epipubic bones on the front of the pelvis. (The epipubic bones are lost later in the evolution of placentals, including all living species.) Epipubic bones project forward into the abdominal wall. They occur in both females and males but tend to be larger in females. It has been suggested that these bones help to support the weight of pouch young in marsupials. Epipubic bones occurred in at least some nonmammalian fossil synapsids called tritylodontids, in which they might have provided a similar function, as well as in locomotion. If they were endothermic and had a high metabolism, these animals would have required prohibitively large eggs, too large for the small mothers to carry until the eggs could be laid. Mammals also have highly specialized teeth that require time for the jaws to grow long enough to accommodate them and for the teeth to develop into an integrated and effective feeding mechanism. A method of parental feeding during this stage of development would have been advantageous. Like other features of their reproduction and development, the common mammalian characteristic of two generations of teeth, deciduous (or milk, or baby) teeth and permanent (or adult) teeth, is also tied to the evolution of lactation.

Mammal skin has several types of glands, typically including sebaceous, eccrine, and apocrine glands. These glands are often associated with hair follicles and produce small amounts of secretions that variously lubricate and cool the skin and hair. Oftedal (2002) suggested that mammary glands most likely evolved from ancestral apocrine-like glands that may have occurred in patches like the nipple-less mammary patches of monotremes. The mammary patch secretions might have initially provided moisture and antimicrobial compounds to the ancestral mammals' soft permeable eggshell and later provided nutrients for the hatchlings. Nipples and breasts would have developed in therians at the common site for the ducts of several milk-secreting lobules of glandular tissue, a more efficient delivery system than the diffuse 100–200 gland openings and hairs in the mammary patch of monotremes. (Apparently, it is not yet known whether egg-moistening by the mammary patch occurs before hatching in modern platypuses.) As the lactation system evolved, the antimicrobial components could have contemporaneously enhanced the nursing infants' chance for survival and evolved to form the immune-defense system bestowed on the infants (Capuco & Akers 2009; Goldman 2002). In general, lactation seems to have evolved to minimize the energy cost to the nursing mother while maximizing the survival of the suckling newborn mammal, thereby enhancing the survival of both (Capuco & Akers 2009). Foods eaten by the mother during nursing can affect taste preferences in the infant mammal later in life and may help to teach the young which foods are safe to eat after weaning.

mammary areas (Burrell 1927; Ewer 1968). Whales, dolphins, and porpoises have muscles that force milk into the mouth of the young, a seemingly necessary adaptation in underwater-feeding animals that also have no lips and are therefore unable to suckle. The number of nipples varies from 2 in many kinds of mammals to about 19 in the opossum *Marmosa* (Tate 1933).

Other types of skin glands are also important in mammals. The watery secretion of the **sweat glands** functions primarily to promote evaporative cooling but also eliminates some waste materials. In humans and some ungulates, sweat glands are broadly distributed over the body surface, but in most mammals, they are more restricted (Folk et al. 1991). In some soricomorphs, rodents, and carnivorans, sweat glands occur on only the feet or the **venter**. The glands are completely lacking in the Cetacea and in some bats and rodents. Hair follicles are supplied with **sebaceous glands** that produce an oily secretion that lubricates the hair and skin (**Fig. 2-2**). Diverse **scent glands** and **musk glands** are found in mammals (e.g., Ewer 1973; Ralls 1971). These glands are variously used for attracting mates, marking territories, communication during social interactions, or protection. The smell of skunk is familiar to all but the most city-bound and has caused the temporary banishment of many a domesticated dog. A musk gland marked by a chevron-shaped patch of dark hairs occurs on the top of the tail of wolves and coyotes, as well as on the tail of many domesticated dogs. The functions of some

mammalian scent glands in connection with social behavior are discussed in Chapter 24.

▪ Hair

The bodies of mammals are typically covered with hair, a unique mammalian feature that has no structural homolog among other vertebrates. Hair was perhaps developed by therapsids before a scaly covering was lost. In modern mammals that possess scaly tails or bony plates (such as armadillos), hairs project from beneath the scales in a regular pattern. A similar pattern of hair distribution, perhaps reflecting the ancestral condition, is also seen in mammals without scales (de Meijere 1894) (e.g., note the pattern of hair projection on your own arm).

A hair consists of dead epidermal cells that are strengthened by keratin, a tough, horny tissue made of proteins. A hair grows from living cells in the hair root. Each hair consists of an outer layer of cells arranged in a scale-like pattern (the **cuticular scale**), a deeper layer of highly packed cells (the **cortex**), and in some cases, a central core of cuboidal cells (the **medulla; Fig. 2-3**). The color of hair depends on pigment in either the medulla or the cortical cells; the cuticular scales are usually transparent (**Box 2-2**).

The coat of hair, collectively termed the **pelage**, functions primarily as insulation. The dissipation of heat from the skin surface to the environment and the absorption of heat from the environment are retarded by the pelage. Seals, sea lions, and walruses, many of which live in extremely

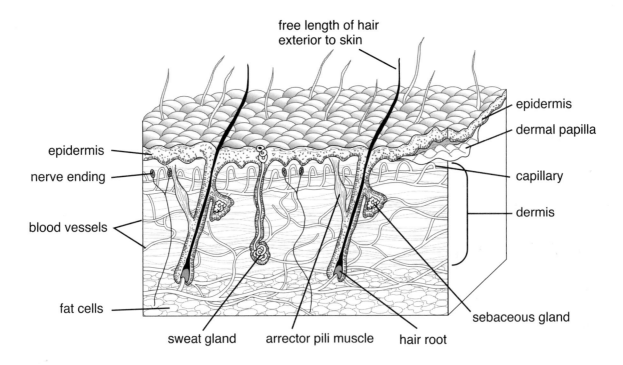

FIGURE 2-2 Generalized section of mammalian skin. The skin on different areas of the body differs in thickness as well as purpose. (Adapted from Romer, A. S., and Parsons, S. T. *The Vertebrate Body*. Saunders, 1977.)

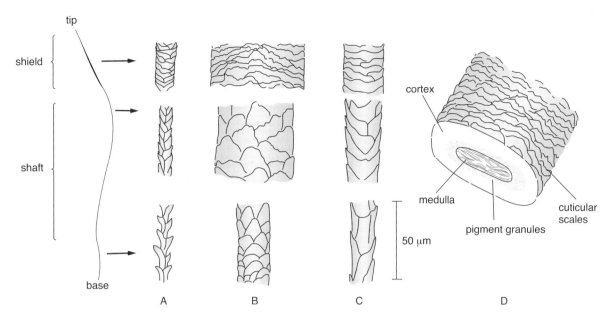

FIGURE 2-3 Structure of a guard hair and cuticular scale patterns of the guard hairs of some mammals. Shown at the left is a diagram of a single guard hair as found on the dorsal surface of a mammal, with regions labeled. Arrows indicate the positions along the hair where the highly magnified scale patterns shown in A, B, and C occur. Note how the scale pattern differs from base to tip. (A) *Pipistrellus nathusii* (pipistrelle bat); (B) *Martes martes* (European pine marten); (C) *Eliomys quercinus* (garden dormouse). (D) Cross-section of hair of *Meles meles* (European badger; same scale as A–C). (Adapted from Teerink, B. J. *Hair of West-European Mammals: Atlas and Identification Key*. Cambridge University Press, 1991.)

cold water, are insulated both by hair and a subcutaneous layer of blubber. Some mammals are hairless, or nearly so. These either live in warm areas or have specialized means of insulation other than hair. Essentially hairless, whales and porpoises have thick layers of blubber that provide insulation. Hair is sparse on elephants, rhinoceroses, and hippopotami. These animals live in warm areas, have thick skins (Jarman 1989) that offer some insulation, and have such favorable mass/surface area ratios for heat conservation because of their large size that retention of body heat is no problem.

Hair, being nonliving material, is subject to considerable wear and bleaching of pigments. Molts occur periodically when a juvenile mammal gains its first adult pelage and thereafter usually once or twice a year. Old hairs are lost, and new ones replace them. These changes often occur in a regular pattern of replacement (**Fig. 2-4**). In many north-temperate species, the molts are in the spring and fall; the summer pelage is generally shorter and has less insulating ability than does the winter pelage. In some species that live in areas with continuous snow cover in the winter, the summer pelage is brown, and the winter coat is white. The

BOX 2-2 Green Mammals?

Numerous kinds of vertebrates—amphibians, reptiles, and birds—are green, but why are there no green mammals? This frequently asked question seems to have no ready answer. Is it because of mammals' nocturnal ancestry and the attendant low level of sunlight in which they are active? A few kinds of mammals actually do appear greenish or olive thanks to chlorophyll pigments in symbiotic algal and cyanobacterial cells. Tree sloths (order Pilosa) achieve a highly camouflaged effect during the rainy season when unicellular algae and cyanobacteria grow on and within the flutings, cuticular scales, and cracks in individual hairs over the entire body, giving the sloths a green appearance (Aiello 1985). Polar bears kept in zoos and monk seals have occasionally been observed to have cyanobacteria and green algae, respectively, growing in their fur, but the advantage of this to the mammals, if any, is unknown and

the phenomenon little studied (Barsanti et al. 2008). Some mammals, particularly those in the rain forests of Australasia (e.g., New Guinean tube-nosed bats of the genera *Nyctimene* and *Paranyctimene*, or the Australian green ringtail possum *Pseudochirops archeri*), that sleep exposed on branches during the day (Flannery 1995) can be olive colored or at least have a strong greenish cast to their wings or body fur. In at least *Nyctimene*, captive animals can host green algae and cyanobacterial cells within the medullas of their hairs (Muller & Byrnes 2006). To our knowledge, no mammals have green fur color that is structural (in which specific wavelengths of light, usually blues and greens, are reflected by the anatomical structure of the hairs, as in the feathers of birds) or from actual pigments in the fur or skin. For the species hosting algae in captivity, it is also unknown whether the algal symbiosis occurs in the wild.

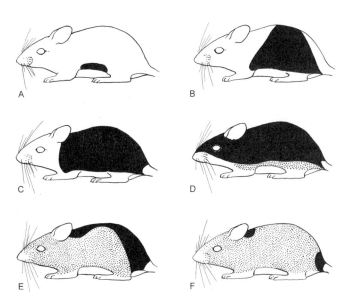

FIGURE 2-4   The pattern of postjuvenile molt in the golden mouse (*Ochrotomys nuttalli*). Black indicates areas in which adult hair is replacing juvenile hair. Stippled areas indicate new adult pelage. (Adapted from Linzey, D. W. and Linzey, A. V., *J. Mammalogy.* 48 (1967): 326–241.)

arctic fox, several species of hares and rodents, and some weasels follow this pattern.

The color of most small terrestrial mammals closely resembles the color of the soil on which they live. In his study of concealing coloration in desert rodents of the Tularosa Basin of New Mexico, Benson (1933) found that white sands were inhabited by nearly white rodents, and adjoining stretches of black lava were home to black rodents. Broadly speaking, nocturnal mammals that are active against dark substrates, such as dark forest soils, are dark colored, whereas those foraging over light-colored soils, such as the soil of deserts, are relatively pale. In large, diurnal, open-country dwellers, such as African antelope, however, coloration may be related to temperature control. The pale bodies of the Arabian oryx (*Oryx leucoryx*) and the addax (*Addax nasomaculatus*), both desert dwellers, may be important in reflecting light and reducing the intake of solar radiation. The color of mammalian hair and skin is determined by the presence of granules of a pigment, melanin, which occurs in the epidermis of the skin and medulla or cortex of hairs. Genes controlling the presence or absence, density, shape, and arrangement of melanin granules determine many of the differences in mammal coat colors (Hoekstra & Nachman 2005).

Countershading is a pattern common to mammals and many other vertebrates. In most lighting conditions, the back of an animal is more brightly illuminated than is the underside. If a mammal were all of a single color, the underside would appear very dark relative to the back, and the form of the animal would be obvious. When the back and sides are darkly colored and the underside and insides of the legs are white (an almost universal color pattern among terrestrial mammals), however, the well-lighted back reflects little light and the shaded white venter tends to reflect light strongly. The result is that the form of the animal becomes obscured to some extent, and the animal becomes less conspicuous.

The color patterns of mammals serve a variety of purposes. Camouflage seems to be the most important factor in the evolution of mammalian coloration, but communication and the regulation of physiological processes also are important (Caro 2005). The pelages of some ungulates and some rodents are marked by white stripes that tend to obliterate the shapes of the animals when they are against broken patterns of light and shade (**Fig. 2-5**). The eye is one of the most conspicuous and unmistakable vertebrate features; some facial markings in mammals may obviate the bold pattern of the eye by superimposing a more dominant and disruptive pattern (**Fig. 2-6**). If these markings only occasionally allow an animal to go unnoticed by a predator or if they cause a predator to be indecisive in its attack for but a fraction of a second, they have adaptive value.

Even in broad daylight, the stripes of zebras cause distant herds to fade into their background, but another potentially adaptive importance has been considered by Cott (1966). The stripes are patterned so as to create an optical illusion. The animal's apparent size is increased. In the dim light in which most predators hunt, this illusion may cause a slight miscalculation of range and an occasional inaccurate leap (Ruxton 2002).

What about the glaringly white rump patch of the pronghorn (see Fig. 18-17), however, or the conspicuous white-next-to-black markings of some African antelope,

FIGURE 2-5   Female nyala (*Tragelaphus angasii*; Bovidae) in South Africa. The stripes and spots conceal this animal in the thickets it inhabits.

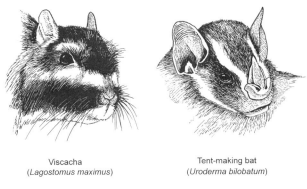

Viscacha
(*Lagostomus maximus*)

Tent-making bat
(*Uroderma bilobatum*)

FIGURE 2-6   The faces of two mammals in which facial masks reduce the conspicuousness of the eye.

or the bold white eye ring of many carnivores, or the white-on-black pattern of skunks (see Fig. 16-27b)? No single explanation can be applied to these diverse patterns, but each makes an animal, or at least part of it, more obvious rather than less so. The pronghorn's markings seemingly function as warning signals to other herd members when an individual begins to run from danger; the bounding gait of some antelope shows off these markings. The eye rings of carnivores are perhaps important as accents for the eyes and face and may emphasize facial expressions used during intraspecific social interactions. The black-and-white coloration of skunks, on the other hand, makes these defensively well-endowed animals conspicuous and unmistakable to their would-be predators. This warning coloration is sometimes termed aposematic.

A poorly studied pelage color phenomenon is the peculiarity that patches of the fur in didelphid opossums reflect ultraviolet light (Pine et al. 1985). Why would ultraviolet reflectance occur in a group of nocturnal mammals that may have lost color vision (Jacobs 1993)? It is difficult to know what the brain of a given mammal actually perceives as colors, but at least a few Australian marsupials and other mammals probably actually can see ultraviolet light (see **Box 2-3**; Sensory Abilities Beyond the Big 5).

BOX 2-3   Sensory Abilities Beyond The Big 5

Many more sensory abilities occur in mammals other than the five or six senses typically thought of (Ackerman 1991). It is important to recognize this not only for our intrinsic curiosity about how other mammals besides us humans perceive their world, but also for our human sensory limitations and potential, and even for the implications for conservation of mammals and other living things (Lim et al. 2008). Smell, sight, hearing, balance, touch, and taste are enhanced and extended by senses of pain, pressure, acceleration, bodily position (proprioception), time, temperature, magnetoreception, electroreception, and others. Cohen (2007) listed 53 different types of inherent natural senses in humans. As examples of other senses in mammals, the "bill" of the duck-billed platypus has electrosensory as well as mechanosensory (touch-sensing) ability. Vampire bats are capable of perceiving infrared radiation (heat) through the pocketings in their noseleaves that can allow them to find a feeding spot on their prey that is well supplied with blood vessels. They also can detect differences and recognize individual humans by the humans' breathing sounds (Gröger & Wiegrebe 2006). Some bats can probably sense barometric pressure changes (Paige 1995). Many kinds of organisms, including at least a few mammals (humans, bats, mice, mole rats, cattle), can orient themselves using the earth's magnetic field (Johnsen & Lohmann 2005); in European noctule bats and American big brown bats, the bats' magnetosensory ability to orient to the landscape and navigate are probably calibrated by visual perception of sunset and the horizon (Holland et al. 2006; Wang et al. 2007). This sense of direction may be provided by a cell or magnetoreceptor, possibly located in the brain, consisting of a chain of magnetite particles. Narwhal males were recently discovered to have in their tusk a density of neurons whose functions are speculative but are actively being investigated (narwhal.org/). Other mammals evolved for an aquatic existence show an abundance of sensory specializations that are functionally adapted to their water environment (Thewissen & Nummela 2008).

Biologists have made remarkable progress recently studying mammals' sense of time, investigating the internal clock that is important in terms of daily, lunar, and seasonal cycles and rhythms. Reproductive cycles, thermoregulatory cycles related to hibernation, migration, and other behavioral–physiological phenomena are under the control of a self-sustaining, autonomous molecular clock in the hypothalamus of the brain (Reppert & Weaver 2002; Sassone-Corsi 1998; Shearman et al. 2000). The clock is capable of being adjusted partly by light perceived by the retina of the eye or by hormonal factors that are triggered or affected by chemical stimuli in the diet or environment (Harrisingh & Nitabach 2008; Hatori et al. 2008; Saper et al. 2005; Tosini & Menaker 1996; Zhang et al. 2006). In mammals, inhabiting arctic regions, where the entire winter day may be dark or the entire summer night may be light, the mammals lose their circadian rhythms (van Oort et al. 2005).

Ultraviolet vision is probably a primitive phylogenetic character among vertebrates that has been lost in many mammals, but ultraviolet photoreceptor cells occur in the retinas of some nocturnal rodents, marsupials, and bats and may be more widespread among mammals (Muller et al. 2009; Peichl 2005; Zhao et al. 2009). Ultraviolet vision is likely to be useful for these species, but we barely fathom its role in their biology at present. A Central and South American nectar- and pollen-feeding bat *Glossophaga* is colorblind but sensitive to ultraviolet light, and some bat-pollinated plants in the tropics bear violet flowers that reflect ultraviolet light (Winter et al. 2003). These bats may be better able to find such flowers during twilight when short wavelengths of light are more perceptible.

Of course, all of a mammal's sensory systems are highly integrated with one another and with the animal's endocrinology, physiology, ecology, and behavior and with the environment.

Fat and Energy Storage

Although fat (adipose tissue) is a feature that is by no means unique to mammals, it is particularly vital in these animals. Fat serves three major functions in mammals: (1) energy storage, (2) a source of heat and water, and (3) thermal insulation. The lives of many species of mammals are punctuated by times of crisis when food is in short supply or energy demands are unusually high. For example, those mammals that hibernate must store enough energy to sustain life through periods when no food is available. Some filter-feeding whales spend their winters in plankton-poor tropical waters where they do little or no feeding. Such mammals survive by metabolizing stored fat. During times when males are competing for mates or defending territories or when females are lactating, stored fat is often the key to survival, and those individuals with the greatest amounts of stored fat have the highest reproductive success. An example of the severity of this energy crisis in a female gray seal is given by Young (1976): during 15 days of lactation, the mother lost 45 kilograms of body weight while her young gained 27 kilograms. Desert dwellers and mammals of temperate areas often have localized fat storage (e.g., in the tail or in the inguinal or abdominal region), whereas boreal and aquatic species typically store fat subcutaneously over much of the body. This subcutaneous layer is important as insulation as well as for food storage.

Circulatory System

In keeping with their active life and their endothermic ability, mammals have a highly efficient circulatory system. A complete separation of the systemic circulation and the pulmonary circulation has been achieved. The four-chambered heart functions as a double pump: The right side of the heart receives venous blood from the body and pumps it to the lungs at low pressure (the pulmonary circulation); the left side receives oxygenated blood from the lungs and pumps it to the body at high pressure (the systemic circulation). The fascinating evolution of the mammalian heart and circulatory pattern is described in detail by Hildebrand (1974).

As might be expected because of the great size difference between the smallest and the largest mammals (the mass of a 2-gram shrew—about the same as a U.S. dime—and that of a 160,000-kilogram whale differ by a factor of 80,000,000), the heart rate is highly variable from species to species. The rate in nonhibernating mammals varies from under 20 beats per minute in seals to over 1,300 in shrews (**Table 2-3**). Especially remarkable is the ability of some mammalian hearts to rapidly alter their rates of beat. As an extreme example, a resting big brown bat (Eptesicus) has a rate of about 400 beats per minute. This rate increases almost instantly to about 1,000 when the bat takes flight and generally returns to the resting rate within 1 second after flight stops (Studier & Howell 1969).

The erythrocytes (red blood cells) of mammals are biconcave disks rather than the ovoid spheres seen in other vertebrates. In all mammals except camels (Camelidae), the erythrocytes extrude their nuclei when they mature, apparently as a means of increasing oxygen-carrying capacity to support the high metabolic rate of mammals.

Respiratory System

In mammals, the lungs are large and, together with the heart, virtually fill the thoracic cavity. Air passes down the trachea, into the bronchi, and through a series of branches of diminishing size into the bronchioles, from which the alveolar ducts branch. Clustered around each alveolar duct is a series of tiny terminal chambers, the alveoli. Gas

TABLE 2-3 **Heart Rates of Selected Mammals**

Species	Common Name	Adult Body Mass	Mean Heart Rate and/or Range (beats/min)
Sorex cinereus	Cinereus shrew	3–4 g	782 (588–1320)
Tamias minimus	Least chipmunk	40 g	684 (660–702)
Sciurus carolinensis	Eastern gray squirrel	500–600 g	390
Neovison vison	American mink	0.7–1.4 kg	272–414
Erinaceus europaeus	European hedgehog	500–900 g	246 (234–264)
Phocoena phocoena	Harbor porpoise	170 kg	40–110
Ovis aries	Domestic sheep	50 kg	70–80
Sus scrofa	Domestic pig	100 kg	60–80
Equus caballus	Domestic horse	380–450 kg	34–55
Elephas maximus	Asiatic elephant	2000–3000 kg	25–50
Phoca vitulina	Harbor seal	20–25 kg	18–25

Data are from Altman and Dittmer (1964: 235); names updated.

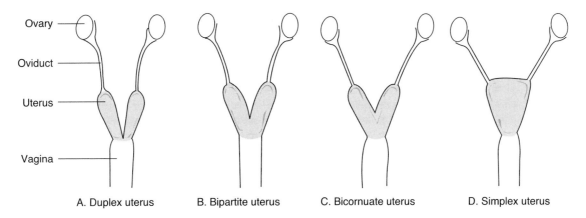

FIGURE 2-7 Several types of uteri (shaded) found in eutherian mammals, showing degrees of fusion of the two "horns" of the uterus. (A) Duplex uterus occurs in the Lagomorpha, Rodentia, Tubulidentata, and Hyracoidea. (B) Bipartite uterus is known in the Cetacea. (C) Bicornuate uterus is found in the Soricomorpha, in some members of the Chiroptera and Primates, and in the Pholidota, Carnivora, Proboscidea, Sirenia, Perissodactyla, and Artiodactyla. (D) Simplex uterus is typical of some members of the Chiroptera, Primates, Cingulata, and Pilosa. (Adapted from Smith, H. M. *Evolution of Chordate Structure*. Holt, Rinehart, and Winston, 1960.)

exchange between inhaled air and the bloodstream occurs in the alveoli; the thin alveolar membranes are surrounded by dense capillary beds. In humans, the lungs contain about 300 million alveoli, which provide a total respiratory surface of about 70 square meters—approximately 40 times the surface area of the body.

Air is forced into the lungs by muscular action that increases the volume of the thoracic cavity and decreases the pressure within the cavity. Some volume increase is gained by the forward and outward movement of the ribs under the control of intercostal muscles, but of greater importance are retraction and depression of the muscular diaphragm (a structure unique to mammals). When relaxed, the diaphragm is bowed forward, but when contracted, its central part moves backward toward the coelomic cavity, thus increasing the volume of the thoracic cavity. Limb and body movements, especially during locomotion, can also strongly affect the movement of air within and between the lungs. Inspiration, expiration, and cross-lung transfer are integrated with the mammal's locomotory style, gait, and body build (e.g., in a quadrupedal dog or rabbit, bipedal human, and flying bat) as well as other physiological and behavioral functions (Bramble & Jenkins 1993; Bramble & Lieberman 2004; Lancaster et al. 1995; Simons 1996).

Reproductive System

In mammals, both ovaries are functional, and the ova are fertilized in the uterine tubes. The embryo develops in the uterus within a fluid-filled amniotic sac. Nourishment for the embryo comes from the maternal bloodstream by way of the placenta. (The female reproductive cycle and the establishment of the placenta are discussed in Chapter 20.) The structure of the uterus is variable (Renfree 1993; Tyndale-Biscoe & Renfree 1987) (**Fig. 2-7**).

The male copulatory organ, the penis, contains erectile tissue and is surrounded by a sheath of skin, the **prepuce**. In many species the penis contains a bone, the **os penis**, or **baculum**, which may differ markedly even between closely related species (**Fig. 2-8**) and therefore may be of considerable use in taxonomic studies. The tip of the penis has an extremely complicated form in some species (Fig. 2-8). The testes of mammals, instead of lying in the coelomic cavity as in other vertebrates, are typically contained in the **scrotum**, a sac-like structure that lies outside the body cavity but is an extension of the coelomic cavity. The testes either descend permanently from the coelomic cavity into the scrotum when the male reaches reproductive maturity or are withdrawn into the body cavity between breeding seasons and descend when the animal again becomes fertile. In most mammals, the maturation of sperm cannot proceed normally at the usual deep-body temperature, and the scrotum functions as a "cooler" for the testes and developing sperm.

Brain

Compared with the brains of other vertebrates, the mammalian brain is unusually large. This greater size is attributable to a tremendous increase in the size of the cerebral hemispheres. These structures were ultimately derived from a part of the brain important in lower vertebrates in receiving and relaying olfactory stimuli.

Most characteristic of the brain of higher mammals is the great development of the **neopallium**, a mantle of gray matter that first appeared as a small area in the front part of the cerebral hemispheres in some reptiles. In mammals, the neopallium has expanded over the surface of the deeper, primitive vertebrate brain. The surface area of the neopallium is vastly increased in many mammals by a complex

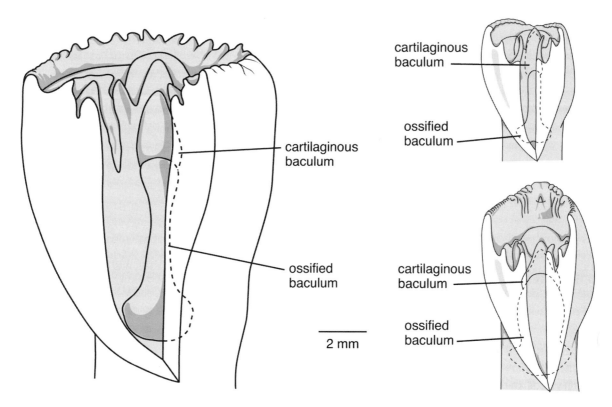

FIGURE 2-8 Ventral view of the penises of several species of New Guinean murid rodents, showing the complex structure of the organ in these mammals. Ossified and cartilaginous components of the baculum are embedded within the penis. (Adapted from Lidicker, W. Z., Jr., *J. Mammalogy* 49 (1968): 609–643.)

pattern of folding (**Fig. 2-9**; Table 2-2, number 12). A new development in eutherian mammals is the **corpus callosum**, a large concentration of nerve fibers that passes between the two halves of the neopallium and provides additional communication between them.

The unique behavior of mammals is largely a result of the development of the neopallium, which functions as a control center that has come to dominate the original brain centers. Sensory stimuli are relayed to the neopallium, where much motor activity originates. Present actions are

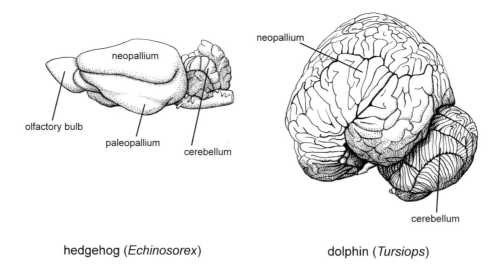

hedgehog (*Echinosorex*)

dolphin (*Tursiops*)

FIGURE 2-9 Left sides of the brains of a hedgehog and a dolphin. The neopallium is relatively small and smooth surfaced in the less derived hedgehog but greatly enlarged and highly convoluted in the dolphin. Anterior is to the left. ((Hedgehog) Adapted from Romer, A. S., and Parsons, S. T. *The Vertebrate Body.* Saunders, 1977; (dolphin) adapted by Norris, K. S. *Whales, Dolphins and Porpoises.* University of California Press, 1966.)

influenced by past experience; learning and "intelligence" are important. The size of the brain relative to total body size is not always a reliable guide to intelligence; brain size apparently need not increase in proportion to increases in body size to maintain intelligence. The degree of development of convolutions on the surface of the neopallium is perhaps a better indication of intelligence. In some groups of mammals (e.g., primates), the increased energy derived from endothermy is consumed in large part by the enlarged brain (Armstrong 1983). In mammals that live in a more complex, three-dimensional world than most terrestrial animals, such as flying bats and swimming-diving cetaceans, relevant parts of the brain and nervous system are modified and optimized accordingly.

■ Sense Organs

The sense of smell is acute in many mammals, probably resulting in part from their nocturnal ancestry. The importance of the sense of smell is reflected in the recent finding—which earned the 2004 Nobel Prize for the discoverers Linda Buck and Richard Axel—that the gene family of olfactory receptors is the largest gene family in the mammalian genome. This gene family is of great interest to researchers in genomics, molecular biology, physiology, and psychology (Gilbert 2008; Zhang & Firestein 2002). The smell sensors are primarily distributed across the mucosal surfaces of the mesethmoid and vomeronasal organ areas (**Fig. 2-10**). The vomeronasal organ or Jacobson's organ in mammals detects pheromones and thus is essential for a variety of physiological and behavioral changes. The organ is involved in the onset of puberty, the female estrous cycle, choosing a mate, defending territory, and other behaviors.

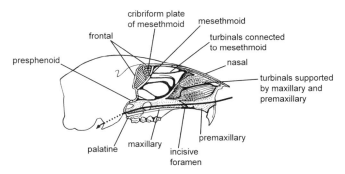

FIGURE 2-10 Cutaway view of the nasal chamber of an Abert's squirrel, showing the complicated arrangement of turbinal bones. The entire right half of the rostral part of the skull is removed, exposing the left side of the nasal chamber. The arrow shows the main air path from the external to the internal nares, but some air circulates through the upper part of the chamber and over the turbinal bones. Branches of the olfactory nerve pass out of the braincase through the cribriform plate of the mesethmoid bone. The incisive foramen transmits the duct to the vomeronasal organ.

It probably allows a mammal to determine the gender, individual identity, kinship, or social dominance of another individual or group of individuals of its species (Bouchard 2001; Holy et al. 2000). The olfactory bulbs and olfactory lobes form a great part of the brain in some moles and shrews and are reasonably large in carnivorans and rodents. The sense of smell is poorly developed, and the olfactory part of the brain is strongly reduced in whales and the higher primates (which are diurnal); the olfactory system is absent in porpoises and dolphins (Kruger 1966).

The sense of hearing is highly developed in mammals; in some 20% of mammalian species, hearing provides an important substitute for vision (see Chapter 22). The acuity of hearing, like olfaction, is probably related to mammals' primitively nocturnal habits. Sound is important for mammals in communication as well as awareness of and orientation to the environment, locating food, and avoiding enemies. Some species such as giraffes and elephants use sounds at frequencies below the range of human hearing ("infrasound"), whereas others such as dolphins and bats use frequencies above our hearing ("ultrasound"), as well as sounds within our hearing range. Mammals alone have an external structure (the **pinna**) to intercept sound waves; the pinnae may be extremely large and elaborate in some mammals, particularly in bats (see Figs. 15-19c and 15-23). Pinnae are missing (presumably secondarily lost) in some insectivorous mammals, phocid seals, and cetaceans. The **external auditory meatus**, the tube leading from the pinna to the tympanic membrane, is typically long in mammals and extremely long in cetaceans. The middle ear is an air-filled chamber that houses the three **ossicles** and is typically enclosed by a bony **bulla** (**Fig. 2-11**). The **cochlea** contains a complex internal array of tiny hairs of different sizes that amplify and are sensitive to different sound frequencies. In some groups such as echolocating bats and dolphins that use sound heavily for orientation and finding food, the ears, cochlea, and parts of the brainstem associated with hearing are highly specialized (Covey 2005).

The mammalian eye resembles that of most amniote vertebrates. In most nocturnal mammals, the **tapetum lucidum** is well-developed. This is a reflective structure within the choroid that improves night vision by reflecting light back to the retina. (This reflection accounts for the shine when headlight beams pick up an animal's eyes at night.) All mammals' retinas have photoreceptor cells called rods and color receptors called cones. The rods enable vision in low light conditions for nocturnal or marine species, and the cones enable color vision in brighter conditions; however, mammals differ greatly in the proportions of various photoreceptors and thus in visual abilities and seem well suited to the light environment they experience. For example, diurnal primates and treeshrews have more

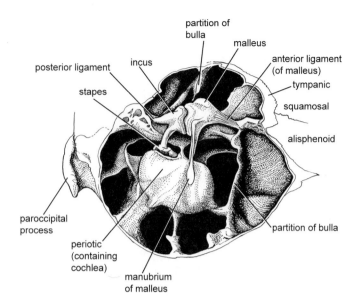

Abert's squirrel
(*Sciurus aberti*)

FIGURE 2-11 Lateral view of the right middle ear chamber (anterior is to the right) of Abert's squirrel, with the auditory bulla largely removed. The complex partitioning of the air-filled bulla, the position of the ossicles of the middle ear, bones of the fluid-filled inner ear, and the ligamentous suspension of the malleus and incus are shown. In life, the manubrium of the malleus is embedded within the tympanic membrane. Much of the stirrup-shaped stapes is hidden behind the periotic in this view.

types of color-sensitive cones than nocturnal carnivores or aquatic whales and seals, whereas semiaquatic mammals such as pygmy hippos and river otters are intermediate (Caro 2005; Peichl 2005).

Although in most mammals the eyes are relatively large, in some insectivorans, cetaceans, and rodents, they are greatly reduced in size and function. In such species, the eyes are able to differentiate only between light and dark and may serve primarily to aid the animal in maintaining the appropriate activity or thermoregulatory cycles (Cooper et al. 1993; Herald et al. 1969; Lund & Lund 1965).

Most mammals have **vibrissae**. These are the whiskers on the muzzle and the long, stiff hairs that are present on other parts of the head and the lower legs of some mammals. The vibrissae are tactile organs, and those on the face enable nocturnal species to detect obstacles near the face. The vibrissae on the muzzle generally arise from a structure termed the **mystacial pad**. They are controlled by a complex of muscles (**Fig. 2-12**), the superficial facial muscles, innervated by cranial nerve VII, the facial nerve. Tactile stimulation of the vibrissae is sensed by barrel structures in the somatosensory cortex of the brain. Bats have tactile-

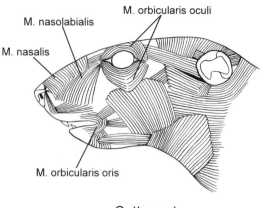

Cotton rat
(*Sigmodon hispidus*)

FIGURE 2-12 The superficial facial muscles of the cotton rat; these muscles almost wholly control facial expression and movement of vibrissae. (Adapted from Rinker, G. C., *Misc. Pubs. Mus. Zool.* 83 (1954): 1–124.)

sensitive hairs on the wing membranes that act somewhat like vibrissae to help sense turbulence and airflow over the wings (Zook 2005).

Digestive System

As in other vertebrates, salivary glands are present in mammals; in some ant-eating species, they are extremely large and specialized for the production of a mucilaginous material that makes the tongue sticky. The esophagus is a simple tube, and the stomach is a single sac-like compartment in most species; however, these structures are complexly

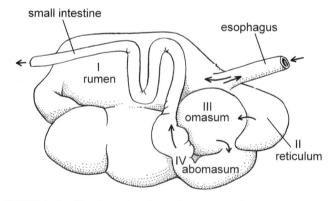

FIGURE 2-13 The four-chambered stomach of a ruminant artiodactyl. As the animal feeds, it swallows vegetation, which is then stored in the large rumen, where digestion is aided by a diverse microbiota. Larger particles of food pass to the reticulum, where the "cud" is formed. While the animal rests, it regurgitates and remasticates the "cud". The reprocessed food is swallowed again and goes through the reticulum to the omasum and finally to the abomasum, where enzymatic digestion occurs. (Adapted from Storer, T. L., and Usinger, R. L. *General Zoology*. McGraw-Hill, 1965.)

elaborated and subdivided in ruminant artiodactyls, cetaceans, sirenians, and some pilosans (**Fig. 2-13**). In herbivorous species, digestion is frequently accomplished partly by microorganisms that inhabit the stomach or the **caecum**, a blind sac that opens into the posterior end of the small intestine (see Fig. 6-29). These microbes digest cellulose from plant cell walls, which no mammals are able to do, and they require an elaborate chamber and internal environment in which to accomplish this. Microbe-hosting mammals (such as rodents and rabbits) derive nutrition from the microbial byproducts, but in these mammals, compared to other small mammals, symbiosis requires a far longer preweaning period before the young are able to handle solid food (Langer 2002).

Muscular System

The mammalian limb and trunk musculature has been highly adaptable. Different evolutionary lines have developed muscular patterns precisely adapted to diverse modes of locomotion. Cetaceans are the fastest marine animals. Certain carnivorans and ungulates are the most rapid runners, and bats as fliers are more maneuverable than most birds. Some muscular specializations favoring specific types of locomotion are described in the ordinal chapters. Intrinsic musculature of the trunk of the body controls posture and contributes to all styles of locomotion. Especially notable in mammals is the great development of dermal musculature, the superficial musculature of the skin. In many mammals, these muscles form a sheath over most of the body and allow the skin to move independently from underlying tissues. Dermal muscles have differentiated and have moved over much of the head (Fig. 2-12), where they control many essential actions. In mammals, there are no more vital voluntary muscles than those that encircle the mouth; these function during suckling and are among the first voluntary muscles to be subjected to heavy use after birth. Facial muscles move the ears, close the eyes, and control the subtle changes in expression that are so important in the social lives of many mammals.

The Skeleton

General Features

The mammalian skeleton differs from that of other amniotes (e.g., in shelled-egg laying vertebrates other than monotremes) in several basic ways, all of which may well be related to the active style of life of mammals. The mammalian skeleton has become simplified and is more completely ossified (**Figs. 2-14** and **2-15**), features perhaps associated with the need for well-braced attachments for muscles. Considerable fusion of bones has also occurred, as, for example, in

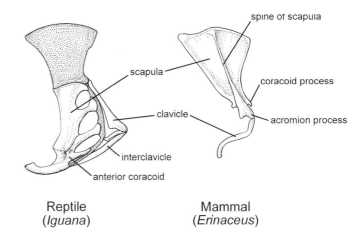

FIGURE 2-14 Lateral view of the right side of the pectoral girdle of a lizard and a hedgehog, showing the more complete ossification and reduction in number of discrete bones in the mammal. The heavily stippled areas are cartilaginous in adults.

the pelvic girdle. There is often great flexibility of the axial skeleton that allows the limbs greater speed and range of movement. The greater range of movement is of particular advantage to arboreal creatures, which many early mammals may have been. The simplification of the skeleton may have been advantageous in terms of metabolic economy—the less bone, the less energy invested in its development and maintenance. Furthermore, selection may have favored a light skeleton in the interest of quick movement with relatively little expenditure of energy.

To an animal as active as a mammal, well-defined articular surfaces on limb bones and solid points of attachment for muscles are highly advantageous during the period of skeleton growth as well as during adult life. Most mammals have evolved a pattern of bone growth very different from that of reptiles. In many reptiles, growth may continue throughout much of life. Growth in nonmammalian amniotes occurs at the ends of limb bones by ossification of the deep parts of a persistently growing cartilaginous cap; such a pattern limits the establishment of a clearly defined joint. In most mammals (except for some marsupials), however, skeletal growth is generally restricted to the early part of life. The articular surfaces and some points of attachment of large muscles become prominent and ossified early, while rapid growth is still under way. Growth continues at a cartilaginous zone where the end of the bone and its articular surface, the **epiphysis**, joins the shaft of the bone, the **diaphysis** (**Fig. 2-16**). When full growth is attained, this cartilaginous zone of growth (the **metaphysis**) becomes ossified, fusing the epiphysis and diaphysis. Because this fusion usually occurs at a certain age within a given species, the degree of closure of the "epiphyseal line" is useful in estimating the age of a mammal.

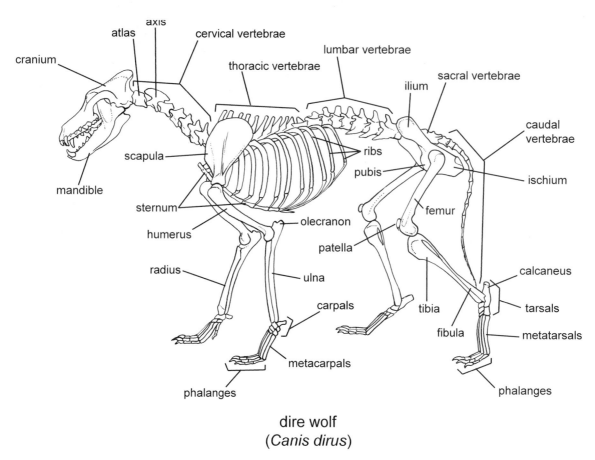

dire wolf
(*Canis dirus*)

FIGURE 2-15   Skeleton of a mammal, the dire wolf (*Canis dirus*), showing the major elements. (Adapted from Stock, C. *Rancho. La Brea: A record of Pleistocene life in California, Science Series, no. 13.* Los Angeles County Museum, 1949.)

Skull

Aspects of the development, structure, diversity, and function of the mammalian skull were summarized by various authors in the useful three-volume series *The Skull* (Hanken & Hall 1993a, 1993b, 1993c). Unlike that of reptiles, the skull of most mammals is akinetic—that is, there are no intrinsic joints within it other than the craniomandibular (jaw) joint and the joints between the middle ear bones. Much of the old, classical literature on mammals has been devoted to the mammalian skull and teeth and is difficult to access or unavailable online, but John Wible and his colleagues (Giannini et al. 2006; Wible 2003, 2008; Wible & Gaudin 2004) have recently produced a number of useful monographs describing in detail the skulls of selected mammals of several orders using a modern phylogenetic perspective. A great deal remains to be learned about mammalian anatomy using the comparative phylogenetic approach (Asher et al. 2008).

The braincase of the mammalian skull (**Fig. 2-17**) is large. In addition to its primary function of protecting the brain, the braincase provides a surface from which the temporal muscles originate. (In many mammals, these are the muscles

that generate the force to close the jaws.) A **sagittal crest** increases the area of origin for the temporal muscles in many mammals; the **lambdoidal crest** gives origin to the temporal muscles and insertion for some cervical muscles. The **zygomatic arch** is usually present as a structure that flares outward from the skull. It protects the eyes, provides origin for the masseteric jaw muscles, and forms the surface with which the condyle of the dentary (lower jaw) bone articulates. The zygomatic arch may be reduced or lost, as in some pilosans, soricomorphs, and cetaceans, or may be enlarged, as in those groups (such as rodents and lagomorphs) in which the masseter muscles largely supplant the temporal muscles as the major jaw muscles (see Fig. 13-3). The skull has a secondary palate (see Fig. 3-4), and there are usually **turbinal bones** within the nasal cavities (Fig. 2-10).

A number of **foramina** (openings) perforate the braincase and allow passage of the cranial nerves and blood vessels (Fig. 2-17). In some rodents, the infraorbital foramen, through which blood vessels and a branch of cranial nerve V (the trigeminal nerve) pass, is greatly enlarged in association with specializations of the masseter muscles (see Fig. 13-3). The incisive foramina, present in the palates of many mammals, form openings for the ducts to

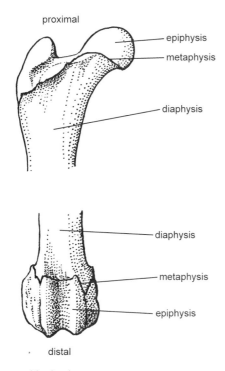

proximal
— epiphysis
— metaphysis
— diaphysis

— diaphysis
— metaphysis
— epiphysis
distal

Hedgehog
(*Erinaceus europaeus*)

FIGURE 2-16 Anterior views of the proximal and distal ends of the right femur of a young hedgehog (*Erinaceus europaeus*), showing the epiphyses, diaphysis, and intervening cartilaginous growth zone or metaphysis.

the vomeronasal organ (Figs. 2-10 and 2-17B). As noted previously, the vomeronasal organ allows a mammal to "smell" the contents of its mouth and the airborne and/or fluid-borne pheromones of potential mates. Vomeronasal organs are widespread among vertebrates; a snake puts the tips of its forked tongue against this part of the palate after "testing" its immediate environment.

Sounds that cause vibration of the tympanic membrane are mechanically transmitted by the three ear ossicles (Fig. 2-11) through the air-filled chamber of the middle ear to a membrane at the wall of the fluid-filled inner ear. The footplate of the stapes is embedded in a membrane that fills an opening into the inner ear and, acting like a piston, transforms the movements of the ossicles to vibrations of the fluid in the cochlea. The inner ear, with the cochlea and semicircular canals, is contained by the periotic (or petrosal) bone, which is generally covered by the squamosal bone but is exposed as the mastoid bone in some mammals (Fig. 2-17). The mammalian cochlea is more or less coiled in most marsupials and placentals; in monotremes, it is angled. The auditory bulla is formed by the expanded tympanic bone or by the tympanic bone plus the entotympanic bone, a bone found only in mammals. The bulbous

tympanic bullae are highly modified in some mammalian species in connection with specialized modes of life.

The lower jaw in Cenozoic mammals is formed exclusively by the **dentary** bone, which typically has a coronoid process on which the temporalis muscle inserts, a coronoid fossa, in which the masseter muscles insert, and an angular process, to which a jaw-opening muscle (the digastricus) attaches. Pterygoid muscles insert on the medial side of the dentary; these muscles originate on the pterygoid bones of the skull and are hypertrophied in mammals. The pterygoid musculature delicately controls lateral movements of the lower jaw for precise occlusion of the teeth. In some herbivores, in which the masseter muscle is enlarged at the expense of the temporalis muscle, the coronoid process is reduced or absent, and the posterior part of the dentary bone becomes dorsoventrally broadened (see Figure 13-4c).

Several skeletal elements in the throat region are highly modified homologs of the gill arches of fish. These elements, the **hyoid apparatus** and laryngeal cartilages, support the trachea, the larynx, and the base of the tongue and are often braced against the auditory bullae.

Teeth

Without doubt, one of the major keys to the success of mammals has been the possession of teeth. Fish, amphibians, crocodilians, reptiles, fossil birds, and mammals all have teeth, but the specialization of the dentition in mammals has exceeded anything found in the other groups. Mainly, it is mammals that have dentitions capable of coping with items so difficult to prepare for digestion as dry grass and large bones. So varied are the dental specializations of mammals and so closely related are they to specific styles of feeding and to patterns of adaptations of the skull, jaws, and jaw musculature, that to know in detail the dentition of a mammal is to understand many aspects of its way of life (Bergqvist 2003). Much of our knowledge of the early evolution of mammals is based on studies of fossil teeth, which because of their extreme hardness are often the only parts of early mammals that are preserved. The earliest known vertebrates, which were jawless, had bodies encased in bony plates. When the visceral arches anterior to those that supported the gill apparatus in primitive vertebrates became modified into jaws, teeth developed on the bony plates that bordered the mouth.

Although the teeth in various dentitions differ widely in number, structure, and function, in most mammals, the dentition is **heterodont**; that is, it consists of teeth that vary in both structure and function. In extant mammals, teeth occur on the premaxillary, maxillary, and dentary bones (Fig. 2-17). The anteriormost teeth, the **incisors** and **canines**, are used to gather or kill food, whereas the more specialized cheek teeth, the **premolars** and **molars**, are used

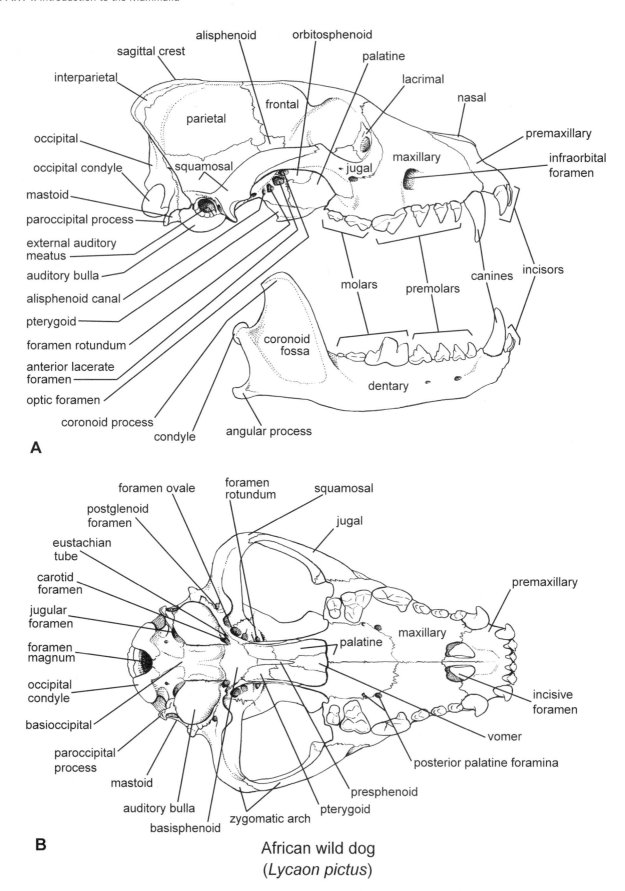

FIGURE 2-17 (A) Side and (B) palatal views of the skull of the African wild dog (*Lycaon pictus*), showing the bones, foramina, and teeth.

to grind or slice food in preparation for digestion. In many mammals, the canines are used in stereotyped displays during social interactions, in addition to their essential food-procuring function. Characteristically, two sets of teeth appear in a mammal's lifetime. The **deciduous** dentition develops early and consists of incisors, canines, and premolars—but no molars. These "milk teeth" are lost and replaced by permanent teeth as the animal matures. The permanent dentition consists of a second set of incisors, canines, and premolars but also includes the molars, which normally have no deciduous counterparts but in some cases may actually represent unreplaced deciduous teeth. The deciduous dentition of some species bears little resemblance to the permanent dentition (**Fig. 2-18**).

The form, function, and origin of the cusp patterns of the cheek teeth, and especially of the molars, are of particular interest. As indicated by fossils, the molars of near-mammals and at least one lineage of early mammals underwent a transition from a primitive, cusps-in-line pattern (as in advanced cynodont synapsids, morganucodontids, and triconodonts; see Chapter 3) to a tritubercular pattern with three cusps in a triangle (as in kuehneotheriids and symmetrodonts). Then a heel called a talonid was added on the lower molars (as in primitive therians, metatherians, and

eutherians) (**Fig. 2-19**). The last type of tooth, in which the protocone of a three-cusped upper molar occludes with a well-developed basin in the talonid of a lower molar (**Figs.** 2-19F and **2-20**), is termed **tribosphenic**. This configuration gives the tooth a crushing function during occlusion, in addition to the shearing function already present. (The occlusal surfaces of teeth are those that contact their counterparts of the opposing jaw—the surfaces the dentist generally treats when putting in a filling.) The majority of living mammals (metatherians and eutherians) either possess tribosphenic molars or originated from groups that primitively possessed them. Indeed, tribosphenic molars probably contributed to the initial Mesozoic radiation and success of therian mammals and may even have evolved twice between different northern hemisphere and southern hemisphere lineages of Mesozoic mammals (but this is controversial) (Woodburne et al. 2003).

A stroke of luck for functional morphologists is that the American opossum (*Didelphis virginiana*) and some other living metatherians have molars that resemble those of certain of the mammals that coexisted with Cretaceous dinosaurs. Opossums are omnivorous, eating insects and other small animals as well as soft plant material; probably many Mesozoic mammals had similar diets. Careful studies of jaw action in the opossum, therefore, can probably indicate how the molars functioned in mammals more than 90 million years ago. The studies by Crompton (1971), Crompton and Jenkins (1968), and Crompton and Hiiemae (1969, 1970) provide much of the basis for the following discussion of the functional morphology of tribosphenic molars.

In the opossum, the molars serve two masticatory functions. For up to 60% of the time involved in chewing and throughout the initial stages of chewing, the high cusps of the upper and lower cheek teeth crush and puncture the food without coming together. After the food is pulped, it is sliced by the six matching shearing surfaces shown in Fig. 2-20. This shearing is facilitated by the way in which food is trapped and steadied by the opposing molars (**Fig. 2-21**). Chewing occurs on but one side of the jaw at a time. During cutting strokes, the jaw action is not one of simple up-and-down movements. Instead, precise lateral adjustments of the jaw during mastication enable opposing molars to slide against each other. As shown in Fig. 2-21, this movement involves a transverse as well as an upward component as the lower molars shear against the uppers. Attrition facets on the molars of Mesozoic mammals indicate that occlusion of tribosphenic teeth during chewing has always had this transverse component (Butler 1972). Major shearing surfaces are those designated in Fig. 2-20 as 1 and 2, but additional cutting occurs when the surfaces on the sides

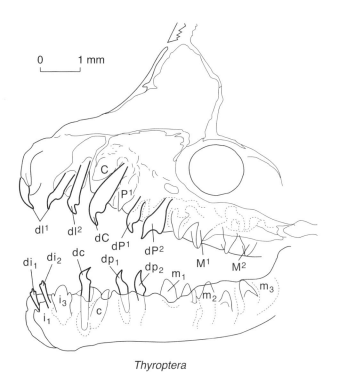

Thyroptera

FIGURE 2-18 Facial portion of the skull of a newborn bat (*Thyroptera tricolor*), showing the deciduous dentition (bold outlines) and developing permanent dentition. Upper teeth are indicated by capital letters and lower teeth by lowercase letters (see text). Deciduous teeth are indicated by d.

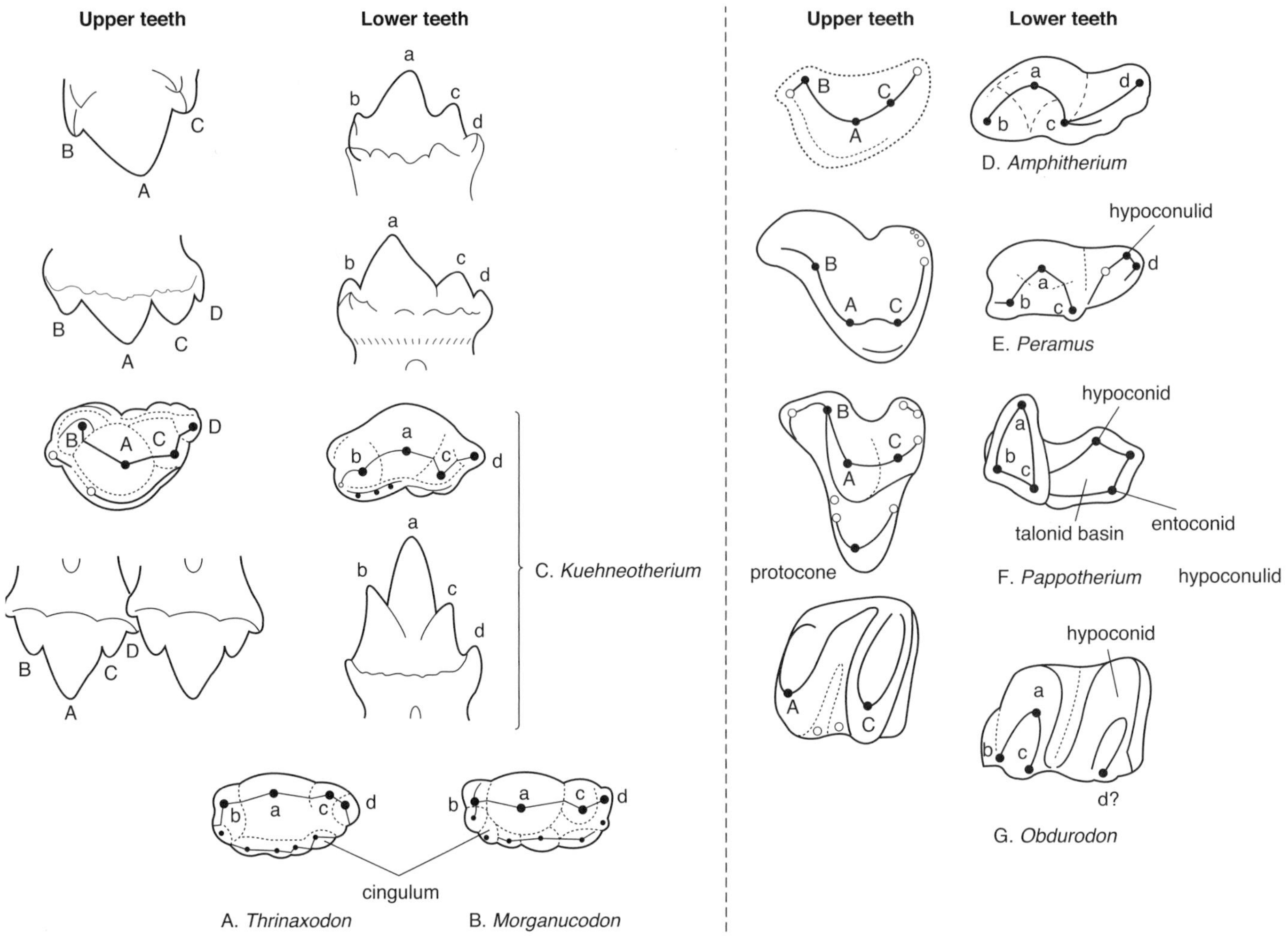

FIGURE 2-19   Cheek teeth of an advanced cynodont synapsid, a basal mammal, and several other primitive mammals, showing important stages in the evolution of tribosphenic molars. In each set of drawings, the left side shows the right upper (maxillary) teeth. Shown to the right side of the upper teeth are left lower (dentary) teeth. (A) *Thrinaxodon* (a cynodont synapsid); (B) *Morganucodon* (a basal mammal); (C) *Kuehneotherium*; (D) *Amphitherium*; (E) *Peramus* (three Mesozoic mammals); (F) *Pappotherium* (mammal with tribosphenic molars); (G) *Obdurodon*, a monotreme (teeth not truly tribosphenic). Cusps of lower teeth are drawn directly below the upper teeth, in the position in which they occlude with the upper teeth. Presumed homologous cusps are indicated by capital letters in upper teeth and lowercase letters in lower teeth: A = paracone; B = stylocone; C = metacone; D = metastyle; a = protoconid; b = paraconid; c = metaconid; d = hypoconulid. Lateral views of the teeth are given in (A), (B), and (C) to show the positions of these teeth as they would be just before they occlude and shear past one another. In addition, occlusal views are shown in all cases, and in (C) through (G), the uppers are shown slightly staggered relative to the lowers, as they would be in their natural positions. In *Amphitherium*, the upper teeth are unknown; this reconstruction is hypothetical. (Adapted from Spencer, R. S. *Major Features in Vertebrate Evolution: Short Courses in Paleontology.* No. 7. Paleontological Society, University of Tennessee, 1994.)

of the cusps of the quadrate posterior part (the talonid) of the lower molars shear against their counterparts on the upper molars. As a result of this complex pattern of occlusion, each time the three (in eutherians) or four (in metatherians) pairs of opposing molars of one side of the jaw come together, 18 or 24 cuts are made in the food, and thus, food already pulped is rapidly sectioned. Natural selection has favored the evolution of efficient dentitions because time spent in masticating food means greater energy

expenditure and less time available for food gathering. In some species, this also means a greater period of vulnerability to predators.

Two major evolutionary trends in tribosphenic molar structure appeared in the late Mesozoic era and became pronounced in the Cenozoic era. In carnivores, portions of some of the cheek teeth became blade like and the vertical shearing function was elaborated. In the interest of powerful sectioning of flesh, transverse jaw action was reduced

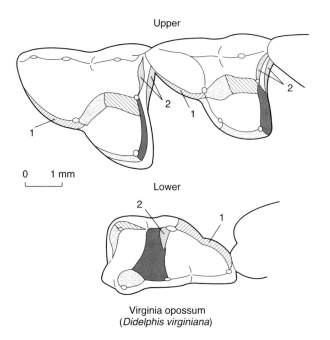

Upper

0   1 mm

Lower

Virginia opossum
(*Didelphis virginiana*)

FIGURE 2-20 Matching shearing planes (numbered patterns) of the occlusal surfaces of the right upper and left lower molars of the Virginia opossum. Anterior is to the right. (Adapted from Crompton, A. W., and Hiiemae, K., *Discovery* 5 (1969): 23.)

material in preparation for digestion, the molars became quadrate; transverse or horizontal jaw action came to be of primary importance, and distinctive features of the skull, jaws, and jaw musculature favoring this action developed. The quadrate configuration of upper molars of mammals arose through the addition of a fourth major cusp, the hypocone, to the posterior lingual corner of the tooth. This kind of configuration is clearly an important innovation related to herbivory in therian mammals; a **hypocone** has evolved convergently more than 20 times among mammals during the Cenozoic era (Hunter & Jernvall 1995). The dentitions of a modern carnivore, a modern herbivore, and a generalized tribosphenic Cretaceous mammal are compared in **Fig. 2-22**. In the carnivore, grasping, slicing, and sometimes bone-cracking functions of the more anterior teeth (premolars and canines) are emphasized, whereas in herbivores, the grinding function of the back teeth is emphasized.

The number of teeth of each type in the dentition is designated by the **dental formula**. This is written as the number of teeth of each kind on one side of the upper jaw over the corresponding number in the lower jaw. Such a formula in carnivorans may be as follows: incisors 3/3, canines 1/1, premolars 4/4, molars 2/3. Because the teeth are always listed in this order, the formula may be shortened to 3/3, 1/1, 4/4, 2/3. (The skull in Fig. 2-17 has this dental formula.) The dental formula lists the teeth of only one side; therefore, the number of teeth in the formula must be doubled to give the total number of teeth in the dentition. As an additional example, the arrangement for humans is

in these animals, and there were a variety of associated changes in the skull, jaws, and jaw musculature. Carnivory and attendant functional changes in the teeth arose convergently in at least seven groups of flesh-eating mammals with tribosphenic teeth (Muizon & Lange-Badre 1997). In herbivores, however, which must finely macerate plant

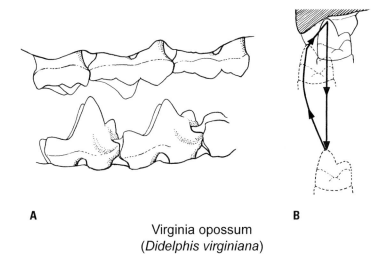

A

B

Virginia opossum
(*Didelphis virginiana*)

FIGURE 2-21 (A) Upper and lower molars of the Virginia opossum, showing the opposing cusps that steady, puncture, and crush food. View from the labial side of the right side of the animal, with anterior to the right. (B) Movement of the lower teeth as they shear against the uppers. These teeth are viewed from the rear on the left side of the animal. See text for description of action. ((B) Adapted from Crompton, A. W., and Hiiemae, K., *Discovery* 5 (1969): 23.)

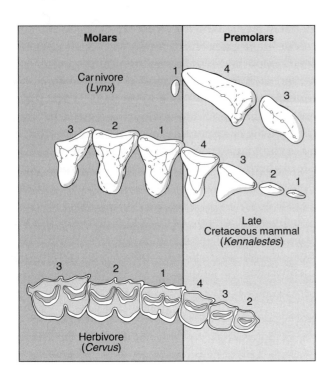

Molars	Premolars

Carnivore
(*Lynx*)

Late
Cretaceous mammal
(*Kennalestes*)

Herbivore
(*Cervus*)

FIGURE 2-22 Comparisons of the occlusal surfaces of the right upper cheek teeth of a carnivore (*Lynx*, at top), a Mesozoic mammal (*Kennalestes*, from Late Cretaceous, at center), and an herbivore (*Cervus*, at bottom). Anterior is to the right. (Adapted from Crompton, A. W., and Hiiemae, K., *Discovery* 5 (1969): 23.)

$2/2, 1/1, 2/2, 3/3 \times 2 = 32$. The basic maximum number of teeth in Cenozoic eutherian mammals is 44 (3/3, 1/1, 4/4, 3/3; higher numbers are known in Cretaceous species) and in metatherians is 50 (5/4, 1/1, 3/3, 4/4). The number of adult teeth is frequently reduced, and a few eutherians lack teeth as adults. Some specialized eutherians, most notably odontocete cetaceans, have more than 44 teeth and have

homodont dentitions (those in which all teeth are structurally alike). Individual teeth are designated by the initial for the type of tooth (I/i for incisor, C/c for canine, P/p for premolar, and M/m for molar) followed by the number for the locus (or position) of the tooth relative to others of its kind, numbered from front to back. Capital letters customarily represent upper teeth and lowercase letters represent lower teeth. Deciduous teeth can be represented by the letter d in front of the tooth type. Thus, i2, dP3, and m4 can be used as shorthand ways of indicating the second lower incisor, the deciduous third upper premolar, and the fourth lower molar, respectively.

The mammalian tooth typically consists of an inner material, **dentine**, covered by a layer of **enamel** (**Fig. 2-23**). Dentine consists largely of hydroxyapatite [$3(Ca_3PO_4)_2$ $Ca(OH)_2$], has an organic fiber content of about 30%, and is harder than bone. Enamel also consists almost entirely of hydroxyapatite, which in the enamel of all living mammals except monotremes is arranged in a prismatic crystalline pattern. Enamel, the hardest mammalian (or vertebrate) tissue, is only 3% organic. The tooth is bound to the jaw by **cementum**, a relatively soft material that may also form part of the tooth crown. Most mammals have teeth that are **brachydont**, or short crowned, and enamel formation ceases when the tooth erupts through the gum. Many herbivores, because their teeth are subject to rapid wear from abrasion by silica crystals in grass and by soil particles that adhere to plants, have **hypsodont**, or high-crowned, teeth (Fig. 2-23). As a further adaptation to abrasive food, in some mammals, some teeth (and in some rodents and lagomorphs all teeth) grow continuously and are termed hypselodont or **ever-growing** teeth. The roots of mammalian teeth are often divided; primitively, the upper molars

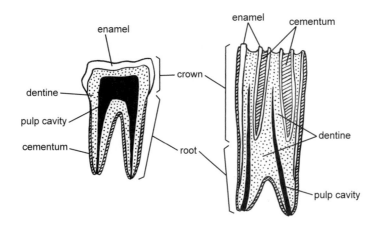

FIGURE 2-23 Generalized sections of mammalian teeth, showing the internal structure. The molar on the left is similar to that of primates and is low crowned; the molar on the right is similar to that of a modern horse and is high crowned.

have three roots and the lower molars have two. Incisors and canines are single rooted in all extant mammals except colugos and elephant shrews, and premolars may have one, two, or three roots. The dentitions of herbivores usually serve only two masticatory functions (but they may have many other functions in grooming, defense, and communication). The incisors, or the incisors and the canines, clip vegetation, and the cheek teeth grind the food. Between these teeth there is usually a space called a **diastema**. This is typical of rodents, for example.

The shape of the molar crown varies in response to the demands of different diets (**Fig. 2-24**). In pigs and in some rodents, carnivorans, and primates, the molars are **bunodont**, which means that the cusps form separate, rounded hillocks that crush and grind food. In herbivores the molars may be **lophodont**, with cusps forming comparatively straight ridges, or **selenodont**, with cusps forming crescents; in these cases the teeth finely section and grind vegetation. In the dental batteries of many insectivores, bats, and carnivores are **sectorial** teeth. These have blade-like cutting edges that section food by slicing against the edges of their counterparts in the opposing jaw. A particularly specialized, functional pair of sectorial teeth in carnivorans is called the **carnassials**; they consist of the last upper premolar (P4) and first lower molar (m1) (see Fig. 2-17). Whether for sectioning, grinding, or slicing, the function of all of these types of teeth is based on interdental shearing and differs only in the relative vectors of movement of the teeth or other details of mechanics.

The generalized, triangular, tribosphenic upper molar (**Fig. 2-25A**) is marked by three major cusps: the **protocone**, the **paracone**, and the **metacone**; the apex of the triangle (the protocone) points medially. The lower molar (Fig. 2-25B) has two main anatomical components: an anterior **trigonid** and a posterior **talonid**. The trigonid is triangular; the apex of the triangle (the **protoconid**) points laterally,

and the **paraconid** and **metaconid** form the medial edge. The talonid typically has three main cusps, from lateral to medial: the **hypoconid**, **hypoconulid**, and **entoconid**.

Axial Skeleton

Compared with that of quadrupedal reptiles, the mammalian vertebral column allows far greater freedom of head movement and powerful dorsoventral, rather than lateral, flexion of the spine. Most of the distinctive structural features of the mammalian vertebral column are related to these functional contrasts. The mammalian vertebral column has five well-differentiated sections: **cervical**, **thoracic**, **lumbar**, **sacral**, and **caudal**. There is often a sharp flexure in the vertebral column in the cervical vertebrae and another bend at the junction of the thoracic and lumbar regions, where a "switch" vertebra or diaphragmatic vertebra sometimes occurs. Only the thoracic vertebrae bear free ribs; elsewhere along the column, the ribs are fused to the vertebrae. The first two cervicals are highly modified; the sacral vertebrae are more or less fused to support the pelvic girdle, and differentiation of the vertebrae of each region is typical (**Fig. 2-26**; see Fig. 2-15). Usually 25 to 35 presacral vertebrae are present. All mammals, with the exception of several xenarthrans, the manatee (order Sirenia), some cetaceans, and possibly the giraffe (order Artiodactyla; Solounias 1997), have seven cervical vertebrae.

The large, ossified, and commonly segmented sternum solidly anchors the ventral ends of the ribs, helping to form a fairly rigid rib cage. The sternum is not highly variable, but in some bats departs strongly from the typical mammalian plan.

Limbs and Girdles

In most terrestrial mammals, the main propulsive movements of the limbs are fore and aft; the toes point forward, and the limb elements are usually greatly angled. In the

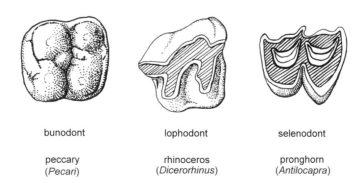

bunodont   lophodont   selenodont

peccary   rhinoceros   pronghorn
(*Pecari*)   (*Dicerorhinus*)   (*Antilocapra*)

FIGURE 2-24 Three major types of right molariform teeth as defined by cusp shapes. Anterior is to the right; the outer (labial) edge of each tooth is toward the top. The cross-hatched parts are dentine.

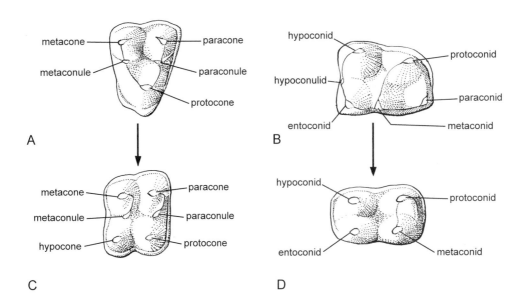

FIGURE 2-25 Basic cusp pattern of mammalian molars: (A) and (C) right upper molars; (B) and (D) left lower molars. Anterior is to the right; labial is at the top. (A) and (B) represent the generalized tribosphenic cusp pattern; this was modified in some evolutionary lines by the addition of a cusp (hypocone) in the upper tooth and the loss of a cusp (paraconid) in the lower tooth, yielding more or less quadrate teeth (C) and (D) adapted to omnivorous or herbivorous diets. (Modified from Romer, A.S. *Vertebrate Paleontology*. University of Chicago Press, 1966.)

most highly **cursorial** species (cursorial mammals are those adapted for running), the joints distal to the hip and shoulder tend to limit movement to a single plane. This allows reduction of whatever musculature does not control flexion and extension and results in lighter limbs. The mammalian pelvic girdle has a characteristic shape, with the **ilium** elongated and projecting forward and the **ischium** and **pubis** extending backward (see Fig. 2-15); these bones are solidly fused in terrestrial mammals, both among themselves and to the sacral vertebrae. In the shoulder girdle, the **coracoid** and **acromion** are usually reduced to small processes on

the scapula and the reptilian interclavicle is gone (see Fig. 2-14); the clavicle is reduced or absent in some cursorial species.

In the **manus** (hand or forefoot) and **pes** (foot or hindfoot) of mammals, there is a standard pattern of bones (**Fig. 2-27**); however, many variations on this basic theme occur among mammals with specialized types of locomotion, such as flight (bats), swimming (cetaceans and pinnipeds), or rapid running (ungulates, rabbits, and some carnivorans). Some of these variations are described in the chapters on orders (Chapters 8–19). The primitive

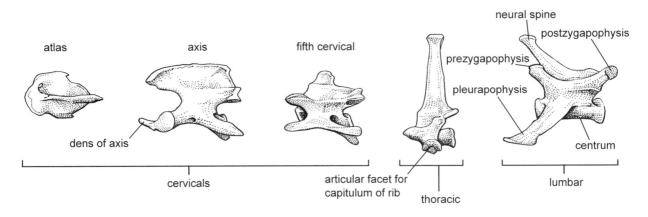

gray fox (*Urocyon cinereoargenteus*)

FIGURE 2-26 Vertebrae of the gray fox, showing the great structural variation in the parts of the vertebral column. The vertebrae are viewed from the left side; anterior is to the left. The atlas is the first cervical vertebra. The axis is the second. The fifth cervical vertebra is also shown. Sacral and caudal vertebrae are not shown.

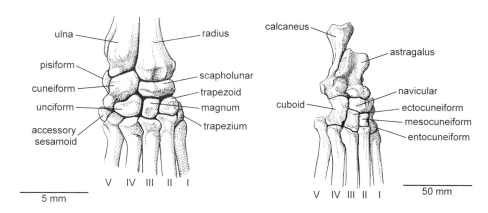

A Right carpus of *Erinaceus europaeus*   **B** right tarsus of *Gulo gulo*

FIGURE 2-27   Patterns of the podials (foot bones) of mammals as seen in anterior views: (A) right carpus (wrist) of a hedgehog; (B) right tarsus (ankle) of a wolverine. The centrale, a carpal element that in some mammals with primitive limbs lies proximal to the trapezoid and magnum, is missing in the hedgehog. The scapholunar in the hedgehog consists of the fused scaphoid and lunar bones. Metacarpals and metatarsals are numbered with Roman numerals from medial to lateral.

mammalian number of digits (five) and the basic phalangeal formula of two phalanges in the thumb (**pollex**) and first digit of the hindlimb (**hallux**) and three phalanges in each of the remaining four digits (2-3-3-3-3) are retained by many mammals. Common specializations involve loss of digits, reduction in the numbers of phalanges, or occasionally, addition of phalanges (**hyperphalangy**), as in the manus of whales and porpoises (see Fig. 19-6).

SUMMARY

Mammals owe their spectacular success to those diagnostic characteristics that enhance their intelligence and sensory ability, promote endothermy, or increase reproductive efficiency or secure and process food. When mammals first appeared in the Triassic period, they retained much of the basic therapsid plan, but they had crossed the nonmammalian–mammalian boundary. Endothermy probably began to develop in the therapsid ancestors of mammals. Mammals have a threefold to sixfold greater capacity for energy production than do reptiles and a standard metabolic rate some 8 to 10 times higher. These improvements in the mammalian capacity for high-energy production were probably part of a suite of anatomical and physiological features that early allowed them to be nocturnal and to expand into niches unoccupied by nonmammals.

Mammalian skin contains several kinds of glands not found in other vertebrates; the most important of these are the mammary glands, which in females provide nourishment for the young during their postnatal period of rapid growth. This extended lactation and suckling period encourages a close social bond between mother and young. For many mammals, especially those with complex foraging and social behavior patterns, this is an extremely important time of rapid learning by the young that prepares them for adult life independent from their mothers. Other types of skin glands are also important in mammals. The sweat glands function primarily in evaporative cooling but also eliminate waste materials. Diverse scent glands and musk glands are used for attracting mates, marking territories, communication during social interactions, or protection.

The bodies of mammals are typically covered with hair, perhaps developed by therapsids before a scaly covering was lost. The pelage functions primarily as insulation. The dissipation of heat from the skin surface to the environment and the absorption of heat from the environment are retarded by the pelage. The color patterns of mammals serve as camouflage and in social communication.

In keeping with their active life and their endothermic ability, mammals have highly efficient circulatory and respiratory systems. As might be expected because of the great size difference between the smallest and the largest mammals, the heart rate is highly variable from species to species. The rate in nonhibernating mammals varies from under 20 beats per minute in seals to over 1,300 beats per minute in shrews. Mammalian lungs are large, and gas exchange between inhaled air and the bloodstream occurs in tiny alveoli surrounded by dense capillaries. Air is brought into the lungs by muscular action under the control of intercostal muscles and the unique muscular diaphragm. Limb and body movements, especially during locomotion, can also strongly affect the movement of air within and between the lungs.

The mammalian brain is unusually large, primarily because of the tremendous increase in the size of the cerebral hemispheres. Most characteristic of the brain of higher mammals is the great development of the neopallium, which has expanded over the surface of the deeper vertebrate brain and is complexly folded. A new development in eutherian mammals is the corpus callosum, a large concentration of nerve fibers that passes between the two halves of the neopallium, providing additional communication between them. The unique behavior of mammals is largely a result of the development of the neopallium, which functions as a control center. Sensory stimuli are relayed to the neopallium, where much motor activity originates. Present actions are influenced by past experience; adaptive learning is important in mammals.

The sense of smell is acute in many mammals, probably resulting in part from nocturnal beginnings. In addition to the smell sensors distributed across the mucosal surfaces of the nose, the vomeronasal organ or Jacobson's organ in mammals detects pheromones and thus is essential for a variety of physiological and behavioral changes. The vomeronasal organ is involved in the onset of puberty, the female estrus cycle, choosing a mate, and defending territory, and it probably allows a mammal to determine the sex, individual identity, kinship, or social dominance of another individual of its species. The sense of hearing is also highly developed in mammals probably because of mammals' primitively nocturnal habits. Sound is important for communicating, locating food, and avoiding enemies. Some mammal species use infrasound (frequencies below the range of human hearing), whereas others use ultrasound (frequencies above our hearing).

The mammalian skeleton has become simplified and more completely ossified, than that of their ancestors allowing for the well-braced attachment for muscles. To active

mammals, epiphyses and well-defined articular surfaces on limb bones, along with solid points of attachment for muscles are highly advantageous during the period of skeleton growth as well as during adult life. Unlike that of other vertebrates, the skull of most mammals is akinetic. The mammalian braincase is large; it protects the brain and provides a surface from which the temporal muscles originate. The zygomatic arch is usually present, flaring outward from the skull. It protects the eyes, provides origin for the masseteric jaw muscles, and forms the surface with which the condyle of the dentary (lower jaw) bone articulates.

One of the major keys to the success of mammals has been the possession of heterodont teeth that are capable of coping with diverse food items such as dry grass and large bones. So varied are mammalian dentitions and so closely related are they to specific styles of feeding that to know in detail the dentition of a mammal is to understand many aspects of its way of life. Much of our knowledge of the early evolution of mammals is based on studies of fossil teeth, which because of their extreme hardness are sometimes the only parts of early mammals that are preserved. As indicated by fossils, the molars of near-mammals underwent a transition from a primitive, cusps-in-line pattern to a tritubercular pattern with three cusps in a triangle. Then a heel called a talonid was added on the lower molars. Finally, a tribosphenic molar was formed when a protocone of a three-cusped upper molar occluded with a well-developed talonid basin of a lower molar. This configuration gives the tooth a crushing function during occlusion, in addition to the shearing crests already present. The majority of living mammals (metatherians and eutherians) either possesses tribosphenic molars or originated from groups that primitively possessed them. In the Cenozoic era, portions of some of the cheek teeth of carnivorous mammals became blade-like, and the vertical shearing function was elaborated. However, in herbivores, which must finely macerate plant material in preparation for digestion, the molars became quadrate; transverse or horizontal jaw action came to be of primary importance, and distinctive features of the skull, jaws, and jaw musculature favoring this action developed.

KEY CHARACTERISTICS OF MAMMALS

- Ancestral accessory jaw bones shifted away from craniomandibular joint and transformed into middle and outer ear bones
- Jaw joint predominantly or solely between squamosal and dentary bones
- Stapes very small relative to skull size
- Atlas intercentrum and neural arches fused to form single, ring-shaped structure
- Epiphyses on the long bones and girdles

- Heart completely divided into four chambers with a thick, compact myocardium
- Heart with atrioventricular node ("pacemaker") and Purkinje fibers
- Single aortic trunk
- Pulmonary artery with three semilunar valves
- Erythrocytes lack nuclei at maturity
- Endothermy
- Central nervous system covered by three meninges

- Cerebellum folded
- Nerve filaments with three polypeptides
- Brain with divided optic lobes
- Facial nerve field strongly represented in brain motor cortex
- Superficial facial musculature expanded and associated with eye, ear, and snout
- Muscular diaphragm encloses pleural (lung) cavities and diaphragmatic breathing

- Complex lung divided into lobes, bronchioles, and alveoli
- Epiglottis
- Skin with erector muscles and dermal papillae
- Hair
- Sebaceous and sweat glands
- Mammary glands
- Loop of Henle in the kidney

KEY TERMS

Acromion
Baculum
Brachydont
Bulla
Bunodont
Caecum (or cecum)
Canine
Carnassial
Caudal
Cementum
Cervical
Cochlea
Coracoid
Corpus callosum
Cortex
Cursorial
Cuticular scale
Deciduous
Dental formula
Dentary
Dentine
Diaphysis
Diastema
Enamel
Entoconid
Epiphysis
Ever-growing
External auditory meatus

Foramina
Hallux
Heterodont
Homodont
Hyoid apparatus
Hyperphalangy
Hypocone
Hypoconid
Hypoconulid
Hypsodont
Ilium
Incisor
Ischium
Lambdoidal crest
Lophodont
Lumbar
Mammary gland
Manus
Medulla
Metacone
Metaconid
Metaphysis
Molar
Musk gland
Mystacial pad
Neopallium
Os penis
Ossicles

Paracone
Paraconid
Pelage
Pes
Pinna
Pollex
Premolar
Prepuce
Protocone
Protoconid
Pubis
Sacral
Sagittal crest
Scent gland
Scrotum
Sebaceous gland
Sectorial
Selenodont
Sweat gland
Talonid
Tapetum lucidum
Thoracic
Tribosphenic
Trigonid
Turbinal bones
Venter
Vibrissae
Zygomatic arch

RECOMMENDED READINGS

Ackerman, D. 1991. *A Natural History of the Senses.* Knopf Doubleday Publishing Group, NY.

Asher, RJ, JH Geisler, & MR Sanchez-Villagra. 2008. Morphology, paleontology, and placental mammal phylogeny. *Systematic Biology,* 57:311–317.

Hildebrand, MH & GE Goslow Jr. 2001. *Analysis of Vertebrate Structure,* 5th ed. John Wiley & Sons, NY.

McKenna, MC & SK Bell. 1997. *Classification of Mammals Above the Species Level.* Columbia University Press, NY.

Rowe, T. 1988. Definition, diagnosis and origin of Mammalia. *Journal of Vertebrate Paleontology,* 8:241–264.

Woodburne, MO, TH Rich, & MS Springer. 2003. The evolution of tribospheny and the antiquity of mammalian clades. *Molecular Phylogenetics and Evolution,* 28:360–385.

Mammalian Origins

Mammals arose from a lineage of vertebrates known as synapsids that had their origin in early **amniotes** over 300 million years ago. After the first vertebrates invaded land, the group known as amphibians remained (and remain) obliged to return to water to reproduce, but a sister group to the amphibians, the Amniota, developed adaptations that allowed reproduction on dry land, including the amniotic or cleidoic egg. This type of egg can be laid on dry land and is characterized by a semipermeable shell and several extra-embryonic membranes. It is capable of gas exchange with the environment, contains food in the form of yolk to nourish the embryo, and stores waste products produced by the embryo until hatching. The first amniotes appeared in the Mississippian period of the Paleozoic era (**Table 3-1**) and soon after their appearance underwent a fundamental split into two clades or phylogenetic lineages: the Sauropsida and the

TABLE 3-1 Geologic Time Scale Since Life Became Abundant

Era	Period	Estimated Time Since Beginning of Each Epoch or Period (Millions of Years)	Epoch	Typical Mammals and Mammalian Ancestors of the Period
		0.00117	Holocene	Modern species and subspecies; extirpation of some mammals by humans
	Quaternary	2.59	Pleistocene	Appearance of modern species or their antecedents; widespread extinction of large mammals
		5.33	Pliocene	Appearance of modern genera
	Neogene	23.03	Miocene	Appearance of modern subfamilies
		33.9	Oligocene	Appearance of modern families
		55.8	Eocene	Appearance of modern orders
Cenozoic	Paleogene	65.5	Paleocene	Adaptive radiation of metatherians and eutherians
	Cretaceous	145		Appearance of metatherians and eutherians
	Jurassic	200		Archaic mammals
Mesozoic	Triassic	251		Therapsids, appearance of mammals
	Permian	299		Appearance of therapsids, cynodonts
	Pennsylvanian	318		Appearance of synapsids
	Mississippian	359		Appearance of amniotes
	Devonian	416		
	Silurian	444		
	Ordovician	488		
Paleozoic	Cambrian	542		

Dates are from 2009 Geological Time Scale of the International Commission on Stratigraphy (stratigraphy.org/).

Synapsida (Benton 2004; **Fig. 3-1**). Whereas the sauropsid lineage eventually gave rise to turtles, lizards, snakes, dinosaurs, and birds, the synapsid lineage eventually led to mammals.

The synapsids dominated terrestrial faunas during the Permian and early Triassic periods. Many groups of early synapsids went extinct at the end of the Permian in one of the largest extinction events of all time, but a few lineages of more advanced synapsids (the Therapsida) survived the Permo-Triassic extinction. By the end of the Triassic, therapsids, too, dwindled in importance but not before giving rise to the cynodonts. From a cynodont ancestry, mammals arose in the late Triassic, slightly after the first appearance of dinosaurs.

Not long ago, the first two-thirds of mammalian history (during the 140 million years from the late Triassic to the late Cretaceous) were regarded as the most poorly known of mammalian history (Lillegraven et al. 1979).

Many exciting discoveries in the last 2 decades have uncovered a plethora of information about Mesozoic mammals, including hundreds of new taxa, their fundamental radiations, global occurrence, and paleobiology (Kemp 2005; Kielan-Jaworowska et al. 2004). Thus, a clearer view of Mesozoic mammals has emerged.

During the vast sweep of the Mesozoic, the dinosaurs dominated the terrestrial scene, and their radiation resulted in a diverse array of herbivorous and carnivorous types, some of which were highly specialized for bipedal locomotion. Dinosaurs were the largest land animals of all time: Most known species were very large, and some species reached weights of almost 100 tons (Gillette 1994). Only a few species were smaller than 10 kilograms as adults. By comparison, Mesozoic mammals were insignificant. Most were mouse-sized, and the occasional "giant," about the size of a domestic cat, was probably not as large as the

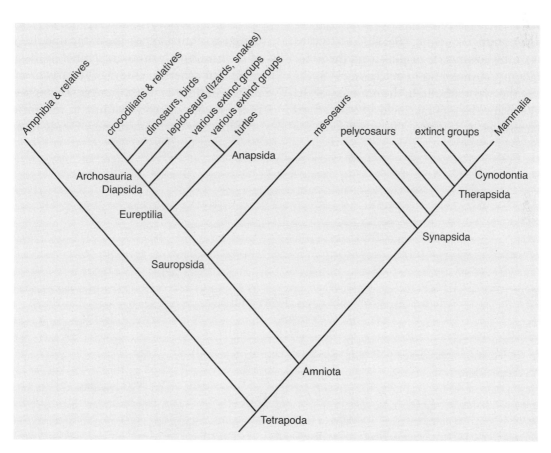

FIGURE 3-1 Phylogeny of Tetrapods, showing the two major branches of Amniotes, the Sauropsida (mesosaurs and reptiles) and the Synapsida (including mammals). More detail of the synapsid branch is shown in Figure 3-3. Among synapsids, the pelycosaurs are a paraphyletic group, artificial because the grouping includes only some of the more primitive descendants of a common ancestor and excludes the more derived descendants of that ancestor. Derived characters of the named clades (nodes or branching points) are given in the text. (Adapted from Gauthier, J, A.G. Kluge and T. Rowe., *Cladistics*, 4 (1988):105–209; adapted from Benton, M.J. *Vertebrate Palaeontology, Second Edition.* Chapman and Hall, 1997; and adapted from Carroll, R.L. *Vertebrate Palaeontology and Evolution.* W. H. Freeman & Co.,1988.)

occasional species of relatively tiny dinosaur. Mesozoic mammals probably hid by day and foraged at night; the fossil record of the postcranial skeleton suggests that they adhered conservatively to a mouse-like body form and to quadrupedal locomotion.

Morphological Features of Synapsids

Amniote groups are characterized in part by various patterns of perforation of the temporal part of the skull (**Fig. 3-2**). The openings, called **temporal fenestrae**, are thought by some researchers to have developed originally to increase the freedom for expansion of the **adductor** muscles of the jaw; these muscles primitively attached inside the solid temporal part of the skull. From the basic split of amniotes, the Sauropsida led to animals with anapsid, diapsid, and other patterns of temporal fenestration.

In the Synapsida, the skull is characterized by a single temporal opening low on the side of the skull, surrounded by the postorbital, squamosal, and jugal bones. A general trend in progressive synapsids was clearly toward the enlargement of the temporal opening (**Fig. 3-3**) and toward the movement of the jaw muscle origins from the inner surface of the temporal shield (as in pelycosaurs) to the braincase and to the zygomatic arch, the remnant of the lower part of the original temporal shield. Other derived features of Synapsida are a contact of the maxilla with the quadratojugal bone, **caniniform** maxillary teeth, and narrow neural arches on the trunk vertebrae. Many Permian synapsids form a series of outgroups to therapsids that are collectively known as pelycosaurs (a **paraphyletic** group; Fig. 3-1). Therapsids themselves seem to be monophyletic (Benton 1997). One way of fitting the synapsids into a classification scheme is given in **Table 3-2**.

Some pelycosaurs, notably the caseids and edaphosaurids, had teeth that seem to have been designed for herbivory (e.g., in the caseid *Cotylorhynchus*; Fig. 3-3). Most pelycosaurs, however, possessed caniniform teeth in the maxillary bone (as in the sphenacodontid *Dimetrodon*; Fig. 3-3) that were clearly designed for predation; these carnivores probably fed largely on fish and amphibians at the water's edge. Many pelycosaurs were relatively large animals compared with contemporary amphibians and sauropsids. *Cotylorhynchus* and *Dimetrodon* each reached a length of approximately 3 meters. Pelycosaurs dominated Early Permian terrestrial faunas but went extinct at the end of the Permian.

Therapsids arose from a common ancestor with the pelycosaurs, first appearing in the middle Permian. Many of the primitive forms were extinct by the end of the Permian, but others, including the early branches of cynodonts, continued into the Jurassic. Therapsids display many morphological trends leading to the basic mammalian anatomical plan. Many of these trends are apparent in the

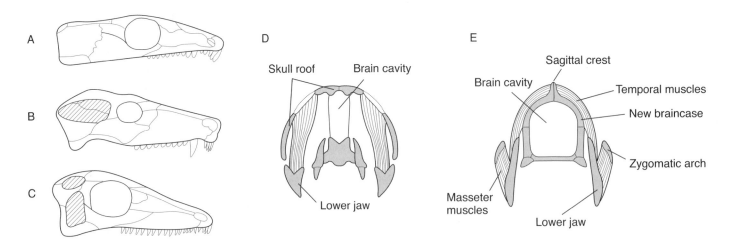

FIGURE 3-2 (A–C) Diagrammatic views of skulls of amniotes showing some of the arrangements of temporal openings. (A) Anapsid condition with no temporal opening. (B) Synapsid condition with postorbital and squamosal bones meeting above a single opening. (C) Diapsid condition with two temporal openings. (D and E) Cross-sections of synapsid skulls showing attachments of jaw muscles. (D) Pelycosaur, with the jaw muscles originating within the remaining parts of the temporal shield ("skull roof"); the sides of the braincase are cartilaginous. (E) Mammal, with the jaw muscles originating on the new and completely ossified braincase ("new braincase," formed partly by extensions of bones that originally formed the skull roof), on the sagittal crest, and on the zygomatic arch, also a remnant of the original skull roof.

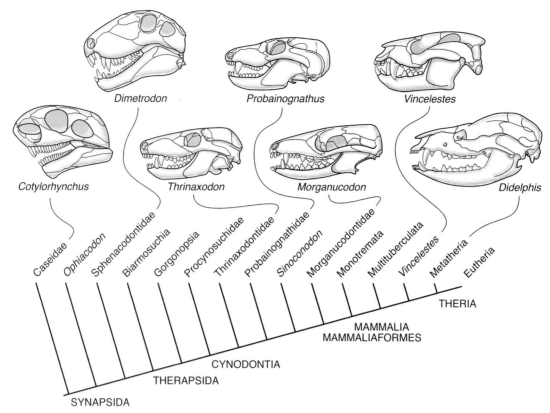

FIGURE 3-3 Phylogeny and representative skulls of synapsids. See text for definitions (synapomorphies) of named nodes. (Phylogeny adapted from Wible, J.R., et al., *American Museum Novitates* 3149 (1995) 1–19; adapted from Rowe, T., *J. of Vertebrate Palaeontology*, 8 (1998): 241–264; adapted from Rowe, T. *Mammal Phylogeny: Mesozoic Differentiation, Multituberculates, Monotremes, Early Therians, and Marsupials* (F.S. Szalay, M.J. Novacek and M.C. McKenna, eds.) Springer-Verlag, 1993 and adapted from Benton, M.J. *Vertebrate Palaeontology, Second edition.* Chapman and Hall, 1997; Skulls adapted from Hopson, J.A. *Major Features in Vertebrate Evolution. Short Courses in Palaeontology* (R.S. Spencer, ed.) Paleontological Society, University Tennessee, 1994.)

TABLE 3-2 A Classification of Synapsids with Emphasis on Nonmammalian Synapsids

Kingdom Animalia

  Phylum Chordata

    Subphylum Vertebrata (Craniata)

     Superclass Tetrapoda

      Series Amniota

       Class Synapsida

        Order Pelycosauria[†]

         Families Eothyrididae, Caseidae, Varanopseidae, Ophiacodontidae, Edaphosauridae, and Sphenacodontida

        Order Therapsida

         Suborder Biarmosuchia

         Suborder Dinocephalia

         Suborder Dicynodontia

         Suborder Gorgonopsia

         Suborder Cynodontia

          Families Procynosuchidae, Galesauridae, Cynognathidae, Diademodontidae, Chiniquodontidae, Tritylodontidae, and Trithelodontidae

      Unnamed rank Mammaliaformes

       Class Mammalia

Data from Benton (1997).
† Denotes a paraphyletic group.

synapomorphies of the successive nodes in Fig. 3-3, and it is instructive to review these characters here. Derived features of the Therapsida include the reduction or loss of the temporal shield as the temporal opening enlarged, revealing the braincase and producing a sagittal crest and zygomatic arches. In connection with these changes to the skull, many of the origins of the jaw muscles moved to the braincase, sagittal crest, and zygomatic arches (Fig. 3-2E). There are septomaxillary bones extensively exposed on the facial part of the skull. The upper canines are enlarged, indicating a capacity for predation. There was a partial abandonment of the primitive, sprawling limb posture by modification of the limb girdles, resulting in a posture in which the limb bones are held more underneath the body. In the pelvic girdle, the **acetabulum** is deep. The feet are shortened. In the skull, the external auditory meatus is formed within the squamosal bone. In the lower jaw, the jaw joint is in line with the occiput, and the anterior coronoid bone is absent. A vertebral notochordal canal is absent in the adult.

The Cynodontia includes an array of advanced, predaceous therapsids. In cynodonts, derived features include the presence of a masseteric fossa on the dentary, development of two occipital condyles (only one ventral, ball-like condyle is present in pelycosaurs), and zygomatic arches that flare laterally. At the back of the lower jaw, the reflected lamina of the angular bone is reduced in size (indicating the incipient emphasis on a hearing function of the postdentary bones; see **Fig. 3-8**). Teeth on the pterygoid bone are absent. The incisors are **spatulate**, and the postcanine teeth have anterior, posterior, and **lingual** accessory cusps (the last two characters result in a strongly heterodont dentition). The maxillary and palatine bones are expanded backward and toward the midline to form a partial, bony **secondary palate** (**Fig. 3-4**). Ribs on the lumbar vertebrae are reduced or fused to the vertebrae as pleurapophyses (**Fig. 3-5**). In the foot there is a distinct calcaneal heel.

Stem mammals first appear in the fossil record in the late Triassic. Many derived features, including numerous characters of the soft anatomy such as mammary glands with nipples, **viviparity** with loss of the eggshell, anal and urogenital openings separate in adults, digastricus muscle used to open the jaw, and other characteristics unite the mammals and near-mammals into a clade (see Table 2-2; McKenna & Bell 1997). In the fossil record, only characters of the bony tissues can be used to define mammals and their nearest sister groups. Mammaliaformes possess a well-developed jaw articulation between the dentary and squamosal bones, double-rooted **cheek teeth**, and expansion of the brain vault in the parietal region. Elsewhere in the skull the tabular bone is absent, and the occipital condyles are large and separated by a notch or groove. The medial wall of the orbit is enclosed by the orbitosphenoid and the

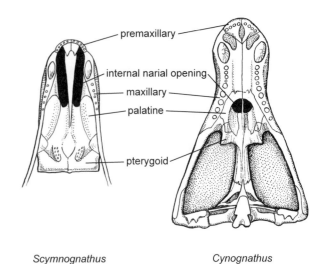

FIGURE 3-4   Palatal views of synapsid skulls. (A) *Scymnognathus* (primitive therapsid); note that the internal nares open into the anterior part of the mouth. (B) *Cynognathus* (more advanced therapsid); note that the maxillaries and palatines have extended medially, forming a shelf that shunts air from the external nares to near the back of the mouth. (Modified from Romer, A.S. *Vertebrate Paleontology*. University of Chicago Press, 1966.)

ascending process of the palatine bone. Mesozoic Mammalia share several derived features of the dentition: The cheek teeth are divided into molars and premolars, there is precise occlusion and a consistent relationship between the upper and lower molars, and the dentition is **diphyodont** (e.g., there are two generations of teeth, juvenile and adult). In addition, the mandibular symphysis is reduced. The clade Theria is diagnosed on the basis of tribosphenic molars, the presence of a supraspinous fossa on the scapula, and an inverted U-shaped area for articulation with the head of the femur is present on the acetabulum. The cochlea is spiraled. The prootic bone lacks an anterior lamina (a specialization in the part of the skull that houses the inner ear).

The characters mentioned previously and many other derived features diagnosing various synapsids (including early mammals) are taken from a flood of recent studies that have resulted from the exciting discoveries around the world of new and informative fossils, with each improving our understanding of the evolution of mammals and their ancestors (see Kemp 2005; Kielan-Jaworowska et al. 2004; Laurin & Reisz 2007).

Through the transition from nonmammalian to mammalian synapsids, many of the derived characters mentioned previously suggest the development of unobservable physiological features in Mesozoic synapsids. For example, the expansion of the dentary bone at the expense of the other jaw elements and the establishment of a dentary-squamosal jaw joint have important implications for the development of the uniquely mammalian middle ear with

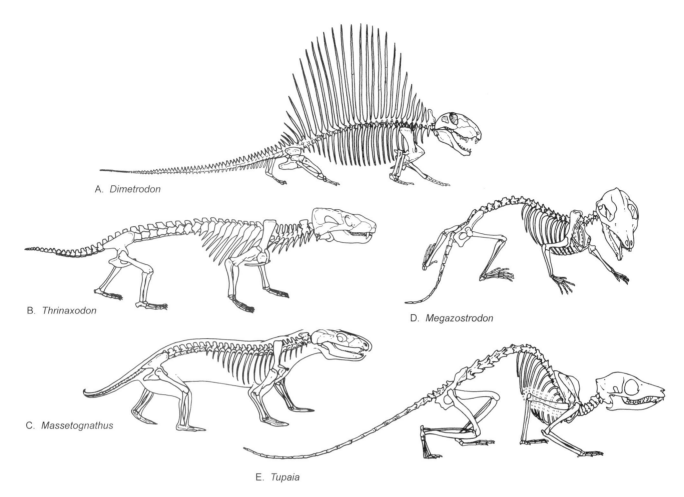

FIGURE 3-5   Reconstructed skeletons of primitive and derived synapsids, showing changes in the postcranial skeleton. (A) *Dimetrodon* (Sphenacodontidae), a pelycosaur that possessed a "sail" supported by elongated neural spines (not all pelycosaurs had sails) and a sprawled limb posture. (B) *Thrinaxodon* (Thrinaxodontidae) and (C) *Massetognathus* (Traversodontidae), cynodonts showing reduction of the lumbar ribs (due to fusion to the vertebrae) and less sprawling limb posture. (D) *Megazostrodon* (Morganucodontidae), in which the forelimbs are sprawled but the hindlimb posture indicates a mammalian stance (in which the femur swings only fore and aft, not out to the side) and the pelvis has a long, rod-like, anterodorsally oriented ilium and a large obturator foramen. (E) *Tupaia* (Tupaiidae), a modern tree shrew, showing considerable flexure in the axial skeleton (cervico-thoracic and thoraco-lumbar regions) and limbs. ((a) Modified from Romer, A.S. *Vertebrate Paleontology.* University of Chicago Press, 1966; (b) adapted from Jenkins, FA Jr., *Journal of Zoology, 165* (1971): 303–315; (c) adapted from Jenkins, FA Jr., *Evolution, 24* (1970): 230-252; (d) adapted from Jenkins, F.A. Jr. and F.R. Parrington, *Transactions of the Royal Society London* B 273 (1976): 387; (e) adapted from Jenkins, FA Jr. *Primate Locomotion.* Academic Press, 1974.)

three ossicles (see **Figs. 3-6, 3-7**, and 3-8) and for improved hearing acuity. Numerous changes in the postcranial skeleton probably affected many aspects of locomotion and agility. These changes include reduction of cervical and lumbar ribs, improved limb posture, alteration of the joints between the occipital condyles, atlas, and axis, and simplification (by fusion of elements) of the **tarsus** and carpus. Reduction of the lumbar ribs in therapsids suggests the presence of a muscular diaphragm. This, together with the development of a secondary palate, possibly indicates the beginnings of higher levels of metabolism in the lineage. Together with the upright posture, increased axial flexion, improved coordination of lung ventilation with locomotion, and other adaptations, these specializations

eventually resulted in cursorial forms of mammals (e.g., cheetah) capable of great running speeds. Precise occlusion of the teeth and rearrangement of the jaw musculature have implications for advanced food-processing (and digestive) abilities. Enlargement of the brain goes hand in hand with many of these adaptations. It is fair to assume that intermediate stages were present in nonmammalian synapsids. Some of these synapsid adaptations are discussed in more detail below.

Important Evolutionary Transformations

A handy and widely used landmark in nonmammalian-to-mammalian synapsid evolution is the structure of the

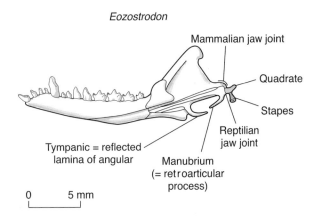

FIGURE 3-6 Parts of the jaw joint of *Eozostrodon*, viewed from the medial side. (Adapted from Lillegraven, J. A., Kielan-Jaworowska, Z., and Clemens, W. A. *Mesozoic Mammals: The First Two-thirds of Mammalian History.* University California Press, 1979.)

jaw articulation. In nonmammals, this joint is typically between the quadrate bone of the skull and the articular bone of the lower jaw, but in mammals, the squamosal and dentary bones form this joint. The earliest mammals are regarded as mammals because of the presence of the dentary-squamosal joint, but the situation is far from simple. In cynodont therapsids, there are several stages in the transformation of the jaw joint. An intermediate stage, appearing in several cynodont families, is the development of a secondary jaw joint—in addition to the quadrate-articular joint—between the surangular bone and the squamosal bone. This secondary joint probably braced the quadrate-

articular joint against backward displacement of the jaw during chewing, but it also might have functioned in sound transmission 100 million years before the appearance of the first mammals (Rowe 1996a). In the cynodont *Probainognathus* (Fig. 3-3), an articular depression (glenoid fossa) developed in the squamosal bone, and into it fit the surangular bone of the lower jaw, braced by an "articular" process of the dentary bone (Fig. 3-6).

In the late Triassic, some cynodonts demonstrated the dentary-squamosal jaw joint for the first time. This involves a posterior extension of the dentary bone into the glenoid fossa of the squamosal bone, with the quadrate-articular joint also present and medial to this "mammalian" jaw joint. In this two-jointed condition, the quadrate and articular bones still formed a jaw joint but functioned together with the stapes to transmit vibrations better from the tympanic membrane to the oval window of the inner ear. Where both joints were present in one animal, the postdentary bones were much reduced in size, and parts of them were loosely fitted into a groove on the medial side of the dentary (Fig. 3-6). There they apparently were capable of more delicate vibrations and more efficient transmission of sound. A part of the angular bone known as the reflected lamina supported a tympanic membrane, and the articular and quadrate bones transmitted vibrations from this membrane (via the stapes) to the inner ear (Allin 1975; Allin & Hopson 1992; Figs. 3-6, 3-7, and 3-8). Accordingly, still further reduction of the size of the articular and quadrate bones improved their sensitivity to vibrations and enhanced the sense of hearing.

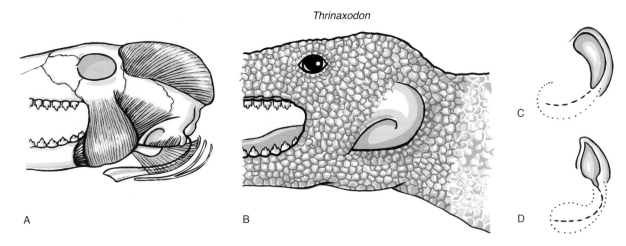

FIGURE 3-7 Reconstructed soft tissues in the early cynodont *Thrinaxodon*. (A) Hypothetical musculature and hypobranchial apparatus. (B) Hypothetical appearance in life. The external auditory meatus is shown as a depression bounded by the masseter and squamosal anteriorly and by a raised soft-tissue fold (primitive pinna) posteriorly. (C, D) Speculative stages in the fusion of the margins of the primitive meatus to form a tubular passage, accompanied by development of a pinna. (Adapted from Webster, D. B., Fay, R. R., and Popper, A. N. *The Evolutionary Biology of Hearing.* Springer-Verlag, 1992.)

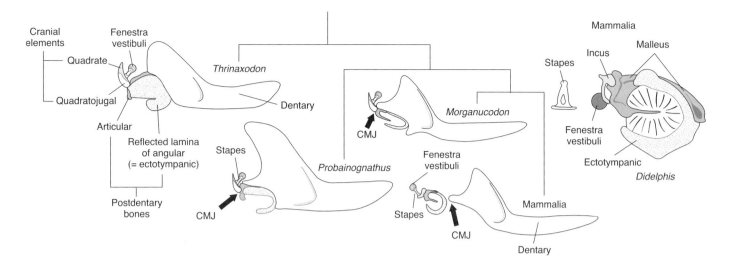

FIGURE 3-8   Selected major stages in the evolution of the mammalian jaw joint and ear region. Homologous bones are shaded similarly from one taxon to the next. The view is of the lateral side. Dentary bones are shown diagrammatically, without teeth. Postdentary bones of nonmammalian synapsids became modified as middle ear bones of Mammalia (reflected lamina of the angular bone = ectotympanic; articular = malleus). Cranial bones (quadrate = incus, quadratojugal, and stapes) are shown without the rest of the cranium for simplicity. Tympanic membrane (eardrum) is shown only in the inset of *Didelphis*. The fenestra vestibuli is the opening into the inner ear in which the footplate of the stapes is normally fitted. CMJ = craniomandibular joint, the joint between the skull and the lower jaw; note the shift in the CMJ from an articular-quadrate joint in *Thrinaxodon* and *Probainognathus* to a condyle on the dentary in Mammalia that articulates with the squamosal of the cranium (not shown). *Morganucodon* has both these joints as jaw joints, but the dentary-squamosal joint is dominant over the reduced articular-quadrate joint. In more derived mammals, the articular-quadrate joint is maintained as a malleus-incus joint, but it is removed from the jaw joint area and serves solely in hearing. Enlarged detail of ear region of *Didelphis* at upper left shows the stapes removed from the fenestra vestibuli. (Adapted from Ghiselin, M. T., and Pinna, G. *New Perspectives on the History of Life*. California Academy of Sciences, 1994.)

Finally, in mammals, the articular, quadrate, and angular bones became completely detached from the lower jaw and became part of the ear apparatus (the malleus, incus, and tympanic ring, respectively). Rowe (1996a, 1996b) proposed that these phylogenetic changes in the ear region were accompanied or caused partly by an increase in brain size and the acquisition of functional jaw muscles during early development. Changes in the ear region are revealed elegantly in the **ontogenetic** stages in the postnatal development of a modern marsupial (**Fig. 3-9**).

The development of masseter muscles with essentially the same attachments as those of mammals was a cynodont innovation that occurred in no other therapsid line. These muscles originate on the zygomatic arch and insert on the lateral surface of the dentary bone and are powerful adductors of the jaws (they close the jaws). The development of these muscles resulted in several important functional refinements (Crompton & Jenkins 1979). First, the masseter muscles formed part of a muscular sling that suspended the jaw and enhanced the precise control of transverse jaw movements. Second, these muscles increased the force of the bite. Third, the forces produced by the bite were "focused" through the point of the bite, and thus, the stress on the jaw joint was reduced.

A shift in the structure and function of the dentition can be traced through the cynodont–early mammal evolutionary line. The front teeth were of primary importance in primitive therapsids and most nonmammalian cynodonts; the incisors and canines were robust and the cheek teeth relatively weak. The reverse was true in early mammals, in which the cheek teeth were the more robust series. The development in cynodonts of masseter muscles and the concentration of jaw-action power through the postcanine teeth attended these changes in function. Most carnivorous nonmammalian cynodonts lacked occlusion of their postcanine teeth so that powerful biting was possible only in the incisors and canines; herbivorous forms had bilateral occlusion of the postcanine teeth (Crompton 1995). Progressively greater precision and breadth of movement of the lower jaw set the stage for the later evolution in mammals of a complex molar cusp pattern. Early mammals uniquely developed unilateral occlusion, in which chewing occurs on only one side of the mouth at a time. The postcanine teeth of some cynodonts were tricuspid and resembled those of some of the earliest known mammals. In other cynodonts, the cheek teeth were complex and double-rooted.

Several other structural features that became well developed in cynodonts are typical of both ancient and modern mammals. One such is the secondary palate, a structure formed by an inward and backward extension of the premaxillary, maxillary, and palatine bones. This bony plate lies beneath the original roof of the mouth and forms a passage that shunts air from the external nares at the front of the snout to the internal narial openings at the back of the mouth (Fig. 3-4). Such a bypass allows mammals to breathe while food is being chewed. Moreover, the presence of a secondary palate may have facilitated **suckling**. Suckling in modern mammals partly depends on the ability to form a seal between the tongue and the front portion of the soft palate. This is achieved with a muscle that originates on the pterygoid bone. The pterygoid bones of *Pachygenelus* (an advanced nonmammalian cynodont) and *Morganucodon* suggest that they had such tensor muscles, supporting the possibility that early mammals suckled their young (Crompton 1995). The incisive foramina were large in cynodonts, a reflection perhaps of the importance of the sense of smell to these animals. The reduction in size of lumbar ribs and the retention of a thoracic rib cage in cynodonts may have been associated with the development of a muscular diaphragm and the respiratory movements typical of mammals.

The limbs and girdles of cynodonts were modified as the sprawling limb posture was partially abandoned in favor of movement in a **parasagittal** plane in the hindlimbs. The ilium shifted forward, and the pubis and ischium moved backward as fore-and-aft limb movement became more important than lateral movement (see Fig. 3-5). The limbs of some cynodonts were slim and adapted to rapid running. A simplistic approach has often been taken in describing the differences between reptilian and mammalian limb postures. The reptilian posture has been characterized as sprawling, with the humerus and femur directed horizontally, whereas mammalian limbs have been described as moving directly fore and aft and being positioned nearly vertically beneath the body. Actually, this latter posture is typical only of cursorial mammals. Jenkins (1971) found that, during locomotion in a group of noncursorial species, the humerus and femur function in postures more horizontal than vertical and at oblique angles relative to the parasagittal plane. The studies of Jenkins further demonstrate that the limb postures of terrestrial mammals are extremely diverse. Certainly a trend toward a vertical limb posture can be detected in cynodonts and in early mammals, but the stereotypical picture of the vertical limb posture shared by all terrestrial mammals should be abandoned. Similarly, in certain extinct forms such as Megazostrodon (Fig. 3-5D), the hindlimbs seem to have had a "mammalian stance," whereas the front limbs were probably sprawled (Jenkins & Parrington 1976). Indeed, a sprawling posture was maintained in the forelimbs in many archaic mammals, such as morganucodontids, symmetrodonts, and multituberculates. The parasagittal posture of the forelimbs first appeared in relatively advanced therians (Gambaryan & Kielan-Jaworowska 1997; Hu et al. 1997).

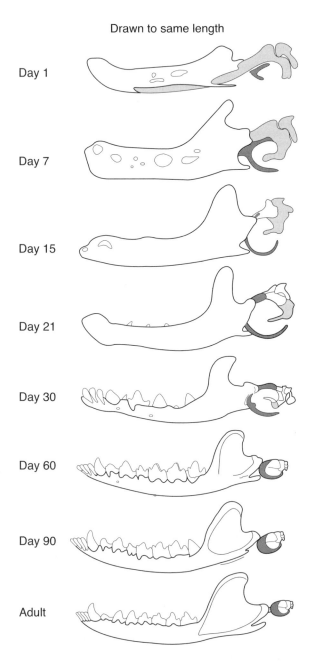

Drawn to same length

Day 1

Day 7

Day 15

Day 21

Day 30

Day 60

Day 90

Adult

FIGURE 3-9 Stages in the development of the lower jaw and ear region in a young opossum *Monodelphis* (Didelphidae), showing how ontogeny recapitulates phylogeny. Compare the ontogenetic stages with Fig. 3-8. The ectotympanic is shown in black, and gray shading indicates cartilage. Jaws are drawn to same length for ease of comparison. (Adapted from Ghiselin, M. T., and Pinna, G. *New Perspectives on the History of Life.* California Academy of Sciences, 1994.)

Cynodonts, and therapsids in general, were active terrestrial synapsids with well-developed senses of hearing and smell, and later species were probably **endotherms** (Feder 1981). Why, after such a long period of dominance, did the progressive cynodonts and other therapsids become extinct? A major cause may have been competition from dinosaurs (Colbert 1982). Early therapsids were often the size of large dogs, whereas the last surviving cynodonts were squirrel-sized and the earliest mammals were no larger than mice. The therapsid–early mammal evolutionary line was apparently under intense selection for small size. As mentioned earlier in relation to Mesozoic mammals, the smaller size of later therapsids may have made available to them more retreats secure from dinosaurs.

Early Mammals

Members of the families Sinoconodontidae and Morganucodontidae (**Figs.** 3-5D, **3-10, 3-13**), from the late Triassic to middle Jurassic period of Asia, Europe, Greenland, Africa, and North America, represent the earliest known mammals (Kielan-Jaworowska et al. 2004). The early Jurassic ones are represented in the fossil record by relatively complete skeletons and dentitions. They were small animals. Their body weight, probably 20 to 30 grams, was an order of magnitude smaller than any middle Triassic cynodont. In morganucodontids, the cheek teeth were differentiated into premolars and molars, and the premolars were probably preceded by deciduous teeth. Chewing was on one side of the jaw at a time, and the lower jaw on the side involved in chewing followed a triangular orbit as viewed from the front (see Fig. 2-20B). During chewing, the outer surface of the lower molars sheared against the inner surface of the upper molars (**Fig. 3-11**). In species of like body size, the sinoconodontid and morganucodontid brain was three or four times larger than that of even the most advanced therapsids, a reflection perhaps of greater neuromuscular coordination and improved auditory and olfactory acuity.

What was the lifestyle of these earliest mammals? Jenkins and Parrington (1976) regarded morganucodonts as insectivores with considerable climbing ability. The apparent ability of the hallux to move independently of the other digits indicates grasping ability, and enlargement of the foramina of the cervical vertebrae through which nerves contributing to the brachial plexus passed suggests refined neuromuscular control of the forelimbs. These mammals were likely secretive, **nocturnal** creatures that depended heavily on their well-developed senses of hearing and smell. Endothermy probably favored nocturnal activity, and the animals must have been covered with hair. It seems unlikely, however, that they had developed the myriad adaptations necessary for coping with the high temperatures encountered during **diurnal** activity.

These earliest mammals possessed a suite of skeletal features that marked them clearly as mammals; but what was their reproductive pattern, and did they have mammary glands? Lillegraven (1979b) stated, "The development of **lactation** was probably a key feature in the origin and later success of mammals in adapting to the changing

FIGURE 3-10 Reconstruction of *Eozostrodon*, a Triassic mammal of the family Morganucodontidae. The length of this animal was about 107 millimeters. (Adapted from Crompton, A.W. and F.A. Jenkins, Jr., *Biological Reviews, 43* (1968): 427.)

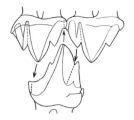

A Late Triassic mammal   B Opossum (*Didelphis virginiana*)

FIGURE 3-11 (A) Shearing planes of opposing molars of a primitive, late Triassic mammal; the shearing surfaces are outlined (Crompton 1974). (B) Occlusal view of the lower molars of an opossum (*Didelphis virginiana*), showing the tongue-in-groove fit of the anterior and posterior surfaces of adjacent teeth.

environments of the Mesozoic and Cenozoic, and was unquestionably fully functional well before the end of the Triassic." One compelling line of histologic evidence supports this view. The mammary tissue of all living mammals is essentially identical, despite the fact that the nontherian and therian evolutionary lines diverged before the end of the Triassic. In all probability, then, the histologic similarities in mammary tissue are due to inheritance by both divisions of mammals from a common late Triassic ancestor that possessed mammary glands. Furthermore, mammary glands and deciduous dentition, which allowed the delay of the growth of the complex adult dentition in the juvenile mammal, probably occurred together. A dentition capable of masticating food can be delayed in a young mammal that is nourished by its mother's milk. During the nursing period, however, a tight social bond between mother and young is essential. Therefore, deciduous teeth, delayed adult dentition, mammary glands, lactation, maternal care, and a tight bond between mother and nursing young must have evolved in concert. When therian mammals abandoned egg laying and began bearing living young (viviparity) is unknown; living monotreme mammals still lay eggs, and some other therian mammals may have kept this pattern long after the Triassic.

Although of great interest in connection with the story of mammalian evolution, these early mammals were diminutive members of a late Triassic terrestrial fauna that was becoming increasingly dominated by dinosaurs. Yet the tiny late Triassic mammals were innovative in unspectacular ways that furthered their survival in the shadow of the dinosaurs. How very different would have been the panorama of post-Triassic vertebrate evolution and how altered would be the face of the earth today if the little late Triassic mammals had proven vulnerable to some contemporary reptilian predator and had relinquished the scene completely to the reptiles.

Mesozoic Mammalian Radiations

Mammals clearly originated monophyletically from cynodont synapsids, and by the late Triassic and early Jurassic, they had diverged into several stocks, but the relationships of these stocks are complex and still unclear because of inconsistencies in the quality and quantity of fossils of each stock. Thus, it is difficult to arrange the various Mesozoic forms and groups into a well-resolved cladogram and even more difficult to "convert" the phylogenetic relationships into a consensus classification. Luo (2007) provided a summary phylogeny (**Fig. 3-12**), but new specimens and data will certainly continue to change this view of mammalian phylogeny. The classification followed here is mainly from Kielan-Jaworowska et al. (2004).

Of course, the better known taxa tend to be included in phylogenetic analyses. One well-known taxon, *Sinoconodon* (Fig. 3-13), from the early Jurassic of Asia, is very near the nonmammal-to-mammal transition and is classified as a mammal because the squamosal and dentary bones make the dominant craniomandibular jaw joint, whereas the quadrate-articular joint is smaller and probably functioned more for hearing (Kielan-Jaworowska et al. 2004). At least some of the Morganucodontidae are very well-known skeletally and are usually considered to be mammals or at least "mammaliaformes" by some authors. The Kuehneotheriidae (Docodonts) of the early Jurassic are probably close to the ancestry of the remainder of the clade and are almost always classified as mammals, as are eutriconodontans, multituberculates, and some other less well-known Mesozoic taxa. The Monotremata are, of course, certainly mammals, but they have a poor Mesozoic fossil record and their phylogenetic ties are uncertain. The South American *Vincelestes* is considered to be a basal member of the radiation of the therians (Metatheria and Eutheria).

As mentioned, the first two-thirds of mammalian history is documented by a spotty fossil record that leaves many geographic areas, time periods, and anatomical changes unrepresented. Nevertheless, Mesozoic mammals are now known from all continents except Antarctica. Many of their fossils have long been known from Europe, Asia, and North America, but recent discoveries in those continents and elsewhere around the world bolster our growing knowledge of early mammals and nonmammalian synapsids. Especially intriguing and to be further sought are Mesozoic mammals from the southern continents that were once a part of the supercontinent Gondwana. Eventually their remains will probably be found in Antarctica, too, as have several kinds of late Cretaceous dinosaurs (Gasparini et al. 1996) and early Cenozoic mammals. (One of the first amniote fossils found in Antarctica was that of a

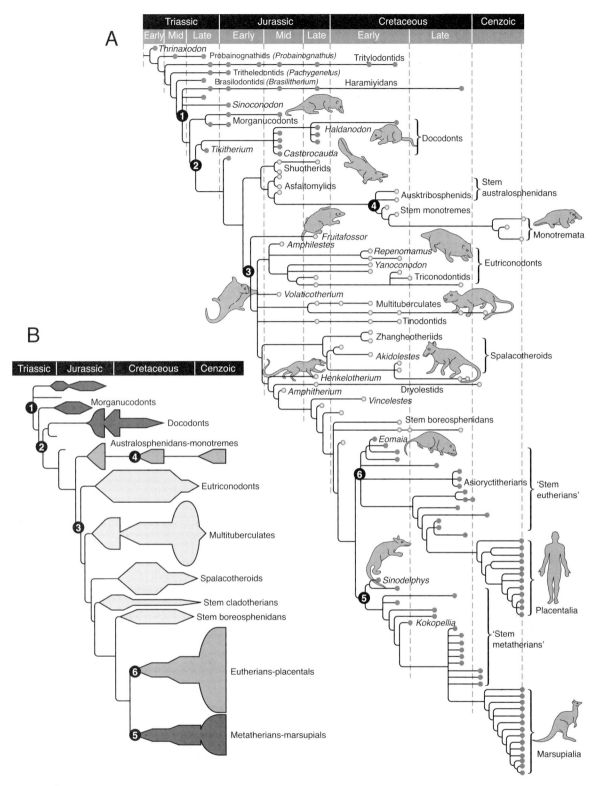

FIGURE 3-12 Hypothesis of relationships of the major groups of mammals and ancestors, with emphasis on Mesozoic types. Only a few of the hundreds of known Mesozoic mammals are shown. (A) Several groups evolved and died out rather quickly, whereas other groups arose, underwent a modest radiation, and persisted for millions of years before eventually becoming extinct, although some (Monotremes, node 4) have persisted with low diversity until today. Metatherians and eutherians (nodes 5 and 6 near the bottom of the phylogeny) arose in the Cretaceous and have continued to undergo pulses of diversification until today. (B) A summary of the same phylogeny in (A) but showing diversity in the higher level groups (orders and families) to which the Mesozoic mammals belong; changes in the depth or thickness of the branches reflect the changes in diversity (number of member species) in each lineage through time. (Modified from Luo, Z.-X., *Nature* 450 (2007): 1011–1019.)

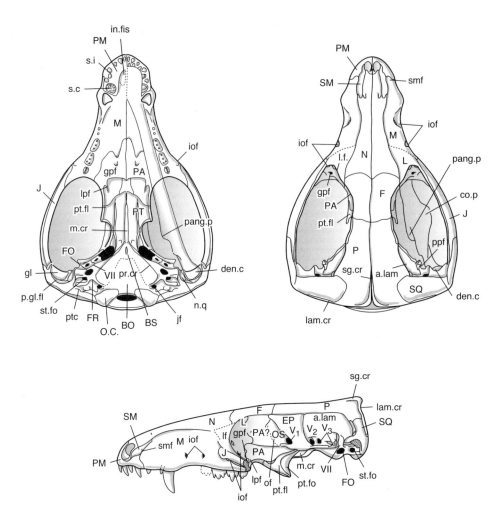

FIGURE 3-13 A reconstruction of the skull of *Sinoconodon* in ventral, dorsal, and lateral views. Ventral view includes the left half of the mandible; dorsal view includes the right half of the mandible. *Sinoconodon* is a derived synapsid considered by some authors and interpretations to be mammalian and by others to be nonmammalian. In either case, it is near the transition, and whether it is considered a mammal depends on semantics and definition of the Mammalia. A, articular; a.lam, anterior lamina of petrosal; BO, basioccipital; BS, basisphenoid; co.p., coronoid process; den.c, dentary condyle; EP, epipterygoid; FO, fenestra ovalis; FR, fenestra rotunda; gl, glenoid fossa for dentary condyle; g.p.f, greater palatine foramen; in.fis, incisive foramen; iof, infraorbital foramen; J, jugal; j.f, jugular foramen; L, lacrimal; lam.cr, lambdoid crest; l.p.f, lesser palatine foramen; M, maxilla; N, nasal; n.q, notch for quadrate (on squamosal); o.c, occipital condyle; P, parietal; PA, palatine; pang.p, pseudangular ("angular") process; p.gl.fl, postglenoid flange; PM, premaxilla; pr., promontorium; pr.cr, promontorium crest; pt.fl, pterygoid flange; Q, quadrate; sc, socket for canine; sg.cr, sagittal crest; si, socket for incisor; SM, septomaxilla; smf, septomaxillary foramen; SQ, squamosal; ST, stapes; st.fo, stapedial muscle fossa; V_1, foramen for the ophthalmic branch of trigeminal nerve; V_2, foramen for the maxillary branch of trigeminal nerve; V_3, foramen for the mandibular branch of trigeminal nerve; VII, foramen for facial nerve. (Adapted from Szalay, F. S., Novacek, M. J., and McKenna, M. C. *Mammal Phylogeny.* Springer-Verlag, 1993.)

Triassic dicynodont therapsid.) Recent improvements in the fossil record of Mesozoic mammals as well as the record of feathered dinosaurs, toothed birds, and many other vertebrates provide interesting glimpses into the terrestrial biota of the Mesozoic world (Dashzeveg et al. 1995; Hu et al. 1997; Sampson et al. 1998). For example, several finds of multituberculate-like mammals known as Gondwanatheria in South America, Madagascar, and peninsular India (all formerly part of Gondwana) indicate their extremely wide southern distribution during the late Cretaceous (Krause et al. 1997). Because Gondwanatherians are known only from fragmentary fossils, their relationships remain unclear (Gurovich & Beck 2009).

Mesozoic mammals are surprisingly diverse. The earliest mammals clearly began to undergo a fundamental radiation that resulted in a complexity of arcane and puzzling forms. The Gondwanatheria just mentioned are one such group, although they appear relatively late in the era. Several other groups, enigmatic because of their poor fossil records, give us tantalizing glimpses into the hidden world of Mesozoic mammals. Other primitive Mesozoic fossil types that are better preserved show convergently evolved structures that are remarkably reminiscent of certain living mammals, and these forms probably filled at least partially similar ecological roles (**Fig. 3-14**; Luo 2007).

Among the earliest archaic therians are the Kuehneotheriidae from the late Triassic or earliest Jurassic of Europe, Greenland, and India. The systematic placement of *Kuehneotherium* and its relatives is uncertain within Mammalia, but *Kuehneotherium*'s cheek teeth show an incipient triangular shape that reflects a fundamental step in the evolution of mammalian molars (**Fig. 3-15**). The molar teeth of more primitive mammals have cusps arranged in a straight line from front to back, as in the morganucodontid *Eozostrodon*, whereas *Kuehneotherium* had up to six molars with cusps arranged in a triangular pattern of three fairly symmetrically situated cusps, called a "symmetrodont" tooth. The upper and lower molars fit together in a series of opposing triangles such that the toothrow provides a zigzag array of efficient shearing surfaces. Another step in this evolution is provided by the Dryolestida, in which the teeth are more

nearly tribosphenic or at least near the ancestral pattern of the truly tribosphenic mammals, the Late Cretaceous Metatheria and Eutheria. *Kuehneotherium* and some of its potential relatives were shrew-sized to rat-sized and were probably insectivorous. The discovery of a nearly complete skeleton of *Zhangeotherium* in the Early Cretaceous of Asia (Hu et al. 1997) gives evidence that these early mammals represent a side branch of Theria before the divergence of the Metatheria and Eutheria. The skeleton of *Zhangeotherium* possesses an uncoiled cochlea, an interclavicle, **epipubic bones**, and a sprawled posture of its forelimbs.

Also stemming from among the earliest mammals were the Eutriconodonta, known from the late Triassic to the early Cretaceous. Eutriconodonts were predatory; the largest genus (*Gobiconodon*) was the size of a domesticated cat. The dentition was heterodont, with as many as 14 teeth in a dentary bone. The canines were large, and typically, the molars had three primary cusps arranged in a front-to-back row. Two early Cretaceous African species of the genus *Ichthyoconodon* were possibly aquatic and fed on fish (Sigogneau-Russell 1995), whereas a North American

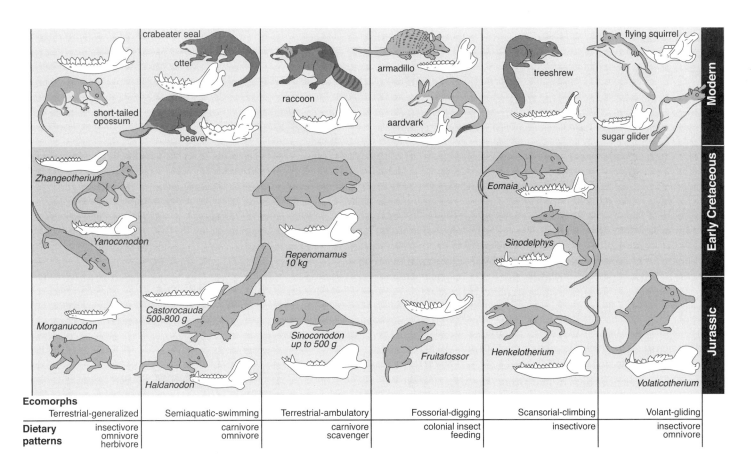

FIGURE 3-14 Life restorations and lower jaws of a number of Mesozoic mammals based on recently discovered and well-preserved skeletal fossils, compared with selected modern mammals. The Jurassic and early Cretaceous mammals (bottom and middle rows) show a variety of convergently evolved body shapes and structures reminiscent of certain modern mammals (top row), and they probably also were partly similar in their lifestyles and some of their ecological relationships. (Modified from Luo, Z.-X., *Nature* 450 (2007): 1011–1019.)

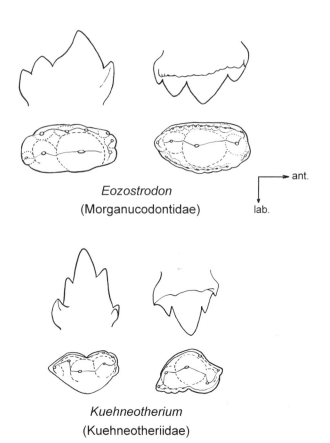

Eozostrodon
(Morganucodontidae)

→ ant.

lab.

Kuehneotherium
(Kuehneotheriidae)

FIGURE 3-15 Diagrams of the molars of Triassic or early Jurassic mammals. In each case, the lower molar is on the left and the upper molar is on the right. Arrows show orientation for all of the occlusal views: ant. = anterior; lab. = labial. (Adapted from Crompton, A.W., *Bulletin British Museum Natural History* (Geology), 24 (1974):399.)

middle Cretaceous eutriconodont *Jugulator* had an estimated body mass of 750 grams and probably was capable of preying on small vertebrates. The true habits of these animals are yet unknown (Cifelli & Madsen 1998).

The order Docodonta is represented by several primitive genera known from the middle Jurassic to early Cretaceous. Members of this group have highly derived, roughly quadrate teeth, with the cusps not aligned anteroposteriorly. The braincase and postcranial skeleton seem to be on a primitive level of development (Martin & Nowotny 2000).

Among other nontherian mammals, the subclass Allotheria, including the orders Haramiyida and Multituberculata, is especially remarkable. These were probably the first mammalian herbivores (or omnivores; Krause 1982), and although the multituberculates finally disappeared in the Paleogene and left no descendants, they were highly successful. Haramiyidans are much less well-known than multituberculates. Unlike all other mammals, allotherians'

dentition allowed them to chew on both sides of the jaw at the same time. Multituberculates appear first in the middle or late Jurassic, and their fossil record spans 100 million years. Their relationships to other mammals are still uncertain; they may have diverged very early from other mammals (Lillegraven & Hahn 1993). These animals were widespread in both the Old World and New World and were the ecological equivalents of rodents in some ways. The strongly built lower jaw provided attachment for powerful jaw muscles. There were usually two (but sometimes three) incisors above and two below, and a diastema was present in front of the premolars (**Fig. 3-16**). Typical of some advanced multituberculates were upper molars with three parallel rows of cusps and remarkably specialized blade-like posterior lower premolars (Fig. 3-16A, B). At least one Paleocene form, Ptilodus, showed **arboreal** specializations of the postcranial skeleton. Climbing adaptations include a hallux (first toe) that could move independently of the other toes, a highly mobile ankle as seen in mammals that can descend trees headfirst, and a possibly **prehensile** tail (Jenkins & Krause 1983). Other multituberculates were **semifossorial** or capable of jumping from a sprawled limb

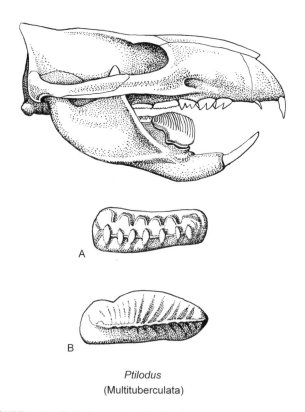

A

B

Ptilodus
(Multituberculata)

FIGURE 3-16 *Ptilodus* (order Multituberculata) skull. Occlusal view of (A) upper and (B) lower cheek teeth of *Ptilodus*. (Modified from Romer, A.S. *Vertebrate Paleontology*. University of Chicago Press, 1966.)

posture (Gambaryan & Kielan-Jaworowska 1997; Kielan-Jaworowska & Gambaryan 1994).

Multituberculates persistently retained several primitive features. Cervical ribs were retained in a few taxa, and throughout their history, they never developed a parasagittal limb posture. Their cranial osteology indicates an unusual musculature and style of chewing that used a posteriorly directed power stroke of the lower jaw, opposite that of rodents and unlike any other mammals (Kielan-Jaworowska 1997; Wall & Krause 1992). The olfactory lobes of the brain were large, the cerebrum was smooth, the incisive foramina were large, and the cochlea of the ear was similar in size and proportion to that of extant small mammals of comparable size, but was uncoiled. The auditory ossicles are strikingly similar in one Paleogene genus, *Lambdopsalis*, to those of living monotremes, which suggests that the ear was ill suited for receiving high-frequency airborne sounds but well suited for low-frequency bone-conducting hearing (Meng & Wyss 1995). In contrast, a late Cretaceous genus *Chulsanbataar* is suggested to have had high-frequency hearing but had a low sensitivity to low-decibel (quiet) sounds (Hurum 1998). Considered together, these features suggest rather primitive mammals that could not remain long in competition with eutherians, but the fossil record indicates otherwise. For over 70 million years, multituberculates and eutherians coexisted. The decline of the multituberculates began in the late Paleocene and spanned 20 million years. The competition probably began with condylarths (ancestors of ungulates) in the late Cretaceous, intensified when primates became common in the Paleocene, and became overwhelming in the Eocene, when rodents became ubiquitous (Van Valen & Sloan 1966). Multituberculates appear last in the early Eocene fossil record of Wyoming, Nebraska, and South Dakota.

Metatherian and eutherian mammals (derived mammals of the subclass Boreosphenida; Luo et al. 2001) evolved from an ancestry within a group of Mesozoic clades (including the Dryolestida, Amphitheriida, and Zatheria). These groups possess an angular process on the dentary bone, a feature shared with members of the Boreosphenida. The Zatheria includes the family Peramuridae, which is known from the early Cretaceous of Europe and Africa and has lower molars with a distinct, posterior "heel" but no true talonid basin. Peramurids or their relatives are dentally closest to the ancestry of boreosphenidans with tribosphenic molars. Another boreosphenidan, *Aegialodon*, of the early Cretaceous of Asia and Europe, and related fossil mammals possessed lower molars considered to be the earliest tribosphenic molars in the fossil record (Kielan-Jaworowska et al. 2004). The shape of the anterior trigonid

section of the lower molar resembles the comparable part of this tooth in many eutherians and metatherians (**Fig. 3-17**). More important, the lower molar has a talonid with a basin into which the protocone of the upper molar fits (Fig. 3-17B), as in eutherians and metatherians.

Interestingly, pseudotribosphenic molars evolved in certain other Mesozoic mammals convergently to those of tribosphenic mammals. One such species is *Shuotherium dongi*, an Asian late Jurassic species in which the lower molars evolved a heel-like structure that is analogous to the talonid of tribosphenic mammals and that functioned in the same manner. In *Shuotherium*, however, this structure occurred on the anterior instead of the posterior end of the molars (Chow & Rich 1982).

Kokopellia represents the earliest known marsupial and is from the middle Cretaceous of North America (Utah; Cifelli 1993; Cifelli & Muizon 1997). Like many fossil vertebrates, *Kokopellia*'s geological age is known by its stratigraphic association with a geochemically dated volcanic ash, in this case 98 million years old. The oldest known eutherian is *Eomaia* from the early Cretaceous of Asia (China; Ji et al. 2002); it is about 125 million years old. The absolute and relative ages of these specimens suggest that metatherians and eutherians had become recognizable as such by perhaps 125 million years ago, although their respective lineages probably split before then. As timed by the molecular clock, genetic evidence suggests that the marsupial

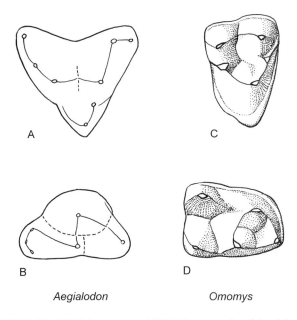

FIGURE 3-17 (A) Right upper and (B) left lower molar of *Aegialodon*, an early Cretaceous tribosphenidan (family Aegialodontidae). The upper molar is a hypothetical reconstruction. (C and D) Comparable teeth of the primitive Eocene eutherian mammal *Omomys*, a tarsier-like primate. (C and D Modified from Romer, A.S. *Vertebrate Paleontology*. University of Chicago Press, 1966.)

lineage split from a therian ancestor earlier (130 to 185 million years ago).

Cretaceous Mammals

A broad view of the Cretaceous—a period of great biotic change—provides a background against which the late Mesozoic evolution of mammals can be viewed. The prior late Jurassic was a time of considerable interchange of biotas between continents, as indicated by the occurrence in western Europe, East Africa, and western North America of identical or closely related species of reptiles (Colbert 1973) and by intercontinental similarities among floras (Vachrameev & Akhmet'yev 1972). Likewise, there are resemblances in the Jurassic mammals and other vertebrates of the southern continents, such as the Australosphenidans of Argentina, Tanzania, and Madagascar, although the fossil record in the Gondwanan continents from that time is still sparse (Rauhut et al. 2002). After the earliest part of the Cretaceous, however, dispersal by mammals became sharply restricted. In the western hemisphere, a series of transgressions of a Cretaceous Sea from the Arctic Ocean to the Gulf of Mexico divided North America for much of the Cretaceous into two separate centers for the evolution of terrestrial plants and animals. In Eurasia also, the dispersal of land animals was restricted: Europe was essentially an archipelago of islands during the Cretaceous, and the "Turgai Strait" seaway separated the land faunas of Europe and Siberia. In the southern hemisphere, South America and Antarctica retained a tenuous connection but had separated from Africa and India 130 million years ago as Gondwana broke apart (Pascual & Ortiz-Jaureguizar 2007). Angiosperms (flowering plants) became established as the dominant plants in terrestrial ecosystems in the middle Cretaceous (Friis & Crepet 1987), an event that strongly affected the evolution of land faunas. Co-adaptive evolution between angiosperm flowers and insects, for example, fostered a Cretaceous insect radiation that must surely have affected the success of mammals.

Most dramatic were Cretaceous changes in the fortunes of the dinosaurs. Throughout the Jurassic and early Cretaceous, they were diverse and abundant and dominated the terrestrial scene. In the late Cretaceous, several herbivorous groups—the ankylosaurs, ceratopsians, and hadrosaurs—diversified in association with the increasing importance of angiosperms and decline of gymnosperms. By the close of the Cretaceous, dinosaurs were gone.

Through much of the early Cretaceous, land dwellers were barred from intercontinental movement (i.e., between the northern and southern continents resulting from the breakups of Laurasia and Gondwana) by oceans and seas, but intracontinental movements would have encouraged relative homogeneity of faunas and floras. By the middle Cretaceous, intracontinental dispersal was restricted, Africa was isolated, but the other portions of Gondwana (South America, Antarctica, India, and Australia) retained increasingly tenuous connections with one another until the late Cretaceous. After the earliest Cretaceous, populations of mammals on different continents, subcontinents, and islands evolved in isolation under different environmental pressures. This isolation of premarsupial and preplacental stocks may well have favored their differentiation. Each group seemingly faced some comparable adaptive problems, but, as in the case of reproduction, each group developed unique solutions to these problems.

Modern mammals (eutherians and marsupials) underwent adaptive radiations in the Cretaceous. Similar significant radiations occurred among dinosaurs (including birds) and multituberculates at this time. These radiations probably reflect the availability of fruits or seeds as an important new food source (Cifelli et al. 1997). Considerable literature (Clemens 1970; Lillegraven 1974; Lillegraven et al. 1979) points to the overriding importance of the early Cretaceous appearance and adaptive burst of flowering plants (angiosperms). The seeds of some angiosperms develop within an edible and nutritious fruit. Angiosperm fruits and seeds are eaten today by many mammals and were probably important to Cretaceous mammals. Mammals probably have a long history of contributing to angiosperm seed dispersal. Krassilov (1973) reported that the seed coats of some of the earliest angiosperms bore hooklets capable of tangling in the fur of foraging mammals. The Lepidoptera (moths and butterflies) appeared in the Cretaceous (MacKay 1970), probably in response to the food offered by the flower nectar and leaves of angiosperms. The Isoptera (termites) also appeared at this time, and the Coleoptera (beetles) underwent an adaptive radiation. These insect groups are important foods for mammals today and were perhaps similarly important in the Cretaceous. The diversification of dentitions may have enabled Cretaceous mammals not only to exploit plant foods but also to profit greatly from the expanding diversity and growing populations of insects.

For mammals, then, the Mesozoic, and especially the Cretaceous, was a time of experimentation. Natural selection, partly in the form of predation by an imposing array of dinosaurian (including avian) and crocodilian predators and partly in the form of competition from dinosaurs, birds, reptiles, and other mammals, affected many changes in mammalian structure and function. Behavioral, physiological, and anatomical changes that increased the efficiency of feeding, reproduction, and thermoregulation may have been critical. Various structural plans evolved and were workable for different lengths of time; some evolutionary side branches proved sterile. During the Mesozoic time of evolutionary trial and error, however, the basic mammalian structural plan was tested, retested, and perfected, and the major taxa were established. Combined fossil and molecular phylogenetic interpretations suggest that several of the major (superordinal) clades, as well as most of the extant orders, of placental mammals appeared during the Late Cretaceous, with first appearances peaking about 89 million years ago before dropping off until another peak during the middle Paleogene. These findings challenge the traditional view that the extinction of the nonavian dinosaurs at the end of the Mesozoic 65 million years ago contributed to the adaptive radiation of mammals early in the Cenozoic (Archibald 2003; Archibald et al. 2001; Bininda-Emonds et al. 2007, 2008; Penny & Phillips 2007).

SUMMARY

Mammals arose within a lineage of vertebrates known as synapsids that dominated terrestrial faunas during the Permian and early Triassic periods. Many groups of early synapsids went extinct at the end of the Permian. However, a few lineages of advanced synapsids (the Therapsida) survived the Permo-Triassic extinction. By the end of the Triassic, therapsids, too, dwindled but not before giving rise to the cynodonts. From a cynodont ancestry, stem mammals arose in the late Triassic, slightly after the first appearance of dinosaurs. Dinosaurs dominated the terrestrial Mesozoic scene. By comparison, Mesozoic mammals were small and probably secretive. Most were mouse-sized, and probably foraged at night.

The synapsid skull is characterized by a single temporal opening low on each side of the cranium, surrounded by the postorbital, squamosal, and jugal bones. A general trend in the synapsids was the enlargement of the temporal opening and movement of the jaw muscle origins to the braincase and the zygomatic arch. Therapsids display many morphological trends leading to the basic mammalian anatomical plan. The upper canines are enlarged, indicating a capacity for predation, and the posture changed to move the limbs underneath the body.

Many derived features, including mammary glands with nipples, viviparity with loss of the eggshell, separate anal and urogenital openings in adults, the digastricus muscle to open the jaw, and other features, unite the mammals and near-mammals into a clade. Mammaliaformes possess a well-developed jaw articulation between the dentary and squamosal bones, double-rooted cheek teeth, and expansion of the brain vault in the parietal region. Mesozoic Mammalia share cheek teeth divided into molars and premolars, precise occlusion between the upper and lower molars, and a diphyodont dentition.

Many of the derived morphological characters mentioned suggest that Mesozoic synapsids were evolving important physiological features as well. For example, the expansion of the dentary bone and the establishment of a dentary-squamosal jaw joint have important implications for the development of the uniquely mammalian middle ear with three ossicles and for improved hearing acuity. Numerous postcranial changes served to improve limb posture and probably affected many aspects of locomotion and agility. Together with the upright posture, increased axial flexion, improved coordination of lung ventilation with locomotion, and other adaptations, these specializations eventually resulted in cursorial forms of mammals. Precise occlusion of the teeth, a secondary palate, and rearrangement of the jaw musculature suggest more efficient food processing abilities. Enlargement of the brain went hand in hand with many of these adaptations.

A spotty fossil record documents the first two-thirds of mammalian history. Nevertheless, Mesozoic mammals are now known from all continents except Antarctica. Mesozoic mammals are surprisingly diverse. Once they appeared, the earliest mammals began a fundamental radiation that resulted in a complexity of arcane and puzzling forms. Eutriconodonts were house cat-sized predators, and the Haramiyida and Multituberculata were probably the first mammalian herbivores or omnivores. Metatherian and eutherian mammals evolved from within a group of clades, including the Dryolestida, Amphitheriida, and Zatheria. *Kokopellia* probably represents the earliest known metatherian and is from the middle Cretaceous of North America, whereas the oldest known eutherian is *Eomaia* from the early Cretaceous of Asia.

The late Jurassic was a time of considerable interchange of biotas between continents. In the early Cretaceous, however, dispersal between continents became sharply restricted. Populations of mammals on different continents, subcontinents, and islands evolved in isolation under different environmental pressures. This isolation of premarsupial and preplacental stocks favored their Cretaceous differentiation. Similar significant radiations occurred among dinosaurs (including birds) and multituberculates at that time. These radiations probably reflect the availability of fruits and seeds (angiosperms) and the appearance of diverse lepidopteran and isopteran insects as important new food sources.

The late Mesozoic, and especially the Cretaceous, was a time of experimentation for mammals; the basic mammalian structural plan was tested, retested, and perfected, and the major taxa were established. Several of the major (superordinal) clades, as well as most of the extant orders, of placental mammals appeared during the late Cretaceous. These findings challenge the traditional view that the extinction of the nonavian dinosaurs at the end of the Mesozoic 65 million years ago contributed decisively to the adaptive radiation of mammals early in the Cenozoic.

KEY TERMS

Acetabulum
Adductor
Amniote
Arboreal
Caniniform
Cheek teeth
Diphyodont
Diurnal

Endotherm
Epipubic bones
Lactation
Lingual
Nocturnal
Ontogenetic
Paraphyletic
Parasagittal

Prehensile
Secondary palate
Semifossorial
Spatulate
Suckling
Tarsus
Temporal fenestrae
Viviparity

RECOMMENDED READINGS

Archibald, JD. 2003. Timing and biogeography of the eutherian radiation: fossils and molecules compared. *Molecular Phylogenetics and Evolution*, 28:350–359.

Benton, MJ. 2004. *Vertebrate Palaeontology*, 3rd ed. Wiley-Blackwell, New York.

Bininda-Emonds, ORP, et al. 2007. The delayed rise of present-day mammals. *Nature*, 446:507–512.

Kemp, TS. 2005. *The Origin and Evolution of Mammals.* Oxford University Press, Oxford, UK.

Kielan-Jaworowska, Z, RL Cifelli, & Z-X Luo. 2004. *Mammals from the Age of Dinosaurs: Origins, Evolution, and Structure.* Columbia University Press, New York.

Luo, Z-X. 2007. Transformation and diversification in early mammal evolution. *Nature*, 450:1011–1019.

Classification of Mammals

Despite their remarkable success, mammals are much less diverse than are most invertebrate groups. This is probably attributable to their far greater individual size, to the high energy requirements of endothermy, and thus to the inability of mammals to exploit great numbers of restricted ecological niches. Wilson and Reeder (2005) recognized about 1,229 genera and 5,416 species of living mammals. Most species of extant mammals have already been described, but approximately 10 to 12 new species continue to be named each year. Many of these are cryptic species, distinct genetically but difficult or impossible to distinguish morphologically from already known species. When fossil mammals are considered, the numbers are more impressive. In the compendium of McKenna and Bell (1997), 5,162 genera of mammals in 425 families—of which 4,079 genera (79%) and 300 families (71%) are extinct—had been named, with additional new fossil taxa named each year as well. Still, the numbers of genera and species are insignificant in comparison with those for invertebrates. There are, for example, an estimated 950,000 named species of insects (perhaps 8 to 100 million undiscovered), 40,000 of protists (perhaps 100,000 to 200,000 undiscovered), and 70,000 of mollusks (perhaps 200,000 undiscovered).

Chapters 5 through 19 consider the taxa of mammals listed in **Table 4-1**. In these chapters, such features as group size, present geographic distribution, time of appearance in the fossil record, structural characteristics, and brief life histories are given for each order and family. When appropriate, morphology is related to function so that the remarkable structural and functional diversity displayed by mammals can be appreciated.

We devote considerable attention to the orders and families of mammals not because we wish to put primary stress on the taxonomic aspect of mammalogy, but rather as an attempt to provide students with sufficient information on the various kinds of mammals to make the subsequent discussions of mammalian biology meaningful. Students' interest is often dulled if they must deal with information about completely unfamiliar kinds of animals. It seems pointless to discuss water regulation in heteromyids, for example, if students have only a vague idea of what a heteromyid is. The chapters on orders, then, should serve as a background for the chapters on selected aspects of the biology of mammals.

Phylogenetic systematics is a major focus of much of the basic research conducted on mammals today. It is also an important feature of many other fields, from conservation biology to molecular genetics. Increasingly, authors attempt to construct or use classification schemes that reflect the presumed phylogeny of mammals and other organisms, as we have done in this book. Reconstructing a phylogeny and producing a classification for a group of organisms are very different goals, however, and the results are often controversial. Although this chapter emphasizes classification, the reader should note that in the chapters of the book dealing with mammalian origins and with the orders of mammals, we have attempted to use published phylogenetic information or at least to cite pertinent references. The classification that follows is largely that of the various authors in Wilson and Reeder (2005). No classification system yet proposed has gained universal acceptance, but Wilson and Reeder's is a useful recent compendium. In keeping with the contemporary approach, categories other than Order and Family are not given for the higher taxa named.

TABLE 4-1 A Classification of Recent Mammals

Classification	Common Name(s)
Prototheria	
Order Monotremata (5 species)	
Family Tachyglossidae	Echidnas, spiny anteaters
Ornithorhynchidae	Duck-billed platypus
Theria	
Metatheria (Marsupialia)	
Order Didelphimorphia (89 species)	
Family Didelphidae	Opossums
Order Paucituberculata (6 species)	
Family Caenolestidae	Rat opossums

Classification	Common Name(s)
Order Microbiotheria (1 species)	
Family Microbiotheriidae	Monito del monte, llaca
Order Dasyuromorphia (71 species)	
Family Thylacinidae (extinct)	Thylacine
Myrmecobiidae	Numbat
Dasyuridae	Dasyures, quolls, antechinuses, dunnarts, devil
Order Peramelemorphia (22 species)	
Family Thylacomyidae	Bilbies
Peramelidae	Bandicoots, echymiperas
Chaeropodidae	Pig-footed bandicoot
Order Notoryctemorphia (2 species)	
Family Notoryctidae	Marsupial "moles," itjaritjaris
Order Diprotodontia (144 species)	
Family Phascolarctidae	Koala
Vombatidae	Wombats
Burramyidae	Pygmy possums
Phalangeridae	Cuscuses, phalangers
Pseudocheiridae	Ring-tailed possums
Petauridae	Gliders, striped possums
Tarsipedidae	Honey possum, noolbenger
Acrobatidae	Feathertail possum, feathertail glider
Hypsiprymnodontidae	Musky rat-kangaroo
Potoroidae	Rat kangaroos, bettongs, potoroos
Macropodidae	Kangaroos, wallabies
Eutheria (Placentalia)	
Order Afrosoricida (51 species)	
Family Tenrecidae	Tenrecs
Chrysochloridae	Golden moles
Order Macroscelidea (15 species)	
Family Macroscelididae	Elephant-shrews
Order Tubulidentata (1 species)	
Family Orycteropodidae	Aardvark
Order Proboscidea (3 species)	
Family Elephantidae	Elephants
Order Sirenia (5 species)	
Family Dugongidae	Dugongs, sea cows
Trichechidae	Manatees
Order Hyracoidea (4 species)	
Family Procaviidae	Hyraxes
Order Cingulata (21 species)	
Family Dasypodidae	Armadillos

(continues)

TABLE 4-1 A Classification of Recent Mammals (continued)

Classification	Common Name(s)
Order Pilosa (10 species)	
Family Bradypodidae	Three-toed tree sloths
Megalonychidae	Two-toed tree sloths
Cyclopedidae	Silky anteater
Myrmecophagidae	Tamanduas and giant anteater
Order Dermoptera (2 species)	
Family Cynocephalidae	Colugos
Order Scandentia (20 species)	
Family Tupaiidae	Tree shrews
Ptilocercidae	Pen-tailed treeshrew
Order Primates (376 species)	
Family Cheirogaleidae	Dwarf lemurs, mouse lemurs
Lemuridae	Lemurs
Lepilemuridae	Sportive lemurs
Indriidae	Wooly lemurs, sifakas
Daubentoniidae	Aye-aye
Lorisidae	Lorises
Galagidae	Bushbabies, galagos
Tarsiidae	Tarsiers
Cebidae	Marmosets, tamarins, capuchins, squirrel monkeys
Aotidae	Night monkeys
Pitheciidae	Titis, uacaris, sakis
Atelidae	Howlers, spider monkeys, wooly monkeys
Cercopithecidae	Old World monkeys
Hylobatidae	Gibbons
Hominidae	Apes, human
Order Rodentia (2,278 species)	
Family Aplodontiidae	Sewellel or mountain beaver
Sciuridae	Squirrels
Gliridae	Dormice
Castoridae	Beavers
Heteromyidae	Kangaroo rats, pocket mice
Geomyidae	Pocket gophers
Dipodidae	Jerboas, birch mice, jumping mice
Platacanthomyidae	Tree mice
Spalacidae	Zokors, bamboo rats, mole rats
Calomyscidae	Calomyscuses
Nesomyidae	Pouched rats and mice, climbing and fat mice, etc.
Cricetidae	Voles, hamsters, New World rats, and mice
Muridae	Rats, mice
Anomaluridae	Scaly-tailed flying squirrels
Pedetidae	Springhaas, springhares

Classification	Common Name(s)
Ctenodactylidae	Gundis
Diatomyidae	Kha-nyou or Laotian rock rat
Bathyergidae	Mole-rats
Hystricidae	African and Asian porcupines
Petromuridae	Dassie rat
Thryonomyidae	Cane rats
Erethizontidae	Bristle-spined rat and New World porcupines
Chinchillidae	Chinchillas, vizcachas
Dinomyidae	Pacarana
Caviidae	Cuis, Guinea pigs, cavies, maras, capybaras
Dasyproctidae	Agoutis, acouchis
Cuniculidae	Pacas
Ctenomyidae	Tuco-tucos
Octodontidae	Degus, rock rats, vizcacha-rats
Abrocomidae	Chinchilla rats
Echimyidae	Spiny rats, tree rats, etc.
Myocastoridae	Coypu or nutria
Capromyidae	Hutias
Heptaxodontidae (extinct)	Giant hutias and key mouse
Order Lagomorpha (92 species)	
Family Ochotonidae	Pikas
Prolagidae (extinct)	Sardinian pika
Leporidae	Rabbits
Order Erinaceomorpha (24 species)	
Family Erinaceidae	Hedgehogs, gymnures
Order Soricomorpha (428 species)	
Family Nesophontidae (extinct)	Nesophontes
Solenodontidae	Solenodons, alamiquis
Soricidae	Shrews
Talpidae	Moles, desmans
Order Chiroptera (1,116 species)	
Family Pteropodidae	Old World fruit bats, flying foxes
Rhinopomatidae	Mouse-tailed bats
Craseonycteridae	Hog-nosed or bumblebee bat
Megadermatidae	False vampire bats
Rhinolophidae	Horseshoe bats
Hipposideridae	Old World leaf-nosed bats
Emballonuridae	Sac-winged bats
Nycteridae	Slit-faced bats
Myzopodidae	Sucker-footed bats
Mystacinidae	New Zealand short-tailed bats
Thyropteridae	Disk-winged bats

(continues)

TABLE 4-1   A Classification of Recent Mammals (continued)

Classification	Common Name(s)
Furipteridae	Smoky bat and thumbless bat
Noctilionidae	Bulldog bats
Mormoopidae	Mustached and ghost-faced bats
Phyllostomidae	New World leaf-nosed bats
Natalidae	Funnel-eared bats
Molossidae	Free-tailed bats
Vespertilionidae	Evening bats, common bats
Miniopteridae	Bent-winged or long-fingered bats
Order Pholidota (8 species)	
Family Manidae	Pangolins
Order Carnivora (286 species)	
Family Felidae	Cats
Viverridae	Civets, genets
Eupleridae	Falanouc, fossa, Madagascaran mongooses
Nandiniidae	African palm civet
Herpestidae	Mongooses
Hyaenidae	Hyenas, aardwolf
Canidae	Wolves, foxes, jackals
Ursidae	Bears, giant panda
Odobenidae	Walrus
Otariidae	Eared seals, fur seals, sea lions
Phocidae	Earless seals
Mustelidae	Weasels, badgers, otters
Mephitidae	Skunks, stink badgers
Procyonidae	Raccoons, ringtails, coatis
Ailuridae	Red panda
Order Perissodactyla (17 species)	
Family Equidae	Horses, asses, zebras
Tapiridae	Tapirs
Rhinocerotidae	Rhinoceroses
Order Artiodactyla (240 species)	
Family Suidae	Hogs, pigs
Tayassuidae	Peccaries
Hippopotamidae	Hippopotamuses
Camelidae	Camels, vicuña, guanaco, llamas
Tragulidae	Chevrotains and mouse deer
Moschidae	Musk deer
Cervidae	Deer
Antilocapridae	Pronghorn
Giraffidae	Giraffe and okapi
Bovidae	Antelope, bison, cattle, duikers, goats, sheep, etc.

Classification	Common Name(s)
Order Cetacea (84 species)	
Family Balaenidae	Right whales
Balaenopteridae	Rorquals
Eschrichtiidae	Gray whale
Neobalaenidae	Pygmy right whale
Delphinidae	Ocean dolphins
Monodontidae	Narwhal and beluga
Phocoenidae	Porpoises
Physeteridae	Sperm whales
Platanistidae	Ganges and Indus river dolphins
Iniidae	Baiji, franciscana, and Amazon river dolphins
Ziphiidae	Beaked whales

SUMMARY

There are approximately 1,229 genera and over 5,420 species of living mammals, with an additional 5 to 12 new species named each year. When fossil mammals are considered, the numbers total over 5,162 genera of mammals in 425 families—of which 79% of genera and 71% of families are extinct. Still, the numbers of genera and species are insignificant in comparison with those for invertebrates. The mammalian classification presented here is largely that of the various authors in Wilson and Reeder (2005) and is based on phylogenetic relationships as currently understood.

RECOMMENDED READINGS

McKenna, MC & SK Bell. 1997. *Classification of Mammals Above the Species Level.* Columbia University Press, New York.

Murphy, MJ, et al. 2001. Molecular phylogenetics and the origins of placental mammals. *Nature, 409*:614–618.

Tree of Life Web Project. 1995–2005. http://www.tolweb.org/tree/phylogeny.html.

Wilson, DE & DM Reeder. 2005. *Mammal Species of the World, A Taxonomic and Geographic Reference,* 3rd ed., 2 vols. Johns Hopkins University Press, Baltimore, MD.

Mammalian Diversity

PART II

Monotremata

The order Monotremata includes the family Tachyglossidae (echidnas, or spiny anteaters), members of which are found in Australia, Tasmania, and New Guinea, and the family Ornithorhynchidae (duck-billed platypuses), restricted to eastern Australia and Tasmania. Represented today by only three genera and five species, monotremes constitute a minor segment of the extant mammalian fauna, but they are of great interest for several reasons. Morphologically, monotremes closely resemble no other living mammals, and they possess several features frequently considered to be more typical of reptiles than of mammals. Monotremes lay eggs and incubate them in bird-like fashion, and yet they have hair and suckle their young. The few surviving species thus retain some primitive features of their synapsid ancestors, yet they are highly derived (advanced) in other features.

One derived feature of monotremes is the rostrum, the "bill" of the platypus or "beak" of the echidnas, which is covered with mechanoreceptors ("push rods" for tactile sensation) and electroreceptors that can detect the weak electrical fields of small invertebrate prey (Augee et al. 2006; Manger et al. 1997). This remarkable electrosensory ability is unique among mammals. Among other vertebrates, electroreception is known only in certain fishes and amphibians, in which a completely different kind of receptor organ has evolved. The sensory importance of the **rhinarium** (the hairless area at the tip of the snout in mammals) is reflected in the cortex of the monotreme brain, which is relatively large and extraordinarily complex (Rowe & Bohringer 1992). Enhancing this sensory capability still further in the echidnas, the snout has a venous cavernous system that can be swollen with blood for better contact of the skin with the soil when the animals probe the soil with their snout; disgorging the blood enables easy removal of the snout from the soil (Augee et al. 2006).

Other specializations, such as the reduction of the dentition (echidnas completely lack teeth, and platypuses retain vestigial teeth only as juveniles) and highly modified skull morphology, make it difficult to determine the monotremes' phylogenetic relationships with other mammals. No phylogenetic analysis within the group seems to have been done, but most authors agree that monotremes themselves form a monophyletic group. The two families are quite divergent and may have gone their separate evolutionary ways since at least the Eocene and possibly since the late Cretaceous (Messer et al. 1998; Westerman & Edwards 1992).

The relationship of monotremes to other mammals has long been debated (Archer et al. 1993). Morphological and paleontological evidence strongly suggests that monotremes diverged from an early mammalian ancestor before the divergence of marsupials and placentals. In this view, monotremes are not therians. Recent molecular genetic data also argue for the monotremes as an independent, primitive group separate from Theria, which consists of the sister groups Metatheria and Eutheria (Kullberg et al. 2008; **Fig. 5-1**). Because of the uniqueness of monotremes and their importance as the few survivors representing a much earlier offshoot of mammalian evolution, the genome of a platypus was one of the first mammalian genomes to be sequenced after that of humans. The platypus genome proved to be as unusual as the animals themselves (Warren et al. 2008).

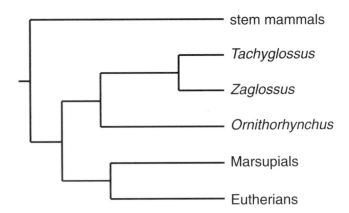

FIGURE 5-1 Phylogenetic relationships of monotremes. (Data from Westerman, M. and Edwards, D. *Platypus and Echidnas* (ML Augee, ed.) Royal Zoology Society of New South Wales, 1992.; Kielan-Jaworowska, Z., Cifelli, R.L. and Luo, Z-X. *Mammals from the Age of Dinosaurs: Origins, Evolution, and Structure.* Columbia University Press, 2004.; and Warren, W.C., et al., *Nature* 453 (2008): 175–184.)

Morphology

Many structural features distinguish monotremes from other mammals. Uniquely bird-like in appearance (**Fig. 5-2**), the monotreme skull is toothless in living forms except in young platypuses, cranial **sutures** disappear early in life, and the elongate and beaklike rostrum is covered by a leathery sheath; this sheath is horny in birds. There are **sclerotic cartilages** in the eyes, but these do not become ossified or form a sclerotic ring as they do in many reptiles and nonmammalian synapsids. The lacrimal and frontal bones are absent, whereas these bones are present in most therian mammals. The skull of monotremes includes a bone, the septomaxilla, that occurs in few other living mammals (Kuhn & Zeller 1987; Wible et al. 1990). There is no auditory bulla, but the chamber of the middle ear is partially surrounded by an oval tympanic ring. In the inner ear, the cochlea is curved but not coiled as in other living mammals.

Monotreme appendages represent excellent examples of "mosaic evolution" (different parts evolving at different rates; Crompton & Jenkins 1973). The shoulder girdle retains a bone pattern typical of therapsids. The forelimb has a rather sprawled posture resulting in part from **fossorial** (digging) specializations, and the pelvis and posture of the hindlimbs are essentially therian. Medially directed spurs occur on the inside of the ankles of adult males; this is a primitive character retained from other early mammal relatives (Musser 2005).

The monotreme pectoral girdle contains an interclavicle, clavicles, precoracoids, coracoids, and scapula and provides a far more rigid connection between the shoulders and the sternum than does the girdle characteristic of therian mammals (**Fig. 5-3**). Large epipubic bones extend forward from the pubes in both sexes. Cervical ribs are present, and the thoracic ribs lack tubercles, processes that occur on the ribs of most other mammals and are braced against the transverse processes of the vertebrae.

As put by Howell (1944), no monotreme "by any strength of the imagination might be considered cursorial [strongly adapted for running]." Monotremes have retained a limb posture that is similar in some ways to that of reptiles, and in monotremes, this posture is associated with limited running ability. In the Australian echidna (*Tachyglossus aculeatus*), the humerus remains roughly horizontal to the substrate during walking (Jenkins 1970). Rotational movement of the humerus, rather than fore-and-aft movement as in most mammals, is largely responsible for propulsion. Because in reptiles the limb posture is splayed, the fore and hind feet touch the ground well to the side of the shoulder and hip joints, respectively. The limb posture in the echidna partially departs from this pattern because the forearm angles medially and the manus (forepaw) is roughly ventral to the shoulder joint; in the hindlimb, the foot is roughly ventral to the knee. The posture of the hindlimb of the echidna resembles that of many generalized therian mammals, but the hind foot is strongly rotated outward and even backward. When the echidna is in motion, its body is elevated well above the ground in nonreptilian fashion. Despite the advances in limb posture in the echidna, locomotion is slow and appears labored and awkward. The ability of monotremes to burrow and the musculoskeletal adaptations associated with burrowing (and in the platypus, swimming)

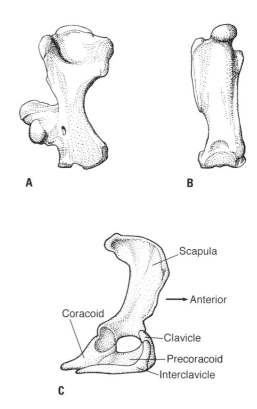

FIGURE 5-3   Bones of monotremes: (A) Left humerus of spiny anteater, (B) right femur of spiny anteater, and (C) pectoral girdle of duck-billed platypus.

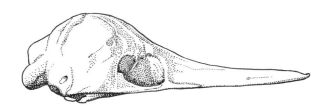

FIGURE 5-2   Skull of the spiny anteater *Tachyglossus aculeatus*. Length of skull 111 millimeters.

confound the distinction between primitive, reptile-like features and specialized adaptations, making phylogenetic interpretation more difficult (Jenkins 1990).

Reproduction

The monotreme reproductive system and pattern are unique among mammals. As in other systems of the body, the reproductive system is a mix, including primitive features shared with amniotes and unique specializations (Renfree 1993). Monotremes differ strongly from therian mammals in having multiple sex chromosomes. They possess 10 X chromosomes (in females) and 5 X and five Y chromosomes (in males), instead of the usual two X or one X and one Y of therians (El-Mogharbel et al. 2007; Grutzner et al. 2004; Rens et al. 2004). Monotremes lay eggs that are **telolecithal** (the yolk is concentrated toward the vegetal pole of the ovum) and **meroblastic** (early cleavages are restricted to a small disk at the animal pole of the ovum) like bird eggs but unlike the reproductive processes of therian mammals. Only the left ovary is functional in the platypus (Asdell 1964), as in most birds, but both ovaries are functional in the echidna. Shell glands are present in the oviducts. Monotremes are **oviparous**: they lay rubbery-shelled eggs that are incubated and hatched outside the mother's body. The fetus employs an "egg tooth" on the egg tooth bone (caruncle) to break out of the shell. There is a **cloaca** (**Fig. 5-4**), and in males, the penis is attached to the ventral wall of this cavity. The testes are abdominal, and seminal vesicles are absent. The female echidna temporarily develops a pouch-like structure to incubate and protect the young, but the platypus never develops one. This structure is not homologous to the pouch developed by some marsupials. The mammae lack nipples, and the young suck milk from two lobules in the pouch in the echidna (**Fig. 5-5**) or from the abdominal fur in the platypus.

Monotremes typically have long periods of lactation and maternal care of the young. The platypus usually lays two eggs in a leaf nest in a burrow, where incubation lasts up to 12 days. The eggs become stuck together as the mother incubates by curling her body around them. The newly hatched young are tiny (11 millimeters in length) and nearly embryonic in appearance. The mother suckles the young and broods them (keeps them warm) for nearly 16 weeks, and they develop slowly. The first growth of hair appears 7 weeks after hatching, and the eyes do not open until about 9 weeks after hatching. Development of the single young echidna is similarly slow. The echidna egg, or sometimes two or three, is laid directly into the mother's pouch. After hatching, the young resides in the pouch until about 12 weeks of age, at which time the eyes open and the juvenile leaves the pouch to live in its mother's burrow. Weaning is at about 20 weeks. Both platypuses and echidnas have low reproductive rates, apparently no more than one clutch a year.

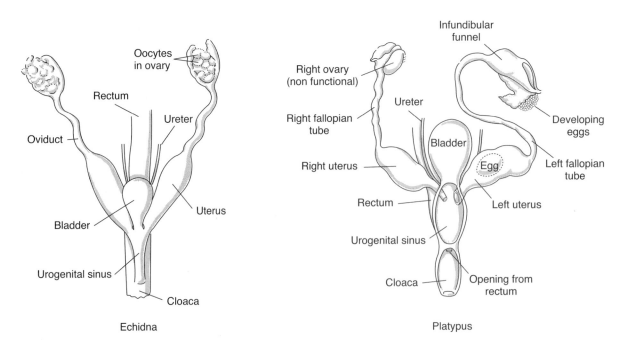

FIGURE 5-4   Anatomy of the female reproductive tract of monotremes. (Adapted from Szalay, F. S., Novacek, M. J., and McKenna, M. C. *Mammal Phylogeny.* Springer-Verlag, 1993.)

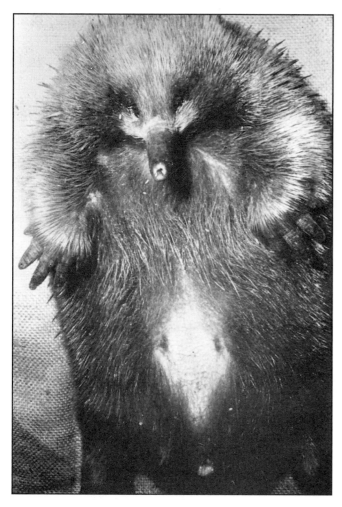

FIGURE 5-5   Ventral view of a live echidna (*Tachyglossus aculeatus*), showing the beaklike rostrum, the poorly developed pouch (typical of the nonbreeding season), and the tufts of hair at the mammary lobules.

Paleontology

The earliest known monotreme fossil is *Teinolophus* from the Australian early Cretaceous, about 112–121 million years ago (Rich et al. 2001). High-resolution X-ray computed tomography (CT or "cat") scans of the fossil jaw of *Teinolophus* indicate that it is a basal platypus (Rowe et al. 2008), but other authors consider it a representative of a separate family (Kielan-Jaworowska et al. 2004). During the time that *Teinolophus* and other early Cretaceous monotremes existed, their portion of Australia was within the Antarctic Circle. Although Antarctica was not covered with ice at that time, an early history in relatively cold climates may partly explain the monotremes' notable thermoregulatory abilities, including **hibernation** (Archer et al. 1993).

Another fossil monotreme was found in early Paleocene sediments, 62–63 million years old, in Patagonian Argentina, and is the only known occurrence of a monotreme outside Australia. Seemingly monotremes were once distributed across the southern continents (Pascual et al. 1992a, 1992b). The Argentine species, *Monotrematum sudamericanum*, is a possible ornithorhynchid, and had molars very similar to those of geologically younger Australian platypuses (Pascual et al. 2002).

Two species of the platypus *Obdurodon* are known in the early Miocene of Australia (Archer et al. 1992; Woodburne & Tedford 1975). One of these, *Obdurodon dicksoni* (**Fig. 5-6**) was large, and its bill included stouter septomaxillary bones and was wider than that of the living *Ornithorhynchus*. If *Obdurodon* foraged in a manner similar to that of *Ornithorhynchus*, these skull differences and its larger size may have enabled it to lift heavier stones underwater in searching for invertebrate prey (Archer et al. 1992, 1993). The extant genus *Ornithorhynchus* first appeared in the middle Miocene in Australia.

An insufficient fossil record and the tendency for reduction of the teeth make it difficult to determine the dental homologies between monotreme teeth and the teeth of other mammals. The dental morphology of fossil monotremes with nonvestigial teeth (*Teinolophus, Steropodon, Monotrematum, Obdurodon,* and others), and the primitive features of the postcranial skeleton of extant species, suggest that monotremes are outside the radiation of mammals with tribosphenic molars (metatherians and eutherians). An early and unusual but poorly known group of Mesozoic mammals, some of which are considered to be near the monotreme stem, are included in the Australosphenida, which includes the living monotremes (Kielan-Jaworowska et al. 2004; Luo et al. 2001). Australosphenidans occurred in the Middle Jurassic to Early Cretaceous on southern continents, including Australia, South America, and Madagascar.

The fossil record of echidnas is much less complete than that of platypuses. Other than the possible Cretaceous echidna humerus mentioned previously here, the oldest known tachyglossid is *Zaglossus robustus* from the late Miocene of Australia. Fossils of *Zaglossus* are also known from Pleistocene deposits in New Guinea, where the genus still occurs. One Pleistocene echidna from southwestern Australia was much larger than any living monotreme; *Zaglossus hacketti* might have weighed 30 kilograms, twice as much as the New Guinea long-beaked echidna.

Family Tachyglossidae

Members of this group have a robust body covered with short, sturdy spines that are controlled by unusually well-developed panniculus carnosus muscles (a sheet of muscles beneath the skin; **Fig. 5-7**). *Zaglossus bruijni*, the New Guinea

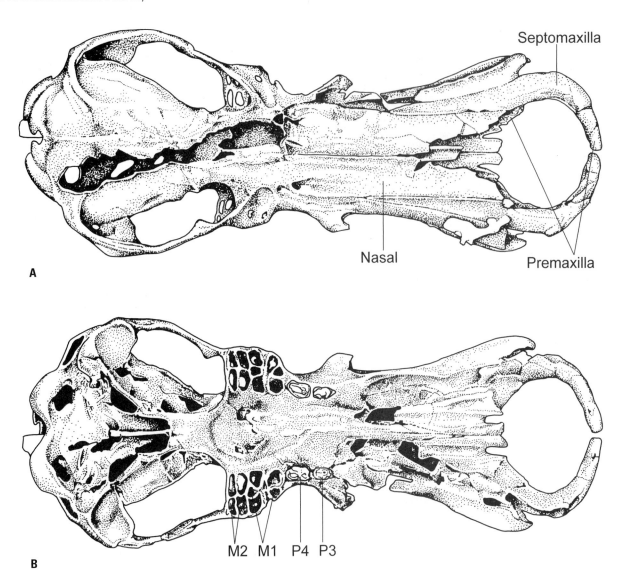

A

B

Septomaxilla

Nasal

Premaxilla

M2 M1 P4 P3

FIGURE 5-6 Well-preserved fossil skull of the Miocene platypus *Obdurodon dicksoni*. Length of skull 137 millimeters (the length of the skull of *Ornithorhynchus* is about 108 millimeters). (A) Dorsal view. (B) Palatal view. M1, M2, sockets for roots of molars; P3, third premolar; P4, fourth premolar. (Reproduced from Archer, M., et al., *Mammal Phylogeny*. 1 (1993): 75–94. With kind permission of Springer Science+Business Media. Courtesy of Michael Archer.)

long-beaked echidna, weighs from 5 to 16 kilograms, and the Australian short-beaked echidna (*Tachyglossus*) ranges from about 2.5 to 6 kilograms. The rostrum is slender and beak like and, at least in the short-beaked echidna (*Tachyglossus*), bears electroreceptors (Augee & Gooden 1992; Gregory et al. 1989). The beak also can function as a snorkel when swimming (Augee et al. 2006). The dentary bones are slender and delicate, and the long, **protrusible** tongue is covered with viscous mucus secreted mostly by the enlarged submaxillary salivary glands. Food is ground between spines at the base of the tongue and adjacent transverse spiny ridges on the palate. The limbs are powerfully built and are adapted for digging. The humerus is highly modified by broad extensions of the medial and lateral

epicondyles that provide unusually large surfaces for the origins of some of the powerful muscles of the forearm (Fig. 5-3). In *Zaglossus*, the number of claws is variable regionally; some animals have only three claws front and rear, whereas others have a full complement of five (Flannery 1995). In *Tachyglossus*, all digits have stout claws. The ankles of all males and some female echidnas bear medially directed spurs, the function of which is not known.

Echidnas have highly specialized modes of life. They are powerful diggers and can rapidly escape predators by burrowing. The food of *Tachyglossus* consists largely of termites and ants, whereas *Zaglossus* eats earthworms and soil arthropods. They forage by turning over stones and digging into termite and ant nests, then capturing the prey with the

FIGURE 5-7 Two species of monotremes: (A) Australian shortbeaked echidna (*Tachyglossus aculeatus*) and (B) New Guinea long-beaked echidna (*Zaglossus bruijni*).

sticky tongue or, in *Zaglossus*, by impaling prey on a barbed tongue (Flannery 1995).

The short-beaked echidna is capable of short-term (daily) **torpor** as well as long-term true hibernation. These animals gain weight during the warmer seasons and enter hibernation for 3 to 4.5 months in winter in the mountains of southeastern Australia (Grigg et al. 1989, 1992). During periods of torpor, the body temperature drops as low as 3.7°C, close to the temperature in the hibernaculum, or winter residence. In laboratory conditions, hibernating echidnas showed various patterns of breathing, from very slow and regular (one breath every 3 to 4 minutes) to periodic, with either constant or varying tidal volume (the volume of air moved into and out of the lungs). Some captive hibernating animals took no breaths for periods from 20 minutes to 2 hours (Nicol et al. 1992), a pattern unknown for any other hibernating mammals. Long-beaked

echidnas (*Zaglossus*) probably do not use torpor (Grigg et al. 2003).

Family Ornithorhynchidae

The duck-billed platypus is smaller than the echidnas, weighing from 0.5 to 2.0 kilograms. Some structural features of the platypus are associated with its semiaquatic mode of life (**Fig. 5-8**). The pelage is dense and velvety, and the underfur is woolly, rather like sea otter (*Enhydra*) or mole (Talpidae) fur. Similarly, these three kinds of mammals have fur that grows straight out of the skin (at right angles to the surface, not laid backward) so that forward and backward movement in water or in a burrow is not impeded. The external auditory meatus is tubular, as in the beaver (*Castor*). The eye and ear openings, which lack pinnae, lie in a furrow that is closed by folds of skin when the animal is submerged. The feet are webbed, but the digits retain claws that are used

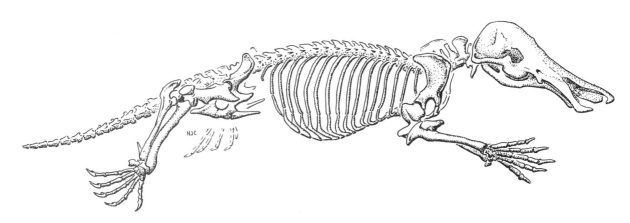

FIGURE 5-8 Skeleton of the duck-billed platypus, *Ornithorhynchus anatinus*.

for burrowing. The web of the forefoot extends beyond the tips of the claws and is folded back against the palm when the animal is digging or when it is on land. The ankles of the male platypus have grooved and medially directed spurs that are connected to venom glands.

The venom produced by platypuses and delivered by the hindfoot spurs of the males shows some structural similarities to snake and lizard venoms, as well as some differences (Warren et al. 2008; Whittington et al. 2008). The convergently evolved venom genes were separately incorporated in platypuses from the same gene "families" as those of reptiles, indicating a holdover from their earlier amniote ancestry. The venom proteins are derived from peptides called defensins, which have antimicrobial properties. The study of defensin-like peptides in the platypus thus may someday lead to the development of novel medical drugs (Whittington et al., in press).

Although the young have teeth, the gums of the adults are toothless and covered by persistently growing, horny plates. Anteriorly, the occlusal surfaces of the plates form ridges that are used to chop food; posteriorly, the plates are flattened crushing surfaces. Some additional **mastication** is accomplished by the flattened tongue, which acts against the palate.

The elongate rostrum bears a flattened, leathery bill that contains densely packed arrays of specialized receptor organs and nerves associated with electroreception (Scheich et al. 1986). The electroreceptors are sensitive to weak electrical currents and are used by the nocturnally active platypus to locate crustacean prey beneath underwater rocks or in turbid water. The sensory importance of the bill is reflected in the cortex of the brain, in which receptive fields for the tactile-electrosensory neurons of the bill are enormously represented (Rowe & Bohringer 1992).

The platypus inhabits a variety of waters, including mountain streams and slow-moving and turbid rivers, lakes, and ponds; it is primarily a bottom feeder. The platypus eats plants, aquatic crustaceans, insect larvae, and a wide variety of other animal material during dives that last for roughly 1 minute. The animals can eat up to half their body weight per day. The platypus takes refuge in burrows dug into banks adjacent to water. Seasonal torpor occurs in some parts of its range. Platypuses were formerly hunted for their plush fur; this helped decimate their populations.

SUMMARY

Represented today by only three genera and a total of five species, monotremes resemble no other living mammals and possess several features typical of reptiles. Monotremes lay and incubate eggs, yet they have hair and suckle their young. Monotremes retain some primitive features of their synapsid ancestors, yet they are highly derived (advanced) in other features. The earliest known fossil monotreme is from the early Cretaceous (112–121 million years ago) of Australia. Available evidence suggests that monotremes (Australosphenida) are an independent, primitive group separate from Theria, which consists of the sister groups Metatheria and Eutheria.

The monotreme reproductive system is a mix of primitive features shared with other amniotes and unique specializations. Monotremes differ strongly from therian mammals in having 10 X chromosomes (in females) and 5 X and 5 Y chromosomes (in males), instead of the usual 2 X or 1 X and 1 Y of therians. Monotremes lay rubbery-shelled eggs and are usually considered oviparous. The fetus employs an "egg tooth" to break out of the shell. The female echidna temporarily develops a pouch-like structure to incubate and protect the young, but the platypus never develops a pouch. The mammae lack nipples, and the young suck milk from two lobules in the pouch in the echidna or from the abdominal fur in the platypus. Monotremes typically have long periods of lactation and maternal care of the young.

Echidnas are powerful diggers and can rapidly escape predators by burrowing. The food consists largely of termites, ants, earthworms and soil arthropods. The short-beaked echidna is capable of short-term (daily) torpor as well as long-term true hibernation.

The platypus is semiaquatic and inhabits waterways from mountain streams to slow-moving and turbid rivers, lakes, and ponds. The platypus eats plants, aquatic crustaceans, insect larvae, and a variety of other animal material during short dives. Their elongate rostrum bears a flattened, leathery bill that contains densely packed arrays of specialized electroreceptors, which are used to detect the weak electrical currents produced by their invertebrate prey. Male platypuses have hindfoot spurs capable of delivering venom that is structurally similar to reptile venoms.

KEY CHARACTERISTICS OF MONOTREMES

- Skull toothless, except in young platypuses
- Cranial sutures disappear early in life
- Elongate and beak-like rostrum covered by leathery sheath
- Electroreceptors present on rostrum
- Sclerotic cartilages in the eyes
- Lacrimal and frontal bones absent
- Septomaxilla present in skull
- Auditory bulla absent, but middle ear partially surrounded by oval tympanic ring
- Cochlea curved, but not coiled
- Sprawled posture of forelimbs, humerus remains roughly horizontal to substrate
- Ankle spurs in adult males, for transporting venom
- Interclavicle, clavicles, precoracoids, coracoids, and scapula retained in pectoral girdle
- Epipubic bones large
- Cervical ribs present
- Ten X chromosomes (in females) and five X and five Y chromosomes (in males)
- Oviparous
- Telolecithal eggs with meroblastic cleavage
- Left ovary functional in the platypus, but both ovaries functional in echidnas
- "Egg tooth" present
- Cloaca present, penis attached to wall of cloaca in males
- Testes abdominal and seminal vesicles absent
- Mammae lack nipples

KEY TERMS

Cloaca	Meroblastic	Sclerotic cartilage
Fossorial	Oviparous	Suture
Hibernation	Protrusible	Telolecithal
Mastication	Rhinarium	Torpor

RECOMMENDED READINGS

Augee, ML, BA Gooden, & AM Musser. 2006. *Echidna—Extraordinary Egg-laying Mammal.* Collingwood, Victoria, Australia.

Augee, ML. 1992. *Platypus and Echidnas.* The Royal Zoological Society of New South Wales, Sydney, Australia.

Grutzner, F, et al. 2004. In the platypus a meiotic chain of ten sex chromosomes shares genes with the bird Z and mammal X chromosomes. *Nature, 432*:913–917.

Manger, PR, R Collins, & JD Pettigrew. 1997. Histological observations on presumed electroreceptors and mechanoreceptors in the beak skin of the long-beaked echidna, *Zaglossus bruijni. Proceedings of the Royal Society B, 264*:165–172.

Moyal, A. 2004. *Platypus: The Extraordinary Story of How a Curious Creature Baffled the World.* The Johns Hopkins University Press, Baltimore, MD.

Musser, AM. 2005. Monotremata. In *Encyclopedia of Life Sciences.* John Wiley and Sons, NY.

Whittington, CM, et al. 2008. Defensins and the convergent evolution of platypus and reptile venom genes. *Genome Research, 18*:986–994.

Metatheria

Metatherians (marsupials and their fossil relatives) and eutherians (placentals and their fossil relatives) represent two evolutionary lines that have been separate since the middle Cretaceous, ca. 98 to 100 million years ago (Cifelli 1993b; Clemens 1968; Lillegraven 1974; Springer et al. 1994). The peak of the diversification of the extant metatherian orders probably did not occur until the latest Cretaceous or early Tertiary period, between 65 and 52 million years ago, several million years later than the peak of the diversification of extant eutherian orders (Springer 1997; Springer et al. 1994, 1996). As a result of their long, independent history, metatherians differ structurally from eutherians in many ways.

Today, only two important strongholds for metatherians remain: the Australian region (Australia, Tasmania, New Guinea, and nearby islands) and the Neotropics (southern Mexico, Central America, and South America). Where they have been isolated from placentals for long periods, metatherians have undergone remarkable radiation. Most metatherians have functional counterparts among eutherians. Living metatherians are commonly called "marsupials," for the **marsupium** or pouch in which young are carried, in order to distinguish them from eutherians, or "placentals." However, several kinds of marsupials do not develop a pouch, and a few kinds (bandicoots) have a type of placenta similar to that of eutherians.

There are many recent hypotheses of phylogenetic relationships among higher level categories for living and extinct metatherians (Archer & Kirsch 2006; Horovitz & Sánchez-Villagra 2003; Meredith et al. 2008; Tyndale-Biscoe 2005). One of these is shown here (**Fig. 6-1**).

Morphology

The metatherian skull, described in detail by Wible (2003), frequently has a small, narrow braincase housing small cerebral hemispheres with simple convolutions. Ossified auditory bullae, when present, are usually formed largely by the alisphenoid bone rather than by the tympanic (both ectotympanic and entotympanic), petrosal, and/or basisphenoid bones, as occurs in most placentals. The metatherian palate characteristically has large vacuities (**Fig. 6-2**), and the angular process of the dentary bone is usually inflected medially. The dentition is unique in that there are never equal numbers of incisors above and below, except in the family Vombatidae. The cheek teeth primitively include 3/3 premolars and 4/4 molars.

Marsupials have a unique pattern of tooth eruption and replacement in which only one tooth in each jaw (the third upper and lower premolars; P3/3) actually replaces a deciduous tooth. The pattern seems to be as follows: (1) The primary (deciduous) incisors and canines begin to develop before their secondary replacement teeth, but they are vestigial and never erupt. Instead, the deciduous precursors are resorbed when the secondary ("permanent") teeth appear. (2) The first two premolars are deciduous teeth that develop late and are not replaced by secondary teeth. (3) The deciduous third premolars are **molariform**

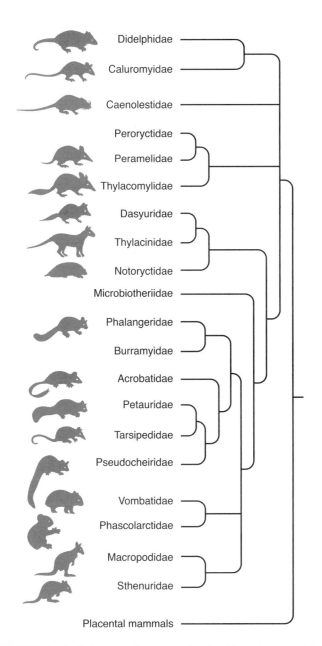

FIGURE 6-1 A phylogeny of present-day families of marsupials. (Adapted from Tyndale-Biscoe, H. *Life of Marsupials.* CSIRO Publishing, 2005.)

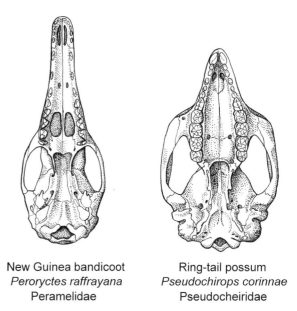

New Guinea bandicoot
Peroryctes raffrayana
Peramelidae

Ring-tail possum
Pseudochirops corinnae
Pseudocheiridae

FIGURE 6-2 Ventral views of two marsupial skulls: a New Guinea bandicoot (length of skull 82 millimeters) and a ringtail possum (length of skull 97 millimeters). (Adapted from Tate, G. H. and Archbold, R., *Bull. Amer. Mus. Nat. Hist.* 73 (1937): 331–476.)

multituberculates, and early therian mammals (including two Cretaceous eutherians; Novacek et al. 1997), epipubic bones extend forward from the pubic bones in both sexes (except in the recently extinct *Thylacinus*, in which the epipubic bones were vestigial, and in the extinct Borhyaenidae, in which they were absent).

Reproduction

The metatherian reproductive pattern differs sharply from that of eutherians (see Chapter 20). The females of about 50% of the metatherians of today have a marsupium (an abdominal pouch) or abdominal folds within which there are nipples (**Table 6-1**). The number of nipples varies from 2 in the family Notoryctidae and some members of the family Dasyuridae to 27 in some members of the family Didelphidae. Individual variation in the number of nipples often occurs within a species. The female reproductive tract is bifid—that is, the vagina and uterus are double (**Fig. 6-4**). In all but the family Notoryctidae, which is adapted for digging, the testes are contained in a scrotum anterior to the penis.

The **gestation** period is characteristically short (8 to 43 days), and the young are tiny and rudimentary at birth. Altricial newborn metatherians probably possess the minimal anatomical development allowing survival outside the uterus. Organogenesis has just begun. The separation of the ventricles of the heart is incomplete. The lungs are vascularized sacs lacking alveoli, and the kidneys lack glomeruli. Also lacking are cranial nerves II to IV and VI, eye pigments,

and are replaced by secondary ones that are premolariform. (Sometimes the deciduous third premolars, dP3/3, are not immediately lost when the P3/3 erupt anterior to them, and the P3/3 and dP3/3 may coexist in the jaws for varying lengths of time.) (4) The four molars are unreplaced primary teeth (Cifelli & Muizon 1998; Cifelli et al. 1996; Luckett 1993).

Metatherians often have highly specialized feet associated with specialized types of locomotion (**Fig. 6-3**). The unusual patterns of specialization of the hind feet are probably a result of an arboreal heritage and the early development of an opposable first digit and an enlarged and powerfully clutching fourth digit. As in the monotremes,

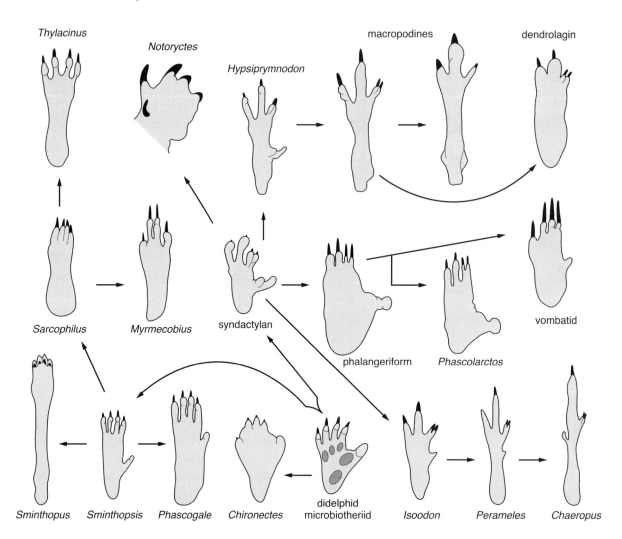

FIGURE 6-3 Ventral views of marsupial right hind feet, showing patterns of specialization associated with various styles of locomotion. At bottom center is the presumed basic arboreal type, represented by the foot of a didelphid or microbiotheriid (with plantar pads shown). The arrows indicate possible evolutionary pathways leading to greater specialization. Didelphidae: *Chironectes* (webbed aquatic foot); Dasyuridae: *Phascogale, Sminthopsis, Antechinomys, Sarcophilus*; Myrmecobiidae: *Myrmecobius*; Thylacinidae: *Thylacinus*; Notoryctidae: *Notoryctes*; Macropodidae: *Hypsiprymnodon*, macropodines, dendrolagin; Phascolarctidae: *Phascolarctos*; Vombatidae: vombatid; and Peramelidae: *Isoodon, Perameles, Choeropus*. (Adapted from Szalay, F. S. *Evolutionary History of the Marsupials and an Analysis of Osteological Characters.* Cambridge University Press, 1994.)

eyelids, and cerebral commissures (nerve fiber bundles connecting the cerebral hemispheres). Despite this minimal development, the naked, blind, and delicate newborn is able to make its way at birth from the vulva to the marsupium (Fig. 20-12). Here it attaches to a nipple and remains there for a period greatly exceeding the gestation period. The weight of the young metatherian when it leaves the pouch and that of the newborn eutherian is roughly the same in species of comparable adult size (Sharman 1970).

Most metatherians, with the exception of the Peramelidae and Peroryctidae, have a **choriovitelline placenta** that lacks villi. These two families of bandicoots are unique among metatherians in possessing a **chorioallantoic placenta** similar to that of eutherians (Hughes et al.

1990; see Chapter 20 for further discussion of metatherian reproduction).

Paleontology

The earliest undoubted metatherian fossils are from the early Cretaceous of Asia (Cifelli & Davis 2003; Kielan-Jaworowska et al. 2004; Luo et al. 2003). In the late Cretaceous, the dinosaurs were still dominant, and surviving groups of Mesozoic mammals included the Monotremata, Multituberculata, Eutheria, and several extinct taxa. The niches filled by primitive metatherians in the late Cretaceous are difficult to ascertain because the fossil record includes few well-preserved postcranial skeletons. The fossil

TABLE 6-1   Marsupium Type and Number of Teats in Some Metatherians

Family	Marsupium	Position of Opening	Number of Teats
Didelphidae	Absent–well developed	—	4–27
Caenolestidae	Absent	—	4–5
Microbiotheriidae	Well developed	Anteroventral	4
Dasyuridae	Usually absent	Posterior	2–12
Thylacinidae	Crescent-shaped flap	Posterior	4
Myrmecobiidae	Absent	—	4
Peramelidae	Well developed	Posterior	6–10
Peroryctinae	Well developed	Posterior	—
Notoryctidae	Small	Posterior	2
Acrobatidae	Well developed	Anterior	1–4
Burramyidae	Well developed	Anterior	4–6
Macropodidae	Well developed	Anterior	4
Petauridae	Well developed	Anterior	2–4
Phalangeridae	Well developed	Anterior	2–4
Phascolarctidae	Well developed	Posterior	2
Potoroidae	Well developed	Anterior	4
Pseudocheiridae	Well developed	Anterior	2–4
Tarsipedidae	Well developed	Anterior	4
Vombatidae	Well developed	Posterior	2

Modified, with kind permission from Springer Science and Business Media, from Marshall, L. G., J. A. Case, and M. O. Woodburne. 1990. Phylogenetic relationships of the families of marsupials, p. 433–505. In H. H. Genoways (ed.), Current Mammalogy. Plenum Press, New York..

record consists mostly of teeth and jaw fragments that are similar in morphology and function to those of modern marsupials with generalized dentitions. An exception is the well-preserved early Cretaceous stem metatherian *Sinodelphys*, which had the same type of specialized feet and ankles known in extant mammals that are agile tree climbers (Luo et al. 2003). Several families of marsupials became extinct at or near the Cretaceous-Paleogene boundary, the same time as the nonavian dinosaurs became extinct.

Members of the Didelphimorphia (opossum-like marsupials) were present in North America during the late Cretaceous and have a nearly continuous fossil record there through the middle Miocene. A few fossil metatherians also have been found in the Cretaceous of Asia and in the Paleogene (early Cenozoic) of Europe and Afro-Arabia (Hooker et al. 2008; Sanchez-Villagra et al. 2007a, 2007b). During most of the Cenozoic, however, the southern continents seem to have served as major centers of metatherian diversification. In Australia, South America, and probably Antarctica when it was not covered by glacial ice, metatherian species radiated in partial or complete isolation from competition with eutherians.

Some experts believe that metatherians arose in North America and moved southward into South America in the Cretaceous and on to Australia, via Antarctica, in the late Cretaceous or earliest Tertiary (Muizon et al. 1997; Woodburne & Case 1996). Late Cretaceous fossils, possibly ancestral to the Didelphimorphia, are known from the interior

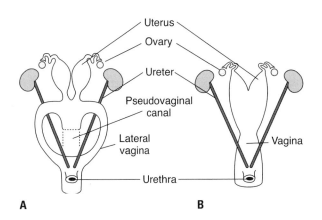

FIGURE 6-4   Generalized diagrams of female reproductive tracts: (A) metatherians and (B) eutherians. (Adapted from Sharman, G.B., *Science* 167 (1970): 1221–1228.)

of North America, and didelphoids and borhyaenoids (dog-like marsupials) are known from Bolivia and Brazil in the Paleocene (Bergqvist et al. 2006; Muizon 1994, 1998; Muizon et al. 1997; Woodburne & Case 1996). The presence of metatherian fossils in Eocene rocks in Antarctica strongly supports a southern route of their dispersal into Australia (Case et al. 1988; Goin & Carlini 1995; Marenssi et al. 1994; Woodburne & Zinsmeister 1982). Of course, some groups could have originated in Antarctica when it was not completely covered in ice (in the Paleocene, Eocene, and part of the Oligocene). Metatherians reached Australia well before eutherians did and underwent a spectacular Cenozoic radiation. Eighteen living metatherian families, including 70 genera, resulted from this radiation, as did several additional extinct families and even orders, many of which were unusual and fascinating (see, for example, Archer & Clayton 1984; Archer et al. 1991).

Anyone seeing the present Australian metatherian fauna for the first time finds the number of species and the structural extremes impressive. By comparison with the fauna of the late Pleistocene, however, the present fauna is severely depleted. Many species of large metatherians became extinct between the late Pleistocene and historic time, and further reductions occurred in historic times. The extinct families Thylacoleonidae and Diprotodontidae are especially noteworthy. *Thylacoleo* was a predaceous Pliocene-Pleistocene genus with some species roughly the size of a mountain lion. The third premolars were greatly elongated shearing blades, and the strongly built front limbs had retractile claws (Archer & Dawson 1982; Keast 1972). The Diprotodontidae were represented in the Pleistocene epoch by several genera. One (*Diprotodon*) was roughly the size of a rhinoceros and is the largest metatherian known. A number of very large kangaroos (family Macropodidae) also became extinct before historic times. Among them was a giant, *Procoptodon goliah*, an extremely short-faced macropodid (**Fig. 6-5**) that stood about 3 meters tall.

Why did these imposing marsupials disappear? This question has been considered carefully by Choquenot and Bowman (1998) and Flannery (2002) who stressed the possible influence of humans. Aborigines had entered Australia by 50,000 years ago, in the late Pleistocene, when the metatherian fauna included the large species just mentioned; however, other workers recently provided evidence in Tasmania supporting a change in climate to a cold, arid period 55,000 years ago that may have caused the extinction of some of the Australian megafaunal marsupials before humans arrived on the continent (Cupper & Duncan 2006; Johnson 2006), whereas additional new dates and evidence from Australia and Tasmania also have been used to renew the argument that humans caused at least some megafaunal

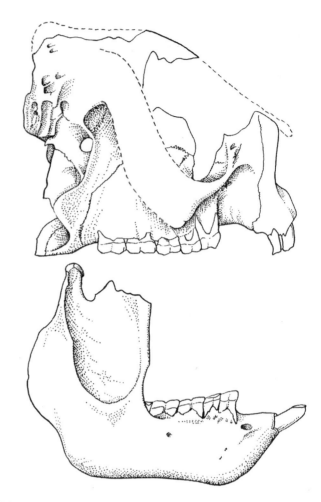

FIGURE 6-5   Fossil skull of *Procoptodon goliah*, a 3-meter-long, grazing, macropodid marsupial from the Pleistocene of Australia. The length of the skull is 218 millimeters. (Adapted from Tedford, R.H., *University of California Publications in Geological Science* 64 (1967): 1–165.)

extinction there (Diamond 2008; Turney et al. 2008). In any case, only modern faunas are found in archeological sites spanning the latest Pleistocene.

Further reductions in the number of species of metatherians began with the coming of European peoples to Australia. The combined effects of heavy grazing by livestock, clearing of land for agriculture, and the introduction of the Old World rabbit (*Oryctolagus cuniculus*), red fox (*Vulpes vulpes*), and feral cat (*Felis catus*) caused widespread declines in the abundance of many species and a number of extinctions (Calaby 1971; Flannery 2002; Lever 1985). As an example, of about 45 species of kangaroos that occupied Australia just before the entry of Europeans, three are extinct and the populations of roughly a dozen have declined drastically, although some species may have become rare before historic times.

The marsupial radiation in South America rivaled that in Australia. There are more than 130 living and fossil genera of metatherians known from South America and more than 130 from Australia (McKenna & Bell 1997). Accompanying the striking structural diversification of the marsupials in response to a wealth of available habitats was a convergence by many types with eutherians. By the middle Tertiary, there were metatherians that structurally (and undoubtedly functionally) resembled shrews, moles, rodents, and carnivorans. In the early Tertiary, several groups of eutherians were in South America. Xenarthrans and various ungulate groups seemingly "owned" the large herbivore adaptive zone throughout much of the Tertiary, to the complete exclusion of metatherians.

All of the South American marsupials presumably evolved from a basal, insectivorous–omnivorous metatherian stock. The occurrence of five orders in the early Paleocene in Bolivia and six orders at a late Paleocene fossil locality in Brazil documents an early Tertiary radiation

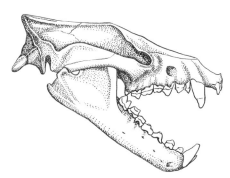

A *Borhyaena*

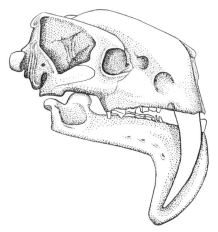

B *Thylacosmilus*

FIGURE 6-6 Skulls of members of the extinct metatherian families (A) Borhyaenidae and (B) Thylacosmilidae. (A) Length of skull 230 millimeters. (B) Length of skull 232 millimeters. (Modified from Romer, A.S. *Vertebrate Paleontology.* University of Chicago Press, 1966.)

(Bergqvist et al. 2006; Muizon & Brito 1993). Perhaps the most spectacular Tertiary metatherians were sparassodonts including the dog-like family Borhyaenidae and the saber-tooth family Thylacosmilidae.

The Borhyaenidae includes a number of metatherians with dentitions that suggest styles of feeding ranging from omnivorous to carnivorous. They are convergently evolved in various features with Australasian thylacinids and placental canids and bears. One borhyaenid, *Stylocynus*, was roughly the size of a bear and presumably omnivorous; a number of small- and medium-sized species were also omnivores. In some borhyaenids, including one of the earliest known (early Paleocene) species *Mayulestes ferox*, numerous skeletal features indicate semiarboreal habits (Muizon 1998). All known borhyaenids are rather short-legged, and the terrestrial types lack marked cursorial specializations. Members of the genus *Borhyaena* had a skull rather like that of a wolf (**Fig. 6-6A**).

The Miocene-Pliocene family Thylacosmilidae includes two genera that had long, recurved, saber-like upper canines. The roots of these teeth extend nearly to the occipital part of the skull, and a long flange on the dentary bone protects the tips of the canines (**Fig. 6-6B**). Although basically well adapted to a carnivorous life, both borhyaenids and thylacosmilids had canines that wore rapidly because they had only a thin layer of enamel. In compensation, the roots in some forms remained open through much of life, permitting continued growth (Patterson & Pascual 1972). The shape of the sabers of *Thylacosmilus* and skeletal features usually associated with powerful neck musculature indicate that the sabers were used like those of eutherian saber-tooth cats (Felidae; Akersten 1985; Goin & Pascual 1987) and saber-toothed barbourofeline carnivores (Nimravidae, not cats). The limbs of thylacosmilids are short, but foot posture was **digitigrade**. Most prey were probably captured by a surprise attack rather than a chase.

The complex story of the decline and eventual extinction of the Borhyaenidae is told by Marshall (1977). Many early Eocene borhyaenids resembled didelphid marsupials and were of moderate size; by the early Oligocene, however, several very large forms had evolved. The largest of all, *Proborhyaena gigantea*, which had a skull roughly 1 meter long, disappeared by the end of the early Oligocene, and with its extinction a trend began toward reduced size in carnivorous borhyaenids. The lack of cursorial adaptations in borhyaenids is unusual for a group of large carnivorous mammals and may have resulted from competition with large carnivorous birds.

During the early Tertiary in South America, the adaptive zones available to carnivores were probably not dominated by any one group. From early Oligocene through

Pliocene times, three families of large carnivorous birds shared with borhyaenids a predatory mode of life. *Phorusrhacos*, a predaceous member of the family Phorusrhacidae (order Gruiformes), stood 1.5 to 3 meters tall and had a heavily built, hooked bill mounted on a skull about the size of that of a horse. The wings were tiny, and the bird was flightless, but the hind legs were long and slim, and the bird must have been a swift runner. Probably the phorusrhacid birds and the borhyaenid carnivores partitioned the resources available to carnivores: the phorusrhacids developed and maintained large size, occupied open savanna areas, and used their cursorial ability in pursuing swift prey; the borhyaenids were of more moderate size, occupied wooded country, and largely killed slow-moving prey or scavenged dead animals. The late Tertiary reduction in the size of borhyaenids may have been part of an adaptive trend toward reduction of competition with the imposing phorusrhacids (Patterson & Pascual 1972).

Further declines in the fortunes of borhyaenids were apparently associated with the mid-Miocene arrival in South America of members of the heretofore entirely North American raccoon family (Procyonidae). The fossil record indicates that replacement of the larger omnivorous borhyaenids by the omnivorous procyonids was complete by the late Pliocene. Extinction of other borhyaenids may have been related to possible competition with marsupials of the opossum family Didelphidae. The small- to medium-sized borhyaenid omnivores declined in the middle Pliocene and became extinct by the late Pliocene, whereas didelphids with similar adaptations appeared in the middle Pliocene and underwent a striking adaptive radiation in the late Pliocene.

The history of the Borhyaenidae was thus intertwined with that of the phorusrhacid birds, the immigrant South American procyonids, and the didelphid metatherians. By the late Pliocene, the borhyaenids were gone, and the three latter groups prevailed.

An event of major importance to the South American mammalian fauna occurred in the late Pliocene. About 2.5 million years ago, the land bridge connecting North and South America was established, and a flood of northern eutherians moved southward; some southern mammals moved northward (see Chapter 25). Although the Borhyaenidae were already extinct and were thus not affected by this collision of faunas, the saber-tooth thylacosmilids were still present and were perhaps decisively affected. The occurrence of eutherian saber-tooth cats in early Pleistocene strata immediately above late Pliocene beds bearing thylacosmilids suggests the possibility of competitive replacement. An early interpretation was that competitively superior North American eutherians entered South

America over the Panamanian **land bridge**, leading to an eventual decline in metatherian diversity. This view has been challenged as overly simplistic by many later authors who showed that the apparent decline in metatherians might not be directly related to the faunal interchange between the two continents. Webb (1991, 2006) further argued that ecological and geographic factors affected the different fates of the land mammals on either side of the land bridge.

The carnivory characterizing a number of the fossil types discussed previously here was just one of a number of feeding patterns of South American metatherians. *Necrolestes* from the middle Miocene was probably insectivorous and fossorial and may have had a mode of life similar to that of the Australian marsupial "moles" (Notoryctidae).

The family Caenolestidae, small shrew-like animals that still persist in relict populations along the Andes Cordillera, appeared first in the Oligocene of South America. Some caenolestids were convergent toward multituberculates in a series of features, including general skull form, enlargement of the anterior incisors, and structure of the **serrate**, lower pair of cheek teeth (**Fig. 6-7**). Both multituberculates and the multituberculate-like caenolestids (subfamily Abderitinae) probably resembled rodents in feeding habits.

The most rodent-like marsupials yet known are two South American species from the early Tertiary. These species compose the family Groeberiidae and are remarkable in having such features as enlarged incisors with enamel only on the anterior surfaces, a simplified pattern on the crowns of the cheek teeth, and a short, deep rostrum and mandible (Pascual et al. 1994).

The metatherian family Argyrolagidae is a supreme example of evolutionary convergence. This unique family is known from the Oligocene to Pliocene of South America (Sánchez-Villagra 2001; Sánchez-Villagra et al. 2000). Argyrolagids did not resemble closely any other group of metatherians, but they possessed a series of morphological characters that are found today in such specialized rodents as kangaroo rats (Heteromyidae) and jerboas (Dipodidae). These rodents occupy mainly sparsely vegetated desert or

FIGURE 6-7   Jaw of a Miocene caenolestid, showing the highly specialized, trenchant cheek tooth. (Modified from Romer, A.S. *Vertebrate Paleontology*. University of Chicago Press, 1966.)

semiarid areas; all are **saltatorial** (use a jumping style of locomotion), and all share certain distinctive morphological features. The hindlimbs are long, and the hind feet are modified by the loss of digits and, in some cases, by the fusion of metatarsal bones (see Fig. 13-18). The long hindlimbs of argyrolagids are highly specialized along similar lines: only the third and fourth digits of the foot are retained, and the metatarsals are closely appressed and resemble to some extent the cannon bone of an artiodactyl (**Fig. 6-8**).

All of these animals have cheek teeth adapted to grinding and incisors suited for gnawing. In kangaroo rats and argyrolagids, the cheek teeth are rootless and have a simple occlusal surface. The cheek teeth have also been considered convergent with those of elephant-shrews (Sánchez-Villagra & Kay 1997).

Condensation of moisture on the cool nasal mucosa during exhalation is a means of reducing pulmonary water loss in kangaroo rats, and the unusual tubular extension of the nasal cavity anterior to the incisors in these animals (see Fig.13-17) is associated with improved water conservation (see Chapter 21). In the argyrolagids there is an even more elongate extension of the nasal cavity. Here, too, this specialization may have facilitated the maintenance of water balance in an arid environment.

The entire form of the skull in kangaroo rats and jerboas is modified by the enormous auditory bullae, and in argyrolagids, the bullae are also inflated. This enlargement of the bullae has been shown to be one of a remarkable series of specializations that allows kangaroo rats to detect faint, low-frequency sounds made by predators such as rattlesnakes (Crotalidae). The enlarged bullae of the argyrolagids probably served a similar end.

It seems, then, that in two lineages that have been separate since at least the middle Cretaceous, but which occupy (or occupied) similar dry habitats, nearly identical suites of characteristics have evolved.

One cannot help but wonder why argyrolagids became extinct for "by all rules of analogy and theories of extinction, they should have survived, as did their close ecological analogs in North America, Asia, Africa, and Australia" (Simpson 1970). Currently, the mystery of the extinction of the argyrolagids remains unsolved. As the fossil record of South American mammals becomes more complete, information on the extinction of the argyrolagids and other metatherian groups will probably come to light.

Metatherians Versus Eutherians: Relative Competitive Abilities

When one views the course of mammalian evolution since metatherians and eutherians diverged from a common ancestor in middle Cretaceous times, the general impression is that the two groups are not of equal adaptive ability. Several lines of evidence can be cited.

FIGURE 6-8 *Microtragulus* (Argyrolagidae), an extinct bipedal metatherian from the Mio-Pliocene of Argentina. (A) Reconstruction of the skeleton. Length of skull 55 millimeters. (B) Enlarged view of partial hind foot. The appressed metatarsals of digits 3 and 4 form a structure resembling the cannon bone of artiodactyls. (Adapted from Simpson, G.G., *Bulletin Museum Comparative Zoology* 139 (1970): 1–86.)

1. Metatherians have not equaled the remarkable functional radiation of eutherians. There are no flying or marine metatherians, and some extremely productive food sources have never been tapped. Marine plankton, utilized by two orders of eutherians (Cetacea and Carnivora), and flying insects, eaten by bats (Chiroptera), have never been part of the marsupial diet.

2. Metatherians have been far more conservative in structural plan. None has modified limbs into fins or wings as eutherians have.

3. Metatherians have not been able to exploit great size. Although there were several large metatherians in the Pleistocene, the largest living metatherian (the red kangaroo) is only 1/1,300 the size of the largest eutherian (the blue whale).

4. Metatherians have never evolved highly social behavior.

5. Metatherians have not developed the systematic diversity of eutherians. Only about 6% (335) of the total number of species of living mammals (5,420) are metatherians.

Although sadly incomplete and equivocal, the fossil record does suggest a competitive edge for eutherians. In North America, where metatherians appeared before eutherians, metatherians had declined seriously by the latest Cretaceous but eutherians had radiated. In South America, similarly, where a Tertiary metatherian radiation occurred in isolation from a balanced assemblage of eutherians, the diversity of metatherians declined late in the Tertiary and in the Pleistocene, perhaps at first because of the entry into South America of only a few eutherians and then (with the emergence of the land bridge connecting North and South America) because of an invasion of eutherians from the north. Especially impressive is the total extinction of the South American metatherian carnivores (of the families Borhyaenidae and Thylacosmilidae) and their ultimate replacement by eutherian carnivores.

Although there is no general consensus on the matter, many scholars consider metatherians adaptively and competitively inferior to eutherians. Here we briefly catalog some basic differences between the two groups to introduce the conflicting views that appear in the literature and to avoid offering a resolution for an unresolved problem.

Each of the following may be associated with adaptive–competitive differences between the two groups:

1. The metatherian mode of reproduction is probably more like that of early mammals than that of eutherians. Metatherians have a brief gestation period and bear almost embryonic young that have precocious forelimbs used to climb the mother's hair to reach the nipple. They undergo most of their basic development during a long period of lactation while attached to the mother's nipple and nourished by milk. Eutherians, in contrast, have long gestation periods, and the young are much more developed at birth; they have a relatively short period of lactation. The metatherian need for precocious grasping forelimbs at birth may constrain their adaptability relative to placentals, precluding the development of wings, flippers, or other specializations of the forelimb.

2. The cerebral cortex develops more rapidly and attains greater volume in eutherians than in marsupials (Muller 1969). The brain and nervous system develop best in the highly nutritive, stable, oxygen-rich environment of the uterus, possibly giving an advantage to eutherians over metatherians. Eisenberg and Wilson (1981), however, reported that overall cranial capacities are similar to eutherians for many didelphid species.

3. Behavioral plasticity is greater in eutherians: social groups with long-term dominance hierarchies and cooperative rearing of young occur only among eutherians. Territoriality, an important aspect of eutherian behavior, is uncommon in metatherians.

4. Antipredator behavior is more highly developed in eutherians: unified herd action, co-operative defense of young, complex vocal and visual communication, and sustained high-speed running are known only among eutherians.

5. The relatively low **diploid number** of chromosomes for marsupials has been mentioned (Hayman 1977; Lillegraven 1975) as possibly being related to their lack of evolutionary flexibility. Marsupials may be capable of far less genetic variability than placentals. The exact reasons for this are unknown but are made more apparent by a lower reproductive rate.

6. The extended gestation of eutherians produces young that are far more endothermic than metatherians and allows for better exploitation of colder climates.

7. The investment of energy by the mother is probably lower in metatherians than in eutherians (Parker 1977), but eutherians are seemingly able to reproduce more rapidly. Unlike eutherians, metatherians show little connection between metabolic rate and rate of reproduction. This is especially evident in small metatherians with high metabolism (Lillegraven et al. 1987). These small metatherians retain a very long lactation period, thereby missing the small-sized advantage that placentals often utilize by producing several litters per year. Metatherians do tend to have larger litter sizes than eutherians, but eutherians dominate by increasing their population sizes much more rapidly.

The view that the metatherian style of reproduction and metatherian biology in general represents an alternative, but not inferior, solution to survival problems has been discussed by a number of researchers (including Kirsch

1977; Lillegraven et al. 1987; Parker 1977; Pond 1977; Tyndale-Biscoe & Renfree 1987). For comparisons of the adaptiveness of the metatherian and eutherian patterns of reproduction, see Lillegraven et al. (1987) and Tyndale-Biscoe and Renfree (1987).

South American Metatherians

Living South American marsupials belong to three separate orders: Didelphimorphia, Paucituberculata, and Microbiotheria. Members of the order Didelphimorphia, opossums and mouse opossums, are the most generalized metatherians and constitute one of the oldest known groups, dating from the early part of the late Cretaceous (Springer 1997). As mentioned, this is a basal group of the metatherian radiation (Muizon et al. 1997); however, the Didelphimorphia may not be a natural group (i.e., not monophyletic; Goin 2003). The order Paucituberculata includes the small rat opossums, whereas the monito del monte (*Dromiciops gliroides*) is the sole living member of the order Microbiotheria.

▪ Order Didelphimorphia

Family Didelphidae

The family Didelphidae includes 17 modern genera comprising 87 species that range from southeastern Canada, with the American opossum (*Didelphis virginiana*), to southern Argentina, with the Patagonian opossum (*Lestodelphys halli*). In these New World opossums, the rostrum is long (**Fig. 6-9A**), the braincase is usually narrow, and the sagittal crest is prominent. The dental formula is 5/4, 1/1, 3/3, 4/4 = 50. The incisors are small and unspecialized, and the canines are large. The upper molars are basically tritubercular with sharp cusps, and the lower molars have a trigonid and a talonid (**Fig. 6-10A, B**).

Except for the opposable and clawless hallux in all species, a feature probably inherited from arboreal ancestral stock, and the webbed hind feet in the yapok or water opossum (*Chironectes minimus*), the feet are unspecialized, with no loss of digits or **syndactyly** (the condition in which two digits are attached by skin, as shown by a number of examples in Fig. 6-3). The foot posture is **plantigrade**. A

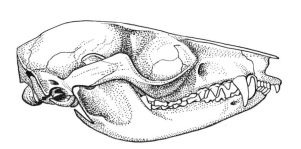

A   Mouse opossum
Tlacuatzin canescens (Didelphidae)

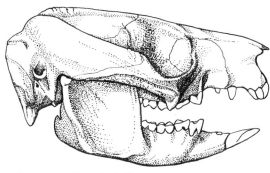

B   Brush-tail possum
Trichosurus vulpecula (Phalangeridae)

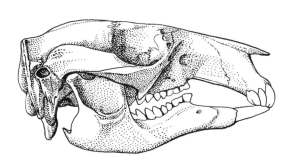

C   Wallaby
Wallabia bicolor (Macropodidae)

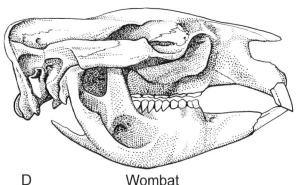

D   Wombat
Vombatus ursinus (Vombatidae)

FIGURE 6-9 Skulls of marsupials: (A) length of skull 35 millimeters. (B) Length of skull 87 millimeters. (C) Length of skull 135 millimeters. (D) Length of skull 180 millimeters.

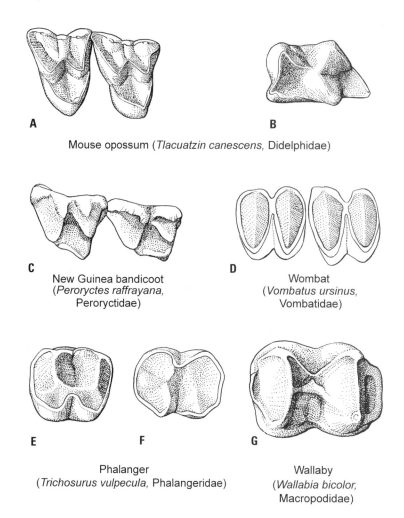

A

B

Mouse opossum (*Tlacuatzin canescens*, Didelphidae)

C

New Guinea bandicoot
(*Peroryctes raffrayana*,
Peroryctidae)

D

Wombat
(*Vombatus ursinus*,
Vombatidae)

E   F

Phalanger
(*Trichosurus vulpecula*, Phalangeridae)

G

Wallaby
(*Wallabia bicolor*,
Macropodidae)

FIGURE 6-10   Occlusal views of metatherian molars: (A) Second and third right upper and (B) third lower left molar of a mouse opossum. (C) Second and third upper right molars of the New Guinea bandicoot. (Adapted from Tate, G. H. and Archbold, R., *Bull. Amer. Mus. Nat. Hist.* 73 (1937): 331–476.) (D) Second and third right upper molars of a wombat. (E) Second upper right and (F) third lower left molar of a phalanger. (G) Second upper right molar of a wallaby.

marsupium is present in some didelphids but is represented by folds of skin protecting the nipples in others and is absent in some. The tail is long and usually prehensile.

Although they occupy a wide range of habitats, didelphids are primarily inhabitants of tropical or subtropical areas, where they are often locally abundant. Most didelphids are partly arboreal and are omnivorous; however, the lutrine opossum *Lutreolina* is predominantly carnivorous, the water opossum is semiaquatic and carnivorous/ insectivorous (Vieira & Astúa de Moraes 2003), and the woolly opossum (*Caluromys*) is largely herbivorous. The small mouse opossum (*Marmosa mexicana*), a widespread Neotropical didelphid, is one of the most abundant small mammals in some parts of Mexico. Mouse-like in general appearance, it seems to be largely insectivorous in some areas, at least during the summer (R. Smith 1971). Poorly defined folds of skin protect the nipples of *Marmosa*, and the young simply hang on to the nipples and the mother's venter as best they can.

■ **Order Paucituberculata**

Family Caenolestidae

Recent members of this family (three genera and six species) bear the common name of shrew-opossum or rat opossum. The three genera have disjunct distributions from one another, each relict population occupying forested areas in a different region of the Andes Mountains of northern and western South America. *Rhyncholestes* and *Lestoros*, each with one species, occupy parts of the southern and central Andes, respectively; *Caenolestes*, with four species, occupies parts of the northern Andes (Albuja & Patterson 1996). The earliest known caenolestids are found in the late Oligocene fossil record of South America. In the Oligocene and Miocene, a diverse group of caenolestids appeared, including some highly specialized types (Fig. 6-7).

Caenolestids resemble shrews because of their elongate heads and small eyes (**Fig. 6-11**). The skull is elongate and the brain primitive; the olfactory bulbs are large and

FIGURE 6-11 An Incan caenolestid (*Lestoros inca*). This individual came from Peru, at an elevation of 3,530 meters.

the cerebrum lacks fissures. The dental formula is 4/3–4, 1/1, 3/3, 4/4 = 46 or 48; the first lower incisors are large and **procumbent**, and the remaining lower incisors, the canine, and the first premolar are **unicuspid**. Kirsch (1977) found that the lower incisors are used like rapiers to stab prey. The atlas bears a movable cervical rib. The feet are unspecialized, the tail is long but not prehensile, and there is no marsupium.

Order Microbiotheria

Family Microbiotheriidae

One species (*Dromiciops gliroides*) that inhabits cool, southern Andean forests is the single living member of the family Microbiotheriidae (Gardner 1993). Although the species occurs in South America, the order to which it belongs is considered by many authors to be a relict of an Australian or Antarctic radiation of metatherians known as the Australidelphia (Luckett 1994; Spotorno et al. 1997; Springer et al. 1996; Woodburne & Case 1996). Fossil microbiotheriids are known from the early Paleocene in South America and the middle Eocene in Antarctica (Carlini et al. 1990; Muizon 1992; Woodburne & Case 1996).

The tiny *Dromiciops* is restricted to south-central Chile. The dental formula is the same as that of the Didelphidae. The tympanic bullae are greatly enlarged and include an entotympanic bone, a component otherwise unknown among metatherians (Hershkovitz 1992). The cloaca opens on the ventral side of the base of the tail, as in monotremes but unlike any other metatherian. The "monito del monte," as it is called, is scansorial/arboreal and nocturnal, preferring habitats that include dense stands of Chilean bamboo (*Chusquea* sp.; Hershkovitz 1999). The species feeds primarily on insects but will eat vegetation. It is the sole known disperser of seeds of a mistletoe species in Chile (Amico & Aizen 2000). In times of food scarcity or cold temperatures, it stores fat in the tail and enters a period of hibernation (Rageot 1978).

Australasian Metatherians

The second great radiation of marsupials occurred primarily in the Australasian region. Often placed in the group Australidelphia to reflect their independent radiation, this assemblage includes five living orders: Dasyuromorphia, Peramelemorphia, Notoryctemorphia, Diprotodontia, and Microbiotheria. The dasyuromorphs, often called carnivorous marsupials, include a large group of unspecialized small carnivores (e.g., quolls, antechinuses, dunnarts, and Tasmanian devil) in the family Dasyuridae, along with the recently extinct Tasmanian thylacine (family Thylacinidae) and the specialized termite-eating numbat (family Myrmecobiidae). Bandicoots and bilbies, divided into two families, comprise the Peramelemorphia. The relationship of the marsupial mole to other metatherians is unclear, and for this reason it is placed in its own order, Notoryctemorphia. Finally, the largest order, Diprotodontia, includes 11 families and 143 species of easily recognizable metatherians: kangaroos, wallabies, wombats, a variety of possums and gliders, and the koala.

Order Dasyuromorphia

Family Dasyuridae

Dasyurids (**Fig. 6-12**) are more progressive than didelphids, both dentally and with regard to limb structure. Although the earliest known dasyurid is from the Australian early or middle Miocene, the family must have arisen at a far earlier time. Recent members of this family include 20 genera and 69 species, and the geographic range includes Australia, New Guinea, Tasmania, the Aru Islands, and Normanby Island (Flannery 1995; Strahan 1995; Van Deusen & Jones 1967).

Many of the major characters of dasyurids are shared by other metatherians, but several features are diagnostic of the former. The dental formula is 4/3, 1/1, 2–3/2–3, 4/4 = 42–46. The incisors are usually small and either pointed or blade like. The canines are large and have a sharp edge, and the upper molars have three sharp cusps adapted to an insectivorous and carnivorous diet. The skulls of some dasyurids resemble rather closely those of didelphids (**Fig. 6-13**). The forefoot has five digits, and the hind foot has four or five digits. The hallux is clawless and usually vestigial and is absent in some cursorial genera (Figs. 6-3 and **6-14**). There is no syndactyly. The foot posture is plantigrade in many species except the long-limbed jumping marsupial,

FIGURE 6-12   Four members of the family Dasyuridae: (A) a Kowari (*Dasyuroides byrnei*), (B) a lesser hairy-footed dunnart (*Sminthopsis youngsoni*), (C) a ningaui (*Ningaui* sp.), (D) a quoll (*Dasyurus viverrinus*).

the kultarr (*Sminthopsis laniger*), and the cursorial and carnivorous eastern quoll (*Dasyurus viverrinus*; Fig. 6-12D). The marsupium is often absent; when present, it is often poorly developed. The tail is long and well-furred, conspicuously tufted in some species, and never prehensile. The size of dasyurids ranges from that of a shrew (*Planigale*) to that of a small dog (*Sarcophilus*).

Dasyurids occupy a wide variety of terrestrial habitats; a few species are arboreal. There is a remarkably diverse array of marsupials in the family Dasyuridae. The smaller species fill the feeding niche occupied in Eurasia and North America by shrews (family Soricidae) and resemble those

animals in the possession of long-snouted heads and unspecialized limbs. A group of rat-sized dasyurids seems adapted to preying on insects and small vertebrates. The desert-dwelling genus *Sminthopsis* has long, slender limbs and a long, tufted tail and uses a rapid, bounding, quadrupedal gait. Another group, the quolls, consists of somewhat civet-like dasyurids that weigh about 0.5 to 3 kilograms and prey on a variety of small vertebrates. Quolls are agile and effective predators, and although primarily terrestrial, they are capable climbers.

The largest dasyurid carnivore, the Tasmanian devil (*Sarcophilus harrisii*; **Fig. 6-15A**), is stocky, short limbed, and

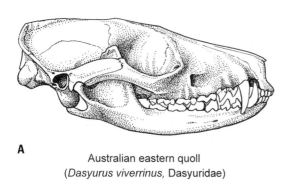

A
Australian eastern quoll
(*Dasyurus viverrinus*, Dasyuridae)

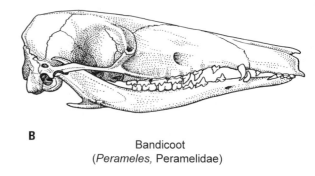

B
Bandicoot
(*Perameles*, Peramelidae)

FIGURE 6-13   Skulls of dasyurid and peramelid marsupials: (A) Australian quoll, length of skull 72 millimeters. (B) Bandicoot, length of skull 81 millimeters. ((b) Adapted from Tate, G. H. H., and Archbold, R., *Bull. Amer. Mus. Nat. Hist.* 73 (1937): 331–476.)

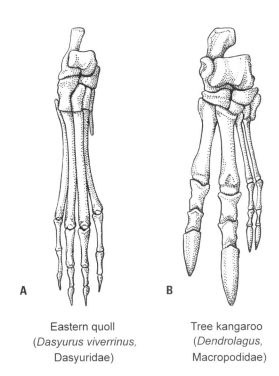

A

Eastern quoll
(*Dasyurus viverrinus*,
Dasyuridae)

B

Tree kangaroo
(*Dendrolagus*,
Macropodidae)

FIGURE 6-14 Feet of two marsupials: (A) a terrestrial species and (B) an arboreal species. (Adapted from Marshall, L.G., *Proceedings of the Royal Society Victoria* 85 (1972): 51–60.)

weighs from 4.5 to 9.5 kilograms. Once widespread over Australia, it is now restricted to Tasmania. Devils occur in a variety of habitats but seem to prefer dry, sclerophyllous forests interspersed with grasslands. They are not territorial, and home ranges of several devils may overlap extensively. The species is a persistent scavenger but will also kill a wide variety of small vertebrates. They tend to forage alone, but groups of up to 22 have been reported feeding on large carcasses. As is typical of scavengers, they have strong jaws (they consume even bones) and can eat up to 40% of their body weight in a single night (Strahan 1995). Tasmanian devils are highly vocal and disputes between neighbors can lead to clashes that result in serious wounds. Recent problems with a deadly and contagious form of facial cancer have decimated devil populations and caused them to be listed as an endangered species.

Family Myrmecobiidae

The most divergent dasyuromorph is the numbat, or banded anteater (*Myrmecobius fasciatus*), a small, long-snouted animal that is the sole representative of the family Myrmecobiidae. The teeth are small and widely spaced in the long tooth row, and the long, protrusible tongue is used in capturing termites. This animal was formerly widespread in eucalyptus forests, in which fallen branches and logs provided lush populations of termites, but the commercial clearing of these forests severely restricted the range of the numbat. Numbats are currently restricted to two isolated

populations in southwestern Australia. Although numbats have several natural predators, the most significant predator is the introduced red fox. Recent efforts to control red foxes and translocation of some numbats to parks and reserves have resulted in an increase in numbat populations (Friend & Thomas 2003).

Family Thylacinidae

Now extinct, the thylacine or Tasmanian "wolf" was doglike in size and general build (Fig. 6-15B); it had long limbs and a digitigrade foot posture. Because of prominent stripes along the back and flanks, it was sometimes referred to as the Tasmanian "tiger." The thylacine was a pounce-pursuit predator that killed 1- to 5-kilogram prey and filled a niche similar to that of some canids (Jones & Stoddart 1998). The last thylacine (*Thylacinus cynocephalus*) died in captivity in 1936. Its demise in Australia probably resulted partly from competition with the dingo (*Canis familiaris*), which was introduced to mainland Australia some 3,500 years ago (Strahan 1995). Populations survived in Tasmania into the early 20th century because Tasmania became isolated from

FIGURE 6-15 (A) The Tasmanian devil (*Sarcophilus harrisii*, Dasyuridae). (B) The thylacine (*Thylacinus cynocephalus*, Thylacinidae), which is now extinct.

the Australian mainland between 8,000 and 10,000 years ago, before introduction of the dingo. On Tasmania, these predators survived until the arrival of Europeans, whose flocks of sheep were thought to be easy prey for the thylacine. Bounties were placed on thylacines, and their populations went into steep decline and final extinction (Owen 2004; Paddle 2000).

Order Notoryctemorphia

Family Notoryctidae

This remarkable family is represented by two species of marsupial "moles," the itjaritjaris, which inhabit sandy soils in arid parts of northwestern and central Australia (*Notoryctes caurinus* and *N. typhlops*, respectively). Most of the diagnostic characters of these mouse-sized animals are adaptations for fossorial life. The eyes are vestigial, covered by skin, lensless, and, as indicated by the specific name *N. typhlops* (typhlops means "blind" in Greek), nonfunctional. The ears lack pinnae. The nose bears a broad, cornified shield, and the nostrils are narrow slits. The dental formula is usually 4–3/3, 1/1, 2/3, 4/4 = 44–42, but the incisors vary in number. The incisors, canines, and all but the last upper premolar are unicuspid; the paracone and metacone of the upper molars form a prominent single cusp, and the lower molars lack a talonid. As an adaptation serving to brace the neck when the animal forces its way through the soil, the five posterior cervical vertebrae are fused. The forelimbs are robust, and the claws of digits three and four are remarkably enlarged and function together as a spade; the other digits are reduced. The central three digits of the hind feet have enlarged claws. The small first digit has a nail, and the fifth digit is vestigial. The rearward-opening marsupium is partially divided into two compartments, each with a single nipple. The fur is long and fine textured, varying in color from silvery white to yellowish red.

Recent aid from aboriginal people of the Pitjantjatjara in central Australia has helped field biologists find and learn more about these elusive but fascinating creatures. Notoryctids use their powerful forelimbs and armored rostrum to force their way through soft, sandy soil, in which they make elaborate labyrinthine underground paths. When the animal forages near the surface, the soil is pushed behind it, and no permanent burrow is formed. The food is predominantly invertebrate larvae. Using new technology with geophones (underground microphones), CyberTracker animal tracking software, DNA samples recovered from itjaritjari predators' scats (foxes, cats, and dingos), and aboriginal knowledge, scientists are finally beginning to study these most fossorial of metatherians (Benshemesh & Johnson 2003).

Order Peramelemorphia

Members of the order Peramelemorphia include the bandicoots and bilbies, characterized in general by an insectivorous dentition (Fig. 6-10C) and a trend toward specialization of the hindlimb for running or hopping. The marsupium is present and opens to the rear, and bandicoots alone among marsupials have a chorioallantoic placenta (Chapter 20). The order historically included 8 genera (one now extinct), and 20 remaining species, mainly from Australia, Tasmania, and New Guinea. Some species of bandicoots have been extirpated or have become uncommon over parts of their former range, apparently because of livestock grazing, brush fires, and the introduction of various eutherian mammals.

Family Peramelidae

This family comprises six genera of bandicoots, including four genera that were until recently placed in a separate family Peroryctidae, but authors disagree about the monophyly of the group. The smaller bandicoots are the size of a rat, with the largest species weighing roughly 2 kilograms (**Fig. 6-16**). The dental formula is 4–5/3, 1/1, 3/3, 4/4 = 46–48. The incisors are small, and the molars are tritubercular or quadritubercular. The rostrum is slender (Fig. 6-13B) and the skull dorsoventrally flattened in cross-section. The ears of some species resemble those of rabbits. Although often long, the tail is not prehensile. The fourth digit of the hind foot is always the largest, and the remaining digits are variously reduced (**Fig. 6-18**). The hind foot posture is usually digitigrade, and the hindlimbs are elongate. The opposable hallux, probably inherited by peramelids from an arboreal ancestral stock, is rudimentary or may be lost. The second and third digits of the hind foot are joined (syndactylous) as far as the distal phalanges by an interdigital membrane,

FIGURE 6-16   Southern brown bandicoot (*Isoodon obesulus; Peramelidae*).

and the muscles of these digits are partially fused, allowing them to act only in unison (F. Jones 1924).

The structure and function of the specialized peramelid hind foot are unique and have been described in detail by Marshall (1972) and Szalay (1994). In mammals, extreme reduction in the number of digits is usually associated with good running ability. Most highly cursorial ungulates have retained only the third digit (as in the horse) or digits 3 and 4 (as in some antelopes). A similar trend occurs in peramelemorphs. Probably partly because of an early development of syndactyly involving the second and third digits and the use of these digits for grooming, the general trend in the cursorial peramelids is toward the reduction of all digits but the fourth, with a great enlargement of this digit (Fig. 6-18). These specializations are accompanied by an alteration in the structure and function of the tarsal bones. The ectocuneiform bone makes broad contact with the proximal end of the fourth metatarsal and partially supports this digit, a character unique to peramelemorphs. The mesocuneiform is lost, and the weight of the body is borne mainly by the cuboid, ectocuneiform, navicular, and astragalus bones. The calcaneum does not serve a major weight-bearing function but, of course, serves as a point of insertion for the extensors of the foot, muscles of great importance in locomotion.

Horses, pronghorns, and peramelemorphs provide beautiful examples of different structural means of solving a similar functional problem, in this case, refining running ability. In the horse, only the third digit is retained, which is supported largely by the ectocuneiform, the navicular, and the astragalus bones (see Fig. 17-8). In the pronghorn, only two digits are retained, and the cannon bone (the fused third and fourth metatarsals) is supported largely by the fused cuboid and navicular bones and the fused mesocuneiform and ectocuneiform bones; the calcaneum is no longer a weight-bearing element. Marshall (1972) pointed out, however, that the structure of the hindlimb of peramelids is not entirely modified for running, perhaps because of the burrowing tendencies of these animals. The fibula is large, and movement at the ankle joint is not restricted to a single plane, as it is in most cursorial mammals. (Cursorial adaptations are discussed in more detail in Chapter 17.)

Bandicoots are largely insectivorous but also eat small vertebrates, a variety of invertebrates, and some vegetable material. Most inhabit relatively dry, open-country habitats. Some species take refuge in nests that they build of plant debris. Fossil peramelids are first known in the Miocene in Australia.

The spiny bandicoots of the subfamily Peroryctinae (formerly family Peroryctidae) probably originated and radiated in New Guinea, although they have no fossil record there. Of the 4 genera and 11 species of peroryctine bandicoots, only 2 species are found outside of New Guinea. The rufous spiny bandicoot (*Echymipera rufescens*) is also found in the tip of Cape York in northeastern Australia, and *Rhynchomeles prattorum* is **endemic** to Seram Island, located between New Guinea and Sulawesi (Nowak & Paradiso 1991). These bandicoots range in size from less than 100 grams (*Microperoryctes murina*) to the giant bandicoot (*Peroryctes broadbenti*), which can exceed 5 kilograms. All are probably insectivorous and/or omnivorous. They occur in many habitats from subalpine grasslands to lowland rain forests; at middle elevations, up to six species may occur **sympatrically** in New Guinea (Flannery 1995).

Family Thylacomyidae

This family was formerly included in Peramelidae, the bandicoots, which they greatly resemble. The family consists of a single genus, *Macrotis*, with two species commonly known as bilbies. One of the species, *Macrotis leucura*, is probably extinct, and the other, *M. lagotis* (**Fig. 6-17**), is endangered. Bilbies feed nocturnally, primarily on invertebrates and some plant material, and dig burrows in which they hide during the day.

Family Chaeropodidae

The Chaeropodidae is comprised of a single genus and species *Chaeropus ecaudatus*, the pig-footed bandicoot, which until recently was classified in the family Peramelidae. The species is extinct, having last been recorded over a century ago in 1907. Analyses of molecular data from preserved specimens show its divergence from other peramelemorphs, hence the reclassification. An extreme degree of cursorial specialization occurred in the pig-footed bandicoot: The forelimb was functionally **didactyl**; the second and third digits of the forelimb were large and had hoof-like claws. The hind foot was functionally **monodactyl**; only the

FIGURE 6-17 The sole remaining thylacomyid species, the bilby *Macrotis lagotis*.

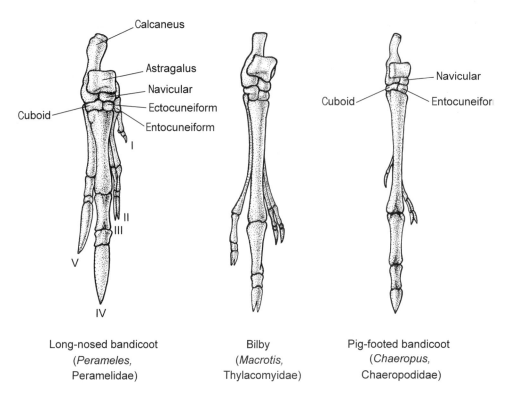

Calcaneus

Astragalus

Navicular

Ectocuneiform

Entocuneiform

Cuboid

I

II

III

V

IV

Cuboid

Navicular

Entocuneiform

Long-nosed bandicoot
(*Perameles*,
Peramelidae)

Bilby
(*Macrotis*,
Thylacomyidae)

Pig-footed bandicoot
(*Chaeropus*,
Chaeropodidae)

FIGURE 6-18   The right feet of three bandicoots of three different families. The least specialized foot is shown on the left and the most specialized is shown on the right. The digits are numbered in *Perameles*. (Adapted from Marshall, L.G., *Proceedings of the Royal Society Victoria*, 85 (1972): 51–60.)

fourth toe was used during running. This toe was greatly enlarged and was supported by the cuboid, navicular, ectocuneiform, and astragalus bones (**FIG. 6-18**).

Order Diprotodontia

Family Vombatidae

This family is represented by two living genera and three species. Known as wombats, these animals are completely herbivorous and show remarkable structural convergence toward rodents. Because of the efforts of humans, wombats have become scarce or absent over much of their former range and now are restricted to parts of eastern and southern Australia, Tasmania, and the islands between Australia and Tasmania.

Wombats are stocky animals with small eyes and rodent-like faces (**Fig. 6-19**); their body weight can exceed 36 kilograms (Strahan 1995). The skull and dentition bear a striking resemblance to those of some rodents (Figs. 6-9D and 6-10D). The skull is flattened. The rostrum is relatively short, and the heavily built zygomatic arches flare strongly to the sides. The area of origin of the anterior part of the masseter muscle is marked by a conspicuous depression in the maxillary and jugal bones that is similar to the comparable depression in the maxillary and premaxillary bones of the beaver (*Castor*; Fig. 13-13). The dental formula is 1/1, 0/0, 1/1, 4/4 = 24; all teeth are rootless and evergrowing. Only the anterior surfaces of the incisors bear enamel, and the incisors and the first premolars are separated by a wide diastema. The molars are **bilophodont** (Fig. 6-10D). As in rodents, the coronoid process of the dentary bone is reduced, and the masseter muscle, rather than the temporalis, is the major muscle of mastication.

The limbs are short and powerful, and the foot posture is plantigrade. The forefeet have five toes; all digits have broad, long claws. The hallux is small and clawless, but the other digits have claws. Digits two and three of the hind feet are syndactylous. The tail is vestigial. The marsupium opens posteriorly and contains one pair of mammae.

This family is first known from the middle Miocene epoch, and both contemporary genera have fossil species. The Pleistocene trend toward large size apparent in many other mammalian groups also occurred in the Vombatidae, as evidenced by the huge Pleistocene "wombat" *Phascolonus*.

Wombats are powerful burrowers. Their burrows are extensive networks of tunnels that are wide enough to admit a small person. Burrow entrances are often clustered to form warrens connected to other **warrens** by well-worn trails and marked by urine and feces. Although one warren system may house up to 10 wombats, they are rarely found together in the same burrow. Young wombats learn

FIGURE 6-19 Hairy-nosed wombat (*Lasiorhinus*, Vombatidae).

to burrow in their mother's burrow system by digging small subsystems, but they abandon the maternal burrows about 4 months after leaving their mother's pouches. Wombats dig burrows in the open or beneath rock piles, as do marmots (*Marmota*), their North American rodent counterparts. Males have a well-developed dominance hierarchy, and females appear to be subordinate to all males.

The southern hairy-nosed wombat (*Lasiorhinus latifrons*) lives in semiarid areas and is able to go for long periods without drinking water. The species copes with the inhospitable environment and low-quality diet, consisting primarily of herbs and grasses, by remaining in their burrows until the cooler evening hours. To conserve energy further, these wombats have a resting metabolic rate two-thirds that of other metatherians. As a result, their high-fiber food may take over a week to pass through the gut (Strahan 1995).

The common wombat (*Vombatus ursinus*) and the southern hairy-nosed wombat (*L. latifrons*) remain common in their limited ranges. The northern hairy-nosed wombat (*Lasiorhinus krefftii*) is endangered; fewer than 70 individuals are restricted to an area less than 300 hectares (750 acres).

Family Phascolarctidae

The familiar koala (*Phascolarctos cinereus*; **Fig. 6-20**) is the sole member of this family. This highly specialized herbivore is restricted to some wooded parts of eastern Australia. The tufted ears, naked nose, and chunky tailless form make the koala one of the most distinctive of Australian marsupials. It is a fairly large marsupial; the adult ranges from 8 to 12 kilograms in weight. The skull is broad and sturdily built, and the dentary bones are deep and robust (**Fig. 6-21A**). The dental formula is 3/1, 1/0, 1/1, 4/4 = 30. The roughly quadrate molars have **crescentic** ridges (Fig. 6-21B), and there is a diastema in both the upper and lower tooth rows between the cheek teeth and the anterior teeth. Tree branches are grasped between the first two and the last three fingers of the hand and between the clawless first digit and the remaining digits of the foot; the long, curved claws help maintain purchase on smooth branches.

FIGURE 6-20 The koala (*Phascolarctos cinereus*, Phascolarctidae).

Koalas are fairly sedentary and feed on only a few species of smooth-barked eucalyptus trees. Dependence on eucalyptus trees, with their high concentration of toxic terpenes and phenolics, low protein content, and high levels of fiber imparts considerable foraging costs. Koalas remain sedentary during most of the day, thereby reducing their energetic needs to a minimum. In addition, their powerful jaws grind leaves into a paste, extracting most of the cellular contents. Undigested fiber is retained in the caecum, where microbial fermentation extracts additional energy from the food. The liver aids in detoxification of plant secondary compounds (Lee & Martin 1988).

Maturation of a koala takes considerable time. A single young is born and is carried in the pouch for 6 months, after which it rides on its mother's back for a few more months. The young koala is dependent on its mother for 1 year, and sexual maturity is not reached until 3 or 4 years of age. Koalas are solitary and have individual home ranges that may overlap if population densities are high. Although they are not territorial, male koalas exhibit a dominance hierarchy, and subordinate males are attacked when encountered.

During the summer breeding season, males advertise their presence to females by low, bellowing calls that can be heard over half a kilometer away. Koalas lived in southwestern Australia during the late Pleistocene but no longer occur there even though suitable habitat appears to be present.

Family Burramyidae

The type genus of this family was known for many years only from Pleistocene fossil material; finally, in 1966, at a ski lodge on Mt. Hotham in Victoria, Australia, a representative of the genus was found alive. More recently, it has been found at other localities. This family contains two genera and five species of small, mouse-like marsupials, called pygmy possums. They lack the feather-like tail of acrobatid possums.

These diminutive marsupials are from 120 to 295 millimeters in total length, are delicately built, and have large eyes and mouselike ears (**Fig. 6-22**). They weigh less than 40 grams and include the world's smallest possum, the 6 to 8 gram little pygmy possum (*Cercartetus lepidus*). Pygmy possums are highly arboreal and have strongly prehensile, nearly naked tails. Members of this family are restricted to wooded areas. They are apparently insectivorous–omnivorous, but the feeding habits of some members of the group are not known.

The mountain pygmy possum (*Burramys parvus*) inhabits cold alpine and subalpine regions and is known to hibernate for long periods. At high elevations (>1,500 meters), where snow cover may last 6 months, mountain pygmy possums double their body weight in fat deposits before the winter snows. Hibernation lasts between 2 and 7 months and is characterized by a body temperature around 2°C for periods of approximately 3 weeks, followed by short bouts at normal body temperature (Geiser & Broome 1993). Because of its restricted mountaintop habitats, this unique mammal lives in several highly disjunct populations totaling less than 2,600 individuals.

Family Phalangeridae

There are 6 genera and 27 species of possums and cuscuses in the family Phalangeridae. Phalangerids are omnivorous and are known to eat a wide variety of plant material, as well as insects, young birds, and bird eggs. The brush-tail possum (*Trichosurus vulpecula*) is one of the most familiar Australian mammals, for it frequently maintains resident populations in suburban areas, where it often seeks shelter in roofs of houses and feeds on cultivated plants. Members of this family mostly inhabit wooded areas, but the adaptable brush-tail possum also occupies treeless areas, where it takes refuge in rocks or the burrows of other mammals. This animal is locally destructive to plantations of introduced pines. The brush-tail possum is solitary and has a sternal scent gland, considerably larger in males than in

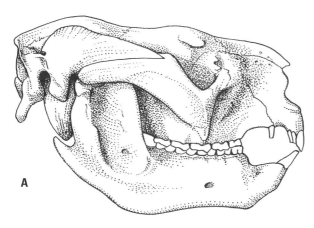
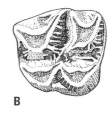

Koala (*Phascolarctos cinereus*, Phascolarctidae)

FIGURE 6-21   (A) Skull of the koala, length of skull 132 millimeters. (B) Occlusal view of the second molar, upper right tooth row, showing the crescentic areas of dentine, exposed by wear, and the complex pattern of furrows.

females, which produces a musky smell that is used in scent marking objects within the animal's territory.

Phalangerids are of moderate size, ranging in weight from 1 to 6 kilograms. The skull is broad and has deep zygomatic arches (Fig. 6-9B). The molars are bilobed with rounded cusps (Fig. 6-10E, F). As adaptations to arboreal life, the hands and feet are large and have a powerful grasp, and the tail is prehensile. The cuscuses have short ears and woolly fur and resemble teddy bears (**Fig. 6-23A**).

Family Tarsipedidae

This family contains but one species, the highly specialized, slender-nosed honey possum, or noolbenger (*Tarsipes rostratus*). This remarkable animal's many specializations obscure its relationships to other marsupials, and its taxonomic position has long been uncertain, although it is certainly a diprotodont.

Tarsipes is small, only about 7 to 12 grams in weight, and has a long, prehensile tail. The pelage is marked by three longitudinal stripes on the back. The rostrum is long and fairly slim, and the dentary bones are extremely

FIGURE 6-22   A pygmy possum (*Cercartetus concinnus*, Burramyidae).

slender and delicate. The cheek teeth are small and degenerate, and only the upper canines and two medial lower incisors are well-developed. The snout is long and slender, and the long tongue tapers to a fine point that has bristles at its tip (these specializations are similar to those of some nectar-feeding bats of the family Phyllostomidae). The muscles of the jaws are reduced and weak; those associated with the tongue are important in these obligate nectar feeders (Rosenberg & Richardson 1995). All digits but the syndactylous second and third digits of the hind feet have expanded terminal pads resembling to some extent those of the primate *Tarsius*. The digits have nail-like structures instead of claws.

Honey possums live only in the southwestern corner of Australia, where a great diversity of plants of the family Proteaceae occurs. Many proteas have spectacular, nectar-rich flowers. The profusion of species flower at various times of the year, providing *Tarsipes* with a steady supply of nectar and pollen. Proteas are an ancient group known from at least the early Eocene in Patagonia (Wilf et al. 2003) and elsewhere in Gondwanan (southern) continents that may have a long history of pollination by *Tarsipes* and its progenitors. Honey possums have no fossil record before the Pleistocene. Like hummingbirds and nectar-feeding bats, honey possums feed on nectar, pollen, and, to some extent, small insects that live in flowers. The long, protrusible tongue is used to probe into flowers. *Tarsipes* can climb delicately over even the insecure footing of clusters of flowers at the ends of branches and often clings upside down to flowers while feeding. Although the animal is still common in some areas today, the expansion of agriculture in southwestern Australia is restricting the honey possum's range, and nonnative foxes and cats prey on the animal.

FIGURE 6-23   Two phalangerids: (A) the common spotted cuscus (*Spilocuscus maculatus*). Note the prehensile tail with the traction-producing ridges on the bare, distal part of the ventral surface. (B) The Australian brushtail possum (*Trichosurus vulpecula*).

Family Acrobatidae

This small family is represented by two living species, each in its own genus: the feathertail glider (*Acrobates pygmaeus*) from Australia and the feathertail possum (*Distoechurus pennatus*) from New Guinea. Both species are characterized by rows of long, stiff hairs along either side of the tail, from which they get the name "feathertail."

The feathertail glider is small, weighing approximately 10 to 15 grams. It has a membrane between the elbows and the knees, and is the smallest gliding mammal. The flattened, feather-like tail is used for steering; the prehensile tail is also used for gripping small branches. The eyes are large and anteriorly placed for improved binocular vision. Expanded pads at the tips of the fingers and toes increase traction between the digits and the trunks and branches of trees; the surfaces of the pads have ridges that further increase the clinging ability of these animals. The molars are typical of insectivores, but the tongue is adapted for extracting nectar from flowers. Feathertail gliders are primarily nocturnal, and groups of up to 20 animals have been found nesting together (or torpid). Females exhibit **embryonic diapause**. The young remain in the pouch for over 2 months, and weaning may not be completed until 100 days. Because of the long developmental period, maternal investment is relatively high, and there are reports of females other than the mother helping with the care of young gliders.

Feathertail possums (*D. pennatus*) are united with the feathertail glider (*A. pygmaeus*) on the basis of three shared characters: the feather-like tail, an unusual middle ear anatomy, and the loss of the last molar (Flannery 1995). Unlike the gliders, *D. pennatus* does not have a gliding membrane. The feathertail possum weighs between 40 and 50 grams, and females are larger than males.

Family Pseudocheiridae

Until recently, the ringtail possums and the greater glider were considered part of the Petauridae, a family to which they are closely related. The pseudocheirids (**Fig. 6-24**) include six genera and 17 species. Ringtails have a strongly prehensile tail, but the tail of the greater glider (*Petauroides volans*) is weakly prehensile and is used mainly as a steering rudder during gliding. All pseudocheirids have molar teeth that bear a series of sharp ridges used in grinding leaves. Microbial fermentation occurs in the large caecum and assists in the breakdown of plant material. The common ringtail (*Pseudocheirus peregrinus*) also practices **coprophagy** (reingestion of feces) to increase its ability to extract nutrients from its food. The largest gliding metatherian, the greater glider (*Petauroides volans*), can glide more than 100 meters and make nearly 90-degree turns. The unusual green ringtail possum (*Pseudochirops archeri*) has fur that appears green but is actually a combination of black, yellow, and white hairs. It has a highly specialized diet; its food is entirely leaves, chiefly those of fig trees (Proctor-Grey 1984). Much of its diet consists of the mature leaves of just four species of trees that provide a continuously available

FIGURE 6-24 Painted ringtail (*Pseudochirulus forbesi*; Pseudocheiridae).

FIGURE 6-25 The sugar glider (*Petaurus breviceps*; Petauridae).

food source that is high in secondary compounds (plant chemicals defensive against herbivores; Jones et al. 2006).

Family Petauridae

The striped possum, trioks, Leadbeater's possum, and six species of lesser glider, so named because of the membrane between their wrists and ankles, are currently recognized in the family Petauridae. Most members of this family are fairly small; weights range from about 100 to 700 grams. The skull is broad, and the molars have low, smooth cusps. The tail is long, bushy, and prehensile. All species have a dark dorsal stripe running from head to tail (**Fig. 6-25**). The gliders (*Petaurus*) have furred membranes that extend between the limbs and function as lifting surfaces for gliding. In these gliders, the claws are sharp and recurved, like those of a cat, and increase the ability of the animal to cling to the smooth trunks and large branches of trees. The petaurids are nocturnal, arboreal creatures that inhabit wooded areas and feed primarily on sap and nectar.

The gliders are strikingly similar to flying squirrels (*Glaucomys*) in gliding style and ability; some can glide more than 100 meters. Sugar gliders (*Petaurus breviceps*) live in family groups, and scent marking plays an important role in the social organization of the group. Each individual has a particular odor recognized by the others. The cohesion of the group is also aided by mutual scent marking, for all members of the group become permeated with the scent of the group's dominant males. Leadbeater's possum (*Gymnobelideus leadbeateri*) exhibits only a vestigial gliding membrane.

The four species of trioks or striped possums (*Dactylopsila*) lack a gliding membrane, and some may be primarily terrestrial. They have a suite of unique characters associated with their specialized insectivorous diet. In *Dactylopsila*, the fourth digit of the hand is elongate and slender, and its claw is recurved, similar to those of the primate *Daubentonia* (**Fig. 6-26**). In addition, the incisors are robust and function roughly as do those of rodents. Striped possums tear away tree bark with their incisors and extract insects from crevices and holes in the wood with the specialized fourth finger and the tongue. *Dactylopsila* wears a conspicuous, striped color pattern that is interesting because it is associated, as in skunks, with a powerful, musky scent.

Family Hypsiprymnodontidae

The musky rat-kangaroo (*Hypsiprymnodon moschatus*) is a muskrat-sized inhabitant of rain forests and riparian

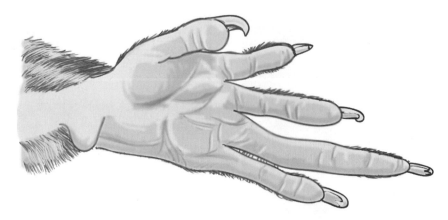

FIGURE 6-26 The hand of the common striped possum (*Dactylopsila*; Petauridae) showing the elongate fourth finger.

situations. This species, the sole member of its family, was recently separated from the Potoroidae and probably is similar to the ancestral stock from which other kangaroos evolved. It has a tail of modest length and retains all of the digits of the hind foot. The hindlimbs are not greatly elongate, and the animal uses quadrupedal rather than saltatorial locomotion. Animal material forms a large share of the omnivorous diet of this seemingly primitive diprotodont. The family first appeared in the late Oligocene of Australia.

Family Potoroidae
The family Potoroidae comprises four genera (one extinct) that depart from the familiar structural pattern of kangaroos and from the grazing or browsing habit. These animals are commonly known as potoroos, rat-kangaroos, and bettongs (**Fig. 6-27A**). They seemingly represent a conservative branch on the diprotodont tree. They retain a slightly prehensile tail and well-developed upper canines but did not evolve the elaborate stomach of the macropodids.

Potoroos and bettongs are small kangaroo-like animals weighing between 0.8 and 3.5 kilograms. The tail is sufficiently prehensile to carry nesting material, and the forelimbs are used in some species for digging the roots, tubers, and fungi that comprise their diets. Male rufous rat kangaroos (*Aepyprymnus rufescens*) probably form loosely defended associations with several females whose ranges overlap their own. Many species dig burrows, but the recently extinct desert rat-kangaroo (*Caloprymnus campestris*) excavated a shallow depression lined with grasses under bushes. Potoroidae fossils are known since the late Oligocene in Australia. Of the nine historically known species of potoroids, two are probably extinct (*C. campestris* and *Potorous platyops*), and seven others are endangered. Competition with introduced rabbits and predation by dingos and introduced foxes and cats are likely responsible for these losses.

Family Macropodidae
Members of the familiar metatherian group Macropodidae, which includes 11 genera of kangaroos, euros, and wallabies

(**Fig. 6-27B**, **C**), are the ecological equivalents of such ungulates as antelope. The present distribution of the approximately 65 modern species of macropodids includes New Guinea, the Bismarck Archipelago, the D'Entrecasteaux Group, Australia, and by introduction, some islands near New Guinea and New Zealand. The family Macropodidae appeared first in the middle Miocene of Australia. Earlier diprotodontians in the late Oligocene and early Miocene included an early side branch of Macropodiformes (family Balbaridae) that had prominent tusks (fangaroos; Archer & Kirsch 2006). Wallabies and kangaroos are known from the Neogene (late Cenozoic), and unusually large macropodids occurred in the Pleistocene.

Living macropodids vary greatly in size and structure. The rock wallaby known as the monjon (*Petrogale burbidgei*) weighs only 900 to 1,400 grams, whereas the great gray kangaroo (*Macropus giganteus*)—the largest living marsupial—reaches 2 meters in height and approximately 90 kilograms in weight. The marsupium is usually large and opens anteriorly. The macropodid skull is moderately long and slender, and the rostrum is usually fairly long (Fig. 6-9C). The dental formula is 3/1, 1–0/0, 2/2, 4/4 = 34–32. The upper incisors have sharp crowns with their long axes oriented more or less front to back. The tips of the procumbent lower incisors are held against a leathery pad just behind the upper incisors when the animals gather vegetation. This specialized arrangement serves a cropping function similar to that of the lower incisors and the premaxillary pad in the artiodactylan ungulates that lack upper incisors. There is a broad diastema between the macropodid incisors and the premolars. The molars are quadritubercular and bilophodont (Fig. 6-10G). In many macropodids, the last molar does not erupt until well after the animal becomes adult. A unique situation occurs in the little rock wallaby (*Petrogale concinna*), in which nine molars may erupt in succession. Usually four or five molars are functional at one time, and replacement is from the rear as the molars are successively lost from the front.

FIGURE 6-27   A potoroid and two kinds of macropodid marsupials: (A) long-nosed potoroo (*Potorous tridactylus*, Potoroidae). (B) Pademelon (*Thylogale billardierii*, Macropodidae). (C) Red-necked wallaby (*Macropus rufogriseus*, Macropodidae).

Macropodids are highly specialized for jumping. The forelimbs are five-toed and are usually small; they are used for slow movement on all fours or for food handling. The hindlimbs are elongate, especially the fourth metatarsal. The hallux is missing in all species. Digits 2 and 3 are small and syndactylous and are used for grooming; the fourth is the largest digit, and the fifth is often robust (**Fig. 6-28**). The unusual pattern of digital reduction and the dominance of the fourth digit in the most highly cursorial Australian metatherians are perhaps due to the arboreal ancestry of these animals. In their ancestors, the foot was five-toed and the hallux opposable; the fourth was the longest remaining

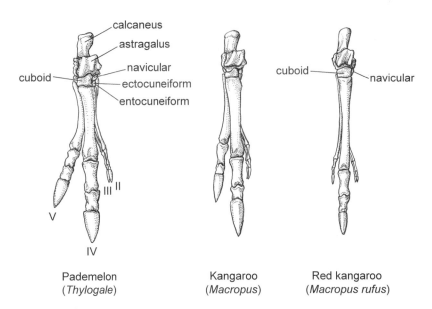

FIGURE 6-28   The right foot bones of some macropodid marsupials; least specialized on the left and most specialized on the right. The digits are numbered in *Thylogale*. (Adapted from Marshall, L.G., *Proceedings of the Royal Society Victoria* 85 (1972): 51–60.)

digit, and the foot was adapted to grasping branches. With specialization of the foot for running or hopping, the hallux was lost, and the longest toe, the fourth, became the most important digit. In most macropodids, the foot is functionally two toed during rapid locomotion, which is characteristically **bipedal**. In the macropodid tarsus, there is no contact between the ectocuneiform and the fourth metatarsal (in contrast to the arrangement in the Peramelidae; Fig. 6-18). Because the hindlimb posture of macropodids is basically plantigrade, the calcaneum is an important weight-bearing element of the tarsus (which it is not in the digitigrade Peramelidae). The macropodid tail is usually long and robust and functions in the more specialized species as a balancing organ and as the posterior "foot" of the tripod formed by the plantigrade hind feet and tail, on which the animal can sit when not in motion.

Macropodids are the only large-bodied mammals to use bipedal hopping for locomotion. Does hopping require more energy than quadrupedal running? When a typical quadrupedal mammal runs, the energetic cost increases linearly with increasing speed, regardless of whether the animal is metatherian or eutherian. Studies by Dawson and Taylor (1973), using kangaroos trained to hop on a treadmill, demonstrated that costs of bipedal hopping are initially high, but that after hopping has begun, it can be maintained at little or no additional cost over a wide range of speeds (Chapter 21). At moderate to high speeds, a hopping kangaroo uses no more energy than a quadrupedal mammal. Part of the explanation for this comes from the energy stored in the huge elastic tendons attached to the hind foot.

The running ability of the larger kangaroos (*Macropus*) is impressive. Speeds on level terrain of roughly 65 to 70 kilometers per hour are attained, and leaps covering distances of 13.5 meters and height of 3.3 meters have been reported (Troughton 1947). The solitary hare wallabies (*Lagorchestes*), which are roughly the size of a large rabbit, also are renowned for their great speed. These animals have a jackrabbit style of escape. They hide beneath bushes or clumps of grass, burst out suddenly when frightened, and run away at high speed. The highly developed jumping ability of macropodids allows these animals to move easily for long distances between scattered sources of water or forage and to escape enemies by erratic leaps. These abilities, rather than the capacity for great speed, are perhaps of primary adaptive importance. Saltation may have been developed by small forms ancestral to kangaroos as a means of erratic escape in open areas. This **ricochetal** style of locomotion is known in a number of desert-dwelling rodents. Rock wallabies (*Petrogale*) can bound at high speed over precipitous, rocky country. According to F. W. Jones (1924), "There seems to be no leap it will not take, no chink between boulders into which it will not hurl itself."

The tree kangaroos (*Dendrolagus*, Macropodinae) spend considerable time on the ground but frequently use their arboreal ability to escape from danger. This mode of life is reflected by the large and robust forelimbs with strong recurved claws, by the hindlimbs, which are not strongly elongate, and by the short, broad hind foot (Fig. 6-14B). Saltation, typical of terrestrial kangaroos, has not been completely abandoned by tree kangaroos; not only are these animals agile climbers, but they also leap from one tree to another and from tree to ground. Their food is large fruit and leaves.

Macropodids resemble eutherian ungulates in their highly specialized limbs and cursoriality. These groups are also alike in being herbivorous and in having skulls and dentitions specialized for this mode of feeding. Even some specializations of the digestive system are similar between these two groups. Like eutherian ruminants, kangaroos have intestinal bacteria that digest the cellulose of plants. These (and probably other) macropodids can thus use the contents of the plant cells, the byproducts of the bacterial digestion of cellulose, and the bacteria themselves (Dellow & Hume 1982; Freudenberger et al. 1989; Hume 1982). The ruminant ungulates also depend on bacteria to increase the efficiency of their utilization of vegetation (Chapter 18). Although both ungulates and macropodids are foregut fermenters (stomach chambers rather than a hindgut caecum contain the microbes responsible for fermentation), the kangaroo stomach is not truly multichambered. Instead, kangaroos have a long tubular stomach with three main sections (**Fig. 6-29**). The esophagus dumps food into an S-shaped blind sac called the sacciform forestomach, where much of the fermentation occurs. Over 60% of the organic matter is digested here, along with most of the sugars. Fiber is broken down in the next section of the stomach, the tubiform forestomach, which forms a coil. Here fiber is processed along the length of the coil, and approximately one-third of it is digested. The remaining fiber is digested in the colon. After passing through the tubiform forestomach, the undigested material passes into the third section, where acid and enzymes are secreted and final processing takes place prior to the food's entering the small intestine. In addition to the breakdown of fiber, the symbiotic microbes also manufacture vitamins, detoxify some plant compounds, and recycle nitrogenous compounds into proteins (Dawson 1995; Hume 2006).

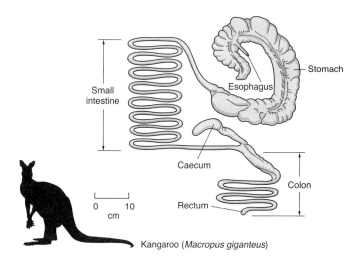

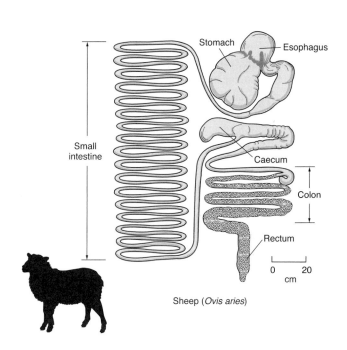

FIGURE 6-29 Comparison of the digestive systems of (A) a kangaroo (*Macropus giganteus*) and (B) a sheep (*Ovis aries*). (Adapted from Dawson, T. J. *Kangaroos: Biology of the largest marsupial.* Cornell University Press, 1995.)

SUMMARY

Metatherians (marsupials) and eutherians (placentals) represent two evolutionary lines that have been separate since the middle Cretaceous, ca. 98 to 100 million years ago. The earliest undoubted metatherian fossils are from the early Cretaceous of Asia. As a result of their long, independent history, metatherians differ structurally from eutherians in many ways. Today, only two important strongholds for metatherians remain: the Australian region (Australia, Tasmania, New Guinea, and nearby islands) and the Neotropics (southern Mexico, Central America, and South America). Where they have been isolated from placentals for long periods, metatherians have undergone remarkable radiation.

Most metatherians have functional counterparts among eutherians. Living metatherians are commonly called marsupials, for the marsupium or pouch in which young are carried; however, several marsupial species do not develop a pouch, and bandicoots have a placenta similar to that in eutherians. The metatherian gestation period is characteristically short (8 to 43 days), and the young are born tiny, naked, blind, and poorly developed. The delicate newborn then makes its way from the vulva to the marsupium, where it attaches to a nipple and remains for a lengthy lactation period that greatly exceeds the gestation period.

South American living marsupials belong to three separate orders: Didelphimorphia, Paucituberculata, and Microbiotheria. Members of the Didelphimorphia, opossums and mouse opossums, are the most generalized metatherians and constitute one of the oldest known groups. The order Paucituberculata includes the small rat opossums, whereas the monito del monte is the sole living member of the order Microbiotheria.

The second great radiation of marsupials survives primarily in the Australasian region. Often placed in the group Australidelphia to reflect their independent radiation, this assemblage includes five orders: Dasyuromorphia, Peramelemorphia, Notoryctemorphia, Diprotodontia, and Microbiotheria. The dasyuromorphs, often called carnivorous marsupials, include a large group of unspecialized small carnivores (e.g., quolls, antechinuses, dunnarts, and Tasmanian devil), along with the recently extinct Tasmanian thylacine and the specialized termite-eating numbat. Bandicoots and bilbies, divided into two families, comprise the Peramelemorphia. The largest order, Diprotodontia, includes 11 families and 143 species of easily recognizable metatherians: kangaroos, wallabies, wombats, a variety of possums and gliders, and the koala. Finally, the relationship of the marsupial moles to other metatherians is unclear, and for this reason, they are placed in their own order, Notoryctemorphia. This remarkable family is represented by two species, the itjaritjaris, which inhabit sandy soils in arid parts of northwestern and central Australia. Itjaritjaris are adapted for fossorial life: their eyes are vestigial, covered by skin, lensless, and nonfunctional.

Macropodidae, which includes 11 genera of kangaroos, euros, and wallabies, are the only large-bodied mammals to use bipedal hopping for locomotion. Macropodids are the ecological equivalents of such ungulates as antelope. The present distribution of the approximately 65 modern species of macropodids includes New Guinea, the Bismarck Archipelago, the D'Entrecasteaux Group of islands, Australia, and by introduction, some islands near New Guinea and New Zealand.

KEY CHARACTERISTICS OF METATHERIANS

- Skull frequently with small, narrow braincase
- Cerebral hemispheres small with simple convolutions
- Auditory bullae ossified, when present usually formed largely by the alisphenoid bone
- Palate characteristically has large vacuities
- Angular process of the dentary bone inflected medially
- Upper and lower incisors never equal in number (except in Vombatidae)
- Only third upper and lower premolars (P3/3) replaced by adult teeth
- Feet often specialized for arboreal or bipedal locomotion
- Epipubic bones extend forward from the pubic bones in both sexes
- Marsupium (an abdominal pouch) or abdominal folds containing nipples present in many species
- Female reproductive tract bifid (vagina and uterus doubled)
- Testes contained in scrotum anterior to penis (except in marsupial "moles")

KEY TERMS

Bilophodont	Embryonic diapause	Ricochetal
Bipedal	Endemic	Saltatorial
Chorioallantoic placenta	Gestation	Serrate
Choriovitelline placenta	Land bridge	Sympatric
Coprophagy	Marsupium	Syndactyly
Crescentic	Molariform	Unicuspid
Didactyl	Monodactyl	Warren
Digitigrade	Plantigrade	
Diploid number	Procumbent	

RECOMMENDED READINGS

Armati, PJ, CR Dickman, & ID Hume. 2006. *Marsupials.* Cambridge University Press, Cambridge, UK.

Cifelli, RL & BM Davis. 2003. Marsupial origins. *Science, 302*:1899–1900.

Dickman, C & RW Ganf. 2007. *A Fragile Balance: The Extraordinary Story of Australian Marsupials.* University of Chicago Press, Chicago, IL.

Flannery, T. 2002. *The Future Eaters: An Ecological History of the Australasian Lands and People.* Grove Press, New York.

Horovitz, I & MR Sánchez-Villagra. 2003. A morphological analysis of marsupial mammal higher-level phgylogenetic relationships. *Cladistics, 19*:181–212.

Jones, M, C Dickman, & M Archer. 2003. *Predators with Pouches: The Biology of Carnivorous Marsupials.* CSIRO Publishing, Collingwood, Victoria, Australia.

Owen, D. 2004. *Tasmanian Tiger: The Tragic Tale of how the World Lost its Most Mysterious Predator.* Johns Hopkins University Press, Baltimore, MD.

Tyndale-Biscoe, H. 2005. *Life of Marsupials.* CSIRO Publishing, Collingwood, Victoria, Australia.

Introduction to Eutherian Mammals

Throughout the Cenozoic era, eutherians have been the most successful group of mammals. During most of the Paleogene and Neogene periods, eutherians dominated the mammalian faunas of all continents but South America and Australia, where marsupials predominated. Now, even on those two continents, placental mammals outnumber marsupials. In South America, the fauna changed about 3 million years ago, along with biogeographical events associated with the Great American Biotic Interchange (see Chapters 6 and 25). In Australia, most of the change occurred during the Holocene and has been partly due to introductions of placental mammals by humans in historic times. Of the over 5,400 mammal species living on Earth today, 5 are monotremes, about 325 are marsupials, and the remaining 5,100 are placentals (Wilson & Reeder 2005).

The story of eutherian mammals begins about 125 million years ago with an exceptional Early Cretaceous fossil named *Eomaia scansoria* from Liaoning Province, China (Ji et al. 2002). *Eomaia* is represented by a complete fossil skeleton with a halo of fur preserved around the body. The remaining record of Cretaceous eutherians, mostly from the Late Cretaceous, includes over 40 genera and is sketchy but improving with better fossils and associated upper and lower teeth (Kielan-Jaworowska et al. 2004; Wible et al. 2009). Most of the fossils are fragments of jaws with partial tooth rows or isolated teeth, and they represent only parts of Asia, western North America, Europe, and South America. As the record of Cretaceous therians improves, it becomes more difficult to distinguish between eutherians and metatherians, based on tooth and jaw features alone (Cifelli 1997; Luckett 1993). These two groups are distinguished mainly by their reproductive systems, which are soft-tissue systems almost never preserved in fossils. We therefore depend partly on molecular-genetic evidence calibrated with known-age fossils to estimate Early Cretaceous as the time of the eutherian–metatherian divergence.

Despite the paucity of complete fossils, some conclusions regarding Cretaceous eutherians can be made. They were small, from the size of a shrew to the size of a marmot, and their structural diversity in the Late Cretaceous reflects radiations during the 79 million years of Cretaceous time. Their skulls lacked an auditory bulla and were typically long, with a narrow braincase and long, narrow snout. The incisors varied in number from 5/4 in some early forms to 3/3. Early Cretaceous eutherians had five premolars (Kielan-Jaworowska et al. 2004), whereas Late Cretaceous eutherians had four. There were three molars that were typically tribosphenic (bearing a protocone on the uppers that fit into a talonid basin surrounded by three cusps on the lowers) with high, sharp cusps (**Fig. 7-1**). The hands and feet of several genera are known, and these lack opposable first digits, indicating that these Cretaceous eutherians were not basically arboreal.

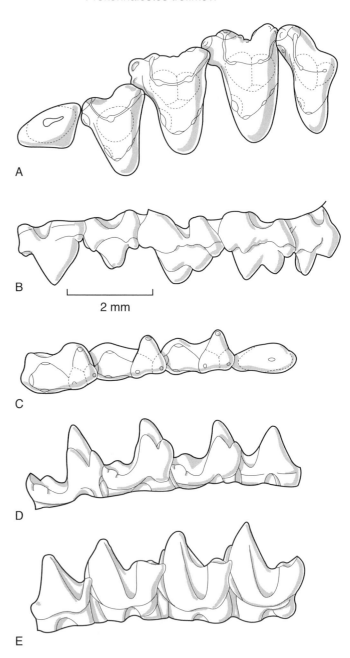

Prokennalestes trofimovi

2 mm

FIGURE 7-1 (A) Occlusal and (B) labial view (view from the outside) of the left upper cheek teeth and (C) occlusal, (D) lingual (from the inside), and (E) labial views of the left lower cheek teeth of *Prokennalestes trofimovi*. The dentition of this eutherian is among the most primitive of those known. (Adapted from Kielan-Jaworowska, Z., and Dashzeveg, D, *Zool. Scr.* 18 (1989): 347–355.)

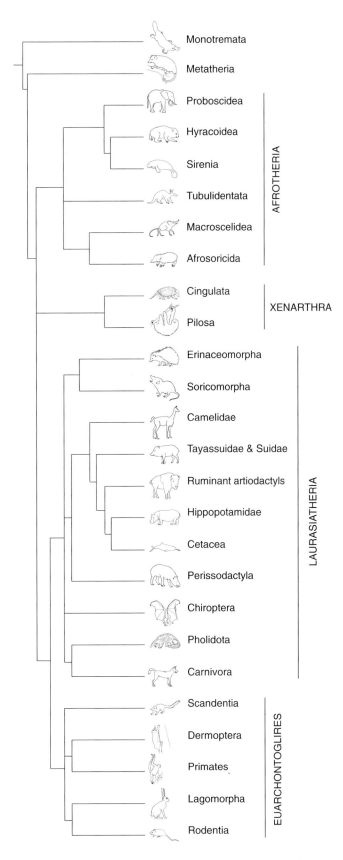

FIGURE 7-2 A phylogenetic tree showing relationships among the major extant orders of placental mammals. Many extinct orders are excluded. In this cladogram, Cetacea are nested within Artiodactyla, and Artiodactyla is a polyphyletic grouping requiring revision. (Adapted from Bininda-Emonds, O.R.P, et al., *Nature* 446 (2007):507–512 and 2008.)

Cretaceous eutherians probably played a variety of ecological roles. The diverse dentitions were seemingly adapted to insectivory, carnivory, and herbivory. As its specific epithet indicates, *Eomaia scansoria* is interpreted as having been scansorial, capable of scrambling or climbing. It would be reasonable to speculate that other Cretaceous eutherians functioned variously as climbers, jumpers, diggers, and those with generalized styles of locomotion, as did other Mesozoic mammals (see Chapter 3). Interpretation of fossil postcranial material is difficult, for we know that some living mammals, such as squirrels (e.g., *Sciurus* spp.), practice several locomotor and dietary lifestyles.

The fossil record documents a Late Cretaceous establishment of the evolutionary lineages leading to several higher level placental groupings including Afrotheria, Xenarthra, Laurasiatheria, Euarchonta, and Glires (**Fig. 7-2**; Archibald et al. 2001; Madsen et al. 2001; Wible et al. 2009). By the latest Cretaceous, some modern orders such as Soricomorpha are recognizable as fossils, with several others quickly following by the early Paleogene (Roca et al. 2004; Rose 2006; Rose & Archibald 2005). A comprehensive "supertree" analysis of mammalian phylogenetic trees by Bininda-Emonds et al. (2007) concluded that there may have been two pulses of diversification that led to modern placental mammals. The first was a Late Cretaceous radiation of superorders and all modern orders between about 100 million and 74 million years ago, and peaking about 89 million years ago. The second diversification pulse occurred well into the Paleogene after the modern orders were already established, peaking near the Paleocene-Eocene boundary about 56 million years ago and resulting in additional families and genera of placental mammals.

The accounts of eutherian mammals in the following chapters include brief comments about when the orders and families appeared and, in some cases, sketches of the evolutionary histories of taxa. The student should regard such comments as summaries of our current knowledge but should bear in mind that the fossil record is sadly incomplete for some groups. Mammals belonging to evolutionary lines that have been restricted to tropical areas, small and delicate species, and arboreal types had relatively little chance of leaving a fossil record. The first appearance of the Chiroptera (bats) in the early Eocene, for example, tells us only that by this time all of the basic chiropteran adaptations had been perfected and that bats were part of the North American fauna. The history of these small, fragile, basically tropical creatures must extend farther back, however, in the Paleocene or perhaps into latest Cretaceous times as the supertree analysis of Bininda-Emonds et al. (2007) suggests. We simply lack documentation of this history and doubtless of the early histories of many other groups, as well, but continue to search for more.

SUMMARY

During most of the Cenozoic era, eutherians dominated the mammalian faunas of all continents except South America and Australia, where marsupials predominated. Now, even on those two continents, placental mammals outnumber marsupials.

The story of eutherian mammals as currently known begins with a 125 million year old fossil named *Eomaia scansoria* from the Early Cretaceous period in China. The remaining record of Cretaceous eutherians is sketchy but improving. Cretaceous eutherians were small, from the size of a shrew to the size of a marmot. They probably played a variety of ecological roles, functioning variously as climbers, jumpers, and diggers, along with forms suited to more generalized styles of locomotion. The diverse dentitions were seemingly adapted to insectivory, carnivory, and herbivory.

A comprehensive phylogenetic analysis suggests that there were two pulses of diversification that led to modern placental mammals. The first radiation led to the establishment of all modern orders between 100 million and 74 million years ago, and a new molecular clock estimate suggests a peak at about 89 million years ago. The second diversification pulse occurred after the modern orders were already established, peaking near the Paleocene-Eocene boundary about 56 million years ago and resulting in additional families and genera of placental mammals.

RECOMMENDED READINGS

Bininda-Emonds, ORP, et al. 2007. The delayed rise of present-day mammals. *Nature, 446*:507–512 (and 2008. *Corrigendum, 456*:274).

Kitazoe, Y, et al. 2007. Robust time estimation reconciles views of the antiquity of placental mammals. *PLoS ONE, April*(4):e384.

Madsen, O, et al. 2001. Parallel adaptive radiations in two major clades of placental mammals. *Nature, 409*:610–614.

Rose, KD. 2006. *The Beginning of the Age of Mammals.* Johns Hopkins University Press, Baltimore, MD.

Wible, JR, et al. 2009. The eutherian mammal *Maelestes gobiensis* from the Late Cretaceous of Mongolia and the phylogeny of Cretaceous Eutheria. *Bulletin of the American Museum of Natural History* 327:1–123.

Afrosoricida, Macroscelidea, and Tubulidentata

For many years, biologists have recognized that certain groups of African mammals are so morphologically unique as to indicate that they diverged early from other mammals and went their own separate evolutionary ways. For decades there has been disagreement as to the taxonomic recognition of these groups. Recent studies, however, demonstrate that these African mammals may belong to a distinct clade, the Afrotheria (**Fig. 8-1**). Within the Afrotheria are two smaller clades (Bininda-Emonds et al. 2007; Seiffert 2007). The orders Afrosoricida, Macroscelidea, and Tubulidentata form one clade, and the elephants, sirenians, and hyraxes comprise a second clade called the Paenungulata (Chapter 9). The Afrosoricida includes the tenrecs of Madagascar, the otter shrews of western and central Africa, and the golden moles of southern Africa. The Macroscelidea comprise the elephant shrews (or sengis), and the aardvark is the sole member of the order Tubulidentata (Fig. 8-1). At first glance, this may seem like an odd assemblage of African mammals. Morphological support for the Afrotheria is almost nonexistent (but see Sanchez-Villagra et al. 2007). Molecular data, however, strongly support the clade Afrotheria (Hedges 2001; Murata et al. 2003; Seiffert 2007; Stanhope et al. 1998).

Order Afrosoricida

The order Afrosoricida includes two suborders, each with a single family (Bronner & Jenkins 2005). The suborder Tenrecomorpha includes the tenrecs and otter shrews (Family Tenrecidae), and the suborder Chrysochloridea contains the golden moles (Family Chrysochloridae). African golden moles, otter shrews, and the Malagasy tenrecs were traditionally considered to be a part of the order Insectivora (or Lipotyphla), which also included shrews, moles, and hedgehogs. The Insectivora, however, was a catchall category that has now been largely abandoned. Instead, the shrews, moles, solenodonts, and hedgehogs are placed in the order Eulipotyphla, and the African radiation of insectivores, the golden moles, otter shrews, and tenrecs, comprise the order Afrosoricida (Asher et al. 2003; Madsen et al. 2001).

Family Tenrecidae

The tenrecs and otter shrews are a distinctive group of primitive insectivores that vary widely in morphology and habits. The family includes 10 genera and 30 species and inhabits Madagascar, the Comoro Islands, and western central Africa. With the exception of the African otter shrews (subfamily Potamogalinae), the major evolution of the group occurred primarily on the island of Madagascar. The largest tenrec, *Tenrec ecaudatus*, has been recently introduced to the Comoro Islands, Mascarene Islands, and the Seychelles. The meager fossil record of tenrecs indicates little about their evolution. Fossils are known from the early Miocene of Africa and the Pleistocene of Madagascar. The ancestral tenrec stock on Madagascar probably dispersed there from Africa at least once during the Miocene epoch (Mouchaty et al. 2000), with a back migration to Africa (Asher & Hofreiter 2006). The exact dispersal mechanism is unknown, but it is clear from molecular data that the colonization of Madagascar occurred after tenrecs diverged from the otter shrews (Potamogalinae) an estimated 55 million years ago (Douady et al. 2002; Rabinowitz et al. 1983). The subsequent isolation of tenrecs on Madagascar resulted in a textbook example of an **adaptive radiation**; tenrecs bear a general resemblance to such diverse mammals as shrews (*Microgale*), hedgehogs (*Echinops* and *Hemicentetes*), desmans, (*Limnogale*), and moles (*Oryzorictes*). Despite their radiation into a wide variety of ecological niches, tenrecs retain a suite of ancestral traits (they lack auditory bullae, zygomatic arches, and have a cloaca).

Tenrecs vary from roughly the size of a shrew to the size of a cottontail rabbit. The snout is frequently long and slender. The jugal bone is absent. The eye is usually small, and the pinnae are conspicuous. The tympanic bone is annular, and the squamosal bone forms part of the roof of the tympanic cavity. The anterior dentition varies from species to species; however, the first upper premolars are never present, and the molars are 3/3 in all but *Tenrec* (4/3) and *Echinops* (2/2). The upper molars have crowns that are triangular in occlusal view, and only in one genus (*Potamogale*) is a W-shaped ectoloph present. The urogenital canal and anus open into a common cloaca. The retractile penis rests in a fold ventral to the anus.

An unusually broad array of adaptive types occurs within the Tenrecidae. *Tenrec* roughly resembles a tailless, coarse-furred, long-snouted opossum and has spines interspersed with soft hairs (**Fig. 8-2A**). It is omnivorous. *Hemicentetes* (Fig. 8-2C), *Echinops* (Fig. 8-2B), and *Setifer*

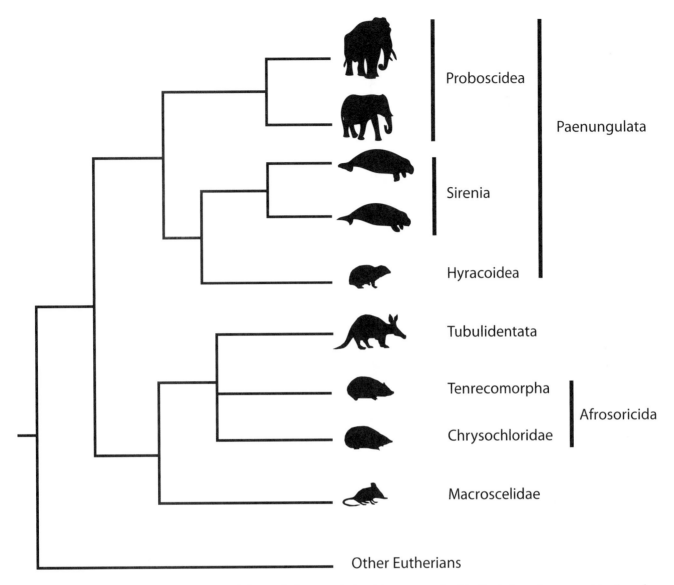

FIGURE 8-1 One phylogentic hypothesis of the evolutionary relationships among the Afrotheria based on a combination of genomic, morphological, and fossil characters. (Adapted from Nishihara, H., et al., *Mol Biol Evol.* 22 (2005): 1823–1833 and Seiffert, E.R., *BMC Evol Biol.* 7 (2007): 1–13.)

are also spiny, and all three closely resemble hedgehogs (Erinaceomorpha, Erinaceidae). In the later two genera, the panniculus carnosus muscle is powerfully developed and enables the animals to erect the spines. It also contributes to the ability of these animals to roll into a protective ball. The feet and head are tucked beneath the body during this protective movement, and the dermal muscles encircling the body allow the spiny dorsal skin to form an impregnable shield of spines (Gould & Eisenberg 1966). *Hemicentetes* has a group of 14 to 16 specialized quills on the middle of the back—the **stridulating organ**—that rub together when underlying dermal muscles are twitched to produce sounds in a variety of repetitive patterns. Differences in these sounds are correlated with differences in associated

behavior of the animals (Gould 1965) and may be used in **intraspecific** communication. In addition, Gould (1965) suggested that some tenrec species may use audible tongue clicks as a rudimentary form of echolocation over short distances. Tongue clicks produced by the four genera studied had frequency ranges of approximately 11–14 hertz and an average duration of between 0.4 and 1.2 milliseconds depending on the species. In contrast, echolocating bats use frequencies of 20,000–100,000 hertz and pulse durations of 0.25 to 100 milliseconds (see Chapter 22).

Some tenrecs are known to be **heterothermic** (e.g., to exhibit fluctuating body temperature or periods of torpor) under natural conditions during seasons of food shortage (Eisenberg & Gould 1970) when conservation of energy is

FIGURE 8-2   (A) A tail-less tenrec (*Tenrec ecaudatus*, Tenrecidae). (B) A lesser hedgehog tenrec (*Echinops telfairi*, Tenrecidae). (C) A lowland streaked tenrec (*Hemicentetes semispinosus*, Tenrecidae).

of vital importance. The ability to maintain a relatively constant body temperature varies widely among Malagasy tenrecs (Olson & Goodman 2003; Racey & Stephenson 1996). The smaller shrew tenrecs in the genus *Microgale* appear capable of maintaining a relatively constant, if somewhat depressed, body temperature. In contrast, the spiny tenrecs (*Tenrec ecaudatus, Hemicentetes nigriceps, H. semispinosus*) undergo daily or seasonal torpor. Others, such as the long-

eared tenrec (*Geogale aurita*) and *Echinops*, appear unable to maintain a constant elevated body temperature under any ambient temperature (Nicoll & Thompson 1987; Racey & Stephenson 1996; Stephenson 1991). *Echinops* typically remains torpid during the day, a condition characterized by low heart rates and body temperatures approaching ambient levels (Scholl 1974), and becomes active at dusk. If environmental temperatures dip below 16°C for a few days, the animal remains in torpor. Waking from torpor involves producing heat by shivering (i.e. thermogenesis); during this process, the heart rate can reach 300 beats per minute, and metabolic rates rise dramatically until the animal reaches a body temperature of about 30°C.

The Malagasy shrew tenrecs of the genus *Microgale* have the most species (18) and are the least specialized tenrecs (Jenkins 2003; Jenkins et al. 1997; MacPhee 1987). Members of this genus lack spines and resemble shrews (**Fig. 8-3**). *Microgale longicaudata*, as its name implies, has 47 caudal vertebrae, the largest number of any mammal except certain pangolins. The tip of the tail in some species of *Microgale* is semiprehensile and is used when climbing thin branches (Ryan, personal observation). The majority of species in this genus inhabit forests, from eastern humid forests at sea level to mossy montane forests at 2,300 meters elevation (Jenkins 2003). They forage along the forest floor for insects, and a few species climb into the understory.

Little is known about the remaining three genera of tenrecs. The large-eared tenrec, *Geogale aurita*, is found in the arid deciduous and spiny forests of southwestern Madagascar. It is unusual in that females have postpartum estrus; this condition allows a second litter to develop in the uterus while the first litter is still nursing. Stephenson and Racey (1993) suggested that postpartum estrus represents an adaptation to maximize reproduction in unpredictable arid environments. The two species of rice tenrecs (genus *Oryzorictes*) are also called mole tenrecs because of

FIGURE 8-3   A shrew tenrec (*Microgale dobsoni*).

their stout limbs, small eyes, and silky fur. They are clearly adapted for digging and are often found in association with rice fields (Stephenson 2003). The most distinctive species is the web-footed tenrec, *Limnogale mergulus*; it is semi-aquatic and lives adjacent to running water where it feeds on aquatic invertebrates. Its forefeet are fringed with stiff hairs, and it has webbed hind feet that aid in swimming. Radio-tracking studies by Benstead et al. (2001) demonstrate that *Limnogale* is strictly nocturnal. Individuals leave their burrows after dusk to forage and may not return until an hour before sunrise. Most of the foraging period is spent in the water along stream channels up to 1,160 meters in length. This is a remarkably large home range for an animal weighing only 60 to 100 grams. Dives of up to 15 seconds are powered by powerful thrusts from the webbed hindfeet. *Limnogale* uses its long vibrissae to locate insect larvae and crayfish, which are snatched in the teeth and brought to the surface for consumption. *Limnogale* deposits feces at specific sites within its home range. Exposed boulders in the middle of channels appear to be preferred latrine sites (Benstead & Olson 2003).

The subfamily Potamogalinae, the otter shrews, includes animals that in many ways are the most remarkable members of the Tenrecidae (**Fig. 8-4**). Members of this relict group live in western and central Africa, are the only living members of a primitive lineage, and have probably survived because of their highly specialized, semiaquatic lifestyle. Although the giant otter shrew (*Potamogale velox*) has been known to scientists since 1860, the genus to which the Nimba and Ruwenzori otter shrews belong (*Micropotamogale*) was not described until 1954 (Heim de Balsac 1954).

The giant otter shrew is large, measures 600 millimeters in length, and weighs up to 1 kilogram. Otter shrews are not shrews at all, but tenrecs that are highly specialized for life as a miniature otter. Like otters, they have a broad, flat rostrum covered with stiff whiskers, small eyes, and reduced external ears. The body is long and streamlined. The limbs are rather short and stocky, and the large tail is laterally compressed. Unlike the web-footed tenrec of Madagascar, otter shrews use their tail for swimming; only the Ruwenzori otter shrew (*Micropotamogale ruwenzorii*) has webbed feet. Propulsion beneath the water is controlled by lateral movements of the flattened tail, and a number of unique features are associated with this locomotor style. The caudal vertebrae have high neural spines and large transverse processes. These unusual caudal vertebrae provide attachment

FIGURE 8-4   A Nimba otter shrew (*Micropotamogale lamottei*).

points for the powerful tail musculature, which is aided by the greatly enlarged gluteal muscles. The posterior parts of the gluteal muscles, which in quadrupedal mammals move the hindlimbs, attach to the muscles overlying roughly the first five caudal vertebrae and move the tail. In the sinuous motion of the back and tail and even in overall body form, *Potamogale* resembles a large salamander. Otter shrews live in permanent streams and rivers and in coastal swamps, and although they rely partly on fish, they seem to prefer freshwater crabs to other food (Kingdon 1984a; Vogel 1983). The habits of otter shrews remain poorly known and provide fascinating opportunities for the resourceful biologist. Unfortunately, habitat destruction continues to limit the geographic range of otter shrews; all three species of otter shrews are currently considered endangered.

Family Chrysochloridae

Golden moles resemble "true" moles (Talpidae), but even more closely resemble, in fossorial adaptations and in function, the marsupial "moles" (Notoryctidae). The 9 genera and 21 species constituting the family Chrysochloridae occur widely in southern Africa, where they occupy forested areas, savannas, and sand dunes. The earliest fossil chrysochlorids from the Miocene of Kenya (*Prochrysochloris*) resemble Recent species.

Golden moles have modes of life similar to those of the fossorial members of the Talpidae and possess some parallel adaptations, as well as some contrasting structural features. The external ear canal of golden moles is covered by dense fur and they lack pinnae altogether. The eyes of chrysochlorids are vestigial and covered with skin and fur, and the optic nerve has degenerated. The pointed snout has a leathery pad at its tip (**Fig. 8-5**). The skull is abruptly conical or wedge-shaped, possibly as an aid in moving loose soil as the animal burrows, and is unlike the flattened and elongate skull of talpid moles. The zygomatic arches are formed by elongate processes of the maxilla, and the occipital area includes bones, the tabulars, not typically found in mammals. An auditory bulla is present that is formed largely by the tympanic bones; the malleus is enormously enlarged. The dental formula is usually 3/3, 1/1, 3/3, 3/3 = 40. The first upper incisor is enlarged, and the molars are basically zalambdodont and lack the stylar cusps and W-shaped ectoloph typical of talpids. The permanent dentition of golden moles emerges fairly late in life. The forelimbs are powerfully built, and the forearm rests against a concavity in the rib cage. The fifth digit of the hand is absent, and digits two and three usually have a huge, pick-like claw. The forelimb has a third bone, which is probably an ossified flexor tendon (**Fig. 8-6**) (Gasc et al. 1986). The forelimbs are not rotated as are those of talpids but more or less retain the usual mammalian posture, with the **palmar** surfaces downward.

FIGURE 8-5 (A) The cape golden mole (*Chrysochloris asiatica*; Chrysochloridae). (B) A Grant's golden mole (*Eremitalpa*) emerging from a dune.

Golden moles are adept burrowers. Gasc et al. (1986) studied golden moles in the laboratory, using x-ray motion pictures, and found that *Eremitalpa granti* swims through loose sand by alternating a forelimb digging phase with up-and-down movements of the head and shoulders that serves to "buttress" the sand above the animal and allow efficient movement through loose substrates (Fig. 8-6). Bateman (1959) also studied golden moles in the laboratory and found that a 60-gram golden mole could push up a 9-kilogram weight covering its cage; this amounts to exerting a force equal to 150 times the animal's weight. When the animal is close to the surface, a ridge marks the course of its progress (**Fig. 8-7**). Both deep and shallow burrows are constructed; the depth of the burrow may depend on the amount of soil moisture. The roofs of shallow burrows in sandy soil frequently collapse, leaving a furrow in the sand as a trace of the former burrow.

Vision is of little use in a subterranean environment. Consequently, golden moles depend on their sense of hearing; however, instead of hearing airborne sounds, several chrysochlorid species use bone conduction to detect seismic vibrations (Mason 2003a, 2003b). The inner ear bones (auditory ossicles) of several genera of golden moles are greatly enlarged. This restricts hearing to low-frequency

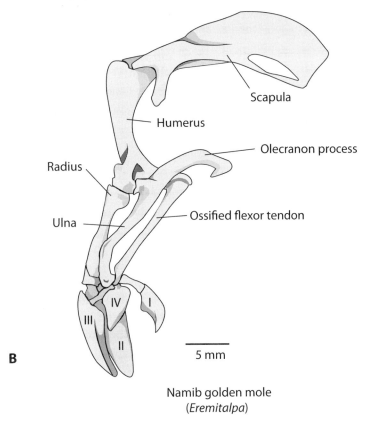

FIGURE 8-6 (A) Schematic drawing of the sand swimming behavior of the Grant's golden mole (*Eremitalpa granti*). The arrows indicate the direction of the thrust of the head and forefeet. (B) Lateral view of the left forelimb skeleton of a golden mole showing the enlarged olecranon process and addition of a third ossified element called the "ossified flexor tendon" in the forearm. The digits are labeled I through IV. (Adapted from Gasc, J.P., et al., *J Zool.* (A) 208 (1986): 9–35.)

sounds, which travel farther in underground tunnels than do high-frequency sounds. Moreover, the massive malleus of *Chrysochloris* can transfer vibrations from the ground through the skull to the oval window, resulting in the detection of seismic vibrations (Mason 2003a, 2003b). Definitive evidence that chrysochloids detect seismic signals is currently lacking, but golden mole behaviors (such as dipping the head into the sand and remaining motionless) are consistent with bone-conduction hearing.

The diet of golden moles consists mostly of invertebrates; two desert-dwelling genera (*Cryptochloris* and *Eremitalpa*) also eat legless lizards. *Eremitalpa* lives in the dunes of the Namib Desert, where it forages mostly on the surface and makes occasional burrowing forays (Fielden et al. 1990). Termites are the primary food. Burrowing activity of *Amblysomus* varies with temperature, rainfall, and prey availability.

Grant's golden mole (*Eremitalpa granti*) inhabits one of the hottest and driest places on Earth, the Namib Desert. These golden moles spend the day buried in the loose dune sand and the nights foraging on the dune surface (Fielden et al. 1990). Daytime surface temperatures can exceed 50°C

FIGURE 8-7 Trail (arrow) left behind a burrowing *Eremitalpa granti*.

at midday during the summer months (November through February in the Namib Desert). Subsurface sand temperatures drop dramatically in the first 10 centimeters and are further reduced beneath clumps of perennial dune grass. These golden moles also choose submerged sites at depths that are within their thermal neutral zone and tend to favor the cooler sites underneath dune grasses. *Eremitalpa* has a metabolic rate only 22% of the predicted rate, a high and narrow thermoneutral zone (31°C to 35°C) and an extremely labile body temperature (19°C to 38°C). The high thermal conductance favors diurnal torpor by allowing the body temperature to assume rapidly the temperature of the surrounding sand. Temperatures below 15°C are lethal. Fielden et al. (1990) suggest that the physiological characteristics of *Eremitalpa* are not adaptations to avoid thermal stress, which is alleviated by behavioral means. Instead, these adaptations evolved in response to a combination of factors, including low oxygen levels and high CO_2 levels encountered while buried in loose sand, the high costs of burrowing, and the limited and often patchy distribution of prey.

Order Macroscelidea

Macroscelidids (elephant shrews or sengis) were traditionally associated with tupaiids and insectivores (Anderson & Jones 1984; Simpson 1945), a view based primarily on shared primitive characters. Others consider macroscelidids to be a sister group of Glires (rodents and lagomorphs) (Gaudin et al. 1996; McKenna 1975; Novacek 1992b; Novacek & Wyss 1986; Szalay 1977; Wible & Covert 1987). Recent studies of both molecular and morphological characters suggest that macroscelids are an ancient African radiation belonging to the Afrotheria (Murphy et al. 2001), although Zack et al. (2005) indicate a North American origin based on the morphology of certain Paleocene-Eocene

fossils. Fossils are known only from Africa, where their sparse fossil record dates far back to the Eocene of Tunisia (Corbert 1971; Corbert & Hanks 1968; Patterson 1965). *Rhynchocyon* is known from the Miocene, and two other extant genera (*Macroscelides* and *Elephantulus*) are present in the Pliocene-Pleistocene of South Africa and Tanzania.

All four genera and 16 living species of elephant shrew are in the single family Macroscelididae and have a disjunct distribution in Africa. A new species of elephant shrew (*R. udzungwensis*) was discovered in 2007 in a remote region of Tanzania (Rovero et al. 2008). The greatest diversity occurs in Central, Southern, and Eastern Africa. *Elephantulus rozeti* lives along the northern coast of Africa (from Morocco, Algeria, Tunisia, and western Libya), separated from the remaining species by the Sahara Desert. Corbet and Hanks (1968) explained this isolated distribution by postulating dispersal westward from eastern Africa along the Mediterranean coast. Contrary evidence (Douady et al. 2003), however, suggests that elephant shrews were once continuously distributed from southern to northern Africa until mid-Miocene climatic changes isolated *E. rozeti* north of the Sahara.

Elephant shrews are remarkable African mammals with lifestyles and morphological features that differ sharply from those of the "true" shrews (family Soricidae). The term elephant shrew refers to the long, mobile snout (**Fig. 8-8**). Elephant shrews vary in size from the diminutive *Macroscelides proboscideus* (30 to 45 grams) to *Rhynchocyon chrysopygus* (approximately 550 grams). Their long, slender limb bones are adapted to rapid cursorial locomotion. The forelimbs and hindlimbs have four or five digits; in *Rhynchocyon*, the forefeet are functionally tridactyl, with the first digit absent and the fifth much reduced, whereas the hind feet have four digits. This more extreme loss of digits in the forelimb than in the hindlimb is unusual in mammals. Elephant shrews have an unusual, but complete auditory bullae (at least seven bones contribute to the bulla). They also have a large jugal bone (complete zygoma) and relatively small olfactory lobes. The large orbits of the skull are never bordered by a complete postorbital bar (**Fig. 8-9**). The dental formula is 1–3/3, 1/1, 4/4, 2/2 = 36–40; the last upper premolar is the largest molariform tooth, and the upper canine is double rooted. The upper molars are quadrate and have four major cusps. Additionally, males have abdominal testes, and an unusual penis morphology (Werdelin & Nilsonne 1999; Woodall 1995).

Many habitats are occupied by elephant shrews, including open plains, savannas, deserts, thornbush, and tropical forests. *Rhynchocyon*, the giant elephant shrew, is a forest dweller and, in some places, such as the coast of Kenya, occupies relict strips of forest. Rathbun (1979) and FitzGibbon (1997) studied *R. chrysopygus* in the coastal

FIGURE 8-8 (A) A rufous elephant shrew (*Elephantulus rufescens*, Macroscelididae). (B) A golden-rumped elephant shrew (*Rhynchocyon chrysopygus*, Macroscelididae).

forests of Kenya, where this diurnal animal feeds on a wide variety of invertebrates, many of which it digs from the leaf litter and soil with the long claws of the front feet. This species is strikingly colored: a dark chestnut brown, with a slightly purplish cast, covers the back and flanks, against which the bright yellow rump patch stands out in sharp contrast. These elephant shrews typically live in pairs occupying stable territories; scent marking by both sexes is

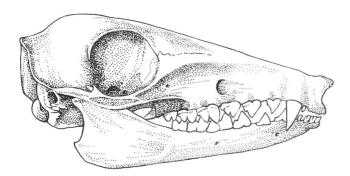

Golden-rumped elephant shrew
(*Rhynchocyon chrysopygus*, Macroscelididae)

FIGURE 8-9 Skull of a golden-rumped elephant shrew (length of skull 67 millimeters).

an important territorial behavior (FitzGibbon 1995, 1997; Rathbun 1979). Foot drumming or tail slapping may also be an important means of communication in some species. A resident pair chases conspecific intruders, with males chasing males and females chasing females. Rathbun (1978) postulated that the conspicuous yellow rump of *R. chrysopygus* functions as a target during aggressive encounters. He found that the skin beneath this patch is thicker than skin over the rest of the body and that scars and cuts are concentrated beneath the patch.

Studies by Rathbun (1979) on the crepuscular rufous elephant shrew have shown that the territories of this brush dweller contain intricate patterns of trails and that 24% of the daylight behavior of a territorial male is devoted to cleaning these trails by using the front feet to meticulously sweep away leaf litter. This male contribution to the common territory of a female and male may help explain why facultative monogamy evolved in this species (Rathbun 1992). This species spends its life above ground, never seeking refuge in burrows, and usually forages within 1 meter of trails. These elephant shrews escape predators by bounding along the trails at amazing speed. Scent marking by feces and sternal glands is concentrated on territorial borders. Rathbun found that the bulk of the diet is termites and ants, although some species supplement this diet with plant material. In coastal Kenya, where the four-toed elephant shrew (*Petrodromus tetradactylus*) and the golden-rumped elephant shrew (*R. chrysopygus*) are sympatric, niche separation is maintained by a combination of different activity patterns, diet preferences, and habitat separation (FitzGibbon 1995).

Elephant shrews in some areas regularly bask (**Fig. 8-10**), probably as a means of reducing energy expended on thermoregulation. In Namibia, for example, an elephant shrew (*Elephantulus rupestris*) basked almost continuously for over 3 hours on a cool (10°C), dry-season afternoon (Vaughan, personal observation). Indeed, several species adjust their physiology to meet environmental demands or to outlast periods of low food availability common in arid environments (Lovegrove et al. 2001; Mzilikazi & Lovegrove 2005; Perrin 1995a). For example, *Elephantulus myurus* exhibits daily torpor, during which minimum body temperatures can drop to 10°C; nocturnal torpor in this species typically lasts about 8 hours.

Like many other mammals inhabiting restricted ranges or fragmented habitats, many species of elephant shrews face an uncertain future. The International Union for Conservation of Nature and Natural Resources red list of threatened animals includes four species of elephant shrews, *Elephantulus revoili* and the three *Rhynchocyon* species. *Rhynchocyon chrysopygus*, for example, is restricted to small forest fragments along the northern coast of Kenya where

FIGURE 8-10   Elephant shrew (*Elephantulus rupestris*) basking in the morning sun.

tourism, development pressures, and bushmeat hunting continue to restrict the habitat of this fascinating species (Rathbun & Kyalo 2000).

Order Tubulidentata

The order Tubulidentata, the aardvarks, includes but one Recent species, *Orycteropus afer*, (family Orycteropodidae; **Fig. 8-11**). The aardvark inhabits much of Africa south of the Sahara Desert. Fossil tubulidentates first appear in the Miocene, and Pliocene fossils appear in Southern Europe, Turkey, Pakistan, and Africa. An extinct member of *Orycteropus* occupied parts of Europe and Asia as recently as the late Pliocene (Holroyd & Mussell 1995; Lehmann et al. 2005 and references therein).

FIGURE 8-11   The aardvark (*Orycteropus afer*, Orycteropodidae).

In the past, tubulidentates often were considered closely related to ungulates. Phylogenetic studies using morphological data cast doubt on a relationship with ungulates but could not otherwise clarify the group's affinities (Novacek & Wyss 1986; Thewissen 1985). Recent data from mitochondrial and nuclear genes, however, support a relationship among the Afrotheria.

Aardvarks are powerful diggers that feed on ants and termites. They weigh up to 65 kilograms, and the thick, sparsely haired skin provides protection from insect bites. The skull is elongate, and the dentary bone is long and slender (**Fig. 8-12**). Incisors and canines are lacking in the dentition of adults (but are present in the deciduous dentition); the cheek teeth are 2/2 premolars and 3/3 molars. Each tooth is rootless and consists of as many as 1,500 hexagonal prisms of dentine, each surrounding a slender, tubular pulp cavity. The columnar teeth lack enamel but are surrounded by cementum. The anteriormost teeth erupt first and are often lost before the posterior molars are fully erupted. The slender tongue is protrusible.

Olfaction is used in finding insects; the olfactory centers of the brain are unusually well developed, and the turbinal bones are remarkably large and complex. The nostrils are highly specialized in a fashion not found in any other mammal. Fleshy tentacles, which presumably have a tactile or other sensory function, occur on the nasal septum (**Fig. 8-13**), and dense hair surrounds the nostrils, which can be sealed when the aardvark digs. Prominent scent glands (resembling a scrotum) in the groin produce a musky secretion used in territorial marking. The pinnae are large and hearing is well-developed but probably used for predator detection rather than for locating insects. The pollex is absent, and the hind feet are five toed; the robust claws are flattened and blunt, intermediate between a hoof and a nail.

Many of the adaptations of aardvarks are specializations for feeding on termites and ants. Some of these, like the elongation of the snout, the highly developed olfactory

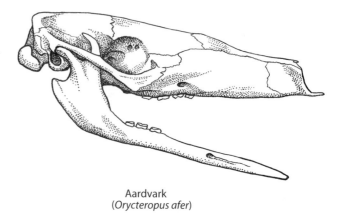

Aardvark
(*Orycteropus afer*)

FIGURE 8-12   The skull of the aardvark (length of skull 240 millimeters). (Adapted from Hatt, R. T., *Bull. Amer. Mus. Nat. Hist.* 66 (1934): 643–672.)

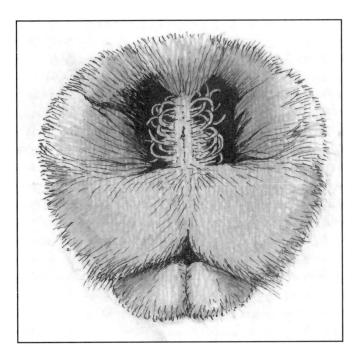

FIGURE 8-13 The complex nose of the aardvark (*Orycteropus afer*). Note the fleshy tentacles on the nasal septum and the dense tracts of hair that can seal the nostrils. (Adapted from Kingdon, J. *The Kingdon Field Guide to African Mammals.* Princeton University Press, 1997.)

system, the long sticky tongue, and the simplification of the dentition, are features that are convergent with those of other ant-eating mammals like the pangolins (order Pholidota) and xenarthran anteaters (order Pilosa). In addition to these structural features, aardvarks are physiologically convergent with other ant-eating mammals. They exhibit a low basal metabolic rate and a low body temperature of about 34.5°C (McNab 1984).

The powerful forelimbs are used in burrowing and in dismantling nearly rock-hard termite mounds and ant nests, and the hindlimbs thrust accumulated soil from burrows. Not only do aardvarks have phenomenal strength, but they can dig astonishingly fast. A meter-long section of burrow the diameter of the animal can be dug in 5 minutes

in appropriate soils. Burrows dug by aardvarks are extensive and numerous in some areas and are used as retreats by a variety of mammals, such as warthogs, hyenas, porcupines, jackals, bat-eared foxes, hares, bats, ground squirrels, and civets, as well as monitor lizards and owls. In Botswana, Smithers (1971) recorded 17 species of mammals, 2 species of reptiles, and a bird using aardvark burrows. One bird, the ant-eating chat (*Myrmecocicla formicivora*), nests in occupied burrows (Taylor & Skinner 2001). In parts of Africa, the abundance of warthogs depends on the availability of abandoned aardvark burrows (Melton 1976).

Although the foot posture is digitigrade, aardvarks are slow runners and can be outrun by a human. They are almost entirely nocturnal and thus are rarely seen by people, even those living in the same area. As a result, their disappearance over much of their historic range has gone largely unnoticed. During their nightly foraging, aardvarks may travel 10 to 30 kilometers. Often their foraging strategy is to move along in a zigzag path about 30 meters wide, sniffing and probing the ground and frequently digging small or large excavations. Although ants and termites are the main foods eaten, other insects such as beetles and grasshoppers are occasionally taken. In some areas they drink regularly, but in parts of southern Africa, they occupy areas that lack surface water in the dry season. During the driest months of the year, aardvarks dig up and eat the moist, fleshy fruits of an unusual species of cucurbit plant (*Cucumis humifructus*; Cucurbitaceae), known as the "aardvark cucumber." The association between the aardvark and this plant may be symbiotic. The geographic distributions of the two species (mammal and plant) are similar, and the cucumber is the only member of its family that produces an underground fruit. Aardvarks pass the ingested seeds intact in their feces, burying and thus replanting and fertilizing them. Aardvarks are the only known agent of dissemination of the seeds; presumably, aardvarks acquire water and nutrients from the fruits (Patterson 1975). Radio-tracking studies reveal that aardvarks have home ranges up to 302 hectares (just over a square mile; Taylor & Skinner 2003).

SUMMARY

The tenrecs and golden moles (Afrosoricida), elephant shrews (Macroscelidea), and aardvark (Tubulidentata) are members of the Afrotheria clade, along with the Paenungulata (elephants, hyraxes, and sirenians; Chapter 9). Among Afrosoricida, the family Afrosoricidae includes the tenrecs of Madagascar and the Central African otter shrews (30 species in 10 genera), and the Chrysochloridae includes the golden moles of southern Africa (21 species in 9 genera). Afrosoricids resemble other widely diverse mammals such as hedgehogs, shrews, mice, otters, and even moles. They

occupy aquatic, arboreal, terrestrial, and fossorial habitats. Golden moles resemble both "true" moles (Talpidae) and marsupial "moles" (Notoryctidae). They are fossorial, have small eyes covered by skin, leathery nose pads, and short, powerful forearms and claws used for burrowing.

The elephant-shrews (Macroscelidea) comprise 16 species in 4 genera. They have long flexible snouts, large eyes, and hind limbs longer than forelimbs, and some species are brightly colored. Elephant-shrews are insectivorous and nocturnal. They escape from predators using a bounding gait over rocks or along the extensive trails they maintain in the undergrowth.

The aardvark (Tubulidentata) is a pig-sized mammal that is specialized for digging and feeding on insects, especially termites. Aardvarks are sparsely haired and have a long, tubular snout, large ears, and a protrusible tongue. The single species of aardvark is evolutionarily distinct and the sole remaining member of its lineage.

KEY CHARACTERISTICS OF AFROSORICIDA

- Cloaca (common opening for urogenital and anal openings)
- Tenrecs and otter shrews:
  - Incomplete zygomatic arch
  - Zalambdodont molars
  - Males that retain abdominal testes (no scrotum)
- Golden moles:
  - Zygomatic arch formed by elongate maxillae
- Zalambdodont molars
- Skull conical
- Pair of tabular bones in occipital area (absent in other mammals)
- Eyes vestigial
- Snout with leathery pad
- Malleus greatly enlarged

KEY CHARACTERISTICS OF MACROSCELIDEA

- Complete zygomatic arches and auditory bullae
- Double-rooted upper canines
- Quadrate molars
- Fenestrated palate
- Snout long and flexible

KEY CHARACTERISTICS OF TUBULIDENTATA

- Elongate, conical skull
- Elaborate turbinal bones
- Complete zygomatic arches
- Incisors and canines absent
- Cheek teeth with hexagonal prisms of dentine coated with cementum
- Cheek teeth rootless and ever-growing
- Milk teeth small and lost before birth
- Stout nails resembling hooves on digits
- Thick, sparsely haired skin
- Long protrusible tongue

KEY TERMS

Adaptive radiation
Heterothermic
Intraspecific

Palmar
Seismic signals
Stridulating organ

Thermal conductance
Thermal neutral zone

RECOMMENDED READINGS

Asher, RJ, MJ Novacek, & JH Geisler. 2003. Relationships of endemic African mammals and their fossil relatives based on morphological and molecular evidence. *Journal of Mammalian Evolution,* 10:131–194.

Garbutt, N. 2007. *Guide to the Mammals of Madagascar.* Christopher Helm Ltd., London

Holroyd, PA & JC Mussell. 1995. Macroscelidea and Tubulidentata. Pp 71-83, in *The Rise of Placental Mammals* (KD Rose & JD Archibald, eds.). Johns Hopkins University Press, Baltimore.

Kingdon, J. 1974. *East African Mammals, Vol. IIA, An Atlas of Evolution in Africa, Insectivores and Bats.* University of Chicago Press, Chicago.

Mason, MJ. 2003. Morphology of the middle ear of golden moles (Chrysochloridae). *Journal of Zoology,* 260:391–403.

Perrin, MR. 1995. The Biology of Elephant-Shrews: A Symposium Held During the 6th International Theriological Congress. Sydney, 5 July 1993. *Mammal Review,* 25(1 and 2).

Sánchez-Villagra, MR, Y Narita, & S Kuratani. 2007. Thoracolumbar vertebral number: the first skeletal synapomorphy for Afrotherian mammals. *Systematics and Biodiversity,* 5(1):1–7.

Seiffert, ER. 2007. A new estimate of afrotherian phylogeny based on simultaneous analysis of genomic, morphological, and fossil evidence. *BMC Evolutionary Biology,* 7:224.

Paenungulata

If general appearance were used as the single criterion for evaluating phylogenetic relationships, the rodent-like hyraxes, massive elephants, and aquatic sirenians (manatees and dugongs) would be judged to be three very distantly related groups of mammals. In this case, however, appearances are deceptive, for the fossil record and a diverse array of anatomical and genomic data suggest that these groups form a monophyletic group, the Paenungulata (characters detailed in Seiffert 2007) (**Fig. 9-1**). Paenungulates ("subungulates"), along with the tenrecs and golden moles (Afrosoricida), elephant shrews (Macroselidea), and aardvark (Tubulidentata) comprise the Afrotheria (Fig. 8-1). Paenungulates share a forked styloglossus muscle in the tongue, nail-like hooves, abdominal testes, microbial fermentation chambers (caecum) in the large intestines, and a number of skeletal characters (Gheerbrant et al. 2005, and references therein). Morphological characters from living and fossil forms originally united hyraxes with perissodactyls (summarized by Gheerbrant et al. 2005). More recent eutherian phylogenies, based on combined morphological and genomic character sets, support a monophyletic clade containing sirenians, proboscideans, and hyracoids (Madsen et al. 2001; Seiffert 2007; Shoshani & Tassy 1996).

The history of paenungulates begins roughly 65 to 60 million years ago in Africa. One lineage of condylarths (primitive hoofed mammals) gave rise to five paenungulate orders (the Embrithopoda and Desmostyla are extinct). Of the three surviving orders, the most basal group were the hyracoids (hyraxes and their ancestors), which appeared in the early Eocene of Africa. Hyracoids diversified in the Eocene and were among the dominant terrestrial herbivores of their time. Competition from ungulates in the Miocene ultimately led to a decline in hyracoid diversity. Some larger hyracoids foraged along the coast of the Tethys Sea and may have been semiaquatic. Sinenians, extinct desmostylians, and proboscideans (elephants and their ancestors) probably evolved from a common hyracoid ancestor near the shores of the Tethys Sea. Early fossils of both proboscideans and sirenians tend to be found in shallow marine or near-shore deposits, and isotope analyses indicate that their common ancestor was semiaquatic (Clementz et al. 2006; Domning 2001; Gheerbrant et al. 2005; Liu et al. 2008).

Order Proboscidea: Elephants

Throughout much of the Cenozoic, some of the largest and most spectacular herbivores were proboscideans; in the Neogene Period, a varied array of these animals occurred widely in North America, Europe, Asia, and Africa. The diversity of proboscideans was reduced in the Pleistocene Epoch, and today, only three species represent this remarkable group. Because elephants now often threaten the interests of humans and because of the great value of their tusks (trade in elephant ivory is now restricted by international treaty), they are being extirpated over wide areas as human populations increase. Today elephants occur only in Africa south of the Sahara Desert (*Loxodonta*) and in parts of southeastern Asia (*Elephas*). Regrettably, we may be witnessing the final stages in the history of one of the most interesting mammalian orders.

Paleontology

The fossil record of proboscideans begins in the early Eocene of North Africa with *Phosphatherium*, *Daouitherium*, and *Numidotherium* (Gheerbrant et al. 1996, 2002, 2005; Mahboubi et al. 1984). Some of these early proboscideans were only the size of a small dog but already had a lophodont dentition (molar teeth with parallel ridges). By the late Eocene, African proboscideans were diverse and already exhibited the typical columnar limbs of today's elephants, indicating that large size and **graviportal** locomotion evolved early in this order (**Fig. 9-2**). Proboscideans expanded out of Africa in the late Oligocene, when Afro-Arabia and Eurasia became contiguous (Kappelman et al. 2003), and reached North America in the middle Miocene. A number of important evolutionary trends occurred during the radiation of proboscideans (Shoshani 1998). The dominant trends include (1) lengthening and thickening of the limb bones to

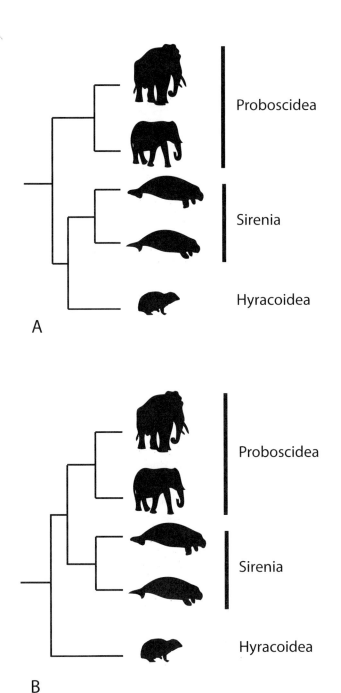

A

B

FIGURE 9-1 Two possible phylogenies of the Paenungulata based on a combination of characters. (Adapted from Seiffert, E.R., *BMC Evol Biol.* 7 (2007): 224 and Nishihara, H., et al., *Mol Biol Evol.* 22 (2005):1823–1833.)

support increased body weight, (2) enlargement of the skull and reduction of the neck, (3) development of an elongate proboscis or trunk, (4) hypertrophy of the incisors to form tusks, (5) increasing numbers of cross lophs on the cheek teeth, and (6) development of forward displacement of the cheek teeth. The rather primitive Eocene proboscideans had

brachydont teeth with few ridges. Most or all the cheek teeth were in place at one time, and both an upper set and a lower set of tusks usually were present (**Fig. 9-3**). By the Miocene, most forms had complexly lophed cheek teeth replaced from behind as worn teeth moved forward along the jaw and eventually fell out.

All three living species belong to the family Elephantidae, which is represented first by the early Miocene Asian genus *Stegolophodon*. Mammoths (*Mammuthus*) evolved in Africa in the Pliocene and radiated outward into Eurasia and later North America, and at least two species of mammoth coexisted in North America with early human populations in the late Pleistocene (**Fig. 9-4**). One Pleistocene species (the woolly mammoth, *Mammuthus primigenius*) was in some ways more specialized than the living elephants. It had a remarkably foreshortened skull and long, upwardly curved tusks that sometimes crossed; the last molar had up to 30 laminae, more than occur in the living elephants. Entire and partial frozen woolly mammoths have been found in Siberia and Alaska. Many recent efforts have been made to sequence the ancient DNA from their remains to learn the species' genome (Gilbert et al. 2008; Krause et al. 2006). Some island populations of mammoths existed off the north coast of Siberia until just 3,700 years ago (Vartanyan et al. 2008).

Morphology

The three extant elephants—the African savanna elephant, *Loxodonta africana*, the forest elephant *L. cyclotis*, and the Indian elephant, *Elephas maximus*—are the largest land mammals, reaching weights of 6,000 kilograms (see Eggert et al. 2002) (**Fig. 9-5**). Central African tropical forests are home to the forest elephant (*L. cyclotis*). In addition to its smaller size and less curved tusks, the forest elephant has five front nails and four on the foot, whereas the savanna elephant has four in front and three in back. Both African species have a long proboscis (trunk) with two finger-like structures at its tip, large ears, and graviportal limbs. As early elephants increased in size, they modified the upper lip and nose into a trunk, which allowed them to feed from the ground and to reach the upper branches of trees. Elephants also use their muscular trunks to smell, manipulate objects, dig water holes, and suck water and squirt it into the mouth or spray it over their bodies.

Asian and African elephants differ in that Asian elephants (Fig. 9-5B) have smaller ears, 19 pairs of ribs instead of 21 pairs, and a flattened forehead; the top of the head is dome shaped and is the highest point on the animal, and there is a single finger-like process at the tip of the trunk. In African elephants, the shoulders are generally the highest point, and there are two finger-like processes at the tip of

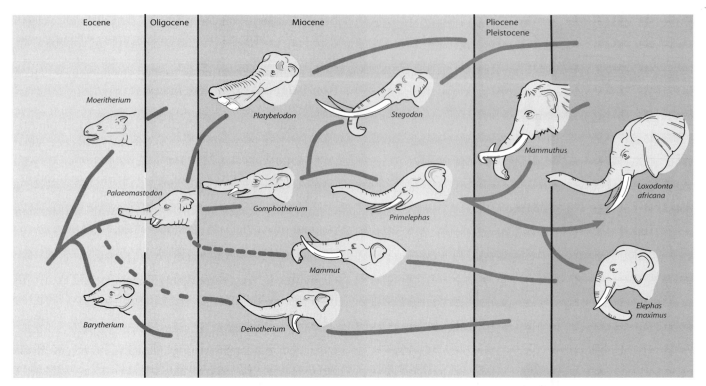

FIGURE 9-2 One family tree of probocideans from the Eocene to the present. (Adapted from Shoshani, J., *Natural History* 106 (1997): 36–47.)

the trunk. African elephants also have much larger, more vascularized ears, which help dissipate excess body heat.

In both genera, the limb bones are heavy, and the proximal segments of the limbs are relatively long; the ulna and fibula are unspecialized, and the bones of the five-toed manus and pes are short and robust with an unusual, spreading, digitigrade posture (**Fig. 9-6**). A heel pad of dense connective tissue braces the toes and largely supports the weight of the animal. An adaptation allowing the efficient support of great weight, the long axis of the pelvic girdle is nearly at right angles to the vertebral column, and the acetabulum faces ventrally. In addition, when the weight of the body is supported by the limbs, the limb segments form a nearly straight line. As described by Howell (1944), an elephant "relies exclusively upon the walk or its more speedy equivalent, the running walk, which permits it to keep at least two feet always on the ground." The gait is unusual, and it has been argued recently that elephants actually can "run" (Hutchinson et al. 2003).

The skull is unusually short and high, perhaps in response to a need for great mechanical advantage for the muscles that attach to the lambdoidal crest and raise the front of the head and the tusks. The skull contains numerous large air cells, particularly in the cranial roof, which lighten this massive structure.

The highly specialized dentition consists of the tusks (each a second upper incisor) and six cheek teeth in each half of each jaw. The pattern of cheek tooth replacement is remarkable. The cheek teeth erupt in sequence from front to rear, but only a single tooth or one tooth and a fragment of another are functional in each half of each jaw at one time (**Fig. 9-7B**). As a tooth becomes seriously worn, it is replaced by the next posterior tooth. The first three cheek teeth erupt during an animal's youth. The fourth erupts at 4 to 5 years of age, the fifth tooth at age 12 to 13, and the final tooth around 25 years of age and lasts for up to 50 more years (Benton 1990; Krumrey & Buss 1968; Laursen & Bekoff 1978). The hypsodont cheek teeth are formed of thin laminae, each consisting of an enamel band surrounding dentine, with cementum filling the spaces between the ridges

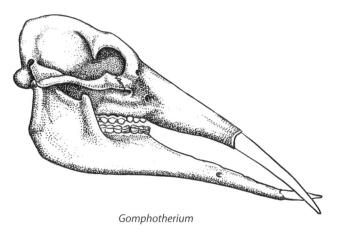

Gomphotherium

FIGURE 9-3 The skull of *Gomphotherium*, a Miocene proboscidean. Length of skull and tusks roughly 1 meter. (Modified from Romer, A.S. *Vertebrate Paleontology*. University of Chicago Press, 1966.)

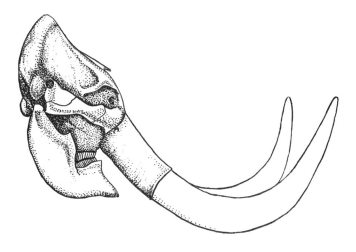

Mammuthus

FIGURE 9-4 The skull of *Mammuthus*, a Pleistocene elephantid. Length of skull and tusks roughly 2.8 meters. (Modified from Romer, A.S. *Vertebrate Paleontology.* University of Chicago Press, 1966.)

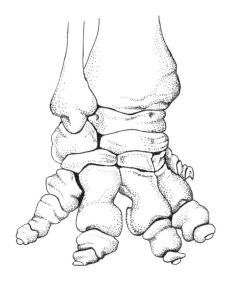

FIGURE 9-6 The right hind foot of *Mammut*, a late Tertiary and Pleistocene proboscidean. (Modified from Romer, A.S. *Vertebrate Paleontology.* University of Chicago Press, 1966.)

(Fig. 9-7A). The last molar, the tooth that must serve for much of the animal's adult life, has the greatest number of laminae. The premolars are considerably smaller, simpler, and less durable than the molars. In old elephants, some of the anterior laminae of the third molar may be lost while the remainder of the tooth is still functional.

◼ Behavior

Elephants occupy forests, semiopen or dense scrub, savanna, and even desert regions in Namibia, but they need access to water. Climate change and the drying out of the Sahara by 2,700 years ago probably drove them out of northern Africa (Kropelin et al. 2008). They feed on a variety of trees, shrubs, grasses, and aquatic plants and characteristically strongly influence their environments (Laws 1970). Each individual eats over 200 kilograms of forage daily. Their great size and strength enable them to "ride down" fairly large trees in order to feed on the leaves (**Fig. 9-8**).

Female elephants are highly social. African elephants live largely in **matriarchal** kinship groups from which adult males are excluded. An elephant family unit consists of a matriarch, several usually related adult females, and their immature offspring (**Fig. 9-9**). The matriarch is the leader of the family unit; she is typically the largest and oldest female in the group (Archie et al. 2006; Douglas-Hamilton & Douglas-Hamilton 1975). African elephant family units may consist of up to 50 individuals that travel together. Female African elephants can use olfactory cues from urine to recognize up to 17 other females and as many as 30 members of their family unit and to keep track of their locations (Bates et al. 2008). Members of the family unit cooperate to raise one another's offspring, acquire food and water, and when necessary defend the group (Moss & Poole 1983). Within a family unit, cohesion is maintained

FIGURE 9-5 (A) A kinship group of African elephants (*Loxodonta africana*) at a waterhole in Etosha National Park. (B) The Indian elephant (*Elephas maximus*) has smaller ears and a flatter forehead.

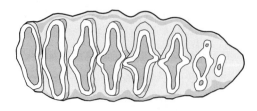

African elephant
(*Loxodonta africana*, Elephantidae)

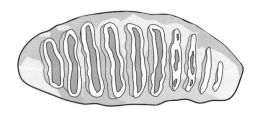

Asian elephant
(*Elephas maximus*, Elephantidae)

A

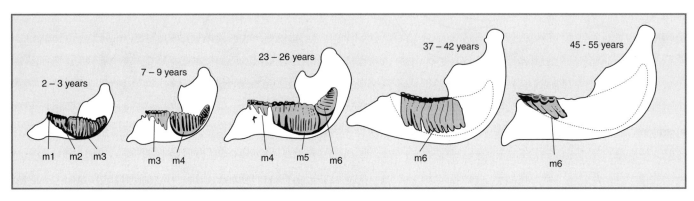

B

African elephant
(*Loxodonta africana*, Elephantidae)

FIGURE 9-7 (A) The occlusal surface of a molar of an African elephant (left) and an Asian elephant (right). (B) The progression of molar toothwear in the African elephant (*Loxodonta africana*) from birth to approximately 55 years of age. Some researchers consider the first three cheek teeth in elephants to represent deciduous premolars (dp2–dp4) and m4–m6 (as numbered here) to represent the permanent m1–m3. (Adapted from Kingdon, J. *The Kingdon Field Guide to African Mammals*. Princeton University Press, 1997.)

by the matriarch via an intricate suite of tactile, visual, and vocal behaviors. When family units become too large for the available resources, the unit splits to form several smaller family units. Thus, in a given area, there may be several genetically related family units; these extended families are called bond groups. Bond groups come together and split apart depending on resource availability (e.g., a fission–fusion social system). Above the level of the bond group is the elephant clan. Clans include several bond groups that all share the same home range when resources are scarce (such as in the dry season or during prolonged droughts). Elephant clans can consist of several hundred elephants in some parts of east Africa. Thus, elephant societies are fluid, with complex relationships forming at several levels from simple mother-offspring relationships to massive assemblages involving several clans.

Male elephants, in contrast, must negotiate two very different social worlds. As youngsters, males live in the tightly bonded family unit. Later, by the age of 10 to 18 years, adolescent males are forced out of the maternal unit (Lee & Moss 1999). This process often takes several years to complete, but eventually, males find themselves at the

bottom of the social ladder in loosely knit bachelor herds. The lives of adult males are characterized by few lasting social bonds and are punctuated by periods of increased aggression and heightened sexual activity called **musth** (Poole & Moss 1981; Rasmussen et al. 1996). Male Asian elephants enter musth only after reaching maturity at approximately 20 years of age. At that time, males begin to compete aggressively for mating opportunities with females (**Box 9-1**).

Long-term field studies of African elephants by Cynthia Moss and others indicate that female elephants in the wild are sexually receptive for only a few days every 3 to 9 years (Laws et al. 1975; Moss 1983, 2001). In addition, elephants range over vast areas, making sexually receptive females a sparsely distributed resource for musth males. How then do musth males and receptive females find each other?

Vocal communications, including the familiar trumpeting sounds, are known to play an important role in maintaining group cohesion in female family units, bond groups, and bachelor herds. Some of these vocalizations include the familiar trumpeting sounds, but elephants are also capable of communicating over distances of several

FIGURE 9-8   A male African elephant (*Loxodonta africana*) using its trunk to reach branches of an acacia tree southern Kenya.

kilometers by using very low-frequency sounds (14 to 24 Hertz) (Garstang 2004; Payne et al. 1986; Poole et al. 1988). These "near-**infrasounds**" are not audible to humans. In addition to their role in general communication, these infrasonic communications may serve as fertility advertisements, allowing musth males to locate sexually receptive females across vast expanses of savanna. In a detailed study examining the relationship between low-frequency calling and the female estrous cycle, Leong et al. (2003), showed that females produced significantly more low-frequency calls during their preovulatory phase, approximately 20 days before ovulation. The implication is that females use low-frequency calls to advertise their reproductive condition with sufficient advance notice to allow competing males to gather from considerable distances. Once within sight of the female herd, bulls use visual and chemical cues to locate the receptive female.

Recent reviews of low-frequency communication in elephants suggest that under ideal atmospheric conditions near-infrasounds carry distances of up to 10 kilometers, but 2.5 kilometers may be the maximum in most situations (Garstang 2004; McComb et al. 2003). At such extreme distances, wind and atmospheric conditions may degrade the information content of the call, but elephants may increase the range of their infrasonic communication by calling at night when atmospheric conditions are best (Larom et al. 1994). Although elephants have massive ears, they are most sensitive to frequencies between 100 and 5,000 Hertz (Heffner & Heffner 1982), well above the infrasonic range of 20 Hertz. Possibly low-frequency calls are received by a combination of seismic vibrations and auditory hearing (O'Connell et al. 1997, 2000). A small area of skin of the forehead vibrates as elephants produce infrasounds. The possibility exists that elephants—like humans who

FIGURE 9-9   A cow-calf family unit of African elephants (*Loxodonta africana*) in Kruger National Park, South Africa.

BOX 9-1   Elephant Musth

Musth, from the Urdu word for intoxicated, is characterized by highly elevated testosterone levels, frequent urination, copious secretions from the temporal gland, and increased aggression toward other males. Rasmussen et al. (2002) have shown that younger, more socially naive males produce sweet-smelling secretions from the temporal glands, perhaps to signal to older males that they are no threat. In contrast, adult males in musth produce malodorous secretions that provide an honest signal of their aggressive state to would-be competitors. Additionally, females may use these chemical signals as well. The main compound in these secretions is frontalin. Interestingly, frontalin is a pheromone also found in bark beetles (Lindgren, 1992). Rasmussen and Greenwood (2003) tested the hypothesis that elephant frontalin acts as a pheromone by presenting varying concentrations of the compound to different age–sex combinations of Asian elephants and recording their behavioral responses. Their results indicate that subadult males exhibited strong avoidance behavior to frontalin, whereas adult males were essentially unresponsive. Moreover, females also responded to frontalin, but their response varied with their reproductive condition. Pregnant females and females about to ovulate showed the strongest response to the compound; initial sniffing was often followed by rumbling sounds or attempts to check a conspecifics temporal gland. In contrast, females in the luteal phase of the estrous cycle (nonreceptive) showed avoidance behaviors.

recognize each others' voices—may recognize the size, age, sex, and sexual and physical status of other elephants from the call sound structure (Garstang 2004). Clearly, much remains to be learned about the role of communication in structuring elephant societies. Asian elephants exhibit behaviors similar to those of their African cousins: they are highly social, long-lived, and communicate using low-frequency sounds. Asian elephants are revered as religious symbols in some parts of Asia and have been domesticated for centuries. Individuals are trained as draft animals, for logging, hunting, and transportation.

Both Asian and African elephants suffer encroachment on their habitat by humans and poaching for the ivory. In Luangwa National Park in Zimbabwe, poaching of elephants for their tusks has increased selection for a genetic condition that results in tuskless elephants. Tuskless female elephants have increased in frequency to 38.2% of the population in 1989, from only 10% two decades ago (Jachmann et al. 1995). African national parks, where settlement by people is prohibited, offer some hope for the survival of wild populations; however, elephants are still killed when they leave the parks and destroy adjoining crops or kill villagers. Both the Asian and the African elephant are currently listed as endangered species in Appendix I of the Convention on International Trade in Endangered Species, except populations of African elephants in four countries in Southern Africa (Blanc et al. 2002). People from Southern

African countries indicated that elephant populations in their areas were stable or increasing because of effective management practices. Consequently, at the 1997 meeting of Convention on International Trade in Endangered Species in Zimbabwe, the African elephant populations in Botswana, Namibia, and Zimbabwe were transferred to Appendix II and listed as not currently threatened with extinction, but likely to become so unless the ivory trade is closely controlled. In 2002, elephant populations from South Africa were also transferred to Appendix II. These four countries are currently allowed to sell elephant ivory and other elephant products according to stringent conditions.

Order Sirenia: Sirenians

The sirenians, or seacows—dugongs and manatees—are the only completely aquatic mammals that are herbivorous and are one of the most anomalous mammalian orders. There are four living species of two genera (*Dugong*, Dugongidae; *Trichechus*, Trichechidae). A fifth species, *Hydrodamalis gigas* or Steller's sea cow (Family Dugongidae), went extinct around 1768. The dugong (*Dugong dugon*) is distributed in tropical waters from eastern Africa, through the Indo-Australian Archipelago, to the western Pacific and Indian Oceans. The Amazonian manatee (*Trichechus inunguis*) lives in the Amazon and Orinoco drainages in South America. *Trichechus manatus*, the West Indian Manatee, inhabits the Caribbean and Atlantic coastlines from Brazil to Virginia, and the African manatee, *T. senagalensis*, occurs in tropical western Africa (Shoshani 2005).

Paleontology

Sirenians share a common ancestry with the proboscideans and hyraxes and are known from Eocene deposits from such scattered points as Europe, Africa, and the West Indies. Nearly all fossil sirenians were tropical and marine. The earliest fossil sirenians, from the early Eocene of Jamaica, had long bodies supported by four short legs (Domning 2001). These early sirenians had four sacral vertebrae and were probably capable of walking on land, but the position of the nasal openings and the massive, heavy ribs suggests an amphibious lifestyle.

Steller's sea cow (*Hydrodamalis gigas*) was the terminal member of a lineage that first appeared in the Pliocene. Some 20,000 years ago, this sea cow had an extensive range around the rim of the North Pacific from Mexico northward and westward, along the coasts, to Japan. Probably as an adaptation to the cold waters it inhabited, it was a giant, reaching some 10 meters in length and 11,000 kilograms in weight. It could not submerge and moved slowly and thus was easily harpooned. The back had thick epidermis. The body had layers of blubber, and the manus was claw-like,

enabling the animal to hold fast to rocks during violent storms with powerful waves. This sea cow ate marine algae, especially kelp, but had no teeth. It crushed its food between upper and lower bony rostral pads.

Our limited knowledge of the natural history of this behemoth comes largely from observations of a relict population of 1,000 to 2,000 animals discovered by Captain Vitus Bering, George Steller, and the crew of the *St. Peter*, which shipwrecked on Bering Island in November of 1741. These sea cows were highly social, lived in family groups, and tried to help injured individuals (Steller 1751). They grazed slowly on nearshore kelp beds.

Near the end of their nearly year-long stay on Bering Island, the crew of the wrecked *St. Peter* began killing sea cows for food. After Steller and the crew (Captain Bering died on the island) returned to Russia, the existence of the remnant population of sea cows became widely known to Russian crews hunting in the North Pacific for sea otters and fur seals. Some crews remained for months each year on the Commander (Komandorskiye) Islands, one of which is Bering Island. In addition to decimating populations of sea otters for their fur, these crews indiscriminately killed sea cows for food. After studying ships' records regarding sea cows, Stejneger (1887) gave 1768 as the date when the last Steller's sea cow was killed. Sadly, humans took but 27 years to drive the species to extinction. Their killing of sea otters, too, may have hastened the extinction of Steller's sea cow by disrupting the ecology of the kelp–sea urchin–otter–sea cow community (Anderson 1995).

The story of the Steller's sea cow requires a postscript. Native peoples along the Pacific Rim as far south as California probably killed sea cows for food long before Europeans entered the New World. Native people likely extirpated the vulnerable sea cows over nearly all of their range. Until the *St. Peter* shipwrecked on Bering Island in 1741, the remote Commander Islands may have been the last refuge of the Steller's sea cow for hundreds or even thousands of years (Domning 1972, 1978; Domning et al. 2007).

Morphology

Living sirenians are large, reaching weights in excess of 1500 kilograms (O'Shea 1994). They are nearly hairless, except for bristles on the snout, and have thick, rough, or finely wrinkled skin (**Fig. 9-10**). The nostrils are **valvular**. The nasal opening extends posterior to the anterior borders of the orbits, and the nasals are reduced or absent. The skull is highly specialized, and the dentary is deep (**Fig. 9-11**). The tympanic bone is semicircular, and the external auditory meatus is small. The periotic bone has a bony attachment to the skull (Berta et al. 2006). The middle ear ossicles are the most massive of any mammal (Domning 2008). The lungs are unlobed, unusually long, oriented horizontally,

FIGURE 9-10 (A) A manatee (*Trichechus manatus latirostris*) from Florida. (B) A dugong (*Dugong dugon*). Note the more rectangular snout, the valvular nostrils, and the pointed tail flukes in the dugong.

and separated from the massive gut by a long, horizontal diaphragm. The orientation of the lungs and dense, heavy bone allow the animal to use minor adjustments in lung volume to maintain a horizontal attitude while feeding at various depths. The heavy bone may counterbalance the added buoyancy from gas production in the gut. Postcranially, sirenians somewhat resemble cetaceans. The five-toed manus is enclosed by skin and forms a flipper-like

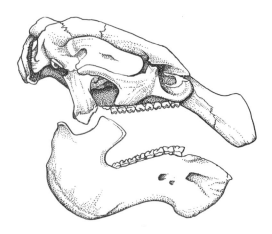

Manatee

(*Trichechus manatus*, Trichechidae)

FIGURE 9-11 The skull of a manatee; length of skull 360 millimeters. (Adapted from Hall, E. R., and K. R. Kelson. *Mammals of North America, Volume 2.* John Wiley & Sons Ltd., 1959.)

structure. The pelvis is vestigial. The hindlimbs are absent, and the tail is a horizontal fluke. There is no clavicle. The ribs are pachyostotic (thickened), and the ribs, limb bones, neural arches of the vertebrae, lower jaws, and some cranial bones are osteosclerotic (made up almost entirely of compact bone; Domning 2008).

The cheek teeth of dugongs (*Dugong*) are large and columnar, lack enamel, and are cementum covered. They have open roots, and the occlusal surfaces are wrinkled and bunodont. Most ancestral dugongids dug seagrass rhizomes with heavy incisor tusks, but the tusks of *Dugong dugon* are relatively degenerate; this Indopacific species specializes on delicate seagrasses (Marsh et al. 1982). In manatees (*Trichechus*), in contrast, there is an indefinite large number of enamel-covered, cementumless teeth, each with two cross ridges and closed roots. As teeth at the front of the cheektooth row wear out, they are replaced by the posterior teeth pushing forward, a pattern similar to that in elephants except that shedding and replacement are not limited to the standard eutherian number of cheek teeth and even more similar to the pattern in a marsupial Australian rock wallaby. This continual replacement in manatees evolved in response to the highly abrasive grasses in the manatee's mainly freshwater habitats and differ from the softer seagrasses eaten by the dugong (Domning 1999). Five to eight teeth in each side of the jaw are functional at one time. Horny plates cover the front of the palate and the adjacent surface of the mandible in all genera. The skull of *Trichechus* is modified by elongation of the nasal cavity, and these animals, virtually alone among mammals, have only six cervical vertebrae. Some differences between the two families of sirenians are shown in **Table 9-1**. Manatees

TABLE 9-1   Comparison of Characteristics of Two Sirenian Families

Dugongidae	Trichechidae
Functional dentition	No functional incisors
1/0, 0/0, 0/0, 2–3/2–3 = 10–14	Indeterminate number of cheek teeth
Cheek teeth columnar, no enamel, cementum-covered; roots single	Teeth with cross ridges, covered with enamel, cementum absent, roots double, continuous tooth replacement
Premaxillaries large; nasals absent; nasal cavity short	Premaxillaries small; nasals present; nasal cavity long
Slender neural spines and ribs	Robust neural spines and ribs
Flippers lack nails	Flippers with nails in two of the three species
Tail notched, as in whales	Tail not notched but spoon shaped

evolved in the Miocene from dugongid ancestral stock and completely replaced the dugongs in the New World.

Behavior

Sirenians are heavy-bodied, slow-moving animals that inhabit coastal seas, large rivers, and lakes and graze while submerged for periods up to about 15 minutes. As an adaptation to a low-nutrient diet, manatees have very slow metabolisms, resulting in the production of relatively little body heat for animals of their size (Gallivan & Best 1986). Water is an excellent conductor of heat, and because manatees have low metabolic rates and their blubber does not provide the insulation seen in other marine mammals, manatees dissipate heat rapidly in cooler waters and are largely restricted to tropical regions (Irvine 1983). Some individuals inhabiting the coasts of Florida move into rivers, springs, and industrial warm-water effluents in winter to avoid cold water.

Sirenians are known to make a variety of sounds underwater (Anderson & Barclay 1995). O'Shea (1994), observing free-ranging West Indian manatees (*Trichechus manatus*), found that these calls form a series of complex, graded signals used to maintain contact between individuals and to communicate basic behavioral information. The underwater calls of dugongs may also be used to maintain exclusive "activity zones" that are patrolled and defended against intruders (Anderson 1997). Manatees, unlike elephants, do not use low frequency communication, but tests by Gerstein et al. (2004) suggest that the short body hairs of manatees may be simple vibrotactile organs.

Dugong are long-lived animals with a low reproductive rate typical of K-selected species. By counting annual growth layers in the tusks, Marsh (1980) estimated the maximum lifespan of dugongs at approximately 70 years. Dugongs reach sexual maturity in 6 to 15 years, with mature females bearing a single calf every 2.5 to 7 years. Gestation

is estimated to be 13 months, and the calf nurses for 18 months. Mating behavior in sirenians appears to vary considerably with species and location. Anderson (2002) reviewed the evolution of mating systems in sirenians. Dugongs from Western Australia may form **leks**, aggregations of males competing for access to females; lekking has not been confirmed in other parts of their range. In contrast, manatees are promiscuous, with males practicing a type of scramble competition for access to females in estrus. Large groups of courting bulls may pursue a single female for several weeks. Anderson believes that a promiscuous mating system evolved in manatees because they frequent rivers in tropical areas where there is limited opportunity to establish lek display areas. Little is known about reproduction in the extinct Steller's sea cow (*Hydrodamalis gigas*), but Anderson (2002) posits that the isolation of sea cows in cold northern waters restricted the breeding season and forced the evolution of the monogamous mating system reported by George Steller in 1751.

Humans have been responsible for the great range reductions of living sirenians and the extinction of the previously discussed Steller sea cow (*Hydrodamalis*). Present serious declines in the populations of dugongs and manatees in some areas are due to persistent hunting by humans. In Florida, where they are stringently protected from hunters, most manatees bear scars inflicted by boat propellers, and collisions with boats and other human-caused factors are responsible for a large proportion of the observed mortality (O'Shea et al. 1995). In addition, **epizootic** infections associated with dinoflagellate blooms (caused by fertilizer runoff) have led to recent die-offs of manatees along the eastern coast of the United States (O'Shea et al. 1991). Because of the rapidly increasing human population in Florida and the low reproductive rate of manatees, serious population decline of manatees due to habitat loss and accidental deaths seems likely in the near future. All four living species are considered endangered or threatened over most of their ranges.

Order Hyracoidea: Hyraxes

Members of this unusual order are small, rodent-like creatures commonly called hyraxes or dassies, and their appearance gives little indication of their relationship to elephants and sirenians (**Fig. 9-12**). This is a small order with a single Recent family, Procaviidae. Recent members include three genera with four species. Hyraxes, which today occupy nearly all of Africa (except the arid northwestern part) and parts of the Middle East, appeared first in early Eocene beds of Morocco. Some early members of the extinct family Pliohyracidae reached the size of a large pig; unlike modern hyraxes, they had elongated feet with unguligrade foot posture (Gheerbrant et al. 2007). Hyraxes were considerably more diverse in the Paleogene of Africa, with approximately 30 species known (Gheerbrant et al. 2005). During the Miocene, Africa's isolation from Eurasia ended and hyrax diversity was greatly reduced as competition from Eurasian ungulates increased.

FIGURE 9-12 Rock hyraxes (*Procavia capensis*) in Mountain Zebra National Park, South Africa.

Morphology

The roughly rabbit-sized procaviids of today have a short skull with a deep lower jaw (**Fig. 9-13**). The dental formula is 1/2, 0/0, 4/4, 3/3 = 34. The incisors are specialized: the pointed, ever-growing uppers are broadly separated, and the flattened posterior surfaces lack enamel; the lowers are chisel shaped and generally tricuspid. Behind the incisors is a broad diastema, and the cheek teeth are either brachydont in yellow-spotted and tree hyraxes, which are mostly browsers, or hypsodont in the grass and leaf-eating rock hyraxes. The molars resemble those of a rhinoceros in miniature: the uppers have an ectoloph and two cross lophs, and the lowers have a pair of V-shaped lophs. Hyraxes have fairly compact bodies with a tiny tail. There are four toes on the forefoot and three toes on the hind foot. The feet are **mesaxonic** (the plane of symmetry goes through the third digit), and the digits are joined by soft tissue (syndactylous) to the bases of the last phalanges. Except for the clawed second digit of the pes, all digits bear flattened nails (**Fig. 9-14**). Hyraxes gain traction on steep rocky slopes by specialized elastic pads on the soles of the feet, which can be "cupped" by muscular contractions, and kept moist by abundant skin glands. Although the clavicle is absent, as in cursorial mammals, the centrale of the carpus is present, a feature decidedly not characteristic of runners. Hyraxes are herbivores; their stomach is simple, but digestion is aided by microbiota in a caecum at the anterior end of the colon and a colonic sac positioned just anterior to the distal colon (Björnhag et al. 1994).

Behavior

Hyraxes are mainly herbivorous and are nimble climbers and jumpers. They occur in a variety of habitats, from forests and scrub country to rock outcrops and lava beds in grasslands

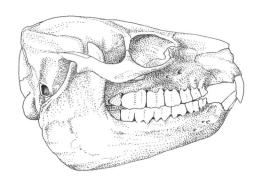

Rock hyrax
(*Procavia capensis*, Procaviidae)

FIGURE 9-13 Skull of a rock hyrax (*Procavia capensis*, Procaviidae) (length of skull 88 millimeters). (Adapted from Hatt, R. T., *Bull. Amer. Mus. Nat. Hist.* 72 (1936): 117–141.)

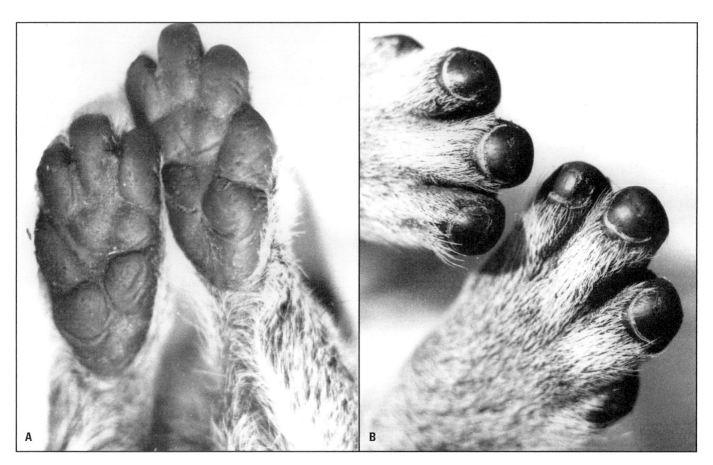

FIGURE 9-14   (A) ventral and (B) dorsal views of the front feet of the rock hyrax (*Procavia capensis*).

and at elevations up to 5,000 meters. The yellow-spotted rock hyrax (*Heterohyrax*) and the rock hyrax (*Procavia*) live in cliffs, ledges, and talus and are adroit at climbing rapidly over steep rock faces and in trees. The two species of tree hyrax (*Dendrohyrax*) are highly arboreal, mostly solitary, nocturnal, and maintain territories in the canopy by loud barking calls (Rahm 1969). Hyraxes are relatively long lived for small mammals, with some individuals surviving more than 10 years in the wild (Hoeck 1982).

The terrestrial species are diurnal and often form polygynous family groups, with one territorial adult male and several genetically related females and their young (Happold 1987; Hoeck et al. 1982). In the Serengeti Plain in Tanzania, *Heterohyrax* and *Procavia* inhabit the same rock outcrops (called kopjes). Combined populations of the two species can reach densities approaching 100 animals per hectare of kopje. Where they co-occur, shared nurseries are common; adults and subadults of both species care for mixed groups of juveniles (Hoeck 1989). Such interspecific cooperation is extremely rare among mammals. In Zimbabwe, both hyrax species share kopjes and suffer heavy juvenile mortality from black eagles and leopards

(**Fig. 9-15**). Barry and Mundy (2002) showed that both species benefit from the increased predator vigilance afforded by mixed-species groups. Moreover, mixed-species basking groups increased in size when offspring were present and declined after weaning. Although both species live in close association, interbreeding is avoided by differences in mating behaviors and penis anatomy.

Mixed-species basking also provides thermoregulatory benefits. The body temperature of the hyrax is quite variable (Bartholomew & Rainy 1971). Behavioral thermoregulation is a conspicuous part of the daily routine of most African rock-dwelling hyraxes. Field studies by Sale (1970) indicate the importance of behavior in adjusting to heat or coolness, and laboratory studies by Bartholomew and Rainy (1971) attest to the unusual system of body temperature regulation in the yellow-spotted rock hyrax (*Heterohyrax brucei*). The body temperature of normally active individuals varies from 35°C to 37°C and is affected by ambient temperatures. The standard metabolic rate is some 20% below that expected on the basis of body mass, and the mean minimum heart rate (118 beats per minute) is 52% below the expected level.

Outside of their nocturnal retreats, hyraxes adjust posture and location to exploit the environment to maintain an appropriate body temperature. When the animals first emerge from their nocturnal retreats in deep rock crevices, they avoid extensive contact between their ventral surfaces and the cool rock, turn broadside to the first rays of the sun, and bask while maintaining a semispherical body form, which presumably prevents excessive heat dissipation. As the air and the rock begin to warm, the hyraxes sprawl on the rock, presenting a large surface area to the sun (Fig. 9-15). Bartholomew and Rainy (1971) found that the lowest body temperatures of hyraxes were reached shortly before sunrise, despite their huddling together during the night. The basking uses solar rather than metabolic heat to raise body temperature. After basking, the hyraxes generally feed.

Of particular importance is the fact that the rocks on which the hyraxes live provide an auxiliary means of adjusting body temperature. The rock outcrops form massive heat sinks with vastly more thermal inertia than air. The hyraxes can use the heat from the rock to compensate for body heat lost to the air. On cool, cloudy days, hyraxes huddle tightly together, thus reducing the surface area exposed to the air.

Hyraxes have very efficient kidneys: they concentrate electrolytes and excrete calcium carbonates (Rübsamen et al. 1982), which forms conspicuous stains on occupied rock outcrops. Like woodrats in the American Southwest, hyraxes in Afro-Arabia accumulate middens (trash heaps) of plant fragments that can indicate the ancient vegetation and potential climatic change of an area (Fall et al. 1990).

FIGURE 9-15 (A) Two *Procavia capensis* (above) and two *Heterohyrax brucei* (below) basking with their bodies broadside to the early morning sun. (B) Hyraxes of both species huddled together on a cool day.

SUMMARY

The Paenungulata are part of the Afrotheria, a diverse assemblage of African mammals. Paenungulates include the Proboscidea (elephants), Sirenia (manatees and dugongs), and Hyracoidea (hyraxes). Hyracoids first appear in the early Eocene of Africa; sirenians and proboscideans probably evolved from a common hyracoid ancestor near the shores of the Tethys Sea.

Proboscideans were more diverse in the Neogene, but today are represented by three living species in two genera: the African savanna elephant, *Loxodonta africana*, the forest elephant *L. cyclotis*, and the Asiatic elephant, *Elephas maximus*. They are the largest extant land mammals, reaching weights of 6,000 kilograms. Elephants are long lived, have complex matriarchal social systems, and communicate via both audible and infrasonic sounds. Both the Asiatic and the African elephant populations suffer from habitat loss and poaching for the ivory in their tusks; all three species are currently listed as endangered.

Sirenians are the only completely aquatic herbivorous mammals. There are four living species of two genera (*Dugong*, Dugongidae; *Trichechus*, Trichechidae). A fifth species, *Hydrodamalis gigas* (Steller's sea cow, Family Dugongidae), was forced to extinction around 1768. Dugongs inhabit tropical waters from eastern Africa, through the Indo-Australian Archipelago, to the western Pacific and Indian Oceans. Three species of manatees live in large South American rivers, coastal areas of the Caribbean Basin, and along the southeastern U.S. coastline, and the African manatee lives in tropical western African river systems. Sirenians are heavy-bodied, slow-moving animals that graze while submerged. They have a relatively slow metabolism, live up to 70 years, and are slow to mature. Because of hunting, boat collisions, and aquatic pollution, all four sirenians are listed as endangered or threatened.

Four extant species of hyraxes occupy forests, scrub country, and rock outcrops in grassland over much of Africa and part of the Middle East. Hyraxes are adroit at climbing rapidly over steep rock faces and in trees. Hyraxes are relatively long lived for small mammals, with some individuals surviving more than 10 years in the wild. Two species may occupy the same rock outcrops and even bask together to help regulate their body temperatures.

KEY CHARACTERISTICS OF PROBOSCIDEANS

- Long, muscular trunk
- Large ears
- Pair of large tusks (upper incisors)
- Sequentially replaced cheek teeth with many transverse lophs
- Skull short and high, with a prominent lambdoidal crest
- Splayed toes cushioned by a dense heel pad

KEY CHARACTERISTICS OF HYRACOIDS

- Rabbit-sized animals with short tails
- Three-toed hind feet and four-toed forefeet
- Digits bear flattened, hoof-like nails, except clawed second digit of the pes
- Feet mesaxonic (the plane of symmetry of the foot goes through the third digit)
- Postorbital bar present
- Small bullae
- Single pair of tusk-like upper incisors, two pairs of lower incisors, with diastema between ever-growing incisors and rooted cheek teeth
- Molars lophodont

KEY CHARACTERISTICS OF SIRENIANS

- Large, nearly hairless body adapted for a fully aquatic lifestyle
- Forelimbs flipper-like, hindlimbs absent
- Pelvis vestigial and without hindlimb bones
- Horizontal tail fluke, no dorsal fin
- Lungs and diaphragm greatly elongated, flattened, and unilobed
- Ear pinnae absent
- Dorsally positioned, valve-like nostrils
- Premaxillae ventrally deflected, with small upper tusks in dugongs
- Nasal bones absent or reduced
- Dentary broad
- Sequentially replaced cheek teeth in manatees
- Middle ear ossicles massive
- Postcranial bones that are dense and heavy (show pachyostosis)

KEY TERMS

Epizootic
Graviportal
Infrasound

Lek
Matriarchal
Mesaxonic

Musth
Valvular

RECOMMENDED READINGS

Berta, A, et al. 2006. *Marine Mammals Evolutionary Biology.* Academic Press, Burlington, MA.

Estes, R & EO Wilson. 1992. *The Behavior Guide to African Mammals: Including Hoofed Mammals, Carnivores, Primates.* University of California Press, Berkeley, CA.

Garstang, M. 2004. Long-distance, low-frequency elephant communication. *Journal of Comparative Physiology A*, 190:791–805.

O'Shea, TJ. 1994. Manatees. *Scientific American*, 271:66–72.

Shoshani, J & P Tassy. 1996. *The Proboscidea Evolution and Palaeoecology of Elephants and Their Relatives.* Oxford University Press, Oxford, UK.

Sukumar, R. 2003. *The Living Elephants: Evolutionary Ecology, Behavior, and Conservation.* Oxford University Press, Oxford, UK.

Cingulata, Pilosa, and Pholidota

These three orders of mammals, Pilosa (sloths and anteaters), Cingulata (armadillos), and Pholidota (pangolins), which include among their extinct relatives some of the most bizarre and improbable of all land-dwelling mammals, are considered together in this chapter as a matter of convenience. Several of their members share features associated with ant eating and termite eating such as elongated crania and tongues, oddly modified dentitions, large, recurved claws, and lowered metabolic rates. As a result, in the past, pilosans, cingulatans, and pangolins were frequently combined in the no-longer-recognized order Edentata. The orders Pilosa and Cingulata (collectively referred to as Xenarthra; **Fig. 10-1**), are closely related and monophyletic, but Pholidota is probably very distantly related to the xenarthrans. Pilosans and cingulatans (or cingulates) underwent an impressive Cenozoic radiation in South America, whereas extant pangolins are an Old World group that has conservatively maintained a single structural plan. Living armadillos are omnivorous or insectivorous. Anteaters are highly specialized ant and termite eaters, and sloths are folivorous (leaf eating). Pangolins are also ant and termite eaters.

There is probably no phylogenetic affinity between pangolins and xenarthrans. Taxonomists have had difficulty distinguishing between features that are shared by these groups because of common ancestry (phylogeny)

versus features that are adaptations to similar modes of life (convergence; Reiss 2001). The relationships of the various groups of xenarthrans to one another are fairly well understood on morphological grounds, and molecular studies have confirmed the close relationship between the Pilosa and Cingulata, as well as the unlikely relationship between those two orders and the Pholidota (Delsuc & Douzery 2008; Gaudin & McDonald 2008). Recent molecular studies of pangolins and xenarthrans seem to indicate a possible relationship between pangolins and carnivorans or with other Laurasiatheria (Delsuc et al. 2002). (Laurasiatheria is a monophyletic clade based on molecular evidence and consists of Erinaceomorpha, Lagomorpha, Rodentia, Primates, Carnivora, Chiroptera, Perissodactyla, Cetartiodactyla, and Pholidota.)

Although the xenarthrans are not represented by a large number of species today, including but 13 genera and 31 species, they are interesting because of their unique structure and unusual ecological roles, their large, aberrant and diverse fossil types, and their remarkable Cenozoic radiation in South America and Caribbean islands.

The living xenarthrans share a series of distinctive morphological features, chief among which are the extra **zygapophysis**-like (**xenarthrous**) articulations that brace at least the lumbar vertebrae (**Fig. 10-2**). The incisors are reduced or absent, and deciduous teeth are absent. The

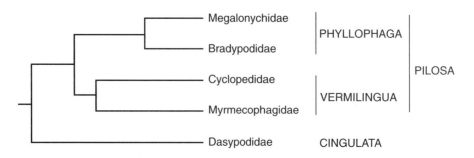

FIGURE 10-1 A phylogeny of the extant families belonging to the xenarthran orders Pilosa and Cingulata, based on molecular and morphological data from several sources.

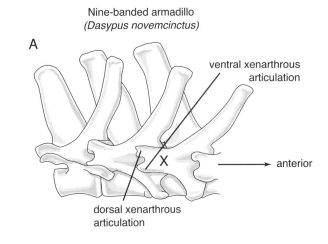

Nine-banded armadillo
(*Dasypus novemcinctus*)

A

ventral xenarthrous
articulation

X

anterior

dorsal xenarthrous
articulation

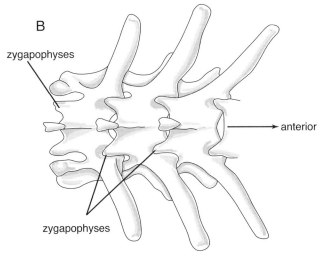

B

zygapophyses

anterior

zygapophyses

FIGURE 10-2   Three lumbar vertebrae of the nine-banded armadillo, showing the dorsal and ventral xenarthrous articulations
upon the xenarthrous process (X). (A) Right lateral view. (B) Dorsal view. The xenarthrous articulations supplement the normal
articulations between zygapophyses, which are visible in the
dorsal view. (Adapted from Szalay, F. S. *Evolutionary History of
the Marsupials and an Analysis of Osteological Characters.* Cambridge University Press, 1994.)

cheek teeth, when present, lack enamel, and each has a
single open root. A septomaxilla, homologous with that of
the monotremes but found in no other living mammals,
is present in the skull of many xenarthrans. The tympanic
bone is annular; the skull is usually long and rather cylindrical or conical in anteaters and armadillos, shorter and
broader in sloths, and short and deep in glyptodonts. The
scapula has a prominent spine and often has a secondary
spine paralleling the acromion process, and the acromion
and coracoid processes are unusually well developed. The
clavicle is present. The ischium is variously expanded and
specialized and usually forms an ischiocaudal, as well as an
ischiosacral, symphysis (**Fig. 10-3**). The hind foot is typically
five-toed, and the forefoot has two or three predominant
toes with large claws. Major xenarthran structural trends
are toward reduction and simplification of the dentition,

specialization of the limbs for such functions as digging
and climbing, and in many extinct species rigidity of the
axial skeleton for the support of great weight or a tripodal
stance on the hind legs and tail.

Paleontology

Xenarthrans, in modern times strictly a New World group
with a modest number of species, include among their extinct relatives many spectacularly unusual creatures. The
earliest fossils are from the late Paleocene of South America
and consist of **osteoderms**, or bony plates, from the armor
of armadillos (Oliveira & Bergqvist 1998; Vizcaíno 1994).
The origins of xenarthrans are confused by several problematic Eocene fossil forms of unclear affinities, such as an
anteater-like form in Europe, sloth-like forms in Asia, and
fragmentary sloth remains in Antarctica, and others (Rose
et al. 2005).

A diverse array of forms resulted from a mid-Cenozoic
radiation of South American armadillos (Dasypodidae).
A large, Pliocene genus (*Macroeuphractus*) had enlarged,
canine-like teeth and was probably a scavenger, and the
Miocene *Stegotherium* was most likely a termite eater. A
bizarre Miocene form (*Peltephilus*) and its relatives had
specialized osteoderms that formed pointed horns on the
top of the snout. Pig-sized armadillos of the family Pam-
patheriidae lived in both North and South America in the
Plio-Pleistocene. The structural diversity of extinct South
American armadillos indicates that they exploited a variety of foods and together formed an important part of the
South American Cenozoic biota.

From Paleocene until late Pliocene times, South America and North America were separated by a seaway, and
dispersal of mammals between the Americas was restricted.
During this interval not only dasypodids but also other
xenarthrans underwent a tremendous radiation in South
America. Several remarkable but now extinct evolutionary
lines arose. Perhaps the most unusual was the Glyptodontidae, which appeared in the middle Eocene epoch and represents one line that probably evolved from armadillo stock.
These often very large and ponderous creatures had unusually deep skulls and highly specialized graviportal limbs
(**Figs. 10-4A** and **10-5**). Many of the unique structural features
of the glyptodonts are associated with their development
of a nearly impregnable, turtle-like carapace composed of
many fused polygonal osteoderms covered by up to 1,800
horny scutes. These are the most completely armored vertebrates known. The pillar-like limbs are distinctive and
highly specialized to support great weight, and most of
the thoracic, lumbar, and sacral vertebrae are fused into
a massive arch that, together with the ilium, supports the
carapace (Fig. 10-5). The pelvic girdle is sometimes fused

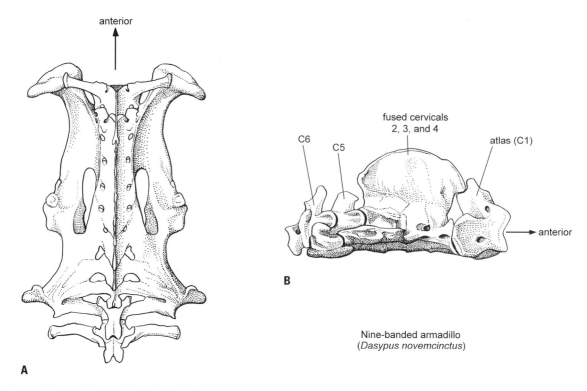

FIGURE 10-3 (A) Dorsal view of the pelvic girdle of the nine-banded armadillo, showing the great degree of fusion of vertebrae with the ilium and ischium (anterior is at the top of the figure). (B) Lateral view of the cervical vertebrae (anterior is to the right). In this individual, the axis and cervicals three and four are fused, and cervical five is partially fused.

to the carapace. Many glyptodonts were massive creatures: the North American Pleistocene species *Glyptotherium arizonae* was 3 meters long and probably weighed at least 1,100 kilograms (1.2 tons). After the Miocene, diversification of the glyptodonts was probably favored by the spread of pampas grassland in South America.

Additional evolutionary lines are represented by sloths of six or seven families, all but two of which are extinct. Sloths first appear in Eocene deposits in South America and Antarctica (Carlini et al. 1990; McDonald & De Iuliis 2008). Traditionally, extinct sloths have been called "ground sloths" because many types were very large and terrestrial. However, many of the smaller fossil species probably were arboreal or at least semiarboreal (White 1997). At least one large Pliocene species was probably a semiaquatic "sea sloth," inhabiting shallow coastal fresh and salt waters (Muizon & McDonald 1995). One of the extant tree sloths (*Choloepus*) has recently been recognized as representing a family of "ground sloths," the Megalonychidae, long believed to be extinct (Webb 1985). The other extant tree sloth, *Bradypus* (Bradypodidae) seems to be phylogenetically basal to other sloths (Gaudin & McDonald 2008). Yet

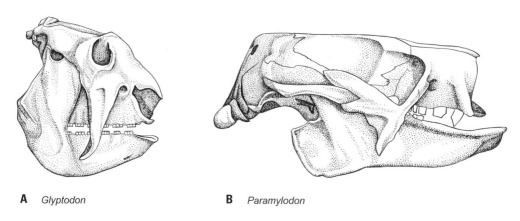

A *Glyptodon*   **B** *Paramylodon*

FIGURE 10-4 Skulls of extinct xenarthrans: (A) *Glyptodon*, length of skull 560 millimeters. (B) *Paramylodon*, length of skull 510 millimeters. (Modified from Romer, A.S. *Vertebrate Paleontology*. University of Chicago Press, 1966.)

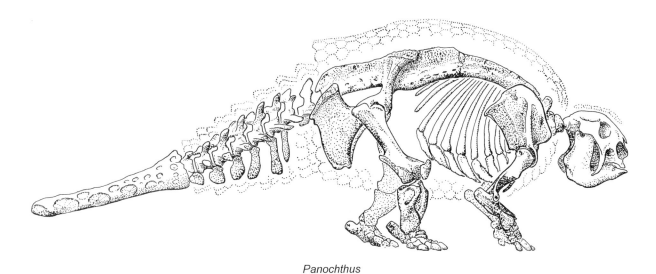

Panochthus

FIGURE 10-5   Skeleton of a glyptodont, *Panochthus*, from the Pleistocene of South America, with a superimposed outline of the carapace and head shield. The caudal armor, which in life probably bore horny spikes on the oval areas, is fused to the caudal vertebrae. Note the extensive fusion of vertebrae and massive graviportal limb bones. This ponderous creature probably weighed about 1,050 kilograms.

another fossil sloth had arboreal adaptations in its skeleton distinct from those of either extant tree sloth (Pujos et al. 2007). All sloths are herbivores with ever-growing teeth that lack enamel. Many fossil sloths are known by well-preserved skulls, which allowed Bargo et al. (2006) to reconstruct their facial musculature and to visualize their appearance (**Fig. 10-6**). The late Pleistocene megatheriid *Eremotherium* was common from Florida to Brazil. It reached the size of an elephant, weighing about 3,500 kilograms (3 to 4 tons), lacked upper canine-like teeth, had graviportal limbs, and walked on the outer edges of the unusually specialized and large hind feet. As in many sloths, it had a massive tail that it used together with the thick hindlimbs for an upright tripodal stance, freeing the forelimbs for gathering food or defending against predators. One can imagine these huge beasts tearing apart large trees or grappling high tree limbs with their heavy forelimbs and claws and using their great weight to pull down the limbs or uproot the trees for the leaves, twigs, and fruits.

The family Megalonychidae (with a single living species) is closely related to the extinct families Megatheriidae and Nothrotheriidae but differs from those groups in having the anterior-most cheek teeth modified into "canines." *Megalonyx*, a Pleistocene genus that reached the size of a cow, was widely distributed in North America. The remains of smaller species of megalonychids have been found in the West Indies in association with human artifacts. Another family, the Mylodontidae, appeared in the Miocene and is characterized in part by the development of upper "canines" (Fig. 10-4B) and remarkably robust limbs. A degree of protection from predators may have been afforded some members of this family by small round dermal

ossicles embedded in the presumably thick skin. The North American mylodont *Paramylodon* (formerly known as *Glossotherium*) was widespread in North America and is the most common xenarthran in the Pleistocene deposits of Rancho La Brea in Los Angeles. This ground sloth had large claws on digits 2 and 3 (**Fig. 10-7B**), an arrangement similar to that in the living armadillo (**Fig. 10-8**). The skeleton was heavy, with the enlarged ischium joining the sacrum and with powerfully developed limbs (**Fig. 10-9**). The strong forelimbs, armed with large claws, were probably defensive weapons that were used against such predators as saber-toothed cats and were used for gathering grasses and browsing for food.

The glyptodonts, megatheriids, megalonychids, mylodonts, and others underwent much of their Cenozoic evolution isolated from the North American mammalian fauna. When a land bridge between the Americas was reestablished in the Pliocene, the glyptodonts and sloths were remarkably successful in invading North America. By some means of chance dispersal, megalonychid sloths reached North America before it was joined with South America, and one genus (*Megalonyx*) seemingly evolved in North America in the Miocene only to reinvade South America in the Pleistocene. *Megalonyx jeffersoni* was named for Thomas Jefferson who, in 1797, spoke to the American Philosophical Society about this sloth. His talk signaled the start of vertebrate paleontology in North America.

After invading Middle America from South America, glyptodonts and the small, glyptodont-like cingulates *Neoglyptatelus* and *Pachyarmatherium* may have evolved in Central America, southern North America, and northern South America in the late Neogene only to spread both

Megatherium americanum

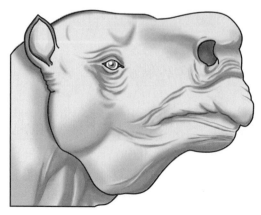

Glossotherium robustum

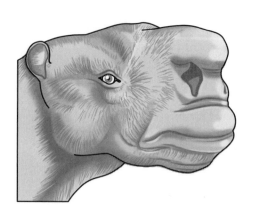

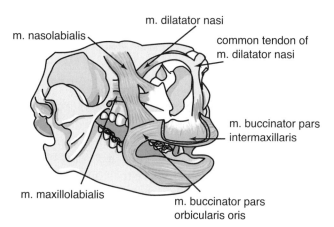

FIGURE 10-6 Restorations of the rostral musculature and facial appearance of two extinct South American Pleistocene ground sloths: the elephant-sized (6.9 ton) *Megatherium americanum*, a selective feeder with a narrow muzzle and prehensile upper lip, and the 1,700 kilogram (1.9 ton) *Glossotherium robustum*, a bulk feeder with a wide muzzle. (Adapted from Bargo, M. S., et al., *J. Morphol.* 267 (2006): 248–263.)

directions northward and southward (Carlini et al. 1997, 2008a,b; Zurita et al. 2008).

The common nothrotheriid *Nothrotheriops* occurred in North America in the Pleistocene, and its remains from Gypsum Cave in Nevada and Aden Crater in New Mexico include bones, skin, and hair (**Fig. 10-10**). *Nothrotheriops* probably weighed about 300 kilograms (660 pounds) and walked on the sides of its highly modified hind feet, as did other large sloths (Fig. 10-7A). The powerful claws of the forelimbs were perhaps used to grasp and tear down vegetation in preparation for ingestion. Dung of this animal is well preserved in dry caves in Nevada and Arizona. Dung from Rampart Cave in Arizona contains such plants as Mormon tea (*Ephedra*) and globe mallow (*Sphaeralcea*) (Hansen 1978; Poinar et al. 1998). These plants remain common today in dry parts of the southwestern United States. This sloth persisted into the late Pleistocene in North America and probably did not disappear until 11,000 years ago. Other sloths probably disappeared from South America by 10,500

years ago and still others from islands in the West Indies as recently as about 4,400 years ago (Steadman et al. 2005). Thus ended a fascinating cycle of xenarthran evolution.

Order Cingulata

Family Dasypodidae

This family includes the armadillos, which differ from the other living xenarthrans in many ways but especially in having protective bony armor. Dasypodidae is the most diverse and widespread living family of xenarthrans. Some 21 modern species, belonging to nine genera, collectively show a distribution from the North American central plains (states of Kansas and Nebraska in the United States) through Central and South America to near the southern end of Argentina. Armadillos occupy many ecological settings from temperate and tropical forests to deserts. Recent morphological and molecular phylogenetic work on this family suggests that armadillos are paraphyletic, with

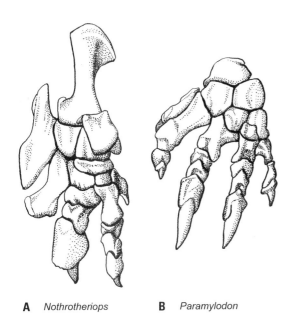

A *Nothrotheriops*   **B** *Paramylodon*

FIGURE 10-7   (A) The right pes of *Nothrotheriops*. (B) The right manus of *Paramylodon*. (Modified from Romer, A.S. *Vertebrate Paleontology*. University of Chicago Press, 1966.)

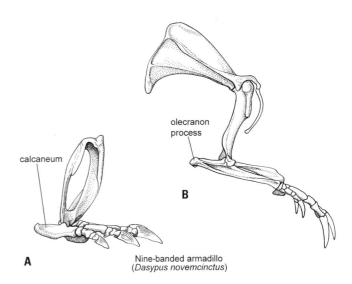

FIGURE 10-8   The right limbs of the nine-banded armadillo: (A) lower part of hindlimb (tibia, fibula, and pes) and (B) forelimb. The flattening of the bones of the forearm and lower leg increases the surface area for attachment of flexor and extensor muscles, and the elongation of the olecranon and the calcaneum gives added mechanical advantage to the muscles that insert on them.

extinct groups such as the pampatheres and glyptodonts as derived members nested deep within the clade Cingulata (Gaudin & McDonald 2008; Gaudin & Wible 2006). If confirmed by additional study, this would require the reclassification of the currently recognized subfamilies within Dasypodidae as families or otherwise.

The most obvious and unique structural feature of armadillos is the jointed armor. This consists of plates, bony osteoderms covered by horny epidermis, which occur in a variety of patterns but always include a head shield and protection for the neck and body (**Fig. 10-11**). Sparse hair usually occurs on the flexible skin between the plates and

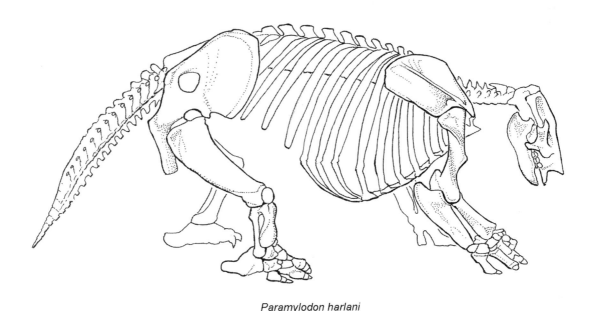

Paramylodon harlani

FIGURE 10-9   Skeleton of a Pleistocene North American sloth, *Paramylodon harlani* (Mylodontidae). Note the powerfully built limbs and the enlarged ilium that joins the sacrum. In life, this creature probably weighed at least 300 kilograms. (Adapted from Stock, C. *Cenozoic gravigrade edentates of western North America with special reference to the Pleistocene Megalonychinae and Mylodontidae of Rancho La Brea. Carnegie, plate 44.* Institution of Washington, 1925.)

FIGURE 10-10 How *Nothrotheriops* might have looked. This extinct megatheriid sloth survived until 11,000 years ago in the southwestern United States.

on the limbs and the ventral surface of the body. Individuals of some species can curl into a ball so that their limbs and vulnerable ventral surfaces are largely protected by the armor. The largest species, the giant armadillo (*Priodontes maximus*), weighs up to 60 kilograms; the smallest, the pygmy armadillo (*Chlamyphorus truncatus*), is roughly the size of a small rat (120 grams).

The skull is often elongate and is dorsoventrally flattened; the zygomatic arch is complete, and the mandible is slim and elongated (**Fig. 10-12C**). With the exception of one species that has premaxillary teeth, the upper teeth are borne only on the maxillary bone. The teeth are homodont and nearly cylindrical and vary, both interspecifically and intraspecifically, from 7/7 to 18/19. Frequently, the teeth are partially lost with advancing age. The axial skeleton is fairly rigid and is partially braced against the carapace. The second and third cervical vertebrae and, in four species, other cervical vertebrae as well, are fused (Fig. 10-3B), and 8 to 13 sacral and caudal vertebrae form an extremely powerfully braced anchor for the pelvis (Fig. 10-3A). Xenarthral articulations between thoracic and lumbar vertebrae (Fig. 10-2) produce a rigid vertebral column. The skeleton braces, but does not come into contact with, the carapace.

In all armadillos, the limbs are powerfully built, and the forefeet and hind feet bear large, heavy claws (Fig. 10-11). The feet are five-toed in all but one genus, and the foot posture is usually plantigrade. The tibia and fibula are fused proximally and distally and are highly modified for the origin of powerful muscles (Fig. 10-8A). Retia mirabilia occur in the limbs and are probably an energy-saving adaptation.

Armadillos are more generalized in their feeding habits and locomotion than other xenarthrans. Most species feed primarily on insects, but a variety of invertebrates, small vertebrates, carrion, and vegetable material are also eaten (McDonough & Loughry 2008). All armadillos are at least

partly adapted for digging, and some species are highly fossorial. One such fossorial form is the pygmy armadillo (*Chlamyphorus truncatus*), which uses a style of digging seemingly unique among mammals. The soil is dug away and pushed beneath the animal by the long claws of the forepaws, and the hind feet rake the soil behind the animal. The pelvic plate is then used to pack the soil behind the body. During the packing, the front limbs push the animal backward, and the hindquarters vibrate rapidly from side to side (Rood 1970). No permanent burrow is formed.

As indicated by their wide range, armadillos seem to be more resistant to cold than are tree sloths. In the northern parts of their range, however, armadillos may suffer 80% mortality during prolonged cold spells (Fitch et al. 1952). Nevertheless, armadillos have extended their range northward in the last 100 years (Humphrey 1974; Taulman & Robbins 1996).

Unusual reproductive strategies occur in the nine-banded armadillo, *Dasypus novemcinctus*. Females may experience an embryonic diapause sometimes lasting over 2

FIGURE 10-11 Living armadillos (Dasypodidae): (A) Six-banded armadillo (*Euphractus sexcinctus*). (B) Nine-banded armadillo (*Dasypus novemcinctus*).

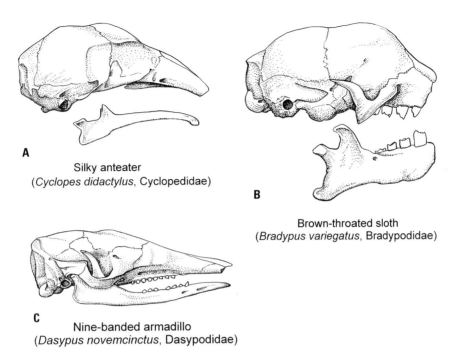

A
Silky anteater
(*Cyclopes didactylus*, Cyclopedidae)

B
Brown-throated sloth
(*Bradypus variegatus*, Bradypodidae)

C
Nine-banded armadillo
(*Dasypus novemcinctus*, Dasypodidae)

FIGURE 10-12   Skulls of living xenarthrans: (A) two-toed anteater (length of skull 46 millimeters), (B) three-toed sloth (length of skull 76 millimeters), and (C) nine-banded armadillo (length of skull 95 millimeters). (Adapted from Hall, E. R., and K. R. Kelson. *Mammals of North America, Volume Two.* John Wiley & Sons, 1959.)

years; others give birth to litters in consecutive years without copulating with a male in between the first and second litters (Storrs et al. 1989). Often, the fertilized egg undergoes two divisions after which each of the four cells develops separately, resulting in identical quadruplets. The nine-banded armadillo is the first xenarthran for which the genome has been sequenced; it was sequenced as part of an effort to provide a diversity of mammalian genome sequences for broad comparisons with the human genome (Chang & Adams 2008).

Order Pilosa

Family Cyclopedidae

Until recently, the single cyclopedid species *Cyclopes didactylus* (two-toed or silky anteater) was included in the family Myrmecophagidae. The Cyclopedidae and Myrmecophagidae are closely related as a monophyletic suborder Vermilingua, but molecular evidence suggests that *Cyclopes* may have diverged from the lineage leading to myrmecophagids way back in the middle Eocene, making it relatively distantly related to the larger anteaters (Delsuc & Douzery 2008). The silky anteater is the size of a squirrel, about 350 grams, and is fully arboreal with a prehensile tail, but unlike most squirrels, *Cyclopes* is nocturnal and forages for insects, mainly ants, high in trees. These small anteaters climb trees using the claws of the manus to hook or to grasp as the animal travels along branches mainly by above-branch quadrupedal walking, unlike the suspensorial locomotion of sloths.

Family Myrmecophagidae

Members of this family, the giant anteater and tamanduas, are highly specialized ant and termite eaters (**Fig. 10-13**). They occur in tropical forests of Central and South America and in South American savanna. There are three living species belonging to two genera.

The most obvious structural features of myrmecophagid anteaters, like those of the silky anteater, are associated with their ability to capture insects and to dig into or tear apart insect nests. The skull is long and roughly conical or cylindrical (Fig. 10-12A). The zygoma are incomplete, and the long rostrum contains complex, double-rolled turbinal bones. Teeth are absent. The dentary bone is long and delicate, and the mandibular rami are unfused. The jaw musculature is reduced, but the tongue musculature is greatly developed. The long, slender tongue is protrusible and covered with sticky saliva secreted by the enlarged and fused submaxillary and parotid salivary glands. The tongue muscles originate from the posterior end of the sternum (**Fig. 10-14**) rather than from the hyoid bones of the throat, as in most mammals. The soft palate is extremely long, extending posteriorly to the level of the fifth cervical vertebra, so that the oropharynx and nasopharynx extend posteriorly into the neck (Naples 1999; Reiss 1997). The forelimbs are powerfully built; the third digit is enlarged and bears a stout, recurved claw, and the remaining digits are reduced. Unlike other anteaters, the giant anteater (*Myrmecophaga*;

FIGURE 10-13 Living anteaters (Myrmecophagidae): (A) a southern tamandua (*Tamandua tetradactyla*) and (B) the giant anteater (*Myrmecophaga tridactyla*).

Fig. 10-13B) is unable to climb and walks on its knuckles with its toes partly flexed, as did some extinct Pleistocene sloths, whereas the two species of *Tamandua* are partly arboreal and walk on the side of the hand with the toes inward. Tamanduas climb in the trees like silky anteaters, by above-branch quadrupedal walking. The plantigrade foot has four or five clawed digits. Myrmecophagid anteaters range in size from that of a marmot (*Tamandua*, 3 to 8 kilograms) to that of a large dog (*Myrmecophaga*, 25 kilograms). *Myrmecophaga* is covered with long, coarse fur, and its laterally compressed, nonprehensile tail has long hairs that hang downward. In the tamanduas, the fur on the body and tail is shorter, and the tail is prehensile.

Anteaters use the powerful forelimbs to expose ants and termites by tearing apart their nests; the insects are captured by the long tongue, swallowed whole, and probably ground up by the thickened pyloric portion of the stomach. Many species of ants are highly venomous. Mucus from the anteater's salivary glands may coat ingested ants and immobilize their mandibles and stingers. The two species of *Tamandua* are individualistically terrestrial and arboreal, nocturnal and diurnal. *Myrmecophaga* is entirely terrestrial and seems largely diurnal. Both *Myrmecophaga* and *Tamandua* eat both ants and termites (Rodrigues et al. 2008). The defensive behavior of these anteaters involves standing bipedally, bracing the body with the tail and hindlimbs, and slashing at the enemy with the forelimb claws.

■ Family Bradypodidae

Together with several families of extinct sloths, the two living families of sloths are monophyletically related. As a result, they are classified together, in a suborder variously called Phyllophaga, Tardigrada, or Folivora by different

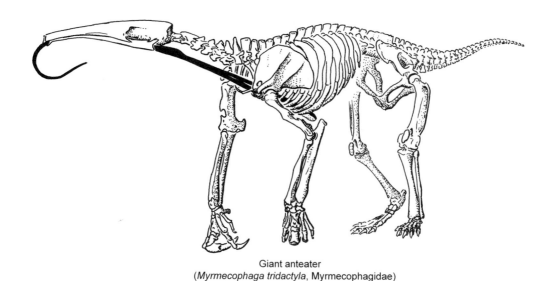

Giant anteater
(*Myrmecophaga tridactyla*, Myrmecophagidae)

FIGURE 10-14 Skeleton of the giant anteater. The tongue and tongue-retracting muscles are shown in black. (Adapted from Paula Couto, C. de. *Tratado de Paleomastozoologia*. Academia Brasileira de Ciencias, 1979.)

authors. The two living families of sloths, Bradypodidae and Megalonychidae, are only distantly related to one another despite their similar lifestyles (Gaudin 2004; Gaudin & McDonald 2008). The members of Bradypodidae, four species called three-toed tree sloths or ais, use a specialized form of arboreal locomotion in an upside-down body posture (called suspensorial or antipronograde). Climbing is done in an upright position by embracing a branch or by hanging upside down and moving along hand over hand. They are so highly modified for this underbranch hanging type of locomotion that they move awkwardly on the ground. They crawl along the ground slowly, using the claws of the forefeet and resting the body on the elbows. The hindlimbs are shorter and move like those in other mammals (R. Timm, personal communication). Interestingly, tree sloths rest in an upright sitting posture that may be necessary for the normal functioning of their multichambered forestomach (Clauss 2004). Sloths are **folivores** (leaf eaters) in which the forestomach acts as a fermentation chamber in some ways similar to certain herbivorous marsupials and ruminant artiodactyls and unlike hindgut-fermenting rodents and lagomorphs. The three modern species of one genus (*Bradypus*) range from Central America (Honduras) through the northern half of South America to northern Argentina. These animals primarily inhabit tropical rain forests.

The bradypodids differ strongly from the anteaters, especially in skull characteristics. The tree sloth skull is short and fairly high, with a strongly reduced rostrum. The zygomatic arch is robust but incomplete, and its jugal portion bears a ventrally projecting jugal process similar to that present in many extinct xenarthrans; part of the superficial masseter muscle originates on the process (Figs. 10-12B and 10-4B). The premaxillary bones are greatly reduced, and the turbinal bones are complexly rolled, as in anteaters. Five maxillary and four or five mandibular teeth are present. The persistently growing teeth are roughly cylindrical and lack enamel. They have a central core of soft dentine surrounded successively by hard dentine and cementum. The teeth erupt as simple cones and acquire features such as "cusps" and "basins" exclusively through wear (Naples 1990).

A departure from the usual mammalian pattern of seven cervical vertebrae occurs in the bradypodids, in which eight or nine occur. Xenarthrism (Fig. 10-2) is developed in the lumbar and possibly thoracic vertebrae, and as in some extinct ground sloths, the coracoid and acromion processes of the scapula are united. The three digits with long and laterally compressed claws are syndactylous (bound together). The tail is short. Three-toed tree sloths are of moderate size, weighing from 4 to 7 kilograms, and are covered with long, coarse hair. This fur provides a habitat for algae, which grow in transverse cracks on the surface of the hairs during the rainy season and tint the fur green. A sample of the fur of the extinct sloth *Nothrotheriops* (Megatheriidae) had ovate bodies resembling the algal cells in fur of the living *Bradypus* (Aiello 1985; Hausman 1929). In addition, the dense pelage of *Bradypus* harbors large numbers of a beetle and the adults of two genera of moths (*Bradypodicola* and *Cryptoses*; Pyralididae, Microlepidoptera).

These remarkably specialized animals eat leaves and descend to the ground only to defecate at 5- to 7-day intervals. The stomach is chambered, and digestion is enhanced by fermentation aided by a **symbiotic** microbiota.

Family Megalonychidae

This once diverse family contains over a dozen extinct genera. The sole surviving members of the family are two species of *Choloepus*, known as two-toed tree sloths or unaus (**Fig. 10-15**). They inhabit Neotropical rain forests from Nicaragua to Bolivia and southern Brazil. There are five upper and four or five lower teeth; the anterior-most teeth are canine-like and are kept sharp by abrasion between the posterior surfaces of the upper teeth and the anterior surfaces of the lower teeth. The number of cervical vertebrae is highly variable, from five to eight, the number differing between the two species and, in some cases, even from one individual to another of the same species. The forelimbs are considerably longer than the hindlimbs. As their name implies, there are two functional digits in the hand, with long grappling claws; the feet have three digits with long claws. These adaptations enable suspensorial locomotion. Green algae and cyanobacteria grow on the hairs, camouflaging the animal, as in *Bradypus*, but in longitudinal flutings rather than in transverse cracks. Like bradypodids, *Choloepus* are arboreal folivores; however, they seem to include nonvegetal matter in their diet occasionally (Chiarello 2008), and the two share low metabolic rates and many other convergently evolved aspects of their ecology, anatomy, and physiology.

Order Pholidota

Family Manidae

The pangolins or scaly anteaters, members of the family Manidae, are represented today by a single genus (*Manis*) with eight species. Pangolins occur in tropical and subtropical parts of the southern half of Africa and in much of southeastern Asia. They are a monophyletic group, with their African and Asian branches arising from three distinct radiations (Gaudin & Wible 1999; **Fig. 10-16**). Accordingly, some authors believe the separate lineages should be afforded taxonomic recognition as different genera. Their fossil record is poor but documents the Eocene occurrence of these animals in Europe, Asia, Africa, and North America (Rose & Emry 1993; Rose et al. 2005).

FIGURE 10-15   The arboreal three-toed sloth (*Bradypus*; Bradypodidae).

Because they are covered by scales, pangolins seem more reptilian than mammalian (**Fig. 10-17**). They are of moderate size, weighing approximately 5 to 35 kilograms. The skull is conical and lacks teeth; the dentary bones are slender and lack angular and coronoid processes but have small laterally directed prongs near the front end. The tongue is extremely long and **vermiform**. Convergent with that of xenarthran anteaters, the tongue musculature of pangolins originates on a short posterior extension of the sternum (the xiphoid process) in Asian species, or on an enormously extended and bifid xiphoid process in African species. This extension passes into the posterior part of the abdominal cavity, curves upward, and ends near the kidneys. The tongue and its extrinsic musculature are therefore longer than the head and body together, allowing the tongue to be extremely protrusible (Kingdon 1971).

The scales are the most distinctive feature of pangolins; they cover the dorsal surface of the body and the tail

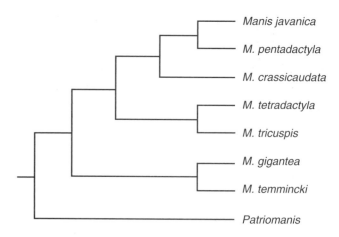

FIGURE 10-16   A phylogeny of pangolins (Pholidota) based on Gaudin and Wible (1999). The Asian species (top three) of *Manis* form a monophyletic clade, while the African species (the other four) are paraphyletic. *Patriomanis* is an extinct pangolin genus from the Eocene of North America. (Adapted from Gaudin, T.J. and J.R. Wible, *Journal of Mammalian Evolution* 6 (1999):39–65.)

FIGURE 10-17   A Sunda pangolin (*Manis javanica*, Pholidota).

one third to one half of the weight of the ground pangolin (*Manis temminckii*). The manus and pes have long, recurved claws borne on unusual, deeply notched ungual phalanges. The pes has five toes, and the manus is functionally tridactyl. The pyloric part of the stomach has thick walls, usually contains small pebbles (that may have been ingested accidentally), and seems to grind food, as does the gizzard of a bird.

The food of pangolins is mostly termites, but ants and other insects are also eaten. The insects are located by smell, as in xenarthrans, and pangolins seem highly selective in their choice of food. Sweeney (1956) found that only rarely would a pangolin dig for the "wrong" species of ant or termite. Like xenarthran anteaters, some pangolins are nocturnal or diurnal. Some are strictly terrestrial. Some are semiarboreal, and two species (one in Java and one in Africa) are quite arboreal and have semiprehensile tails. Also, like xenarthrans, pangolins have low metabolic rates. Pangolins roll up into a ball when disturbed, erect the scales, flail the tail, or move the sharp scales in a cutting motion. Some species spray foul-smelling fluid from the anal glands.

and are composed of agglutinated hair (keratin). The skin and scales account for a large share of the weight of these animals. Kingdon (1971) reports that these parts constitute

SUMMARY

The orders Pilosa (sloths and anteaters), Cingulata (armadillos), and Pholidota (pangolins) include among their extinct relatives some of the most bizarre and improbable of all land-dwelling mammals. They are considered together in this chapter as a matter of convenience. Several of their members share features associated with ant- and termite-eating such as elongated crania and tongues, oddly modified dentitions, large, recurved claws, and lowered metabolic rates. As a result, in the past they were frequently combined in an order Edentata. The Pilosa and Cingulata (collectively referred to as Xenarthra) are closely related and monophyletic, but Pholidota is probably very distantly related to the xenarthrans. Sloths, anteaters, and armadillos underwent an impressive Cenozoic radiation in South America, whereas extant pangolins are an Old World group.

Perhaps the most unusual xenarthrans, glyptodonts, appeared in the middle Eocene epoch, probably from armadillo stock. They were very large, ponderous creatures with a nearly impregnable, turtle-like carapace composed

of many fused polygonal osteoderms covered by up to 1,800 horny scutes. Sloths first appear in Eocene deposits in South America and Antarctica. Traditionally, extinct sloths have been called "ground sloths" because many types were very large and terrestrial, some reaching elephant size. Many of the smaller fossil species, however, were probably arboreal or semiarboreal, and at least one large Pliocene species was seemingly a semiaquatic "sea sloth." The extant tree sloths in the genus *Choloepus* have recently been recognized as representing a family of otherwise extinct "ground sloths," the Megalonychidae.

Today xenarthrans are represented by 13 genera and 31 species. They are interesting because of their unique structure and unusual ecological roles. Living armadillos are omnivorous or insectivorous. Anteaters are highly specialized ant and termite eaters, and sloths are folivorous (leaf eating). Pangolins are also ant and termite eaters; they bear a skin covering of imbricated keratinous scales, unique among mammals.

KEY CHARACTERISTICS OF XENARTHRANS

- Lumbar vertebrae with xenarthrous articulations
- Incisors reduced or absent
- Deciduous teeth absent
- Cheek teeth, when present, lack enamel
- Septomaxilla in skull
- Tympanic bone annular

Armadillos

- Plates or bony osteoderms covered by horny epidermis, present on head and body
- Hair sparse and usually occurs between plates, on limbs, and ventral body surface
- Skull elongate and dorsoventrally flattened
- Zygomatic arch complete
- Mandible slim and elongate
- Teeth homodont and nearly cylindrical
- Second and third cervical vertebrae fused
- Eight to 13 sacral and caudal vertebrae anchor pelvis
- Forefeet and hind feet bear large, heavy claws

Anteaters

- Zygoma incomplete
- Teeth absent
- Dentary bone long and delicate, and mandibular rami unfused
- Tongue long, slender, and protrusible and covered with sticky saliva
- Tongue muscles originate from posterior sternum
- Soft palate extremely long, extending posteriorly to the level of the fifth cervical vertebra
- Third digit enlarged, bearing a stout, recurved claw

Sloths

- Skull short with reduced rostrum
- Zygomatic arch robust but incomplete and bearing a ventral jugal process
- Premaxilla greatly reduced
- Teeth, 5 maxillary and 4–5 mandibular teeth present
- Teeth cylindrical with a central core of soft dentine surrounded by hard dentine and cementum
- Eight to nine cervical vertebrae
- Three syndactylous (fused) digits with long and laterally compressed claws

KEY CHARACTERISTICS OF PANGOLINS

- Keratinous scales cover dorsal surface of body and tail
- Skull conical and dentary bones slender
- Teeth absent
- Tongue extremely long and vermiform and originates on xiphoid process of sternum
- Manus and pes with long, recurved claws borne on unusual, deeply notched ungual phalanges

KEY TERMS

Folivore

Osteoderms

Symbiotic

Vermiform

Xenarthrous

Zygapophysis

RECOMMENDED READINGS

Clauss, M. 2004. The potential interplay of posture, digestive anatomy, density of ingesta, and gravity in mammalian herbivores: why sloths do not rest upside down. *Mammal Reviews, 34*:241–245.

Gaudin, TJ. 1999. The morphology of xenarthrous vertebrae (Mammalia, Xenarthra). *Fieldiana (Geology), 41*:1–38.

Gaudin, TJ. 2004. Phylogenetic relationships among sloths (Mammalia, Xenarthra, Tardigrada): the craniodental evidence. *Zoological Journal of the Linnean Society, 140*:255–305.

Montgomery, GG. 1985. *The Evolution and Ecology of Armadillos, Sloths, and Vermilinguas.* Smithsonian Institution Press, Washington, DC.

Reiss, KZ. 2001. Using phylogenies to study convergence: the case of the ant-eating mammals. *American Zoologist, 41*:507–525.

Rose, KD, et al. 2005. Xenarthra and Pholidota. Pp. 106–126 in *The Rise of Placental Mammals: Origins and Relationships of the Major Extant Clades* (KD Rose & JD Archibald, eds.). Johns Hopkins University Press, Baltimore, MD.

Vizcaíno, SF & WJ Loughry. 2008. *The Biology of the Xenarthra.* University Press of Florida, Gainesville, FL.

Dermoptera and Scandentia

Order Dermoptera

Family Cynocephalidae

Members of the Dermoptera are generally called colugos or, sometimes, "flying lemurs" (a poor name choice, as they neither fly nor are lemurs). They have lemur-like faces and are able to glide long distances between trees. One family, Cynocephalidae, with two genera (*Cynocephalus* and *Galeopterus*) each with one species (*C. volans* and *G. variegatus*), represents the order today. Recent molecular evidence, however, shows that one species, the Sunda colugo (*G. variegatus*), is actually several cryptic species, but these have not yet been formally recognized nor studied in detail. The distribution includes tropical forests from southern Myanmar and southern Indochina, Malaya, Sumatra, Java, Borneo, and nearby islands to southern Mindanao and some of the other southern islands of the Philippine group.

In the past, numerous early Paleogene fossils from various parts of Europe and North America were considered to represent extinct dermopterans. None of these are true dermopterans related to the extant colugos, but they are convergent with them in dental features. The first true dermopteran fossil, *Dermotherium*, is known from the late Eocene in Thailand (Ducrocq et al. 1992). The fossil, although only a jaw fragment, differs little from modern colugos despite being 34 million years old. Most recent phylogenetic hypotheses place the Dermoptera in a monophyletic group called Euarchonta, together with treeshrews and primates (Janecka et al. 2007; Schmitz et al. 2005; Silcox et al. 2005).

The two living species of colugos are modest in size, weighing roughly 0.5 to 1.75 kilograms, and have large eyes and faces that resemble those of Old World fruit bats and some lemurid primates (**Fig. 11-1**). The dorsal pelage color varies between the sexes and is very cryptic against tree bark; females are grayish and males brownish or chestnut, irregularly blotched with white. Although the molars have retained a basically three-cusped insectivore pattern, the broad cheek teeth have a shearing action that includes a large transverse component. This action and the crenulated enamel of the molars provide for efficient mastication of plant material (Rose & Simons 1977). The anterior dentition is highly specialized: the lateral upper incisor (I2) is caniniform, and the first two lower incisors are broad and pectinate (comb-like). The unusual lower incisors are used to groom the fur but may also be used to scrape leaves when the animal feeds (Rose et al. 1981). Unique among extant mammals, the canines are double rooted. The dental formula is 2/3, 1/1, 2/2, 3/3 = 34. Differences in the cranial and dental functional morphology between the two genera of colugos suggest that feeding and dietary differences will be found in them (Stafford & Szalay 2000).

A broad, furred membrane extends from the neck to near the ends of the fingers, between the fingers and toes, between the limbs, and from the hind foot to the tip of the tail. The hands and feet retain five digits that bear needle-sharp, strongly curved claws for clutching branches. As in bats, the neural spines of the thoracic vertebrae are short, the sternum is slightly keeled, the ribs are broad, the radius is long, and the distal part of the ulna is strongly reduced (**Fig. 11-2**). A striking feature of colugos is their long, soft, and luxurious fur, similar to that of marsupial sugar gliders and flying squirrels. In addition to conferring excellent camouflage, the impressive fur may add an important aerodynamic component to gliding in that it appears to dampen turbulence (R. M. Timm, personal communication).

Colugos are **crepuscular** and nocturnal, and they seek refuge during the day in holes in trees; several individuals may occupy the same den (Francis 2008; Wharton 1950). These animals cling to the vertical surfaces of tree trunks or undersides of branches while traveling and feeding. Colugos are slow but skillful climbers, but they are unable to stand upright and are virtually helpless on the ground. They can glide distances well over 100 meters in traveling to and from feeding places. Their spectacular glides are facilitated in part by taking off from some of the tallest trees in the world's rain forests, with few lianas (vines) to block the way (Dudley & DeVries 1990; Emmons & Gentry 1983). Laser scanning of southeast Asian tropical forest canopies

The diet includes leaves, buds, flowers, fruit, and sap from a variety of tree species. The enlarged tongue and specialized lower incisors are used in cow-like fashion in picking leaves (Winge 1941). The great lengthening of the intestine typical of herbivorous mammals is well illustrated by colugos. *Cynocephalus*, which has a head plus body length of only about 410 millimeters, has an intestinal tract approaching 4 meters in length, nine times its head and body length (Wharton 1950). The caecum, a blind diverticulum at the proximal end of the colon, is greatly enlarged (to about 48 centimeters in length) and is divided into compartments. This chamber harbors microorganisms that help break down cellulose and other relatively indigestible carbohydrates. Caecal enlargement is often associated with an herbivorous diet (as in many rodents).

The clearing of forests for agriculture is restricting the distribution of colugos, and in some regions the animals are hunted for their meat and their fur. Because colugos are challenging to study, knowledge of their natural history remains rudimentary. Nevertheless, the study of colugos will make for fascinating fieldwork if it can be completed before their habitat is destroyed and they disappear. Lim (2007) has led the way with field research and a captivating and important account of colugo biology, and has provided ideas for future research and conservation of these curious mammals.

Order Scandentia

Two families, Ptilocercidae and Tupaiidae, which collectively consist of 5 extant genera and 20 species, represent the order Scandentia. All of these are referred to as treeshrews. Members of this order resemble small long-snouted squirrels and occur from India through Burma to the islands of Sumatran Borneo and the Philippines (**Fig. 11-3**). The first definitive fossil tupaiids appear in middle Eocene deposits from Asia, but these fossils are morphologically similar to extant genera and add little to our understanding of the phylogenetic position of treeshrews relative to other mammalian orders. Most modern systematists exclude treeshrews from the Primates and place them in their own order, but there is strong disagreement about their phylogenetic relationships. Morphologists link them with Primates, Dermoptera, and Chiroptera in the group Archonta (Kay et al. 1997b; Sargis 2002, Silcox et al. 2005; Wible & Covert 1987), but molecular geneticists have variously allied them with the Dermoptera and Primates (as Euarchonta), or Lagomorpha, and/or Macroscelidea (Adkins & Honeycutt 1993; Bailey et al. 1992; Murphy et al. 2001; Stanhope et al. 1993). Until recently, the families Ptilocercidae and Tupaiidae were recognized as subfamilies

FIGURE 11-1 (A) Malayan colugo (*Galeopterus variegatus*) with young clinging to a tree branch. (B) Colugo in mid-glide with all patagial membranes extended.

also shows that these forests have more and larger vertical gaps between treetops than do forest canopies in Africa and the Americas; this has likely favored the evolution of a greater diversity of southeast Asian vertebrates that glide and brachiate (Dudley et al. 2007; Gibbons & Culotta 2008). Using a custom-made accelerometry system similar to the motion-detection technology used in some video games, Byrnes et al. (2008) showed that five Malayan colugos (*Galeopterus*) fitted with the devices made 4 to 29 glides each night for distances from 2.5 to 150 meters per glide. They usually took off and landed on vertical or inclined surfaces of trees, and each glide lasted between about 0.6 and 15 seconds (average 3.5 seconds) at a glide speed of about 10 meters per second. The gliding membrane allows colugos to conserve energy and travel quickly and efficiently across gaps between trees relatively safe from arboreal and terrestrial predators. The animals also use it as an air brake on long glides, allowing them to reduce their airspeed to 4 meters per second before landing and to land with lower impact force, typically making quadrupedal landings.

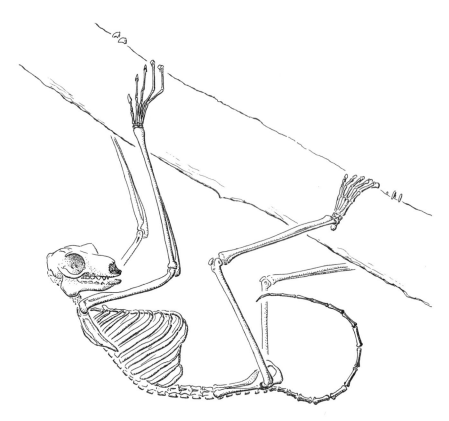

FIGURE 11-2   Skeleton of a colugo shown in a natural posture, hanging from a tree branch. In life, the terminal phalanges of the fingers and toes would bear very sharp, strongly curved claws for grasping.

FIGURE 11-3   A treeshrew (*Tupaia longipes,* Tupaiidae).

of Tupaiidae, but these subfamilies were raised to full family rank by Helgen (2005); a preliminary molecular analysis confirmed this change (Olson et al. 2005). Related to the possibility of the close relationship of treeshrews to Primates is their increasing use in biomedical research (Cao et al. 2003).

The dental formula is 2/3, 1/1, 3/3, 3/3 = 38; the upper incisors are caniniform, and the upper canine is reduced. Treeshrews have a tooth comb consisting of the middle four lower incisors, but unlike lemurs, which also possess a tooth comb, the treeshrew tooth comb does not include the canines (Rose et al. 1981). The upper molars have **trenchant**, W-shaped ectolophs, and the lower molars retain the basic insectivore pattern (**Fig. 11-4**). There is no loss of digits, and all digits have strongly recurved claws. The long tail is usually heavily furred. One unique feature of treeshrews is a prominent hole in the zygomatic arch (**Fig. 11-5**). Treeshrews have well-developed postorbital processes that join the zygoma, a feature long used as evidence of a close relationship with primates. Extant treeshrews, however, lack virtually all of the characteristics that define the order Primates, and the superficial similarities with primates may stem from the shared arboreal and diurnal niche.

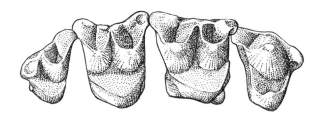

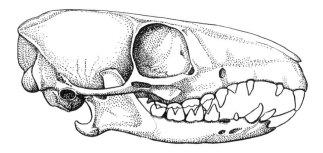

FIGURE 11-5 Skull of a treeshrew (length of skull 52 millimeters).

FIGURE 11-4 Fourth premolars and molars of a treeshrew (*Tupaia* sp., Tupaiidae).

Treeshrews occupy deciduous forests and forage both in the trees and on the ground. They are opportunistic feeders and use a variety of foods, including insects, small vertebrates, fruits, seeds, nectar, and bird eggs (Emmons 2000; Smith & Xie 2008). Emmons (1991) demonstrated that four species in Malaysia are mostly frugivorous. Her studies reveal that treeshrews eat a wide variety of fruit, which passes rapidly (in 13 to 29 minutes) through their simple digestive tracts. In their rapid food transit times and simple gut morphology, tupaiids resemble frugivorous bats and may be important seed dispersers of some tropical forest trees. Many aspects of the ecology and behavior of these poorly known and peculiar mammals were described through the exemplary fieldwork by Emmons (2000).

Family Ptilocercidae

Unlike the tupaiid treeshrews, the sole species of the Ptilocercidae is the arboreal and nocturnal feather-tailed or pen-tailed treeshrew (*Ptilocercus lowii*) of the Malay Peninsula. The eyes of these treeshrews are even larger than those of tupaiid treeshrews and are better adapted for night vision, with a well-developed tapedum lucidum and a brilliant white eyeshine. Their facial vibrissae are more prominent, and they pay close attention to insect sounds when foraging (Emmons 2000). They move quickly and with agility and have extraordinarily strong grip on tree trunks and lianas with their claws. They nest high in the forest canopy in tree hollows lined with leaves and fibers. Many osteological characters in the postcranial skeletons of treeshrews hint at an arboreal ancestry, especially in *Ptilocercus*, whereas the Tupaiidae have a more derived skeleton favoring their stronger tendency for terrestrial

locomotion (Sargis 2001, 2002a, 2002b). Recent fieldwork in Malaysia with radio-collared *Ptilocercus* showed that while flower buds of the bertam palm (*Eugeissona tristis*) are available these treeshrews nightly consume large amounts of alcohol from the fermenting nectar while also pollinating the flowers. Despite having continuously high blood alcohol levels that would intoxicate humans, the treeshrews showed no signs of intoxication and are somehow able to tolerate the risk of high blood alcohol concentrations (Wiens et al. 2008).

Family Tupaiidae

Tupaiid treeshrews are characteristically quick moving and highly vocal. They are active at any time of day or night, but are mainly diurnal or crepuscular. Some species, especially *Tupaia tana* and *Urogale everetti*, are more terrestrial than others, and *Urogale* even has some adaptations in its forelimbs for scratch digging (Sargis 2002a). The mountain treeshrew (*Tupaia montana*) lives in social groups in which a rigid dominance hierarchy is apparent, whereas in parts of Borneo, other species occupying lowland areas do not form social groups (Sorenson & Conaway 1968). The common treeshrew (*Tupaia glis*) is territorial, and small family groups, made up of a male and female and their offspring, patrol the hectare-sized territory regularly, frequently scent marking the boundaries. Parental care appears to be rudimentary, and mothers visit the highly **altricial** young in the nest only to suckle them every other day (Emmons 2000; Emmons & Biun 1991; Martin 1968). The milk is rich in protein, however, and the short suckling bouts are enough to sustain the developing young. This form of maternal behavior, called "absentee parental care," probably serves as an antipredator strategy because the nestlings remain silent and mothers approach the nest by different pathways on each visit so as not to leave a conspicuous odor trail for predators to follow (Emmons 2000; Emmons & Biun 1991).

SUMMARY

Members of the Dermoptera are called colugos or, "flying lemurs" (they neither fly nor are lemurs). They have lemur-like faces and are able to glide long distances between trees. One family, Cynocephalidae, with two genera (*Cynocephalus* and *Galeopterus*) each with one species (*C. volans* and *G. variegatus*), represents the order today. The distribution includes tropical forests from southern Myanmar and southern Indochina, Malaya, Sumatra, Java, Borneo, and nearby islands to the southern Philippines. Most recent phylogenetic hypotheses place the Dermoptera in a monophyletic group called Euarchonta, together with treeshrews and primates.

Colugos are crepuscular and nocturnal and they seek refuge during the day in tree holes. Colugos are slow but skillful climbers, clinging to tree trunks or the undersides of branches while traveling, but they are virtually helpless on the ground. Using skin membranes more extensive than those of any mammals other than bats, they can glide distances well over 100 meters in traveling to and from feeding places. The diet includes leaves, buds, flowers, fruit, and sap from a variety of tree species. The enlarged tongue and specialized lower incisors are used in cow-like fashion in picking leaves.

The treeshrews, order Scandentia, include two families with 5 extant genera and 20 species. Treeshrews resemble small, long-snouted squirrels and occur from India through Burma to the islands of Borneo and the Philippines. Most modern systematists place treeshrews in an order separate from primates, but there is strong disagreement about their phylogenetic relationships. Treeshrews occupy deciduous forests and forage both in the trees and on the ground. They are opportunistic feeders and use a variety of foods including insects, small vertebrates, fruits, seeds, nectar, and bird eggs.

KEY CHARACTERISTICS OF COLUGOS

- Face lemur-like and eyes large
- Dorsal pelage cryptic against tree bark
- Lateral upper incisor caniniform, and first two lower incisors are broad and comb-like
- Canine double rooted
- Broad, furred gliding membranes present from neck to fingers, between limbs, and from hind foot to tail tip

KEY CHARACTERISTICS OF TREESHREWS

- Upper incisors caniniform and upper canine reduced
- Tooth comb consisting of middle four lower incisors present
- Zygomatic arch with prominent hole
- Postorbital processes join zygoma to form postorbital bar

KEY TERMS

Altricial   Crepuscular   Trenchant

RECOMMENDED READINGS

Byrnes, G, NT-L Lim, & AJ Spence. 2008. Take-off and landing kinetics of a free-ranging gliding mammal, the Malayan colugo (*Galeopterus variegatus*). *Proceedings of the Royal Society B*, 275:1007–1013.

Emmons, LH. 2000. *Tupai: A Field Study of Bornean Treeshrews*. University of California Press, Berkeley, CA.

Lim, N. 2007. *Colugo: The Flying Lemur of South-East Asia*. Draco Publishing and National University of Singapore, Singapore.

Silcox, MT et al. 2005. Euarchonta (Dermoptera, Scandentia, Primates). Pp. 127-144 in *The Rise of Placental Mammals*. (KD Rose and JD Archibald, eds.). Johns Hopkins University Press, Baltimore, MD.

Primates

The order Primates is of particular interest not only because it includes ourselves and our closest relatives but also because its members display a fascinating breadth of structural and behavioral adaptations to their environments. The primate radiation can be viewed as an exploitation of arboreal herbivory, a mosaic of adaptations including arboreal locomotion, manual dexterity, stereoscopic vision, and complex social behavior and communication. The immediate ancestors of humans are among the relatively few primate lineages that adopted a largely terrestrial mode of life.

Primates have been most successful in tropical and subtropical areas, where today most species are arboreal. Some primates, such as baboons and chimpanzees, have become partly terrestrial, but only humans have become fully bipedal. Sixty-nine genera and 376 species of primates live today, of which 16 genera and 128 species occur in the New World (Groves 2005). Recent discoveries in remote areas of the world, coupled with molecular research, have resulted in the recognition of 24 new species in the last decade (Wilson & Reeder 2005).

A great deal of literature on primates has appeared during the past two decades, and at least six journals are devoted exclusively to primatology. The Wisconsin Primate Research Center maintains a computerized primate bibliographic service available on the Internet (http://primatelit. library.wisc.edu/).

Primate Origins

The sister group of primates remains controversial, although many believe that primates, colugos (Dermoptera), and treeshrews (Scandentia) form a monophyletic clade, the Euarchonta. Primates constitute one of the oldest eutherian orders, dating from the early Paleocene of North America. Primates of modern aspect (often called Euprimates to exclude a number of extinct primitive primate-like groups) first appeared about 55 million years ago, in the early Eocene (Ni et al. 2003). The evolution of primates is still the subject of heated debate because there are few shared derived characters uniting early primates with members of the other mammalian orders. There is little doubt, however, that primates are a monophyletic group. Primate characters are a mosaic of ancestral and derived traits with the derived traits largely associated with adaptations to an arboreal lifestyle. Among these traits are those associated with locomotion, stereoscopic vision, increased size and complexity of the brain, increased manual dexterity, small litter size with prolonged gestation, modifications of the auditory complex, and dentition (**Table 12-1**). Although primates are largely herbivorous, many are **omnivorous** and seem to be primarily opportunistic feeders. As a result, the molars of primates are largely bunodont and brachydont and have the quadrate form typical of molars of many herbivores or generalized feeders (Kay 1975). Early in the evolution of primates, a hypocone was added to the upper molar and the paraconid of the lower molar disappeared, leaving a basically four-cusped pattern (**Fig. 12-1**). The rostrum of most primates is proportionately shorter than in most mammals, resulting in a shorter tooth row. This is probably related to the importance of stereoscopic vision and a reduced dependence on olfaction.

Plesiadapiforms, mammals with primate-like characteristics, first appear in the late Cretaceous. Recent fossil discoveries from Wyoming, Montana, and China suggest that plesiadapiforms are ancestral to Euprimates (primates of modern aspect) (Bloch & Boyer 2002, 2003), but the relationship between plesiadapiforms and Euprimates remains controversial (Kirk et al. 2003). Plesiadapiformes include nine distinct families that share enlarged and procumbent incisors, low-crowned molars, and other dental characters (Silcox et al. 2005). In addition, many plesiadapiforms show postcranial adaptations for an arboreal lifestyle. A recently discovered 56 million year old plesiadapiform, *Carpolestes simpsoni*, from Wyoming, exhibits a mosaic of ancestral

TABLE 12-1 Shared Derived Features of the Order Primates

Feature	Primates
Locomotor characters	Grasping hands and feet with opposable thumbs and toes Hallux bearing a nail Nails present on all or most digits (may be lost secondarily) Elongation of the calcaneum Hindlimb dominance during locomotion Center of gravity shifted toward hindlimbs
Stereoscopic vision	Some degree of forward rotation of the orbits and narrowing of the interorbital distance Enlargement of the orbital cavity Exposure of the ethmoid bone on the inner orbital wall Stereoscopic vision in which approximately half of the retinal axons project to the ipsilateral side of the brain and half to the contralateral side
Brain characters	Increased fetal brain size compared with fetal body weight, retained in neonates Sylvian sulcus and triradiate calcarine sulcus both present on brain
Reproductive and developmental features	Descent of testes early in life Urogenital sinus absent in females Long gestation times relative to body size, resulting in small litters Sexual maturity relatively late in life Long lifespans (typically)
Auditory characters	Auditory bulla bony, the floor derived from the petrosal bone Extension of the ectotympanic into the auditory meatus
Dentition	Loss of one incisor and one premolar from the ancestral eutherian condition

pleisadapiform and derived euprimate characters (Bloch & Boyer 2002).

Euprimates share a suite of characters, including grasping hands and feet, digits with nails instead of claws, enlargement of the orbits and some degree of orbital convergence, the presence of a postorbital bar or a postorbital wall, and an enlarged brain (Table 12-1; Cartmill 1992). Ancestral Euprimates also showed a number of adaptations for arboreal leaping. It is hypothesized that these traits evolved as primate ancestors exploited food resources available only on slender, terminal branches. The plesiadapiform *C. simpsoni* had an opposable grasping toe (hallux) with a nail, but claws on the other digits. It also lacked orbital convergence and was not a specialized leaper. Based on this specimen, Bloch and Boyer (2003) suggest that grasping evolved before the orbits rotated forward, thus allowing stereoscopic vision, a hypothesis that remains controversial (Kirk et al. 2003).

Euprimates appear in the early Eocene, when the Adapidae and Omomyidae were common on the northern continents (**Fig. 12-2**). Adapidae from the Eocene-Oligocene are believed to be the ancestors of (or sister group to) Lemuriformes (Lemurs, Lorises, and Galagos) and therefore belong to the Strepsirrhini (**Fig. 12-3**). A second Paleogene group, the Omomyidae, are considered to be the sister group of the Haplorrhini (Silcox et al. 2005).

Historically, the Primates were divided taxonomically into the Prosimii and Anthropoidea (Simpson 1945). The prosimians included the lemurs of Madagascar, along with the Old World lorises, galagos, and tarsiers. The Anthropoidea contained the remaining monkeys, apes, and humans. As Groves (2001) points out, however, this arrangement is, "universally admitted to be based on grades, not clades." Cladistic analyses supports the two primate clades noted previously: the Strepsirrhini and Haplorrhini (reviewed by Groves 2001, 2005).

Suborder Strepsirrhini

The extant strepsirrhine primates include the five families of lemurs endemic to Madagascar, the Galagidae of Africa, and the Lorisidae of Africa and Southeast Asia. The suborder derives its name from the features of the nose, with its naked rhinarium, unfused nasal prominences, and comma-shaped nostrils. The strepsirrhine nose is of little use, however, in defining this assemblage of primates because this feature is shared by many nonprimate mammals (Andrews 1988). Strepsirrhine primates are better defined by presence of a toothcomb (**Fig. 12-4**), composed of lower incisors and canine teeth, and a grooming claw on the second digit of the foot. The toothcomb (secondarily lost in *Daubentonia*) projects forward from the lower jaw and is used during grooming. Treeshrews (Scandentia) have a similar toothcomb, but it does not include the canine teeth (Rose et al. 1981). The extant dermopteran *Cynocephalus* also has elaborate toothcombs with up to 15 tines, but these "combs" evolved independently and are used to crop leaves and not for grooming.

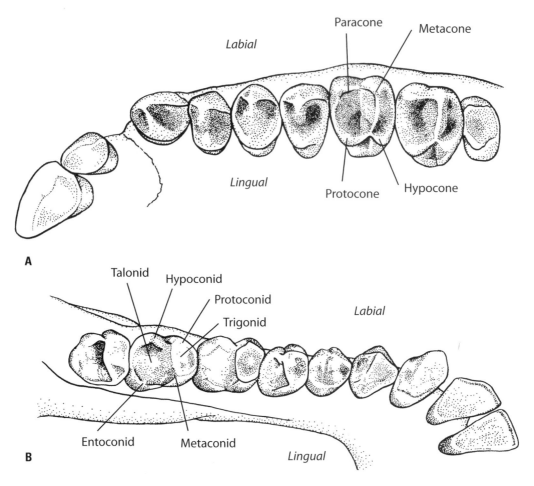

FIGURE 12-1 A diagrammatic representation of the maxillary and mandibular tooth row of the platyrrhine primate *Aotus*: (A) left upper tooth row and (B) left lower tooth row. Lingual refers to the side of the tooth row adjacent to the tongue. (Modfied from Swindler, D. R., and Erwin, J. *Comparative Primate Biology, Volume 1.* Alan R. Liss, Inc., 1986.)

The origin of strepsirrhine primates remains an enigma. Until relatively recently, the oldest fossils clearly attributable to Strepsirrhini were from the early Miocene of Kenya and Uganda. More ancient fossils unearthed in Egypt now suggest that the strepsirrhine primates arose by the middle Eocene in Afro-Arabia (Seiffert et al. 2003, 2005). Molecular evidence suggests that the clade comprising the lorises and galagos diverged from the lemuriform primates during the Paleocene, over 62 million years ago, in Africa (Seiffert et al. 2005; Yoder & Yang 2004). Later (47 to 54 million years ago), ancestral lemuriforms dispersed from Africa to Madagascar (Yoder 2003). In isolation on the

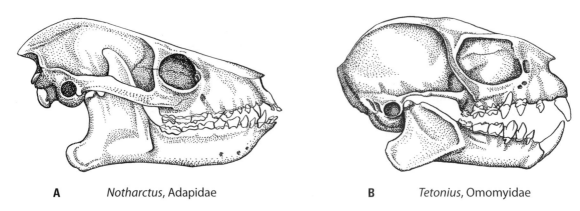

| A | *Notharctus*, Adapidae | B | *Tetonius*, Omomyidae |

FIGURE 12-2 Skull of a fossil Eocene lemuroid (A) with a skull length of 75 millimeters and an Eocene tarsier-like primate (B), length of skull 46 millimeters. (Modified from Romer, A.S. *Vertebrate Paleontology.* University of Chicago Press, 1966.)

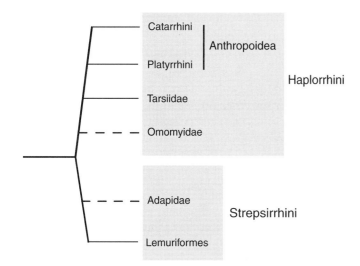

FIGURE 12-3 One hypothesis of primate origins. Note that the Tarsiidae are more closely related to anthropoids than to strepsirrhine primates. The dashed lines indicate fossil groups. (Adapted from Poux, C. and Douzery, E. J. P., *Am. J. Phys. Anthropol.* 124 (2004): 1–16; Schmitz, J., Roos, C., and Zischler, H., *Cytogenet. Genome Res.* 108 (2005): 26–37; and Ross, C. F. and Kay, R. *Anthropoid Origins: New Visions.* Kluwer/Plenum Publishing, 2004.)

World's fourth largest island, they radiated into the spectacular diversity we see today. There are 60 living lemur species and at least 16 recently extinct species of subfossil (partially fossilized) lemurs. Meanwhile, on the African continent, ancestral lorises and galagos diverged along separate evolutionary pathways in the Eocene, approximately 40 million years ago (Seiffert et al. 2003) (**Fig. 12-5**).

Family Cheirogaleidae

One of the five families of primates commonly called lemurs that inhabit the island of Madagascar, the cheirogaleids, includes 5 genera and 21 species commonly known as mouse and dwarf lemurs (**Fig. 12-6**). The family Cheirogaleidae lacks a fossil record. In the past, cheirogaleids were considered a subfamily of the Lemuridae (Simpson 1945) but are now considered a separate family (Groves 2005). Cheirogaleids, like other Malagasy primates, have a toothcomb comprising the incisors and canines.

These small lemurs are nocturnal and weigh less than 500 grams. Mouse lemurs (*Microcebus*) and the hairy-eared dwarf lemur (*Allocebus*) are among the smallest living primates: adult *Microcebus* may weigh as little as 60 grams. Field surveys beginning in the 1990s, and more recent molecular studies, have supported the recognition of eight rather than two species of mouse lemur (Kappler & Rasoloarison 2003). They are arboreal and move by means of quadrupedal walking or by a series of bipedal leaps between branches of trees. Mouse lemurs feed primarily on fruits, nectar, leaves, insects, and even an occasional vertebrate,

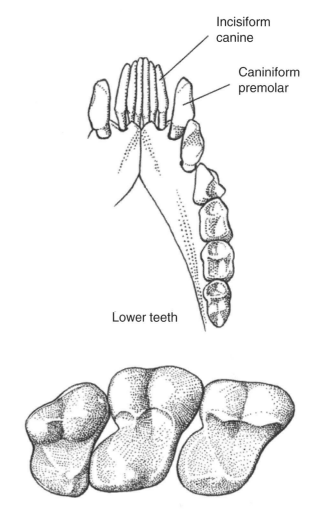

FIGURE 12-4 Teeth of a galago (strepsirrhine) showing the modified canine and premolars that contribute to the tooth comb. (Adapted from Clark, W. E. L. *The Antecedents of Man.* Quadrangle Books, 1971.)

including tree frogs and chameleons (Hladik et al. 1980). *Microcebus* tends to forage relatively close to ground level (less than 10 meters) and inhabits a wide variety of forest types, from desert forests of southern Madagascar to the montane rain forests along the eastern escarpment (Schmid & Kappeler 1994; Tattersall 1982). Mouse lemurs can achieve densities of up to 700 individuals per square kilometer where food resources and adequate shelter are abundant. When food resources and temperatures decline during the austral winter, mouse lemurs enter periods of torpor. The duration of torpor ranges from several hours to several weeks. *Microcebus* conserves energy during torpor by reducing body temperature (to as low as 7°C) and

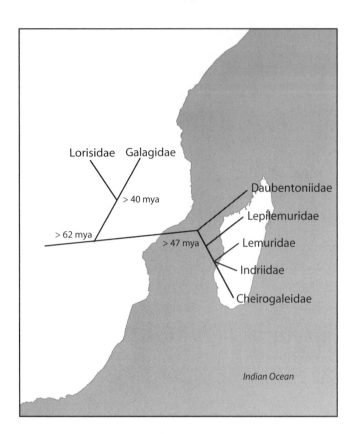

FIGURE 12-5 The timing of dispersal and subsequent diversification of strepsirrhines on Madagascar. (Adapted from Yoder A.D., et al., *Proc. Natl. Acad. Sci.* 93 (1996):5122–5126.)

metabolic rate (Schmid 2000). Additional energy savings come from sharing tree holes with conspecifics. During prolonged torpor or hibernation, mouse lemurs can lose one third of their body weight. Males typically are active several weeks before females re-emerge from hibernation, and during that time, males increase their chances of encountering receptive females by dramatically expanding their home ranges.

The mating system in the gray mouse lemur (*M. murinus*) appears to be promiscuous, but little is known about the behavior of the other species. The gray mouse lemur occurs in dispersed "population nuclei" (Martin 1973; Petter 1962). The proportion of females to males in these nuclei is four to one, and as many as 15 females may occupy the same nest. Surplus males not accompanying groups of females often occupy nests on the periphery of the area. Although mouse lemurs occupy nests together, perhaps because nest sites are at a premium, there is no organized social life, and animals forage alone. This primate, therefore, must be regarded as a basically solitary animal that is flexible enough to occupy nests communally.

The tiny hairy-eared dwarf lemur (*Allocebus trichotis*, weight 80 grams) was considered extinct, but was rediscovered in northern Madagascar in 1989. This small lemur and the closely related *Microcebus* and *Mirza coquereli* (Yoder

2003) have dentitions similar to those of some other lemurs that eat mostly tree gums.

The dwarf lemurs, *Cheirogaleus*, have pointed snouts and long bodies. They are arboreal quadrupeds and rarely resort to bipedal leaping. *Cheirogaleus* species tend to be more frugivorous than other lemurs. *Cheirogaleus* are obligate hibernators (Fietz 2003). Hibernation in the fat-tailed dwarf lemur (*C. medius*) may last for up to 8 months, during which fat reserves stored in the lemur's tails provide energy and body weights may drop nearly 50% (Hladik et al. 1980). Dwarf lemurs hibernate in tree holes where daily temperatures may fluctuate widely, and fluctuations in body temperatures match changes in ambient temperatures. During austral summer (November to April) dwarf lemurs live in family groups consisting of a pair of monogamous adults and their offspring. These family groups defend one to two hectare territories and sleep together in common tree holes. Estrus lasts for a single day (usually in December), and both parents participate in raising the young (Fietz 2003).

The four species of fork-crowned lemurs (genus *Phaner*) are the largest members of the family Cheirogaleidae. Their name derives from the dark rings around their eyes that join to form a single stripe down the middle of the back. These lemurs forage almost exclusively on gums and sap secreted from wounds (made by the lemurs) in the outer bark of certain species of trees. They use their procumbent lower incisors to gouge into the inner gum-producing layers of the tree, and their large caecum contains microbes used to break down the gum (Charles-Dominique & Petter 1980). Because gum and sap production is not seasonal, these lemurs are active throughout the year.

■ Family Lemuridae

Lemurids inhabit Madagascar and the nearby Comoro Islands. Among the 19 Recent species belonging to 5 genera, some are arboreal, some are semiarboreal, and one species, *Lemur catta*, is largely terrestrial. The 11 species of *Eulemur* are widely distributed in forests across the island of Madagascar. The Lemuridae also includes four species of bamboo lemur (*Hapalemur*), two species of ruffed lemurs (*Varecia*), and two monotypic genera, *Lemur catta* and *Prolemur simus*.

In contrast to most primates, the cranium of lemurs is elongate, and the rostrum is usually of moderate length, giving the face a fox-like appearance. The dental formula is 0–2/2, 1/1, 3/3, 3/3 = 32–36. There is a broad diastema and the lower incisors form a toothcomb; the upper incisors are reduced. The pollex and hallux are more or less enlarged and are opposable in all genera (**Fig. 12-7**). The pelage is woolly. The tail is long and heavily furred. The limbs are usually slim, and the tarsal bones are not greatly

FIGURE 12-6   A mouse lemur (*Microcebus rufus*) from Ranomafana National Park, Madagascar.

elongated as in unrelated tarsiers (**Fig. 12-8**). Conspicuous color patterns occur in some species.

Lemurs are variously herbivorous or **frugivorous** and, depending on the species, are primarily diurnal. They are agile climbers, and the hands are used both for climbing and for handling food. Bamboo lemurs (*Hapalemur* and *Prolemur)* inhabit restricted areas of the rain forest. As their common name implies, they feed almost entirely on bamboo shoots and leaves. Until the 1970s, only *H. griseus* and the much larger *P. simus* were known, but a third species (*H. aureus*) was later discovered in isolated patches of primary rain forest (Meier & Rumpler 1987; Meier et al. 1987). Remarkably, the three species coexist by partitioning their food resources. *H. griseus* eats a wide range of plant species and eats young bamboo leaves and tender bases of mature leaves, whereas *P. simus* eats mature bamboo leaves and the densely lignified main stems. The golden bamboo lemur, *H. aureus*, specializes on the young shoots of the giant bamboo (*Cathariostachys madagascariensis*), which are rich in cyanide, a poisonous defensive compound (Mutschler & Tan 2003.

Ring-tailed lemurs (*Lemur catta*) are by far the best known lemurs; they are largely terrestrial and live in large social groups of 20 or more individuals. Interestingly, ring-tailed lemurs, like most lemur societies, are female dominance hierarchies (Kappeler & Ganzhorn 1993). Lemurid social systems are less complex than those of many other primates. The ring-tailed lemur is perhaps the most social of all lemuroid primates (Jolly 1972, 2003). Adult males and females are equally represented in the troop, and the total number of young usually equals the number of adults. Troops occupy exclusive areas, and there is little intertroop contact. Social organization within a troop is based on dominance patterns. All adult females in the troop are dominant over all males, a reversal of the usual primate system. A male dominance hierarchy is established among males, but (surprisingly) the dominant male does not always have first access to estrus females. A given female comes into estrus for only a few hours each year, although the breeding season for the entire troop may last several weeks. During the breeding season, males compete with one another for female attention; male rivals jump toward each other, each attempting to slash the other with the upper canines. These violent clashes often lead to injuries, yet the winner is not always chosen by the female (Jolly 2003).

Ring-tailed lemurs make wide use of olfactory signals. Both sexes mark branches with secretions from the genitals, and using the palms of the hands, males mark branches

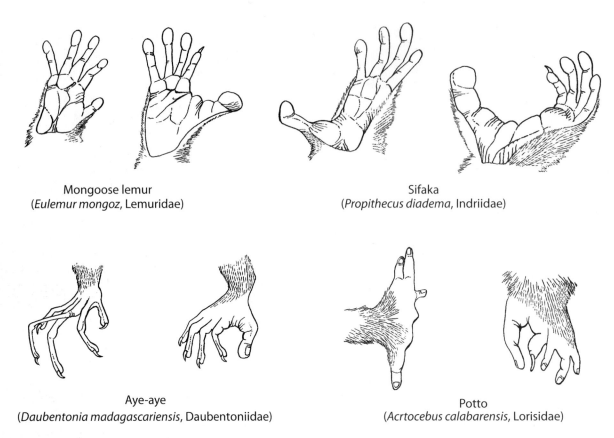

Mongoose lemur
(*Eulemur mongoz*, Lemuridae)

Sifaka
(*Propithecus diadema*, Indriidae)

Aye-aye
(*Daubentonia madagascariensis*, Daubentoniidae)

Potto
(*Acrtocebus calabarensis*, Lorisidae)

FIGURE 12-7 Hands and feet of some strepsirrhine primates (the hand is on the left in each pair).

with scent from glands on the chest and forearms. During aggressive confrontations, the tail is pulled between the forearms, anointed with scent, and then lifted high and waved to disperse the scent. Males indulge in "stink fights," which involve palmar marking, tail marking and waving,

and often displacement of one animal by the other. The animals face each other when performing the scent marking, and the visual displays by the two animals, each using the conspicuously banded tail, seem to be mirror images of each other. The dominant animal moves forward, whereas

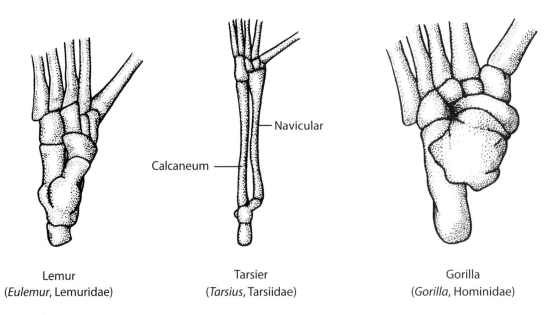

Lemur
(*Eulemur*, Lemuridae)

Tarsier
(*Tarsius*, Tarsiidae)

Gorilla
(*Gorilla*, Hominidae)

Navicular

Calcaneum

FIGURE 12-8 Dorsal views of the left foot bones of several primates, showing the tarsal bones. Note the remarkable elongation of the calcaneum and the navicular in the tarsier. (Adapted from Clark, W. E. L. *The Antecedents of Man.* Quadrangle Books, 1971.)

the other retreats. Vocalizations are also important during social interactions, and a variety of vocal signals are used. Kappeler (1998) showed that female scents probably function in mate attraction and in maintaining female dominance hierarchies, whereas male scents are primarily used in male–male competition.

Family Lepilemuridae

The sportive lemurs, genus *Lepilemur*, were previously grouped in the Megaladapidae or the Lemuridae. They are now placed in a separate family, Lepilemuridae (Groves 2005). Eight living species of sportive lemurs were recognized by Wilson and Reeder (2005), but a recent reanalysis of morphological and molecular characters reveals up to 11 previously unrecognized species (Louis et al. 2006). Sportive lemurs are nocturnal folivores and rely on bacteria in their digestive tracts, especially in the large caecum, to digest plant cellulose. Leaves are poor in nutritional content and often contain defensive chemicals that reduce digestibility; sportive lemurs conserve energy by ingesting their feces and redigesting the material to extract the remaining nutrients. Indeed, field metabolic rates of sportive lemurs are 33% lower than would be predicted of herbivorous mammals of similar body size. Schmid and Stephenson (2003) hypothesized that such a low metabolic rate allows *Lepilemur* species to survive on a diet low in energy and rich in toxic plant compounds.

Family Indriidae

The Malagasy indrids include three species of woolly lemur (*Avahi*) (**Fig. 12-9**), seven species of sifaka (*Propithecus*) (**Fig. 12-10**), and the largest living lemur, *Indri indri*. Sub-Recent indrid fossils from Madagascar include a huge terrestrial lemur, *Archeoindris*, with limb dimensions similar to the extinct ground sloths of the New World and an estimated weight of 200 kilograms (Mittermeier et al. 1994).

Indrids are fairly large, measuring up to 90 centimeters in head-and-body length and weighing about 10 kilograms. They have a shortened rostrum and a monkey-like face. They have a reduced dental formula of 2/2, 1/0, 2/2, 3/3 = 30, with only four incisors making up the tooth comb. The hands and feet are highly modified for grasping branches during climbing (Fig. 12-7). The limbs are relatively long, and the tail is approximately equal to the head-and-body length, except in *Indri*, which has a short stub of a tail. The pelage is mottled brown in the nocturnal avahi, either black and white or orange-brown and white in *Propithecus* and black and white in *Indri*.

These primates are folivorous. Their diets resemble those of the Neotropical howler monkey (*Alouatta*; Cebidae) and the African colobus monkey (*Colobus*; Cercopithecidae). Indrids are typically fairly slow, deliberate

FIGURE 12-9   A nocturnal woolly lemur or avahi (*Avahi laniger*, Indridae) from Ranomafana, Madagascar.

climbers, but can move rapidly through the canopy by using a series of bipedal leaps. The hindlimbs are long relative to the front limbs; when traveling on the ground, these primarily arboreal animals proceed by a series of hops. The hands are used for climbing and for handling food; however, manual dexterity seems limited, and food is often picked up in the mouth.

The nocturnal avahi lives in **monogamous** pairs, forages through the rain forests for leaves, and sleeps huddled in small family groups during the day. Unlike most nocturnal prosimians, avahi forage as a family group; females carry the young during foraging instead of parking them. The family group visits specific trees that often are widely dispersed within the home range. It has been suggested that monogamy is an optimal strategy when food or mates are difficult to defend. *Propithecus* and *Indri* live in small family groups and forage during the day. The indri is also monogamous and often pairs for life. The diurnal sifakas (*Propithecus*), however, is promiscuous. Like most other lemur species, females are dominant in troops of sifakas; males lose 95% of conflicts with females. Males also compete with one another, often violently, for access to females during the short breeding season (Pochron & Wright 2005).

FIGURE 12-10 A Coquerel's sifaka (*Propithecus coquereli*) from northwestern Madagascar.

All indrids are vocal to some extent. The specialized larynx enables indri to produce loud, resonant calls. These howls are given with greatest frequency in the morning and evening, as are the calls of the howler monkey, and may help maintain territorial boundaries between neighboring bands. Contagious calling, involving neighboring family groups responding to each others calls, occurs on an almost daily basis.

■ Family Daubentoniidae

This family is represented by one highly specialized living species, the aye-aye (*Daubentonia madagascariensis*). This secretive nocturnal animal occurs locally in dense forests throughout Madagascar (Sterling 2003). The fossil record of the Daubentoniidae consists of sub-Recent fossils from Madagascar of an extinct species that was slightly larger than the surviving aye-aye.

Aye-ayes weigh approximately 2 kilograms and have prominent ears and a long, bushy tail, and the skull and dentition depart strongly from the usual primate plan (**Fig. 12-11**). The skull is short and moderately high. The orbit is

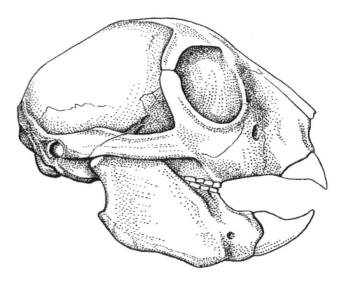

Aye-aye
(*Daubentonia madagascariensis*, Daubentoniidae)

FIGURE 12-11 Skull of an aye-aye (length of skull 110 millimeters).

prominent and faces largely forward; the postorbital bar and zygomatic arch are robust, and the rostrum is short and deep. The dentition differs from the basic primate type both in the extensive loss of teeth and in the strong specialization of the teeth that are retained. The dental formula is 1/1, 0-1/0, 1/0, 3/3 = 18 or 20. The canine is often absent, and the cheek teeth have flattened crowns with no clear cusp pattern. The laterally compressed, ever-growing incisors are greatly enlarged and wear to a sharply beveled edge because only the anterior surfaces are covered with enamel (as in rodents). Because of the shape of the teeth and the presence of a diastema between the incisors and the cheek teeth, *D. madagascariensis* was first described as a rodent. The hand is unique among primates; the digits are clawed, all but the nonopposable pollex are long and slender, and the third digit is remarkably slender (Oxnard 1981; Fig. 12-7). In the hind foot, the hallux is opposable and bears a nail, but the other digits are clawed.

Aye-ayes are arboreal, nocturnal, and mainly insectivorous. Their foraging technique is unique. The elongate third finger is used to tap on wood harboring wood-boring insects (**Fig. 12-12**). The aye-aye then listens carefully for insects within the wood, and the remarkable third digit is used for removing adult and larval insects from holes or fissures in the wood (Erickson 1991). When necessary, the powerful incisors tear away wood to enable the third digit to reach insects in deep burrows. Interestingly, this strange mode of foraging is shared (convergently) by metatherians of the family Petauridae from Australia and New Guinea (Flannery 1995; Strahan 1995). In *Dactylopsila*, the most

FIGURE 12-12   An aye-aye (*Daubentonia madagascariensis*) from Madagascar.

specialized of these metatherians, the front incisors are modified, and the manus is specialized along lines parallel to those in the hand of *Daubentonia*, except that the fourth rather than the third digit is the probing finger (see Figure 6-26). Aye-ayes also use their procumbent incisors to scrape waxy substances from tree trunks and have been observed to use their elongate fingers to ladle nectar from flowers. During the day, aye-ayes sleep in nests constructed of branches high in the trees; at night, they often travel on the ground covering as many as 4.3 kilometers a night (Sterling 2003).

As is the case with many specialized mammals that occupy limited geographic areas, the future of the aye-aye and many other species of Malagasy primates depends on preservation of areas undisturbed by humans. In 1927, the government of Madagascar protected all lemur species, but habitat loss continues to reduce many lemur populations. Today, most Malagasy lemurs have restricted ranges, and most species are endangered or critically endangered.

▪ Family Lorisidae

The lorisids are more widely distributed than are the lemurs of Madagascar and can be common at some localities.

Lorisids occur in Africa south of the Sahara, India, Sri Lanka, and southeastern Asia and in the East Indies. The fossil record of lorisids is scanty but suggests that these animals evolved in the Afro-Arabia area and diverged from galagos by the Middle Eocene (Seiffert et al. 2003). There are five living genera and nine species.

The eyes face forward in lorisids (**Fig. 12-13**), rather than more or less to the side as in lemurs, and the rostrum is short. Lorisids are arboreal, and their locomotion usually involves methodical hand-over-hand climbing (called slow climbing). They vary from the size of a rat to the size of a large squirrel. The braincase is globular. The facial part of the skull is often short and ventrally placed, and the anteriorly directed orbits are separated by a thin interorbital septum. The dental formula is 1-2/2, 1/1, 3/3, 3/3 = 34 or 36. The upper incisors are small. The lower canine is incisiform, and the molars are basically quadritubercular. Tails are either absent or short, less than 15% of head-and-body length.

The manus and pes are specialized in a variety of ways for clutching branches. In the genus *Arctocebus*, a pincerlike hand has been developed by the reduction of digits two and three and a change in the posture of the remaining digits; the first digit of the pes is opposable and frequently

FIGURE 12-13   A slow loris (*Nycticebus coucang*) from Borneo.

greatly enlarged (Fig. 12-7). Circulatory adaptations in the appendages provide for an increased blood supply to the digital flexor muscles, which are used in gripping branches during extended periods of contraction. Similar circulatory modifications, involving the formation of a rete mirabile, are also important in this and many other mammals in conserving body heat (Chapter 21).

These nocturnal primates are omnivorous, including gum and nectar in their diet, or are frugivorous. Prey is usually captured by the hands after a stealthy approach. The African pottos (*Perodicticus* and *Arctocebus*) are slow climbers that prefer a continuous canopy. *Perodicticus* is generally found high in the canopy and restricts its movements to larger branches, whereas *Arctocebus* forages in the understory, below 5 meters, where it climbs on very small branches and vines. Diet depends on the region; however, generally, *Perodicticus* prefers fruits and gums, and *Arctocebus* is primarily insectivorous (Fleagle 1988; Oates 1984; Walker 1969). The two Asian lorisids are omnivores and exhibit two distinct body types: the slender loris (*Loris*) of southern India and Sri Lanka is lightly built, whereas the slow loris (*Nycticebus*) of southeastern Asia is stocky (Fig. 12-13).

The lorises present a paradox: their diets include high-energy items, such as sap, nectar, and fruit, that are available year round. However, all lorises that have been studied have metabolic rates less than 60% of that predicted on the basis of size, and all move slowly. This slow lifestyle is seemingly associated with the expenditure of considerable energy to detoxify defensive compounds in their food (Weins et al. 2006).

Family Galagidae

Galagos vary from the size of a rat to that of a large squirrel, have very large eyes and expressive ears that resemble those of some bats, and have a remarkable ability to make prodigious arboreal leaps (**Fig. 12-14**). There are 19 species of galagos in three genera (Groves 2005). The tail is long and well furred and is used as a balancing organ during leaping. Galagos have unusually long hindlimbs with powerful thigh muscles. The skull has a long rostrum (relative to those of other primates), and the dental formula is 2/2, 1/1, 3/3, 3/3 = 36. Specialized lower incisors and canines are procumbent (**Fig. 12-15**) and form a tooth comb (Fig. 12-4) used in grooming the fur and in feeding on tree resin (gum). The specialized hands and feet are well adapted to

FIGURE 12-14   A galago (*Galago* sp., Galagidae).

grasping: both the thumb and hallux are large and opposable. The fourth digits are unusually long, and the distal pads of the digits have well-developed traction ridges (**Fig. 12-16**) that are important during climbing. The second digit of the hind foot is short and bears a claw used for grooming, but all other digits bear flattened nails. The long foot segment of the hindlimb is associated with leaping ability; the elongation involves the tarsus but is not as extreme as that in *Tarsius*.

Galagos are common in many sections of Africa, and their arboreal leaps are often a fascinating part of the twilight and nocturnal scene. By day, this species usually takes refuge in family groups in holes in trees. The evening dispersal from these retreats often involves the use of favorite pathways through the trees. Leaps of up to 7 meters between adjacent trees have been reported (Kingdon 1971).

Galagos have an extremely varied diet but seem to prefer insects. When insects are abundant during the rainy season, galagos depend primarily on this food source. A precisely timed leap terminating in a quick grab with one hand is a common style of capturing insects. Kingdon (1971, 1997) reports that *Galago senegalensis* feeds also on seeds and small vertebrates and that it takes nectar from the large flowers of baobab trees. In some areas, during long dry seasons and periodic food shortages, the diet of galagos is mostly tree resins. These contain polymers of hexose and pentose sugars, which galagoes can digest, and are the key to survival (Charles-Dominique 1971; Martin & Bearder 1979). Although known to contain defensive secondary compounds, in many areas, resins are year-long dietary staples.

Galagos have a variety of vocalizations that serve as warnings, as communication signals between mother and young, and perhaps as appeasement during intraspecific encounters. *G. senegalensis* has a "vocabulary" of about 10 basic sounds (Andersson 1969). The loud, raucous calls of the greater galago (*Otolemur crassicaudatus*) are an impressive addition to the chorus of night sounds in wooded parts of southern and eastern Africa.

Suborder Haplorrhini

The haplorrhine primates include the tarsiers (family Tarsiidae), the New and Old World anthropoid primates, and several fossil families including Omomyidae and Eosimiidae

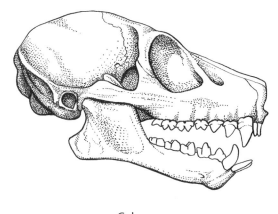

Galago
(*Galago* sp., Galagidae)

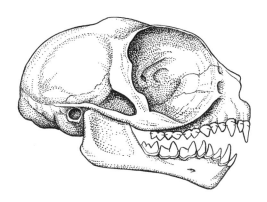

Tarsier
(*Tarsius spectrum*, Tarsiidae)

FIGURE 12-15   Skulls of a galago (*Galago*, Galagidae), length of skull 65 millimeters and a tarsier (*Tarsius spectrum*, Tarsiidae), length of skull 36 millimeters. (Adapted from Clark, W. E. L. *The Antecedents of Man*. Quadrangle Books, 1971.)

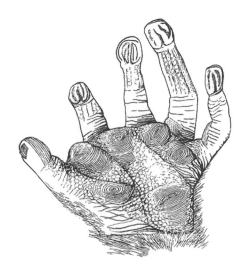

FIGURE 12-16 Traction patterns on the palm of the hand of the greater galago (*Otolemur crassicaudatus*).

(Kay et al. 1997b, 2004; Groves 2005). Haplorrhine primates are phylogenetically united by shared dental and cranial features, as well as synapomorphies of the soft anatomy, molecular characters, a **hemochorial placenta**, and a **fovea centralis** in the retina, that is associated with improved visual acuity (Kay & Williams 1994; Kay et al. 1997b; Porter et al. 1994). Haplorrhines are further divided into the Platyrrhini (New World) and the Catarrhini (Old World). Platyrrhines have flat noses, with the nasal openings facing outward, whereas the catarrhines have downward-directed nasal openings.

The origins of haplorrhine primates remain controversial. Ancestral tarsiers and anthropoids may have entered Africa from Asia in the Early to Middle Eocene (more than 45 million years ago). Furthermore, molecular divergence time data posit a 34 to 36 million year old (Eocene) split between platyrrhine and catarrhine primates. Platyrrhines first appear in the late Oligocene of Bolivia. According to one hypothesis, primitive primates ancestral to the platyrrhines may have entered South America from Central America on logs or debris that floated across the stretch of water that separated these land masses or by island hopping along an early Antillean archipelago present during much of the Tertiary Period (Gingerich 1980; Marshall & Sempere 1993; Simpson, 1945). Alternatively, they may have arrived from Africa, at a time when Africa and South America were closer together and ocean currents favored dispersal from Africa (Ciochon & Chiarelli 1980; Flynn et al. 1995). Once established in the New World, platyrrhines became a diverse and common component of the South and Central American fauna (Fleagle et al. 1997; Tejedor 1998).

The taxonomic position of the tarsiers has also been controversial (Fig. 12-3), but most primatologists now agree that a suite of shared cranial features indicates a close tarsier–anthropoid relationship (Kay et al. 1997b; Poux & Douzert 2004). Kay et al. (2004) proposed that the earliest anthropoids were small, primarily insectivorous, nocturnal, and arboreal quadrupeds. They hypothesized that tarsiers diverged from this ancestral plan by switching to a bipedal leaping style of arboreal locomotion. Stem anthropoids, they suggested, became more herbivorous, increased in body size, developed a quadrupedal form of arboreal locomotion, became diurnal, and evolved greater visual acuity.

In addition to the tarsiers, the suborder Haplorrhini includes the four extant New World families and three Old World families (**Fig. 12-17**). New World anthropoids are a monophyletic group, usually called the Platyrrhini, that includes the marmosets, tamarins, capuchins, and squirrel monkeys (family Cebidae) along with three recently recognized families: Aotidae (night monkeys), Pitheciidae (titi, uakari, and saki monkeys), and Atelidae (howler, spider, and woolly monkeys). The Catarrhini are an Old World assemblage, including the family Cercopithecidae (macaques, baboons, mangabeys, and many other monkeys), as well as

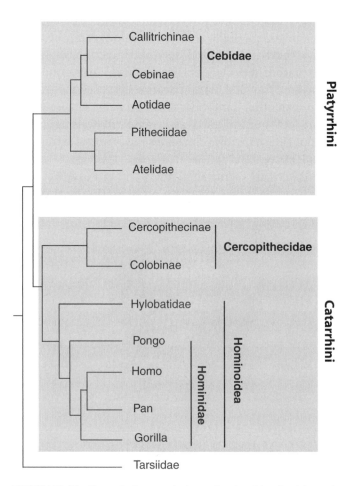

FIGURE 12-17 One phylogenetic hypothesis of haplorrhine primates. (Adapted from *Tree of life,* University of Arizona.)

the Hylobatidae (gibbons and siamang) and the Hominidae (orangutan, gorilla, chimpanzee, bonobo, and humans).

■ Family Tarsiidae

This family is represented today by seven species in a single genus, *Tarsius*, and occurs in jungles and secondary growth in Borneo, southern Sumatra, some East Indian islands, including Sulawesi, and some of the Philippine Islands. Tarsiids are first known from the middle Eocene of China (Rossie et al. 2006).

The tarsier is roughly the size of a small rat and, with its large head, huge eyes, long limbs, and long tail, has a distinctive appearance (**Fig. 12-18**). The most conspicuous cranial features are the enormous orbits, which face forward and have expanded rims and a thin interorbital septum (Fig. 12-15). The eye of the tarsier is apparently adapted entirely to night vision, for it lacks cones in the retina; however, tarsiers also lack a reflecting tapetum lucidum in the retina, suggesting that they evolved from a diurnal ancestor. Tarsiers and anthropoids are the only mammals that possess a fovea centralis for improved visual acuity. The dental formula 2/1, 1/1, 3/3, 3/3 = 34 is unique among primates. The medial upper incisors are enlarged. The premolars are simple. The crowns of the upper molars are roughly triangular, and the lower molars have large talonids. The neck is short, a characteristic of many saltatorial (jumping) vertebrates. All but the clawed second and third pedal digits have flat nails, and all digits have disk-like pads (**Fig. 12-19A**). The limbs, especially the hind ones, are elongate; the tibia and fibula are fused. Tarsiers share with other haplorrhines a hemochorial placenta (capillaries of the chorion burrow into the uterine wall, making direct contact with the maternal blood), unlike the **epitheliochorial placenta** of lemurs, where the capillaries are separated from maternal blood by the uterine lining (Fleagle 1988).

The trend toward jumping ability is developed to an extreme degree in the family Tarsiidae. As in all highly specialized jumpers, the hind foot is elongate, but in the tarsier, the elongation is unique. It involves two tarsal bones (hence the name *Tarsius*) rather than metatarsals, as in such jumpers as jerboas and kangaroo rats. In *Tarsius*, the calcaneum and navicular are greatly elongate (Fig. 12-8), whereas the metatarsals are not unusually long relative to the phalanges (Fig. 12-19A). An important functional end is achieved by this unusual arrangement: The elongation of the tarsus has not sacrificed the dexterity and grasping ability of the digits themselves (the metatarsals and phalanges). Elongation of the metatarsals would have caused a reduction of dexterity. In elephant shrews and kangaroo rats, dexterity and gripping ability of the hind foot are not important, and a more "direct" means of elongation—lengthening of the already somewhat elongate metatarsals—occurred.

Tarsiers are primarily arboreal and nocturnal and feed largely on insects, which they pounce on and grasp with the hands. Fogden (1974) observed tarsiers quietly watching and waiting on a low perch and then leaping to the ground to capture insects. Although more highly adapted to leaping than any other primate, tarsiers can walk and climb quadrupedally, hop or run on their hind legs on the ground, and slide down branches (Sprankel 1965). Tarsiers and some species of galagos share the ability to leap long distances with great precision, and in both of these types of primates, the landing from a leap is largely bipedal. In association with jumping ability, much of the weight of the tarsier is concentrated in the hindlimbs, which together constitute 21% of the total weight of the animal; the musculature of the thighs alone equals 12% of the body weight, largely because of the great enlargement of the quadriceps femoris, a powerful extensor of the shank (Grand & Lorenz 1968). Horsfield's tarsiers (*T. bancanus*) usually live in pairs, but single animals are most frequently observed, suggesting that only large, dominant males associate with females

FIGURE 12-18   A Philippine tarsier (*Tarsius syrichta,* Tarsiidae).

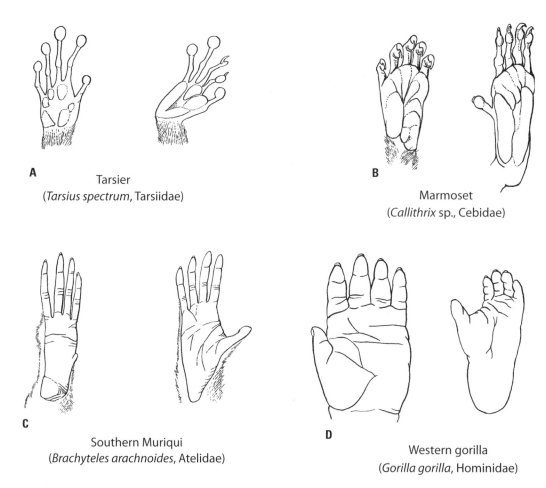

A
Tarsier
(*Tarsius spectrum*, Tarsiidae)

B
Marmoset
(*Callithrix* sp., Cebidae)

C
Southern Muriqui
(*Brachyteles arachnoides*, Atelidae)

D
Western gorilla
(*Gorilla gorilla*, Hominidae)

FIGURE 12-19 Hands (left) and feet (right) of four primates. (Adapted from Clark, W. E. L. *The Antecedents of Man.* Quadrangle Books, 1971.)

(Fogden, 1974). Spectal tarsiers (*T. tarsier*) are more gregarious and often sleep in small family units (Gursky 2000, 2002). Studies of captive animals have demonstrated considerable differences in social structure between the species of tarsiers (Crompton & Andau 1987). Tarsiers have an unusually long gestation period (up to 6 months) for their size and give birth to a single offspring that may weigh 30% of the mother's weight.

Platyrrhini

Family Cebidae

This New World family includes 6 genera and 56 species, including the tamarins, marmosets, capuchins, and squirrel monkeys (Groves 2005). Marmosets and tamarins belong to the subfamily Callitrichinae. They are the smallest of New World primates, weighing between 100 and 750 grams, depending on the species. Considerable debate remains about the evolutionary history of these tiny primates. Hershkovitz (1977) considered marmosets and tamarins to be relatively primitive groups, whereas several other researchers (Rosenberger 1984; Sussman & Kinzey 1984)

argued that they represent derived forms specialized to unique insectivorous and resin-feeding niches.

The dentitions of marmosets and tamarins differ from those of other New World primates. Marmosets and tamarins have chisel-shaped medial incisors, triangular upper molars, and lack the third molar, except in Goeldi's monkey (*Callimico goeldii*) where a third molar is retained. The typical dental formula is 2/2, 1/1, 3/3, 2/2 = 32. Several species have heads adorned with manes or conspicuous tufts of fur (**Fig. 12-20A**). The hands and feet lack opposable thumbs and big toes; except for the hallux, all of the digits have claws instead of the flattened nails of other primates (Fig. 12-19B). These sharp claws provide purchase and allow these tiny primates to cling vertically to the bark of trees.

Except for *Callimico*, marmosets and tamarin females usually give birth to fraternal twins that share a common placental circulation (Haig 1999). This remarkable arrangement results in **chimerae**, where blood cells from both twins are present in each individual (Gengozian & Merritt 1970). After a gestation period of up to 150 days, the mother and one or more of the adult males care for the twins (Terborgh 1983).

FIGURE 12-20   (A) A pair of cottontop tamarins (*Saguinus oedipus*, Cebidae) and (B) a squirrel monkey (*Saimiri* sp., Cebidae).

The recent discoveries of several new species of callitrichines in South America demonstrate that much remains to be learned about tropical forests threatened by deforestation (Lorini & Persson 1990; Mittermeier et al. 1992; Van Roosmalen et al. 1998).

The Cebidae also includes the capuchins (subfamily Cebinae) and the squirrel monkeys (subfamily Saimiriinae) (Fig. 12-20B). They range from southern Mexico through Central America to southern Brazil. They have elongate limbs and digits that bear curved nails, and the furry tail is typically long. In capuchins, the tail is semiprehensile. The size ranges from about 750 grams to 4.5 kilograms. The skull is more or less globular, with a high braincase and short rostrum (**Fig. 12-21**). The orbits face forward, and the nostrils are separated by a broad internarial pad and face to the side. Bare, brightly colored patches of skin on the rump (**ischial callosities**) often present in Old World monkeys (Cercopithecidae) do not occur in platyrrhines.

Capuchins and squirrel monkeys typically occur in tropical forests, and most are diurnal. They are primarily vegetarians; fruit is often preferred, but a wide variety of plant and animal material is eaten. They are active, intelligent animals and adroit climbers, and some species move with amazing speed through the trees. Most species are gregarious, the most common social aggregation consisting of a family group. Some species form larger troops: the squirrel monkey (*Saimiri*) occurs in bands of up to 200 animals, which break into smaller bands in a fission–fusion type of social system. Capuchins, on the other hand, live in troops of six to 35 animals. A typical capuchin troop consists of several related females, their offspring, and several adult males. A single large male dominates other members of the troop and usually fathers most of the young. Squirrel monkeys apparently do not defend regular territories, but some capuchins (*Cebus albifrons*) are strongly territorial and defend their core areas against other troops. The array of capuchin and squirrel monkey calls may serve to keep troop members together, to reinforce group dynamics, or to warn of the presence of predators.

Capuchin and squirrel monkey females give birth to single young after a gestation period of from 150 to 180 days. Mothers carry their young with them as they forage.

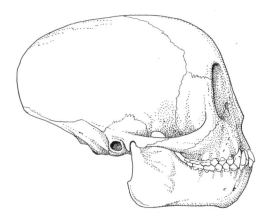

Squirrel monkey
(*Saimiri oerstedii*, Cebidae)

FIGURE 12-21 Skull of a squirrel monkey (*Saimiri oerstedii*, Cebidae). Note the large braincase. This neotropical monkey has a complex social system (length of skull 59 millimeters).

Males play virtually no role in raising the young. Females reach sexual maturity at 4 years of age, males at 8 years.

The social organization of a group of arboreal primates affects their foraging behavior. As an example, a consistent spacing is maintained by a group of Neotropical capuchin monkeys (*Cebus olivaceus*) moving through the trees. The front and center positions are occupied by the dominant male and female and by individuals tolerated by them, whereas peripheral positions toward the rear are occupied by individuals of lower social rank (Robinson 1981). The monkeys near the center of the group can afford to be less alert for predators than those monkeys with peripheral positions and can therefore eat fruit more rapidly.

Remarkably, capuchins (*C. olivaceus*) in Venezuela use crushed millipedes that are rubbed into their fur as a natural insect repellent (Valderrama et al. 2000). These millipedes contain the potent mosquito repellent benzoquinone.

Some species of the family Cebidae are locally common, but most are considered endangered, threatened, or vulnerable to extinction, primarily because of habitat loss (International Union for Conservation of Nature and Natural Resources).

Family Aotidae

The night monkeys, or douroucoulis (*Aotus*), include eight species that inhabit forests from Panama south to northern Argentina and Paraguay. They are the only truly nocturnal anthropoids. Night monkeys have only recently been regarded as a family (Aotidae) separate from the Cebidae (Groves 2005). *Aotus* species are small (450 to 950 grams in body weight), quadrupedal primates that lack a prehensile tail. There is no sexual dimorphism. Night monkeys have large eyes and good nocturnal vision despite their retention of traits normally associated with diurnal habits, such as the retinal fovea and the lack of the reflective tapetum lucidum (Wright 1996). Night monkeys are monochromatic and have lost the ability to see color (Jacobs et al. 1993; **Box 12-1**).

Aotus social groups typically include an adult male and female and their offspring (Wright 1994). At only 133 days, *Aotus* has one of the shortest gestation periods of any platyrrhine primate. *Aotus* infants weight 90 to 105 grams at birth and are covered in fur (Dixson 1994). A unique feature of *Aotus* family groups is that the adult male is the primary caregiver; the mother only carries the infant during the first week after birth and while the infant is nursing. Between bouts of nursing, it is the foraging male who carries and guards the young until it is about 18 weeks old (Dixson 1994).

Aotus family groups defend small territories (approximately 10 hectares) from neighboring groups. Neighboring groups rarely come into contact except at fruiting trees on territory borders (Wright 1994). Territories are defended by vocalizations, scent marking with urine and

BOX 12-1 Primate Color Vision

An unusual form of color vision occurs in some species of New World monkeys; males and females of the same species may differ in their ability to see color. In contrast, humans and Old World primates are **trichromatic**. They see colors by comparing the signals from three types of color-sensing cone cells in the retina of the eye. Each type of cone cell contains a slightly different type of light-sensitive pigment, which is sensitive to specific wavelengths of light. Most nonprimate mammals have only two types of color-sensing cells and are **dichromatic** (analogous to certain forms of color blindness in humans). In many species of New World primates, however, males are dichromats, and about 60% of females are trichromats (Ja-

cob & Deegan 2001). This unusual system occurs because one of the light-sensitive pigment genes is located on the X chromosome and has three possible alleles. Males inherit only one X chromosome and are dichromatic, but females have two X chromosomes and can have several combinations of the three alleles. The result is that heterozygous females are trichromatic and homozygous females are dichromatic. Not all New World monkeys share this condition; the nocturnal night monkeys (*Aotus*) are monochromats, and both sexes of howler monkeys are trichromats because they lack the polymorphism in this allele.

glandular secretions, and when necessary, aggressive behavior (Moynihan 1964). The diet includes fruit, flowers, resins, insects, and small vertebrates.

Family Pitheciidae

The family Pitheciidae includes the New World titi monkeys, saki monkeys, and uacaris. There are 28 species in the genus *Callicebus* (titis), 2 *Cacajao* species (uacari), and 10 species of saki monkeys in two genera (*Chiropotes* and *Pithecia*). Titi monkeys inhabit much of South America, from southern Colombia and Venezuela to Ecuador, Brazil, and northern Paraguay. The number of known *Callicebus* species has increased in recent years, thanks to discoveries in the Amazon basin. Groves (2005) recognized 28 species, not including the recently discovered species *Callicebus aureipalatii* from western Bolivia. (An online casino won an auction for the naming rights to this species, resulting in the unfortunate common name GoldenPalace.com Monkey. The $650,000 paid by the casino was used to help maintain Madidi National Park where the species was discovered.)

Titi monkeys have long reddish to black fur. The tail is long, furry, and not prehensile. Uacaris have tails (15–18 cm) less than half of their head and body length (40–45 cm; Nowak 1999). Their bodies are also covered with long hair, but their heads and faces are bald. Coloration varies with geography from the whitish pelage and shockingly scarlet-red face of *Cacajao calvus* in northwestern Brazil to the dark brown fur and black face of the Venezuelan *C. melanocephalus* (Emmons 1990). Saki monkeys of genus *Pithecia* are smaller than uacaris and have long, bushy tails. Their fur ranges from black to reddish brown depending on the species. The faces may be naked or ringed with fur, and there is a prominent hood of fur around the head. Bearded sakis (*Chiropotes*) have a pronounced beard of long hair on the jaws, throat, and neck. This is most prominent in males.

Titis, sakis, and uacaris are all diurnal and arboreal. Most are agile climbers capable of spectacular jumps between trees. They forage in the rain forest canopy for fruits, leaves, flowers, nuts, and the occasional vertebrate prey. Sakis, for example, are known to search tree holes for rodents and bats (Nowak 1999).

Titis live in small family groups comprising parents and their offspring. They defend territories vigorously, often chasing off intruders. Titi monkeys are believed to by monogamous and may mate for life. Frequent grooming and vocalizations help maintain group cohesion. Saki monkeys have similar family groups, but bearded sakis (*Chiropotes*) may live in groups of up to 30 members. Typically, a single infant is born after a gestation period of 150 to 180 days. Titi monkeys are weaned after 5 months, reach adult size after 1 year, and leave the family by the end of the third year.

Members of the family Pitheciidae are relatively long lived (18 to 35 years in captivity), but their low birth rate means populations are slow to recover. Throughout their range, populations are in decline from continued habitat destruction and from bushmeat hunting as human populations increase in remote areas. Consequently, both species of uacaris and the white-nosed saki (*C. albinasus*) are listed as endangered.

Family Atelidae

The family Atelidae includes the howler monkeys, spider monkeys, and woolly monkeys (**Fig. 12-22**). Howler monkeys (*Alouatta*) include 10 species previously included in the family Cebidae. They range in size from 4 to 10 kilograms; males are usually larger than females (Emmons 1990). The tail is long, prehensile, and naked on the ventral surface of the tip. Digits one and two on each hand are separate from and opposable to the remaining digits (Nowak 1999). The long, coarse fur ranges from light brown to black depending on the species. Howler monkeys inhabit a wide variety of forest types from mangroves to cloud forests from southern Mexico to northern Argentina.

Spider monkeys (*Ateles*), the closely related muriqui (*Brachyteles*), and the woolly monkeys (*Lagothrix*) all have long, thin limbs and highly prehensile tails that are much longer than the head and body length. Like howler monkeys, the tip of the tail is naked on the underside and covered in grooves reminiscent of fingerprints. The hands are long and narrow, and the thumbs are highly reduced or absent (Fig. 12-19C). Adult spider monkeys reach an average body weight of 6 kilograms. Woolly monkeys have short fur with a dense under fur layer, whereas spider monkeys tend to have long, coarse hair with no underfur. Coloration varies from a yellowish gray to jet back; the small head and the face is naked. In *Lagothrix* species, the head is large, and the face is black, whereas the two species of muriqui (also called woolly spider monkeys) have bright red faces.

The southern muriqui (*Brachyteles arachnoides*) and the northern muriqui (*Brachyteles hypoxanthus*) are the two largest species of neotropical monkeys (weights are between 12 and 15 kilograms). Both have highly restricted ranges; the northern muriqui is one of the world's most endangered monkeys, with an estimated population size of fewer than 400 individuals (Mittermeier 1987). They occur only in Brazil's southeastern Atlantic coast forests. The seven species of spider monkey (*Ateles*) live in forested regions from Mexico to Bolivia.

Howler monkeys live in groups ranging in size from 4 to 45 individuals. They are diurnal and highly arboreal, moving quadrupedally through the canopy in search of leaves, fruits, and flowers. For dazzling arboreal ability, the Neotropical spider monkey is probably surpassed only

FIGURE 12-22   A howler monkey (*Alouatta* sp., Atelidae).

by the Old World gibbons (Hylobatidae). Most species in the family Atelidae are vocal to some extent, and several species have loud, penetrating calls. Outstanding among these are the howler monkeys, in which the hyoid apparatus is enlarged into a resonating chamber. The males emit loud roaring sounds that carry for long distances through the tropical rain forest. These sounds seem important in maintaining the cohesiveness of the troop. In addition, the territories of howler troops are probably announced and partly maintained by the loud vocalizations.

Spider monkeys live in large groups; up to 100 individuals may congregate at certain times, but groups of fewer than 25 spider monkeys are more common. Even these groups may further subdivide in a fission–fusion social system similar to that found in chimpanzees (*Pan troglodytes*). Symington (1988) proposed that the unusual fission–fusion social system of red-faced spider monkeys (*Ateles paniscus*) and chimpanzees was the result of spatially and temporally patchy food resources. In most species of the Atelidae, males are **philopatric** (tending to stay in their natal area), and females disperse to join new groups when

they reach adolescence (5 to 7 years old). Spider monkeys mate year round. A single offspring is born after a gestation period of up to 232 days for spider monkeys (approximately 180 days for howler monkeys). Until they are able to keep up with the group on their own, at about 1 year of age, the young are carried by their mothers. Males do not provide parental care. Older males of the Venezuelan red howler (*Alouatta seniculus*) troops kill infants (infanticide) when these males newly enter the group or when the male dominance hierarchy is reshuffled (Crockett & Sekulic 1984; Crockett & Rudan 1987).

Spider monkeys may move as much as 5 kilometers in a day in search of fruits and have home ranges of up to 4 square kilometers. Because they require such large areas of forest, their populations are declining in many parts of South America in response to habitat destruction. In addition, in many areas, spider monkeys and howler monkeys are hunted for food and for export in the pet trade. Three species of howler monkey, one species of spider monkey, both species of muriqui, and the yellow-tailed woolly monkey (*Oreonax flavidauda*) are endangered.

Catarrhini

The Old World catarrhines are the most successful primates in terms of number of species (153 Recent species in 29 genera). They include the Old World monkeys (family Cercopithecidae), the gibbons and siamang (Hylobatidae), and the Hominidae (orangutans, gorillas, bonobo, chipanzees, and humans). The name **catarrhine** derives from the structure of the nose, with its narrow septum and downward facing nostrils. Living nonhuman catarrhines occupy much of Africa and Asia, including Gibraltar and northeastern Africa, sub-Saharan Africa, the southern Arabian Peninsula, and much of southern and southeastern Asia. Catarrhine primates diverged from the New World platyrrhine primates approximately 38 to 40 million years ago.

Family Cercopithecidae

Harrison (1987) listed 21 cranial, postcranial, and dental characters that define the monophyletic cercopithecid clade. Cercopithecids first appear as fossils in the Miocene of Africa, but older relatives (extinct cercopithecoid families) date back to the Eocene. As in many other groups, some Pleistocene cercopithecids reached large sizes; an extinct South African baboon reached the size of a gorilla. There are two distinct subfamilies: Cercopithecinae and Colobinae. Cercopithecines include 73 species in 11 genera, including mangabeys, guenons, vervet monkeys (**Fig. 12-23A**), macaques, and baboons. Members of this subfamily are largely omnivorous and have simple stomachs, cheek pouches, and ischial callosities (patches of bare, thickened skin on the rump). In contrast, the 59 species in the subfamily Colobinae lack both ischial callosities and cheek pouches. Colobines include colobus monkeys (Fig. 12-23B), snub-nosed monkeys, and langurs. Among non-hominid primates, some cercopithecids have the greatest tolerance for cold climates where winter snows occur; some occupy high forests in Tibet, and others live in northern Honshu, Japan.

Many cercopithecids show **sexual dimorphism** in body size, coloration, or canine size. In weight, cercopithecids

FIGURE 12-23   (A) A vervet monkey (*Chlorocebus pygerythrus*, Cercopithecidae) from East Africa and (B) a colobus monkey (*Colobus* sp., Cercopithecidae).

range from 1.5 to over 50 kilograms, and some species are stocky in build. The braincase is large, and the rostrum is often short (**Fig. 12-24**); in some species the skull is robust and heavily ridged, and in the baboons, the rostrum is long (**Fig. 12-25**). The dental formula is 2/2, 1/1, 2/2, 3/3 = 32, as in the hylobatids and hominids. The medial upper incisors are often broad and roughly spoon shaped; the upper canines are usually large and, in many species, tusk like. When the jaws are closed, the lower canine rests in a diastema between the upper canine and the last incisor (Fig. 12-25). The first lower premolar is enlarged and forms a shearing blade that rides against the sharp posterior edge of the upper canine. Most of the molars have four cusps, the outer pair connected to the inner pair by two transverse ridges, producing a bilophodont tooth. The last lower molar has an additional posterior cusp, the hypoconulid (**Fig. 12-26**).

All of the digits have nails, and the pollex and hallux are opposable, except in the strongly arboreal, leaf-eating genus *Colobus*, in which the pollex is vestigial or absent. The tail is vestigial in some species but long in others. Ischial callosities are well developed in many species, are often red, and are used in conjunction with ritualized postures as a means of communication between members of a social group. Bare facial skin may also be red but is bright blue and red in the mandrill (*Mandrillus sphinx*). These patches of skin are more brightly colored in the male than in the female in most species. The olfactory epithelium is greatly reduced in cercopithecids, and apparently their sense of smell is rudimentary. The facial muscles are well developed and produce a wide variety of facial expressions. Some cercopithecids are brightly or conspicuously marked. For example, the variegated langur of Indochina (*Pygathrix nemaeus*) has a bright yellow face, a chestnut strip beneath

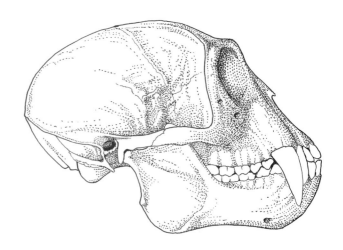

Vervet monkey
(*Cercopithecus aethiops*, Cercopithecidae)

FIGURE 12-24   Skull of a male vervet monkey (*Cercopithecus aethiops*, Cercopithecidae). Note the large canines (length of skull 110 millimeters).

the ears, black and chestnut limbs, a gray body, and a white rump and tail.

Although most cercopithecids are probably largely omnivorous, some are adapted to an herbivorous diet. Members of the subfamily Colobinae (the arboreal langurs and colobus monkeys) are herbivorous and frugivorous, and some species feed primarily on leaves. Colobus monkeys (*Colobus* and *Procolobus*) rely on a multichambered stomach containing colonies of cellulose digesting bacteria, allowing them to forage exclusively on leaves (Bauchop 1978). The baboons (*Papio*, *Theropithecus*) are the most successful terrestrial cercopithecids, and one species, in

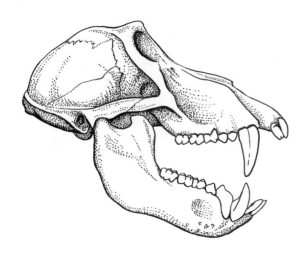

Baboon
(*Papio* sp., Cercopithecidae)

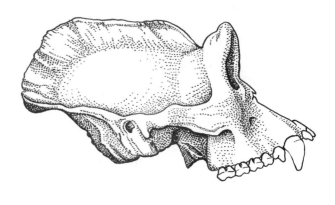

Gorilla
(*Gorilla gorilla*, Hominidae

FIGURE 12-25   Skulls of a baboon (length of skull 200 millimeters) and a gorilla (length of skull 320 millimeters).

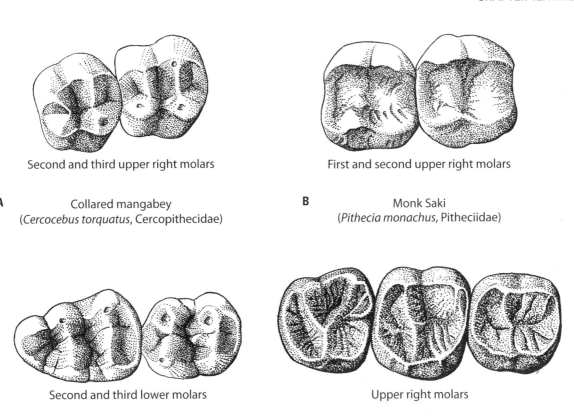

Second and third upper right molars

A   Collared mangabey
(*Cercocebus torquatus*, Cercopithecidae)

First and second upper right molars

B   Monk Saki
(*Pithecia monachus*, Pitheciidae)

Second and third lower molars

C   Hamadryas baboon
(*Papio hamadryas*, Cercopithecidae)

Upper right molars

D   Orangutan
(*Pongo pygmaeus*, Hominidae)

FIGURE 12-26   Cheek teeth of anthropoid primates. Note the cross lophs on the teeth of the mangabey and the baboon and the extra posterior cusp (hypoconulid) on the third lower molar of the baboon. (Orangutan adapted from Clark, W. E. L. *The Antecedents of Man*. Quadrangle Books, 1971.)

areas where suitable trees are not available, assembles on cliffs in large groups of up to 750 individuals (Kummer 1968). Baboons are known to kill and eat small mammals opportunistically, including young gazelles, hares, and even vervet monkeys (Strum 1981). These predatory events are relatively uncommon, and the bulk of their diet consists of fruits, roots, tubers, seeds, and leaves.

Interesting contrasts between the behavior of the baboons and that of the equally terrestrial patas monkey (*Erythrocebus patas*) were discussed by Hall (1968). Savanna baboons are highly vocal, live in fairly large troops controlled by several dominant males, and are prone to noisy, rough, aggressive interactions. In contrast, the patas monkey usually maintains "adaptive silence" but has a repertoire of soft calls, lives in small troops, each with a single adult male that serves as a sentry, is rarely aggressive, and never fights. The patas monkey has a slim greyhound-like build and is the fastest runner of all primates, having been timed at a speed of 55 kilometers per hour (more than twice as fast as Olympic sprinters). Adaptations for speed in this animal include elongation of the limbs, carpals, and tarsals; shortening of the digits; reduction of the pollex and

hallux; and the development of palmar and plantar pads. This remarkably cursorial primate has a quiet mode of life. It usually attempts to escape detection and depends on its speed to escape danger. In these respects, the patas monkey is the primate counterpart of the small antelopes. The noisy baboon troop, however, frequently depends on its aggressive dominant males to confront and discourage a predator, and terrestrial locomotion is less important than in the patas as a means of escape.

Although they remain excellent climbers, baboons that live in savannas and pursue an almost entirely terrestrial life are often vulnerable to attack by large predators. Food is frequently scattered, and a troop must forage over wide areas, thus increasing the chance of encounters with predators. A large and tightly organized social group has evolved in these baboons, perhaps largely in response to this pressure. These groups include from about a dozen to over 150 individuals. Each group occupies a largely exclusive home range; although home range boundaries are usually respected, they are seemingly not defended. When a group is moving, males quickly respond to threats from any quarter, and their united action provides the primary defense of the troop. Baboons are aggressive thieves and can do severe damage to unguarded safari camps (**Box 12-2**).

Sexual dimorphism is pronounced in baboons and enhances the male's intimidating appearance, as well as his fighting ability. Male baboons are about twice as large as females, are more powerfully built, and have comparatively huge canines. A mane of long fur over the crown, neck, and shoulders of the male accentuates the impression of size (**Fig. 12-27**).

The mating pattern of baboons seems to have a consistent relationship to dominance ranking. Whereas subadult, juvenile, and less dominant males copulate with females in the early stages of the estrous cycle, dominant males have exclusive rights to females during the period of maximal sexual swelling (the time when ovulation occurs). In some groups, only the highest ranking male copulated with females during the height of the swelling, and in a group observed by DeVore (1965) in Kenya, not one dominant male attempted copulation until the swelling was at its peak.

◼ Family Hylobatidae

This family includes 14 species of Southeast Asian gibbons in four genera: *Bunopithecus*, *Hylobates*, *Nomascus*, and *Symphalangus* (Groves 2005). Sometimes called lesser apes, gibbons share several features with apes (family Hominidae) and with cercopithecid monkeys. The hylobatid fossil record is scanty, and fossils are known only from the Miocene to the Recent of southeastern Asia (Thorington & Anderson 1984). There is no consensus concerning the ancestry of modern gibbons (Szalay & Delson 1979). Andrews and Martin (1987) listed 11 morphological characters shared by extant gibbons and apes, and molecular data also support a close relationship between hylobatids and hominids.

Gibbons are small, weighing 4 to 11 kilograms, and lack the sexual dimorphism in size found in some other primates. They retain the same dental formula as cercopithecids and have bunodont molars and prominent canines. The most notable feature of gibbon morphology is the limb proportions (**Fig. 12-28**). They have extremely elongate forelimbs, modified for **brachiation** (arm-over-arm swinging), with forelimbs over twice the length of the body. Hindlimbs are also elongate and are nearly one and a half times the body length. The digits on the hand and feet are long and slender, and there is a deep cleft between the first and second digits. The pollex and hallux are strongly opposable. Ischial callosities are present, but gibbons lack an external tail and cheek pouches.

Gibbons occur throughout tropical regions of southeastern Asia from Bangladesh to Vietnam and south to Malaysia, Sumatra, Borneo, Java, and neighboring islands. They prefer tropical forests with well-developed canopies that facilitate brachiation. Gibbons feed principally on fruit, leaves, and flowers (Chiver 1974; Fleagle 1976; Leighton 1986). Gibbons live in small family groups that defend territories by loud vocalizations. Once believed to mate for life, recent evidence suggests that gibbons form long-term pair bonds, but extra-pair copulations may occur (Palombit 1994). Gibbon family groups average four individuals, usually including an adult male and female pair, an infant, and an adolescent or subadult.

The calm of the predawn hours of many Southeast Asian tropical forests is often shattered by piercing gibbon choruses (Fig. 12-28). So loud are these dawn songs that they can be heard several kilometers away. Typically, the male begins his song before sunrise, and the song becomes increasingly more complex toward sunrise. Females often wait until after sunrise to begin singing. Female songs begin with long notes that become increasing shorter and closer together until they climax in a series of short barks that eventually trail off (Geissmann 2002). In some species, males and females sing while swinging rapidly back and forth between canopy branches. Some gibbon species sing only solos, whereas monogamous pairs of other species

BOX 12-2   Habitat and Baboon Social Behavior

Habitat differences between savanna-dwelling baboons and desert-dwelling baboons are also associated with certain differences in social behaviors. Group size, for example, differs markedly. Savanna baboons depend on food patches that are typically rich enough to support large social groups: in Amboseli National Park of Kenya, groups contained an average of 51 individuals (Altmann & Altmann 1970). Desert-dwelling baboons, on the other hand, have a relatively unpredictable and sparse food supply, and foraging groups are small, from several animals to perhaps 20. Seemingly because the two habitats differ markedly in resources, social organization differs between these baboons. Savanna baboons form only one type of social unit, the group, which includes many females, their young, and multiple mature males. The male–female pair bond is brief, lasting only a few hours or days, during the female's estrous period. Desert baboons, in sharp contrast, have four levels of social organization (Kummer 1968, 1984). The smallest and most tightly knit unit is the family group, consisting of a single mature male, one to several adult females, their young, and often a young adult bachelor male "follower." Several such family groups band together to form a less tightly knit foraging unit called the clan, and a number of these clans form a fairly stable traveling unit called a band, which usually includes some 60 baboons. Many bands tolerate each other in order to sleep in safety on the same cliff at night; this loose aggregation is called a troop and contains several hundred baboons.

FIGURE 12-27 An olive baboon (*Papio anubis*, Cercopithecidae) from Kenya.

combine specific song phrases into coordinated duets. Geissmann (2002) suggested that the common ancestors of living gibbons sang male–female duets, but that temporally separated solo songs evolved secondarily.

Gibbon songs may serve to maintain spacing among groups, attract or retain mates, defend territories or food resources, and strengthen the monogamous pair bond. Field studies show that the song frequency of mated males increases when population densities are high. Where densities are low and there are few other unmated males nearby, males sing only rarely or not at all (MacDonald 2001).

Gibbon populations throughout southern Asia have declined due to habitat loss from agriculture and logging and to the illegal pet trade. Currently, all 14 species are listed as endangered or critically endangered (International Union for Conservation of Nature and Natural Resources 2008).

Family Hominidae

The family Hominidae has until recently included only one living member, *Homo sapiens*, with the great apes being assigned to a separate family, Pongidae; however, recent data indicate a relationship between humans and chimpanzees so close that a redefinition of the Hominidae is warranted (Groves 2005). Humans now occur worldwide, but the great apes are restricted to isolated habitats in equatorial Africa (*Pan* and *Gorilla*) and in Borneo and Sumatra (*Pongo*).

Living hominids include two species of *Gorilla* (**Fig. 12-29**), two species in the genus *Pan* (chimpanzee and bonobo), two species of orangutan (*Pongo*), and *Homo sapiens*.

The oldest fossil anthropoids likely to be near the ancestry of hominids are from the Early and Middle Eocene of northern Africa (*Algeripithecus*; Algeria) and southern Asia (*Amphipithecus*, *Pondaungia*, and *Eosimias*; Myanmar, China, Oman) (McKenna & Bell 1997; Takai & Shigehara 2004). Fossils of primitive hominids include *Kamoyapithecus* from the Late Oligocene and *Dryopithecus* from the Early Miocene in Africa; the closest fossil relative of *Pongo* is *Lufengpithecus* from the Middle Miocene of Thailand (Chaimanee et al. 2003). *Gorilla* as yet has no fossil record, although recently the 10 to 11 million year old *Chororapithecus* from Ethiopia has been proposed as a member of the gorilla clade (Suwa et al. 2007). The oldest hominids (*Sahelanthropus* in Chad, *Orrorin* in Kenya, and *Ardipithecus* in Ethiopia) appeared in the Late Miocene in northern Africa 5 to 7 million years ago; of these, *Ardipithecus* probably gave rise to *Australopithecus* by the Pliocene, 4 million years ago (Semaw et al. 2005; White et al. 2005). In East Africa, the oldest tools and cut marks on bone made by early humans (possibly *Australopithecus*) are 2.6 million years old, just slightly older than the oldest known fossils of *Homo*, descendant of *Australopithecus*, at 2.3 to 2.4 million years ago (Semaw et al. 2003). The oldest chimpanzee (*Pan*) fossils were recently discovered in the middle Pleistocene of Kenya (where they were associated with an extinct species of *Homo*) and are between 545,000 and 284,000 years old (McBrearty & Jablonski 2005). Anatomically modern human fossils, *Homo sapiens*, are first known in Ethiopia about 154,000 to 160,000 years ago (Clark et al. 2003; White et al. 2003). It is generally believed that chimpanzees and humans began to diverge from one another approximately 7 million years ago. Molecular evidence suggests that the *Pan-Homo* (Tribe Hominini) split may have taken 4 million years to complete (Patterson et al. 2006).

The living great apes (excluding humans) vary from 48 to 270 kilograms in weight and have robust bodies and powerful arms. The hands and feet are similar to those of humans, but the hallux is opposable (Fig. 12-19D). The elongate skull is typically robust and in older animals is marked by bony crests and ridges (Fig. 12-25). The dental formula is 2/2, 1/1, 2/2, 3/3 = 32, as in cercopithecids and humans. The incisors are broad, and the premaxillae and anterior parts of the dentary bones are broadened to accommodate them. The canines are large and stoutly built. The upper molars are quadrangular and usually four cusped, and the lower molars have an additional posterior cusp (hypoconulid). In contrast to cercopithecids, a trend toward elongation of the molars does not occur in great apes, and the molars lack well-defined cross ridges (Fig. 12-26). The

FIGURE 12-28 A pair of siamang gibbons (*Symphalangus syndactylus*, Hylobatidae) giving their dawn chorus.

tooth rows are parallel, and the mandibular symphysis is braced by a bony shelf (the "simian shelf").

The forelimbs are longer than the hindlimbs in living great apes and they lack tails. The hands are longer than the feet, and all digits bear nails. The thorax is wide, and the scapula has an elongate vertebral border. Adaptations allowing advantageous muscle attachments during erect or semierect stances include lengthening of the pelvis and enlargement and lateral flaring of the ilium. Regarding structural details, locomotor ability, molecular genetics, brain size, and level of intelligence, the great apes are closer to humans than are any other mammals.

The great apes are largely vegetarian, but some chimpanzees (*Pan*) are occasionally carnivorous. Arboreal locomotion in chimpanzees and *Pongo* (orangutan) involves brachiation. Although the gorilla (*Gorilla*) and chimpanzee are able climbers, both are capable of a bipedal stance and limited bipedal locomotion. The behavior of hominids and cercopithecids has been studied intensively in recent years (Fleagle 1998; Fleagle et al. 1999; Kappeler & Pereira 2003; Strier 2002).

The great apes (Hominidae) have social systems that display no radical departures from basic primate patterns, but there are some unique features. Groups of the mountain gorilla (*Gorilla gorilla*) include from 2 to 30 animals. Social interplay between individuals is generally amiable, and assertions of dominance are modest (Fossey 1972; Schaller 1963, 1965a, 1965b). Particularly notable is the age-graded troop, with the nucleus of the group consisting of the dominant silver-backed male (10 years of age or older) and adult females and their young. Additional males, including less dominant silver-backs and black-backed males, attach themselves to the periphery of the group.

The chimpanzee (*Pan troglodytes*) has been the subject of considerable field observation by a number of workers, including Izawa and Itani (1966), van Lawick-Goodall (1968, 1973), Izawa (1970), Nishida and Kawanaka (1972), Sugiyama (1973), Goodall (1983, 1986), and Stanford (1995). The basic social unit of chimpanzees is an often dispersed group of 30 to 80 animals that show considerable fidelity to a large home range. Particularly unusual is the looseness of the social organization, with intricate

FIGURE 12-29   A gorilla (*Gorilla gorilla*, Hominidae).

patterns of establishment and dissolution of small parties. Highly evolved visual, tactile, and vocal communications are used. When a party of chimpanzees finds trees bearing fruit, their almost manic vocalizations and actions attract other parties to the bonanza. Male chimpanzees are capable of quick and coordinated hunting behavior (Stanford 1995; Stanford et al. 1994; Teleki 1973). Typically, chimps hunt opportunistically, such as when they find a baboon or colobus monkey that has become separated from its troop. In this situation, some males will act to block its main path back to the troop while others block off other escape routes. Chimpanzee hunting may be more common than was previously thought. Stanford (1995) estimated that Gombe chimps consume approximately 1 ton of meat annually. Lower ranking males appear to barter meat for sex. Chimpanzees are clearly not always gentle and benign: On several occasions males of one group have been observed systematically killing males of another group. Also, chimps occasionally kill and eat baby chimpanzees, but this behavior is considered aberrant (Goodall 1986).

The closely related bonobo (*Pan paniscus*) has evolved a highly unusual complex of social behaviors (de Waal 1995). Bonobo social systems are based on sexual contacts between individuals and include such behaviors as mouth-to-mouth "kissing," genito-genital rubbing between adult females, and pseudocopulation between adult males (de Waal 1989, 1995). These behaviors appear to have evolved to minimize conflicts among group members. Bonobos also spend more time using bipedal gaits than do chimpanzees. They inhabit the lowland swamp forests of the Democratic Republic of Congo, where they feed on fruits and other plant material; unlike chimpanzees, they are not known to hunt other mammals (White 1996).

Orangutans (*Pongo*) include two species native to the rainforests of Borneo (*Pongo pymaeus*) and Sumatra (*Pongo abelii*). The orangutans of Sumatra differ from the Bornean species in several characteristics; Sumatran orangutans tend to be thinner and have paler orange pelage, and adult males have smaller cheek pads than the orangutans of Borneo. Both orangutan species have coarse, long hair and long arms. Infants are born with pink faces, but the pigmentation darkens with age. Males have large, pendulous throat pouches and cheek pads. Orangutans are highly sexually dimorphic; males weigh, on average, 87 kg (192 lb) compared with an average of only 37 kg (81.6 lb) for females (Markham & Groves 1990). Orangutans are highly arboreal and only reluctantly come to the ground.

Humans (*Homo sapiens*) are the only living members of the genus *Homo*. In humans, the skull has a greatly inflated cranium, housing a large cerebrum, and the rostral part of the skull is virtually absent. The foramen magnum is beneath the skull, a feature associated with an upright stance. The dentition is not as robust as in their fellow hominids: the incisors are less broad, the canines typically rise but slightly above adjacent teeth, and the cheek teeth are less heavily built. The premolars are usually bicuspid. The upper molars have four cusps, the first lower molar has five cusps, the second has four, and the third has five. The dental formula 2/2, 1/1, 2/2, 3/3 = 32 occurs in most individuals, but one or more of the posterior most molars (the "wisdom teeth") may not appear. The tooth rows are not parallel, as they are in great apes, nor is the simian shelf present in the mandible. The pollex is opposable, but the hallux is not. With a change in posture and use of the forelimbs, the thorax has become broad, and the scapulae have come to lie dorsal (or posterior) to the rib cage, as in bats, rather than lateral to the rib cage, as in most other mammals. As in many primates, human males are significantly larger than the females. Molecular data indicate that humans are more closely related to chimpanzees than they are to either orangutans or gorillas (Miyamoto et al. 1987).

Primarily because of the efforts of humans, the great apes are dangerously close to extinction. Destruction of habitat and killing of animals for food has led to the serious reduction of some populations. The International Union for Conservation of Nature and Natural Resources classifies all species of great apes as either endangered or critically endangered. Wild populations are in decline from bushmeat hunting, deforestation for agriculture and timber, and disease. In recent years, chimpanzee and gorillas, like many other African primates, have the added misfortune of living in areas ravaged by civil war (Plumptre 2003). The sudden influx of soldiers or human refugees increases habitat destruction and poaching in war-torn areas.

SUMMARY

Primates are a diverse group, with 69 genera and 376 living species worldwide, of which 16 genera and 128 species occur in the western hemisphere. Most are tropical and arboreal. Primates range in size from the 50- to 70-gram mouse lemurs to the 175-kilogram mountain gorilla.

Primates, colugos (Dermoptera) and treeshrews (Scandentia) are believed to form a monophyletic clade, the Euarchonta. Ancestral primate fossils date from the early Paleocene of North America. Primates of modern aspect (Euprimates) first appeared in the early Eocene.

Living primates are divided into the Strepsirrhini and the Haplorrhini. Extant strepsirrhine primates include the five lemur families endemic to Madagascar, the Galagidae of Africa, and the Lorisidae of Africa and Southeast Asia. Malagasy lemurs are very diverse; they include insectivorous and herbivorous species, nocturnal and diurnal species, and species that are solitary and highly social. The most unusual lemur is the aye-aye.

Haplorrhine primates are more diverse and widely distributed. In addition to the tarsiers, the living haplorrhines include the four New World and three Old World families. The New World Platyrrhini include the marmosets, tamarins, capuchins, squirrel monkeys, night monkeys, howler monkeys, and spider monkeys, among others. The Old World Catarrhini assemblage includes three families. The Cercopithecidae comprise the macaques, baboons, mangabeys, and many others. The Hylobatidae include the gibbons and siamang, whereas the Hominidae includes the orangutans, gorillas, chimpanzees, bonobos, and humans.

Most primates inhabit tropical or subtropical regions. They are primarily folivorous, frugivorous, or insectivorous. Some species are omnivorous, and chimpanzees, baboons and humans hunt and consume meat. Most primate species retain an arboreal lifestyle characterized by adaptations such as stereoscopic vision, opposable hallux and pollex, and mobile radius and ulna in the forelimb for a dextrous hand and movement in the trees. Several species, including humans and some baboons, are primarily terrestrial.

KEY CHARACTERISTICS OF PRIMATES

Primates radiated in arboreal habitats, and most of the characteristics of the group are arboreal adaptations.
- Grasping hands and feet with opposable thumbs and toes
- Nails present on all or most digits (may be lost secondarily)
- Calcaneum elongated
- Orbits rotated forward, interorbital narrowed, and orbital cavity enlarged
- Ethmoid bone exposed on the inner orbital wall
- Stereoscopic vision (retinal axons project to both ipsilateral and contralateral sides of the brain)
- Increased fetal brain size compared to fetal body weight, retained in neonates
- Sylvian sulcus and triradiate calcarine sulcus both present on brain surface
- Testes descend early in life
- Urogenital sinus absent in females
- Long gestation times relative to body size
- Sexual maturity relatively late in life
- Auditory bulla bony, the floor derived from petrosal bone
- Extension of the ectotympanic into the auditory meatus
- Loss of one incisor and one premolar from the ancestral eutherian condition

KEY CHARACTERISTICS OF STREPSIRRHINI

- Lower incisors that form a tooth comb used in grooming (may be secondarily modified)
- Rhinarium naked and moist, with unfused nasal prominences, median cleft, and comma-shaped nostrils

KEY CHARACTERISTICS OF HAPLORRHINI

- Rhinarium dry, nostrils oval, and no median cleft
- Hemochorial placenta
- Fovea centralis in the retina
- Postorbital plate present

KEY TERMS

Brachiation
Catarrhine
Chimerae
Dichromatic
Emigration

Epitheliochorial placenta
Fovea centralis
Frugivorous
Hemochorial placenta
Ischial callosity

Monogamous
Omnivorous
Philopatric
Sexual dimorphism
Trichromatic

RECOMMENDED READINGS

Fleagle, JG. 1999. *Primate Adaptation and Evolution.* Academic Press, NY.

Fleagle, JG, C Janson, & K Reed. 1999. *Primate Communities.* Cambridge University Press, Cambridge, UK.

Goodall, J. 1986. *Chimpanzees of Gombe.* Harvard University Press, Cambridge, MA.

Groves, C. 2001. *Primate Taxonomy.* Smithsonian Institution Press, Washington, DC.

Hershkovitz, P. 1977. *Living New World monkeys (Platyrrhini), With an Introduction to the Primates,* vol. 1. University of Chicago Press, Chicago, IL.

Kappeler, PM & ME Pereira. 2003. *Primate Life Histories and Socioecology.* University of Chicago Press, Chicago, IL.

Kay, RF, et al. 2004. Anthropoid origins: a phylogenetic analysis. Pp 91-135, in *Anthropoid Origins: New Visions* (CF Ross & RF Kay, eds.). Kluwer/Plenum, NY.

Ross, CF & RF Kay. 2004. *Anthropoid Origins: New Visions.* Kluwer/Plenum Publishing, NY.

Silcox, MT, et al. 2005. Euarchonta (Dermoptera, Scandentia, Primates). Pp. 127–144, in *The Rise of Placental Mammals* (KD Rose & JD Archibald, eds.). Johns Hopkins University Press, Baltimore, MD.

Yoder, AD. 2003b. Phylogeny of the lemurs. Pp. 1242–1247, in *The Natural History of Madagascar* (SM Goodman & JP Benstead, eds.). University of Chicago Press, Chicago, IL.

Rodentia and Lagomorpha

Order Rodentia

Because plants are the most abundant food source for terrestrial mammals, it is not surprising that most members of the largest mammalian order, Rodentia, are herbivorous. Approximately 42% of all mammals are rodents, with some 33 living families, 481 genera, and roughly 2,277 species (Wilson & Reeder 2005). Rodents have been, and remain today, spectacularly successful: They are nearly cosmopolitan in distribution, exploit a broad spectrum of foods, and outnumber all other mammals in terrestrial faunas (both as species and as individuals). Repeated rodent radiations have occurred at many times in many places. As a result of convergent evolution, many rodents have lifestyles and morphological features that are similar to those of members of other orders. Among rodents that radiated in South America, for example, are species that resemble rabbits or small antelopes and one species that fills the ecological role of a miniature hippopotamus.

Rodents are an unusually complex group with respect to morphological diversity, lines of descent, and parallel evolution of similar features in different evolutionary lines. Wood (1935) regarded parallelism as "the evolutionary motto of the rodents in general." Because of these complexities, disagreement among zoologists as to relationships among rodent taxa has been the rule (Carleton & Musser 2005; Luckett & Hartenberger 1985). For example, the terms "sciuromorph," "myomorph," and "hystricomorph" refer to basic patterns in the arrangement of the masseter muscles, the skull, and the zygomatic arch and have been used repeatedly to designate major divisions within the order Rodentia. There has been little agreement, however, regarding the use of these terms in the formal taxonomy of rodents.

The classification used here follows that of Carelton and Musser (2005), who recognized five main suborders. The suborder Sciuromorpha includes the mountain beaver, squirrels, and dormice. The beavers, pocket gophers, and kangaroo rats and their relatives are placed in the suborder Castorimorpha. The Myomorpha includes the jerboas and a diverse group of muroid rodents, whereas the scaly-tailed flying squirrels and the springhare are placed in the Anomaluromorpha. Finally, the suborder Hystricomorpha includes porcupines, cavies, and mole-rats, among others (**Fig. 13-1**).

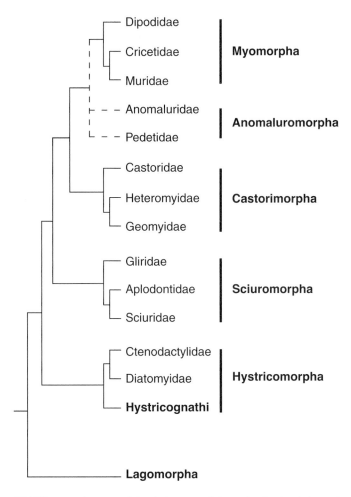

FIGURE 13-1 One possible phylogeny of the order Rodentia based on combined morphological and molecular evidence. Dashed lines indicate uncertain relationships. (Data from Huchon, D, et al., *Proc Natl Acad Sci USA* 104 (2007):7495–7499.)

Morphology

Rodents range in size from about 5 grams to 50 kilograms, and Recent members of the order Rodentia share a series of distinctive cranial features (**Table 13-1**). The upper and lower jaws each bear a single pair of persistently growing incisors derived from the deciduous second incisors (dI2/2) (Carleton & Musser 2005). Enlarged, ever-growing incisors developed early in the evolution of rodents and committed them to an essentially herbivorous mode of feeding, while still permitting the exploitation of such abundant foods as insects. Because only the anterior surfaces are covered with enamel, the incisors assume a characteristic beveled tip as a result of wear. Gnawing is a unique behavior that requires the separation of the upper and lower cheek teeth while the incisors are engaged and the disengagement of the upper and lower incisors when the cheek teeth are occluded. The occlusal surfaces of the cheek teeth are often complex and allow for effective sectioning and grinding of plant material. In some rodents, the cheek teeth are ever growing. The dental formula seldom exceeds 1/1, 0/0, 2/1, 3/3 = 22 (except in the families Bathyergidae and Ctenodactyidae), and a diastema is always present between the incisors and the premolars. The incisors and canines are always 1/1, 0/0.

The glenoid fossa of the squamosal bone is elongate and allows anteroposterior and transverse jaw action. The mandibular symphysis has sufficient give in many species to enable the transverse mandibular muscles to pull the ventral borders of the rami together and spread the tips of the incisors. The masseter muscles are large and complexly subdivided, provide most of the power for mastication and gnawing, and in all but one species (*Aplodontia rufa*), have at least one division that originates on the rostrum. The temporal muscles are usually smaller than the masseters, and their point of insertion, the coronoid process, is usually reduced. In general, rodents have undergone little postcranial specialization. There are notable exceptions, however, among saltatorial, fossorial, and gliding rodents.

Rodents are fascinating, partly because of the very features that make them difficult to classify. Their complex patterns of evolution, different morphological solutions to similar basic functional problems, intricate systems of resource allocation, and finely tuned adaptations to such extreme environments as those in arctic and desert areas make rodents a rewarding group to study.

TABLE 13-1 Uniquely Derived Characters Used to Diagnose the Monophyletic Rodentia

- One pair of upper and lower incisors; each tooth is enlarged, sharply beveled, and ever growing
- Broad diastema (space) between incisors and premolars of both upper and lower jaws resulting from the loss of canines and some cheek teeth
- Incisor enamel restricted to the anterior surface only
- Paraconid lost on lower cheek teeth
- Orbital cavity lying just dorsal to cheek teeth
- Ramus of the zygoma lies anterior to the first cheek tooth
- Glenoid fossa is an anterior–posterior trough allowing fore and aft movement of the mandible

Data from Luckett and Hartenberger (1985).

Paleontology

Rodents are an old group, dating back to the late Paleocene of North America and Asia. The earliest fossils are jaws and scattered teeth, about 57 million years old, representing the primitive families Alagomyidae and Ischyromyidae (Dawson 2003; Dawson & Beard 1996; Meng 2004; Meng et al. 1994; Rose 2006). The structure of the skull of primitive rodents (**Fig. 13-2**) indicates that the temporalis muscle was large, and the masseter muscles were not highly specialized and originated entirely from the zygomatic arch (**Fig. 13-3**). Rodents with a **hystricognathous** masseter morphology first appear as fossils in the late Eocene or early Oligocene in Africa, Asia, and South America (Meng & Wyss 2005). The primitive dental formula is 1/1, 0/0, 2/1, 3/3 = 22, and the cheek teeth were brachydont.

Rodents may have diverged early into two or three major groups. Three suborders (Sciuromorpha, Myomorpha, and Hystricomorpha) are defined by some taxonomists based on the anatomical position of the masseter muscles relative to the infraorbital foramen (described later here). Alternatively, other taxonomists recognize the suborders Sciurognathi and Hystricognathi, names that refer to contrasting types of mandibles. In the **sciurognathous** type (**Fig. 13-4A**), the angular process of the dentary bone originates in the vertical plane that passes through the alveolus of the incisor and is ventral to the alveolus. In the hystricognathous type (Fig. 13-4B), the angular process is not in the same plane as the horizontal ramus (Korth, 1994). In contrast to the sciurognathous dentary, the hystricognathous dentary tends to have a more strongly reduced coronoid process, and its lower border is generally marked by a more prominent projection at the base of the

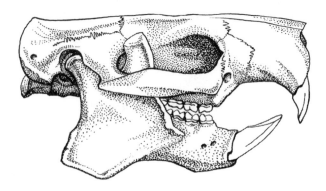

(*Paramys*, Paramyidae)

FIGURE 13-2 Skull of a primitive late Paleocene and early Eocene sciuromorphous rodent (*Paramys* Paramyidae; length of skull 89 millimeters). (Modified from Romer, A.S. *Vertebrate Paleontology*. University of Chicago Press, 1966.)

incisor root. These characters, however, vary widely within each suborder.

The early success of rodents is indicated by their Eocene abundance in Eurasia and North America and by their rapid radiation. In the Eocene, when they were seemingly replacing the ancient and formerly highly successful multituberculates, rodents were abandoning the primitive zygomasseteric arrangement. The time between the late Eocene and middle Oligocene was one of accelerated rodent evolution (Korth 1994; Wible et al. 2005; Wilson 1972), and it was then that the major patterns of jaw muscle specialization typical of modern rodents (Figs. 13-3 and **13-5**) were established. Over half of the living families of rodents appeared by the end of the Oligocene. The Miocene spread of grasslands or savannas in both the Old World and the New World provided new adaptive zones for rodents. The evolution of the jerboas (Dipodidae) in the Old World and the kangaroo rats and pocket mice (Heteromyidae) in the New World was probably decisively affected by increasing aridity in the late Miocene and the appearance of deserts in the Pliocene.

Because of the remarkable diversity and degree of evolutionary parallelism among rodents, phylogenetic relationships are difficult to discern and are controversial. Numerous authors, using both morphological and molecular data, are investigating the relationship among rodents with the result that some broad patterns are emerging: (1) the Rodentia is a monophyletic order (Luckett & Hartenberger 1993; Musser & Carelton 2005; Philippe 1997; Sullivan & Swofford 1997). (2) Lagomorphs are the most likely living sister group of the Rodentia (Douzery & Huchon 2004; Luckett & Hartenberger 1985; Meng & Wyss 2005). (3) Alagomyids including *Alagomys* and *Tribosphenomys* are among the most primitive rodents morphologically and represent a probable sister group to all other rodents (Dashzeveg 1990; Dawson & Beard 1996; Meng 2004). (4) Ctenodactyloids also represent an early, primitive line of rodent evolution (Wible et al. 2005). (5) Similarities among early geomyoids (represented today by the Heteromyidae and Geomyidae) and the Castoridae indicate that these groups may share a common ancestral stock (McKenna & Bell 1997). (6) The Sciuridae, Aplodontiidae, and possibly the Gliridae are each other's closest living relatives (Musser & Carelton 2005). (7) The relationships of the Anomaluridae and Pedetidae to other families are unresolved. (8) The Dipodidae and members of the superfamily Muroidea, including the Cricetidae and Muridae, comprise the suborder Myomorpha (Musser & Carelton 2005). (9) The African gundis, Ctenodactylidae, are related to the Hystricognathi, which originated in Asia and includes a number of closely related groups (Douzery & Huchon 2004; Huchon & Douzery 2001).

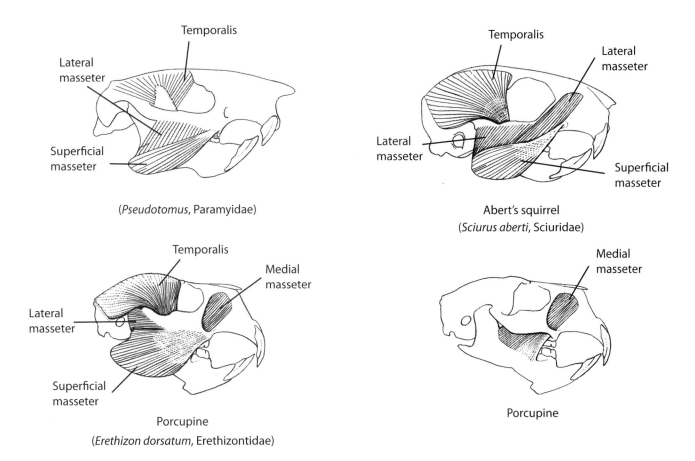

FIGURE 13-3 Patterns of specialization of the skull, jaws, and jaw musculature of rodents. The jaw muscles are restored in *Pseudotomus* (Paramyidae), a primitive Eocene rodent. The masseter muscles originate entirely on the zygomatic arch. In Abert's squirrel (*Sciurus aberti*, Sciuridae), the anterior part of the lateral masseter originates on the rostrum and the zygomatic plate. In the porcupine (*Erethizon dorsatum*, Erethizontidae), the anterior part of the medial masseter originates largely on the rostrum and passes through the greatly enlarged infraorbital foramen. The temporalis muscle, typically reduced in size in hystricomorphous rodents, is unusually large in porcupines.

Jaw Muscle and Skull Specializations

Despite their diversity and success in adapting to contrasting environments and lifestyles, rodents have rather consistently followed certain basic trends in the evolution of the jaw muscles, the bones from which these muscles take origin or insertion, and the teeth. Even in the early stages of their evolution, selective pressures apparently favored forward migration of the jaw muscles.

Since their appearance in late Paleocene times, rodents have had a dentition featuring a division of labor between incisors and cheek teeth. The incisors serve as chisels that gnaw food, clip vegetation, or in some fossorial forms, dig soil and move rocks. These teeth are subject to heavy wear, and they became ever-growing early in the evolution of rodents. The cheek teeth, separated from the incisors by a broad diastema, perform a different function, mastication of food. A complicated jaw movement allows the lower cheek teeth to move transversely or anteroposteriorly against the upper teeth, producing a crushing and grinding action. Not

only are the gnawing and grinding functions performed by different teeth, but they must be performed separately. When the cheek teeth are in position for grinding, the tips of the incisors do not meet; the lower jaw must therefore be moved forward for the incisors to be in position for gnawing.

This division of labor between incisors and cheek teeth clearly "guided" the evolution of the rodent jaw musculature. During gnawing, muscles that attach far forward on the jaw and skull are advantageous because they confer great power on the jaw action through increased mechanical advantage. Furthermore, because forward movement of the lower jaw is required for gnawing, selection probably favored the attachment of some of the jaw muscles far forward on the rostrum; contraction of these muscles caused a forward shift of the jaw. During the grinding of food by the cheek teeth, jaw muscles with mechanical advantage for power are also important, and the complex jaw action associated with grinding demands jaw musculature with precise control over forward, backward, and transverse jaw movement. The jaw musculature and skull specializations

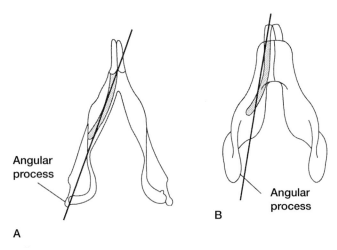

FIGURE 13-4   The sciurognathous and hystricognathous dentary bones. (A) Ventral view of the sciurognathous pattern of *Sciurus niger* with angular process in line with the ramus of the dentary bone (line). (B) Ventral view of a hystricognathous dentary of the porcupine (*Erethizon dorsatum*), showing the angular process located off the plane of the dentary.

to be considered can best be put in functional perspective if the importance of complex grinding movements and powerful forward movements of the rodent jaw are kept in mind.

Not every rodent can readily be classified as **sciuromorphous**, **hystricomorphous**, or **myomorphous**, and as mentioned previously, experts do not always agree on how the types evolved. The myomorphous pattern, for example, may have evolved through a hystricomorphous ancestry (Klingener 1964). The broad trend in rodents was away from the primitive condition, in which the masseter muscles originated entirely on the zygomatic arch, toward

the placement of the origin of at least one division of the masseter on the rostrum. The primitive condition, termed **protrogomorphous** by A. E. Wood (1965), is retained today by only one rodent (*Aplodontia rufa*, Aplodontiidae).

Presumably, as a result of competition among the rapidly diversifying rodents, in the late Eocene, the skull and jaw musculature were altered in some phylogenetic lines in a way that increased the effectiveness of gnawing and grinding. These specializations involved primarily the lateral and medial masseter muscles and their areas of attachment. In some rodents, the insertion of the anterior part of the lateral masseter shifted onto the anterior surface of the zygomatic arch and the adjacent part of the rostrum (Fig. 13-3). This pattern is termed sciuromorphous and occurs in the Sciuridae and Castoridae. In these families, the temporalis muscle is relatively large, and the coronoid process is moderately well developed.

A second pattern of zygomasseteric specialization involves the shift of the origin of the medial masseter muscle from the zygomatic arch to an extensive area on the side of the rostrum. This muscle passes through the often greatly enlarged infraorbital foramen, an arrangement termed hystricomorphous (Fig. 13-3). It occurs in the families Dipodidae, Ctenodactylidae, and Pedetidae and in most families in the suborder Hystricognathi. Although included in the Hystricognathi, members of the African family Bathyergidae are not hystricomorphous as adults: in this family, the infraorbital foramen is reduced and transmits a small slip of the medial masseter only in the embryo (Maier & Schrenk 1987). Lavocat (1973) believed that ancestral bathyergids were hystricomorphous but that they secondarily reduced the infraorbital foramen. In living bathyergids, the large

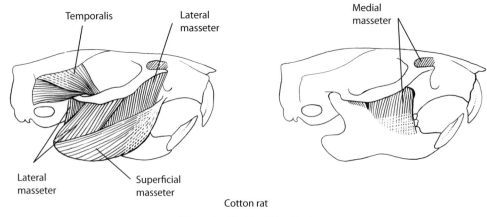

Cotton rat

(*Sigmodon hispidus*, Cricetidae)

FIGURE 13-5   Zygomasseteric pattern in a myomorphous rodent, the cotton rat (*Sigmodon hispidus*, Cricetidae). The superficial masseter originates on the rostrum and the anterior part of the lateral masseter originates on the anterior extension of the zygomatic arch. The superficial muscles have been removed on the right to expose the medial masseter, which originates partly on the rostrum and passes through the slightly enlarged infraorbital foramen. (Adapted from Rinker, G. C., *Misc. Pubs. Mus. Zool.* 83 (1954): 1–124.)

anterior part of the medial masseter originates in the orbit, a condition perhaps permitted by the reduction of the eye in these highly fossorial rodents.

Most rodents use a third type of zygomasseteric specialization, termed myomorphous. In such rodents, the anterior part of the lateral masseter originates on the highly modified anterior extension of the zygomatic arch (the zygomatic plate and zygomatic spine; Fig. 13-5), and the anterior part of the medial masseter originates on the rostrum and passes through the somewhat enlarged infraorbital foramen. The temporalis is typically reduced, and the coronoid process ranges from well developed to vestigial (Fig. 13-5). The suborder Myomorpha includes the huge families Muridae and Cricetidae. Maier et al. (2002) demonstrated that dormice, family Gliridae, have convergently evolved "myomorphy" and that the masseter arrangement in dormice is better described as "pseudomyomorphy."

Suborder Sciuromorpha

Family Aplodontiidae

This family (formerly spelled Aplodontidae) is interesting because of the unique ancestral morphological features that characterize its one extant member, *Aplodontia rufa*, the sewellel or mountain "beaver," an animal restricted to parts of California and the Pacific Northwest. Mountain beavers are roughly the size of a small rabbit and have a robust, short-legged form (**Fig. 13-6**). *Aplodontia* is the only living rodent in which the masseters have an entirely zygomatic origin. The skull is flat, and the coronoid process of the dentary bone is large (**Fig. 13-7**). The cheek teeth are ever growing (a specialized feature) and have a unique crown pattern (**Fig. 13-8**). The dental formula is 1/1, 0/0, 2/1, 3/3 = 22.

The earliest records of aplodontids are from the late Eocene of western North America. The closely related

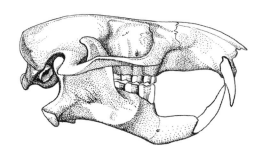

Mountain beaver
(*Aplodontia rufa*, Aplodontiidae)

FIGURE 13-7   Skull of a mountain beaver (*Aplodontia rufa*, Aplodontiidae; length of skull 68 millimeters).

family Mylagaulidae appeared in late Oligocene. These extinct marmot-sized rodents were fossorial, and one or two forms are notable for having prominent nasal "horns." The aplodontids spread later to Europe and Asia but since the middle Pliocene have lived only in the moist, forested parts of the Pacific slope of North America, where their relict distribution today is from central California to southern British Columbia. Widespread late tertiary aridity may have restricted the aplodontids to their present range.

Aplodontia favors moist areas supporting lush growths of forbs and often builds its burrows next to streams. The tunnel systems have multiple exits, and this rodent seldom forages far from its burrows. The diet includes a variety of forbs and the buds, twigs, and bark of such **riparian** plants as willow (*Salix*) and dogwood (*Cornus*). On occasion, *Aplodontia* builds "hay piles" of cut sections of forbs (Grinnell & Storer, 1924). During winter months, mountain beavers may climb shrubs or small trees to cut off branches and twigs.

Family Sciuridae

The successful and widespread family Sciuridae includes 278 living species representing 51 genera. Tree squirrels, chipmunks, marmots (**Fig. 13-9**), and prairie dogs belong to

FIGURE 13-6   A mountain beaver (*Aplodontia rufa*) from Oregon.

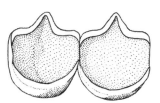

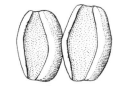

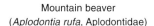

Mountain beaver
(*Aplodontia rufa*, Aplodontidae)

Merriam's kangaroo rat
(*Dipodomys merriami*, Heteromyidae)

FIGURE 13-8   Crowns of first two right upper molars of a mountain beaver (note the simplified and unique crown pattern) and a Merriam's kangaroo rat (note the highly simplified crown pattern). The labial border of the tooth is above; anterior is to the right. The unshaded part is enamel, and the stippled part is dentine.

FIGURE 13-9 (A) Least chipmunk (*Tamias minimus*) in Wyoming. (B) Yellow-bellied marmot (*Marmota flaviventris*) in Colorado.

this family. Sciurids probably evolved from ischyromyid ancestors in the late Eocene of North America. Ground squirrels and tree squirrels are distinguishable from each other by the end of the Oligocene. Sciurids remain widespread today, absent only from the Australian region, Madagascar, the polar regions, southern South America, and certain Old World desert areas.

Sciurids are structurally distinctive. The skull is usually arched in profile (**Fig. 13-10**), and the front of the zygomatic arch is flattened where the anterior part of the lateral masseter rests against it. The dental formula is 1/1, 0/0, 1–2/1, 3/3 = 20–22. The cheek teeth are rooted and usually have a crown pattern that features transverse ridges. Sciurids have relatively unspecialized bodies: a long tail is usually retained, and the limbs seldom have a loss of digits

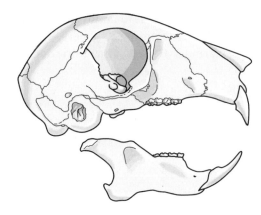

Pygmy squirrel
(*Exilisciurus whiteheadi*, Sciuridae)

FIGURE 13-10 Lateral view of the skull and mandible of a pygmy squirrel (length of skull 28 millimeters). (Modified from Heaney, L. R., *Misc. Pubs. Mus. Zool.* 170 (1985): 1–58.)

or reduction of freedom of movement at the elbow, wrist, and ankle joints. Several semifossorial types, including ground squirrels (*Spermophilus*), prairie dogs (*Cynomys*), and marmots (*Marmota*), have variously departed from this plan in the direction of greater power in the forelimbs and, in some cases, reduction of the tail.

Sciurids are basically diurnal herbivores, but a great variety of food is used. Tree squirrels occasionally eat young birds and eggs; chipmunks (*Tamias*) and the antelope ground squirrels (*Ammospermophilus*) are seasonally partly insectivorous in some areas. Sciurids are tolerant of a great range of environmental conditions. Some, such as marmots, some prairie dogs, chipmunks, and some ground squirrels, hibernate during cold parts of the year.

Styles of locomotion vary among sciurids; the most specialized style occurs in the "flying" squirrels. This group of 15 genera, constituting the tribe Petromyini, is characterized by gliding surfaces formed by broad folds of skin between the forefoot and hind foot (Johnson-Murray 1977; Thorington et al. 1998). These animals usually glide distances of only 10 to 20 meters, but some can glide considerable distances. The giant flying squirrel (*Petaurista*) of southeastern Asia, for instance, can glide up to 450 meters and can bank in midair (Nowak & Paradiso 1991).

■ Family Gliridae

The Gliridae includes 28 species of dormice (formerly the Myoxidae) in nine genera. Dormice are squirrel-like Old World rodents known first from the early Eocene of Europe and Asia. Glirids are an old and distinctive group that was one of the first to branch from the primitive and extinct family Ischyromyidae. Dormice are entirely Old World in distribution, occurring in much of Africa south of the Sahara, England and Wales, Europe from southern Scandinavia southward, Asia Minor, southwestern Russia, southern

India, southern China, and Japan. The distinctive desert dormouse (*Selevinia betpakdalaensis*) is restricted to the Betpak-dala Desert of Russia (and until recently was placed in its own family).

Dormice are small (up to 325 millimeters in length), and most genera have bushy or well-furred tails (**Fig. 13-11**). The skull has a smooth, rounded braincase, a short rostrum, and large orbits. The dental formula is 1/1, 0/0, 1/1, 3/3 = 20. The crowns of the brachydont molars have parallel cross-ridges of enamel; in some cases, the ridges are reduced, and the crowns have basins. The infraorbital foramen is somewhat enlarged and transmits part of the medial masseter muscle, a condition called "pseudomyomorphy" to distinguish it from murid and cricetid myomorphy (Maier et al. 2002). The limbs and digits are fairly short, and the sharp claws are used in climbing. The manus has four toes, and the pes has five. The desert dormouse eats primarily insects and spiders and, as is typical of many desert mammals, has greatly enlarged auditory bullae. Its dental formula is 1/1, 0/0, 0/0, 3/3 = 16; the cheek teeth are small and short crowned, and the much-simplified crown pattern features smooth, concave surfaces.

Glirids are swift and agile climbers that occupy trees and shrubs, rock piles, or rock outcrops. These rodents are omnivorous but are capable little predators that kill small birds and large insects. In temperate areas, the animals hibernate in winter and may have an above-ground activity season of only 4 to 6 months. In autumns with low food availability, males of *Glis glis* enter hibernation with minimal fat reserves and may forego reproducing in the spring in order to replenish their energy stores (Bieber 1998). A

FIGURE 13-11   A forest dormouse (*Dryomys nitedula,* Gliridae).

unique feature of glirids is their ability to lose and regenerate the tail (Mohr 1941).

Suborder Castorimorpha

Family Castoridae

The family Castoridae is represented today by only two species: *Castor canadensis* of the United States and Canada and *Castor fiber* of northern Europe and northern Asia. Beavers had an important role in the history of the North America, for much of the early European exploration of many major river systems in Canada and the western United States was done by trappers in quest of valuable beaver pelts.

The fossil record of beavers dates to the late Eocene of Asia, and beavers reached North America and Europe in the early Oligocene. Several lines of descent developed in the Tertiary. One line developed fossorial adaptations; the fossil remains of the North American Miocene beaver *Palaeocastor* have been found in spectacular corkscrew-shape burrows dug by these animals. Another evolutionary line led to the bear-sized giant beaver (*Castoroides*) of the North American Pleistocene. Throughout their history, castorids have been restricted to the Northern Hemisphere.

Extant beavers are semiaquatic, and several of their distinctive structural features are adaptations to this mode of life (**Fig. 13-12**). The animals are large, reaching over 30 kilograms in weight. Their large size produces a low surface-to-volume ratio that is more advantageous in terms of heat conservation than that of smaller rodents. In addition, the body is insulated by fine underfur protected by long guard hairs. These are important adaptations in animals that frequently swim and dive for long periods in icy water. The large hind feet are webbed. The small eyes have **nictitating membranes** (membranes that arise at the inner angle of the eye and can be drawn across the eyeball), and the nostrils and ear openings are valvular and thus can be closed during submersion.

Two structural specializations allow beavers to open their mouths when gnawing under water, and while swimming, they can carry branches in the submerged open mouth without danger of taking water into the lungs (Cole 1970). The epiglottis is internarial (it lies above the soft palate), an arrangement allowing efficient transfer of air from the nasal passages to the trachea but not allowing mouth breathing or panting. Also, the middorsal surface of the posterior part of the tongue is elevated and fits tightly against the palate and, except when the animal is swallowing, blocks the passage to the pharynx (Cole 1970).

The tail is broad, flat, and largely hairless. The skull is robust. The zygomasseteric structure is specialized in that the rostrum is marked by a conspicuous lateral depression from which a large part of the lateral masseter muscle

FIGURE 13-12 A beaver (*Castor canadensis*) dragging a branch to a dam.

originates (**Fig. 13-13**). The jugal is broad dorsoventrally, and the external auditory meatus is long and surrounded by a tubular extension of the auditory bulla. The dental formula is 1/1, 0/0, 1/1, 3/3 = 20. The premolars are molariform, and the complex crown pattern features transverse enamel folds (**Fig. 13-14**).

Beavers are always found along waterways. Although they are most typical of regions supporting coniferous or deciduous forests, they also live in some hot desert regions, as, for example, along the lower Colorado River of Arizona and California. Along major rivers of the southwestern and middle-western United States, beavers dig burrows in the river banks, but on smaller streams and in northern and mountainous regions, they build lodges of sticks and mud in ponds formed behind their dams. Beavers remain active beneath the ice throughout the winter, feeding on the cambium of aspen and willow branches that they have stuck in the mud bottoms of the ponds. Numerous mounds of mud marked with **castoreum** (a urine-based secretion of the castor glands) and scattered within the territory serve to advertise the family unit's territorial boundaries (Schulte 1998).

Beavers are remarkable in their ability to modify their environment by felling trees and building dams. Many high valleys in the Rocky Mountains have been transformed by beaver dams from a series of meadows through which a

narrow, willow-lined stream meandered to a terraced series of broad ponds bordered by extensive willow thickets and soil saturated with water. A valley suitable for grazing by cattle before occupancy by beavers may be more suitable afterward for moose, trout, and waterfowl.

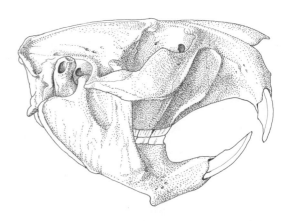

Beaver

(*Castor canadensis*, Castoridae)

FIGURE 13-13 Skull of a beaver (note the channel-like depression in the side of the rostrum from which the anterior part of the lateral masseter originates; length of skull 139 millimeters).

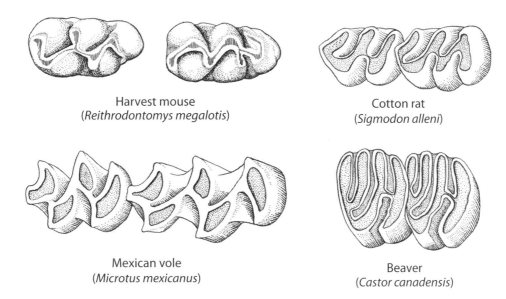

FIGURE 13-14 Crown patterns of rodent molars. The first right upper and first left lower molar of a harvest mouse (Cricetidae) and the first two right upper molars of a cotton rat (Cricetidae) and a Mexican vole (Cricetidae) are shown. The first two right upper molars of a beaver (Castoridae) are shown. Unshaded areas on occlusal surfaces are enamel; stippled areas within the enamel folds are dentine.

■ Family Geomyidae

Members of family Geomyidae, the pocket gophers, are the most highly fossorial North American rodents. They are distributed from Saskatchewan to northern Colombia. The family includes 40 Recent species in six genera. Pocket gophers appeared first in the early Oligocene of North America. Although they are not restricted to semiarid habitats today, many of their most characteristic specializations probably evolved in response to the soil conditions and vegetation of the semiarid environments that developed in the Miocene. Geomyids, along with their sister group the Heteromyidae, form a monophyletic clade and share external, fur-lined cheek pouches (Genoways & Brown 1993; Ryan 1989; Wahlert 1985).

The most obvious structural characteristics of pocket gophers were developed in response to fossorial life. These animals are fairly small (weighing from 100 to 900 grams) have small pinnae, small eyes, and a short tail. The head is large and broad, and the body is stout. External fur-lined cheek pouches are used for carrying food and nesting material. The dorsal profile of the geomyid skull is usually nearly straight. The zygomatic arches flare widely, and in the larger species, the skull is angular and features prominent ridges for muscle attachment. The rostrum is broad, robust, and marked laterally by depressions from which the lateral masseter muscles take origin. The large incisors often protrude forward, in some species beyond the anterior-most parts of the nasals and premaxillae; the lips close behind the incisors, which are therefore outside the mouth. The dental formula is 1/1, 0/0, 1/1, 3/3 = 20. The cheek teeth are ever growing and have a highly simplified crown pattern.

There is no loss of digits. The forelimbs are powerfully built and bear large, curved claws; the toes of the forefoot have fringes of hair that presumably increase the effectiveness of this foot during digging (**Fig. 13-15**).

Pocket gophers occupy friable soils in environments ranging from tropical to boreal, eat a variety of aboveground and underground parts of forbs, grasses, shrubs, and trees, and strongly affect their environment. Pocket gopher burrows provide channels that allow deep penetration of water and decrease surface erosion during periods of snow melt

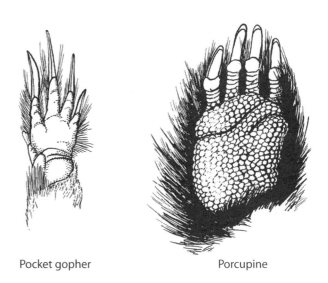

Pocket gopher   Porcupine

FIGURE 13-15 Ventral views of the left manus of the pocket gopher (*Thomomys bottae*, Geomyidae). Note the fringes of hairs on the toes. A right manus of the porcupine (*Erethizon dorsatum*, Erethizontidae) is shown. Note the pattern of tubercles on the pads, a design that increases traction.

in mountainous areas. The disturbance of the soil and the mounds of soil created by pocket gophers strongly influence vegetation by favoring pioneer plants. In some mountain meadows, roughly 20% of the ground surface is covered with mounds, and in such meadows in Utah, pocket gophers may eat more than 30% of the annual underground productivity of forbs (Andersen & MacMahon 1981). Because their preference for alfalfa and some other cultivated plants results in great crop damage, large amounts of money have been spent by farmers and federal agencies to control pocket gophers on cultivated land in the western United States.

The evolutionary history of pocket gophers has featured morphological variation among local populations. Pocket gophers typically occupy sandy or loamy soils, because burrowing, energetically costly as best, is insupportably costly in hard soils. Furthermore, pocket gophers live subterranean lives and are quite sedentary. Because areas of friable soils are commonly isolated from one another by hard soils, pocket gopher populations are also isolated, a situation favoring the proliferation of species. Texas offers an example: six of the *Geomys* species that live there are restricted to isolated areas of sandy soil. For the two most widespread western species of pocket gopher (*T. bottae* and *T. talpoides*), over 245 subspecies have been recognized in the literature. Although the proper resolution of this taxonomic mess is uncertain, these "subspecies" clearly reflect morphological differentiation of local populations. Along with other differences, the populations exhibit wide color variation: those living in pale-colored soils are pale. Those living in the darkest soils are equally dark, with a range of intermediate colors in populations living in intermediate soils. This matching of fur color to soil color is probably due to selection of conspicuous pocket gophers by predators.

Family Heteromyidae

Members of family Heteromyidae are the North American rodents most strongly adapted to desert life. Heteromyids are restricted to the New World, where they range from southern Canada through the western United States to Ecuador, Colombia, and Venezuela. Although they occupy areas ranging from temperate to tropical, they reach their greatest diversity and density in arid and semiarid regions. This family contains 60 Recent species in six genera.

Heteromyids first appeared in the Oligocene of North America. The kangaroo rats (*Dipodomys*; **Fig. 13-16A**) are known from the late Miocene, when the deserts and semiarid brushlands that heteromyids now frequently inhabit became widespread in western North America. Large eyes, small ears, silky fur, and handsome facial markings are typical of kangaroo rats. Certain diagnostic characteristics of kangaroo rats, such as the greatly enlarged auditory bullae (**Fig. 13-17**) and features of the hindlimbs that allow saltation, probably evolved under constraints associated with desert or semidesert conditions (Genoways & Brown 1993; Grinnell 1922; Nikolai & Bramble 1983).

Most heteromyids are specialized for bipedal jumping (richochetal locomotion). Such adaptations are most strongly developed in the kangaroo rats (*Dipodomys*) and kangaroo mice (*Microdipodops*). In these heteromyids, the forelimbs are small, the neck is short, and the tail is long and serves as a balancing organ. The hindlimbs are elongate, and the thigh musculature is powerful (Howell 1932; Ryan 1989). The hind foot is elongate; except for the almost complete loss of the first digit in some kangaroo rats (**Fig. 13-18E**), there is no loss of digits. The cervical vertebrae are largely fused in *Microdipodops*, and in *Dipodomys*, they are strongly compressed and partly fused, producing a short, rigid neck (Fig. 13-18C). These species are bipedal when

FIGURE 13-16 (A) A Merriam's kangaroo rat (*Dipodomys merriami*, Heteromyidae) in Arizona. (B) An Egyptian jerboa (*Jaculus jaculus*, Dipodidae) from North Africa.

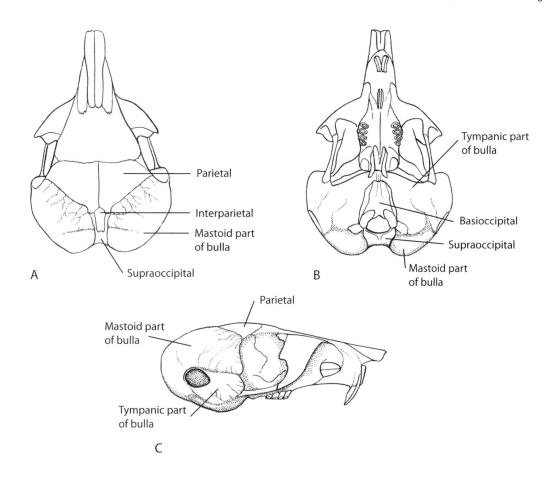

Merriam's kangaroo rat
(*Dipodomys merriami*, Heteromyidae)

FIGURE 13-17 (A) Dorsal, (B) ventral, and (C) lateral views of the skull of Merriam's kangaroo rat (Heteromyidae). Note the great enlargement of the auditory bulla, the chamber surrounding the middle ear (length of skull 45 millimeters). (Adapted from Grinnell, J, *University of California Publications in Zoology* 24 (1922):1–124.)

moving rapidly; when frightened, they move by a series of erratic hops.

As in the pocket gophers, external fur-lined cheek pouches are present in heteromyids (see Ryan 1986, for a discussion of the evolution of cheek pouches in rodents). The skull is delicately built, with thin, semitransparent bones; the zygomatic arch is slender. The auditory bullae are usually large and in some genera are enormous, being formed largely by the mastoid and tympanic bones (Fig. 13-17). The enlargement of the bullae in heteromyids (and convergently in jerboas, Dipodidae) greatly increases auditory sensitivity. The nasals are slender and usually extend well forward of the slender upper incisors. The dental formula is 1/1, 0/0, 1/1, 3/3 = 20. The ever-growing cheek teeth have a strongly simplified crown pattern (Fig. 13-8) resembling that of pocket gophers, to which heteromyids are closely related (Ryan 1989; Wahlert 1985).

Small annual plants in deserts are able to make the most of irregular moisture by germinating, growing, and flowering rapidly and by producing abundant seeds that remain dormant for various periods. This enormously abundant seed crop is the major food source of heteromyids. Individuals of some species develop conspicuous trails from their burrows to their foraging areas (**Fig. 13-19**). Perhaps the most remarkable heteromyid adaptation is the ability to survive for long periods on a diet of dry seeds with no free water (MacMillen 1983a, 1983b). This capability does not occur in all heteromyids, nor is it developed to the same degree in all species adapted to dry climates. In the species of kangaroo rats (*Dipodomys*) and pocket mice (*Chaetodipus* and *Perognathus*) that occupy deserts, however, this ability is well-developed (Chapter 21).

Suborder Myomorpha

Family Dipodidae

This family includes the jerboas (strongly saltatorial), the jumping mice (moderately saltatorial), and the birch mice

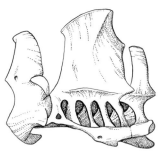

A. Jerboa

B. Springhare

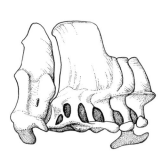

C. Kangaroo rat

D. Jerboa

E. Kangaroo rat

FIGURE 13-18 The cervical vertebrae and hind feet of several saltatorial rodents. Cervical vertebrae (anterior is to the left) of (A) a jerboa (*Jaculus* sp., Dipodidae), (B) the springhare (*Pedetes capensis*, Pedetidae), and (C) Heermann's kangaroo rat (*Dipodomys heermanni*, Heteromyidae). (D) Dorsal view of hind foot of a jerboa (*Jaculus* sp., Dipodidae); note the reduction of digits and the cannon bone formed by metatarsals two, three, and four. (E) Left hind foot of the desert kangaroo rat (*Dipodomys deserti*, Heteromyidae); note the near loss of the first digit and the elongation of the foot. (A, B, and C adapted from Hatt, R. T., *Bull. Amer. Mus. Nat. Hist.* 63 (1932): 599–738.; D adapted from Howell, A.B. *Speed in Animals*. University of Chicago Press, 1965.; E, adapted Grinnell, J., *University of California Publications in Zoology* 24 (1922): 1–124.)

(nonsaltatorial). Jerboas occur in arid and semiarid areas in northern Africa, Arabia, and Asia Minor and in southern Russia eastward to Mongolia and northeastern China. The Nearctic jumping mice occupy Alaska and much of Canada and in the United States, living as far south as New Mexico and Georgia. Palearctic jumping mice and birch mice are found from Scandinavia into central Europe and in Russia, Mongolia, and China. The family first appeared in the early Eocene of Asia and is represented today by 16 genera and 51 species.

Jerboas have a compact body, large head, reduced forelimbs, and elongate hindlimbs—features associated with saltatorial locomotion (**Figs. 13-20A** and 13-16B). The tail is long and usually tufted, and, as in the New World kangaroo rats (Heteromyidae), the tuft is frequently conspicuously black and white. The posterior part of the skull is broad (**Fig. 13-21A**), mostly because of the enlargement of the auditory bullae, which are huge in some species. The rostrum is usually short. The orbits are large, and most of the anterior part of the medial masseter, which originates largely on the side of the rostrum passes through the enlarged infraorbital

canal (Fig. 13-21B). The zygomatic plate is narrow and below the infraorbital canal. The dental formula is 1/1, 0/0, 0–1/0, 3/3 = 16–18, the cheek teeth are hypsodont, and the crown pattern usually features re-entrant enamel folds.

The hindlimbs are elongate in all species, but varying stages of specialization for saltation are represented (Stein 1990). In members of the subfamily Cardiocraniinae, the toes vary in number from three to five, and the metatarsals are not fused. At the other extreme are members of such genera as *Dipus* and *Jaculus* (subfamily Dipodinae), which represent the greatest degree of hindlimb specialization for saltation in rodents. In these genera, only three toes (digits 2, 3, and 4) remain and the elongate metatarsals are fused into a cannon bone (Fig. 13-18D). An additional specialization that occurs in some species is a brush of stiff hairs on the ventral surfaces of the hind toes. The ears of jerboas vary from short and rounded to long and rabbit like.

Jumping mice and birch mice are small and graceful, weighing approximately 10 to 25 grams, with a long tail and, in all genera but *Sicista*, elongate hindlimbs (Fig. 13-20B). The coloration in most species is striking: the belly is white,

FIGURE 13-19 Trails made by Merriam's kangaroo rat (*Dipodomys merriami*) near Yuma, Arizona.

and the dorsum is bright yellowish or reddish brown. Much of the anterior part of the medial masseter muscle originates on the side of the rostrum and passes through the enlarged infraorbital foramen (Klingener 1964). Although all living dipodids have an hystricomorphous zygomasseteric arrangement, they are regarded as related to the myomorphous Muridae. Both Wilson (1949) and Klingener (1964) concluded that the murid myomorphous masseter was derived from a dipodoid-like ancestor. The dental formula is 1/1, 0/0, 0–1/0, 3/3 = 16–18; the cheek teeth are brachydont or semihypsodont and have quadritubercular crown patterns with re-entrant enamel folds. The hindlimbs in jumping mice are elongate and somewhat adapted for hopping, but unlike the situation in more specialized saltatorial rodents, all digits are retained. As an additional contrast with jerboas and other specialized saltators, the cervical vertebrae of jumping mice are unfused.

Jerboas occupy arid areas and lead lives that resemble in some ways those of members of the New World family Heteromyidae (Fig. 13-16A) (Mares 1980). They live in burrows that are frequently plugged during the day, a habit that favors water conservation by keeping the humidity in the burrow as high as possible. They are nocturnal, and many species sift seeds from sand or loose soil with the forefeet, although some species depend largely on insects for food. Unlike kangaroo rats, jerboas hibernate during the winter in fairly deep burrows. Locomotion in jerboas is chiefly bipedal, but when they are moving slowly, the forefeet may be used to some extent. When frightened, jerboas move

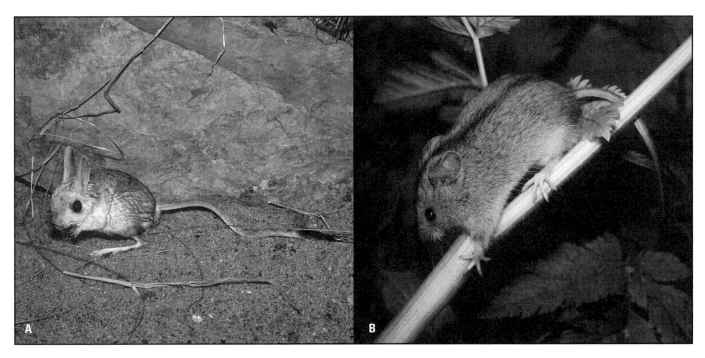

FIGURE 13-20 (A) A small five-toed jerboa (*Allactaga elater*, Dipodidae) from Asia, showing the elongate tail and enlarged hind feet. (B) A northern birch mouse (*Sicista betulina*, Dipodidae) from Eastern Europe.

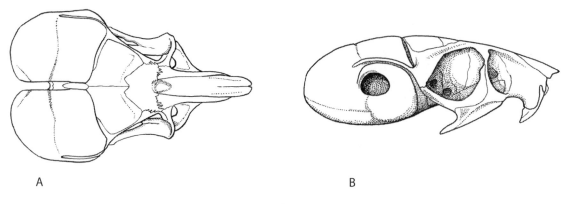

A   B

Jerboa

(*Salpingotus kozlovi*, Dipodidae)

FIGURE 13-21   (A) Dorsal and (B) lateral views of the skull of a Kozlov's pygmy jerboa (*Salpingotus kozlovi*, Dipodidae); the length of the skull is 27 millimeters. Note the greatly enlarged auditory bullae and the general resemblance to the skull of the kangaroo rat (Figure 13-17). (Adapted from Allen, G. M. *The Mammals of China and Mongolia, Part II.* American Museum Natural History, 1940.)

rapidly in a series of long leaps, each of which may cover 3 meters. Such a rapid and, more important, erratic mode of escape from predation is especially effective in the barren terrain that jerboas occupy (Schröpfer et al. 1985).

Jumping mice and birch mice usually inhabit boreal forests. Some species occur typically in coniferous forests, whereas others appear in birch stands or in mixed deciduous forests. Usually, jumping mice favor moist situations, and *Zapus princeps* of the western United States is most abundant in dense cover adjacent to streams or in wet meadows. These mice hibernate in the winter and emerge during or after snow melt. Food consists of a variety of seeds and other plant material, and fungus (*Endogone*), but insect larvae and other animal material made up approximately half of the food of *Z. hudsonius* in New York (Whitaker 1963) and roughly one third of the diet of *Z. princeps* in Colorado (Vaughan & Weil 1980).

Family Platacanthomyidae

This family includes two species of tree mice: *Platycanthomys lasiurus* and *Typhlomys cinereus* (Musser & Carelton 2005). Originally described as dormice (Gliridae) based on superficial similarities in dentition, platycanthomyids likely evolved from an Asian muroid some time in the Eocene or Oligocene (Carleton & Musser 1984). To avoid confusion with dormice, Musser and Carleton advocate the use of the name "tree mice" for the two platycanthomyid species. Tree mice first appear in the Miocene of China and Pakistan and now inhabit southwestern India (*P. lasiurus*) and montane forests of southern China and northwestern Vietnam (*T. cinereus*).

The dental formula is 1/1, 0/0, 0/0, 3/3 = 16. The spiny tree mouse, *P. lasiurus*, weighs approximately 75 grams. They have spiny hairs intermingled with the underfur on the back and sides, with fewer, more delicate spines on the belly. The

base of the tail is scaly and thinly haired, whereas the terminal half is covered with long dense fur (Nowak 1999). Spiny tree mice have light brown to reddish pelage dorsally, with gray to white fur on the underparts. The soft-furred tree mouse, *T. cinereus*, is considerably smaller and has short spineless fur. The scaly tail is sparsely haired and ends in a terminal brush of long thin hairs.

Little is known of the ecology and behavior of tree mice. Spiny tree mice inhabit forests at elevations up to 900 meters, where they live in tree holes and rock crevices. *T. cinereus* lives in dense, mossy forests above 1,200 meters elevation. The diet of *P. lasiurus* apparently consists of grains, seeds, and fruits, and this species is considered an agricultural pest in some areas of India (Nowak 1999).

Family Spalacidae

This family contains four subfamilies of fossorial rodents: Myospalacinae (zokors), Rhizomyinae (bamboo rats), Spalacinae (blind mole rats), and Tachyoryctinae (African root rats). There are 36 Recent species in six genera. Musser and Carleton (2005) provided evidence that these rodents form a clade that diverged from other muroid rodents in the late Oligocene. The oldest known member of the Spalacidae is *Debruijnia*, an early Miocene mole rat from Turkey (Mein et al. 2000). Today zokors occur in Russia and China, and bamboo rats occur in Nepal and southern China southward into India and eastward to Vietnam and the Malay Peninsula. Blind mole rats occur, in suitable habitat, from eastern Europe, southern Russia and Ukraine, to the Middle East and northern Africa. African mole rats live in eastern Africa.

Members of the Spalacidae are compact, stoutly built rodents with short, muscular limbs. The forelimbs of zokors and bamboo rats have long robust claws used for digging, whereas blind mole rats and root rats use strongly protruding incisors for digging tunnels (**Fig. 13-22**). The skull is robust, and the cheek teeth are cylindrical; the dental formula is 1/1, 0/0, 0/0, 3/3 = 16. Ear pinna are absent or reduced, and the tail is often vestigial. The eyes are small and partially hidden in the fur in most species. Blind mole rats (*Spalax*) have small eyes covered with skin. The eye muscles and optic nerve are degenerate, but the eyes have photopigments that may enable the animals to establish circadian rhythms of activity (David-Gray et al. 1998; Sanyal et al. 1990). Unlike many fossorial rodents, *Spalax* digs mostly with its large incisors and bulldozes soil with its blunt head. As adaptations to this style of digging, the neck and jaw muscles are powerful. The incisors are robust, and the nose is protected by a broad, horny pad. The feet, surprisingly, are not unusually large; the claws have been described as blunt, round nubbins. These nocturnal rodents burrow in both alluvial and stony soils and eat both aboveground and underground parts of plants. *Spalax* often lives in burrows in water-saturated or snow-covered soil; these burrows have extremely high levels of carbon dioxide and extremely low levels of oxygen. In most mammals, these conditions interfere seriously with heart action, but *Spalax* can raise its heart rate and maintain a stable pulse at low oxygen levels (Arieli & Ar 1981a). As an adaptation to improve oxygen delivery to tissues under these conditions, the capillary density in the heart and skeletal muscles of *Spalax* is nearly twice as high as in the laboratory rat (Arieli & Ar 1981b).

Members of the subfamily Rhizomyinae range from 0.5 kilogram to over 4 kilograms and have short, robust limbs and a compact body. The procumbent incisors and long powerful claws are both used to dig the extensive burrow systems. Bamboo rats live primarily in areas with at least 500 millimeters of annual precipitation and occupy a variety of habitats, including dense bamboo thickets at elevations up to 4,000 meters.

The 13 species of African root rats (Tachyoryctinae) favor savannahs or agricultural land with fairly high rainfall (**Fig. 13-23**). *Tachyoryctes* burrows by slicing away the soil with powerful upward sweeps of the protruding lower incisors (Jarvis and Sale 1971). The dislodged soil is moved behind the animal by synchronous thrusts with the hindlimbs. When the burrow becomes blocked with freshly dug soil, *Tachyoryctes* turns and pushes the load to the surface with the side of its head and one forefoot. The conspicuous mounds (up to 6 meters in diameter) associated with the activities of *Tachyoryctes* on Mount Kenya resemble the Mima mounds that are formed by pocket gophers in some parts of the western United States (see Fig. 23-10b). *Tachyoryctes* is solitary and aggressive and eats a variety of below- and above-ground parts of plants.

■ **Family Calomyscidae**

The eight species in the family Calomyscidae were once associated with the cricetine hamsters (or with New World *Peromyscus*), but Musser and Carleton (2005) found little evidence for such a union. Instead, they tentatively

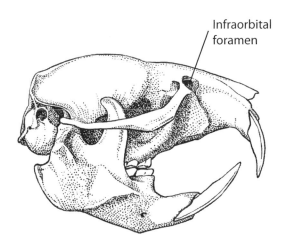

Northeast African mole rat
(*Tachyoryctes splendens*, Spalacidae)

FIGURE 13-22 Skull of a Northeast African mole rat (Spalacidae, Tachyoryctinae). Note the procumbent incisors, which are used for digging and the dorsal position of the infraorbital foramen (length of skull 41 millimeters).

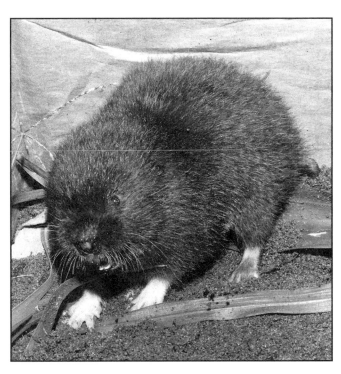

FIGURE 13-23 A Northeast African mole rat (*Tachyoryctes splendens*, Spalacidae, Tachyoryctinae) from Kenya.

suggested family-level status and, to avoid confusion with hamsters, advocated the use of calomyscus as part of the common names. Calomyscids lack the cheek pouches found in cricetid hamsters. The fossil record of *Calomyscus* begins in the Miocene of Europe and Turkey. Today various species occupy suitable habitats from Pakistan to the Middle East.

Calomyscus weighs between 15 and 30 grams, and its tail is slightly longer than the head and body. The tail is densely furred and ends in a tuft of longer hairs. The ears are large and the pelage soft. The dental formula is 1/1, 0/0, 0/0, 3/3 = 16. In Turkmenistan, *Calomyscus mystax* inhabits rocky talus slopes where it feeds on seeds, flowers, and other plant material (Sapargeldyev 1984). Females from Turkmenistan produce two litters between March and June, but in Iran, the breeding season extends to December (Lay 1967). Hotson's calomyscus (*Calomyscus hotsoni*) is listed as endangered and three other species are considered at risk.

◼ Family Nesomyidae

This family includes 6 subfamilies, 21 genera, and 61 species of obscure African and Malagasy rodents (Musser & Carleton 2005). The Nesomyidae, as currently defined, includes the subfamilies Cricetomyinae (pouched rats), Delanymyinae (swamp mouse), Dendromurinae (climbing mice, desert mouse, fat mice), Mystromyinae (white-tailed rat), Nesomyinae (Malagasy rodents), and Petromyscinae (rock mice). Several of these subfamilies were once thought to be members of the Muridae or Cricetidae, but evidence from molecular genetic studies suggest that the Nesomyidae evolved before the Cricetidae and Muridae (Jansa et al. 1999; Michaux et al. 2001).

Molecular characters unite the six families in the Nesomyidae and suggest that this morphologically diverse group diverged from the muroid rodents approximately 25 million years ago (Steppan et al. 2004). The morphological diversity of nesoymids, however, reflects the variety of their habitats. Nesomyids include species that resemble rats (*Cricetomys* and the Mystromyinae), mice (*Dendromus*), voles (*Brachyuromys*), and gerbils (*Macrotarsomys*). The dental formula is 1/1, 0/0, 0/0, 3/3 =16. Body size ranges from a bulky 3,000 grams in *Cricetomys* to a petite 6 grams for Delany's swamp mouse (*Delanymys brooksi*). Tails are variously long or short, prehensile or not, furred, naked, or with a tuft on the tip. Coloration also varies widely among members of the Nesomyidae. Interestingly, one subfamily, the Cricetomyinae, evolved cheek pouches superficially similar to those found in cricetid hamsters, but the pouch retractor muscle is cricetomyines is different from those in hamsters (Ryan 1989).

The three genera of pouched rats (Cricetomyinae) are found throughout much of sub-Saharan Africa. Male giant pouched rats (*Cricetomys*) can reach 3 kilograms (females are half the weight of males), whereas the hamster-like *Saccostomus* weighs less than 85 grams. All cricetomyines have large internal cheek pouches used to transport food and nesting materials to den sites. The subfamily Petromyscinae contains four species of rock mice (*Petromyscus*) from southwestern Africa.

Delany's swamp mouse (*Delanymys brooksi*) is the only member of the subfamily Delanymyinae, and lives in montane sedge swamps where Uganda, Rwanda, Burundi, and the Democratic Republic of Congo border one another. The Dendromurinae (21 species in 6 genera) include small- to medium-sized mice from sub-Saharan Africa, some with isolated geographic ranges. The sole member of the Mystromyinae, the white-tailed rat (*Mystromys albicaudatus*), inhabits the savannah grasslands of South Africa. White-tailed rats resemble small hamsters. They are nocturnal and spend the day in burrows or deep cracks in the dry soil; they are also known to use abandoned burrows of meerkats (*Suricata*) on occasion.

An obscure, but interesting subfamily is the Nesomyinae of Madagascar. This probably monophyletic group of rodents appeared in the Miocene in Africa, dispersed to Madagascar approximately 20 million years ago, and subsequently radiated in isolation from their African ancestors (Poux et al. 2005). Released from competition with other rodents, ancestral nesomyines quickly evolved to fill the vacant niches on Madagascar. Living nesomyines include 22 species that resemble rodents as diverse as pocketmice, voles, arboreal rats, and even rabbits. Until the 1990s, 7 genera and only 10 species of nesomyine rodents were known (Musser & Carleton 1993). Detailed surveys of the island in the past several years, coupled with taxonomic revisions of several genera, indicate that the nesomyine fauna is much more diverse (Carleton 1994; Carleton & Goodman 1996, 1998; Carleton & Schmidt 1990; Musser & Carleton 1993; summarized in Goodman & Benstead 2003). In fact, surveys of the previously unexplored high mountain forests in Madagascar have resulted in the addition of two new genera and species of nesomyine rodents: *Monticolomys koopmani* and *Voalavo gymnocaudus* (Carleton & Goodman 1996, 1998).

◼ Family Cricetidae

This, the second largest family of mammals, includes 681 species of 130 genera. Within the family are the Old World hamsters (Cricetinae), the African maned rat (Lophiomyinae), a diverse assemblage of New World rats and mice (Neotominae, Sigmodontinae, and Tylomyinae) and voles and lemmings (Arvicolinae). Members of the Cricetidae occupy much of the globe, including the entire Western Hemisphere, nearly all of Eurasia, and parts of East Africa.

In the extreme north of the Western Hemisphere, the collared lemming (*Dicrostonyx groenlandicus*), a cricetid, lives on the bleak tundra of the north coast of Ellesmere Island. This is about 83° north latitude, only 7° south of the North Pole, the northernmost occurrence of any land mammal. At the southernmost tip of the Western Hemisphere, the chinchilla mouse (*Euneomys chinchilloides*), another cricetid, lives on cold and windswept Tierra del Fuego, at over 55° south latitude, only 12° north of the Antarctic Circle. Between these latitudinal extremes, a diverse array of cricetids inhabit virtually every terrestrial environment, from tundra, taiga, coniferous and deciduous forests, alpine areas (including those of the high Andes Mountains), grasslands, tropical and subtropical areas, to the driest North American and South American deserts.

Hamsters (18 species and 7 genera) have a Palearctic distribution. They are compact rodents with small ears, short legs, thick fur, and short tails (**Fig. 13-24**). The dental formula is 1/1, 0/0, 0/0, 3/3 = 16. The skull has a relatively long and robust rostrum with a characteristic hourglass-shaped area between the orbit (as viewed from above). Hamsters prefer grains, leaves, and roots, but some species are more omnivorous. Food is transported to the extensive burrow systems in large internal cheek pouches. Ninety kilograms of winter food stores have been found in burrows of *Cricetus cricetus* (Nowak 1999). Hamsters live largely solitary lives, and some species are highly aggressive toward conspecifics. Most species are either nocturnal or crepuscular. During winter, hamsters enter long periods of torpor. The golden hamster (*Mesocricetus auratus*) is a common pet and laboratory research animal. Interestingly, all golden hamsters sold in the pet and research animal

trade are thought to have descended from a single litter captured in Syria in 1930 (Murphy 1985).

The subfamily Arvicolinae includes the voles, lemmings, and muskrat, a group of 28 genera and 151 species of rodents distributed throughout the Northern Hemisphere, where some species occupy harsh Arctic or alpine environments. These rodents frequently have short tails, ear openings that are partially covered by fur, and a chunky, short-legged appearance (**Fig. 13-25**). The cheek teeth often feature complex crown patterns (Fig. 13-14), and some are ever growing, adapted to masticating abrasive grasses. The collared lemming (*Dicrostonyx*) of the far north are the only North American rodents that turn entirely white in winter. Some Inuit people recognize the brown summer lemmings as one species and the white winter ones as another. Inuit children use the white lemming skins to make doll clothes. Digging claws, unique to the three species of collared lemmings, develop in the autumn and are used to dig through snow and ice. The voles and lemmings generally have high reproductive rates, and several species undergo spectacular population fluctuations in some areas (Chapter 23).

The Neotominae includes 124 species and 16 genera of New World woodrats, deer mice, and harvest mice (**Fig. 13-26**). They occupy a wide variety of habitats from Alaska to Panama. In many parts of the western United States, the large dens of woodrats (*Neotoma*) are conspicuous features. These dens may be a meter high or more and variously include sticks, leaves, cactus, small skulls and bones, shotgun shells, or anything else the animals can carry. These rodents are commonly called packrats in reference to their collecting habits. Dens are built at the base of small trees or shrubs, in patches of cactus or yucca, in rocky places, or in the canopy in tall chaparral. A woodrat can build a

FIGURE 13-24 Striped dwarf hamster (*Cricetulus barabensis*).

FIGURE 13-25 A field vole (*Microtus agrestis*, Cricetidae, Arvicolinae).

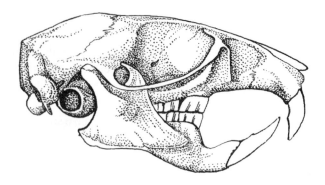

Stephen's woodrat
(*Neotoma stephensi*, Neotominae)

FIGURE 13-26 Skull of myomorphous rodent, a Stephens' woodrat (Cricetidae, Neotominae; length of skull 42 millimeters).

den under the hood of a truck within 4 days (Vaughan, personal observation). Dens offer some protection from predators. The white-throated woodrat (*Neotoma albigula*) builds the most well-armored dens (**Fig. 13-27**), which are typically festooned with cactus, often the formidable teddy-bear cholla (*Opuntia bigelovii*). Most importantly, however, dens insulate the animals against extremes of temperature and humidity (Brown 1968). This is especially critical for a group of rodents that occur from cold alpine areas to hot deserts, places where they could not live without the protection of dens (Chapter 21). Woodrats collect local plants and deposit them on dens. Ancient dens that are preserved in rock fissures or caves therefore provide samples of past plant communities. These samples document biotic and climatic changes over the past 40,000 years (Betancourt et al. 1990).

Deer mice (*Peromyscus*; some 53 species) seem to occupy nearly every terrestrial habitat in the Western Hemisphere. *Peromyscus*-like rodents (*Copemys*) entered North America from Eurasia in the early Miocene but underwent little radiation until late Miocene and Pliocene, when such neotomine genera as *Peromyscus*, *Neotoma*, *Onychomys*, and *Reithrodontomys* appeared (Korth 1994; Musser & Carleton 2005).

The Sigmodontinae is the second largest cricetid subfamily (377 species) and occupies South America and most of North America. Neotropical sigmodontines underwent rapid radiation in the late Miocene and Pliocene; by the early Pliocene, the extant genus *Sigmodon* appeared. Sigmodontines possibly entered South America when the Panamanian land bridge formed in late Pliocene but could have entered earlier by **waif dispersal** across the seaway separating the Americas (Engel et al. 1998; Smith & Patton 1999). Experts disagree as to whether the present diversity of South American sigmodontines resulted from a tropical North American radiation before access to South America was gained or whether this radiation occurred in South America after the land bridge formed or both, but the investigation of this radiation will be a fruitful area of research (Pardiñas et al. 2002). In any case, of the 74 living sigmodontine genera, the vast majority live in South America or no farther north than tropical Central America. Today, sigmodontines occupy alpine, tropical, and desert habitats and are variously terrestrial, amphibious, fossorial, or arboreal. The Argentine gerbil mouse (*Eligmodontia*, also known as laucha) is the only mammal that can live in some of the driest South American deserts. This sigmodontine can excrete highly concentrated urine and may maintain water balance by eating the succulent but salty leaves of halophytic plants (Mares 1977; Mares et al. 1997; Chapter 21).

FIGURE 13-27 (A) A white-throated woodrat (*Neotoma albigula*, Cricetidae) in the Sonoran Desert of Arizona. (B) Den of a white-throated woodrat in the Sonoran Desert. This den was over 1 meter high and was covered with joints of teddy-bear cholla (*Opuntia bigelovii*).

Family Muridae

This is the largest family of mammals, including some 32% of the living species of rodents (roughly 730 species and 150 genera) and is nearly worldwide in distribution; its members occupy environments ranging from alpine tundra to tropical forests to desert sand dunes. The Muridae includes five subfamilies: Deomyinae (spiny mice), Gerbillinae (gerbils and their relatives), Leimacomyinae (groove-toothed forest mouse), Murinae (Old World rats and mice), and Otomyinae (vlei rats and whistling rats). This incredible diversity was reached quickly, within the last 15 to 20 million years, after an initial appearance in the middle Miocene.

Most murids retain a "standard" mouse-like form, with a long tail, generalized limb structure, and no loss of digits. Murids range in size from about 5 grams, as in *Micromys* (**Fig. 13-28A**), to 2 kilograms, as in the New Guinean rat *Mallomys*. The skull varies widely in shape, but the infraorbital foramen is always above the zygomatic plate and is enlarged dorsally for the transmission of part of the medial masseter, which originates on the side of the rostrum. Through the narrowed ventral part of this foramen pass blood vessels and a branch of the trigeminal nerve. The maxillary root of the zygomatic arch is plate-like and provides surface for the origin of part of the lateral masseter. This myomorphous zygomasseteric arrangement was perhaps derived through an hystricomorphous ancestry. The dental formula is generally 1/1, 0/0, 0/0, 3/3 = 16; in some species, the molars are reduced in number. Molars range from brachydont to hypsodont and are ever growing. The basic cusp pattern involves transverse crests (**Fig. 13-29**); however, crests are absent in some species, and molar crown patterns vary widely.

Within the array of murid species, a variety of modes of life and morphological and behavioral specializations are represented. The following paragraphs discuss the subfamilies that display some of this broad structural and functional variety.

The Murinae is the largest murid subfamily (561 species), occurs nearly worldwide, and includes a wide diversity of species adapted to terrestrial, fossorial, semiaquatic, or arboreal life. Some murines (mainly the house mouse, *Mus musculus*, and brown rat, *Rattus norvegicus*) live in close association with humans in situations ranging from isolated farms to the world's largest cities. As a result of introductions by humans, these animals have become nearly cosmopolitan in distribution and are probably the rodents most familiar to most of us. Murines that are not commensal with humans occur in much of southeastern Asia, Europe, Africa, Australia, Tasmania, and Micronesia. Tropical and subtropical areas are centers of murine abundance; however, these animals have occupied a wide variety of habitats, and some genera are highly adapted to specialized modes of life. Murines range in size from about 10 grams to about 2 kilograms. Although the tail is usually more or less naked and scaly, it is occasionally heavily furred. The molars are rooted or ever growing and usually have crowns with two cusps on the first molar that form v-shaped transverse lamina; great simplification of

FIGURE 13-28 (A) Eurasian harvest mouse (*Micromys minutus*). (B) Roof rat (*Rattus rattus*).

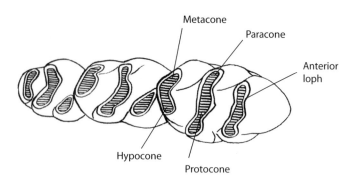

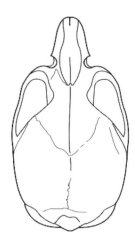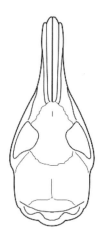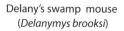

Rattus sp.

FIGURE 13-29 Occlusal surfaces of the right upper molars of a murid rodent (*Rattus*). With wear, the cross lophs become lakes of dentine (cross-hatched areas) rimmed with enamel.

Delany's swamp mouse
(*Delanymys brooksi*)

Northern Luzon rhynchomys
(*Rhynchomys soricoides*)

FIGURE 13-30 Extreme skull shape in rodents of the family Muridae. The Delany's swamp mouse is a rock-dwelling omnivore, whereas the shrew-like northern Luzon rhynchomys is a rare species that apparently feeds on invertebrates.

the crown pattern occasionally occurs (Musser & Carleton 2005). The dental formula is usually 1/1, 0/0, 0/0, 3/3 = 16. In some murines, the reduction of the cheek teeth has become extreme. The greatest reduction occurs in *Pseudohydromys ellermani*, a rare mouse from New Guinea, in which only one molar is retained on each side of each jaw. In all murines, the feet retain all of the digits, but the pollex is rudimentary.

Murines appear fairly late in the fossil record (late Miocene), but the subfamily has been remarkably plastic from an evolutionary point of view (Carleton & Musser 1984). In both Africa and the Australian faunal region, murines have undergone impressive radiations and are variously amphibious, terrestrial, semifossorial, arboreal, and saltatorial. The water rats (*Crossomys*) have greatly reduced ears, large, webbed hind feet, and nearly waterproof fur. These animals live along waterways in New Guinea. At the other extreme is the hopping mouse (*Notomys alexis*), a saltatorial inhabitant of extremely arid Australian deserts. This rodent needs no drinking water and has the greatest ability to concentrate urine as a means of conserving water of any animal in which water metabolism has been studied (MacMillen & Lee 1969). Murines feed on a variety of plant material and on invertebrate and vertebrate animals. In association with the great diversity of feeding habits of murines, the skull form varies widely within the subfamily, with a shrew-like elongation of the rostrum occurring in some insectivorous genera (**Fig. 13-30**).

Extremely high population densities have been recorded for feral populations of some murines that are often commensal with humans. A 35-acre area near Berkeley, California, which had only an occasional house mouse for a number of years after population studies began in 1948, supported 7,000 *Mus musculus* in June 1961 (Pearson 1964). Among other factors, the high reproductive rate and year-round breeding contribute to its ability to reach high

densities quickly. Murines that live with humans have great economic impact. Not only do they spread such serious diseases as bubonic plague and typhus, but the damage they do to stored grains and other foods is so severe that in many countries species of *Rattus* (Fig. 13-28B) and *Mus* compete effectively and devastatingly with humans for food.

The subfamily Gerbillinae, the gerbils, includes 16 genera and 103 species and is a group of rodents that resembles jerboas (Dipodidae) and kangaroo rats (Heteromyidae) in being semifossorial, more or less saltatorial, and inhabiting mainly desert regions. Gerbils occur in arid parts of Asia, in the Middle East, and in Africa. The hindlimbs are large. The central three digits are larger than the lateral ones, and the tail is often long and functions as a balancing organ (**Fig. 13-31**). The skull does not depart strongly from the general murid plan (**Fig. 13-32**). Gerbils maintain water balance in hot, arid conditions, partly by eating food with high water content and partly by concentrating urine, as do heteromyids.

The Deomyinae is an Afro-Asian subfamily diagnosed solely by molecular characters. It includes the link rat (*Deomys ferugineus*; also known as the Congo forest rat), 19 species of spiny mice (*Acomys*), 21 species of brush-furred rats (*Lophuromys*), and Rudd's bristle-furred rat (*Uranomys ruddi*). All deomyines have stiff guard hairs over at least part of the body. They are best developed in spiny mice. Perhaps the most unusual deomyine is the link rat (*D. ferugineus*). It has relatively long legs, large ears, a long bicolored tail, and stiff guard hairs on the rump. Until recently, *Deomys* was included in the Dendromurinae or placed in its own subfamily, the Deomyinae (reviewed by Musser &

FIGURE 13-31 Mongolian Jird (*Meriones unguiculatus*, Gerbillinae).

Carleton 2005). A suite of molecular characters now unites *Deomys* with *Acomys*, *Lophuromys*, and *Uranomys* in the Deomyinae. They are believed to share a common ancestor with members of the Gerbillinae some 16 to 20 million years ago, probably in Africa (Steppan et al. 2004, 2005).

Büttner's African forest mouse (*Leimacomys buettneri*) is the sole member of the subfamily Leimacomyinae. It is known only from two specimens collected in Togo in 1890; it has not been seen since and thus is considered by some to be extinct and by others as critically endangered.

The Otomyinae includes 3 genera and 23 species of sub-Saharan rodents (Musser & Carleton 2005). Some molecular evidence suggests that this group should be included in the subfamily Murinae (Jansa & Weksler 2004). Otomyines include 2 species of Karroo rats (*Myotomys*) from South Africa, 2 species of whistling rats (*Parotomys*)

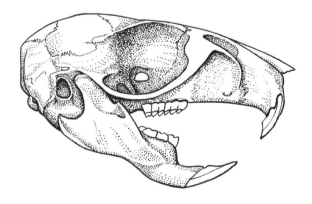

Gerbil
(*Tatera humpatensis*, Gerbillinae)

FIGURE 13-32 Skull of a gerbil (Muridae, Gerbillinae); length of skull 38 millimeters. (Adapted from Hill, J. E., and Carter, T. D., *Bull. Amer. Mus. Nat. Hist.* 78 (1941): 1–211.)

from southern Africa, and 19 species of Vlei rats (*Otomys*) from much of sub-Saharan Africa. Otomyines are vole-like in appearance, with short limbs, small eyes and ears, and dense, soft fur. Like voles, otomyines are largely herbivorous and terrestrial.

Suborder Anomaluromorpha

Family Anomaluridae

The Anomaluridae, composed of seven Recent species of three genera, includes the scaly-tailed squirrels. These animals occupy forested, tropical parts of western and central Africa. They resemble the flying squirrels (Sciuridae) in some structural features and in gliding ability. The relationship of anomalurids to other rodents groups is poorly understood.

Increased surface area for gliding is provided in anomalurids by a fold of skin that extends between the wrist and the hind foot and is supported and extended during gliding by a long cartilaginous rod, roughly the length of the forearm, that originates on the posterior part of the elbow (**Fig. 13-33**; Kingdon 1984b). In gliding sciurids, in contrast, a short cartilaginous brace arises from the wrist. In anomalurids, folds of skin similar to the uropatagium of bats extend between the ankles and the tail a short distance distal to its base (Johnson-Murray 1987). The fur over most of the membrane is fine and soft, but a tract of stiff hairs occurs along the outer edge of the membrane behind the cartilaginous elbow strut. This tract may improve the efficiency of gliding by controlling the flow of the boundary layer of air sweeping over the membrane (Johnson-Murray 1987). Relative to those of most rodents, the limbs of anomalurids are unusually long and lightly built; they provide for a wide spreading of the gliding membranes. One genus (*Zenkerella*) does not have a gliding membrane. The anomalurid tail is usually tufted and has a bare ventral area near its base that has two rows of keeled scales (Fig. 13-33). These scales seemingly keep the animals from losing traction when they cling to the trunks of trees. The feet are strong and bear sharp, recurved claws. In the anomalurid skull, the infraorbital canal is enlarged and transmits part of the medial masseter muscle. The dental formula is 1/1, 0/0, 1/1, 3/3 = 20, and the cheek teeth are rooted.

Anomalurids are handsome animals that are beautifully adapted to an entirely arboreal life. Their diet includes leaves, sap, and insects (Julliot et al. 1998). Lord Derby's scaly-tailed squirrel (*Anomalurus derbianus*) is a graceful and highly maneuverable glider, capable of glides of over 100 meters and of midair turns. Giant flying squirrels (*Petaurista*) are capable of glides up to 450 meters long (Nowak 1999). Groups of anomalurids often take shelter during the day in cavities in trees. Group size varies from six to eight

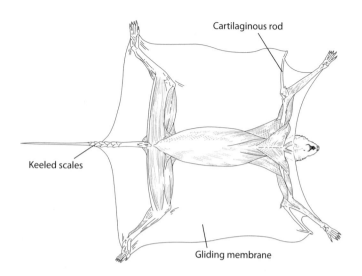

FIGURE 13-33   Drawing of the ventral surface of a scaly-tailed flying squirrel (Anomaluridae) with the skin and fur removed to highlight the gliding membrane and the cartilaginous rod supporting the membrane near the elbow. As their name implies, the tail has several rows of keeled scales on the ventral surface. (Adapted from Kingdon, J. *The Kingdon Field Guide to African Mammals.* Princeton University Press, 1997.)

animals to colonies of over 100 anomalurids of several species. Rosevear (1969) reports four species occupying the same hole in a tree. On occasion, these rodents share a hollow tree with dormice (*Graphiurus*) and with several species of bats (Julliot et al. 1998). Such mixed species roosting is highly unusual in mammals.

Family Pedetidae

This family includes two species of the genus *Pedetes*, called springhares (Matthee & Robinson 1997). Their distribution includes East Africa and southern Africa, where they inhabit sandy soils in semiarid regions. Sparsely vegetated areas or places where the vegetation has been heavily grazed by ungulates are preferred. *Megapedetes*, an early-mid Miocene genus from eastern Africa, is the earliest member of this family. The phylogenetic position of this family within the Rodentia is controversial because pedetids share both sciurognathous and hystricognathous features.

Springhares are saltatorial and roughly the size of a large rabbit, weighing up to about 4 kilograms (**Fig. 13-34**). The eyes are extremely large, suggesting a reliance on vision for detecting predators. The forelimbs are short but robust and bear long claws that are used in digging (**Fig. 13-35**). The hindlimbs are long and powerfully built. The fibula is reduced and fused distally to the tibia, and the feet have only four toes. The long tail is heavily furred throughout its length. Through the enormous infraorbital foramen (**Fig. 13-36**) passes the large anterior division of the medial

masseter. As in a number of saltatorial rodents, the cervical vertebrae are partly fused (Fig. 13-18B). The dental formula is 1/1, 0/0, 1/1, 3/3 = 20, and the cheek teeth are ever growing with a simplified crown pattern. A tragus fits against the ear opening and keeps out sand and debris when the animal digs.

Springhares dig fairly elaborate burrows. Because of their restriction to friable soils, these animals are not evenly distributed and appear to occur in colonies. When frightened, springhares can make tremendous bipedal leaps of over 6 meters, but when foraging and moving slowly, they are quadrupedal. A variety of plant material is eaten, including bulbs, seeds, and leaves. Springhares probably maintain water balance during some seasons by eating succulent vegetation or insects. The springhare has an unusually low reproductive rate for a rodent. There are only two pectoral mammae, and typically, the female bears only one young. Newborn young are large—roughly one third the size of the adult (Coe 1967; Hediger 1950)—and well-developed and remain in the maternal burrows until they weigh at least half as much as adults.

Suborder Hystricomorpha

Carleton and Musser (2005) review the considerable morphological and molecular support for uniting ctenodactylid and hystricognath rodents in the suborder Hystricomorpha.

Family Ctenodactylidae

Members of family Ctenodactylidae, commonly called "gundis," inhabit arid parts of northern Africa from Senegal, Chad, Niger, and Mali on the west to Somalila on the east. There are four Recent genera and five Recent species. The earliest known ctenodactylids are from the early Oligocene of Asia. Ctenodactylids share many derived characters with the Hystricognathi, but ctenodactylids retain many primitive features and are considered (along with the Diatomyidae) a possible sister group to hystricognath rodents (Dieterlen 2005; Huchon et al. 2007).

These are small (about 175 grams), compact, short-tailed rodents with long, soft fur. The ears are round and short and in some species are protected from wind-blown debris by a fringe of hair around the inner margin of the pinnae. The infraorbital canal is enlarged, and through it passes part of the medial masseter muscle. The skull is flattened, and the auditory bullae and external auditory meatus are enlarged. The cheek teeth are ever growing. The crown pattern is simple, and the premolars are nonmolariform. The dental formula is 1/1, 0/0, 1–2/1–2, 3/3 = 20–24. The limbs are short; the manus and pes each have four digits.

FIGURE 13-34 A springhare (*Pedetes capensis*).

These herbivorous rodents occur in arid and semiarid areas, where they are restricted to rocky situations. They are diurnal and crepuscular and scurry into jumbles of rock or fissures when threatened. Gundis spend much time basking and do not emerge on cold or stormy days, habits suggesting that they conserve energy by the use of behavioral thermoregulation.

Family Diatomyidae

In 2004, a perceptive scientist of the Wildlife Conservation Society recognized the uniqueness of a mammal on a skewer in a Laotian market. This mammal (kha-nyou to Laotians) was subsequently described as a new genus and species (*Laonastes aenigmamus*) representing a new family (Laonastidae), based on additional specimens from rural markets (Jenkins et al. 2005).

The discovery of a new family of living rodents was properly heralded as remarkable, but the story became ever more fascinating. A group of paleontologists familiar with Asian fossil rodents compared the living *Laonastes* with several species of rodents of the extinct family Diatomyidae (Dawson et al. 2006). They concluded that *Laonastes* was a living member of the family Diatomyidae. "*Laonastes* is a particularly striking example of the 'Lazarus Effect' in Recent mammals, whereby a taxon that was formerly thought to be extinct is rediscovered in the extant biota, in this case after a temporal gap of roughly 11 million years" (Dawson et al. 2006).

Laonastes resembles a long-nosed squirrel. The limbs are generalized and not adapted to climbing or digging, and the well-furred tail is long. The skull has a very large infraorbital foramen and a slightly elongate rostrum. The dentary lacks a coronoid process; the masseteric fossa (the depression in the lateral surface of the dentary into which the masseter muscles insert) is horizontally subdivided and extends forward to the anteriormost cheek tooth. The incisors are short, and the cheek teeth are bilophodont. The lower molars each have four roots, a derived character that contrasts with the primitive condition in rodents of two roots.

The natural history of *Laonastes* is unknown. It has been trapped along limestone outcrops. As the lone surviving

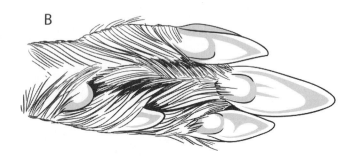

FIGURE 13-35 Ventral surfaces of the feet of a springhare (*Pedetes capensis*): (A) forefoot, (B) the heavily-furred and functionally three-toed hindfoot. (Adapted from Skinner, J. D., and Chimimba, C. T. *The Mammals of the Southern African Subregion.* Cambridge University Press, 2006.)

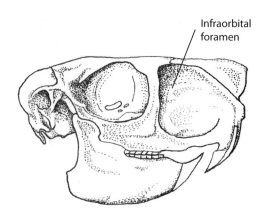

Springhare
(*Pedetes capensis*, Pedetidae)

FIGURE 13-36 The skull of the springhare (Pedetidae). Note the enormously enlarged infraorbital foramen.

member of a family with a history stretching back some 30 million years, *Laonastes* is of great importance to paleontologists, paleoecologists, and biologists.

Infraorder Hystricognathi

Although the African hystricognaths (Bathyergidae, Hystricidae, Petromuridae, Thryonomyidae) and those of South America (termed caviomorphs) share many morphological characters (**Table 13-2**) and are regarded by most authorities to be closely related, their origins and relationships were long a source of controversy and speculation. The controversy centers in part on the geographic origins of

these groups. One school of thought (Wood 1980) held that the South American caviomorphs were derived from early members of the Tertiary Franimorpha, a basal stock from North America. Lavocat (1976, 1980), and others, on the other hand, believed that Africa was the source for South American caviomorphs. Recent discovery of a new fossil rodent from Chile pushes the origin of caviomorph-like rodents in South American back to 31.5 million years ago (Flynn & Wyss 1998). Although more data would be welcome, it is now generally agreed that the South American hystricognaths (caviomorphs) arose from an African ancestry (MacFadden 2006; Martin 1994; Wyss et al. 1993).

Competition from other orders of South American mammals was apparently not intense, for the caviomorphs rapidly radiated; 7 of the 13 living caviomorph families can be distinguished among fossils from the late Oligocene (Deseadan). The Oligo-Miocene climatic changes in South America were accompanied by an expansion of grasslands that strongly affected the fortunes of the caviomorphs. The

TABLE 13-2   Shared Derived Characters in Hystricognathous Rodents

- Mandible hystricognathous
- Infraorbital foramen enlarged and hystricomorphous
- Deepened pterygoid fossa opening into the orbit
- Enlarged slip of the superficial masseter muscle attached to the medial mandible
- Malleus and incus closely appressed or fused
- Internal carotid artery and canal absent
- Stapedial artery absent
- Molars four or five crested

Data from Luckett and Hartenberger 1985.

Octodontidae increased its range in the Pliocene, whereas the range of the Echimyidae shrank. The Dasyproctidae became less common in the Pliocene, and the Chinchillidae became less diverse. From South America or Central America, the hutias (Capromyidae) probably reached the Lesser Antilles in the Oligocene by rafting between islands or along an ancient cross-Caribbean landmass (Iturralde-Vinent & MacPhee 1999) and had varying degrees of success there. The New World porcupines (Erethizontidae) have been successful in moving northward from their ancestral home in South America. They are now widespread in tropical North America, and one species (*Erethizon dorsatum*) is widely distributed in North American forests and woodlands to the Arctic Slope of Canada and Alaska.

Family Bathyergidae

This family contains the African mole-rats, a fascinating group of highly specialized fossorial rodents of special interest because of the unique social behavior of two of its members (Chapter 24). The family includes 6 genera with 17 living species. Mole-rats occupy much of Africa from Ghana, Sudan, Ethiopia, and Somaliland southward. The earliest fossil records of bathyergids are from the early Miocene of Africa. Bathyergids are allied with the Old World hystricognaths (Luckett 1980b; Maier & Schrenk 1987; Sarich & Cronin 1980; Sherman et al. 1991).

Bathyergids are from 120 to 330 millimeters in total length, and they possess a number of unique structural features associated with their fossorial life. The eyes are small in all species. Their vision apparently is poorly developed, and the visual centers in the brain are reduced (Jarvis &

Bennett 1991). The ears lack or nearly lack pinnae. The skull is usually robust. The powerful incisors are procumbent in all species (**Fig. 13-37**), and the roots of the upper incisors usually extend behind the molars. The lips close tightly behind the incisors so that dirt does not enter the mouth when the animal is burrowing (**Fig. 13-38**). The cheek teeth are hypsodont but rooted and typically have a simplified crown pattern. The dental formula is variable (1/1, 0/0, 2–3/2–3, 0–3/0–3 = 12–28). In *Heliophobius*, there are six cheek teeth, but not all are functional simultaneously. The zygomasseteric structure is distinctive. The infraorbital foramen transmits little or no muscle. The masseter muscles, however, are highly specialized: The large anterior part of the medial masseter originates from the upper part of the medial wall of the orbit, and the superficial part of the lateral masseter originates partly on the anterior face of the zygoma. The mandibular fossa and angular part of the dentary bone are greatly enlarged and provide an extensive area for the insertion of the masseter muscles (Figs. 13-4 and 13-37). The limbs are fairly robust in all species but are apparently used for digging only in *Bathyergus*. The hind feet are broad, and the animals back up against a load of soil and push it from the burrow with the hind feet. The tail is short and is used as a tactile organ. Most species have a complete pelage, but in *Heterocephalus glaber* the skin is nearly naked with only a sparse sprinkling of long hairs (Fig. 13-38A).

The African mole-rats are herbivorous and eat primarily underground storage organs of plants (**geophytes**), such as tubers, corms, and bulbs, which the mole-rats find by burrowing. In some areas, mole-rats depend on a single

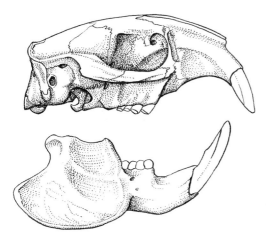

A Giant mole-rat
(*Fukomys mechowi*, Bathyergidae)

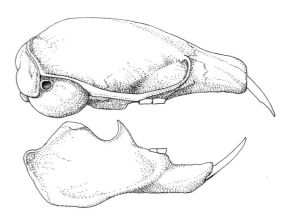

B Naked mole-rat
(*Heterocephalus glaber*, Bathyergidae)

FIGURE 13-37 (A) Skull of a giant mole-rat (*Fukomys mechowi*; note the large, procumbent incisors, used for digging, and the enormously enlarged angular process of the dentary; length of skull 57 millimeters). (B) Skull of the naked mole-rat (*Heterocephalus glaber*). Note the procumbent incisors. (A, adapted from Hill, J. E., and Carter, T. D., *Bull. Amer. Mus. Nat. Hist.* 78 (1941): 1–211.)

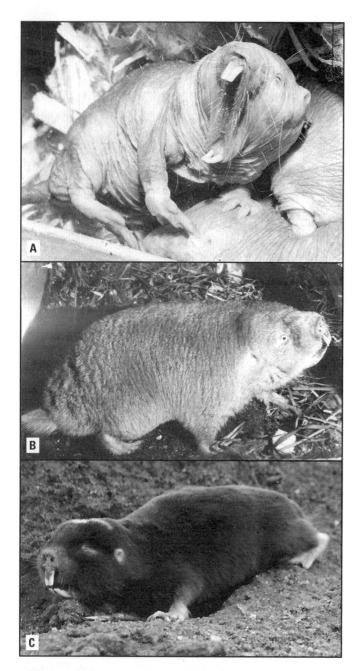

FIGURE 13-38 (A) The naked mole-rat (*Heterocephalus glaber*, Bathyergidae) of East Africa. The lips close behind the incisors, and a fold of skin guards the nostrils. These adaptations keep soil from being inhaled and ingested when the incisors are involved in digging. (B) The Cape dune mole-rat (*Bathyergus suillus*; Bathyergidae) of South Africa. (C) A Damara mole-rat (*Fukomys damarensis*, Bathyergidae) in the Kalahari Desert of Namibia.

plant species. Such simple food chains can be unstable. In the Kalahari Desert of Namibia, a large colony of *Fukomys damarensis* (formerly *Cryptomys*; Kock et al. 2006) disappeared when their staple food (wild onion, *Dipcada glaucum*) died out (J. Jarvis, personal communication). Mole-rats seldom appear above ground and usually occupy soft loamy or sandy soils in desert and savanna areas. The

huge incisors are the major digging tools in all species except *Bathyergus suillus* (Fig. 13-38B). Three of the six bathyergid genera are solitary; however, *Heterocephalus*, *Cryptomys*, and *Fukomys* are colonial, and their burrow systems are far more extensive than those of solitary species. Burrowing *Heterocephalus* workers form "digging chains" with an organized division of labor. *Heterocephalus glaber* (Fig. 13-38A) and *F. damarensis* are of special interest because they are the only known **eusocial** mammals, indeed, the only eusocial vertebrates (Jarvis & Bennett 1993; O'Riain et al. 2000; Scantlebury et al. 2006) (see Chapter 24). All the other *Cryptomys* and *Fukomys* species are social with division of labor among colony members and cooperative care of the young, but only *Heterocephalus* and *F. damarensis* have true overlapping generations (with replacement of the breeder occurring from within the colony) and other features of eusociality. The other social mole-rats seem to be obligate outbreeders, and the colony appears to break up with the death of a breeder.

Family Hystricidae

The Old World porcupines are a widely distributed group of rodents (3 genera with 11 Recent species) that resemble the New World porcupines (Erethizontidae) in having quills for protection. Hystricids occur throughout Africa (except Madagascar), in Sicily, Italy, Albania, and northern Greece (the Romans perhaps introduced porcupines into this area), in southern Asia and South China, and in Borneo, southern Sulawesi, Flores, and the Philippines. Hystricids first appear in the Miocene of Europe.

These large, stocky, and imposing rodents weigh up to 27 kilograms. The occipital region of the skull is unusually strongly built and provides attachment for powerful neck muscles. The zygomasseteric arrangement is hystricomorphous, with the large anterior part of the medial masseter originating on the deep rostrum. The skull is strongly domed in some African species because the nasoturbinal, lacrimal, and frontal bones are highly pneumatic (filled with air cavities; **Fig. 13-39**). The dental formula is 1/1, 0/0, 1/1, 3/3 = 20. The hypsodont cheek teeth have re-entrant enamel folds that, with wear, become islands on the occlusal surfaces. Some of the hairs are stiff, sharp spines that reach at least 40 centimeters in length; in some species, open-ended, hollow spines make a noise when rattled that appears to have a warning function. One genus (*Trichys*, of Borneo, the Malay Peninsula, and Sumatra) lacks stiff spines. The large, plantigrade feet of hystricids are five toed, and the soles are smooth.

Hystricids are herbivorous but, in contrast to New World porcupines, are terrestrial rather than partly arboreal and often dig fairly extensive burrows that are used as dens. The quills of some species are conspicuously marked

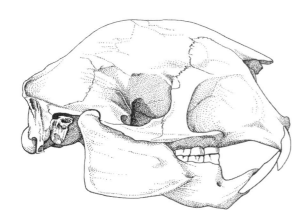

African crested procupine
(*Hystrix cristata*, Hystricidae)

FIGURE 13-39   Skull of the African crested porcupine (Hystricidae). Note the greatly inflated rostrum and frontal part of the skull.

with black and white bands (**Fig. 13-40A**). The two African species of *Hystrix* have an impressive intimidation display, involving fanning and "whirring" the quills, stamping the feet, and rushing rump-first toward the adversary (Kingdon 1984b). Aside from humans, the larger cats are the main predators of hystricids.

Family Petromuridae

The relict family Petromuridae includes but a single species, *Petromus typicus*, the dassie rat (Fig. 13-40B). This animal lives in parts of southwestern Africa, including the extremely dry and barren Namib Desert. The family is known by fossils only in the Pliocene of Africa (Sénégas 2004).

The dassie rat is a small rodent (about 200 grams) with a squirrel-like appearance. The infraorbital foramen is enlarged and transmits part of the medial masseter. The rooted, hypsodont cheek teeth have a simplified crown pattern; the dental formula is 1/1, 0/0, 1/1, 3/3 = 20. Structurally, these animals are most remarkable for specializations enabling them to seek shelter in narrow crevices. Such specializations include a strongly flattened skull, flexible ribs that allow the body to be dorsoventrally flattened without injury, and mammae situated laterally at the level of the scapulae, where the young can suckle while the female is wedged in a rock crevice.

Dassie rats are diurnal and feed largely on leaves and flowers. They are restricted to dry, rocky areas (**Fig. 13-41**). They bask in the early morning sun on cool mornings, a well-known energy-conservation behavior of another leaf eater, the hyrax (Hyracoidea). The daily energy expenditure of the dassie rat is much lower than the average for mammals its size (Withers et al. 1979), perhaps partly as a result of the use of behavioral thermoregulation.

Family Thryonomyidae

One genus with only two species makes up the small family Thryonomyidae, the cane rats. These animals are broadly distributed in Africa south of the Sahara. The earliest records of cane rats are from the late Eocene of Africa. Fossil cane rats from the Miocene of southern Asia and an extinct species of *Thryonomys* from the central Sahara indicate that cane rats and marshy areas were once more widespread than they are now (López et al. 2004).

Cane rats are large rodents, from 4 to 6 kilograms in weight, with a coarse, grizzled pelage. The snout is blunt, and the ears and tail are short. The robust skull has prominent ridges, a heavily built occipital region, and a large infraorbital foramen. The cheek teeth are hypsodont, and the large upper incisors are marked by three longitudinal grooves. The dental formula is 1/1, 0/0, 1/1, 3/3 = 20. The fifth digit of the forefoot is small, and the claws are strong and adapted to digging.

FIGURE 13-40   (A) African porcupine (*Hystrix* sp.) (B) A dassie rat (*Petromus typicus*; Petromuridae).

FIGURE 13-41   The rocky habitat of the dassie rat (*Petromis typicus*, Petromuridae) in Namibia.

Cane rats are capable swimmers and divers and are largely restricted to the vicinity of water, where they take shelter in matted vegetation or in burrows. Males indulge in ritualized snout-to-snout pushing contests, and the blunt shape of the snout seems to enable the animals to avoid damage during these bouts. They are herbivorous and do considerable local damage to crops, particularly cassava and sugar cane. Cane rats are prized for food in many parts of Africa. The animals are often taken by snares, during organized drives using dogs or are driven from their hiding places and captured when hunters set fire to reeds.

Family Erethizontidae

To this small family belong the New World porcupines, a group including five Recent genera with 16 living species. These animals occur from the Arctic south through much of the forested part of Canada and the United States into Sonora, Mexico, in the case of *Erethizon* and from southern Mexico through much of the northern half of South America in the case of the other genera. Porcupines are of interest because of their remarkable coat of quills and because the animals often have little apparent fear of humans and thus can be observed easily.

New World porcupines are large, heavily built rodents, weighing up to 16 kilograms. All species have quills on at least part of the body. The stiff quills are usually conspicuously marked by dark- and light-colored bands, and the sharp tips have small, proximally directed barbs. These barbs make the quills difficult to remove from flesh and aid in their penetration, which progresses at the rate of 1 millimeter or more per hour. The skull is robust. The rostrum is deep, and the greatly enlarged infraorbital foramen is nearly circular in some species (**Fig. 13-42**) and

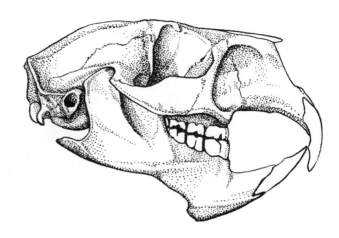

Porcupine
(*Erethizon dorsatum*, Erethizontidae

FIGURE 13-42   Skull of a porcupine. Note the large infraorbital foramen (length of skull 115 millimeters).

accommodates the highly developed medial masseter. The dental formula is 1/1, 0/0, 1/1, 3/3 = 20; the rooted cheek teeth have occlusal patterns dominated by re-entrant enamel folds (**Fig. 13-43**). New World porcupines have some arboreal adaptations that are lacking in their more terrestrial Old World counterparts. The feet of erethizontids have broad soles marked by a pattern of tubercles that increase traction (Fig. 13-15); in some species, the hallux is replaced by a large, movable pad. The toes bear long, curved claws, and the limbs are functionally four-toed. In *Coendou*, the long tail is prehensile and curls dorsally to grasp a branch. Beginning in the Oligocene, this family underwent its early evolution in South America, becoming established in North America only after the emergence of the previously inundated Isthmus of Panama in the Pliocene.

New World porcupines eat a variety of plant material. Cambium is a staple winter food for *Erethizon*, and in many timberline areas, trees missing large sections of bark and cambium give evidence of long-term winter occupancy by porcupines. In summer, porcupines eat a variety of plants and often "graze" on sedges (*Carex*) along the borders of meadows. Most species of porcupines are able climbers, and *Coendou* spends most of its life in trees. *Erethizon* is inoffensive and at times almost oblivious to humans. When in danger, however, the animal directs its long dorsal hairs forward, exposing the quills; it erects the quills and arches its back. The tail is flailed against an attacker as a last resort. Surprisingly, *Erethizon* is killed by a variety of carnivores. Some mountain lions learn to flip porcupines on their backs and kill them by attacking the unprotected belly. Occasionally, however, dead or dying carnivores are found with masses of quills penetrating the mouth and face, indicating that learning to prey on porcupines may be a fatal undertaking. *Erethizon* characteristically takes shelter in rock piles, beneath overhanging rocks, or in hollow logs but (as other New World porcupines) does not dig burrows as do Old World porcupines.

 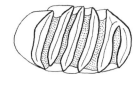

Porcupine   Capybara

FIGURE 13-43 Crowns of hystricognath molars. The upper right molars one and two of the porcupine (*Erethizon dorsatum*, Erethizontidae) and the third lower left molar of the capybara (*Hydrochoerus hydrochaeris*, Caviidae). The cross-hatched areas on the porcupine teeth are dentine; the stippled areas on the capybara tooth are cementum. (Capybara adapted from Ellerman, J. R. *The Families and Genera of Living Rodents*. British Museum of Natural History, 1940.)

Family Chinchillidae

One member of the family Chinchillidae, the chinchilla (*Chinchilla*), is somewhat familiar to many because of the publicity given to chinchilla fur farming some years ago. The family also includes the viscachas (*Lagidium* and *Lagostomus*). Three genera with seven Recent species represent the family, which occurs in roughly the southern half of South America in the high country of Peru and Bolivia and throughout much of Argentina to near its southern tip. The fossil record of this group is entirely South American and extends from the early Oligocene to the Recent; one of the oldest known rodents in South America is a chinchillid (Flynn et al. 2003).

Chinchillids are densely furred and of moderately large size (1 to 9 kilograms), with a long, well-furred tail. Mountain viscachas (*Lagidium*) and chinchillas have fairly large ears and a somewhat rabbit-like appearance, whereas the plains viscacha (*Lagostomus*) has short ears. The cheek teeth are ever growing, and the occlusal surfaces are formed by transverse enamel laminae with intervening cementum. The dental formula is 1/1, 0/0, 1/1, 3/3 = 20. There are some cursorial adaptations, but the clavicle is retained. The forelimbs are fairly short and tetradactyl. The hindlimbs are long, however, and the elongate feet have four toes in the chinchillas and mountain viscachas or three toes in the plains viscacha.

Chinchillids are herbivorous and occupy a variety of habitats, including open plains (pampas), brushlands, and barren, rocky slopes at elevations ranging from 800 to 6,000 meters. The mountain viscachas and chinchillas are diurnal and seek shelter in burrows or rock crevices. Although adept at moving rapidly over rocks and broken terrain, they seem not to depend on speed in the open to escape enemies. The plains viscacha, in contrast, occurs in open pampas areas with little cover, where colonies live in extensive burrow systems marked by low mounds of earth and accumulations of such debris as bones, livestock droppings, and plant fragments. The habit of collecting items is displayed even by captive animals. Colonies may occupy large areas; one such area measured 20 by 300 meters, and this colony was known to have been in existence for at least 70 years (Weir 1974). In *Lagostomus*, cursorial ability is highly developed. These animals are able to make long leaps and to evade a pursuer by abrupt turns. They have considerable endurance and can run at speeds up to 40 kilometers per hour.

Family Dinomyidae

The Dinomyidae includes a single South American species, *Dinomys branickii*, the pacarana. This rare animal inhabits the foothills of the Andes and adjacent remote valleys in Peru, Colombia, Ecuador, and Bolivia. Several extinct members of this family were huge, including one that is the largest known rodent. The late Miocene and Pliocene species *Josephoartigasia monesi* weighed over 1,000 kg (Rinderknecht & Blanco 2008). *Phoberomys pattersoni* probably weighed 700 kg (Sanchez-Villagra et al. 2003), and *Telicomys* was yet another giant. Although dinomyids appear to be near extinction today, they were more successful and diverse in the past with up to 58 species, the oldest of which appears in the Miocene of South America (Mones 1981).

The pacarana weighs up to 15 kilograms; the dark-brown pelage is marked by longitudinal white stripes and spots. Pacaranas lack the cursorial adaptations of the Caviidae and Hydrochaeridae. Instead, the broad tetradactyl feet of pacaranas have long, stout claws seemingly adapted to digging, and the foot posture is plantigrade. The clavicle is complete, another departure from the conventional cursorial morphology. The unusually hypsodont cheek teeth consist of a series of transverse plates. The dental formula is 1/1, 0/0, 1/1, 3/3 = 20.

This unusual rodent is herbivorous, slow moving, and docile in captivity. It is probably nocturnal and communicates by foot stamping and an elaborate series of vocalizations (Collins & Eisenberg 1972). It is extremely rare in the wild and is currently listed as endangered.

■ Family Caviidae

This family includes just 6 genera and 18 species and contains the familiar guinea pig (*Cavia*, Caviinae), as well as several similar types, the cursorial Patagonian mara (*Dolichotis*, Dolichotinae; Fig. 13-44) and the semiaquatic capybara (*Hydrochoerus*, Hydrochoerinae). Caviids occur nearly throughout South America, except in Chile and parts of eastern Brazil, and first appeared in the Miocene of South America.

The guinea pig-like caviids (subfamily Caviinae) are chunky and moderately short-limbed and weigh from 400 to 700 grams. *Dolichotis* (Dolichotinae), in contrast, has long, slender legs and feet and weighs up to approximately 16 kilograms. All caviids have ever-growing cheek teeth with occlusal patterns consisting basically of two prisms. The dental formula is 1/1, 0/0, 1/1, 3/3 = 20. The dentary bone has a conspicuous lateral groove into which insert the temporal muscle and the anterior part of the medial masseter. Although only *Dolichotis*, the Patagonian mara, is strongly cursorial, all caviids have certain features typical of cursorial mammals: The clavicle is vestigial, the tibia and fibula are partly fused, and the digits are reduced to four on the manus and three on the pes. Members of the subfamily Caviinae, despite these cursorial adaptations, have a plantigrade foot stance and scuttle about in mouse-like fashion. *Dolichotis*, however, is a swift runner and

FIGURE 13-44 Patagonian mara (*Dolichotis* sp.)

makes long bounds. The foot posture of *Dolichotis* during running is digitigrade, and specialized pads beneath the digits (Fig. 13-45D) cushion the impact when the feet strike the ground. With its deep, somewhat laterally compressed skull and large ears, *Dolichotis* resembles a large rabbit and is sometimes called (mistakenly) Patagonian hare. *Dolichotis* is an inhabitant of open, arid regions. It is diurnal and forms large, loosely knit groups on occasion, but the basic social unit is a monogamous pair that mates for life.

Caviids are herbivorous, and some have complex systems of social behavior and vocal communication (Lacher 1981; Rood 1970b, 1972). They occupy habitats ranging from grassland to brushy and rocky areas and forest edges. They are nocturnal, diurnal, or crepuscular (active at dawn and dusk) and often live in large colonies with distinct social hierarchies.

The subfamily Hydrochoerinae contains the largest living rodent, the capybara (*Hydrochoerus*). The single species occupies Panama and roughly the northern half of South America east of the Andes. Extinct species of capybara are known from the Pliocene and Pleistocene in North America and the Miocene to Recent in South America. An extinct Pliocene giant (*Protohydrochoerus*) probably weighed over 200 kilograms.

Capybaras are large (up to 79 kilograms in weight), robust, rather short-limbed rodents with a coarse pelage (Fig. 13-46). The head is large and has a deep rostrum and truncate snout. The skull and dentary bone are similar to other members of the Caviidae, but the paroccipital processes are unusually long. The teeth are ever growing. Both upper and lower third molars are much larger than any other cheek tooth in their respective rows and are formed by transverse

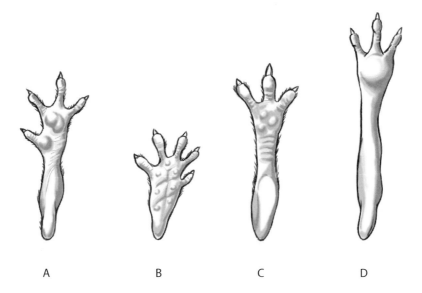

FIGURE 13-45 Ventral views of the left hind foot of some South American rodents. (A) Chinchilla (*Chinchilla* sp., Chinchillidae), (B) degu (*Octodon* sp., Octodontidae), (C) *Dasyprocta* sp. (Dasyproctidae), (D) Patagonian mara (*Dolichotis* sp., Caviidae). (Adapted from Howell, A. B. *Speed in Animals*. University of Chicago Press, 1965.)

lamellae united by cementum (Fig. 13-43). The dental formula is 1/1, 0/0, 1/1, 3/3 = 20. The tail is vestigial, and the same cursorial features listed for the caviids occur in the hydrochoerids. The digits are partly webbed and unusually strongly built, adaptations that probably allow the support of the considerable body weight on marshy ground.

Capybaras are semiaquatic and occur along the borders of marshes or the banks of streams, where they forage on succulent herbage (Quintana et al. 1998). They are largely crepuscular and, although they can run fairly rapidly, usually seek shelter in the water. They swim and dive well and can remain submerged beneath water plants with only the nostrils above water. In both form and function, a capybara roughly resembles a miniature hippopotamus. In parts of Venezuela, capybaras are raised commercially as food for human consumption.

▪ Family Dasyproctidae

Members of family Dasyproctidae and their relatives, called agoutis (**Fig. 13-47A**), occur in the Neotropics from southern Mexico south to Ecuador, Bolivia, Paraguay, and northeastern Argentina. Two genera and 13 species (some of which are of doubtful validity) are included in this family. One of the earliest fossils of rodents in South America (earliest Oligocene) is a possible dasyproctid from Chile (Flynn et al. 2003).

These rodents are medium-size, up to 2 kilograms in weight. The tail is short. The skull is robust. The incisors are fairly thin, and the crowns of the hypsodont cheek teeth are flat and bear five crests. The dental formula is 1/1, 0/0, 1/1, 3/3 = 20. Although these rodents are compactly built, the limbs are slim and have many cursorial adaptations. The forefeet are tetradactyl, and the plane of symmetry passes between digits three and four (as in the Artiodactyla). The hind feet have three toes, and the plane of symmetry passes through digit three (Fig. 13-45C), as in the Perissodactyla. The clavicle is vestigial, and the claws are hoof like.

These herbivorous rodents typically inhabit tropical forests, where they are largely diurnal. *Dasyprocta punctata* is territorial and scatterhoards fruit and nuts in times of plenty to be used during lean times. These animals are probably important seed dispersers. A male *D. punctata* sprays a female with urine during courtship; the female then goes into a "frenzy dance" and allows the male to copulate (Smythe 1978). Some species take refuge in burrows that they dig in the banks of arroyos, beneath roots, or among boulders. Agoutis are rapid and agile runners and usually travel along well-worn trails. Agoutis have a habit of remaining still when approached by a predator and then bursting from cover and running away after the fashion of a small antelope.

▪ Family Cuniculidae

Family Cuniculidae includes two species of a single genus (*Cuniculus*, formerly Agouti). These rodents live in tropical forests from central Mexico to southern Brazil and are known from the Oligocene to Recent. Often called pacas, they are large (weighing up to about 12 kilograms) nearly tailless and have a conspicuous pattern of white spots and stripes on the body (Fig. 13-47B). They have an exceptionally ungraceful form, with short legs and a blunt head. There are four digits on the forefeet and five digits on the hind feet. The cheek teeth are high-crowned, and the dental formula is 1/1, 0/0, 1/1, 3/3 = 20. Resonating chambers are formed by concavities in the maxillaries and by greatly broadened

FIGURE 13-46   South American capybara (*Hydrochoeris hydrochaeris*, Caviidae).

zygomatic arches (**Fig. 13-48**); air is forced through associated pouches, producing a resonant, rumbling sound. The massively enlarged zygoma are unique to these animals.

These terrestrial rodents live in tropical forests along streams and rivers, where they dig burrows in banks. The diet consists of a variety of plant material, including fallen fruit. They are nocturnal and not particularly swift on land but are good swimmers and often escape enemies by fleeing into the water. Intensive hunting for their highly prized meat, coupled with the fact that pacas seldom bear more than one young at a time, has lead to dramatic declines in their populations in many areas.

FIGURE 13-47   (A) An agouti (*Dasyprocta* sp., Dasyproctidae) and (B) a lowland paca, *Cuniculus paca* (Cuniculidae).

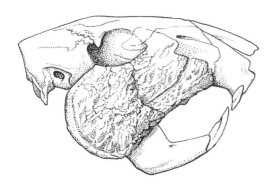

Paca
(*Cuniculus paca*, Cuniculidae)

FIGURE 13-48 Skull of a paca. Note the great enlargement of the zygomatic arch in this hystricognath rodent (length of skull 150 millimeters). (Adapted from Hall, E. R., and K. R. Kelson. *Mammals of North America, Volume 2*. John Wiley & Sons Ltd., 1959.)

Family Ctenomyidae

Members of family Ctenomyidae, called tuco-tucos, are fossorial and superficially resemble pocket gophers (Geomyidae). They occupy much of the southern two thirds of South America, from Peru to Tierra del Fuego, including the Andes Mountains to elevations of 4,000 meters. There is a single genus (*Ctenomys*) with 60 species. Ctenomyids first appear in the late Miocene of South America.

These rodents range in size from 100 to 700 grams and are unusual in having simplified cheek teeth that are roughly kidney shaped; the dental formula is 1/1, 0/0, 1/1, 3/3 = 20, and the third molar is vestigial. The skull is broad and dorsoventrally flattened. Among South American hystricognathous rodents, only in the Ctenomyidae and in one species of the Octodontidae (*Spalacopus cyanus*) are fossorial adaptations strongly developed. The head of the tuco-tuco is large and broad, and the stout incisors protrude permanently from the lips. The eyes and ears are small. The neck is short and powerfully built. The forelimbs are powerful. The manus has long claws, and the tail is short and stout. In contrast to pocket gophers, tuco-tucos have greatly enlarged hind feet with powerful claws, and they lack external cheek pouches. Fringes of hair on the toes of the fore and hind feet in tuco-tucos are presumably an aid to the animals when they are moving soil.

Tuco-tucos are herbivorous and eat such underground parts of plants as roots, tubers, and rhizomes. They dig extensive burrow systems in open, often barren areas, and most species live in "colonies" composed of many solitary individuals, each with its burrow systems spaced widely apart from those of its neighbors. An animal typically occupies a given burrow system permanently but periodically seeks adjacent foraging areas by digging new burrows.

Tuco-tucos occasionally make short forays from their burrows to gather leaves and stems (Pearson 1959). In contrast to pocket gophers (Geomyidae), tuco-tucos are quite vocal and give distinctive cries from burrow entrances. Tuco-tucos share **karyotypic** similarities with the Octodontidae (George & Weir 1972). Because *Ctenomys* is restricted to sandy soils, the isolation of these areas by hardpan or rocky terrain has seemingly led to the rapid proliferation of species.

Family Octodontidae

Although mostly burrow dwellers, octodontids are rat like in general appearance, and most species lack the fossorial specializations typical of the Ctenomyidae (**Fig. 13-49**). Octodontids, variously called degus, cururos, or rock rats, have a restricted range near the west coast of South America from southwestern Peru south to northern Argentina and northern Chile. There are 8 genera and 13 species. The earliest fossil records are from the late Oligocene of South America. These small rodents (200 to 300 grams) derive their family name from the "eight-shaped" crown pattern of the cheek teeth; the dental formula is 1/1, 0/0, 1/1, 3/3 = 20. Most species have large ears, large eyes, and long vibrissae. The forefeet have four digits, and the hind feet have five; the tail varies from long to rather short.

Octodontids occupy a variety of habitats, from grassy areas to high Andean forests to dry cactus and acacia slopes. *Octodon* is an able climber that takes shelter in rocks or the burrows of other animals, and *Octodontomys* lives in burrows and in rock crevices or caves and feeds on acacia pods and cactus. Neither of these genera is specialized for fossorial life, nor is *Octomys*. Of the remaining genera,

FIGURE 13-49 A rock rat or degu (*Octodon degus*, Octodontidae).

Aconaemys is somewhat modified for fossorial life, and *Spalacopus* is strongly so. In Chile, *Spalacopus cyanus* occupies sandy coastal areas where it occurs in colonies, all members of which occupy a common burrow system. The animals feed entirely below ground, and the tubers and underground stems of huilli, a species of lily (*Leucoryne ixiodes*), form the bulk of the diet. *Spalacopus* is nomadic, an exceptional mode of life for a rodent. When a colony exhausts the supply of huilli roots at one place, the animals abandon this foraging site and move to a nearby undisturbed area (Reig 1970). *Spalacopus* is unusually vocal for a rodent and gives distinctive calls at burrow openings. It uses its forelimbs and teeth to loosen soil and its large hind feet to throw dirt from the mouth of the burrow. The ranges of the tuco-tucos (Ctenomyidae), and the similarly adapted *Spalacopus* do not overlap. The restricted diet and nomadic nature of *Spalacopus* have not favored the rapid speciation found in the more sedentary Ctenomyidae (Reig 1970).

Family Abrocomidae

Members of family Abrocomidae, the "chinchilla rats," occur in parts of west-central South America. Their range includes southern Peru, Bolivia, and northwestern Argentina and Chile. The family is represented today by 2 genera with 10 species. This family appeared first in the South American Miocene.

Chinchilla rats look roughly like large woodrats (*Neotoma*, Cricetidae), reach over 400 millimeters in total length, and resemble octodontids in many ways. The pelage is long and dense. The skull has a long, narrow rostrum, and the bullae are enlarged. The cheek teeth are ever growing; the upper teeth have an internal and an external enamel fold, whereas the lowers have two internal folds. The dental formula is 1/1, 0/0, 1/1, 3/3 = 20. The limbs are short and have short, weak nails. The pollex is absent. *Abrocoma bennetti* is unusual among rodents in having 17 pairs of ribs.

These herbivorous rodents are poorly known. They are seemingly colonial, climb well, and usually seek shelter beneath or among rocks or in tunnels under rocks or bushes. *A. cinerea* lives in cold, bleak, rocky areas in the Andes at elevations between 3,700 and 5,000 meters. Its extremely long digestive tract (2.5 meters), and voluminous caecum suggest a diet of leaves.

Family Echimyidae

Members of the important Neotropical family Echimyidae, which includes a variety of roughly rat-sized rodents, are called spiny rats. Most of the living species have flattened, spine-like hairs with sharp points and slender basal portions. Approximately 87 living species of 21 genera are recognized. Spiny rats are widely distributed in the Neotropics, occurring from Nicaragua southward through the northern half of South America to Paraguay and southeastern Brazil.

Echimyids have prominent eyes and ears. Some species are fairly large for rodents, weighing over 600 grams. The tail, which in some genera is longer than the head and body, is lost readily, a feature perhaps of value in aiding escape from predators. The point of weakness is at the centrum of the fifth caudal vertebra. Among 637 *Proechimys* taken in Panama, 18% were tailless (Fleming 1970). Typically, the skull has a deep rostrum and a large infraorbital foramen; the coronoid process of the dentary is reduced, and the angular process is prominent (**Fig. 13-50**). The cheek teeth are rooted, and the occlusal surfaces in most species are marked by transverse re-entrant folds. The dental formula is 1/1, 0/0, 1/1, 3/3 = 20. The feet are not highly specialized in most genera. In the arboreal tree rats (*Echimys*), however, the digits are elongate and partially syndactylous. When an animal is climbing, the first two digits grasp one side of a branch in opposition to the remaining digits, which grasp the other side. Tree rats inhabit the canopy zone, and as a result, their ecology and behavior are poorly known.

Echimyids are an old group, appearing first in the late Oligocene of South America. Two extinct genera are known from skeletal material found in Indian kitchen middens in Cuba and Haiti. These genera seemingly became extinct fairly recently. In the case of the genus from Haiti (*Brotomys*), extinction may have resulted from the introduction of predators by Europeans.

As far as is known, spiny rats are completely herbivorous. They are variously semifossorial, terrestrial, or arboreal. In Panama, fruit was the primary food found in the stomachs of *Proechimys semispinosus* (Fleming 1970). *Kannabateomys* apparently prefers the young shoots of bamboo and inhabits dense bamboo thickets near waterways (Emmons 1990). Several arboreal species take shelter in tree holes. The bamboo rat (*Dactylomys dactylinus*) of South

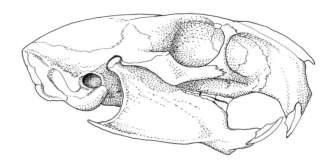

Spiny rat

(*Diplomys labilis*, Echimyidae)

FIGURE 13-50 Skull of a spiny rat (*Diplomys labilis*, Echimyidae). Note the prominent angular process of the dentary and the greatly enlarged infraorbital foramen.

America is completely nocturnal and arboreal, eats leaves and buds, and gives explosive calls that presumably play a role in territoriality (Emmons 1981). *Proechimys* is terrestrial and is among the most abundant lowland rainforest rodents (Eisenberg 1989; Emmons 1990).

Family Myocastoridae

The nutria or coypu (*Myocastor coypus*), the only living member of the family Myocastoridae, is familiar to many people in North America, Europe, and Asia because this South American rodent has been introduced widely and has thrived in certain areas. In some places, it has become a serious pest because of its destruction of aquatic vegetation and crops and its disruption of irrigation systems. This animal is native to southern South America, from Paraguay and southern Brazil southward, but now also occurs in some 15 states in the United States, as well as in some countries in Europe and East Africa. The family is also represented by nine extinct genera ranging back to the early Miocene in South America.

The nutria is large, up to roughly 8 kilograms, and looks like a rat-tailed beaver (*Castor*; Castoridae). The skull is heavily ridged and has a deep rostrum. The zygomasseteric structure is hystricomorphous; in association with the reduction of the temporal muscles, the coronoid process of the dentary bone has nearly disappeared. The hypsodont cheek teeth well illustrate changes in crown pattern that occur with increasing age and wear (**Fig. 13-51**). The dental formula is 1/1, 0/0, 1/1, 3/3 = 20. The feet have heavy claws, and a web joins all but the fifth toe of the pes.

Nutrias resemble beavers in some of their habits. They dig burrows in banks, use cleared trails through vegetation, are extremely destructive to plants, and are capable swimmers and divers. Because of their dense, fine underfur, they have been raised in some fur farms in the United States and are trapped for their fur in some states. In the trapping season of 1975 to 1976, the nutria sold to the fur trade in Louisiana yielded $8 million (Woods 1984). Most biologists strongly oppose the indiscriminate introductions of such nonnative animals (referred to as "exotics") as the nutria. The activities of nonnative species often result in the alteration of the vegetation, with the resultant disappearance of native species and the destruction, perhaps irretrievably, of the original biotic community.

Family Capromyidae

Members of family Capromyidae are known locally as hutias and are restricted to the West Indies. Of the 20 species in eight genera listed by Woods (1993), 6 species are now extinct and 2 others are probably extinct (Woods et al. 1985). The 12 living species occupy the Bahamas, Cuba, Isle of Pines, Hispaniola, Puerto Rico, and Jamaica. The oldest fossils are from the early Miocene. These mostly herbivorous rodents weigh up to about 7 kilograms and look like unusually large, thick-furred rats (**Fig. 13-52**). They closely resemble the nutria (Myocastoridae) structurally and are often included in the same family. Hutias have hypsodont and flat-crowned molars; the dental formula is 1/1, 0/0, 1/1, 3/3 = 20. They are primarily herbivorous but are known to take small lizards. Although most species are solitary, *Geocapromys ingrahami* lives in colonies.

These rodents, adapted to the insular conditions of the West Indies before the coming of Europeans, were unable to cope with predation by the introduced mongoose (*Herpestes*) or by humans and their dogs. Of the 35 Recent species of capromyids, 23 are extinct; the remaining ones are restricted to steep or inaccessible areas (Woods 1982; Woods et al. 1985). One living species (*Mesocapromys nanus*) was first described from bones found in a cave but was later found alive. (Members of the West Indian family Heptaxodontidae have gone extinct in Recent or sub-Recent times and are not considered here.)

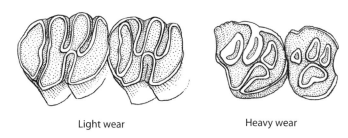

Light wear   Heavy wear

Nutria
(*Myocastor coypus*, Myocastoridae)

FIGURE 13-51 First and second upper right molars of the nutria. Note the tremendous changes in the crown pattern caused by wear. Stippled areas on the occlusal surfaces surrounded by enamel (unshaded) are dentine.

FIGURE 13-52 A Desmarest's hutia (*Capromys pilorides*) from Cuba.

Antelope jackrabbit
(*Lepus alleni*, Leporidae)

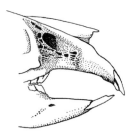

Arctic hare
(*Lepus arcticus*, Leporidae)

FIGURE 13-53 Skull of the antelope jackrabbit (*Lepus alleni*); note the highly fenestrated maxillary and occipital bones. On the right is the anterior part of the skull of the arctic hare (*L. arcticus*), showing the procumbent incisors and the receding nasals, specializations associated with this animal's habit of using the incisors to scrape away ice and snow to reach food. (Artic hare adapted from Hall, E. R., and K. R. Kelson. *Mammals of North America, Vol 2*. John Wiley & Sons, 1959.)

Order Lagomorpha

The taxonomic position of the Lagomorpha (pikas and rabbits) with respect to the other mammalian orders has been vigorously debated over the past century. Lagomorphs were traditionally united with rodents in the cohort Glires (Gregory 1910; Simpson 1945); however, such groups as primates, insectivorans, and artiodactyls have also been considered the sister group of the lagomorphs (Wood 1957). Cranial characteristics (Gaudin et al. 1996; Novacek 1985), placental membranes (Luckett 1985), molecular genetic data (Douzery & Huchon 2004), and new fossil evidence (Li & Ting 1985; Li et al. 1987; Meng & Wyss 2001; Meng et al. 2003) support a sister group relationship between rodents and lagomorphs (Glires).

Although lagomorphs—the rabbits (Leporidae) and pikas (Ochotonidae)—are not a diverse group, including but 13 genera with 92 Recent species, they are important members of many terrestrial communities, are nearly worldwide in distribution, and are familiar to most people. Considering large land masses only, lagomorphs were absent only from Antarctica, the Australian region, and southern South America (before recent introductions by humans). Lagomorphs occupy diverse terrestrial habitats from the Arctic to the tropics. In some temperate and boreal regions, rabbits and hares are subject to striking population cycles marked by periods of great abundance alternating with periods of extreme scarcity (Chapter 23). In such regions, the population cycles of some carnivores are influenced strongly by changes in rabbit population densities.

Many important diagnostic features of Recent lagomorphs are related to their herbivorous habits and, in the case of leporids, to their cursorial locomotion. Lagomorphs have a **fenestrated** skull (having areas of thin, lattice-like bone), a feature highly developed in some leporids (**Fig. 13-53**). The anterior dentition resembles that of a rodent, but whereas rodents have 1/1 incisors, rabbits and pikas have 2/1 incisors; the second upper incisor is small and peg like and lies immediately posterior to the first (Fig. 13-53). As in rodents, the lagomorph incisors are ever growing. A long postincisor diastema is present, and the canines are absent. The cheek teeth are hypsodont and rootless, and the crown pattern features transverse ridges and basins (**Fig. 13-54**). The distance between the two upper tooth rows is greater than that between the two lower rows, restricting occlusion of upper and lower cheek teeth on only one side at a time and requiring a lateral or oblique jaw action. The masseter muscle is large, and the pterygoideus muscles are well developed and help control transverse jaw movements. The temporalis is small, and the coronoid process, its point of insertion, is rudimentary. The skull of leporids is unique among mammals in having a clearly defined joint at which slight movement occurs. This joint fully encircles the skull just anterior to the occipital and otic bones. This unusual cranial specialization appears first in Miocene leporids and is a mechanism that absorbs shock to the skull while the animal is bounding at high speeds (Bramble 1989). The clavicle is either well developed (Ochotonidae) or rudimentary

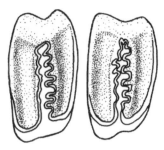

A. Antelope jackrabbit   B. Pika

FIGURE 13-54 (A) Occlusal view of upper right premolars three and four of the antelope jackrabbit. (B) Occlusal view of upper right premolars three and four of the pika.

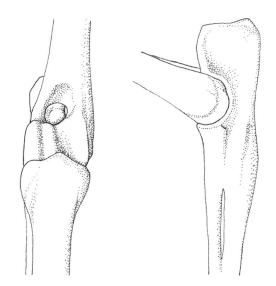

FIGURE 13-55 Unique elbow joint of *Lepus.*

(Leporidae), and the elbow joint limits movement to a single anteroposterior plane (**Fig. 13-55**). The tibia and fibula are fused distally; the front foot has five digits, and the hind foot has four or five digits. The soles of the feet, except for the distalmost toe pads in *Ochotona*, are covered with hair. The foot posture is digitigrade during running but plantigrade during slow movement. The tail is short and in pikas is not externally evident. All leporids have extremely large ears, and those of leporids occupying hot areas are important as heat-dissipating structures (Fig. 21-19).

The first fossil record of mammals with lagomorph-like characters is from the Paleocene of China. The family Leporidae probably originated in Asia in the Eocene and underwent most of its early (Oligocene and Miocene) evolution there and in North America. The pikas appeared first in the Eocene of Asia and spread in the Oligocene to Europe and North America. The Recent genus *Ochotona* is known from the late Miocene. In contrast to the leporids, which have remained widespread since the Pliocene, the ochotonids reached their greatest diversity and widest distribution in the Miocene, when they occupied Europe, Asia, Africa, and North America; they have declined since. In North America, pikas are now mostly restricted to high mountains north of Mexico. They occur more widely and are more diverse in the Old World, where they inhabit eastern Europe and much of northern and central Asia.

Why, although they are an old and thriving group, have lagomorphs lost much of their earlier adaptive radiation? Perhaps their conservatism is related to the limitations of their functional position as "miniature ungulates." Competition with members of the larger and more diverse order Artiodactyla, a group highly adapted to an herbivorous diet and cursorial locomotion, may have limited lagomorphs to the exploitation of but a single, limited adaptive

zone, although this zone was occupied with great success over broad areas. Of interest in this regard are the scarcity and local occurrence of lagomorphs in many parts of east and southern Africa, where there is an extremely rich ungulate fauna.

■ Family Ochotonidae

The pikas are represented today by 1 genus with 30 species. Pikas are less specialized with regard to cursorial adaptations than are rabbits and usually venture only short distances from shelter. Pikas occur in the mountains of the western United States and south central Alaska and over a wide area in the Old World, including eastern Europe and much of Asia southward to northern Iran, Pakistan, India, and Burma.

Pikas are smaller than rabbits, weighing about 100 to 150 grams. They have short, rounded ears, short limbs, and no externally visible tail (**Fig. 13-56**). The ear opening is guarded by large valvular flaps of skin that may provide protection during severe weather. The skull is strongly constricted between the orbits and lacks a supraorbital process; the rostrum is short and narrow. The skull is less strongly arched in ochotonids than in leporids (**Fig. 13-57**), and the angle between the basicranial and palatal axes is smaller. The maxilla has a large fenestra. The dental formula is 2/1, 0/0, 3/2, 2/3 = 26. The third lower premolar has more than one re-entrant angle, and the re-entrant enamel ridges of the upper cheek teeth are straight (Fig. 13-54B). The anal and genital openings are enclosed by a common sphincter, and males have no scrotum.

In North America, pikas usually inhabit talus slopes in boreal or alpine situations and occur from near sea level in Alaska to near the treeless tops of some of the highest peaks in the Rocky Mountains and Sierra Nevada–Cascade chain. When frightened, pikas seek shelter in the labyrinth of chambers and crevices between rocks and seldom forage

FIGURE 13-56 A pika (*Ochotona princeps*) at a lookout point on a rock.

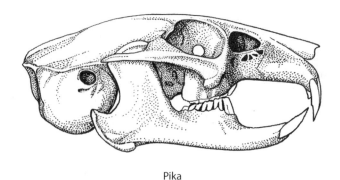

Pika

(*Ochotona princeps*, Ochotonidae)

FIGURE 13-57 Skull of the pika (*Ochotona princeps*, length of skull 41 millimeters).

far from such shelter. Large "hay piles" are built each summer in the shelter of large, usually flat-bottomed boulders; these provide food during times of winter food shortages. In Eurasia, pikas occupy an extensive geographic range and a wide variety of habitats, including talus, forests, rock-strewn terrain, open plains, and desert-steppe areas. Unusually large hay piles, weighing up to 20 kilograms, are made by pikas inhabiting dry areas in southern Russia (Formozov 1966). The Sardinian pika (*Prolagus sardus*) was native to the Mediterranean isles of Sardinia and Corsica until its extinction in the early 1800s.

Steppe-dwelling pikas of Europe have a much higher reproductive potential than do talus-dwelling North American species. Female *Ochotona pusilla* of Russia mature reproductively by 5 weeks of age and have up to three litters before winter; adult females average over eight young per litter and have up to five litters a year. Female talus-dwelling *Ochotona princeps* in California, in contrast, reach sexual maturity at 1 year of age, usually raise but one litter a year, and have only three young per litter (Smith 1978).

Ochotona princeps of North America has an unusually high metabolic rate, low thermal conductance (rate of heat loss from the skin to the environment), and little ability to dissipate heat at even moderately high ambient temperatures (MacArthur & Wang 1973). This pika avoids activity on warm or hot days. These responses reflect a preference for living in cold places.

For most of their history, pikas in North America were not restricted to montane rocky areas or talus. In the Miocene, *Ochotona* and other extinct pikas lived in riparian and marshy habitats in the Great Plains (Dalquest et al. 1996; Wilson 1960). Similarly, pikas occupied valleys and grasslands in the Pleistocene, but by 7,500 years ago, warming temperatures caused the northward or upslope withdrawal of boreal plants. Pikas, too, retreated to the higher mountains and became talus dwellers (Mead 1987); their

dwindling subalpine retreats in North America continue to be threatened by climate change today (Grayson 2005).

Family Leporidae

The rabbits and hares are a remarkably successful group in terms of ability to occupy a variety of environments over broad areas; they are now nearly cosmopolitan. Their distribution before introduction by humans included most of the New and Old Worlds, and rabbits have been introduced into New Zealand, Australia, parts of southern South America, and various oceanic islands in both the Atlantic and the Pacific Ocean. Eleven Recent genera represented by 61 recent species are known.

Several major leporid evolutionary structural trends were recognized by Dawson (1958). The cheek teeth have become hypsodont. Some of the premolars have become molariform, and the primitive crown pattern has been modified into a simple arrangement in which most traces of the primitive cusp pattern have been lost. These changes resemble those in some groups of strictly herbivorous rodents. The skull has become arched, and the angle between the basicranial and palatal axes has increased. The changes are associated with a posture involving a greater angle between the long axis of the skull and the cervical vertebrae than that typical of primitive leporids. Trends in limb structure leading to increased cursorial ability include elongation of the limbs and specializations of the articulation.

The leporid skull (Fig. 13-53) is arched in profile, and the rostral portion is fairly broad. The maxillae and often the squamosal, occipital, and parietal bones are highly fenestrated, and a prominent supraorbital process is always present. The auditory bullae are globular, and the external auditory meatus is tubular. The dental formula is usually 2/1, 0/0, 3/2, 3/3 = 28; the re-entrant enamel ridges of the upper cheek teeth are usually crenulated (Fig. 13-54A). The clavicle is rudimentary and does not serve as a brace between the scapula and the sternum. The limbs, especially the hindlimbs, are more or less elongate; movement at the elbow joint is limited to the anteroposterior plane (Fig. 13-55). The tail is short. The ears have a characteristic shape: The proximal part is tubular, and the lower part of the opening is well above the skull (**Fig. 13-58A**). The testes become scrotal during the mating season. In some species that inhabit regions with snowy winters, the animals molt into a white winter pelage in the fall and into a brown summer pelage in the spring. Wild leporids weigh from 0.3 to 5 kilograms.

Leporids inhabit a tremendous array of habitats, from Arctic tundra and treeless and barren situations on high mountain peaks to coniferous, deciduous, and tropical forests, open grassland, savanna, and deserts. Some species,

FIGURE 13-58   (A) A black-tailed jackrabbit (*Lepus californicus*) and (B) a desert cottontail (*Sylvilagus auduboni*) from Arizona.

such as the marsh rabbit (*Sylvilagus palustris*) and the swamp rabbit (*S. aquaticus*) of the southeastern United States, are excellent swimmers and lead semiaquatic lives. Leporids are entirely herbivorous and eat a wide variety of grasses, forbs, and shrubs. Several species are known to reingest fecal pellets and are thought to obtain essential nutrients (proteins and some vitamins) from material as it passes through the alimentary canal a second time.

Habitat preference and cursorial ability differ markedly from species to species and are strongly interrelated. Broadly speaking, species with less cursorial ability, such as *Brachylagus idahoensis* and *S. bachmani* of the western United States, scamper short distances to the safety of burrows or dense vegetation when disturbed. Cottontails, such as *S. floridanus* of the eastern and *S. audubonii* of the western United States, are intermediate in cursorial ability and typically inhabit areas with scattered brush, rocks, or other cover and do not run long distances to reach a hiding place. Representing the extreme in cursorial specialization among lagomorphs are some members of the genus *Lepus*, such as the New World jackrabbits (*L. californicus, L. townsendii,* and *L. alleni* and their relatives) and some hares of the Old World (such as *L. capensis* of Africa). These animals, which have greatly elongate hindlimbs, have adopted a bounding gait and occupy areas with limited shelter, such as deserts, grasslands, or meadows, where they take shelter in shallow depressions (forms) in the grass or under bushes. Instead of taking cover at the approach of danger, they depend for escape on their running ability. Jackrabbits and other similarly adapted members of the genus *Lepus* are extremely rapid runners for their size; some attain speeds up to 70

kilometers per hour. This speed allows them to occupy areas with little cover, where they can often outrun predators. The arctic hare (*L. arcticus*) of the North American Arctic uses bipedal locomotion at times and can stand and jump using only its hind legs.

Rabbits are strong competitors and are remarkably adaptable. In some parts of Australia, the extinction or near extinction of certain marsupials is perhaps due primarily to competition from introduced European rabbits (*Oryctolagus*). In addition, these prolific rabbits have caused great damage to crops and rangeland and at various times have been a primary agricultural pest in many parts of Australia as well as in New Zealand, where they were also introduced. Leporids have adapted in various ways to a wide range of environmental conditions. Along the coasts of Greenland, *L. arcticus* uses its protruding incisors (Fig. 13-52) to scrape through snow and ice to reach plants during the long Arctic winters. Far to the south, in the deserts of northern Mexico, jackrabbits (*L. alleni*) maintain their water balance through hot, dry periods by eating cactus and yucca.

Over their broad range, rabbits are the "universal prey." Humans have captured and eaten rabbits for centuries, and a diversity of predators finds rabbits to be prime fare. From the deserts of the southwestern United States to the Great Basin and Great Plains, black-tailed jackrabbits (*L. californicus*), for example, are eaten by eagles, large hawks, owls, coyotes, foxes, and bobcats. Gopher snakes (*Pituophis melanoleucus*) and even roadrunners (*Geococcyx californianus*) eat baby jackrabbits. Feldhamer (1979) found jackrabbit survival in Idaho to be only 9% after the first year of life.

Not surprisingly, populations are comprised mostly (over 70%) of young of the year.

There are several important keys to the success of various species of *Lepus*. They can subsist on a great variety of plant species, including some, such as the creosote bush (*Larrea tridentata*) of the deserts, that concentrate defensive chemicals in their leaves. *Lepus* species that inhabit dry areas get sufficient water from their food to maintain water balance without access to free water (Chaper 21). Those that live in deserts have physiological and behavioral specializations that allow survival in searing summer heat. Of major importance is a high reproductive potential: *L. californicus* females average 14 young per year (North & Marsh 1999); females of the tundra-dwelling Alaskan hare (*L. othos*) average six young per year (Anderson & Lent 1977).

SUMMARY

These two orders together probably form a monophyletic group called Glires. Rodents are the most diverse mammals but lagomorphs are far less so. Approximately 42% of all mammals are rodents; there are roughly 2,277 species in 481 genera among the 33 living rodent families. Rodents have undergone repeated radiations, resulting in a nearly cosmopolitan distribution. Convergent and parallel evolution in many rodent groups leads to morphologically and ecologically similar species in comparable habitats.

Rodents are an ancient group, dating to the late Paleocene epoch of North America and Asia. Between the late Eocene and middle Oligocene, rodent evolution accelerated. Over half of the living families of rodents appeared by the end of the Oligocene.

The classification used here recognizes five main suborders. The suborder Sciuromorpha includes the mountain beaver, squirrels, and dormice. The "true" beavers, pocket gophers, along with the kangaroo rats and their relatives are placed in the suborder Castorimorpha. The Myomorpha includes the jerboas and a diverse group of muroid rodents, whereas the scaly-tailed flying squirrels and the springhare are placed in the Anomaluromorpha. Finally, the suborder Hystricomorpha includes porcupines, cavies, and mole-rats, among others.

Some highly arboreal rodent species live in rainforest canopies. Others are highly fossorial and rarely emerge from underground tunnels, and a few species are highly aquatic. Rodents may be found above the Arctic Circle and in some of the world's harshest deserts. Most rodents are herbivorous, but a few species are omnivorous. Rodent social systems are highly diverse; they range from solitary species to species with complex social communities (e.g., prairie dog towns) and from monogamous species to promiscuous species and even to eusocial species where a single female produces all of the offspring for the colony.

Lagomorphs (hares, rabbits, and pikas) comprise 13 genera with 92 living species. Lagomorph-like mammals first appear in the Paleocene of China. The family Leporidae probably originated in Asia in the Eocene and underwent most of its early evolution there and in North America. Pikas appeared first in the Eocene of Asia and spread in the Oligocene to Europe and North America.

Lagomorphs are nearly worldwide in distribution and occupy terrestrial habitats from the Arctic to the tropics. In some regions, rabbits and hares are subject to striking population cycles. Leporids have a very high reproductive potential, with up to 14 offspring per year in some species. Rabbits and hares are highly cursorial, have elongate hindlegs, and adopt a bounding gait. Pikas are smaller, less cursorial, and inhabit a variety of habitats in Asia and talus slopes in North America. They construct "hay piles" among the rocks for winter food. Their high-mountain habitats in North America are shrinking fast because of climate change.

KEY CHARACTERISTICS OF RODENTS

- One pair of enlarged upper and lower incisors, each tooth sharply beveled and ever growing
- Broad diastema (space) between incisors and cheek teeth of both upper and lower jaws
- Canines and some cheek teeth absent
- Incisor enamel restricted to anterior surface
- Paraconid lost on lower cheek teeth
- Orbital cavity lying just dorsal to cheek teeth
- Anterior ramus of zygoma anterior to first cheek tooth
- Glenoid fossa is an anterior–posterior trough, allowing fore-and-aft movement of the mandible
- Well-developed pterygoid region
- Clavicle usually present
- Portions of the masseter muscles originate from rostrum anterior to zygomatic arch

KEY CHARACTERISTICS OF LAGOMORPHS

- Skull fenestrated (having areas of thin, lattice-like bone)
- Two pairs of upper incisors, with the second, posterior, pair small, and peg-like
- Incisors ever growing
- Canines absent, long diastema between incisors and cheek teeth
- Cheek teeth hypsodont and rootless
- Skull of leporids unique in having a joint at which slight movement occurs
- Clavicle well-developed (Ochotonidae) or rudimentary (Leporidae)
- Tibia and fibula fused distally
- Tail short
- Ears extremely large (Leporidae only)

KEY TERMS

Castoreum
Eusocial
Fenestrated
Geophyte
Hystricognathous

Hystricomorphous
Karyotype
Myomorphous
Nictitating membrane
Protrogomorphous

Riparian
Sciurognathous
Sciuromorphous
Waif dispersal

RECOMMENDED READINGS

Bramble, DM. 1989. Cranial specialization and locomotor habit in the Lagomorpha. *American Zoologist, 29*:303–317.

Carleton, MD & GG Musser. 2005. Order Rodentia. Pp. 745–751 in *Mammal Species of the World* (D Wilson & D Reeder, eds.), vol. 2. Johns Hopkins University Press, Baltimore, MD.

Chapman, JA & JEC Flux. 1990. *Rabbits, Hares and Pikas: Status Survey and Conservation Action Plan.* IUCN Lagomorph Specialist Group, International Union for Conservation of Nature and Natural Resources, Gland, Switzerland.

Genoways, HH & JH Brown. 1993. *Biology of the Heteromyidae.* Special Publication No. 10, American Society of Mammalogists.

Luckett, WP & JL Hartenberger. 1985. *Evolutionary Relationships Among Rodents: A Multidisciplinary Analysis.* Plenum Press, New York.

Mares, MA. 1980. Convergent evolution among desert rodents: a global perspective. *Bulletin of the Carnegie Museum of Natural History, 16*:1–51.

Meng, J. 2004. Phylogeny and divergence of basal Glires. *Bulletin of the American Museum of Natural History, 285*:93–109.

Meng, J & AR Wyss. 2005. Glires (Lagomorpha, Rodentia). Pp. 145-158 in *The Rise of Placental Mammals* (KD Rose & JD Archibald, eds.). Johns Hopkins University Press, Baltimore, MD.

Rose, KD. 2006. Anagalida: rodents, lagomorphs, and their relatives. Pp. 306-334 in *The Beginning of the Age of Mammals* (KD Rose, ed.). Johns Hopkins University Press, Baltimore, MD.

Rowlands, IW & BJ Weir. 1974. *The Biology of the Hystricomorph Rodents.* Academic Press, New York.

Sherman, PW, JUM Jarvis, & RD Alexander. 1991. *The Biology of the Naked Mole-Rat.* Princeton University Press, Princeton, NJ.

Wolff, J & PW Sherman. 2007. *Rodent Societies, An Ecological and Evolutionary Perspective.* University of Chicago Press, Chicago, IL.

Erinaceomorpha and Soricomorpha

The orders Erinaceomorpha and Soricomorpha include the hedgehogs, solenodons, moles, and shrews (**Table 14-1**). Because the most ancestral eutherian mammals were insectivorous and because their descendants have often retained dentitions that remain adapted to an insect diet (although frequently highly specialized), the tendency has been to include some ancestral types together with all modern descendants in the order Insectivora. This order has thus long been used as a catch-all repository for taxa of doubtful affinities. As an example, until recently, treeshrews (Scandentia), elephant-shrews (Macroscelidea), and golden moles and tenrecs (Afrosoricida) were included in the Insectivora but are now considered separate orders.

Even among the remaining insect eaters, the relationships are still not completely understood. Genetic and morphological evidence now indicates that the families Solenodontidae (solenodons), Soricidae (shrews), Talpidae (moles), and Erinaceidae (hedgehogs and allies) form a monophyletic group (Eulipotyphla) having a North American origin (Roca et al. 2004). There is disagreement, however, over the arrangement of these four families. One hypothesis supports solenodontids as the sister group to soricids plus talpids in the order Soricomorpha, with the hedgehogs in a separate order Erinaceomorpha (**Fig. 14-1A**; Corneli 2002; Hutterer 2005; Wilson & Reeder 2005). Other molecular and genetic studies suggest that shrews and hedgehogs are more closely related to each other than either group is to moles, with the solenodons even more

distantly related to all three of these groups but nevertheless included with them in a single order called Eulipotyphla (Fig. 14-1B; Douady & Douzery 2003; Douady et al. 2002, 2004). Until there is greater consensus, we follow Wilson and Reeder (2005) in provisionally recognizing the Orders Erinaceomorpha and Soricomorpha.

Taxonomic assignment of numerous but incomplete fossils has also been highly tentative and arbitrary (Kielan-Jaworowska et al. 2004). An insectivorous lifestyle was common to many Cretaceous eutherians, and several major groups of Cretaceous and Paleogene mammals radiated from the eutherian stem (Asioryctitherians, Leptictidae, Palaeoryctidae, Apternodontidae, Nyctitheriidae). Some of these have been proposed as ancestral to modern "Insectivora" (reviewed by Asher 2005). Nevertheless, the modest fossil record coupled with few demonstrable shared derived characters makes the classification of insectivoran-grade mammals difficult, and the system used here is provisional.

Order Erinaceomorpha

Family Erinaceidae

Members of the family Erinaceidae, the hedgehogs, moon rats, and gymnures, are morphologically primitive but remain successful even in areas highly modified by humans. The family is represented today by 10 genera and 24 species. They occur in Africa, Eurasia, and southeastern Asia and on the island of Borneo. Erinaceids are first known from

TABLE 14-1 Classification and Number of Species of Recent Erinaceomorpha and Soricomorpha

Family	Genera	Species	Distribution
Erinaceidae	10	24	Africa, Eurasia, southeastern Asia, and Borneo
Solenodontidae	1	4[a]	Haiti and Cuba
Nesophontidae[b]	1	9[b]	West Indies
Soricidae	26	376	Worldwide except Australia and most of South America
Talpidae	17	39	North American, Europe, and Asia

[a] Includes two extant species and two extinct species.
[b] Includes nine late Holocene extinct species.

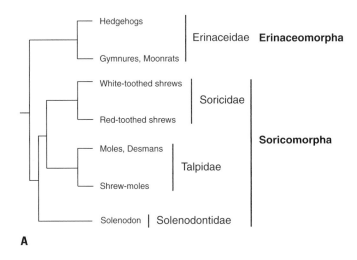

A

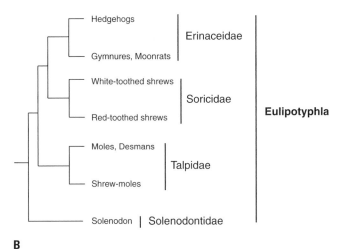

B

FIGURE 14-1 Two possible phylogenetic hypotheses of the evolutionary relationships among the families Erinaceidae, Soricidae, Talpidae, and Solenodontidae. (A) A phylogenetic hypothesis that supports a sister group relationship between the Orders Erinaceomorpha and Soricomorpha and (B) a phylogeny of the Order Eulipotyphla.

the Paleocene of North America (*Litolestes;* Novacek et al. 1985), and fossil material is known from the Eocene of Europe and Asia and the early Miocene of Africa. Other extinct families containing primitive relatives of the hedgehogs occurred in the Paleogene in the northern hemisphere.

Erinaceids include two subfamilies: the Erinaceinae, or hedgehogs, and the Galericinae, the moon rats and gymnures. They vary from the size of a mouse to the size of a small rabbit (1.4 kilograms). The eyes and pinnae are moderately large, and the snout is usually long (Corbet 1988). The zygomatic arches are complete. The dental formula is 2–3/2–3, 1/1, 3–4/2–4, 3/3 = 34–44. The first upper and, in some species, the first lower incisors are enlarged, but the front teeth never reach the degree of specialization typical of shrews (**Fig. 14-2**). In hedgehogs, the upper

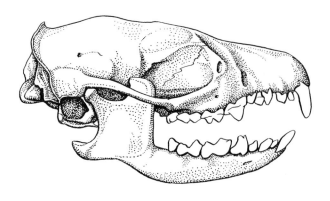

Hedgehog
(*Erinaceus*, Erinaceidae)

FIGURE 14-2 Skull of a hedgehog (length of skull 32 millimeters).

molars have simple nonsectorial cusps, with the paracone and metacone near the outer edge; the hypocone completes the quadrate form of the upper molars (**Fig. 14-3A**). The talonids of the lower molars are somewhat simplified (Fig. 14-3B). The molars are thus better adapted to an omnivorous than to an insectivorous diet. The feet retain five digits in all but one genus, and the foot posture is plantigrade (walking on the entire soles of the hands and feet). An obvious pelage specialization is spines in the hedgehogs (subfamily Erinaceinae; **Fig. 14-4**). In these animals, the panniculus carnosus muscle is greatly enlarged and controls the pulling of the skin around the body and the erection of

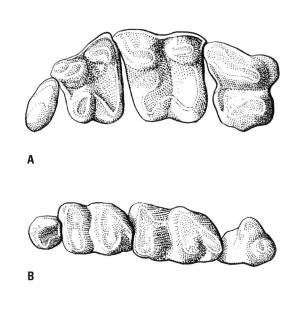

A

B

Hedgehog
(*Erinaceus*, Erinaceidae)

FIGURE 14-3 Cheek teeth of a hedgehog. Only the fourth premolar and three molars of the (A) upper right and (B) lower left tooth rows are shown.

FIGURE 14-4   The southern white-breasted hedgehog (*Erinaceus concolor*).

the spines when the animal rolls into a tight ball to defend itself from predators.

In various parts of their wide range, hedgehogs occupy deciduous woodlands, cultivated lands, and tropical and desert areas. They are omnivorous. However, animal food seems to be preferred, and a wide variety of invertebrates are eaten (Reeve 1994). Kingdon (1984a) reported that hedgehogs attack and kill small snakes and, during the attack, direct their spines forward, leaving only a small part of the body exposed to the strikes of the snake. Hedgehogs seem remarkably resistant to snake venom.

Members of the subfamily Erinaceinae are probably heterothermic. (Heterothermic animals can regulate their body temperature physiologically, but temperature is not regulated as precisely or at the same levels at all times.) Hibernation occurs in the widespread genus *Erinaceus*, and the desert species *Hemiechinus aethiopicus* enters a state of warm-season torpor called estivation. A related species from India, *H. micropus*, has survived in captivity for periods of 4 to 6 weeks without food or water (Nowak & Paradiso 1991). This species and *Hemiechinus auritus* undergo winter periods of dormancy in India. In Kenya, *Erinaceus albiventris* disappears and apparently estivates through the long dry season from May to September or October. In this instance, the animals are probably responding primarily to food shortages, for temperatures remain moderate through this period.

The subfamily Galericinae includes five genera and seven species restricted to southeastern Asia, the Malay Peninsula, Borneo, and the Philippines. Moon rats and gymnures lack spines and instead have a thick coat of coarse hairs and a long, nearly naked tail (**Fig. 14-5**). They are largely nocturnal and solitary foragers (Gould 1978), but little is known about their ecology or behavior. Several species are listed as endangered, primarily because of habitat loss.

Order Soricomorpha

Family Solenodontidae

Represented today by only one genus and three living species, the solenodons are a relict group that is unable to prosper in competition with other placentals recently introduced into their ranges. Solenodons occurred in late Holocene and historic times in Cuba, Puerto Rico, and Hispaniola but are now restricted to Haiti (*Solenodon paradoxus*) and Cuba, where a declining population of *Solenodon cubanus* occurs. The introduction by humans of rats (*Rattus*), the mongoose (*Herpestes*), and dogs and cats into the West Indies and the extensive clearing of land for agriculture combined to cause the rapid decline of the solenodons. These animals are now listed as endangered, and hopes for their survival under natural conditions are slim (Morgan & Woods 1986). Solenodons evolved in the late Mesozoic or Paleogene from their close relatives, the extinct Apternodontidae of North America (Asher et al. 2002; Roca et al. 2004).

Solenodons are roughly the size of a muskrat and have the form of an unusually large and big-footed shrew (**Fig. 14-6**). The five-toed feet and the moderately long tail are nearly hairless. The snout is long, slender, and highly flexible. The cartilaginous snout in *S. paradoxus* attaches to the cranium via a type of ball-and-socket joint (MacDonald 1984). The eyes are small, and the pinnae are prominent. The zygomatic arch is incomplete. No auditory bulla is present, and the dorsal profile of the skull is nearly flat. The dentition is 3/3, 1/1, 3/3, 3/3 = 40. The first upper incisor is greatly enlarged and points backward slightly; the second lower incisor has a deep lingual groove that may function to transport the toxic saliva that empties from a duct at the base of this tooth. (*Solenodon* is from Greek *solen*, or

FIGURE 14-5   A Malaysian moon rat (*Echinosorex gymnura*).

FIGURE 14-6 *Solenodon paradoxus* (Solenodontidae) is now restricted to Haiti, where populations of this large (800 gram) soricomorph are in decline.

"channel," and *odons*, or "tooth"). The upper molars lack a W-shaped ectoloph and are basically tritubercular with a morphology called **zalambdodont**. A sharp, blade-like (trenchant) ridge is formed by a high crest at the outer edge of each molar.

Solenodons are generalized omnivores that prefer animal material. They are nocturnal and often find food by rooting with their snouts or by uncovering animals with their large claws. The submaxillary salivary glands produce toxic saliva that is carried by ducts to the lower incisors and presumably enters a prey's wound by capillary action. The saliva of this soricomorphan is similar in effect to that of the venomous shrews. Solenodons prey primarily on invertebrates but may also scavenge vertebrate remains. Solenodons also have highly developed senses of touch, smell, and hearing. Vocalizations include high frequency "clicks," similar to those made by some shrews, that may be a crude form of echolocation (Nowak & Paradiso 1991). Solenodons are relatively long-lived species (over 11 years in captivity) but have low reproductive rates. Being archaic animals that seem to have little competitive ability, their isolation on islands is probably the key to their continued, if tenuous survival.

Family Nesophontidae

This West Indian family is represented by nine fossil species in the genus *Nesophontes*. Of the nine species that survived the Pleistocene, most seem to have become extinct by the 1600s, and all are now thought to be extinct (MaPhee et al. 1999). Most of the information on this genus comes from skulls and skeletal remains from fossil sites and owl pellets found in 1930 (Morgan & Woods 1986; Nowak & Paradiso 1991). The skulls of *Nesophontes* lack a jugal bone, zygomatic arch, and probably an auditory bulla. These mouse- to rat-sized animals probably resembled solenodons in having a long, flexible snout and small eyes.

Family Soricidae

Members of this family, the shrews, are among the smallest and least conspicuous of mammals. In many areas, they are the most numerous soricomorphans; they have the widest distribution and are the most familiar. The family Soricidae is represented today by 26 genera and 376 species in three subfamilies (Crocidurinae, Myosoricinae, and Soricinae). Shrews represent 6.9% of all extant mammal species and occur throughout the world except in the Australian area, most of South America, and the polar areas. Undoubted soricid fossils first appear in the Eocene of North America and Eurasia and the Miocene of Africa (Wojcik & Wolsan 1998). Because soricids are rare as fossils, their early evolution is obscure; they may have evolved from Eocene Parapternodontidae or Nyctitheriidae (Asher et al. 2002).

Shrews are small: The smallest weighs 2.5 grams, and the largest weighs roughly 180 grams, the weight of a small rat. The snout is long and slim. The eyes are small, and the pinnae are usually visible (**Fig. 14-7**). The feet are five toed and unspecialized, except for fringes of stiff hairs on the digits in semiaquatic species and enlarged claws in semifossorial forms. The foot posture is plantigrade. The narrow and elongate skull usually has a flat dorsal profile (**Fig. 14-8A**). There is no zygomatic arch or tympanic bulla, and the tympanic bone is annular (Fig. 14-8B). The specialized dentition consists of 26 to 32 teeth; the dental formula of *Sorex* is 3/1, 1/1, 3/1, 3/3 = 32. In the subfamily Soricinae, the teeth contain the reddish pigment goethite, an iron-containing mineral (Akersten et al. 2002; Strait & Smith 2006). The first upper incisor is large and hooked and bears a notch and projection resembling those on the beak of a

FIGURE 14-7   (A) The European pygmy shrew (*Sorex minutus*) and (B) the Eurasian water shrew (*Neomys fodiens*).

falcon (Fig. 14-8A). Behind the first upper incisor is a series of small unicuspid teeth (presumably incisors, a canine, and premolars). The fourth upper premolar is large and has a trenchant ridge, and the upper molars have W-shaped ectolophs (**Fig. 14-9A**). Both the trigonid and talonid of the lower molars are well developed (Fig. 14-9B), and the first lower incisor is greatly enlarged and procumbent (leaning forward), often with extra cusps (Carraway 1995).

Perhaps the most unusual skeleton is that of the hero shrew (*Scutisorex somereni*) of West Africa. The vertebral column of this animal consists of vertebrae with numerous interlocking processes that presumably provide added strength to the vertebral column and additional attachment sites for the complex back muscles (Cullinane et al. 1998a, 1998b). These shrews are thought to have incredible strength, but almost nothing is known of their ecology or the function of this unique vertebral specialization.

Because they are unusually small, shrews can exploit a unique mode of foraging. Many patrol for insects beneath logs, fallen leaves, and other plant debris and in the narrow spaces and crevices beneath rocks. Rodents' surface runways and burrows may also be used as feeding routes. Because of this style of foraging, shrews are seldom observed, even in areas where they are common. Although shrews are

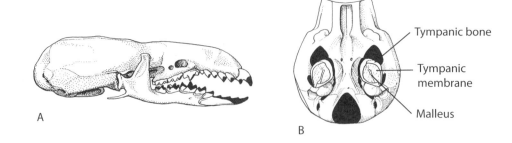

Vagrant shrew
(*Sorex vagrans*, Soricidae)

FIGURE 14-8   Skull of the vagrant shrew. (A) Side view, showing the pincer-like anterior incisors. The red-pigmented parts of the teeth are shown in black. (B) Ventral view of the basicranial region, showing the annular (ring-like) tympanic bone, the tympanic membrane, and the malleus (length of skull 17 millimeters).

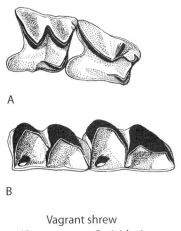

Vagrant shrew
(*Sorex vagrans*, Soricidae)

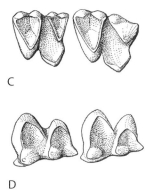

Eastern mole
(*Scalopus aquaticus*, Talpidae)

FIGURE 14-9   Cheek teeth of Soricomorpha (not to scale). (A) The fourth upper right premolar and first molar and (B) the first two lower left molars of the vagrant shrew (Soricidae). The red-pigmented parts of the teeth are shown in black. (C) The first and second upper right and (D) the comparable lower left molars of the eastern mole (Talpidae).

typically associated with moist conditions, some species, such as the gray shrew (*Notiosorex crawfordi*) of the southwestern United States and the piebald shrew (*Diplomesodon pulchellum*) of southern Russia, inhabit desert areas. Aquatic adaptations in some species allow them to dive and swim and feed mainly on aquatic invertebrates (see Fig. 24-10). One of the most aquatic species is the Tibetan water shrew (*Nectogale elegans*), which inhabits mountain streams and feeds primarily on fish. In this species, the streamlined shape is enhanced by the strong reduction of the pinnae, and the digits and feet have fringes of stiff hairs that greatly increase their effectiveness as paddles. The distal part of the tail is laterally compressed, and the edges bear lines of stiff hairs to aid in swimming. The North American water shrew (*Sorex palustris*) is similar in morphology and lifestyle.

Except for the duck-billed platypus (order Monotremata), the only mammals known to be venomous are shrews and solenodons. Over 350 years ago, there were reports on the symptoms that developed when humans were bitten by the short-tailed shrews of North America (*Blarina* species), and work in the present century has confirmed that short-tailed shrews have venomous saliva (Pournelle 1968; Tomasi 1978) that is toxic to small mammals (Kita et al. 2004). It has also been demonstrated that the European water shrew (*Neomys fodiens*; Fig. 14-7B) and the Haitian solenodon (*Solenodon paradoxus*) are venomous (Pournelle 1968; Pucek 1968). The salivas of other soricomorphans closely related to these two shrews have been studied but are apparently not toxic. In some people, the bites of the musk shrew, *Suncus murinus*, cause minor aches and hypersensitivity and reddening of the skin, especially at the finger joints (G. L. Dryden, personal communication).

Both *N. fodiens* and *Blarina* species have similar adaptations for delivering venom, and the effects of the venoms are similar (Pournelle 1968; Pucek 1968; Tomasi 1978). In these shrews, the first lower incisors have concave posterior surfaces, forming a crude channel, and the ducts from the venom-producing submaxillary glands open near the base of these teeth. *Neomys fodiens* salivates copiously during attacks on prey, and saliva is seemingly channeled to wounds via the two first lower incisors. Frogs bitten by *N. fodiens* were partially immobilized and, when forced to move, were uncoordinated. Laboratory mice injected with a homogenate of these salivary glands immediately developed paralysis of the hindlimbs. Kita et al. (2004) showed that mice injected with extracts of the submaxillary glands of *Blarina brevicauda* were killed. What is the functional importance of venom to shrews? *Blarina* species can kill mice considerably larger than themselves, and Eadie (1952) and Getz et al. (1992) reported that meadow voles (*Microtus pennsylvanicus*) were an important fall and winter food of these shrews. Frogs and small fish are known to be preferred foods of *N. fodiens*. Both of these shrews attack prey from behind and direct their bite at the neck and base of the skull, an area where neurotoxic venom might readily be introduced into the central nervous system. The adaptive importance to a very small predator of making its relatively large prey helpless seems to be great, and one wonders why more shrews are not venomous.

Shrews forage day and night and are not known to hibernate. They have a reputation as ravenous predators. Survival depends on their ability to consume nearly twice their body weight in food every day. The need for such large amounts of food can be explained by their small body size. In mammals, the metabolic rate is inversely proportional to body size, in part because the great surface area to volume ratio of small mammals fosters rapid dissipation of heat.

Because metabolic heat is produced by the conversion of food into usable energy for cellular metabolism, to maintain constant body temperatures small mammals must consume proportionally larger amounts of food than do larger mammals (Chapter 21). In winter, demands for food increase precisely when invertebrate food supply declines. Shrews of the **Holarctic** genus *Sorex* adapt to winters of the north by doubling their **nonshivering thermogenesis** (a means of heat production using specialized brown adipose tissue), while their body weight drops by up to 53%. This loss of weight does not help winter survival but reflects low food supply and loss of energy/fat reserves. The increased shivering does aid winter survival, however. Old World tropical *Crocidura* adapt to winter by undergoing daily torpor (Genoud 1985; Merritt 1995; Merritt et al. 1994). The high metabolic rate of shrews is associated with the highest blood oxygen content of any mammal measured, a heart rate between 900 and 1,400 beats per minute in some species, and more than three times the number of red blood cells per cubic milliliter of blood than that found in humans (Grzimek 1990).

Some shrews (*Sorex* and *Blarina*) are known to use a rudimentary form of **echolocation** to find prey. These shrews emit 30- to 60-kilohertz pulses as they explore their environment (Gould et al. 1964; see Chapter 22). Observations of shrew behavior in the wild are few, and what little is known comes from relatively few laboratory studies. One interesting behavior is the innate following behavior unique to some shrews of the subfamily Crocidurinae, which results in the formation of "caravans." When a female and her young are moving, the first young grabs the base of her tail with its mouth. The next young grasps the base of the tail of the first young in its mouth, and so on, forming a chain of young that, under some conditions, is dragged by the mother (**Fig. 14-10**).

■ Family Talpidae

This family includes 17 genera and 39 species of small rat- to mouse-sized animals usually referred to as moles. These predominantly burrowing invertebrate feeders occur in parts of North America, Europe, and Asia. The European fossil record of talpids begins in the Eocene (*Eotalpa*; Sigé et al. 1977); talpids are known first in Asia and the New World from the Oligocene. Apparently, the anatomical modifications typical of Recent fossorial genera were attained within the first 10 or 15 million years of their existence, and the living European genus *Talpa* is first known from the early Miocene. The fossil family Proscalopidae was once thought to be closely related to talpids, but discovery of postcranial material from the Miocene of Montana indicates that members of this fossil group did not use the same digging technique observed in modern moles, and consequently,

FIGURE 14-10 Photo of a mother *Crocidura* and her young forming a caravan.

they represent a distinct family of fossil moles (Barnosky 1982). Modern talpids occupy a wide range of habitats ranging from terrestrial, semiaquatic, fossorial/aquatic, semifossorial, to fossorial (Shinohara et al. 2003).

The head and forelimbs of most talpids are strongly modified for fossorial life (**Fig. 14-11**). The zygomatic arch is complete. The tympanic cavity is not fully enclosed by bone, and the eyes are small and often lie beneath the skin. The snout is long and slender. The ears usually lack pinnae, and the fur is characteristically lustrous and velvety. Mole fur "lays" in any direction, allowing the mole to reverse direction easily within the tunnel without getting its pelage dirty (their reduced pelvis also allows them to turn around in a tunnel barely wider than their bodies). The dental formula is 2–3/1–3, 1/0–1, 3–4/3–4, 3/3 = 32–44. The first upper incisors are usually inclined backward (**Fig. 14-12**), and the upper molars have W-shaped ectolophs (Fig. 14-9C and D). In the fossorial species, the forelimbs are more or less rotated from the usual orientation typical of terrestrial mammals in such a way that the digits point to the side, the palms face backward, and the elbows point upward (**Fig. 14-13A**). In addition, the phalanges are short. The claws are long, and the clavicle and humerus are unusually short and robust. Unlike most mammals, the scapula is long and slender, and serves both to anchor the forelimb solidly against the axial skeleton and to provide advantageous attachments for some of the powerful muscles that pull the forelimb backward. The anteriormost segment of the sternum (the manubrium) is greatly enlarged and extends forward beneath the base of the skull (Fig. 14-13B). These specializations increase the area for attachment of the large pectoralis muscles and move the shoulder joint forward, allowing the forepaws to remove or loosen soil beside the snout.

The short clavicle provides a large secondary articular surface for the humerus. The double articulation of the shoulder joint, with articular contacts between the humerus and the scapula and clavicle, provides an unusually strong bracing for this joint during the powerful rotation of the humerus that accompanies the digging stroke of the forelimb. In some genera, the falciform bone is large and increases the breadth of the hand and braces the first digit (Fig. 14-13A). Unlike the unrelated golden moles (Chrysochloridae), talpids live in areas where it is possible to dig more or less permanent tunnel systems. As a result, talpids have evolved a different digging technique involving an alternating fore and aft scraping against the tunnel walls with their laterally oriented, spade-like hands (**Fig. 14-14**) (Skoczen 1958; Yalden 1966).

Fossorial talpids occur typically in moist and friable soils in forested, meadow, or stream-side areas and feed largely on animal material including earthworms, slugs, snails, and arthropods. A species that occurs in the eastern United States (*Scalopus aquaticus*), however, locally

FIGURE 14-11   The European mole (*Talpa europaea*).

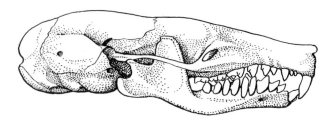

Eastern mole
(*Scalopus aquaticus*, Talpidae)

FIGURE 14-12   Skull of the eastern mole (length of skull 37 millimeters).

penetrates the moderately dry sandhill prairies of Nebraska and eastern Colorado, where the characteristic ridges of soil made by the animals appear only during wet weather. In most areas, these ridges are a common evidence of the presence of moles and are made by the animals as they travel

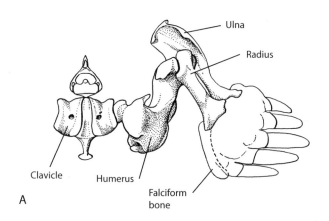

A

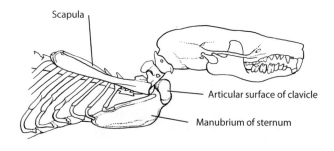

B

Eastern mole
(*Scalopus aquaticus*, Talpidae)

FIGURE 14-13   The pectoral girdle and forelimb of the eastern mole. (A) Side view of the skeleton showing the pectoral girdle with forelimb bones removed. (B) Anterior view of part of the pectoral girdle and the forelimb, with the shoulder joint slightly disarticulated to show the head of the humerus and its secondary articular surface. The head of the humerus articulates with the glenoid fossa of the scapula, and a second articulation (involving considerably larger surfaces) occurs between the secondary articular surface of the humerus and an articular surface of the clavicle.

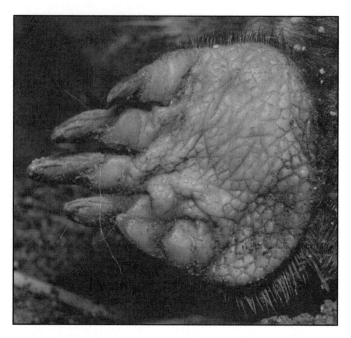

FIGURE 14-14   Spade-like hand of the European mole (*Talpa europaea*).

just beneath the surface by forcing their way through the soil (Yalden 1966). Soil from deep burrows is deposited on the surface in more or less conical mole hills. Some species of mole appear to be territorial, although territories of males may overlap those of females (Loy et al. 1994). Moles are not known to hibernate or estivate.

Moles forage along their tunnel systems and probe the walls of the tunnel with the aid of tens of thousands of touch receptors located on the snout in the **Eimer's organ** (Quilliam 1966). These organs also are found in desmans (Richard 1982). *Condylura cristata*, the star-nosed mole, has a ring of 22 fleshy "tentacles" on the tip of its snout (**Fig. 14-15**). This species is semiaquatic and uses these sensory "tentacles" when trying to locate prey. Recent experiments on the neurobiology of the nose of star-nosed moles demonstrated that the tentacles of the star contain an elaborate system of Eimer's organs. With approximately 25,000 Eimer's organs, the nose of *Condylura* contains six times as many touch receptors (mechanoreceptors) as the entire human hand (**Fig. 14-16**; Catania & Kaas 1996). Thus, the nose of *Condylura* is used primarily for touch reception and not for olfaction.

The behavior of star-nosed moles suggests a second function of the sensory tentacles. Star-nosed moles frequently submerge all or part of their tentacles in the water before entering the water themselves. During this time, the proboscis scans the underwater environment and extremely quickly (within 200 milliseconds) allows the mole to identify and capture tiny prey (Catania & Remple 2005). This behavior, along with the density and complexity of

FIGURE 14-15   (A) Head and forelimbs of a star-nosed mole (*Condylura cristata*). The unique "star" of finger-like structures is found in no other mammal. (B) Scanning electron micrograph of the nose of a star-nosed mole showing the 22 fleshy appendages that ring the nostrils. Each appendage is covered with small epidermal papillae called Eimer's organs that are very sensitive to touch. The scale bar equals 2 millimeters. About half of the mole's brain cortex that processes the sense of touch is dedicated to the star, and one fourth of that is associated with the two rays (number 11 in B) that are immediately in front of the first upper incisors.

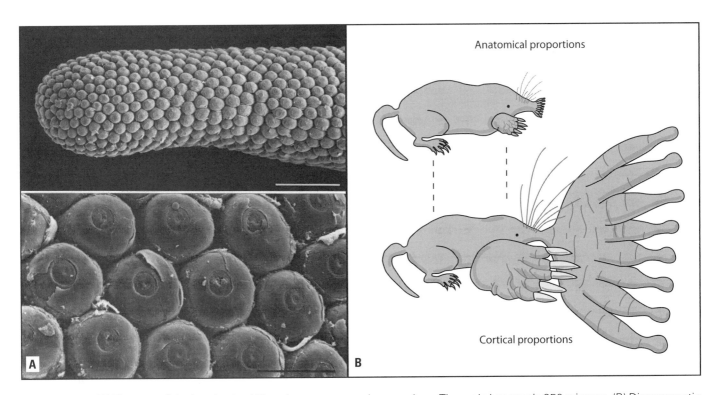

FIGURE 14-16   (A) Close-up of the hundreds of Eimer's organs on each appendage. The scale bar equals 250 microns. (B) Diagrammatic drawing of the body parts of a star-nosed mole in their normal anatomical proportions (above) and as they would be shown relative to their proportional representation in the somatosensory cortex of the brain (below), highlighting the relatively large somatosensory field for the nose of the mole. (Modified from Catania, K. C., and Kass, J. H., "The unusual nose and brain of the star-nosed mole," *BioScience* 46 (1996): 578–586. © 1996 American Institute of Biological Sciences.)

the Eimer's organs, suggests that the tentacles may also serve not only as touch receptors but also possibly as **electroreceptors** analogous to the function of the receptor pits in the duck-billed platypus (*Ornithorhyncus*; Grand et al. 1998). Intriguing as this idea may be, it has not yet been tested experimentally.

The evolution of Eimer's organs in talpids appears to be correlated with habitat preference (Catania 1995). A gradient from wet to dry soils exists among three North American species, with the eastern mole (*S. aquaticus*) preferring drier soils, star-nosed moles (*C. cristata*) living in very wet habitats, and hairy-tailed moles (*P. breweri*) preferring intermediate soil types. Eimer's organs in eastern moles are degenerate, probably because of the thicker skin necessary to combat the considerable wear and tear of burrowing through dry, sandy soils. Conversely, Catania (1995) suggested that the relatively wet, muddy soils inhabited by star-nosed moles allowed for the elaboration of the more delicate sensory structure containing thousands of sensitive mechanoreceptors. The extreme touch sensitivity of the star and the millisecond speed with which it can identify prey makes star-nosed moles the fastest known predators at processing tactile information (Catania & Remple 2005).

The Russian desman (*Desmana moschata*) and the Pyrenean desman (*Galemys pyrenaicus*) are two unusual and poorly known members of the Talpidae. Pyrenean desmans weigh between 35 and 80 grams and Russian desmans up to 220 grams (Niethammer 1970; Nowak 1999). Desmans are similar in appearance to moles, but with long tails and huge hindfeet. They have conical heads with long, naked snouts, which are dorsoventrally compressed. The snorkel-like snout terminates in a pair of nostrils. Desmans have small eyes and no external pinna. They have thick, glossy fur, large, webbed hind feet, and a **fusiform** body shape—all adaptations for a semiaquatic lifestyle.

The elongate snout is highly mobile; *Galemys* continually presses the snout against the stream bottom when foraging underwater. Like the tentacles on the snout of the star-nosed mole, the desman snout is covered with a dense array of Eimer's organs. The combination of Eimer's organ receptors and numerous vibrissae on the snout suggests that desmans feed using touch to detect prey (Palmeirim & Hoffmann 1983).

According to Palmeirim and Hoffmann (1983), Pyrenean desmans have permanent home ranges along streams of up to 200 meters in length. Males and females appear to be monogamous, with male territories completely encompassing those of the female. Territories are marked with a musky scent from a gland at the base of the tail. These desmans are largely nocturnal and prey on aquatic insects and crustaceans captured underwater.

SUMMARY

Genetic and morphological evidence indicates that the solenodons (Solenodontidae), shrews (Soricidae), moles (Talpidae), and the hedgehogs and moonrats (Erinaceidae) form a monophyletic group (Eulipotyphla) with North American origins. A meager fossil record and few demonstrable shared derived characters make this classification provisional.

The family Erinaceidae includes the hedgehogs, moonrats, and gymnures, represented today by 10 genera and 24 species from Africa, Eurasia, and southeastern Asia. Erinaceids are first known from the Paleocene in North America, from the Eocene in Europe and Asia and the early Miocene in Africa. Hedgehogs are small, spiny mammals that prey on invertebrates, and they are probably heterothermic. Moonrats and gymnures lack spines and instead have a thick coat of coarse hair and a nearly naked tail.

The Soricomorpha (formerly part of the polyphyletic Insectivora) includes three extant families: the Soricidae (shrews), Talpidae (moles), and Solenodontidae (solenodons). The family Soricidae is represented today by 26 genera and 376 species. Undoubted soricid fossils first appear in the Eocene of North America and Eurasia and in the Miocene in Africa. Shrews are small (weighing 2.5 to 180 grams) and patrol for insects beneath logs and fallen leaves and in narrow crevices beneath rocks. Water shrews forage underwater using their hind feet as paddles. Several species produce venom from saliva that helps immobilize larger prey. Some shrews (*Sorex* and *Blarina*) are known to use a rudimentary form of echolocation to find prey.

There are 17 genera and 39 species of moles and desmans (Talpidae). Moles are strongly modified for fossorial life. They occur typically in moist and friable soils in forested, meadow, or streamside areas and feed largely on earthworms, slugs, snails, and arthropods. Moles forage along tunnel systems by probing the tunnel walls with tens of thousands of touch receptors (Eimer's organs) located on their snouts. Desmans resemble moles, but have long tails and huge hind feet. They have conical heads with long, snorkel-like snouts that terminate in a pair of nostrils. Desmans have small eyes and no external pinna. They have thick, glossy fur, large, webbed hind feet, and a generally fusiform body shape—all adaptations for a semiaquatic lifestyle.

Solenodons (Solenodontidae) are represented today by two surviving species restricted to Haiti and Cuba. Formerly two additional solenodons and nine species in a second genus and family (*Nesophontes*; Nesophontidae) occurred in the West Indies but became extinct in recent centuries. Introduced rats (*Rattus*), mongoose (*Herpestes*), dogs and cats, and extensive land clearing caused the rapid decline of the solenodons; they are now listed as endangered species. Solenodons weigh up to 1 kilogram and resemble very large shrews. They are nocturnal omnivores that find food by rooting with their long, flexible snouts or by uncovering animals with their large claws. *Solenodon* produces toxic saliva from submaxillary salivary glands.

KEY CHARACTERISTICS OF HEDGEHOGS, MOON RATS, AND GYMNURES (ERINACEIDAE)

- Complete zygomatic arch
- Upper molars quadritubercular and bunodont
- Lower molars with well-developed trigonids and talonid basins
- Hair modified into sharp spines (hedgehogs)
- Panniculus carnosus muscle extensive (hedgehogs)

KEY CHARACTERISTICS OF SHREWS (SORICIDAE)

- Skull long and narrow
- Zygomatic arches, auditory bullae, and postorbital processes absent
- Tympanic bone annular (ring shaped)
- Doubled condyloid processes on dentary form double jaw articulations
- Enlarged first incisor "falciform" with forward projecting main cusp and small secondary cusp
- Remaining incisors, canines, and premolars (except P4) small and peg-like (unicuspid)
- Upper molars dilambdodont, with W-shaped ectoloph
- Milk teeth shed before birth

KEY CHARACTERISTICS OF MOLES AND DESMANS (TALPIDAE)

- Body fusiform
- Eyes tiny, sometimes covered by skin
- Legs short and powerful, forelimbs rotated so palms face posteriorly
- Forelimb claws robust
- Pinnae absent
- Fur velvety (moles), allowing easy movement in the burrow
- Skull long, flattened with narrow rostrum
- Zygomatic arches complete
- Auditory bullae present
- Humerus broader than long
- Clavicle short and broad
- Molars dilambdodont
- Snout snorkel-like (desmans)

KEY CHARACTERISTICS OF SOLENODONS (SOLENODONTIDAE)

- Snout long and flexible
- Feet large and clawed
- Tail long, nearly naked
- Zygomatic arch incomplete
- Auditory bulla absent
- Os proboscidis bone supports tip of rostrum
- Second lower incisor with deep groove for transferring toxic saliva

KEY TERMS

Echolocation
Eimer's organ
Electroreceptor

Fusiform
Holarctic

Nonshivering thermogenesis
Zalambdodont

RECOMMENDED READINGS

Asher, RJ, et al. 2002. Morphology and relationships of *Apternodus* and other extinct, zalambdodont, placental mammals. *Bulletin of the American Museum of Natural History, 273*:1–117.

Catania, KC. 1995. A comparison of the Eimer's organs of three North American moles: the hairy-tailed mole (*Parascalops breweri*), the star-nosed mole (*Condylura cristata*), and the eastern mole (*Scalopus aquaticus*). *Journal of Comparative Neurology, 354*:150–160.

Catania, KC, JF Hare, & KL Campbell. 2008. Water shrews detect movement, shape, and smell to find prey underwater. *Proceedings of the National Academy of Sciences, 105*:571–576.

Churchfield, S. 1990. *The Natural History of Shrews.* Comstock, Cornell University Press, Ithaca, NY.

Douady, CJ & EJP Douzery. 2003. Molecular estimation of eulipotyphlan divergence times and the evolution of "Insectivora." *Molecular Phylogenetics and Evolution, 28*:285–296.

Gorman, ML & RD Stone. 1990. *The Natural History of Moles.* Cornell University Press, Ithaca, NY.

Kita, M, et al. 2004. *Blarina* toxin, a mammalian lethal venom from the short-tailed shrew *Blarina brevicauda*: isolation and characterization. *Proceedings of the National Academy of Sciences USA, 101*:7542–7547.

Merritt, JF, GL Kirkland, Jr, & RK Rose. 1994. *Advances in the Biology of Shrews.* Special Publications No. 18, Carnegie Museum of Natural History, Pittsburgh, PA.

Shionhara, A, KL Campbell, & H Suzuki. 2003. Molecular phylogenetic relationships of moles, shrew moles, and desmans from the New and Old Worlds. *Molecular Phylogenetics and Evolution, 27*:247–258.

Chiroptera

Because of their secretiveness, their appearance in gathering darkness during Shakespeare's "very witching time of night," their ability to fly, and their unusual form, which for the casual observer sets them apart from the more familiar groups of mammals, bats have long been central figures in superstition and have provided shadowy forms for poetry:

> *Bats, and an uneasy creeping in one's scalp*
> *As the bats swoop overhead!*
> *Flying madly.*
> *Pipistrello!*
> *Black piper on an infinitesimal pipe.*
> *Little lumps that fly in air and have voices indefinite,*
>   *wildly vindictive;*
> *Wings like bits of umbrella.*
> *Bats!*

This is the bat of D. H. Lawrence, but the biologist's bat has taken on substance. Accelerated research on bats in recent years has revealed fascinating aspects of their lives, including complex social behavior, intricate communication signals, and magnetic sensory ability (Covey 2005; Wang et al. 2007). Of dominant importance in the lives of most bats is **echolocation**, involving a coordinated assemblage of sensory, neurological, neuromuscular, and behavioral adaptations allowing bats to perceive in detail their prey and their environment by their extremely sophisticated use of sound (see Chapter 22). Bats are also unsurpassed by other mammals in the ability to survive through periods of stress or to conserve energy daily by drastic reductions in the metabolic rate and by tracking barometric pressure (Paige 1995).

In recent years, as more bat species become endangered, even as new genera and species are still being discovered, public awareness about bats and their habitats and concern for their conservation have increased dramatically. Biologists and a growing segment of the general public are coming to realize not only that bats deserve respect as remarkably specialized products of at least 53 million years of evolution but also that they merit our protection for their importance in terrestrial ecosystems as efficient predators of insects and as important pollinators of flowering plants.

Bats are a remarkably successful group today and constitute the second largest mammalian order (behind Rodentia); approximately 202 genera and over 1,120 species of living bats in 19 families are known. They are nearly cosmopolitan in distribution, being absent only from arctic and polar regions and from some isolated oceanic islands. Bats exploit an impressive diversity of foods, from insects, spiders, scorpions, centipedes, crustaceans, mammals, reptiles, birds, frogs, and blood to fruit, nectar, pollen, and leaves. Although bats are frequently abundant members of temperate faunas, they reach their highest densities and greatest diversity in tropical and subtropical areas. In certain Neotropical localities, for example, there are more species of bats than of all other kinds of mammals combined. Bats occupy a number of terrestrial environments, including temperate, boreal, and tropical forests, grasslands, chaparral, and deserts. In some areas, the molesting of bat colonies by humans has caused alarming declines in populations, and the heavy use of insecticides continues to be a threat.

A great deal of effort has been made in late years to clarify the phylogenetic relationships within bats and between the Chiroptera and other groups of mammals. These studies have drawn evidence from fossils and diverse kinds of morphological and molecular data from living bats. Phylogenetically, bats seem to be a part of a radiation of mammals called Laurasiatheria, a group that includes the Chiroptera, Carnivora, Cetartiodactyla, Perissodactyla, Erinaceomorpha, Soricomorpha, and Pholidota, but bats' particular relationship among these other groups is still unresolved. Within the bats themselves, however, the resolution of relationships among the families is improving (**Fig. 15-1**; Simmons 2008; Simmons et al. 2008; Teeling et al. 2005).

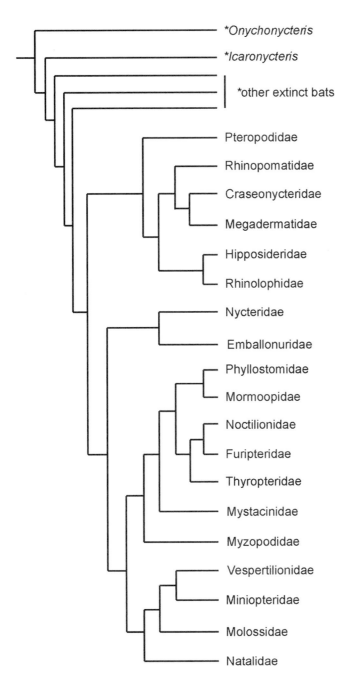

FIGURE 15-1 Hypothesis of phylogenetic relationships of bats. (Adapted from Simmons, N. B., et al., *Nature*. 451 (2008):181–821 and Teeling, E. C., et al., *Science*. 307 (2005): 580–584.)

Some authors name the two main branches of living bats as separate entities: "Yinpterochiroptera" (consisting of Pteropodidae, Rhinopomatidae, Craseonycteridae, Megadermatidae, Hipposideridae, and Rhinolophidae) and "Yangochiroptera" (consisting of the remaining 13 families).

Among the Yinpterochiroptera one family, Pteropodidae, formerly the sole members of a suborder known as Megachiroptera or megabats, is distinct from all other living bat families. Pteropodids also are not known to hibernate, although some nectar-feeding species enter hypothermia with lowered rates of metabolism (Bonaccorso & McNab 1997; McNab & Bonaccorso 1995). In contrast, many other families are heterothermic, and some, especially in temperate zones, hibernate for long periods. Unlike all other bat families, pteropodids use vision and not echolocation as a primary mode of orientation. Two exceptions are the pteropodids *Rousettus* and *Stenonycteris*, in which the ability to echolocate perhaps re-evolved independently after being lost early in the evolution of the Yinpterochiroptera, or else echolocation evolved twice in bats, in the other yinpterochiropteran convergently with the Yangochiroptera. *Rousettus* and *Stenonycteris* use clicks made by the tongue as the basis for their acoustical orientation (Novick 1958a). All other bat families use sound produced by the larynx. Often these pulses are ultrasonic, that is, above the range of human hearing. In most families, the pulses are emitted through the open mouth, but in those with elaborate nose-leaves (Megadermatidae, Nycteridae, Hipposideridae, Rhinolophidae, and Phyllostomidae), the sound pulses are emitted through the nostrils while the mouth is kept closed. The differences are reflected in underlying anatomy of the skull and vocal apparatus (see Fig. 22-8; Freeman 1984; Pedersen 1998). Nose-leaves vary from simple to elaborate in structure and are related in some yet poorly known ways to echolocation and foraging style and help shape sonar beams (Bogdanowicz et al. 1997; Zhuang & Müller 2006).

Echolocation, a means of perceiving the environment even in darkness or varied lighting conditions (see Chapter 22), and flight, a means of great motility, have been two major keys to the success of bats. These abilities enable nocturnal bats to occupy many of the niches filled by birds during the day. The abilities for echolocation and relevant portions of the brain in many bats are incredibly complex and fascinating, allowing bats to "see" three-dimensional acoustic images processed from information gleaned from echoes (Covey 2005; Simmons 1989; Ulanovsky & Moss 2008). In addition, the remarkably maneuverable flight of bats facilitates a mode of foraging for insects that birds have never exploited. Heterothermy, allowing bats to hibernate or to operate at a lowered metabolic output during part of the daily torpor cycle, has enabled these animals to occupy areas only seasonally productive of adequate food and to use an activity cycle involving only nocturnal or crepuscular foraging periods. The metabolic economy resulting from hibernation and from lowered metabolism during part of the daily cycle has affected the longevity of some bats. For their size, some bats are remarkably long lived. Several taxa are known to have reached 30 years of age, and a *Myotis lucifugus*, a small bat weighing roughly 10 grams, is known to have lived 34 years (Arlettaz et al. 2002).

Morphology

Many of the most important diagnostic features of bats are adaptations to flight and involve exquisitely integrated functional and structural modifications of the body. Many such adaptations are morphophysiological, such as the demand for synchronized respiration and echolocation in flight. Flying is strenuous and highly aerobic; echolocation sounds made by many bats are precise, forcefully produced through the larynx, and may be extremely loud. Wingbeats must be coordinated with breathing, and breathing must be synchronized with sound production. As a result of these interrelated functions, bats have evolved the largest hearts and lungs relative to body size of all mammals (Canals et al. 2005; Lancaster & Speakman 2001). In the most extreme example known, the South American big-eared bats *Histiotus* have a heart mass that is 2.18% of their body mass, a value that is about 63% above the average for mammals. The remarkably large lungs and related optimal circulatory parameters afford bats more extensive lung alveolar surface area, large capillary blood volume, a thin blood–gas barrier, and high blood and hemoglobin concentrations (Maina 2000).

In terms of the skeleton, the bones of the arm and hand (with the exception of the thumb) are elongate and slender (**Fig. 15-2**). Flight membranes extend from the body and the hindlimbs to the arm and the fifth digit (**plagiopatagium**), between the fingers (**chiropatagium**), from the hindlimbs to the tail (**uropatagium**), and from the arm to the occipitopollicalis muscle (**propatagium**) (**Fig. 15-3B**). In some species, the uropatagium is present even when the tail is absent. The muscles bracing the wing membranes are often well developed and anchor a complex network of elastic fibers (Fig. 15-3A). Rigidity of the outstretched wing during flight is partly controlled by the specialized elbow and wrist joints, at which movement is limited to the anteroposterior plane.

In most bats, the enlarged greater tuberosity of the humerus locks against the scapula at the top of the upstroke (**Fig. 15-4**), allowing the posterior division of the serratus ventralis thoracis muscle to tip the lateral border of the scapula downward to help power the downstroke of the wing (Vaughan 1959; **Fig. 15-5**). The adductor and abductor muscles of the forelimb raise and lower the wings and are therefore the major muscles of locomotion. (In the contrasting arrangement found in terrestrial mammals, the flexors and extensors provide most of the power for locomotion.) The distal part of the ulna is reduced in bats, and the proximal section usually forms an important part of the articular surface of the elbow joint (**Fig. 15-6**). The clavicle is present and articulates proximally with the enlarged manubrium and distally with the enlarged acromion process and enlarged base of the coracoid process (see Fig. 15-5). The hindlimbs are either rotated to the side 90 degrees from the typical mammalian position and have a reptilian posture during quadrupedal locomotion or else they are rotated 180 degrees, have a spider-like posture, and are used primarily to suspend the animal upside down from a horizontal

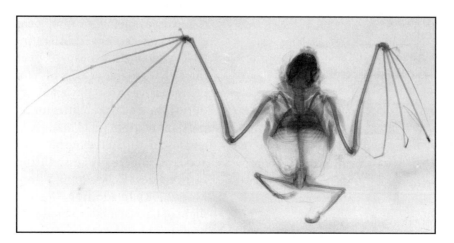

FIGURE 15-2 An X-ray photograph of the big fruit-eating bat (*Artibeus lituratus*, Phyllostomidae), showing the great elongation of the bones of the arm and hand.

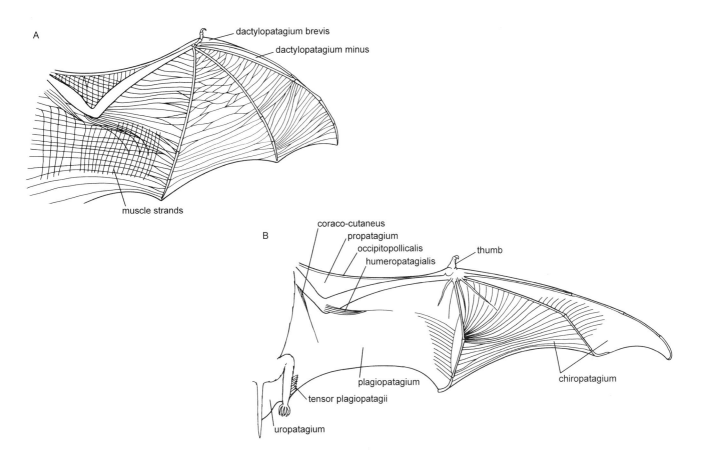

FIGURE 15-3   Ventral views of the wings of two bats, showing the parts of the wing and muscles and the elastic fibers that brace the membranes: (A) The big fruit-eating bat (*Artibeus lituratus,* Phyllostomidae); note the muscle strands that reinforce the plagiopatagium and the system of elastic fibers. This broad-winged bat does not remain on the wing for long periods. (B) The western mastiff bat (*Eumops perotis*, Molossidae). This narrow-winged bat is a fast and enduring flier. ((A) Adapted from Slaughter, B. H., and Walton, D. W. *About Bats.* Southern Methodist University Press, 1970; (B) Adapted from Wimsatt, W. A. *Biology of Bats.* Academic Press, 1970.)

support. The fibula is usually reduced, and support for the uropatagium, in the form of the **uropatagial spur** or **calcar**, is usually present (**Fig. 15-7**) (Schutt & Simmons 1998).

The evolution of the muscular control pattern of the wing-beat cycle typical of bats has seemingly been strongly influenced by their use of echolocation. Highly maneuverable flight is essential for these bats because objects are perceived in detail only at fairly close range by echolocation (Aldridge & Rautenbach 1987; Fenton 1994, 1995; Fenton et al. 1995). In contrast, birds use vision for more long-range perception of their environment and thus have relatively little need for extremely maneuverable flight. In both groups, similar trends toward rigidity of the axial skeleton and lightening of the wings occur, but many of the muscular and skeletal specializations that enable these animals to control their wings differ in the two groups. The pectoral girdle in birds is braced solidly by a tripod formed by the clavicles and coracoids anchored to the sternum and by the nearly blade-like scapula, which rests almost immovably against the rib cage. The pectoralis and supracoracoideus muscles, both of which originate on the sternum, supply nearly all the power for the wing beat (see Fig. 15-5).

In bats, nearly the reverse mechanical arrangement for flight occurs: the scapula is braced against the axial skeleton by the clavicle alone, movements of the clavicle during flight increase flight efficiency (Hermanson 1981), and the job of powering the wing beat is shared by many muscles (Vaughan 1959; see Fig. 15-5). This division of labor is made possible partly because the scapula is free to rotate on its long axis. The pectoralis, the posterior division of the serratus ventralis thoracis, and the clavodeltoideus muscles control the downstroke of the wings; only the pectoralis originates on the sternum. The muscles of the deltoideus and trapezius groups and the supraspinatus and infraspinatus muscles largely power the upstroke. The subscapularis is responsible for fine control of the wings during the entire wing-beat cycle (Altenbach & Hermanson 1987; Hermanson & Altenbach 1981). The thoracic skeleton in non-echolocating pteropodids is relatively similar to that in non-volant mammals, but in echolocating bats, it is modified often for greater rigidity, sometimes extremely so in a few bats (DesRoche et al. 2007). For example, in *Hipposideros gigas* (Hipposideridae), the first few ribs, sternum, and associated vertebrae are fused into a pectoral "ring"

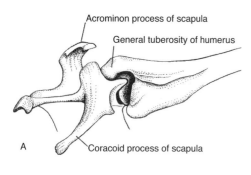

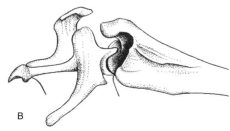

FIGURE 15-4 Anterior view of the left shoulder joint of a free-tailed bat (*Molossus ater*) at (A) the top of the upstroke and (B) during the downstroke. The greater tuberosity of the humerus locks against the scapula at the top of the upstroke, transferring the responsibility for stopping this stroke to the muscles binding the scapula to the axial skeleton. During the downstroke, the greater tuberosity of the humerus moves away from its locked position. This type of action and this type of shoulder joint also occur in the Vespertilionidae and other derived families of bats. (Adapted from Slaughter, B. H., and Walton, D. W. *About Bats*. Southern Methodist University Press, 1970.)

(Vaughan 1970), and in *Chilonatalus tumidifrons*, (Natalidae) all of the ribs, sternum, all thoracic and all but two lumbar vertebrae form an extremely rigid trunk skeleton (**Figs. 15-8** and **15-26**).

In terms of mechanical and physical properties, bat wings deform complexly and extensively during flight and

act completely unlike most human-engineered aircraft (Swartz et al. 2006). Even the fur of a bat's body and wings is aerodynamically involved in flight (Bullen & McKenzie 2008; Vaughan 1980). A morphological trend of critical importance to bats and all other flying animals is toward the reduction of wing weight. Propulsion is obtained in all flying animals by movement of the wings, and the kinetic energy produced by such movement depends on the speed and weight of the wing. The amplitude of a stroke and its speed are progressively greater toward the wing tip. Consequently, reduction of the weight of the distal parts of the wing results in a reduction of the kinetic energy developed during a wing stroke. A considerable advantage in metabolic economy is thus gained; as less kinetic energy is developed during each stroke, less energy is necessary to control the wings. In addition, light wings can be controlled with speed and precision during extremely rapid maneuvers when bats chase flying insects. Equally important, reduction in thickness of cortical bone in the humerus and radius of the bat wing is an adaptation to resist the large torsional stresses generated on these bones during flight (Swartz et al. 1992). Such torsion is not normally encountered during locomotion by terrestrial mammals, whose bones are relatively thick walled, but bird and pterosaur forelimb bones are thin walled like the humerus and radius of bats. Interestingly, the more distal elements in the wings (metacarpals and phalanges) of bats show the opposite trend, with greatly increased cortical thickness (Swartz et al. 1992).

Reduction of the wing weight has been furthered in bats by many specializations. Movement at the elbow and wrist joints is limited to one plane, thus eliminating musculature involved in rotation and bracing at these

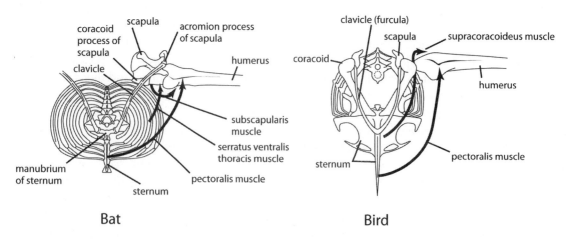

Bat   Bird

FIGURE 15-5 Anterior views of the thorax and part of the left forelimb of a bird and a bat, with some of the major muscles controlling the wing-beat cycle shown diagrammatically. In the bird, the supracoracoideus muscle raises the wing, and the pectoralis muscle powers the downstroke; both muscles originate on the sternum. In the bat, three muscles primarily control the downstroke: the subscapularis, the serratus ventralis thoracis, and the pectoralis. Only the pectoralis originates on the sternum. Many muscles power the upstroke in bats. (Adapted from Slaughter, B. H., and Walton, D. W. *About Bats*. Southern Methodist University Press, 1970.)

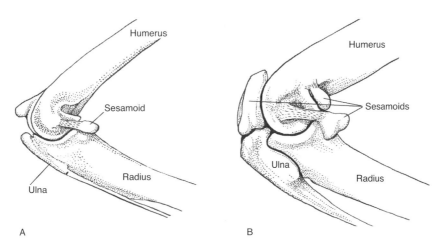

FIGURE 15-6 Lateral view of the right elbow of (A) a myotis (*Myotis volans*, Vespertilionidae) and (B) a free-tailed bat (*Molossus ater*, Molossidae). (Adapted from Slaughter, B. H., and Walton, D. W. *About Bats*. Southern Methodist University Press, 1970.)

joints. In addition, the work of extending and flexing the wings is transferred from distal muscles (of the forearm and hand) to large proximal muscles (pectoralis, biceps, and triceps), thereby allowing a reduction in the size of the distal musculature (Vaughan 1959). Certain forearm muscles are made nearly inelastic by investing connective tissue. Because of this modification and specializations of their attachments, these muscles "automatically" extend the chiropatagium with extension at the elbow joint or flex the chiropatagium with flexion at this joint (Vaughan & Bateman 1970; **Fig. 15-9**).

The hindlimbs of bats are generally quite thin but are not drastically reduced in length because of their importance in supporting the trailing edge of the plagiopatagium and the lateral edge of the uropatagium. The thinness of the hind legs and the fact that bats generally hang upside down to roost probably evolved under selective pressures favoring reductions in weight and the advantage of a quick takeoff to escape predators. As an adaptation to hanging, the normally delicate bat femur is suited to tensional stresses rather than compressional stresses associated with the femur of terrestrial mammals. In addition, many bats possess a tendon-locking mechanism on the tendons of the digital flexors that, when locked during hanging, reduces the muscular activity (Quinn & Baumel 1993). In a few kinds of bats, the hindlimb bones are relatively stout and are used where terrestrial locomotion is more frequently required. Examples are several members of Molossidae, including one genus *Cheiromeles* with an opposable hallux capable of grasping twigs, the unique New Zealand mystacinid *Mystacina*, the bulldog bats Noctilionidae, and vampire bats. Vampire bats must sometimes stalk up to their prey and avoid being stepped on and have been run on treadmills to analyze their locomotion (Riskin et al. 2005, 2006; Schutt & Simmons 2006).

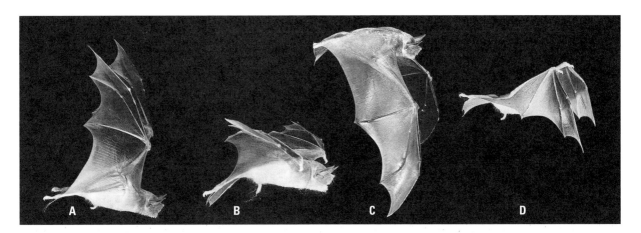

FIGURE 15-7 A fishing bat (*Noctilio albiventris*) in flight, showing several stages in the wing-beat cycle. A bone (the calcar) braces the uropatagium next to the ankle. (A) Top of the upstroke. (B) Midway through the downstroke. (C) End of the downstroke. (D) Midway through the upstroke.

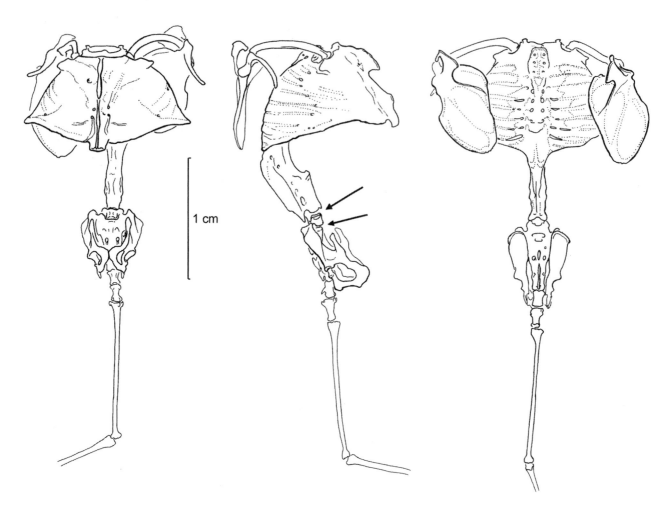

FIGURE 15-8 Anterior, lateral, and posterior views of part of the skeleton of *Chilonatalus tumidifrons* (Natalidae) showing extreme fusion of many of the central parts of the axial skeleton, sternum, ribs, thoracic vertebrae, and most lumbar vertebrae. The arrows indicate the two remaining joints on either side of the single unfused lumbar vertebra.

Flight

The three modern groups of flying animals—insects, birds, and bats—are all highly successful. Viewing the terrestrial scene, there are more flying than nonflying species of animals, but each flying group has evolved a different type of wing: bird wings are formed of feathers braced by a simplified forelimb skeleton along the leading edge; insect wings are membranous sheets of chitin braced by intricate patterns of chitinous veins, and bat wings are sheets of skin braced by the five-digited forelimb and elastic connective tissue. Flight styles also differ. Birds usually depend on relatively fast and not especially maneuverable flight. Insects usually use extremely rapid wing beats, a variety of flight speeds, and often a remarkable ability to hover. Most bats, however, use slow, highly maneuverable flight, generating lift and thrust very differently from birds and using complex and dynamic wing shape and motion changes (Muijres et al. 2008; Tian et al. 2006). As might be expected, these groups of fliers face diverse and complex mechanical and

aerodynamic problems, and animal flight remains incompletely understood. Inasmuch as 20% of all mammal species are bats, it is important to understand the phenomenon of flight in bats. Competing hypotheses about the evolution of flight and echolocation in bats have been proposed or discussed by Fenton et al. (1995), Speakman (1993), Norberg (1994), Arita and Fenton (1997), Simmons and Geisler (1998), with more recent genetic and developmental data about the potentially rapid evolution of elongated wing digits by Cooper and Tabin (2008), Dudley et al. (2007), and Sears et al. (2006).

Most students have been introduced at least once to the basic aspects of aerodynamics; this topic can therefore be treated briefly. Because the wings of animals usually provide both the thrust and the lift necessary for sustained flight—whereas in aircraft the wings provide only the lift—flight in animals presents special problems.

Lift is generated when an airstream sweeps over a wing with an asymmetrical cross-section. The profile of the cross-section of a wing (the airfoil) varies widely from one species

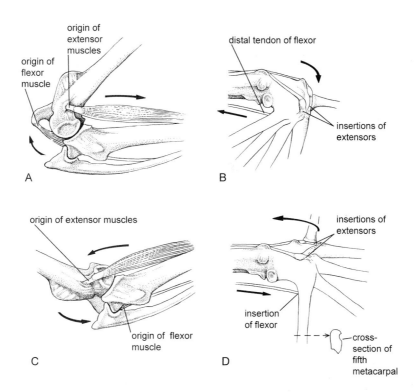

FIGURE 15-9 Lateral views of the elbow joint of the leaf-chinned bat (*Mormoops megalophylla*), showing the "automatic" flexion and extension of the fingers caused by certain forearm muscles in many advanced bats. (A) Flexion of the elbow joint moves the origins of the extensor muscles toward the wrist and the origin of one flexor muscle away from the wrist. (B) Because the flexor muscle is largely inelastic, with flexion at the elbow the distal tendon of the flexor pulls on the fifth digit and tends to flex the fingers. (C) With extension at the elbow joint, the origin of the extensor muscles is moved away from the wrist, and the origin of the flexor muscle moves toward the wrist. (D) This action pulls the extensor tendon toward the elbow and releases tension on the flexor tendon, thus extending the fingers. In D, the complex cross-sectional shape of the fifth metacarpal is shown. (Adapted from Vaughan, T. A. and Bateman, *J. Mammal.* 51 (1970): 217–235.)

of flying animals to another, but characteristically in birds and bats, it has an arched dorsal surface and a concave ventral surface (**Fig. 15-10B**). The tendency is for the parts of the airstream flowing over the opposite surfaces of the wing to arrive at the trailing edge simultaneously; this necessitates faster movement of air over the dorsal surface than over the ventral surface. The more rapidly the air moves over a surface, the less pressure it exerts, a relationship described by Bernoulli in 1738 and exploited by flying vertebrates for over 150 million years. The unequal pressure on opposing wing surfaces creates lift, a force opposing the force of gravity on a flying animal. Recent wind-tunnel experiments with a nectar-feeding phyllostomid bat, *Glossophaga*, show that during the downstroke in low-speed forward flight the leading edge of a bat's wings creates a vortex on top of the wing that greatly increases lift (Muijres et al. 2008). Lift is also created when a surface is presented at an **angle of attack** to the airstream (the angle of attack is the angle that the chord line of the airfoil makes with the plane of motion of the wing; Fig. 15-10A). Within limits, raising the angle of attack can increase lift. When, however, the angle of attack becomes so great that the air moving over the upper surface breaks away from the wing and forms turbulent eddies, the

lift produced by the wing abruptly falls as the **drag** sharply rises and stalling occurs. (Drag is the force exerted by air on an object in motion and in a direction opposite that of the motion.)

Wing performance is influenced by a series of variables. Lift increases directly as the surface area of the wing, but so does drag. Lift increases (within limits) as the **camber** of an airfoil increases (camber is the curvature, or arching, of an airfoil), but this also increases drag. Lift increases as the square of the speed, as does drag. Intuitively, then, one might expect that some of the constraints forcing modifications of wing design on fast fliers are of relatively little importance in slow-flying bats. This seems to be true and leads us to a consideration of the unique structure and function of the chiropteran wing.

The wings of bats form very thin airfoils of high camber. Several important features enhance the performance of these wings in the low-speed flight typical of most bats (Norberg 1969, 1972, 1981; Swartz et al. 2006). Thin airfoils, essentially cambered membranes, are more effective in producing high lift at low speeds than are conventional airfoils with some thickness. Of further importance is the ability of the bat to vary the camber of the wing in the

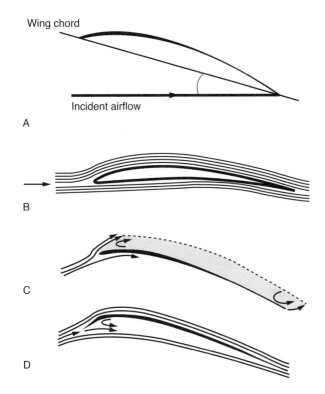

FIGURE 15-10 Cross-sections of wings (airfoils) and air flow over the wings. (A) Thin airfoil, showing the angle of attack, the angle between the wing chord and the incident airflow. (B) Flow of air over a thick airfoil. (C) Turbulence and separated air flow over a wing at a high angle of attack. (D) The addition of a leading-edge flap keeps the air flowing smoothly over the surface. (Adapted from Norberg, U. M., *Zoomorphology*. 73 (1972): 45–61.)

interest of producing high lift at low speeds. Camber of the bat wing is largely under the control of the occipitopollicalis muscles, the flexors of the thumb, the inclination of the dactylopatagium minus (see Fig. 15-3), the fifth digit, and the hindlimbs.

Compared with birds, bats have low **wing loadings** (Norberg 1981). (Wing loading is the ratio of body weight to wing area [W/S]. In general, the lower the wing loading, the slower an animal can fly and still maintain adequate lift to remain airborne). Most bats also have broad wings with a low **aspect ratio**, which is the relationship of the length of a wing to its mean breadth that for wings of irregular shape is expressed as the ratio of the span squared to the wing area (b^2/S). Some bats that fly rapidly and remain in flight for long periods have long, narrow, high-aspect ratio wings (see Fig. 15-3B). Broad wings suffer some loss of lift owing to air spillage from the high-pressure area on the ventral surface to the low-pressure area on the dorsal surface of the wing tip. Wings that are strongly tapered toward the tip minimize this spillage and loss of lift and are typical of fast-flying bats. Some bats are capable of advanced flight maneuvers such as sideslips and rolls through 180 degrees in order to lose altitude rapidly.

To produce lift, an airfoil must move through the air, and this requires a means of propulsion. In animals, movements of the wings create propulsion, and photographs of the wing-beat cycle in bats in level flight indicate that the downstroke is the power stroke and the upstroke is largely a recovery stroke (see Fig. 15-7). During the downstroke, the wings are fully extended and the powerfully braced fifth digit and the hindlimbs maintain the plagiopatagium at a fairly constant angle of attack, but the air pressure against the membranes becomes progressively greater toward the wing tip as the speed of the wing increases. This increase in pressure, coupled with the elasticity of the membranes between the digits, causes the trailing edges of the chiropatagium to lag behind the well-braced leading edge. In effect, the wing tip is twisted into a propeller-like shape and serves a propeller-like function. As the wing tip sweeps rapidly downward, it tends to force air backward, resulting in forward thrust of the animal. The membrane between the third and fourth digits (dactylopatagium longus) is probably of primary importance in producing thrust. During the upstroke, or recovery stroke, the wing is partly flexed, the stroke is directed upward and, to some extent, backward, and the force of the air stream partially aids the movement. Judging in part from the large muscles that power the downstroke and the relatively small muscles that control the upstroke, one would expect that the latter demands relatively little power and energy.

Some bats can fly very slowly, and some can hover; during these types of flight, the action of the wings is different from that used in level flight. When the nectar-feeding bat *Leptonycteris curasoae* hovers, the downstroke is directed largely forward, and the upstroke is directed backward (**Fig. 15-11**). The posture of the wings during the downstroke is similar to that in level flight, but because the stroke is largely horizontal, vertical thrust is developed. The upstroke, however, is complicated by a reversal of the usual posture of the wing tip: the tip turns over in such a way that the dorsal surface of the chiropatagium faces downward, and the leading edge of the wing still leads in this stroke but is posterior to the trailing edge (Fig. 15-11). Toward the end of the upstroke, the reversed wing tip is flipped rapidly backward and produces considerable upward thrust; at the start of the downstroke, the wing tip swings into its normal posture. This powerful flip probably demands considerable energy, but the vertical thrust that it develops strongly augments the thrust resulting from the downstroke and enables the bat to remain nearly stationary in the air (Altenbach 1977). Probably because of the high-energy cost of hovering, bats generally use it only briefly.

During the early evolution of the bat wing, selection seemingly favored refinements in design that allowed the development of high lift at low speeds. Later, however, in the Eocene epoch, bats underwent an adaptive radiation involving, in part, exploitation of various styles of flight (Habersetzer & Storch 1989; Habersetzer et al. 1994). The wings of some bats (members of the family Molossidae and some members of the family Emballonuridae, for example) developed characteristics advantageous during rapid flight. Because lift varies as the square of the speed of an airfoil, it would seem that rapid-flying bats could afford the luxury of higher wing loadings because of the greater lift developed per unit of wing area at higher speeds. Because drag also increases as the square of the speed, however, a reduction in wing surface area, angle of attack, and camber during rapid flight would be highly advantageous. Wing design in rapid-flying bats is clearly the result of a series of evolutionary compromises, and not all of these animals have wings that are alike; nonetheless, a number of bats have evolved roughly the same type of high-speed wing. High speed is used here only in a relative sense, for probably few bats achieve speeds in level flight above 80 kilometers per hour (50 miles per hour). Those bats actually measured to date showed flight speeds ranging from slow 1 meter per second (2.2 miles per hour; *Glossophaga*, Phyllostomidae) and 2.39 meters per second (5.3 miles per hour; *Natalus*, Natalidae) to 13.5 meters per second (30.2 miles per hour; *Nyctalus*, Vespertilionidae) (Akins et al. 2007; Bruderer & Popa-Lisseanu 2005; Muijres et al. 2008).

Paleontology

Because of their small size, delicate structure, and greatest abundance in tropical areas (where fossils are less often found), bats are relatively rare as fossils. Consequently, the evolution of bats is poorly known. Ironically, however, the best bat fossils known also happen to be some of the earliest known (Simmons & Geisler 1998; Simmons et al. 2008). The earliest undoubted fossil bats, *Onychonycteris finneyi* of the extinct family Onychonycteridae and *Icaronycteris index* of the extinct family Icaronycteridae, are from early Eocene beds in Wyoming and India (**Fig. 15-12**; Simmons et al. 2008; Smith et al. 2007). The recently discovered *Onychonycteris* is the most primitive known bat. It has relatively small wings, well-developed claws on all five wing digits, and a calcar (a partially ossified cartilaginous spur off the ankle that can help spread the tail membrane) and lacks specializations of the bony parts of the ears and hyoid apparatus. These features indicate that it could fly (perhaps rather weakly), retained the claws of an arboreal ancestry, but was unable to echolocate, solving a debate about whether bat flight evolved before, after, or at the same time as, echolocation (Simmons et al. 2008). The contemporaneous *Icaronycteris* is somewhat more advanced but retains several primitive features, such as claws on the first two digits and tiny

FIGURE 15-11 A nectar-feeding bat (*Leptonycteris curasoae*, Phyllostomidae) in flight, showing positions of the wings during slow or hovering flight.

FIGURE 15-12   A skeleton of the most primitive known bat, *Onychonycteris finneyi* (Onychonycteridae), from the early Eocene of Wyoming.

remnant nubbins of the ungual phalanges on the other digits of the hand, fairly short, broad wings, and bony indications of the ears and hyoids that suggest that it could echolocate. Its basic limb structure is that of modern bats. The upper molars of both *Onychonycteris* and *Icaronycteris* have teeth typical of most insectivorous bats. Equally well-preserved specimens of European Eocene bats from Messel, Bavaria have been found to have moth scales (Microlepidoptera and Macrolepidoptera) and parts of the chitinous exoskeletons of beetles (Coleoptera), cockroaches (Blattoidea), mosquitoes (Diptera: Culicidae), and caddis flies (Trichoptera) among their fossilized gut contents (Habersetzer et al. 1994). Eocene and early Oligocene deposits in Europe, Asia, Africa, and North America have yielded the earliest records of the modern families Rhinopomatidae, Emballonuridae, Megadermatidae, Nycteridae, Rhinolophidae, Natalidae, Molossidae, and Vespertilionidae (Gunnell et al. 2008; Simmons 2005). Pteropodidae appear first in the fossil record in the late Eocene of Thailand (Ducrocq et al. 1992, 1993).

The early Eocene appearance of bats clearly adapted for flight and of a diversity of fossil bats belonging to modern families by the late Eocene and Oligocene (Simmons 2005) have long been used to infer a much earlier origin of bats, in the Paleocene or perhaps Cretaceous, assuming a necessarily slow evolution of complex features required for flight (Gunnell & Simmons 2005). However, until fossils of bats or bat ancestors older than early Eocene are recognized or discovered, a recent finding suggests that bat wings and flight may have evolved much faster than was previously thought possible. Sears et al. (2006) showed that during ontogenetic development the timing of elongation of the embryonic digits of the forelimbs of bats is under genetic control of a single gene. This gene allows the digital bones to grow in length quickly relative to the same bones in the hindlimbs or the forelimb bones in mouse digits. This same ontogenetic development likely was key in the phylogenetic evolution of bat wings and, once it evolved, has remained relatively static in the numerous kinds of bats that have existed during the last 50 million years.

Non-Echolocating Bats

Family Pteropodidae

Pteropodids are called Old World fruit bats or flying foxes because of their fox-like faces and large size. Most are fruit eaters; others are nectar and pollen feeders. These bats are abundant and often conspicuous members of many tropical biotas in the eastern hemisphere. This family is represented in historic times by 42 genera and about 186 species. Pteropodids occur widely in tropical and subtropical regions from Africa and southern Eurasia to Australia and on many South Pacific islands eastward to Samoa and the Carolines. Members of this family are often relatively large, including the largest known bats, up to 1.5 kilograms in weight and 1.2 meters in wingspan, but some are small (13 grams in weight and 245 millimeters in wingspan). Based on molecular data, the Pteropodidae appear to be a monophyletic family that probably originated in Southeast Asia and Melanesia and invaded Africa several times (Giannini & Simmons 2003; Hollar & Springer 1997). Unlike other bats, pteropodids do not echolocate, except for two genera (*Stenonycteris* and *Rousettus*) that use tongue-click echolocation that is very different from the echolocation involving the larynx as the sound-producing organ.

The Pteropodidae have the following features that set them apart from other families of bats. The face is usually fox-like or lemur-like, with large eyes, usually a moderately long snout with a simple, unspecialized nose pad, and simple ears lacking a **tragus** (**Fig. 15-13**). (The tragus, a fleshy projection of the anterior border of the ear opening, may be seen in many other families such as the Vespertilionidae; see Fig. 15-19C.) The orbits are large and are bordered posteriorly by well-developed postorbital processes that

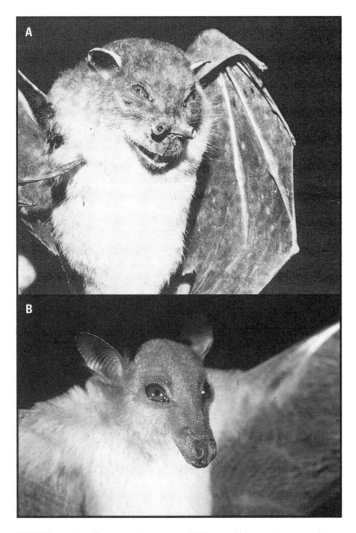

FIGURE 15-13 Faces of pteropodid bats: (A) a tube-nosed bat (*Paranyctimene raptor*) from New Guinea and (B) a nectar-feeding pteropodid (*Macroglossus minimus*) from Sabah, Borneo.

may meet to form a postorbital bar. The rostrum is never highly modified (**Fig. 15-14**). The dental formula is 1–2/0–2, 1/1, 3/3, 1–2/2–3 = 24–34 (Giannini & Simmons 2007). A benchmark monograph of the genus *Pteropus* thoroughly describes the cranial osteology (Giannini et al. 2006). The molars are never **tuberculosectorial** with W-shaped ectolophs, as in most echolocating bats, but instead are low, moderately flat-crowned, more or less quadrate, and lacking in stylar cusps (**Fig. 15-15**). The teeth are adapted basically to crushing fruit. The wing is primitive in having two-clawed digits, except in the genera *Eonycteris* and *Dobsonia* (in which only the thumb has a claw); the greater tuberosity of the humerus is not enlarged to make contact with the scapula at the top of the upstroke (**Fig. 15-16**). The tail typically is short or rudimentary. A cartilaginous uropatagial spur that projects from the tendon of the gastrocnemius muscle supports the uropatagium posteriorly (Schutt & Simmons 1998).

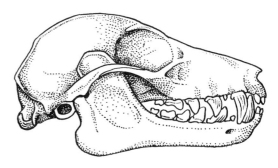

FIGURE 15-14 The skull of a megachiropteran bat (Pteropodidae), length of skull 62 millimeters.

Broadly speaking, pteropodids eat two types of food. One group is fruit eaters, whereas members of the other eat mostly nectar and pollen. Nectar-feeding probably evolved several times among pteropodids (Giannini & Simmons 2003; Kirsch et al. 1995). Few types of pteropodids (some of the genus *Nyctimene*) eat insects. The fruit eaters have fairly robust or moderately reduced dentitions. The jaws in these species are usually fairly long; in some species that presumably eat hard fruit, the jaws are shorter, and the teeth and dentary bones are unusually robust. The fruit bats often roost in trees in large colonies; in the case of the Australian species *Pteropus scapulatus*, as many as 100,000 have been observed roosting together. Fruit bats occasionally travel long distances during their nocturnal foraging, and *Pteropus alecto* regularly flies at least 70 kilometers from roosting sites to feeding areas. The fruit eaters usually are not particularly maneuverable fliers but have a steady, direct style of flight. Pteropodids are adroit at climbing about in vegetation, where the clawed first and second digits of the wing come into play (**Fig. 15-17**).

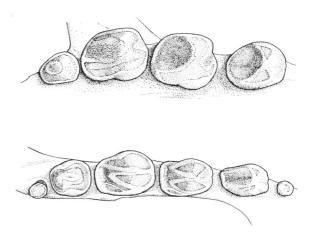

FIGURE 15-15 The cheek teeth of a pteropodid bat, *Pteropus* sp.: (top) right upper tooth row, showing two molars and two premolars, (bottom) lower left tooth row, showing three premolars and three molars. (Adapted from Wimsatt, W. A. *Biology of Bats*. Academic Press, 1970.)

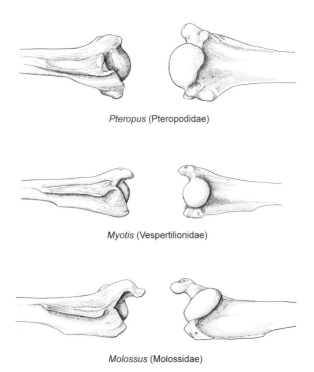

Pteropus (Pteropodidae)

Myotis (Vespertilionidae)

Molossus (Molossidae)

FIGURE 15-16   The proximal end of the right humerus in three bats. Anterior views are on the left and posterior views on the right. (Adapted from Wimsatt, W. A. *Biology of Bats*. Academic Press, 1970.)

Hypsignathus monstrosus, the hammer-headed bat, is unique among mammals in the fantastic degree to which the vocal apparatus is specialized in the males. This large, frugivorous pteropodid, with a wing spread approaching 1 meter, occupies tropical forests in much of central Africa.

FIGURE 15-17   A male Dyak bat (*Dyacopterus spadiceus*) hanging from a branch by its feet and thumb claws.

Communal displays by males in courtship areas are important in the breeding cycle of *Hypsignathus*. The males on the courtship arena, which is called a lek, use a penetrating call, described by Kingdon (1984a) as "guttural, explosive, and blaring," to attract females. The remarkable specializations of the vocal apparatus clearly evolved in association with the loud vocalizations employed during breeding displays.

Externally, the most striking feature of the male is the strange hammer-head appearance (**Fig. 15-18**). This is due in part to the enlarged and elevated nasal bones but is accentuated by a large pouch that encloses the rostrum and extends back over the cranium. These features enhance the resonance of the calls, and pharyngeal sacs in the throat are probably also resonators. Equally impressive are internal features attending the massive enlargement of the larynx. This structure, which contains huge vocal cords, has moved into the thorax, where it occupies most of the space filled by the heart and lungs in other mammals. As a result of this migration, the large trachea lies against the diaphragm and curves sharply craniad to the lungs, which are also forced against the diaphragm. The thoracic cavity thus serves largely as a container for the huge larynx in male *Hypsignathus*, with a drastic sacrifice in lung capacity. Kingdon has called this animal a flying loudspeaker; this

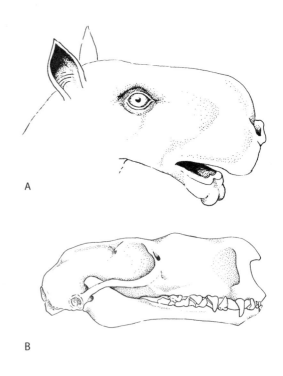

Hypsignathus monstrosus
Pteropodidae

FIGURE 15-18   (A) A drawing of the head of a male hammer-head bat (*Hypsignathus monstrosus*, Pteropodidae) showing the enlarged rostrum and elaborate lips. (B) Skull of a hammer-head bat. Length of skull 65 millimeters.

characterization seems especially apt when one considers the enlarged lips of the males, which can be formed into almost perfect megaphones. Schneider et al. (1967) described the anatomy of the larynx, and Matschie (1899) and Lang and Chapin (1917) carried out early studies on the morphology of the vocal apparatus of this bat.

The pteropodids that eat nectar and pollen have long, slender rostra, strongly reduced cheek teeth, and delicate dentary bones. The tongue is long and protrusible and has hair-like structures at its tip to which pollen and nectar adhere. Pollen, which adheres to the fur and is ingested when the bats groom themselves, is probably an essential source of protein to nectar-feeding bats. Some species roost in groups in caves, and some roost solitarily or in huge groups called "camps" in trees or other vegetation. Flight is slow and maneuverable.

Echolocating Bats

Eighteen families form the remainder of Chiroptera other than Pteropodidae. They reflect great structural diversity and widely contrasting modes of life. In these 18 families, echolocation universally involves sound production by the larynx, but echolocation styles vary. Bats of families other than the Pteropodidae are usually small, with body weights ranging from 2 to 196 grams. The eyes are often small. The rostrum is usually specialized, and the nose pad and lower lips may be modified in a variety of ways (**Fig. 15-19**). The ears have a tragus, a small flap of skin at the base of the pinna in all except members of the families Rhinolophidae and Hipposideridae. The postorbital process is usually small. Dentitions vary tremendously, but most families (except some members of the Phyllostomidae) have tuberculosectorial molars; the upper molars have a W-shaped ectoloph with strongly developed stylar cusps, and in the lower molars, the trigonid and talonid are roughly equal in size (**Fig. 15-20A**). In many insectivorous species and in some frugivorous members of the family Phyllostomidae, one or more premolars above and below are caniniform, and in some insectivorous species, the premaxillae are separate (**Fig. 15-21**).

The flight apparatus of the echolocating bats is more derived than that of the pteropodids. The second digit does not bear a claw and lacks a full complement of phalanges, and its tip is connected by a ligament to the joint between the first and second phalanges of the third digit. During flight, this connection allows the second digit to brace the third digit, which forms much of the leading edge of the distal part of the wing, against the force of the airstream. The greater tuberosity of the humerus is usually enlarged and locks against a facet on the scapula at the top of the upstroke of the wings (see Fig. 15-6). The size of the tail and uropatagium varies. The uropatagium is supported posteriorly by a cartilaginous or bony calcar that articulates directly with the calcaneus of the ankle (see Fig. 15-7), unlike the uropatagial spur of pteropodids (Schutt & Simmons 1998). The shape of the wing varies according to foraging pattern and style of flight. In general, slow, maneuverable fliers have short, broad wings, whereas rapid, enduring fliers have long, narrow wings (see Fig. 15-3).

Family Rhinopomatidae

Members of this small family, containing but one genus with five species (possibly more; Hulva et al. 2007), occur in northern Africa and southern Asia east to Sumatra. These animals are called mouse-tail bats because of the long tail that is largely free from the uropatagium (**Fig. 15-22A**). The oldest known fossil representatives of the family are from the late Eocene of Egypt (Gunnell et al. 2008).

These small bats (10 to 15 grams in body weight) have premaxillaries that are separate from one another, and their palatal portions are much reduced. The second digit of the hand, in contrast to the arrangement in all other echolocating bats, retains two well-developed phalanges. Perhaps the clearest indication of the primitiveness of these bats is the structure of the shoulder joint. In contrast to the situation in most bats, the greater tuberosity of the humerus is small and does not lock against the scapula at any point in the wing-beat cycle. Other rhinopomatid features include laterally expanded nasal chambers, no fusion of cervical, thoracic, or lumbar vertebrae, and a complete fibula.

The dentition is adapted to an insectivorous diet. The molars are tuberculosectorial; the upper molars have W-shaped ectolophs of the usual type. The dental formula is 1/2, 1/1, 1/2, 3/3 = 28. These are fairly small bats (the length of the head and body is up to 80 millimeters) with slender tails nearly the length of the head and body. The eyes are large, and a fold of skin joins the anterior bases of the large ears across the forehead. The nostrils are slit-like.

Mouse-tail bats are insectivorous and typically occupy hot, arid areas. They roost in a wide variety of situations, including fissures in rocks, houses, ruins, and caves; one species roosts in large colonies in some Egyptian pyramids. Although locally common, mouse-tailed bats are outnumbered by other types of bats over much of their range and are not as important today as other families. Rhinopomatids perhaps hibernate in some areas. Large deposits of subcutaneous fat occur in the abdominal area and around the base of the tail in individuals from some localities. These bats tolerate body temperatures as low as 22°C and can spontaneously rewarm themselves (Kulzer 1965).

Family Craseonycteridae

This family is found only in Thailand and Myanmar (Ramos Pereira et al. 2006). As far as is known, only one species

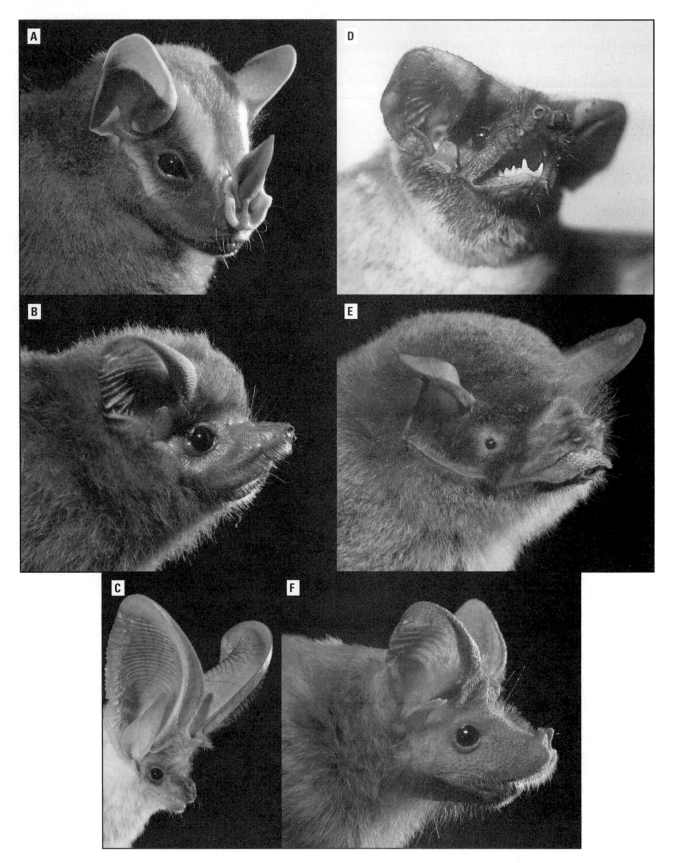

FIGURE 15-19 Faces of some bats: (A) common tent-making bat (*Uroderma bilobatum*; Phyllostomidae). (B) Gray sac-winged bat (*Balantiopteryx plicata*, Emballonuridae). (C) Brown long-eared bat (*Plecotus auritus*, Vespertilionidae). (D) African little free-tailed bat (*Chaerephon pumilus*; Molossidae) from Kenya. (E) Common mustached bat (*Pteronotus parnellii*, Mormoopidae). (F) Lesser mouse-tailed bat (*Rhinopoma hardwickii*, Rhinopomatidae).

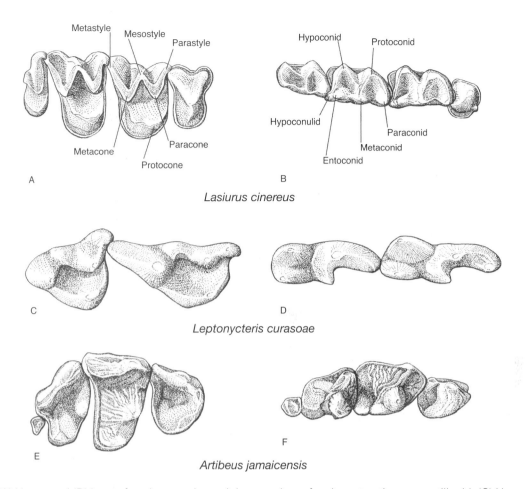

FIGURE 15-20   (A) Upper and (B) lower fourth premolar and three molars of an insect-eating vespertilionid. (C) Upper and (D) lower second and third molars of a nectar-eating phyllostomid. (E) Upper and (F) lower fourth premolar and three molars of a fruit-eating phyllostomid.

(*Craseonycteris thonglongyai*) represents the family. The common name is bumblebee bat or Kitti's hog-nosed bat. The species (and hence the entire family) represents a distinct evolutionary lineage, and its populations are thought to be declining; because of this, it is considered one of the top 50 most highly endangered mammal species (Isaac et al. 2007; Puechmaille et al. 2008).

Craseonycteris is delicately built and is one of the smallest living mammals; adults weigh about 2 grams (Fig. 15-22B). It has small eyes and large ears. The premaxillae are not fused to adjacent bones, a feature that may increase the mobility of the upper lip, and the much-reduced coronoid process of the dentary bone probably allows a wide gape of the jaws. The dental formula is 1/2,

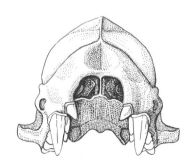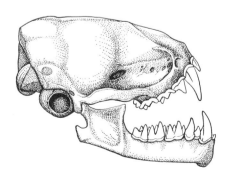

FIGURE 15-21   The skull of the hoary bat *Lasiurus cinereus* (Vespertilionidae): (left) anterior view, showing the emarginate front of the palate, and (right) side view, showing the shortened rostrum typical of some insect-feeding bats. The length of the skull is 17 millimeters.

1/1, 1/2, 3/3 = 28 and is of the usual insectivorous type with W-shaped ectolophs on the upper molars. The greater tuberosity of the humerus extends beyond the head of the humerus and may serve as a locking device; the second digit of the wing has only one very short phalanx, and the wing is broad. The pelvis and axial skeleton are highly specialized: The last three thoracic vertebrae and all but the last two lumbars are fused, and the sacral vertebrae are fused, whereas the pelvis is delicately built. The hindlimbs are slender, and the fibula is thread-like.

Bumblebee bats share molecular genetic similarities with members of the clade known as the Yinpterochiroptera (including the families Megadermatidae, Rhinopomatidae Hipposideridae, and Rhinolophidae; Hulva & Horacek 2002; Simmons et al. 2008; Teeling et al. 2005). These families share behavioral similarities. For example, in certain members of all these families, while the mother is roosting head downward, the nursing young clings to her head upward with its hind legs around the mother's neck and its mouth clinging to non–milk-producing pubic nipples (Duangkhae 1990; Simmons 1993). Bumblebee bats roost by day in caves and eat small arthropods that are probably gleaned from leaf surfaces (Hill & Smith 1981).

■ Family Megadermatidae

This is a small family, consisting of but four genera and five species. These bats are known as false vampires, an inappropriate title as they neither resemble vampires nor feed on blood. They occur in tropical areas in East Africa, southeastern Asia including Indonesia, the Philippines, and Australia. The earliest fossil megadermatid is from the late Eocene of Europe (Simmons 2005). No phylogenetic analysis of the family has been attempted, but would be quite interesting given the megadermatids' unique biology and wide distribution.

These are fairly large, broad-winged bats. The largest species has a wing spread approaching 1 meter and weighs up to nearly 200 grams. Smaller species have wing spreads of about 320 millimeters and weigh about 25 grams. The ears are large and are connected across the forehead by a

FIGURE 15-22 (A) Mouse-tailed bat, *Rhinopoma hardwickii* (Rhinopomatidae). (B) The endangered Kitti's hog-nosed bat *Craseonycteris thonglongyai* (Craseonycteridae), one of the world's smallest mammals. Note its short uropatagium.

ridge of skin. The tragus is bifurcated. The snout bears a conspicuous "nose leaf," and the eyes are large and prominent (**Fig. 15-23**). The premaxillae and upper incisors are absent; the upper canines project forward and have a large secondary cusp (**Fig. 15-24B**). The molars are tuberculosectorial; the dental formula is 0/2, 1/1, 1–2/2, 3/3 = 26–28. In *Megaderma*, and to a still greater extent in *Macroderma*, the W-shaped ectoloph of the upper molars is modified by the partial loss of the commissures connecting the mesostyle to the paracone and metacone. This trend is toward the development of an anteroposteriorly aligned cutting blade and is associated with the carnivorous habits of these genera. The shoulder and elbow joints are primitive; the second digit of the hand has one phalanx, and the third has two phalanges. The pectoral girdle has specializations similar to those of the nycterids (discussed later here), but the strengthening of the pectoral girdle is carried further in megadermatids. The manubrium of the sternum is broader in megadermatids than in the nycterids and is fused with the first rib and the last cervical and first thoracic vertebrae into a robust ring of bone. The megadermatid sternum is moderately keeled. The tail is very short or absent.

These bats occur in tropical forests and savannas, often near water, and eat a variety of foods. Of the five species of megadermatids, three are known to be carnivorous, and two are mostly insectivorous. The Australian ghost bat (*Macroderma gigas*), an unusually large, pale megadermatid, feeds on a variety of small vertebrates. In some areas, it seems to feed largely on other bats. The ghost bat and related species in southeastern Asia frequently consume their prey while hanging from the ceilings of spacious covered porches or verandas of large homes and detract from the gracious atmosphere by littering the floors with feet, tails, and other discarded fragments of frogs, birds, lizards, fish, bats, and rodents. Prakash (1959) observed *Megaderma lyra* of India eating bats of the genera *Rhinopoma* and *Taphozous*, a gecko, and a large insect. In the stomach contents of these bats, he found bones of amphibians and fishes. In captivity, *Megaderma lyra* begins feeding vertebrate prey

FIGURE 15-23 The bat *Megaderma spasma* (Megadermatidae) from Laos. Note the very large nose leaf, huge ears, and bifurcate tragus.

to their young during the late lactation period and weaning (Raghuram & Marimuthu 2007). The megadermatids may hunt partly by sight. One species, the partially diurnal, insectivorous African *Lavia frons*, uses a style of foraging similar to that of a flycatcher, hanging from a branch and making short flights to capture passing insects.

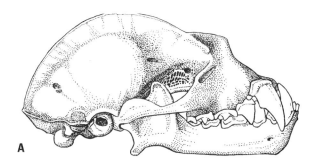
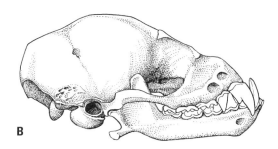

FIGURE 15-24 Skulls of two African bats that eat large beetles: giant leaf-nosed bat *Hipposideros gigas* [Hipposideridae; (A)], length of skull 32 millimeters, and African heart-nosed bat *Cardioderma cor* [Megadermatidae (B)], length of skull 27 millimeters.

Cardioderma cor of Africa has the most sedentary style of foraging known for any insectivorous bat (Vaughan 1976). This entirely nocturnal bat hangs from a perch in low vegetation when foraging and at some seasons regularly gives loud, humanly audible calls that seem to be territorial announcements. The body revolves through approximately 360 degrees as the hanging bat meticulously "scans" the ground, listening for sounds made by terrestrial invertebrates, such as large beetles and centipedes. When prey is detected, the bat flies directly to the ground, snatches up the food, and returns to the same perch to consume it. When insects are abundant during the wet seasons, *Cardioderma* spends very little time in flight: on some nights, it perches for periods averaging nearly 11 minutes and spends less than 1% of its foraging time in flight. In the dry season, however, when prey abundance declines, flights from perch to perch are more frequent and more time is spent in flight, although flights after prey average only 3 seconds. Considering all seasons, flights after prey average only 5 seconds. In *Cardioderma*, and perhaps in all megadermatids, the technique of searching for prey has departed markedly from that of most insectivorous bats, whereas the style of flight and the morphology of the forelimb have remained generalized.

Megadermatids roost in many types of places, from hollow trees, caves, and buildings in the case of most species, to sparse, occasionally sunlit vegetation in the case of *Lavia*. Wickler and Uhrig (1969) found this bat to occupy fairly small foraging territories and to have several humanly audible calls during social interactions.

■ Family Hipposideridae

This is a large and successful Old World family, called Old World leaf-nosed bats, with 9 genera and approximately 81 species, 67 of which are in the genus *Hipposideros*. This family has often been included as a subfamily in the Rhinolophidae; the two families are clearly sister taxa. The oldest fossil record of the Hipposideridae is from the middle Eocene in Europe, but the family might have originated and radiated early in the southern hemisphere, possibly in Australia (Hand & Kirsch 1998, 2003). One member of the family, *Hipposideros gigas* of Africa, is one of the largest insectivorous bats (Fig. 15-24A). Some individuals reach body weights of over 150 grams and wingspans of over 500 millimeters. Hipposiderids range from Africa and Madagascar across southern Asia to Japan, New Guinea, northern Australia, Indonesia, Borneo, the Philippines, and other islands.

In contrast to many bats that emit pulses from the open mouth in echolocation, hipposiderids keep the mouth closed during flight; the ultrasonic pulses used in echolocation are emitted through the nostrils and are beamed by the complex nasal apparatus (Möhres 1953). Because of the unique and complex face, hipposiderids are one of the most unmistakable groups of bats (**Fig. 15-25A**). In males of one species, the shield-faced leaf-nosed bat, *Hipposideros lylei*, the huge noseleaf completely blocks the bat's forward visual field (Francis 2008). The ears are usually large but lack a tragus, and the eyes are relatively small. The tail is of moderate length in some species but is small or rudimentary in others. The pectoral girdle is remarkable because it represents the extreme development of the trend (that occurs also in the Nycteridae and Megadermatidae) toward powerful bracing and enlargement of the sternum. In the most extreme manifestation of this trend, the seventh cervical vertebra, the first and second thoracic vertebrae, the first and most of the second rib, and the enormously enlarged and shield-like manubrium of the sternum are fused into a powerfully braced ring of bone (**Fig. 15-26**). The shoulder joint has a moderately well-developed locking device.

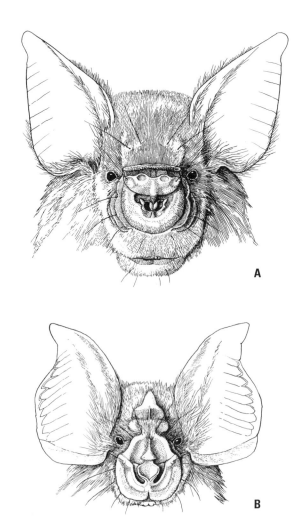

FIGURE 15-25 Faces of bats of the families Hipposideridae and Rhinolophidae. (A) Commerson's leaf-nosed bat *Hipposideros gigas*. (B) Lander's horseshoe bat *Rhinolophus landeri*.

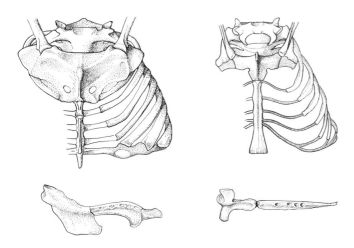

FIGURE 15-26 Ventral views of the thorax and lateral views of the sternum of a hipposiderid *Hipposideros gigas* (left) and a vespertilionid *Myotis yumanensis* (right). Note the highly specialized sternum of *Hipposideros*, to which the first two ribs are fused. (Adapted from Wimsatt, W. A. *Biology of Bats*. Academic Press, 1970.)

In some hipposiderids, all but the last two lumbar vertebrae are fused; a similar specialization occurs in the Natalidae (see Fig. 15-8). The pelvis is uniquely modified by enlargement of the anterior parts and by an accessory connection between the ischium and the pubis. These unusual pelvic specializations may be responses to the mechanical stresses imposed on the hindlimbs and pelvis by the repeated takeoffs and landings that occur during foraging in some of these bats. When they roost, these bats often hang upside down, and the hindlimbs are rotated 180 degrees from the usual mammalian posture so that the plantar surfaces of the feet face forward.

Most Old World leaf-nosed bats are cave dwelling, but some kinds roost in trees. All are insectivorous, with some species specializing to a high degree on certain insect prey thanks to their sophisticated biosonar and highly maneuverable flight. One, the African trident-nosed bat, *Cloeotis percivali*, feeds mainly on small adult moths year round, irrespective of the relative abundance or seasonality of these insects. This same species uses echolocation calls dominated by sounds of extremely high frequencies over 200 kilohertz, but like most hipposiderids, the calls are made at low intensity (loudness). The high frequencies allow the bats to feed on moths that have ears sensitive to the sonar frequencies used by less specialized bats (Bogdanowicz et al. 1999). When foraging, *Hipposideros gigas* of Africa hangs fairly high in trees, uses echolocation to detect large and straight-flying beetles at distances up to 20 meters, and makes brief and precise interception flights that last an average of 5.1 seconds (Vaughan 1977). This bat returns to the perch to consume prey, which consists of very large beetles (up to 60 millimeters in length).

Family Rhinolophidae

This family, commonly called horseshoe bats, contains only a single genus, *Rhinolophus*, but, remarkably, over 78 species. The horseshoe bats were previously often combined with hipposiderids as two subfamilies (Rhinolophinae and Hipposiderinae) of the Rhinolophidae, but virtually all systematists today treat these two groups as separate families. Uncertainty remains, however, as to the phylogenetic relationships among the species within Rhinolophidae (Bogdanowicz & Owen 1992; Sun et al. 2008).

The term "horseshoe bats" refers to the complex, basically horseshoe-shaped, cutaneous ridges and depressions on the nose (Fig. 15-25, botttom). Some rhinolophids are quite small, with body weights of nearly 6 grams and wing spans of 250 millimeters, whereas large species can weigh up to 45 grams. The geographic distribution of rhinolophids includes much of the Old World from western Europe and Africa to Japan, the Philippines, Indonesia, New Guinea and adjacent islands, and eastern Australia. The earliest known fossils of rhinolophids are from the middle Eocene of Europe and are very similar to some early hipposiderids (Hand & Kirsch 1998).

In keeping with the elaborate nose-leaves and style of echolocation, the rhinolophid skull is long and narrow with enlarged nasal chambers and a hump on the rostrum. The auditory bullae are small, but the cochleae are very large. The pinnae are medium-sized to enormous and lack a tragus but have a prominent antitragus at the lateral border of the pinna that probably has a similar function in hearing. The eyes are small, reflecting the dominance of hearing as a primary sense. The tail is fairly long and is nearly completely enclosed in the uropatagium. Some species have long, woolly fur, whereas in others, the fur is shorter. Horseshoe bats are common in many areas, and in Germany, the "Hufeisennase" is a familiar inhabitant of attics and church steeples that can be entered by uninterrupted flight. These bats have wide environmental tolerances. Various species inhabit temperate, subtropical, tropical, and desert regions. Rhinolophids hibernate in caves in some parts of their range and characteristically rest or hibernate with the body enshrouded by the wing membranes. Several species in East Africa are migratory. The food is largely arthropods, and the style of foraging resembles that of the hipposiderids, nycterids, and some megadermatids. Horseshoe bats pick spiders and insects from vegetation or capture flying insects in midair, and *Rhinolophus ferrumequinum* was observed to alight on the ground and capture flightless arthropods (Southern 1964). Like some hipposiderids, a number of rhinolophids make short foraging flights and do not remain continuously on the wing while foraging. Perhaps the wing membranes are important in some species in aiding in the capture of insects. Webster and Griffin (1962) demonstrated photographically that rhinolophids are able to capture insects in the chiropatagium, as do some vespertilionids and some other bats.

The specialized and intricately varied nose-leaves dominate the face and function to aid nasal emission of rhinolophids' biosonar. Zhuang and Müller (2006) showed that furrows in the nose-leaves act as resonance chambers helping to shape the sonar beam. Rhinolophids produce echolocation calls with a long, constant-frequency component, with a short, frequency-modulated downsweep at the end; the constant-frequency component is usually characteristic of a species in a particular region. A remarkable series of coordinated behaviors is associated with the highly specialized rhinolophid style of echolocation (see Chapter 22).

Most horseshoe bats are colonial, but some are solitary. Many kinds of roosting sites are used; caves, buildings, and hollow trees are generally preferred, but rock crevices, foliage of trees and shrubs, root systems of trees, and the burrows of large rodents are used by some species. Although the sexes roost separately in some species, a long-term breeding study of *Rhinolophus ferrumequinum* by Rossiter et al. (2005) showed that most females breed with specific males for several years, and their female offspring may also share the same mating partners with their mothers.

Family Emballonuridae

This family contains a variety of bats that are frequently called sac-winged or sheath-tailed bats. They range in size from small (about 4 grams) to large (up to 105 grams). *Saccolaimus peli*, an African emballonurid, is among the largest of the insectivorous echolocating bats, with a wing spread of nearly 70 centimeters. The small emballonurids have a wingspan of about 240 millimeters. Thirteen genera and about 51 species are currently recognized. Phylogenetic relationships within the family have been investigated using molecular data together with an interesting analysis of morphological and behavioral characteristics (Lim & Dunlop 2008; Lim et al. 2008). The wide geographic range of emballonurids includes the Neotropics (much of southern Mexico, Central America, and northern South America), most of Africa, Madagascar, southern Asia, most of Australia, and the Pacific Islands east to Samoa. Emballonurids are known from the early Eocene of Europe.

These bats combine a number of primitive features with several noteworthy specializations. In the possession of postorbital processes and reduced premaxillaries that are not in contact with one another, emballonurids resemble pteropodids. In addition, the shoulder and elbow joints are primitive. In contrast to the rhinopomatids, emballonurids retain only the metacarpal in the second digit; the flexion

of the proximal phalanges of the third digit onto the dorsal surface of the third metacarpal is a specialization also found in some advanced families of bats. Obvious external specializations include a glandular sac in the propatagium in some genera and the emergence of the tail from the dorsal surface of the uropatagium. The nose is simple in that it lacks leaf-like structures or complex patterns of ridges and depressions (Fig. 15-19B). In addition to the more common gray and dark brown species of emballonurids, some species of two genera (*Rhynchonycteris* and *Saccopteryx*) have whitish stripes on the back, and members of the genus *Diclidurus* are entirely white.

These insectivorous bats inhabit tropical or subtropical areas, where they use a great variety of roosting sites. Emballonurids occupy houses, caves, culverts, rock fissures, hollow trees, vegetation, or the undersides of rocks and dead trees for daytime retreats and usually roost in colonies. They are often fairly tolerant of well-lighted situations. Proboscis bats of the neotropics, *Rhynchonycteris naso*, often roost on exposed surfaces of the basal buttresses of large closed-canopy rainforest trees, tree trunks leaning over water, or fallen logs where they are well-camouflaged by their grizzled and striped fur and by small hair tracts across their forearms. In East Africa, *Taphozous mauritianus* often roosts on the trunks of large trees such as baobab trees (*Adansonia digitata*). In some areas, emballonurids probably forage mainly over water. Some members of the genus *Taphozous* have long, narrow wings, are swift and dashing fliers and often forage in clearings and above the canopies of tropical forests. One emballonurid (*T. mauritianus*) was recorded foraging at an altitude of 600 meters in Zimbabwe (Fenton & Griffin 1997).

A distinctive feature of some emballonurids is a sac in the propatagium that plays an important role in social communication. Study of one Neotropical species (*Saccopteryx bilineata*) has shown that this sac, especially well-developed in males, is used in ritualized displays to females during the breeding season. The sacs themselves are not glandular, but males fill their wing sacs with urine and secretions from the genital area and a gular gland (Voigt & von Helversen 1999). The sac odors are then "salted" (the wing is shaken at the female) and fanned toward females during a stereotyped complex hovering display.

Family Nycteridae

Members of this small family (16 species of one genus) are called slit-faced bats. These bats occur in Madagascar, Africa, the western Arabian Peninsula, the Malay Peninsula, and parts of Indonesia, including Sumatra, Java, and Borneo. The phylogenetic study of Griffiths (1997) suggests that the family originated in Africa, where it is most broadly distributed today. A fossil nycterid *Chibanycteris* is known from the early Oligocene of Oman, but an even older, middle Eocene tooth of an indeterminate species of nycterid is known from Tunisia, Africa (Sigé et al. 1994).

These fairly small- to moderate-sized bats weigh from about 6 to 35 grams and have wingspans ranging from 250 to 350 millimeters. They can be recognized by their large, separate ears, very small eyes, distinctive "hollow" face with a nose leaf and fringe of skin flaps, and a T- or Y-shaped cartilage at the end of the tail. The skull has a conspicuous interorbital concavity that is probably associated with the beaming of the sound pulses used in echolocation (**Fig. 15-27**). The extreme downward tilt of the rostrum relative to the basicranium strongly supports this suggestion (Freeman 1984). The interorbital concavity is connected to the outside by a slit in the facial skin. The dental formula is 2/3, 1/1, 1/2, 3/3 = 32, and the molars are tuberculosectorial. Postcranially, these bats combine primitive and specialized features. The shoulder and elbow joints are fairly primitive, but the retention of only the metacarpal of the second digit of the hand and the reduction of the number of phalanges of the third digit to two are obvious specializations. The pectoral girdle is modified in the direction of enlargement and strengthening of the bracing of the sternum, a pattern parallel to the trend in birds. The sternum in nycterids is robust, and the mesosternum is strongly keeled. The manubrium is broad. The first rib is unusually strongly built, and the seventh cervical and first thoracic vertebrae are fused. This general pattern also occurs in the family Megadermatidae and reaches its most extreme development in the families Hipposideridae and Natalidae (see Figs. 15-26 and 15-8). Because the specializations of the pectoral girdle in these bats parallel to some extent the roughly similar modifications in birds, they could be associated with a progressive structural trend in bats. Actually, it is doubtful that this is the case. Some of the most advanced and successful families of bats have less bird-like pectoral girdles than those in the families listed herein but have modifications of the shoulder and elbow joints and forelimb musculature that provide for efficient flight. Perhaps the nycterid-megadermatid-hipposiderid pectoral girdle is associated with a foraging style typified by short intervals of flight. In any case, this style of pectoral girdle seems to be a divergent type and does not represent a progressive morphological trend common to most "advanced" bats.

Slit-face bats inhabit tropical forests and savanna areas and seem to feed largely on arthropods picked from vegetation or from the ground. Nycterids are amazingly delicate and maneuverable fliers. When foraging, they often seem to drift effortlessly around the trunks of large trees and near foliage. Flying insects form part of the diet, but orthopterans and flightless arachnids, such as spiders and scorpions,

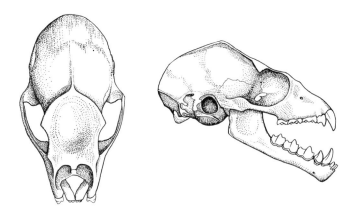

FIGURE 15-27 The skull of a slit-faced bat, *Nycteris thebaica* (Nycteridae): dorsal view, showing the depression in the forehead; side view, showing the flattened profile. The length of the skull is 19 millimeters. (Adapted from Hill, J. E., and Carter, T. D., *Bull. Amer. Mus. Nat. Hist.* 78 (1941): 1–211.)

are also important food items. The large species *Nycteris grandis* of Africa eats a remarkably varied assortment of animals, including orthopteran and lepidopteran insects, fish, frogs, birds, and small bats (Fenton et al. 1990, 1993). Nycterids show a similar variability in their approach to foraging; some individuals hunt by short flights from a perch, whereas others hunt while continuously on the wing.

Nycterids roost in a variety of situations, and some are even known to occupy burrows made by porcupines and aardvarks. In Kenya, in some remote safari camps, the pits dug for privies ("longdrops" in local parlance) are occasionally used as daytime retreats by *Nycteris thebaica*, to the consternation of the uninitiated users of these toilets.

Family Myzopodidae

Until recently the only known species representing this family was *Myzopoda aurita*, the sucker-footed bat, but in 2007 a second species, *M. schliemanni*, was described (Goodman et al. 2007). Both species are restricted to Madagascar. Fossils of this family are known from the early Pleistocene in East Africa, but the species certainly had a much longer history than the few fossils indicate. Many recent molecular genetic studies place Myzopodidae as a basal member of the Noctilionoidea, otherwise including the New Zealand family Mystacinidae and the Neotropical families Thyropteridae, Furipteridae, Noctilionidae, Phyllostomidae, and Mormoopidae.

The structure of the shoulder joint of *Myzopoda* is similar to that in the Natalidae, Furipteridae, and Thyropteridae. The lumbar vertebrae are not fused as they are in natalids. The dental formula is 2/3, 1/1, 3/3, 3/3 = 38. The ears are very large, and the ear opening is partly covered by an unusual mushroom-shaped structure found in no other bat. The claw of the thumb is rudimentary, and the thumb bears a sucker disk. Only the metacarpal of the second digit is bony; the third digit has three ossified phalanges. The foot bears a sucker disk on its sole, and as in thyropterids, each digit has only two phalanges; however, unlike thyropterids, the disks of *Myzopoda* seem to function by gluing instead of active suction (Thewissen & Etnier 1995). In *Myzopoda*, the metatarsals are fused, and all of the toes fit tightly against one another.

Myzopoda aurita appears to be rare, and its life history is poorly known, but *M. schliemanni* seems to be more common. Both species are insectivorous, with *M. aurita* feeding on small moths and *M. schliemanni* taking mostly small moths and roaches as well as lesser numbers of wasps and beetles (Rajemison & Goodman 2007). *Myzopoda schliemanni* inhabits dry deciduous forest in western central Madagascar, whereas *M. aurita* occupies the moister lowlands of eastern Madagascar. One individual of *M. aurita* was captured roosting in a traveler's palm (*Ravenala madagascariensis*), and a captive specimen readily roosted in *Ravenala* fronds by using its thumb and foot disks, orienting itself head upward, and bracing itself with its stiff tail (**Fig. 15-28**; Göpfert & Wasserthal 1995).

◼ Family Mystacinidae

This rather aberrant family until recently consisted of two species in the genus *Mystacina*, restricted to New Zealand (**Fig. 15-29**). One of the species is now extinct, not having been reported since 1965 (Hill & Daniel 1985), and the surviving species is threatened. Late Pleistocene fossils of mystacinids are known in New Zealand, but extinct, primi-

FIGURE 15-29   Portrait of a New Zealand lesser short-tailed bat, *Mystacina tuberculata* (Mystacinidae).

tive types were present in Australia in the Miocene (Hand et al. 2001).

The phylogenetic relationship between *Mystacina* and other bats has long defied explanation. In the past, based on various morphological features, it has been considered to have at least some affinities with practically every other echolocating bat family (Daniel 1979). Most recent studies using molecular evidence (e.g., Teeling et al. 2003) and hyoid morphology (Griffiths 1997) suggest that it is a noctilionoid, that is, a member of the southern hemisphere radiation of the endemic Neotropical families Noctilionidae, Mormoopidae, Phyllostomidae, Furipteridae, and Thyropteridae and possibly also the Madagascaran family Myzopodidae. The recently discovered fossils of mystacinids in Australia indicate that the family arose by at least the Miocene. These fossils and molecular data suggest that mystacinids colonized New Zealand from Australia (Hand et al. 2001).

Mystacina has an advanced locking shoulder joint, one phalanx in the second digit and two in the third, and no fusion of presacral vertebrae. These small bats weigh up to 22 grams, with a wingspan of 280 to 300 millimeters. The skull is relatively robust with a rather globose braincase and conical rostrum, and there is no anterior palatal emargination. The lower jaw symphysis is fused, an unusual characteristic for insectivorous bats but often found among nectar-feeding bats (Freeman 1988, 1995). The teeth are tuberculosectorial, and the dental formula is 1/1, 1/1, 2/2, 3/3 = 28. The tongue is partly protrusible and bears a brush of fine papillae on its tip, also found in nectar-feeding bats (Freeman 1995). The limbs resemble, in some ways, those of molossids: The wing membranes and uropatagium are

FIGURE 15-28   Golden bat, or sucker-footed bat, *Myzopoda aurita* from Madagascar. The animal is clinging by its sucker discs to a frond of the palm *Ravenala*.

tough and leathery. The first phalanx of the third digit folds back on the dorsal surface of the metacarpal. The distal end of the humerus has a long spinous process, and the hind foot is unusually broad; the fibula is complete, and the hindlimb is robust. The tail is short and protrudes from the dorsal surface of the uropatagium. Unique among bats, each of the claws of the thumb and foot has a secondary talon at its ventral base.

These unusual bats have many distinctive and flexible aspects of their natural history that are probably related, in large part, to the fact that, until historic introductions of exotic species to New Zealand by humans, no other ground-dwelling small mammals occurred on the islands. In New Zealand, the two species of *Mystacina* tended to fill this niche (Daniel 1990; Worthy et al. 1996). The limbs are well adapted to quadrupedal locomotion, and the bats are quite agile on the ground. In experiments with the bats moving on a treadmill, they seem to have a single type of locomotory gait that is used for both walking and running (Riskin et al. 2006). The wing can be folded compactly because of the unique flexion pattern of the third digit, and during quadrupedal locomotion, it is partially protected by the leathery proximal part of the plagiopatagium. The claws, and possibly the incisors, are used in excavating roosting tunnels in decaying kauri (*Agathis*) trees, in which several *Mystacina* will roost head to tail. These bats roost in a wide variety of places, including abandoned seabird burrows, holes in volcanic pumice, and hollows in other types of trees (Sedgeley 2006).

Mystacina tuberculata uses aerial foraging and surface gleaning but feeds primarily on the ground, where it burrows through leaf litter. This species feeds on large ground-dwelling insects and other arthropods and on very small vertebrates, including nocturnal frogs and lizards (Worthy et al. 1996). However, *Mystacina* is not strictly predatory, for it eats fruit, nectar, and pollen (Daniel 1979) and pollinates the endangered endemic plant *Dactylanthus* while crawling on the ground (Meyer-Rochow & Stringer 1997). *Mystacina* enters torpor but apparently does not hibernate during the austral winter. It has been observed foraging at ambient temperatures as low as 22°C (Daniel 1990). Another unusual aspect of its biology is that, unlike most other bats, *Mystacina* breeds in leks.

Family Thyropteridae

This family includes three small Neotropical species of bats that are known as disk-winged bats because of the remarkable sucker disks that occur on the thumbs and feet (**Fig. 15-30**). These animals are the only bats and the only mammals that have true suction cups. Disk-winged bats occur in southern Mexico, Central America, and South America as far as Peru and southern Brazil. This family's evolutionary relationships to other bat families are not completely resolved; several studies place them with vespertilionoid bats but recently more often with noctilionoid bats. Solari et al. (2004) gave relationships within the family. The earliest known fossil thyropterids have been recorded in the middle Miocene of Colombia.

In general appearance and in many skeletal details, these small (weight about 4 grams; wing span 225 millimeters), delicately formed bats resemble natalids, but the lumbar vertebrae are not fused as in the latter. The skulls of natalids and thyropterids are similar, and the dental formulas are the same. The thumb is reduced but retains a small claw, and its first phalanx has a sucker disk. The second digit is short, being represented by only a rudimentary metacarpal, and, as a result, the membrane between digits 2 and 3 is unusually small. The third digit has three bony phalanges. The digits of the feet have only two phalanges each. The third and fourth digits are fused, and the metatarsals bear suction disks, which have a complex structure that allows them to act as suction cups. The bats can cling to smooth surfaces and can even climb a vertical glass surface. A fibrocartilaginous framework braces each disk; the rim of the disk consists of 60 to 80 chambers, each supplied by a sudoriferous (sweat) gland, which improves the tightness of contact with the substrate by ensuring that the face of the disk is constantly moistened. The disk itself lacks muscles, but specialized

FIGURE 15-30 Spix's disk-winged bat, *Thyroptera tricolor*, from La Selva, Costa Rica, clinging by its suction discs to a smooth-surfaced leaf.

forearm muscles produce suction by cupping the middle of the face of the disk and release suction by lifting a section of the disk rim (Wimsatt & Villa-Ramirez 1970).

Disk-winged bats are insectivorous and restricted to tropical forests. Their roosting habits are highly specialized. *Thyroptera tricolor* roosts in the young, slightly unfurled or furled dead leaves of certain tropical plants that are partially or completely shaded by larger trees (Findley & Wilson 1974). Such a roosting site is provided by heliconias or "platanillos" (*Heliconia* spp.) and *Phenakospermum* plants, which resemble the banana plant (Simmons & Voss 1998). When a young leaf of one of these plants is beginning to unroll, it forms a tube roughly 1.3 meters long and 25 millimeters in diameter with a small opening at its tip. Several disk-winged bats may occupy such a tubular leaf in a head-to-tail row, heads upward, with the sucker disks anchoring them to the slippery surface of the smooth leaf. Because the leaf soon unfurls, it is suitable for occupancy for only about 24 hours, and the bats move periodically to new and more suitable leaves. Findley and Wilson (1974) found that these bats usually roost in social groups of six or seven, that the bats of a given group always roost together, and that each group occupies an exclusive area within which it roosts in the daytime. Another species, *T. lavali*, is thought to roost in palm trees (Solari et al. 2004). Disk-winged bats seem to forage mainly by surface gleaning on extremely small, mostly nonflying invertebrates such as jumping spiders, leafhoppers, and insect larvae (Dechmann et al. 2006).

■ Family Furipteridae

This small family contains but two genera, each with one species. These bats occur from Costa Rica south to southern Brazil and northern Chile and in Trinidad. Furipterids, known as smoky bats because of their gray pelage, are seemingly closely related to the Natalidae, Thyropteridae, and Myzopodidae. All of these groups share certain structural similarities and might be related, but some recent phylogenetic studies suggest that they are related to noctilionoids. They have almost no fossil record; one extant genus *Furipterus* is known in the late Pleistocene of Brazil, and the other, *Amorphochilus*, is known from the late Quaternary (Pleistocene or Holocene) of Peru. The family's roots, however, surely go much farther back in time.

Externally, furipterids resemble natalids in the structure of the ears and in the slender build. The shoulder joint and the fused lumbar vertebrae are also similar in these families. Furipterids differ from natalids in minor features of the skull and dentition, such as partially cartilaginous premaxillaries and reduced canines. The furipterid dental formula is 2/3, 1/1, 2/3, 3/3 = 36. The thumb of smoky bats is greatly reduced.

These bats are not common, and their habits are poorly known. They are insectivorous and have been found in caves and buildings. In French Guiana, Simmons and Voss (1998) found *Furipterus* roosting in or under fallen rainforest trees or logs. Most of the area inhabited by smoky bats is tropical, but the two genera differ in their habitat selection: *Furipterus* occurs mainly in moist forests, whereas *Amorphochilus* occurs in the hyperarid coastal sections (Atacama Desert) of northwestern South America.

■ Family Noctilionidae

Although this family is not large in terms of numbers of species (it contains only two species in one genus), it is of particular interest because one species is highly specialized both structurally and behaviorally for eating fish. Noctilionid bats are often referred to as bulldog bats or fishing bats. Evolutionarily, their closest relatives are members of the other noctilionoid families, especially Mormoopidae and Phyllostomidae. Bulldog bats inhabit the Neotropics from Sinaloa, Mexico, and the West Indies to northern Argentina. They are known as fossils from the middle Miocene of South America (Czaplewski 1997).

Noctilionids are morphologically distinctive. They are fairly large (from roughly 30 to 60 grams in weight and up to 585 millimeters in wing spread), and the heavy lips, somewhat resembling those of a bulldog, pointed ears, and simple nose make the face unmistakable (**Figs. 15-31** and 15-7). The dorsal pelage varies in color from orange to dull brown, and a whitish or yellowish stripe is usually present from the interscapular area to the base of the tail. The hindlimbs and feet are remarkably large, especially in *Noctilio leporinus*, and the feet have sharp, recurved claws. The premaxillae are complete, and in adults, the two maxillae are fused together and are fused with the premaxillae, forming a strongly braced support for the enlarged upper medial incisors. The dental formula is 2/1, 1/1, 1/2, 3/3 = 28. The teeth are robust, and the molars are tuberculosectorial. *Noctilio* has unusually long canines and a low coronoid process for a bat. The seventh cervical vertebra is not fused to the first thoracic. The shoulder and elbow joints are primitive, and the second digit of the hand has a long metacarpal and a tiny vestigial phalanx. The pelvis is powerfully built, with the ischia strongly fused together and fused to the posterior part of the laterally compressed, keel-like sacrum. The tibia and hind foot of *N. leporinus* have a series of unusual specializations. The calcar in noctilionids is uniquely ossified, rather than cartilaginous as in other bats.

The two species of *Noctilio* primarily eat insects, but are quite flexible in foraging style and diet. *Noctilio leporinus* is well-known for its habit of catching fish and occasionally taking crustaceans. The smaller species, *Noctilio*

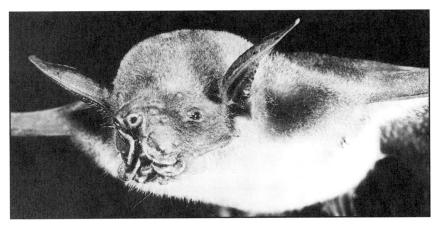

FIGURE 15-31   The face of the fishing bat *Noctilio albiventris* (Noctilionidae).

albiventris, too, sometimes eats fish and occasionally even fruits, which are often taken from the surface of the water. These bats use a mode of echolocation that is well-suited to this trawling style of foraging (Kalko et al. 1998; Lewis-Oritt et al. 2001). *Noctilio leporinus* is able to catch flying insects in midair using the tail or wing membranes, but also catches fish, shrimp, terrestrial insects, and other invertebrates by using its feet (Brooke 1994; Crasso & Wagner 2008). The style of foraging of *N. leporinus* involves the use of the hind claws as gaffs (Bloedel 1955; Brooke 1994). This bat recognizes concentrations of small fish or single fish immediately beneath the surface of the water by detecting (by echolocation) the ripples or breaks in the surface that these fish create (Suthers 1965, 1967). *Noctilio* spp. are the loudest known bats, producing cries with a sound pressure level up to 140 decibels, but above the range of human hearing (Surlykke & Kalko 2008). The bat skims low and drags its feet in the water, with the limbs rotated so that the hook-like claws are directed forward. (This involves rotation of the hindlimbs 180 degrees from the typical mammalian position.) When a small fish is "gaffed," it is brought quickly from the water and grasped by the teeth. From 30 to 40 small fish were captured in this fashion per night by one *N. leporinus* under laboratory conditions. Terrestrial invertebrates, such as large insects, scorpions, and small crabs, are gaffed from the surface of the ground by the hind feet (Brooke 1994).

A series of modifications of the hindlimb are clearly advantageous in allowing this animal to pursue efficiently its specialized style of foraging. The long bony calcar, which is roughly as long as the tibia, the calcaneum, the digits and claws, and the distal part of the tibia are all strongly compressed so that they are streamlined with respect to their direction of movement when they are dragged through the water. During foraging sweeps, the short tail is raised, and the blade-like calcar is pulled craniad and clamped against the flattened side of the tibia. In this way, the large uropatagium is brought clear of the water and the streamlined calcar and tibia knife through the water, producing a minimum of drag.

Noctilionids roost during the day in groups in hollow trees and rock fissures, caves, and occasionally buildings. *Noctilio leporinus* is seemingly most common in tropical lowland areas, frequently occurring along coasts, where it forages along rivers or streams, over mangrove-lined marshes and ponds, or over the sea. In western Mexico in the dry season, individuals often forage over small, disconnected ponds in nearly dry streambeds. Such ponds have concentrations of small fish. In French Guiana, a *N. albiventris* was caught over a roadside puddle next to primary rainforest (Simmons & Voss 1998).

Family Mormoopidae

The two genera, *Mormoops* and *Pteronotus*, and 10 to 13 species that make up this family can appropriately be called leaf-chinned bats because in all species a conspicuous, leaf-like flap of skin occurs on the lower lip (Fig. 15-19E). These bats are largely tropical in distribution and occur from the southwestern United States and the West Indies south to Brazil. Phylogenetically, mormoopids are most closely related to other noctilionoid families, especially Phyllostomidae; within-family relationships have been described by Dávalos (2006), who suggested that 13 species might exist within the family instead of the 10 widely recognized. Their fossil record extends to the early Oligocene in Florida, United States (Morgan & Czaplewski, unpublished manuscript).

Leaf-chinned bats are fairly small, weighing between 7 and 20 grams, and have several distinctive external specializations. The snout and chin always have cutaneous flaps or ridges (that reach their most extreme form in *Mormoops*), but a nose leaf is not present. The ears are moderately large, have a tragus, and vary in shape but always have large ventral extensions that curve beneath the fairly small eyes. The tail is short and protrudes from the dorsal surface of the fairly large uropatagium. The rostrum is tilted more or less upward (this feature is most extremely developed in *Mormoops*), and the floor of the braincase is elevated. The coronoid process of the dentary bone is reduced, allowing the jaws to gape widely. The teeth are of the basic insectivorous type; the dental formula is 2/2, 1/1, 2/3, 3/3 = 34.

The second digit in the hand has one phalanx and the third, three, as in the related family Phyllostomidae, but the shoulder and elbow joints differ markedly from the phyllostomid pattern. The greater tuberosity of the humerus in mormoopids does not form a well-developed locking device with the scapula; the head of the humerus is more or less elliptical, perhaps favoring a specialized wing-beat cycle. The elbow joint is specialized in all species, and in *Mormoops*, modifications of the distal end of the humerus and the forearm musculature provide for a highly efficient "automatic" flexion and extension of the hand. The musculature of the hand is reduced and simplified; this lightens the hand and probably favors maneuverability and endurance. The hindlimbs do not have the spider-like posture typical of phyllostomids but instead have a "reptilian" posture that allows the animals to crawl on the walls of caves with considerable agility.

Leaf-chinned bats are among the most abundant bats in many tropical localities, where they are seemingly the major chiropteran insectivores. They are most common in tropical forests but occur also in some desert areas. Some species appear early in the evening; their insect-catching maneuvers resemble those of their temperate zone counterparts, the vespertilionids. Leaf-chinned bats usually roost in caves or deserted mine shafts and may concentrate in large numbers. They often select the hottest, most humid parts of tropical caves (Rodríguez-Durán & Soto-Centeno 2003). A colony of *Mormoops* observed by Villa-Ramirez (1966) in Nuevo León, Mexico contained more than 50,000 bats, and a colony of four species of mormoopids in Sinaloa, Mexico was estimated to contain 400,000 to 800,000 bats (Bateman & Vaughan 1974). When the bats from the latter colony emerged in the evening, they swept down the nearby arroyos and trails in such numbers and at such speeds that one hesitated to move across their path. When they form large colonies, these bats disperse many kilometers from their roosting site to forage at night and to remain continuously on the wing for several hours. Their impact on tropical ecosystems must be great, for the bats in the Sinaloan colony probably consume over 1,400 kilograms of insects per night. It is not surprising that the bats must disperse over a wide area to forage.

Family Phyllostomidae

This is the most diverse family of bats with respect to structural variation and contains more genera than any other chiropteran family. Fifty-five genera and 161 modern species are included in the family. These "Neotropical leaf-nosed bats" are so named because of the biogeographic region they occupy and the conspicuous leaf-like structure that is nearly always present on the nose (see Figs. 15-19A and 22-10A). These bats have exploited the widest variety of foods used by any family of bats. Some leaf-nosed bats have retained insectivorous feeding habits, but some are partly carnivorous and eat small vertebrates, including other bats, rodents, birds, frogs, and lizards. Some eat nectar and pollen, some are frugivorous, and some feed solely on the blood of other vertebrates. In recent years, leaves have been recognized as important items in the diet of some frugivorous species (Kunz & Diaz 1995; Zortéa & Mendes 1993). Phyllostomids are the most important bats in the Neotropics and occur from the southwestern United States and the West Indies south to northern Argentina. The oldest known fossil of this family is from the early Miocene of Argentina (Czaplewski 2010).

The great structural variation that occurs in the phyllostomids is largely associated with an adaptive radiation into a wide variety of feeding niches (Freeman 2000). Phyllostomids form a monophyletic clade with the other families of Noctilionoidea, especially Mormoopidae, but the extensive functional specializations have made it difficult to determine the phylogenetic relationships within the family. The most inclusive recent attempts are those of Baker et al. (2003), who proposed 11 subfamilies based on mitochondrial DNA sequencing, and Wetterer et al. (2000), who analyzed a combination of morphological and molecular data and proposed 7 subfamilies. Some phyllostomids are fairly small bats (*Choeroniscus* has a wing spread of roughly 220 millimeters and weighs 8 grams), and one species is the largest western hemisphere bat (*Vampyrum*, with a wing spread of over 1 meter and a body weight of up to 190 grams).

In most species, the nose leaf is conspicuous and spear shaped, but in a few species, the nose leaf is rudimentary or highly modified. The ears vary from extremely large to small, and a tragus is present. The tail and uropatagium are long in some species, with many stages of reduction and the absence of the tail and uropatagium in various species. Some species have a uropatagium but lack a tail; only *Sturnira* has completely lost the uropatagium. The wings are typically broad; the second digit has one phalanx, and the third has three phalanges. The shoulder joint has a moderately well-developed locking device formed between the greater tuberosity of the humerus and the scapula.

However, the elbow joint and forearm musculature are primitive, and the forelimb is, for a bat, generalized. Probably all phyllostomids, whatever their feeding habits, do not remain on the wing for long periods during foraging.

The forelimbs are not used only for flight but are important in many species in handling food, as well as in climbing over and clinging to vegetation (especially in the fruit-eating species). The importance of such use of the forelimbs has probably favored the retention in phyllostomids of limbs more generalized than those of many strictly insectivorous bats. The seventh cervical and first thoracic vertebrae are not fused, and in phyllostomids there is no fusion of elements to form the sturdy pectoral ring as is characteristic of rhinolophoid bats. In some leaf-nosed bats, however, the sternum is strongly keeled. The ventral parts of the pelvis are lightly built in most species, but the ilia are robust and more or less fused to the sacral vertebrae. These vertebrae are fused into a solid mass that becomes laterally compressed posteriorly. The acetabulum is characteristically directed dorsolaterally; the hindlimbs are rotated 180 degrees from the usual mammalian orientation and have a spider-like posture. Because of this position of the hindlimbs, some phyllostomids are unable to walk on a horizontal surface and use the hindlimbs only for hanging upside down.

All of the Holocene leaf-nosed bats probably evolved from an ancestral type that had tuberculosectorial teeth adapted to a diet of insects. Only the subfamily Phyllostominae has retained this dentition, however, and in some species there is no trace of the ancestral pattern. The noteworthy adaptive radiation of phyllostomids will be traced by considering the dentitions and foraging habits of each subfamily.

The subfamily Phyllostominae deviates least from the ancestral structural plan, and some species retain insectivorous feeding habits. This subfamily contains all the leaf-nosed bats with tuberculosectorial teeth of the ancestral type; however, in some species (*Chrotopterus* and *Vampyrum*, for example), the W-shaped ectoloph of the upper molars is distorted by the reduction of the stylar cusps and the closeness of the protocone, paracone, and metacone. Most members of this subfamily are insectivorous, and some species are known to pick insects either from vegetation or from the ground. A few of the largest phyllostomines resemble their Old World look-alikes and ecological counterparts, the megadermatids, in their partly carnivorous habits. The large phyllostomine species—*Phyllostomus hastatus, Trachops cirrhosus, Chrotopterus auritus,* and *Vampyrum spectrum*—are known to include small vertebrates in their diet. *Trachops* primarily eats insects, but in Panama, it specializes in frogs, which it locates using ears specially attuned to the low-frequency calls of the frogs

(Bruns et al. 1989; Tuttle & Ryan 1981). Beneath the roosts of *V. spectrum*, feathers and the tails of rodents and geckos give indications of feeding preferences (Vehrencamp et al. 1977), and stomach contents reflect a preference for small birds, marsupials, bats, and rodents (Bonato et al. 2004). *Phyllostomus hastatus* and *C. auritus* also opportunistically prey upon small bats (Oprea et al. 2006; Rodrigues et al. 2006). The means by which these carnivorous–omnivorous bats perceive small vertebrates is not known. They may well hear the faint sounds made as their prey moves, and the large eyes of these bats indicate that hunting also involves vision. Bats of this type generally have large ears, however, suggesting highly discriminatory echolocation. These and other phyllostomids exhibit much diversity in their modes of echolocation, just as they do in other aspects of their anatomy, behavior, and ecology (e.g., Mora & Macías 2006; Weinbeer & Kalko 2007).

Nectar feeding is common among tropical vertebrates (as indicated by the presence of more than 300 species of hummingbirds in the American tropics) and has also been adopted by bats of the subfamilies Glossophaginae and Lonchophyllinae of Mexico and Central and South America and by bats of the subfamily Phyllonycterinae of the West Indies. Like hummingbirds, these bats feed on the nectar and pollen of a great variety of plants and have many physiological and structural features associated with this mode of life, such as the ability to metabolize sugars at a fast rate in at east some species (Voigt & Speakman 2007). At least one nectar-feeding bat, *Glossophaga soricina*, may use ultraviolet vision to locate flowers at night (Winter et al. 2003). The rostrum of the skull is often elongated, and in highly specialized forms like *Choeronycteris mexicana* and *Musonycteris harrisoni*, it is extremely elongated and the dentaries are slender and fused together (**Fig. 15-32**; Freeman 1995; Tschapka et al. 2008). The tongue is long and protrusible and has a brush-like tip; in a recently described species (*Anoura fistulata*), the tongue is an extraordinary 150% of the head and body length (Muchhala 2006). The cheek teeth have largely lost the tuberculosectorial pattern (see Fig. 15-20C and 20D). The hairs of at least some of these nectar feeders have divergent scales that catch pollen as the bat feeds on nectar. This pollen is swallowed when the bats groom their fur and provides a protein supplement without which the animals could not survive. In nectar-feeding bats, the wings are usually broad and the uropatagium is reduced.

Some nectar feeders migrate long distances with the seasons, following the flowering of the plants that form their main food supply. Nectar feeders can maneuver delicately through dense tropical vegetation or around spiny desert cacti and can hover. Flowers seem to be located by the senses of smell and vision, and these bats feed by hovering

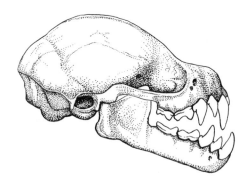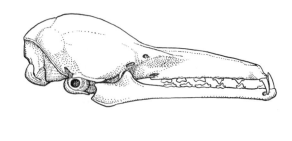

FIGURE 15-32   Skulls of leaf-nosed bats (Phyllostomidae): a fruit eater (*Artibeus phaeotis*; left), length of skull 19 millimeters; a nectar feeder (*Choeronycteris mexicana*; right), length of skull 30 millimeters.

briefly and thrusting the long tongue into the flowers. The pollination of many night-blooming Neotropical plants is accomplished by nectar-feeding phyllostomids, just as many plants in the Old World tropics are pollinated by nectar-feeding pteropodids. Selective forces determined by this method of pollination have probably been important in the evolution of flower structure and in the timing of pollen and nectar production in these plants. A dietary continuum from heavy reliance on insects (in *Glossophaga*) to dependence on nectar, pollen, and fruit (in *Leptonycteris*) occurs among nectar-feeding bats.

Fleming and his colleagues intensively studied the biology of nectar-feeding phyllostomids, especially *Leptonycteris* (Fleming et al. 1993; Fleming 1995). Some of their studies employed a method widely used by geochemists, plant physiologists, paleoecologists, and others. These scientists use stable isotopes of certain elements, especially carbon and nitrogen, to study various biological and physical processes. Stable isotopes extracted from the tissues of these bats indirectly show the general type of plants on which the bats were feeding. Plants with different photosynthetic pathways (called C3, C4, and CAM pathways) store characteristic proportions of these isotopes in their tissues, which can be analyzed using a mass spectrometer. The poorly known Peruvian bat, *Platalina genovensium*, proved to be a specialist on CAM plants (columnar cacti of the genus *Weberbauerocereus*; Sahley & Baraybar 1994). Biochemical studies have shown that some bats are migratory (*Leptonycteris*) and others are not (*Glossophaga*), and seasonal differences in diet among different populations of *Leptonycteris* have been detected.

The members of three subgroups of phyllostomids—Carolliinae, Brachyphyllinae, and Stenodermatinae—are frugivorous. The success of these groups and the richness of this food source in the Neotropics is indicated by the fact that within the family Phyllostomidae, the largest and most abundant group of Neotropical bats, nearly one-half of the species (approximately 78 out of 161), is basically fruit eaters. Several variations on this fruit-eating theme can be recognized. Members of the subfamily Carolliinae have reduced molars with the original tuberculosectorial pattern largely obliterated. These bats apparently prefer ripe, soft fruit and are known to eat a great variety of it. Stenodermatines of the genus *Sturnira* are fairly small, often brightly colored bats that have robust molars with no trace of the basic tuberculosectorial pattern. Indeed, their molars strongly resemble those of Neotropical monkeys (Cebidae). *Sturnira* eats small and often hard fruit, such as the fruits of low-growing species of nightshade (*Solanum* sp.).

Other stenodermatines have robust teeth that are highly adapted to crushing fruit. The upper molars have lost the stylar cusps, and the inner portion is much enlarged and marked by complex rugosities (Fig. 15-20E and 20F). The skulls are variously specialized in having a short rostrum (Fig. 15-32), and fairly high coronoid process of the dentary bone in many species, conferring considerable mechanical advantage for powerful jaw action to the large temporal muscles (Dumont 2006). Large species of the stenodermatine genus *Artibeus* are remarkably abundant in some Neotropical areas, and their piercing calls are characteristic sounds of the tropical nights. Often many *Artibeus* of several species concentrate on a single fig tree (*Ficus* sp.) with abundant fruit. In central Sinaloa, Mexico, two students and the senior author camped beneath such a fig tree—but only for one night. The activities of dozens of *A. lituratus*, *A. hirsutus*, and *A. jamaicensis* caused a nearly continuous rain of fruit and bat excrement throughout much of the night, and with sunrise came herds of aggressive local pigs to gather the night's fallout of figs. Stenodermatines often eat unripe and extremely hard fruit, and it is perhaps as an adaptation to this type of food that the robust teeth and powerful jaws evolved.

Recently it was learned that some fruit-eating phyllostomids eat leaves and drink at pools of murky water formed on clay mineral licks that are used by larger mammals. Geophagy (earth-eating) is thought to help large mammals acquire certain mineral nutrients or is done as self-medication. In fruit-eating phyllostomids, the murky water on clay licks was mostly consumed by female bats and might help them to acquire calcium in the diet or to produce milk (Bravo et al. 2008).

Interestingly, many species of Stenodermatinae construct "tents" in which to roost by chewing the veins of leaves of various species of tropical plants (Rodríguez-Herrera et al. 2008; Timm & Lewis 1991).

The subfamily Desmodontinae contains the vampire bats, the only mammals that feed solely on blood. Only three genera and three species constitute this group, but they are widely distributed from northern Mexico southward to northern Argentina, Uruguay, and central Chile.

Vampire bats are fairly small. Mexican specimens of the common vampire bat (*Desmodus rotundus*) usually weigh about 30 grams and have an average wing span of 365 millimeters. The skull and dentition are highly specialized. The rostrum is short. The braincase is high (**Fig. 15-33**), and in all species the cheek teeth are reduced in both size and number. In *D. rotundus*, the most specialized species, the dental formula is 1/2, 1/1, 2/3, 0/0 = 20. The upper incisors are unusually large and are compressed and blade-like, as are the upper canines. These teeth have remarkably sharp cutting edges. The cheek teeth are tiny. Except for the canine, the lower teeth are small.

The flight of vampires is strong and direct and not highly maneuverable, but the agility of vampires during quadrupedal locomotion is remarkable compared with most bats (Altenbach 1979; Riskin et al. 2005). The thumb in *Desmodus* is unusually long and sturdy and contributes an additional segment with three joints to the forelimb during quadrupedal locomotion. The hindlimbs are large and robust, and the fibula is not reduced. The proximal part of the femur and the tibia and fibula is flattened and ridged; this irregularity provides large surfaces for the attachment of the powerful hindlimb musculature. *Desmodus* can run

rapidly and easily on the ground and can even jump short distances (**Fig. 15-34**). Their agility in terrestrial locomotion is reflected in the histochemistry of the primary flight muscle, the pectoralis. The histochemical composition of this muscle is complex, with four different fiber types similar to those in terrestrial mammals and unlike the pectoralis of other bats (Hermanson et al. 1993). *Diphylla* feeds entirely on the blood of birds; in order to do this, it must climb and feed arboreally. During arboreal locomotion, *Diphylla* climbs head upward, gripping branches with opposed thumbs and hind feet, and uses the short, stout calcar on its hindlimbs like an opposable "sixth digit" to facilitate grasping (Schutt & Altenbach 1997).

Vampire bats can detect local temperature differences of the skin of their "prey," an ability probably of importance when they are seeking a place to bite. Temperature stimuli are probably detected in three pits in the hairless skin surrounding the nose (Kurten & Schmidt 1982). The surface temperature of this skin is up to 9°C lower than that of nearby parts of the face.

The feeding habits of vampire bats are of particular interest (Schutt 2008). Vampires begin foraging after dark and have one foraging period per night. *Diaemus* and *Diphylla* prefer the blood of birds, but *Desmodus* is more generalized, feeding on both mammals and birds. Indeed, *Desmodus* is a major pest of livestock and poultry, such as chickens and turkeys kept in poorly enclosed pens in Latin America, and there are reports of it feeding on penguins and sea lions along the coast of South America. *Desmodus* alights on the ground near its chosen host, often a cow, horse, or mule, and climbs up the foreleg to the shoulder or neck. The bat uses its upper incisors and canines to make an incision several millimeters wide from which it "laps" blood with its tongue. Vampire bats occasionally feed on the feet of cattle, at which time their ability to jump quickly may enable them to avoid injury when the host animal moves its feet. In *Desmodus*, the ingestion of blood is facilitated by an **anticoagulant** in the saliva that retards clotting. It has been estimated that each bat takes a meal of blood each night that amounts to over 50% of the fasting weight of the bat; a vampire bat weighing 34 grams, then, takes roughly 18 grams of blood per night (Wimsatt 1969a). Because of this nightly drain of blood from cattle in certain localities and because vampire bats transmit rabies and other diseases, they have great economic impact in many Neotropical areas. Occasionally, vampire bats feed on humans. Female *Desmodus* and *Diphylla* that have fed will regurgitate blood to feed their young and closely related roost mates, a form of altruistic behavior (see Chapter 24; Elizalde-Arellano et al. 2007; Wilkinson 1987). Related bats and those that share the same roost will also share the same feeding sites on their prey.

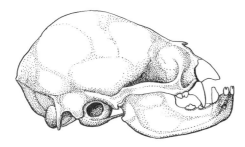

FIGURE 15-33 The skull of a vampire bat (Phyllostomidae). The length of the skull is 24 millimeters.

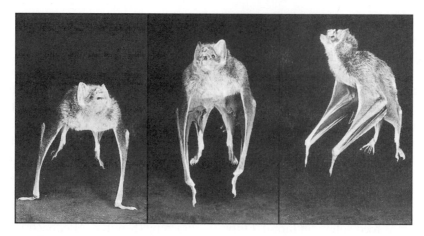

FIGURE 15-34 A vampire bat (*Desmodus rotundus*) leaping. Note the use of the long, robust thumbs.

Family Natalidae

This small Neotropical family includes three genera with nine species currently recognized, but probably at least two more species are unrecognized (Dávalos 2005; Tejedor 2005, 2006). These bats are commonly referred to as funnel-eared bats and occur from Baja California, northern Mexico, and the West Indies southward to Colombia, Venezuela, and Brazil. Their evolutionary relationships to other bat families are controversial, but most recent molecular phylogenies of bats show that natalids may be basal members of a superfamilial group Vespertilionoidea, which also includes Vespertilionidae, Miniopteridae, and Molossidae (e.g., Teeling et al. 2005). Family relationships show the distinctive species *Nyctiellus lepidus* as the most basal member of the family and *Natalus* as the most derived genus, with *Chilonatalus* and *Natalus* as sister genera including several species (Dávalos 2005). Early fossil natalids are known by a radius fragment from the Oligocene of Florida, United States, and the extinct genus *Primonatalus* from the Miocene there (Morgan & Czaplewski 2003).

These small bats weigh from roughly 5 to 10 grams and have slender, delicate-looking limbs, broad wing membranes, and a large uropatagium that encloses the long tail. The funnel-shaped ears with a tragus, the simple nose lacking any sort of nose leaf, and the long, soft pelage that is frequently yellowish or reddish are characteristic (**Fig. 15-35**). The skull has a long, wide rostrum with complete premaxillaries, and the braincase is high. The teeth are tuberculosectorial; the dental formula is 2/3, 1/1, 3/3, 3/3 = 38. The humeroscapular locking device is well developed. Reduction of the phalanges of the hand is well advanced (the second digit lacks a phalanx and the third has two), and the manubrium of the sternum is unusually broad and has a well-developed keel. Some of the most distinctive natalid features, however, reduce the flexibility of the axial skeleton:

In the most extremely specialized skeleton (of *Chilonatalus tumidifrons*), all of the thoracic vertebrae and all but the last lumbar vertebra are fused together with the ribcage and sternum into a solid bony structure (see Fig. 15-8), and the sacral vertebrae are mostly fused. As a result of these specializations, the strongly arched thoracolumbar section of the vertebral column is nearly rigid, with movement between this and the sacral section of the column allowed only by the joints at either end of the last lumbar vertebra.

Funnel-eared bats are insectivorous, and their foraging flight is slow, delicate, and maneuverable. Released in dense vegetation, they are amazingly adroit at flying slowly through small openings between the interlacing branches of trees and shrubs. These bats inhabit tropical

FIGURE 15-35 Portrait of a Mexican funnel-eared bat, *Natalus stramineus* (Natalidae).

and semitropical lowlands and foothills and typically roost in groups in hot, moist, and deep caves or mines. These are handsome little bats; groups of *Natalus stramineus* scattered over the ceiling of a cave look like bright orange jewels in the beam of a flashlight. Many of the species are threatened or in danger of extinction, even before the actual number of species is known (Dávalos 2005; Silva Taboada et al. 2007; Taddei & Uieda 2001).

Family Molossidae

Members of this family, the free-tailed bats, are important components of tropical and subtropical chiropteran faunas throughout much of the world. The family comprises 16 genera and 100 species. Phylogenetically, their closest relatives are probably the Vespertilionidae, although the two families have followed separate evolutionary trajectories for 50 million years. At least two distinct subfamilies, Tomopeatinae and Molossinae, are recognized (Sudman et al. 1994). Freeman (1981) examined relationships of the genera within the family using morphological characteristics, but no comprehensive molecular phylogeny has been completed. Molossids occupy the warmer parts of the Old World, from southern Europe and southern Asia southward, as well as Australia and the Fiji Islands. In the western hemisphere, they occasionally occur as far north as Canada, but the main range begins in the southern and southwestern United States and the West Indies and extends southward through all but the southern half of Chile and Argentina. Fossils of molossids are known from as early as the middle Eocene of Canada.

Structurally, this is a peripheral group of bats; the most extreme manifestations of many of the typically chiropteran adaptations for flight occur in the Molossidae. These bats weigh from 8 to 196 grams, and wingspans range from 240 to at least 516 millimeters. The greater tuberosity of the humerus is large (see Figs. 15-4 and 15-16), and the locking device between it and the scapula is highly developed. The origins of the extensor carpi radialis longus and brevis and flexor carpi ulnaris muscles are well away from the center of rotation of the elbow joint and probably act more effectively than in any other bats as "automatic" extensors and flexors of the hand. The wing is typically long and narrow (Fig. 15-3B), with the fifth digit no longer than the radius, and the membranes are leathery because they are reinforced by numerous bundles of elastic fibers. In many species, there are structural refinements that favor high-speed flight. In many molossids, for example, the radius and forearm muscles are flattened, and the arrangement of specialized hairs (**Fig. 15-36**) is such that the forearm is streamlined with respect to the airstream during flight (Vaughan & Bateman 1980). Adding rigidity to the outstretched wing during flight, movement at the wrist and elbow joint is strictly

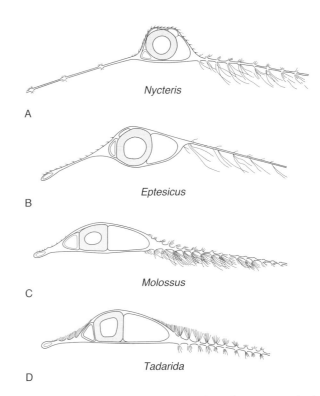

FIGURE 15-36 Cross-sectional views of the forearms and wing membranes of four species of bats, showing in (C) and (D) the pronounced streamlining of the wing by hair tracts and the flattening of the forearm in molossids. Two slow fliers: (A) *Nycteris thebaica* and (B) *Eptesicus fuscus*. Two fast fliers: (C) *Molossus ater* and (D) *Tadarida condylura*. (Adapted from Wilson, D. and Gardner, A. *Proceedings of the Fifth International Bat Research Conference*. Texas Tech Press, 1980.)

limited to one plane. The muscles that brace the fifth digit and maintain an advantageous plagiopatagium attack angle during the downstroke of the wings are large and unusually highly specialized. Except for fusion of the last cervical and first thoracic vertebrae, the presacral vertebrae are unfused. The body of the sternum is not keeled.

The general appearance of molossids is distinctive. The tail extends well beyond the posterior border of the uropatagium when the bats are not in flight, and the fur is usually short and velvety. (In one genus, *Cheiromeles*, hair is virtually absent.) Typically, the ears are broad, project to the side, and are like short canard winglets. As viewed from the side, the pinnae are arched and resemble an airfoil of high camber. The ears are frequently braced by thickened borders and are connected by a fold of skin across the forehead. Because of their unique design, the ears in most species do not directly face the force of the airstream during flight, an adaptation probably of considerable importance to these fast-flying bats. The muzzle is broad and truncate, and the thick lips are wrinkled in some species (Fig. 15-19D). In *Eumops*, the lips are finely wrinkled and almost prehensile when used in feeding (Freeman 1981).

The molossid skull is broad. The teeth are tuberculosectorial, and the dental formula is 1/1–3, 1/1, 1–2/2, 3/3 = 26–32 (**Fig. 15-37**). Several characteristically molossid features are associated with the well-developed quadrupedal locomotion typical of these bats. The first phalanges of digits 3 and 4 flex against the posterodorsal surfaces of their respective metacarpals, allowing the chiropatagium to be folded into a compact bundle, no longer than the forearm, that is manageable when the animals run. The feet are broad and have sensory hairs along the outer edges of the first and fifth toes. The fibula is not reduced, and the short hindlimbs are stoutly built. Within the structural limits of the basic chiropteran plan, these bats have seemingly made the better of two types of locomotion. The highly specialized wings are clearly adapted to fast, efficient flight, and the primitive hindlimbs have not lost their ability to serve in rapid quadrupedal locomotion. In *Cheiromeles*, the big toe is opposable and has a nail instead of a claw.

These insectivorous bats are remarkable for their speedy and enduring flight. Whereas most bats fly fairly close to the ground or to vegetation when foraging, many molossids fly high and may move long distances during their nightly foraging. Some populations of Brazilian free-tailed bats (*Tadarida brasiliensis*), the bats that occur in great numbers in Carlsbad Caverns and other large caves in the southwestern United States, fly at least 90 kilometers to their foraging areas each night (Davis et al. 1962). These bats were observed in Texas with radar and helicopters, and dispersal flights were tracked (Williams et al. 1973). Dispersing bats were recorded at elevations of over 3,000 meters, and masses of bats moved at an average speed of 40 kilometers per hour. The western mastiff bat (*Eumops perotis*) forages over broad areas and in southern California may on occasion fly more than 650 meters above the ground (Vaughan 1959). Because of the temperature inversions that frequently prevail for many nights in this area, in the winter, these high-flying bats may be surrounded by air warmer than that at the ground and may be catching insects that are flying in the warm air strata. Using bat detectors on helium-filled balloons, Fenton and Griffin (1997) recorded several species of African molossids foraging 600 meters above the ground. Some molossids remain in flight for much of the night; foraging periods of at least 6 hours have been recorded for some species.

Some molossids make spectacular dives when returning to their roosts. The western mastiff bat (shown on the title page of this chapter), for example, often makes repeated high-speed dives and half loops past the roosting site. It returns to its roost in a cliff by diving toward the cliff base, pulling sharply upward at the last instant, and entering the crevice with momentum to spare. Several other molossids are known to return to their roosting places by similar maneuvers. Because the wings of many molossids are narrow and have relatively small surface areas relative to the weights of the bats, these animals must attain considerable speed before they can sustain level flight. As a result, some species roost high above the ground in cliffs, buildings, or palm trees, in situations where they can dive steeply downward for some distance in order to gain appropriate flight speed. These species are unable to take flight from the ground.

Most molossids inhabit warm areas. Migration, therefore, is not generally characteristic of these bats. The Brazilian free-tailed bat, however, is known to make extensive migrations from the United States to as far south as southern Mexico (Villa-Ramirez & Cockrum 1962). The tremendous deposits of **guano** in some large caves inhabited by molossids attest to the effect that large colonies of these bats must have on insect populations in some areas. One such species in the Americas, *Tadarida brasiliensis*, forms aggregations numbering millions of individuals that when foraging can be seen on radar to spread out across hundreds of square miles after leaving their cave roosts. This species performs a huge ecological service to humankind by con-

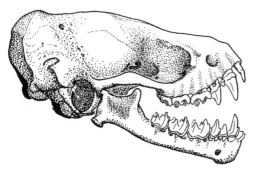

Brazilian free-tailed bat
(*Tadarida brasiliensis*, Molossidae)

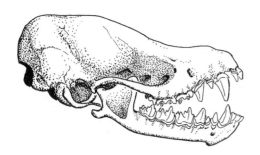

Long-eared myotis
(*Myotis evotis*, Vespertilionidae)

FIGURE 15-37 Skulls of two bats, a common molossid *Tadarida brasiliensis* (length of skull 16 millimeters) and a vespertilionid *Myotis evotis* (length of skull 16 millimeters).

trolling several kinds of agricultural-pest insects (Cleveland et al. 2006).

Family Vespertilionidae

This is the largest family of bats in terms of numbers of species and the most widely distributed. Forty-seven genera and approximately 395 species are included in this family, and in temperate parts of the world, these are usually by far the most common bats. In the Americas, vespertilionids occur from the tree line in Alaska and Canada southward throughout the United States, Mexico, and Central and South America. All of the Old World is inhabited north to the tree line in northern Europe and Asia. Most islands, with the exception of some that are remote from large land masses, support vespertilionids. As can be inferred from their geographic distribution, these bats occupy a wide variety of habitats, from boreal coniferous forests to barren sandy deserts. In the Neotropics, however, they are greatly outnumbered by bats of other families, particularly by leaf-nosed bats (Phyllostomidae).

Perhaps because of the diversity of habits and structure represented within the Vespertilionidae, no common name for this group is in general use; they are usually called simply "vespertilionid bats." For the same reasons, interpreting the phylogenetic relationships within the family has been extremely problematic; several subgroups (subfamilies or tribes) are often recognized. The most inclusive phylogenetic analysis of vespertilionids to date is that of Hoofer and Van Den Bussche (2003). The vespertilionid genus *Myotis* is remarkable for its long fossil record, which begins in the middle Eocene of Europe, and for its broad geographic distribution. *Myotis* has the widest range of any mammalian genus except *Homo* (and our domesticates).

Vespertilionids are rather plain-looking bats that lack the distinctive facial features characteristic of many families (**Fig. 15-38**). A nose leaf is rarely present (only in the subfamily Nyctophilinae, best developed in *Pharotis*, which is now extinct; Flannery 1995), and complex flaps or pads do not occur on the lower lips. The eyes are usually small. The ears are of moderate to enormous size, and the tragus is present but differs in shape markedly from one species to another (Fig. 15-19C). These bats are usually small, weighing from 4 to 83 grams, with wingspans ranging from 200 to 400 millimeters. The wings are typically broad, and the uropatagium is large and encloses the tail. The shoulder joint is of a derived type and provides for a locking of the large greater tuberosity of the humerus (see Fig. 15-16) against the scapula at the top of the upstroke of the wing. The elbow joint is also derived, and the spinous process of the medial epicondyle, which is well developed in many species, enables certain forearm muscles to "automatically" extend and flex the hand (see Fig. 15-19). The shaft of the

FIGURE 15-38   An eastern small-footed myotis (*Myotis leibii*; Vespertilionidae) in flight.

ulna is vestigial, but the proximal portion forms an essential part of the elbow joint. The second digit of the hand has two bony phalanges, and the third digit has three. The fibula is rudimentary. The manubrium of the sternum has a keel, but the body of the sternum has at best a slight keel. In all species, the presacral vertebrae are unfused.

The teeth are tuberculosectorial, and the W-shaped ectoloph of the upper molars is always well developed. The dental formula is 1–2/2–3, 1/1, 1–3/2–3, 3/3 = 28–38. The skull lacks postorbital processes (Fig. 15-37); the palatal parts of the premaxillaries are missing, and the front of the palate has an open gap (see Fig. 15-21). In general, vespertilionids are mostly small, plain bats that are characterized by refinements of echolocation and of the flight apparatus that enable them to forage and maneuver efficiently.

Most vespertilionids are insectivorous, and in their ability to capture flying insects they are expert. At least two large species (*Nyctalus lasiopterus* and *Ia io*) occasionally prey on small nocturnally migrating birds in Europe and Asia, respectively (Popa-Lisseanu et al. 2007; Thabah et al. 2007). Most children in Europe and North America gain their first experience with bats by watching vespertilionid bats, silhouetted against the twilight sky, making abrupt turns and sudden dives while pursuing insects. The most commonly used vespertilionid foraging technique is probably the most demanding: It involves the pursuit and capture of flying insects, which means that the bats must remain on the wing throughout most of their foraging periods. The insects, detected by echolocation, are usually followed in their erratic flight by a series of intricate maneuvers by the bat and are either captured in the mouth or, in some species of bats, trapped by a wingtip or by the uropatagium (Webster & Griffin 1962). This type of foraging demands highly maneuverable flight, and this is the type of flight to which vespertilionids seem best adapted.

Styles of foraging vary from one type of vespertilionid to another. The tree-roosting bats (*Lasiurus*) remain on the wing throughout their foraging, whereas some other bats alight to eat large prey. Some vespertilionids capture insects from the surface of the water, and several species of *Myotis* also capture fish or crustaceans from the water, probably by gaffing the prey with the large claws of the enlarged feet. Still other kinds of vespertilionids glean insects from the surfaces of vegetation or the ground. The pallid bat (*Antrozous pallidus*), a common species in the southwestern United States, feeds on such large terrestrial arthropods as scorpions, Jerusalem crickets (*Stenopelmatus*), and sphinx moths (Sphingidae). After localizing these prey items mostly by listening for the sounds they make, the bats snatch them from vegetation or pounce on them on the ground. Occasionally, pallid bats eat small vertebrates, such as horned lizards (*Phrynosoma*). This bat pollinates cactus and *Agave* flowers in southwestern North American deserts while capturing and eating insects that are in the flowers (Fleming 1995). The feeding habits of one of the tube-nosed bats, *Harpiocephalus harpia* of India and Southeast Asia, are poorly known and include at least beetles as a food item. This species has evolved some morphological features, such as a broad rostrum and strongly modified teeth, that seem convergent with those of certain fruit-eating bats.

A wide variety of roosting places are used by vespertilionids. They adapt well to urban life and frequently roost during the day in attics, in spaces between rafters, or behind shutters or loose boards. Crevices in rocks, spaces beneath rocks or behind loose bark, caves, mines, holes in trees, and foliage are also used. Often these bats are colonial, frequently with maternity colonies of females with young occupying one roost and adult males using another. Many species, however, such as the foliage-roosting bats, roost singly or in small groups. Some species rest for part of the night beneath bridges, in grottoes in cliffs, or in porches or buildings, often in places never used as daytime retreats.

In temperate regions, many vespertilionids hibernate. Although the hibernation sites of many species are not known, some well-known species hibernate in caves and mines or buildings, and some species migrate various distances to reach "optimal" hibernacula.

Family Miniopteridae

Long included as a subfamily in the Vespertilionidae, molecular phylogenetic data indicate that miniopterids are divergent enough to be considered a separate family (Hoofer & Van Den Bussche 2003; Miller-Butterworth et al. 2007). Commonly known in English as long-fingered bats or bent-wing bats, the family includes one genus, *Miniopterus*, with over 20 species, some of which are not yet formally named (**Fig. 15-39**). The similar appearance among many members of the family has led to a confused taxonomy and difficulty in distinguishing species, some of which probably comprise complexes of several cryptic species. Phylogenetically, miniopterids are vespertilionoids and appear to be most closely related to the Vespertilionidae. Long-fingered bats are small, weighing less than 20 grams and have very long wingspans for their size, up to about 135 mm. They are widespread in the Old World from southern Europe, Africa, Madagascar, southern and southeast Asia to Japan, the Philippines Islands, Australia, New Guinea, and many islands. Their fossil record extends to the late Oligocene of Germany.

Bent-winged bats have a distinctive wing morphology in which the third digit bears a very short first phalanx and

FIGURE 15-39 A little long-fingered bat, *Miniopterus australis* (Miniopteridae).

a very long second phalanx that is folded forward onto the palmar (ventral) surface of the wing when the bat is not flying. The fur color is usually black or dark brown. The ears are short and broad with a narrow tragus, and the tail is fully enclosed in the uropatagium. The skull has a high and very rounded braincase and a short, low, upturned rostrum.

Long-fingered bats roost mainly in caves, but also in rock crevices, mines, tree hollows, culverts, and buildings. The long third finger supports a long, narrow wing; like the narrow-winged molossids, some miniopterids forage for high-flying insects above the forest canopy. They tend to fly fast and erratically. The most widespread species in this family is *Miniopterus schreibersii*, whose distribution extends into cool north temperate climates in China, where some populations migrate seasonally and hibernate in caves in the winter (Smith & Xie 2008). *Miniopterus* roosts in groups sometimes numbering in the tens of thousands, where during emergence individuals of such large populations as those of *M. australis* in northeastern Australia are preyed on by ghost bats *Macroderma gigas* (Megadermatidae) and pythons.

SUMMARY

Echolocation, a means of perceiving the environment, and flight, a means to great motility, have been two major keys to the success of bats. Bats constitute the second largest mammalian order (behind Rodentia); approximately 202 genera and over 1,120 species of living bats in 19 families are known. They are nearly cosmopolitan in distribution, being absent only from arctic and polar regions and from some isolated oceanic islands. Bats exploit an impressive diversity of foods, from insects, spiders, scorpions, centipedes, crustaceans, mammals, reptiles, birds, frogs, and blood to fruit, nectar, pollen, and leaves.

Bats seem to be part of a radiation of mammals called Laurasiatheria, a group that includes the Chiroptera, Carnivora, Cetartiodactyla, Perissodactyla, Erinaceomorpha, Soricomorpha, and Pholidota, but bats' particular relationship among these groups remains unresolved. Some authors separate the two main branches of living bats as the Yinpterochiroptera (consisting of Pteropodidae, Rhinopomatidae, Craseonycteridae, Megadermatidae, Hipposideridae, and Rhinolophidae) and Yangochiroptera (consisting of the remaining 13 families).

The earliest undoubted fossil bats (Onychonycteridae and Icaronycteridae) are from early Eocene beds in Wyoming and India. The recently discovered *Onychonycteris* is the most primitive known bat. It has relatively small wings, well-developed claws on all five wing digits, and a calcar, suggesting that it could fly. *Onychonycteris*, however, lacks bony specializations of the ears and hyoid apparatus, suggesting that it was unable to echolocate; this suggests that bat flight evolved before bat echolocation.

Among the Yinpterochiroptera one family, Pteropodidae, formerly the sole members of a suborder known as Megachiroptera or megabats, is distinct from all other living bat families. Pteropodids use vision and not echolocation as a primary mode of orientation. Two exceptions are *Rousettus* and *Stenonycteris*, which may have evolved echolocation independently; they use tongue-clicks for their acoustical orientation. All other bat families use sound produced by the larynx. Often these pulses are ultrasonic, that is, above the range of human hearing.

Most bats use slow, highly maneuverable flight, generating lift and thrust very differently from birds. Bat wings form very thin airfoils of high camber. Thin airfoils, essentially cambered membranes, are more effective in producing high lift at low speeds than are conventional airfoils. Additionally, bats vary the camber of the wing during flight to generate added lift at low speeds. The wings of some bats (e.g., Molossidae and Emballonuridae) have become specialized for rapid flight. A wide range of wing loadings and aspect ratios has evolved in bats and is coupled with foraging mode.

Pteropodids (Old World fruit bats or flying foxes) have fox-like faces and a large body size. Most are fruit eaters; others are nectar and pollen feeders. These bats are abundant members of many tropical biotas in the eastern hemisphere. Some species roost in groups in caves, and some roost solitarily or in huge groups called "camps" in trees or other vegetation. Flight is slow and maneuverable.

By the current classification scheme, 18 families form the remainder of Chiroptera other than Pteropodidae. They reflect great structural diversity and widely contrasting modes of life. In these 18 families, echolocation universally involves sound production by the larynx, but echolocation styles vary. Echolocating bats are usually small, with reduced eyes and specialized nose pads or lower lips. The ears have a tragus in all but members of the Rhinolophidae and Hipposideridae. The wings of the echolocating bats are more derived than those of the pteropodids. The shape of the wing varies according to foraging pattern and style of flight. In general, slow, maneuverable fliers have short, broad wings, whereas rapid, enduring fliers have long, narrow wings.

KEY CHARACTERISTICS OF BATS

- Wings present, capable of sustained self-powered flight
- Wing surface covered by skin, supported by four digits, and attached to hindlegs
- Uropatagium usually present between hindlimbs
- Humerus and radius greatly elongated
- Hindlimbs short and rotated 90 to 180 degrees from typical mammalian posture
- Uropatagial spur supports uropatagium

Pteropodids

- Face fox-like or lemur-like
- Eyes large
- Nose usually unspecialized, without nose leaf
- Tragus absent
- Postorbital process well-developed
- Wing with two clawed digits (I and II)

Nonpteropodids

- Capable of echolocation using sounds produced in larynx
- Eyes often reduced
- Tragus present (except in Rhinolophidae and Hipposideridae)
- Wing with one clawed digit (digit I)
- Calcar supports uropatagium

KEY TERMS

Angle of attack
Anticoagulant
Aspect ratio
Calcar
Camber
Chiropatagium

Drag
Echolocation
Guano
Plagiopatagium
Propatagium

Tragus
Tuberculosectorial
Uropatagial spur
Uropatagium
Wing loading

RECOMMENDED READINGS

Fenton, MB. 1992. *Bats.* Facts of File, New York.

Fleming, TH. 1988. *The Short-tailed Fruit Bat: A Study in Plant-Animal Interactions.* University of Chicago Press, Chicago, IL.

Kunz, TH & MB Fenton. 2003. *Bat Ecology.* University of Chicago Press, Chicago, IL.

Kunz, TH & PA Racey. 1998. *Bat Biology and Conservation.* Smithsonian Institution Press, Washington, DC.

Maina, JN. 2000. What it takes to fly: the structural and functional respiratory refinements in birds and bats. *Journal of Experimental Biology, 203:*3045–3064.

Neuweiler, G. 2000. *The Biology of Bats.* Oxford University Press, Oxford, UK.

Nowak, RM. 1994. *Walker's Bats of the World.* Johns Hopkins University Press, Baltimore, MD.

Simmons, NB, et al. 2008. Primitive early Eocene bat from Wyoming and the evolution of flight and echolocation. *Nature, 451:*818–822.

Simmons, NB. 2008. Taking wing. *Scientific American, 299:*96–103.

Simmons, NB. 2005. Chiroptera. Pp. 159–174 in *The Rise of Placental Mammals* (KD Rose & JD Archibald, eds.). Johns Hopkins University Press, Baltimore, MD.

Wilson, DE. 1997. *Bats in Question.* Smithsonian Institution Press, Washington, DC.

Carnivora

Predation in mammals is an ancient practice. Primitive carnivorous mammals (creodonts) appeared in the Paleocene before most of the Recent mammalian orders. Mammalian carnivores probably evolved in response to the food source offered by an expanding array of herbivorous mammals and underwent adaptive radiation as herbivores diversified. However, not all modern carnivorans are carnivores.

The classification of carnivorans remains a subject of considerable debate (Dragoo & Honeycutt 1997; Flynn & Wesley-Hunt 2005; Hunt & Tedford 1993; Wayne et al. 1989; Wozencraft 1989; Wyss & Flynn 1993). Living members of the order Carnivora include terrestrial carnivores (such as dogs, cats, weasels, and bears) as well as aquatic carnivores (seals, sea lions, and walruses). Flynn and Wesley-Hunt (2005) list several morphologic features that characterize extant carnivorans: (1) an expanded braincase in which the frontoparietal suture is located anteriorly, (2) a flange on the lateral margin of the basioccipital (entotympanic attachment), (3) fused scaphoid and lunar bones in the carpals, and (4) the loss of the third molar.

No consensus on the classification of the Carnivora has been reached. However, numerous morphologic and molecular studies reveal several broad patterns as follows: (1) the Carnivora is composed of two clades: the Feliformia, which includes felids, hyaenids, viverrids, herpestids, and the Malagasy Eupleridae, and the Caniformia, including the canids, ursids, procyonids, mustelids, mephitids, ailurids, odobenids, otariids, and phocids (**Fig. 16-1**; Flynn et al. 2005; Wesley-Hunt & Flynn 2005; Wozencraft 2005). (2) The African palm civet, *Nandinia binotata*, is the sole member of the family Nandiniidae and is considered basal to the remaining Feliformia. (3) The Malagasy carnivores (Family Eupleridae) descended from a single invasion from Africa during the late Oligocene or early Miocene (Yoder & Flynn 2003). (4) Four clades comprise the living Caniformia: Canidae, Ursidae, Pinnipedia, and Musteloidea (Flynn et al. 2005). (5) The seals, sea lions, and walruses (Pinnipedia) are closely related to the superfamily Musteloidea and along with ursids form the superfamily Arctoidea (Higdon et al. 2007; Yu & Zhang 2006). (6) The superfamily Musteloidea includes the red panda (*Ailurus*), skunks (family Mephitidae), procyonids, and mustelids. (7) The red panda is neither an ursid nor a procyonid. Rather, it is the sole member of the Ailuridae, a basal member of the superfamily Musteloidea (Flynn et al. 2005a). (8) There is strong support that skunks (Mephitidae) form a clade distinct from mustelids (Flynn et al. 2005).

The oldest carnivorous mammals, order Creodonta, appeared in the Paleocene. They were the typical carnivores of the Paleocene and Eocene and persisted in Old World tropical **refugia** into the Miocene. As currently recognized, the creodonts include the families Oxyaenidae and Hyaenodontidae. Although creodonts share with Carnivora the presence of primitive carnassial teeth, they are no longer considered ancestral to the Carnivora (Flynn & Wesley-Hunt 2005). The Paleocene and Eocene Viverravidae and "Miacidae" (Miacidae is not a monophyletic group) are considered to be basal members of the Carnivora. Modern carnivorans probably share a common ancestor with one or both of these two groups, but few fossils representing modern carnivoran families are known until the Oligocene, by which time many of the modern families were already present (Carroll 1988; Heinrich & Rose 1995; Hunt & Tedford 1993; Wyss & Flynn 1993).

Most Recent carnivorans are predaceous and have an acute sense of smell. Cursorial ability may be limited, as in the Ursidae and Procyonidae, or may be strongly developed, as in the cheetah and some canids. The braincase is large with the orbit usually confluent with the temporal fossa. The turbinal bones are usually large, and their complex form provides for a large surface area. There are usually 3/3 incisors (3/2 in the sea otter, *Enhydra lutris*), and the canines are large and usually conical. All teeth are rooted, and the cheek teeth vary from 4/4 premolars and 2/3 molars in long-faced carnivores, such as the Canidae and Ursidae, to 2/2 premolars and 1/1 molars, as in some cats.

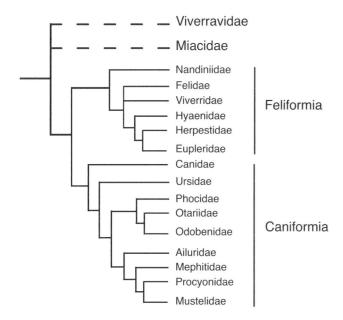

FIGURE 16-1 One phylogeny of the Carnivora. The Carnivora include two major clades: feliforms and caniforms. The pinnipeds are a monophyletic group within the caniform clade. (Adapted from Flynn, J. J., et al., *Syst. Biol.* 54 (2005): 317–337.)

In most species, the fourth upper premolar and the first lower molar are carnassials (specialized shearing blades). The condyle of the dentary bone and the glenoid fossa of the squamosal bone are elongated transversely, allowing only limited transverse movement. This tight joint allows the jaws to act like scissors with the shearing carnassials as the slicing blades.

Cursorial adaptations evident in the carpus include the fusion of the scaphoid and lunar bones and the loss of the centrale (**Fig. 16-2**). The foot posture is plantigrade, as in ursids and procyonids, or digitigrade, as in canids, hyaenids, and felids. Little reduction from the ancestral number of five digits has occurred. The greatest reduction occurs in the hyenas and in the African wild dog (*Lycaon pictus*), in which the manus and pes have four toes.

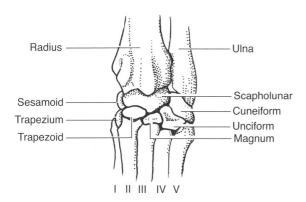

FIGURE 16-2 Anterior view of the left carpus of the gray fox (*Urocyon cinereoargenteus*). Roman numerals at the bottom indicate the digit number.

Feliformia

The feliform clade includes six families of extant carnivorans: the African palm civet (Nandiniidae), cats (Felidae), civets and genets (Viverridae), hyenas and the aardwolf (Hyaenidae), mongooses (Herpestidae), and the Malagasy civets and mongooses (Eupleridae). A number of morphological and biochemical features characterize this clade (Flynn & Wesley-Hunt 2005; Wyss & Flynn 1993). Among the most important characters defining the Feliformia is the two-chambered auditory bullae. In feliforms, two bones (ectotympanic and entotympanic) unite to form the bullae with a septum separating the two resulting chambers. In contrast, caniforms have an auditory bullae composed of a single entotympanic bone.

Family Nandiniidae

The African palm civet (*Nandinia binotata*) is the only species in the family Nandiniidae. Palm civets are considered basal to all feliforms because they are genetically distinct from other feliforms and retain primitive auditory bullae lacking a septum (Flynn & Nedbal 1998). *Nandinia* occurs throughout much of sub-Saharan Africa as far south as Zimbabwe. It weighs between 2 and 5 kilograms and has short, woolly fur. *Nandinia* is omnivorous and feeds primarily on fruit or other plant material (Nowak 1999).

Family Felidae

The cats are highly proficient predators, with some species regularly killing prey as large as themselves and occasionally overcoming prey several times their own weight (as in the case of the African lion and the giraffe). Throughout the history of the order Carnivora, the cats have been the carnivores most highly specialized morphologically for predation (see Turner 1997 for a discussion of felid paleontology). Included in the Felidae are 14 extant genera and 40 living species. All cats, from the pampered tabby to the tiger, bear a strong family resemblance (**Fig. 16-3**). This family occurs nearly worldwide, with the exception of Antarctica, Australia, Madagascar, and various isolated islands.

In members of the family Felidae, the rostrum is short, an adaptation furthering a powerful bite, and the orbits in most species are large (**Fig. 16-4**). The number of teeth is reduced. The typical dental formula is 3/3, 1/1, 3/2, 1/1 = 30, and the anterior-most upper premolar is strongly reduced or lost (as in *Lynx*). The carnassials are well developed and have specializations that enhance their shearing ability (**Fig. 16-5**). The foot posture is digitigrade. The forelimbs are strongly built, and the manus can be rotated so that the soles face upward. The claws are sharp, recurved, and completely retractile, except in the cheetah (*Acinonyx*). These features of the forelimbs allow cats to clutch and grapple

FIGURE 16-3 (A) Caracal (*Caracal caracal*), in Augrabies National Park, South Africa. This small cat (up to 19 kilograms) occurs widely in Africa, the Middle East, and as far east as India. (B) A margay (*Leopardus wiedii*) from Central and South America. (C) A leopard (*Panthera pardus*) from Okavango Delta, Botswana. (D) A mountain lion (*Puma concolor*) from the western United States.

with prey. Some species have spotted or striped color patterns that enable them to conceal themselves effectively. One fundamental difference between most large cats, such as members of the genus *Panthera*, and most smaller cats (*Felis*, *Prionailurus*, etc.) is in the structure of their hyoid, which in the larger cats is partially replaced with a flexible

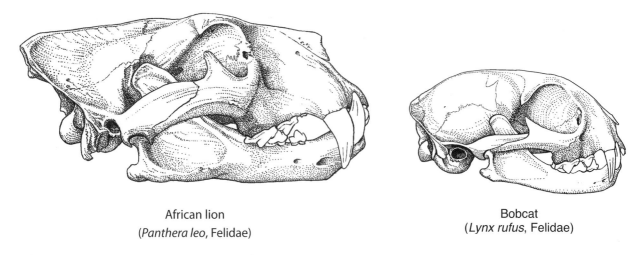

African lion
(*Panthera leo*, Felidae)

Bobcat
(*Lynx rufus*, Felidae)

FIGURE 16-4 Skulls of an African lion (length of skull 366 millimeters) and a bobcat (length of skull 120 millimeters).

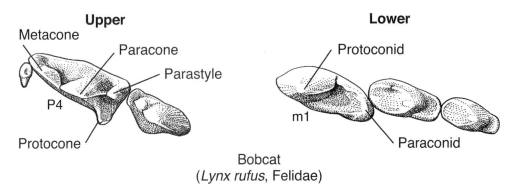

Upper

Metacone

Paracone

Parastyle

P4

Protocone

Lower

Protoconid

m1

Paraconid

Bobcat
(*Lynx rufus*, Felidae)

FIGURE 16-5 Occlusal views of the cheek teeth of the bobcat. Note the lack of crushing molars. The cusps of the carnassials (P4 and m1) are labeled. Only sectorial teeth are present; the parastyle of P4 increases the length of its shearing blade, and the loss of the talonid of m1 makes this tooth entirely blade-like.

cartilage, allowing them to roar instead of purr. The weight of cats varies from that of the black-footed cat of Africa (*Felis nigripes*), at 1 to 2 kilograms, to that of the tiger (*Panthera tigris*) at 275 kilograms.

The groundwork for an understanding of the phylogeny of cats was laid by Matthew (1910), and new fossil material described in the last decade has improved our knowledge of the felid fossil record. The earliest cat-like mammals, which appeared in North America in the Eocene, were already strongly differentiated from other carnivores. All had retractile claws, sectorial carnassials, reduced cheek teeth anterior and posterior to the carnassials, and some

had saber-like upper canines (**Fig. 16-6**). These fossil "cats," known as false sabertooths, were separated from the Felidae and placed in the family Nimravidae, based on characters of the auditory region, dentition, and skull (Bryant 1991; Neff 1983). The phylogenetic position of the Nimravidae, with respect to the rest of the Carnivora, remains a subject of considerable debate. For example, the Nimravidae have been allied with the canids (Flynn & Galiano 1982) or separated from both the canids and felids and placed in a more basal position as a sister group to the Carnivora (Neff 1983). Morlo et al. (2004) suggested that one group, the Barbourofelidae, was distinct from nimravids and may be the sister

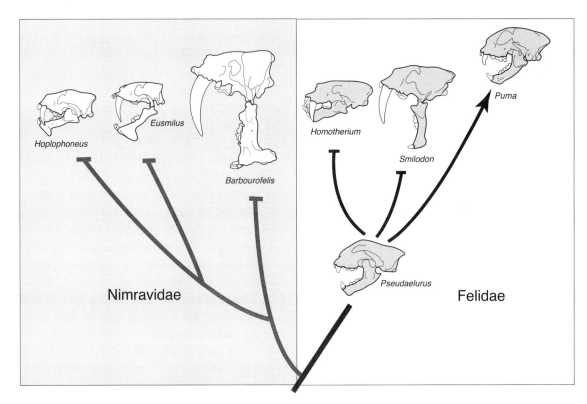

Hoplophoneus

Eusmilus

Barbourofelis

Homotherium

Smilodon

Puma

Pseudaelurus

Nimravidae

Felidae

FIGURE 16-6 A hypothetical phylogeny of the sabertooth "cats". (Adapted from Martin, L.D., *Transactions of the Nebraska Academy of Sciences*, 7 (1980):141–154. 1980.)

taxon of felids. Taxonomic placement of the Nimravidae aside, their radiation culminated in the Miocene with one of the most highly specialized of all sabertooths, *Barbourofelis fricki* (Fig. 16-6). The nimravids are not known after the Miocene. Features that distinguish nimravids from felids include the lack of an auditory bulla in most species and the absence of a cruciate sulcus (a conspicuous, deep groove) on the brain.

The family to which all living cats belong, Felidae, appeared in the early Oligocene. The basal felid *Proailurus* had relatively short upper canines, but a Miocene and Pliocene radiation resulted in both sabertooth types and cats with short upper canines (Fig 16-6). The Pleistocene felid *Smilodon* (Machairodontinae), although not as extreme in sabertooth specialization as *Barbourofelis*, was an imposing predator that survived until the end of the Pleistocene and coexisted with humans (Miller 1969). Using DNA extracted from fossil bones (called ancient DNA), Barnett et al. (2005) compared DNA of two sabertooths (*Smilodon* and *Homotherium*) to DNA from modern felids. Their analysis indicates that sabertooth cats and modern cats shared a common ancestor but diverged into two separate lineages relatively early on. The extinction of the sabertooth cats is difficult to explain. Saber-like upper canines evolved independently (convergent evolution) four times among mammals (in Eocene creodonts, nimravids, Miocene-Pleistocene felids, and the Miocene-Pliocene marsupial family Thylacosmilidae), each time, perhaps, in response to the abundance of large herbivores. Probably, in each case, the extinction of the sabertooth carnivores was linked to the decline or extinction of their preferred, large prey.

From their origins in the Miocene, the 40 species of living felids have radiated widely and now live on all continents except Antarctica. Modern cats began their diversification approximately 11 million years ago in the Miocene of Asia, according to Johnson et al. (2006). These authors constructed a felid phylogenetic tree that, when compared with sea level changes over the past 10 million years, suggested that early cats exploited land bridges to move between continents. As they moved, cats diversified. According to this study, eight felid lineages evolved, beginning with the *Panthera* lineage in Eurasia 11 million years ago. Subsequent sea level changes exposed land bridges that allowed intercontinental migrations of a caracal lineage into Africa and several felid lineages into North and South America. One of the last lineages to appear was a group of small Eurasian and African cats (approximately 6 million years ago), including the ancestor of the domestic cat (*Felis catus*). Archaeological finds, including a cat carefully buried with other artifacts in a human grave in Cyprus, suggest that domestication occurred at least 9,500 years ago (Vigne et al. 2004). Further support for a Near Eastern origin of cat domestication comes from the molecular genetic studies of both wild and domestic cats by Driscoll et al. (2007).

Cats usually catch prey by a stealthy stalk followed by a brief burst of speed (Chapter 24). They are typically sight hunters, and some species spend considerable time watching for prey and waiting for it to move into striking distance. Many kinds of animals are eaten, from fish, mollusks, and small rodents to ungulates as large as African buffalo. The fishing cat (*Prionailurus planiceps*) of southeastern Asia is unusual in that it forages near streams and rivers for fish and frogs (Nowak & Paradiso 1991).

Felids have well-developed senses of smell, sight, and hearing. Many species hunt nocturnally, despite having good color vision. A tapetum lucidum enhances night vision. Most cats are agile climbers, and leopards (*Panthera pardus*) often drag their kills up into the branches of trees to protect the carcass from scavengers. Cheetahs (*Acinonyx jubatus*) are poor climbers, but their ability to sprint is unsurpassed (**Fig. 16-7**). Cheetahs make short dashes after prey and can attain speeds in excess of 90 kilometers per hour (about 60 miles per hour). Cheetahs have relatively long limbs and a slender build. Cheetahs can accelerate from a standstill to nearly 90 kilometers per hour in only a few seconds. In most chases, however, the average speed is less than 65 kilometers per hour and does not cover more than 300 meters. After being knocked off its feet by the cheetah, the prey is usually killed by a suffocating bite to the throat (Estes 1991). Despite the cheetah's speed, chases result in kills only 40% of the time.

Most felids are solitary predators, but lions (*Panthera leo*) live in large family groups, called prides. Prides consist of one or more males (usually brothers), a group of related females, and their cubs. Both males and females contribute to hunting, and both will defend the pride against intruders. However, females do most of the hunting, and males are the primary defenders of the pride. Subadult males are expelled from the pride as they approach sexual maturity, at around 3 years of age. The outcasts become nomadic for several years, but as they reach their prime, a nomadic male (or a coalition of males) takes over a pride and ousts the previous pride holder (Chapter 24). The new male often kills any remaining cubs in order to induce females to enter estrus sooner.

Many of the larger cats, particularly those with spotted or striped coats, are hunted because their fur is valuable in the fur trade. Others, most notably the tiger, are hunted for their body parts, which are sold at high prices to supply the ever-increasing demands for home remedies in the Asian traditional medicine markets. Habitat loss has also contributed to the decline of many cat populations worldwide. In Australia and many islands, the spread of introduced feral cats is exacerbating the decline of native mammals.

FIGURE 16-7   The cheetah (*Acinonyx jubatus*), the fastest cursorial mammal. Note how the extension of the spine contributes to an increase in stride length.

■ Family Viverridae

The family Viverridae includes the Old World civets and genets. Fifteen genera and 35 species are recognized, but the taxonomy of this family is still uncertain. Viverrids inhabit much of the Old World: The center of their distribution, however, is in tropical and southern temperate areas, and they are absent from Madagascar, northern Europe, and all but southern Asia, as well as from New Guinea and Australia. This is an old group; it appeared in Europe in the early Oligocene. The first African records are from the Miocene (Martin 1989).

Viverrids are small to medium sized, short-legged, long-tailed carnivores (**Fig. 16-8**). They vary in size from about 600 grams in the spotted linsang (*Prionodon pardicolor*) to roughly 20 kilograms in the African civet (*Civettictis civetta*). The viverrid skull has a moderately long rostrum (**Fig. 16-9**). The premolars are large, and the carnassials are usually trenchant (**Fig. 16-10**). The upper molars are tritubercular and are wider than they are long; the lower molars have well-developed talonids. The dental formula is generally 3/3, 1/1, 3–4/3–4, 2/2 = 36–40. The five toes on each foot include a much-reduced pollex or hallux. The foot posture is plantigrade or digitigrade, and the claws are semiretractile and covered by a fleshy sheath in some genera. The tail is typically long and bushy. Some species are banded. Others are spotted, and still others are striped. Most species have well-developed perineal (near the anus) scent glands, used in intraspecific and interspecific communication.

Viverrids have many lifestyles. Palm civets are agile climbers, and with the aid of traction pads on the hind feet and hook-like claws on its medial toes, they can crawl headfirst down a nearly smooth tree trunk. Several viverrids, including the aquatic genet (*Genetta piscivora*) and the otter civet (*Cynogale bennettii*), show adaptations for a semiaquatic lifestyle. Aquatic genets, for example, have naked palms on the forefeet that allow a better grip on slippery fish or frogs. As the name implies, otter civets have an otter-like face, complete with nostrils that open on the top of the rhinarium and close by special flaps that prevent water from entering the nostrils. Long stiff vibrissae (whiskers) that are located on the snout and under the ears serve

FIGURE 16-8 (A) A Maylayan civet (*Viverra tangalunga*). (B) Large-spotted genets (*Genetta tigrina*), agile climbers that often feed on birds.

as tactile hairs in the water. The feet of otter civets are only slightly webbed, and the animals are probably relatively slow swimmers. Nevertheless, they are good climbers and often seek the safety of trees (Nowak & Paradiso 1991).

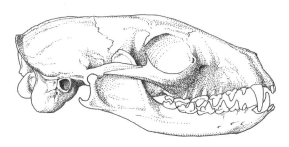

White-tailed mongoose
(Ichneumia albicauda, Herpestidae)

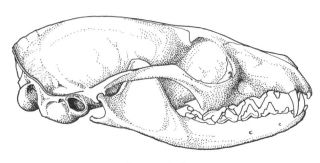

Large-spotted genet
(*Genetta tigrina*, Viverridae)

FIGURE 16-9 Skull of a large-spotted genet (Viverridae, length of skull 87 millimeters). Skull of a white-tailed mongoose (Herpestidae, length of skull 106 millimeters). Note the large auditory bulla of the mongoose.

Viverrids are primarily carnivorous and eat small vertebrates or insects. Most are nocturnal ambush predators, and many are excellent climbers. The binturong (*Arctictis binturong*) of Asia and Indochina is a slow deliberate climber and has a bushy tail that is prehensile at its tip. Binturongs are highly vocal and live in small groups consisting of adults and their offspring. The Malabar civet of India (*Viverra civettina*) is considered critically endangered, and the southeast Asian otter civet and the central African crested genet (*Genetta cristata*) are considered endangered. The reason for the decline in these populations appears to be the continued destruction of appropriate habitat and their use as food. In 2004, the United States banned importation of civets because they are suspected of being a host for the SARS (severe acute respiratory syndrome) virus.

Family Herpestidae

Members of the family Herpestidae, commonly known as mongooses, include 33 Recent species in 14 genera. The fossil record of mongooses dates back to the late Oligocene of Europe (Martin 1989). They are strictly an Old World group with a poor fossil record. Long considered a subfamily of the Viverridae, the herpestids are now considered a distinct lineage (Hunt 1987; Neff 1983; Wozencraft 1989, 2005). Herpestids inhabit most of Africa (except the Sahara Desert), southeastern Europe, the Middle East, India, Sri Lanka, and much of southeastern Asia to the Philippine Islands and Borneo. Mongooses were introduced to several of the Caribbean and Hawaiian Islands in a failed attempt to control rats (*Rattus*). Herpestids occupy habitats ranging from deserts to tropical forests and in many places are the most common carnivores.

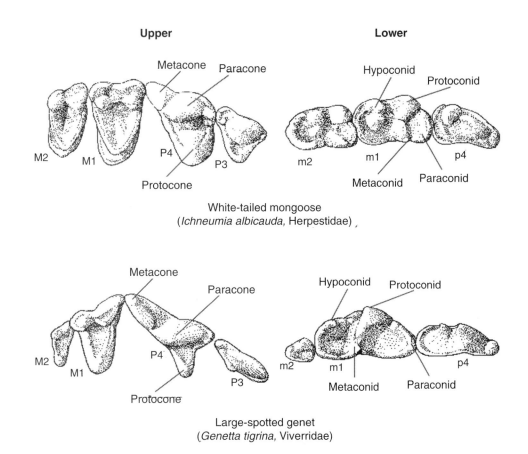

Upper

Metacone   Paracone

M2   M1   P4   P3

Protocone

Lower

Hypoconid   Protoconid

m2   m1   p4

Metaconid   Paraconid

White-tailed mongoose
(*Ichneumia albicauda*, Herpestidae)

Metacone

Paracone

M2   M1   P4   P3

Protocone

Hypoconid   Protoconid

m2   m1   p4

Metaconid   Paraconid

Large-spotted genet
(*Genetta tigrina*, Viverridae)

FIGURE 16-10   Occlusal views of the cheek teeth of a large-spotted genet and a white-tailed mongoose. Note the large inner lobes (bearing the protocone) of P4 (carnassials) of both species and the large inner lobe of P3 of the mongoose.

Mongooses are small, typically long-bodied and long-tailed carnivores (**Fig. 16-11**). They range in size from about 270 grams (the dwarf mongoose, *Helogale parvula*) to 5 kilograms (the white-tailed mongoose, *Ichneumia albicauda*). The postorbital part of the skull is relatively long (Fig. 16-9), and the complex structure of the large auditory bullae, featuring an expanded ectotympanic and a circular external auditory tube, is unique among carnivores (van der Klaauw 1931). The dental formula is 3/3, 1/1, 3–4/3–4, 2/2 = 36–40. Usually all but the first premolar are large. The upper molars are tritubercular and are broader than long, and the third upper premolar has an internal cusp in most mongooses (Fig. 16-10). The upper carnassial is usually trenchant and has a large inner, protocone-bearing lobe. In most herpestids, the front and hind feet have five toes. The feet are digitigrade to semiplantigrade, and the forefeet bear long claws in many species. One African species, the marsh mongoose (*Atilax paludinosus*), has long-toed feet similar to those of a raccoon. The marsh mongoose, like the raccoon, forages by feeding along the bottom of a pool with its dextrous "hands." The herpestid anal scent glands, situated adjacent to the anus in an anal pouch, are highly specialized sacs that secrete carboxylic acid, a byproduct of bacterial metabolism (Gorman et al. 1974). Herpestids also have facial glands that are used in scent marking.

Complex social life, characterized by highly structured social behavior and the lavish use of scent marking, has evolved in roughly half the genera of mongooses. Social mongooses are diurnal, live in clans, and are primarily insectivorous. Sociality has enabled such species (e.g., the dwarf mongoose, the banded mongoose, *Mungos mungo*, the meerkat, *Suricata suricatta*) to forage by day in open situations where threats from aerial and terrestrial predators abound (Fig. 16-11B). By concerted group action, these species can deter some predators.

Although herpestids eat a wide range of food, from various invertebrates and small vertebrates to eggs and even fruit, insects are the mainstay. Of the 14 genera of mongooses, most are basically insectivorous. Adult and larval beetles (Coleoptera), termites (Isoptera), and grasshoppers and their kin (Orthoptera) are most important. One species, Meller's mongoose (*Rhynchogale melleri*), is seemingly

FIGURE 16-11 Two herpestid carnivores: (A) Yellow mongoose (*Cynictis penicillata*), in Kalahari Desert of South Africa. This animal is diurnal, social (up to 20 in a colony) and often lives in association with colonies of cape ground squirrels. (B) A group of meerkats (*Suricata suricatta*) on lookout duty in the Kalahari Desert of Botswana.

a termite specialist. Because in Africa some termites inhabit ungulate dung, as do the larvae of the diverse and numerous dung beetles, the abundance of African ungulates may influence population levels of mongooses.

A number of herpestids break the shells or exoskeletons of such items as eggs or crabs by throwing them backward between their hind legs against a hard object. Only among the mongooses and their ecological counterparts, the spotted skunks (*Spilogale*), is this behavior known.

Family Eupleridae

Malagasy carnivores include three species of civet-like carnivores and five species of mongoose-like animals that, until recently, have been placed in the Viverridae or Herpestidae. Yoder et al. (2003) demonstrated that living Malagasy carnivores descended from an African herpestid ancestor that colonized Madagascar sometime during the late Oligocene or early Miocene.

The largest Malagasy carnivore, *Cryptoprocta ferox*, resembles a small mountain lion (**Fig. 16-12A**). It is nocturnal and arboreal with sharp retractile claws. Its diet includes many small vertebrates, and it is thought to be a principal predator of small lemurs. The smaller, terrestrial Malagasy euplerid (*Fossa fossana*) is nocturnal and has a varied diet that includes small mammals, frogs, crabs, and insects. The rare Malagasy falanouc (*Eupleres goudotii*) has a reduced

dentition of short conical teeth used to grasp earthworms, snails, and other invertebrates. There are five species of Malagasy euplerids ranging is size from 500 grams to nearly 2 kilograms in weight. The diurnal ring-tailed euplerid (*Galidia elegans*; Fig. 16-12B) forages for small vertebrates and invertebrates in small family groups. Short, high-pitched whistles are used by *Galidia* to keep family members together as they move through the dense rainforest understory (Ryan, personal observation).

Family Hyaenidae

Many carnivorans will eat **carrion** if the opportunity arises, but most members of the family Hyaenidae have become more or less specialized for carrion feeding. This is a small family, with but four Recent species in three genera. The distribution includes Africa, Turkey, and the Middle East to parts of India. The Hyaenidae, probably derived from viverrid stock, appeared in Eurasia and Africa in the early Miocene and, except for *Chasmaporthetes* (which probably crossed the Bering Strait land bridge and is known from the Pliocene-Pleistocene in North America), has been an entirely Old World family.

Except for the aardwolf (*Proteles cristatus*), hyenas are characterized by rather heavy builds with the forelimbs longer than the hindlimbs (**Fig. 16-13**). The skull is strongly built. The carnassials are well developed (**Fig. 16-14**). The dental formula is 3/3, 1/1, 4/3, 0–1/1 = 32–34. The feet are digitigrade, and both forepaws and the hind paws have four toes that bear blunt, nonretractile claws. The pelage is either

FIGURE 16-12 Several euplerid carnivores: (A) The fossa, *Cryptoprocta ferox*, is the largest living carnivore in Madagascar. (B) The *Galidia elegans*, the Malagasy ring-tailed "mongoose," is diurnal and lives in small family groups. (C) *Galidictis fasciata*, the Malagasy broad-striped "mongoose," is nocturnal and solitary.

spotted (*Crocuta*) or variously striped (*Hyaena*). Members of the family Hyaenidae weigh up to 80 kilograms.

Hyenas in some areas specialize in scavenging on the kills of lions and other large carnivores and are able to drive cheetahs (*Acinonyx*) from their kills. They may also forage in villages at night for edible refuse. In their ability to crush large bones, they are unsurpassed. Kruuk (1972) and Mills (1989) have found that spotted hyenas are also powerful predators. Often hunting in packs of up to 30 animals, these mostly nocturnal hunters can bring down animals as large as zebras. In the Kalahari Desert of southern Africa, spotted hyenas kill 72% of the food they consume, and nearly half of those kills are of large ungulates such as gemsbok (*Oryx gazella*) and wildebeest (*Connochaetes taurinus*; Mills 1989). Indeed, in the Ngorongoro Crater in Tanzania, a reversal of the usual pattern of interactions between lions and hyenas has occurred. Spotted hyenas are better able than the lions to make regular kills, and lions live partly by driving hyenas from their kills and eating the carrion (Ewer 1968). In contrast to spotted hyenas, Kalahari brown hyenas (*Hyaena brunnea*) are primarily scavengers, killing and scavenging mainly small animals, such as springhares, with kills making up only 5.8% of the diet in the Kalahari (Mills 1989). Striped hyenas (*Hyaena hyaena*) are more omnivorous and include fruit and insects in their diet (Kruuk 1976). The differences in diet between hyena species may explain the differences in their hunting behavior. Brown hyenas and striped hyena are mostly solitary, whereas spotted hyenas usually hunt in groups of at least three individuals.

Spotted hyenas live in "clans" of up to 80 individuals, and they scent mark the borders of their territories, which may exceed 30 square kilometers (Kruuk 1972). These loosely defined clans are usually made up of smaller hunting groups spread out across the territory. Hyena clans in Kruger National Park spent one third of their activity time on territorial patrols (Henschel & Skinner 1990). Fierce fighting can break out among adjacent clans when territorial borders are violated (Chapter 24). Care of the young in spotted hyenas is the sole responsibility of the mother, although all young from the same clan may live in a communal den. Young spotted hyenas suckle from their mothers for up to 18 months because food is not brought back to the den by the mother (MacDonald 1984). In contrast, brown hyena mothers will suckle any infant in the clan, and adults frequently bring food back to the den for the young to consume (Skinner et al. 1980).

The aardwolf is more lightly built than the hyenas and has a more delicate skull and much reduced teeth (Fig. 16-14). All teeth except the canines are small, and the cheek teeth are simple and conical. The dental formula is generally 3/3, 1/1, 3/1–2, 1/1–2 = 28–32, but frequently some

FIGURE 16-13 Spotted hyena (*Crocuta crocuta*) in Etosha National Park, Namibia.

of these teeth are lost (as in the skull shown in Fig. 16-14). The forefeet have five toes, and the hind feet have four. The animal is striped and has a mane of long hair from neck to rump. The tail is quite bushy. When the animal is threatened and adopts a defensive posture, the hair of the mane and tail is erected, but the mouth remains closed. *Proteles* has abandoned the open-mouthed threat used by most carnivores—which in its case would merely advertise the weakness of the dentition—in favor of extensive erection

of the long hair. The aardwolf also releases a pungent fluid from the well-developed anal glands when attacked. The aardwolf eats mostly termites, to which it is attracted largely by the sounds they make (Kruuk & Sands 1972). Its unusually large auditory bullae are probably associated with an enhanced sense of hearing. In contrast to many termite feeders, the aardwolf does not dig for termites but laps them from the surface of the ground with its large, flat tongue.

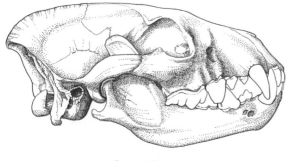

Spotted hyena
(*Crocuta crocuta*, Hyaenidae)

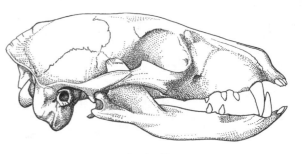

Aardwolf
(*Proteles cristatus*, Hyaenidae)

FIGURE 16-14 Skulls of a spotted hyena (length of skull 248 millimeters) and an aardwolf (length of skull 148 millimeters).

Caniformia

The remaining carnivorans form the caniform clade, which includes the families Canidae, Ursidae, Ailuridae, Mephitidae, Procyonidae, Mustelidae, and three families of aquatic carnivorans (Fig. 16-1). The aquatic members of the Carnivora (Pinnipedia) include the earless seals (Phocidae), the eared seals (Otariidae), and the walrus (Odobenidae). Basing their conclusions on features of the skull and on genetic evidence, many researchers regard the pinnipeds as a monophyletic group related to either ursids or members of the superfamily Musteloidea (Flynn & Wesley-Hunt 2005; Wozencraft 2005). The superfamily Musteloidea includes four families. The red panda (Ailuridae) and the skunks (Mephitidae) are now considered separate from the Procyonidae and Mustelidae (Wozencraft 2005).

Family Canidae

Thirteen genera and about 35 Recent species make up the family Canidae. Canids (**Fig. 16-15**)—wolves, jackals, foxes, and dogs—occupy a great array of environments from arctic to tropical. Before their domestication and dispersal with humans, canids occurred nearly worldwide, except on most oceanic islands. The dingo (*Canis*) was probably brought to Australia by early humans between 11,000 and 3,500 years ago (Menkhorst 1995). The fossil record of canids dates from the Eocene of North America (*Hesperocyon*). The African and South American fossil records are restricted to the Pliocene and Pleistocene (see Wang & Tedford 2008, for a complete discussion of canid paleontology).

Canids are broadly adapted carnivores; this is reflected in their morphology. The canid skull typically has a long

FIGURE 16-15 (A) An African wild dog (*Lycaon pictus*). (B) Gray wolf (*Canis lupus*) in Arizona. (C) A black-backed jackal (*Canis mesomelas*) in Namibia. (D) A bat-eared fox (*Otocyon megalotis*) in South Africa, showing the large ears.

rostrum that houses a large nasal chamber with complex turbinal bones, a feature associated with a remarkable sense of smell (**Fig. 16-16**). Most canids have a nearly complete eutherian complement of teeth (3/3, 1/1, 4/4, 2/3 = 42). The bat-eared fox (*Otocyon megalotis*) is unique among canids in having an "extra" molar above and below (3/3, 1/1, 4/4, 3/4 = 46; Fig. 16-16). The canines in canids are generally long and strongly built, and the carnassials retain

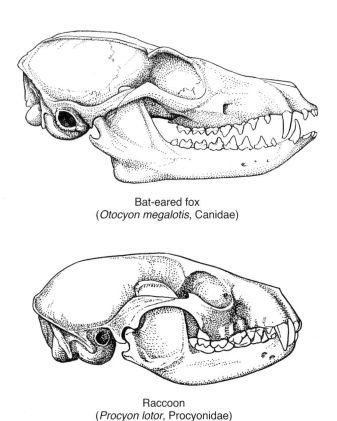

Bat-eared fox
(*Otocyon megalotis*, Canidae)

Raccoon
(*Procyon lotor*, Procyonidae)

FIGURE 16-16 Skulls of a bat-eared fox (length of skull 111 millimeters) and a raccoon (length of skull 115 millimeters).

the shearing blades (**Fig. 16-17**). The postcarnassial teeth have crushing surfaces, indicating a more flexible diet than that of the more strictly carnivorous cat family (Felidae). In cursorial species, the limbs are long, and rotation at the joints distal to the shoulder and hip joints is reduced in the interest of cursorial ability. The clavicle is absent. The feet are digitigrade, and the well-developed but blunt claws are nonretractile. The forefoot usually has five toes, and the hind foot has four. The weight of canids ranges from 1 to 75 kilograms.

Some canids forage tirelessly over large areas, and lengthy pursuit is frequently part of the hunting technique. The coyote (*Canis latrans*), probably one of the swiftest canids, can run at speeds of about 65 kilometers per hour. The fact that coyotes in many areas depend partly on jackrabbits for food is an impressive testimonial to this carnivore's speed. Canids often hunt in open country. Wolves (*Canis lupus;* Fig. 16-15B) and the African wild dog (*Lycaon pictus*; Fig. 16-15A) seem to rely more on endurance than on speed when hunting. These canids and the eastern Asian dholes (*Cuon alpinus*) habitually hunt in packs and kill larger prey than could be overcome by a solitary hunter. The gray fox (*Urocyon cinereoargenteus*), in contrast, does not generally forage in open areas but is amazingly agile and can run rapidly through the maze of stems beneath a canopy of chaparral. The foods of canids include vertebrates, arthropods, mollusks, carrion, and many types of plant material. Blackbacked jackals (*Canis mesomelas*; Fig. 16-15C) once were a problem in parts of South Africa because of their extensive feeding on pineapples (Ewer 1968). Coyotes in parts of the western United States feed heavily on such cultivated crops as melons and grapes and such uncultivated plant material as juniper berries and pricklypear cactus fruit. The average canid is clearly an opportunist; this may in large part account for the great success of this family.

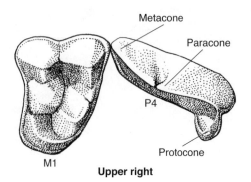

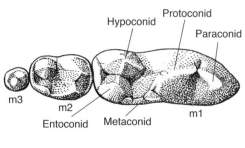

FIGURE 16-17 Occlusal view of selected cheek teeth of a coyote (*Canis latrans,* Canidae). The cusps of the carnassials (P4 and m1) are labeled. Notice that the cheek teeth both shear and crush.

Canid hunting behavior is correlated loosely with body size, whereby small canids, such as arctic foxes (*Vulpes lagopus*), bat-eared foxes (*Otocyon megalotis*; Fig. 16-15D), and kit foxes (*Vulpes macrotis*), are solitary hunters. Medium-sized canids (coyotes) hunt alone or in small family groups, and large canids (wolves, African wild dogs, and dholes) hunt in large cooperative groups (Moehlman 1989). One exception is the large maned wolf (*Chrysocyon brachyurus*), which is a solitary hunter. Social canids can have a major impact on the biota and dynamics of an ecosystem. Before recent declines, African wild dogs formed large packs of up to 50 dogs. Members of these packs hunt cooperatively, and although the young are usually the progeny of only one dominant breeding pair, all adult pack members provision the young (Estes 1991). Hunting success varies with prey species, season, and presence of potential competitors (Fuller & Kat 1993).

Family Ursidae

Bears are notable for their large size and their departure from a strictly carnivorous mode of life (**Fig. 16-18**). The family Ursidae contains five genera and eight Recent species. Morphologic and molecular evidence suggests that the giant panda, *Ailuropoda melanoleuca*, is a bear, and it is included here in the Ursidae (Chorn & Hoffmann 1978; Goldman et al. 1989; O'Brien et al. 1985; Wozencraft 2005). The red panda (*Ailurus fulgens*) is not an ursid and based on molecular and chromosomal evidence is placed in the monotypic family Ailuridae (Goldman et al. 1989; Wozencraft 2005).

The distribution of ursids includes most of North America and Eurasia, the Malay Peninsula, the South American Andes, the Atlas Mountains of extreme northwestern Africa, and for the giant panda, parts of China. Bears inhabit diverse habitats, from drifting ice in the Arctic to the tropics, but are most important in boreal and temperate areas.

Bears first appeared in the northern hemisphere (Europe, Asia, North America) in the early Miocene and reached Africa in the late Miocene. The small dog-like *Cephalogale*, a primitive bear-like carnivoran, first appeared during the late Eocene in Europe. Early bears of the genus *Ursavus* are probably derived from a *Cephalogale*-like ancestor. *Ursavus* radiated in Asia and ultimately gave rise to bears of the genus *Ursus* in Europe approximately 5 million years ago. Bears reached North America in the early Miocene (Zhanxiang 2003) and South America in the early Pleistocene (McKenna & Bell 1997).

The bear skull retains the long rostrum typical of the canids, but the orbits are generally smaller. The nasals are short, and the dentition is very different (**Fig. 16-19**). The postcarnassial teeth are greatly enlarged, and the occlusal surfaces are "wrinkled" and adapted to crushing. On the other hand, the first three premolars are usually rudimentary or may be lost, and a diastema usually occurs between the canines and the premolars. The upper carnassial is roughly triangular because of the posterior migration of the protocone and is much smaller than the neighboring molars (Fig. 16-19B); both upper and lower carnassials no longer have a shearing function. The dental formula is usually 3/3, 1/1, 4/4, 2/3 = 42, but premolars may be lost

FIGURE 16-18 (A) A Kodiak brown bear (*Ursus arctos*), also known as the grizzly bear, in Alaska. (B) A giant panda (*Ailuropoda melanoleuca*) from China.

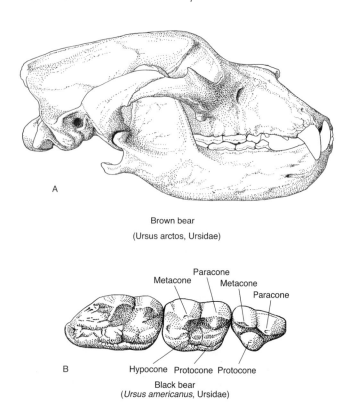

Brown bear
(*Ursus arctos*, Ursidae)

Black bear
(*Ursus americanus*, Ursidae)

FIGURE 16-19 (A) Skull of a brown bear (*Ursus arctos*) from Kodiak Island, Alaska (length of skull 355 millimeters). In this adult female, all of the sutures of the cranium and the rostrum are obliterated because the bones have grown together. (B) Right upper carnassial and two molars of the black bear (*Ursus americanus*); note the small, non–blade-like carnassial and the greatly lengthened molars.

with advancing age. The limbs, especially the forelimbs, are strongly built. The plantigrade feet have long, nonretractile claws. There are five toes on each foot. The ears are small, and the tail is extremely short. In size, bears range from 30 kilograms in the Malayan sun bear (*Helarctos malayanus*) to over 800 kilograms in the polar bear (*Ursus maritimus*).

The abandonment of cursorial ability in favor of power in the limbs and the loss of the shearing function of the cheek teeth in favor of a crushing battery have accompanied the adoption of omnivorous feeding habits among extant bears. The strong forelimbs can aid in the search for food by rolling stones or tearing apart logs, and the crushing surfaces of the molars can cope with many kinds of food, from insects and vertebrates to berries, grass, and pine nuts. Berries are of prime importance for northern populations of brown bears. Carrion is also avidly sought. The polar bear has a more restricted diet, consisting mainly of seals, and the giant panda eats mostly bamboo shoots (Schaller et al. 1985).

In areas with cold winters, some bear species retreat for much of the winter to caves or other retreats protected from drastic temperature fluctuations. Here they become lethargic or dormant, and they live off fat reserves accumulated

during the fall (Chapter 21). Whether this is "true," hibernation is a matter of semantics. Some ecophysiologists do not consider this hibernation because the bear's body temperature does not drop substantially and bears can arouse easily. However, Folk et al. (1976) maintain that bears do hibernate because their pulse rate drops by nearly half.

Pinnipedia

Pinnipeds are a monophyletic group of aquatic carnivorans, including the families Odobenidae (walrus), Otariidae (eared seals), and Phocidae (earless seals) (Arnason et al. 2006). Pinnipeds evolved from a bear-like ancestor during the late Eocene or early Oligocene in the northern hemisphere. Molecular studies suggest that during the late Oligocene pinnipeds split into clades. Arnason et al. (2006) proposed that the otariid-odobenid clade originated in the northern Pacific, whereas the phocids originated and diversified along the coast of the southeastern United States. Only later did these early phocids disperse to the colder waters of the North Atlantic, Arctic, and Antarctic Oceans.

Many of the distinctive morphologic features of aquatic carnivorans are adaptations to marine life. They are larger than land carnivores, ranging from 45 kilograms in the ringed seal (*Phoca hispida*) to 3,600 kilograms in the southern elephant seal (*Mirounga leonina*). Large size saves energy in cold environments because of the favorable surface area-to-volume ratio of large animals (Chapter 21). According to Scheffer (1958), large size in pinnipeds is primarily an adaptation to a cold environment. The body is insulated by thick layers of blubber and, in otariids, by insulating fur as well. The pinnae are either small or absent. The external genitalia and mammary nipples are withdrawn beneath the body surface. The tail is rudimentary, and only the parts of the limbs distal to the elbow and knee protrude from the body surface. As a result, the torpedo-shaped body has smooth contours and creates little drag during swimming. The slit-like nostrils are normally closed and opened by voluntary effort.

The skull is partially telescoped, with the supraoccipital partially overlapping the parietals. The rostrum is usually shortened, and the orbits are usually large and encroach on the narrow interorbital area (**Fig. 16-20**). Either one or two pairs of lower incisors are present. The canines are conical. The cheek teeth are homodont, two rooted, and usually simple and conical. The teeth vary in total number from 12 to 24. In some pinnipeds, cheek teeth are characteristically lost with advancing age. The limbs and girdles are highly specialized. The clavicle is absent, and the humerus, radius, and ulna are short and heavily built (**Fig. 16-21**). The pollex is the longest and most robust of the five digits and forms the leading edge of the wing-like fore

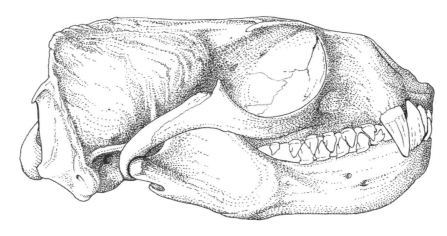

Cape fur seal
(*Arctocephalus pusillus*)

FIGURE 16-20   Skull of a female Cape fur seal from South Africa (length of skull 228 millimeters).

flipper. The pelvic girdle is small and nearly parallel to the vertebral column in phocids (Fig. 16-21). The femur is broad and flattened. The first and fifth are the longest digits of the pes, and both the manus and pes are fully webbed.

The reduction of the vertebral zygapophyses and the absence of the clavicle allow the vertebral column and the forelimbs considerable flexibility and freedom of movement. These features may favor rapid maneuvering during the pursuit of prey. Although terrestrial locomotion is characteristically slow and laborious in most species, the importance of terrestrial locomotion when the animals haul themselves out on rocks or ice or are on breeding grounds has probably limited the extent to which the limbs of aquatic carnivores have become specialized for swimming.

Aquatic carnivorans are extremely capable divers, with some species surpassing most cetaceans in this skill. The diving performance of the eared seals (Otariidae) is probably similar to that of many dolphins. The most spectacular diving occurs among earless seals (Phocidae). The elephant seal (*Mirounga angustirostris*) of the North Atlantic reaches depths of at least 1,500 meters and can stay beneath the surface for up to 90 minutes (Le Boeuf et al. 1993; Stewart 1997). The Weddell seal (*Leptonychotes weddellii*) commonly reaches depths of 300 to 400 meters and is known to dive to 600 meters (Kooyman 1981). Dives often last more than 40 minutes, and a dive of 70 minutes was recorded (Kooyman et al. 1971). This duration surpasses that recorded for such exceptional divers as the sperm whale (*Physeter catodon*), bowhead whale (*Balaena mysticetus*), and bottle-nosed whale (*Hyperoodon ampullatus*). Distances traveled during dives by Weddell seals are also remarkable. Kooyman (1968) found that during a single dive

they can swim 5 kilometers from their breathing holes in the ice and return.

An integrated array of specializations is associated with the diving ability of seals (King 1983; Kooyman 1968, 1975, 1981; Kooyman et al. 1976, 1980). The lungs of aquatic carnivores tend to be larger than those of terrestrial mammals of comparable size, a character important primarily in contributing to buoyancy and allowing the animals to rest at sea. The respiratory airways, even the terminal segments supplying the alveoli, are made rigid by cartilage and muscle. This feature, shared by cetaceans, allows for free passage of air from the alveoli when the lungs collapse under the great pressures to which the body is subjected during deep dives. In addition, this rigidity contributes to the animal's ability to expire air from the lungs extremely rapidly and provides for very quick exchanges of large gas volumes. Another adaptation for prolonged diving is the large blood volume of seals. Weddell seals usually have approximately 14% of their body weight in the form of blood compared with only around 7% for terrestrial animals of similar size. During extended dives, there is a redistribution of blood: blood flow to the major muscle masses is restricted to maintain adequate oxygen supply to the brain and heart. It has been estimated that the metabolic rate of a seal during a long dive is less than 20% of the basal rate.

Family Odobenidae

The family Odobenidae contains only one extant species, *Odobenus rosmarus*, the walrus (**Fig. 16-22**). This species occurs near shorelines in arctic waters of the Atlantic and Pacific Oceans but may stray southward along the coastlines. Odobenids first appeared in the North Pacific in the early Miocene and later diversified into over 20 subfamilies

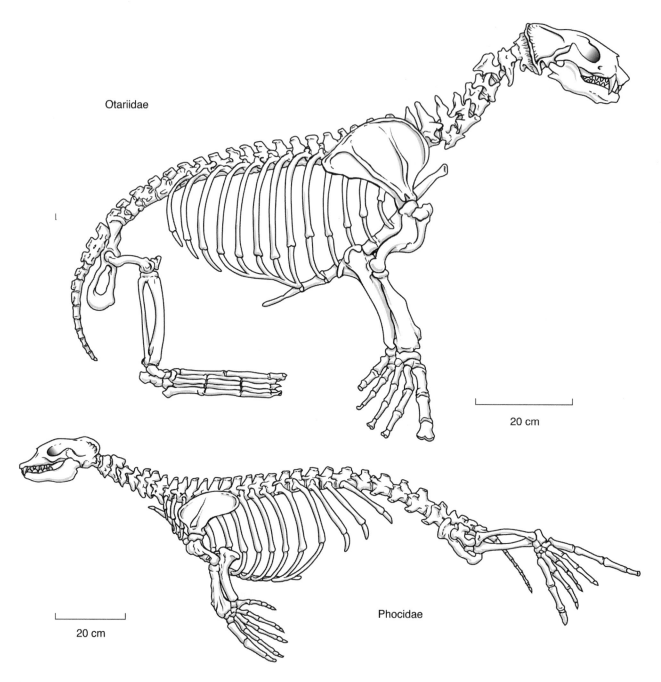

Otariidae

Phocidae

20 cm

20 cm

FIGURE 16-21 Drawings of the skeletons of an otariid seal and a phocid seal. (Adapted from King, J. E. *Seals of the World*. Cornell University Press, 1983.)

(Kohno 2006). The sole surviving species, *Odobenus rosmarus*, is believed to have originated in subtropical Atlantic waters, gradually moved into North Atlantic and Arctic waters, and later re-colonized the Northern Pacific in the Pleistocene (Arnason et al. 2006; Repenning et al. 1979).

The walrus is a large pinniped (up to 1,270 kilograms) with a robust build, a nearly hairless skin, and no external ears. The hind flippers can be brought beneath the body and are used for terrestrial locomotion, which is ponderous and slow. In both sexes, the upper canines are modified into long tusks (**Fig. 16-23**), which in the adult lack enamel. There are no lower incisors in adults, and 12 cheek teeth are usually present. The dental formula is 1–2/0, 1/1, 3–4/3–4, 0/0 = 18–24. On the huge mastoid processes attach the powerful neck muscles that pull the head downward.

Walruses feed on mollusks, which they take from the sea floor by means of their lips, opening the shells with tongue suction. Their huge tusks are used during fighting but are seemingly not used for digging (Fay 1981). Walruses are **gregarious** and **polygynous** and frequently assemble

FIGURE 16-22   A walrus (*Odobenus rosmarus*; Odobenidae).

in large groups of more than 1,000 individuals. They are migratory to some extent, moving southward in winter. Walruses make a variety of loud noises when out of the water and make a church-bell sound and rasps and clicks underwater (Schevill et al. 1963, 1966). The rasps and clicks made during swimming suggest their use in echolocation, but this has not been confirmed.

Family Otariidae

The family Otariidae, containing seven genera and 16 Recent species, includes the eared seals and the sea lions (**Fig. 16-24**). These animals inhabit many of the coastlines of the Pacific Ocean and parts of the South Atlantic and Indian Oceans, including the coasts of southern Australia and New Zealand. They are common along the Pacific coast of North America. The earliest otariids are known from the middle Miocene (*Pithanotaria* and *Thalassoleon*; Carroll 1988) and may have originated in the food-rich kelp reefs of the North Pacific (Arnason et al. 2006; Repenning 1976; Scheffer 1958).

Otariids differ from phocids in being less highly modified for aquatic life and better able to move on land. The hind flippers can be brought beneath the body and used in terrestrial locomotion (Fig. 16-24B). Well-developed nails occur on the three middle digits. A small external ear is present. Males are much larger than females. In the northern fur seal (*Callorhinus ursinus*), males weigh 4.5 times as much as females. Considerable sexual dimorphism in the shape of the skull occurs in some species. In males, the skull becomes larger and more heavily ridged with advancing age. The dental formula is 3/2, 1/1, 4/4, 1–3/1 = 34–38. The body is covered with uniformly dark fur. Weights of otariids range from roughly 60 to 1,000 kilograms.

These seals are generally highly vocal and utter a great variety of sounds. They tend to be gregarious all year round and are social during the breeding season, when they assemble in huge breeding rookeries (**Fig. 16-25**). Propulsion in the water is accomplished by powerful downward and backward strokes of the forelimbs; speeds up to 27 kilometers per hour have been recorded (Scheffer 1958). Otariids eat mostly squid and small fish that occur in schools, and they maneuver rapidly in pursuit of prey. Given the opportunity, Cape fur seals (*Arctocephalus pusillus*) of southern Africa prey on juvenile cormorants (*Phalacrocorax capensis*).

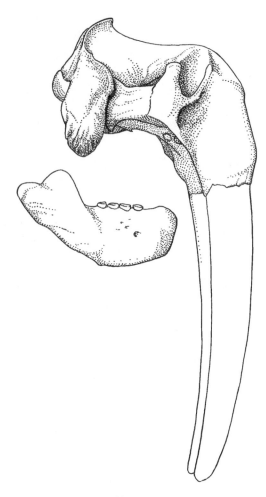

Walrus
(*Odobenus rosmarus*)

FIGURE 16-23 The skull of a walrus (length of skull approximately 355 millimeters). The tusks are enlarged upper canines.

Family Phocidae

Earless seals include 19 species and 13 genera, making the Phocidae the most species-rich family of aquatic carnivorans. They occur along most northern (above 30 degrees northern latitude) and most southern (below 50 degrees southern latitude) coastlines and in some intermediate areas. The Caspian seal (*Pusa caspica*) is restricted to the Caspian Sea, and the Baikal seal (*P. siberica*) lives only in freshwater Lake Baikal in eastern Russia. Phocids appear in the early Miocene and probably arose in the Atlantic Ocean (Muizon 1982).

The earless seals are more highly specialized for aquatic life than are the other aquatic carnivorans. As the vernacular name implies, there is no external ear. The hind flippers are useless on land, but, as a result of lateral undulatory movements, are the primary propulsive organ in the water (Fig. 16-24A). The fore flippers are short and well furred. The structure of the cheek teeth is highly variable but is usually fairly simple (homodont). In the crab-eater seal (*Lobodon carcinophagus*), however, the cheek teeth have complex cusps (**Fig. 16-26**). The pelage of most phocids is spotted, banded, or mottled. These seals frequently have extremely heavy layers of subcutaneous blubber that give the body smooth contours and, in some cases, a nearly perfect fusiform shape. Most species weigh from about 80 to 450 kilograms, but male elephant seals (*Mirounga*) occasionally weigh as much as 3,600 kilograms. The considerable size of most phocids is seemingly an adaptation to the cold arctic or Antarctic waters they inhabit.

Many phocids are monogamous and form small, loose groups in which no social hierarchy is evident, but some, such as the elephant seal, are gregarious and polygynous and have a dominance hierarchy. The monogamous species are quiet, whereas the polygynous species are highly vocal

FIGURE 16-24 (A) Harbor seal (*Phoca vitulina*; Phocidae) on the California coast. (B) A New Zealand sealion (*Phocaractos hookeri*; Otariidae).

FIGURE 16-25   Part of a colony of many thousands of Cape fur seals (*Arctocephalus pusillus*) at Cape Cross, Namibia, southern Africa. The cacophony of vocalizations at such a colony is almost deafening.

(Evans & Bastian 1969). The sole function of the **proboscis** of the male elephant seal is the production of vocal threats (Bartholomew & Collias 1962).

The usual foods of phocids are fish, cephalopods, and other mollusks. Large prey may be taken. Sterling (1969) reports a fish weighing 29.5 kilograms removed from the stomach of a Weddell seal, and the powerful leopard seal (*Hydrurga leptonyx*) eats penguins and small seals. Two other species of phocids are filter feeders and use the complex cheek teeth to filter crustaceans and other plankton from the water. So abundant is the filter-feeding Antarctic crab-eater seal

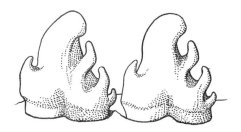

FIGURE 16-26   Medial view of two right lower cheek teeth of the crab-eater seal (*Lobodon carcinophagus*). These complex teeth enable this animal to depend on filter-feeding.

that it constitutes a major share of the world population of phocid seals. Dehnhardt et al. (1998) report that harbor seals (*Phoca vitulina*) use their vibrissae as hydrodynamic receptors to detect the water movements of nearby prey.

Many Weddell seals of the Antarctic are year-round residents as far south as 79 degrees, where broad areas are covered all year with stationary sea ice and the seals must depend on scattered breathing holes. The seals use the upper canines and incisors to ream these holes open by using violent side-to-side thrashing of the head (Kooyman 1975). The ice is about 1 meter thick away from the holes, and the survival of seals depends on their ability at the end of a dive to find their way back to the original breathing hole or to locate a new one. These animals must be skilled navigators in darkness, for many spend the winter where the sun does not appear above the horizon (polar night) for up to 3.5 months (Born et al. 2002).

Musteloidea

Family Ailuridae

The red panda (*Ailurus fulgens*) is the sole member of the family Ailuridae (**Fig. 16-27A**). The name red panda is

FIGURE 16-27 (A) A red panda (*Ailurus fulgens*). (B) Striped skunk (*Mephitis mephitis*) in Arizona.

unfortunate because genetic and morphological evidence suggest that they are not bears whereas the giant panda is considered to be a bear (Flynn et al. 2000; Slattery & O'Brien 1995). Instead, recent evidence supports including the red panda within the superfamily Musteloidea along with skunks, procyonids, and mustelids (Flynn et al. 2000; Wozencraft 2005).

The Miocene *Simocyon* and other extinct genera form a sister-group to red pandas. The cougar-sized *Simocyon* was semiarboreal and possessed an enlarged radial sesmoid bone or "false thumb" similar to those in *Ailurus* and the giant panda (*Ailuropoda melanoleuca*, Ursidae). The fossil evidence suggests that the "false thumb" of the red panda evolved to facilitate grasping branches as it moved through the trees and only later became useful for handling food (Salesa et al. 2006). Thus, it appears that the "false thumbs" of red pandas and giant pandas represent a striking example of convergent evolution.

Today the geographic range of *Ailurus* is restricted to isolated forests in northern Burma (Myanmar) and the districts of Sichuan, Xizang, and Yunnan in China. Red pandas weigh approximately 3 to 6 kilograms, about the size of a raccoon. Their reddish fur is long and they have

long, furry tails with alternating rings (Fig. 16-27A). The head is round with dark marks under the eyes. The pelage on the legs and feet is black. The feet are plantigrade. Red pandas are usually solitary and most active at dawn, at dusk, and during the night. They are largely arboreal and descend trees headfirst. On the ground, their gait is slower and somewhat awkward.

Red pandas eat berries, flowers, leaves, and the occasional bird's egg. Bamboo leaves are the red pandas' primary food in many parts of their range. They eat bamboo leaves in a manner reminiscent of the giant panda (*Ailuropoda*) by grasping the bamboo with the manus, inserting the stalk into the mouth, and stripping the leaves from the stalk (Glatston 1994; Roberts & Gittleman 1984). Red pandas are considered endangered throughout their range due primarily to habitat loss.

■ Family Mephitidae

Skunks were formerly considered a subfamily of the Mustelidae, but genetic evidence now suggests that skunks (family Mephitidae) are a separate clade. There are 4 genera and 12 species of living skunks. The two species of stink badgers (*Mydaus*) live in Indonesia, Malaysia, and the island of Palawan in the Philippines. The remaining species all inhabit the Western hemisphere from Canada to South America. The oldest fossil mephitids are from the early Miocene of Europe.

Skunks are moderately small mammals, ranging in body weight from 0.5 kilogram (*Spilogale*) to more than 4 kilograms (*Conepatus*). Most species are conspicuously colored with black and white spots or stripes that serve to warn predators (Fig. 16-27B). Skunks are well-adapted for digging, with short muscular legs and long claws on the forepaws. They are nocturnal and shelter during the day in burrows, among rocks, under buildings, or in hollow logs.

Skunks are well-known for spraying a foul-smelling secretion from their anal glands as a defense against predators. The paired anal glands produce a noxious mixture of sulfurous chemicals that are ejected by muscle contractions. The spray is accurate to 3 meters and causes intense irritation to any predator that gets within range. The pungent odor may linger in the area for days. A combination of hisses and the skunk's warning coloration is its first line of defense. Skunks are reluctant to use their chemical defenses because after several sprays it takes several days to refill the anal glands.

Except during the breeding season, skunks are solitary animals. The skunks, with no claim to remarkable agility or killing ability, seem to feed on whatever animal material is readily available, which during the summer is generally insects. They also eat a variety of plant material, small vertebrates, and readily seek human garbage. They have an excellent sense of smell. Their vision, however, is limited to a few meters, and roadkills are common in settled areas. Skunks can be endearingly trusting creatures that at night will bumble along within inches of a person.

Family Procyonidae

The family Procyonidae includes the raccoons (*Procyon*), ringtails (*Bassariscus astutus*), and their relatives. As with the bears, omnivorous feeding habits have become predominant in procyonids. Six genera and 14 species are known. Procyonids occupy much of the temperate and tropical parts of the New World, from southern Canada through much of South America. Procyonids chiefly inhabit forested areas, but the range of one species of ringtails includes arid desert mountains and foothills. Procyonids are known from the late Eocene of North America and early Oligocene of Europe and from the late Miocene in Asia. They reached South America from North America at the end of the Miocene (Baskin 1982, 1989; Koepfli et al. 2007).

The structural and functional departure of procyonids from the carnivorous norm has included adaptations favoring both omnivorous feeding habits and climbing ability (**Fig. 16-28**). Associated with the omnivorous trend has been a specialization of the cheek teeth. The premolars are not reduced, as in the bears, but the shearing action of the carnassials is nearly lost. Instead, the carnassials are low-cusped crushing teeth. A hypocone is retained in the upper and in the lower the talonid is enlarged and broadened (**Fig. 16-29**).

FIGURE 16-28   (A) A raccoon (*Procyon lotor*) from North America. (B) A South American coati (*Nasua nasua*).

Upper

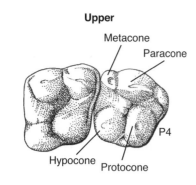

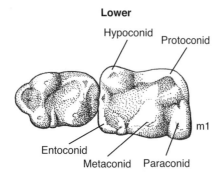

Raccoon
(*Procyon lotor*, Procyonidae)

FIGURE 16-29 Occlusal view of selected cheek teeth of the raccoon. Note that the upper carnassial has a hypocone and that all the teeth are adapted to crushing.

In contrast to the elongate upper molars of bears, those of procyonids are usually broader than they are long. The coati (*Nasua narica*) has flattened, blade-like canines that are formidable defensive weapons. The dental formula is usually 3/3, 1/1, 4/4, 2/2 = 40. Five toes are on each foot. The foot posture is usually plantigrade, and the claws are nonretractile or semiretractile. The limbs are fairly long. The toes are separate, and the forefoot has considerable dexterity in some species and is used in handling food. Tracks left by the human-like hand of the raccoon are familiar to many people. The tail is long, generally marked by dark rings, and is prehensile in the arboreal kinkajou (*Potos flavus*). Procyonids are of modest size, weighing from less than 1 to about 20 kilograms.

The familiar raccoon often takes advantage of cultivated crops. Corn is a staple food item for raccoons living in the middle and western United States, and they eat grapes, figs, and melons in parts of California (Grinnell et al. 1937). In addition, they prey on a variety of small vertebrates and some invertebrates, especially crustaceans. In many parts of the United States, raccoons have become urban pests. They are garbage-can raiders, adroitly catch goldfish and koi in backyard fishpools, and occasionally reside in the attic of an urban home. Some tropical procyonids are largely vegetarians. Hall and Dalquest (1963) reported that in Veracruz the coati eats corn, bananas, and fruit of the coyol palm and that kinkajous eat mostly fruit. The ringtails, on the other hand, eat mostly small rodents. Procyonids reach their greatest diversity and greatest densities in the Neotropics, where they are largely arboreal. In tropical forests, several species may occur together. In such areas, the quavering cries of kinkajous liven the nocturnal scene. Coatis are social animals and assemble in female-young tribes of from 5 to 20 or so animals. These animals are highly vocal and have a varied repertoire of communication calls.

■ Family Mustelidae

The family Mustelidae is large, with 25 genera and some 65 Recent species, including the weasels, badgers, otters, and the wolverine (**Fig. 16-30**). Mustelids occupy virtually every type of terrestrial habitat, from arctic tundra to tropical rain forests, and live in rivers, lakes, and oceans. The distribution is nearly cosmopolitan, but they do not inhabit Madagascar, Australia, or oceanic islands. Mustelids first appear in the fossil records of Europe and Asia in the early Oligocene.

Mustelids are typically fairly small, long-bodied carnivores with short limbs and a pushed-in face. The skull generally has a long braincase and a short rostrum (**Fig. 16-31**), and the postglenoid process partially encloses the glenoid fossa so that in some species the condyle of the dentary bone is difficult to disengage from the fossa. Obviously, little lateral and no rotary jaw action is possible. The dentition is quite variable but is generally 3/3, 1/1, 3/3, 1/2 = 34. The carnassials are blade-like in many species (**Fig. 16-32**) but have been modified into crushing teeth in others. In the sea otter (*Enhydra lutris*), for example, none of the cheek teeth are trenchant. The carnassials have rounded cusps adapted to crushing, and the postcarnassial teeth, M1 and m2, are broader than they are long (Fig. 16-32). In mustelids, the first upper molar is frequently hourglass-shaped in occlusal view (Fig. 16-32).

The limbs are usually short. The five-toed feet are either plantigrade or digitigrade, and the claws are never completely retractile. Anal scent glands are usually well-developed. The tail is generally long, and the pelage may be conspicuously marked, as in badgers. Some mustelids have beautiful, glossy fur that has considerable value in the fur trade. In size, mustelids range from the smallest member of the order Carnivora, a circumboreal weasel (*Mustela nivalis*) that weighs 35 to 250 grams, to the wolverine (*Gulo*) and the sea otter at 32 and 45 kilograms, respectively.

Mustelids, although basically carnivorous, pursue many styles of feeding. Most aggressively search for prey in

FIGURE 16-30 (A) A North American river otter (*Lontra canadensis*). (B) A wolverine (*Gulo gulo*).

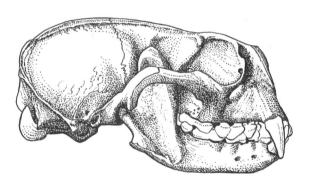

Sea otter
(*Enhydra lutris*, Mustelidae)

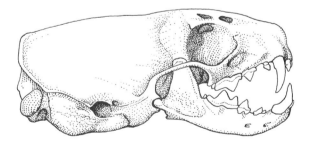

Least weasel
(*Mustela nivalis*, Mustelidae)

FIGURE 16-31 The skull of the sea otter, *Enhydra lutris* (length of skull 152 millimeters), with heavy cheek teeth adapted to crushing marine invertebrates, and a least weasel, *Mustela nivalis* (length of skull 31 millimeters).

burrows, crevices, or dense cover, and many are able killers. A male long-tailed weasel (*M. frenata*) can kill young cottontail rabbits (*Sylvilagus audubonii*) roughly twice its own weight. The weasel typically kills by repeatedly biting the back of the rabbit's skull. Some mustelids, such as the beautiful and graceful marten (*Martes americana*), are swift and agile climbers that occasionally catch arboreal squirrels. Otters are semiaquatic, or almost completely aquatic, in the case of the sea otter, and feed on a wide variety of vertebrates and invertebrates. Sea otters use rock "tools," brought up from the bottom of the ocean, against which the otters crack open the hard shells of crabs and mollusks. Badgers are also tool users on occasion. A badger that was excavating ground squirrel burrows in Alberta, Canada plugged the squirrels' escape holes with blocks of wood (Michener 2004). Badgers have long been known to plug burrows with soil or chunks of sod as a hunting technique.

Mustelids are unusual in several features of their reproduction. Apparently, all mustelids must prolong copulation to induce the female to **ovulate** (release an ovum for fertilization). As a result, copulation may last for several hours in some species. After fertilization has taken place, the embryo may or may not immediately implant in the lining of the uterus. **Delayed implantation** occurs is some, but not all, mustelids (Chapter 20). For example, in the Old World badger (*Meles meles*) implantation may be delayed for up to 10 months, and implantation of the **blastocyst** takes place in response to environmental cues, such as temperature and/or day length (Canivenc & Bonnin 1979; Mead 1989).

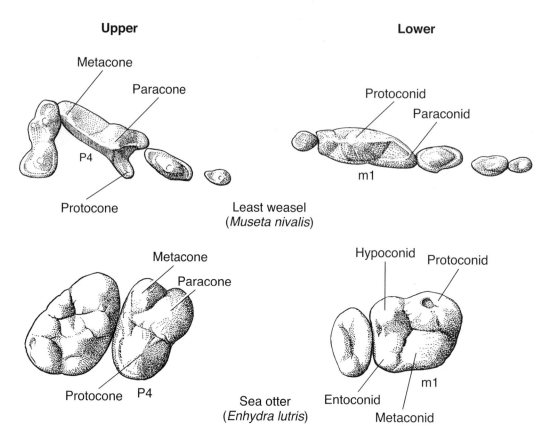

Upper

Metacone

Paracone

P4

Protocone

Least weasel
(*Museta nivalis*)

Lower

Protoconid

Paraconid

m1

Metacone

Paracone

Protocone   P4

Sea otter
(*Enhydra lutris*)

Hypoconid   Protoconid

m1

Entoconid

Metaconid

FIGURE 16-32   Occlusal views of the cheek teeth of two carnivores. The crushing teeth, M1 and m2, of the least weasel are reduced, and the shearing function of the cheek teeth is of major importance. The bunodont teeth of the sea otter have rounded cusps and are adapted to crushing; the carnassials retain no shearing function.

SUMMARY

Carnivorans probably evolved in the Paleocene and underwent adaptive radiation as herbivores, their primary prey, diversified. Some modern carnivorans are omnivorous. Living members of the Carnivora include terrestrial carnivores (such as dogs, cats, weasels, and bears), as well as aquatic carnivores (seals, sea lions, and walruses). Carnivoran classification remains a subject of considerable debate. The Carnivora is composed of two basic clades, the Feliformia and Caniformia. The seals, sea lions, and walruses (Pinnipedia) are closely related to the superfamily Musteloidea, and along with ursids form the superfamily Arctoidea.

The suborder Feliformia includes 121 species in 54 genera distributed worldwide except in Australia (where they were introduced by humans) and Antarctica. The feliform clade includes six families of extant carnivorans: the African palm civet (Nandiniidae), cats (Felidae), civets and genets (Viverridae), hyenas and the aardwolf (Hyaenidae), mongooses (Herpestidae), and the Malagasy civets and mongooses (Eupleridae). Cats usually catch prey by a stealthy stalk followed by a brief burst of speed. Felids have well-developed senses of smell, sight, and hearing, and

many are agile climbers. Most members of the Feliformia are solitary predators, but lions and spotted hyenas live in large family groups. Lion prides are dominated by a coalition of males, but spotted hyenas form matrilineal clans. Viverrids, herpestids, nandiniids, and euplerids are smaller carnivores with diverse diets and lifestyles. Some species, such as the meerkat and dwarf mongoose, have complex social lives.

The caniform clade includes 165 species in 72 genera, the families Canidae, Ursidae, Ailuridae, Mephitidae, Procyonidae, Mustelidae, and three families of aquatic carnivorans. Canids are broadly adapted carnivores, often foraging over large areas. Small canids such as foxes are solitary hunters, whereas medium-sized canids (e.g., coyotes) hunt alone or in small family groups, and large canids (wolves and African wild dogs) hunt in large cooperative groups. Bears (Ursidae) are large-bodied animals that abandoned cursorial ability and trenchant carnassial cheek teeth in favor of powerful limbs and crushing molars in keeping with their more omnivorous feeding habits. In temperate and boreal regions, some bear species retreat to protected dens where they become lethargic or dormant and live off fat reserves during the cold winter months.

The remaining terrestrial members of the Caniformia include the mustelids, skunks, and the red panda. Skunks were long considered a subfamily of the Mustelidae, but evidence now suggests that skunks and stink badgers form a separate clade (family Mephitidae). Most skunks are conspicuously colored with black-and-white spots or stripes, which serve to warn predators about their defensive and foul-smelling anal secretions. The family Mustelidae includes weasels, polecats, badgers, otters, martens, the tayra, grisons, and the wolverine. Mustelids occupy virtually every type of terrestrial habitat, from arctic tundra to tropical rain forests, and live in rivers, lakes, and oceans. They are typically fairly long-bodied carnivores with short limbs and a blunt face and large brain. Mustelids are unusual in several features of their reproduction; all mustelids prolong copulation to induce the female to ovulate. After fertilization the embryo may not immediately implant in the lining of the uterus (delayed implantation) in some species. The family Procyonidae includes the raccoons, ringtails, coatis,

and their relatives. Procyonids are mostly omnivorous and arboreal; some such as the olingos and kinkajou are rather primate-like.

Pinnipeds are a monophyletic group of aquatic carnivorans, including the families Odobenidae (walrus), Otariidae (eared seals), and Phocidae (earless seals). Pinnipeds evolved from a bear-like ancestor during the late Eocene or early Oligocene in the northern hemisphere. The large size of pinnipeds is an adaptation to a cold environment. Their bodies are insulated by thick layers of blubber and, in otariids, by thick insulating fur as well. The external genitalia and mammary nipples are withdrawn beneath the body surface. The tail is rudimentary, and only the distal portions of the limbs protrude from the body surface. As a result, their torpedo-shaped body has smooth contours and creates little drag during swimming. Aquatic carnivorans are extremely capable swimmers and some are deep divers. Otariids differ from phocids in being less highly modified for aquatic life and better able to move on land. Otariid hind flippers can be brought beneath the body and used in terrestrial locomotion.

KEY CHARACTERISTICS OF CARNIVORA

- Enlarged P4 and m1 form shearing carnassial pair (secondarily reduced in bears, raccoons, and seals)
- Canines large and conical
- Third molar lost
- Skulls heavily built, with strong zygomatic arch
- Braincase enlarged and frontoparietal suture located anteriorly
- Well-defined, transverse glenoid fossa restricts jaw motion to dorsal-ventral plane
- Sagittal crest prominent (secondarily reduced in some members)
- Turbinals relatively large and complex
- Fused scaphoid and lunar bones in carpals
- Simple stomach

Caniformia
- Auditory bullae composed of a single bone and not divided or only partially chambered
- Claws nonretractile
- Baculum well developed

Pinnipedia
- Body insulated with thick layer of blubber and may also be covered with hair
- Body fusiform and adapted for swimming
- Digits fused together and covered with skin to form flippers
- Forelimbs and hindlimbs paddle like
- Pinnae highly reduced or absent
- Vibrissae well developed
- Molariform teeth homodont
- Tail very short or absent
- External genitalia hidden within slits or grooves
- Baculum present

Feliformia
- Auditory bullae two-chambered, joined by septum
- Rostrum shorter, teeth fewer, and carnassial teeth more sectorial than caniforms
- Claws strong and sharp, retractile or semiretractile

KEY TERMS

Blastocyst
Carrion
Delayed implantation

Gregarious
Ovulate
Polygynous

Proboscis
Refugia

RECOMMENDED READINGS

Arnason, U, et al. 2006. Pinniped phylogeny and a new hypothesis for their origin and dispersal. *Molecular Phylogenetics and Evolution, 41*:345–354.

Flynn, JJ & GD Wesley-Hunt. 2005. Carnivora. Pp. 175–198 in *The Rise of Placental Mammals* (KD Rose & JD Archibald, eds.). Johns Hopkins University Press, Baltimore, MD.

Flynn, JJ, et al. 2005. Molecular phylogeny of the Carnivora (Mammalia): assessing the impact of increased sampling on resolving enigmatic relationships. *Systematic Biology, 54(2)*:317–337.

Gittleman, J. 1993. *Carnivore Behavior, Ecology and Evolution.* Cornell University Press, Ithaca, NY.

King, JE. 1983. *Seals of the World.* Cornell University Press, Ithaca, NY.

Kruuk, H. 1972. *The Spotted Hyena. A Study of Predation and Social Behavior.* University Chicago Press, Chicago, IL.

Ridgway, SH & RJ Harrison. 1981. *Handbook of Marine Mammals.* Academic Press, London.

Schaller, GB. 1972. *The Serengeti lion, a study of Predator–Prey Relations.* University of Chicago Press, Chicago, IL.

Turner, A. 1997. *The Big Cats and Their Fossil Relatives.* Columbia University Press, New York.

Wang, X & RH Tedford. 2008. *Dogs, Their Fossil Relatives and Evolutionary History.* Columbia University Press, New York.

Yoder, AD & JJ Flynn. 2003. Origin of Malagasy Carnivora. Pp. 1253-1256 in *The Natural History of Madagascar* (SM Goodman & JP Benstead, eds.). University of Chicago Press, Chicago. IL.

Perissodactyla

Since Eocene times, some of the most specialized and spectacular cursorial mammals have been perissodactyls (horses, rhinos, tapirs, and their extinct relatives). In the early Tertiary, these were the most abundant **ungulates**, but their diversity declined in the Oligocene, perhaps because of changing environments. With the diversification and "modernization" of the artiodactyls (pigs, camels, antelopes, and their relatives) in the Miocene, the fortunes of perissodactyls further declined. The surviving perissodactylan fauna (consisting of 6 genera and 16 species) is but an insignificant remnant of this once important and diverse group and is vastly overshadowed by an impressive living artiodactylan assemblage (consisting of 220 species). Perissodactyls occur today largely in tropical and subtropical areas, including Africa, parts of central and southern Asia, and tropical parts of North America and northern South America.

There is universal agreement that the Perissodactyla is a monophyletic group (**Fig. 17-1**). A suite of morphological characters defines perissodactyls, including that the axis of symmetry of the foot passes through the large middle digit (the mesaxonic condition; **Fig. 17-2**). The result is that most of the body weight is borne by the third digit (in artiodactyls, the axis of symmetry passes between the third and fourth digits [**paraxonic**], and the weight is shared by both digits). Most perissodactyls have three digits on the hindfoot and three or four on the forefoot. The most cursorial species are the horses (Equidae), which retain only a single third digit. Among living perissodactyls, only rhinos have horns, but unlike the paired, bony horns of artiodactyls, rhino horns are derived from dermal tissues and are located on the nasals or frontals of the skull. Perissodactyls also have an elongated rostrum that houses a large battery of molariform cheek teeth. These teeth are hypsodont and lophodont (having complex cross-lophs) in the grazing equids and brachydont in browsers such as tapirs. Modern artiodactyls, in contrast, have selenodont or bunodont cheek teeth. Perissodactyls have a simple stomach and a very large cecum where bacterial fermentation of cellulose occurs (artiodactyls have multichambered stomachs for fermentation).

Perissodactyl Evolution

Perissodactyls probably arose in the late Paleocene in Asia. By the early Eocene, ancestral perissodactyls had diversified and spread to Europe and North America (Hooker 2005). This Eocene radiation included the massive brontotheres (titanotheres), paleotheres (the likely ancestors of horses), the strange, knuckle-walking chalicotheres, and the enormous hyracodontid *Paraceratherium* (also known as *Indricotherium* and *Baluchitherium*). Although early perissodactyls were the dominant browsers of the Eocene and Oligocene, climatic changes during the Miocene and probably competition from artiodactyls resulted in the decline of perssiodactyls; only three families survive today.

Fossil evidence suggests that tapir-like animals, lacking a proboscis, evolved in North America in the Eocene. Unlike modern tapirs, with their tropical distribution, Eocene tapiroids were found as far north as Ellesmere Island on the Arctic Circle at a time when the climate of that region was probably subtropical (Eberle 2005). Roughly 20 million years ago the ancestral tapirs of North America diverged and spread to Asia. A second wave of migration took place approximately 3 million years ago, as some North and Central American species moved into South America. Temperate North American species died out 10,000 years ago.

The fossil record of the rhinoceroses (Rhinocerotidae) and their relatives is remarkably complex and parallels that of the horses. Rhinoceroses diverged from other ancestral perissodactyls in the early Eocene. Two fossil groups are traditionally allied with modern rhinos: the Hyracodontidae and Amynodontidae. Amynodontids were large, semiaquatic animals that ranged across Eurasia and North America. Two hyracodontid genera that illustrate well the diversity of early Tertiary rhinocerotoids are *Hyracodon* and *Paraceratherium*. *Hyracodon*, a small North American Eocene–Oligocene "running rhinoceros," had slender legs

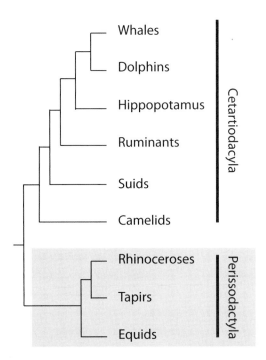

FIGURE 17-1 One phylogenetic hypothesis of the Perissodactyla and their sister groups.

and tridactyl feet and was probably similar in cursorial ability to Oligocene horses. Hyracodonts became extinct in early Miocene. A contemporary of *Hyracodon* in the Oligocene was the Eurasian form *Paraceratherium*, the largest known land mammal. This giant was nearly 5 meters high at the shoulder and probably weighed on average about 11 metric tons (11,000 kg) with a potential maximum up to 15 to 20 metric tons (Fortelius & Kappelman 1993), and the skull (small in proportion to the great size of the rest of the animal) was 1.2 meters long. The neck was long, and they may have browsed on high vegetation in giraffe-like fashion.

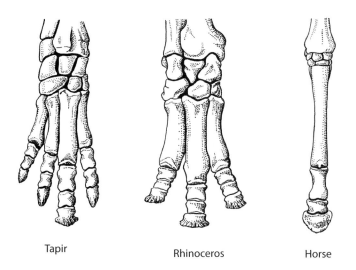

FIGURE 17-2 Front feet of three perissodactyls: a tapir; a rhinoceros, and a horse. (Adapted from Howell, A. B. *Speed in Animals.* University of Chicago Press, 1965.)

The limbs were long and graviportal, but the tridactyl feet were unique in that the central digit was greatly enlarged and terminated in a broad hoof, whereas the lateral digits were much smaller than in any other rhinocerotoid (Lucas & Sobus 1989). Rhinoceroses died out in the New World in the earliest Pliocene (Prothero 2005), but remained common and diverse in Eurasia through the Pleistocene. The Pleistocene woolly rhinoceros (*Coelodonta*) was apparently adapted to cold climates. Entire preserved specimens of this animal have been found in an oil seep in Poland. The reasons for the demise of the woolly rhinoceros in Eurasia are debated, but climate change or human predation may have been responsible. Sumatran rhinos (*Dicerorhinus sumatrensis*) have a fossil record dating back to the early Miocene and are considered the sister group to the Asian rhinos (*Rhinoceros*) and African rhinos (*Ceratotherium* and *Diceros*; Lacombat 2005).

The evolution of horses is well-documented by an excellent and largely New World fossil record discussed by MacFadden (2005, 1992), Radinsky (1984), and Simpson (1951). For many years, equids were thought to be first represented by *Hyracotherium* from the early Eocene record of Europe and North America. However, recent phylogenetic analysis shows a much more complex early history, with *Hyracotherium* being a paleothere (an extinct early sister group to horses) and a half-dozen genera known from that time period in North America alone (Froehlich 1999, 2002). The oldest of these are *Arenahippus* and *Sifrihippus* from the earliest Eocene (Janis et al. 2008). These primitive horses had a generalized skull with 44 teeth (**Fig. 17-3**). The upper and lower molars were brachydont and basically four cusped (**Fig. 17-4**). The upper molars bore a protoconule and a metaconule, and the paraconid of the lower molars was reduced. The premolars were not molariform. The limb structure reflected considerable running ability: The limbs were fairly long and slender, the front foot had four toes, and the hind foot had three toes. The animal, however, was functionally tridactyl. Each digit terminated in a small hoof. Eocene horses included several genera that ranged in size from 25 to 50 centimeters at the shoulder and presumably browsed on low-growing vegetation in forested or semiforested areas (MacFadden 2005).

Cenozoic changes in climate and in the flora of North America may have had a critical influence on the evolution of horses, as well as on other mammals on other continents. Accompanying the climate changes were the Miocene development of grasslands and savannas over much rolling or nearly level land in the present Great Plains, the Great Basin, and the southwestern deserts of the United States. Many of the most progressive equid skull and dental features probably arose in response to the shift to a grazing habit. Grass, at least at certain times of the year, has low nutritional value

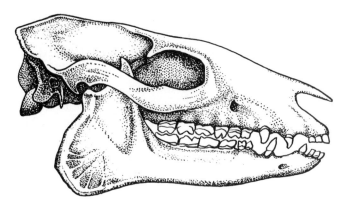

Hyracotherium

FIGURE 17-3 Skull of *Hyracotherium* (*Eohippus*), one of the first known equids (length of skull 134 millimeters). (Modified from Romer, A.S. *Vertebrate Paleontology*. University of Chicago Press, 1966.)

and must be eaten in large quantities to sustain life. High-crowned, persistently growing teeth evolved in horses, and many other lineages of mammals to cope with grasses made highly abrasive by silica in the leaves and by particles of soil deposited on leaves by wind and the splash effect of rain (Jacobs et al. 1999; Janis & Fortelius 1988). A similar trend toward hypsodonty occurred during the Miocene for mammals in Europe (Jernvall & Fortelius 2002; Theodor 2002), but the same trend occurred much earlier, in the Oligocene, in South American native ungulates, including some convergently evolved horse-like types (extinct litopterns and notoungulates; MacFadden 2000). Also of great adaptive value for horses were the highly cursorial limbs with single-toed feet, which facilitated rapid and efficient locomotion on the firm, level footing of the grasslands. Cursorial ability was perhaps as advantageous for traveling between widely scattered concentrations of food and distant water holes in semiarid regions as for escaping from predators.

Side branches from the main stem of equid evolution developed at various times. During part of the late Miocene,

some localities in North American savannas supported up to 12 species of horses. The main evolutionary line leading to *Equus* can be traced from Eocene ancestors through such intermediate genera as *Orohippus*, *Mesohippus*, *Merychippus*, and Miocene *Dinohippus* to the Pliocene-Pleistocene and Recent *Equus* species (**Fig. 17-5**). Late Eocene-Oligocene equids such as *Mesohippus* became slightly taller and sported a longer face. By the Miocene, equids had split into several lineages: the three-toed browsing forms that spread to the Old World, a short-lived line of pygmy horses, and a third lineage of cursorial horses that specialized on grasses (MacFadden 2005). *Merychippus*, a group of pony-sized horses that were functionally tridactyl but retained short lateral digits, belonged to this third lineage. *Merychippus* had a number of cursorial adaptations, including an **unguligrade** stance, a fused radius and ulna, and strong ligaments running under the fetlock to store elastic energy during running. The dentary bone was deep. The face was long, and the orbit was fully enclosed. The cheek teeth were high crowned, covered with cementum, and had an occlusal pattern similar to that of *Equus* (Fig. 17-4).

By the middle to late Miocene, horses had reached their greatest diversity. *Pliohippus* occurred in the late Miocene; it had the skull features of its progenitor, *Merychippus*, but was more progressive in having higher crowned teeth and lateral digits usually reduced to splint-like vestiges. About 12 million years ago, *Dinohippus*, a genus of one-toed horses resembling modern *Equus* evolved in North America. *Equus*, the genus to which all living horses belong, differs little from *Dinohippus*. Approximately 4 million years ago, the first *Equus* species appeared in North America. *Equus* underwent a rapid burst of speciation, forming several species that coexisted with other Pliocene equids. Late Pliocene-Pleistocene intercontinental land bridges associated with glaciation events allowed some *Equus* species to migrate to the Old World, where they subsequently spread across Asia and entered Africa. The African migrants gave

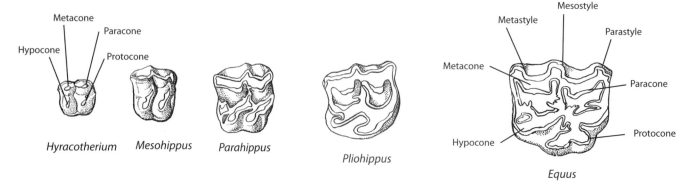

FIGURE 17-4 Right upper molars of four fossil equids and the extant *Equus*. These teeth illustrate stages in the evolution of the equid molars. (Modified from Romer, A.S. *Vertebrate Paleontology*. University of Chicago Press, 1966.)

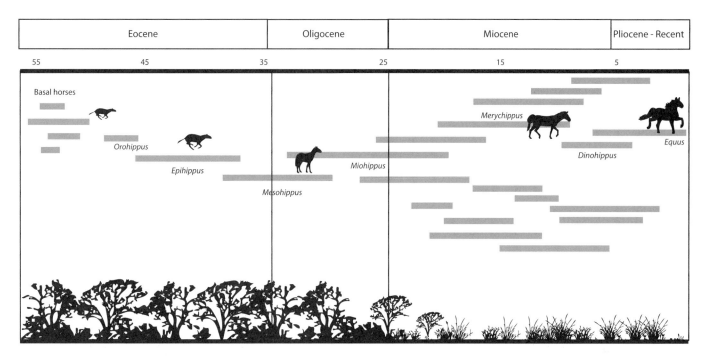

FIGURE 17-5 A time chart of the North American equids over the past 55 million years. The early history, from 55 to 25 million years ago, was characterized by a diversity of small browsing species. In the Miocene, the climate of North America shifted from predominantly woodlands to vast grassland areas. An adaptive radiation of equids occurred in the Miocene in response to changes in habitat and food. The gray bars represent different lineages, and the numbers represent millions of years. (Adapted from MacFadden, B. J., *Science.* 307 (2005): 1728–1730.)

rise to zebras; Asian and European forms gave rise to the modern horse, *Equus caballus*. For some unknown reason, horses disappeared from the New World, their place of origin and the primary center of their evolution, at the end of the Pleistocene. Their decline began toward the end of the Miocene Epoch, when savannas were being replaced by cooler and drier steppe conditions. In the early 16th century, Europeans reintroduced horses into their ancestral New World range (Luis et al. 2006).

Textbooks have long portrayed the major evolutionary trends of horse evolution as a series of transitions from the diminutive *Hyracotherium*, through a series of intermediate-sized three-toed horses, to modern *Equus*. MacFadden (2005), however, cautioned that such oversimplifications misrepresent the complexity of equid evolution. As MacFadden pointed out, the available fossil evidence suggests that "rather than a linear progression toward larger body size, fossil horse macroevolution is characterized by two distinctly different phases." In the first 30 million years, primitive equids remained relatively small (less than 50 kg). From the early Miocene to the present, some horse lineages became larger (e.g., the lineage that gave rise to *Equus*), whereas others either became smaller or changed little in body size.

From the late Eocene through the Oligocene, perissodactyls were the dominant herbivores in the Northern Hemisphere. For most of this period, perissodactyls co-existed with the less diverse artiodactyls (even-toed

ungulates). Competition with perissodactyls possibly forced artiodactyls to occupy marginal habitats and subsist on coarser, low-energy plants. However, as the climate of North America and Eurasia became drier beginning in the Oligocene and continuing to the Pliocene-Pleistocene ice ages, woodlands gave way to more open savannas, grasslands, and even deserts, the mix of perissodactyls and artiodactyls continued to change. The key to understanding how climatic changes profoundly altered the course of perissodactyl and artiodactyl evolution is knowledge of their cursorial and feeding adaptations.

Cursorial Specialization

Exceptional running ability has evolved independently in a number of mammalian groups. It provides a means of escaping predators (as in ungulates, rabbits, and some rodents) or of capturing prey (as in carnivores). An entire series of specializations for cursoriality is unique to mammals and involves the integration of locomotor and respiratory functions. Galloping mammals synchronize their breathing and stride cycles, using the inertia of their intestines as a "visceral piston" to help ventilate the lungs as the liver "sloshes" back and forth against the diaphragm with each running stride (Bramble & Jenkins 1993, 1998). As an aid to this process, a phenomenon known as "pneumatic stabilization" involves tracheal valving to shunt air from side to side between lungs and help to control inhalation and exhalation (Simons 1996). Thus pressurized, the lungs not only function in respiration but also help to stabilize the shoulder and chest wall on alternating sides as the leading forelimb strikes the ground (Bramble & Jenkins 1998).

Cursorial adaptations appear in the early Cenozoic history of mammals. The earliest known perissodactyls, of the early Eocene Epoch, were not highly specialized (Radinsky 1966). The early Eocene genus *Diacodexis*, one of the earliest known artiodactyls, had slim, elongate limbs and was highly cursorial (Carroll 1988; Rose 1982, 1996). The refinement of cursorial adaptations in ungulates was favored by the Miocene expansion of grasslands. Not only was speed the primary means of escaping predators in this open country, but long daily or seasonal movements to seek water or nutritious food probably became an important part of the ungulate mode of life.

Running speed is determined basically by two factors: stride length and stride rate (the number of strides per unit of time). Cursorial specializations lengthen the stride or increase its rate. Perhaps the most universal cursorial adaptation that lengthens the stride is lengthening of the limbs. In generalized mammals or in many powerful diggers, the limbs are fairly short and the segments are all roughly the same length (**Fig. 17-6**). In cursorial species, however, the limbs are long. In the most specialized runners, the metacarpals and metatarsals are elongate, and the manus and pes are the longest segments. Limb joints become more deeply excavated in cursorial mammals, further restricting limb movement to the parasagittal plane. Loss or reduction of the clavicle contributes further to the length of the stride. This occurs in carnivorans, leporids, and ungulates. With the loss of the clavicle, the scapula and shoulder joint are freed from a bony connection with the sternum, and the scapula is free to rotate to some extent about a pivot point approximately at its center.

Substantial lengthening of the stride also results from an inchworm-like **flexion** and **extension** of the spine (**Fig. 17-7**). In small- or moderate-sized runners, the flexors and extensors of the vertebral column are powerfully developed. The vertebral column extends as the forelimbs reach forward and the hindlimbs are driving against the ground, then it flexes when the front feet move backward while braced against the ground as the hindlimbs swing forward. Hildebrand (1959) estimated that such movements of the cheetah vertebral column added nearly 10 kilometers per hour to the cheetah's running speed.

The speed of limb movements and thus stride rate are increased by a combination of structural modifications. The total speed of the foot, which drives against the ground and propels the animal, depends on the speed of movement of the distal joints of the limb. If another movable joint is added to the distal limb, the speed of the limb will

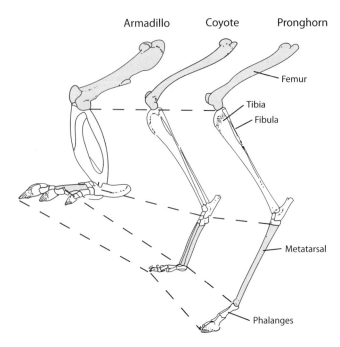

FIGURE 17-6   The hindlimbs of three mammals: armadillo (*Dasypus novemcinctus*), a powerful digger with plantigrade feet; coyote (*Canis latrans*), a good runner with digitigrade feet; pronghorn (*Antilocapra americana*), an extremely speedy runner with unguligrade feet. Note the lengthening of the shank and foot in the coyote and pronghorn especially; the metatarsals have undergone the greatest lengthening. The limbs are not drawn to scale, but the femur is the same length in each drawing.

be increased by the speed of movement at the new joint. The greater the number of joints that move in the same direction simultaneously, the greater is the speed of the limb. One way to add a joint to the distal limb is to change the foot posture such that only the hoof-bearing tips of the digits contact the ground. This lifting of the heel from the ground allows another limb joint, that between the metapodials and the phalanges, to contribute to limb speed. Not surprisingly, nearly all cursorial mammals have abandoned a plantigrade foot posture in favor of one that is a digitigrade or unguligrade. In addition, the movable scapula and the flexion of the vertebral column (in some cursorial species), which help to increase stride length, also contribute their motion to total foot speed.

Specializations of the musculature also add importantly to limb speed and hence to running speed. A trend in many cursorial mammals is toward a lengthening of the tendons of some limb muscles (in association with the elongation of the distal segments of the limbs) and, in some cases, a migration of the insertion points of these muscles toward the body. Generally, the nearer the insertion point of a muscle approaches the joint it spans and at which it causes motion, the greater is the advantage for speed.

Speedy limb movements are further facilitated by a reduction of the mass of the distal parts of the limbs and the

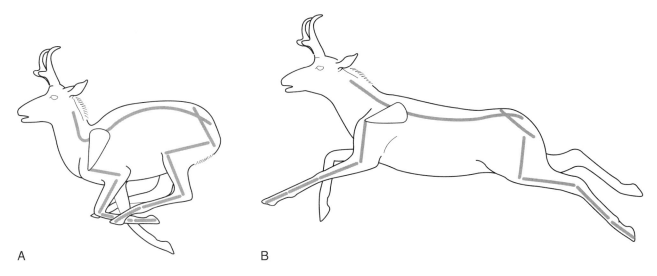

FIGURE 17-7   Two positions of a running pronghorn showing the flexion and extension of the vertebral column and the changing position of the scapula. (A) The forelimbs have just left the ground; the hindlimbs are reaching forward and will touch the ground as the forelimbs swing forward. (B) The animal is bounding ahead after the limbs have driven against the ground; the forelimbs are reaching forward.

resultant reduction in the amount of inertia that must be overcome at the end of one limb movement and the start of another. Because the distal part of the limb moves more rapidly than the proximal part during a stride, reduction of mass of the distal parts is especially advantageous. Several specializations commonly serve this end. The most obvious is the loss of certain digits. Also, the heaviest muscles are mostly in the proximal segment of the limb, thus keeping the center of gravity of the limb near the body. The combined effect of these modifications that reduce and redistribute weight is to favor rapid limb movement and to reduce the outlay of energy associated with that movement. For excellent discussions of cursorial adaptations in mammals, see Gambaryan (1974), Hildebrand (1959, 1960, 1965, 1985, 1987), and Kardong (1998).

In the ungulate ankle joint, the calcaneum is pushed aside, so to speak. In mammals in which no drastic reduction of digit number has occurred, the distal surface of the astragalus articulates with the navicular, the calcaneum articulates with the cuboid (see Figure 2-27, and the weight of the body is transferred through the digits, the distal carpals, both the astragalus and the calcaneum, and the tibia and fibula. In ungulates, a different arrangement occurs in association with the loss of some digits: The astragalus rests more or less directly on the distal tarsal bones, which may be highly modified by fusion and loss of elements (**Fig. 17-8**, see Fig. 18-6, and the weight of the body is borne by the central digits (or digit, in the case of equids), the distal tarsals, and the astragalus. The astragalus thus becomes the main weight-bearing bone of the two proximal tarsals. The calcaneum remains important as a point of insertion for extensors of the foot, but it no longer is a major weight-bearing bone of the tarsus. A similar bypassing of

the calcaneum occurs in the cursorial peramelid marsupials (see Fig. 6-17).

Two distinctive ungulate specializations involve connective tissue. The nuchal ligament is a heavy band of elastin (an elastic protein found in vertebrates) that is anchored posteriorly to the tops of the neural spines of some of the anterior-most thoracic vertebrae and attaches anteriorly high on the occipital part of the skull (**Fig. 17-9**). This ligament, especially robust in large, heavy-headed ungulates such as

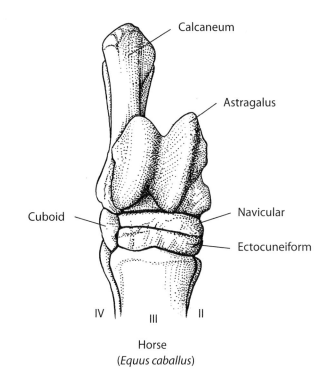

FIGURE 17-8   The tarsus of the horse (*Equus caballus*). The metatarsals are numbered.

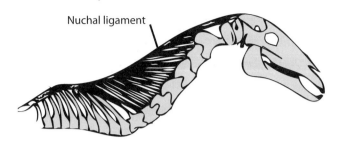

Nuchal ligament

FIGURE 17-9 The nuchal ligament of the horse (*Equus caballus*) illustrating the extent of the insertion on the skull and vertebral spines.

the horse and moose, helps support the head so that the burden on the muscles that lift the head is greatly lightened. The elasticity of the ligament allows the animal to lower its head when it is eating or drinking. A second specialized ligament, the springing ligament, occurs in the front and hind feet of ungulates; it evolved from muscles that flexed the digits (Camp & Smith 1942; Hildebrand 1985). In the hind foot of the pronghorn (*Antilocapra*), for example, the springing ligament arises from the proximal third of the back of the **cannon bone** and inserts distally on the sides of the first phalanges of digits three and four (see Figure 18-5). When the foot supports the weight of the body, the phalanges are extended, thereby stretching the springing ligament. As the foot begins to be relieved of the weight of the body toward the end of the propulsion stroke of the stride, however, the elastic ligament begins to rebound. When the foot is leaving the ground, the phalanges snap toward the flexed position. The familiar backward flip of the horse's foot just as it leaves the ground is controlled by the springing ligament. This flip gives a final increase in speed and thrust to the stride and increases the ungulate's speed without the use of muscular effort.

Feeding Specialization

The herbivorous diet characteristic of most ungulates has favored the development of cheek teeth with large and complex occlusal surfaces that finely section plant material as an aid to digestion. Premolars tend to become molariform and thus to increase the extent of the grinding battery, and the anterior dentition becomes variously modified. In advanced types, there is a diastema between the anterior dentition and the cheek teeth.

A diet of vegetation puts unusual demands on the digestive systems of ungulates. Vegetation yields far less fat and protein than meat, is more difficult to digest, and is often protected by defensive secondary compounds. In addition, an herbivore must break down the cell wall, a fairly rigid structure formed largely of cellulose, not so much for the energy it yields as to gain access to the proteins,

carbohydrates, and lipids within the cells. This process is difficult, however, for mammals lack enzymes that digest cellulose. All herbivores must therefore have an alimentary canal that can digest cellulose by means other than the herbivore's own enzymatic action. Both perissodactyls (horses, rhinoceroses, and tapirs) and **ruminant** artiodactyls (camels, deer, antelope, sheep, goats, and cattle) use a fermentation process that breaks down cellulose with the cellulitic enzymes of microorganisms living in the alimentary canal. **Microbial fermentation** is a relatively slow process requiring specialized fermentation chambers. Microbial digestion of cellulose in a modified stomach is called gastric (foregut) fermentation and is typical of artiodactyls (Chapter 18).

In perissodactyls, protein is digested and absorbed in the relatively small and simple stomach, whereas microbial fermentation takes place in the large intestine and enlarged cecum (intestinal or hindgut fermentation—**intestinal fermentation** also occurs in rodents, lagomorphs, elephants, and hyraxes). The horse cecum is an expanded portion of the ascending colon with a blind sac at one end (**Fig. 17-10**). As a result of microbial fermentation, cellulose is broken down and sugars and fatty acids are released and absorbed across the intestinal walls.

Perissodactyls and ruminant artiodactyls have thus evolved contrasting nutritional strategies. Ruminant artiodactyls can satisfy their nutritional needs with relatively unnutritious food. In areas with marked seasonality, the only areas where ruminants are abundant, they can remain widespread through times when forage is nutritionally poor. During comparable times, perissodactyls must seek sites that support the greatest quantity of vegetation and the most nutritious vegetation, and their range is correspondingly restricted. When ruminant artiodactyls compete with perissodactyls for nutritious food, however, ruminants are the losers. Ruminants are less efficient in using nutritious food because proteins are used inefficiently by the ruminant system of microbial fermentation (Reid 1970; Smith 1975). Ruminants have reduced their efficiency in transforming forage into animal biomass while enhancing their ability to survive on low-quality food (Kinnear et al. 1979).

An extremely interesting African grazing succession, strongly influenced by differences between digestive efficiencies and food requirements, has been described by Bell (1971). He studied primarily the most abundant ungulates: the zebra (*Equus burchellii*), a nonruminant, and the wildebeest (*Connochaetes taurinus*) and Thompson's gazelle (*Gazella thompsonii*), both ruminants. He found that, on the Serengeti Plains, the zebra was the first of these ungulates to be forced by food shortages to move from the preferred short-grass area down into the longer, coarser grasses of the lowlands. After the zebras' feeding and trampling in the lowlands had removed the coarse upper parts of the grass

Foregut Fermentation

Hindgut Fermentation

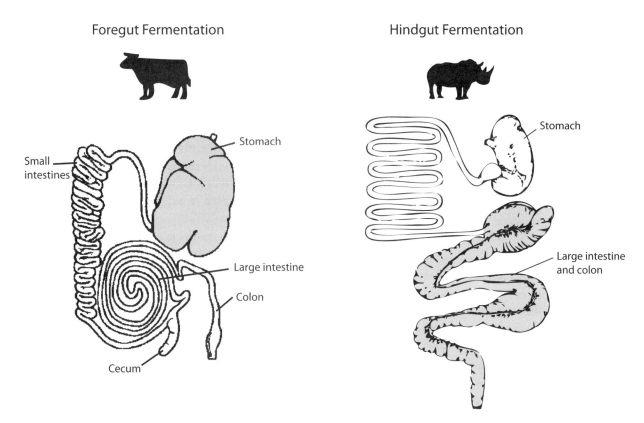

FIGURE 17-10 A comparison of the anatomy of the gastrointestinal tract of an artiodactyl (cow), which relies on a multichambered stomach for foregut fermentation, and a perissodactyl (white rhino), which uses an enlarged cecum and colon for hindgut fermentation.

and had made the lower, more nutritious parts more readily available, the wildebeest, a more selective feeder, moved in. By this time, the zebras were becoming less able to get sufficient quantities of forage and were moving to new tall-grass pastures. A similar replacement of wildebeest by Thompson's gazelles occurred after the wildebeest had removed still more grass and had made available to the small, highly selective gazelles the fruits and leaves of low-growing forbs. Competition between these abundant ungulates is minimized by this grazing pattern, and the activities of the early members of the grazing succession were highly advantageous to the later, more selective members. Bell's study clearly illustrates that differences in the digestive systems of ungulates have pronounced effects on food preferences, migratory patterns, and many basic interactions within a grazing ecosystem.

Living Perissodactyls

The features of several important perissodactylan families illustrate the considerable structural and functional diversity within the group. The dentition and cranial morphology are developed in response to herbivorous feeding habits. Living perissodactyls have elongate skulls because of an enlargement of the facial region to accommodate a full complement of 44 teeth. The teeth are usually lophodont and are either hypsodont in grazing types (all equids and *Ceratotherium* of the Rhinocerotidae) or brachydont in browsers (all tapirids and *Rhinoceros* and *Dicerorhinus* of the Rhinocerotidae). Many postcranial specializations further cursorial ability. The clavicle is absent, and usually the manus has three or four digits and the pes, three digits. In the equids, however, only one functional digit is retained on each foot (Fig. 17-2).

Family Equidae

Horses, the most highly cursorial and graceful of perissodactyls, now occur wild only in Africa, the Middle East, and parts of western and central Asia. In addition, feral populations of domestic horses and burros live in various places. There is but one genus with eight living species.

Wild horses in general are not as large as domesticated breeds. The average weight of a female zebra (*Equus burchellii*) is given by Bell (1971) as 219 kilograms, but some domestic breeds of horses weigh over 1,000 kilograms. The skull has a fairly level profile, and the rostrum is long and deep (**Fig. 17-11**); the dental formula is 3/3, 0–1/0–1, 3–4/3, 3/3 = 36–42. The cheek teeth are hypsodont and have complex patterns on the occlusal surfaces (Fig. 17-4). The limbs are of a highly cursorial type: Only the third digit is

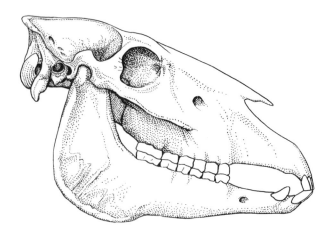

Horse
(*Equus caballus*, Equidae)

FIGURE 17-11   The skull of a horse (length of skull 530 millimeters).

functional; all but the proximal joints largely restrict movement to one plane, and the foot is greatly elongate. In the tarsus, the main weight-bearing bones are the ectocuneiform, navicular, and astragalus. The calcaneum is mostly posterior to the astragalus (Fig. 17-8).

In addition to the horse, living equids include the ass (*E. asinus*; donkey), the onager (*E. hemionus*), the kiang (*E. kiang*), and three species of zebras (Grubb 2005). Although wild horses now occupy only Africa and parts of Asia, within historic times they occurred throughout much of Eurasia. Wild equids inhabit grasslands in areas ranging from tropical to subarctic.

Zebras are native to central and southern Africa. The zebra's stripes (**Fig. 17-12**) have been hypothesized to act as camouflage against the tall grasses of the savannah, to allow individual recognition, and to serve as disruptive coloration, thereby confusing predators or disrupting the visual system of biting tsetse flies (Ruxton 2002).

Zebras are thick bodied, have relatively short legs, and stand approximately 1.5 meters at the shoulder. They are highly social. Like most equids, zebras are polygynous animals (one male controls access to more than one breeding female) that often form large herds consisting of extended family groups or "clans." Within these groups, a social hierarchy exists, usually led by a stallion. Status within the group is maintained by complex behavioral and vocal communication. Asses (*E. asinus*), onagers (*E. hemionus*) and Grevy's zebra (*E. grevyi*) range over vast areas in small family groups that come together and split apart regularly (fission–fusion social systems; Rubenstein & Hack 2004). These species live in xeric regions where patches of grass and water are widely distributed. Furthermore, females do not give birth synchronously, and therefore, lactating or pregnant females must remain closer to water supplies than nonbreeding females. Consequently, females range widely over vast areas so as to reduce competition with one another for the limited grazing and watering sites. Stable female groups do not form in these species. Rather, females live in small fision–fussion social groups. With females dispersed widely across such large home ranges, males cannot defend harems of females. Instead, they must defend resources needed by females, such as waterholes or prime grazing areas (Rubenstein 1989). In contrast, some populations of mountain zebras (*E. zebra*) and wild horses (*E. caballus*) live in more mesic environments where grasses and water supplies are abundant. In these environments, competition among females for food and water is reduced.

FIGURE 17-12   (A) A mountain zebra (*Equus zebra*) in Mountain Zebra National Park, South Africa. (B) A Przewalski's horse (*Equus caballus przewalskii*) stallion from Mongolia.

Predator densities also tend to be higher in these areas, and herds benefit from increased vigilance. Therefore, selection favors the formation of stable female herds that are more easily defended by dominant stallions. Under these conditions, stallions defend groups of females and not their resources.

On the Mongolian steppes, stallions of Przewalski's horse (*E. caballus przewalskii*; Fig. 17-12B) maintain harems in part by scent marking (King & Gurnell 2007). Stallions form stud piles by repeatedly defecating in the same spot within their home range. Harem-holders communicate their status to other stallions by ritually urinating on their dung piles. Stallions also urinate on the dung piles of females within their harem in an apparent attempt to communicate to interlopers that the mare is a member of his harem. On occasion, these female defense signals are ignored and fights occur between stallions.

In polygynous mating systems, where a single male controls access to several females, many males have limited breeding opportunities. These non–harem-holding males group together in small herds of up to 40 "bachelor" males, probably as a defense against predators. In the 25,000 square kilometer Serengeti-Mara ecosystem of Africa, zebra family groups and bachelor herds coalesce into massive herds (often mixed with wildebeest and other ungulates) that move to better grazing areas seasonally.

The genus *Equus* evolved in North America, and crossed into Asia during the late Pliocene glaciation events (beginning around 2.6 million years ago). Thus, since the early Pleistocene, horses and humans coexisted in Eurasia. Horses appear in association with human remains in archaeological sites as far back as 12,000 years ago, but the first archaeological evidence for domestication is from 4,500 year old sites from the Central Asian steppes and the Iberian Peninsula (Jansen et al. 2002; Vila et al. 2001). Analysis of maternally inherited mitochondrial DNA from modern horse breeds, Przewalski's horse, and DNA extracted from horse remains found in the permafrost of Alaska (dating back 12,000 to 28,000 years ago), suggest that horses were domesticated several times. Since their domestication, horses have been an integral part of human culture, providing transportation, serving as draft animals, mounts for military conquest, and companions.

Family Tapiridae

Tapirs occupy tropical parts of the New World and the Malayan area. The family includes one living genus (*Tapirus*) and four species. Structurally, tapirs are notably conservative and share many features with the common ancestors of all perissodactyls.

Tapirs have a stocky build and weigh up to about 320 kilograms (**Fig. 17-13**). The limbs are short, and both the ulna and fibula are large and separate from the radius and tibia, respectively. The front feet have four toes (Fig. 17-2) and vestiges of the fifth (the pollex), and the hind feet have three toes. Tapirs retain a full placental complement of 44 teeth. Three premolars are molariform, and the brachydont cheek teeth retain a simple pattern of cross lophs. The short proboscis and reduced nasals are among the few specializations of tapirs (**Fig. 17-14**). The short, trunk-like proboscis is formed from tissues of the nose and lips along with the elaboration of several facial muscles (Witmer et al. 1999). Retraction of the nasal bones to a position above the orbital cavity (**Fig. 17-15**), and loss of several cartilaginous elements, results in a fleshy proboscis lacking bony or cartilaginous support. Consequently, the proboscis is a muscular hydrostat composed of longitudinal, transverse, and helical muscles that provide a wide range of possible movements (Witmer et al. 1999).

Today, tapirs occupy moist forests, where their primitive feet serve well on the soft soil and their teeth are adequate for masticating plant material that is not highly abrasive. Tapirs are usually found near water. They are rapid swimmers and often take refuge from predators in the water. Tapirs are solitary and nocturnal, and their presence is frequently made known chiefly by their systems of well-worn trails between feeding areas, resting places, and water. Their food is largely succulent plant material, including fruit. Tapir populations have declined in recent years because of deforestation and bushmeat hunting, and all four species are classified as endangered.

Family Rhinocerotidae

The family Rhinocerotidae is represented today by four genera and five species and is restricted to parts of tropical and subtropical Africa and southeastern Asia. These ponderous creatures—the armored tanks of the mammal world—are surviving members of the spectacular late Tertiary and Pleistocene ungulate fauna. Although they had an illustrious past, rhinoceroses are now a declining group.

All Recent rhinoceroses are large, stout-bodied herbivores with fairly short, graviportal limbs (**Fig. 17-16**). Weights range up to about 2,800 kilograms. The front foot has three or four toes (Fig. 17-2), and the hind foot is tridactyl. The nasal bones are thickened and enlarged, often extend beyond the premaxillary bones, and support a horn (**Fig. 17-17**). Where there are two horns, the posterior one is on the frontals. The horns are of dermal origin and lack a bony core; they are convergent with the papillary cornified epidermis of artiodactylan horn sheaths, cetacean baleen, and bird beaks (Hieronymus et al. 2006). The occipital part of the skull is unusually high and yields good mechanical advantage for neck muscles that insert on the lambdoidal crest and raise the heavy head (Fig. 17-17). The incisors and canines

FIGURE 17-13 A Brazilian tapir (*Tapirus terrestris*) from South America. The upper lip is elongate, forming a down-curved proboscis.

are absent in some rhinoceroses and are reduced in number in others; the dental formula is 0–1/0–2, 0/0–1, 3–4/3–4, 3/3 = 24–36. The cheek teeth have a primitive pattern of cross lophs far simpler than that of equids (**Fig. 17-18**).

Rhinoceroses inhabit grasslands, semideserts, savannas, brushlands, forests, and marshes in tropical and subtropical areas. Some species are usually solitary (black rhinoceros, *Diceros bicornis*), whereas others occur in

FIGURE 17-14 A Malayan tapir (*Tapirus indicus*) showing the flexibility of the proboscis.

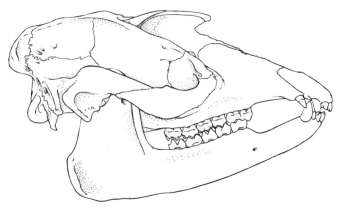

Baird's tapir
(*Taprirus bairdii*)

FIGURE 17-15 Skull of a Baird's tapir (*Taprirus bairdii*) (length of skull 420 millimeters).

FIGURE 17-16   (A) The white rhinoceros (*Ceratotherium simum*) of South Africa. This species had been decimated throughout its range by the beginning of the 20th century. Because of strenuous conservation efforts in South Africa, the species is now well-represented in a number of reserves and parks in southern Africa. The northern subspecies (*C. s. cottoni*) of East Africa, however, is critically endangered. (B) An Indian rhino (*Rhinoceros unicornis*). Note the single horn and the thick, plate-like skin in this species.

family groups (white rhinoceros, *Ceratotherium simum*) or even in groups including up to 24 animals (Heppes 1958). Rhinoceroses are territorial and practice scent marking by establishing dunghills along well-worn trails. A variety of plant material is eaten. White rhinos have a square lip for cropping grasses, and others have a pointed prehensile upper lip adapted to browsing.

Sumatran rhinos (*Dicerorhinus sumatrensis*) are the smallest living rhinos, and their skin is covered with coarse hair. They retain many ancestral characters and are believed to be the oldest of extant rhino species. Sumatran rhinos live largely solitary lives in the dense forests of Borneo and

Sumatra. They are the most vocal of rhinos, and their calls may serve as a means of communication between individuals. Analysis of captive Sumatran rhino vocalizations reveals that they are similar to those of humpback whales (von Muggenthaler at al. 2003). There are three main types of vocalizations; short duration "eeps," complex sounds reminiscent of whale songs, and "whistle-blows." These components are strung together to produce "songs." Whistle-blows have infrasonic qualities and may be useful for communicating in the dense forests (von Muggenthaler et al. 2003).

Adults are nearly invulnerable to predation, except by humans, but young rhinoceroses are occasionally killed

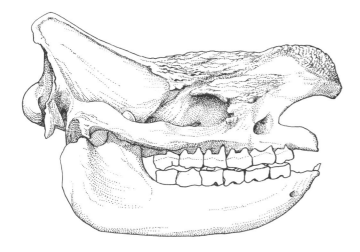

Black rhinoceros
(*Diceros bicornis*, Rhinocerotidae)

FIGURE 17-17   Skull of a black rhinoceros (*Dinoceros bicornis*; length of skull 692 millimeters).

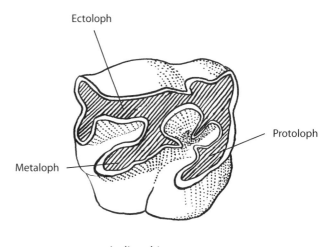

Indian rhinoceros
(*Rhinoceros*, Rhinocerotidae)

FIGURE 17-18   The right upper molar of a rhinoceros.

by lions, spotted hyenas, or tigers. The Asian rhinoceroses (*Rhinoceros* and *Dicerorhinus*) and the African black rhinoceros are facing possible extinction. Because of the supposed medicinal properties of the horn and other parts, rhinoceroses have been hunted persistently for at least 1,000 years. There are thought to be fewer than 300 Sumatran rhinos (*Dicerorhinus*) surviving in highly fragmented populations in Indonesia and Malaysia. Because of strict protection from Indian and Nepalese governments, Indian rhino (*Rhinoceros*) populations have recovered from fewer than 200 to over 2,600 today. The African black rhinoceros continues to decline over broad areas despite efforts to protect it from poaching. It has suffered an estimated 85% decline in the past 2 decades, but recently its numbers have stabilized at just over 3,000 individuals; a thriving population lives in Etosha National Park, Namibia. Regrettably, the future of rhinoceroses seems dim. All five species are currently listed as endangered.

SUMMARY

Perissodactyls (odd-toed ungulates) were the dominant herbivores in the Eocene, Oligocene, and Miocene epochs, with 14 families and many species present, including a rhinoceros (*Indricotherium* = *Baluchitherium*) that stood over 5 meters tall at the shoulder and weighed roughly 30,000 kilograms. Perissodactyls began to decline during the Miocene as artiodactyls diversified. Today there are three living families, with 17 species. They include the lithe and graceful horses, the long-snouted tapirs, and the huge, armored rhinoceroses.

Perissodactyls are browsing and grazing mammals. Their herbivorous diet favored the development of large cheek teeth with complex occlusal surfaces that finely section plant material. Diet puts unusual demands on the digestive systems of ungulates. Vegetation yields little fat and protein, is difficult to digest, and is often protected by defensive secondary compounds. In addition, mammals lack enzymes that digest cellulose; therefore, herbivores must use a fermentation process that breaks down cellulose with the cellulytic enzymes of microorganisms. Microbial fermentation is a relatively slow process requiring specialized fermentation chambers. In perissodactyls, microbial fermentation takes place in the large intestine and enlarged cecum (intestinal or hindgut fermentation).

Perissodactyls and ruminant artiodactyls have evolved contrasting nutritional strategies. Ruminant artiodactyls can satisfy their nutritional needs with relatively unnutritious food and can remain widespread through times when forage is nutritionally poor. During comparable times, perissodactyls must seek sites that support the greatest quantity of vegetation and the most nutritious vegetation, and their range is correspondingly restricted.

Members of the family Equidae are swift runners that inhabit open areas. They tend to form herds of extended family members. Equids are fairly vocal, but much of the communication among the herd is visual (ear twitches, tail positions, and body postures).

Tapirs occupy tropical parts of the western hemisphere and the Malayan area. The family includes one living genus and four species. Structurally, tapirs are notably conservative with the exception of their short, trunk-like proboscis formed from tissues of the nose and lips. Tapirs are solitary and nocturnal. They occupy moist forests, usually near water: They are rapid swimmers and often take refuge from predators in the water.

Four genera and five species of rhinoceroses survive today in restricted parts of tropical and subtropical Africa and southeastern Asia. They are stout-bodied herbivores with short, graviportal limbs. Rhinoceroses are territorial and practice scent marking by establishing dunghills along well-worn trails. One or two horns of cornified epidermis may be present on the dorsal midline of the rostrum. Regrettably, rhinos are often killed solely for their horns and the future of rhinoceroses is uncertain. All five species are currently listed as endangered.

KEY CHARACTERISTICS OF PERISSODACTYLS

- Skull thick, body stocky
- Foot structure mesaxonic (with symmetry of the foot passing through the central, or third, digit)
- Digits bear hooves
- Skull elongated and houses a full set of cheek teeth
- Molars and premolars hypsodont in grazing forms (brachydont in browsers)
- Dermal horns without bony cores located along midline of nasals and frontals in some species
- Stomach simple; cecum enlarged and sacculate for microbial fermentation

KEY TERMS

Cannon bone	Intestinal fermentation	Ruminant
Extension	Microbial fermentation	Ungulate
Flexion	Paraxonic	Unguligrade

RECOMMENDED READINGS

Hieronymous, TL, LM Witmer, & RC Ridgely. 2006. Structure of whole rhinoceros (*Ceratotherium simum*) horn investigated by X-ray computed tomography and histology with implications for growth and external form. *Journal of Morphology, 267*:1172–1176.

Hildebrand, M. 1960. How animals run. *Scientific American, 202*:148–156.

Hildebrand, M. 1987. The mechanics of horse legs. *American Scientist, 75*:594–601.

Prothero, DR & RM Schoch. 1989. *The Evolution of Perissodactyls.* Oxford University Press, New York.

MacFadden, BJ. 2008. *Fossil Horses: Systematics, Paleobiology and Evolution of the Family Equidae.* Cambridge University Press, New York.

Prothero, DR. 2005. *The Evolution of North American Rhinoceroses.* Cambridge University Press, Cambridge, UK.

Artiodactyla

For thousands of years humankind has regarded members of this order as symbols of beauty, grace, strength, and fertility. Domesticated cattle, sheep, and goats have long been close associates of people of many cultures. In some parts of the world today, a family's prestige and social stature depend on the size of its herd of cattle. Among wild artiodactyls are many of the fastest and most enduring runners, the most impressive herd animals, and species variously able to live in most terrestrial environments, including harsh deserts and the bleakest arctic tundra.

Evolution

The earliest fossil artiodactyls were rabbit-sized members of the Diacodexidae, a family that includes *Diacodexis* from the early Eocene of North America (Theodor et al. 2007). Newly discovered fossils of *Diacodexis* (and closely related taxa) also place these early artiodactyls in Europe and Asia during the early Eocene (Theodor et al. 2005, 2007). *Indohyus* (Raoellidae), another early artiodactyl from the Eocene of Asia, was a raccoon-sized animal that was ecologically similar to chevrotains (*Hyemoschus*). *Indohyus* shares several ear and tooth characteristics with ancestral whales and is thought by some researchers to be the sister group to cetaceans (Thewissen et al. 2007). Artiodactyl origins are further complicated by the discovery of early fossil whales with double-trochlea astragali and paraxonic (even-toed) foot symmetry (Gingerich et al. 2001; Thewissen et al. 2001), characters once used to define artiodactyls. Conversely, deciduous teeth of an Eocene artiodactyl show a morphology that was formerly thought to be restricted to Eocene whales (Theodor & Foss 2005).

Just a decade ago, it would have been difficult to convince most mammalogists that a hippopotamus was more closely related to a whale than it was to a horse, but given the new evidence available today, it would be difficult to convince them that this is not true! Molecular genetic data strongly support the hypothesis that whales evolved from within the Artiodactyla, with hippos (Hippopotamidae) as the sister group to cetaceans and the remaining even-toed ungulates in a separate clade (**Fig. 18-1**; Gatsey et al. 1999; Nikaido et al. 1999; among others). The term Cetartiodactyla is used to indicate the shared ancestry of artiodactyls and cetaceans. Morphological and paleontological evidence also recently favors a monophyletic Artiodactyla, with cetaceans as either a sister group to Artiodactyla (Geisler 2001; Thewissen et al. 2007; **Fig. 18-2**) or included within Artiodactyla (Geisler et al. 2007).

The original evidence for the evolution of whales from within the Artiodactyla came from analyses of DNA and protein sequences in these taxa. More recently, molecular systematists using SINEs (short interspersed nuclear elements) provided evidence for a hippo-cetacean clade, with ruminant artiodactyls as a sister-clade (Nikaido et al. 1999; Shedlock et al. 2000). SINEs are short ($<$ 500 bases), repetitive DNA sequences that are occasionally copied and inserted elsewhere in the genome. It is extremely unlikely that identical SINEs will be inserted in the same location by chance. Therefore, taxa that share identical SINEs at the same insertion point in the genome are thought to have acquired this character from a common ancestor (that originally inserted the SINE). Interestingly, hippos, cetaceans, and ruminants share several SINEs not present in suids and camels, suggesting that hippos and suids (pigs) are not as closely related as hypothesized based on morphological evidence.

Combined analyses using morphological, fossil, and molecular data sets also support the hippo-cetacean clade (Geisler et al. 2007; O'Leary & Gatesy 2008; Price et al. 2005). With the exception of Boisserie et al. (2005), morphological studies support two separate clades, the Cetacea and the Artiodactyla (including hippos). Morphologists and paleontologists point out that whale fossils from the Eocene predate the first fossil hippos by approximately 30 million years. Nevertheless, most mammalogists now support the hypothesis of a Cetartiodactyla clade that includes the Cetacea and Artiodactyla as separate clades. Disagreement continues over whether or not cetaceans evolved from

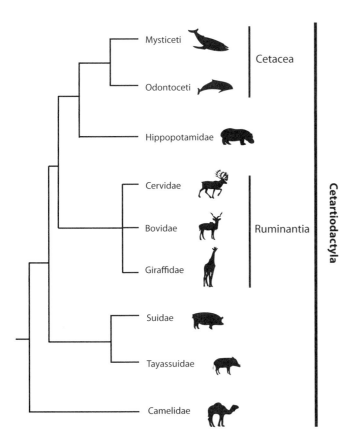

FIGURE 18-1 A phylogenetic hypothesis of the Cetartiodactyla based on molecular data from Price et al. (2005) and O'Leary and Gatesy (2008). Hippos are the sister group of cetaceans. The Artiodactyla, as originally described, is a paraphyletic group.

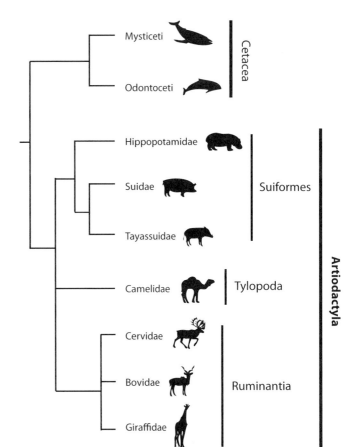

FIGURE 18-2 Traditional phylogenetic hypothesis of the orders Artiodacyla and Cetacea. Three suborders of artiodactyls are shown. Hippos are considered members of the suborder Suiformes, and the Artiodactyla is a monophyletic group.

within artiodactyls, with hippos as the sister group to cetaceans. For now, we retain the traditional classification of the Artiodactyla proposed by Grubb (2005) but acknowledge the possible sister-group relationship of the Hippopotamidae to the Cetacea.

Today, artiodactyls (pigs, camels, deer, antelope, cattle, and their kind) far overshadow perissodactyls in diversity and abundance. Although the perissodactyls appeared in the late Paleocene and reached their greatest diversity in the Eocene, the first artiodactyl is from the early Eocene (Rose 1996; Theodor et al. 2007), and the major artiodactyl radiation occurred in the Miocene (Gentry et al. 1999; Janis 2007; Prothero & Foss 2007). Since the Miocene, the perissodactyls have steadily declined, but the artiodactyls have remained diverse and successful. Of the 36 artiodactyl families present in the Cenozoic, 10 families (and 89 genera) survive to the present. In contrast, the perissodactyls, represented by 14 Cenozoic families, are reduced today to 3 families and only 6 genera. It is tempting to relate the decline of the perissodactyls to the rise of the artiodactyls and to regard the latter as the more effective competitors, but the true story is certainly far more complex than this. Although many structural differences between perissodactyls

and artiodactyls are apparent, the functional advantages conferred by many of the features of the latter are not easily recognized.

Morphology

The food of most artiodactyls is leaves, which are generally high in cellulose, low in protein, and contain defensive compounds, but in ruminant artiodactyls, the enlarged and multichambered stomach harbors microorganisms that break down cellulose. Initially, food passes into the large sac-like **rumen** that serves as a holding chamber and a microbial fermentation vat (see Fig. 2-13). Large, undigested food particles float on top of the fluid in the rumen. During **regurgitation**, these larger plant fibers are forced into the esophagus by a combination of contractions of the diaphragm and peristaltic contractions in the esophagus itself; the fibers finally reach the mouth, where they are rechewed. The animals "chew their cud." Following the remastication of the cud, it is swallowed a second time. The well-digested, finer particles of food are drawn from the **reticulum** into the **omasum** by the negative pressure caused by a relaxation of the omasum walls. Contractions of the omasum then force the slurry into the **abomasum**. The abomasum is glandular and functionally equivalent to the true stomach of other mammals. Gastric fermentation (also called foregut fermentation) has evolved independently in many leaf-eaters, including sloths, some leaf-eating primates, peccaries, camels, hippopotami, many rodents, hyraxes, and kangaroos.

The complex system of recycling and reconstituting food in the stomach enhances the nutritional yield of poor-quality food in four ways. First, gastric fermentation releases proteins, carbohydrates, and lipids early in the digestive process so that they can be efficiently absorbed by the intestines. Second, remastication results in more complete breakdown of cell walls. Third, nitrogen, a byproduct of cellulose digestion, is used by the microbes to make their own proteins. The ruminant periodically flushes these microbes into its intestines and thereby digests the proteins formed by the microbes. Finally, ruminants can quickly gather large volumes of plant material into the huge stomach and digest it later. As might be expected, ruminants can subsist on remarkably low-quality food.

There are disadvantages to gastric fermentation, however. The ruminant's food is digested slowly, and its rate of passage through the gut is slow. Food takes 70 to 100 hours to pass through the gut of a cow compared with only 30 to 45 hours for a nonruminant horse. Thus, when food is abundant, nonruminants, such as horses, consume greater quantities of food each day, process the easily digestible portion, and excrete the rest.

In the order Artiodactyla, the structure of the foot is especially diagnostic. The foot is paraxonic; that is, the plane of symmetry passes between digits three and four (**Fig. 18-3**). The weight of the animal is borne primarily by these digits: The first digit is always absent in living members, and the lateral digits (two and five) are always more or less reduced in size. Four complete and functional digits occur in the families Suidae, Hippopotamidae, and Tragulidae and in the forelimb of the Tayassuidae (the hindlimb has the medial digit suppressed). Two complete toes, with the lateral digits absent (Camelidae, some Bovidae, Antilocapridae, and Giraffidae) or with incomplete remnants of the lateral digits (Cervidae and some Bovidae), occur in the more cursorial families. The cannon bone (fused third and fourth metapodials) is present in the families Camelidae, Cervidae, Giraffidae, Antilocapridae, and Bovidae. Typically, the terminal phalanges are encased in pointed hoofs. The limbs have springing ligaments (**Figs. 18-4** and **18-5C**), and the astragalus has a "double pulley" or "double-trochlea" arrangement of articular surfaces (**FIG, 18-6**) that completely restricts lateral movement. The proximal articulation (with the tibia) and the distal articulation (with the navicular and cuboid, which are fused in many advanced types) of the astragalus are critical in allowing great latitude of foot and digit flexion and extension, as is the extension of the articular surfaces and keels on the distal ends of the cannon bones to the anterior surfaces (Fig. 18-5A, B). The distinctive artiodactyl astragalus is regarded by some as a key to the success of the group.

The limbs of artiodactyls, especially the distal segments, are usually elongate and fairly slim. The femur lacks a third

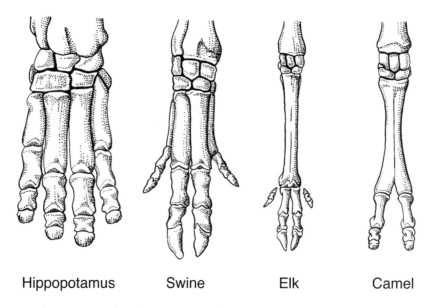

Hippopotamus   Swine   Elk   Camel

FIGURE 18-3 Right front feet of four artiodactyls: a hippopotamus (*Hippopotamus amphibius*), swine (*Sus scrofa*), elk (*Cervus elaphus*), and camel (*Camelus dromedarius*). (Adapted from Howell, A. B. *Speed in Animals*. University of Chicago Press, 1965.)

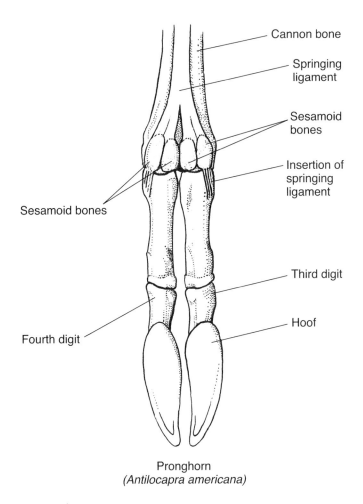

Cannon bone

Springing ligament

Sesamoid bones

Insertion of springing ligament

Sesamoid bones

Third digit

Hoof

Fourth digit

Pronghorn
(Antilocapra americana)

FIGURE 18-4 Posterior view of the left hind foot of a pronghorn (*Antilocapra americana*), showing the position of the springing ligament.

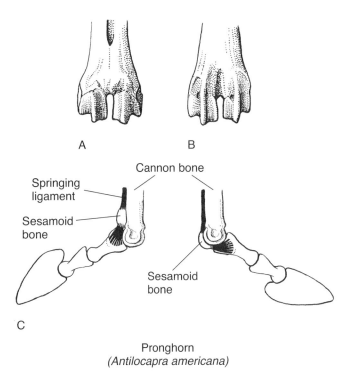

A   B

Springing ligament

Cannon bone

Sesamoid bone

Sesamoid bone

C

Pronghorn
(Antilocapra americana)

FIGURE 18-5 Left hind foot of the pronghorn (*Antilocapra americana*). (A) Anterior view of the distal end of the cannon bone. (B) Posterior view of this bone. (C) Position of the phalanges: when the foot is supporting the weight of the body and the springing ligament (shown in black) is stretched and when the foot leaves the ground and the springing ligament flexes the phalanges.

trochanter. Whereas this prominence serves as a point of insertion of gluteal muscles in perissodactyls, in artiodactyls, these muscles insert more distally, on the tibia. The distal parts of the ulna and the fibula are usually reduced and may fuse with the radius and tibia, respectively; this fusion is associated with the restriction of limb movement to one plane. The clavicle is seldom present. The intrinsic muscles of the feet (those that both originate and insert on the feet) are usually absent, being replaced by specialized tendons and ligaments.

The skull usually has a long preorbital section, and a postorbital bar or process is always present. Horns, always of bone or with a bony core, are most often borne on the frontal bones, which are enlarged at the expense of the parietal bones. The teeth are brachydont or hypsodont and vary from 30 to 44 in number. The crown pattern is bunodont or, more often, selenodont. The premolars are not fully molariform, in contrast to the perissodactyl situation, and considerable specialization of the anterior dentition occurs in derived artiodactyls.

The classification used here, that of Grubb (2005), recognizes eight living families, all of which except Hippopotamidae are first known from the Eocene.

Family Camelidae

Camelids are ruminants that are restricted to arid and semiarid regions. *Camelus*, with two species, occupies the Old World, and wild populations persist in the Gobi Desert of Asia. A single species of *Lama* (guanaco) occurs in South America from Peru through Bolivia, Chile, Argentina, and Tierra del Fuego. The genus *Vicugna* contains one species restricted to the Andes Mountains south of Peru (**Fig. 18-7**). Camelids first appear in the middle Eocene in North America and remained there for 35 million years until migrating to Eurasia in the late Miocene and reaching South America relatively recently, in the late Pliocene (2 or 3 million years ago).

Of special interest, as an example of a reversal of a well-established evolutionary trend, is the development of the camelid foot. By the Oligocene, the camel's foot was already highly specialized. It was nearly unguligrade in posture and didactyl, and the distal-most phalanges probably bore hoofs. The distinctive distal divergence of the metapodials (Fig. 18-3), however, was already

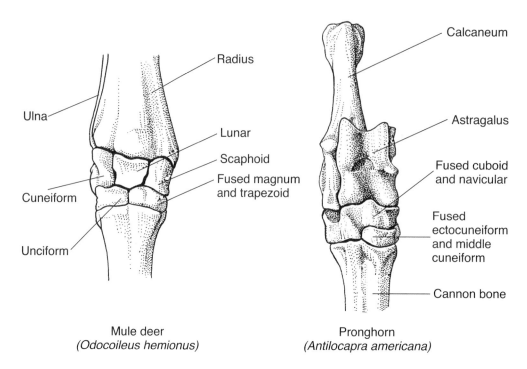

FIGURE 18-6 The right carpus of a mule deer (*Odecoileus hemionus*) and the right tarsus of a pronghorn (*Antilocarpa americana*).

recognizable. In the Miocene, the central metapodials fused to form a cannon bone. However, at this same time, a retrograde trend toward the secondary development of a digitigrade foot posture began, and from Pliocene times onward, camels were digitigrade. They are the only extant fully digitigrade ungulates. Because semiarid conditions developed and became widespread in the Miocene, it is tempting to relate the changes in the camelid foot posture to changing soil conditions. In any case, the camelid foot clearly provides effective support on soft, sandy soil, into which the feet of "conventional" unguligrade artiodactyls sink deeply.

FIGURE 18-7 (A) A one-humped camel (*Camelus dromedarius*) roaming the deserts of Qatar. (B) The vicugna (*Vicugna vicugna*), a camelid that inhabits the central Andes of South America.

Although highly specialized in foot structure, camelids are the least derived living ruminants. They are large mammals, ranging in weight from about 60 to 650 kilograms, and have long necks and long limbs. The dentition has evolved less toward herbivorous specialization than has that of the Ruminantia. In camelids, only the lateral upper incisor is present, but it is caniniform (**Fig. 18-8**); the lower canines are retained and are little modified. The lower incisors are inclined forward and occlude with a hardened section of the gums on the premaxillary bones. A broad diastema is present, and the premolars are reduced in number (to 3/1–2 in *Camelus* and 2/1 in *Lama*). As in other ruminants, the limbs are long and the ulna and fibula are reduced. The trapezium is absent in the carpus, and the mesocuneiform and ectocuneiform are fused in the tarsus. The digitigrade feet are didactyl, but the cannon bone is distinctive in that the distal ends of the metapodials remain separate and flare outward (Fig. 18-3). The toes are separate, and each is supported by a broad cutaneous pad that largely encases the second phalanx and greatly increases the surface area of the foot. The short ungual phalanges do not bear hoofs but have nails on the dorsal surfaces.

Camels are remarkably well adapted to arid areas. They have the ability to go without drinking free water for months (Chapter 21). They conserve water by producing dry feces and little urine. They allow their body temperature to rise 6°C during the heat of the day. Camels have specialized nasal cavities that reduce water loss via evaporation. In addition, camels can store fat in their humps for use during food shortages (Fig. 18-7A; MacDonald 1984). They are grazers and can survive in regions with only sparse vegetation. Wild Bactrian camels (*Camelus bactrianus*) still roam the inhospitable deserts of northwestern China and Mongolia, including a remote area that was formerly a nuclear test range. Their numbers have declined dramati-

cally in recent years, and they are now listed as critically endangered.

The guanaco (*Lama glama*) and vicugna (*Vicugna vicugna*) of South America are gregarious and live in small social groups dominated by an adult male. Guanacos are fairly speedy runners but are especially adroit at moving rapidly over extremely rough terrain (Fig. 18-7B). South American camelids live on high-altitude, grassy plains where evening temperatures routinely drop below freezing. Their wooly fur consists of hollow hairs that interlock to provide excellent insulation. Domesticated lamas and alpacas are thought to be the descendants of wild guanacos and vicugnas (Kadwell et al. 2001; Marín et al. 2008). The Inca first domesticated these animals for fiber production. Hunting was prohibited by Inca royalty, but after Spanish conquest, hunting for food and fiber decimated vicugna populations. By the early 1970s, when they were formally listed as endangered, only a few thousand vicugnas remained. Thanks to the establishment of protected areas, vicugna populations have recovered; there are now an estimated 125,000 animals in the wild (Torres 1992).

▪ Family Suidae

Swine are omnivorous and lack many structural modifications typical of more specialized artiodactyls. The Suidae is an Old World family whose present distribution includes much of Eurasia and Africa south of the Sahara. There are 18 Recent species in five genera. Suids appeared in the late Eocene of Asia (Ducrocq et al. 1998; Harris & Liu 2007; Liu 2001). The entelodonts (Entelodontidae), an early branch on the artiodactyl evolutionary line, were huge, somewhat pig-like creatures with skulls up to 1 meter in length.

Most suids resemble the domestic swine (*Sus*). Adults may weigh as much as 275 kilograms and typically have thick, sparsely haired skin. The skull is long and low and usually has a high occipital area and a concave or flat profile (**Fig. 18-9**). The large canines are ever growing, and the upper canines form conspicuous tusks that protrude from the lips and curve upward. In male *Babyrousa*, an Indonesian suid, the upper canines are rotated upward to protrude from the top of the snout (**Fig. 18-10**). Suid molars are bunodont, and the last molars are often elongate, with many cusps and a complexly wrinkled crown surface (**Fig. 18-11**). The dental formula is variable even within a species; the total number of teeth ranges from 34 to 44. The limbs are usually fairly short, and their four-toed feet lack cannon bones (Fig. 18-3).

Swine inhabit chiefly forested or brushy areas, but the warthog (*Phacochoerus*) favors savanna or open grassland and is almost entirely herbivorous (**Fig. 18-12**). Most suids are gregarious, and some assemble in groups of up to 50 individuals. Most species eat a broad array of plant food and carrion and, given the opportunity, kill and eat such

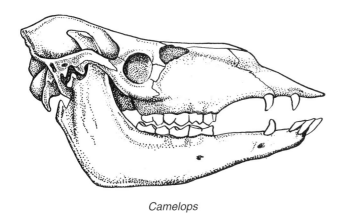

Camelops

FIGURE 18-8 Skull of an extinct Pleistocene New World giant llama (*Camelops*); length of skull 565 millimeters. (Modified from Romer, A.S. *Vertebrate Paleontology*. University of Chicago Press, 1966.)

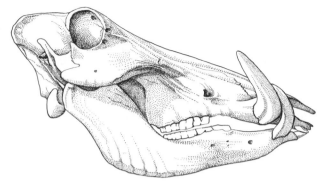

Warthog
(*Phacochoerus aethiopicus*)

FIGURE 18-9 Skull of the African warthog (*Phacochoerus africanus*; length of skull 376 millimeters).

animals as small rodents and snakes. By comparison with ruminant artiodactyls, cursorial ability in suids is modest. Warthogs are fairly swift, however, and escape their predators by speed and by taking refuge in burrows.

FIGURE 18-10 A male babirusa (*Babyrousa celebensis*) showing the massive tusks.

Swine
(*Sus scrofa*, Suidae)

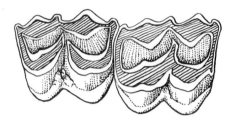

Elk
(*Cervus elaphus*, Cervidae)

FIGURE 18-11 Second and third right upper molars of the swine and the comparable teeth of an elk (*Cervus elaphus*), with the enamel ridges unshaded, the enamel-lined depressions stippled, and the dentine cross-hatched. The molars of *Sus scrofa* are bunodont; those of *Cervus* are selenodont.

Domestication of wild suids probably took place between 8,000 and 10,000 years ago. Two centers of domestication have been proposed based on archaeological evidence—one in western Eurasia and one in eastern Asia. Recent phylogeographic analyses of genetic data, however, suggest multiple centers of pig domestication. According to this hypothesis, wild boars were domesticated independently in Turkey (or Iran), Europe, India, and China (Larsen et al. 2007).

Family Tayassuidae

Tayassuids, usually called peccaries or javelinas (**Fig. 18-13**), are restricted to the New World, where they occur from the southwestern United States to central Argentina. The fossil record of tayassuids begins in the Eocene. Presumably, peccaries evolved from Old World suids, but they are not known as fossils from the Old World (Harris & Liu 2007). Tayassuids colonized South America when the Isthmus of Panama united North and South America about 2.5 million years ago.

There are but three currently recognized species of three genera (*Pecari*, *Tayassu*, and *Catagonus*). The Chacoan peccary (*Catagonus wagneri*) was first known from Pleistocene fossils and was for many years thought to be extinct, but Wetzel et al. (1975) reported a surviving

FIGURE 18-12 A warthog (*Phacochoerus africanus*) showing the prominent tusks and broad snout.

population in the biologically poorly known Gran Chaco area of Paraguay.

Tayassuids are much smaller than suids; the weight of peccaries ranges up to about 30 kilograms. The skull has a nearly straight dorsal profile, and the zygomatic arches are unusually robust (**Fig. 18-14**). The canines are sharp and long and are directed slightly outward; the upper canines never turn upward, however. These opposing canines slide against one another, and the anterior surface of the upper and posterior surface of the lower are planed flat by this contact. These interlocking canines form "occlusal guides" that, together with the hinge-like jaw joint, stabilize the jaw joint against forces generated when hard nuts or seeds are cracked by the rear teeth (Kiltie 1981). The molars are roughly square and have four cusps; they have thick enamel and lack the complex wrinkled and multicusped pattern typical of suids. The dental formula is 2/3, 1/1, 3/3, 3/3 = 38.

Peccaries are more progressive in limb structure than are suids and are less carnivorous. The feet of peccaries are slender and appear delicate, and the side toes are small relative to those of suids and usually do not reach the ground. Of the three living genera of peccaries, *Catagonus* is the most cursorial. All three genera have four toes on the front foot. Although in *Catagonus* digits two and five of the hind foot are vestigial (they lack phalanges and hoofs), digit two is complete, and digit five is vestigial in *Tayassu*. In all genera, the medial metatarsals are partly fused. An additional difference is the more elongate distal segments of the limbs in *Catagonus* (Wetzel 1977). Modern peccaries are not as cursorial as was *Mylohyus*, an extinct Pleistocene species of North America in which the side toes of the forefoot were very strongly reduced and the didactyl hind foot had a fully developed cannon bone.

Peccaries occupy diverse habitats, from deserts and pinyon-juniper foothills in Arizona to dense tropical forests and thorn scrub in southern Mexico, Central America, and South America. They are highly social, with each social group containing up to 12 individuals. Peccaries are omnivorous but seem to rely more heavily on plant material than do suids. The Chacoan peccary lives in small groups that range over 2 kilometers per day in search of ground cacti (Mayer & Brandt 1982; Taber et al. 1993). The presence of peccaries is often indicated by shallow excavations where roots have been exposed beneath bushes or patches of prickly-pear cactus. Despite their chunky build, peccaries are fairly rapid and extremely agile runners.

FIGURE 18-13 A collared peccary (*Pecari tajacu*; Tayassuidae).

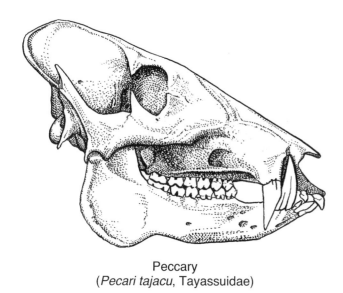

Peccary
(*Pecari tajacu*, Tayassuidae)

FIGURE 18-14 Skull of a collared peccary, *Pecari tajacu* (length of skull 225 millimeters).

Ruminants

Suborder Ruminantia includes giraffes, deer, antelope, sheep, goats, and cattle. Members of this most derived artiodactylan suborder have been in the past and remain the dominant artiodactyls. In general, these animals are committed strictly to an herbivorous diet and to highly cursorial locomotion. Ruminants chew their cud; the stomach has three or four chambers (see Fig. 2-13) and supports microorganisms that have cellulolytic enzymes. Ruminants have selenodont molars (Fig. 18-11), and the anterior dentition is variously specialized by loss or reduction of the upper incisors, by the development of incisiform lower canines, and commonly by the loss of upper canines. The skull differs from those of suids in the exposure of the mastoid bone between the squamosal and exoccipital bones. Antlers or horns, often large and complex structures, are present in some families. In the limbs, there is a pronounced trend toward elongation of the distal segments, fusion of the carpals and tarsals, and perfection of the two-toed foot. A diagnostic feature of the ruminants is the fusion of the

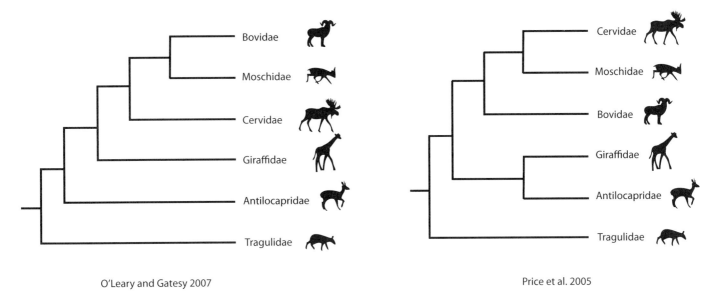

O'Leary and Gatesy 2007

Price et al. 2005

FIGURE 18-15 Two phylogenetic hypotheses of ruminant artiodactyls based on the work of O'Leary and Gatesy, 2008 (left), and Price et al., 2005 (right). Both hypotheses suggest that chevrotains and mouse deer (Tragulidae) are the sister group to other ruminant artiodactyls.

navicular and cuboid bones, over which the astragalus is nearly centered (Fig. 18-6).

The most derived artiodactyls (Cervidae, Giraffidae, Antilocapridae, and Bovidae) share a series of progressive features. The upper incisors are absent. The upper canines are usually absent. The lower canines are incisiform, and the cheek teeth are selenodont. The dental formula is typically 0/3, 0/1, 3/3, 3/3 = 32. The cannon bone is present in fore and hindlimbs. Its distal articular surfaces are extensive, and the lateral digits are always incomplete (Fig. 18-3) and are often lacking. Movement of the foot is strongly limited to a single anterior-posterior plane by the tongue-in-groove contacts between the astragalus and the bones with which it articulates and by the specialized articular surfaces of the joint between the cannon bone and the first phalanges (Fig. 18-5). Some fusion of carpal elements always occurs and further restricts movement to a single plane of motion. The navicular and cuboid bones are always fused (Fig. 18-6), and a variety of fusion patterns occur in the other elements. The four-chambered stomach is of a ruminant type. Although all ruminants but the tragulids share this basic structural plan, each family has distinctive features, usually related to diet and degree of cursorial ability. A number of phylogenetic hypotheses have been offered to explain the evolution of ruminant artiodactyls (**Fig. 18-15**; see also Métais & Vislobokova 2007, for the numerous extinct ruminants), but there is little consensus.

■ Family Tragulidae

The family Tragulidae, which contains the chevrotains (mouse deer), has only three living genera with eight species, but is of interest because these animals are the least derived extant ruminants and probably resemble in many ways the ancestors of other ruminants (**Fig. 18-16**; Métais & Vislobokova 2007; Rössner 2007). Chevrotains and mouse-deer are small, delicate creatures, weighing from 2.3 to 4.6 kilograms, that occur in tropical forests in central Africa (*Hyemoschus*) and in parts of southeastern Asia (*Tragulus*). The tragulids are long-term survivors of the basal radiation of ruminants. Their fossil record begins with *Krabitherium* in the late Eocene of southeast Asia (Metais et al. 2007). Tragulids were widespread and successful until the Miocene; the few living species remain relatively unchanged since that time.

Although apparently related to more derived ruminants, chevrotains combine a unique complex of features. The skull never bears antlers, but seemingly in compensation, the upper canines are tusk-like and are used by males in intraspecific combat. Otherwise, the dentition resembles that of higher ruminants: The upper incisors are lost, the lower canine is incisiform, and the cheek teeth are selenodont. Unique to tragulids is an ossified plate, derived from an **aponeurosis** (a membranous sheet of tendon) to which the sacral vertebrae attach. The limb structure is a mosaic of ancestral and derived features. Although the limbs are long and slender and the carpus is highly specialized in having the navicular, cuboid, and ectocuneiform bones fused, the lateral digits are complete, a condition never present in the more derived ruminants. In addition, although a cannon bone occurs in the hindlimb, the metacarpals of the central digits are separate in the African tragulid and are partly fused in the Asian form, whereas the cannon bone is represented by fully fused metapodials in all other ruminants.

FIGURE 18-16 A mouse-deer (*Tragulus*) from Southeast Asia.

Tragulids are secretive, chiefly solitary, nocturnal creatures that inhabit forests and underbrush and thick growth along water courses. They escape predators by darting along diminutive trails into dense vegetation or into water in the case of the water chevrotain (*Hyemoschus aquaticus*). Their food consists of grass, the leaves of shrubs and forbs, and some fruit. They have a four-chambered stomach that is not as specialized as in other ruminants.

■ Family Antilocapridae

This family is represented today by one species, the pronghorn (*Antilocapra americana*; **Fig. 18-17**), which occupies open country from central Canada to north central Mexico. Antilocaprids probably arose from an ancestry among pecorans in Asia in the late Oligocene, but the fossil record of these animals is entirely North American and begins in the early Miocene (Davis 2007). The horns of the earliest antilocaprids were cervid-like in form; that is, they were forked and most had a basal burr. In contrast to cervids, however, the cores were never shed, and both sexes had horns. In addition, the unusually prominent orbits were situated high and far back in the skull, as in *Antilocapra*. The horns of fossil pronghorns were generally more complex

FIGURE 18-17 The pronghorn (*Antilocapra americana*), one of the fastest cursorial mammals. The long, slender limbs are typical of the more cursorial ungulates.

than those of *Antilocapra*, with two-, four-, and six-horned, and even spirally horned genera during the Miocene and Pliocene, but the limbs and teeth of fossil forms were similar to those of the present species. The pronghorn fauna was at one time more diverse than it is now. From the middle Pleistocene Tacubaya Formation in central Mexico, which contains numerous fossils of mammals that lived in a small area at one time, Mooser and Dalquest (1975) list four species of extinct pronghorns, ranging in size from a tiny species to one at least as large as the living species.

The pronghorn is unique in being the only mammal that sheds its horn sheaths annually. The sheaths are of keratinized skin. The old sheath, beneath which a new sheath is beginning to develop, is shed annually in early winter, and the new sheath is fully grown by July (O'Gara et al. 1971). Whereas the mature sheath is forked, the bony core is a single, laterally compressed blade. Both sexes have horns, but those of the females are small and inconspicuous.

The dental formula is that typical of ruminants, and, although the animals are largely browsers on low shrubs and forbs, the cheek teeth are high crowned. The abrasive soil particles adhering to the low-growing vegetation they eat probably make high-crowned teeth advantageous. Perhaps as an adaptation allowing a pronghorn to watch for danger while its head is close to the ground, the orbits are unusually far back in the skull (**Fig. 18-18**). The legs are long and slender, and all vestiges of lateral digits are gone. The tarsus distal to the astragalus and calcaneum consists of only three bones (Fig. 18-6).

Pronghorns inhabit open prairies and deserts that support at least fair densities of low grasses, shrubs, and forbs. The numbers of pronghorn were seriously reduced during the pioneering period of the western United States. Today pronghorns are common in a number of the western states (O'Gara 1978), but they have been all but eliminated from much of their original range in states such as Kansas. They are among the fastest of cursorial mammals; even the very young ones run for miles learning a crucial survival skill (Byers 2003). At full speed on level footing, adults can attain a speed of at least 85 kilometers per hour (53 mph; Einarsen 1948; Kitchen 1974; McLean 1944). Interestingly, the need for this speed probably evolved under selection pressure from the pursuit predator *Miracinonyx*, the "American cheetah" during the Plio-Pleistocene. *Miracinonyx* became extinct at the end of the Pleistocene, but the high cursorial ability of the pronghorn continues in the predators' absence (Byers 1997).

Although it may yield top speed to the cheetah, in high-speed endurance running, the pronghorn is probably unsurpassed by any other mammal. Pronghorns can sustain high speeds (about 65 kilometers per hour or 40 mph) over distances greater than 10 kilometers (McKean & Walker

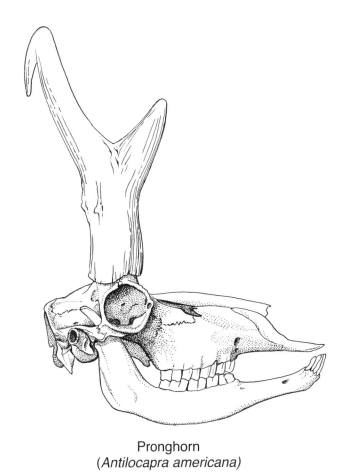

Pronghorn
(*Antilocapra americana*)

FIGURE 18-18   Skull of a pronghorn, *Antilocapra americana* (length of skull 292 millimeters).

1974). Hence, they can cover 11 kilometers in only 10 minutes. This performance demands a rate of oxygen uptake over three times that predicted for a similar-sized cursor. This aerobic ability is not facilitated by unusual locomotor efficiency (energy cost per unit of distance covered) or unusually large muscles, but by an integrated suite of muscular and respiratory adaptations that provide for extremely rapid and prolonged oxygen uptake (Lindstedt et al. 1991). Compared with other similar-sized mammals, pronghorns have enlarged airways to the lungs (trachea and bronchi), greater lung surface areas, higher concentrations of hemoglobin, and greater densities of muscle capillaries and mitochondria.

Family Giraffidae

This family is represented today by but two monotypic genera, *Giraffa* and *Okapia*. The family occurs in much of Africa south of the Sahara.

The robust cheek teeth of giraffids are brachydont and are marked with rugosities. Short horns covered with furred skin are borne on the front part of the parietals, and a medial thickening of the bone in the area where the nasals and the frontals join is conspicuous (**Fig. 18-19**). The prominent horns (called ossicones) are formed from ossified cartilage. In some populations from north of the equator in East Africa, this thickening produces a median horn. Horns occur in both sexes and are never shed. The lateral digits of the elongate limbs are entirely gone, and the tarsus is highly specialized. Distal to the astragalus and calcaneum, only two tarsal bones are present. One is formed by the fusion of the navicular and cuboid bones and the other by the fusion of the three cuneiform bones. The okapi lacks the extreme elongation of the neck and legs that is typical of giraffes but has an even more specialized tarsus in which all bones distal to the astragalus and calcaneum are fused. The fossil record of the giraffids begins in the Miocene Epoch of Africa, with many extinct forms having occurred also in Eurasia (Solounias 2007a). The okapi is remarkable in its close resemblance to primitive late Miocene and early Pliocene giraffids long known to paleontologists (Palaeotraginae).

Giraffes occur in savannas, semideserts, and lightly wooded areas, where their exceptional height enables them to browse on branches of leguminous trees up to 6 meters (19 feet) above the ground (**Fig. 18-20**). Some of the species of *Acacia* that these animals feed on bear long thorns, but giraffes adroitly use their long tongue (up to 45 centimeters long or 18 inches) and prehensile upper lip to gather leaves from even the thorniest acacias. *Acacia* leaves are high in protein and are the most preferred and important food of giraffes (Sauer et al. 1982). Giraffes have functionally modified their last cervical and first thoracic vertebrae in a

FIGURE 18-20 A giraffe (*Giraffa camelopardalis*) among umbrella trees (*Acacia tortilis*) in Tsavo West National Park, Kenya. The markings on the body enable this huge animal to blend inconspicuously into savanna vegetation.

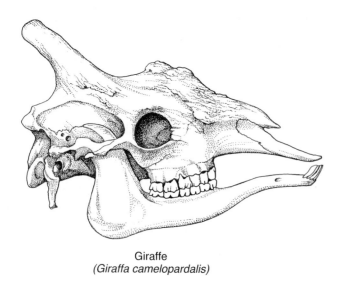

Giraffe
(*Giraffa camelopardalis*)

FIGURE 18-19 Skull of a male giraffe, *Giraffa camelopardalis* (length of skull 708 millimeters). The heavy deposits of bone on the frontal and nasal bones seemingly protect the skull when the head is used as a weapon in fights between males.

unique manner, repositioning the base of the neck farther back on the torso so that the forelimbs appear to protrude forward (Solounias 1999) and shifting the center of balance for running. Despite their considerable weight (to nearly 1,820 kilograms in males), giraffes can gallop for short distances up to 60 kilometers per hour. Relative to those of lighter, shorter limbed artiodactyls, the limbs of giraffes are flexed little during each stride, producing a stiff-legged gait. When walking, giraffes swing both legs from the same side forward at nearly the same time, but when galloping, the hind legs swing forward together. When the animal is walking or galloping, its center of gravity is partly controlled by fore-and-aft movements of the head and neck (Dagg 1962). Ritualized fighting by males involves powerful blows by the head against the opponent's head, neck, and body.

Okapis are secretive forest animals first described in 1901. They live in dense tropical forests of central Africa. They resemble a dark brown horse with a long neck and horizontal white stripes on the legs. Males have hair-covered horns up to 15 cm in length that project rearward from the frontal bones. Females lack horns and are generally taller than males. Okapis have long tongues that are used for grasping vegetation from shrubs and trees. Okapis generally live solitary lives, coming together for mating or to share mineral deposits at specific sites in the forest. Newborn okapis are highly precocial and begin following the mother within hours after birth. After a few days, the young bed down in a hiding spot where they will spend up to 80% of each day during the next two months (Bodmer & Rabb 1992).

Family Cervidae

Members of family Cervidae, which includes the muntjacs, deer, elk, caribou, and moose, occur throughout most of the New World and in Europe, Asia, and northwestern Africa; they have been introduced widely elsewhere. Grubb (2005) recognized 19 genera and 51 species. Cervids appeared in the early Miocene in Eurasia, reached North America in the late Miocene, and entered South America in the late Pliocene or early Pleistocene over the Panama land bridge (Janis & Scott 1987).

Antlers are the most widely recognized characteristic of members of the family Cervidae (**Fig. 18-21**). Antlers are branched, horn-like structures that are lost and grow anew annually. Antlers attain spectacularly large size in some species and vary widely from one species to another. People have long been fascinated by their complexity, variety, and symmetry. Antlers occur only in males, except in caribou (*Rangifer*; Fig. 18-21C). In some antlered cervids, the upper canines are retained but reduced (as in the elk, *Cervus elaphus*). Two cervids with short antlers have enlarged canines (*Elaphodus* and *Muntiacus*; Fig. 18-21B). In one deer, there are no antlers (*Hydropotes inermis*), but the canines are enlarged sabers. Deer antlers undergo annual cycles of regeneration, and grow from **pedicles** (permanent extension of the frontal bone) into large branched structures used for

FIGURE 18-21 Four members of the Cervidae. (A) A moose (*Alces americanus*) from Wyoming. (B) A Reeve's muntjac (*Muntiacus reevesi*) from China. (C) A caribou (*Rangifer tarandus*; also known as reindeer) from Alaska. (D) A North American elk (*Cervus elaphus*) from Yellowstone National Park, Wyoming.

fighting and/or display (**Fig. 18-22**; Price & Allen 2004). Of particular interest is the annual cycle of rapid growth of the antlers, their use during the breeding season in ritualized social interactions, and their subsequent loss.

This annual cycle has been thoroughly studied in the white-tail deer (*Odocoileus virginianus*) of North America (Goss 1983; Wislocki et al. 1947) and in red deer (called elk in North America, *Cervus elaphus*; Price & Allen 2004). The cycle is primarily under the control of testicular and pituitary hormones. In the Northern Hemisphere, secretions from the pituitary, activated by increasing day length in the spring, initiate antler growth in April or May. Some time later, pituitary gonadotropin stimulates growth of the testes and subsequent production of androgens. Until recently, the assumption was that testosterone levels were largely responsible for the cycle of regeneration. Testosterone's effects may be indirect, however, for the effect may occur after testosterone has been converted to estrogen by aromatase (summarized in Price & Allen 2004). In response to circulating hormones, antler growth begins in the spring and early summer. The antlers of elk and moose may grow at a rate of up to 2 centimeters per day during this phase (Price & Allen 2004). The growing antlers are covered by "velvet," a fur-covered skin that carries blood vessels and nerves. In the fall, circulating hormones inhibit the action of the pituitary antler-growth hormone, leading to the drying and loss of the velvet. Then the animals rub and thrash their antlers against vegetation, and as the velvet is removed, the antlers are stained by resins and take on a brown, polished look. In the fall and early winter, androgens maintain the connection between the dead bone of the antlers and the live frontal bones, and during the fall breeding season, the antlers are used in clashes between males competing for females. In winter, pituitary stimulation of the testes declines as day length is reduced, and androgen secretion diminishes. This results in decalcification in the pedicel, weakness at the point of connection between the antler and the pedicel, and shedding of the antlers. For several months in late winter, before reinitiation of antler growth, the males are antlerless.

The cheek teeth of cervids are brachydont, reflecting a browsing habit. Body sizes of cervids range widely: The pudu deer (*Pudu mephistophiles*) weighs from 7 to 10 kilograms, whereas the moose (*Alces alces*) weighs up to roughly 800 kilograms. The feet are always four-toed, but the lateral toes are often greatly reduced. Distal to the astragalus and calcaneum, the tarsus is usually composed of three bones: the fused navicular and cuboid, the fused ectocuneiform and mesocuneiform, and the internal cuneiform (Fig. 18-6).

Cervids occur from the Arctic to the tropics. Many are well-adapted to boreal regions and occupy mountainous or subarctic areas with severely cold winters. Effective insulation is provided in many cervids by the long, hollow hairs of the pelage. Some species are gregarious for much of the year and may assemble in large herds during the winter and during migratory movements. Annual migrations from high elevations with deep snow to lower elevations are common among North American deer (*Odocoileus*) and elk.

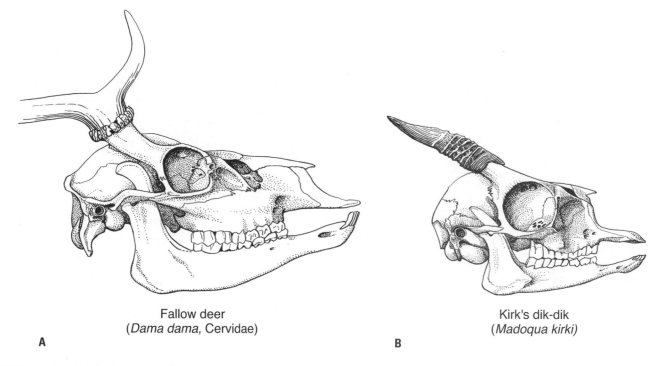

Fallow deer
(*Dama dama*, Cervidae)

A

Kirk's dik-dik
(*Madoqua kirki*)

B

FIGURE 18-22   (A) Skull of a male fallow deer, *Dama dama* (length of skull 265 millimeters). The bony antlers are shed yearly. (B) The skull of a Kirk's dik-dik, *Madoqua kirki* (length of skull 108 millimeters). The receding nasal bones of the dik-dik are an adaptation allowing mobility of the short proboscis.

Family Moschidae

The musk deer were traditionally placed in the family Cervidae but are now placed in a separate family (Groves & Grubb 1987; Janis & Scott 1987; Webb & Taylor 1980). This small family contains one genus and seven species that occupy an area including northern Afghanistan, Pakistan, India, the Himalayan plateau of southern China, Nepal, Tibet, and North Vietnam. The earliest musk deer are known from the Oligocene of Europe (Dawson & Krishtalka 1984). Musk deer have coarse fur, and their hindlimbs are longer than the forelimbs. They lack antlers, but have saber-like upper canines. When pursued, these animals make huge leaps with radical changes of direction and are so agile as to be able to leap into trees where they can balance on thin branches to forage (Smith & Xie 2008). As their name implies, males have a musk gland in the abdomen that secretes a waxy substance that is used as a base for expensive perfumes, soaps, and medicine. In 2007, the musk was valued at US$45,000 per kilogram, higher than gold, cocaine, or rhinoceros horn (Prothero 2007). Because of extensive hunting, musk deer populations have declined rapidly, and all species are listed as threatened or vulnerable. The Chinese have succeeded in breeding musk deer in captivity for the production of musk, and this may take the pressure off some wild populations (MacDonald 1984).

Family Bovidae

The family Bovidae, which includes African and Asian antelope, bison, sheep, goats, and cattle, is the most commercially important and most diverse living group of ungulates (**Table 18-1**). The family includes 50 genera and 143 species, and wild species occur throughout Africa, in much of Europe and Asia, and in parts of North America. A new bovid species, the saola (*Pseudoryx nghetinhensis*) was recently discovered in the forests of Laos and Vietnam (Dung et al. 1993). The domestication of bovids began in Asia roughly 8,000 to 10,000 years ago (Achilli et al. 2008, 2009; Edwards et al. 2007), and domesticated bovids are nearly as cosmopolitan in distribution as are humans.

The systematic relationships of genera and species within the Bovidae remain controversial (Allard et al. 1992; Essop et al. 1997; Gentry 1992), and several authors even disagree about the monophyly of the family (Allard et al. 1992; Gatesy et al. 1992). Bovids seemingly derived from traguloid ancestry in the Old World and first appeared in the Oligocene of Asia (*Palaeohypsodontus* and *Hanhaicerus* may be among the oldest; Solunias 2007b). Africa was the center of the early and rapid radiation that gave rise to a diverse bovid fauna, judging from the many kinds of bovids known from the Miocene. Far more extinct genera than Recent genera of bovids are known (158 versus 50; Grubb 2005; McKenna & Bell 1997).

Toward the end of the Pleistocene, most bovids were driven from Europe by the southward advance of the cold climate. A few reached the New World in the Pleistocene via the Bering Strait land bridge. Because this boreal avenue of dispersal was under the influence of cold climates, it functioned as a "filter bridge" (Simpson 1965b), and only animals adapted to the cold dispersed across it. As a consequence, the New World received from Asia such bovids as the bighorn sheep (*Ovis*; **Fig. 18-23A**), the mountain "goat" (*Oreamnos*; Fig. 18-23B), the musk ox (*Ovibos*; **Fig. 18-24A**), and the bison (*Bison*; Fig. 18-24B). Bovids less able to withstand boreal conditions—the Old World antelopes and the gazelles are prime examples—were forced from the northern parts of Europe and Asia in the Pleistocene back to their present strongholds in Africa and Asia and hence did not disperse across the Bering bridge to North America. An exception is the saiga antelope (*Saiga*), which now inhabits arid parts of Asia but occurred in Alaska and Canada in the Pleistocene. Bison were extremely abundant members of grassland faunas in the Recent in North and Central America, where they occurred as far south as El Salvador. Some structural divergence occurred in Pleistocene bison; in some areas, several species may have occupied common ground. Some Pleistocene bison were considerably larger than the present *Bison bison*. Specimens of the Pleistocene species *Bison latifrons* from North America indicate that this animal was over 2 meters high at the shoulder and had horns that in larger individuals spanned more than 2.5 meters.

Bovids characteristically inhabit grasslands, and their derived dentition and limbs probably developed in association with grazing habits. The cheek teeth are high crowned, and the upper canines are reduced or absent. **Preorbital vacuities** in the skull are present in some bovids and absent in others. The lateral digits are reduced or totally absent. The ulna is reduced distally and is fused with the radius, and only a distal nodule remains as a vestige of the fibula. Horns, formed of a bony core and a keratinized sheath, are present in males of all wild species, and females often bear horns also. The entire horns (including both sheath and core) are never shed and in some species grow throughout the life of the animal. Bovid horns are never branched but are often large and occur in a variety of forms (Fig. 18-22B). Males of the Indian four-horned antelope (*Tetracerus quadricornis*) are unique in having four short, dagger-like horns. The horns are frequently used in fights between males during the breeding season, but in many species, the horns are used in ritualized tests of body strength in such a way as to minimize injury. Some bovids, such as oryx and gemsbok (*Oryx*), can use their horns as awesome defensive weapons, respected even by lions.

Horns occur in both sexes of many antelope, but the females of some species are hornless. Females have horns in 75% of the genera in which the average female weight is more than 40 kilograms, whereas females nearly always

TABLE 18-1 Classification of Bovidae and Distribution of Recent Genera

Genera (number of species)	Common Name and Continent(s)	Genera (number of species)	Common Name and Continent(s)
Subfamily Aepycerotinae		*Tetracerus* (1)	Four-horned antelope (Asia)
Aepyceros (1)	Impala (Africa)	*Tragelaphus* (7)	Bushbuck, nyala, kudu, bongo (Africa)
Subfamily Alcelaphinae		**Subfamily Caprinae**	
Alcelaphus (3)	Hartebeests (Africa)	*Ammotragus* (1)	Barbary sheep (North Africa)
Beatragus (1)	Hunter's Hartebeest (Africa)	*Budorcas* (1)	Takin (Asia)
Connochaetes (2)	Wildebeest (Africa)	*Capra* (8)	Ibex, goat (Asia, Europe, North America)
Damaliscus (4)	Hartebeest, topi, blesbok (Africa)	*Capricornis* (6)	Serows (Asia)
Subfamily Antilopinae		*Hemitragus* (3)	Tahr (Asia)
Ammodorcas (1)	Dibatag (Africa)	*Naemorhedus* (4)	Goral (Asia)
Antidorcas (1)	Springbok (Africa)	*Oreamnos* (1)	Rocky mountain "goat" (North America)
Antilope (1)	Blackbuck (Asia)	*Ovibos* (1)	Musk ox (North America, Greenland)
Dorcatragus (1)	Beira antelope (Africa)	*Ovis* (5)	Mouflon, argali, bighorn sheep, sheep (Asia, Europe, North America, North Africa)
Eudorcas (3)	Gazelles (Africa)		
Gazella (10)	Gazelles (Africa)		
Litocranius (1)	Gerenuk (Africa)		
Madoqua (4)	Dik-dik (Africa)	*Pantholops* (1)	Chiru (China, Tibet, Northern India)
Nanger (3)	Gazelles (Africa)		
Neotragus (3)	Royal, pygmy, and suni antelope (Africa)	*Pseudois* (2)	Bharal (Asia)
Oreotragus (1)	Klipspringer (Africa)	*Rupicapra* (2)	Chamois (Southwest Asia)
Ourebia (1)	Oribi (Africa)	**Subfamily Cephalophinae**	
Procapra (3)	Black-tailed gazelle (Asia)	*Cephalophus* (15)	Duiker (Africa)
Raphicerus (3)	Steinbuck, grysbuck (Africa)	*Philantomba* (2)	Maxwell's and blue duikers (Africa)
Saiga (2)	Saiga (Asia, Europe)		
Subfamily Bovinae		*Sylvicapra* (1)	Bush duiker (Africa)
Bison (2)	Bison (Europe, North America)	**Subfamily Hippotraginae**	
		Addax (1)	Addax (Africa)
Bos (5)	Cattle (worldwide)	*Hippotragus* (3)	Roan, sable (Africa)
Boselaphus (1)	Nilgai (Asia)	*Oryx* (4)	Oryx (Africa)
Bubalus (4)	Asiatic buffalo (Asia)	**Subfamily Reduncinae**	
Pseudoryx (1)	Siola (Vietnam)	*Kobus* (5)	Waterbuck, kob, lechwe (Africa)
Syncerus (1)	African buffalo (Africa)		
Taurotragus (2)	Eland (Africa)	*Pelea* (1)	Rhebuck (Africa)
		Redunca (3)	Reedbuck (Africa)

From Grubb 2005.

lack horns in genera in which females weigh less than 25 kilograms. Furthermore, when both sexes have horns, those of the males are thicker at the base, more complex in shape, and are adapted to withstanding the forces encountered during intraspecific combat (Lundrigan 1996). The horns of the female are straighter and thinner, better adapted to stabbing, and thus more effective defensive weapons. The relationship in females between body weight and the possession of horns is probably correlated with antipredator behavior. Larger species depend on direct defense, but small species flee or use concealment. Large antelope offer especially effective antipredator defense because they are larger than most of their predators and are much larger than the predators of their young.

The last great strongholds of bovids are the grasslands and savannas of East Africa, but a diversity of bovids still

FIGURE 18-23   (A) A male desert bighorn sheep (*Ovis canadensis*) from southern Arizona. (B) A mountain goat (*Oreamnos americanus*) from North America.

occurs in parts of central and southern Africa (**Fig. 18-25**). Seemingly, every conceivable bovid niche has been occupied. Some antelope, such as the Bohor reedbuck (*Redunca redunca*) and the lechwe (*Kobus leche*), inhabit river borders and swampy ground, while at the other extreme the oryx (*Oryx*) and springbok (*Antidorcas marsupialis*) live in arid plains and deserts with limited access to drinking water. The protection afforded game in some parts of Africa should allow the survival of many species of this diverse and handsome group.

Family Hippopotamidae

The family Hippopotamidae is represented today by the genera *Hippopotamus* and *Hexaprotodon* (the pigmy hippopotamus), each with one species. Hippos first appeared in the middle Miocene in East Africa and their evolution has been centered mainly in Africa. They are most likely derived from anthracotheres (Boisserie et al. 2005), and their close relationship with Cetacea is supported by molecular and paleontological-morphological evidence (Boisserie 2007). When considered as sister taxa, hippopotamids plus cetaceans have been called Cetancodonta or whippomorpha (Arnason et al. 2000). By late Miocene, the family Hippopotamidae occurred widely in the southern parts of the Old World (including southern Asia) and even reached southern Europe by the Pleistocene. Although subfossils of two species are known from Madagascar, hippopotami now occur only in Africa. In North Africa, they

FIGURE 18-24   (A) A group of musk oxen (*Ovibos moschatus*) from northern Canada. (B) A pair of North American bison (*Bison bison*) from Wyoming. Note the considerable size dimorphism between the large bull (right) and the cow (left).

FIGURE 18-25 Some members of the diverse bovid fauna of Africa: (A) greater kudu (*Tragelaphus strepsiceros*) from Etosha National Park, Namibia. (B) A group of gemsbok (*Oryx gazella*). (C) An adult and juvenile bontebok (*Damaliscus pygargus*). (D) A group of African buffalo (*Syncerus caffer*).

are restricted to the Nile River drainage, but they occur widely in the southern two thirds of the continent.

Hippopotami are bulky creatures with huge heads and short limbs (**Fig. 18-26**). They are large: *Hippopotamus* weighs 510 to 3,200 kilograms and *Hexaprotodon* 180 to 275 kilograms (Kingdon 1997). Some of the distinctive features of these animals probably evolved in association with their amphibious mode of life. Specialized skin glands secrete an oily substance that protects the sparsely haired body from the sun. These secretions include hipposudoric acid, a red pigment, and norhipposudoric acid, an orange pigment that change from colorless to pink and then to brown with exposure to the sun. These compounds act as sunscreens and have antibiotic properties that provide protection from microbes (Hasimoto et al. 2007; Saikawa et al. 2004). The highly specialized skull has elevated orbits and enlarged and tusk-like canines and incisors (**Fig. 18-27**). The bunodont molars are basically four cusped; the dental formula is 2–3/1–3, 1/1, 4/4, 3/3 = 38–44. The limbs are robust, and the feet are four toed (Fig. 18-3). The foot posture is semidigitigrade;

only the distal phalanx of each toe touches the ground. The broad foot is braced by a sturdy "heel" pad of connective tissue, and the central digits are nearly horizontal.

FIGURE 18-26 A hippopotamus (*Hippopotamus amphibius*) from Swaziland, southern Africa, with its mouth agape.

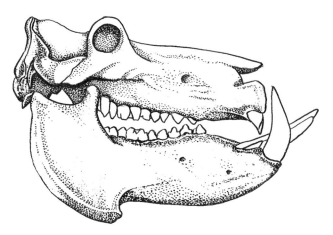

Hippopotamus
(*Hippopotamus amphibius*, Hippopotamidae)

FIGURE 18-27 Skull of a hippopotamus, *Hippopotamus amphibius* (length of skull 600 millimeters). (Modified from Romer, A.S. *Vertebrate Paleontology*. University of Chicago Press, 1966.)

Hippopotami are gregarious, and groups spend much of the day in the water. When bodies of water are scarce during the dry season, hippopotami often concentrate in stagnant ponds, which they churn into muddy morasses. They are good swimmers and divers and, when submerged, are able to walk on the bottom of rivers or lakes using a slow-motion gait. Mating, birth, and much of calf rearing, including suckling, takes place in the water. Hippos can profoundly change the local environment. In the Okavango Delta of Botswana, for example, hippos spend the day in deep water lakes, and at night they move to grazing areas along specific pathways. Local overgrazing and the repeated use of limited pathways create deep channels that divert water through the swamp, link lakes, or lead to lake closure (McCarthy et al. 1998).

Hippos are known to communicate by producing a wide variety of sounds; some of their underwater sounds resemble the sonar clicks of dolphins. Grunts and other sounds are produced while the throat is submerged, and only the eyes and nostrils are above the surface. Consequently, these low-frequency grunts are transmitted simultaneously through the air and the water. Submerged hippos surface in response to these grunts and add their voices to those of hippos already on the surface in a chorus that can spread from territory to territory for several kilometers (Barklow 2004). The pygmy hippo is solitary or occurs in pairs and inhabits forested areas (**Fig. 18-28**). Instead of seeking shelter in the water when disturbed, as is characteristic of *Hippopotamus*, the pygmy hippo seeks refuge in dense vegetation.

FIGURE 18-28 A pygmy hippopotamus (*Hexaprotodon liberiensis*) from West Africa.

SUMMARY

The order Artiodactyla includes 240 species in 89 genera. The earliest fossil artiodactyls were rabbit-sized members of the Diacodexeidae and Raoellidae from the early Eocene of North America, Europe, and Asia. Today artiodactyls (pigs, camels, deer, antelope, cattle, and their kind) far overshadow perissodactyls in diversity and abundance. Forests gave way to grasslands in the late Eocene to Miocene. Since the Miocene, the perissodactyls have steadily declined, but the artiodactyls, with their highly specialized stomachs and ability to digest a low-nutrition diet of grasses, have remained diverse and successful. Of the 36 artiodactyl families present in the Cenozoic, 10 families survive to the present. It is tempting to relate the decline of the perissodactyls to the rise of the artiodactyls and to regard the latter as the more effective competitors, but the true story is certainly far more complex.

Recent discoveries of early fossil whales with double-trochlea astragali and paraxonic (even-toed) foot symmetry, an Eocene artiodactyl with tooth morphology similar to Eocene whales, and molecular evidence all support the hypothesis that cetaceans (whales and dolphins) and artiodactyls are closely related. The term Cetartiodactyla is used to indicate their shared ancestry. Morphological and paleontological evidence favors a monophyletic Artiodactyla, with cetaceans as either a sister group to Artiodactyla, or when combined with molecular data, cetaceans are included within Artiodactyla in a hippo-cetacean clade. For now, we retain the traditional classification of the Artiodactyla but acknowledge the possible sister-group relationship of the Hippopotamidae with the Cetacea.

Living artiodactyls are often grouped into suborders. The suborder Suiformes includes the pigs, peccaries, and hippos. Suiformes have relatively simple stomachs and do not ruminate. The Tylopoda includes the Old World camels and the New World guanaco and vicuña. These are ruminants with a three-chambered stomach. They also have a unique Y-shaped cannon bone (fused third and fourth metapodials). The third and largest suborder, the Ruminantia, includes chevrotains and mouse deer (Tragulidae), musk deer (Moschidae), deer (Cervidae), pronghorn (Antilocapridae), giraffes (Giraffidae), and a diverse group of antelopes, sheep, goats, and cattle (Bovidae). All members of the Ruminantia have 4-chambered stomachs and most are highly cursorial.

Artiodactyls are native to every continent except Australia (where they were introduced) and Antarctica. They include species adapted for semiaquatic lives (hippos and chevrotains), for life in the arctic (musk ox and caribou), and for survival in some of the driest and hottest deserts on Earth (oryx and camels). Some species, such as wildebeest and caribou, form vast herds that must undertake lengthy annual migrations. Others, such as the Philippine mouse deer and the siola have restricted distributions and are exceedingly rare. In addition, humans have domesticated several species (pigs, camels, llamas, sheep, goats, and cattle) for meat, wool, and milk production and as pack animals.

KEY CHARACTERISTICS OF ARTIODACTYLS

- Feet paraxonic, the plane of symmetry passes between the third and fourth digits
- First digit absent
- Toes even in number, only two digits in extreme cases
- Ankle bones reduced in number, and astragalus bears most of weight
- Astragalus with double-trochlea (restricting lateral movement)
- Femur lacks a third trochanter
- Horns with bony core or antlers often present
- Postorbital bar or process present
- Cheek teeth bunodont or more commonly selenodont
- Stomach often multichambered (four chambers in extreme cases) for microbial fermentation

KEY TERMS

Abomasum

Aponeurosis

Omasum

Pedicel

Preorbital vacuities

Regurgitation

Reticulum

Rumen

RECOMMENDED READINGS

Boissere, J-R, F Lihoreau, & M Brunet. 2005. The position of Hippopotamidae within Cetartiodactyla. *Proceedings of the National Academy of Sciences USA, 102*:1537–1541.

Byers, JA. 2003. *Built for Speed: A Year in the Life of a Pronghorn.* Harvard University Press, Cambridge, MA.

Lundrigan, B. 1996. Morphology of horns and fighting behavior in the family Bovidae. *Journal of Mammalogy, 77*:462–475.

Mayer, JJ & PN Brandt. 1982. Identity, distribution, and natural history of the peccaries, Tayassuidae. Pp. 433–455 in *Mammalian Biology in South America* (MA Mares & HH Genoways, eds.). Special Publication Series, Pymatuning Laboratory of Ecology Vol. 6. University of Pittsburgh Press, Pittsburgh, PA.

O'Leary, MA & J Gatesy. 2008. Impact of increased character sampling on the phylogeny of Cetartiodactyla (Mammalia): combined analysis including fossils. *Cladistics, 23*:1–46.

Price, J & S Allen. 2004. Exploring the mechanisms regulating regeneration of deer antlers. *Philosophical Transactions of the Royal Society, London B, 359*:809–822.

Prothero, DR & SE Foss. 2007. *The Evolution of Artiodactyls.* Johns Hopkins University Press, Baltimore, MD.

Theodor, JM, KD Rose, & J Erfurt. 2005. Artiodactyla. Pp. 215–233 in *The Rise of Placental Mammals* (KD Rose & JD Archibald, eds.). Johns Hopkins University Press, Baltimore, MD.

Thewissen, JG, et al. 2007. Whales originated from aquatic artiodactyls in the Eocene epoch of India. *Nature, 450*:1190–1194.

Cetacea

Cetaceans (order Cetacea) are notable for being the mammals most fully adapted to aquatic life. Baleen whales (suborder **Mysticeti**), the largest living or fossil mammals known, are mainly plankton feeders (although some eat fish) and thus draw from the tremendous productivity near the middle of the marine food web. Some of the large toothed whales (suborder **Odontoceti**), in contrast, exploit the top of the marine food web. Sperm whales, for example, eat giant squid, large sharks, and bony fish. The food of killer whales includes fish, seals, porpoises, and baleen whales. Remarkable swimming and diving ability, the capability of many (perhaps all odontocetes) to echolocate, considerable intelligence, and complex social behavior have all contributed to the great success of the cetaceans.

Cetaceans have long intrigued and inspired humans. Leaping dolphins have been the embodiment of beauty and exuberance to seafarers for centuries: 4,000 years ago Minoan artists included graceful drawings of dolphins in their frescoes at the palace of Knossos on Crete. It has remained for modern humans to deplete the cetacean populations that once seemed limitless. All species of cetaceans are now listed in appendices I or II of the Convention on International Trade in Endangered Species.

Paleontology

In recent years, paleontologists have unearthed a remarkable series of Eocene cetaceans from coastal deposits of the former Tethys Sea. These early cetaceans comprise six families, which together illustrate a transformation from terrestrial quadrupeds to fully aquatic marine mammals (Fordyce & Muizon 2001; Perrin et al. 2002; Rose 2006; Thewissen & Williams 2002; Thewissen et al. 2007). The earliest known cetaceans (*Pakicetus*, *Ichthyolestes*, and *Nalacetus*; Pakicetidae) are known from the early to middle Eocene of Pakistan and India. Their fossils are found in association with various land mammals in river sediments deposited at the border of the shallow Tethys Sea, which previously had separated the northern supercontinent

Laurasia and the southern supercontinent Gondwana (including India, which at about the time was beginning its slow collision with Asia; Gatesy & O'Leary 2001). These and other primitive whales belong to the extinct suborder Archaeoceti and were once believed to have evolved from the mesonychid carnivores. However, an improved fossil record along with recent molecular studies shows that the Archaeoceti probably evolved from ancestral artiodactyls (Cetartiodactyla; Berta et al. 2006; Geisler & Uhen 2005; Thewissen et al. 2007). The middle Eocene artiodactyl family Raoellidae, including the small, semiaquatic *Indohyus* from India are believed to be the sister group to Archaeocete cetaceans (Geisler & Theodor 2009; Thewissen at al. 2007). The raccoon-sized *Indohyus* has a number of aquatic adaptations, including heavy, osteosclerotic bones (bones with a thickened cortex) that provided ballast while wading. Presumably, *Indohyus* spent considerable time in the water, perhaps foraging in the shallows for aquatic vegetation, or coming onto land to feed as modern hippos do. Thewissen and colleagues hypothesize that cetaceans evolved from an *Indohyus*-like ancestor that was already semi-aquatic. Further specializations for aquatic feeding in archaeocetes led to the morphological and physiological adaptations we see in extant cetaceans.

The hippos (Hippopotamidae) are now considered by many researchers to be the sister group to a combined Raoellidae-Cetacea clade (Geisler & Theodor 2009). Indeed, as derivatives of the artiodactyls, a strict adherence to phylogeny in the taxonomy of mammals would require us to include Cetacea in the Artiodactyla chapter in future editions of this textbook.

The ancestors of archaeocetes probably diverged from archaic artiodactyls at the end of the Cretaceous. Early Eocene archaeocetes were quadrupedal and mostly terrestrial (**Fig. 19-1**; Thewissen & Bajpai 2001). One remarkably preserved 47.5 million year old female archaeocete whale contained a near-term fetus positioned for head-first delivery. Modern cetaceans give birth tail first. Head-first births are characteristic of terrestrial mammals, suggesting that

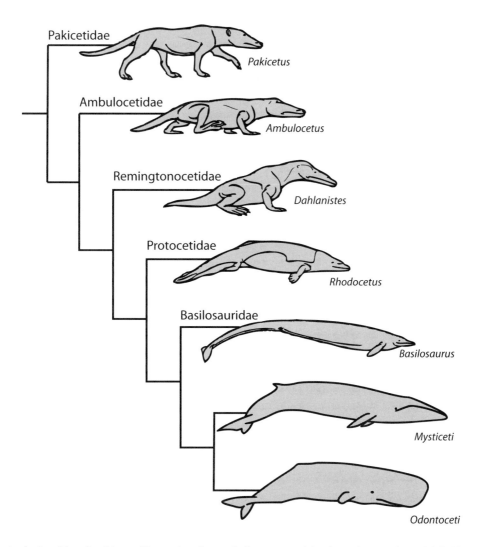

FIGURE 19-1 Morphological and fossil evidence illustrating the evolutionary transition from the quadrupedal *Pakicetus* to the semiaquatic *Ambulocetus* and *Dahlanistes*, to the fully aquatic Protocetidae. Members of the Basilosauridae are thought to be the ancestors of modern mysticete and odontocete whales. (Adapted from Zimmer, C. *At the Water's Edge*. Simon & Schuster, 1998.)

this quadrupedal whale (*Maiacetus inuus*) returned to land to give birth (Gingerich et al. 2009). Later archaeocetes (Ambulodetidae, Remingtonocetidae, and Protocetidae) evolved reduced limbs, thickened tails, and more elongate bodies and were either semiaquatic or fully aquatic. In fact, fossil archaeocetes provide remarkable examples of a major macroevolutionary transition and even offer evidence (from isotope studies) for changes in osmoregulatory abilities as the animals moved from fresh to salt water (Roe et al. 1998; Thewissen et al. 1996).

By the end of the Eocene, surviving archaeocetes had long, streamlined bodies, tiny hindlimbs, and horizontal tail **flukes**. Some archaeocetes attained large size. For example, *Basilosaurus* (Basilosauridae), a late Eocene type from Egypt, was elongate and slender (Fig. 19-1). Its body was roughly 17 meters long, with a skull 1.5 meters long. In primitive cetaceans, the external nares had not migrated so far toward the back of the skull as they have in modern

cetaceans (**Fig. 19-2**). Early modern cetaceans are believed to have evolved from a basilosaurid ancestor (O'Leary & Uhen 1999; Thewissen & William 2002).

Cetaceans in general appear to form a monophyletic group, but considerable debate remains over the relationships within the order (Price et al. 2005). Controversies over the phylogeny of odontocetes have been especially intense (Árnason & Gullberg 1994; Messenger & McGuire 1998; Milinkovitch & Thewissen 1997).

Early mysticetes (baleen whales) first appeared in the early Oligocene, lacked baleen, and instead possessed teeth inherited from their ancestors (Fitzgerald 2006; Fordyce 1992). The Oligocene species *Janjucetus hunderi* from Australia was a mysticete whale with sharp teeth and large eyes, suggesting that it hunted fish and squid as well as potentially larger prey; it was clearly not directly ancestral to modern baleen whales. Early mysticetes were relatively small compared with modern mysticetes and had large

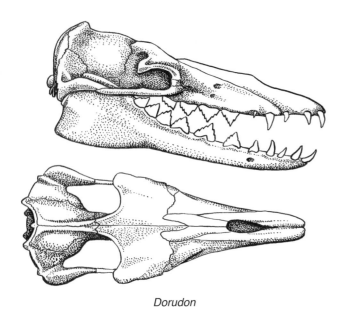

Dorudon

FIGURE 19-2 Skull of a primitive cetacean (*Dorudon*), a fossil archaeocete from the Eocene. The length of the skull is approximately 600 millimeters. Note the lack of telescoping of the skull and the heterodont dentition. (Adapted from Romer, A.S. *Vertebrate Paleontology.* University of Chicago Press, 1966.)

orbits and sharp teeth. They probably relied on vision to capture prey, exhibiting a predatory lifestyle similar to that of extant leopard seals (*Hydrurga leptonyx*). Other Oligocene whales (e.g., *Chonecetus* and *Aetiocetus*) had more elongated rostra and probably possessed both teeth and baleen, suggesting that they were capable of limited bulk filter-feeding (Fitzgerald 2006). By the middle Miocene, mysticetes had lost teeth, reduced the size of the orbits, and evolved longer rostra to support more baleen (**Box 19-1**). These whales were bulk filter-feeders that entrapped large quantities of small prey rather than relying on visual acuity to hunt larger prey. They seem to have evolved after the initiation of the circum-Antarctic current in the Oligocene, at about the same time as there was a flourishing of plankton in the Southern Ocean, which possibly led to an explosion of krill that the whales could exploit (Barnes 1976; Barnes

& Sanders 1996; Emlong 1966; Fordyce 1992; Fordyce & Barnes 1994). The transition from visually oriented predators to bulk filter-feeders allowed early mysticetes to exploit a huge untapped energy source, which ultimately led to an increase in body size. Modern mysticetes, for example, can consume up to 600,000 kilograms of krill and other marine food in a single year (Gaskin 1982).

The suborder Odontoceti, the toothed whales, probably evolved from a basilosaurid ancestor in the late Eocene. Early odontocetes appeared in the Oligocene record of Australia, Europe, and New Zealand. Advanced odontocetes with highly telescoped skulls, homodont dentitions, and many more teeth than the primitive eutherian complement are known from the Miocene (**Fig. 19-3**).

In summary, there were two major radiations of cetaceans. The first occurred in the middle Eocene in the shallow waters of the Tethys Sea and led to the diversification of the stem cetaceans (Archaeoceti). A second Oligocene-Miocene radiation led to modern cetaceans; both odontocetes and mysticetes radiated rapidly in response to dramatic changes in global climate and continental tectonics. The movement of continental plates during the final breakup of Gondwana created new oceanic circulation patterns, which in turn increased marine productivity (Scher & Martin 2006). The stage was set for the evolution of filter-feeding mysticetes and the refinement of echolocation-based hunting in odontocetes (Fordyce 2003; Lindberg & Pyenson 2007).

Morphology

All cetaceans are completely aquatic, and their anatomy reflects this mode of life. The body is fusiform (cigar-shaped), nearly hairless, insulated by thick blubber, and lacks sebaceous glands. To aid in streamlining the body, the single pair of mammae lies flat along the abdomen and the teats are enclosed within slits adjacent to the urogenital opening. In males, the testes remain abdominal, and the penis is retractile (Rice 1984). Most vertebrae have high neural spines (**Fig. 19-4**), and the cervical vertebrae are highly compressed

BOX 19-1 From Teeth to Baleen

The transition from tooth-based predation to baleen-based filter feeding was apparently a stepwise process with intermediate forms possessing both teeth and baleen (Deméré et al. 2008). This evolutionary transformation parallels the developmental trajectory of modern mysticetes in that rudimentary teeth are present in fetal baleen whales, but the tooth buds degenerate before birth. This degeneration coincides with the formation of baleen-forming papillae. Thus, fetal mysticetes pass through a toothed stage, followed by a stage with rudimentary teeth and baleen, and finally a baleen only stage before birth. Deméré et al. (2008) suggested that a similar sequence of events oc-

curred in the evolution of mysticete whales. Nutrient foramina and sulci, which contain blood vessels that supply the baleen, are present on the palates of several species of toothed Oligocene mysticetes, suggesting that these species had at least rudimentary baleen. In addition, these authors demonstrate that two enamel-specific genes (enamelin and ameloblastin) are present, but nonfunctional in modern mysticetes, leading to the degeneration of tooth primordia in the fetus. Taken together, the evidence records an evolutionary transition from toothed ancestors, to an intermediate with both teeth and baleen, to modern baleen whales.

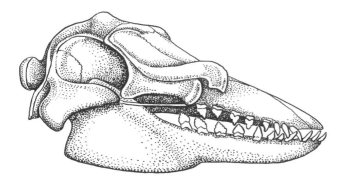

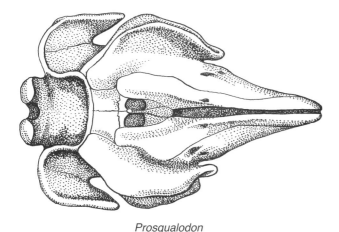

Prosqualodon

FIGURE 19-3 Skull of *Prosqualodon*, a fossil porpoise from the Miocene. The length of the skull is approximately 450 millimeters. The skull is highly telescoped, and the maxillary and frontal bones form a roof over the temporal fossa. (Adapted from Romer, A.S. *Vertebrate Paleontology*. University of Chicago Press, 1966.)

(**Fig. 19-5**). The clavicle is absent. The forelimbs (flippers) are paddle shaped, and no external digits or claws are present. Little movement is possible between the joints distal to the shoulder. The proximal segments of the forelimb are short, whereas the digits frequently are unusually long because of the development of more phalanges per digit than the basic eutherian number (**Fig. 19-6**). The hindlimbs are vestigial, do not attach to the axial skeleton and are not visible externally.

The flukes (tail fins) are horizontally oriented. The cetacean fossil record documents hindlimb reduction and the expansion of tail flukes for propulsion (Fig. 19-1). Nevertheless, modern cetaceans retain reduced hindlimbs (usually the innominate, femur, and tibia) within the abdominal cavity. Dolphin embryos (Carnegie stages 12–16) possess both forelimb and hindlimb buds, but the hindlimbs fail to complete normal development. Thewissen et al. (2006) suggest that reduced gene expression in at least one key gene that acts to maintain limb development occurred some 41 million years ago prior to the origin of basilosaurids and resulted in the loss of the distal hindlimb elements. Later, prior to the origin of modern cetaceans (approximately 34 million years ago), this gene became completely nonfunctional, leading to further hindlimb reduction in modern whales, dolphins, and porpoises.

The cetacean skull is typically highly modified as a result of the posterior migration of the external nares (**Fig. 19-7**). The premaxillary and maxillary bones form most of the roof of the skull, and the occipitals form the back. The nasals and parietals are telescoped between these bones and form only a minor part of the skull roof. Large frontal bones are mostly covered by the maxillaries and premaxillaries. The tympanoperiotic bone (the bone that houses the middle and inner ear) is not braced against adjacent bones of the skull in most cetaceans and is partly insulated from the rest of the cranium by surrounding air sinuses (**Fig. 19-8**). Members of the families Ziphiidae and Physeteridae have a pneumatic bony strut that braces the tympanoperiotic bone against the skull.

Experimental work on the bottlenose dolphin (*Tursiops truncatus*) by Herman et al. (1975) showed that visual acuity is similar above and below the water. The above-water acuity may be due to the "pinhole camera" effect of the pupil. Mechanisms for fine adjustment of lens shape or displacement are seemingly absent in cetaceans, but in bright light, the central part of the pupil closes completely, leaving two tiny apertures that yield great depth of field, as

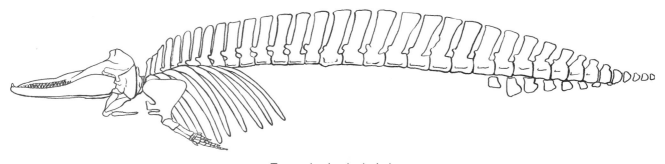

Tasmanian beaked whale
(*Tasmacetus shepherdi*, Ziphiidae)

FIGURE 19-4 The skeleton of the Tasmanian beaked whale. Total body length is approximately 6.6 meters.

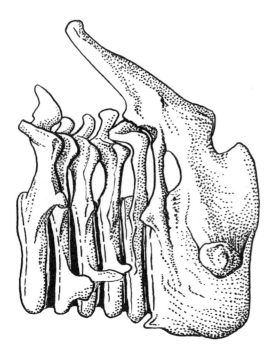

FIGURE 19-5 Cervical vertebrae of a dolphin (*Delphinus delphis*, Delphinidae). Only the axis and atlas are fused, whereas most of the series are fused in some cetaceans.

does a pinhole camera. This allows the dolphin to receive a sharp image from distant objects when ambient light levels are relatively high (Dawson 1980; Dawson et al. 1979).

Cetacean Adaptations

The vulnerable point of cetaceans is their need to breathe atmospheric air, but unlike most other mammals, many cetaceans are able to alternate between periods of **eupnea** (normal breathing) and long periods of **apnea** (cessation of breathing). Some whales, such as sperm whales (*Physeter catodon*), remain submerged for over 70 minutes. Most delphinids can hold their breath 4 to 5 minutes, but they often surface to breathe several times a minute. The ability of cetaceans to remain active during periods

of apnea probably depends on many adaptations. Rapid gas exchange is enhanced by two layers of capillaries in the interalveolar septa of the lungs. During expiration, most of the air can be exhausted from the lungs, and up to 12% of the oxygen from inhaled air is used (the corresponding figure for terrestrial mammals is only 4%). Compared with terrestrial mammals, cetaceans have at least twice as many erythrocytes per volume of blood and about two to nine times as much myoglobin (a molecule able to store oxygen and release it to tissue) in the muscles. During deep dives, the heart rate drops to roughly half the surface rate, and vascular specializations allow blood to bypass certain muscle masses while maintaining flow to the brain. Important physiological adaptations to prolonged submersion include tolerance to high levels of lactic acid and carbon dioxide. Good discussions of cetacean adaptations to deep diving are given by Irving (1966), Elsner (1969), Kooyman and Anderson (1969), and Rommel et al. (2006).

Some large odontocetes and some mysticetes can perform deep dives (**Table 19-1**). Beaked whales (*Ziphius cavirostris*) fitted with depth recording instruments routinely dove to depths of over 1,000 meters, with a maximum reported depth of 1,885 meters (Tyack et al. 2006). Short-finned pilot whales (*Globicephala macrorhynchus*) dive deep very quickly, sprinting after squid, their main prey, at speeds up to 9 meters/second and depths down to 1,000 meters (Aguilar Soto et al. 2008). During deep dives, cetaceans are subjected to tremendous pressures, for with every 10-meter increase in depth an additional 1 atmosphere of pressure is exerted on the body. A cetacean swimming at a depth of only 200 meters, probably a common depth for many species, is subjected to 20 atmospheres of pressure, or 294 pounds per square inch. Any gases that remain within the body cavities are therefore subjected to great pressure, resulting in a decrease in their volume and an increase in the amount of gases that dissolve in the body solvents, such as blood. One serious result of similar activities in humans is a condition known as "the bends," or decompression sickness. When humans use equipment that allows them to breathe

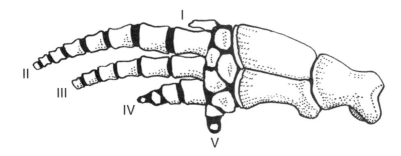

FIGURE 19-6 Dorsal view of the right forelimb of the bottlenose dolphin (*Tursiops truncatus*, Delphinidae).

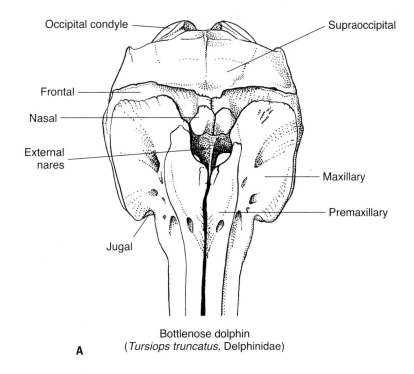

Bottlenose dolphin
(*Tursiops truncatus*, Delphinidae)

A

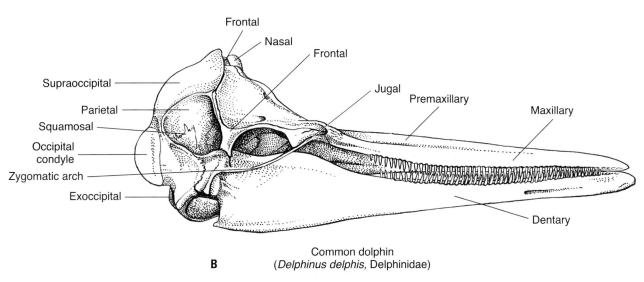

Common dolphin
(*Delphinus delphis*, Delphinidae)

B

FIGURE 19-7 (A) Dorsal view of part of the skull of a bottlenose dolphin. Note the asymmetry of the bones surrounding the external nares. The function of this remarkable asymmetry is not known. (B) Skull of a common dolphin. Note the highly telescoped skull with the maxillary and frontal bones roofing the small temporal fossa. The frontal bone is barely exposed on the skull roof. The length of the skull is 475 millimeters.

underwater and undergo prolonged exposure to high pressures during diving, greater than normal amounts of gases are dissolved in the tissues and the blood. If decompression is too rapid, these gases cannot be carried to the lungs rapidly enough to be removed from the body; instead, gases quickly leave solution and appear as bubbles in the tissues. Intravascular bubbles may occlude capillaries and result in injury to tissues or even death.

Whether cetaceans are subject to the bends is controversial. Cetacean lungs collapse in the first 100 meters of

a dive, and they have evolved the following anatomical specializations for deep diving: (1) A large proportion of the ribs lack an attachment to the sternum or other ribs, which allows rapid lung collapse. (2) The lungs are dorsally situated above the oblique diaphragm. (3) In deep divers, the lungs are small and the volume of the nonvascular air passages is relatively large. (4) The trachea is short and often of large diameter, and the cartilaginous rings bracing the trachea are nearly complete, have small intermittent breaks, or are fused (Slijper 1962). (5) The bronchioles are

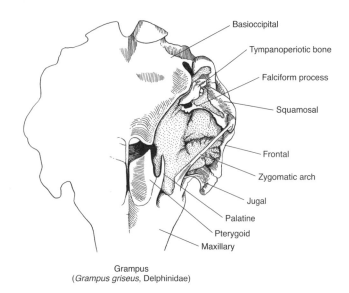

FIGURE 19-8 Ventral view of part of the skull of the grampus (*Grampus griseus*), showing the large air sinuses (stippled) anterior to the tympanoperiotic bone and partly surrounding it. (Reproduced from Norris, K. S. *Whales, Dolphins and Porpoises*. © 1966 by the Regents of University of the University of California. Published by the University of California Press.)

reduced in length, and the entire system of bronchioles, to the very origin of the alveolar ducts, is braced by muscles and cartilaginous rings. (6) The lungs, especially the walls of the alveolar ducts and the septa, contain unusually high concentrations of elastic fibers. (7) Finally, in some of the small odontocetes, a series of myoelastic sphincters occurs in the terminal sections of the bronchioles.

The adaptive importance of all these features is not yet completely understood. Clearly, the specializations of the ribs and the placement of the lungs relative to the diaphragm permit the lungs to collapse and air to move from space where gas exchange occurs to space where it does not occur. Bracing the respiratory passages may facilitate alveolar collapse and rapid inhalation and exhalation at the surface. All of these cetacean adaptations were thought to help prevent the bends. Moore and Early (2004) investigated bone tissues from the carcasses of stranded sperm whales and concluded that the necrosis in the bone and marrow tissue was consistent with the bends. Others, however, have suggested that the abnormal bone tissue in these whales and reports of gas bubble formation in the tissue of stranded beaked whales are instead due to the use of powerful naval sonar tests (Fernandez et al. 2005; Tyack et al. 2006).

Cetaceans are fast swimmers. Powerful dorsoventral movements of the tail provide propulsion, and the flippers are used for steering. Dolphins can swim up to 36 kilometers per hour, and speeds of 55 and 27 kilometers per hour have been reported for killer whales and a pilot whale, respectively (Lang 1966). Gawn (1948) reports the huge blue whales' speed as 37 kilometers per hour for 10

TABLE 19-1 Depths at Which Cetaceans Have Been Recorded

Species	Depth (m)	Method of Observation
Balaenopteridae	500	Harpooned and collided with bottom
Fin whale, *Balaenoptera physalus*	355	Depth manometer on harpoon
Physeteridae	900	Entangled in deep-sea cable
Sperm whale, *Physeter catodon*	1134	Entangled in deep-sea cable
	520	Echo-sounder
Ziphiidae*		
Cuvier's beaked whale, *Ziphius cavirostris*	1885	Attached depth recorder
Blainville's beaked whale, *Mesoplodon densirostri*	1251	Atttached depth recorder
Delphinidae		
Rough-toothed dolphin, *Steno bredanensis*	30	Attached depth recorder
North Pacific pilot whale, *Globicephala macrorhynchus*	366	Inferred from feeding behavior
Bottlenose dolphin, *Tursiops truncatus*	92	Visual observations from underwater craft
	185	Vocalizations near underwater craft
	170	Trained to activate buzzer

Data are from Kooyman and Andersen (1969), who cite the sources of the observations.
* Data are from Tyack et al. (2006).

minutes. This remarkable swimming performance of cetaceans has proved difficult to explain. Recent studies have demonstrated that their speed is due to a combination of subdermal "springs" that store and release energy during muscle contraction and to specializations that greatly reduce resistance (drag) as the animals swim. In dolphins, for example, thrust is produced by the up and down movement of the tail flukes and powered by the muscles in the tailstock. Surrounding the tailstock muscles is a layer of blubber and a connective tissue sheath. Dolphin blubber is reinforced with a network of collagen fibers giving it stiffness and flexibility that, coupled with the underlying subdermal sheath, act as a biological spring (Pabst 1990, 1993). As the tailstock muscles act to raise the fluke, the blubber and sheath "spring" on the dorsal surface of the tailstock are compressed and simultaneously stretched on the ventral surface. This stores elastic energy, which is released at the top of the stroke and helps to push the tail back downward. The opposite process occurs when the tail fluke moves in the downward direction. The combination of these dorsal and ventral "springs" increases thrust while saving energy. At the same time, the dolphin's body is superbly designed to reduce drag. The amount of resistance depends on the type of water flow over the body surface. If the flow is smooth, parallel to the surface, it is said to be laminar. When such smooth flow is interrupted by water movements that are not consistently parallel to the body surface, however, turbulent flow occurs. All other conditions being equal, laminar flow creates much less resistance than does turbulent flow. If the bodies of small dolphins were subjected to turbulent flow, swimming at 38 kilometers per hour would require their muscles to be five times as powerful as those of humans (Lang 1966). Assuming flow to be nearly laminar, however, this speed is approximately that expected if their power output were that of a well-trained human athlete. Scientists for many years have attempted without success to design bodies shaped so that air or water flow over the surface is laminar or nearly so. What is the cetacean solution to this problem?

Several factors seemingly contribute importantly to reducing resistance as a porpoise swims rapidly (Hertel 1969). The body is hairless, and no obstructions except the streamlined appendages break the extremely smooth surface. In addition, the body form of dolphins is approximately parabolic; this form creates even less resistance than the rounded (elliptical) head end and tapered body of such rapid swimmers as a trout. In addition to being streamlined, dolphins have a compliant spongy layer in their outer skin that reduces drag by dampening local pressure fluctuations that cause turbulence (McNeill Alexander 1968). Body size also plays an important role in swimming speed. Power output is directly proportional to muscle mass, and muscle mass generally increases with overall body mass. Simply stated, larger animals generate more power. At the same time, drag as a function of body mass decreases with body mass (at a given speed). In other words, the most important components of drag, surface area and cross-sectional area, increase with the square of a linear dimension, such as body length, whereas body mass, which determines power output, increases with the cube of body length. Thus, a larger aquatic animal will always be able to swim faster than a smaller animal of identical body shape (Eckert et al. 1988).

Large body size is also important in thermoregulation. Many cetaceans spend considerable time in polar waters, where water temperatures often approach −2°C (the freezing point of sea water). Smaller cetaceans use a great deal of metabolic energy to maintain a body temperature of approximately 35°C. Heat loss is minimized by retia mirabilia, complexes of blood vessels that accomplish countercurrent heat exchange in the flippers and tail flukes (see Chapter 21). In addition, a thick layer of blubber under the skin acts as a layer of insulation. Larger whales have a low surface area to volume ratio and thus have little problem keeping warm in polar seas. Gray whales, which filter feed in near-freezing Arctic waters using their huge (1.6 meter long), uninsulated tongues, even have a retia mirabilia in the tongue to conserve heat (Heyning & Mead 1997).

The ability of cetaceans to use the Earth's geomagnetic fields for navigation is poorly understood. Evidence from geomagnetic anomaly maps and records of cetacean strandings points to the use of some sort of geomagnetic detection system in whales. Klinowska (1990) and Kirschvink (1990) hypothesized that cetaceans use the flux density of the earth's magnetic field as a map. The evidence comes from the fact that cetaceans generally migrate parallel to the geomagnetic fields along the sea bottom. Furthermore, this geomagnetic field exhibits regular fluctuations that allow cetaceans to mark their progress along the geomagnetic map. Although they do not use the geomagnetic information like a compass, cetaceans can apparently use the information for local and long-distance migrations. Further support for this theory comes from data on strandings of live whales along coastal beaches. Dates of mass strandings coincide with unusually large fluctuations in the normal geomagnetic flux (Kirschvink et al. 1986; Klinowska 1985a, 1985b), suggesting that whales blunder onto beaches while trying to navigate along contours of geomagnetic minima (Kirschvink 1990; Walker et al. 1986, 1992).

Suborder Mysticeti

The suborder Mysticeti contains the huge baleen whales, which inhabit all oceans. There are 13 Recent species, grouped in six genera and four families. Before intensive whaling decimated their populations, the mysticetes were perhaps more important than the odontocetes in terms of biomass, although there are fewer species of mysticetes. The phylogenetic relationships of baleen whales have proven difficult to reconstruct from morphological evidence alone, but studies using molecular dating and SINE insertion patterns have improved understanding mysticete relationships (**Fig. 19-9**). Seemingly ancestral baleen whales experienced a period of rapid diversification between the early and middle Miocene (Nikaido et al. 2006). Bowhead and right whales (Balaenidae) diverged early in this radiation, followed by the pygmy right whale (Neobalaenidae). The rorquals (Balaenopteridae) and gray whales (Eschrichtiidae) diverged from a clade that radiated rapidly in the early Miocene.

Although all baleen whales are filter feeders, they use three rather distinct feeding styles (Pivorunas 1979). The first is typical of the right whales (Balaenidae). These are large-headed animals with long **baleen plates** and conspicuous lips (cheek flaps; **Fig. 19-10**). Right whales generally graze on small plankton, usually copepods less than 1 centimeter in length, that concentrate in layers at or near the water surface. The whales swim slowly through these concentrations. The water and plankton flow in the front of the mouth, and the water passes along the side of the tongue and out through the baleen, which traps the plankton (**Fig. 19-11**). The trapped plankton is then removed from the baleen plates by the action of the tongue. This sifting of the water can involve skimming a surface layer of plankton or moving through plankton swarms at greater depths.

The second method is that of the rorquals (Balaenopteridae). These whales have huge mouths and heads,

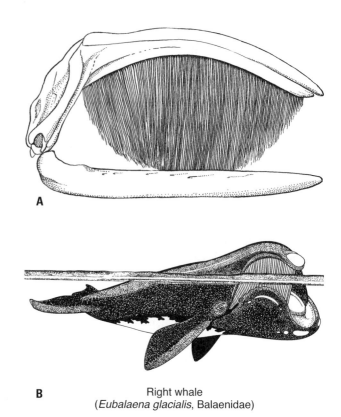

A

B   Right whale
          (*Eubalaena glacialis*, Balaenidae)

FIGURE 19-10   (A) Skull of the Atlantic right whale; length of skull roughly 4 meters. Note the baleen plates attached to the maxilla. (B) Right whale feeding on plankton at the surface of the sea. (Modified from Pivorunas, *American Scientist* 67 (1979): 432–440.)

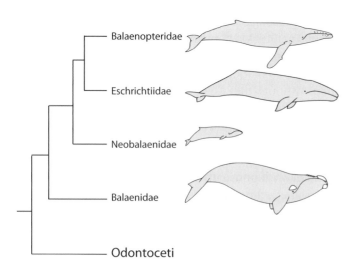

FIGURE 19-9   One phylogenetic hypothesis of the families of extant baleen whales (Mysticeti) based on molecular data from 36 SINE insertion patterns. (Modified from Nikaido, M, et al., *Molecular Biology and Evolution* 23 (2006):866–873, by permission of Oxford University Press.)

relatively short baleen (Fig. 19-11B), and extensive furrowing of the blubber of the throat and anterior abdomen that forms a highly distensible pouch (Lambertsen et al. 1995; **Fig. 19-12**). Rorquals engulf food occurring in dense swarms, usually krill or fish. One of the rorquals, the blue whale (*Balaenoptera musculus*), has been observed to swim up to its prey and engulf it, along with huge amounts of water (up to 6,400 kilograms). The food and water are contained briefly in the capacious ventral pouch. The pouch is then contracted. The water passes through the baleen, and the food trapped by the baleen is swallowed. During feeding, the mandible is lowered until it forms a 90-degree angle with the plane of movement through the water. Lambertsen et al. (1995) described a frontomandibular stay apparatus consisting of a strong fibrous tendon that prevents hyperextension of the mandible as the whale moves through the water (**Fig. 19-13**). A number of variations on the basic feeding pattern are used (Leighton et al. 2007). On occasion, for example, the humpback whale (*Megaptera novaeangliae*) blows a ring of bubbles, a bubble "net," near the surface and, by rising upward within the net, engulfs animals that hesitate to pass through the net and are trapped at the surface (**Fig. 19-14** and 24-9).

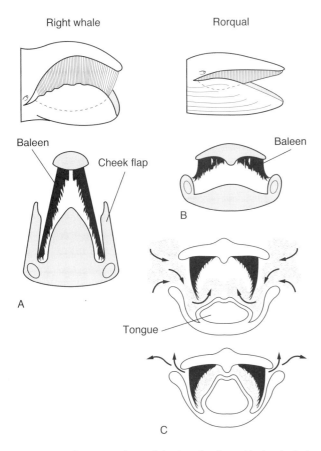

Right whale   Rorqual

Baleen

Cheek flap

Baleen

A

B

Tongue

C

FIGURE 19-11   Cross-sections of the heads of two kinds of whales, showing contrasting arrangements of baleen: (A) right whale and (B) rorqual. (C) The pattern of water flow into and out of the mouth during feeding. (Modified from Pivorunas, *American Scientist* 67 (1979): 432–440.)

The third feeding style is that of the gray whale (*Eschrichtius robustus*, Eschrichtiidae), which feeds by scooping or suctioning material from the bottom and filtering out bottom-dwelling organisms. Gray whales dive to the bottom and, turning on their sides, plow their open mouths through the muddy sediments, leaving shallow trenches in the bottom. They filter out the crustaceans, polychaete worms, and other bottom-dwelling organisms on their baleen plates (MacDonald 1984).

Family Balaenidae

Members of this family, the right whales and the bowhead whale (Fig. 19-10B), were killed in such great numbers during the height of whaling activities that they are rare today and are protected by international treaty. The family includes four Recent species of two genera that inhabit most marine waters except tropical and south polar seas. The earliest fossil balaenids are from the early Miocene of Argentina (McLeod et al. 1993).

Balaenids are large, robust whales that reach about 18 meters in length and weigh over 67,000 kilograms. The

head and tongue are huge; the head amounts to nearly one third of the total length. There are more than 350 long baleen plates on each side of the upper jaw; these plates fold on the floor of the mouth when the jaws are closed. No furrows are present on the skin of the throat or chest. The cervical vertebrae are fused, and the skull is telescoped to the extent that the nasal bones are small and the frontal bones are barely exposed on the top of the skull. The rostrum is arched to accommodate the long baleen plates (Fig. 19-10A). The flippers are short and rounded, and the dorsal fin is usually absent.

Right whales feed largely on copepods and are most common near coastlines or near pack ice. The bowhead whale (*Balaena mysticetus*) never ventures far south of the Arctic Circle (Burns et al. 1993). Eskimos claim it can break through ice nearly 1 meter thick to reach air (Rice 1984). Southern right whales (*Eubalaena australis*) of the southern oceans make long annual migration from temperate or tropical waters to spend the austral summer in Antarctic waters. Northern right whales remain in North Atlantic (*E. glacialis*) or North Pacific (*E. japonica*) waters.

Family Balaenopteridae

Known as rorquals, this family includes seven species and two genera. A possibly new species of balaenopterid (*Balaenoptera omurai*) was recently described by three Japanese researchers (Sasaki et al. 2006; Wada et al. 2003; but see Mead & Brownell 2005). The fossil record of this family dates back to the middle Miocene of the western South Pacific (Bearlin 1988). The distribution includes all oceans. These whales vary in size from the relatively small Minke whale (*Balaenoptera acutorostrata*), at approximately 11 meters long and 4,000 kilograms in weight, to the extremely large blue whale (*B. musculus*), at approximately 31 meters and 160,000 kilograms. In some species of rorquals, the body is slender and streamlined, but it is chunky in others (**Fig. 19-15**). The baleen plates are short and broad, and the skin of the throat and chest is marked by numerous longitudinal furrows (Fig. 19-12). The nasals are small, and the frontals are either not exposed or only barely exposed on the skull roof.

Some of these whales feed in cold waters near the edges of the ice where upwelling, nutrient-rich water results in great growths of plankton in summer. Planktonic crustaceans and small schooling fish are eaten (Tershy 1992). During the northern winter, the Northern Hemisphere populations move southward toward the equatorial areas, and during the southern winter, southern populations move northward. Wintering adults do not feed but instead live off stored blubber. Breeding occurs in the wintering areas. However, because the southern and northern winters are 6 months out of phase, no interbreeding between

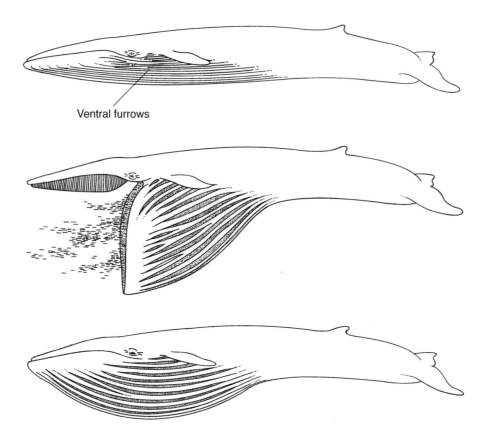

Ventral furrows

FIGURE 19-12 The style of feeding of rorquals and the use of the furrowed, expandable pouch. (Modified from Pivorunas, *American Scientist* 67 (1979): 432–440.)

populations occurs. The humpback whale (**Fig. 19-16**), an animal given to spectacular leaps, makes remarkably melodious and varied underwater sounds (Payne 1970; Payne & McVay 1971). Southern humpback whales (*Megaptera novaeangliae*) make annual migrations of roughly 8,300 kilometers from the Pacific coast of Central America to Antarctica (Rasmussen et al. 2007). Mothers are accompanied by their calves in this movement, and they try to migrate along warm waters between about 21°C and 28°C.

Excessive commercial exploitation has resulted in a tremendous decline in the populations of rorquals such as fin whales (*B. physalus*), humpbacks, and blue whales. Over broad areas, blue and humpback whales are so scarce as to be "commercially extinct" (Rice 1984), and some species of smaller rorquals are also in decline due to overhunting. In the 120 years before their protection, it has been estimated that 338,000 blue whales were killed (Baskin 1993); at up to 175 tons each, this equals 59 million tons of blue whales! Although blue whales are possibly beginning to increase since being legally protected in 1968, it has been difficult to estimate their numbers, a problem exacerbated by the intentional coverup and severe underreporting of numbers taken by some whaling countries (Yablokov 1994). Despite a moratorium on commercial whaling established by the International Whaling Commission in 1982, Iceland, Norway, and Japan, still harvest whales. Iceland and Norway withdrew from the International Whaling Commission and are not bound by its decisions. Japan retained membership, but harvests approximately 1,300 whales annually for "scientific" purposes, thereby circumventing the moratorium. Finally, a small number of whales are hunted each year by indigenous populations in Canada, Greenland, the United States, several Caribbean islands, the Faroe islands, two small Indonesian islands, and the Russian far east.

Family Eschrichtiidae

This family is represented today only by the gray whale (*Eschrichtius robustus*), which occupies parts of the North Pacific. The fossil record for this species dates back only as far as the late Pleistocene (McLeod et al. 1993).

The gray whale is fairly large, weighing up to 31,500 kilograms and measuring 15 meters in length, and has a slender body with no dorsal fin. It has a relatively small head, with a narrow, arched rostrum. The baleen plates are short, and the telescoping of the skull is not extreme. The nasal bones are large, and the frontals are broadly visible on the roof of the skull. The throat usually has two longitudinal furrows in the skin.

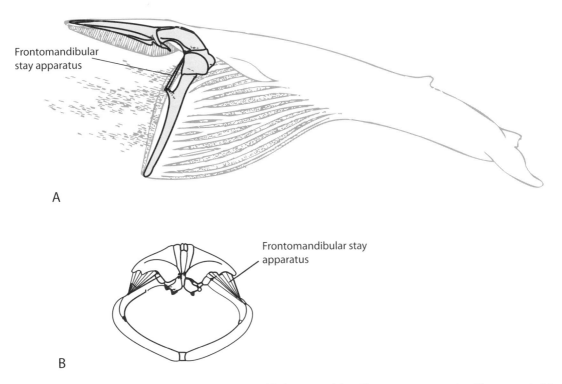

FIGURE 19-13 Drawing of the frontomandibular stay apparatus of Balaenopteridae. The stay apparatus is a fibrous part of the temporalis muscle that mechanically links the frontal bone of the skull with the coronoid process of the lower jaw, thereby limiting hyperdepression of the lower jaw during feeding. The lateral view of the skull shows the maximum gape. The anterior view of the skull shows a moderate gape angle. In both figures, the uniformly shaded areas represent connective tissues covering the temporomandibular joint and the mandibular symphysis. (Adapted from Lambertsen, R., et al., *J. Mammalogy.* 76 (1995): 877–899.)

Gray whales migrate extremely long distances but probably no farther than blue whales. The round trip distance of the migration is from 10,000 to 22,000 kilometers. Gray whales occupy parts of the North Pacific (the Bering, Chukchi, and Okhotsk seas) in the summer. Here they feed largely on bottom-dwelling crustaceans (amphipods), which they take by stirring up the sediments with their snouts. In late autumn, gray whales migrate southward along the coastlines. The western Pacific population winters along the coast of the Korean Peninsula, and the eastern Pacific gray whales winter along the coast of Baja California and Sonora. Young are born in shallow coastal lagoons in the wintering areas. Many people each year watch migrating gray whales from a vantage point at Cabrillo National Monument, near San Diego, and from numerous whale-watching vessels.

The future of the gray whale today seems reasonably bright. Driven nearly to extinction by whaling activities between 1850 and 1925, they are now protected by the International Convention for the Regulation of Whaling, and their numbers have increased greatly in recent years (recent estimates range from 18,000 to 29,000 individuals; Alter et al. 2007). Future populations may decline, however, because of changes in the Bering Sea ecosystem caused by global warming.

Family Neobalaenidae

This family consists of one living species, the pygmy right whale (*Caperea marginata*), which inhabits the marine waters of the Southern Hemisphere. The species is unusual in that it is small, typically between 5 to 6 meters in length, and has more ribs (34) than any other whale. It is also slender, like some rorquals, has narrow flippers, and a small dorsal fin. Pygmy right whales feed mainly on copepods. These whales are not thought to make long-distance migrations but may instead form small, localized populations. Unlike the true right whales, pygmy right whales do not breach or "lob-tail" (slap the water surface with the tail fluke). Little is known about these whales, but they are thought to swim and feed primarily near the surface (Mitchell 1975; Ross et al. 1975).

Suborder Odontoceti

The odontocetes—the toothed whales, porpoises, and dolphins—form the largest suborder of cetaceans in terms of abundance and species diversity, and they are the most widely distributed. Odontocetes include 69 Recent species within 34 genera and seven families, and occur in all oceans and seas connected to oceans. Some members of three families inhabit some rivers and lakes in North America, South

FIGURE 19-14 Two humpback whales (*Megaptera novaeangliae,* Balaenopteridae) breaching the surface during feeding. Note that a flipper is visible in the foreground.

America, Asia, and Africa. Odontocetes are readily observed. They frequently forage close to shore, often make spectacular leaps and roll repeatedly out of the water. Some ride the bow waves of ships much as humans ride shore waves.

Primitive cetaceans, from the early Eocene, generally had laterally compressed, serrated teeth that possibly functioned as do the similar teeth of today's crab-eater seal (see Fig. 16-26), allowing water to escape but trapping small prey when the jaws are closed. As odontocetes radiated in the Miocene, the serrated teeth were progressively replaced by many simple, conical teeth borne on slim, elongate jaws. Such pincer-type jaws probably enabled Miocene odontocetes to capture small, underwater prey by quick snaps and thrusts of the jaws. Many living odontocetes retain such teeth and beak-like jaws. In living odontocetes, engulfment of prey is caused by a rapid retraction of the piston-like tongue, with the resultant rush of water and prey into the mouth. In many living odontocetes, the beak pincers are lost or reduced, perhaps in association with increasing perfection of the ability to stun prey acoustically.

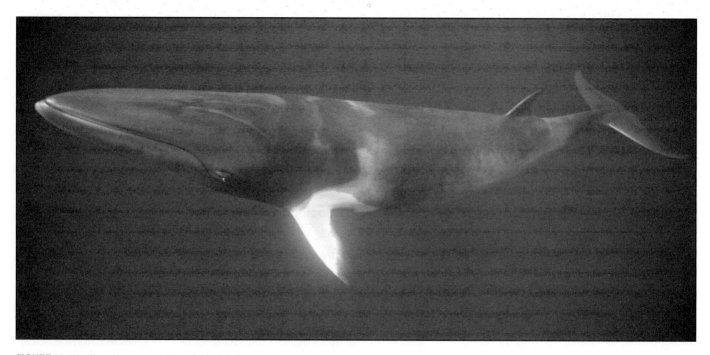

FIGURE 19-15 A common minke whale (*Balaenoptera acutorostrata*).

FIGURE 19-16   A mother and young humpback whale (*Megaptera novaeangliae*; Balaenopteridae).

The evolutionary history of extant odontocetes has been difficult to elucidate. Traditionally, the seven families of toothed whales are grouped into four categories: (1) sperm whales (Physeteridae); (2) beaked whales (Ziphiidae); (3) oceanic dolphins, porpoises, narwhals, and bulugas (Delphinidae, Monodontidae, and Phocoenidae); and (4) river dolphins (Platanistidae and Iniidae). Sperm whales are considered the most basal lineage (**Fig. 19-17**). Until relatively recently, the four genera of river dolphins were grouped together despite their vastly separated geographic distributions. River dolphins are probably not a monophyletic group (Hamilton et al. 2001; Nikaido et al. 2001). Rather, the Amazon River dolphin (*Inia*), LaPlata River dolphin (*Pontoporia*), and Yangtze River dolphin (*Lipotes*) form a clade separate from the Ganges River and Indus River dolphins (*Platanista*). *Platanista* species appear to be basal to all extant odontocetes except the sperm whales (Phyeteridae). Hamilton et al. (2001) suggested that ancestral river dolphins invaded the vast shallow, marine deltas of China, the Indian subcontinent, and South America during the middle Miocene when sea levels were much higher than today. As the epicontinental seas receded in the late Miocene, ancestral odontocetes survived in these isolated river systems.

Just as the mysticetes are characterized by a unique filter-feeding mode of life and by using sound primarily for communication, odontocetes are characterized by

their ability to echolocate (see Chapter 22). An echolocating animal emits acoustic signals and detects objects by receiving and interpreting echoes of these signals. Much has

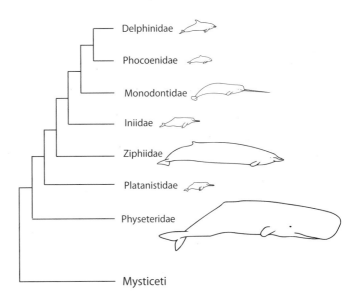

FIGURE 19-17   One phylogenetic hypothesis of toothed whales (Odontoceti) based on molecular and morphological data. The river dolphins are not a monophyletic group; instead, the Platanistidae is a sister group to all odontocetes except the sperm whales. (Adapted from Price, S. A., Bininda-Emonds, O. R. P., and Gittleman, J. L., *Biol. Rev. Camb. Philos. Soc.* 80 (2005): 445–473, and Nikaido, M., et al., *Mol. Biol. Evol.*, 23 (2006): 866–873.)

been learned about cetacean echolocation since the original discovery by McBride in 1947 that dolphins (*Tursiops truncatus*) could avoid nets made invisible by turbid water (Au 1993; McBride 1956; Thomas et al. 2004). Odontocetes emit a great variety of sounds, broadly grouped into two categories: narrow-band continuous tones, such as whistles and squeaks used for intraspecific communication, and broad-band clicks used for echolocation (Au 1993; Herman & Tavolga 1980).

In addition to the use of sound for prey detection, odontocetes may acoustically stun their prey. Bel'kovich and Yablokov (1963) were the first to suggest that an intense, high-frequency sound of the sort that could be made by even a small dolphin could produce a shock sufficient to stun prey. Berzin (1971), who carefully studied sperm whales, was impressed by the fact that large squid and sharks found in the stomachs of sperm whales bore no teeth marks and that even sperm whales with deformed or injured jaws, incapable of grasping prey, had normally full stomachs. Berzin proposed that sperm whales acoustically stun their prey and that this reduced the importance of jaws for feeding. More recently, Hult (1982) observed disorientation among schooling fish and hypothesized that this resulted from the high-intensity sounds made by approaching bottlenose dolphins.

Two additional important and obvious evolutionary trends in odontocetes may also be related to this ability. The first is a trend among advanced odontocetes (such as the bottlenose dolphin) toward the focusing of sound energy into a concentrated beam. Such a specialization furthers long-range echolocation ability, and prey debilitation could be its byproduct. A second suggestive trend involves the loss of teeth, a reduction in their size, or the development of specialized teeth unsuitable for capturing prey (such as the long narwhal tusk). A specialized means of debilitating prey would open the door to such dental changes.

Experimental trials with trained bottlenose dolphins showed that they can produce sounds of intensities equal to the lethal threshold of some marine fish and close to those that have been observed to kill moderately large squid. These intense sounds were not harmful to the dolphins but were some five orders of magnitude above levels usually reported for wild or captive dolphins. In addition, sperm whales (*Physeter catodon*) are capable of producing clicks lasting only one ten-thousandth of a second, but at a deafening 160 decibels (the loudest biologically produced sound; Brownell 2003). Playing recordings of beaked whale sounds to squid (*Loligo*), however, did not debilitate the squid nor cause them to show predator avoidance behavior (Wilson et al. 2007). Although acoustic debilitation of prey by odontocetes has not yet been proved nor disproved, it seems probable and could have evolved by a series of

entirely plausible steps. Highly specialized sound-producing structures evolved in response to sociality and the importance of communication. Sound production later became used for echolocating obstacles and prey. With an increasing ability to produce intense and focused sounds for long-range echolocation, a remarkable new ability may have appeared; prey subjected to these sounds could be stunned. This innovation would free the jaws and dentition from many selective pressures associated with killing or grappling with prey and could allow the shortening of beaks, reduction of dentition, or development of teeth primarily adapted to ritualized combat, as in the case of the male narwhal.

Family Delphinidae

Dolphins are by far the largest and most diverse group of cetaceans. Because some species come close to shore and roll and jump conspicuously, they are the most frequently observed cetaceans. The word "dolphin" is commonly applied to any small cetacean having a beak-like snout, whereas the term "porpoise" generally refers to small cetaceans with a blunt snout and a less streamlined body form. These terms can be misleading, however, as not all delphinids have a beak-like snout (**Fig. 19-18**). Thirty-four Recent species representing 17 genera are known. Delphinids inhabit all oceans and some large rivers and estuaries in southern Asia, Africa, and South America. Fossil delphinids appear first in the Oligocene record of Europe.

Small delphinids are roughly 1.5 meters in length and 50 kilograms in weight, but the killer whale (*Orcinus orca*) reaches 9.5 meters and at least 7,000 kilograms (Fig. 19-18A). The facial depression of the skull is large, and the frontal and maxillary bones roof over the reduced temporal fossa (Fig. 19-7).

The "melon," a lens-shaped fatty deposit that lies in the facial depression, is well-developed and gives many delphinids a forehead that bulges prominently behind the beak-like snout. Some delphinids, such as the killer whale, lack a beak and have a rounded profile. The number of teeth varies from 65/58 to 0/2. From two to six cervical vertebrae are fused (Fig. 19-5). Males are typically larger than females, and in some species, there is considerable sexual dimorphism in the shape of the flippers and dorsal fin. Coloration is varied: Some species are uniformly black or gray, some have striking patterns of black and white, and still others have yellowish or tan stripes or spots.

Delphinids characteristically feed by making shallow dives and surfacing several times a minute. They are rapid swimmers, and some species regularly leap from the water during feeding and traveling. In the Gulf of California, bottlenose dolphins have been observed leaping completely out of the water and catching mullet in midair. Most small

FIGURE 19-18   (A) A killer whale (*Orcinus orca*) breaching. (B) A Pacific white-sided dolphin (*Lagenorhynchus obliquidens*) swimming in the North Pacific.

delphinids eat fish and squid, but the killer whale is known to take a great variety of items, including large bony fish, sharks, seabirds, sea otters, seals and sea lions, porpoises, dolphins, and whales.

Delphinids are typically highly gregarious, and assemblages of approximately 100,000 individuals have been observed. Some groups of delphinids kept in large tanks establish a dominance hierarchy, with an adult male having the highest position (Bateson & Gilbert 1966). Most dolphins spend their entire lives in schools (Norris & Dohl 1980; Pryor & Norris 1991). Schooling behavior enhances the effectiveness of food searching, prey capture, and predator avoidance and may increase reproductive synchrony and efficiency. Dolphins have highly evolved systems of communication and social behavior (Herman & Tavolga 1980). Many recent studies have indicated that cetaceans are remarkably intelligent, inventive, and capable of "higher-order" learning (Pack & Herman 1995). Herman (1980) compares the learning ability of odontocetes with that of primates: "Both taxa have advanced capabilities for classifying, remembering, and discovering relationships among events." In cetaceans, these capabilities are based on auditory perception, whereas primates use mostly visual perception. Remarkably inventive behavior was observed by Hoese (1971), who watched two bottlenose dolphins cooperatively pushing waves onto a muddy shore and stranding small fish. The dolphins rushed up the bank, snatched the fish from the mud, and then slid back into the water. More recently, Visser et al. (2007) reported and videorecorded orcas (*Orcinus orca*) cooperatively and synchronously swimming up to four abreast to create a broad wave that would wash a seal off a small ice floe so that they could catch it.

Family Monodontidae

Family Monodontidae contains two species: the narwhal (*Monodon monoceros*), remarkable for its long, straight, forward-directed tusk (**Fig. 19-19**), and the beluga (*Delphinapterus leucas*), also called the white whale. These two species occur in the Arctic Ocean and the Bering and Okhotsk seas, in Hudson Bay, in the St. Lawrence River in Canada, and in some large rivers in Siberia and Alaska. The earliest monodontids are from the early Miocene of Europe.

These are small to medium-sized cetaceans. Belugas reach about 6 meters in length and 2,000 kilograms in weight, and narwhals, without the tusk, are of similar length. In both species, the facial depression in the skull is large and the maxillary and frontal bones roof over the reduced temporal fossa. The zygomatic process of the squamosal bone is strongly reduced, and the cervical vertebrae are not fused. The beluga has 11/11 teeth, and the narwhal has 1/0. The single upper tooth of the narwhal (usually the left) forms a straight, spirally grooved tusk up to 2.7 meters long; the corresponding tooth in the other upper jaw is normally rudimentary (occasionally the right incisor also develops a tusk).

The gregarious belugas and narwhals are characteristic of northern seas, where in winter they assemble in openwater areas. In summer, belugas move far up large rivers. They feed largely on fish, both benthic (bottom-dwelling) kinds and those that live at intermediate depths, and squid. Narwhals are seemingly largely pelagic (open-sea dwellers). Male narwhals fence with their long tusks, and a tusk occasionally becomes imbedded and broken off in the head of one of the combatants (Silverman & Dunbar 1980). Electron micrographs of the tusk surface reveal millions of tiny pits that apparently connect to sensory neurons,

FIGURE 19-19 A male (foreground) and female (background) narwhal (Monodon monoceros) in the Arctic Ocean. Note the distinctly shaped tail fluke and the long tusk on the male.

leading some researchers to suggest that it may serve to detect temperature or salinity (Milius & Nweeia 2006). Both belugas and narwhals are quite vocal, and the trilling sounds made by belugas account for their common name of "sea canary."

Family Phocoenidae

Members of family Phocoenidae are generally called porpoises. Six Recent species of three genera are recognized. They occur widely in coastal waters of all oceans and connected seas of the Northern Hemisphere, as well as in some coastal waters of South America and some rivers in southeastern Asia. The earliest fossil records of phocoenids are from the late Miocene of Asia, South America, and North America.

Phocoenids are small, from about 1.5 to 2.1 meters in length and 90 to 118 kilograms in weight, and have fairly short jaws and no beak. The dorsal fin is either low or absent. The skull resembles that of the Delphinidae but has conspicuous prominences anterior to the nares. The teeth of most phocoenids are distinctive in being laterally compressed and spade-like; the crowns have two or three weakly developed cusps. *Phocoenoides*, however, has conical teeth. The number of teeth varies from 15/15 to 30/30. From three to seven cervical vertebrae are fused.

Some phocoenids (*Phocoena* and *Neophocaena*) inhabit inshore waters, such as bays and estuaries, whereas the swift Dall's porpoises (*Phocoenoides*) generally inhabit deeper water. Schools of at least 100 phocoenids may assemble, and crescent-shaped formations associated with feeding have been noted (Fink 1959). A variety of food is taken, including cuttlefish and squid, crustaceans, and fish. The small, shy porpoise known as the vaquita (*Phocoena sinus*) is one of the two most seriously endangered cetaceans. Vaquitas inhabit only the northern Gulf of California, and Mexican officials are desperately working to conserve them (Rojas-Bracho & Jaramillo-Legorreta 2000).

Family Physeteridae

The sperm whales occur in all but Arctic oceans, and the giant sperm whale (*Physeter catodon*) of *Moby Dick* fame has long been an important species to the whaling industry. There are two genera and three species of living sperm whales, and numerous fossil forms are known from as far back as the early Oligocene.

Physeter is large, attaining a length of over 18 meters and a weight in excess of 53,000 kilograms. The pygmy sperm whales (*Kogia*) are relatively small, reaching about 4 meters in length and 320 kilograms in weight. The head is huge in *Physeter*, accounting for about one third of the total length. In both genera, the rostrum is truncate, broad, and flat. The facial region of *Physeter* contains two massive oil-filled organs, which together account for approximately 25% of the whales total body mass (Carrier et al. 2002). The **spermaceti organ** is the dorsal oil sac (so called because the semisolidified oil turns milky white and resembles semen).

The ventral sac is called the junk because it contains denser, less valuable oil separated by connective tissue baffles. In addition to playing an important role in echolocation (Chapter 21), these oil-filled organs may serve as battering rams during aggressive encounters between males (Carrier et al. 2002). The blowhole is toward the end of the left side of the snout, and the nasal passages are highly specialized. The posterior skull forms a concave depression that supports the posterior end of the spermaceti organ (Fig. 22-14). The upper jaw (except in occasional individuals) lacks functional teeth; the lower jaw has 25 functional teeth on each side in *Physeter* and from 8 to 16 in *Kogia*. All of the cervicals are fused in *Kogia*, and all but the atlas are fused in *Physeter*.

The habits of *Kogia* are not well-known, but those of *Physeter* are better understood, probably because humans have persistently hunted this animal for many years. *Physeter* is social and assembles in groups with occasionally as many as 1,000 individuals. Schools of females with their calves, together with male and female subadults, are overseen by one or more large adult males, whereas younger males congregate in "bachelor schools." Some adult males are solitary. Sperm whales generally forage in the open sea at depths where little or no light penetrates (the use of echolocation by *Physeter* is discussed in Chaper 22). Dives to depths of 1,000 meters are probably usual, and dives of 1,130 meters have been recorded (Heezen 1957). *Physeter* feeds largely on deep-water squid, including giant squid, sharks and skates, and such bony fish as tuna and barracuda. Males commonly migrate north in summer and are occasionally seen in the Bering Sea, but females remain in temperate and tropical waters. *Kogia* occurs in small schools and feeds largely on cephalopods such as squid and cuttlefish. Sperm whales were one of the most important species to the whaling industry until the early 1980s, when hunting of these animals was prohibited by the International Whaling Commission. Sperm whales were prized for the spermaceti oil and used to produce smokeless candles, lubricating oil, and soaps. Before 1750, most sperm whaling took place along the coastlines because each whale had to be towed to shore for processing. After 1750, on-board processing and market demand for whale oil products lead to a rapid increase in sperm whale hunting on the open ocean (**Table 19-2**; Gosho et al. 1984).

Family Platanistidae

This family includes two species of long-snouted river dolphins, the Ganges River dolphin (*Platanista gangetica*), and the Indus River dolphin (*P. minor*). Small populations of both species persist in the Ganges, Brahmaputra, and Indus Rivers of India, Nepal, Bangladesh, and Pakistan. Until relatively recently *Platanista*, were grouped with the other species of river dolphins in a single family. However, molecular data indicate that *Platanista* is not closely related to Chinese or Amazon River dolphins (Hamilton et al. 2001; Nikaido et al. 2001).

These are small cetaceans, from 2.0 to 2.9 meters in length and roughly 90 kilograms in weight. The jaws are unusually long and narrow and bear numerous teeth (28–29 on either side of the jaws). The forehead rises abruptly and is rounded, giving the head an almost bird-like aspect. In *Platanista*, there are large crests that are probably of sesamoid origin. The large temporal fossa is not roofed by the maxillary and frontal bones. None of the cervical vertebrae is fused. The eyes of all members of the family are reduced, and presumably food and obstacles are detected

TABLE 19-2   Sperm Whale Harvest Records for the United States Whaling Vessels from 1835 to 1885

Year	Number of U.S. Whaling Ships	Estimated Sperm Whale Harvest	Estimated Barrels of Sperm Whale Oil
1835	500	7,598	172,683
1840	559	6,943	157,791
1845	696	6,949	157,917
1850	543	4,088	92,892
1855	638	3,197	72,649
1860	569	3,243	73,708
1865	276	1,463	33,242
1870	321	2,428	55,183
1875	—	—	—
1880	173	1,656	37,614
1885	133	1,065	24,203

The whaling effort declined in the 1860s because of the American Civil War and newer sources of oil and natural gas. Data are from Gosho et al. (1984).

by echolocation. *Platanista*, which usually swims on its side, lacks eye lenses and can perhaps detect only light and dark.

These cetaceans often inhabit rivers that are made nearly opaque by suspended sediments. Under these conditions, echolocation may completely supplant vision. A variety of fish and crustaceans are captured by probing the muddy river bottom.

■ Family Iniidae

This family includes three monotypic genera; the Amazon River dolphin or boto (*Inia geoffrensis*), the Chinese river dolphin or baiji (*Lipotes vexillifer*), and the La Plata River dolphin or franciscana (*Pontoporia blainvillei*). Amazon dolphins are found in the larger tributaries of the Amazon and Orinoco Rivers of South America. *Inia* feed largely on fish and, during the rainy seasons, may move deep into flooded tropical forests (Humbolt & Bonpland 1852). Observations by Layne (1958) suggest that river dolphins are not as social as many other cetaceans, and they may have fairly acute vision above water. Individuals approaching a narrow channel used their eyes above water, presumably to scan the banks for danger.

Chinese river dolphins (*Lipotes*) inhabit the middle and lower reaches of the Yangtze (Changjiang) River. Adults reach lengths of up to 2.7 meters and may weigh 230 kilograms. These dolphins have suffered considerable habitat loss from the construction of hydroelectric dams and from habitat degradation from river pollution brought about by the rapid expansion of the Chinese economy. In 1992, the Chinese government established five protected reserves along sections of the Yangtze River. Sadly, these efforts may be too late; the last confirmed sighting of a baiji in the wild was in 2004. Since then, concerted efforts by researchers to locate individuals have failed, and baijis may be extinct in the wild (Turvey et al. 2007).

The franciscana (*Pontoporia*) is found in coastal rivers and estuaries from Brazil to Argentina; it is the only river dolphin that can live in saltwater environments. *Pontoporia* is small, ranging from 1.2 to 1.8 meters in length and up to 50 kilograms in weight. This species has an unusually long, thin beak studded with up to 242 teeth. It hunts along the river bottoms for small fish and invertebrates, which it locates by echolocation. Little is known about the conservation status of this species, but hundreds are accidentally killed each year in nets set for sharks.

■ Family Ziphiidae

The beaked whales are widely distributed—they occupy all oceans—but are rather poorly known. Some species have never been seen alive. Nineteen Recent species of six genera are recognized. The earliest fossil record of ziphiids is from the early Miocene of Australia.

These are medium-sized cetaceans with fairly slender bodies. The length varies from 4 to over 12 meters, and the weight reaches 11,500 kilograms. The snout is usually long and narrow, and in some species, the forehead bulges prominently. One species (*Tasmacetus shepherdi*) has a large number of teeth; in the others, the dentition is strongly reduced. Only two lower teeth on each side occur in the two species of *Berardius*. In all remaining ziphiids, there is only a single functional tooth, a lower one, on each side (**Fig. 19-20**). In some species, the lower jaw extends beyond the upper jaw, and the teeth are visible outside the mouth. Two to seven cervical vertebrae are fused. The stomach is divided into 4 to 14 chambers, an unusual morphology for a mammal that eats largely squid and fish.

Beaked whales are deep divers able to remain submerged for long periods. The northern bottlenose whale (*Hyperoodon ampullatus*) can dive for periods well over 1 hour and Cuvier's beaked whale (*Ziphius cavirostris*) can reach depths of 1,885 meters. Some species forage in the open ocean far from land. Most beaked whales are highly social and travel in schools in which all members surface and dive in synchrony. The teeth in those species with reduced dentition may be used primarily during intraspecific social interactions and may be of little use during feeding. The primary food is squid, but deep-sea fish are also taken. The North Atlantic bottlenose whale is known to make annual migrations, and other species are probably migratory as well.

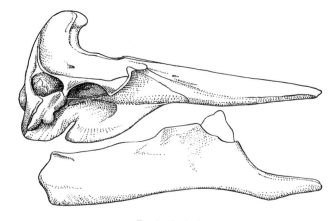

Beaked whale
(*Mesoplodon* sp., Ziphiidae)

FIGURE 19-20 Skull of a beaked whale (Ziphiidae). Note the single large tooth in the dentary bone. The length of the skull is about 590 millimeters.

SUMMARY

Cetaceans comprise the extant clades Mysticeti (baleen whales) and Odontoceti (toothed whales) as well as the extinct Archaeoceti (ancient whales). There are 84 living species in 40 genera.

A remarkable series of Eocene cetaceans from coastal deposits of the ancient Tethys Sea have been discovered recently. These early cetaceans illustrate a transformation from terrestrial quadrupeds to fully aquatic marine mammals. The earliest fossil cetaceans (*Pakicetus*, *Ichthyolestes*, and *Nalacetus*; Pakicetidae) are known from the early to middle Eocene of Pakistan and India. These and other primitive whales belong to the Archaeoceti and probably evolved from ancestral artiodactyls. The hippos (Hippopotamidae) are now considered by many researchers to be the sister group to the Archaeoceti-Cetacea clade.

The first radiation of cetaceans occurred in the middle Eocene in the shallow waters of the former Tethys Sea and led to the diversification of the stem cetaceans (Archaeoceti). By the end of the Eocene, surviving archaeocetes had long, streamlined bodies, tiny hindlimbs, and horizontal tail flukes. A second radiation in the Oligocene-Miocene led to modern cetaceans; both odontocetes and mysticetes radiated rapidly in response to dramatic changes in global climate and continental tectonics. The movement of continental plates during the final breakup of Gondwana created new oceanic circulation patterns, which in turn increased marine productivity. The stage was set for the evolution of filter-feeding mysticetes and the refinement of echolocation-based hunting in odontocetes.

Early odontocetes appeared in the Oligocene, and advanced odontocetes with highly telescoped skulls, homodont dentitions, and many more than the primitive eutherian complement of teeth are known from the Miocene. The transition from tooth-based predation to baleen-based filter feeding was apparently a stepwise process, with intermediate forms possessing both teeth and baleen.

The baleen whales (Mysticeti) inhabit all oceans. Although all baleen whales are filter feeders, they use distinct feeding styles. Some whale species filter-feed by swimming slowly through concentrations of plankton; the water and plankton flow in the front of the mouth, and the water passes along the side of the tongue and out through the baleen, which traps the plankton. In contrast, rorquals are lunge feeders. They engulf huge amounts of water along with krill or fish occurring in dense swarms. The food and water are contained briefly in the capacious ventral pouch, which is then contracted, forcing the water through the baleen and trapping the food. Humpback whales modify this technique by blowing a ring of bubbles near the surface that acts like a net and, by rising upward within the net, engulfing prey trapped within the net at the surface. The gray whale exhibits a third filter-feeding style; it scoops or suctions material from the bottom and filters out bottom-dwelling organisms.

The odontocetes—the toothed whales, porpoises, and dolphins—form the largest suborder of cetaceans in terms of abundance and species diversity. Odontocetes occur in all oceans and seas connected to oceans. Members of three families inhabit rivers and lakes primarily in South America and Asia. Just as the mysticetes are characterized by a unique filter-feeding mode of life and use sound primarily for communication, odontocetes are characterized by their ability to echolocate.

Cetaceans are notable for being the mammals most fully adapted to aquatic life. Baleen whales (Mysticeti), the largest living or fossil mammals known, are mainly plankton feeders (although some eat fish), and thus, they draw from the tremendous productivity near the middle of the marine food web. Some of the large-toothed whales (Odontoceti), in contrast, exploit the top of the marine food web. For example, food of killer whales includes fish, seals, porpoises, and baleen whales. Remarkable swimming and diving ability, the capability of odontocetes to echolocate, considerable intelligence, and complex social behavior have all contributed to the great success of the cetaceans. Unfortunately, excessive commercial exploitation (whaling) has resulted in tremendous declines in many populations.

KEY CHARACTERISTICS OF CETACEANS

- Body fusiform
- Skin almost hairless and sebaceous glands absent
- Thick layer of blubber under skin surrounding body
- Rostrum elongated, cervical vertebrae compressed
- External ear pinnae absent, auditory meatus closed
- Nares located dorsally
- Tail long, muscular, terminating in horizontal flukes
- Forelimbs flipper-like and lacking claws or nails
- Clavicle absent
- Elbow and wrist joints immobile
- External hind limbs absent
- Pelvic girdle vestigial, not attached to vertebral column

- Sacral vertebrae separate
- Diaphragm oblique
- Gall bladder absent
- Testes abdominal
- Uterus bipartite
- Placenta epitheliochorial
- Mammae extrudable, elongated, in narrow cavities flanking urogenital opening

Mysticeti

- Teeth absent in adults
- Rows of baleen plates composed of flexible keratin plates

- External nares (blowholes) paired
- Skull symmetrical
- Auditory bullae attached to skull

Odontoceti

- Teeth homodont and monophyodont
- Echolocate using complex nasal passages
- Skull asymmetrical, with concave profile
- External nare (blowhole) single
- Fatty melon present

KEY TERMS

Altruistic behavior	Eupnea	Odontoceti
Apnea	Flukes	Spermaceti organ
Baleen plates	Mysticeti	

RECOMMENDED READINGS

Berta, A, JL Sumich, & KM Kovacs. 2006. *Marine Mammals: Evolutionary Biology*, 2nd ed. Elsevier/Academic Press, Boston, MA.

Burns, WCG & A Gillespie. 2003. *The Future of Cetaceans in a Changing World*. Transnational Publishers, New York.

Deméré, TA, et al. 2008. Morphological and molecular evidence for a stepwise evolutionary transition from teeth to baleen in mysticete whales. *Systematic Biology, 57*:15–37.

Estes, JA, DP Demaster, & DF Doak. 2006. *Whales, Whaling, and Ocean Ecosystems*. University of California Press, Berkeley, CA.

Geisler, JH & MD Uhen. 2005. Phylogenetic relationships of extinct cetartiodactyls: results of simultaneous analyses of molecular, morphological, and stratigraphic data. *Journal of Mammalian Evolution, 12*:145–160.

Mann, J & PL Tyack. 2000. *Cetacean Societies: Field Studies of Dolphins and Whales*. University of Chicago Press, Chicago, IL.

Perrin, WF, B Würsig, & JGM Thewissen. 2002. *Encyclopedia of Marine Mammals*. Academic Press, San Diego, CA.

Thewissen, JGM. 1998. *The Emergence of Whales: Evolutionary Patterns in the Origin of Cetacea*. Plenum Press, New York.

Thomas, JA, CF Moss, & M Vater. 2004. *Echolocation in Bats and Dolphins*. University of Chicago Press, Chicago, IL.

Zimmer, C. 1998. *At the Water's Edge: Fish Without Fingers, Whales With Legs, and How Life Came Ashore But Then Went Back to Sea*. Touchstone Books, New York.

Mammalian Structure and Function

PART III

Reproduction

Because of its primary importance to all life, reproduction is tied to virtually every structural, physiological, and behavioral adaptation of an individual or a species. The unique mammalian pattern of reproduction must be of great antiquity: Mammary glands, nourishment of the newborn or newly hatched with milk, and a close mother–young bond probably evolved together with the diphyodont dentition in the late Triassic, although the bearing of live young (viviparity) most likely appeared later.

Of all the vertebrate classes, the reproductive pattern typical of mammals departs most from that of other vertebrates. Primitive ancestral vertebrates presumably were egg layers, and this style of reproduction, or some fairly modest variation on this theme, is typical of all classes of vertebrates but the Mammalia. Monotreme mammals retain the amniotic egg of their ancestors, but at ovulation the ova are smaller and contain less yolk than those of similarly sized reptiles or birds. Other shared primitive features of monotremes include the presence of an **egg tooth**, a leathery eggshell, and incubation of the egg that continues outside the mother's body. However, unlike amniotes, after the egg hatches, the young monotreme is nourished by milky secretions produced by the mother's mammary glands. According to one scenario proposed by Lynch and Wagner (2005), ancestral therians evolved the ability to retain the developing eggs in the oviducts for longer periods, thereby protecting the eggs from changing weather patterns and food supplies and freeing the mother from incubating the eggs after laying. Retention of the eggs within the mother's reproductive tract was likely accompanied by loss of the leathery eggshell. For example, in living metatherians, the shell membrane is thin and highly permeable, permitting the embryo to receive additional nourishment from the uterus during gestation. At this point, selection likely favored traits that enhanced *in utero* embryonic development, such as the modification of embryonic membranes to form a more intimate contact with the uterus, resulting in a placenta. These dramatic changes in development may have taken place rapidly, in as few as 3 million years, but would have profound consequences for the evolution of therian mammals (Lynch & Wagner 2005).

In therians, there are important differences in the relative lengths of gestation and lactation periods. The young remain within the uterus, and it is here that many embryonic tissues and organs differentiate and the fetus grows. Nourishment and protection for the intrauterine young are provided by the mother, and under most conditions, fetal survival rates are high. After birth, all young mammals are nourished by milk from the mother, and parental care, or in most cases maternal care, lasts until the young are reasonably capable of caring for themselves. The young of some mammals stay with their parents through an additional period in which they learn complex foraging and social behavior. In sharp contrast, in most nonmammalian vertebrates (birds are an exception), the young have little or no parental care after hatching or, in the case of **ovoviviparous** animals (those that produce eggs that are incubated and hatch within the parent's body), after birth.

Most amniote vertebrates lay great numbers of eggs at tremendous metabolic cost, and the success of the species depends on the survival of an extremely small percentage of young. For any given young of most lower vertebrates, survival is unlikely. In mammals, in contrast, relatively few young are produced, but extensive **postpartum** care results in high-energy expenditure per young and an increased likelihood each offspring will survive to sexual maturity.

Reproductive System Anatomy

Male Anatomy

In mammals, the reproductive system includes the gonads, which produce gametes and hormones, the structures that transport gametes, and in females, the structures that house and nourish the embryos after fertilization. The paired testes of males produce sperm (gametes) and testosterone (steroid hormones). During the breeding season, special cells within the testes, called spermatogonia, undergo spermatogenesis, the process by which mature spermatozoa (sperm cells) develop (**Fig. 20-1**). Spermatogenesis begins when a diploid spermatogonium (germ cell) near the basal lamina of seminiferous tubules, undergoes mitosis, yielding two diploid primary spermatocytes. Typically, one primary spermatocyte remains near the basal lamina and generates additional spermatogonia. The other primary spermatocyte duplicates its DNA and undergoes the first of two meiotic divisions (meiosis I) to produce two secondary spermatocytes and then four haploid spermatids. The meiotic divisions introduce genetic variation by the random assignment of maternal and paternal chromosomes and by crossing-over events. The result is that no two spermatids will be genetically identical. The haploid spermatids now begin a process of maturation called spermiogenesis (Fig. 20-1). Each spermatid begins to grow a tail. The energy-producing mitochondria congregate in a thickened midpiece, and DNA is tightly packaged into the head region. These maturation events are controlled by testosterone, which removes unnecessary cytoplasm and organelles, which are then engulfed by sertoli cells. Mature spermatozoa are released into the lumen of the seminiferous tubule (spermiation) and transported to the epididymis in fluid secreted by the sertoli cells. It is in the epididymis (a tightly coiled tube connecting the testis to the vas deferens) where sperm finally acquire motility. Spermatozoa are stored in the epididymis until ejaculation, when smooth muscle in the walls of the vas deferens contract propelling the sperm forward. The sperm enters the urethra, along with secretions from the seminal vesicles, prostate gland, and the bulbourethral glands, which form the semen. Muscular contractions of the penis expel sperm during ejaculation. The final stage in the maturation of sperm (capacitance) takes place in the female reproductive tract.

The testes are retained within the abdomen in monotremes, edentates, elephants, sirenians, and cetaceans, but in most mammals, they descend into an external pouch, the scrotum. In many mammals, the testes are in a scrotal position only during the breeding season and are withdrawn into the body cavity at other times. The penis, or intromittent organ, transfers sperm to the interior of the female reproductive tract during copulation. When not in use, the penis may be retracted into a sheath. Many mammals, including rodents, bats, carnivores, and nonhuman primates, have a bone or baculum embedded within the penis (Fig. 2-8). Bacula are often highly distinct and are useful in taxonomy. During copulation, the sinuses within the penis become engorged with blood, stiffening and erecting the penis. The monotreme penis has four rosettes (two are functional) at its tip. Most marsupials, except the gray and red kangaroos, have a bifid penis (Dawson 1995). Penile morphology varies widely among eutherian mammals. In rodents, the glans penis may be cup shaped (Fig. 2-8), and in many rodent and bats, the exterior is covered with spines (**Fig. 20-2**), which are believed to aid in the removal of copulatory plugs (Ryan 1991a, 1991b, and references therein). The variation in mammalian penile morphology is likely the result of sexual selection. In mammal species where females frequently mate with multiple males, competition between males extends

BOX 20-1   *Hox* Genes and the Evolution of Therian Placentas

The ability to retain and nourish embryos in the uterus was a key evolutionary innovation in therians. The shift from egg laying to internal gestation required a suite of morphological and physiological adaptations. How did this complex reproductive transition evolve? New evidence from the field of evolutionary developmental biology (evo-devo for short) is beginning to provide some answers.

One class of genes, the so-called master control genes, or *Hox* genes, are believed to have played a key role in reshaping the female reproductive tract to allow therian embryos to invade the uterine lining and establish a placenta. *Hox* genes exert control over other genes and are therefore responsible for organizing the proper tissues in the correct location at a specific time in development. Changes in the timing or location of *Hox* gene expression may lead to novel morphologies. For example, *HoxA-9* is expressed in tissues destined to become the fallopian tubes, *HoxA-10* and *HoxA-11* in regions that will form the uterus, and *HoxA-13* in a novel structure in therians, the vagina.

Hox genes are not only important during development. They play different roles in adult tissues. *HoxA-10* and *HoxA-*

11 are required for uterus formation during development, but they also are required in adult females for the embryo to implant successfully in the uterine lining. Likewise, *HoxA-13* is required for vaginal formation during development and for the formation of umbilical arteries during pregnancy. Lynch and Wagner (2005) sequenced these genes in placental mammals, marsupials, monotremes, reptiles, and amphibians and studied those sequences for evidence of positive selection. They found specific mutations in *Hox* genes that were shared by all placental and marsupial mammals but which were lacking in monotremes and nonmammalian vertebrates. These mutations led to the evolution of a uterus and vagina in therian mammals. Later, additional gene mutations led to the formation of a placenta and umbilical vessels in support of longer gestation periods. These results indicated that the evolution of new genes is not required for evolution of novel structures or physiologies. Rather, subtle changes in existing genes (or transcription factors that regulate those genes) under positive selection are sufficient to generate new morphologies (see Carroll 2005).

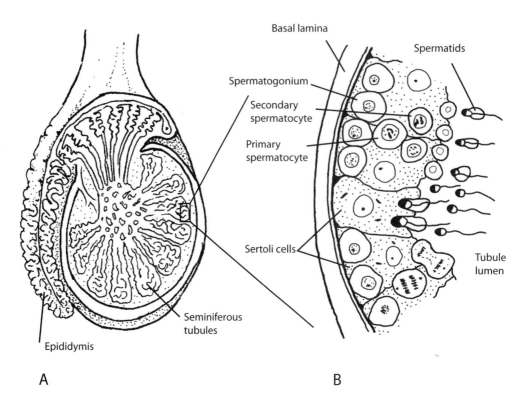

Basal lamina

Spermatids

Spermatogonium

Secondary
spermatocyte

Primary
spermatocyte

Sertoli cells

Tubule
lumen

Seminiferous
tubules

Epididymis

A

B

FIGURE 20-1 Sectional view of the testis and epididymis (A) with an enlargement of a region of the testis showing spermatogenesis within the walls of the seminiferous tubules (B). Spermatogonia along the tubule walls divide to form more spermatogonia and primary spermatocytes. The primary spermatocytes undergo meiosis to form secondary spermatocytes, which then are reorganized into spermatozoa. Sertoli cells house the developing spermatozoa until they are ready to be released into the lumen of the tubules. From the lumen, they travel to the epididymis where they are stored.

beyond copulation, to include sperm competition; competition is between rival ejaculates for fertilization of ova. Sperm competition may result in increase ejaculate volume and the formation of copulatory plugs, coagulated accessory gland secretions that plug the female reproductive tract. Accessory glands are particularly well-developed in mammals that produce copulatory plugs (some rodents, primates, marsupials, bats, and insectivorans; Bedford 2004). Selection for larger plugs may explain the variation in accessory gland size in mammals (Ramm et al. 2005).

Female Anatomy

Females have a more complicated role in reproduction than males. Like males, they manufacture gametes (ova), but unlike males, females must receive and transport sperm to the ova, provide nourishment to the developing young during gestation, are responsible for the laying of an egg or in the live birth of young (parturition), supply continued nourishment for the young in the form of energy rich milk (lactation), and often offer protection and support for the young after lactation ends.

The gametes (ova) of mammalian females mature in paired ovaries. The process of oogenesis, however, differs markedly from spermatogenesis in males. Although male

spermatogenesis continues throughout the lifetime of the individual, oogenesis in females takes place in the fetus during gestation. Primordial germ cells called oogonia in the fetal ovary divide repeatedly (via mitosis), forming many primary **oocytes**. These primary oocytes begin, but do not complete, the first of two meiotic divisions; meiosis is arrested at this stage until just before ovulation. Just before birth, many primary oocytes in the fetal ovary are surrounded by follicle cells forming primary follicles. Those that do not form primary follicles disintegrate. At birth, only a subset of the original oocytes remains, serving as a final reservoir from which ova are drawn for continued development and subsequent ovulation. After reaching sexual maturity, one or more primary follicles develop into secondary follicles and may be ovulated (in humans 99.8% of all primary follicles degenerate; Sherwood et al. 2005). Just before ovulation, the arrested primary oocyte enlarges and completes the meiotic division (except in carnivores and equids where primary oocytes are ovulated) forming two haploid cells: One is the secondary oocyte, which contains most of the cytoplasm and will go on to form a mature ovum, and the other forms the first polar body and degenerates. It is this secondary oocyte that is ovulated and passes into the fallopian tube, where, if viable sperm are present,

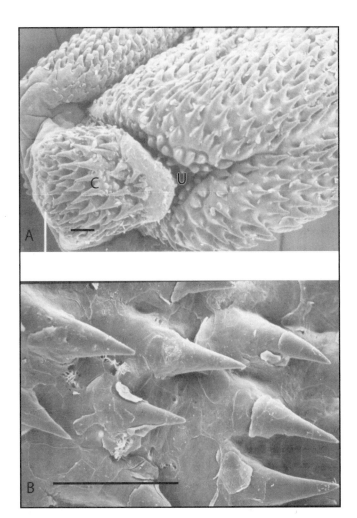

FIGURE 20-2 Scanning electron micrograph of (A) the ventral view of the glans penis of the wrinkle-lipped bat (*Mormopterus jugularis*; Molossidae) showing the urethral cap, C; urethral opening, U; and epithelial spines. (B) A close-up of the epithelial spines on the penis in black mastiff bat (*Molossus rufus;* Molossidae). Scale bars equal 100 microns.

it is fertilized. Sperm entry is required for the secondary oocyte to complete the second meiotic division forming a mature ovum.

The female reproductive tract varies considerably among the three major mammalian groups. In monotremes, each ovary is surrounded by an **infundibulum** connected to a narrow uterine tube (fallopian tube), which continues to a widened uterus. The right and left uteri independently enter a common urogenital sinus (**Fig. 20-3A**). Excretory products from the urinary bladder and rectum also enter the urogenital sinus before passing out of the body through the cloaca. In female marsupials, the posterior portion of each oviduct also forms a pair of uteri that enter separately into the vaginal sinus (Fig. 20-3B). Although there is a common vaginal sinus, each uterus retains a cervix (duplex condition). In the opossum (*Didelphis*), the vaginal sinus connects to separate right and left

vaginas, which enter a common urogenital sinus. Before birth, the vaginal sinus expands posteriorly to connect with the urogenital sinus forming a median birth canal called a pseudovaginal canal. In many marsupials (e.g., kangaroos), the pseudovaginal canal is retained permanently after the first birth, but in others, it closes after each birth (Dawson 1995). The bifid penis of many marsupials is thought to position semen at the base of the paired vaginas: Sperm travels up the right and left vaginas, and the pseudovaginal canal serves as the birth canal.

Eutherians retain paired ovaries, infundibula, and fallopian tubes, and in such groups as elephants, lagomorphs, and many rodents, the uteri and cervix are paired (duplex uterus). In other species, including many carnivores and ungulates, the two uterine horns are fused near the base (bipartite uterus or bicornuate uterus): Primates and armadillos have a single uterus (simplex uterus). Regardless of the degree of uterine fusion, eutherians always have a single vagina (Fig. 20-3C).

The features of gametogenesis are similar for all mammals, but the reproductive patterns occurring after fertilization are vastly different among monotremes, marsupials, and placentals. Montremes lay eggs and share a number of other features with reptiles and birds. Both metaterians and eutherians bear live young but have diverged from one another with respect to the relative length of the gestation and lactation periods.

Monotreme Reproduction

The five living monotremes are of great interest because of their combination of reptilian and mammalian characteristics. Among the ancestral characteristics retained by monotremes are a yolky egg, a fetal egg tooth, a urethra that enters the urogenital sinus dorsally, somite stage embryos at the time of egg laying, eggs that are incubated outside the mother's body, and a lack of teats (Renfree 1993).

Male echidnas (*Tachyglossus acelatus*) exhibit several unusual reproductive traits. Urine and semen exit from separate openings. The urine passes into the cloaca, whereas semen passes through the urethra to the tip of a uniquely shaped glans penis (Johnston et al. 2007). The short-beaked echinda's penis, when erect, is nearly a quarter of the animal's body length. Initially, the penis has four protruberences (rosettes), two on each side of the bifid glans penis. However, it appears that one side of the glans retracts after erection leaving only two functional rosettes on one side of the erect penis. These two lateral rosettes are positioned to deliver semen into the female's paired oviducts. Another unusual feature is that echidna sperm are bundled together in packets of up to 100 spermatozoa. The packets are fused

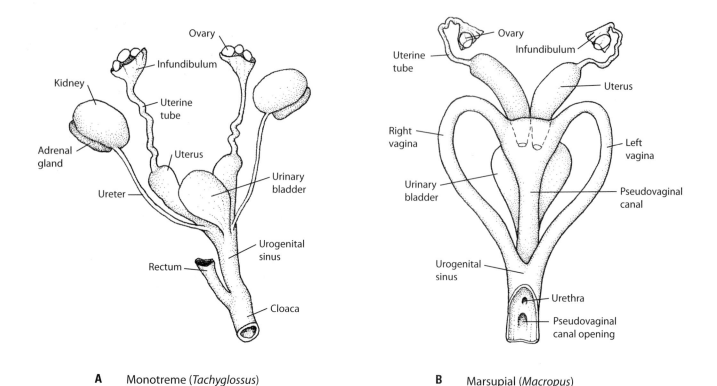

A Monotreme (*Tachyglossus*)

B Marsupial (*Macropus*)

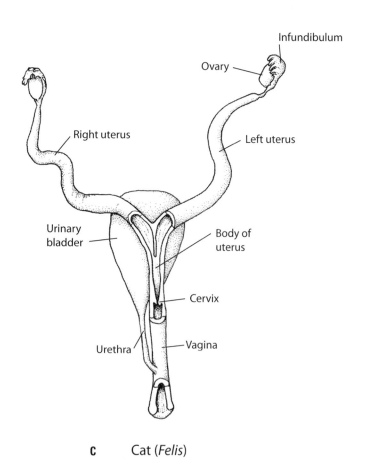

C Cat (*Felis*)

FIGURE 20-3 (A) The female reproductive tract in a monotreme (*Tachyglossus*). (B) A wallaby (*Macropus*). (C) A placental mammal (*Felis*). The kidneys and ureters are shown in the monotreme as they enter the urogenital sinus.

together at their tips, leaving the tails free to beat rhythmically as the packet advances. Whether these sperm packets serve in sperm competition or aid in sperm storage is not known (Johnston et al. 2007).

One characteristic that truly sets monotremes apart from other mammals is that females are oviparous (lay eggs). The reproductive tract of female monotremes (Fig. 20-3A) resembles that of reptiles. Relative to placental mammals, the ovaries of monotremes are much larger because of the greater amounts of yolk in each egg. Mature oocytes (approximately 3 to 4 millimeters in diameter) are shed into the infundibulum, where they are fertilized before entering the oviduct. Although monotremes retain the paired oviducts of their reptilian ancestors, only the left oviduct is functional in the platypus. During passage of the fertilized egg down the fallopian tube, a mucoid coat (consisting of glycoprotein) is secreted around the egg (Hughes et al. 1975). Farther down the fallopian tube, in the tubal gland region, a basal layer of ovokeratin is secreted, forming the thin shell (Griffiths 1978). After the egg passes into the anterior uterus, a second shell membrane is added (Hill 1941). In the uterus, development proceeds via meroblastic cleavage (the cleavage furrow does not pass through the yolk), forming a **blastodisc** on top of the yolk, as in reptiles. Eventually, the blastodisc completely envelops the yolk, and embryogenesis continues. Uterine secretions provide additional nutrition at this stage. After the embryo reaches about 10 to 12 millimeters in diameter, the third, more porous, layer of the shell is applied, and the eggs are laid soon afterward.

Very little is known about the manner in which eggs are laid and their subsequent incubation by the female, but the following comes from a variety of observations of the echidna (*Tachyglossus aculeatus*), summarized by Griffiths (1978). Apparently the female curls up, and in this position, the cloaca is everted and nearly enters the pouch. The egg moves through the cloaca directly to the pouch, where further incubation occurs. The pouch forms a shallow depression with swollen lips that forms the incubation chamber. Newly laid eggs are covered with a sticky coating that keeps the egg attached when the female must move about. Approximately 10 days after the egg is laid, the tiny young uses its egg tooth to emerge from the shell (**Fig. 20-4**). After hatching, the young finds its way to a mammary lobule inside the pouch and begins to suckle the rich milk. Young *Tachyglossus* remain in the pouch for approximately 55 days. When their spines begin to emerge, they are dropped from the pouch by the mother. Evicted young can continue to suckle from their mother for several more months (Griffiths 1978).

It is a mistake to assume that because monotremes lay eggs their reproduction is similar to that of reptiles.

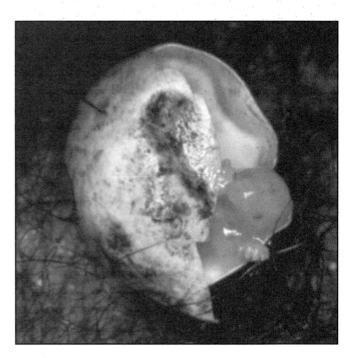

FIGURE 20-4   Young echidna (*Tachyglossus* sp.) emerging from the egg.

Although oviparity is undoubtedly the ancestral condition, monotreme eggs are much smaller at ovulation and contain considerably less yolk, and for most of the intrauterine development, growth of the embryo depends on absorption of **endometrial gland** secretions across the yolk sac. Furthermore, monotreme young hatch at a very early stage of development (as do metatherian young) and depend on lactation for continued development.

Therian Reproduction

The recognition of a "metatherian-eutherian dichotomy" is based on a number of biological differences; primary among these are the contrasting reproductive patterns of the two groups. Metatherians bear virtually embryonic young after a brief gestation period. In contrast, many eutherians bear anatomically complete, often highly precocious young after a relatively long gestation period (**Fig. 20-5**). These contrasting reproductive patterns are driven by differences in estrous cycles, rates of embryogenesis, degree of placentation, sequence of hormonal events, as well as unique environmental cues and behavioral traits. The assumption has been that metatherian reproduction was somehow inferior to reproduction in placental mammals. Part of the basis for this, now antiquated assumption, was that placental mammals tend to outcompete marsupials, except in Australia, where many species evolved in the absence of eutherian competitors. A more complete understanding of marsupial reproduction reveals that their reproductive patterns are likely derived, albeit divergent,

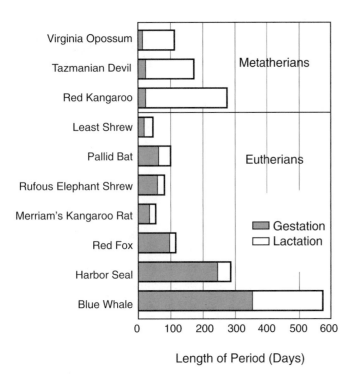

FIGURE 20-5 A comparison of the gestation and lactation periods in metatherian and placental mammals. Metatherians invest more heavily in the lactation period. Placental mammals show a wide range of variation but generally invest in a longer gestation period. (Data from Dawson, T. *Kangaroos: Biology of the Largest Marsupial.* Cornell University Press, 1995; data from Eisenberg, J. *The Mammalian Radiations: An Analysis of Trends in Evolution, Adaptation, and Behavior.* University of Chicago Press, 1981.)

reproductive modes (**Table 20-1**; Hayssen et al. 1985; Renfree 1993). Tyndale-Biscoe and Renfree (1987) propose that eutherians evolved reproductive characters associated with longer gestation, whereas metatherians evolved along a separate trajectory with an emphasis on lactation for early growth of the young. In general, eutherians have faster reproductive rates than marsupials. This may be the main reason that metatherian diversity is declining in areas where both groups occur.

■ The Estrous Cycle

In female mammals, reproduction is characterized by a series of cyclic events that are under nervous and hormonal control. As with many complex functions of the vertebrate body, the regulation of the reproductive cycle is maintained by environmental and social cues and by reciprocal controls between endocrine organs and their secretions. The events characterizing the stages in the mammalian reproductive cycle are well known, but details of the hormonal regulation of these events are not completely understood for many mammal species. The ovarian cycle results in the development of ova, their release from the ovary, and their passage into the uterus; the uterine cycle involves a series of cyclic changes in the uterus. These changes result in a period of

heightened sexual receptivity called estrus or heat, and the estrous cycle is the time from one entry into estrus and the next. Mammals may be monoestrous, exhibiting a single estrous cycle per year, or **polyestrous**, having more than one estrous cycle annually (**Table 20-2**). Many primates have a menstrual cycle, in which the endometrium (uterine lining) is shed at **menstruation** instead of being reabsorbed if conception does not occur. Primates with menstrual cycles remain sexually receptive throughout the cycle. Thus, the estrous and menstrual cycles are named after the most obvious period of the cycle, behavioral estrus and menstruation, respectively.

The mammalian estrous cycle includes four phases: **proestrus**, **estrus**, **metestrus**, and **diestrus** (**Fig. 20-6**). In species with a nonbreeding period (anestrus), the cycle begins with proestrus as the female enters a short period of sexual receptivity (heat). Proestrus is characterized by the growth of the follicle (the ovum, or egg cell, surrounded by specialized cells of the ovary) and its secretion of estradiol and inhibin. Ovulation typically occurs during estrus and is triggered by a massive surge in luteinizing hormone (LH). After the release of the ova, the ruptured follicle develops into the corpus luteum (a glandular structure). LH maintains the corpus luteum and stimulates it to secrete the hormones progesterone and to a lesser extent estrogen. Progesterone (and estrogen) prepares the **endometrium** (inner lining of the uterus) for the implantation of the blastocyst (a hollow sphere resulting from early cleavage of the zygote), mobilizes the endometrial glands to produce glycogen, increases vasculature in the uterine lining, and suppresses uterine contractions. These events set the stage for implantation and placenta formation (described later here). If **fertilization** of the ova does not occur and there is no implantation, however, the endometrium secretes prostaglandins. Declining progesterone secretion, coupled with increasing prostaglandin levels, leads to regression of the corpus luteum during metestrus. If fertilization and implantation occurs, the corpus luteum continues to produce progesterone for the duration of gestation in most mammals.

A period of variable length called anestrus occurs before the next proestrus. In **monestrous** mammals, anestrus lasts most of the year and females are receptive to males for only a short period during the year. In contrast, polyestrous mammals have a short diestrus and therefore are receptive on a regular schedule throughout the year, or part of the year (**Fig. 20-7**).

Spontaneous ovulation without copulation occurs widely among mammals, but some deviations are common. In several carnivores, some rodents, some insectivorans, and most lagomorphs, the follicles develop, but ovulation does not occur until shortly after copulation (**induced ovulation**). In rabbits, the follicles do not develop fully before

TABLE 20-1   Reproductive Characters in Monotremes, Metatherians, and Eutherians

Feature	Monotremes	Metatherians	Eutherians
Sperm head	Elongate	Short	Short
Glans penis	Bifid	Bifid or single	Single
Testes	Abdominal	Inguinal or scrotal	Abdominal, inguinal, or scrotal
Seminal vesicles	Absent	Absent	Present
Prostate gland	Present	Present	Present
Endometrium	Secretory	Secretory	Secretory
Ovum	Large, yolk-filled	Small, some yolk	Very small, no yolk
Cleavage pattern	Meroblastic	Holoblastic	Holoblastic
Mucoid coat	Present	Present	Absent in most species
Shell membrane	Present	Present	Absent
Shell	Present	Absent	Absent
"Birth" of young	Altricial from egg	Altricial from uterus	Altricial to precocial from uterus
Mammary glands	Present	Present	Present
Mammary hairs	Present	Present	Absent
Teats	Absent	Present	Present
Pouch	Present/absent	Present/absent	Absent
Lactation period	Lengthy	Lengthy	Short

Data are from Tyndale-Biscoe and Renfree (1987) and Renfree (1993).

copulation, and a long estrus may occur; ripening of the follicles and ovulation are initiated by copulation. In some rodents, pheromonal and tactile stimuli from the male may initiate ovulation. However, not all species within the same family show the same pattern of ovulation. It may therefore be more useful to regard ovulation patterns along a continuum with different selection pressures resulting in variation among species (Lariviere & Ferguson 2003).

When copulation occurs, the sperm reach the oviducts in a matter of minutes in some species, and fertilization of

TABLE 20-2   Characteristics of the Estrous Cycle in Mammals

Species	Type of Cycle	Estrous Cycle Duration (Days)	Estrus Period	Ovulation Type
Vulpes vulpes (red fox)	Seasonal, Monestrous	90	1–5 days	Spontaneous
Canis lupus familiaris (dog)	Seasonal, Monestrous	60	7–9 days	Spontaneous
Felis catus (cat)	Seasonal, Polyestrous	14	4 days	Induced
Ovis aries (sheep)	Seasonal, Polyestrous	15–18	30–36 hours	Spontaneous
Equus caballus (horse)	Seasonal, Polyestrous	19–23	4–7 days	Spontaneous
Bos taurus (cow)	Continuous, Polyestrous	17–27	13-14 hours	Spontaneous
Sus scrofa (pig)	Continuous, Polyestrous	18–23	2–3 days	Spontaneous
Rattus norvegicus (rat)	Continuous, Polyestrous	4–5	13–15 hours	Spontaneous
Cavia porcellus (guinea pig)	Continuous, Polyestrous	16	6–11 hours	Spontaneous
Macaca mulatta (rhesus monkey)	Continuous, Menstral	28	None	Spontaneous
Homo sapiens (human)	Continuous, Menstral	28	None	Spontaneous
Mustela putorius (ferret)	Seasonal	—	Continuous	Induced
Oryctolagus cuniculus (rabbit)	Continuous	—	Continuous	Induced

Data are from Sherwood et al. (2005).

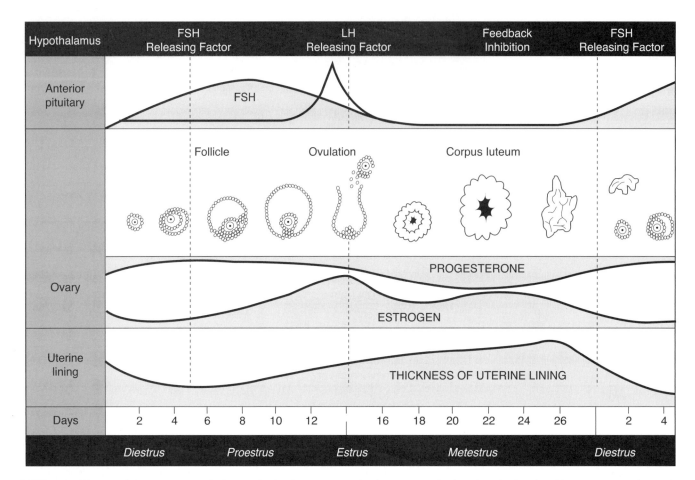

FIGURE 20-6 Timing of the estrous cycle in a typical polyestrous eutherian mammal, showing the levels of important regulatory hormones, state of the uterus, and the transition from the follicle to the corpus luteum. (Modified from Withers. *Comparative Animal Physiology, 1E.* © 1992 Brooks/Cole, a part of Cengage Learning, Inc. Reproduced by permission. www.cengage.com/permissions)

the ova usually occurs within 24 hours of ovulation. The **zygotes** move down the oviducts, aided by **peristaltic** contractions of the muscles of the oviducts and movement of cilia, and usually reach the uterus and implant within a few days.

The estrous cycle of metatherians and eutherians is similar with respect to the changes in the uterus and ovary during the follicular and luteal phases. In addition, many of the eutherian steroid hormones and gonadotropins also occur in marsupials (although their functions may differ; Tyndale-Biscoe & Renfree 1987). Important differences between the two groups are that the corpus luteum in marsupials is generally not prolonged by the implantation of the embryo and the marsupial estrous cycle is typically longer than the period of pregnancy. Tyndale-Biscoe (1984) divided marsupial reproductive patterns into several categories:

1. Species with a gestation period considerably shorter than estrous cycle and in which postpartum estrus ovulations are suppressed during lactation. This basic pattern seems

to be common in the Petauridae, Phalangeridae, and Peramelidae (among others).
2. Species with a gestation period roughly equal in length to the estrous cycle and in which postpartum estrus and ovulation occur. If fertilization occurs as a result of postpartum ovulation during lactation (for the previously born young), development of the resulting embryo is arrested at the blastocyst stage (embryonic diapause). This pattern is typical of most of the Macropodidae.
3. Species with a gestation period roughly equal in duration to the estrous cycle and with postpartum estrus and ovulation but where further development is not controlled by lactation (it may be seasonal). Honey opossums (*Tarsipes rostratus*; Renfree 1981) and the feathertail gliders (*Acrobates pygmaeus*; Ward & Renfree 1986) may exhibit this unusual pattern.

The estrous cycle differs among species, but typically, smaller mammals have more frequent cycles. Species such as golden hamsters (*Mesocricetus auratus*) have estrous cycles lasting only 4 days. The cycle may last 16 weeks in

Phyllostomus hastatus

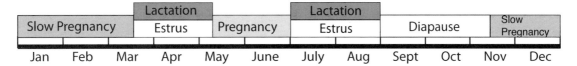

Estrus	Pregnancy				Lactation			Anestrus			Estrus	
Jan	Feb	Mar	Apr	May	June	July	Aug	Sept	Oct	Nov	Dec

Artibeus jamaicensis

Slow Pregnancy		Lactation		Pregnancy	Lactation		Diapause		Slow Pregnancy
		Estrus			Estrus				
Jan	Feb	Mar	Apr	May	June	July	Aug	Sept	Oct	Nov	Dec

FIGURE 20-7 Drawing of the estrous cycle of the greater spear-nosed bat, *Phyllostomus hastatus* (Phyllostomidae), exhibiting a typical pattern for a monestrous bat, and the estrous cycle of the Jamaican fruit-eating bat, *Artibeus jamaicensis* (Phyllostomidae), a polyestrous bat species. (Adapted from Altringham, J. *Bats, Biology and Behaviour*. Oxford University Press, 1996.)

African elephants (Sherwood et al. 2005; Hermes et al. 2000). Regardless of the pattern, estrous cycles result in behavioral and physiological changes, including ovulation of one or more ova.

Embryonic Development and Placentation

In eutherians, fertilization normally takes place in the oviducts shortly after ovulation. The fertilized egg, or zygote, begins a series of cell divisions as it is passed down the oviduct by peristaltic muscular contractions. By the time the blastocyst reaches the uterus, it is a hollow ball of approximately 64 cells (called a **morula**). The journey usually takes 3 to 4 days from ovulation to arrival in the uterus (**Fig. 20-8**).

Ovulation is spontaneous in metatherians, usually occurring 10 hours after copulation in the opossum (*Didelphis*). Sperm transport in marsupials is quite rapid; fertilized eggs have been found in the oviducts five hours after copulation and in the uterus only 22 hours after copulation (Rodger & Bedford 1982a, 1982b). Shortly after fertilization, the fertilized egg divides. As it passes down the oviduct, a mucoid layer forms around the fertilized egg, trapping any remaining sperm. Upon entering the uterus (roughly 48 hours after fertilization), a thin coat of keratin fibers, secreted by the uterus, forms a thin shell membrane homologous to those found in snakes and monotremes (Tyndale-Biscoe 2005). These membranes are absent in placental mammals, although lagomorphs retain the mucoid coat.

Early cleavage stages in metatherians and eutherians eventually result in a hollow sphere, the blastocyst. In metatherians, the blastocyst remains within its shell membranes and absorbs nutritious uterine secretions directly across these membranes. At a similar stage in eutherian development, the outer layer or **zona pellucida** surrounding the eutherian blastocyst ruptures, freeing the outer cells of the blastocyst (called the **trophoblast**) to invade the uterine lining. In many metatherian and some placental mammals, the blastocyst may enter a dormant stage (embryonic diapause) until reactivated by hormones or other chemical signals from the mother.

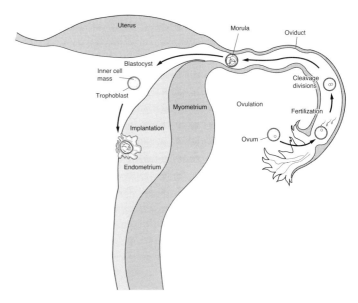

FIGURE 20-8 Drawing of early embryonic development in a human from ovulation to implantation of the trophoblast into the endometrium of the uterus. Fertilization occurs in the upper oviduct, and it takes three to four days for the embryo to reach the uterus.

By the time the blastocyst reaches the uterus, the endometrium has been conditioned by the combined action of follicle-stimulating hormones and LH hormones from the anterior pituitary and progesterone from the corpus luteum. At implantation, the eutherian embryo, enclosed within a sphere of trophoblast cells, adheres to the surface of the endometrium, triggering a round of proliferation in the trophoblast cells. These cells apparently secrete enzymes that erode the lining of the endometrium, permitting the trophoblast (and the enclosed embryo) to invade the highly vascular lining of the uterus. The trophoblast grows rapidly at this stage, extending thousands of **chorionic villi** into the surrounding endometrium and eventually establishing the placenta. In contrast, the metatherian embryo remains within the shell membranes for nearly 75% of the gestation period, and a few days before birth, these outer membranes are enzymatically degraded, allowing the embryonic membranes to contact the uterine lining to form a yolk-sac placenta (choriovitelline placenta). Thus, the metatherian placenta is established only 4 to 5 days before birth in bandicoots and up to 10 days in kangaroos (Tyndale-Biscoe 2005).

One of the most distinctive and important structures associated with reproduction in therian mammals is the placenta. Differences among the major placental types have been used to distinguish some higher taxonomic categories of mammals (subclasses and infraclasses), and some primary contrasts between reproductive patterns in mammals relate to placental differences.

A functional connection between embryo and uterus is necessary in therian mammals in which the fetus develops within the uterus and takes nutrients from the uterus rather than from yolk stored in the ovum. This connecting structure, the placenta, allows for nutritional, respiratory, and excretory interchange of material by diffusion between the embryonic and maternal circulatory systems. The placenta achieves varying degrees of contact between embryonic and uterine tissues. Of critical importance, the placenta also functions as a barrier that excludes bacteria and many large molecules from the embryonic circulation. In addition, the eutherian placenta produces certain food materials and functions as an endocrine organ by producing a variety of hormones. Amino acids supplied by the maternal circulation, for example, provide the placenta with the raw materials for protein synthesis for fetal growth. In some mammal species, the placenta produces progesterone and estrogen to help maintain pregnancy. Hormones, such as placental lactogen, are also secreted to promote mammary development (Talamantes 1975). Mammals are not unique in having a placenta, for certain fishes and reptiles establish placenta-like connections that allow diffusion of materials between the vascularized oviduct and the embryo. Among mammals, the major types of placentae differ sharply in structure and in the efficiency with which they facilitate the nourishment of the embryo.

The choriovitelline (yolk-sac) placenta occurs to varying degrees in all metatherians and in the early stages pregnancy in some eutherians (Selwood & Johnson 2006). In metatherians, after the shell membrane degenerates, the degree to which the blastocyst invades the endometrium varies widely. In some species, the blastocyst does not implant deep in the **uterine mucosa**, as it does in eutherians, but merely sinks into a shallow depression in the mucosa. The contact may be strengthened by the wrinkling of the blastocyst wall that lies against the uterus; this wrinkling increases the absorptive surface area of the blastocyst. The embryo absorbs nutritive substances secreted by the uterine mucosa and also derives nourishment from limited diffusion of substances between the maternal blood in the eroded depression in the mucosa and the blood vessels within the large yolk sac of the embryo. The yolk sacs of older embryos of opossums (*Didelphis virginiana*) are fused to each other and to the endometrium such that they cannot be pulled apart (New et al. 1977). In this species, microvilli increase absorption from the endometrium (Krause & Cutts 1984).

Metatherian placentae come in four basic types, based on the degree of association between fetal and maternal tissues and the structure of the allantois relative to the chorion (Hughes 1984; Tyndale-Biscoe & Renfree 1987). The quokka (*Setonix* sp.; **Fig. 20-9**), a small marcropodid, exhibits the most common of these, in which a highly vascularized portion of the yolk sac makes an intimate connection with the endometrium of the uterus (**Fig. 20-10**). Similar placentae occur in some members of the Didelphimorphia and other Diprotodontia. The most derived type occurs in members of the Peramelemorphia where a true chorioallantoic placenta is formed in addition to the choriovitelline placenta (Fig. 20-10). In peramelids (banidcoots), the allantois is fairly large and becomes highly vascularized; the embryo rests against the endometrium on the side where the allantois contacts the chorion. At the point of contact with the embryo, the uterus is highly vascularized, and the part of the chorion against the vascularized endometrium is more or less lost. Because of the close apposition of the maternal bloodstream and the allantois, exchange of materials occurs across the allantoic membranes. The peramelid allantois, however, lacks villi, and only its corrugations increase its absorptive surface, limiting the surface area available for maternal–fetal exchange. Supplementary nutrition may be supplied by uterine secretions absorbed across the choriovitelline surfaces, which in macropodids continues until birth (Freyer et al. 2002). Peramelids have one of the shortest gestation periods of any marsupial (12.5 days), yet their

FIGURE 20-9 A quokka (*Setonix brachyurus*) from Australia.

offspring are born relatively well-developed. The eastern barred bandicoot (*Parameles gunnii*) and the eastern quoll (*Dasyurus viverrinus*) are similar in adult size, yet the newborn bandicoot weighs 20 times that of a newborn quoll (Tyndale-Biscoe 2005). The bandicoot's chorioallantoic placenta forms on day nine and, therefore, provides nourishment for only 3.5 days in utero. Nevertheless, the more efficient exchange of the chorioallantoic placenta probably accounts for the larger newborns.

In eutherian mammals, the chorioallantoic placenta reaches its most advanced condition with regard to facilitating rapid diffusion of materials between the fetal and uterine bloodstreams. In eutherians, the blastocyst first adheres to the uterus and then sinks into the endometrium (Fig. 20-8). As implantation proceeds, chorionic villi grow rapidly and push farther into the endometrium as local breakdown of uterine tissue occurs. The resulting tissue "debris," often called **embryotroph**, is absorbed by the blastocyst and nourishes the embryo until the villi are fully developed and the embryonic vascular system becomes functional. In response to the presence of the blastocyst, the uterus becomes highly vascularized at the site of implantation. When the eutherian placenta is fully formed, the complex and highly vascularized villi provide a remarkably large surface area through which rapid interchange of materials between the maternal and fetal circulations can occur (Fig. 20-10). The extent to which the villi increase the

surface area available for diffusion is difficult to imagine but is suggested by the fact that the total surface area of villi in the human placenta is approximately 90 square meters at term (Blackburn 2007).

Among eutherians, the degree to which the maternal and fetal bloodstreams are separated in the placenta varies widely (Fig. 20-11). Lemurs, some ungulates (suids and equids), and cetaceans have an epitheliochorial placenta, in which the epithelium of the fetal chorion is in contact with the uterine epithelium and the villi rest in pockets in the endometrium. Under these structural conditions, oxygen, nutrients, and **immunoglobins** (in many species) must pass through the walls of the uterine blood vessels and through layers of connective tissue and epithelium before entering the fetal bloodstream. In carnivorans, erosion of the endometrium is carried further, and the epithelium of the chorion is in contact with the endothelial lining of the uterine capillaries. This is called an **endotheliochorial placenta**. Destruction of the endometrium in some mammals may involve even the endothelium of the uterine blood vessels, allowing blood sinuses to develop in the endometrium; the chorionic villi may then be in direct contact with maternal blood (Fig. 20-11). This most invasive hemochorial placenta occurs in some insectivorans, rabbits, bats, higher primates, and some rodents. In rabbits and some rodents, the destruction of placental tissue is so extreme that only the endothelial lining of the blood vessels in the

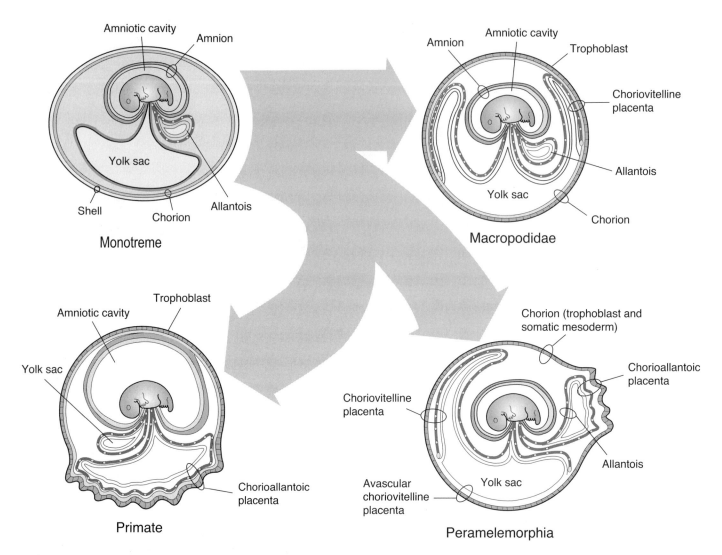

FIGURE 20-10 Drawing of the arrangements of fetal membranes in mammals. Monotremes have an amniotic egg. Kangaroos and wallabies (Macropodidae) have a relatively simple yolk sac (choriovitelline) placenta. Bandicoots (Peramelemorphia) have both chorioallantoic and choriovitelline placentas. The chorioallantoic placenta of a typical primate shows the penetration of the endometrium by thousands of chorionic villi. (Adapted from Dawson, T. J. *Kangaroos: Biology of the Largest Marsupial.* Cornell University Press, 1995 and Tyndale-Biscoe, H. *Life of Marsupials.* CSIRO Publishing, 2005.)

villi separates the fetal blood from the surrounding maternal blood sinuses (Arey 1974).

Recent studies (Carter & Mess 2008; Wildman et al. 2006) contradict the long-held assumption that the highly invasive hemochorial placenta is a derived condition in mammals. They suggest instead that the ancestral eutherian placenta was discoidal in shape with a hemochorial interface, where the fetal trophoblast deeply erodes the endometrium until the trophoblast directly contacts the maternal circulation. They propose that the less invasive edotheliochorial and epitheliochorial placentas are derived conditions that evolved in a number of mammalian lineages.

The rate at which substances move from the maternal to the fetal bloodstream in the placenta is increased, of course, when the number of interposed membrane barriers is reduced. Because of the difference between the number of such barriers in the human placenta (hemochorial) and the pig's epitheliochorial placenta, sodium is transferred 250 times more efficiently by the human placenta than by the pig placenta (Flexner et al. 1948). The remarkable absorptive ability of the allantoic placentae of such mammals as insectivorans, bats, primates, rabbits, and rodents is due largely to the great surface area afforded by the complex system of villi, to the extensive erosion of uterine mucosa and the resulting development

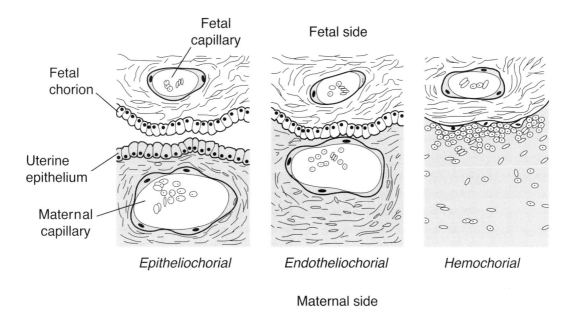

Fetal capillary

Fetal chorion

Uterine epithelium

Maternal capillary

Fetal side

Epitheliochorial   Endotheliochorial   Hemochorial

Maternal side

FIGURE 20-11   Placental types based on proximity of fetal and maternal circulations. Fetal tissues are above and maternal tissues below. Maternal and fetal blood supply is separated by multiple layers in the epitheliochorial placenta. At the opposite extreme, the hemochorial interface is more deeply invasive and the fetal blood supply is separated from the maternal blood by only a single layer of cells. (Modified from Arey, L. B. *Developmental Anatomy*. W.B. Saunders Company, 1965.)

of blood sinuses into which the villi extend, and to the loss of nearly all of the membranes separating uterine from fetal blood.

The shape of the placenta is governed by the distribution of villi, and several different distributions of villi occur in mammals. The lemurs, some artiodactyls, and perissodactyls have a **diffuse placenta**, which has a large surface area because the villi occur over the entire chorion. Ruminant artiodactyls have a **cotyledonary placenta** consisting of more or less evenly spaced groups of villi scattered over the mostly avillous chorion. Carnivorans have a **zonary placenta**, in which a continuous band of villi encircles the equator of the chorion. Insectivorans, bats, some primates, rabbits, and rodents have a **discoidal placenta**, in which villi occupy one or two disk-shaped areas on the chorion.

■ Parturition

The maternal **corpus luteum** and the fetus both play critical roles in stimulating **parturition** (birth) in metatherians. The following series of events are summarized by Tyndale-Biscoe (2005) for the tammar wallaby. As parturition approaches, the maternal corpus luteum releases progesterone and relaxin, which are believed to soften the cervix and pseudovaginal canal (**Fig. 20-12**). Simultaneously, the fetal pituitary gland in the brain releases ACTH

(adrenocorticotrophic hormone), which acts on the fetal adrenal glands causing them to release cortisol. Fetal cortisol stimulates the release of placental prostaglandins (PGF) and stimulates uterine contractions. Rising PGF metabolite levels in the maternal blood stream causes the mother's pituitary to (1) release mesotocin, further enhancing uterine contractions, and (2) release a pulse of prolactin, which shuts down progesterone secretion in the corpus luteum and probably also causes mammary glands to prepare for lactation. As described previously, fetal marsupials are directly involved in initiating parturition, even those such as the honey possum (*Tarsipies*), which can weigh as little as 4 milligrams at birth.

Just before birth, a female kangaroo adopts a characteristic birth posture: she sits with the hind legs extended so that the cloaca is oriented upward and begins licking the interior of the pouch and the cloacal region. At birth, the yolk-sac ruptures, and the female licks the fluids that leak from the cloaca. Typically, the allantois is shed next, followed by the appearance of the tiny young still partially wrapped in the amnion. After freeing itself from the amnion, the newborn begins crawling with its forelimbs toward the pouch. In less than 5 minutes, the newborn kangaroo enters the pouch (Tyndale-Biscoe 2005). This short journey is remarkable because of the extremely altrical state of the newborn. Tyndale-Biscoe (2005) describes the newborn

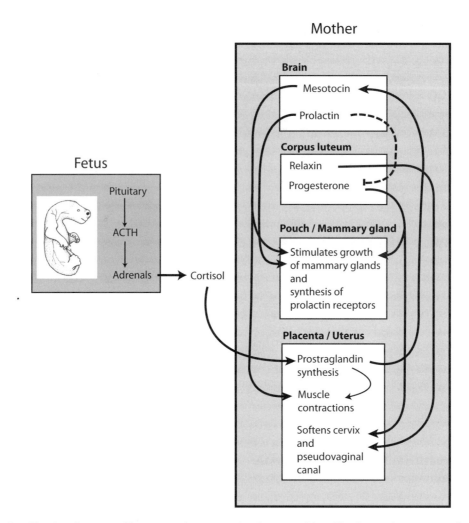

FIGURE 20-12   The role of fetal and maternal hormones in preparation for parturition. The fetus plays a central role in initiating parturition by secreting cortisol. The mother secretes a variety of hormones important in preparation of the mammary glands for lactation and in preparing the uterus and cervix for birth. (Adapted from Tyndale-Biscoe, H. *Life of Marsupials.* CSIRO Publishing, 2005.)

marsupial as "a marvelous composite of embryonic structures and **precociously** developed functional organs." Among the later are well-developed head and forelimbs: The forelimbs are innervated by a large brachial nerve, have sharp claws, and exhibit an alternating right and left, grasp and pull pattern (crawling). The mouth and tongue are large, allowing the newborn to attach to the bud at the tip of the teat. The teat then expands inside the mouth and the lips and tongue create a firm seal. The trigeminal nerve is robust and supplies innervation to the jaw and hyoid muscles, which create the negative pressure needed to suck milk from the teat. Newborns have an unusual arrangement of the epiglottis that allows the young to suckle without interrupting breathing (Tyndale-Biscoe 2005). Newborn marsupials have rudimentary lungs, and at this stage, gas exchange occurs primarily across the skin surface (cutaneous respiration). In one species of dunnart (*Sminthopis douglasi*), the young are born after a 12-day gestation period weighing only 15 milligrams. For the first few days in

the marsupium, nearly 100% of gas exchange takes place across the skin; cutaneous respiration declines to one third by week 3 as the developing lungs take on a larger role in respiration (Frappell & Mortola 2000). The senses of smell and touch are also believed to be well developed at birth (Gemmell & Rose 1989). The posterior portion of the body, in contrast, is rudimentary: The hindlimbs are simple paddles, and the small intestines and colon are poorly developed. Once securely attached to the teat, the lactation phase begins (**Fig. 20-13**).

In eutherians, the delicate hormonal control of pregnancy is also exerted by interactions between hormones produced by the pituitary, the ovary, and the uterus. During early pregnancy, the corpus luteum, because of its production of progesterone, is important in maintaining pregnancy by keeping the endometrium in a thickened and highly vascularized condition and by preventing coordinated contractions of the myometrium that might expel the embryo. In some species, progesterone sensitizes the

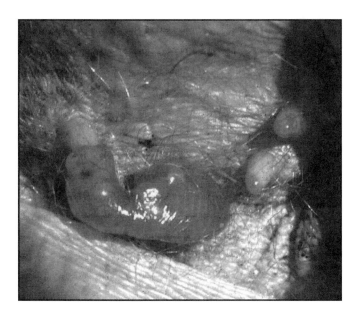

FIGURE 20-13 Newborn kangaroo (*Macropus* sp.) attached to a nipple inside the mother's marsupium.

endometrium and increases the efficiency of blastocyst implantation. In the early part of pregnancy, chorionic gonadotropin is of critical importance in preserving the corpora lutea and in preventing regression of the endometrium. This hormone is produced first by the outer layer of the blastocyst (trophoblast) during its implantation in the endometrium and then by the chorion, which develops from the trophoblast.

The maintenance of pregnancy in many mammals is not entirely under the control of progesterone from the corpus luteum. Instead, as pregnancy continues, hormones are produced progressively more by the placenta and less by the ovary. In humans, the placenta produces estrogens and is probably the most important source of progesterone after the third month. This is a time of great risk of miscarriage, when corpus luteum progesterone declines and placental progesterone commences. During the latter stages of pregnancy, the placenta seems to be a nearly independent endocrine gland that, in humans at least, takes over the functions of the pituitary gland and the corpus luteum. Near parturition, the fetal placenta produces corticotropin-releasing hormone into both fetal and maternal blood streams, which stimulates ACTH production in the pituitary. ACTH stimulates production of cortisol and estrogen from the placenta. Cortisol promotes the maturation of the lungs in preparation for breathing. The combination of rising estrogen and declining progesterone levels results in (1) rhythmic and powerful contractions of the uterine myometrium, (2) increased sensitivity of the uterus to oxytocin, and (3) increased prostaglandin production and subsequent softening of the cervix. Continued contractions force the fetus against the cervix, creating a positive

feedback system whereby both prostaglandin and oxytocin levels continue to increase (Sherwood et al. 2005). With uterine contractions continuing and the cervix fully dilated, the offspring is delivered through the vaginal opening. In bats, a full-term fetus is larger than the birth canal opening in the pelvis. At birth the interpubic ligament stretches to over 15 times its normal length to allow the passage of the huge fetus (Crelin 1969).

At birth, the fetal contribution to the eutherian placenta is always expelled as part of the "afterbirth," but the maternal part may or may not be lost at this time. In mammals with an epitheliochorial placenta, the villi pull out of the uterine pits in which they fit. None of the endometrium is pulled away, and no bleeding occurs at birth. This placenta is termed **nondeciduous**. In mammals with placentae allowing more intimate approximation of uterine and fetal bloodstreams, there is extensive erosion of the uterine tissue and extensive intermingling of uterine and chorionic tissue, and the uterine part of the placenta is torn away at birth, resulting in some bleeding. This type of placenta is **deciduous**. The hemorrhaging after birth is soon stopped by the collapse of the uterus, by contractions of the **myometrium** (the smooth muscle layer of the uterus), which tend to constrict the blood vessels, and by clotting of blood. The loss of the placenta suddenly removes a major source of progesterone, an inhibitor of prolactin, which is critical for the release of milk.

In summary, the hormonal events leading to parturition are similar in metatherian and eutherian mammals, but the length of gestation/lactation periods, developmental stage of the newborn, and events subsequent to birth differ dramatically.

▪ Lactation

Lactation, the synthesis and secretion of milk to nourish the young, is one of the defining characteristics of mammals. Exactly how and when lactation evolved is not known (see Box 2-1). Pond (1977) proposed that lactation and suckling must have evolved in concert with precise dental occlusion and diphyodonty (replacement of milk teeth with a single set of adult teeth). According to this scenario, lactation evolved first. Newborns obtained nourishment from milk as they grew, and without the need for chewing, permanent tooth eruption was delayed until near weaning when the jaw was approximately adult size. Many hypotheses have been offered to explain the evolution of mammary glands, lactation, and suckling in ancestral mammals. Mammary glands probably evolved from apocrine-like glands associated with hair follicles (Oftedal 2002a). Thus, secretions of proto-milk probably originally occurred directly onto the hair as it does in extant monotremes. Accordingly, Blackburn et al. (1989) proposed that early mammals initially secreted milk-like substances with antimicrobial properties to protect their thin, parchment-shelled eggs before hatching. Selection later favored the production of more copious, nutritive secretions (proto-milk), which supplemented the energy provided by the yolk. Eventually, reliance on embryonic yolk was supplanted by longer periods of lactation. Oftedal (2002a, 2002b), however, thought that mammary secretions evolved initially in synapsids to keep the eggs from desiccating and proposed that synapsids laid eggs covered in thin, "parchment-like" shells (typical of some modern reptiles and living monotremes), not the hard, calcified shells of modern birds. Additionally, he suggested that these proto-mammals were likely endothermic and incubated their highly permeable eggs at above-ambient body temperatures. High incubation temperature speeds desiccation but also increases embryonic growth rates, thereby reducing incubation time. Oftedal believes that the eggs were left in a nest and rehydrated with mammary-patch secretions when the mother returned from foraging. These secretions were co-opted by Triassic therapsids to provide energy-rich milk to hatchlings, resulting in "a progressive decline in egg size and an increasingly altricial state of the young at hatching." Support for an early transition to lactation comes from genomic studies of egg yolk genes and genes involved in milk protein production (caseins). Analysis of the platypus genome by Warren et al. (2008), and by Brawand et al. (2008) showed that ancestral genes for egg yolk proteins are retained in the platypus, an egg-laying mammal, whereas these same genes were lost during the evolution of metatherian and eutherian mammals. These authors further showed that genes called caseins related to milk production, which function rather like egg yolk gene proteins in nourishing a developing embryo, evolved in a common mammalian ancestor in the Mesozoic. Their interpretation suggests that the development of lactation and placentation in metatherain and eutherian mammals allowed for the evolutionary loss of egg and yolk genes found in monotremes and other primitive mammals.

Today all newborn mammals are nourished by milk produced by the mother's mammary glands. Under the influence of estrogen, progesterone, insulin, and placental lactogen, mammary glands undergo considerable growth during pregnancy. Milk production is stimulated by suckling of the young and regulated by prolactin, produced by the anterior lobe of the pituitary. Prolactin is secreted in progressively larger amounts during the latter part of pregnancy and after parturition; it is inhibited by progesterone. When the inhibition of prolactin by progesterone is removed after birth, milk secretion can begin. Oxytocin, released under the stimulus of suckling, causes contraction of the myofibrils surrounding the milk-containing alveoli of the mammary glands, inducing a release of milk (milk

letdown) for the young. Milk production is partly under neural control and continues only as long as the suckling stimulus persists.

Lactation lasts far longer in metatherains than in similar-sized eutherians. Three stages occur in metatherians: mammogenesis (preparation of the mammary glands), early lactogenesis (lactation during the permanent attachment of the young to the teat), and late lactogenesis (after detachment of the young from the teat). Late-stage lactogenesis in marsupials is characterized by rapid growth of the young and is equivalent to the entire lactation phase of placental mammals.

During pregnancy, mammogenesis occurs when the milk-secreting cells within the mammary gland swell and the ducts branch in response to progesterone secreted by the corpus luteum. Interestingly, mammogenesis also occurs in nonpregnant marsupials at an equivalent stage in the estrous cycle (Sharman 1962). This feature is currently being used to conserve rare marsupial species by transferring the young of rare species into the pouches of common species (Tyndale-Biscoe 2005). Milk synthesis begins when prolactin is secreted from the mother's pituitary (Nicholas & Tyndale-Biscoe 1985). In placentals, the hormones insulin, thyroxin, and cortisol are required in addition to prolactin. Lactogenesis in placental mammals is maintained in multiple teats via the repeated stimulus of suckling and the removal of milk from the glands. Suckling triggers secretion of prolactin, which furthers milk synthesis. In contrast, milk is supplied only to teats with attached young in metatherians. Stewart (1984) showed that prolactin receptors increase on cells within the gland after a newborn wallaby attaches to the teat. Teats without attached young shrink, and the cells of these glands show vastly reduced prolactin receptor densities. Fostering a young onto a shrunken teat increases prolactin receptors on those cells, making them responsive to circulating prolactin and initiating milk synthesis. In the tammar wallaby (*Macrpus eugenii*), stage 2 lactogenesis lasts roughly 180 to 200 days (**Fig. 20-14**).

In metatherians, a newborn young may be attached to a nipple and suckling while a much older young is returning periodically to suckle from a separate nipple. During double suckling in kangaroos, a remarkable thing occurs; separate mammary glands concurrently produce vastly different milks. The gland supporting the pouch young produces dilute milk containing little fat, and the gland supporting the advanced young produces milk with three times as much fat.

Late-stage lactation in metatherians is equivalent to the entire lactation period in eutherian mammals. At this point, the young metatherian suckles intermittently, and its growth rate increases markedly (Fig. 20-14). The young develops a functioning immune system. Its eyes and ears

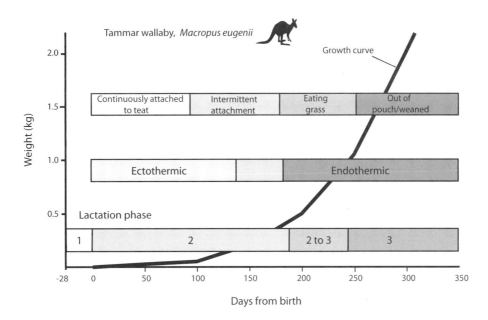

FIGURE 20-14 Events during the development of a Tammar wallaby (*Macropus eugenii*) from conception to weaning. Phase 2 lactation occurs over the first 180 days in the pouch and is characterized by relatively slow growth of the young. The growth curve shows rapid weight gain at the end of phase 2 and beginning of phase 3 as the young becomes able to regulate its own body temperature and makes forays outside the pouch to feed on grass. (Data from Tyndale-Biscoe, H. *Life of Marsupials*. CSIRO Publishing, 2005.)

open. The nervous system matures, and it makes its first forays outside the pouch. Shortly after, the young develops the capacity to regulate its own metabolism and becomes endothermic (Tyndale-Biscoe 2005).

Two major differences in milk composition occur between metatherian and eutherian mammals. First, the composition of milk carbohydrates, lipids, and proteins in eutherian mammals remains fairly uniform throughout the lactation period (except the secretion of colostrum during the first few days). In metatherians, milk composition changes during the long lactation period. In the early stages, when the newborn marsupial is tiny, the milk is dilute and contains mostly simple sugars. Later, during the transition to stage 3 lactation, the composition changes dramatically (Tyndale-Biscoe 2005). By day 200 in the tammar wallaby, carbohydrate composition changes from simple to more complex sugars. Protein concentrations continue to increase, and milk fat levels (mostly in the form of triglycerides) rise dramatically (**Fig. 20-15**). Second, milk composition varies little among marsupial species at similar stages in lactation. In placental mammals, however, some species produce relatively dilute milk (10% solids) and other species, such as seals, produce concentrated, fat-rich milk (up to 78% solids) throughout the lactation period.

The period of lactation is one of the most critical times in the life of a female mammal. The survival and vigor of her young, and thus her own fitness, depend on a female's ability to meet the energetic demands of lactation, which requires far more energy than does pregnancy (**Table 20-3**).

This is due to postnatal costs in the developing young of growth, thermoregulation, and activity. In a laboratory study, food intake by pregnant cotton rats (*Sigmodon hispidus*) increased 25% over the intake of nonreproductive females; during lactation, the increase was 66% (Randolph et al. 1977). This disparity is actually greater than it seems, for about two thirds of the energy accumulated by a pregnant female was not invested in young but was stored and used during lactation, when the mother could not assimilate energy fast enough to support her rapidly growing young. Even with ample food, lactating cotton rats lost about 11% of their body weight (Mattingly & McClure 1985).

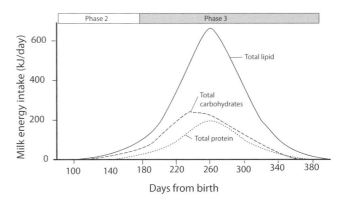

FIGURE 20-15 Milk is relatively dilute early in lactation, but milk composition changes dramatically during phase 3 lactation coinciding with rapid weight gain of the joey. (Data from Tyndale-Biscoe, H. *Life of Marsupials*. CSIRO Publishing, 2005.)

TABLE 20-3   Increase (in Percentage) in Dietary Energy Used by Pregnant or Lactating Animals Relative to Energy Used by Nonreproductive Individuals

Species	Pregnancy	Lactation	Reference
Homo sapiens (humans)	+10%	+50%	Nat. Res. Council Food and Nutrition Board (1968)
Ovis aries (sheep)	+81%	+146%	NRC (1974)
Rattus norvegicus (Norway rat)	+46%	+152%	NRC (1972)
Mus musculus (house mouse)	+33%	+111.3%	Myrcha et al. (1969)
Microtus arvalis (European vole)	+32%	+133%	Migula (1969)
Myodes glareolus (bank vole)	+24%	+92%	Kaczmarski (1966)
Sigmodon hispidus (cotton rat)	+25%	+66%	Randolph et al. (1977)

Adapted from Randolph et al. (1977).

Lactation is especially costly for metatherians because of their long lactation periods. Female eastern quolls (*Dasyurus viverrinus*) provided 11%, 33%, and 55% of their total digestible energy intake to litters of one, three, and five young, respectively. Consequently, mothers with large litters lost weight during the lactation period even though young from large litters were smaller at weaning than those from smaller litters (Green et al. 1997).

The energetic costs of pregnancy and lactation have now been assessed in several marsupial and placental mammals (**Fig. 20-16**). The general pattern is for placental mammals to invest heavily in a relatively short gestation period, whereas marsupials invest little in gestation and the early stages of lactation, reserving the bulk of their energetic investment for late stage lactation. When controlled for variation in body mass, the total energetic investment (expressed as megajoules per kilogram body weight, MJ/kg) for gestation and lactation is remarkably similar in eutherians and metatherians: A 50 kilogram ewe (sheep) invests approximately 22.5 MJ/kg, and a 5 kilogram tammar wallaby invests nearly 21 MJ/kg. The relatively low early investment in wallabies is thought to be an adaptation to uncertain environmental conditions, whereby females can end pregnancies early with little energy invested if conditions turn harsh. For placental mammals, terminating pregnancy often results in considerable loss of the energy invested in the fetus.

Female ungulates of several species respond to the cost of gestation and lactation and poor nutrition by periodic infertility. In northern Alaska, for example, female caribou (*Rangifer tarandus*) have a reproductive pause once every 4 years; this may increase reproductive performance over a female's lifetime (Cameron 1994).

Although lactation alone is energetically costly, many species of rodents, rabbits, and otariid seals have a postpartum estrus, and pregnancy and lactation may be concurrent (Asdell 1964). The energetics of concurrent pregnancy and lactation in laboratory colonies of cotton rats and eastern woodrats (*Neotoma floridana*) were studied by Oswald and McClure (1990). Compared with females that were only lactating, females of both species that were concurrently pregnant and lactating had higher resting metabolic rates (26% higher for cotton rats and 14% for woodrats). In contrast to concurrently pregnant and lactating cotton rats, woodrats in this condition increased food intake, delayed implantation of the second litter, and had smaller second litters. The length of delay of implantation was positively correlated with the number of suckling young, an effect also observed in some other rodents. The woodrat's period of lactation was twice as long as that of the cotton rat (24 versus 14 days). For species with prolonged lactation, delayed implantation postpones the progressively increasing costs of pregnancy until after the litter in the nest is weaned (Oswald & McClure 1990). Probably as a result of energetic stress, concurrent pregnancy and lactation in wild rabbits (*Oryctolagus cuniculus*) in Wales frequently resulted in the death and resorption of the embryos in the uterus (Asdell 1964). Although concurrent pregnancy and lactation are an energetic strain that may prejudice a female's postbreeding survival, selection seemingly favors this strategy in some mammals (such as some rodents, shrews, and rabbits) that may live only long enough to breed during a single season.

Female mammals meet the cost of reproduction in several ways. Probably all species increase food intake during reproduction, and this increase can be drastic. In addition, some females may save energy during reproduction by decreasing activity. Energy stored in the body during pregnancy is used by some species during lactation. Mother white-footed mice (*Peromyscus leucopus*) metabolized body fat during lactation to augment food intake. When food was restricted, however, they abandoned or killed (and ate) their litters rather than threatening their own survival (Millar 1975). With restricted food, lactating cotton rats drew

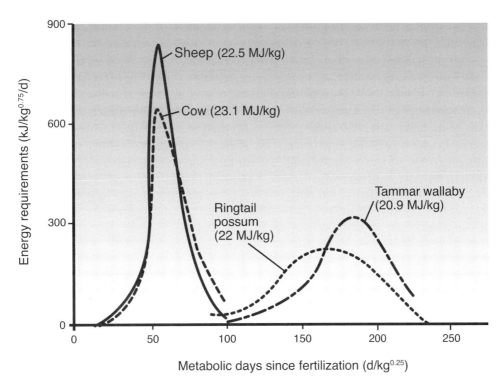

FIGURE 20-16 Investment in gestation and lactation in metatherian and eutherian mammals. Body size differences have been accounted for by conversion to "metabolic days" since conception. The total energy invested (MJ/kg), the area under each curve, is similar for metatherians and eutherians; however, eutherian investment occurs over a relatively short period, whereas that in metatherians is distributed over a much longer time, reflecting their investment in the later stages of lactation. (Modified from Tyndale-Biscoe, H. *Life of Marsupials*. CSIRO Publishing, 2005; Data for cattle, sheep and tammar from Cork and Dove (1989, *Journal of Zoology*, London 219); for the koala from Krockenberger (1993, PhD Thesis, University of Sydney); and for the ringtail possum from Munks and Green (1997, *Physiological Zoology* 70).)

heavily on their own energy stores, losing 5.2 grams of body weight per day (Mattingly & McClure 1985).

Some mammals make dietary shifts in an attempt to compensate for the heavy demands of lactation (Brody 1945). A particularly interesting example involves spotted dolphins (*Stenella attenuata*) in the eastern Pacific. The diet of the general population of these dolphins, including pregnant females, was squid, but lactating females switched to a diet of flying fish (Bernard & Holn 1989). Analysis of these foods showed that the fish yielded more energy than did squid (per 100 grams of muscle: 420 kilojoules versus 310 kilojoules) and higher levels of protein (21.2% versus 17.6%).

The high energy demands of pregnancy, and even higher demands of lactation, are known to be met in some mammals by anatomical changes that favor rapid absorption of nutrients. These adaptive changes compensate for the sharp increase in food consumption during lactation and involve increases in the volume and capacity of the alimentary canal and increased blood flow to the digestive system (Fell et al. 1963; Souders & Morgan 1957). The scale of these changes in one small mammal is shown by data on the common shrew (*Sorex araneus*; Jaroszewska & Wilczynska 2006). Compared with the esophagus of nonpregnant and

nonlactating females, that of lactating females was 24.3% longer. There were comparable changes for other organs: The duodenum was 39.8% longer; the mesenteric intestine was 12.1% longer; and area of the mucosa of the small intestine was 49.1% greater. The greatest lengthening of villi was in the duodenum and proximal part of the mesenteric intestine, where the rate of absorption of proteins and fats is highest. Female common shrews typically have two litters in close succession and then die, and thus, the fate of the alimentary canal after reproduction is not known. In the laboratory rat (*Rattus norvegicus*), however, all of the changes in the reproductive female's alimentary canal are temporary and reversible. Future research may demonstrate for most female mammals the occurrence of changes during lactation similar to those described above.

Postnatal Growth

In general, small mammals have relatively higher postnatal growth rates than do large mammals. That is to say, small mammals attain adult weight, or a given percentage thereof, at an earlier age than do large mammals. Because of the higher metabolic rate of the small mammal, it can presumably mobilize energy for growth more rapidly than can a large mammal. Even among mammals of similar size, however, both metabolic rates and postnatal growth rates vary.

Low growth rates in certain small mammals may be associated with unusually low metabolic rates (as is probably the case with the tenrec, *Hemicentetes semispinosus*) or with an adaptively long mother–young association. Consider the following case discussed by Eisenberg (1981). Relative to body weight of the adult, postnatal growth is more rapid in vespertilionid bats than in phyllostomid bats (Kleiman & Davis 1978), and the former apparently have richer milk. The Neotropical phyllostomids can seemingly afford a long suckling period because the young need not prepare for hibernation.

Postnatal growth rates reflect the life history of a mammal. High rates have evolved under such demanding conditions as stressful environments and short seasons for preparing for hibernation. The need for high growth rates in young bats that must store fat for a long hibernation has been mentioned. Equally demanding conditions have selected for extremely rapid growth in some marine mammals whose young must rapidly prepare for life at sea. This is true for the northern and southern elephant seals (*Mirounga* spp.). These are huge animals; males reach weights of over 3,000 kilograms. An average pup weighs about 35 kilograms at birth, and this weight doubles by about 11 days of age. Pups often triple their weights by 28 days of age, when they are weaned (Condy 1978). The young Weddell seal (*Leptonychotes weddellii*) doubles its weight within 2 weeks after birth (Bertram 1940).

The rapid growth of pinnipeds is facilitated by the high-energy milk these animals produce (up to 53% fat and 78% solids). In contrast, the suckling period is long in some pinnipeds, up to 1.5 years in the walrus and 1 year in some California sea lions (*Zalophus californianus*; Peterson & Bartholomew 1967). In these species, preparation for life at sea involves both high-calorie milk and a long suckling period.

Because pinnipeds forage at sea but must haul out on land or ice floes to give birth and nurse their young, otariid and phocid seals have a different suite of maternal strategies than those of other mammals. Female elephant seals (Phocidae) typically give birth to a single pup per year, and the duration of maternal care varies from 27 days to 3 years. Males do not contribute to parental care. Therefore, the lifetime reproductive success of those males able to gain access to **harems** greatly exceeds that of females. A harem bull may mate with over 100 females in a single season (LeBoeuf & Reiter 1988), whereas a successful female has a relatively low lifetime reproductive output. Female northern elephant seals wean an average of 10 pups in a lifetime.

Pinnipeds exhibit three basic maternal strategies with respect to lactation and maternal care of pups: aquatic nursing, foraging-cycle, and fasting-cycle strategies (Boness & Bowen 1996). Walruses are unique among pinnipeds in exhibiting an aquatic-nursing strategy (**Fig. 20-17**). Female walruses accumulate blubber prior to parturition and fast for the first few days after giving birth. Young follow their mothers out to sea and nurse in the open water (Miller & Boness 1983).

A foraging-cycle strategy is used by many female otariids, which rely on stored blubber to support lactation for the first 2 weeks postpartum. After this initial fasting period, females make foraging trips to sea (from 3 to 5 days for Antarctic fur seals and from 5 to 14 days for northern fur seals) before returning to the beach to nurse their pups. The milk is high in fat, and lactation can last from 4 months to 3 years.

The fasting-strategy is typical of phocids and involves females arriving at birthing sites with large stores of blubber

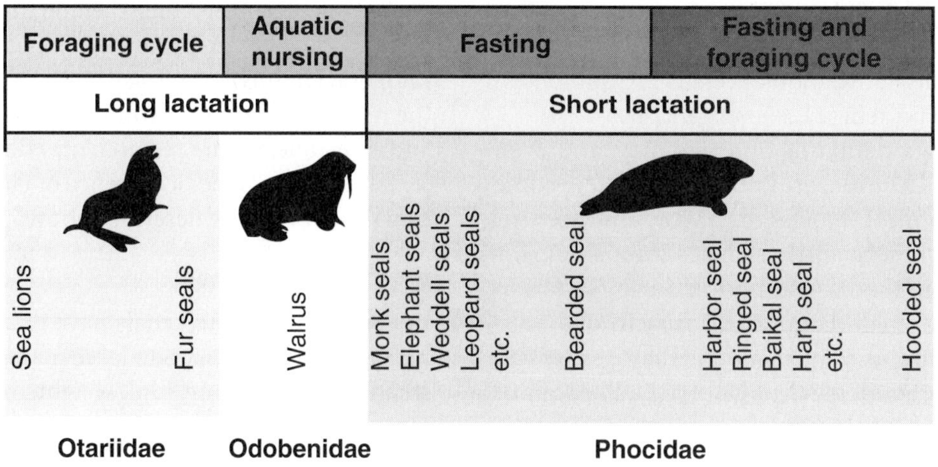

FIGURE 20-17 Lactation and feeding strategies of pinnipeds. Phocid seals typically have short lactation periods relative to otariids and the walrus. (Adapted from Boness, D. J., and Bowen, W. D., *BioScience* 46 (1996): 645–654.)

and fasting throughout lactation. Lactation periods are often short but vary widely among species (from 4 to 50 days). Lactation lasts only 4 days in the hooded seal, but the milk is extremely high in fat, and the young ingest 250 megajoules of energy per day. At the other extreme, lactation lasts 27 days in the northern elephant seal and the daily intake of energy by the young is 98 megajoules per day (**Table 20-4**; Boness & Bowen 1996; Oftedal et al. 1993).

Until recently, it has been assumed that otariids used a foraging-cycle strategy and that phocids used the fasting strategy. Recent advances in radio telemetry and isotope labeling have revealed that several small phocids, including harbor seals (*Phoca vitulina*), use the foraging-cycle strategy but retain a short lactation (Boness et al. 1994; Bowen et al. 1992). Boness and Bowen (1996) concluded that the energetic constraints of small body size favored the evolution of a maternal foraging-cycle strategy in harbor seals. According to their hypothesis, small-bodied seals cannot store sufficient energy in the form of blubber to sustain their 24-day lactation and must replenish these stores with short foraging bouts, especially toward the end of lactation.

Major Reproductive Patterns

In eutherian mammals, the length of time from fertilization to implantation is typically considerably shorter than the period between implantation and birth. Fertilization usually occurs shortly after ovulation, and the development of the embryo from fertilization to birth is an uninterrupted process. Perhaps in response to specialized activity cycles or seasonal food shortages, some mammals have abandoned this usual pattern of continuous development. One departure involves a delay of ovulation and fertilization until long after copulation (**delayed fertilization**). Another is typified by normal fertilization and early cell cleavages but arrested embryonic development at the blastocyst stage (delayed implantation). Another strategy involves a long delay in the development of the blastocyst after it has implanted (**delayed development**).

■ Delayed Fertilization

This pattern of development occurs in a number of bats inhabiting northern temperate regions. As early as 1879, Fries recognized that the males of some species of the families Rhinolophidae and Vespertilionidae could store viable sperm through the winter, long after **spermatogenesis** had ceased; later studies detailed the reproductive cycles of the females of these species (Guthrie 1933; Hartman 1933; Wimsatt 1944, 1945). These remarkable reproductive tactics are seemingly adaptations to continuous or periodic winter dormancy and occur in a number of New World and Old World species in the genera *Rhinolophus*, *Myotis*, *Pipistrellus*, *Eptesicus*, *Nycticeius*, *Lasiurus*, *Plecotus*, *Miniopterus*, and *Antrozous*. Delayed fertilization may be the typical pattern in all but the tropical members of the family Vespertilionidae. Papers by Wimsatt (1944, 1945),

TABLE 20-4 Characteristics of Maternal Support and Lactation in Pinnipeds

Species	Female Mass (kg)	Body Fat (%)	Lactation Period (d)	Milk Fat (%)	Daily Energy (MJ/d)
Phocidae	Elongate	Short	Short		
Mirounga leonina Southern elephant seal	515	23	24	47	184
M. angustirostris Northern elephant seal	504	40	27	54	98
Cystophora cristata Hooded seal	179	40	4	61	250
Phoca vitulina Harbor seal	84	25	24	50	31
Otariidae					
Eumetopias jubatus Steller sea lion	273	—	330	24	21
Zalophus californianus California sealion	88	—	300	44	10
Callorhinus ursinus Northern fur seal	37	—	125	42	6

Adapted from Boness and Bowen (1996).

Racey (1982), Uchida and Mori (1987), and Bernard and Cumming (1997) describe delayed fertilization as it occurs in Microchiroptera, and the following remarks are based largely on those studies.

The reproductive cycle of the little brown bat (*Myotis lucifugus*) follows a timetable similar to that of many temperate-zone vespertilionids. The testes descend into the scrotum in the spring. This descent is caused mostly by increased production of testosterone, which is cyclical in bats. The testes begin to enlarge in the spring, and spermatogenesis peaks in the spring and ends by September. Accessory reproductive organs remain enlarged throughout the winter. A female may be inseminated repeatedly in the fall and winter, and males frequently copulate with hibernating females, although usually all females are inseminated by the end of November. In males of this species, mating occurs when circulating testosterone levels are at their lowest (**Fig. 20-18**; Gustafson & Damassa,1985). In the females, a single follicle enlarges in the autumn but remains in the ovary throughout the winter. Ovulation and fertilization occur at the end of hibernation. The most typical vespertilionid pattern is for copulation to occur before hibernation. The sperm are stored in the uterus, where they remain motile for at least 198 days in *Nyctalus noctula* (Racey 1973). Sperm remain viable for such long periods because the sperm heads are embedded in the microvilli of the uterus where secretions from the uterine lining nourish them (Uchida & Mori 1987). The gestation period in this species is highly variable, probably because of regional differences in ambient temperatures and hence to the different body-temperature

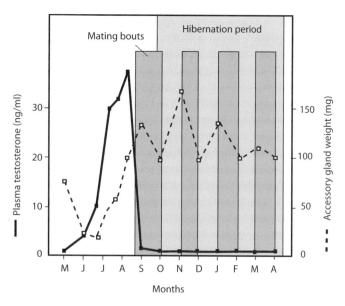

FIGURE 20-18 Graph of the reproductive cycle in male little brown bats (*Myotis lucifugus*). Circulating testosterone levels peak in September and fall to very low levels during the mating season, which begins just before hibernation. (Data from Gustafson, A.W. and Damassa, D.A. *Biol Reprod.* 33 (1985): 1126–1137.)

routines that occur in bats of widely separated colonies. Periodic torpor or low body temperature after the beginning of gestation slows the development of the embryo.

Several features of this unique reproductive cycle are especially noteworthy. The development of the male reproductive organs is out of phase; that is, the testes have regressed when the caudal epididymides and accessory organs are most enlarged and when breeding activity is at its peak. Males retain viable sperm in the caudal epididymides long after spermatogenesis has ceased. Females do not ovulate until long after they have been inseminated but are able to store viable sperm for several months. Because of differing metabolic routines in different individuals, the rate of development of the embryo is highly variable.

Delayed fertilization is seemingly a highly advantageous adaptation in mammals with long periods of dormancy. Spermatogenesis, enlargement of reproductive organs, and copulation require considerable energy. In species that practice delayed fertilization, these activities occur in the late summer and autumn, when males are in excellent condition and have abundant food, rather than in spring, when the animals are in their poorest condition and when food (insects) may not yet be abundant. Ovulation, fertilization, and zygote formation occur almost immediately on emergence from dormancy, rather than being delayed until after males attain breeding condition and copulation occurs. The female can therefore channel more energy into nourishment of the embryo than would be available if copulation occurred after hibernation. Perhaps the major advantage is that the time of parturition is hastened; thus, young have the longest possible time to develop before the next period of winter dormancy.

Delayed Implantation

This deviation from the normal reproductive pattern occurs in a variety of mammals representing the orders Chiroptera, Pilosa, Carnivora, and Artiodactyla (**Tables 20-5** and **20-6**). These mammals obviously do not share a direct common ancestor, and they occupy a wide variety of habitats and pursue different modes of life. Delayed implantation in each group, therefore, has probably evolved independently in response to different selective pressures. Delayed implantation is either obligate and constitutes a consistent part of the reproductive cycle, or facultative, and provides for a delay of implantation on occasions when an animal is nursing a large litter. Good discussions of delayed implantation are given by Daniel (1970) and Mead (1989).

In mammals with obligate delayed implantation, ovulation, fertilization, and early cleavages up to the blastocyst stage occur normally, but further development of the blastocyst is arrested, as is implantation in the uterine endometrium. The blastocyst remains dormant in the uterus

TABLE 20-5 Periods During Which Blastocysts Remain Dormant in Some Mammals with Obligate Delayed Implantation or Delayed Development

Species	Dormancy of Blastocyst (Months)
Order Chiroptera	
Equatorial fruit bat (*Eidolon helvum*)	3+
Jamaican fruit bat (*Artibeus jamaicensis*)	2.5
Order Cingulata	
Nine-banded armadillo (*Dasypus novemcinctus*)	3.5–4.5
Order Carnivora	
Grizzly bear (*Ursus arctos*)	6+
Polar bear (*U. maritimus*)	8+
River otter (*Lontra canadensis*)	9–11
Harbor seal (*Phoca vitulina*)	2–3
Gray seal (*Halichoerus grypus*)	5–6
Walrus (*Odobenus rosmarus*)	3–4
Order Artiodactyla	
Roe deer (*Capreolus capreolus*)	4–5

Data from Daniel (1970), except data on *A. jamaicensis* from Fleming (1971).

for periods from 12 days to 11 months. Little blastocyst growth occurs during dormancy, which begins generally when the embryo consists of approximately 100 to 400 cells. The western spotted skunk (*Spilogale gracilis*), of North America west of the continental divide, follows a reproductive pattern fairly typical of mammals with delayed implantation (Mead 1968a, 1968b). Males become fertile in the summer, and copulation and fertilization of the ova occur in September. The zygote undergoes normal cleavage but stops at the blastocyst stage; the blastocysts float freely in the uterus for 180 to 200 days. After implantation, the gestation period is about 30 days, and the young are usually born in May. During dormancy, each blastocyst is covered by a thick and durable zona pellucida. In the eastern

TABLE 20-6 Reproductive Cycles of Some North American Mammals with Delayed Implantation

Species	Breeding Season	Time of Implantation	Length of Delay Period (Months)	Time of Parturition	Litter Size	Gestation Period (Months)
Long-tailed weasel (*Mustela frenata*)	July	March	8	April–May	6-10	9
Ermine (*Mustela erminea*)	June–July	March	8.5–9	April–May	6–10	9.5–10
Mink (*Mustela vison*)	Feb–March	March	0–1	April–May	3–8	1.3–2.3
Marten (*Martes americana*)	July–August	March–April	8	May	2–3	9
Fisher (*Martes pennanti*)	March–April	Feb–March	11	March–April	2–4	11.5–12
Wolverine (*Gulo gulo*)	Spring–Summer	Jan–Feb	5+	March–April	2–4	8–9+
Badger (*Taxidea taxus*)	July–Aug	Feb	6	March–April	2–3	8
Western spotted skunk (*Spilogale gracilis*)	Sept	April	6–7	May–June	4–7	8
Black bear (*Ursus americanus*)	June	Nov	6	Jan–Feb	1–4	7
Northern fur seal (*Callorhinus ursinus*)	Late July	Nov–Dec	3.5–4.5	Late July	1	12

From PL Wright.

spotted skunk (*S. putorius*), of North America east of the continental divide, delayed implantation does not occur. A number of carnivorans practice delayed implantation, but the timing of the cycle varies from species to species (Table 20-6).

All pinnipeds probably practice delayed implantation, which in these animals enhances the survival of young by providing for the optimal timing of births (Boyd 1991). The timing of life-history events relative to the reproductive cycle in the northern fur seal (*Callorhinus ursinus*) was described by York and Scheffer (1997). These seals are widely dispersed in the North Pacific for most of the year but assemble each summer at breeding colonies (rookeries) on islands off the coasts of California, Alaska, and Russia. In the Pribilof Islands, females arrive from mid-June through early August. They bear a single young within 2 days of arrival and mate within the next week. Cell division in the blastocyst accelerates just before implantation, which occurs mostly in November, 123 days after copulation and 237 days before birth. This schedule allows males and females to disperse widely and feed in separate areas for most of the year and to assemble at rookeries to give birth and to breed.

Facultative delayed implantation occurs in some species in which the female is inseminated soon after the birth of a litter. This type of delay is known in some metatherians, some insectivorans, and many rodents. In rodents that have postpartum estrus, implantation of blastocysts is delayed when the female is suckling her first litter.

Our understanding of the factors controlling normal blastocyst development or dormancy in eutherian mammals is incomplete. Present evidence suggests that estrogen causes the uterine endometrium to form proteins essential for rapid growth of the blastocyst and that a deficiency of these proteins results in blastocyst dormancy (Daniel 1970; Heideman & Powell 1998). Experimentally administered doses of estrogen or progesterone, or both, have been used in an attempt to reinitiate growth of a dormant blastocyst in mammals with obligate delayed implantation. These procedures have not been successful in renewing growth of the blastocyst (Daniel 1970). McLaren (1970) has proposed that during lactation in house mice (*Mus musculus*) implantation is delayed by an initial inability of the blastocyst to "hatch" from the zona pellucida, which must be shed before implantation can occur.

Delayed implantation is an important part of the reproductive cycles of many metatherians. In most macropodids for which embryonic diapause is known, the mother undergoes postpartum estrus. A female with a newborn in the pouch may copulate and produce a fertilized zygote. In this case, the blastocysts cease growing at the 70- to 100-cell stage. The metatherian blastocyst is surrounded by protective membranes consisting of an albumin layer and a shell membrane. When the young leaves the pouch, development of the corpus luteum and growth of the blastocyst resume, the blastocyst implants, and rapid growth of the embryo resumes. The resumption of blastocyst growth may be due to the secretion of progesterone, platelet-activating factor, or some other growth factor (Shaw 1996). Embryonic diapause is common in the red kangaroo (*Macropus rufus*), a large herbivore living in arid regions of Australia. When conditions are good, a female *M. rufus* may have a blastocyst in diapause, a neonate suckling in the pouch, and a **joey** at her side. When drought occurs, the joey and the neonate may both die, in which case the corpus luteum reinitiates the development of the blastocyst (Tyndale-Biscoe 1984, 2005). This system favors survival of the mother at the expense of the young during stressful times.

Delayed Development

The California leaf-nosed bat (*Macrotus californicus*; Phyllostomidae) does not hibernate, mates in the fall, and females give birth the following summer. Fertilization occurs in the fall shortly after mating and development proceeds until the embryo reaches the blastocyst stage, at which time the blastocyst implants in the uterine lining, where embryonic development is delayed for about 4 months, to resume again in the spring (Bradshaw 1962).

Delayed development also occurs in the Jamaican fruit bat (*Artibeus jamaicensis*; Fleming 1971). Mating occurs after the birth of young in July or August. The blastocyst soon implants in the uterus, as in most mammals, but then becomes dormant, and further development is delayed until mid-November. This delayed development allows the young resulting from late summer matings to be born in early spring, when food (fruit) is abundant. In the wild, the ability to delay pregnancies would be highly adaptive when resources are scarce or unpredictable. Short-tailed fruit bats (*Carollia perspicillata*) can extend gestation up to 229 days if conditions are particularly stressful (Rasweiler & Badwaik 1997).

Delayed development also occurs in several species of pteropodid bats (Heideman 1989; Heideman & Powell 1998; Heideman et al. 1993; Mutere 1967). In all reported cases of delayed development in bats, the delay occurs at a very early stage, before gastrulation of the blastocyst. Postimplantation delays in tropical species, including phyllostomids and pteropodids, are not the result of lowered body temperatures as they are in some temperate zone vespertilionids (Heideman & Powell 1998; Racey 1982). This strategy in tropical bats may allow females flexibility in responding to poor environmental conditions or to synchronize the birthing period so that all young are born within a short time period (which may be important in species with nursery colonies). The mechanism producing these delays is not known.

Control of Reproductive Timing

A major factor affecting reproductive success in mammals is the precise timing of reproduction to coincide with favorable environmental conditions. This timing is mediated by interactions between environmental, behavioral, and physiological stimuli. These complex interactions defy simple explanation but have become better understood (Bronson 1989; Jameson 1988).

Many environmental factors have some effect on the timing of reproduction in mammals, but **photoperiod** (the period of light during the daily light–dark cycle), temperature, energy, and nutrition are probably of prime importance. Seasonal breeding in many temperate-zone mammals is partly regulated by photoperiod (Bronson 1989; Elliott 1976; Kenagy & Bartholomew 1981; Reiter 1983). In mammals, the pineal gland is photosensitive and can transmit information about the photoperiod to the endocrine system. The pineal secretes the hormone melatonin. When the photoperiod is short (short day length), melatonin levels increase, causing a depression of gonadal activity in the Syrian hamster (*Mesocricetus auratus*; Bartness & Wade 1984). The white-footed mouse (*Peromyscus leucopus*) is found over a great range of latitudes, and, as expected, the number of hours of daylight that it needs to preserve testicular function varies with latitude (Lynch et al. 1981). Laboratory studies of tammar wallabies (*M. eugenii*) show that their parturition times are determined by daylength (Sadlier & Tyndale-Biscoe 1977). Further evidence for a photoperiod cue in these wallabies comes from experiments in which the pineal was denervated, which abolished the pattern of delayed development of the blastocyst (Renfree et al. 1981). In some species, reproduction is tied to food abundance and not to photoperiod. Arctic-dwelling lemmings are reproductively unresponsive to changes in photoperiod (Hasler et al. 1976), although the molting of the collard lemmings (*Dicrstonyx* spp.) into their white winter pelage is triggered by decreasing photoperiod (Nagy et al. 1993).

The effects of temperature on reproduction are often more difficult to interpret. The testes of male pocket mice (*Chaetodipus formosus*) exposed experimentally to high temperatures were small relative to those of mice exposed to low temperatures. At high temperatures (35°C), reproduction in house mice was depressed (Pennycuik 1969). This inverse relationship is unusual, however, for low temperatures usually inhibit testicular growth (Clarke & Kennedy 1967; Kenagy 1981). Temperature may have little effect on gonadal activity, but changes in ambient temperature can deplete energy stores, resulting in changes in testicular activity.

Two prime factors affecting the timing of reproduction are energy and nutrition. Gestation and lactation demand large amounts of energy, and it is axiomatic that in mammals faced with seasonal variations in food availability breeding coincides with the time (or times) of food abundance. Unpredictable food resources may cause irregular reproduction. In Nevada deserts, when the annual seed crop fails to appear because precipitation has been scant, Merriam's kangaroo rats (*Dipodomys merriami*) do not breed (Beatley 1969). In the Great Basin Desert of southwestern Idaho, the Townsend ground squirrel (*Spermophilus townsendii*) has a single pulse of reproduction in early spring when green forage is available but will suspend reproduction in response to inadequate food supply. Availability of free water may be an important factor in controlling reproductive timing in desert environments. Giving supplemental free water to rodents in the Namib Desert caused a marginal improvement in reproductive success (Christian 1979).

The nutritional quality of the food supply also has an effect on reproductive performance. A lack of sufficient protein or a particular vitamin can negatively affect **gametogenesis**. Specific factors, such as plant secondary compounds, have been known to influence reproduction in such herbivores as voles. Both male and female montane voles (*Microtus montanus*) become reproductive rapidly when they ingest 6-methoxybenzoxazolinone (6-MBOA), a compound derived from young, actively growing plants (Berger et al. 1981; Sanders et al. 1981). This chemical cue allows the voles to initiate reproduction when the growing season has begun—when survival of the young of this short-lived rodent would be high—and synchronizes breeding with abundant plant growth in a fluctuating environment. Although still subject to experimental verification, there is evidence that 6-MBOA is one cue for reproduction in a number of vertebrates and may affect population dynamics in arvicoline rodents. Other plant compounds have the opposite effect. At the end of the growing season in August, salt grass (*Distichlis* sp.) becomes dry and brown and accumulates high levels of the phenol 4-vinylguaiacol, which is known to suppress reproduction in female montane voles (Berger et al. 1977). A wide variety of **phytoestrogens** (plant compounds that mimic animal estrogens) occur in nature, and the role they play in regulating mammal reproduction is likely to be complex.

Behavioral and physiological regulation of reproduction is clearly of great importance, and the importance of **pheromones** has been established by laboratory studies. Mammals deposit urine at various places in the course of their activities, and this urine serves as an individual's olfactory "signature" or "fingerprint" (Caroom & Bronson 1971; Jones & Nowell 1973a, 1973b) that provides information on the species, sex, reproductive condition, and social status. Such olfactory cues can regulate reproduction by triggering endocrine responses or modifying behavior. In association with tactile stimuli, chemicals in male urine called "priming pheromones" regulate the reproductive maturation and timing of ovulation in female house mice (*Mus musculus*) and deer mice (*Peromyscus maniculatus*) by inducing a series of hormonal responses (Bronson 1971, 1989; Bronson & Maruniak 1975). Males are also sensitive to urinary cues. The male endocrine system responds to female urine by secreting hormones that may increase the effect that the male's urine has on the female's reproductive system (Bronson 1979). Among prepubertal females, female urine suppresses reproductive maturation and overrides the effect of male urine for a period during development. In the presence of a male, female **puberty** can occur at 25 days of age but may not occur until 50 days in the absence of a male. Tactile stimuli are important at various stages of the reproductive cycle, and domination of one mouse by another can drastically reduce the secretion of sex hormones in the subordinant. Socially induced reproductive repression is known to occur in a eusocial mole-rat (Bennett et al. 1994a) and yellow-bellied marmots (Armitage & Schwartz 2000).

Bronson (1989) describes "three dimensions of complexity" in mammalian reproductive patterns. First, "mammals have evolved in ways that allow them to surmount environmental complexity." In other words, mammals can respond appropriately to environmental complexity, and no single pattern of responses exists in mammals. The second dimension of complexity "relates to the fact that the actions and interactions of environmental factors can vary greatly with the stage of the mammal's life cycle during which they are perceived." Responses to a particular set of factors may change dramatically after a female becomes pregnant, for example. The third dimension corresponds to "the ongoing interplay between a mammal's need to reproduce and its need to survive." In other words, reproductive processes are not independent of other demands on the organism. Bronson (1989) illustrates this point by describing reproduction as a process that competes for calories with thermoregulation. Calories allocated to reproduction are given low priority in nonpregnant females, but the balance shifts when the female becomes pregnant or is lactating.

Infanticide and Pregnancy Termination

Male-induced termination of pregnancy is a remarkable phenomenon mediated by social, olfactory, and endocrine factors (Bruce 1966; Hausfater & Hrdy 1984). Such termination is called the **Bruce effect** and is initiated in the laboratory by the replacement of the original male by an unfamiliar male or exposure of a pregnant female to the odor of an unfamiliar male. In response to a new male, females of several species of voles will abort their fetuses, enter estrus, and breed with the new male within a few days (Stehn & Jannett 1981). The appearance of the new male induces estrus regardless of the reproductive condition of the female. The Bruce effect is thought to result from a drop in prolactin secretion in the female induced by nervous inputs to the brain from the olfactory lobes. Male-induced abortion is to be expected in many arvicolines and could influence the population dynamics of wild populations.

Infanticide is an additional factor affecting the timing of reproduction in many mammals. Under some conditions, male langur monkeys (*Semnopithecus entellus*) kill young during a troop takeover (Hrdy 1977), and infanticide is recorded among several other primates (Goodall 1977; Rudran 1973, 1979; Struhsaker 1977). In a 4-year study of a troop of Namib Desert baboons living under harsh conditions, only 3 of 22 infants survived more than 6 months (Brain 1992; Brain & Bohrmann 1992). Infant deaths were due to unusually heavy tick (*Rhipicephalus* sp.) infestations around the nose and mouth that resulted in an inability to suckle and to infant kidnapping by high-ranking, non-lactating females (called "aunts"). Hrdy (1976) also details infant-kidnapping by females, which was referred to as "aunting-to-death" in other primate species. Infanticide in male collared lemmings (*Dicrostonyx groenlandicus*) is directed toward unfamiliar young, but males show little or no infanticidal behavior toward their own young (Mallory & Brooks 1978). Young Asian rhinoceroses (*Rhinoceros unicornis*) that are unprotected by their mothers are commonly killed by adult males (Dinerstein et al. 1988). Infanticide by males occurs in the puma (*Puma concolor*; Logan & Sweanor 2001) and is especially well-understood in lions (Chapter 24). A new male or group of male lions taking over a pride often kill the cubs, which hastens the onset of estrus in the females and seemingly increases the reproductive fitness of the infanticidal males.

Reproductive Cycles and Life-History Strategies

Underlying the tremendous variation in mammalian reproductive cycles is a broad pattern. Mammals can be segregated into two groups, those that bear altricial young (helpless, naked young in which the eye and ear openings are covered by membranes and locomotion and thermoregulation are undeveloped) and those that bear precocial young (fur-bearing young in which the eyes and ears are functional and thermoregulation and locomotion are well-developed). These patterns fall along a continuum, and each is typically associated with a different life-history strategy.

Mammals with altricial young typically live under unstable conditions with seasonal or unpredictable food abundance. They are small and subject to heavy predation pressure. Litters are large (often seven young or more). The young are born in a nest, and the gestation and suckling periods are short. The young grow rapidly and reach sexual maturity early. Life spans are short. The brain size is relatively small. The mother–young bond is brief, and social behavior is minimal. Estrus is short and frequently triggered in part by male–female interactions. Under favorable environmental conditions, breeding may occur repeatedly throughout the year. These high reproductive rates favor reproductive opportunists that take advantage of even brief periods of food abundance. With such opportunism goes high population turnover and population densities that are unstable seasonally and from year to year. An array of mammals, including tree shrews, shrews, tenrecs, many kinds of rodents, and small carnivorans, fit this pattern. Examples of altricial mammals are the European lemming (*Lemmus lemmus*) and the montane vole (*Microtus montanus*). Females of both these species can breed at a remarkably early age (15 and 21 days, respectively; Kalela 1961), and the gestation period is only some 21 days. The polyestrous females have postpartum estrus and may have several litters during the summer growing season. Females rarely survive to bear young during a second reproductive season. In extreme cases, reproduction may continue nearly all year, as in some arvicoline rodents (Baker & Ranson 1933; Greenwald 1956).

Semelparity, the life history strategy involving the death of males and the decline of fecundity of females after a single breeding season, occurs in many metatherians (Leiner et al. 2008). Among Australian marsupials, dasyurids (**Fig. 20-19**) have a brief, highly synchronized breeding season, after which all adult males die, and females undergo "exaggerated senescence," with much reduced fecundity in their second reproductive season (Braithwaite & Lee 1979; Wood 1970; Woolly 1966). Semelparity is seemingly common also among small Neotropical didelphids. The Brazilian slender opossum (*Marmosops paulensis*) offers an example (Leiner et al. 2008). All adults die soon after the September to March breeding season. The population includes

FIGURE 20-19   A yellow-footed antechinus (*Antechinus flavipes,* Dasyuridae) from Australia. Males of this species are semelparous.

exclusively juveniles and subadults from April to August, a period when food is in short supply. The death of post-breeding male dasyurids, and perhaps other semelparous marsupials as well, probably has several causes. The mating season is short but energetically costly, and during this time males lose half their body weight and much of their fur. Stress and high cortisol levels at this critical time initiate the degradation of protein and suppression of the immune system, resulting in the inability to mount **immune responses** to pathogens and parasites and a fatal decline in condition. Selective pressures associated with dry-season food shortages, with resulting high mortality of juveniles caused by intense competition between young and adults for food, may have favored the evolution of semelparity.

Mammals with precocial young, including many ungulates, cetaceans, primates, hyraxes, and some hystricognath rodents, typically live in relatively stable environments with a fairly predictable food base. These mammals are often large, reach sexual maturity late, and some are not subject to intense predation pressure. The estrous cycle is long. Ovulation is usually spontaneous, and gestation is prolonged. Usually the single young is not born in a nest but accompanies (or clings to) the female virtually from birth. Lactation is also prolonged, with an enduring mother–young social bond in many species. The brain of the precocial young mammal is usually large. Social behavior is complex, and an individual may spend its entire life as a member of a social group. These mammals have a low reproductive rate (exceptions include *Lepus* spp.), but the survival rate of young is usually high because of the extended period of maternal care. Population stability, a low reproductive rate, low population turnover, and dependence on a fairly stable environment make these animals vulnerable to habitat alteration by humans. An example of a mammal with precocial young is the South American woolly monkey (*Lagothrix lagotricha*). This 5.4-kilogram primate mates first at 8 years of age, has a long gestation (255 days), and one large (1 kilogram) young, and births are widely spaced (1.5 to 2.0 years). The young can cling to its mother shortly after birth. Lactation lasts from 9 to 12 months. The mother–young bond is tight, and social behavior is complex.

The length of the gestation period in eutherian mammals is, in general, positively correlated with body weight and with the degree of development of the newborn. The larger the mammal, the longer is the gestation period; for mammals of equal weight, the species with the heaviest neonate has the longest gestation (Huggett & Widdas 1951). Striking departures from this pattern occur among elephant-shrews, sloths, hystricognathous rodents, cetaceans, some pinnipeds, and primates. These reproductive departures probably evolved under selection associated with certain life-history strategies.

Elephant-shrews have unusually long gestation periods and bear highly precocial young. The mouse-sized (45 grams) *Macroscelides proboscideus*, for example, has a gestation period nearly 2 weeks longer (76 versus 63 days) than the wolf (*Canis lupus*), an animal over 600 times larger. Some species of elephant-shrews do not use burrows or nests, but rest on the surface of the ground, remain perpetually alert for predators, and escape by rapid bounds along well-known trails. Survival of young elephant-shrews depends on their precociously developed sensory and locomotor abilities and thus indirectly on their long gestation period.

Members of the Pilosa, which have a very specialized lifestyle (see Chapter 10), have a long gestation period, but there is no consistent progression toward longer gestation with increasing body size. McNab (1979, 1980) has found that pilosans have unusually low metabolic rates and has related this to their myrmecophagous (ant-eating) or folivorous (leaf-eating) habits. The case of the two-toed sloth (*Choloepus hoffmanni*) is of particular interest. This modest-sized (9 kilograms) animal has a gestation period of 332 days, some 3.5 months longer than the American elk or wapiti (*Cervus elaphus*), which is 22 times larger (200 kilograms). Both of these mammals are herbivorous and bear a single, precocious young at long intervals (12 months for the elk and 18 months for the sloth). The elk is highly mobile and ranges widely, feeding selectively on a wide variety of plants. It has relatively modest protein and energy needs per unit of weight because of its large size. In contrast, the arboreal sloth is sedentary, occupies a small home range (1.96 hectares), and feeds on the leaves of but a few tree species. Not only are the leaves often low in energy and protein, but they also contain defensive secondary compounds (such as tannins and terpenoids) that retard digestion. This sloth became adapted to a dependable and ubiquitous, but energetically marginal, food by adopting a very low metabolic rate, heavy insulation, and a counter-current heat exchange system in the limbs. The reproductive pattern is also the result of selection pressure to reduce energy needs: The extremely long gestation period avoids the rapid mobilization of energy, allows for the development of a precocial young able to cling to its mother, and reduces the energetically costly lactation period.

Baleen whales stand out as remarkable exceptions to the general body-weight, gestation-period trend. These whales, the largest animals of all time, have gestation periods similar to or shorter than those of camels or horses. Thus, the blue whale (*Balaenoptera musculus*), 250 times heavier than the camel (*Camelus bactrianus*), has a shorter gestation period (360 versus 406 days). The reason for this amazing rate of fetal growth is unknown. One possibility is that, by being extremely efficient harvesters of plankton, the most abundant marine food source, and by spending summers in boreal or austral waters where plankton reaches peak productivity, female whales may be able to invest great amounts of energy in the fetus and still store the reservoir of blubber necessary for migration and early lactation (young are usually born in tropical waters). Another possibility is that carrying a fetus through more than one migratory cycle and continuing pregnancy through tropical fasting periods is unfeasible energetically; the very rapid development of a fetus may be more economical.

The Australian sea lion (*Neophoca cinerea*) is unique in having an extended, nonannual, and nonseasonal breeding cycle. The pupping period is 5 months. Lactation is prolonged (15 to 18 months). The females stay near the breeding grounds all year, and breeding events shift forward 13.8 days earlier every 18 months (Higgins 1993). The only other pinniped with a nonannual cycle is the walrus (*Odobenus rosmarus*), in which individual females breed about every 24 months. Breeding occurs annually in the population but does not involve all reproductive females (Fay 1981).

The primates as a group have long gestation periods, and those of the small primates are most remarkable. Although most primate young are precocial (the human young is an exception), they depend on a long period of maternal and sometime paternal care and often on indoctrination into a complex social system. This is reflected by the typically long lactation period. The evolution of an early-primate reproductive pattern involving long gestation and precocial young may have been critical in setting the stage for the highly social lives of higher primates.

Inbreeding Avoidance

Seemingly, no mammalian species is immune to the deleterious effects of **inbreeding** (Lacy 1997). Inbreeding (breeding between close relatives) increases the probability that an individual will have two identical **alleles** at a **locus** due to inheritance from an ancestor shared by the individual's parents. Observed effects of inbreeding include heightened mortality, susceptibility to disease, and increased frequency of developmental defects, as well as reduced **fecundity**, lowered growth rates, an inability to withstand stress, and a lowered competitive ability (Ralls et al. 1988; Wright 1977). A reduced ability to adapt to environmental change is often cited as an especially adverse result of depletion of **heterozygosity**. Inbreeding in mammals is avoided or reduced in many ways, including dispersal by one or both sexes from the **natal** area before sexual maturity in some species, in others by male dispersal from natal social groups, and by recognition of **kin** and avoidance of breeding with them. In *Microtus*, the degree of inbreeding avoidance differs among species, perhaps being highest in monogamous species (Berger et al. 1997). Avoidance in *Microtus* may be based entirely on kin recognition due to

familiarity of parents and siblings before weaning (Ferkin 1989, 1990). Delayed reproductive development was observed by Rissman and Johnson (1985) in male California voles that were raised in bedding impregnated with "family" odor. This was probably a response to a unique family scent. These authors regarded this delay as a response that avoided inbreeding. Similarly, young white-footed mice (*P. leucopus*) were reproductively inhibited by air-borne chemicals present in their mother's urine (Terman 1992). In this species, movement away from the preweaning social environment may trigger reproduction (Wolff et al. 1988). Belding ground squirrels (*Spermophilus beldingi*), in contrast, can recognize kin without prior contact (Holmes & Sherman 1983). Clearly, avoidance of inbreeding in mammals is mediated by a variety of physiological and behavioral mechanisms.

Litter Size and Seasonal Timing

Because the metabolic cost of raising large, well-nourished litters is paid by a lowering of future reproduction, litter size represents the best reproductive investment for the environmental situation in which any population is living (Williams 1967). This best investment may differ within a species from area to area (Spencer & Steinhoff 1968). Within a species, large litters occur at far northern latitudes and high elevations, where severe winters and brief growing seasons limit the number of litters per year. The general pattern is for the mammals of a boreal community to have a few large litters each year, for those of a less severe temperate area to have smaller but more frequent litters, and for those in tropical communities to have many small litters each year. Within any area, however, strategies typically differ from one species to another. Thus, in Panama, most rodents breed throughout the year, but some species breed seasonally (Fleming 1970). In a subalpine community in Colorado, some rodents have several litters each summer and some have only one (Vaughan 1969).

In the view of Pianka (1976), optimal reproductive tactics involve maximizing an individual's reproductive fitness (the sum of all present plus future offspring) at every age. Reproductive effort, therefore, should vary inversely with residual reproductive value (expectation of offspring). The oldfield mouse (*Peromyscus polionotus*) follows these tactics (Dapson 1979). In one population of this species, the mice lived about 2 years, reproduced at a moderate rate, and had few differences in reproductive pattern among the **age cohorts** (segments of the population). A contrasting strategy was followed by another population of this same species. One cohort, made up of younger mice that had a chance of surviving to reproduce the following year, had a moderate reproductive effort, whereas another cohort, consisting of older animals with low residual reproductive value (animals that would not survive to reproduce the following season), had litters in rapid succession and had double the number of young per female.

The timing of reproduction is a vital factor influencing reproductive success. During gestation, and especially during lactation, the mother's energy needs increase tremendously. During the period when young are weaned and are becoming independent, their survival depends on adequate food. All phases of reproduction may be within the seasonal period of food abundance if it is long. If this period is short, however, gestation may occur at a stressful time, with lactation and weaning (or just weaning) occurring when food is most abundant.

Even in tropical regions, there is typically seasonality in food abundance, and mammalian reproductive patterns are responsive to this. In many species of Neotropical bats, the weaning of young coincides with peak food abundance just after the start of the rainy season (Bernard & Cumming 1997; Wilson 1979), but among phyllostomid bats, there are three common reproductive patterns: aseasonal polyestry, bimodal polyestry, and monestry. Similarly, there are several reproductive patterns among African bats (**Table 20-7**). The timing of reproduction in some species varies geographically. In Gabon, Africa, colonies of the bat *Hipposideros caffer* usually give birth in October in southern latitudes and in March north of the equator. The majority of tropical rainforest mammals breed seasonally (Bourliere 1973). This is the case for rodents in Africa (Delany 1971; Dubost 1968; Rahm 1970) and the typical pattern for monkeys of the genus *Cercopithecus* (Bourliere 1973).

In boreal and montane areas, brief growing seasons and long, severe winters compress breeding seasons and select for optimal allocation of energy between reproduction and hibernation. Long-term field studies by Michener and her coworkers (Michener 1998; Michener & Locklear 1990a, 1990b; Michener & Michener 1977) have revealed that the extremely tight activity schedules for the Richardson's ground squirrel (*Spermophilus richardsonii*) differ between the sexes. Typically, males awake from hibernation in early February but remain in their burrows for a week, during which they eat seeds stored the previous summer, renew fat deposits, and begin spermatogenesis. Males emerge from their burrows in late February, 2 to 3 weeks before the appearance of females. Females emerge soon after awakening, over a 2- or 3-week period in early March. Each female enters estrus within 4 days after emergence and is usually receptive to males for only a part of a single afternoon.

Male *S. richardsonii* eat little while competing intensely for females, lose weight, sustain injuries from fighting, and suffer high mortality. In poor physical condition, surviving males eat, replenish fat deposits, store seeds in their burrows, and enter hibernation in late June. Females bear young in mid April, and young emerge from burrows a month later. Adult females then spend several weeks laying on fat deposits and enter hibernation in early July. After emergence, juvenile ground squirrels gain weight rapidly and begin hibernation from about 7 weeks (for females) to approximately 15 weeks (for males) after the adults (**Fig. 20-20**). Adults of both sexes are active for only about 110 days a year. Males spend this time preparing for mating (by storing fat), mating, and preparing for hibernation and next year's mating (by storing body fat and caching seeds). Females spend about half their active time in pregnancy (23 days) and in nursing young (30 days) and half in eating and storing fat for hibernation. Thus, in a given activity season,

TABLE 20-7   Reproductive Patterns in Selected Paleotropical (African) Bats

Polyestrous: Year Long Asynchronous Breeding	Polyestrous: Asynchronous Breeding for Part of Year	Diestrus: Two Synchronous Breeding Periods	Monestrous: One Synchronous Breeding Period
Rousettus aegyptiacus (P)	*Epomophorus wahlbergi* (P)	*Rousettus aegyptiacus* (P) *Myonycteris torquata* (P)	*Eidolon helvum*[a] (P) *Myonycteris torquata* (P)
R. lanosus (P)		*Taphozous mauritianus* (E)	*Rhinopoma hardwickei* (RM)
Epomophorus labiatus (P)		*Nycteris hispida* (N) *N. thebaica* (N)	*Coelura afra* (E) *Taphozous nudiventris* (E)
Epomops franqueti (P)		*Cardioderma cor* (MG)	*Hipposideros commersoni* (R)
		Rhinolophus landeri (R) *Mops condylurus* (M) *Lavia frons* (MG)	*H. cyclops* (R) *H. caffer* (R) *Pipistrellus nanus* (V) *Eptesicus somalicus* (V) *Chaerephon pumila* (M)

[a] This species has delayed fertilization; copulation and parturition seem not to be synchronous within a population. Abbreviations: E, Emballonuridae; M, Molossidae; MG, Megadermatidae; N, Nycteridae; P, Pteropodidae; R, Rhinolophidae; RM, Rhinopomatidae; V, Vespertilionidae. Data are mostly from Kingdon (1974); some data were taken by TJ O'Shea and TA Vaughan in Kenya.

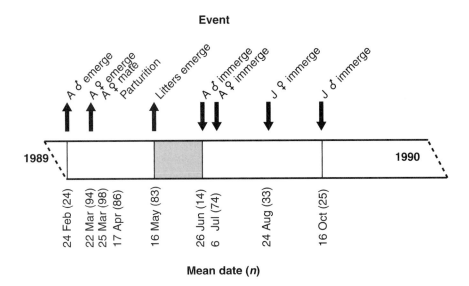

FIGURE 20-20 The annual cycle of Richardson's ground squirrels in Alberta, Canada, as indicated by major events in parts of 1989 and 1990: emergence from hibernation for adult males (A ♂) and adult females (A ♀); mating, parturition, emergence of litters from natal burrows, and emergence (entry into hibernation burrows) for adults and juvenile males (J ♂) and juvenile females (J ♀). Arrows pointing right designate mean emergence dates; arrows pointing left designate mean emergence dates. The cross-hatched segment (16 May to 26 June) is the only period when all age and sex classes are active above ground at the same time. The figures in parentheses are sample sizes. (Modified from Michener, G.R., J. *Mammalogy* 79 (1998): 1–19. American Society of Mammalogists, Allen Press, Inc.)

the fitness of a male depends primarily on his ability to gain body mass and competitive condition in the brief time (about 5 weeks) between awakening from hibernation and the availability of estrus females. A female's fitness, however, depends on her ability to assimilate sufficient energy after emergence for pregnancy and lactation. (This pattern, typical of mammals, involves males competing for females and females competing for energy to invest in young.)

The reproductive cycles of desert rodents are timed to take advantage of seasonal bursts in the growth of ephemeral forbs (Beatley 1976). In Arizona, both sexes of Merriam's kangaroo rat ate more green plant material during the semi-annual periods of plant growth than during the rest of the year, and there were surges in kangaroo rat reproduction immediately after these periods (Reichman & Van DeGraaf 1975). Two Arizona sites only 147 kilometers apart differed markedly in rainfall. One site received three times more rain in the autumn than the other and had vastly more green vegetation. At the wetter site, 90% of the adult kangaroo rats were reproductively active, whereas only 14% of the males and none of the females sampled at the drier site were in reproductive condition (Van De Graaff & Balda 1973).

SUMMARY

The unique mammalian reproductive pattern, with mammary glands providing milk for newborns or newly hatched young and a close mother–infant bond, probably evolved together with the diphyodont dentition in the late Triassic. Viviparity probably appeared later. Monotremes retain the amniotic egg of their ancestors, an egg tooth, a leathery eggshell, and incubation of the egg outside the mother's body. Unlike other egg-laying amniotes, however, after the egg hatches, the young monotreme is nourished by milk from the mother's mammary glands. Ancestral therians likely evolved the ability to retain the developing eggs in the oviducts for longer periods, resulting in loss of the leathery eggshell. This protected the eggs from environmental changes and freed the mother from incubating the eggs after laying. Living metatherians have a thin and highly permeable shell membrane, permitting additional nourishment from the uterus during gestation. In eutherians, selection favored lengthy *in utero* embryonic development and modification of embryonic membranes to form a placenta.

Metatherians bear virtually embryonic young after a brief gestation period. In contrast, many eutherians bear anatomically complete young after a relatively long gestation period. These contrasting reproductive patterns are driven by differences in estrous cycles, rates of embryogenesis, degree of placentation, sequence of hormonal events, as well as unique environmental cues and behavioral traits. Metatherian and eutherian reproductive patterns are derived, albeit divergent, reproductive modes.

The regulation of the reproductive cycle is maintained by environmental and social cues and by reciprocal controls between endocrine organs and their secretions. The events in the mammalian reproductive cycle are well-known, but details of these events are not completely understood for many mammal species. The ovarian cycle results in the development of ova, their release from the ovary, and their passage into the uterus; the uterine cycle involves a series of cyclic changes in the uterus. These changes result in a period of heightened sexual receptivity called estrus or heat, and the estrous cycle is the time from one entry into estrus and the next. Mammals may exhibit a single estrous cycle per year (monoestrous) or have more than one estrous cycle annually (polyestrous).

Eutherian fertilization normally takes place in the oviducts, after which the zygote begins a series of cell divisions as it passes down the oviduct. In metatherians, the resulting blastocyst remains within its shell membranes and absorbs nutritious uterine secretions directly across these membranes. A few days before birth, these outer membranes disappear, and embryonic membranes contact the uterine lining forming a choriovitelline placenta. Metatherian placentae come in four basic types, based on the degree of association between fetal and maternal tissues and the structure of the allantois relative to the chorion.

In contrast, the eutherian embryo, enclosed within a sphere of trophoblast cells, adheres to the uterine surface and erodes the endometrium, permitting the trophoblast (and the enclosed embryo) to invade the highly vascular uterine lining. The trophoblast grows rapidly, extending thousands of chorionic villi into the surrounding endometrium and eventually establishes a chorioallantoic placenta (a highly vascularized placenta with a remarkably large surface area through which rapid interchange of materials occurs). The degree to which the maternal and fetal bloodstreams are separated in the chorioallantoic placenta varies widely, as does the shape of the placental surface.

Lactation, the synthesis and secretion of milk to nourish the young, is a defining characteristic of mammals. Many hypotheses seek to explain the evolution of mammary glands, lactation, and suckling. Mammary glands may have evolved from apocrine-like glands associated with hair follicles. Lactation and suckling probably evolved together with precise dental occlusion and diphyodonty (replacement of milk teeth with adult teeth). According to this scenario, newborns obtained nourishment from milk as they grew, and without the need for chewing, permanent tooth eruption was delayed until near weaning when the jaw was approximately adult size.

Today all newborn mammals are nourished by milk produced by the mother's mammary glands. Under the influence of hormones, mammary gland growth during pregnancy and milk production is stimulated by suckling of the young. Lactation lasts far longer in metatherians than in similar-sized eutherians. Three stages occur in metatherians: mammogenesis (mammary gland preparation), early lactogenesis (while young are permanently attached to teats), and late lactogenesis (after detachment of young). Late-stage lactogenesis in marsupials is characterized by rapid growth of the young and is equivalent to the entire lactation phase of placental mammals.

In metatherians, milk composition changes during the long lactation period. In the early stages, the milk is dilute and contains mostly simple sugars. Later, the composition changes dramatically to include complex sugars and increased protein and fat concentrations. In placental mammals, the composition of the milk remains relatively constant throughout lactation, but some species produce relatively dilute milk and other species, such as seals, produce concentrated, fat-rich milk.

In eutherian mammals, the length of time from fertilization to implantation is typically short, but some mammals have abandoned this pattern. One departure involves a delay of ovulation and fertilization until long after copulation (delayed fertilization); another is typified by normal fertilization but arrested embryonic development at the blastocyst stage (delayed implantation). In yet another, there is a long delay in the development of the blastocyst after it has implanted (delayed development).

Placental mammals can bear altricial young (helpless, naked young in which the eye and ear openings are covered by membranes and locomotion and thermoregulation are undeveloped) or precocial young (fur-bearing young in which the eyes and ears are functional and thermoregulation and locomotion are well-developed). These patterns fall along a continuum, each associated with different life-history strategies. Mammals bearing altricial young usually live under unstable conditions with seasonal or unpredictable food abundance. Altricial young are small and subject to heavy predation pressure. Litters are large, and the gestation and suckling periods are short. The young grow rapidly and reach sexual maturity early. Life spans are short. The mother–young bond is brief, and social behavior is minimal.

In contrast, mammals with precocial young typically live in stable environments with predictable food supplies. These mammals are often large and reach sexual maturity late, and some are not subject to intense predation. The estrous cycle is long. Ovulation is usually spontaneous, and gestation is prolonged. Usually a single young is born, lactation is prolonged, an enduring mother-young social bond is formed, and social behavior is more complex. These mammals have a low reproductive rate, but the survival rate of young is usually high because of the extended period of maternal care.

KEY TERMS

Age cohort
Alleles
Blastodisc
Bruce effect
Chorionic villi
Copulation
Corpus luteum
Cotyledonary placenta
Deciduous
Delayed development
Delayed fertilization
Didelphis
Diestrus
Diffuse placenta
Discoidal placenta
Egg tooth
Embryotroph
Endometrial gland
Endometrium
Endotheliochorial placenta
Estrus

Fecundity
Fertilization
Gametogenesis
Harem
Heterozygosity
Immune response
Immunoglobins
Inbreeding
Induced ovulation
Infanticide
Infundibulum
Joey
Kin
Locus
Menstruation
Metestrus
Monestrus
Morula
Myometrium
Natal
Nondeciduous

Oocytes
Ovoviviparous
Parturition
Peristaltic
Pheromones
Photoperiod
Phytoestrogens
Polyestrus
Postpartum
Precocious
Proestrus
Puberty
Spermatogenesis
Spontaneous ovulation
Trophoblast
Uterine mucosa
Vivipacity
Zona pellucida
Zonary placenta
Zygote

RECOMMENDED READINGS

Bronson, FH. 1989. *Mammalian Reproductive Biology.* University of Chicago Press, Chicago, IL.

Eisenberg, JF. 1981. *The Mammalian Radiations.* University of Chicago Press, Chicago, IL.

Johnston, SD, et al. 2007. One-sided ejaculation of echidna sperm bundles. *American Naturalist, 170*:E162–E164.

Hayssen, VD, et al. 1993. *Asdell's Patterns of Mammalian Reproduction: A Compendium of Species-specific Data.* Cornell University Press, Ithaca, NY.

Oftedal, OT. 2002. The mammary gland and its origin during synapsid evolution. *Journal of Mammary Gland Biology and Neoplasia, 7*:225–252.

Oftedal, OT. 2002. The origin of lactation as a water source for parchment-shelled eggs. *Journal of Mammary Gland Biology and Neoplasia, 7*:253–266.

Saunders, NR & LA Hinds. 1997. *Marsupial Biology: Recent Research, New Perspectives.* University of New South Wales Press, Sydney, Australia.

Tyndale-Biscoe, CH. 2005. *Life of Marsupials.* CSIRO Publishing, Victoria, Australia.

Tyndale-Biscoe, H & MB Renfree. 2005. *Reproductive Physiology of Marsupials.* Cambridge University Press, Cambridge, UK.

Aspects of Physiology

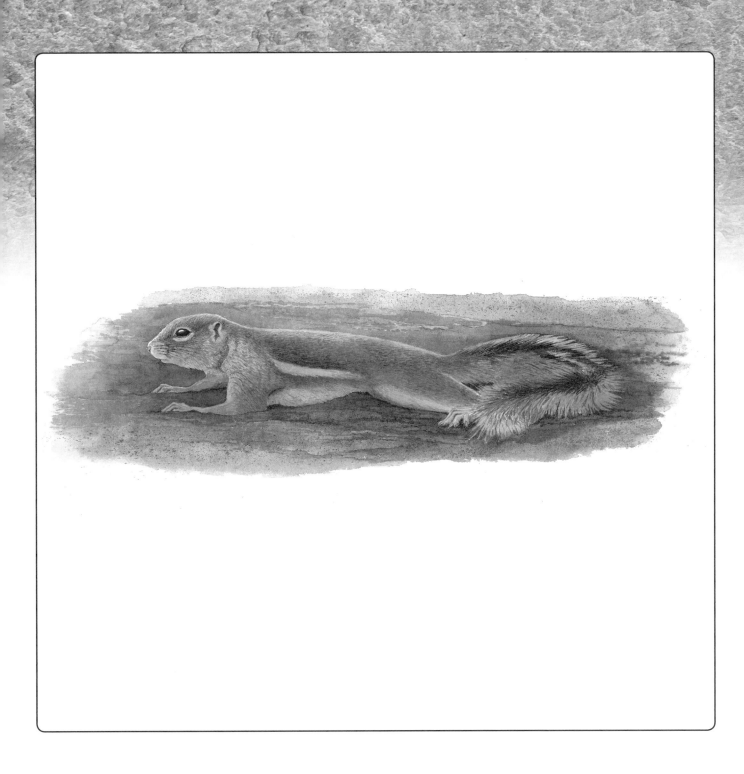

Some barriers to mammalian distribution are easily recognized. Bodies of water, ice fields, or mountains may be absolute barriers to dispersal, depending on the environmental tolerances of the specific mammals. Equally limiting, however, are environmental temperatures and the availability of water. The distributions of some mammals—Neotropical sloths, for example—might be described most precisely by reference to the extremes of temperature and to the seasonal patterns of temperature change that can be tolerated. Air temperatures from –50°C to 50°C may be encountered at various times and places on the earth, but mammals can only survive body temperatures of approximately 0°C to 45°C and can be normally active only within the narrow range of body temperatures between approximately 30°C and 42°C. Just as some mammals are adapted to a few food sources or to a restricted type of habitat, some can live only within a narrow range of temperatures. In mammals, interspecific differences in the ability to withstand temperature extremes or scarcity of water occur even among closely related species, and it is not surprising that no one species is adapted to facing the full range of environmental extremes known for mammals as a group. Knowledge of various aspects of mammalian physiology is essential to understanding how mammals adapt to the great array of ecological settings they occupy.

Endothermy: Benefits and Costs

Most animals are **ectothermic**; their body temperature is regulated behaviorally by heat gained from the environment rather than by metabolically produced heat. Mammals and birds, however, are **endothermic** (from the Greek *endo* for "within" and *therm* for heat). Their body temperatures are controlled largely by a combination of internal metabolic activity and physiological regulation of heat exchange with the environment, with **behavioral thermoregulation** of lesser importance for most mammals (**Fig. 21-1**). Instead of endotherm, many physiologists prefer the term **homeotherm** (from the Greek *homeo* for "same" and *therm* for "heat"), which refers to thermoregulation that maintains a stable internal body temperature regardless of external influence. The main difference is that endotherm refers to the location (internal) of heat production, whereas homeotherm refers to the constancy of the heat produced. Most homeotherms maintain a high and fairly constant body temperature throughout life, but for this they pay an extremely high cost in energy. Typically, a mammal expends 5 to 10 times more energy for maintenance than a reptile of equal size and equal body temperature (38°C to 40°C). At lower temperatures, the cost of maintaining a high body temperature rises abruptly: a mammal uses 33 times more energy than a reptile at 20°C and 100 times more at 10°C. A foraging mouse uses 20 to 30 times more energy than a foraging lizard of equal weight. In small mammals, such as most rodents, 80% to 90% of the total energy budget is spent on thermoregulation. Consequently, some mammals become either regionally or temporally heterothermic (from the Greek *hetero* for "other"). **Regional heterothermy** occurs when core body temperature is maintained well above that of other body regions (e.g., limbs), whereas temporal heterothermy occurs when mammals modify their body temperature in response to fluctuations in daily or seasonal

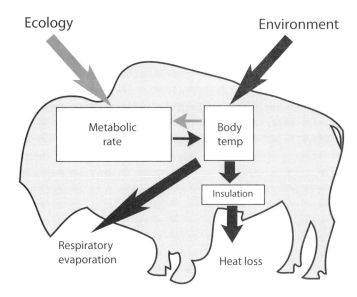

FIGURE 21-1 Factors influencing body temperature in mammals. The thin arrows represent influence—ecology and body temperature both influence metabolic rate. The thick arrows denote heat flow. (Adapted from Clarke, A., and Rothery, P., *Funct. Ecol.* 22 (2008): 58–67.)

environmental temperatures. Regardless, the costs of homeothermy (endothermy) are clearly high. What are the benefits?

A variety of benefits have been recognized. In mammals (or in mammalian ancestors), endothermy probably evolved in the late Triassic under selective pressure for sustained activity and reduced body size. As a costly byproduct of these capabilities, the resting metabolic rate was raised (McNab 1978). The primary advantage of endothermy is in greatly enhancing the ability to sustain high levels of activity (Bennett & Ruben 1979). High body temperatures are supported by high oxygen-transport capabilities and high rates of enzymatic action. The maximum capacity of endotherms to use aerobic metabolism to produce power surpasses that of ectotherms by a factor of roughly 10. Homeotherms can be active under an imposing array of temperature extremes, ranging from intense desert heat to extreme arctic cold. Moreover, they are freed from dependence on the warmer sunlit part of the daily light–dark cycle and from becoming inactive during cold seasons. Most mammals have responded to this freedom by being nocturnal, and many are active through all seasons.

Although core body temperatures are held relatively constant within groups, there is considerable variation in mammalian body temperature (T_b) between groups. Monotremes, armadillos (Cingulata), and sloths (Pilosa) have relatively low body temperatures (**Fig. 21-2**). With the exception of shrews, many insectivous mammals and metatherians also have lower T_b than many other groups of mammals. Most eutherian mammals attempt to maintain

T_b between 36°C and 38°C. However, their ability to do so depends largely on the environmental temperature. Every endotherm has a **thermal neutral zone** (TNZ) within which little or no metabolic energy is expended on temperature regulation (**Fig. 21-3**). Within this zone, the fluffing or compressing of the fur, local vascular changes, or shifts in posture suffice to maintain thermal homeostasis.

Outside the TNZ, metabolic costs go up (Fig. 21-3). To see why this is so, consider the relationship between the amount of metabolic heat (VO_2) needed to offset heat lost to the environment and ambient temperature (T_a).

$$VO_2 = C(T_b - T_a)$$

Because T_b is held constant in mammals and **thermal conductance** (C; the rate of heat loss from an animal to its environment) is determined by the degree of insulation (i.e., the better the insulation, the lower the C is), metabolic heat production (VO_2) varies linearly with ambient temperature (T_a) outside the TNZ. At the lower limit of the TNZ is the **lower critical temperature**, the point below which the balance between metabolic heat production and heat lost to the environment cannot be maintained (except by

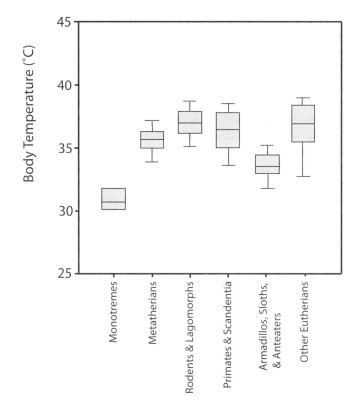

FIGURE 21-2 Box plots of body temperature for various mammalian groups. Members of the Monotremata, Cingulata, and Pilosa have lower body temperatures than other mammalian groups, but there is considerable variation in body temperature within all taxonomic groups. The box encloses the median and the upper and lower 25% of data values. The "whiskers" represent the maximum and minimum values excluding outlier points. (Data from Clarke and Rothery, *Funct. Ecol.* 22 (2008): 58–67.)

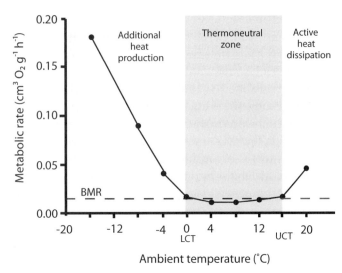

FIGURE 21-3 The effect of environmental temperature on metabolic rate (oxygen consumption) in Arctic ground squirrels (*Spermophilus parryii*). No additional energy is required across the thermoneutral zone (from 0°C to 16°C). Above the upper critical temperature (UCT) additional energy is required to actively dissipate excess heat. Below the lower critical temperature (LCT) additional energy is needed to maintain a constant body temperature. (Data from Buck, C. L., and Barnes, B. M., *Am. J. Physiol. Regul. Integr. Comp. Physiol.*, 279 (2000): R255–R262.)

variations in thermal conductance). Below the lower critical temperature, oxidative metabolism must be increased to keep the body temperature constant. The rate at which metabolism must be increased (denoted by the slope of the line) is determined by thermal conductance. The better the insulation, the lower the slope and the less metabolic heat that must be produced to keep the body temperature constant for an animal of given body size. Thermal conductance also depends on body size, because the surface area to volume ratio is an important determinant of metabolic rate (**Fig. 21-4**). Obviously, if a constant body temperature is to be maintained over a wide range of ambient temperatures, adjustments of both thermal conductance (through changes in insulation) and heat production (through metabolic changes) are necessary.

The **upper critical temperature** is the point above which metabolic work (greater than resting level) is needed in order to dissipate heat and thus maintain constant body temperature (Fig. 21-3). This temperature is far less variable than the lower critical temperature but is of great importance to desert mammals, which usually do not have access to drinking water and must strictly minimize water loss. Many mammals faced with temperatures above the upper critical temperature dissipate heat by **evaporative cooling**, which involves considerable water loss. Because such loss in desert species is extremely disadvantageous, these animals try to avoid temperatures above the upper critical temperature, often by spending the day in cooler

shelters and being active on the surface only at night. Flying foxes in Australia deal with increasingly excessive heat by first wing-fanning, then shade-seeking, **panting**, and saliva-spreading (licking the wings for evaporative cooling). Yet they can still suffer tremendous mortality during heat waves; in a single night in 2002, environmental temperatures over 42° C (108° F) killed over 3,500 bats in 9 mixed-species colonies (Welbergen et al. 2007).

Selection in mammals has favored body temperatures and metabolic rates that save energy or facilitate the exploitation of a particular environment. There are many variations on the endothermic theme among mammals, and the survival of a species depends just as much on matching its thermoregulation pattern to its lifestyle and environment as it does on foraging efficiency or predator avoidance. Many large mammals have sufficient body mass, and therefore enough **thermal inertia**, to maintain their body temperature within narrow limits. For small mammals, however, in which the ratio of mass to surface area favors rapid heat loss, wider fluctuations of body temperature are common. Indeed, for small mammals, a regular pattern of body temperature change through the daily temperature cycle may be the rule rather than the exception (Kenagy & Vleck 1982).

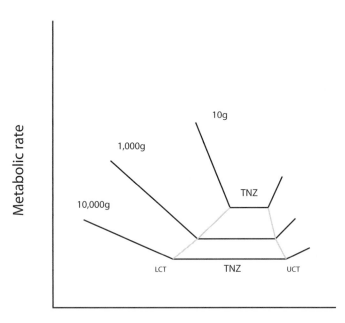

FIGURE 21-4 Relationship between metabolic rate and ambient temperature for mammals of various body sizes ranging from 10 grams to 10 kilograms. Metabolic rate stays relatively constant within the thermal neutral zone (TNZ). Above the upper critical temperature (UCT), the animal must expend additional energy to dissipate heat. Below the lower critical temperature (LCT), metabolic rate increases to keep the animal warm. Larger mammals have greater thermal inertia and require less metabolic energy to keep body temperature constant below the LCT.

Even within a species, body temperatures and metabolic rates are strongly influenced by climate (Careau et al. 2007). For example, coyotes (*Canis latrans*) of the hot Sonoran Desert have **basal metabolic rates** (BMR; the minimum metabolic rate necessary for simply maintaining life in a resting organism in thermal neutrality) 79% of that predicted for its body size (Golightly & Ohmart 1983), whereas Shield (1972) found that for cold-adapted Alaskan coyotes the metabolic rate was 126% of that predicted. Of all members of the Carnivora, red foxes (*Vulpes vulpes*) have the broadest geographic range, including hot North African deserts and the cold arctic coast of Canada (**Fig. 21-5**). The desert red foxes have roughly 20% lower metabolic rates than do their arctic conspecifics. In desert climates, high temperatures and water scarcity put a premium on reducing use of energy and on limiting water loss. By decreasing heat production through a reduced metabolic rate, the need for evaporative cooling is reduced, and precious water is conserved. In arctic climes, there are different physiological problems: Temperatures there are often far below an animal's thermoneutral zone, and the ability to maintain high levels of endogenous heat production is critical. This ability is associated with a high BMR (but usually also with behavioral adjustments, larger body size, and more insulative fur).

Metatherian Versus Eutherian Metabolic Rates

Of the metatherians that have been studied, all but one (the water opossum, *Chironectes minimus* of the American tropics) have low BMRs relative to eutherians. Whereas in eutherians such aspects of reproduction as gestation period, litter size, and rate of postnatal development are positively correlated with metabolic rate, in metatherians there is no such relationship (McNab 1983). Metatherians in general are as good at thermoregulation as eutherians, and differences in metabolic rates between these infraclasses are interpreted by McNab as demonstrating that the high metabolic rates of eutherians resulted from selection for high reproductive rates rather than for more effective thermoregulation (see Chapter 20 for a discussion of the metatherian–eutherian reproductive dichotomy). Of interest here is that the water opossum has the highest known rate of metabolism of any metatherian and also has an unusually rapid growth rate of the young (Rosenthal 1975).

The view that metatherians are characterized by lower metabolic rates than those of eutherians may be an oversimplification. Dasyurid metatherians, when resting at temperatures within their thermoneutral zones, have metabolic rates 32% below those of comparable-sized eutherians, but with lower ambient temperatures and locomotor activity simulating foraging, the metabolic rates of dasyurids and

FIGURE 21-5 A red fox (*Vulpes vulpes*) from Canada.

eutherians are indistinguishable (Baudinette 1982; Hinds & MacMillen 1984). Future research may show that active metatherians and eutherians have similar metabolic rates.

Among eutherians, factors independent of body mass may strongly influence metabolic rate as well as other aspects of life history. Consider the following comparisons, based largely on data from George et al. (1986) and Perrin and Fielden (1999). The short-tailed shrew (*Blarina brevicauda*) and Grant's golden mole (*Eremitalpa granti*; Fig. 8-5B) are miniature carnivores/insectivores of about 20 grams, but their metabolic physiology and most aspects of their life histories differ radically. The short-tailed shrew lives in the **mesic** northcentral and northeastern United States, where it forages in runways or burrows beneath heavy grass or leaf groundcover. Grant's golden mole, however, lives in the arid Namib Desert of southwestern Africa, where it occupies nearly barren sand dunes (**Fig. 21-6**) and forages on the surface or by burrowing. The metabolic rate of the short-tailed shrew is more than 12 times that of the golden mole, and food consumption is at least 3.7 times greater. These differences reflect adaptations to very different environments, the short-tailed shrew's habitat offering relatively abundant food plus cover for diurnal foraging, whereas the golden mole's environment provides limited food that is patchily distributed, and no cover for daytime foraging. The short-tailed shrew's food supply can sustain a high metabolic rate, large litter size (average of 4.5 young per litter), and high population density (1.6 to 121 animals per hectare), but the inhospitable environment of the golden mole has selected for conservation of energy, a low reproductive rate (1 young per litter), and low population densities (0.14 to 1.19 individuals per hectare).

The metabolic rates for most mammals that have been studied lie between 90% and 110% of the expected value. Although body mass has the primary influence on metabolic rate, food habits, independent of body size or taxonomic status, are of great importance (McNab 1980, 1983). Desert seed-eating rodents (Heteromyidae), fossorial herbivorous rodents (Geomyidae, Ctenomyidae, Rhizomyidae, Bathyergidae), and arboreal omnivores (various primates) have low metabolic rates (between 60% and 89% of that expected). Even lower rates (less than 60% of that expected) are those of many insectivorous bats, ant and termite specialists (anteaters, Myrmecophagidae; pangolins, Manidae; aardvark, Orycteropodidae), leaf eaters (sloths, Bradypodidae), and soil and litter omnivores (armadillos, Dasypodidae; hedgehogs, Erenaceidae). These mammals face a suite of problems: Their food is either widely dispersed, is only seasonally available or abundant, is deficient in energy, or contains defensive secondary compounds (see Chapter 23. These mammals, then, have been forced to lower their rates of energy output to a level sustainable by their low energy intake. Many of these mammals have low body temperatures, poor thermoregulatory abilities, and some can undergo torpor. At the opposite end of the scale,

FIGURE 21-6 The sand dune habitat of Grant's golden mole (*Eremitalpa granti*) in Namibia.

high metabolic rates (111% to 140% of expected) occur among some terrestrial carnivores, insectivores (crocidurine shrews), some small carnivores (weasels, mink, wolverine, marten, Mustelidae), some small herbivores (voles, Arvicolinae; rabbits, Leporidae), horse (Equidae), and cattle (Bovidae). Extremely high rates are found among some shrews (Soricidae), some semiaquatic carnivores (otters, Mustelidae), and some aquatic carnivores (seals, Phocidae; dolphins, Delphinidae; porpoises, Phocoenidae). McNab (1983) stressed that because of the positive relationship between their metabolic rate and reproductive rate, eutherians tend to have the highest possible metabolic rates.

Probably all mammals make metabolic adjustments throughout the daily cycle of activity in response to their levels of activity or to environmental changes (temporal heterothermia). For some groups of mammals that live in temperate areas (e.g., insect-eating bats and squirrels), major seasonal changes in metabolic rate are a key to survival. Even in a resting mammal, different parts of the body have contrasting metabolic rates: Brain tissue has a higher rate than muscle tissue, and one organ system may have a higher rate than another (regional heterothermia). Considering these complexities, it is apparent that the BMR indicates the "average" metabolic rate for an organism at a given time but does not further our understanding of the metabolic division of labor among organs or organ systems (Eisenberg 1981).

As yet, our knowledge of mammalian daily or seasonal adjustments in metabolism is incomplete, but there is no doubt as to its importance. Just as the capacity for drastic morphological change has been a keynote of mammalian evolution for over 250 million years, so also has metabolic plasticity almost certainly been important. Finely tuned metabolic responses to internal and external stimuli are now and must long have been a critical part of the mammalian adaptive repertoire.

Coping with Cold

There are essentially four mammalian strategies for surviving in cold environments. Mammals inhabiting cold climates can (1) over many generations, evolve large body size, resulting in a more favorable surface-area-to-volume ratio and consequently reduced heat loss; (2) decrease their rate of heat loss through increased insulation or behavioral thermoregulation; (3) increase their rate of metabolic heat production; or (4) abandon their normal body temperature and allow it to decline to a level closer to ambient temperatures (**hypothermia**). Obviously, the optimal strategy will depend on the environmental conditions and physiological constraints of the individual at any given time.

◼ Consequences of Body Size

Boreal mammals are generally larger than their ecological counterparts and close relatives in warmer areas (referred to as Bergmann's Rule). For example, the collared lemming (*Dicrostonyx groenlandicus*) of the Greenland tundra, a giant among nonaquatic arvicoline rodents, is more than twice as large (76 versus 32 grams) as its related eastern North American counterpart, the meadow vole (*Microtus pennsylvanicus*). Other cold-climate mammals that outsize their warm-climate relatives are the Arctic hare, the Alaskan wolf, and the Alaskan moose. The uniformly large size of marine mammals (which is discussed later) is an adaptation to living in cold water. Bergmann's Rule was later modified to include only within-species comparisons across latitudinal gradients (Mayr 1956). Under these more restricted conditions, the validity of Bergmann's Rule has been challenged (McNab 1971; but see Ashton et al. 2000). An intraspecific increase in body size with increasing latitude suggests that there is an advantage to being large in colder regions. In fact, large size is a common and effective adaptation to cold. In general, large size favors heat conservation, and small size favors heat dissipation. To understand why this is true, consider how surface area and volume (or mass) change as an animal doubles in height (or length). As height doubles, the surface area (SA) increases fourfold (as the square of the linear dimension), but the volume (V; or mass) increases eightfold (as the cube of the linear dimension). Thus, the larger the animal, the greater is the volume or mass relative to surface area, and the smaller the animal, the greater is the surface area relative to mass (**Fig. 21-7**). The fact that surface area and mass (or volume) scale at different rates has profound consequences for mammalian physiology. In mammals, the amount of metabolic heat produced is roughly proportional to the number of cells in the body (volume). The rate at which that heat is lost is proportional to the surface area of the skin. Consequently, large mammals (with low SA:V ratios) have difficulty staying cool (by dissipating the heat they produce), and small mammals (with their high SA:V ratios) have difficulty staying warm.

The empirical relationship between body temperature (T_b), lower critical temperature (T_{lc}), and body weight (W) in mammals is represented by the expression $T_b - T_{lc} = 4W(0.25)$. Because body temperature is usually held constant in mammals, as weight decreases, T_{lc} approaches T_b (Gordon et al. 1977) (Fig. 21-4). BMR, lower critical temperature, and thermal conductance all vary inversely with body size and are intimately interrelated. Mass-specific metabolic rate, as measured by oxygen consumption per gram of body weight per hour, rises so precipitously with decreasing body weight that the smallest mammals (Etruscan shrew, *Suncus etruscus*, and Kitti's hog-nosed

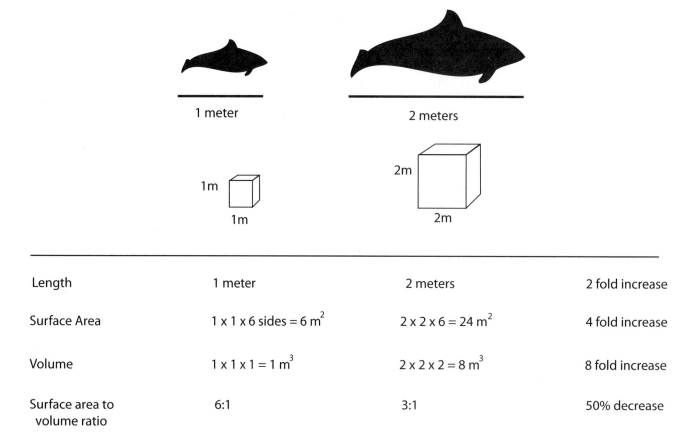

FIGURE 21-7   The effects of increasing body length on surface area and volume. A Dall's porpoise doubles in length from 1 meters to 2 meters over its lifetime. The square boxes represent what happens to the area and volume of an object when its length increases twofold. As the linear dimension doubles, the surface area increases fourfold, and the volume increases eightfold.

bat, *Craseonycteris thonglongyai*), each weighing about 2 grams, probably represent the lower limit of mammalian body size (newborns are considered in the following discussion). The mass specific metabolic rate of a mammal the size of the Etruscan shrew, for example, is approximately 141 J/g/hr (joules/gram/hour) compared with only about 2 J/g/hr for an elephant (Withers 1992). Rates of oxygen consumption differ markedly even among small mammals (**Fig. 21-8**): The tiny masked shrew consumes oxygen at a rate over four times that of the larger deer mouse. Carrying the comparison further, the mouse consumes oxygen 10 times faster than the horse (Krebs 1950).

The smallest mammals are often **neonates** (newborns) (**Fig. 21-9**). Until relatively recently, studies of neonatal thermoregulation involved measurements made on individual young removed from the nest. Under these highly artificial conditions, altricial neonates appeared to be devoid of physiological regulation of body temperature and were erroneously referred to as ectothermic (Case 1978). As Hill (1992) pointed out, in nature, neonates huddle in groups in well-insulated nests and are often warmed by the close proximity of their parents and siblings. Measurements of neonatal and weanling thermoregulation under these more

natural conditions revealed that even altricial young begin to thermoregulate at an early age. Thus, neonates rely on shared body warmth from litter mates and parents during much of their early development. Hill (1992) hypothesized that neonates suspend their own thermoregulation facultatively when the mother is present but begin endogenous **thermogenesis** (heat production) when she is out foraging.

BMRs are typically measured under controlled laboratory conditions far removed from those experienced by the animal in the field. Applying BMR data from the laboratory to free-living mammals responding to a wide variety of environmental and behavioral conditions is difficult. Field metabolic rates (FMR) can now be measured with accuracy using the doubly labeled water technique. FMR is the total energy cost (including BMR, thermoregulation, locomotion, feeding, digestion, reproduction, growth, and so forth) experienced by a free-ranging animal during the course of 24 hours. The doubly labeled water technique involves measuring the relative turnover of radioactively labeled hydrogen and oxygen injected into the animal as water. The hydrogen isotope measures the rate of water loss, and the oxygen isotope measures the sum of water and

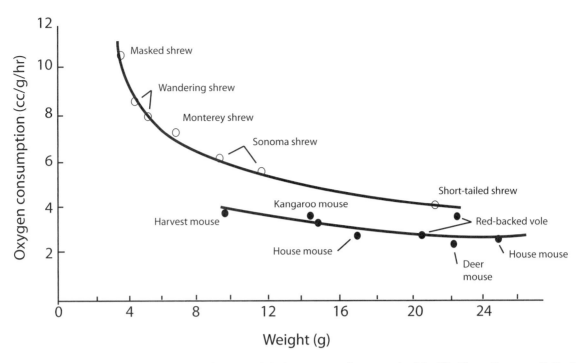

FIGURE 21-8 Oxygen consumption as a function of body weight in some small mammals. (Modified from Pearson, O. P., *Science.* 108 (1948): 44. Reprinted with permission from AAAS.)

CO_2 loss (Kunz & Nagy 1988; Nagy 1987). The difference between the two turnover rates can be converted to FMR.

Using FMR data from many doubly labeled water experiments on a variety of vertebrate taxa, Nagy (1987) showed that FMR is strongly correlated with body mass (**Fig. 21-10**). The slope of the relationship between FMR and body mass for 23 species of eutherians (0.81) was significantly higher than that measured for BMR (0.75), but the opposite was true for 13 metatherian species (slope = 0.58). These data showed that FMRs of medium-sized mammals and birds (200 to 500 grams) are similar, suggesting that it costs the same amount of energy to live one day. Small eutherians

(less than 200 grams) had lower FMRs than small metatherians, and large eutherians have relatively high FMRs. The FMR of a 67-kilogram mule deer (40,000 kilojoules per day [kJ/d]) was nearly four times higher than a 62-kilogram

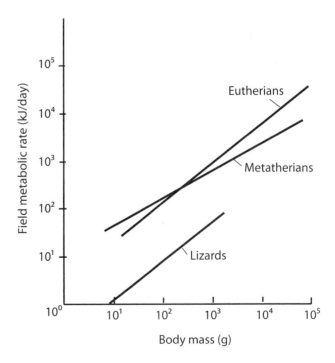

FIGURE 21-10 Scaling of FMRs with body size in eutherian (23 species) and metatherian (13 species) mammals along with comparable data for lizards (25 species). FMRs of endotherms are approximately 17 times that of similar-sized ectotherms. (Modified from Nagy, K. A., *Ecol. Monogr.* 57 (1987): 111–128.)

FIGURE 21-9 A group of rat (*Rattus rattus*) neonates.

gray kangaroo (11,734 kJ/d; Nagy 1987). Compared with a lizard (ectotherm) of similar size, a mammal expends approximately 17 times as much energy. The energetic difference is due to the increased cost of maintaining a high body temperature for the mammal. Nagy (1987) also showed that FMRs scaled differently with diet, season, and habitat. Desert-dwelling mammals typically have 30% lower FMRs than nondesert mammals. The relationship between body size and metabolic costs is further complicated by the ability to store fat and the type of insulation. Large size

enhances fat storage, which may serve both as insulation and an effective metabolic fuel.

Insulation

Individual mammals generally lack the ability to increase their body mass sufficiently for this to be an effective response to cold stress. A more useful strategy is to lower their thermal conductance. Effective insulation is, therefore, an important feature of cold-adapted mammals (**Fig. 21-11**). The insulative value of fur increases with its thickness. A

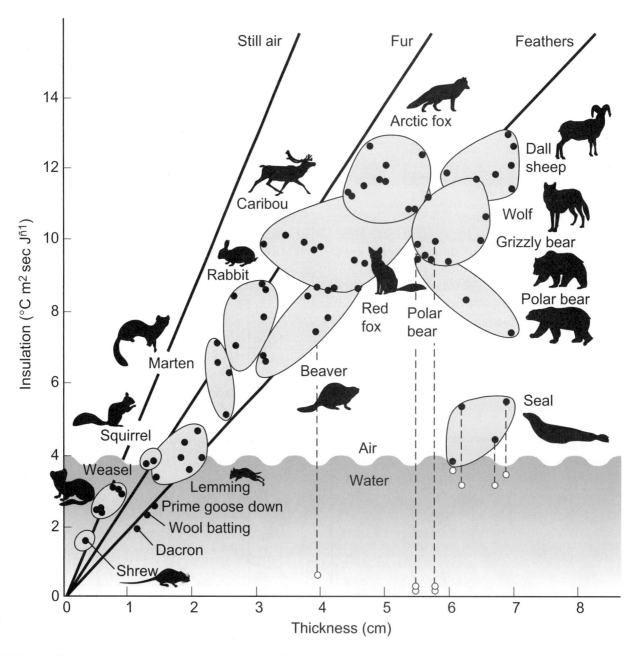

FIGURE 21-11 The insulative value of various thicknesses of mammalian fur. The dotted vertical lines illustrate the loss of insulation when fur is wet. Seal fur provides less insulation in air than the fur of other mammals, but retains most of its insulative value when wet. (Modified from Scholander, P. F., et al., *Biol. Bull.* 99 (1950): 225-236. Reprinted with permission from the Marine Biological Laboratory, Woods Hole, MA.)

half-centimeter-thick layer of shrew fur would have approximately the same insulative capacity as a half-centimeter layer of caribou fur. Shrew fur, however, is less than 0.3 centimeters thick. Caribou fur is often 4 or 5 centimeters thick, and winter pelage can be as thick as 15 centimeters. Fur is so remarkably effective in some species that the TNZ may extend down to −30°C, as in the case of the arctic fox (*Alopex lagopus*). In many mammals active in the cold, the length of the woolly underfur and of the longer guard hairs varies seasonally. The summer pelage, which is acquired in spring, is short and has reduced insulating ability, but the winter coat, which replaces the summer pelage in autumn, is long and has great insulating ability. The hollow hair of some ungulates (such as the pronghorn) is remarkable insulation that allows winter activity under extreme conditions. In water, fur loses much of its insulative value because it is compressed by water (less thickness) and because water has higher thermal conductivity than air. The insulation provided by beaver fur drops from a value near seven to less than one when immersed in water. Seal fur has considerably less insulative value than an equivalent thickness of polar bear fur, yet seals inhabit some of the coldest waters on Earth. Seal pelage serves to trap a layer of still water rather than trapping a layer of air. This still water layer probably does not contribute much to insulation because skin temperatures of harp seals (*Phoca groenlandica*) are approximately the same as water temperatures. A thick layer of subcutaneous blubber, rather than fur, provides insulation in most aquatic mammals.

In some cetaceans and pinnipeds, up to 50% of the animal's body mass may be from blubber. A layer of lipid-rich blubber surrounds much of the body and acts as a thermal insulator. Blubber is superior to fur in aquatic environments because blubber does not compress under pressure. Fur traps air in pockets among the hairs, but when compressed, the air is forced out of the fur and insulation is lost. Blubber is also highly vascularized. In cold arctic waters, the surface vessels in the blubber of whales constrict to prevent heat loss (discussed below). For large pinnipeds that spend considerable time on ice floes or beaches, heat dissipation is vital. During periods of heat stress, tiny vessels in the blubber that bypass capillary beds (arteriovenous anastomoses) increase blood flow at the periphery and serve to shed excess heat to the environment. In addition to its role in thermoregulation, blubber serves as a storage site for fat-based energy, aids in buoyancy, contributes to streamlining, and may even act as a biological spring during swimming (see Chapter 19).

■ Regional Heterothermia

Despite the adaptations mentioned previously, at very low ambient temperatures, the costs of endothermy may be unsupportably high. To offset these costs, some cold-adapted mammals practice regional heterothermy; the temperatures of the skin or extremities drop well below the core body temperature. Extremities, such as legs and ears, which are poorly insulated and dissipate heat rapidly, are allowed to become cool, thereby reducing heat loss by minimizing the temperature differential between these parts and the environment. In an Eskimo dog exposed to cold, the deep-body temperature was 38°C. The toe pads were 0°C, and the tops of the feet 8°C (Irving 1966). This cooling of the extremities is due to **vasoconstriction** or to a **countercurrent heat exchange** system (**Fig. 21-12**). Arterial blood leaving the heart is at the core body temperature, whereas venous blood returning from the extremities is substantially colder. In a countercurrent heat exchange system, as warm arterial blood passes next to veins containing cool blood returning from the periphery, the warm arterial blood gives up heat to the venous blood. Consequently, the arterial blood is precooled before it reaches the skin and has little heat to lose to the environment. At the same time, the cold venous blood is warmed before it returns to the body core, minimizing a drop in core body temperature.

Regional heterothermy is a major cold adaptation of beavers (*Castor canadensis*). In fact, beavers employ two types of countercurrent heat exchangers, a rete system of small arteries and veins in the tail and a venae comitantes (a pair of veins running parallel to and surrounding an artery) in the hindlimbs (Cutright & McKean 1979). In the beaver's hindlimbs, the venae comitantes consist of a central artery surrounded by a series of veins. The rete system of the tail is even more specialized for heat exchange and consists of a series of interwoven arteries and veins. During the warm summer months, blood is shunted through a bypass to allow warm arterial blood to reach the skin and dissipate heat to the environment.

A similar heat exchanger is found in the dorsal fin of bottlenose dolphins (*Tursiops truncates*; Meagher et al. 2002). The dorsal fin contains blood vessels that function either to conserve body heat during diving or to dissipate heat during exercise (**Fig. 21-13**). The dorsal fin contains a series of superficial veins along the periphery and a row of countercurrent heat exchangers, called periarterial venous rete, at the fins center. Researchers recorded heart rate, respiration, and the heat loss and skin temperature at three positions on the dorsal fins of wild bottlenose dolphins. When submerged, heat loss values were highest across the superficial veins of the dorsal fin, but in air there was no such relationship. This study suggests that the dorsal fin acts as a counter-current heat exchanger capable of both heat conservation and heat dissipation.

A unique twist on heat dissipation occurs in the testis of bottlenose dolphins. Viable sperm production in many

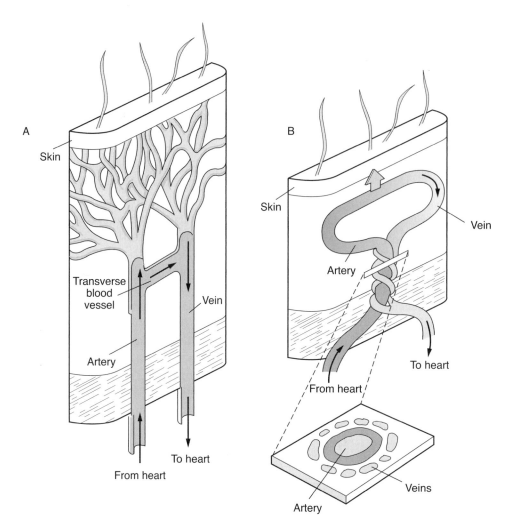

FIGURE 21-12 Diagram of an idealized heat exchange system. (A) Warm arterial blood moves into a network of capillaries near the skin surface, where heat can be lost through the skin. The cooled blood returns to the body core through the veins. In some mammals, this heat loss can be reduced by shunting the warm arterial blood through transverse blood vessels, thereby preventing heat from reaching the skin. (B) In mammals that live in very cold climates, a countercurrent heat exchange system may be used to conserve heat. Here, warm arterial blood is used to warm the cooler venous blood; the venous blood serves to precool the arterial blood before it reaches the skin. In this system, an artery surrounded by veins gives up some of its heat to the venous blood returning to the body core.

mammals requires temperatures below core body temperature. Consequently, many mammals have evolved a mechanism for testis descent from the abdominal cavity to an external position in a scrotal sac, where optimal temperatures for sperm production can be modulated below core body temperature (Bedford 1977). However, dolphins possess intra-abdominal testes surrounded by robust swimming muscles capable of producing temperatures above the core body temperature during exercise. To keep the testis cool, dolphins evolved a vascular countercurrent heat exchanger that functions to regulate the temperature of their intra-abdominal testes (Pabst et al. 1995). Surprisingly, the testes receive cooled blood from the surfaces of the dorsal fin and tail flukes through a venous plexus that juxtaposes cool venous blood from the extremities with warm spermatic arterial blood (Fig. 21-13). Thus, cooled blood from

the dorsal fin and tail fluke is brought deep into the abdominal cavity to cool the dolphin's abdominal testis.

Temporal Hypothermia

Maintaining a relatively high and constant body temperature (endothermy) by endogenous heat production is metabolically expensive. Endothermy requires a relatively constant supply of high-quality food for fuel. Under natural conditions energy resources (food and water) in the environment vary in time and location. The patchy distribution of food coupled with seasonal and daily fluctuations in ambient temperatures have selected for patterns of energy use that increase survival when food (or water) is scarce or when environmental temperatures are very high or very low.

Temporal hypothermia refers to a continuum of responses that allow energy to be saved by temporarily

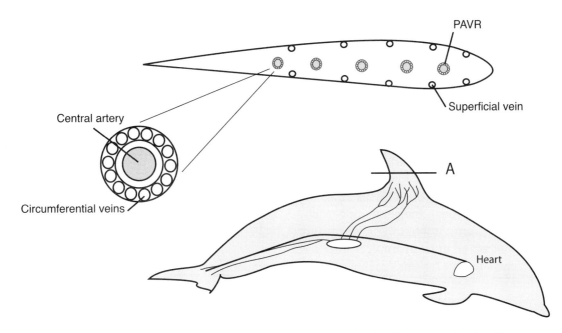

FIGURE 21-13 Drawing of the countercurrent heat exchangers in a bottlenose dolphin. The upper drawing illustrates a cross-section through the dorsal fin at A and reveals a series of superficial veins near the periphery and a central row of deep periarterial venous retia (PAVR) surrounding central arteries. The drawing of the dolphin shows the location of the countercurrent heat exchanger at the dolphin testis. Cooled blood in the superficial veins of the dorsal fin and the tail flukes drains via large veins toward the abdominal testes. Near the testes an arteriovenous plexus forms a countercurrent heat exchanger that regulates the temperature of the testes. (Adapted from Pabst, DA, et al., *J Exp Biol.* 198 (1995): 221–226 and Meagher, EM, et al., *J Exp Biol.* 205 (2002):3475–3486.)

abandoning homeothermy (maintaining a constant T_b). Temporal hypothermia occurs in monotremes, some metatherians, and some members of the orders Soricomorpha, Erinaceomorpha, Afroscoricida, Chiroptera, Primates, Carnivora, and Rodentia.

At one end of the continuum is daily torpor, in which body temperature, metabolic rate, respiration, and heart rate are lowered for a portion of the 24-hour daily cycle. Body temperature typically declines to only 10°C to 20°C in daily torpor. When energy savings from daily torpor are insufficient and migration to a more favorable climate is not an option, some species enter prolonged periods of torpor lasting from a few days to several months. These extended periods of hypothermia are typically seasonal and are referred to as hibernation (in response to cold) or estivation (in response to heat). The length and degree of seasonal torpor varies widely among mammals. Larger mammals, with access to protected hibernacula (dens) and the thermal advantage of large body size, exhibit shallow hibernation (or winter lethargy), in which body temperature drops and is regulated within 10°C of normal (**Fig. 21-14**). At the opposite end of the continuum is profound hibernation. Profound hibernation is defined by a suite of characteristics, including a T_b falling to within 1°C of ambient temperature, reduced oxygen consumption, prolonged periods of apnea (suspended respiration), a markedly reduced heart rate, and the ability to arouse by mobilizing endogenous

heat production (Gordon et al. 1977). Except for occasional periods of arousal, hibernating mammals remain inactive and without food or water for weeks to months. Therefore, major changes occur in the cellular and tissue physiology of the brain, muscles, and blood.

It is clear that different animals respond to different stimuli for torpor, and many respond differently to the same stimulus. Chipmunks (*Tamias*) and some species of hamster (*Phodopus sungorus*) enter hibernation in response to environmental cues and are referred to as facultative hibernators. Such cues include diminishing autumn photoperiods and falling temperatures and, in some small mammals, declining food resources (Nestler et al. 1996). Marmots (*Marmota*) and some species of ground squirrels (*Spermophilus*) are obligate hibernators because they

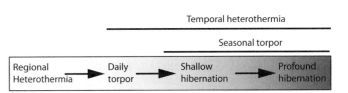

FIGURE 21-14 The continuum of responses for reducing body temperature to save energy. For some mammals, cooling only certain body parts (regional heterothemia) is sufficient to conserve energy. Other species must enter daily or seasonal torpor to save enough energy for survival.

enter hibernation in response to endogenous physiological cues. Several species of pocket mice and a kangaroo mouse (*Microdipodops pallidus*) also enter torpor spontaneously (Brown & Bartholomew 1969; French 1977, 1993; Kenagy 1973a; Meehan 1976).

Body size is an important factor in the degree of torpor. Although small mammals can exploit small food items (such as seeds) that are usually unavailable to larger animals and small mammals have access to a nearly limitless array of retreats, small size is a liability energetically. The high metabolic rates of small mammals must be sustained by high intakes of food, and seasonal changes in food availability present severe problems. Winters in the north and dry seasons in the deserts and in many tropical areas are times of potential food shortage for small mammals, and these are also periods when temperature or lack of moisture may limit activity. It is not surprising, therefore, that some small species have evolved means of surviving periods of food shortage and temperature stress and of using times of moderate temperatures and high food productivity to reproduce and to store food or fat.

Many small mammals (up to approximately 5 kilograms or less) periodically conserve energy by allowing the body temperature to drop to near that of the environment. This is not a manifestation of some ancestral inability to sustain a steady temperature at all times, but is instead a highly adaptive ability. Such adaptive hypothermia may well have been a factor important in furthering the success of the two largest mammalian orders, Rodentia and Chiroptera. Many small bats would not be able to forage only at night and fast throughout the day if they could not conserve energy in the day by means of daily torpor. Similarly, seasonally hostile areas would not be inhabited by some small rodents if these animals retained constant thermal homeostasis.

Typically, torpor (daily and seasonal) has three phases: a rapid entry phase, a prolonged period of torpor, and a relatively rapid arousal period (**Fig. 21-15**). In hibernators, periods of torpor become progressively longer with intervening arousals. The rate of entry into torpor is proportional to thermal conductance and therefore also to body mass. Consequently, small mammals enter torpor much more rapidly than large mammals (**Table 21-1**).

During torpor, body temperature drops to a minimum critical level (T_{crit}) and is maintained at this level by endogenous heat production. Species differ in their lower critical body temperature. This temperature is fairly high in some heteromyids that undergo shallow torpor (12°C in *Chaetodipus hispidus*) but, as would be expected, is low in mammals using deep torpor. Critical body temperature is close to freezing in some bats of the genus *Myotis*, 2.8°C in the golden-mantled ground squirrel, and 4°C in the European hedgehog (*Erinaceus europaeus*). The minimum critical temperature is generally above freezing, not to prevent tissue freezing, as torpid arctic ground squirrels can supercool (discussed later here), but to reduce the costs of arousal from torpor.

Daily Torpor

Torpor cycles can be seasonal (hibernation and estivation) or daily. Torpor in many small mammals is a circadian phenomenon. These animals are torpid by day and active and homeothermic by night. Daily torpor cycles are characterized by declines in body temperature and metabolic rates but usually not to the extent seen in hibernating mammals. There are exceptions. The eastern rock elephant shrew (*Elephantulus myurus*) lives in southern Africa (a region with moderate temperature extremes), yet this species has the lowest reported T_b (5°C) of any placental mammal in daily torpor (Lovegrove et al. 2001; Mzilikazi et al. 2002). Rock elephant shrews have bouts of torpor averaging between 7 and 9 hours in duration, which is consistent with those expected for daily heterotherms (less than 24 hours). However, their unusually low body temperature (T_b), oxygen consumption (VO_2), and decline in BMR are consistent with those of hibernators (Geiser & Ruf 1995; Mzilikazi et al. 2002). One torpor bout exceeded 24 hours, suggesting that rock elephant shrews may be capable of modest hibernation. Interestingly, rock elephant shrews arouse from daily torpor around sunrise, emerge from rock crevices, and bask in the morning sun (Fig. 8-10). Mzilikazi et al. (2002) reported that passive warming (basking) occurred in over 80% of observations (see behavioral thermoregulation). Mouse lemurs (*Microcebus myoxinus*) also experience daily torpor in the cool dry season of western Madagascar (Schmid et al. 2000). These tiny primates have daily torpor bouts averaging 9.6 hours, in which their metabolic rates are reduced by 86%. *Microcebus* also uses passive warming in the early morning to help return T_b to normal levels.

Mammals that exclusively exhibit daily torpor in response to energy stress are most common in Afrotropical and Australasian biogeographic regions (Lovegrove 2000). In both regions, species using daily torpor occur in both mesic and desert habitats and during both summer and winter months. Selection seemingly favored the evolution of daily torpor in desert regions and in mesic regions where energy resources are unpredictable and not primarily as a response to cold. In contrast, seasonal torpor (hibernation) is favored in regions with strongly seasonal but predictable climates (i.e., those with long cold winters) and in mammals capable of storing fat.

Seasonal Torpor

Hibernation or seasonal torpor is believed to be triggered by food deprivation, declining day length, and/or low

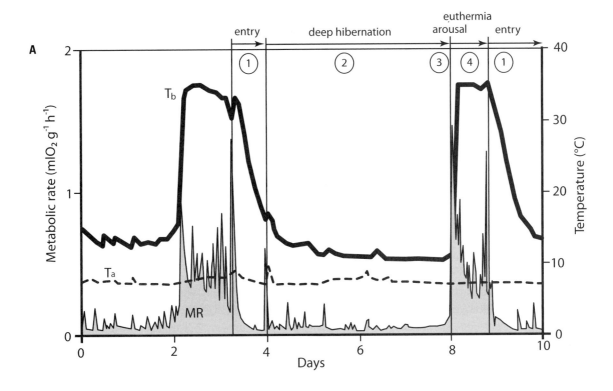

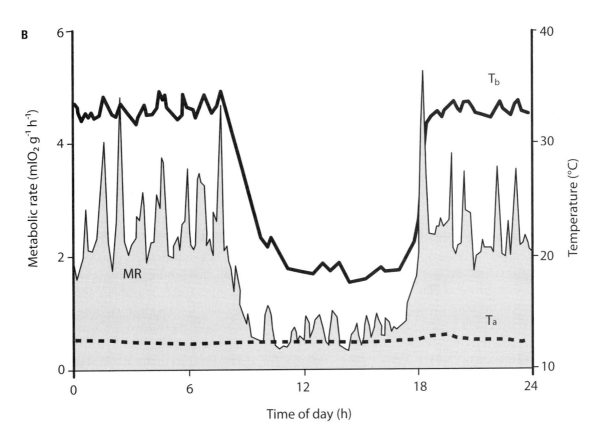

FIGURE 21-15   Graphs of the metabolic rate and body temperatures of (A) an alpine marmot (*Marmota marmota*) during hibernation and (B) a Djungarian hamster (*Phodopus sungorus*) during a bout of daily torpor. In both cases, the entry, torpor phase, and arousal phases are similar physiologically. The main differences are in the degree of body temperature decline and the length of the torpor period. (Reprinted from *Respiratory Physiology & Neurobiology*, Vol 141, G. Heldmaier, et al., "Natural hypometabolism during hibernation and daily torpor in mammals," pp. 317-329, Copyright 2004, with permission from Elsevier.)

TABLE 21-1 Time Required for Mammals of Various Body Masses for Entry into and Arousal from Torpor at an Ambient Temperature of 15°C

Species	Body Mass (g)	Entry Time (min)	Arousal Time (min)
Shrew (*Suncus*)	2	35	13
Honey possum (*Tarsipes*)	4	59	17
Echidna (*Tachyglossus*)	3,500	1,648	226
Marmot (*Marmota*)	4,000	1,766	237
Badger (*Taxidea*)	9,000	2,685	323
Bear (*Ursus*)	80,000	8,307	741

Adapted from Withers (1992).

environmental temperatures. The specific cue (or cues) involved varies among species. Preparation for hibernation often (but not always) involves great increases in body weight resulting from fat storage. This gain ranges in sciurids from 80% of the fat-free weight in the golden-mantled ground squirrel (*Spermophilus lateralis*) to 30% of this weight in the yellow-pine chipmunk (*Tamias amoenus;* Jameson & Mead 1964). Black bears (*Ursus americanus*) fatten rapidly before becoming dormant for weeks or months in the winter. Although described as a hibernator by many biologists, the black bear actually undergoes shallow hypothermia. In many areas, this animal retires to a protected place into which it has carried insulative nesting material and remains there from October to April. Rogers (1981) found that bears hibernating in Minnesota maintained body temperatures above 31°C, some 7°C below their normal temperature. A wild, hibernating Alaskan black bear had a heart rate of 8 beats/minute when sleeping soundly in December, whereas the heart rate of active bears in summer is from 50 to 80 beats/minute. Dormant black bears do not eat, drink, urinate, or defecate for prolonged periods. Surprisingly, levels of protein synthesis remain normal during dormancy, but protein breakdown (especially muscle protein) slows dramatically. The recent discovery of a compound in the blood of dormant bears that prevents protein degradation helps explain why bears lose fat but not muscle mass during dormancy (Argiles et al. 2007). Relatively high levels of lipid and protein metabolism generate metabolic heat which, combined with their large body size and thick fur, helps explain their relatively high body temperature of black bears during dormancy.

Most profound hibernators select or construct chambers (hibernacula) below ground where variation in external temperatures is minimized. Hibernacula of arctic ground squirrels (*Spermophilus parryii*) varied less than 10°C over the 6-month winter, whereas temperatures above ground ranged from 15°C in October to a minimum of −40°C in late December. Nevertheless, den temperatures were at or below freezing for much of the winter. Barnes

(1989) exposed free-ranging arctic ground squirrels to ambient temperatures of −4.3°C for prolonged periods. In response, the animal's core body temperatures dropped to as low as −1.3°C. These squirrels were able to avoid freezing by supercooling for brief periods. Supercooling occurs when temperatures are below freezing, but ice crystals are unable to form. In an extreme case of profound hibernation, Geiser (2007) showed that captive pygmy possums (*Cercartetus nanus*) were capable of hibernating for over one year (367 days) using only body fat as fuel. A single gram of body fat supplied sufficient energy for approximately half a day in a nonhibernating pygmy possum, but supported over 16 days of metabolism in a hibernating possum. This astonishing capacity for prolonged hibernation in a tropical marsupial illustrates that seasonal hibernation is not restricted to temperate regions with cold winters.

Profound hibernation is not a uniform state. Instead, it is a dynamic state in which hibernators remain sensitive to environmental changes and are capable of responding appropriately. Experiments on hibernating mammals reveal that they are capable of increasing metabolic rate in response to sharp drops of ambient temperature below T_{crit}. An extreme example of this ability is shown by a small pocket mouse (*Perognathus parvus;* MacMillen 1983a). When hibernating at ambient temperatures down to 2°C, this mouse maintains a T_b about 1°C above ambient temperature. At ambient temperatures between 2° and −5°C, it increases its metabolic rate just enough to maintain body temperature at 2°C. Thus, despite a body and brain temperature near freezing, this mouse resets its T_b thermostat to maintain body temperature above lethal limits.

Seasonal hibernators frequently exhibit a series of hibernation bouts over the course of the entire hibernation season. For example, alpine marmots hibernate from early October to late March. During this 6-month period, individual bouts of torpor average 11 to 12 days, and there are as many as 20 separate hibernation bouts during the winter (**Fig. 21-16**; Ortmann & Heldmaier 2000). Each bout

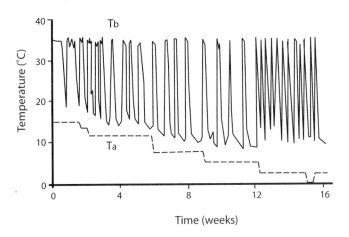

FIGURE 21-16 Body temperature (Tb) of an alpine marmot (*Marmota marmota*) over a 4 month hibernation period. (Modified from Ortmann, S., and Heldmaier, *American Journal of Physiology* 278 (2000): R698–R704.)

terminates in an arousal phase (a period of re-warming) followed by a 1- to 2-day period of normal body temperature (euthermic period; Fig. 21-15A). These arousal periods are short, but energetically costly. Alpine marmots (*Marmota marmota*) consume approximately 70% of their energy reserves for hibernation on arousals (Heldmaier et al. 2004). Arousal rates for Etruscan shrews averaged 0.83°C per minute, which is similar to rates reported for *Sorex cinereus* (Morrison et al. 1959) and among the highest rates found in mammals (Stone & Purvis 1992). This high rate of arousal is costly and suggests that the most advantageous strategy for this hibernator would involve continuous deep torpor. Nonetheless, periodic arousal during the hibernation period is the rule among many species studied. The broad pattern is one of progressively increasing periods of torpor through the early stages of hibernation and decreasing periods in the late stages.

Among several rodents, the maximum periods of torpor were from 12 to 33 days, but a period of 80 days was recorded for the little brown bat (*Myotis lucifugus*). The European hedgehog was found to be torpid 31% of the time at 10°C and 80% of the time at 4.5°C. Small mammals that maintain very long hibernation bouts with few, brief arousal periods may conserve over 90% of their energy reserves compared to nonhibernating animals (Heldmaier et al. 2004). By comparison, maximum energy savings from daily torpor are only about 40%. Ruf and Heldmaier (1992) showed, however, that metabolic economy is only part of the story for hamsters. Additional energy savings come from the foraging behavior of hamsters. Foraging occurs outside the protection of the den and requires considerable energy. Hamsters emerging from torpor require less food and therefore spend less time foraging (and digesting) than hamsters not in torpor. Taken together, the energy saved

from reduced metabolism during torpor and the savings from reduced foraging yield a total energy savings of 67% for hamsters using daily torpor.

Hibernation clearly saves more energy than does daily torpor, but the latter may have certain advantages. For example, daily torpor does not require the long preparation periods needed to store fat or cache food that is required of hibernators. The different patterns exhibited by bats, shrews, and rodents may reflect the fact that rodents can store energy (food or fat) to use during periodic arousals, and shrews and bats generally do not store food. Furthermore, the degree and duration of daily torpor can be varied day to day in response to changes in food availability or temperature changes (Heldmaier et al. 2004). Viewed broadly, daily torpor reduces use of energy by 40% to 60% and provides the flexibility to maintain social interactions (for part of the day) and to adjust to changes in resources and environmental conditions.

Arousal

Arousal from torpor occurs when a rapid drop in environmental temperature beyond the animal's ability to compensate triggers a rapid re-warming (alarm arousal). Arousal also occurs in the absence of external signals. Such periodic arousal varies in duration and frequency among species (Fig. 21-16). After final arousal, the animal maintains its normal body temperature.

All forms of arousal occur rapidly and without external heat sources. In some ground squirrels, body temperatures increase from 7°C to 37°C in less than 1 hour (Mayer 1960). In some small mammals, arousal is accompanied by violent contractions of skeletal muscles (shivering), with attending heat production. Mammals can arouse from hibernation even when injected with curare to block the neuromuscular signaling and prevent skeletal muscle contraction. These experiments indicate that shivering is not the only means of re-warming.

Production of heat by metabolizing fat is called nonshivering thermogenesis. Mammalian nonshivering thermogenesis typically involves the oxidation of a special type of fat called brown adipose tissue (BAT). BAT is found in limited deposits around the neck and between the shoulders in many cold-adapted eutherian mammals and some newborn mammals. Loudon et al. (1985) reported a similar type of fat tissue in a metatherian. Arousal is associated with the metabolism of energy-rich BAT in some species and with shivering in others (Chaffee & Roberts 1971). Etruscan shrews, the smallest mammals studied, use both mechanisms during arousal from daily torpor. During arousal in some mammals, heat production is primarily from BAT, augmented with shivering at body temperatures above 17°C (Fons et al. 1997).

Although all tissues in endotherms are capable of generating some heat, BATs are specialized for rapid, nonshivering thermogenesis. The individual cells in BAT contain high densities of mitochondria, and the entire fat body contains numerous capillaries. It is the degree of vascularization and mitochondrial density that give these adipocytes their brown color. Normal fat cells (white adipocytes) store lipids that can be mobilized and transported to other tissues (i.e., skeletal muscle) for use as fuel. In contrast cold stress induces cell division in BAT and vasodilation in its capillaries. A number of cellular events also occur (reviewed by Cannon & Nedergaard 2003). In short, the brain responds to information on body temperature and energy reserves by producing norepinephrine, which in turn stimulates lipid breakdown within the brown adipocytes. The fatty acids released as a result serve as substrates for thermogenesis in the mitochondria. In non-BAT mitochondria, these substrates are converted to ATP via a complex series of reactions on the mitochondrial inner membranes. This process releases heat as a byproduct of ATP synthesis. In contrast, BAT mitochondria have high concentrations of a protein called thermogenin (or uncoupling protein-1) in the inner membrane, which bypasses ATP production and increases the rate of heat production substantially (Cannon & Nedergaard 2003).

BAT can produce heat at 10 times the rate possible for active skeletal muscle contraction (Withers 1992). Not surprisingly, nonshivering thermogenesis via BAT is especially important in infants (which often lack the insulation provided by a rich coat of fur) and during arousal from torpor or hibernation in adults. In soricid shrews nearly all body fat is in the form of BAT, which can account for up to 20% of the animal's body weight (Hyvarien 1994). The rapid heat production from these fat stores via nonshivering thermogenesis is partially responsible for the high degree of overwinter survivorship of shrews (Merritt 1995).

Temperature Regulation in Bats

Studies of a variety of bats have both clarified and complicated the picture of temperature regulation in these animals. Among different species, contrasting reactions to temperature changes occur, and within the Chiroptera most mammalian styles of temperature regulation are represented. Seemingly, the larger pteropodids are homeotherms. Those that have been studied are able to maintain body temperature within fairly narrow limits (35°C to 40°C) over a range of ambient temperatures from approximately 0°C to 40°C. Many pteropodid bats react to cold stress by shivering and by enveloping the body with the wings, which serve as blankets that provide considerable insulation for the body (**Fig. 21-17**; Bartholomew et al. 1964). Shallow diurnal torpor does occur in a few small

pteropodids, including two tube-nosed bats (*Nyctimene*) from New Guinea (Bartholomew et al. 1970).

Compared with pteropodids, other bats are highly variable in their responses to temperature extremes. The Australian species *Macroderma gigas* (Megadermatidae), probably because of its large size (100 to 140 grams compared with less than 40 grams for most bats), is able to maintain a stable body temperature in the face of ambient temperatures as low as 0°C, and many of the reactions to temperature extremes in this bat are similar to those of non-pteropodids (Leitner & Nelson 1967). The Neotropical species *Desmodus rotundus*, a vampire bat (15 to 50 grams), was originally reported to be unable to regulate its body temperature in response to moderate changes in ambient temperature (Lyman & Wimsatt 1966). More recent studies of the three genera of vampire bats, however, suggest that under natural conditions they are able to regulate their body temperature down to ambient temperatures approaching 0°C, and *Diphylla* and *Diaemus* may even become heterothermic under certain conditions (Hill & Smith 1992).

FIGURE 21-17 A pteropodid fruit bat with its wing membranes enshrouding its body.

Many tropical non-pteropodids bats from the Old World and from the Neotropics are active at night and inactive during the day, and this activity cycle is reflected by their temperature cycle. In the Neotropical phyllostomid bats that have been studied, body temperatures are from 37°C to 39°C during the night and 2°C to 3°C lower during the day (Morrison & McNab 1967). In general, these bats are able to maintain a high body temperature despite moderately low ambient temperatures. Broadly speaking, cold stress can usually be tolerated by tropical bats for only fairly short periods, after which the body temperature falls uncontrollably. Core body temperatures below 20°C are often fatal.

Adaptive hypothermia, often involving (at different seasons) both daily torpor and hibernation, occurs in many vespertilionids of north temperate areas and seems to be the key to the survival of some species in cool or cold regions. During the summer, some temperate zone bats undergo daily torpor at low ambient temperatures. Tremendous metabolic savings are realized by bats that become hypothermic at low temperatures. Under experimental conditions, the average metabolic rate of six little brown bats (*Myotis*) kept at an ambient temperature of 35°C was 33 times that of these same bats when kept at 5°C (Henshaw 1970). In addition, strikingly abrupt rises in metabolic rate occur during flight. The metabolic rate of the phyllostomid bat *Phyllostomus hastatus* was about 30 times greater in a flying individual than in those resting at a body temperature of 36.5°C (Thomas & Suthers 1972). In the interest of saving energy, bats in temperate areas have the briefest possible periods of flight and, in bats with the ability, the longest possible daily periods of hypothermia.

In summer, at the northern limit of their range in southern Canada, male pallid bats (*Antrozous pallidus*) are heterothermic: They choose roosts with high and constant temperatures and use shallow daytime torpor, thus minimizing pulmocutaneous water loss and the use of energy (Rambaldini & Brigham 2008). At times, bats flew before their temperatures reached those of normally active bats. The flight muscles probably generated the heat that caused full rewarming to normal activity levels.

Some bats that are homeothermic in summer abandon this pattern well before winter. Fat deposition is known to occur in the late summer or early fall in some species of vespertilionids that hibernate (Baker et al. 1968; Ewing et al. 1970; Krzanowski 1961; Weber & Findley 1970), and three species of *Myotis* became hypothermic during this time (O'Farrell & Studier 1970). In *M. thysanodes*, the metabolic rate for homeothermic individuals at an ambient temperature of 20.5°C is 6.93 ml³ O_2/g/h (cubic milliliters of oxygen per gram per hour) but drops to 0.59 ml³ O_2/g/h in hypothermic bats. This decrease results in the saving of 2.81 kilocalories per day as a bat becomes hypothermic. Fat is deposited in preparation for hibernation at the rate of 0.17 grams per day in the period of maximum fat accumulation. This requires an extra 1.60 kilocalories per day, which is available primarily because of the late summer–autumn shift to daily hypothermia (Ewing et al. 1970; Krzanowski 1961).

Winter hibernation in bats differs from short-term torpor largely in the length of dormancy and in the levels to which the metabolic rate and temperature drop. The duration of hibernation for bats differs widely among species and within a species, depending on the area. In the northeastern United States, *M. lucifugus* remains in hibernation for 6 or 7 months, from September or October to April or May (Davis & Hitchcock 1965). Periods of hibernation for bats in warmer areas are probably considerably shorter. At ambient temperatures near 5°C, bats in deep hibernation maintain a body temperature about 1°C above ambient temperature. These bats are responsive to certain stimuli and will begin arousal when handled or when subjected to unusual air movement. As a defense against freezing to death, bats spontaneously raise the metabolic rate at dangerously low ambient temperatures (below roughly 5°C) and either arouse fully or regulate body temperature and remain in hibernation.

The red bat (*Lasiurus borealis*) of the eastern United States has a geographically variable system of balancing its winter energy budget. Although most red bats are seemingly migratory, many (mostly males) remain through winter in areas such as southern Illinois, southern Indiana, and Missouri, where periodic freezing temperatures occur (Davis & Lidicker 1956). These bats roost in trees and forage in warmer times, but during cold spells they hibernate in the leaf litter of the forest floor. Torpid red bats in the laboratory had body temperatures close to ambient temperatures down to the "defended" level of 5°C, but responded to lower temperatures by increasing metabolism and keeping body temperatures above this level (Dunbar & Tomasi 2006). Leaf litter provides some insulation from temperature extremes, while allowing bats to respond to warmer temperatures when foraging may be profitable. Perhaps for red bats in winter there is a continuum between lengthy periods of hibernation in the north and daily torpor in the south.

Other Ways of Coping with Cold

Not all north temperate bat species enter torpor in response to declining fall temperatures. At least some species of *Nyctalus*, *Vespertilio*, *Lasiurus*, *Lasionycteris*, *Pipistrellus*, and *Tadarida* migrate from north to south as winter approaches. Although the cost of flight is high (4.6 kilocalories per 100 kilometers), many species of tree-roosting bats (for example, *Lasiurus* and *Nyctalus*) make seasonal migratory flights of over 1,500 kilometers (McNab 1982). A number of temperate zone species of *Myotis* make shorter (200 to 500 kilometers) seasonal migratory flights between their hibernation caves and summer roosts or maternity colonies. These species use both migration and torpor to save on energetic expenses.

For many years the overwintering habits of most bats of the western United States have remained a mystery. These bats, unlike those of the Midwest or east, do not congregate in large numbers in caves or buildings in winter. A study in Colorado, involving the radio-tagging of big brown bats (*Eptesicus fuscus*), made progress toward solving this mystery. Neubaum et al. (2006) found that 24 *E. fuscus*, originally captured in Fort Collins at the eastern base of the Rocky Mountains, migrated west in autumn, up the Poudre River drainage. They were found in crevices in rock from 26 to 87 kilometers away from Fort Collins. Presumed hibernacula were crevices where radio-tagged bats stayed for at least a week before their transmitter batteries expired. These crevices were deep enough that the bats could not be seen. Temperatures within these crevices varied little compared with temperatures at the mouths of the crevices (**Fig. 21-18**). Crevices used as hibernacula differed from a large sample of randomly chosen crevices in being deeper, higher above the ground, and in having a warmer average winter temperature (1.9°C). These crevices were on northwest-facing hillsides shaded from the winter sun. The bats could conserve energy by maintaining minimal metabolic rates while avoiding long periods of subfreezing temperatures. Short autumn migrations from warm maternity roosts to higher elevations, and the use of deep rock crevices as hibernacula, may be strategies used by bats in the many mountainous areas of western North America.

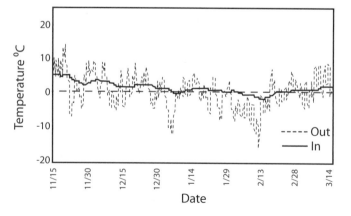

FIGURE 21-18 Temperatures inside and outside the roost sites of big brown bats (*Eptesicus fuscus*) in Colorado. Roosts were at 2,206 meters elevation. (Adapted from Neubaum, D. J., et al., *J. Mammalogy*. 87 (2006): 470–479. American Society of Mammalogists, Allen Press, Inc.)

All animals that maintain a reasonably constant body temperature must balance heat gains and losses. Whereas lizards and many other ectotherms do this by practicing behavioral thermoregulation, mammals rely primarily on metabolic adjustments. Mammals use behavioral thermoregulation too, but their high resting metabolic rate and their activity under temperature extremes impose special thermoregulatory burdens. Nevertheless, behavior plays a critical role in reducing cold stress.

Among the 31 species of northern hemisphere ground squirrels (*Spermophilus*), the arctic ground squirrel (*S. parryii*) is unique in occurring widely north of the Arctic circle and along thousands of kilometers of tundra bordering the Arctic Ocean, where its burrows are commonly less than a meter above prermafrost. Most ground squirrels hibernate. All are diurnal, and of central importance to them is the storage of enough fat when active to fuel the long periods of torpor. Probably more than for any other species, maximizing foraging time and minimizing the cost of thermoregulation are survival imperatives for the arctic ground squirrel.

The arctic-slope populations of arctic ground squirrels emerge from hibernation and immerge into hibernation during roughly the same 2 to 3 weeks each year (late April and early September, respectively), as do most ground squirrels of temperate areas. Conditions facing the arctic squirrels during emergence, however, are far more hostile. In a study of these squirrels on the north slope of the Brooks Range in Alaska (Buck & Barnes 1999), the first emergence of a male was on April 12 in 1994. The mean temperature that day was –41°C, and snow depth was 24 centimeters. The following year the first emergence was on the same date, with similarly harsh conditions (–24°C and 37 centimeters of snow). Continuous snow cover lasted into May each year.

Thermoregulation in arctic ground squirrels was studied in northern Alaska by Long et al. (2005). In early May, when breeding dominated above-ground activity, temperatures were often below freezing, and the squirrels used physiological thermoregulation. On cold and wet days, they stayed in nests in their burrows. These bulky, grass nests probably provide enough insulation in summer to enable the BMR to maintain high body temperatures (Chappell 1981). In June and July, as the days warmed, surface activity increased. The squirrels' morning emergence coincided with the time the ambient temperature reached the squirrels' thermoneutral zone (18°C to 36°C), and they foraged almost continuously from 5:00 a.m. until 10:00 p.m., when temperatures stayed in that zone. Body temperatures in active squirrels averaged 39.4°C and varied 3°C to 5°C during the 24-hour cycle. When ambient temperatures rose above 33°C, most squirrels reduced surface activity. Instead of panting or sweating, overheated squirrels retreated below ground, where the differential between body temperature (39°C to 42°C) and burrow temperature (less than 10°C) allowed rapid dissipation of heat. On the warmest days, squirrels alternated between brief stays in the burrows and long times spent foraging. The squirrels seemed to remain in their burrows when the energy gained by foraging was less than that spent on thermoregulation. Winter soil temperatures recorded near arctic ground squirrel hibernacula were very cold, averaging –8.9°C (Buck & Barnes 1999), whereas soil temperatures for lower-latitude species seldom reach 0°C. Arctic ground squirrels must continue thermogenesis all winter, and thus, for them, hibernation must be especially costly.

Long et al. (2005) regard the thermal environment as the major selective force affecting the foraging behavior of arctic ground squirrels and conclude that they respond behaviorally to the potentially severe costs of thermoregulation during the active season. By selective use of burrows, they minimize the cost of thermoregulation on cold days and maximize foraging times on hot days.

Nest-building behavior is important for small mammals. Nests, usually built in protected places, provide insulation that greatly augments that provided by the pelage. Group thermoregulation, involving several animals huddling together (West 1977), reduces the exposed surface area of each animal and is apparently widespread in mammals. Social behavior may be an important energy-conserving strategy. Taiga voles (*Microtus xanthognathus*), which are active through severe interior Alaskan winters, enhance winter survival by living in groups of five to ten animals (Wolff & Lidicker 1981). The communal nest of these voles is always occupied by one or several animals, and nest temperature is thus kept well above ambient temperature. Morton (1978) observed nest sharing in a small marsupial (*Sminthopsis crassicaudata*) and regarded it as an energy-saving behavior. The response of seeking shelter or a favorable place is also of importance. An animal foraging at ground level beneath a deep snowpack faces temperatures near 0°C, whereas ambient temperatures above the snow may be many degrees below zero.

Coping with Heat

Some of the most severe problems in thermoregulation are those faced by mammals living in hot regions. In many low-latitude deserts, daytime surface and air temperatures in the summer rise well above the body temperature of most mammals. Under such conditions, heat from the environment is absorbed while the animals are producing considerable metabolic heat. In order to maintain thermal homeostasis, these animals must avoid the absorption of heat from the environment, dissipate such heat as it is absorbed, and lose endogenous heat. For passive heat loss to occur, the body surface temperature must exceed the ambient temperature. Several mechanisms for dumping excess heat were observed in bats under heat stress (Bartholomew et al. 1970). Vasodilation occurred in surfaces such as the scrotum, wing membranes, and ears. Naked surfaces are seemingly efficient heat dissipators (**Fig. 21-19**). Other reactions to high temperatures in bats were extension of the wings, fanning of the wings, and panting. Under intense heat stress, the animals facilitated heat transfer via evaporative cooling. They salivated copiously and licked their bodies. Evaporative cooling is common, but it is a luxury that most desert mammals cannot afford because they live in regions where water is in critically short supply.

▪ Evaporative Cooling

Evaporative cooling involves the evaporation of a fluid (typically sweat or saliva) from a body surface. Typical panting, which involves rapid and shallow respiration, is used entirely for heat dissipation and is an effective aid to temperature regulation. Laboratory studies of dogs, for example, indicated a tolerance of an ambient temperature of 43°C for at least 7 hours (Robinson & Lee 1941). Panting uses evaporative cooling of the mouth, tongue, and probably most important, the nasal mucosa (Schmidt-Nielsen et al. 1970). In the dog and in many other mammals with a long snout and excellent olfaction, the turbinal bones of the nasal cavity are intricately rolled and provide a large surface area of nasal mucosa. This moist surface is ideal for evaporative dissipation of heat. The tongue is probably also important as a site of heat loss during panting. Blood flow to it increases sharply (vasodilation) at the onset of panting and during heat stress increases six times over normal.

The resting respiratory rate of a dog is roughly 30 breaths per minute, but this rate rises abruptly, with virtually no intermediate rate, to over 300 per minute during panting. The lateral nasal glands, which open some 2 centimeters inside the opening of each nostril, supply a major share of the water used in evaporative cooling during panting in dogs (Blatt et al. 1972). Under experimental conditions, the rate of secretion of one of these glands in a dog rose from no secretion at 10°C to 9.6 grams per hour at 50°C. Between 20% and 40% of the evaporative cooling during panting at high temperatures results from evaporation of the fluid from these glands. Because the glands are situated anterior to the turbinals, they tend to keep the nasal mucosa moist when air is drawn rapidly in through the nostrils during panting, enhancing evaporation.

Panting has several advantages over sweating. There is minimal loss of salt during panting, whereas salt loss during sweating (except probably in donkeys and camels) is always appreciable. In addition, adequate ventilation of

FIGURE 21-19 A jack rabbit (*Lepus californicus*) from the Kofa Mountains of Arizona. The large ears of these jackrabbits contain networks of blood vessels that dilate to dissipate heat to the environment.

evaporative surfaces always occurs during panting. In still air, however, sweating is seemingly not equally efficient. One potential disadvantage of panting is that the increased activity increases metabolism, thereby contributing more heat to be dissipated. Studies of respiratory frequency in dogs (Crawford 1962) indicated that these animals pant at the resonant frequency of oscillation of the diaphragm (the natural frequency of vibration of this structure) and may therefore economize energy output.

Considering water loss relative to total body surface area of a mammal, the amounts of water lost in sweating and panting are probably similar in many mammals. One apparent exception to this occurs in heteromyid rodents. Approximately 84% of the water lost by *Dipodomys merriami* is lost via the respiratory tract compared with only 16% from **cutaneous evaporation** (Chew & Dammann 1961; French 1993). However, because desert heteromyids have lower metabolic rates (up to 25% lower) than many other rodent species and consequently pass less air over the respiratory surfaces per unit time, respiratory water

loss is reduced. Both panting and sweating are obviously not effective means of cooling at high humidities. Because they spend the days in cool, humid burrows, the necessity of evaporative cooling is seldom experienced by *D. merriami*.

Thermoregulation at high temperatures in the rat kangaroo (*Potorous tridactylus*), a small and rather generalized macropodid, involves a specialized system of evaporative cooling (Hudson & Dawson 1975). This metatherian weighs about 1 kilogram, and its metabolic rate and body temperature (36°C) are low relative to those of eutherian mammals of similar size. Thermal conductance from the well-furred body is low. Studies by Dawson and Hulbert (1970) and MacMillen and Nelson (1969) have shown that a number of metatherians have body temperatures equivalent to those of eutherians but have metabolic rates that are about two-thirds those of placentals of comparable size. During exercise, at ambient temperatures below body temperature, heat is dissipated by the rat kangaroo primarily by panting, but at ambient temperatures approaching and exceeding body temperature, the bare tail, which contributes 9.4% of the total surface area, is a major route for heat loss. Vasodilation in the skin of the tail allows for increased nonevaporative heat loss, and at temperatures near and above body temperature, profuse sweating of the tail, but not of the body, produces rapid evaporative cooling. Constant side-to-side movement of the tail further facilitates evaporation. The maximum rate of sweating in the tail is extremely high, reaching 620 to 650 grams/meter^2/hour, roughly double the highest measured rates in eutherians such as horses and cows.

Evaporative cooling is not always the optimal strategy. For example, the very adaptations that enable seals or sea lions to reduce heat loss in cold water make them unable to stand high temperatures on their basking beaches. Whittow (1974) and his coworkers at the University of Hawaii studied the terrestrial thermal budget of California sea lions (*Zalophus californianus*), which inhabit some arid coasts of Mexico and the Galapagos Islands, where high temperatures occur regularly. At air temperatures of about 30°C, sea lions were unable to maintain a constant body temperature after the seawater had evaporated from their body surfaces. With continued exposure under experimental conditions, body temperature rose to slightly over 40°C. When the sea lions slept, their heat production dropped 24%, an obvious advantage for energy conservation and temperature regulation, but they were unable to dissipate sufficient heat in direct sunlight to avoid heat stress. Under heat stress, the sea lions fanned their flippers, which are known to sweat, thus increasing evaporative cooling. They also urinated and wet the underside of the body, thus further increasing evaporation. Under experimental conditions, however, these behaviors were inadequate, and the animals were increasingly hyperthermic. Only when they were able to wet their bodies in the sea did body temperatures drop and stabilize.

Terrestrial heat production by a sea lion is dissipated approximately as follows: Two percent is lost by respiratory evaporative cooling, 12% by evaporation from the skin, 52% by nonevaporative heat loss (conduction and convection) from the skin, and 15% by conduction from the parts of the body against the sand. Nineteen percent of the metabolic heat is stored, leading eventually to an elevation of body temperature. It is obvious why sea lions and other pinnipeds have difficulty staying out of the water for long periods on a warm day, and increased activity on land at night is understandable. Although their physiologic makeup limits the amount of time they can spend on land on a warm or hot day, by choosing the windy side of an island and by basking at sites where spray from breaking waves repeatedly wets them and increases evaporative cooling, sea lions can considerably extend their resting time on land.

■ Adaptive Hyperthermia

Some mammals have evolved physiological, anatomical, and behavioral strategies for tolerating long exposure to air temperatures higher than body temperatures. Large size itself is advantageous to mammals that must tolerate high temperatures. Because of the volume-to-surface area ratio discussed earlier, the larger the animal the greater its ability to withstand exposure to high temperatures because of a relatively reduced surface area for heat gain. Stated differently, large animals have greater thermal inertia than do small ones. Of additional importance, just as insulation in the form of thick pelage slows the loss of body heat in low ambient temperatures, fur slows the penetration of heat to the body surface when temperatures are high.

Studies of temperature regulation in captive camels (*Camelus dromedarius*) by Schmidt-Nielsen (1959) revealed a carefully regulated and highly adaptive circadian cycle of changes in body temperature. Fully hydrated camels in the Sahara Desert in winter, when cool temperatures (from roughly 0°C to 20°C) prevailed, had fairly constant body temperatures that varied between 36°C and 38°C. The fluctuations in body temperature were not random but followed the same pattern day after day, regardless of weather. In the summer, variations in body temperature were considerably greater. Generally, body temperature was between 34°C and 35°C in the morning and reached a peak of approximately 40°C late in the day (**Fig. 21-20**). The camels seemed to be able to regulate their temperature but did so only above or below these extremes. When body temperatures reached 40.7°C, evaporative cooling in the form of sweating was used to dissipate heat and stabilize body temperatures. Thus, during the day the camel accepted a heat load that sharply elevated its temperature. During the relative coolness of the desert night, however, the heat stored during the day was passively dissipated and the body temperature dropped. Schmidt-Nielsen (1964) estimated that 5 liters of water would be required to dissipate the camel's load by evaporative cooling during a hot day. For an animal that does not have frequent access to water, such daily water loss would lead to fairly rapid dehydration. An additional advantage of high body temperature during the day results from narrowing the gap between environmental and body temperature; the smaller this temperature differential, the lower the rate of heat flow from the environment to the body.

Recent studies using miniature data loggers to record core body temperature and carotid arterial temperatures in free-ranging African artiodactyls have challenged the view that desert animals are capable of conserving water by adaptive heterothermy. Mitchell et al. (2002) listed four requirements that must be met to qualify as adaptive heterothermy: (1) Variation in core T_b must be large enough

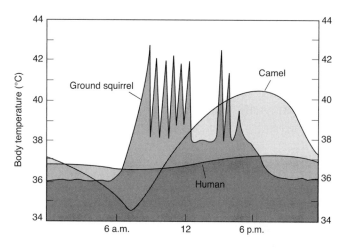

FIGURE 21-20 Daily patterns of body temperature change in three mammals subjected to desert heat. The antelope ground squirrel (*Ammospermophilus leucurus*) goes through a series of heating-cooling cycles during the day, whereas the camel slowly becomes hyperthermic as the day progresses. (Data from Bartholomew, G. A., *Symp. Soc. Exp. Biol.* 18 (1964): 7–29.)

to serve as heat storage, (2) T_b variations must be in phase with daily changes in T_a, (3) T_b should increase with increases in ambient heat stress, and (4) should be linked

to preventing water loss. These authors reported T_b data in free-ranging eland (*Tragelaphus oryx*), impala (*Aepyceros melampus*), oryx (*Oryx gazella*), and black wildebeest (*Connachaetes gnou*; **Fig. 21-21**) in their natural habitats in Namibia and South Africa. In all of the species studied, T_b varied only 2.3°C on average (maximum of 3.7°C) over the daily cycle and none met all the criteria necessary for adaptive heterothermy. Mitchell et al. (2002) caution that adaptive heterothermy in desert ungulates may be less common than once believed because early studies used captive animals that were unable to employ their full range of thermoregulatory behaviors.

Arabian oryx (*Oryx leucoryx*), however, which live under some of the most extreme desert conditions faced by any ungulate, appear capable of adaptive heterothermy in the wild. Ostrowski et al. (2003) used implanted radiotelemetry to measure core body temperature in free-ranging Arabian oryx over a 2-year period in Saudi Arabia. Daily changes in T_b averaged 4.1°C, and maximum T_b averaged 40.5°C over the summer. Furthermore, daily changes in T_b appeared to track thermal loads. These oryx also practiced behavioral thermoregulation by seeking shade trees in the early morning and remaining in the shade for an average of

FIGURE 21-21 A black wildebeest (*Connachaetes gnou*) from South Africa.

9 hours each day. Remaining in the shade and their ability to store excess heat during the day resulted in an estimated savings of 0.5 liter of water per day that would otherwise have been used in evaporative cooling. Arabian oryx can live without drinking water and instead acquire all of their water from their diet. Consequently, even a meager half a liter savings in water each day can be critical for survival.

Exertion and Heat Stress

During heavy exercise, mammals produce heat much faster than it can be dissipated. Indeed, for moderate-sized mammals (5 to 200 kilograms), the most intense thermal stress encountered is during heavy exertion. The rise in deep-body temperature during exertion in mammals is far more rapid than that in resting mammals in desert heat. Taylor (1974) calculated that the rate of excess heat production in the domestic dog during heavy exercise is 10 times the highest possible heat gain the dog could face in the hottest desert.

During running, when muscular effort increases drastically, metabolic heat production may greatly exceed the ability of the animal to dissipate heat to the environment. Under these conditions, the ability to store heat by elevating body temperature can be highly advantageous. Cheetahs (*Acinonyx jubatus*), the fastest land mammals, inhabit many hot parts of Africa. During a sprint to capture a gazelle, a cheetah produces heat at nearly 50 times the rate produced at rest (Gordon et al. 1977). This tremendous heat load cannot be dissipated by evaporative cooling alone, as it would be at rest and must therefore be temporarily stored, causing body temperature to rise. Consequently, cheetahs can pursue prey only over short distances before they must stop to prevent overheating. Similarly, the gazelle fleeing from a cheetah stores heat. Gazelles running at low speeds store relatively little heat because they can dissipate much of the heat produced by muscular work by evaporative cooling (sweating) while they run. At speeds of at least 80 kilometers per hour, such as those encountered when a gazelle is pursued by a cheetah, over 70% of the metabolic heat is temporarily stored and body temperatures rise.

A bizarre escape behavior is pertinent here. On hot days (over 40°C), when approached by a person, desert-dwelling black-tailed jackrabbits (*Lepus californicus*; Fig. 21-19) will at times replace their familiar high-speed, bounding dash from danger with a slow-motion, ears down, creeping (Vaughan, personal observation). On such days, when ambient temperatures exceed the jackrabbit's body temperature of 39.2°C, the large surface area of the ears tends to absorb heat and the animals depend on evaporative cooling, which expends water. Desert jackrabbits rely on plants as their sole source of water. For the animals to maintain **water balance** the water content of their diet must be at least 68% (Nagy et al. 1976). In summer, the animals are on tight water budgets. They cannot afford needless exertion and overheating (unless a coyote is involved). A reasonable speculation, then, is that the prime importance of this slow-escape behavior is to avoid overheating and consequent water loss.

Mammalian brain tissue is very sensitive to temperature fluctuations. Bovids appear to tolerate elevated body temperatures in part because they regulate their brain temperatures at a lower set point than their body temperature. In the oryx, in Thompson's gazelles (*Gazella thompsonii*), and probably in many other antelopes, the brain is provided with a specialized countercurrent cooling system of its own in the sinus cavernosus (Taylor 1969a). The external carotid artery, on its way to the brain, divides into many branches in this sinus, and these branches are in close proximity to veins returning from the nasal passages (**Fig. 21-22**). This system is called a **carotid rete**. These veins carry relatively cool blood because evaporative cooling of the nasal mucosa cools the blood supplying these surfaces. Countercurrent heat exchange in the carotid rete assures that the blood supply of the brain is cooler than that of most of the rest of the body. A similar countercurrent system involving a carotid rete occurs in the domestic sheep, goat, cat, and dog and probably occurs widely. In free-ranging wildebeest (*Connochaetes gnou*) and springbok (*Antidorcas marsupialis*), selective brain cooling helps modulate thermoregulation but may not play a significant role in protecting the brain from thermal stress (Mitchell et al. 1997). Selective brain cooling occurred in these animals only during moderate increases in body temperature and was abandoned entirely when physical activity was highest. These results from free-ranging animals suggest that selective brain cooling does not protect the brain during strenuous activity (Mitchell et al. 2002).

Hyperthermia during running may be an important adaptation in a number of mammals. Taylor et al. (1971) found that the body temperature of a running African hunting dog (*Lycaon pictus*) is higher than that of the domestic dog (41.2°C versus 39.2°C) and that the percentage of heat lost by respiratory evaporation produced by running is much lower (25.1% versus 49.7%). Taylor et al. suggested that hyperthermia and the greatly reduced pulmonary water loss of the African hunting dog allow this animal to conserve water and maximize the distance that it can chase prey.

Behavioral Thermoregulation

Most desert animals, wherever they occur in the world, are neither subjected to extremely high daytime summer

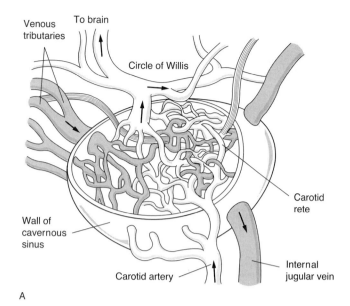

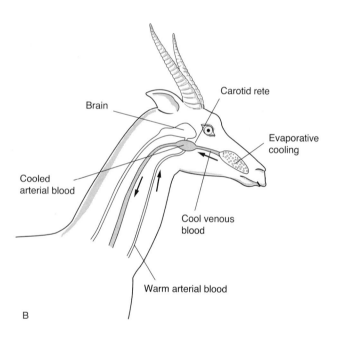

FIGURE 21-22 Cutaway diagram of the carotid rete of a gazelle (A) and the head of a gazelle (B) showing the location of the carotid rete used in cooling the blood to the brain. (Adapted from Taylor, C. R., *Sci. Am.* 220 (1969): 89–95 and Maloiy, G. M. O. *Comparative Physiology of Desert Animals*. Academic Press, 1972.)

temperatures nor are they able to survive them. Their success is based on the ability to avoid such temperatures rather than to cope with them. The saving grace of the desert is the great daily and seasonal fluctuation in temperature. Temperatures typically drop markedly at night, and winters are usually cool or cold. As a result, soil temperatures below the surface are never high, even in the summer, and nearly all desert rodents retreat to this refuge of coolness and relatively high humidity during the day. All but a very few desert rodents are strictly nocturnal, and all are more

or less fossorial. These mammals are active above ground in the part of the circadian cycle when temperatures are lowest.

Various means of avoiding daytime heat are used by desert mammals. Stick and debris dens that insulate woodrats (*Neotoma*, Cricetidae) from ambient temperature extremes have allowed three species (*N. albigula*, *N. devia*, and *N. lepida*) to occupy the hottest and driest North American deserts. The Karoo bush rat (*Otomys unisulcatus*, Muridae) of South Africa, another desert dweller, builds large stick dens at the bases of bushes and, as in some species of woodrats, digs burrows beneath the dens (Vermeulen & Nei 1987). Bighorn sheep (*Ovis canadensis*) and javelinas (*Pecari tajacu*) often take shelter in rock grottos or in the shade of steep rock outcrops, where for much of the day their body temperature is above air temperature and they can dissipate heat. Bovid horns are richly vascularized, and considerable heat can be lost across horn surfaces during hot weather (Bubenik & Bubenik 1990). Horns also lose heat during cold months, and Picard et al. (1994) estimate that Barbary sheep (*Ammotragus lervia*) can suffer energy losses that exceed by 20% and 29% the energy output of a resting female and male, respectively.

Behavioral thermoregulation is a conspicuous part of the daily routine of several mammals with energy-poor or seasonally restricted diets. For folivorous, diurnal rock-dwelling hyraxes (*Heterohyrax brucei*, *Procavia capensis*), such behavior allows them to adjust to temperature extremes with minimal output of energy (Bartholomew & Rainy 1971; Sale 1970). These animals have low metabolic rates (20% below the expected level), low heart rates (52% below the expected), variable body temperatures, and high thermal conductance, yet live in many places with searing summer temperatures and subfreezing nights in winter (**Fig. 21-23**).

Behavioral thermoregulation is an important part of the lives of some other rock dwellers with energy-poor diets. The dassie rat (*Petromus typicus*, Petromuridae) lives in the extremely arid Namib Desert, where it eats primarily leaves. On cool mornings this diurnal rodent basks on rock ledges, probably an energy-saving way to raise its body temperature from a reduced nocturnal level. Even nocturnal mammals may bask. On cold mornings Stephen's woodrat (*Neotoma stephensi*), known to specialize on eating juniper leaves and to have a low metabolic rate (Sorensen et al. 2005), will occasionally bask in a sunlit rock crevice (Vaughan, personal observation). Rock-dwelling species of elephant shrews (Macroscelididae) bask on cold mornings and alternatively bask and forage on winter days when their insect food is in short supply and conserving energy is vital.

Passive rewarming (basking in the sun) during arousal from daily torpor is also practiced in several dasyurid marsupials.

FIGURE 21-23 Hyraxes behaviorally thermoregulating.

In the winter, false antechinus (*Pseudantechinus*) typically seek shelter in rock crevices and enter daily torpor when ambient temperatures are low. In the morning, while still in partial torpor, they crawl from their crevices to sunlit basking sites to rewarm (Geiser & Pavey 2007). Another small (10 g) dasyurid, the fat-tailed dunnart (*Smminthopsis crassicaudata*), also uses radiant heat from the sun to reduce arousal costs from torpor. These arid zone animals sheltered in deep cracks in the soil during the night. When they slowly emerged in the morning their body temperature may be as low as 14.6°C. Solar radiation is used to complete arousal from torpor. Fat-tailed dunnarts also use solar radiation to maintain **normothermia** throughout the day. Warnecke et al. (2008) hypothesized that basking-assisted arousal saves an estimated 25% of their daily energy budget. Faster arousal rates may also have considerable survival benefits in reducing predation.

Fossorial Mammals: Physiological Problems

Several families of insectivorous mammals (Notoryctidae, Chrysochloridae, Talpidae) and a number of rodent families (Geomyidae, Spalacidae, Muridae, Octodontidae, Ctenomyidae, Bathyergidae) have fossorial members. Fossorial life in sealed burrows offers several advantages, including stable (and often moderate) temperatures, high humidities, and safety from predators. There are major liabilities, however, such as the high energetic cost of burrowing. Vleck (1979) estimated that burrowing a given distance consumes 360 to 3,400 times as much energy as moving the same distance on the surface. A North American mole (*Scapanus townsendii*) has been estimated to expend over 5,000 joules of energy to tunnel one meter but only 9 joules to walk a comparable distance (Gorman & Stone 1990). Also, microclimate factors can be more variable in a burrow than above ground: In burrows, rapid changes in oxygen levels (6% to 21%) and carbon dioxide concentrations (0.5% to 4.8%) can occur after heavy rain or rapid digging (Withers 1978). Schaefer and Sadleir (1979) found CO_2 levels in burrows to be 10 times atmospheric levels. A dispersed or scarce food supply may pose added problems. In mammals, the demands of burrowing have favored small size, a compact, fusiform shape, specialized digging structures, and modified sensory organs. Less obvious, but of decisive importance, are physiological and behavioral adaptations.

A fascinating example of such adaptations is offered by the naked mole-rat (*Heterocephalus glaber*, Bathyergidae). This eusocial mammal (see Fig. 13-38A) lives in large colonies of usually 70 to 80 animals in semideserts of East Africa. The workers average only 32 grams. These rodents have high thermal conductance (because they are naked), a BMR that is less than 60% of the expected rate, and low and labile body temperatures. Their narrow thermoneutral zone (31°C to 34°C) is within the range of dry-season burrow temperatures. Jarvis (1978) found that the metabolic rate of naked mole-rats increases modestly in response to temperatures below 30°C (**Fig. 21-24**). The decline in body temperatures and metabolic rates of individual mole-rats resting at ambient temperatures between 20°C and 25°C indicates that these animals abandoned their attempts at physiological thermoregulation.

Jarvis (1978) viewed this animal's behavioral and physiological peculiarities as part of a strategy for surviving under high temperatures and limited food. The low metabolic rate and high thermal conductance may be prerequisites for energy-saving behavioral thermoregulation. When heat stressed from burrowing, the mole-rats can passively unload heat from their naked bodies by moving to a cool section of the burrow. When cold, they can warm passively by "basking" in a warmer section. Mole-rats also reduce the cost of thermoregulation by huddling together when resting.

One type of "subterranean rodent syndrome" (low BMR, high and extremely narrow thermoneutral zone, high thermal conductance, limited thermoregulatory ability) is well-illustrated by some members of the bathyergid genera *Cryptomys* and *Fukomys* (Bennett et al. 1993, 1994b). These social mole-rats range from equatorial Africa to 35 degrees south latitude, occupy semiarid deserts and mesic tropics, and differ markedly in size between species (60 to 272 grams). Nevertheless, most share the above suite of adaptations (**Table 21-2**). Such characteristics also occur among a diverse group of other rodent families or subfamilies (Spalacinae, Rhizomyinae, Geomyidae, Ctenomyidae; McNab 1966; Reig 1970).

Viewed broadly, then, the trend is toward fossorial rodents of warmer areas displaying the "typical" subterranean rodent physiology described above for *Cryptomys*, but at least some of those from colder areas departing variously from this pattern. Yet even among warm-area species there are departures; some of the social bathyergid mole-rats have high resting metabolic rates (Bennett et al. 1993; Buffenstein & Yahau 1991). The frequent deviations from the central pattern suggest that there is no universal subterranean rodent syndrome.

Fossorial insectivores display diverse adaptations. The talpid moles that are fossorial occupy temperate areas and have high metabolic rates and low thermal conductance. They are homeotherms that neither hibernate nor estivate, but the southern African golden moles (Chrysochloridae) are at the opposite thermoregulatory pole. The Namib golden mole has a metabolic rate only 22% of the predicted rate, a high and narrow thermoneutral zone (31°C to 35°C), and an extremely labile body temperature (19°C to 38°C). The high thermal conductance favors diurnal torpor by allowing the body temperature to assume rapidly the temperature of the surrounding sand. Temperatures below 15°C are lethal. Low oxygen concentrations in the sand and a sparse and clumped food supply (mainly termites) have favored energy conservation in this golden mole (Fielden et al. 1990). In the temperate talpids, in contrast, a dependable, year-long supply of soil invertebrates seems to pay the higher energetic costs of homeothermy.

Many fossorial mammals are convergent regarding respiratory and tissue adaptations (Nevo 1979). Typically, the blood has a high affinity for oxygen. In the blind mole-rats (**Fig. 21-25**; *Spalax*, Muridae), this is due to a high erythrocyte

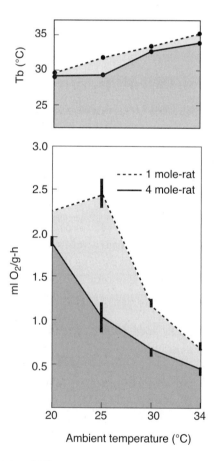

FIGURE 21-24 Differences in metabolic rates and body temperatures of naked mole-rats (*Heterocephalus glaber*), resting alone or huddling together in groups of four, at different ambient temperatures. (Adapted from Jarvis, *Bulletin of the Carnegie Museum of Natural History* 6 (1978).)

TABLE 21-2 Thermoregulatory Characteristics of Mole-Rats of the Genus *Cryptomys* **and** *Fukomys*

Species or Subspecies	Mean Body Mass (g)	Mean Body Temperature (°C)	RMR $(cm^3O_2g^{-1}h^{-1})$	TNZ[a] (°C)	Conductance $(cm^3O_2g^{-1}h^{-1}°C^{-1})$	Social Status	Habitat
C. hottentotus	75	34	0.90	27–30	0.13	Social	Semiarid
C. h. amatus	77	33.8	0.63	28–32	0.12	Social	Mesic
C. h. natalensis	102	33.8	0.80	30–31.5	0.13	Social	Mesic
C. bocagei	94	33.7	0.74	31.5–32.5	0.12	Social	Mesic
F. darlingi	60	33	0.98	28–31.5	0.19	Social	Mesic
F. damarensis	131	35	0.66	28–31	0.065	Social	Arid, semiarid
F. mechowi	272	33.7	0.60	29–30	0.09	?	Mesic

Modified with kind permission from Springer Science and Business Media: Bennett, N.C., Aguilar, G.H., Jarvis, J.U.M. & Faulkes, C.G. 1994. Thermoregulation in three species of Afrotropical subterranean mole-rats (Rodentia: Bathyergidae) from Zambia and Angola and scaling within the genus *Cryptomys*. *Oecologia* 97: 222–228. RMR = resting metabolic rate; TNZ = thermal neutral zone.
[a] Note the extremely narrow thermal neutral zones.

count and small corpuscle volume (Ar et al. 1977). In both the North African *Spalax* and the North American pocket gophers (Geomyidae), the myoglobin content of skeletal muscles is high (as in marine mammals), favoring rapid oxygen diffusion from capillaries to mitochondria (Lechner 1978). Respiratory rates of *Spalax* and *Talpa* (a mole) are 40% below expected levels, and both experience greatly increased rates at high carbon dioxide concentrations (Stahl 1967). The resting heart rates of these animals are also low, but rise sharply under low oxygen levels. These respiratory and tissue adaptations are essential to mammals that regularly experience adverse concentrations of low oxygen and

FIGURE 21-25 A Middle East blind mole-rat (*Spalax ehrenbergi*) from central Israel.

high carbon dioxide and are probably common to most fossorial mammals.

Aquatic and Semiaquatic Mammals: Physiological Problems

Temperature regulation is a demanding problem for mammals that inhabit cold water. The rate of heat loss by an endotherm in water is some 10 to 100 times as great as the rate of loss in air of the same temperature (Kanwisher & Sundnes 1966). Arctic and Antarctic waters are near 0°C year round, and high-latitude lakes and rivers approach this temperature in winter. Consequently, a temperature differential of about 35°C between deep-body temperature and ambient temperature is common in mammals swimming in these waters. Despite the thermal inhospitability of this environment, cold waters are permanently inhabited by some cetaceans, and some pinnipeds spend much of their lives in such waters.

In addition, the muskrat (*Ondatra zibethicus*), beaver, some shrews (*Sorex* spp.), some otters (*Lontra canadensis*, *Enhydra lutris*), and the mink (*Neovison vison*) spend considerable time in cold water. Although these semiaquatic mammals lose heat to the water most rapidly from the foot pads, the nose, and other bare surfaces, most of the body is insulated by a layer of air entrapped by the fur. Nonetheless, heat is lost far more rapidly during immersion in water than when the animal is in air. Calder (1969) found that in two species of shrews (*Sorex palustris* and *S. cinereus*) and two species of mice (*Zapus princeps* and *Peromyscus maniculatus*) thermal conductance in the water when the fur had entrapped air was 4.5 times that in air. When the fur was wet to the skin, the conductance rose to nine times that in air. Calder also measured heat loss in the water shrew (*S. palustris*), the smallest homeothermic diver. The body temperature of water shrews with air entrapped in the fur dropped an average of 1.4°C in 30 seconds during dives beneath the surface of the water, whereas shrews with fur wet to the skin had a temperature drop of 4.5°C in the same time. The meticulous grooming and drying of the fur by a shrew after a dive is clearly highly adaptive because it removes water trapped in the fur.

Hinds et al. (1993) showed that for a given body mass there is no difference in cold-induced metabolism between metatherians and eutherians. Prolonged immersion in cold water, however, is especially difficult for all small mammals. In some species, heat loss from the extremities is reduced in water by countercurrent heat exchange. When a muskrat swims, vasoconstriction and countercurrent systems keep its limbs at ambient temperature in cool and cold water, but vasodilation in the limbs allows for heat dissipation at a water temperature of 30°C or above (Fish 1979). Because of the very high rate of thermal conductance in the limbs

of these semiaquatic mammals, they would lose heat extremely rapidly when swimming if the limbs were kept near body-core temperature.

Some nearly permanent inhabitants of the sea, such as otariid seals, use entrapped air as insulation, but many marine mammals (cetaceans, phocid seals, and walruses) lack insulative fur, and their bodies are in contact with water that may in extreme cases be 40°C below their deep-body temperature. How they maintain a constant body temperature under such demanding conditions is of considerable interest.

These marine mammals have a thick layer of subcutaneous blubber that forms an insulating envelope around the deep, vital parts of the body. A substantial amount of the weight of a marine mammal may be contributed by blubber. For example, in the small (75 kilograms) harbor porpoise (*Phocoena*), 40% to 45% of the weight is blubber, and only 20% to 25% is muscle (Kanwisher & Sundnes 1966; Koopman 1998). Studies of seals by Irving and Hart (1957) have shown that the skin temperature varies directly with the water temperature down to 0°C. The cooled surface of the body and the thick blubber are an effective insulation, as indicated by the fact that the lower critical temperature of some seals is 0°C.

The young of seals and polar bears (*Ursus maritimus*) face especially severe thermoregulatory problems. The young must face extreme cold but are far smaller than the adults and thus lack the heat-conserving advantages of large size. The baby harp seal (*Phoca groenlandicus*) is born on drifting ice in the North Atlantic in winter and must survive air temperatures of −20°C or below. The pup weighs only 11 kilograms, whereas its mother may weigh as much as 140 kilograms. The pup has long fur, which offers better insulation than the short fur of the adult, and thermogenic adipose tissue yields energy during shivering and helps maintain the core body temperature (Blix et al. 1979). This adipose tissue is transformed to insulative blubber when the pup is several days old. Young harp seals use hypothermia (lowered body temperature) to conserve energy during extreme cold, wind, or rain. Their overall tolerance to cold is largely due to a high metabolic rate supported by rich (high fat, high energy) milk, shivering thermogenesis, and vasoconstriction in the skin (Blix et al. 1979).

Some of the most extreme thermal demands faced by endotherms are those met by cetaceans. Whales and porpoises live their entire lives in the water, and some species continuously occupy water at or near the freezing point. All cetaceans have insulating layers of blubber, but an extreme situation is faced by a small porpoise, which must maintain a deep-body temperature some 40°C higher than that of the sea, from which it is insulated by only 2 centimeters of blubber. An inflexible pattern of thermoregulation is inadequate even in inhabitants of the sea, which offers a relatively constant thermal environment. Some cetaceans migrate seasonally from cold waters to warm tropical seas. Because of the high thermal conductivity of water, skin temperatures generally equal water temperatures, and variations in water and skin temperatures of roughly 20°C to 30°C may occur seasonally. The temperature of the body core, however, remains constant, and insulation requirements therefore may vary fivefold.

Gigantic differences in the ability of cetaceans to keep warm result from differences in body size and in thickness of blubber. The biggest whale is 10,000 times as heavy as the smallest porpoise, has roughly a 10-fold greater mass-to-surface area ratio that favors heat retention, and has a much thicker shell of blubber. Because of these differences, the whale has approximately a 100-fold advantage over the small porpoise in its ability to keep warm. The very factors working in favor of heat retention in the large cetaceans, however, are obviously disadvantageous under conditions of great activity or warm water. Because of the vast bulk of these animals, dissipation of heat is an acute problem.

Although much remains to be learned, several points seem well established. First, metabolic rates of cetaceans differ markedly from species to species. The small porpoises have much higher BMRs than do large whales, far higher, in fact, than what would be predicted on the basis of weight. The harbor porpoise (*Phocoena phocoena*), for example, metabolizes at about 1.6 times the predicted rate.

Second, blood flow through the well-developed vascular system in the flippers, dorsal fin, and flukes of cetaceans allows these structures to function effectively as heat dissipators under conditions of heat stress. The flow can apparently be shut down during cold stress, allowing for a minimum of heat loss from these surfaces.

Third, a remarkable series of vascular specializations allows for great variation in the thermal resistance offered by the blubber (**Fig. 21–26**). A system of countercurrent heat exchangers in the vascular network supplying the blubber minimizes heat loss to the blubber and skin and hence to the environment (Fig. 21-12). In cetaceans, a second venous system in the blubber bypasses the countercurrent system during heat stress and allows considerable heat loss to the environment when heat dissipation is of prime importance. Similar countercurrent and bypass systems occur in the flippers and fins. The extremities and much of the surface of the body can thus dissipate heat or when necessary can conserve heat by using vascular adaptations.

The great quantities of blubber on large whales (up to 20 centimeters thick) seem not primarily useful as insulation. Because of their size, these animals could probably maintain a constant deep-body temperature with much less insulation. These fat deposits may be useful primarily

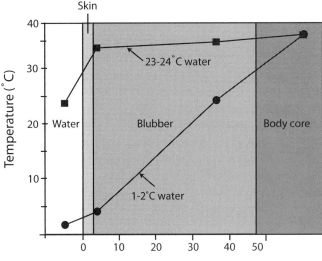

FIGURE 21-26 Mean blubber and core body temperatures for a harp seal (*Phoca groenlandica*) resting in 1–2°C water and in water at 23–24°C. In cold water, subcutaneous temperature is only 1–2°C warmer than that of the surrounding water, but deep blubber temperature (at 40mm tissue depth) measures approximately 23°C. (Adapted from Kvadsheim, P.H. and Folkow, L.P., *Acta Physiol Scand.* 161 (1997): 385–95.)

as food stores that can support an animal during periods of migration and fasting. It has been estimated that consumption of only half of a whale's blubber could fuel the BMR for 4 to 6 months (Parry 1949). Koopman (1998) showed that blubber in the harbor porpoise can be divided into two compartments, based on blubber thickness. The thoracic-abdominal blubber varies little in thickness around the girth of the animal and probably serves as insulation and energy storage. Blubber posterior to the anus, however, forms thick dorsal and ventral ridges but is very thin laterally. This unusual distribution suggests that postanal blubber may serve primarily to maintain a favorable hydrodynamic shape of the peduncle.

Energy Costs of Locomotion

A 10-kilogram coyote and a 400-kilogram horse have similar top speeds, but at such speeds, how do these animals compare with regard to the expenditure of energy? Some time ago, Hill (1950) made a series of predictions as to how energy use during running would change with the size of the runner. (Energy consumption by the muscles of a running animal is generally regarded as the result of the transformation of chemical energy to mechanical energy.) Hill reasoned that, although large and small runners could often reach similar top speeds, the rates of work and energy use at these speeds would be higher in the small runners. His logic was that, whether an animal is small or large, each gram of muscle performs the same amount of work and consumes the same amount of energy during a stride, but the short legs of the small animal have to take many strides to cover the same distance covered in one stride by a large animal. When large and small mammals run at the same speed, then, the small ones should have the higher stride rates and should consume more energy per unit of body weight.

These proposed relationships have been studied experimentally by Taylor and his associates (Fedak et al. 1982; Heglund et al. 1982; Taylor et al. 1970, 1982), who used mammals ranging in weight from 21 grams (a house mouse, *Mus musculus*) to 254 kilograms (African cattle, *Bos taurus*). Several important relationships were demonstrated by these studies:

- At the trot-gallop transition speed, the amount of energy used per stride per gram of muscle is nearly constant over a wide range of body sizes (**Table 21-3**). Mouse, baboon, and horse all expend nearly the same amount of energy per gram of muscle during a stride.
- The metabolic cost of muscle action in running animals increases linearly with speed. As shown in **Fig. 21-27**, the amount of energy a mammal expends increases as running speed increases.
- The mass-specific (per gram) use of energy by a running animal decreases as a function of weight (Fig. 21-27), varying as the –0.3 power of body mass. Thus, when a chipmunk and a horse are running at the same speed, each gram of chipmunk uses 15 times more energy than each gram of horse.

Several probable explanations of these relationships are available. Seemingly, muscular force is generated and dissipated more rapidly as an animal runs faster. With increasing speed, more muscle fibers that have a rapid contraction–relaxation cycle are brought into play. Each cycle uses a unit of energy, and the increase in the cost of rapid locomotion perhaps results from the increased use of rapid-cycling muscle fibers.

Muscular force must be generated and dissipated more rapidly in small mammals than in larger ones because, at comparable speeds, small mammals have higher stride rates. The muscles of small mammals contain higher percentages

of "faster" fibers that have rapid contraction–relaxation cycles and use energy at a high rate. The decrease in the use of "fast" muscle fibers accompanying increased body size may partially account for the mass-specific decrease in the cost of running in large mammals.

Fedak et al. (1982) found that, during high-speed running by large mammals, the energy expended by the muscles was not sufficient to provide the work necessary for the total kinetic energy (energy of motion) developed. The authors concluded that the storage of energy by muscles and tendons and its release by elastic recoil provide a significant part of the total kinetic energy. In large mammals, elastic recoil is of considerable importance in locomotion. Its contribution to locomotor efficiency in small mammals is unknown (Alexander 1992).

These relationships provide at least a partial explanation for the scaling of size in cursorial mammals. Artiodactyls, perissodactyls, and cursorial carnivorans generally weigh at least 10 kilograms and thus expend less energy per gram of body weight during running than smaller mammals. The mechanical problems associated with the support and propulsion of great weight may set upper limits on the size of runners, but some quite heavy mammals (such as 500-kilogram horses) are rapid and enduring runners. The swiftest cursors, however, generally weigh from about 50 to 125 kilograms (e.g., antelope), and the cheetah, arguably the fastest of all runners, weighs between 50 and 65 kilograms.

Bipedal locomotion, involving leaping or bounding, has evolved independently many times in the class Mammalia. Questions as to the relative energy cost of bipedal versus quadrupedal locomotion have inspired much controversy. Some researchers have claimed that, over a considerable range of hopping speeds in bipeds (such as kangaroos), there is no increase in energy cost attending increasing speed (**Fig. 21-28**). This aerobic plateau is thought to represent an energetically more efficient mode of locomotion in bipeds, involving elastic storage in the hindlimb tendons. Research shows that kangaroos and wallabies can store and recover up to 60% of the energy required to hop via elastic recoil of the large Achilles tendons in the hindlimbs (Alexander, 1982; Biewener & Baudinette 1995).

TABLE 21-3   Speed, Stride Frequency, and Metabolic Energy Consumed at the Trot-Gallop Transition by Mammals of Three Size Classes

Body Mass (kg)	Speed at Trot-Gallop Transition (m/s)	Stride Frequency at Trot-Gallop Transition (stride/s)	Energy Used (J/kg/stride)
0.01	0.51	8.54	5.59
1.0	1.53	4.48	5.00
100.0	4.61	2.35	5.53

Values calculated by Taylor et al. (1982) from equations given by Heglund et al. (1974).

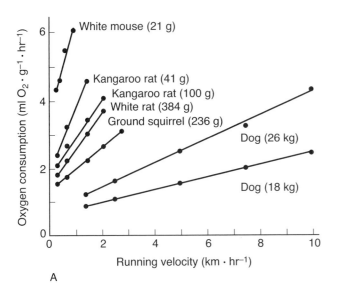

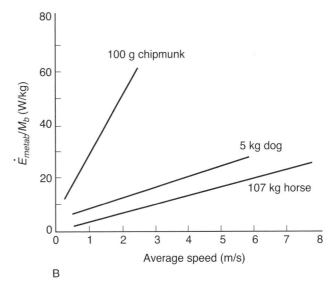

A

B

FIGURE 21-27 (A) Oxygen consumption by mammals of different sizes at various running speeds. (B) Mass-specific use of energy (use of energy per unit of weight) at various speeds by animals of different sizes. The notation E_{metab}/M_b (W/kg) = metabolic energy consumed in watts per kilogram of body weight. A watt is a unit of power equal to about 1/746 of an English horsepower. ((A) Adapted from Taylor, C. R., et al., *Am. J. Physiol.* 219 (1970): 1104-1107, (B) Adapted from Fedak, M. A., et al., *J. Exp. Biol.* 97 (1982): 23-40.)

A female kangaroo has the added burden of transporting a young joey in her pouch for several months. At the time the young permanently exits the pouch, the joey can weigh approximately 4.5 kilograms or about 20% of the mother's weight. The smaller Tammar wallaby (*Macropus eugenii*) carry pouch young equivalent to 15% of the mother's body weight. When 15% of each mother's body weight was artificially added to the pouch and the females were trained to run on a treadmill, they showed no increase in metabolic rate over speeds up to 4.5 meters per second (Baudinette & Biewener 1998). Thus kangaroos and wallabies carry their young at no extra energetic cost because the added weight is recovered by the elastic storage of the tendons.

Seemingly no such energy savings is associated with hopping in small (less than 3 kilogram) "bipedal" mammals (Thompson et al. 1980; MacMillen & Hinds 1992). Although referred to here as bipedal, these rodents are bipedal only when hopping. Unlike their quadrupedal kin, bipedal heteromyids show a plateau in oxygen consumption at higher running speeds (over 3 to 4 km/h), but this plateau is an anaerobic plateau accompanied by high blood lactate levels and does not contribute to energetic savings (MacMillen 1983; MacMillen & Hinds 1992). In addition, there is no energetic cost difference among running bipedal and quadrupedal heteromyids. Bipedal hoppers (*Dipodomys* and *Microdipodops*), however, do appear much more "willing" to run anaerobically at higher speeds, thereby creating an artificial "bipedal plateau." Although hopping is no less costly than running, the erratic turns made by hopping small mammals to avoid predators probably adds to the energy costs of hopping.

One of the most energetically expensive forms of locomotion is burrowing. Rodents that burrow through compacted soils (such as *Thomomys*, *Cryptomys*, or *Heterocephalus*) expend considerable energy to scrape away the soil and remove it from the burrow. The net cost of transport for the sand-swimming Namib mole (*Eremitalpa granti*), on the other hand, is much less than that for burrowing rodents (Seymour et al. 1998). Sand swimming involves pushing through loose sand, which collapses behind

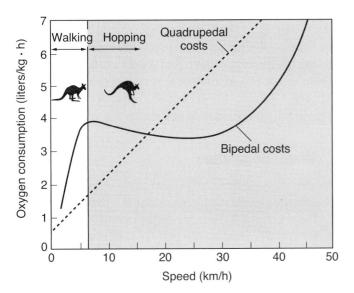

FIGURE 21-28 The relationship between metabolic rate and hopping speed in red kangaroos. The dotted line shows the relationship in a similarly-sized quadrupedal mammal. (Data from Dawson, T. J. *Kangaroos: Biology of the Largest Marsupials.* Cornell University Press, 1995.)

the animal, leaving no tunnel. Sand-swimming uses approximately 80 times the energy used by running on the surface (**Fig. 21-29**). Nevertheless, sand swimming in the Namib mole is still an order of magnitude less expensive than is burrowing through compacted soil for *Thomomys bottae* (Seymour et al. 1998; Vleck 1979).

The metabolic costs of aquatic and aerial locomotion depend on very different principles. Aquatic mammals swim in an environment that is 800 times as dense as air. The fluid environment, therefore, provides sufficient buoyancy to offset much of the gravitational force. Because of the high density of water, however, drag in water is far higher than that in air (drag is the fluid force acting opposite to the direction of thrust or forward movement). Because metabolic rate must be nearly doubled every time swimming speed increases by one body length per second, there are metabolic limits to swimming speed. A swimming mammal experiences both inertial drag and viscous drag. Viscous drag results from the friction between the water and the mammal' body surface. This type of drag is relatively constant over a wide range of swimming speeds and is reduced when the body surface is smooth (free of projecting appendages such as limbs and ear pinna). Inertial drag, on the other hand, results from pressure differences as the mammal displaces water as it swims. Inertial drag is minimized at slow swimming speeds, but at high swimming speeds, its effects rapidly increase (Vogel 1988). Both types of drag are reduced when the body form is **streamlined**. Streamlined bodies are roughly teardrop shaped, are

circular in cross section, tapered at both ends, have width-to-length ratios (w:l) of approximately 0.25, and have the maximum width located about one third of the way back from the tip of the head. Seals, sea lions, dolphins, and cetaceans all have streamlined body shapes.

The energetic cost of swimming also depends on the mode of swimming and the position of the body in the water column. Mammals such as the platypus swim using a rowing motion of the forelimbs, whereas muskrats and beavers paddle with their hindlimbs underneath their bodies (Fish 1993). Both rowing and paddling are relatively inefficient because thrust is generated only during half the stroke and the appendages increase drag. Sea lions also use rowing motions of the fore-flippers to generate thrust, resulting in maximum mechanical efficiencies of 80% (Feldkamp 1987b). The surface area of the platypus forefeet is over 13% of its total surface area, a value comparable to sea lion fore-flippers and considerably larger than the value of 4% to 6% for the hind feet of semiaquatic rodents (Feldkamp 1987a; Fish et al. 1997). In addition to the greater thrust generated by the platypus' forefeet, swimming platypuses spend much of the time submerged, which reduces drag (Evans et al. 1994). Bow waves associated with surface swimming increase drag up to fivefold, resulting in considerably higher metabolic costs. Williams (1989), for example, reported a 41% savings in oxygen consumption and a 35% reduction in drag for a sea otter swimming submerged compared to one swimming at the surface.

Aerial locomotion involves some of the same physical principles described for aquatic locomotion, except that the major cost is due to gravity and not to the density of the medium. Velocity for gliding flight, such as that exhibited by flying squirrels (Rodentia), honey opossums (Metatheria), and colugos (Dermoptera), is produced by gravity. Unlike flapping flight, there is relatively little metabolic cost to gliding because muscular effort is only needed to hold the gliding surface rigid and the animal loses altitude at rates between 1 and 2.5 meters per second (Alexander 1992). Flapping flight is discussed in Chapter 15.

Water Regulation

Roughly 35% of the earth's land surface is desert, where water is the primary limiting factor for plant and animal life. Desert areas are characterized by intense solar radiation by day and maximal heat loss by night, resulting in large daily variation in air temperature (commonly up to 30°C) in the summer, extremely low humidity through most of the year, and small amounts of precipitation, often at irregular intervals. On a summer day, the searing dry winds, radiation, and reflection of heat from the hot, pale soil add to the harshness of the desert environment. Few environments

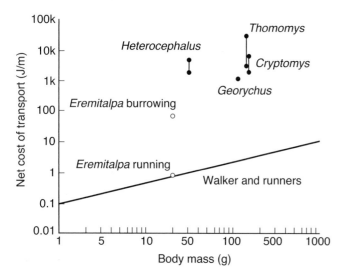

FIGURE 21-29 Net cost of transport of burrowing via sand-swimming and running on the dune surface in the Namib mole (*Eremitalpa*). Similar costs for several species of mole-rat and one species of pocket gopher (*Thomomys*), which burrow through more compacted soils, are shown for comparison. The solid line denotes the cost of transport for mammals in general. (Adapted from Seymour, R. S., et al., *J. Zool.*, London. 244 (1998): 107–117. Reproduced with permission of Blackwell Publishing Ltd.)

on earth are as hostile to life; to the casual observer the desert gives the impression of overwhelming sterility. This impression is deceptive, however, for, in reality, the desert supports a great variety of life. Most mammals who live in the desert remain hidden in shelters by day and forage in the relative cool of night.

Water is absolutely essential to life. To all mammals, life depends on the maintenance of an internal water balance within fairly narrow limits. (Water balance results when water intake, through drinking, eating, and production of **metabolic water**, equals water loss by evaporation from skin and lungs, defecation, and urination.) Mammals are approximately two-thirds water by mass. Most mammals are stressed when water loss reduces their body weight by as little as 10% or 15%, and death occurs in many mammals when such loss reduces the body weight by 20%. Loss of water occurs rapidly on the desert; water loss in a human on a hot summer day in the southwestern deserts of the United States has been recorded as 1.41% of body weight per hour. Comparable figures for the donkey and dog are 1.24% and 2.62%, respectively (Schmidt-Nielsen 1964). Deprived of drinking water, a human or a dog can survive only a day or two of exposure in the summer. Nonetheless, some small desert rodents live without drinking water and must satisfy their water needs by using water in their food and water derived from the metabolism of food. Similarly, some large desert mammals must maintain water balance with only occasional access to drinking water. Although more remains to be learned about mammalian adaptations for water conservation in arid environments, excellent studies have provided a solid base of knowledge.

A number of solutions to the problem of maintaining water balance are used by desert mammals. These solutions depend on seasonal weather patterns, size of the animal, timing of activity cycles, diet, and a variety of behavioral, structural, and physiological features. The following discussions do not cover the subject of water conservation in mammals exhaustively but consider the adaptations that permit some mammals to maintain water balance in dry environments.

Periodic Drinkers

In many arid or semiarid regions, scattered water holes or widely separated rivers offer water to mammals that can move long distances. The extensive grasslands of Africa form such an area, as did the North American Great Plains before the coming of settlers. Most large mammals in such areas probably drink every day or two in hot weather and seemingly are unable to survive for long periods without drinking. A few ungulates, such as the camel, however, occupy an intermediate position with regard to water needs. Although they can go for moderate periods without drinking, these mammals are not independent of drinking water, as are some desert rodents, and must drink water periodically.

Our present knowledge of the water metabolism of the camel is largely a result of the work of Schmidt-Nielsen et al. (1956, 1957). Their work, done in the northwestern Sahara on local domesticated camels, substantiated the popular idea that camels can tolerate long periods without drinking water, but more importantly, Schmidt-Nielsen and his group explained the adaptations allowing this tolerance. The ability of their experimental animals to tolerate dehydration was remarkable. One camel went without water for 17 days in the winter on a diet of dry food. During this period, it lost 16.2% of its body weight. In some areas, camels that foraged on native vegetation in the winter were never watered. Two camels kept without water for 7 days in the heat of the summer lost slightly over 25% of their body weight. All of these animals drank tremendous amounts of water after their periods of dehydration, and none showed ill effects.

The camel economizes on water in several ways. Its body temperature drops sharply at night and then rises slowly during the heat of the day (Fig. 21-20). It is able to tolerate considerable hyperthermia, and typically the day is largely over before the animal's body temperature rises to levels at which evaporative cooling, in the form of sweating, must occur. Thus, relative to humans under similar conditions, very little moisture is expended each day in cooling the camel. Excess heat gained by day is lost passively at night. Further water saving results from the modest ability of the kidneys to concentrate urine and from the absorption of water from fecal material. Despite these important water-saving adaptations, however, the camel loses water steadily through evaporation from lungs and skin and in the urine and feces. Its ability to tolerate tremendous water loss (up to 27% of body weight) during these periods of dehydration is striking.

Apparently, the proportions of water lost from various parts of the body differ in humans and camels. When a person in the desert has lost water equal to about 12% of body weight, the blood becomes viscous. As a result, the heart has difficulty moving the blood and the rate of blood circulation decreases. This leads to a marked reduction in the rate of dissipation of metabolic heat, to a sudden rise in body temperature, and to death. In a camel that has lost 20% of its body weight because of water loss, water content of the plasma remains nearly normal, but large amounts of water are lost from **interstitial fluid** and intracellular water. In a camel deprived of water for 8 days, interstitial fluid volume decreased 38%, and intracellular water volume fell 24%; plasma volume, however, decreased only 10% (Schmidt-Nielsen 1964). Although the camel becomes strikingly dehydrated during periods without water, the blood apparently retains its fluidity and its ability to contribute to heat dissipation without straining the circulatory system. The donkey, which was also studied by Schmidt-Nielsen (1964), proved to be as capable as the camel of tolerating dehydration. The donkey lost water 2.5 times faster than the camel, however, and could not be independent of water for more than a few days.

■ Dietary Moisture

A number of mammals that occupy deserts or semiarid areas are no better adapted to surviving without considerable moisture in their diet than are mammals of fairly moist areas. Even in some areas with fairly high precipitation, small mammals do not have regular access to drinking water and, as in the case of some desert rodents, satisfy their water requirements by eating moist food.

Succulent plants provide water for some desert rodents, such as the white-throated woodrat (*Neotoma albigula*), which occupies the hot deserts of the southwestern United States and northern and central Mexico. Paradoxically, this rodent needs large amounts of water, which it obtains largely from cactus. The desert woodrat (*N. lepida*) and the cactus mouse (*Peromyscus eremicus*) also use large quantities of cactus (*Opuntia*) as a source of both food and water (MacMillen 1964a). These mammals have evolved the ability to cope metabolically with oxalic acid, a compound abundant in cactus and toxic to some mammals (Schmidt-Nielsen 1964). The ability to obtain water from cacti and to deal with oxalic acid is not limited to the rodents mentioned above, all of which belong to the family Cricetidae, but also occurs in the rodent family Geomyidae, the pocket gophers. The northern pocket gopher (*Thomomys talpoides*), inhabiting fairly dry short-grass prairies of Colorado, also obtains water by eating prickly pear cactus (Vaughan 1967).

Some desert rodents obtain water from succulent plants that contain high salt concentrations. These rodents have kidneys that are able to produce highly concentrated urine (urine that has little water relative to the contained solutes). The North African sand rat (*Psammomys obesus*, Muridae) is such an animal. The sand rat obtains water from the fleshy leaves of **halophytic** plants (plants that grow in salty soil), which grow along dry river beds in the desert (Mares et al. 1997; Schmidt-Nielsen 1964). These leaves are 80% to 90% water but contain higher concentrations of salt than seawater and also have large amounts of oxalic acid. In order to use this water source, the sand rat produces urine with extremely high concentrations of salt and metabolizes large quantities of oxalic acid. The Australian hopping mouse, *Notomys cervinus* (Muridae), and a South American desert-dwelling rodent, *Eligmodontia typus* (Cricetidae), have remarkably

well-developed abilities to concentrate salts in their urine and probably use the succulent but highly saline leaves of halophytic plants as a water source (MacMillen & Lee 1969; Mares 1977).

Most deserts support a number of carnivorous and insectivorous mammals whose moisture requirements are seemingly met by the water in their food. The grasshopper mouse (*Onychomys*), a small rodent widely distributed in the deserts and semiarid sections of the western United States and Mexico, is almost exclusively insectivorous at some times of the year. This mouse has thrived in the laboratory on an entirely meat diet, with no drinking water (Schmidt-Nielsen 1964). Similarly, the desert hedgehog (*Hemiechinus auritus*) and the fennec (a fox; *Vulpes zerda*), both inhabitants of North African deserts, could get adequate water from a predominantly carnivorous diet, as could the mulgara (*Dasycercus cristicauda*), an Australian dasyurid metatherian (Schmidt-Nielsen & Newsome 1962). The fennec can maintain water balance for at least 100 days on a diet of mice and no drinking water. This small animal has an unusually low rate of evaporative water loss (EWL) and equals water-independent desert rodents in its ability to concentrate urine (Noll-Banholzer 1979b; **Table 21-4**), a capacity that enables it to excrete little water with the large amounts of urea produced by its high-protein diet. Ruppell's foxes (*Vulpes rueppellii*) from Saudi Arabia obtain all of their water from metabolic water (water generated via the mammal's own metabolism) and from preformed water (water available from solid foods in their diet; Williams et al. 2002). On average, Ruppell's foxes took in 123 mililiters of water per day, 26 mililiters from metabolic production, and the remainder from their diet of rodents and invertebrates. Ruppell's foxes have a basal metabolic rate similar to other foxes of the same size. Nevertheless, their mass-specific field metabolic rate in the winter is the lowest reported in foxes, and their total evaporative water loss is 50% less than predicted for animals of their body size.

Few large ungulates inhabit barren deserts where no drinking water or little cover is available. One notable exception is the oryx, or gemsbok (*Oryx gazella*; **Fig. 21-30**), a large antelope that occurs in arid and semiarid sections of Africa including the extremely dry Namib Desert. More remarkable than the amazing ability of the oryx to withstand intense desert heat is the animal's lack of dependence on drinking water. Careful studies by Taylor (1969a) showed that the water needs of the oryx are probably satisfied by its food, which consists of grasses and leaves of shrubs that by day may contain as little as 1% water. After nightfall, as the temperature drops and the humidity rises, these parched leaves absorb moisture from the air and probably contain approximately 30% water during much of the night. By

TABLE 21-4 **Relative Urine-Concentrating Ability of Some Mammals as Indicated by Osmotic Concentration of Urine from Dehydrated Animals**

Species	Common Name	Urine Osmolality (mOsm/liter)	Diet
*Leggadina forresti**	Sandy inland mouse	4,710	Granivorous
*Notomys alexis**	Spinifex hopping mouse	6,550	Granivorous
*Notomys cervinus**	Fawn hopping mouse	3720	Granivorous
Dipodomys merriami	Merriam's kangaroo rat	3,936	Granivorous
Peromyscus crinitus	Canyon mouse	3,047	Omnivorous
Perognathus longimembris	Little pocket mouse	3,716	Granivorous
Onychomys torridus	Grasshopper mouse	3,259	Insectivorous
Neotoma lepida	Desert woodrat	1,484	Herbivorous
Lepus californicus	Black-tail jackrabbit	3,600	Herbivorous
Vulpes zerda	Fennec	4,022	Carnivorous
Canis lupus	Wolf	2,608	Carnivorous
Felis catus	Cat	3,118	Carnivorous
Madoqua sp.	Dik-dik antelope	4,300	Herbivorous
Camelus dromedarius	One-humped camel	3,100	Herbivorous
Oryx gazella	Oryx	2,900	Herbivorous
Equus asinus	Donkey	1,500	Herbivorous
Bos taurus	Zebu cattle	1,400	Herbivorous

Data are from MacMillen (1972), MacMillen et al. (1972), MacMillen and Grubbs (1976), MacMillen and Lee (1969), Nagy et al. (1976), Noll-Banholzer (1979a, 1979b), and Maloiy (1973).
*Data from Tyndale-Biscoe and Renfree (1987) and Renfree (1993)

feeding at night, therefore, the oryx can manage a nightly intake of some 5 liters of water with its forage. This is a minimal amount of water for a 200-kilogram mammal living in shelterless desert, and is sufficient for the oryx only because of a combination of mechanisms that favor water conservation. When water is available, however, oryx are avid drinkers.

■ Metabolic Water

Many rodents that inhabit deserts, seasonally dry chaparral, or woodlands must survive for extended periods without access to preformed water. Among the rodent members of these habitats are species that primarily eat seeds. These rodents, some of which are saltatorial and can move rapidly over considerable distances in search of seeds, occupy even the most barren and inhospitable deserts of the world. They represent the families Heteromyidae, Dipodidae, and Muridae. These rodents share two life-history features: They are nocturnal, and they are semifossorial.

As a basis for further discussion, the routes of water intake and loss in desert rodents must be reviewed. Sources of water intake include succulent foods, metabolic water (seeds high in carbohydrates have high yields of water released as a byproduct of metabolism), and drinking water.

FIGURE 21-30 An oryx (*Oryx gazella*) from Etosha National Park, Namibia.

Water is lost by lactation, defecation, urination, and **pulmocutaneous evaporation**. Many seed-eating desert rodents do not have regular access to drinking water and eat little succulent food. Their major water source is metabolic water. Water lost in lactation is important periodically to females. Urinary water loss is typically reduced in these rodents by the concentration of urine, and water is absorbed from fecal material. The primary channel for water loss is pulmocutaneous evaporation. Such loss may account for 90% of total water loss, and MacMillen and Grubbs (1976) and MacMillen (1972) demonstrated that there is no difference in the rate of such water loss between desert and nondesert rodents. Nagy and Peterson (1980) confirmed these results for free-living mammals. Because these desert rodents do not sweat and may have reduced cutaneous water loss, pulmonary evaporation is of greatest importance.

Water intake in kangaroo rats and other seed-eating desert rodents can be accounted for fairly easily. Many seeds are high in carbohydrates, which yield large amounts of water when they are oxidized. For example, for every 100 grams of dry barley metabolized, 53.7 grams of water is produced. This may be augmented by preformed water in the food: Seeds in the soil or on the surface at night absorb moisture, and seeds stored in nests in burrows may contain as much as 20% water (Morton & MacMillen, 1982). Merriam's kangaroo rat (*Dipodomys merriami*) consumes significant amounts of insects and succulent vegetation during the hottest part of the Sonoran Desert summer and need not rely on metabolic water from dry seeds for survival (Tracy & Walsberg 2002).

Of central importance to desert rodents, then, is the balance between water lost via evaporation (the principal route of water loss) and water gained from the metabolism of food (the major source of water). This balance must be maintained through a wide range of daily and seasonal temperatures. MacMillen and his associates (MacMillen 1972; MacMillen & Christopher 1975; MacMillen & Grubbs 1976; MacMillen & Hinds 1983a, 1983b) studied water balance in the laboratory for a range of heteromyid rodents from the southwestern United States. Because heteromyid rodents are nocturnal, they are active in the coolest part of the daily cycle when pulmonary water loss is lowest. Except in summer, temperatures in the desert drop dramatically at night. For much of the year, foraging desert rodents are faced with temperatures below their TNZ and must raise their metabolic rate accordingly. As metabolic rate rises, so does production of metabolic water. To desert rodents, therefore, the relationship between evaporative water loss (EWL), production of metabolic water, and ambient temperature is critically important.

Through careful analyses of data on simultaneous measurements of EWL and oxygen consumption (an indicator of the level of production of metabolic water), MacMillen and Hinds (1983a, 1983b) showed that, at ambient temperatures below 16.6°C, many rodents produce more water metabolically than they lose by evaporation. At even lower temperatures, metabolic water production (MWP) still further exceeds EWL (**Fig. 21-31**). There is, however, considerable interspecific variation in the ability to limit EWL. Merriam's kangaroo rat, for example, far surpasses average performance (MacMillen & Hinds 1983a). MacMillen and coworkers concluded that nocturnality provides desert rodents with a favorable relationship between EWL and MWP.

The classical view of kangaroo rats (derived largely from laboratory studies) is that they are relatively intolerant of high ambient temperatures ($> 30°C$), and by being semifossorial escape the heat of the day in their cool, humid burrows (**Fig. 21-32**). Some species are active on the surface for as little as 1 hour each night (Kenagy 1973a). These rodents were, therefore, believed to live mostly under conditions of moderate humidities and temperatures. Field studies in the Sonoran Desert refute this view. Temperatures rarely fell below 30°C even in tunnels 2 meters below the surface, and the kangaroo rats spent the bulk of their time within 1 meter of the surface where burrow temperatures averaged over 35°C during the day; they emerged just after sundown, during the hottest part of the night (Tracy & Walsberg 2002). Field measurements also revealed that

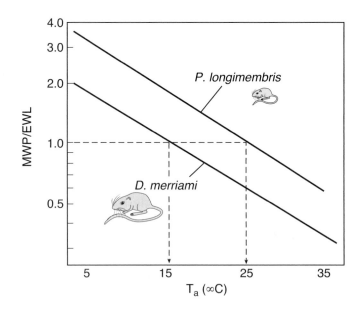

FIGURE 21-31 Relationship between metabolic water production and evaporative water loss (MWP/EWL) in *Dipodomys merriami* and *Perognathus longimembris* while on a diet of dry millet seeds. An extension of the horizontal line at MWP/EWL = 1.0 indicates the temperature at which MWP = EWL (dotted lines). This equality occurs at a much higher temperature in *P. longimembris* (about 25°C) than in *D. merriami* (about 17°C). (Adapted from MacMillen, R. E., and Hinds, D. S., *Ecology* 64 (1983): 156.)

FIGURE 21-32 A giant kangaroo rat (*Dipodomys ingens*) emerging from its burrow in California to begin foraging.

kangaroo rat burrows were not particularly humid. At typical summer temperatures of 35°C, a resting *D. merriami* fed only seeds would lose 2.2% of body mass per day as water and would survive less than a week on such a diet. The key to kangaroo rat survival during the summer months may be tolerance of higher temperatures and the ingestion of significant quantities of succulent vegetation and insects (**Fig. 21-33**).

Although under laboratory conditions most species of kangaroo rats (*Dipodomys*) and all species of pocket mice (*Chaetodipus* and *Perognathus*) remain healthy on a dry-seed diet, the urine of pocket mice has roughly one-half the osmotic concentration of that of kangaroo rats. Pocket mice, then, should be in a more favorable state of water balance than kangaroo rats. MacMillen and Hinds (1983) tested this hypothesis by measuring the primary route of water input (MWP) and output (EWL) in *Dipodomys merriami* (36 grams) and *Perognathus longimembris* (8.0 grams) fed millet seeds and subjected to a wide temperature range. In these heteromyids, EWL is independent of ambient temperature at and below the rodent's temperatures of thermoneutrality. MWP is inversely related to ambient temperature: The lower the temperature the higher the metabolic rate, hence the greater the MWP. In both species, the ratio MWP:EWL (an expression of the state of water balance) is inversely related to ambient temperature (Fig. 21-31). Because of its smaller size and associated higher metabolic rates, *P. longimembris* is always in a more favorable state of water balance than *D. merriami*, as indicated by the more dilute urine of *P. longimembris*.

MacMillen and Hinds (1983) regarded urine concentration, the traditional criterion of water regulatory efficiency in desert rodents, to be misleading and proposed that a more meaningful indicator for granivorous rodents is the ambient temperature at which MWP = EWL. Greater water regulatory efficiency of *P. longimembris* at higher temperatures is demonstrated by the fact that the ambient temperature at which MWP = EWL is 10°C higher in *P. longimembris* than in *D. merriami*. The demonstrated relationship between metabolism and EWL led MacMillen and Hinds to hypothesize that water regulatory efficiency in heteromyids should vary inversely with body weight. This hypothesis was confirmed by experiments on five genera and 13 species of heteromyids.

MacMillen and Hinds speculated that under selective pressures associated with increasing aridity, heteromyids became granivorous, improved water regulatory ability, minimized energy needs by reducing metabolic rate, and became smaller. At a body weight threshold of 35 to 40 grams, a divergence in locomotor style occurred: The larger

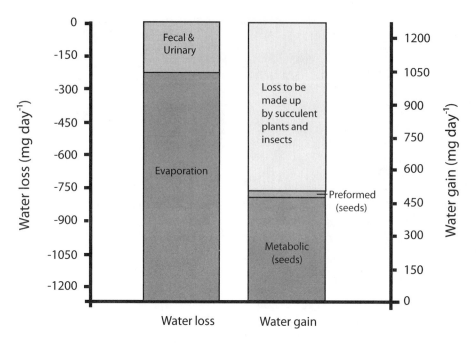

FIGURE 21-33 Daily water budget for a Merriam's kangaroo rat (*Dipodomys merrriami*) during the summer in the Sonoran Desert. Water losses exceed gains when the diet is restricted to dry seeds. Field studies show that *D. merriami* supplement their diet with succulent plants and insects during the summer months to help balance water losses and gains. (Adapted from Tracy, R. L., and Walsberg, G. E., *Oecologia*. 133 (2002): 449–457.)

kangaroo rats (more than about 40 grams) became bipedal hoppers, and the smaller pocket mice (less than 40 grams) maintained quadrupedal locomotion. They proposed that the ecological importance of the mass-related differences in water regulatory efficiency favors survival in the small species that have limited locomotor ability. Whereas bipedal hopping provides kangaroo rats with a rapid means of locomotion that allows erratic escape from predators in open situations, the slower quadrupedal locomotion of pocket mice leaves these animals more vulnerable to predators on open ground and restricts the foraging area. Thus, kangaroo rats can forage widely and selectively and can choose seeds high in carbohydrates (seeds that have high yields of metabolic water), and their water regulatory ability became fixed at an intermediate level. Pocket mice, in contrast, must forage less widely and eat a variety of seeds, some of which, because of relatively low carbohydrate content, yield relatively less metabolic water and more urea to eliminate via the kidneys.

■ Urine-Concentrating Ability

An important factor in reducing water loss is the ability of the kidneys of water-independent rodents to concentrate urine. Kangaroo rat urine is roughly five times more concentrated than that of a human. Therefore, in excreting comparable amounts of urea, the kangaroo rat uses one fifth as much water as do humans. The concentration of dissolved compounds in the urine of kangaroo rats may be roughly twice that of seawater; in the laboratory,

these animals have maintained water balance by drinking seawater. The urine of humans, on the other hand, has a concentration of dissolved compounds lower than that of seawater. When a human drinks seawater, the excretion of the dissolved salts requires the withdrawal of water from body tissues, resulting in severe dehydration.

Among desert rodents of the Old World, jerboas (*Jaculus jaculus*) and several gerbils of two genera (*Meriones* and *Gerbillus*) can live on dry food, and some surpass kangaroo rats in their ability to concentrate urine. Adaptations to a dry diet have clearly evolved independently in several rodent families (Heteromyidae, Dipodidae, Muridae). Striking convergent evolution in these families has led to saltatorial adaptations in some members of each family, as well as to similar specializations favoring water conservation.

Two species of spiny mice (**Fig. 21-34**; *Acomys*, Muridae) studied by Shkolnik and Borut (1969) in the desert of Israel are remarkable in their unusual pattern of adaptation to arid conditions. These animals have highly specialized kidneys that can concentrate urine to a greater degree than can the kangaroo rat kidney, but the spiny mice have an EWL two to three times as great as that in Merriam's kangaroo rat. Probably because of high water loss through the skin, spiny mice are unable to subsist on a diet of dry seeds (**Table 21-5**). Seemingly, the high cutaneous water loss is important as a means of dissipating heat in a hot climate, and the great ability of the kidney to concentrate urine, coupled with a diet high in land snails (which have a high water content), compensates for their extravagant use of water in thermoregulation.

FIGURE 21-34 A spiny mouse (*Acomys cahirinus*) from the Fujairah Mountains, United Arab Emirates.

A different combination of adaptations have evolved in several rodents of the arid center of Australia, where rain may not fall for years or may come in a deluge. Aridity has prevailed here for about 6 million years (Bowler 1982), and monsoon patterns date back some 65,000 years (Johnson et al. 1999), model conditions for the evolution of desert-adapted animals. Although the three well-known murid rodents of this area—the sandy mouse, *Leggadina forresti*, spinifex hopping mouse, *Notomys alexis*, and fawn hopping mouse, *N. cervinus*—share certain life-history features with many other desert rodents (they are nocturnal, rest in burrows by day, and are granivorous or omnivorous), they are physiologically unique (MacMillen & Lee 1967,

1969, 1970; MacMillen et al. 1972). Like some kangaroo rats, *L. forresti* and *N. alexis* are independent of drinking water, but they differ in having higher pulmocutaneous and fecal water losses. They compensate by having extremely efficient kidneys suited to concentrating urea and are able to produce the most concentrated urine known among mammals (Table 21-5). The third species, *N. cervinus*, is less able to concentrate urine but is better able to concentrate electrolytes in saline solutions, allowing it to obtain water from the saline sap of halophytic plants. In response to high temperatures and low humidities in the laboratory, the *Notomys* species reduced their metabolic rates and went into hyperthermia but never used evaporative cooling, which, in the interest of water conservation, is not within their physiological repertoire. The adaptations of *N. cervinus* must be closely matched to its environment, for despite periodic flooding of its clay pan habitats, this species was fairly abundant through a period (1966–1992) when the two sandhill species were scarce (MacMillen & MacMillen 2007).

The kidneys of some bats are specialized to concentrate urine, but these animals are seemingly not independent of drinking water. Carpenter (1969) found that two desert-dwelling insectivorous bats produced concentrated urine. Their need for water was increased by high EWLs during flight and when they were not torpid. He estimated that these bats lost 3.1% of their body weight through evaporation per hour of flight. Carpenter concluded that they were not independent of drinking water but that their ability to

TABLE 21-5   Physiological Characteristics of Four Desert Rodents on a Natural Diet

Characteristic	*Dipodomys merriami*	*Acomys cahirinus*	*Ammospermophilus leucurus*	*Psammomys obesus*
Diet	Granivore	Omnivore	Omnivore	Herbivore
Body mass (g)	35.0	49.0	88.0	135.0
Dry matter intake (g/day)	3.57	3.14	6.55	12.11
Water influx (ml/day)	2.13	4.99	14.21	46.47
Metabolic water (ml/day)	1.91	1.29	2.42	3.45
(% water influx)	89.7	25.9	17.0	7.4
Preformed water (ml/day)	0.22	3.7	11.8	42.8
(% water influx)	10.3	74.1	83.0	92.2
Water efflux (ml/day)	2.13	4.99	14.21	46.47
Fecal water (ml/day)	0.07	0.44	1.30	7.57
(% water efflux)	3.2	8.8	9.2	16.3
Urinary water (ml/day)	0.49	2.11	6.73	18.9
(% water efflux)	23.0	42.3	47.4	41.6
EWL (ml/day)	1.57	2.44	6.26	20.0
(% water efflux)	73.7	48.9	44.0	43.1

Data are from Degen (1996).

fly long distances to drink water enabled them to maintain water balance in desert areas. Urine concentrations in the little brown bat (*Myotis lucifugus*) reach peak levels during high evaporative cooling and just after feeding, but by drinking after feeding, this species avoids water stress (Geluso & Studier 1979). A marine fish- and crustacean-eating bat (*Myotis vivesi*) that inhabits the arid islands and coasts of the Gulf of California has the ability to concentrate urine to the extent that it can use seawater as a water source (Carpenter 1968). Because of high EWLs, particularly during flight, the water gained from this bat's food probably is not sufficient to meet its water requirements, and, presumably, it must drink seawater.

Desert insectivores, such as the desert shrew (*Notiosorex crawfordi*), grasshopper mice (*Onychomys*), and golden moles (Chrysochloridae), face slightly different problems. Insectivores compared with granivores obtain more preformed water from their food, but insectivores too must conserve water. The Namib Desert golden mole (*Eremitalpa granti*) obtains sufficient water from its diet of termites and insect larvae, which contain 60% to 80% water (Fielden et al. 1990a, 1990b; Redford & Dorea 1984). Water loss is minimized by the golden mole's nocturnal foraging habits and by a low metabolic rate during diurnal torpor deep in the sand, where humidities are high. Although insects have a high moisture content, they also are high in protein. Insectivores, therefore, excrete wastes from protein catabolism as urea in the urine, generally resulting in higher urinary water losses for these insectivores than for granivores (Lindstedt 1980).

The osmotic concentration of urine in desert insectivores far exceeds that of more mesic-adapted insectivores (**Table 21-6**). Namib golden moles, for example, lose only 11% of their total water budget via their urine compared with over 30% in the similar-sized short-tailed shrew (*Blarina brevicauda*; Deavers & Hudson 1979; Fielden et al. 1990b).

■ Nasal Countercurrent Heat and Water Exchange

The observation by Schmidt-Nielsen that kangaroo rats exhale air that is cooler than body temperature led to studies showing that the nasal passages of many mammals serve as heat exchange systems (Langman et al. 1979; Schmidt-Nielsen et al. 1970, 1980). These systems result in a significant reduction of pulmonary water loss.

The nasal passages of rodents function as heat exchangers with alternating flow in opposite directions in a single tube rather than steady flow in opposite directions in adjacent tubes (as in retia mirabilia). Inhaled air that is below body temperature cools the moist nasal mucosa, which is further cooled when water evaporates from it. Inhaled air then becomes warmed and saturated with water in the airways and lungs. During expiration, this humid, relatively warm air passes back through the narrow nasal passages and over the cool mucosa. The expired air is cooled, and thus moisture condenses on the mucosa. This moisture is subsequently absorbed back into the animal's system. This pattern, repeated with every respiratory cycle, results in expired air that is far below body temperature and substantially below the temperature of the inhaled (or ambient)

TABLE 21-6 **Average Osmotic Concentration of Urine from Insectivorous Mammals Exposed to Water Stress in the Laboratory**

Species	Common Name	Urine Osmolarity (mosmol/kg)	Source
Arid Zone			
Parastrellus hesperus	Western pipistrelle	4,340	Geluso 1978
Hemiechinus auritus	Long-eared hedgehog	4,010	Yaakobi & Sholkni 1974
Antrozous pallidus	Pallid bat	3,980	Geluso 1975
Eremitalpa granti	Grant's golden mole	3,820	Fielden et al. 1990
Macrotis lagotis	Greater bilby	3,566	Hulbert & Dawson 1974
Onychomys torridus	Southern grasshopper mouse	3,180	Schmidt-Neilsen & Haines 1964
Mesic Zone			
Erinaceus europaeus	European hedgehog	3,062	Yaakobi & Sholkni 1974
Myotis volans	Long-legged myotis	2,910	Geluso 1978
Planigale maculata	Pygmy planigale	2,317	Morton 1980
Blarina brevicauda	Short-tailed shrew	1,820	Deavers & Hudson 1979

Data from Tyndale-Biscoe and Renfree 1987 and Renfree 1993

air. Although the exhaled air is saturated with water vapor, it is far cooler than the air in the lungs and thus contains considerably less water. A kangaroo rat with a body temperature near 38°C, resting in air at 30°C and 25% relative humidity, exhaled air that was 27°C. In this case, 54% of the water used to humidify the inhaled air was recovered from the exhaled air by condensation on the mucosa. Ambient temperature influences the rate of water recovery: The cooler the inhaled air, the more the nasal passages are cooled. Consequently, the more the air being exhaled is cooled, the more water is recovered. At an ambient temperature of 15°C and 25% relative humidity, the kangaroo rat mentioned above would recover up to 88% of the water used to humidify the inhaled air. Moreover, in rodents in general, EWL is independent of ambient temperature below thermal neutrality, likely because of the influences of nasal mucosal cooling of expired air (Hinds & MacMillen 1985; MacMillen & Hinds 1983).

A dehydrated camel, however, not only cools exhaled air but also desaturates it. This is due to a **hygroscopic** (water-absorbent) layer of dried mucous and cellular debris that coats the nasal passages. This layer absorbs water rapidly during exhalation, and as a result, the exhaled air is dried. Schmidt-Nielsen (1981) estimated that if a camel in an air temperature of 28°C exhaled air that was at this same temperature but at only 75% relative humidity, it would lose by pulmonary evaporation only one-third as much water as it would be forming as a byproduct of metabolism. These estimates are not intended to be full accounts of the camel's water intake and loss, but they illustrate the effectiveness of the coupling of a heat exchange system and hygroscopic moisture exchange.

Lactation and Water Balance

The loss of water by a lactating female is substantial. For example, lactating female bats in arid western North America drink at desert water holes 13 times as often as non-reproductive adult females (Adams & Hayes 2008). Thus it is not surprising that some lactating mammals recycle such water. Cat breeders have long known that a lactating domestic cat stimulates urination and defecation in her young by licking their genital areas. She then ingests the urine and feces. This keeps the nest sanitary and recycles to the mother much of the water lost in the milk. Similar behavior occurs in several Australian murid rodents, the dingo (*Canis familiaris*), and two species of kangaroos (Baverstock & Green 1975). These species recovered about 30% of the water lost in milk. A laboratory study indicated that most of the water lost by female house mice (*Mus musculus*) during lactation, with the exception of that dissipated by EWL from the young, was recovered in a similar manner by the mother (Baverstock et al. 1979).

Recycling of water is far more important for desert dwellers than for mesic species. The importance of maternal ingestion of urine was studied by Oswald et al. (1993) in lactating females of mesic red-backed voles (*Myodes gapperi*), white-footed mice (*Peromyscus leucopus*) that occupy a variety of habitats, and two species of desert-dwelling gerbils (*Gerbillus*). For all these species, water requirements during lactation increased more than 100%, and all species recycled comparable absolute amounts of their young's urine. Although recycled water contributed only 3.2% to the total water budget for the mesic-adapted vole, this figure was 11.5% for the widespread white-footed mice and 39% for the xeric-adapted gerbils. In contrast to the other species, the gerbils' highly developed ability to minimize water loss in urine and feces results in a minimal water budget. Thus, recycled water is a major part of this budget during lactation. The milk of the desert-dwelling Merriam's kangaroo rat contains only 50% water, making it one of the most concentrated types of milk known and comparable to those of pinnipeds and cetaceans (Boness & Bowen 1996; Kooyman 1963). Behaviors that result in recycling water lost in milk are probably widespread among mammals that bear altricial young or have limited access to fresh water.

SUMMARY

Knowledge of mammalian physiology is essential to understanding how mammals adapt to their environments. Mammals are endothermic as well as homeothermic. The main difference between these terms is that endothermy refers to internal heat production, whereas homeothermy refers to the constancy of the heat produced. Most homeotherms maintain a high and fairly constant body temperature throughout life, but for this they pay an extremely high cost in energy. Consequently, some mammals become regionally or temporally heterothermic. Regional heterothermy occurs when core body temperature is maintained above that of limbs or other body regions, whereas temporal heterothermy occurs when

mammals modify their body temperature in response to fluctuations in daily or seasonal environmental temperatures.

Endothermy probably evolved in synapsids in the late Triassic under selective pressure for sustained activity and reduced body size. The primary advantage of endothermy is the ability to sustain high levels of activity under an imposing array of temperature extremes. Therefore, endotherms are free to exploit nocturnal niches and many are active through all seasons.

Every endotherm has a thermal neutral zone within which little or no metabolic energy is expended on temperature regulation. Within this zone, the fluffing or compressing of the fur, local vascular changes, or shifts in posture suffice to maintain thermal homeostasis. Outside this zone, metabolic costs increase to either lower or raise body temperature.

Mammals inhabiting cold climates can (1) over many generations, evolve large body size, resulting in a more favorable surface-area-to-volume ratio and consequently reduced heat loss; (2) decrease their rate of heat loss through increased insulation or behavioral thermoregulation; (3) increase their rate of metabolic heat production; or (4) abandon their normal body temperature and allow it to decline to a level closer to ambient temperatures (hypothermia). Obviously, the optimal strategy will depend on the environmental conditions and physiological constraints of the individual at any given time.

At very low ambient temperatures, the costs of endothermy may be unsupportably high. To offset these costs, some cold-adapted mammals practice regional heterothermy: extremities, such as legs and ears, which are poorly insulated and dissipate heat rapidly, are allowed to become cool, thereby reducing heat loss by minimizing the temperature differential between these body parts and the environment.

Temporal hypothermia refers to responses that save energy by temporarily abandoning homeothermy. At one end of the continuum is daily torpor, in which body temperature, metabolic rate, respiration, and heart rate are lowered for a portion of the 24-hour daily cycle. When energy savings from daily torpor are insufficient and migration to a more favorable climate is not an option, some species enter prolonged periods of torpor lasting from a few days to several months. These extended periods of hypothermia are typically seasonal and are referred to as hibernation (in response to cold) or estivation (in response to heat).

The length and degree of seasonal torpor vary widely among mammals. Seasonal hibernators frequently exhibit a series of hibernation bouts over the course of the entire hibernation season. Each bout terminates in an arousal phase followed by a 1- to 2-day period of normal body temperature (euthermic period). These arousal periods are short, but energetically costly. Arousal is associated with the metabolism of energy-rich brown adipose tissue in some species and with shivering in others.

Mammals inhabiting hot regions also face severe thermoregulatory problems. In many deserts, daytime air temperatures in the summer rise well above the body temperature of most mammals. Consequently, heat from the environment is absorbed while the mammals are producing considerable metabolic heat of their own. In order to maintain thermal homeostasis, these animals must avoid the absorption of heat from the environment, dissipate such heat as it is absorbed, and lose endogenous heat.

For passive heat loss to occur, the body surface temperature must exceed the ambient temperature. Mechanisms for dumping excess heat include vasodilation of naked skin surfaces, such as the scrotum, wing membranes, and ears, to dissipate heat. Other mechanisms include fanning of the wings in bats, panting, and evaporative cooling. Evaporative cooling is common, but it is a luxury that most desert mammals cannot afford because they must also conserve water. Some mammals have evolved physiological, anatomical, and behavioral strategies for tolerating long exposure to heat. Behavioral thermoregulation is a conspicuous part of the daily routine of several mammals with energy-poor or seasonally restricted diets.

Fossorial life in sealed burrows offers the advantages of stable (and often moderate) temperatures, high humidity, and safety from predators, but there are major liabilities, such as the high energetic cost of burrowing. Burrowing a given distance consumes 360 to 3,400 times as much energy as moving the same distance on the surface. Microclimate factors may also present physiological challenges. Carbon dioxide levels in burrows can be ten times atmospheric levels. Moreover, dispersed or scarce food supply may pose added problems. Fossorial mammals have behavioral and physiological strategies for surviving underground. Such a "subterranean syndrome" includes low basal metabolic rate, high and extremely narrow thermoneutral zone, high thermal conductance, and limited thermoregulatory ability.

Aquatic lifestyles also present physiological challenges for mammals. The rate of heat loss by a mammal in water is 10 to 100 times greater than the rate of loss in air of the same temperature. Arctic and Antarctic waters are near 0°C year round, and high-latitude lakes and rivers approach this temperature in winter. Consequently, a temperature differential of about 35°C between deep-body temperature and ambient temperature is common in mammals swimming in these waters. Yet cold waters are permanently inhabited by some cetaceans, and pinnipeds spend much of their lives in such waters. Some form of insulation is critical in these environments. Air trapped in the fur serves as insulation for many mammals, but others (cetaceans, phocid seals, and walruses) have a thick layer of subcutaneous blubber that forms an insulating envelope around the deep, vital parts of the body. The ability of cetaceans to keep warm also results from the large body size. Because of these differences, the whale has approximately a 100-fold advantage over the small porpoise in its ability to keep warm. Second, blood flow through the well-developed vascular system in the flippers, dorsal fin, and flukes of cetaceans allows these structures to function effectively as heat dissipators under conditions of heat stress. The flow can apparently be shut down during cold stress, allowing for a minimum of heat loss from these surfaces. Third, a remarkable series of vascular specializations allows for great variations in the thermal resistance offered by the blubber. A system of countercurrent heat exchangers in the vascular network supplying the blubber minimizes heat loss to the blubber and skin, and hence to the environment.

Water is absolutely essential to all mammals. Survival depends on the maintenance of an internal water balance within fairly narrow limits. Water balance results when water intake, through drinking, eating, and production of metabolic water, equals water loss by evaporation from skin and lungs, defecation, and urination. Mammals employ a number of solutions to the problem of maintaining water balance. These solutions depend on seasonal weather patterns, size of the animal, timing of activity cycles, diet, and a variety of behavioral, structural, and physiological features.

In many arid or semiarid regions, scattered water holes or widely separated rivers offer water to mammals that can move long distances. Most large mammals in such areas are periodic drinkers, drinking every day or two in hot weather. Although they can go for moderate periods without drinking, these mammals are not independent of drinking water. A number of mammals that occupy deserts or semiarid areas do not have regular access to drinking water and satisfy their water requirements by eating moist food (dietary moisture). Other species, including some desert rodents, must survive for extended periods without access to preformed water (water already in the form of H_2O). They appear to do so by forming water as a byproduct of chemical reactions, such as the oxidation of starches from the dry seeds they consume (i.e., forming metabolic water).

The routes of water intake from drinking, dietary water, and metabolic water must exceed water loss by lactation, defecation, urination, and pulmocutaneous evaporation. Water lost in lactation is important periodically to females. Some lactating mammals lick their young to stimulate urination and defecation, then ingest the urine and feces to recover much of the water lost in the milk. In other desert-adapted mammals, water loss is typically reduced by concentrating the urine and by absorbing additional water from fecal material. The primary channel for water loss in many desert mammals is pulmocutaneous evaporation. Some, such as kangaroo rats, exhale air that is cooler than body temperature by using their nasal passages as heat exchange systems, resulting in a significant reduction of pulmonary water loss.

KEY TERMS

Basal metabolic rate
Behavioral thermoregulation
Carotid rete
Countercurrent heat exchange
Cutaneous evaporation
Ectothermic
Endothermic
Evaporative cooling
Halophytic
Homeotherm

Hygroscopic
Hyperthermia
Hypothermia
Interstitial fluid
Lower critical temperature
Mesic
Metabolic water
Neonate
Normothermia
Panting

Pulmocutaneous evaporation
Regional heterothermy
Streamlined
Thermal conductance
Thermal inertia
Thermal neutral zone
Thermogenesis
Upper critical temperature
Vasoconstriction
Water balance

RECOMMENDED READINGS

Degen, AA. 1996. *Ecophysiology of Small Desert Mammals.* Springer-Verlag, Berlin.

French, AR. 1993. Physiological ecology of the Heteromyidae: economics of energy and water utilization. Pp 509-538, in *Biology of the Heteromyidae* (HH Genoways & JH Brown, eds.) Special Publication No. 10, American Society of Mammalogists.

Kanwisher, J & G Sundnes. 1966. Thermal regulation in cetaceans. Pp 397-409, in *Whales, Dolphins and Porpoises* (KS Norris, ed.). University of California Press, Berkeley, CA.

Kenagy, GJ. 1973. Daily and seasonal patterns of activity and energetics in a heteromyid rodent community. *Ecology,* 54:1201–1219.

MacMillen, RE. 1983. The adaptive physiology of heteromyid rodents. *Great Basin Naturalist, 7*:65–76.

Marshall, PT & GM Hughes. 1980. *Physiology of Mammals and Other Vertebrates*, 2nd ed. Cambridge University Press, Cambridge, UK.

McNab, BK. 2002. *The Physiological Ecology of Vertebrates: A View From Energetics.* Cornell University Press, Ithaca, NY.

Tomasi, TE & TH Horton. 1992. *Mammalian Energetics: Interdisciplinary Views of Metabolism and Reproduction.* Cornell University Press, Ithaca, NY.

Withers, PC. 1992. *Comparative Animal Physiology.* Saunders College Publishing, Philadelphia, PA.

Echolocation

Animals that echolocate use echoes of sounds they produce to locate objects in their path. Because the mammals most familiar to us depend largely on vision for perceiving their environment, it is surprising to note that at least 18% of the known species of mammals (1,014 species of bats and cetaceans out of a total of 5,416 mammals) probably use echolocation as their primary means, or at least as important secondary means, of "viewing" their surroundings. Most bats, some members of the orders Soricomorpha and Afrosoricida, and probably all odontocete cetaceans echolocate. Future research may demonstrate that the use of echolocation among mammals is even more widespread.

An accurate picture of the bat's use of acoustical orientation was long in emerging. As early as 1793, Lazzaro Spallanzani performed experiments which suggested that bats use acoustical rather than visual perception when avoiding obstacles and when feeding. A few years later (1798), Louis Jurine showed that bats could not avoid obstacles if their ear canals were plugged with wax, suggesting that hearing played an important role in bat orientation and prey capture. Not until the early 1940s, however, was the use of echolocation by bats conclusively demonstrated by the careful laboratory experiments of Griffin and Galambos (1940, 1941) and by the observations of Dijkgraaf (1943, 1946). Continued research, aided by electronic equipment, has contributed to our present detailed, if incomplete, knowledge of echolocation. Detailed reviews of many aspects of echolocation in bats and cetaceans can be found in Thomas et al. (2004).

Introducing Echolocation

The Realities of Echolocation Sounds

Mammals produce two types of echolocation calls. Odontocetes produce clicks in the nasal passage, and two species of bats make clicks with the tongue, whereas the signals of the vast majority of echolocating bats and shrews are made in the larynx (Novick 1955). The vocal signals of bats are more intense than clicks, a factor of key importance because the more intense the signal, the greater the echolocation range and thus the more time to react to prey or to obstacles. Echolocation calls can be described in terms of time (duration, repetition rate), frequency (pitch), and intensity (an expression of signal strength). The following discussions deal primarily with the signals of bats.

Echolocation calls are brief pulses of sound that vary in duration from about 100 milliseconds (a millisecond is 1/1000 of a second) to 0.25 milliseconds. Some bats (such as vespertilionids) typically give short signals; others (such as rhinolophids) emit longer signals (**Fig. 22-1**; see also Chapter 15). An individual changes the duration and the rate (number per unit of time) of its calls during a search and capture sequence (**Fig. 22-2**).

The **frequency** (pitch) of echolocation calls varies widely among species and seems to be controlled by varying tension on the vocal cords (Suthers & Fattu 1982). The echolocation signals of most bats are **ultrasonic**, that is to say, above the range of human hearing, which reaches about 20 kilohertz (a kilohertz [kHz] equals 1000 cycles per second). Most bats emit echolocation signals within the range of about 12 to 100 kilohertz, but some use frequencies above 200 kilohertz. Although conventional wisdom holds that echolocation is based on ultrasonics, a number of bats use echolocation signals that are below 20 kilohertz and are therefore audible to many humans (Fenton 1982; Fenton & Griffin 1997; Rydell & Arlettaz 1994). These bats represent three families and were observed at such widely scattered places as Arizona, British Columbia, Colombia, Europe, and Africa. Odontocete cetaceans also use echolocation signals that are audible to humans. Therefore, echolocation clearly does not depend on ultrasonics. Many echolocation calls are complex, consisting of a **fundamental frequency**

Yinpterochiroptera

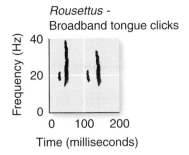

Rousettus -
Broadband tongue clicks

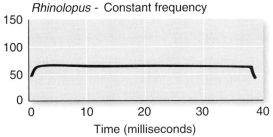

Rhinopoma - Narrowband, multiharmonic

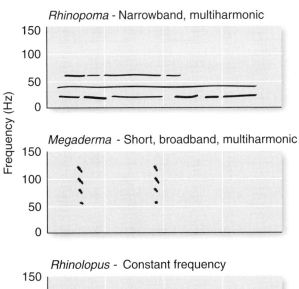

Megaderma - Short, broadband, multiharmonic

Rhinolopus - Constant frequency

Yangochiroptera

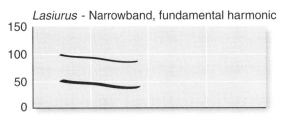

Lasiurus - Narrowband, fundamental harmonic

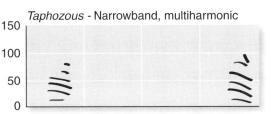

Taphozous - Narrowband, multiharmonic

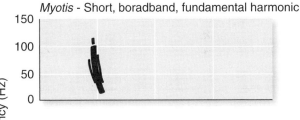

Myotis - Short, boradband, fundamental harmonic

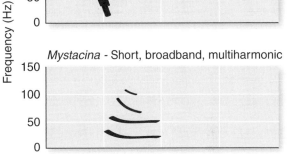

Mystacina - Short, broadband, multiharmonic

Myzopoda - Long, broadband, multiharmonic

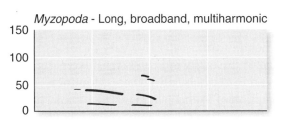

Pteronotus - Constant frequency

FIGURE 22-1 Duration and pattern of frequency change through time of characteristic types of echolocation calls from members of the Yinpterochiroptera (Pteropodidae, Rhinopomatidae, Megadermatidae, and Rhinolophidae) and Yangochiroptera (including the remaining 13 families). Among the Pteropodidae only *Rousettus* echolocates (using tongue clicks). The diversity of call types is greater among members of the Yangochiroptera. (Adapted from Jones, G., and Teeling, E. C., *Trends Ecol. Evol.* 21 (2006): 149–156.)

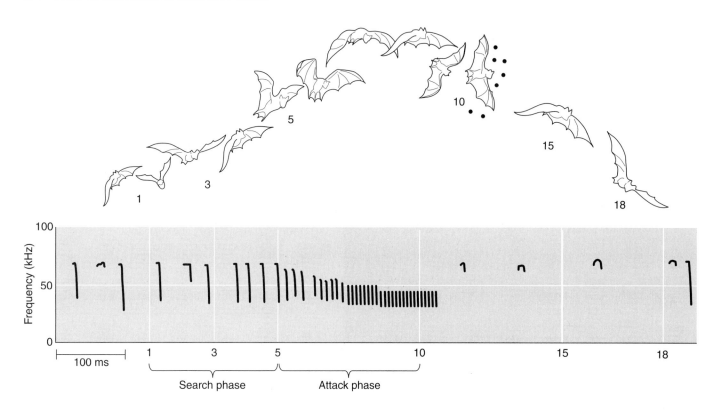

FIGURE 22-2 (A) Drawings of 18 stages in the flight path of *Noctilio albiventris* during an attack on a flying insect. The insect is denoted by a small dot. The bat attempts to capture the insect at stage 10. (B) The corresponding sequence of echolocation pulses (sonograms) recorded as frequency versus time during the attack. The numbers correspond to the stages illustrated in (A). Notice that the characteristics of the echolocation pulses changed as the bat switched from search phase to attack phase at stage 5. The pulse duration and pulse intervals are reduced during the attack phase. (Modified from Kalko, E. K., et al., *Behav. Ecol. Sociobiol.* 42 (1998): 305-319. With kind permission of Springer Science & Business Media.)

(the lowest, or root, tone of a chord) and several **harmonics** (frequencies that are integral multiples of the fundamental frequency).

Echolocation calls also vary in intensity of sound (**Fig. 22-3**). Intensity indicates signal strength, usually expressed in decibels (dB), whereas loudness refers to our perception of sound. The signal of a smoke detector and the echolocation call of a little brown bat have the same intensity (110 decibels measured at 10 centimeters from the source), but to us the smoke detector is loud and we hear no sound from the bat. "Whispering bats" and "loud bats" were recognized by Griffin (1958), and other bats have signals of intermediate intensity. Using broadband microphones, Simmons et al. (1979) found that whereas the intense signals of some bats could be detected at 30 meters or more, those of others were only detectable at less than 0.5 meters.

Among different species of bats, echolocation signals differ in **bandwidth** (breadth of frequencies produced) and in the information they provide. **Narrowband** calls (those that span less than 10 kilohertz) are often designated as

constant frequency (CF) signals, in contrast to **broadband, frequency-modulated** (FM) calls (those that span more than 10 kilohertz). Narrowband signals are useful for detecting a target, but do not provide details as to its position. Many bats use such narrowband calls (search-phase calls) when searching for prey. By increasing the bandwidth of their signals, bats increase the precision with which they can pinpoint a target. Some broadband calls are shallow, covering a narrow range of frequencies relatively slowly, and others are steep, covering a wider range of frequencies more rapidly (Fig. 22-1). Although the shallow broadband (FM) calls are excellent for target detection and are used as search-phase signals by many bats, such calls are sensitive to distortion by **Doppler shift** and do not provide precise information on target location. By contrast, steep broadband (FM) signals provide information for accurately localizing a target and are used when a bat attacks its prey (the attack phase). Typically, a bat detects prey with narrowband (CF) or shallow broadband (shallow FM) calls, but switches to steep broadband (steep FM) calls during the last instant of the attack (Fig. 22-2).

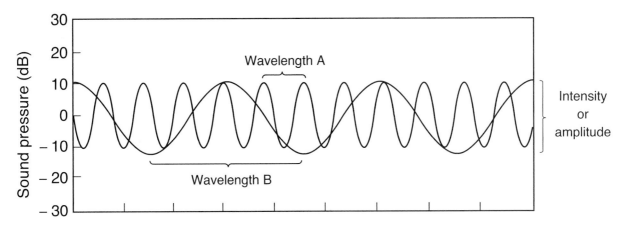

FIGURE 22-3 The details of a sound wave: (A) a high-frequency (short-wavelength) sound; (B) A low-frequency (long-wavelength), sound. The frequency of a sound is the number of cycles per unit time. The sound in part A completes approximately 12 cycles in the same time it takes the sound in part B to complete 3 cycles.

Range Limits

Mention of the "last instant" introduces the realities of time and distance. Compared with vision, echolocation in air is an extremely short-range system. Flying bats must react rapidly to echoes from objects that are close and rapidly getting closer. Spheres 10 millimeters in diameter were first detected by echolocating *Eptesicus fuscus* (a bat with intense, long-range calls) at 5 meters (Kick 1982). Using this figure and the observed foraging-flight speed for *E. fuscus* of 3.5 meters per second, Fenton (1990) estimates that these targets were first detected only 1.5 seconds in front of the bat. If the target is an edible flying insect, then this bat has only 1.5 seconds to locate the insect precisely, to track its trajectory, and to maneuver for the capture. Novick (1970) found that the pursuit and capture of an insect take a mustached bat (*Pteronotus parnellii*, **Fig. 22-4**) only 0.25 to 0.33 seconds. **Table 22-1** shows the estimated maximum ranges of target detection for some bats. Most insectivorous bats are small, have low wing loadings, fly at slow or moderate speeds, and are highly maneuverable, features that accord with the demands of short-range perception.

Echolocation in Water and Air

Because water and air differ drastically in density, they transmit sound differently. Sound travels more than four times faster in water than in air (1541 meters per second versus 361 meters per second). The intensity of a given signal is greater in water than in air, and sound is **attenuated** (reduced in intensity) less rapidly in water. Sound waves in air spread from a source and rapidly lose intensity according to the inverse square law (Lawrence & Simmons 1982; see **Fig. 22-5**). Relative to an echolocating mammal in air, then, one in water receives information from echoes much faster, uses less energy to produce a signal of given intensity, and transmits a signal farther. An additional complication

for terrestrial echolocators is that temperature and humidity strongly influence the ability of air to transmit sound (Harris 1996; Knudsen 1931). Because of the relatively poor sound transmission qualities of air, high-intensity tonal signals are important for bats. The less intense clicks of odontocetes are adequate because of the superior sound-transmission qualities of water.

For an aquatic mammal, echolocation can provide long-range information. Goold and Jones (1995) estimate that the maximum echolocation range of the sperm whale (*Physeter catodon*) was at least 1500 meters. But for terrestrial mammals, echolocation yields only short-range information. Fenton (1990) estimates maximum range of target

FIGURE 22-4 Closeup of the face of a mustached bat (*Pteronotus parnellii*), showing the lips formed like a megaphone during echolocation.

TABLE 22-1 **Estimated Maximum Ranges of Target Detection for Some Bats**

	Interpulse Interval (ms)	Range[a] (m)	Source
Rhinopoma hardwickei	100	17.0	Simmons et al. 1984
Taphozous mauritianus	110	18.7	Fenton et al. 1980
Cormura brevirostris	81	13.8	Barclay 1983
Saccopteryx bilineata	53	9.0	Barclay 1983
Nycteris grandis	20	2.4	Fenton et al. 1983
Nycteris thebaica	22	3.7	Fenton et al. 1983
Noctilio leporinus	100	17.0	Suthers and Fattu 1973
Trachops cirrhosus	24	4.1	Barclay et al. 1981
Myotis adversus	90	15.3	Thompson and Fenton 1982
Myotis daubentoni	15	2.6	Jones and Rayner 1988
Lasionycteris noctivagans	167	28.4	Barclay 1986
Eptesicus fuscus	100	17.0	Simmons et al. 1979
Chalinolobus variegatus	128	21.2	Obrist 1989
Lasiurus cinereus	303	51.5	Barclay 1986
Euderma maculatum	365	62.1	Leonard and Fenton 1984

[a] The estimates are based on intervals between signals, assuming that echolocating bats do not tolerate overlap between the echo from one signal and the next signal.
Source: Reproduced from Fenton, M.B. 1990. The foraging behaviour and ecology of animal-eating bats. *Canadian Journal of Zoology*, 68:411–422. © 2008 NRC Canada or its licensors. Reproduced with permission.

detection for 16 species of bats (Table 22-1): The greatest range was 62 meters. The mean was only 19 meters, and the range for some bats was only about 2.5 meters.

High-Frequency Sound

Because high frequencies are more rapidly attenuated in air than are low frequencies (Fig. 22-5), one might wonder why bats typically use high-frequency echolocation calls. Perhaps of prime importance is the relationship between prey size and the wavelength of echolocation pulses. The higher the frequency of a sound, the shorter is its wavelength. Frequencies of roughly 30 kilohertz have a wavelength of approximately 11.5 millimeters, roughly the size of a small moth; this balance between prey size and wavelength is ideal because objects approximately the size of a given wavelength reflect that wavelength particularly well. Low-frequency sounds have long wavelengths and tend to bend around small objects without producing an echo. Some species of bats can detect wires with a diameter as small as 0.08 millimeters (1/30 of a wavelength), but in general, the wavelengths of the pulses emitted by bats are in the range that is most efficient for the detection of small to medium-sized insects.

Frequency also influences directionality of hearing. All echolocating bats depend on accurately localizing the source of sound (echoes). In all bats studied by Obrist et al. (1993), hearing became increasingly directional with increasing frequency of echolocation calls.

Self-Deafening

As a protection for the inner ear against outgoing echolocation sounds, bats have several "self-deafening" adaptations. In bats, as in all mammals, sound waves (vibrations in air) strike the tympanic membrane, where they are converted to mechanical vibrations that are amplified and transmitted by the ear ossicles to the oval window of the inner ear. Here vibrations are set up in the fluid in the inner ear, are transmitted to the basilar membrane, and are converted to nerve impulses that pass to the brain. Two muscles of the mammalian middle ear dampen the ability of the ossicles to transmit vibrations when an individual is subjected to unusually loud sounds or when it is vocalizing. These muscles—the tensor tympani, which changes the tension on the tympanic membrane, and the stapedius, which changes the angle at which the stapes contacts the oval window—are extremely well-developed in bats, and their contractions reduce the bat's sensitivity to its own pulses (**Fig. 22-6**). Jen and Suga (1976), using electronic equipment, found that action potentials of the cricothyroid laryngeal muscles were followed 3 milliseconds later by action potentials of the middle ear muscles. This coordination of laryngeal and middle ear muscles ensures that the latter muscles contract just prior to vocalization and attenuate (weakens in intensity) the auditory self-stimulation by 25% (Suga & Shimozawa 1974). In Brazilian free-tailed bats (*Tadarida brasiliensis*), the stapedius muscle contracts approximately 10 milliseconds before pulse generation, pulling the stapes away from the

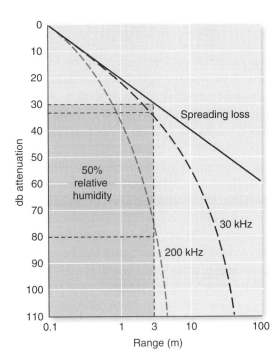

FIGURE 22-5   Rate of attenuation of echolocation calls in air. The dotted lines show the attenuation of calls at 30 kilohertz and 200 kilohertz. Echoes from a target 3 meters in front of an echolocating bat are weaker by 30 decibels (spreading loss line). Because bats listen for the echo, sounds must make a roundtrip of 6 meters and are attenuated by nearly 60 decibels. In addition to spreading losses, the atmosphere also absorbs sound (curved lines) and thereby adds to the overall attenuation. Therefore, a bat 3 meters from a target echolocating at 30 kilohertz will perceive an echo that is 64 decibels weaker than when the sound was emitted (60 decibels for roundtrip spreading loss and 4 decibels for atmospheric attenuation). A bat echolocating at 200 kilohertz will perceive a total attenuation of about 110 decibels (with atmospheric losses accounting for 50 decibels) at 50% relative humidity (RH). (Reprinted with permission from Lawrence, B. D., and Simmons, J. A., *J. Acoust. Soc. Am.* 71 (1982): 585–590. Copyright 1982, American Institute of Physics.)

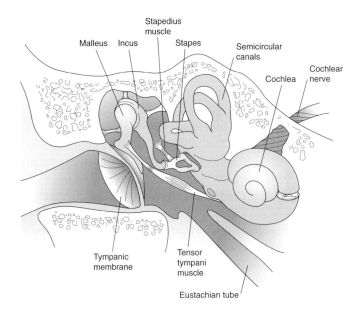

FIGURE 22-6   Cutaway drawing of the middle and inner ear of a typical mammal. Sound waves pass down the external auditory canal to the tympanic membrane. Vibrations in the tympanic membrane are amplified and transferred to the oval window of the cochlea by the three middle ear ossicles. Note the two middle ear muscles, tensor tympani and stapedius, which facilitate self-deafening in bats .

oval window and partly preventing emitted sounds from reaching the inner ear. A few milliseconds later, the stapedius relaxes, re-engaging the stapes with the oval window to allow detection of the echo (Hill & Smith 1992).

Additional neural attenuation of the direct reception of echolocation signals occurs in the brain. Nerve impulses arising from direct reception and passing from the cochlea to the inferior colliculus of the brain are attenuated by the neurons of the lateral lemniscus of the brain. This change, plus that effected by the middle ear muscles, attenuates the neural events by 40%. Suga and & Shimozawa (1974) suggest that similar attenuating mechanisms occur in humans and keep the sounds of our own speech from becoming disturbingly loud. Direct reception of echolocation calls in bats is doubtless reduced also by the beaming of sounds by the lips and the complex noses and noseleaves.

As additional structural refinement in bats, the bones housing the middle and inner ear are insulated from the rest of the skull. This bony otic capsule does not contact other bones of the skull (**Fig. 22-7**) and is insulated from the skull by blood-filled sinuses or fatty tissue. During the emission of signals, the conduction of sound from the larynx and the respiratory passages through the bones of the skull is thus greatly reduced.

Oral and Nasal Pulse Emission

Echolocation signals of bats are emitted through either the mouth (oral emitters) or the nose (nasal emitters). Most bats are oral emitters; only members of the Nycteridae, Megadermatidae, Rhinolophidae, and Phyllostomidae are

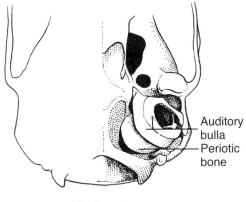

Myotis volans

FIGURE 22-7   Ventral view of the posterior part of the skull of a vespertilionid bat, showing that the periotic bone and auditory bulla are not directly attached to the skull.

nasal emitters. A basic dichotomy in skull shape is associated with these two modes of sound emission (Freeman 1984; Pedersen 1998). The skulls of oral-emitting bats are similar to those of most terrestrial mammals, but in nasal-emitting bats the rostral part of the skull is rotated ventrally, below the level of the braincase. This configuration results in the alignment of the nasal cavity, instead of the oral cavity, with the direction of flight (**Fig. 22-8**). Oral emitters keep their mouths open when echolocating and typically produce high-intensity pulses (Fig. 22-4). Nasal emitters, on the other hand, usually keep their mouths closed when echolocating; some (Rhinolophidae) give high-intensity pulses, whereas others produce relatively low-intensity signals ("whispering bats"; Griffin 1958). Viewed broadly, bats that emit high-intensity signals, whether nasal or oral emitters, usually catch flying insects, whereas those nasal emitters that have low-intensity calls feed on fruit, nectar, small animals taken from the ground or from vegetation, or combinations of these foods.

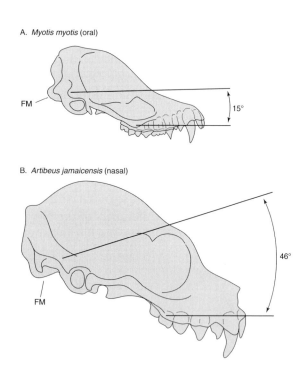

A. *Myotis myotis* (oral)

FM

15°

B. *Artibeus jamaicensis* (nasal)

46°

FM

FIGURE 22-8 Comparison of the cranial characteristics in (A) *Myotis myotis*, an oral emitter, and (B) *Artibeus jamaicensis*, a nasal emitter. The angle between the plane of the semicircular canals in the ear and the palate is much greater in nasal emitters. The foramen magnum (FM) opens ventrally, resulting in a head posture during flight that, coupled with the displacement of the rostrum below the level of the basicranium, aligns the nasal area, rather than the mouth, with the direction of flight in nasal emitters such as *A. jamaicensis*. (Adapted from Freeman, P. W., *Biol. J. Linn. Soc. Lond.* 21 (1984): 387–408 and Pedersen, S. C., *J. Mammal* 79 (1998): 91–103.)

Two Contrasting Approaches to Echolocation

Most bats avoid self-deafening resulting from signal–echo overlap by separating the signal and its echo in time. They produce signals a small percentage of the overall time and do not broadcast signals and receive echoes at the same time. This is the **low duty cycle** approach to echolocation. Duty cycle refers to the percentage of time that a signal is produced: A low duty cycle bat emits signals less than 20% of the time, less than 200 milliseconds out of 1000 milliseconds. Because the time between a signal and its echo becomes progressively shorter as a bat approaches its target, a bat must drastically shorten its signals during an attack sequence to continue to avoid signal–echo overlap. During the final pursuit of an insect, a low duty cycle bat not only shortens the signals but also increases the repetition rate (Fig. 22-2), which probably enables the bat to track the final-instant movements of the prey (Fenton & Bell 1979). This drastic increase in signal rate produces the "feeding buzz."

A fundamentally different echolocation strategy enables some bats to detect the fluttering wings of insects and to track precisely the trajectories of prey by perceiving Doppler-shifted echoes (the change in frequency of a sound produced by a moving object). Echolocation in these bats—*Pteronotus parnellii* (Fig. 22-4), a mormoopid, and probably all species of the Rhinolophidae—is characterized by **high duty cycles** (greater than 60%). That is to say, these bats emit signals for more than 600 milliseconds out of 1000 milliseconds. These bats can produce pulses and receive echoes at the same time. The signals are dominated by long, constant-frequency components, and the bats separate signal from echo by using frequency (Doppler-shift compensation). These bats perceive Doppler shifts in the echoes from targets approaching or moving away (Simmons 1974) and shifts generated by fluttering insect wings in echoes of the constant-frequency part of the echolocation signal (Schnitzler 1987; Schnitzler et al. 1983). Some flutter-detecting bats studied by Schnitzler (1987) could discriminate between the wing-beat signatures of different insects (**Fig. 22-9**). They thus minimize self-deafening during signal emission by tuning the inner ear to different frequencies than those of emitted signals. In *Rhinolophus ferrumequinum*, for example, the organ of Corti, which transduces vibrations into neural impulses to the brain, is most sensitive to frequencies just above and below those of the outgoing signal (Vater 1987, 1998). This style of echolocation, seemingly present as early as the Eocene (Habersetzer & Storch 1987), enables bats to detect fluttering insects against a complex but nonfluttering background (Bell & Fenton 1984; Vater 1987). The reliance on flutter detection by high duty cycle bats is suggested by observations in Italy of foraging *R. ferrumequinum* that only

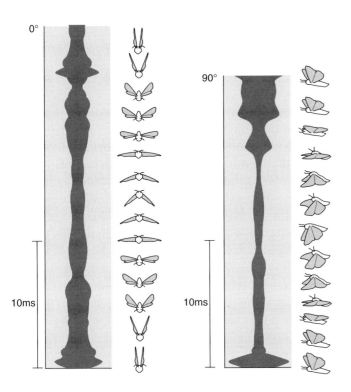

FIGURE 22-9 Changes in echo intensity of a CF pulse during the wing-beat cycle of a flying moth. The bat can use the amplitude changes to determine the wing-beat frequency of the prey, as well as its orientation with respect to the bat. Large, high-intensity glints are produced when the moth's wing surface area is greatest. (Adapted from Schnitzler et al. 1983. With kind permission of Springer Science & Business Media.)

1995; Watkins 1977). But the importance of echolocation to odontocetes and its specific use during foraging are poorly known.

Regarding the importance of echolocation to bats, however, there is little uncertainty. Echolocation plays a central role in the lives of all non-pteropodid bats that have been studied. The foraging strategies of bats that eat flying insects are built around the ability to echolocate. Echolocation in these bats has become highly specialized, allowing them not only to orient themselves and to locate obstacles but to detect, assess (as to size, wing-beat frequency, and so on), and track flying insects. These bats can discriminate between types of insects (e.g., moth versus beetle) and can track flying insects against a close and complex background of vegetation. Some bats that do not concentrate on flying insects but glean insects or small vertebrates from vegetation can perceive details of the texture of prey (Schmidt 1988).

Some insectivorous bats may use other means than echolocation for locating prey. Sounds made by their prey are used by some bats, and some bats may use smells as cues (Fiedler 1979; Kolb 1976; Rentz 1975; Thies et al. 1998). Little is known about the use of vision in prey detection by insectivorous bats. Bell (1982a) found that the California leaf-nosed bat (*Macrotus californicus*) can use vision to find prey, and the African yellow-winged bat (*Lavia frons*) seems to use vision to detect large, high-flying insects silhouetted against a twilight sky (Vaughan & Vaughan 1986).

attacked insects whose wings were beating (Griffin & Simmons 1974) and by comparable observations of *P. parnellii* in the laboratory (Goldman & Henson 1977).

The Role of Echolocation in the Lives of Mammals

Echolocation in land mammals probably evolved from communication calls and first became a means of detecting obstacles in dim or unpredictable lighting (Gould 1970, 1971). This seemingly remains the sole function of echolocation in shrews. Tactile and olfactory cues and sounds made by prey are used by shrews for the recognition and tracking of prey. Odontocetes clearly use echolocation for general obstacle avoidance, and some may use echolocation for locating and tracking prey. The periodically increased click rates of deep-diving sperm whales are probably associated with the pursuit of prey (Goold & Jones

Echolocation Performance

One aspect of echolocation performance in air is its strictly short-range usefulness. Another aspect is acuity, considered here as the smallest detectable differences in location, size, and texture of a target. Simmons (1973) tested the range-detecting ability of one species each from the families Vespertilionidae, Phyllostomidae, Mormoopidae, and Rhinolophidae. All of the bats could discriminate range differences of from 1 to 2 centimeters, and this remarkable performance was apparently achieved by cross-correlation of the transmitted pulse with the returning echo. The essential variable from which the bats estimate distance is the time it takes for a pulse to reach a target and the echo to return. By comparing the relative **target-ranging** abilities of the bats in relationship to the bandwidths of their echolocation pulses, Simmons determined that the FM components of the pulses are used for target ranging. Laboratory studies on a limited number of species have further demonstrated the ability to discriminate minor differences in target size, shape, and direction (Pollak & Casseday 1989). In the laboratory, *E. fuscus* detected differences of less than 1 millimeter in the depth of small holes in a target (Simmons et al. 1974). The holes modified the spectrum of the echo from the target by absorbing sound energy at certain frequencies in the bandwidth of the bat's FM sweep; with changes in hole depth the absorption peaks shifted to different frequencies. Schmidt (1988) did laboratory experiments on texture discrimination by echolocating *Megaderma lyra*, an Asian bat that gleans prey from vegetation or the ground. This bat distinguished depth differences of only 0.2 millimeter, an ability seemingly based on the perception of frequency modifications in the spectra of echoes. This ability would be extremely useful in detecting differences in textures of mice or frogs, common prey of this bat. Apparently, bats associate features of the echo spectrum with target shape and texture when selecting prey.

Bats that use the high-duty-cycle approach to echolocation (*P. parnellii* and rhinolophids) precisely track the flight trajectories of insects. The longer the CF component of the pulse, the greater is the sensitivity to target velocities. Computed velocity resolutions listed by Simmons et al. (1975) indicate that the European *R. ferrumequinum*, with its very long narrowband signal (up to 60 milliseconds), can perceive relative target velocities of less than 0.04 meters per second; *P. parnellii*, with its fairly long signals (up to 28 milliseconds) can detect relative target velocities as low as 0.10 meters per second. These bats concentrate on the trajectory of prey while remaining aware of clutter (echoes returning from objects other than the target of interest). The ability to track prey with precision enables some high

duty cycle bats to intercept rather than pursue prey, thus reducing energy expended on foraging.

Faces and Ears and Echolocation

The strange faces of bats, always a source of amazement to those unfamiliar with these animals, have an important function in connection with echolocation. When mormoopids are echolocating, their lips are formed into megaphones that seemingly focus echolocation calls (Fig. 22-4). Möhres (1953) showed that the complex horseshoe-like structure surrounding the nostrils of rhinolophids serves as a diminutive megaphone to focus the short-wavelength pulses emitted by these bats into a beam; the 80- to 100-kilohertz pulses have wavelengths of only 3 or 4 millimeters. In addition, because the nostrils are situated almost exactly 0.5 wavelength apart, the pulses emitted through the nostrils undergo interference and reinforcement that tends to beam the pulses (see also Pye 1988; Schnitzler & Grinnell 1977). Zhuang and Muller (2006) showed that the rhinolophid noseleaf furrows act as resonance cavities to shape the outgoing beam. Beam shaping may allow rhinolophids to maintain an appropriate height and avoid collisions as they cruise low above the ground. Prominent noseleaves occur in five chiropteran families: Rhinolophidae, Hipposideridae, Nycteridae, Megadermatidae, and Phyllostomidae (**Fig. 22-10**). In nasal emitters, noseleaves enhance the directionality of sound emissions (Hartley & Suthers 1987). The facial patterns of many bats may well function similarly to direct pulses such that some species can scan their

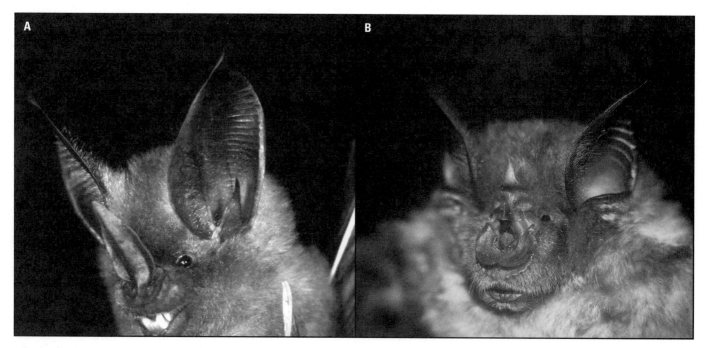

FIGURE 22-10   (A) The large, spear-shaped noseleaf of a *Mimon bennettii*, Phyllostomidae; (B) The distinctive noseleaf of a horseshoe bat, *Rhinolophus ferrumequinum*, Rhinolophidae.

surroundings with a concentrated beam of sound, much as we probe the darkness with a flashlight beam. Arita (1990) surveyed noseleaf morphology of 46 phyllostomid species and hypothesizes that noseleaf morphology is correlated with foraging behavior. Bogdanowicz et al. (1997) showed that noseleaf morphology correlates with diet (fruit versus nectar) but not with the type of echolocation calls.

The tragus (Fig. 22-10A), an often bladelike structure that is a prominent part of the external ear in most echolocating bats (but lacking in the Rhinolophidae), plays an important role in localizing sound (Lawrence & Simmons 1982). Laboratory experiments using *E. fuscus* demonstrated that sounds reflected from the pinna to the tragus produce a secondary echo of sounds entering the ear canal. The time between this echo and the sound directly entering the ear canal encodes the vertical direction of a sound. Bats with the tragus intact could perceive changes of 3° of arc in vertical angles separating targets; discrimination in bats with the tragus folded down was reduced to 12° to 14°.

Obrist et al. (1993), who studied the external ears of 47 species of bats, found that, in general, the pinnae serve to focus and amplify sound and to increase the directionality of hearing. In many species the pinnae are "tuned" to the bats' echolocation calls. That is to say, for a given species, the frequencies of the echolocation calls with the most sound energy are those most strongly amplified by the pinnae. The structure of the pinnae in relationship to echolocation calls is discussed in a later section here on echolocation calls and foraging strategies.

Some high duty cycle bats, adapted to flutter detection and to tracking trajectories of prey, enhance their ability to scan their surroundings by coordinating body, head, and ear movements. For example, when the African species *Hipposideros gigas* hangs from an acacia branch and scans for insects, its pendant body revolves continuously back and forth through an arc of approximately 180°; the head is in constant motion up and down, and the tips of the ears vibrate forward and backward. These movements seem to allow the bat to use its beamed signals to scan its surroundings meticulously (Vaughan 1977). But why the ear movements? Simmons et al. (1975) proposed that, by moving the direction the ears aim, these bats scan the vertical plane. The rapid ear movements are out of phase: The tip of one ear moves forward whereas the other moves backward. These movements are in approximate synchrony with the signals emitted by the bat. These ear movements, alternately toward and away from the target, may heighten the Doppler shift of echoes from moving targets and improve the bat's discrimination of movement (Simmons et al. 1975; Valentine & Moss 1998). One cannot help but marvel at the elaborate neural coordination of stereotyped body, head, and ear movements with signal emission rate.

Evolution of Echolocation

The two distinguishing functional features of bats are their capacity for powered flight and their ability to use echolocation for obstacle avoidance and to detect and capture prey (except Pteropodidae). There is no doubt that both flight and echolocation contributed significantly to the evolution and diversity of bats (more than 20% of extant mammals are bats), but which evolved first? Several competing hypotheses have emerged.

Gould (1970, 1971) hypothesized that the sonar pulses of bats were derived originally from vocalizations that established or maintained spacing or contact between individuals. This "echolocation first" hypothesis maintains that the repetitive communication sounds used by infant bats and similar pulses, perhaps used originally during flight to maintain adequate spacing of foraging individuals, may have become important secondarily for detecting prey and avoiding obstacles. According to Gould (1971), "The prominence with which continuous, graded signals pervade the lives of such social and nocturnal mammals as bats suggests that echolocation is an inextricable and integral part of a communication system." This author suggested that some of the vocalizations used by early bats may have been inherited from their insectivorous ancestors, in which auditory communication was perhaps as important as it has been shown to be in some living shrews and tenrecs (Eisenberg and Gould 1970; Gould 1969).

Fenton et al. (1995) augmented the echolocation first hypothesis by suggesting that the ancestors of bats were small, nocturnal gliders that produced low duty cycle clicks (short pulses separated from the returning echoes by relatively long interpulse intervals). These clicks were used primarily for orientation as the animals glided from tree to tree. Later, the ancestors of bats replaced these simple orientation clicks with tonal signals of greater intensity. Tonal signals increased the distance at which flying insects or obstacles could be detected. Fenton and his colleagues hypothesized that only after echolocation was optimized for the detection of flying prey did flapping flight evolve. The evolution of flight, coupled with sophisticated echolocation, led to the rapid diversification of bats in the Eocene. Further advances in echolocation design, such as high duty cycle pulses (long pulses separated by short interpulse intervals, overlapping with the returning echoes, and separated from them in frequency) and the ability to achieve self-deafening during pulse emission, allowed bats to specialize in a variety of food sources.

Simmons (Simmons 1994; Simmons & Geisler 1998) offered an alternative "flight first" hypothesis. She suggested that the common ancestor of all modern bats could already fly but showed no specializations for echolocation. These ancestral bats may have evolved flight to reduce energy and increase foraging efficiency. Only later did these proto-bats evolve echolocation to improve detection and tracking of aerial prey.

Finally, according to the "tandem-evolution" hypothesis, echolocation and flight evolved together (Arita & Fenton 1997; Norberg 1985a, 1985b, 1989; Rayner 1991a, 1991b; Speakman1993). The basis for this hypothesis is that the cost of echolocation is high in stationary bats but significantly reduced during flight. Contraction of powerful flight muscles assists in lung ventilation and generates the forceful airflow needed to produce intense laryngeal calls. Thus, flight and echolocation may have evolved in tandem.

The recent discovery of the early Eocene bat *Onychonycteris finneyi* coupled with new molecular phylogenies now sheds light on the evolution of echolocation and flight in bats (Jones & Teeling 2006; Simmons 2008; Simmons et al. 2008). Unlike *Icaronycteris*, which is morphologically similar to modern bats, *Onychonycteris* is a mosaic of ancestral and modern characters (Fig. 15-12). *Onychonycteris* retains claws on all five digits and has proportionately shorter forearms and longer hindlimbs than modern bats. Nevertheless, its modestly elongate digits supported a flight membrane and its well-developed clavicle and keeled sternum suggest it was capable of rudimentary powered flight. Simmons et al. (2008) believe that *Onychonycteris* flew using a combination of gliding and fluttering flight that represented a transitional form of powered flight. Although capable of flight, *Onychonycteris* lacked "all three of the bony correlates of echolocation," which are an expanded stylohyal bone (one of the hyoid bones supporting the enlarged larynx in bats), an enlarged cochlea, and an enlarged head on the malleus bone. Thus, flight seems to have preceded echolocation during the evolution of bats.

By mapping traits associated with flight and echolocation onto the chiropteran family tree, scientists hope to establish when and how many times echolocation evolved during the rapid radiation of bats in the Eocene. Molecular data from huge datasets for all chiropteran families are now rewriting the evolution of the Chiroptera (Eick et al. 2005; Jones & Teeling 2006; Springer et al. 2001). Emerging evidence indicates that bats formerly included in a suborder called Microchiroptera are paraphyletic. Accordingly, the Rhinolophidae, Hipposideridae, Megadermatidae, Craseonycteridae, and Rhinopomatidae, all of which echolocate, are united with the Pteropodidae, which do not echolocate (except via tongue clicks in *Rousettus and Stenonycteris*), in a monophyletic clade called Yinpterochiroptera. The remaining families of echolocating "microbats" form the clade Yangochiroptera (Fig. 15-1).

The new molecular phylogenies have profound implications for the evolution of echolocation. One possibility is

that echolocation evolved in the common ancestor of all bats, was lost in the lineage leading to the Pteropodidae, and secondarily evolved the very different tongue-clicking method in the genus *Rousettus* (Springer et al. 2001). Alternatively, Eick et al. (2005) provide evidence that echolocation evolved twice, once in the ancestor of Yangochiroptera and once in the Yinpterochiroptera after the split between the Pteropodidae and the remaining five families. The debate over how many times echolocation evolved in the Chiroptera remains unresolved, but if the division of the Chiroptera into Yangochiroptera and Yinpterochiroptera is valid, then it suggests that certain call types have evolved independently in several families of bats. For example, the long CF signals of rhinolophid bats (Yinpterochiroptera) and mormoopid bats (Yangochiroptera) must have evolved independently in the two lineages (Fig. 22-1; Jones & Holderied 2007; Teeling 2009). Similarly, short, broadband, multiharmonic calls evolved separately in members of the Megadermatidae (Yinpterochiroptera) and Nycteridae (Yangochiroptera). Obviously, there is considerable diversity and plasticity in call structure in extant bats (Fig. 22-1). Perhaps Jones and Teeling (2006) are correct in suggesting that "the animal's habitat is often more important in shaping its call design than is its evolutionary history."

Echolocation Calls and Foraging Strategies

Adaptive radiation in foraging styles among insectivorous bats has been accompanied by the evolution of a variety of echolocation signal types. The following discussion considers how different echolocation strategies enable bats to exploit contrasting modes of foraging and the degree to which bats that forage similarly and occupy similar habitats share similar styles of echolocation and certain morphological features.

Open-Habitat Bats

A number of bats within the Molossidae, Emballonuridae, and Vespertilionidae forage for insects in open situations such as meadows, above canopies of vegetation, or in some cases, up to 600 meters above the ground (Fenton & Griffin 1997). These bats are typically fast fliers and, when foraging, encounter no obstacles except insects, other bats, and wind-carried objects such as parachute seeds of cottonwood trees and "ballooning" spiders or their web strands. Most of these bats have narrow wings with pointed tips (high aspect ratio wings; **Fig. 22-11**). The energetic cost of flight in these bats is low (Norberg & Rayner 1987), and all seem to remain in flight while foraging. Some are remarkably enduring fliers, remaining on the wing continuously for up to 6 or 7 hours (**Table 22-2**).

Most of these bats use long, high-intensity, low-frequency, narrowband search-phase calls (a, i, and l in **Fig. 22-12**) that

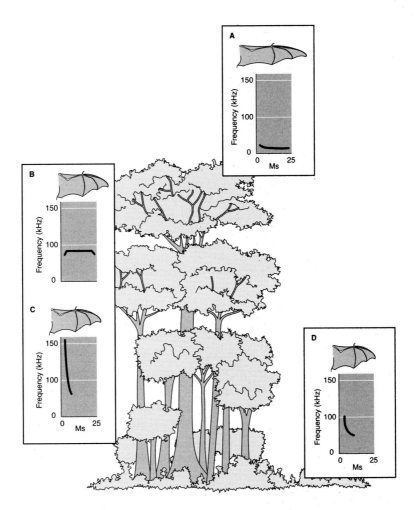

FIGURE 22-11 Habitat partitioning in a bat community. Bats foraging in open spaces above the canopy typically have high aspect ratio wings and produce low-frequency FM calls (A). Those species that forage in dense, cluttered spaces within the forest tend to have low aspect ratio wings and produce either long CF pulses or steep FM pulses of relatively high frequency (B and C). Bats that forage along the edges of vegetation often have long, rounded wings and emit steep FM calls of intermediate frequency (D). (Adapted from Schnitzler and Kalko. *Bat Biology and Conservation* (Kunz and Racey, eds). Smithsonian Institution Press, 1998.)

have low duty cycles (less than 20%). Low frequencies are relatively immune to atmospheric attenuation and are therefore ideal for maximizing the range of target detection. In general, these search-phase calls enhance long-range (for echolocation) detection of large prey. All six species of high-flying molossids studied in Zimbabwe had search-phase signals that conformed to this pattern (Fenton & Bell 1981; Fenton & Griffin 1997). The calls were long (10 to 20 milliseconds), of low frequency (mostly below 20 kilohertz), and had fairly narrow bandwidths (5 to 10 kilohertz). The calls of five of these species were within the range of human hearing. Among eight sympatric Neotropical emballonurids studied by Kalko (1995), the search-phase calls with the lowest frequencies and longest durations were those of the two species that foraged above the forest canopy (**Table 22-3**). Because smaller insects are less detectable by low-frequency signals, open-habitat bats probably capture mostly large insects. But some small vespertilionids and emballonurids that forage in the open

use moderately high-frequency signals and can detect smaller insects. During the attack phase of foraging, open-habitat bats (like other bats that capture flying insects) increase the rate and bandwidth of their echolocation signals.

Many open-habitat bats are flexible with regard to echolocation calls (Obrist 1995). One such species is the northern bat (*Eptesicus nilssoni*) of Europe. In Sweden this bat used higher frequency calls when foraging close to the ground than when foraging high above the ground (Rydell 1993).

Maximizing the range of target detection is of prime importance to rapid-flying bats. This is done in part by increasing echo intensity. This is one function of the pinnae, which act like acoustic funnels. By focusing (amplifying) sound, the pinnae increase the sound pressure arriving at the tympanic membranes (**pinna gain**). In open-habitat bats, the pinnae are "tuned" (in shape and size) to enhance reception of the main frequency of the search-phase call and to improve the localization of targets (Obrist et al.

TABLE 22-2   Aspects of Foraging Behavior of Some Animal-Eating Bats

	Number of Bats	Longest Flight (min)	One-Way Distance (km)	Foraging Strategy	Source
Megaderma lyra	17	45	4	CF, SF	Audet et al. 1988
Nycteris grandis	4	2	27	CF, SF	Fenton et al. 1987
Rhinolophus hildebrandti	10	3	113	CF, SF	Fenton and Rautenbach 1986
Myotis myotis	21	9	307	CF	Audet 1990
Eptesicus fuscus	78	5	120	CF	Brigham and Fenton 1986
Scotophilus borbonicus	9	5	62	CF	Fenton and Rautenbach 1986
Nyctalus noctula	32	3	270	CF	Kronwitter 1988
Lasiurus cinereus	23	20	436	CF	Barclay 1989
L. borealis	4	0.6	206	CF	Hickey 1988
Euderma maculatum	3	12	400	CF	Wai-Ping and Fenton 1989
Mops midas	10	15	100	CF	Fenton and Rautenbach 1986

Note: Data from field observations of known individuals carrying radio transmitters. CF = constant flight; SF = short flights from perches.
Source: After Fenton 1990.

1993). In *Miniopterus schreibersii*, for example, the area of best hearing is at the lowest range of the echolocation call, where most of the signal energy is concentrated (Obrist et al. 1993). Large ears increase low-frequency sound pressure gain but are not similarly effective with high frequencies. Thus, the high-flying molossids that use low-frequency search-phase signals tend to have large ears. The hoary bat (*Lasiurus cinereus*), a vespertilionid, has such a signal but has rather small ears. Obrist et al. (1993) found that the pinnae of this bat are tuned to the second harmonic of its call, and the authors speculated that because this bat occurs seasonally at high latitudes and roosts in foliage small ears may be important in reducing heat loss.

The range of target detection in open-habitat bats with narrowband search-phase signals may be further extended by the detection of acoustic "glints." These are produced by an echolocation signal that hits a flying insect at the instant in its wing-beat cycle when its wings are perpendicular to impinging sound waves (Fig. 22-9). The sound pressure levels of echoes from flying insects can thus be increased up to 20 decibels relative to echoes from nonflying insects (Kober & Schnitzler 1990). Kalko (1995) estimated that some Neotropical emballonurids with high duty cycle, narrowband signals could detect acoustic glints at the rate of 9 to 15 per second.

To summarize, open-habitat bats optimize long-range target perception by using intense, long, low-frequency,

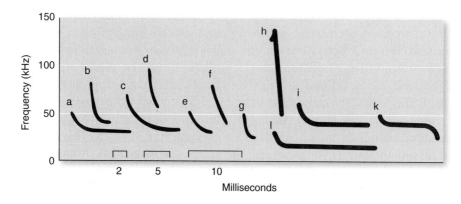

FIGURE 22-12   Diagnostic echolocation calls of members of a bat community near Portal, Arizona. a, *Lasiurus cinereus*; b, *Myotis volans*; c, *Eptesicus fuscus*; d, *Pipistrellus hesperus*; e, *Myotis thysanodes*; f, *Myotis californicus*; g, *Antrozous pallidus*; h, *Myotis auriculus*; i and k, *Tadarida brasiliensis* (i, an unmodified echolocation call; k, a honk); and l, *Nyctinomops macrotis*. (Modified from Fenton, M. B., *Q. Rev. Bio.* 59 (1984): 33–53. University of Chicago Press.)

TABLE 22-3 **Characteristics of Echolocation Search-Phase Signals of Six Neotropical Emballonurids**

Species	CF Components (kHz)	Sound Duration (ms)	Pulse Interval (ms)
Diclidurus albus[a]	24.3 (20)[b]	11.3 (17)	183.3 (17)
Peropteryx sp.[a]	25.4 (28)	9.3 (17)	84.0 (17)
Cormura brevirostris	42.0 (60)	5.2 (54)	102.3 (48)
Saccopteryx bilineata	45.1 (60)	9.4 (41)	85.8 (41)
S. leptura	52.7 (41)	5.3 (31)	60.1 (25)
Rhynchonycteris naso	102.5 (70)	5.0 (90)	53.5 (82)

[a] The first two species forage in the open, whereas the others forage near vegetation.
[b] Sample sizes are in parentheses.
Source: Data from Kalko 1995.

narrowband calls and by having pinnae tuned to the main frequency of the search-phase call. Some species may enhance long-range echolocation by the perception of acoustic glints. These bats tend to have narrow wings and are rapid, enduring fliers that may forage in edge habitats but never forage in closed habitats.

Edge-Habitat Bats

Many, perhaps most echolocating bats forage in edge situations (Fig. 22-11). These bats forage in such places as the borders of woodlands, along streams, or along cliff faces or the edges of arroyos. In a typical perceptual setting faced by these bats, obstacles and clutter are on one side and largely uncluttered space is on the other. Insectivorous bats using edges often capture prey near vegetation but usually do not fly within its clutter. Flight in these bats is typically slow, highly maneuverable, and energetically economical. The wings are usually long, with short, rounded wingtips, low wing loadings, and moderately high aspect ratios (Norberg & Rayner 1987). Many of these bats forage while continuously in flight, but probably most flights are no longer than 120 minutes. At least some members of the Mormoopidae, Phyllostomidae, Rhinolophidae, and Vespertilionidae are edge-habitat bats.

The search-phase signals of these bats tend to be short and intense, with a combination of broadband and narrowband components. Many of these bats use shallow FM search-phase calls that allow for medium- or short-range detection of targets and obstacles (Fig. 22-12B-G). By increasing the bandwidth of their signals, bats can increase the precision of target localization (Simmons & Stein 1980). Some bats increase bandwidth by adding harmonics; other bats do this by increasing the range of frequencies through which the signals sweep. The latter strategy is used by many vespertilionids. The echolocation call of *Myotis auriculus*, an extreme example, sweeps steeply through a range of roughly 100 kilohertz (h in Fig. 22-12). Some edge-habitat species, such as the mormoopid *P. parnellii* and some rhinolophids, use Doppler-shift flutter detection; their calls are dominated by constant frequency components. The low-flying vespertilionid *Antrozous pallidus* is remarkable

because it often stops giving echolocation signals when hunting insects and concentrates on sounds made by its prey (Bell 1982a).

Many species of bats, including many that forage along edges, adapt their echolocation behavior to suit different situations (Obrist 1989). Such a bat is *Lasiurus borealis*, which uses narrowband search-phase and broadband attack-phase calls. In this bat, and perhaps many other vespertilionids, the pinnae enhance perception of the dominant frequency of the search-phase call (Obrist et al. 1993). Among rhinolophids, many of which forage along edges, matching of external ear characteristics with the dominant echolocation frequency is especially striking. The pinna gain in some species is as high as 30 decibels.

Many species that use echolocation calls adapted primarily to one habitat can forage in others. Phyllostomids that eat fruit or nectar, such as *Artibeus* and *Glossophaga*, are good examples. These bats have low-intensity echolocation calls and can forage amid clutter but also forage along edges.

Closed-Habitat Bats

A number of bats, particularly those of tropical forests, forage amid or close to the clutter of branches and foliage. Thies et al. (1998) observed that for many such bats "the returning echoes of the food are often buried in a multitude of clutter—echoes from leaves, branches, and surfaces on which the food rests." Among these closed-habitat bats are at least some members of the families Emballonuridae, Nycteridae, Megadermatidae, Rhinolophidae, Phyllostomidae, and Vespertilionidae. Their diets range from invertebrates and small vertebrates to fruit and nectar. Flight is slow and highly maneuverable. The wings are broad with rounded wingtips and have low wing loadings and low aspect ratios (Fig. 22-11); this design is associated with energetically expensive flight (Norberg & Rayner 1987). High energetic cost is the price these bats pay for the ability to hover and to make complex, split-second maneuvers. Although details of foraging behavior for most species are unknown, most seem not to remain in flight for long periods. Some species are sit-and-wait predators with brief foraging flights. The

African megadermatid *Cardioderma cor*, during a night of foraging, made flights lasting an average of less than 5 seconds and was in flight for a total of less than 11 minutes (Vaughan 1976). Some closed-habitat species make flights lasting up to 113 minutes (e.g., *Rhinolophus hildebrandti*; Table 22-2).

Most of these bats use short (less than 2 milliseconds), broadband, steep FM calls of low intensity (less than 70 decibels). This signal design yields precise, close-range information of obstacles and targets (Simmons & Stein 1980). Low-intensity calls minimize echoes from all but nearby clutter and may also partially avoid alerting prey (Fullard 1987). Among closed-habitat bats are many nasal emitters, including phyllostomids, rhinolophids, nycterids, and megadermatids. Rhinolophids that use closed habitats have narrowband echolocation calls and use flutter detection.

Of particular interest is the dependence by some of these bats on olfaction or prey-generated sounds, rather than echolocation, for finding food. Thies et al. (1998) found that in flight cages the Neotropical frugivores *Carollia perspicillata* and *C. castanea* flew slowly (2 to 3 meters per second) and used echolocation continuously while foraging but located fruit by olfaction (see also Kalko & Condon 1998). Realistically shaped artificial fruit that would have been indistinguishable by echolocation from real fruit was ignored. The high-frequency, multiharmonic calls of these bats provide detailed, close-range information. *Megaderma spasma* of India uses echolocation for perceiving its surroundings but can locate flying insects by using prey-generated sounds (Tyrrell 1988). The Neotropical *Trachops cirrhosus* listens for mating calls of male frogs in order to locate the frogs as prey; however, during most attacks on frogs, the bat continues to give echolocation calls (Tuttle & Ryan 1981). This bat attacked a tape recorder broadcasting frog calls, indicating that echolocation was not used for target detection. Two African species of *Nycteris* use this combination of echolocation and prey-generated sounds during foraging (Fenton et al. 1983). Both *Megaderma lyra* (Fiedler 1979) and *Macrotus californicus* (Bell 1982a) can locate nonflying prey by using prey-generated sounds; under light conditions equaling bright starlight, the latter species uses vision rather than echolocation when searching for nonflying prey (Bell 1982b).

Many closed-habitat bats are gleaners, taking prey from the surfaces of vegetation, rocks, or the ground. These species tend to have large ears that amplify low-frequency sounds (below 15 kilohertz), such as those made by prey rustling through leaves. The pinnae of these bats are remarkable in their ability to localize low-frequency sounds (Obrist et al. 1993). Among species that forage at short range and use broadband echolocation calls with no dominant frequency, such as fruit- or nectar-eating phyllostomids,

pinna gain and ability to localize sounds are only moderately developed.

■ Detection of Prey by Fishing Bats

The Neotropical fishing bat *Noctilio leporinus* locates prey by detecting ripples caused by small fish or by recognizing parts of the fish that break the water surface. During attacks on prey, the sequence is from short CF/FM to FM signals and only one harmonic is used. This fairly simple, single-harmonic pattern seems well adapted to this bat's unique foraging style. Of central importance is the detection of a disturbance on a fairly uniform background, a problem similar to that facing a bat flying high above obstacles. Their calls are the most intense known among bats, at a sound pressure level up to 140 decibels.

■ Variability and Multiple Uses of Echolocation Calls

Echolocation signals are used by bats for communication, as well as for orientation and locating prey. In addition, modifications in signal design allow some bats to exploit a variety of habitats. It is to be expected, then, that echolocation is flexible.

Echolocation calls display considerable geographic variation within a species. Thomas et al. (1987) found (in an extreme case) that *L. cinereus* in Arizona had calls with a minimum frequency 53.8% higher than the minimum frequency of calls in Manitoba (26.0 kilohertz versus 16.9 kilohertz). Eight of 12 species showed geographic variation of at least 3 kilohertz in minimum echolocation frequency.

Individual variation in echolocation calls is known for some species. For example, a male *Euderma maculatum* foraging not far from a female of this species had calls of lower frequencies and longer intervals (437 milliseconds versus 352 milliseconds; Obrist 1995). In British Columbia, *Myotis evotis* calls varied in frequency range from 30 to 86 kilohertz to 54 to 97 kilohertz (Thomas et al. 1987).

Much of the variability in call design is probably a response to conspecifics. When conspecifics flew by, all four species of vespertilionids studied by Obrist (1995) decreased call duration and increased the intervals between calls. Signals were "personalized" by three of these species by increasing frequency, and one species drastically increased the intensity of its calls when foraging near other lasiurines. One of these species (*Lasiurus borealis*) sharply altered its signal design in the presence of conspecifics. Under certain circumstances, some bats add an extra pulse to the echolocation calls. On a collision course with another bat, an *N. leporinus* lowered its call frequency and added a warning "honk" to the signal (Suthers, 1965). Fenton and Bell (1981) recorded a similar honk given by a *T. brasiliensis* when near other bats. *Rhinopoma hardwickei* consistently maintains

CF signals when flying alone but changes frequencies when flying in a group (Habersetzer 1981).

Echolocation calls communicate several kinds of information. Barclay (1982) reported that the gregarious *Myotis lucifugus* relied on signals of conspecifics to locate day roosts, mating sites, hibernation sites, and feeding areas. The solitary forager *E. maculatum*, in contrast, reacted aggressively to playbacks of calls of a conspecific individual. This bat either attacked the speaker or abruptly moved away (Leonard & Fenton 1984). Apparently this species' low-frequency calls, with most of the sound energy at about 10 kilohertz, provide fairly long-range warnings that maintain spacing between individuals.

Echolocation calls also provide vocal signatures. This is the basis for mother–young recognition in a number of species (e.g., *R. ferrumequinum*, Matsumura 1981; *A. pallidus*, Brown 1976). In a captive colony of *R. ferrumequinum*, individuals recognized each other by listening to echolocation calls and had clear roostmate preferences (Möhres 1967). Vocal signatures and individual recognition may be important also during interactions among foraging bats (Obrist 1995).

An additional kind of flexibility allows some bats to optimize echolocation performance in different habitats (Simmons et al. 1978). A number of species are known to alter their signal design (by adding steep FM components) when changing from foraging in open areas to those closer to obstacles (e.g., *Eptesicus nilssoni*, Rydell 1990; *Lasiurus borealis*, Obrist 1995; *Nyctalus noctula*, Zbinden 1989).

Bats Versus Moths

Several studies have revealed a remarkable series of adaptations by certain moths in response to predation by bats (Dunning 1968; Dunning & Roeder 1965; Roeder 1965; Roeder & Treat 1961; summarized by Fullard 1987). Nocturnal moths of the families Noctuidae, Ctenuchidae, Geometridae, and Arctiidae have an ear on each side of the rear part of the thorax. Each ear is a small cavity within which is a transparent membrane. The ears are sensitive to a wide range of frequencies and allow the insects to detect the ultrasonic pulses of foraging bats. Upon detecting the approach of a bat, the moths alter their level flight and adopt various erratic flight patterns. Some members of these families of moths have carried the business of evading bats to an even greater extreme and have a noise-making organ on each side of the thorax. When the moths are disturbed, these organs produce trains of clicks with prominent ultrasonic components.

Under laboratory conditions, flying bats about to capture mealworms tossed into the air regularly turned away from their targets when confronted with recorded trains of moth-produced pulses. These pulses apparently protect moths from bats; probably these signals interfere with the echoes returning to the bat and deafen them temporarily. In addition, some moths jam a bat's echolocation system by producing sounds that resemble the terminal buzz of a bat closing in on an insect (Fullard et al. 1979).

Investigations of the strategies bats use to hear and feed on moths have shown that predator–prey interactions between these animals are complex, with considerable fine-tuning of adaptations favoring prey detection, predator avoidance, and inconspicuousness on the part of both predator and prey (Fenton & Fullard 1981). In moth–bat confrontations, the relative times or distances at which each animal detects the other are critical. Bats using loud signals in the 20- to 50-kilohertz range can be detected by some moths at distances up to 40 meters, whereas bats probably detect moths at no more than roughly 20 meters. Neither bats nor moths do consistently well, however, for some bats detect insects at ranges of only 1 meter, and experiments using an African slit-faced bat (*Nycteris macrotis*) showed that moths cannot detect this bat at distances greater than 0.2 meter. Nevertheless, the fact that ears are present in more than 95% of the moths of Ontario and in more than 85% of those in southern Africa suggests that sound detection is important to moths.

Many moth species have apparently become tuned to bat signals (Fullard et al. 2008). One Australian moth (*Speiredonia*) roosts in the very caves inhabited by its main predators, *Rhinolophus megaphyllus* and *Miniopterus australis*. These moths survive by listening for the bats' echolocation calls and taking evasive action. In most areas, the echolocation frequencies used by bats are those to which the ears of local moths are most highly sensitive. Most bats of North America have echolocation signals at frequencies between 15 and 60 kilohertz, and the ears of moths of this region are tuned to these frequencies. Because some bats obviously eat moths, however, we might well ask how such bats avoid the early warning systems of moths.

One major bat strategy is to use frequencies to which moths are relatively insensitive. Bats that use echolocation signals of low intensity and those that have extremely high-frequency signals can escape detection except at close range. Frequencies above and below the range used for detection by moths are called **allotonic frequencies**. In the Afrotropical, Indo-Malayan, and Australasian tropics, roughly one-third of the species of insectivorous bats uses allotonic signals and thereby presumably partially avoids detection by moths. Fenton and his associates have found that, in many cases, bats with such signals tend to specialize on moths. Each of three moth-eating rhinolophids was found to use very high frequencies (100, 139, and 210 kilohertz; Fenton 1984; Fenton & Fullard 1981). There are costs to using allotonic frequencies, however, because very high frequencies are severely attenuated, and low frequencies provide poor target discrimination. The benefits of remaining undetected by the prey probably outweigh the costs for many bats, including some North American vespertilionids. *Myotis evotis* and *M. keenii* both emit very high-frequency pulses that are not detected by their main prey, noctuid moths (*Catocala* spp.; Faure et al. 1990, 1993). On the opposite side of the auditory spectrum, the vespertilionid *Euderma maculatum* uses extremely low-frequency pulses (9 to 12 kilohertz) while foraging for moths (Leonard & Fenton 1984; Obrist 1995; Woodsworth et al. 1981). Fullard and Dawson (1997) tested the hypothesis that the use of allotonic frequencies by *E. maculatum* allows them to remain "acoustically inconspicuous" to sympatric eared moths. Their results indicated that the short-duration, low-frequency calls of *E. maculatum* are not easily detected by the moths. Fullard and Dawson estimated that eared moths would be unable to detect the presence of *E. maculatum* until the bats were within 1 meter of the moth, whereas *Eptesicus fuscus* would be detected at a distance of more than 20 meters.

Another strategy that allows bats to remain inconspicuous to moths is that of partially abandoning echolocation. Some bats, such as some megadermatids, do not use echolocation for detecting prey but instead listen for sounds made by the prey while remaining silent themselves.

Bat–moth predator–prey interactions provide an interesting example of evolutionary gamesmanship. Echolocation is a major key to the great success of bats, allowing them to perceive and capture insects in darkness. Bat echolocation signals provide moths (and other insects) with a means of detecting foraging bats, and moths have developed antipredator behaviors that depend on this ability. A third countermove is seemingly in progress. The shift to allotonic echolocation among some moth-specialists, insectivorous bats may be a response to early detection by eared moths (Fenton & Fullard 1981). Why, then, haven't eared moths responded by tuning their ears to these allotonic frequencies? Fullard (1982) and Fullard and Belwood (1988) suggested that eared moths have not adapted because their ears are tuned to the average "echolocation frequencies of all of the bats that present a significant risk." Common bats, such as *E. fuscus*, present a disproportionately greater risk to moths than the relatively rare *E. maculatum*.

Echolocation in Cetaceans

Just as bats must cope with darkness, cetaceans frequently must perceive their underwater environment under conditions that render vision difficult, if not impossible. In some waters inhabited by cetaceans, suspended material such as soil particles or plankton limits visibility to a few meters or even a few centimeters. Water transmits light poorly, and even under ideal conditions, visibility under water is limited. Also, some cetaceans forage at considerable depths, where there is not a trace of light. It is not surprising, then, that some cetaceans have developed echolocation.

Probably all odontocetes (toothed whales and dolphins) use echolocation for detecting obstacles and prey. Mysticetes (the baleen whales), however, are not known to echolocate. A tremendous variety of sounds, some having a fascinating musical quality, is made by both mysticetes and odontocetes. The wailing, creaking, and squealing noises of cetaceans have become commonplace to sailors operating sonar equipment at sea (**Table 22-4**; Payne 1970). Some of these underwater sounds may have a communication function (Clark & Clark 1980). Biosonar emissions by odontocetes occur in two basic types: narrowband continuous tones called whistles, used for intraspecific communication, and broadband clicks used for echolocation. Acoustical energy moves through water very efficiently, making sonar ideal for the aquatic environment.

Since the account by Schevill and Lawrence (1949) of the underwater noises made by the white whale (*Delphinapterus*), considerable research has been done on the vocalizations of cetaceans. Much research has dealt with a common dolphin, *Tursiops truncatus* (Au et al. 1974; Kellogg et al. 1953; Lilly 1962, 1963; Norris et al. 1961; Schevill & Lawrence 1953; Wood 1959). *Tursiops* is able to detect obstacles and recognize food by means of echolocation; it uses short pulses resembling those of bats. *Tursiops* is capable of producing a great variety of sounds, but of primary importance for echolocation are the trains of clicks that it emits. The clicks are audible to humans but cover a wide spectrum of frequencies. The pulse rate rises as a dolphin approaches a target, and *T. truncatus* can distinguish between a piece of fish and a substitute water-filled capsule with a similar shape (Norris et al. 1961), or even between sheets of different thicknesses of the same metal (Evans & Powell 1967). Pack and Herman (1995) presented bottlenose dolphins with a variety of complex shapes and showed that the dolphins were capable of immediately recognizing these shapes by echolocation.

Peak frequencies recorded in free-ranging dolphins by Au et al. (1974) were between 120 and 130 kilohertz, whereas those reported by Evans (1973) in an aquarium were close to 60 kilohertz. Au and his colleagues attributed the lower peak values in captive dolphins to possible interference reverberations from aquarium walls. Noisy environments may also affect the sonar signals used by cetaceans. The peak frequencies of click emission varied with location in the false killer whale (*Pseudorca*) from only 20 kilohertz to more than 110 kilohertz in a noisy environment. These

TABLE 22-4   Characteristics of Sounds Produced by Representative Cetaceans

Suborder Species	Common Name	Sound Type	Frequency Range (kHz)	Frequency at Max. Energy (kHz)
Odontoceti				
Inia geoffrensis	River dolphin	Click	25–200	95–105
Phocoena phocoena	Harbor porpoise	Pulse	100–160	110–150
Delphinus delphis	Common dolphin	Whistle	0.2–150	4–9
		Click	0.2–150	30–60
Orcinus orca	Killer whale	Scream	0.25–35	12
Stenella longirostris	Long-beaked spinner	Click	1–160	60
		Whistle	1–20	8–12
Tursiops truncatus	Bottlenose dolphin	Click	0.2–150	60–80
		Whistle	2–20	—
Physeter catodon	Sperm whale	Coda	16–30	—
Mysticeti				
Balaenoptera musculus	Blue whale	Moan	0.2–0.02	0.012–0.018
B. physalus	Fin whale	Call	0.16–0.75	0.02
Balaena mysticetus	Bowhead whale	Call	0.1–0.58	0.14–0.34
Megaptera novaeangliae	Humpback whale	Song	0.05–10	,4.0

Source: From data summarized by Ketten 1997.

studies suggest that caution should be used in interpreting data gathered from animals held in tanks.

Taylor et al. (1997) recently advanced an intriguing alternative hypothesis concerning cetacean sonar. These authors suggested that some marine mammals may not use active sonar (hearing echoes of sound pulses, such as bat echolocation or the ping of a submarine) or passive sonar (noisy objects are detected without the production of sound) at all. Active sonar advertises the position of the cetacean to the prey species. Indeed, there is evidence that dolphin echolocation clicks tend to alarm prey fish. Evidence that captive dolphins use something other than active or passive sonar comes from experiments in which eye cups were used to temporarily blind bottlenose dolphins (*Tursiops truncatus*) as they pursued live prey. Simultaneous recordings by the listening devices secured to the dolphins' heads or by hydrophones suspended at various locations in the pool indicated that these dolphins did not use underwater clicks or other sounds to locate the prey, yet were able to follow and capture 100% of prey fish in all trials (Taylor et al. 1997). These observations lead to the proposal that dolphins may be able to use ambient noise imaging to see their environments at close ranges.

Ambient noise imaging is fundamentally different from both active and passive sonar systems and refers to the use of sound to "see" underwater. The physics of this type of underwater sound imaging are well beyond the scope of this discussion, but computer-generated models predict that dolphins could use ambient noise imaging to "see" useful images for tens of meters underwater (Taylor et al. 1997). If verified in living dolphins, this suggests that dolphins, and perhaps other marine mammals, have an entirely new way of seeing with sound.

Where do the cetacean biosonar clicks and buzzes originate? Echolocation signals are generated in the upper nasal passages just ventral to the blowhole (Berta et al. 2005; Cranford et al. 1996, 1997). Cranford et al. (1996) using such techniques as magnetic resonance imaging scans, demonstrated that these biosonar signals are produced by vibrations of the nasal sac system (**Fig. 22-13**). With the exception of sperm whales, odontocetes have bilateral phonic lip/dorsal bursae complexes (PLDBs) positioned below the vestibular air sacs and blowhole. Each PLDB complex consists of a pair of muscular phonic lips with a slit between adjacent lips through which air passes. The phonic lips are embedded within fatty bursae and the entire complex is suspended by muscles and ligaments below the blowhole and posterior to the melon. Cranford et al. (1997) demonstrated that when air forces the slit open and it subsequently slaps shut, the entire PLDB vibrates (much as the lips of a trumpet player vibrate against the mouthpiece). It is not the air passing through the lips that generates the sound, but the vibrations in the PLDB tissues. The tissue-borne sounds are propagated forward through the melon, which

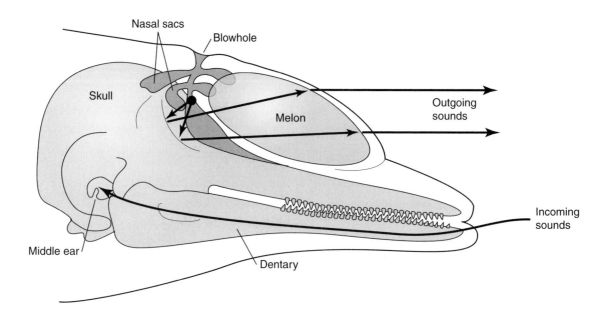

FIGURE 22-13   A dolphin generates clicks when air is forced through the slit in the phonic lips of the phonic lip/dorsal bursae complexes causing the entire PLDB to vibrate. These vibrations are then reflected off the skull bones and focused forward by the melon on the dolphin's forehead. Returning echoes from the target are transmitted to the middle ear via the dentary. (Adapted from Thomas, J. and Kastelein, R., eds. *Sensory Abilities of Cetaceans: Laboratory and Field Evidence*. Plenum Press, 1990.)

focuses the sound waves into a beam that is then directed forward into the water. The position of the melon in front of the skull and the PLDB complex and its composition (primarily low-density lipids) create an acoustic lens that shapes the beam (Fig. 22-13; Cranford 2000). Importantly, smaller odontocetes have a pair of PLDB complexes that can generate sound pulses independently or together. Bottlenose dolphins (*T. truncatus*) have asymmetric PLDB complexes and asymmetric bones around the external nares (Fig. 19-7). The right complex is nearly twice the size of the left. This asymmetry suggests that these dolphins can generate pulses at two different frequencies and with two distinct rhythms simultaneously (Cranford 2000; Cranford et al. 1996).

The basic sound generation system is probably homologous among all odontocetes, but compared to other odontocetes sperm whales (*Physeter catodon*) have larger, more conified PLDB complexes, which are anterior to the spermaceti organ and the junk (a region in the sperm whale head, below the spermaceti organ, which contains spermaceti oil and connective tissues; **Fig. 22-14**). Cranford and colleagues (1996) proposed that the spermaceti organ of sperm whales is homologous to the right posterior bursae in smaller odontocetes and that the junk in *Physeter* is homologous to the melon in other odontocetes.

The sperm whale (*Physeter catodon*) presents an interesting case of echolocation among cetaceans because it uses a unique mode of foraging that is probably made possible by echo ranging. The click of a sperm whale is known to consist of a series of brief pulses each lasting roughly 20 milliseconds (Backus & Schevill 1966). The clicks are repeated at rates from less than 1 click per second to 40 per second. One remarkable feature of sperm whale clicks is that a single click results in evenly spaced pulses of declining amplitude. Hydrophones aligned with the whale's echolocation beam by Mohl et al. (2003) showed only a single, high-decibel pulse (at roughly 235 decibels, it is the loudest biologically produced sound). In sperm whales, there is a single pair of phonic lips at the front of the head. A sound pulse generated by the phonic lips is not directed forward into the surrounding water. Cranford (1999) concluded that the sound pulse is initially transmitted back through the spermaceti organ to the frontal air sac just in front of the skull. The pulse is then reflected forward through the junk (homologous to the melon), where a series of parallel fatty lenses focuses the sound into a beam that then enters the water in front of the head (Cranford 1999; Mohl et al. 2003).

Sperm whales feed largely on squid that they take at depths (down to at least 2000 meters; Watkins et al. 1993) at

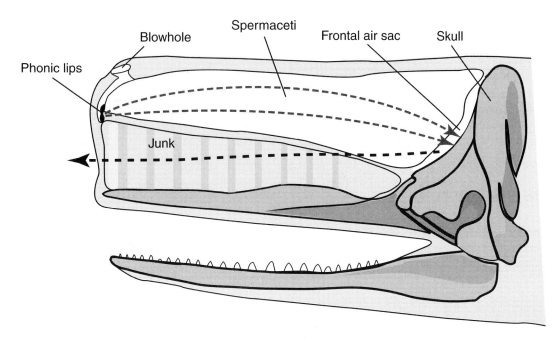

FIGURE 22-14   Cutaway view of a sperm whale head showing the anterior location of the phonic lips. Sound produced in the tissues of the phonic lips is radiated backward through the spermaceti organ to the frontal air sac near the skull. From there, the sound is reflected forward through the fatty lenses of the junk and out into the water in front of the head. (Adapted from Wahlberg, P.T., M. and Mohl, B., *Behav Ecol Sociobiol*, 53 (2002): 31– 41.)

which prey is scarce and light is virtually absent. Seemingly, the sperm whale is able to forage at great depths largely because it is able to use echo scanning to locate food under conditions that require efficient long-range echolocation. Goold and Jones (1995) estimated that sperm whales, by using click rates of 0.5 to 1 per second, should be capable of detecting prey 750 meters away, but Madsen et al. (2002) calculated that sperm whale clicks may be able to detect targets up to 16 kilometers away. As the whales approach the potential prey, click rates must increase to provide information on target position and velocity (Au & Benoit-Bird 2003). It has also been suggested that the rapid bursts of clicks referred to as "creaks" (up to 200 pulses per second) produced by sperm whales may be analogous to the terminal "buzz" of a flying bat in giving greater target resolution. Miller et al. (2004) attached digital recorders to free-living sperm whales and recorded body position, depth, and sound. These whales switched from typical echolocation clicks used for range finding to creaks at the bottom of the dive. Creaks were also associated with rapid turns and body rolls, suggesting that foraging sperm whales were pursuing prey. Several researchers suggest that the intense blasts of sound, referred to as bangs, produced by sperm whales may serve to disorient or incapacitate prey (Bel'kovich & Yablokov 1963; Berzin 1971). Evidence for acoustic prey debilitation is primarily circumstantial; proponents point out that beaked whales, narwhals (without functional teeth), and toothless immature sperm whales are still able to capture fast-moving prey. These observations suggest that these whales may stun their prey with sound before engulfing them. Definitive evidence, such as acoustical recordings of bangs with sufficient intensity (more than 240 decibels) to stun prey, is currently lacking.

■ Cetacean Hearing

Each ear of a cetacean functions as a separate hydrophone, allowing the animal to localize a sound source by discriminating (as we do) between the times the sound is received by each ear. The pressure that sound transmitted through water exerts on the bones of the entire skull causes vibrations to be transmitted by the skull. When the bone that houses the middle and inner ear is attached rigidly to the skull, as it is in most mammals, vibrations from water are transmitted through the bones of the skull and reach the ear from various directions. As a consequence, when a mammal with this type of skull is submerged, it is unable to localize accurately the source of a sound. Because sound localization is of great importance to cetaceans that use echolocation, these animals have evolved several structural features that insulate the bone surrounding the middle and inner ear (tympanoperiotic bone) from the rest of the skull (Ketten 1997).

First, the tympanic bullae (tympanoperiotic bone; the fused auditory bulla and cochlea) are not fused to the skull in any cetacean, and in the specialized porpoises and dolphins, the bullae are separated by an appreciable gap from adjacent bones of the skull. In addition, the bullae are insulated by an extensive system of sinuses unique to cetaceans. These sinuses surround the bullae and extend forward into the enlarged pterygoid fossae, and each sinus is connected by the eustachian tube to the cavity of the middle ear. The sinuses are filled with an oil–mucus emulsion, foamed with air, and are surrounded by fibrous connective tissue and venous networks. These sinuses can apparently retain air even when subjected to pressures of 100 atmospheres, pressures higher than those to which cetaceans are subjected during deep dives. The foam in the air sinuses apparently forms a layer around the bullae that retains remarkably constant sound-reflecting and sound-insulating qualities through a wide range of pressures and effectively insulates the bullae from the rest of the skull.

Norris (1964, 1968) regarded the extremely thin back part of the lower jaw of delphinids (pan bone) as an acoustical window. He held that sound passes into the skin and blubber overlying the dentary, through the thin part of this bone, which at its thinnest may be only 0.1 millimeter thick, to the intramandibular fat body, which leads directly to the wall of the auditory bulla, into which the sound presumably passes. Weight is given to Norris' hypothesis by experiments done by Bullock et al. (1968), who found that the jaw is the most acoustically sensitive area of the dolphin's head.

Echolocation by Other Mammals

Echolocation is especially useful for nocturnal mammals or aquatic mammals that frequent depths below which light cannot penetrate. Not surprisingly, echolocation abilities are best developed in members of the Chiroptera and the Odontoceti. Ultrasonic sounds have also been reported for certain species of shrews, tenrecs, rodents, pinnipeds, and sirenians (Schusterman et al. 2004; Thomas & Jalili 2004). In some cases these sounds may represent a rudimentary form of echolocation, but in others, these calls appear to be used for communication and not echolocation. Echolocating mammals are able to produce and detect ultrasonic sounds, and they must be able to use those sounds to detect targets (prey or objects).

Some shrews have these abilities (Buchler 1976; Crowcroft 1957; Komarek 1932; Reed 1944; Swinhoe 1870; Thomas & Jalili 2004). In a series of carefully controlled laboratory experiments, Gould et al. (1964) demonstrated that three species of *Sorex* could echolocate. These shrews searched around an elevated disk, found a lower platform, and jumped to it, all without the use of tactile, visual, or olfactory senses. While the shrews searched their environment, they emitted pulses with frequencies between 30 and 60 kilohertz; the pulse duration was from 5 to more than 33 milliseconds. The shrews were unable to find the disk when their ears were plugged. The familiar short-tailed shrew of the eastern United States, *Blarina brevicauda*, produces similar pulses (Thomas & Jalili 2004). Gould (1969) provided evidence that least shrews (*Cryptotis parva*) emit "twitters" and "putts" that were used when exploring unfamiliar areas. Laboratory trials revealed that three species of tenrecs (Tenrecidae) from Madagascar also echolocate (*Hemicentetes semispinosus*, *Echinops telfari*, and *Microgale dobsoni*; Gould 1965; Novick & Gould 1964). Tenrecs produce pulses by clicking the tongue, and the pulses are of frequencies audible to humans (from 5 to 17 kilohertz). In addition, streaked tenrecs (*H. semispinosus*) produce ultrasonic sounds between 10 and 70 kilohertz by rubbing a set of specialized spines together (stridulation); but these sounds are used for communication and not for echolocation. Hedgehogs (*Hemiechinus auritus*; Erinaceidae) have also been reported to orient toward ultrasonic sounds (Sales & Pye 1974), but there is no evidence that they can produce ultrasonic sounds themselves.

Several groups of rodents have also been reported to produce ultrasonic sounds, including dormice (*Glis glis*), golden hamsters (*Mesocricetus auratus*), rats (*Rattus norvegicus*), and several species of voles (*Microtus*) (Floody 1979; Kahmann & Osterman 1951; reviewed in Thomas & Jalili 2004). Although platform location studies similar to those described previously for shrews seemed to indicate that several species were capable of echolocation, these studies failed to properly control for the use of tactile and olfactory cues. Thus, several rodent species are known to produce and detect ultrasonic calls for communication, but there is little definitive evidence that rodents use echolocation for target detection and discrimination.

Pinnipeds and sirenians produce a number of underwater sounds related to social communication. Among the most interesting are the "knocking" sounds of male walruses (Fay et al. 1984; Miller 1985; Schevill et al. 1966; Stirling et al. 1987), which are used to establish and maintain dominance. Underwater clicks have also been recorded from harbor, ringed, harp, grey, hooded, and leopard seals (Phocidae; Ballard & Kovacs 1995; Schusterman et al. 2004). Renouf and Davis (1982) hypothesized that these clicks are used for echolocation, and Evans et al. (2004) speculated that the clicks produced by Weddell seals (*Leptonychotes weddellii*) are used to navigate under the ice. Experimental evidence, including fitting seals with video cameras (Davis et al. 1999), suggests that the underwater clicks of seals are not used when chasing fish, nor do Weddell seals appear to use echolocation to locate breathing holes from under the ice. Instead, Schusterman et al. (2000) argued that the amphibious lifestyle of pinnipeds excluded the evolution of a specialized underwater biosonar system; pinnipeds retain hearing tuned to the airborne vocalizations of their rookeries and have only modestly adapted their auditory systems for hearing underwater. Currently, there is no unequivocal evidence of echolocation in pinnipeds (Evans et al. 2004; Schusterman et al. 2004).

The only other group of fully aquatic marine mammals is the Sirenia. Studies of underwater sound communication in Florida manatees by O'Shea and Poche (2006) revealed complex, single-note calls with multiple harmonics that are used to maintain group cohesion. These calls also appear to provide information on the size and identity of the caller, but there is no evidence that sirenians use echolocation. Indeed, manatee calls provide little directional information, which may explain why large numbers of manatees are killed by collisions with boats each year in the southeastern United States (Novacek et al. 2004).

SUMMARY

At least 18% of the known mammal species use echolocation as a means of "viewing" their surroundings. Most bats, some members of the orders Soricomorpha and Afrosoricida, and probably all odontocete cetaceans echolocate. Echolocation involves extracting information from echoes of sounds produced by the animal.

Mammals produce two basic types of echolocation calls, with or without the vocal cords. Without involving the vocal cords, odontocetes produce clicks and squeals in the nasal passage, and two genera of bats make clicks with the tongue. The vast majority of echolocating bats and shrews produce signals using the vocal cords in the larynx. Bats emit these echolocation signals through either the mouth (oral emitters) or the nose (nasal emitters). Most bats are oral emitters; members of the Nycteridae, Megadermatidae, Rhinolophidae, Hipposideridae, and Phyllostomidae are nasal emitters. Bat signals are more intense than clicks and therefore provide a greater range of detection of objects in air. Echolocation calls are described in terms of time (duration, repetition rate), frequency (pitch), and intensity (an expression of signal strength). Bat echolocation signals are typically ultrasonic, above about 20 kilohertz, with most signals between 12 to 100 kilohertz, but some bat signals exceed 200 kilohertz.

Bat echolocation signals also differ in bandwidth (breadth of frequencies produced). Narrowband calls (spanning less than 10 kilohertz) are often called CF signals; they are useful for target detection at greater distances but provide fewer details about target position or movement. Many bats use such CF calls as search-phase calls when foraging. By increasing the bandwidth of their signals, bats increase the precision with which they can pinpoint a target. Such broadband, FM calls sweep through a greater range of frequencies and provide greater detail about the target but with reduced detection distance because of attenuation of the sound intensity.

Most bats avoid self-deafening that could result from signal-echo overlap by separating the signal and its echo in time. They produce signals a small percentage of the overall time and do not broadcast signals and receive echoes at the same time (low duty-cycle approach). During the final pursuit of an insect, a low duty cycle bat drastically increases its signal rate (making a "feeding buzz"), enabling the bat to track the final-instant movements of the prey. A fundamentally different echolocation strategy enables some bats to detect the fluttering wings of insects and to track precisely the trajectories of prey by perceiving Doppler-shifted echoes (the change in frequency of a sound produced by a moving object). These bats use a high duty cycle approach in which they separate the pulse they make from the echo they hear by frequency instead of by time; they tend to make long CF signals whose echoes overlap the bat's calls in time.

Both flight and echolocation contributed significantly to the success and diversity of bats. The "echolocation first" hypothesis suggests that the ancestors of bats were small, nocturnal gliders that produced low duty cycle clicks used primarily for orientation as the animals glided from tree to tree. Later, tonal signals, which increase detection distance, replaced simple orientation clicks and set the stage for detecting flying prey, and flapping flight evolved. The alternative "flight first" hypothesis proposes that the common ancestor of all modern bats could already fly, but showed no specializations for echolocation. The recent discovery of the early Eocene bat *Onychonycteris finneyi* coupled with new molecular phylogenies sheds light on the evolution of echolocation and flight in bats. *Onychonycteris* was likely capable of flight but lacked characteristics of the hyoid and ear region that are correlated with echolocation. Thus, flight appears to have evolved before echolocation in bats. The recent discovery and sequencing of a mammalian hearing gene important in echolocating bats may well lead to further consideration of the hypothesis.

Bats exhibit many foraging styles in habitats that are open, closed, or transitional between the two, and these styles are accompanied by a variety of echolocation strategies. Most open-habitat bats encounter few obstacles during open-air flight; they are fast fliers, with narrow, pointed wings, and use long, high-intensity, low-frequency, narrowband search-phase calls. Edge-habitat bats forage along transition areas such as the borders of woodlands and streams. These bats fly slowly, are highly maneuverable, and have wings with low wing loadings and moderately high aspect ratios. The search-phase signals of these bats tend to be short and intense, with a combination of broadband and narrowband components. Many of these bats use shallow FM search-phase calls that allow for medium- or short-range detection of targets and obstacles. Closed-habitat bats forage amid or close to the clutter of branches and foliage. Flight is slow and highly maneuverable. Their wings are broad with rounded wingtips and have low wing loadings and low aspect ratios. Most of these bats use short, broadband, steep FM calls of low intensity.

Nocturnal moths are a favored prey of many insectivorous bats. Over long evolutionary time spans, many moth species have become tuned to bat signals, whereas bats have responded with improved predatory behaviors. Several families of nocturnal moths have ears on the thorax. These ears are sensitive to a wide range of frequencies and allow the insects to detect the ultrasonic pulses of foraging bats. Some bats evolved countermeasures: they use frequencies to which moths are relatively insensitive (allotonic frequencies) to escape detection by moths. Other bats partially abandon echolocation to remain inconspicuous to moths.

Probably all odontocetes (toothed whales and dolphins) use echolocation for detecting obstacles and prey. Mysticetes (baleen whales), however, are not known to echolocate. After decades of research, the mechanism of sound production in odontocetes has been discovered in the upper nasal passages just ventral to the blowhole. The basic sound-production system includes a pair of phonic lips and surrounding fat-filled bursae suspended by muscles and ligaments below the blowhole and posterior to the melon. Echolocation clicks are produced when air forces the slit open and it subsequently slaps shut, causing vibrations in the tissues. The sounds are propagated forward through the melon, which focuses the sound waves into a beam that is then directed forward into the water. The position of the melon in front of the skull and PLDB complex and its composition create an acoustic lens that concentrates the beam.

Cetacean ears serve as separate hydrophones, allowing the animal to localize an echo by discriminating (as we do) between the times the sound is received by each ear; however, sound transmitted through water causes vibrations to be transmitted by the entire skull. If the middle and inner ear bones were attached rigidly to the skull, as they are in most mammals, vibrations would reach the ear from many directions simultaneously. To allow localization of the sound, the tympanic bullae are not fused to the skull in any cetacean. Foam-filled sinuses surround the bullae, effectively insulating the bullae from the rest of the skull. In addition, the extremely thin posterior part of the lower jaw (pan bone) acts as an acoustical window. Sound passes through the thin part of this bone to the intramandibular fat body, which leads directly to the wall of the auditory bulla, into which the sound passes. The jaw is the most acoustically sensitive area of the dolphin's head.

Ultrasonic sounds are also produced by certain species of shrews, tenrecs, rodents, pinnipeds, and sirenians. Tenrecs produce audible pulses by clicking the tongue, and streaked tenrecs produce ultrasound (10 and 70 kilohertz) by rubbing specialized dorsal spines together (stridulation); these sounds are used for communication, not echolocation. Several groups of rodents produce and detect ultrasonic calls for communication, but there is little definitive evidence that rodents use echolocation for target detection and discrimination. Harbor, ringed, harp, grey, hooded, and leopard seals make underwater clicks, but there is yet no unequivocal evidence of echolocation in these pinnipeds. The calls of Florida manatees are used to maintain group cohesion, and there is no evidence that sirenians use echolocation.

KEY TERMS

Allotonic frequency	Duty cycle	Low duty cycle
Attenuated	Frequency	Narrowband
Bandwidth	Frequency-modulated	Pinna gain
Broadband	Fundamental frequency	Target ranging
Constant frequency	Harmonics	Ultrasonic
Doppler shift	High duty cycle	

RECOMMENDED READINGS

Arita, HT & MB Fenton. 1997. Flight and echolocation in the ecology and evolution of bats. *Trends in Ecology and Evolution, 12*:53–58.

Berta, A, JL Sumich, & KM Kovacs. 2005. *Marine Mammals: An Evolutionary Approach*, 2nd ed. Academic Press. NY.

Cranford, TW & ME Amundin. 2003. Biosonar pulse production in odontocetes: the state of our knowledge. Pp. 27–35, in *Echolocation in Bats and Dolphins* (JA Thomas, CF Moss, & M Vater, eds.). The University of Chicago Press, Chicago, IL.

Cranford, TW. 2000. In search of impulse sound sources in odontocetes. Pp. 109–156, in *Hearing by Whales and Dolphins* (WWL Au, AN Popper, & RR Fay, eds.). Springer-Verlag, NY.

Fenton, MB. 1984. Echolocation: Implications for ecology and evolution of bats. *Quarterly Review of Biology, 59*:33–53.

Jones, G & EC Teeling. 2006. The evolution of echolocation in bats. *Trends in Ecology and Evolution, 21*:149–156.

Li, G, et al. 2008. The hearing gene *Prestin* reunites the echolocating bats. *Proceedings of the National Academy of Sciences USA, 105*:13,959–13,964.

Schusterman, RJ, et al. 2004. Pinniped sensory systems and the echolocation issue. Pp. 531–535, in *Echolocation in Bats and Dolphins* (JA Thomas, CF Moss, & M Vater, eds.). University of Chicago Press, Chicago, IL.

Simmons, NB. 2008. Taking wing. *Scientific American, December,* 96–103.

Simmons, NB & TM Conway. 2003. Evolution of ecological diversity in bats. Pp. 493–535, in *Bat Ecology* (TH Kunz & MB Fenton, eds.). University of Chicago Press, Chicago, IL.

Simmons, NB, et al. 2008. Primitive early Eocene bat from Wyoming and the evolution of flight and echolocation. *Nature, 451*:818–822.

Teeling, E. 2009. Hear, hear: the convergent evolution of echolocation in bats? *Trends in Ecology and Evolution, 24*(7):351–354.

Thomas, JA, CF Moss, & M Vater. 2004. *Echolocation in bats and dolphins.* University of Chicago Press, Chicago, IL.

Ecology, Behavior, and Conservation

Ecology

One of the most remarkable attributes of humans is our ability to recognize relationships between disparate phenomena or events—to discover order, pattern, symmetry, predictability, and beauty in an apparently disordered world. The study of ecology demands such ability, for of central interest to ecologists is an understanding of the often complex relationships between living things. Kendeigh (1961) described ecology as "a study of animals and plants in their relations to each other and to their environment." Early ecological work featured descriptive field studies, and later investigations added controlled field or laboratory experiments. These provide a base of knowledge on which the modern theoretical ecologists depend.

It is difficult to overemphasize the value of an ecological approach to the study of mammals. An understanding of mammalian ecology has been long in emerging, however, not because mammalogists lack interest in ecology but because a study of the ecology of even a single species involves detailed knowledge of many aspects of that species' biology, of its physical environment, and of the biology of species with which it is associated. Consequently, ecology overlaps with the disciplines of physiology, genetics, behavior, and evolution.

Mammalian ecologists are primarily interested in interactions at three levels: populations, communities, and entire ecosystems. Populations consist of individuals of the same species living in the same area at the same time. The many lion prides in the Serengeti ecosystem or the vast herds of caribou in the Canadian arctic are both examples of populations (a group or groups of potentially

interbreeding individuals). Populations grow or decline over time, depending on the rate at which new members are added by birth or immigration and lost by mortality and emigration. Populations have limited geographic boundaries, determined by the available resources and barriers to dispersal. Populations are dynamic and potentially immortal, in that those members alive today are the descendants of those alive in past generations. Genetic structure, life history patterns, sex ratio, and age structure all play important roles in shaping populations of mammals.

Ecological communities, another level of organization, are structured by interactions between populations of different species. Such multispecies interactions include predation, competition, and mutualism. With respect to mammals, it is here that many interesting and exciting interactions take place in the natural world. For example, competition for food among lions, hyenas, and hunting dogs in the Serengeti are important factors affecting community structure and stability. Also, energy flow between the trophic levels (e.g., primary producers to primary consumers) of a community has a major influence on community organization and biodiversity.

Ecosystems form the largest and most complex level because they often include hundreds or thousands of species (both plant and animal) over broad geographic areas. Ecologists often resort to studying common "currencies," such as energy, when comparing ecosystems. Clearly, the scope of ecology is extremely broad. This chapter concentrates on ecological relationships and principles basic to an understanding of mammalian biology and is, therefore, necessarily incomplete and selective in its coverage.

Global Climate Patterns

The environments of animals can be characterized in terms of physical and biotic factors. Physical factors include temperature, humidity, climatic patterns, precipitation, and soil types; biotic factors are those associated with interactions between organisms.

Solar radiation provides the energy, in the form of heat and light, on which living organisms depend. The intensity of solar radiation at the Earth's surface is influenced largely by the directness with which the sun's rays strike the Earth. The angle of these rays decreases and the climate becomes progressively cooler the farther north or south of the equator an area is situated.

Warm air holds more moisture than does cool air, and equatorial areas, especially areas near 25°N and 25°S latitude, receive relatively heavy precipitation. In addition, major global patterns of air circulation occur as warm air rises from equatorial regions and moves northward or southward. In a belt centered at 30°N or 30°S of the equator, the equatorial air masses reach a stage of cooling that causes them to sink. Cool air also carries less moisture. Thus, some of the major deserts of the world, such as those in the southwestern United States and North Africa, are roughly 30 degrees from the equator and are under the influence of this system of descending air (**Fig. 23-1**).

Over most of the world, even in tropical areas, rainfall is seasonal, and animals and plants must adapt to times of relative scarcity of food that grows only during periods of abundant rainfall. Migrations of some tropical bats coincide with seasonal shifts in the abundance of insects, fruits, or flowers, and the dramatic migrations of wildebeest in East Africa are in response to seasonal changes in the availability of nutritious forage.

Superimposed on global climatic patterns are regional or local variations. For example, periodic increases in Pacific Ocean surface temperatures, referred to as El Niño Southern Oscillations, have pronounced effects on marine and terrestrial mammal populations. The 1991 to 1992 El Niño resulted in three times the normal annual rainfall in semiarid areas of northern Chile, superabundant food resources, and rapid increases in small mammal populations (Meserve et al. 1995). El Niño-caused increases in rodent densities may have been partly responsible for an outbreak of rodent-borne hemorrhagic fever viruses (Hantavirus) in the southwestern United States (Childs et al. 1995).

Topographic features may affect precipitation and local distributions of plants and animals. When storms sweeping in from the west move over north-south-oriented mountain ranges, the western slopes receive high precipitation, the eastern slopes receive lower precipitation, and the eastern basins of the mountains are often deserts (**Fig. 23-2**). This "rain-shadow" effect strongly influences the distribution of both plants and animals (**Fig. 23-3**). When fog funnels through the passes in the Santa Ana Mountains of southern

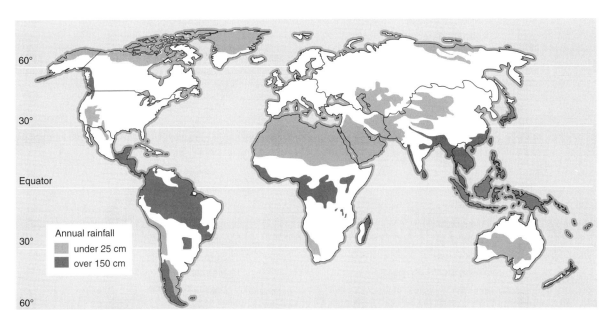

FIGURE 23-1   The important desert regions (those with less than 25 centimeters of precipitation annually) and the major wet regions (those with greater than 150 centimeters of precipitation annually). (Adapted from Espenshade, E. B. *Goode's World Atlas, Thirteenth edition*. Rand-McNally, 1971.)

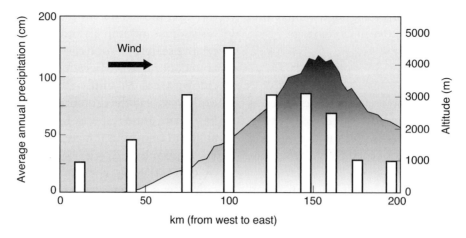

FIGURE 23-2 Effects of mountain ranges on local climate are illustrated by the rainfall patterns from the Pacific coast to the Sierra Nevada in California. Prevailing winds sweep moist air masses from the Pacific Ocean inland, where they are deflected upward by the Sierra Nevada. As the moist air rises, it cools and condenses, forming rain on the windward side of the mountain range and leaving the leeward side dry.

California, condensation dripping from the needles of the knob-cone pine (*Pinus attenuata*) can total a remarkable 10.2 centimeters of precipitation per month. This supplies sufficient moisture to allow the knob-cone pine to extend its growth into rainless periods (Vogl 1973) and affects the distributions of small mammals.

Local topography may also strongly affect the amount of heat the surface of the Earth receives. The main axes of most mountain ranges lie north and south. The drainage systems are oriented more or less east and west, and the canyon walls face roughly north or south. In northern latitudes, because the sun's rays strike a south-facing slope more directly than a north-facing slope, south-facing slopes are considerably drier and warmer and have different biotas than do nearby north-facing slopes. In the precipitous, chaparral-covered mountains of southern California, for example, contrasting biotas occupy adjacent north- and south-facing slopes (Vaughan 1954; Fig. 23-3).

Vertical temperature gradients are encountered as one ascends a mountain (or descends in an aquatic environment). Temperatures decline with increased elevation at a rate of approximately 1°C for every 150 meters. This effect,

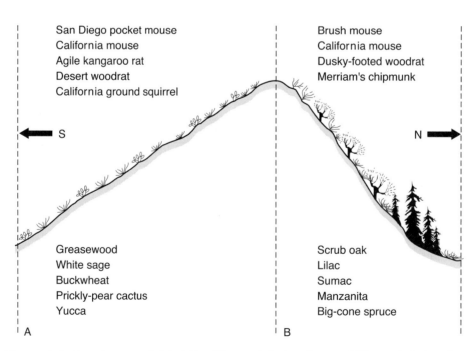

FIGURE 23-3 Assemblages of plants and mammals inhabiting (A) a south-facing slope and (B) a north-facing slope in lower San Antonio Canyon, San Gabriel Mountains, Los Angeles County, California. (Data from Vaughan, T. A., *Univ. Kans. Publ., Mus. Nat. Hist.* 7 (1954): 513–582.)

coupled with increased precipitation at higher elevations, shorter growing seasons for plants, and drastic diurnal-nocturnal fluctuations in temperature, is associated with a distinct separation of climatic zones in high mountains throughout the world. In some areas of the western United States, an assemblage of "desert" mammals resembling those typical of arid lands as far south as central Mexico may occupy the arid or semiarid land at the foot of a mountain range, whereas the crests of the mountains a few miles away may support boreal genera that occur as far north as northern Canada or Alaska (**Fig. 23-4**).

Human-induced climate change is beginning to have an impact on mammal populations. Over the last 20 years scientists have documented significant loss of sea-ice cover in several arctic regions and changes in breakup and freeze-up dates of the sea ice over broad areas of the circumpolar Arctic. These changes are likely a consequence of climate warming (Comiso & Parkinson 2004; Gough et al. 2004; Parkinson et al. 1999). Polar bears (*Ursus maritimus*) live in isolated subpopulations throughout the Arctic and depend on sea ice for seal hunting and travel. In Western Hudson Bay and Baffin Island, the ice is melting earlier, and polar bear populations are declining (Stirling & Parkinson 2006; Stirling et al., 1999). Stirling and Parkinson (2006) reported that five polar bear populations in the Canadian Arctic must hunt on shore because there is no sea ice for several months. Reduced sea ice may be responsible for a cascade of impacts, including lowered reproductive rates because female bears will have less energy to invest in

cubs, increased walking and swimming costs as ice thins and becomes more fractured, reduced seal hunting success rates, and more human–bear encounters (Derocher et al. 2004).

In other cases, global warming may disrupt entire communities. Population cycles in voles have been fading out in northern Europe (Ims et al. 2007; Kausrud et al. 2008). Researchers suspect that global warming is the underlying cause of the collapse of vole, lemming, and snowshoe hare cycles in boreal ecosystems. The unprecedented late winter and early spring temperatures probably reduce subnivean microhabitats and expose lemming to an increased risk of predation and increased energy costs (Kausrud et al. 2008). The resulting scarcity of lemmings results in predators, such as arctic fox and snowy owls, switching to other prey. Thus, climatic changes may alter boreal **food webs** and change the dynamics of the ecosystem, further limiting lemming cycles.

■ Biomes

Climate obviously has a dramatic effect on the distribution of plants and animals across the Earth's surface. Wet regions support dense vegetation, whereas very dry areas support little or no vegetation. The short growing season and low temperatures of alpine zones support unique communities of dwarf or slow-growing plants. Biological communities, although never exactly alike in plant and animal species, can be grouped into categories based on the dominant vegetation. These large-scale biological communities are called

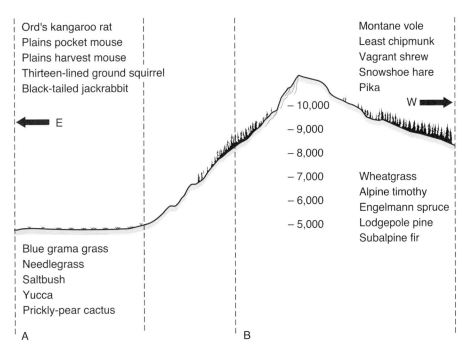

FIGURE 23-4 Assemblages of plants and mammals inhabiting (A) short-grass prairie and (B) subalpine habitats in northern Colorado (Larimer County).

biomes. The world's major biomes include tropical forests, savanna, deserts, polar regions, chaparral, temperate grasslands, temperate deciduous forests, coniferous forests, and tundra (arctic and alpine; **Fig. 23-5**). Biomes can, of course, be more finely subdivided. Temperate grasslands can be divided into short- or tall-grass prairies.

Biomes are also dynamic in both space and time. As the world's climate has changed over the millennia, the boundaries between adjacent biomes have shifted position (e.g., during ice ages). The dynamic nature of the biomes has led to changes in the opportunities for speciation and reinvasion. The result is that groups of species living in the Sahara Desert of Africa are not identical to the species inhabiting the Gobi Desert of Asia. Both are desert biomes, but evolution has led to similar traits in independently evolved species (convergent evolution).

Biomes typically grade into each other over fairly large areas, forming **ecotones**, and within each biome, there is considerable local variation, resulting in a patchy appearance. Fire or floods may create openings in certain areas. **Ecological succession** may result in one community being replaced by another, and human activity adds to the patchy appearance of many landscapes and biomes.

Succession occurs in places denuded or mostly denuded of life, such as newly formed river sandbars or land cleared by humans. Succession in a place that has never before supported life, such as a lava flow, is **primary succession**. Where the area previously supported a community, such as abandoned cropland or burned-over brushland, **secondary succession** occurs. Secondary succession is responsible for the mosaic of habitat patches that one commonly observes in agricultural areas or in many chaparral-dominated foothills of California.

The initial successional community on land typically consists of sun-tolerant, annual herbs with seeds adapted to wide dispersal. This **pioneer community** is usually short lived and is replaced by perennial herbs, which maintain themselves by spreading vegetatively. Succession might then proceed by shrubs replacing perennial herbs and shrubs eventually giving way to trees. In the course of tens or hundreds of years, a relatively stable **climax community** is established. Plants in this community tolerate or are favored by their own effects on their environment (shading of the ground, leaf litter, and added nitrate in the soil, in the case of forest trees). The climax is dynamic, but its species interactions, structure, and energy flow tend to perpetuate the community. The different communities in the succession are called **seral stages**. Succession in different areas obviously involves different seral stages and different climax communities; the climax community in the Chihuahuan Desert consists of creosote bush (*Larrea*) and other xeric shrubs. On the high plains of eastern Colorado,

FIGURE 23-5 (A) Tropical forest in the Chyulu Range, Tsavo West National Park, Kenya. (B) Sandy dunes in the Namib Desert of Namibia, southern Africa. (C) Subalpine coniferous forest near Rabbit Ears Pass, Routt County, Colorado.

short grasses form the climax; the climax community in the mountains of Colorado is typically coniferous forest.

Succession has pronounced effects on the distribution and densities of mammals. White-tailed deer favor early successional stages. In Pennsylvania, in areas in the "brush" stage of succession, 24 species of plants yielded more than 91 kilograms of deer food per acre, but in later successional stages, 7 species of plants produced only 16 kilograms of food per acre (Gerstell 1938). In Massachusetts, an upsurge in the production of deer food followed the abandonment of cultivated land; when a mature hardwood forest became established, however, production of deer food declined to nearly zero (Gould 1937). As might be expected, white-tailed deer are uncommon in mature northern forests, and the main factor allowing their spread and increased abundance in parts of northeastern United States was widespread logging (Hosley 1956). Most small mammals are restricted to specific climax communities or successional stages and are absent or scarce in others.

Populations

A **population** is a group of potentially interbreeding individuals that occupy a given locality (Mayr 1970). The boundaries of the "locality" may depend on the presence of barriers to dispersal and the range of the species. Local populations may occupy very small habitat patches. In contrast, a **metapopulation** is a series of local populations connected by dispersing individuals (gene flow). At the largest scale, the entire species can be considered a single population. Populations are characterized by such features as density, spacing patterns, age structure (ratio of one age class to another), rates of growth or decline, and genetic structure. These characteristics change over time and space as a result of changing selective pressures. Natural selection shapes the life history strategies of populations and plays a vital role in regulating the distribution and abundance of populations.

Population characteristics and processes emerge from the characteristic processes of individuals (**Fig. 23-6**). Such properties of populations include spatial distribution, density, age structure, sex ratio, and growth rate. The characteristics of populations are determined by interactions between and among individuals and their environment over both ecological and evolutionary time scales.

As detailed in Chapter 21, all mammals must maintain a nearly constant internal environment in an ever-changing physical environment. Local environmental conditions vary spatially and temporally. For example, temperature, light, rainfall, and a host of other physical variables vary seasonally or daily. The very adaptations that allow an animal to flourish under certain conditions constrain its survival under a different set of conditions. This environmental variation is reflected in the animal's distribution across the landscape and its abundance within the habitat.

Habitat Requirements

Just as no two species of animals are structurally identical, each species requires a specific environment—a particular combination of physical and biotic factors—and each is functionally unique, pursuing a particular mode of life within its environment. The very morphological, physiological, and behavioral characters that determine the distinctness of a species also determine the distinctness of its habitat requirements. The specific environmental setting a species occupies and the functional "role" it plays in this habitat comprise the animal's ecological **niche**.

Many factors contribute to the selection of appropriate habitats, and each species has physical and behavioral traits suited to its habitat. Habitats change constantly, occasionally abruptly. Animals either adapt to these changes or disappear. Degrees of habitat tolerance differ among species, and local populations may adapt to local conditions. In the western Sierra Nevada of central California, four forest communities are delineated by elevation (**Fig. 23-7**). Each of four species of shrew (*Sorex*) occupies a different

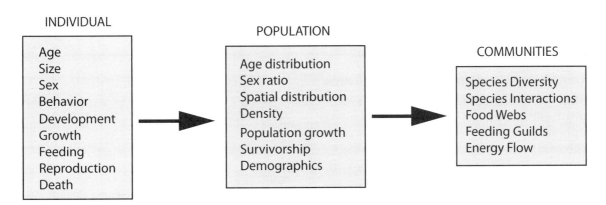

FIGURE 23-6 Diagram of the properties of individuals, populations, and communities. Notice that unique properties emerge at each level.

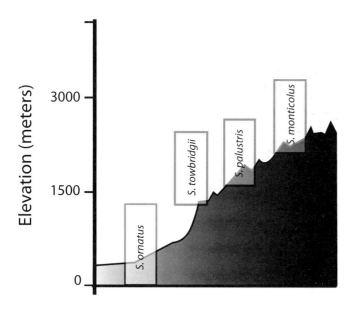

FIGURE 23-7 Distribution of four species of shrews belonging to the genus *Sorex* along an elevational gradient in the Sierra Nevada Mountains. (Data from Findley, J. S. and Yates, T. L., *Mus. Southwestern Biol. Spec. Publ.* 1 (1991): 1–14.)

major forest community (Williams 1991). *Sorex ornatus* occupies the low elevation (500 to 1,000 meters) gray pine/blue oak woodlands and ponderosa pine forests and is replaced by *S. trowbridgii* in intermediate elevation (1,500 to 2,000 meters) mixed coniferous forests. As mixed coniferous forests give way to high elevation (> 2,000 meters) red fir and lodgepole pine forests, *S. trowbridgii* is replaced by *S. monticolus*. The water shrew (*S. palustris*) was associated with riparian habitats at elevations over 1,500 meters. Although sympatry occurs, habitat segregation between species occurs on a local scale such that interspecific contacts are probably rare in shrews (Kirkland 1991; Sheftel 1994).

The environment at a terrestrial locality is not uniform but consists of a complex mosaic of microenvironments. As a general rule, few terrestrial mammals can withstand the most extreme temperatures (or other conditions) that occur in the habitats they occupy, but many are able to select microenvironments in which temperature extremes are avoided or moderated. Rather than being a source of winter hardship for small mammals, snow is actually a boon. It forms an insulating mantle that provides a microenvironment at the surface of the ground where activity, including breeding in some species, continues through the winter. To these small mammals, such as shrews (*Sorex* and *Blarina*), pocket gophers (*Thomomys*), voles (*Microtus, Myodes, Phenacomys*), and lemmings (*Lemmus, Dicrostonyx*), the most stressful periods are in the fall, when intense cold descends but a snow cover has not yet

moderated temperatures at the surface of the ground (Formozov 1946), and in the spring, when rapid melting of deep snow often results in local flooding (Ingles 1949; Jenkins 1948; Vaughan 1969). As mentioned previously, when such **subnivean** microhabitats are lost (because of climate change), small mammal populations suffer (Ims et al. 2007, and references therein).

A group of beavers occupying a beaver lodge in the winter is not subjected to the extreme air temperatures outside the lodge (**Fig. 23-8**). Similarly, when shrews forage beneath litter, under logs or rocks, or beneath dense foliage, not only is their food abundant but temperature and humidity are moderated. During the dead of winter, short-tailed shrews (*Blarina brevicauda*) maintain a relatively constant core body temperature of approximately 38°C, despite ambient temperatures above the snow as low as –21°C, by using subsurface runways where temperatures hover around 1°C to 2°C (Merritt & Adamerovich 1991; Merritt & Bozinovic 1994). These animals cannot tolerate the general climatic conditions of the regions they occupy but are instead adapted to a limited set of conditions that occur in their chosen microenvironment.

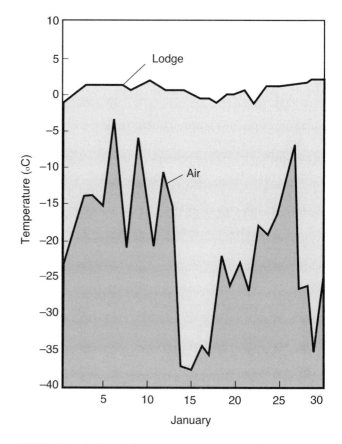

FIGURE 23-8 Daily minimum temperatures inside and outside a beaver lodge in Algonquin Park, Ontario, Canada. (Modified from Stephenson, A. B., *J. Mammalogy* 50 (1969): 134–136. American Society of Mammalogists, Allen Press, Inc.)

Although often depicted as continuous in field guides, the distribution of a species is not uniform throughout its geographic range. Rather, it is a mosaic of the distributions of the local populations. Neither are individuals in the population equally abundant in all regions. The natural world is heterogeneous, resulting in a patchwork distribution of habitats of varying quality.

Some landscape–ecology research has dealt with "vegetation mosaics" as they influence population regulation. This term refers to the spatial arrangement of habitat patches of different quality. In heterogeneous environments, habitat patches often differ sharply in plant composition and cover. For a species of vole, some patches may afford optimal habitat, whereas some may be uninhabitable. High-quality patches harbor "source" populations (where emigration exceeds immigration), whereas poor quality patches often function as "sinks," where immigration exceeds emigration. An example of this situation involves the California vole (*Microtus californicus*) in an area with discrete vegetation patches. Relative to poor-quality (sparsely populated) patches, the high-quality (densely populated) patches were characterized by female-biased sex ratios, longer persistence of females, higher reproductive and emigration rates, and lower immigration rates (Ostfeld & Klosterman 1986). The density of males was similar in all patches, but the female density varied threefold. Laboratory experiments showed that vegetation from the densely populated patches supported rapid growth by voles, whereas voles lost weight on vegetation from sparsely populated patches. Patch quality seemingly depended primarily on food quality.

One of the best examples of habitat selection is in the desert rodent communities of the American Southwest. Merriam's kangaroo rat, *Dipodomy merriami*, prefers to forage in more open areas, where its long hind legs and richochetal locomotion give it an advantage. In contrast, the smaller pocket mouse, *Chaeodipus penicillatus*, favors closed habitats with good cover where it can hide from predators. To test the habitat preferences of these two rodents, Price (1978) created additional open habitat by removing half the shrubs from one of her study grids. Within 6 weeks, the density of the kangaroo rat population had increased in the more open study grid (**Fig. 23-9**). Later, in a similar experiment, after Thompson (1982) added artificial cover (cardboard shelters) to another desert study site, the density of kangaroo rats declined as predicted, whereas pocket mouse populations increased as more cover became available.

Clearly, each species has a preferred habitat type, but are these preferences learned by experience or inherited? North American deer mice, *Peromyscus maniculatus*, include a short-eared prairie form and a long-eared woodland form. Research has shown that the prairie form is capable of living

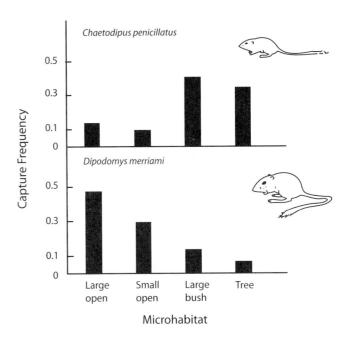

FIGURE 23-9 Summer microhabitat distributions of two heteromyid rodents. *Chaetodipus penicillatus* spend more time under large bushes and trees, whereas *Dipodomys merriami* prefer open areas. (Data from Price, M. V., *Ecology* 59 (1978): 910–921.)

in woodlands but "prefers" grasslands. Wild-caught and lab-reared prairie deermice were tested by Wecker (1963) in an enclosure that straddled a grassy field and an oak-hickory forest. Wecker recorded the time spent in each habitat type. Not surprisingly, wild-caught adults spent over 80% of the time in the grassland side of the enclosure. Interestingly, offspring of the wild-caught prairie forms that had been reared in a woodland habitat still preferred the grassland habitat. The lab-reared mice, bred in the lab for 12 to 20 generations, spent only half as much time in the grassland habitat; apparently they had lost some of their innate preference for grasslands. Wecker concluded that habitat selection in prairie deer mice is largely under genetic control. Natural selection may shape the behaviors that determine habitat choice because individuals that select, or are forced to accept, marginal habitats raise fewer offspring.

In theory, individuals should always select the best possible habitat, but because environments vary, high-quality habitats may be in short supply when population density is high. Fretwell (1972) developed a simple model of habitat selection involving only three habitats (high-, medium-, and low-quality habitats). In this model, all individuals are free to occupy any of the three habitat types (i.e., an **ideal free distribution**). When population density is low, all members of the population may occupy high-quality habitats because there are plenty to go around. If densities exceed the carrying capacities of high-quality habitats, however, competition and crowding force some individuals into medium- or low-quality habitats. At very high densities, it may be advantageous for some individuals to choose poorer but less crowded habitats.

Territorial mammals, however, are not free to occupy all available habitats because they are constrained by the defensive behavior of current territory holders. Fretwell (1972) calls this the **ideal despotic distribution** and predicted that high-quality habitats would be occupied by dominant individuals, with subordinate animals (i.e., those with lower fitness) forced into more marginal habitats. Under these conditions, the density of animals in marginal habitats may exceed that of animals in high-quality habitats. If the stress of crowding becomes intense, some individuals may attempt to disperse to less crowded habitats.

Dispersal

Dispersal is one of the key features of populations, for it provides a genetic connection between subpopulations (equivalent to "local population" in metapopulation terminology) and, given enough time, between isolated populations over the species' entire geographic range. Two categories of dispersal are usually recognized. **Natal dispersal** occurs when an individual moves permanently away from its birth place to a new location where it attempts to reproduce. Breeding dispersal, in contrast, arises when adults relocate between breeding attempts ("transfer" dispersal in Cockburn 1992). Male spotted hyenas (*Crocuta crocuta*) exhibit both natal dispersal at about 24 months of age and may transfer between clans (breeding dispersal) several times in their lifetimes (Van Horn et al. 2003). In both types of dispersal, movements are one way and (depending on the species) may be as short as a few dozen meters or as long as several hundred kilometers.

Dispersal tends to be male biased in polygynous mammals because females gain fitness by breeding in the natal area (**philopatry**). A female provides more parental investment than does the male and should benefit from knowledge of the local resources and the presence of cooperative kin in the vicinity. In contrast, males experience local competition for mates (or other resources) and may suffer inbreeding if they remain philopatric. Because male reproductive success is tied to the number of females impregnated, males benefit by dispersing to find more mates. Dispersal is usually costly (in terms of both energetics and survival), and only one sex need disperse to reduce competition or avoid inbreeding.

The extent of female philopatry and male dispersal in polygynous mammals should increase with increasing complexity of the social system (Perrin & Goudet 2001). If the species is relatively nonsocial, few kin are likely to remain in the natal area, and males need not travel far to avoid inbreeding or competition. On the other hand, in highly social species, males should disperse farther from the natal area. Support for this prediction comes from an analysis of 11 species of ground squirrels (Devillard et al. 2004). Male-biased dispersal increased with increasing social complexity.

Female-biased dispersal may evolve in monogamous mammals because monogamous males that secure and defend resources are more likely to attract females and thus to exhibit parental care (Greenwood 1980). Female dispersal has been demonstrated in several monogamous mammals, including the white-toothed shrew, *Crocidura russula* (Favre et al. 1997). In the polygynous greater sac-winged bat (*Saccopteryx bilineata*) males defend small harems of females and females disperse (Voigt et al. 2008).

Ultimately, dispersal by either sex will be selected for only when the individual's fitness is increased by moving to a new habitat. A dispersing individual's fitness may increase if it avoids inbreeding, reduces local mate competition, and/or reduces competition for resources other than mates. Individuals in a few mammal species may even be innately programmed to disperse under certain conditions. Evidence for each of the hypothetical causes of dispersal is presented in the following paragraphs.

Inbreeding Avoidance

According to the inbreeding avoidance hypothesis, mammals evolved dispersal mechanisms to reduce the possibility of breeding with closely related kin in the natal area. Close inbreeding increases homozygosity and often leads to fewer, less fit offspring, because harmful recessive alleles are more likely to occur in the homozygous condition. The loss of fitness in inbred populations (called **inbreeding depression**) leads to the evolution of such inbreeding avoidance mechanisms as single-sex dispersal (Ralls et al. 1986).

Early research on naked mole-rats (*Heterocephalus glaber*) indicated that the underground colonies were highly inbred and only formed by fission from existing colonies (Sherman et al. 1991). More recent studies on wild populations suggest that dispersal and outbreeding are more common than previously realized (Braude 2000; O'Riain et al. 1996). The discovery of a rare "disperser morph" in naked mole-rat colonies provides a mechanism to reduce inbreeding and increase gene flow between colonies (O'Riain et al. 1996). Dispersers are phenotypically and physiologically unlike other colony members: They are larger, more obese, and have elevated luteinizing hormone levels. Their urge to disperse is strong, and they show little interest in mating with the resident "queen."

Experiments on root voles (*Microtus oeconomus*) also seem to support the inbreeding avoidance hypothesis (Gundersen & Andreassen 1998). A study of fifty-three mothers released with newly weaned litters (i.e., matrilines) into separate enclosures with access to three empty habitat patches demonstrate that (1) long-distance dispersal of offspring was male biased; (2) male offspring that dispersed into adjacent habitats tended to avoid their mothers; and (3) mothers were apparently able to suppress reproduction in sons that failed to disperse. The authors conclude that inbreeding avoidance was driving the male-biased dispersal in root voles.

Local Mate Competition

Philopatric males are likely to suffer competition for mates among their male kin. In addition, because males tend to have higher **reproductive potential** than females do, they suffer more under such competition. The cost of dispersal is predicted to be less than the loss of fitness associated with remaining in the natal area. This may be particularly true for subadults or competitively inferior members of the population: Subadult male lions are always driven from the pride before they pose a serious threat to the dominant male (Pusey & Packer 1987), and dispersing common shrews (*Sorex araneus*) are usually small and lower-ranking (Hanski et al. 1991).

Resource Competition

Individuals may disperse to increase their access to food, territories, or unrelated mates. Nunes et al. (1997) found that female Belding ground squirrels (*Spermophilus beldingi*) disperse to improve access to food or territories. Seemingly, the quantity and distribution of resources shape dispersal in females. In contrast, male arctic ground squirrels dispersed independent of food supply or population density, probably to avoid inbreeding (Byrom & Krebs 1999).

Innate Dispersal

Innate dispersal is not well studied in mammals, but there is evidence that some species are genetically predisposed to disperse. An ontogenetic switch is triggered when young male Belding ground squirrels reach a particular body mass or condition. They disperse during their juvenile summer only if they attain sufficient body mass or fat stores; otherwise, fat stores are used for hibernation and dispersal is delayed until the next summer (Holekamp 1986; Nunes et al. 1998). An internal circannual timing mechanism, seemingly tied to day length and food supply, regulates body mass and fat deposition in these squirrels. In house mice (*Mus musculus*), a number of risk-taking behaviors, including exploratory forays into unfamiliar habitats and dispersal, increase at the onset of sexual maturity (Macri et

al. 2002). Mammalian dispersal patterns are complex, and an array of proximate and ultimate factors act on the sexes in various ways to influence dispersal.

Territoriality and Home Range

Because individuals of the same species are potentially competing for the same environmental resources at the same place and time, intraspecific competition should be moderate to intense, depending on population density. Indeed, this seems to be the case, with the result that individuals of the same species often occupy separate, or nearly separate, home ranges. Burt (1943) described the **home range** of a mammal as "that area traversed by the individual in its normal activities of food gathering, mating, and caring for the young." Home ranges may have irregular shapes and may partially overlap. Within the home range of some mammals is an area that is defended against other members of the species. This area, which usually does not include the peripheral parts of the home range, is called the **territory**, and species that apportion space in this fashion are termed territorial. A home range or a territory may be occupied by one individual, by a pair, by a family group, or by a social group consisting of a number of families.

To solitary mammals or to members of a group, the occupancy of a home range has several important advantages. Each home range provides all of the necessities of life for an individual or group, permitting self-sufficiency within as small an area as possible. The less extensively the individual must range, the less chance there is of encounters with predators. Because the home range quickly becomes familiar to the individual, it can then find food and shelter with the least possible expenditure of energy and can escape predators more effectively because escape routes and retreats are familiar and no time or movement is lost in seeking shelter. Some mammals, such as some rabbits and meadow voles (*Microtus*), maintain trails that serve as routes to food and as avenues of escape. To some male mammals that guard harems, the territory is an exclusive mating area.

A mammal's reproductive success may be increased by its knowledge of areas adjoining its home range that are occupied by other individuals of the same species (in the case of solitary species), or by familiarity with individuals sharing its home range (in the case of social species). During early life, young can develop under parental care largely free from interference by other individuals of their own species. Infanticide that involves killing of unrelated offspring may also play an important role in the evolution of female territoriality (Wolff 1997). According to Wolff's hypothesis, females that have helpless, nonmobile young that are reared in a burrow should evolve territoriality to reduce the risks of infanticide from strangers and to ensure an exclusive food supply.

Home range size varies tremendously (**Table 23-1**). Many mammals in the orders Soricomorpha, Chiroptera, Primates, Rodentia, Lagomorpha, Carnivora, Perissodactyla, and Artiodactyla are known to be territorial. The recognition of territorial boundaries in some species depends on scent marking and other means of territorial marking, and much remarkable behavior is associated with the maintenance of territories (some of this behavior is discussed in Chapter 24).

Some territorial species such as pocket gophers (**Fig. 23-10A**) are distributed according to a pattern of home ranges that may persist through many generations. Hansen (1962) found such a pattern to be typical of northern pocket gophers (*Thomomys talpoides*) in Colorado. Each animal occupies an area of raised ground called a mima mound (Fig. 23-10B), which is some 10 meters in diameter. The mima mounds are more productive of food than are the relatively narrow intermound areas, which usually have shallow soil. Except in the winter, the intermound areas are used little by pocket gophers, and the chances of survival are slim for an animal that is unable to establish itself in a mima mound. Likewise, sites for woodrat dens are at a premium and ideal sites may be used over periods of thousands of years (Fig. 13-27 Betancourt et al. 1990; Wells & Jorgensen 1964), as indicated by the presence of plant fragments in the dens that no longer occur in the area but did thousands of years ago.

Migration

Migratory behavior evolved primarily as a way to exploit food abundance at one time of year and to avoid undesirable conditions at another. Migrations are energetically expensive and expose the participants to increased risks along the route, but the benefits clearly exceed the costs. Migration in some species brings large numbers of individuals together during the mating season and may enhance the survival of the newborn.

Many mammals make regular seasonal migrations: Gray whales (*Eschrichtius robustus*) travel annually more than 10,000 kilometers, from northern Pacific summer feeding areas to Mexican winter breeding areas and back. Northern elephant seals (*Mirounga angustirostris*) migrate biannually from the northern Pacific to California's Channel Islands and back. These seals are sexually dimorphic with respect to their foraging areas and migratory routes, and males tend to migrate greater distances (about 21,000

TABLE 23-1 Sizes of Home Ranges of Some Mammals

Species	Home Range (Acres)	Source
Common shrew (*Sorex araneus*)	0.7	Buckner 1969
Varying hare (*Lepus americanus*)	14.5	O'Farrell 1965
Mountain beaver (*Aplodontia rufa*)	0.3	P. Martin 1971
Least chipmunk (*Tamias minimus*)	2.1–4.7 (summer only)	Martinson 1968
Yellow-pine chipmunk (*T. amoenus*)	3.89 (males); 2.49 (females)	Broadbooks 1970
White-footed mouse (*Peromyscus*)	0.08–10.66	Redman and Sealander 1958; Blair 1951
Red-backed vole (*Myodes gapperi*)	0.25 (winter only)	Beer 1961
Prairie vole (*Microtus ochrogaster*)	0.11 (males); 0.02 (females)	Harvey and Barbour 1965
Gray wolf (*Canis lupus*)	23,040 (pack of 2) 345,600 (pack of 8)	Stenlund 1955 Rowan 1950
Red fox (*Vulpes vulpes*)	1,280	Ables 1969
Grizzly bear (*Ursus arctos*)	50,240 (1 mother & 3 yearlings)	Murie 1944
Russian brown bear (*U. arctos*)	6,400–8,320	Bourliere 1956
Raccoon (*Procyon lotor*)	13.3–83.4	Shirer and Fitch 1970
Badger (*Taxidea taxus*)	2100	Sargent and Warner 1972
Mountain lion (*Puma concolor*)	9,600–19,200 (males) 3,200–16,000 (females)	Hornocker 1970b
Lynx (*Lynx canadensis*)	3,840–5,120	Saunders 1963
Blacktail deer (*Odocoileus hemionus*)	90 (winter); 180 (summer)	Leopold, et al. 1951
Mule deer (*O. hemionus*)	502–2,534	Swank 1958
White-tail deer (*O. virginianus*)	126–282	Ruff 1938
Pronghorn (*Antilocapra americana*)	160–480	Bromley 1969

FIGURE 23-10 (A) A pocket gopher (*Thomomys bottae*) from Campbell, California, and (B) mima mounds in Mima Prairie, Thurston County, Washington. These mounds, some 10 meters in diameter, are probably formed by the burrowing of the Mazama pocket gopher (*T. mazama*).

kilometers) than do females (**Fig. 23-11**; Stewart 1997; Stewart & DeLong 1995).

The migratory journeys of terrestrial mammals are generally shorter than those of marine mammals, but

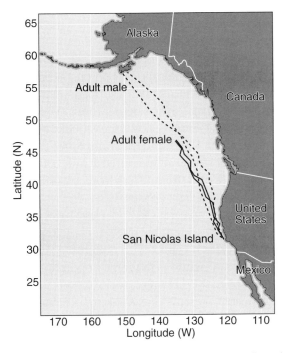

FIGURE 23-11 Different migration patterns of adult male and adult female northern elephant seals (*Mirounga angustirostris*). (Modified from Stewart, B. S., *J. Mammalogy* 78 (1997): 1101–1116. American Society of Mammalogists, Allen Press, Inc.)

some are impressive (**Fig. 23-12**). The ungulate migrations in the Serengeti ecosystem are one spectacular example. The complex, circuitous migration of the Serengeti wildebeest (*Connochaetes taurinus*) in 1960 was estimated to have covered at least 1,700 kilometers (Talbot & Talbot 1963). These migrations vary in length and route from year to year and involve the movement of millions of animals of several species including: wildebeest (**Fig. 23-13**), zebra (*Equus burchelli*), and Thomson's gazelle (*Eudorcas thomsonii*), among others. Rainfall patterns are tightly linked with the patterns of ungulate migrations in the Serengeti. American pronghorn (*Antilocapra americana*) in Wyoming move 200 to 500 kilometers seasonally along the Wind River Mountains to areas in Grand Teton National Park (Sawyer & Lindzey 2000). In Canada and Alaska, caribou (*Rangifer tarandus*) follow well-established routes between their calving and winter feeding grounds. Wolves (*Canis lupus*) follow these caribou herds, accomplishing one of the longest seasonal migrations for terrestrial carnivores (Walton et al. 2001). Migration distances in bats vary from less than 100 kilometers in some populations of *Eptesicus fuscus* (Neubaum et al. 2006) to more than 1,500 kilometers in some populations of *Tadarida* (Hill & Smith 1992).

Much remains to be learned about the physiological, behavioral, and genetic adaptations for mammalian migration. Reviews of the evolution of migration (Alerstam et al. 2003; Berthold 2001) suggest that migration apparently requires (1) innate instructions about the circannual timing

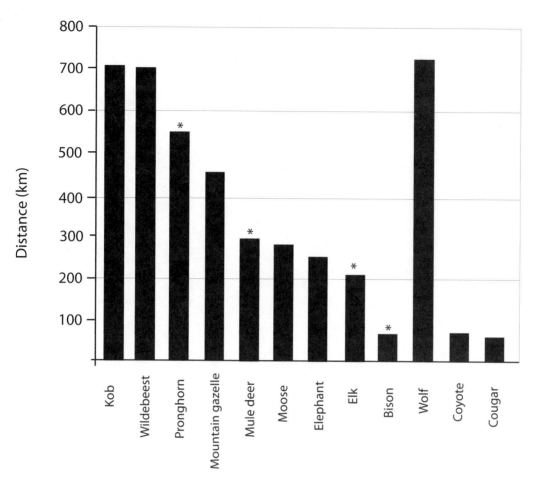

FIGURE 23-12 Maximum round-trip migration distances for selected large mammals. (Data from Berger, Joel, *Conservation Biology* 18 (2004): 320–331.)

and duration of movement, (2) physiological adaptations supporting energy storage and use, (3) behavioral adaptations that allow flexibility in response to varying weather or currents en route, and (4) awareness of navigational cues.

Life History Patterns

Life tables, which recognize age-specific fecundity and age-specific survivorship for each age class (cohort), help us predict how populations will change over time. In his study of Alaskan wolves, Adolph Murie (1944) developed a life table for Dall sheep (*Ovis dalli*) to understand their population dynamics as affected by wolf predation. Consider Fleming's (1988) life table data for female short-tailed fruit bats (*Carollia perspicillata*; **Fig. 23-14** and **Table 23-2**). Mortality is expressed in a number of ways: survivorship, age-specific mortality, and life expectancy. A newborn *Carollia* has a 53.7% chance of surviving to its first birthday (l_x). Put another way, a cohort of newborns would be expected to suffer approximately 50% mortality in their first year of life (d_x). Age-specific mortality is designated by q. A 5-year-old bat, for example, has a 20.6% chance of dying in its fifth year of life. Life expectancy (e_x) indicates the number of years a bat of a given age (x) is expected to live. A newborn, for

example, is expected to live 2.56 years. Because this is a life table for females, the fecundity is also represented by m_x (the *m* derives from maternity). From these data, it is possible to calculate the expected number of offspring a single female would produce in her lifetime (R_o). In the case of *Carollia*, a female will replace herself with a single daughter over her lifetime. For comparison, the net reproductive rate (R_o) of the killer whale (*Orcinus orca*) was measured in one study as 2.17 (Olesiuk et al. 1990), and the chimpanzee (*Pan troglodytes*) in another study was measured at only 0.44 (Goodall 1986). No value of R_o that differs from 1.0 is sustainable in the long run; if it were greater than 1.0, it would lead to unchecked exponential growth, and if less than 1.0, it would lead to decline or extinction. For example, a population at **carrying capacity** is expected to manifest a very low net reproductive rate, whereas the same population with low density relative to the carrying capacity is expected to have a very high net reproductive rate. A summary of life table data for a variety of mammalian species is given in **Table 23-3**.

In mammals, the contribution made to a population by reproduction (R_o) clearly depends on a variety of factors and is seldom constant within a species from year to year. As mentioned, variation in litter size, number of litters, length

FIGURE 23-13   Wildebeest crossing the Mara River in Kenya during the annual migration.

of breeding season, age at which young animals breed, and survival rates of young are all important variables. In addition, the litter size and the percentage of females that becomes pregnant change with the age distribution of a species; the age composition of a population may therefore have a marked effect on its reproductive performance.

Reproductive performance varies in response to environmental conditions. Whereas well-nourished female white-tailed deer may breed first at 17 months of age, those that occupy poor ranges may wait until 41 months of age (Taber & Dasmann 1957). Similarly, Stevenson-Hamilton (1947) reported that the litter size of the African lion dropped when food was scarce.

Survival rates of young also strongly affect population levels. Young are clearly a vulnerable part of a population and show the greatest fluctuations during population changes. A 74% decline in pocket gopher density in western Colorado in 1958 was associated with an extraordinary drop in the survival rate of young (Hansen & Ward 1966), and in southern Colorado, high survival rates of young pocket gophers were characteristic of periods of high densities, whereas low survival rates of young were associated with a declining population (Hansen 1962). Low

survival rates of young have also been found in declining vole populations.

Survivorship curves, such as those in **Fig. 23-15**, graphically illustrate the generally high juvenile mortality. In this figure, Dall sheep have low juvenile mortality relative to the eastern gray squirrel (*Sciurus carolinensis*). In the case of the African buffalo (*Syncerus caffer*), the life expectancy at 1 year of age is considerably higher than at birth. Generally, life expectancy increases with age in large mammals because after they attain a certain percentage of their adult size they are more difficult for all but a few predators to kill. Although they are small, some bats that hibernate also have long lives (up to 30 or more years), and their survivorship curves resemble those of much larger species.

A population's life history traits may change over time (via selection) as environmental conditions change. Environments in which resources are abundant favor rapid population growth. Because the offspring will not be competing with one another for food, selection favors production of larger numbers of smaller offspring. In contrast, in crowded environments, competition for limited resources is severe and selection favors allocating more resources to fewer but larger offspring.

FIGURE 23-14   A Seba's short-tailed bat (*Carollia perspicillata*) in Jurua, Brazil.

■ Population Growth and Regulation

In 1908, a small group of moose colonized Isle Royale in Lake Superior from the adjacent mainland. The moose found ample food, no competitors, and no effective predators because wolves did not become established on the island until the 1940s (Mech 1966; Peterson 1977). Under these nearly ideal conditions, the moose population soon increased to nearly 3,000 by 1929. As moose numbers grew, browse became depleted and starvation ensued. A second census in the mid 1930s revealed that the moose population had plummeted to only a few hundred animals (Hickie 1936). As this example illustrates, each environment is capable of supporting only a limited number of animals at any given time (carrying capacity).

Humans are aware of the tremendous capacity of populations to increase, for our own population has grown exponentially in recent history. The result of **exponential growth** is a continuously accelerating curve whose slope increases with increasing population size. Exponential growth may occur when small populations are recovering from natural catastrophes or when they colonize new areas that have ample food resources and few predators or competitors. For example, grey seal (*Halichoerus grypus*) populations in the northwest Atlantic were reduced to only a few thousand individuals by the early 1960s; however, detailed monitoring of grey seal pups on Sable Island, Nova Scotia, between 1976 and the 1990s, indicated

TABLE 23-2   Life History Table for Female *Carollia perspicillata* from Costa Rica

Age (x Years)	Probability of Survival at Year x l_x	Proportion Dying During xth Age Interval d_x	Probability of an x Year Old Dying During Age Interval q_x	Number of Females Produced by Each Female m_x	$l_x m_x$	Life Expectancy e_x
0	1.000	0.537	0.537	0.0	—	2.56
1	0.537	0.281	0.523	0.650	0.349	2.90
2	0.256	0.046	0.180	0.650	0.166	
3	0.210	0.047	0.224	0.650	0.137	$T = 3.14 \text{ yr}^a$
4	0.163	0.037	0.227	0.650	0.106	
5	0.126	0.026	0.206	0.650	0.082	
6	0.100	0.028	0.280	0.650	0.065	
7	0.072	0.026	0.361	0.650	0.047	
8	0.046	0.015	0.326	0.650	0.030	
9	0.031	0.013	0.419	0.650	0.020	
10	0.018	—	—	0.650	0.012	
				6.500	$R_o = 1.013^b$	

[a] T = generation time in years.
[b] R_o = reproductive rate.

TABLE 23-3   Life History Table for Representative Mammals

Common Name	Species Name	Age at First Reproduction (yr)	Maximum Age (yr)	Mean Female Offspring per Female	Mean Fertility (F)	Mean Adult Survival (P)	R_0	T_c	l_α
Red deer[1]	*Cervus elaphus*	2	15	0.34	0.30	0.80	1.32	4.86	0.78
Wildebeest [2]	*Connochaetes taurinus*	2	16	0.38	0.15	0.84	0.78	6.70	0.34
Zebra[3]	*Equus burchellii*	3	20	0.24	0.20	0.89	1.45	8.12	0.67
African elephant[4]	*Loxodonta africana*	14	60	0.10	0.07	0.94	1.01	26.62	0.57
Lion[5]	*Panthera leo*	2	17	0.39	0.19	0.90	0.75	7.21	0.40
Harbor seal[6]	*Phoca vitulina*	4	35	0.41	0.16	0.89	0.94	12.18	0.27
Orca[7]	*Orcinus orca*	13	60	0.07	0.07	0.99	2.17	27.58	0.79
Chimpanzee[8]	*Pan troglodytes*	14	50	0.09	0.08	0.95	0.44	23.98	0.32
Short-tailed fruit bat[9]	*Carollia perspicillata*	1	10	0.65	0.35	0.66	1.01	3.14	0.54
Little brown bat[10]	*Myotis lucifugus*	1	12	0.50	0.16	0,86	0.92	4.76	0.31
Belding's ground squirrel[11]	*Spermophilus beldingi*	1	9	1.70	0.66	0.52	1.36	2.44	0.39
Red squirrel[12]	*Tamiasciurus hudsonicus*	1	6	1.90	0.63	0.52	1.30	1.83	0.33
Snowshoe hare[13]	*Lepus americanus*	1	4	9.30	0.95	0.20	1.19	1.25	0.10

Abbreviations F, P, R_0, T_c, and l_α are explained in the text.
[1] Lowe (1969), [2] Attwell (1982), [3] Spinge (1972) and Smuts (1976), [4] Laws (1966), [5] Packer at al. (1988), [6] Reijnders (1978), [7] Olesiuk et al. (1990), [8] Goodall (1986), [9] Fleming (1988), [10] Humphrey and Cope (1976), [11] Sherman and Morton (1984), [12] Kemp and Keith (1970), [13] Meslow and Keith (1968).
Modified from Fleming, T. H. *The Short-Tailed Fruit Bat: A Study in Plant-Animal Interactions.* Chicago: University of Chicago Press, 1988.

that pup populations have been increasing exponentially at an annual rate of approximately 13% for more than 4 decades (**Fig. 23-16A**) (Bowen et al. 2003). This rapid rate of seal population growth occurred despite periods of dramatic environmental fluctuations. Bowen et al. conclude that exponential growth was possible because of increases in abundance of small prey fish in the region and a reduction in the seal's competitors. How long the Sable Island seal population can continue to increase at this rate is not known.

Although virtually all populations have the capacity for rapid rates of growth under ideal conditions (known as their **biotic potential**), populations rarely attain this potential because of many factors (Fig. 23-16B). Collectively, these factors are referred to as environmental resistance or carrying capacity. Food shortages, a lack of suitable cover, disease, and predation all serve to limit a population's biotic potential. As a population approaches the carrying capacity of its environment, the population typically grows at an ever-decreasing rate (**logistic population growth**; Fig. 23-16B).

The abundance of a mammal species at a given time and a given locality depends on the carrying capacity of the habitat and the relationship between the rate at which the animals are added to the population (by reproduction or immigration) and the rate at which they are lost from the population (by death or emigration). Over the long run, population size is controlled by **density-dependent** and **density-independent** processes. Density-dependent processes, such as competition, reproductive rate, predation, dispersal, and disease, have increasing effects as population density increases. For example, disease can spread more rapidly in a dense population because increased crowding favors rapid transmission. Density-independent factors affect a population in ways and at rates unrelated to population size. For example, a series of large fires burned out of control in Yellowstone National Park in 1988 destroying 1.4 million acres of habitat, and the catastrophic eruption of the Mount Saint Helens volcano in Oregon in 1980 devastated an area of 600 square kilometers. Both natural disasters acted to reduce or eliminate mammal populations regardless of their density.

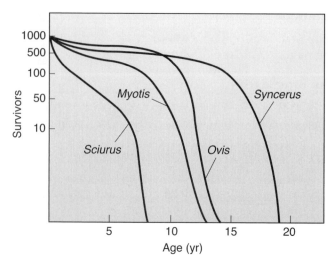

FIGURE 23-15 Survivorship curves for four mammals: Gray squirrel (*Sciurus carolinensis*), greater mouse-eared bat (*Myotis myotis*), Dall sheep (*Ovis dalli*), and the African buffalo (*Syncerus caffer*). Juvenile mortality is high in all species but remains relatively high throughout the lifespan in *Sciurus*. (Gray squirrel data from Barkalow, F.S., Jr., *Journal of Wildlife Management* 34 (1970): 489–500; Greater mouse-eared bat data from Gaisler, J. *Ecology of Small Mammals* (DM Stoddart, ed.) Chapman and Hall, 1979; Dall sheep data from Deevey, E.S., *Q Rev Biol.* 22 (1947): 283–314 and African buffalo data from Sinclair, A.R.E. *The African Buffalo, A Study in Resource Limitation of Populations.* University Chicago Press, 1977.)

Density-Dependent Factors Limiting Populations

When population densities are high, density-dependent factors depress the birth rate, increase mortality, or force migration to new areas. **Competition** for increasingly limited resources at high population densities and high levels of predation or parasitism is thought to be the most important density-dependent factors controlling mammalian population size.

Competition Competition occurs when two or more individuals occupying the same habitat at the same time are using some environmental resource in short supply. Competition can be direct or indirect. Individuals competing indirectly may never come in contact—a chipmunk may eat so many cutworms during the morning that it becomes unprofitable for other chipmunks to search the same area for cutworms in the afternoon. Individuals competing directly, on the other hand, are in direct confrontation for a given resource. Competition may also be intraspecific (between members of the same species, as in the preceding chipmunk example) or interspecific (between members of different species), as when a pride of lions takes over a freshly killed wildebeest from a group of spotted hyenas. Typically, interspecific competition has one of the following outcomes: One species becomes extinct and is replaced by the other (an evolutionary process repeated countless times

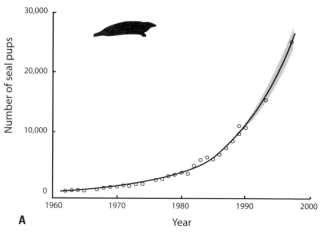

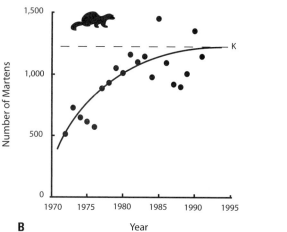

FIGURE 23-16 (A) Exponential population growth in gray seal pups (*Halichoerus grypus*) on Sable Island over 4 decades. The gray zone to the right is the 95% confidence limits for census data based on incomplete tagging. (B) Logistic growth over 25 years in a marten (*Martes americana*) population from Algonquin Provincial Park, Ontario, Canada. The population stabilized around 1,200 individuals beginning in 1985. ([A] Data from Bowen, W.D., et al., *Journal of Marine Science* 60 (2003): 1265–1274; [B] Data from Fryxell, J. M., et al., *Ecology* 80 (1999): 1311–1321.)

during the history of life on Earth); one species emigrates to another area; or one or both species change with regard to their use of the disputed resource (the latter two occur on an ecological time scale). Interspecific competition is discussed later under community-level species interactions.

Often competition is most intense among **conspecifics** (members of the same species). Resource depletion by conspecifics at high population density can reduce fecundity because females produce fewer offspring when food is in short supply. Stewart et al. (2005) manipulated elk densities in the wild to test whether density-dependent mechanisms alter physical condition and fecundity. They predicted, and confirmed, that body condition and fecundity of females is lower in an area of high population density than in a low-density area.

The action of conspecifics may also directly or indirectly increase mortality. Individuals that cannot gain access to sufficient food are likely to starve, or their poor condition increases their likelihood of dying from predation or disease. Individuals unable to find sufficient resources in one area may be forced to risk dispersal to new areas. There is a strong trade-off between food availability and safety in high-density house mouse (*Mus domesticus*) populations (Ylonen et al. 2002). Mice moved from recently mowed pastures, which offered little food and almost no cover from predators, to the vegetation along the fence rows. Here cover was better but food was still scarce. When mouse population densities were highest, the mice took greater risks in all habitats.

Intraspecific strife is probably an important cause of mortality in some large and powerful predators. In a 10-year study of pumas (*Puma concolor*) in New Mexico, Logan and Sweanor (2001) found that the primary source of natural mortality was conspecific killing. Of 27 puma cubs whose causes of death were known, 12 were killed (and in some cases, eaten) by adult males. In the case of 30 adults, intraspecific strife, involving fighting over food or territory, or predation and cannibalism, caused 46% of the male deaths and 53% of the female deaths. In all cases, males were the killers and in nearly all cases the killer was much larger than the victim. Logan and Sweanor stressed that adult females are under strong selective pressure to avoid males.

The question "Why do male pumas kill other pumas?" was considered by Logan and Sweanor (2001). Breeding success, and therefore fitness in males, is strongly linked to longevity. A male's taking over and eating prey of a smaller puma improves the physical condition of the aggressor, may enhance his survival, and can potentially increase his longevity. Furthermore, males that kill males trespassing on their territories are reducing the threat of infanticide for cubs the killer has sired. If a male is hungry, killing and eating an unfamiliar female provides nourishment, but also reduces competition for food and could reduce competition with the mother of the male's cubs and his matrilineal daughters (daughters that have remained as adults within their mother's territory). A male's killing and eating an unrelated litter of cubs (infanticide) provides nourishment but probably more importantly causes the mother to cycle into breeding condition far sooner than if she had raised the litter; this gives the male an earlier chance to sire her litter. In the Logan and Sweanor study, both of the males that killed and ate entire litters later sired litters with the mothers of the cannibalized cubs. Similar situations are known for African lions (Bertram 1975) and leopards (Packer et al. 1988).

Intraspecific competition may also lead to space depletion. When resources are in short supply, individuals may compensate by increasing territory size and defending the territory against conspecifics. Individuals without a territory may not breed and are exposed to increased predation risk. As population density increases, even successful territory holders will have to devote more time to defense and less time to foraging. Some species reduce intraspecific competition by reducing competition between sexes. Sexual dimorphism (differences in size between the sexes) in some predators may enable males and females to exploit different sizes or kinds of prey. This dimorphism is well-illustrated by the long-tailed weasel (*Mustela frenata*) and the ermine (*M. erminea*), in which males weigh approximately twice as much as females. Male long-tailed weasels kill larger prey than do females, and this is probably also true for the ermine.

Disease and Parasitism Disease and parasitism are also density-dependent population regulating mechanisms; the transmission rate of disease increases with crowding in the population. Parasitism and disease are known to be significant causes of mortality among mammals and may occasionally cause dramatic population crashes. Such was the case in a die-off of prairie dogs (*Cynomys gunnisoni*) in Colorado caused by bubonic plague (Lechleitner et al. 1962). Recently, canine distemper virus (CDV) was identified in several carnivore species in the Serengeti-Mara ecosystem of East Africa. Domestic dogs living adjacent to the park are probably the source of CDV, which is thought to have caused several fatal epidemics in canids, including black-backed jackals (*Canis mesomelas*), bat-eared foxes (*Otocyon megalotis*), and African wild dogs (*Lycaon pictus*). Although the Serengeti lion population was not affected by CDV, an epidemic of the closely related morbillivirus emerged in the lion populations in early 1994. By 1997, nearly 1,000 of the 3,000 lions in the ecosystem had died (Troyer et al. 2004). In many mammal populations, disease as the single cause of death may be relatively unimportant but may be important in contributing to the vulnerability of an animal to predation or to stressful environmental conditions.

In extreme cases, disease may lead to a species' extinction. In 1996, Tasmanian devils (*Sarcophilus harrisii*) were observed with large facial tumors. By 2005, Tasmanian devil populations had fallen precipitously and individuals with tumors were reported over 51% of Tasmania (Hawkins et al. 2006, and references therein). Devil facial tumor disease (DFTD, as it is now called) is lethal and responsible for population declines of up to 90% in some areas (see Chapter 28, online). The Tasmanian devil has been placed on the endangered species list (IUCN 2008), and there is a real possibility that this species may become extinct.

Parasitic diseases may also cause species extinction. Wyatt et al. (2008) reported the first case of mammalian extinction brought about by an infectious parasite (a trypanosome). The endemic rat, *Rattus macleari*, disappeared from Christmas Island around 1908, after the arrival of humans and black rats (*Rattus rattus*) in 1857. By extracting ancient DNA from museum specimens of both rat species and searching for the molecular signal of trypanosome infection, these researchers showed that endemic rats lacked trypanosome infection before the arrival of black rats. They also ruled out hybridization between the two rat species. Instead, their data indicate that a parasitic pathogen, *Trypanosoma lewisi*, caused the extinction of the "immunologically naïve" endemic rat.

Parasitism occasionally is an important cause of mortality, but careful observation indicates that otherwise healthy animals can often tolerate a moderately heavy parasite load. Heavy infestations of botflies (*Cuterebra*) in white-footed mice are known to affect reproduction. Botfly larvae burrow beneath the skin and take up residence, breathing through an air hole to the outside. They mature within a month and exit via the air hole to pupate in the soil (Whitaker 1968). These botflies often infest the inguinal and scrotal regions of white-footed mice, and up to 65% of the mouse population can be infested. Botfly-infested females produced fewer litters and fewer offspring than did their uninfested counterparts (Burns et al. 2005). Thus, botfly parasitism has an indirect effect on population density in white-footed mice populations.

Predation   Although predation is clearly a source of mortality among mammals, our understanding of the ability of predators to control or influence densities of prey species or to regulate populations remains incomplete. Predator populations may exert density dependent pressure on prey populations. Solomon (1949) and Holling (1959) concluded that (1) higher prey density resulted in increased prey consumption rate by each predator (functional response) and (2) predator density increased with increasing prey density (numerical response). When prey densities are low, a predator spends most of its time searching for prey, whereas at high prey densities, a predator captures and consumes more prey in a given period. Predators become more abundant as prey density increases (numerical response) because predators reproduce at higher rates when ample energy can be allocated to reproduction and because predators aggregate at sites of high prey density (aggregation response).

The results of 27 studies of wolf (*Canis lupus*) and moose (*Alces alces*) interactions in North America indicated that wolves exerted a strong functional response on moose populations, and that the killing rate was strongly correlated with moose density (Messier 1994; **Fig. 23-17A**). Wolves became satiated at a killing rate of approximately 3.4 moose/wolf/100 days. Additionally, there was a clear numerical response as wolf density increased with moose density (Fig. 23-17B).

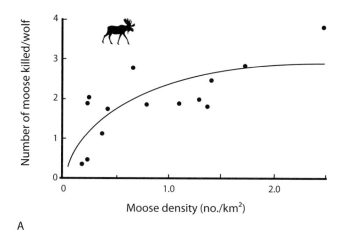

A

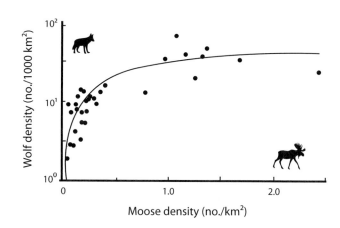

B

FIGURE 23-17   (A) The functional response of wolves preying on moose. The per capita kill rate by wolves was strongly related with moose density. (B) The numerical response of wolves to increasing moose density. (Data from Messier, F., *Ecology*. 75 (1994): 478–488.)

Predation has also been assumed to be a prime factor regulating snowshoe hare (*Lepus americanus*) cycles. In the boreal forests of the Yukon Territory, Canada, snowshoe hare populations cycle with peak densities occurring every 8 to 11 years. O'Donoghue et al. (1997) demonstrated a strong numerical response by both coyote (*Canis latrans*) and lynx (*Lynx canadensis*) populations to swings in snowshoe hare density. Both predator populations increased several fold a year after hare populations peaked (i.e., a delayed response). As hare populations declined, lynx populations had lower pregnancy rates, smaller litters, and poor offspring survival (Mowat et al. 1996).

One implication of functional, numerical, and aggregate responses for predator–prey systems is that predators may exert larger effects on prey populations when prey densities are low, but may be less important at high prey densities. Generalist predators (such a coyotes) consume a wide variety of prey, whereas specialist predators (such as lynx) depend almost exclusively on one or two prey species. The general result is that generalist predators tend to stabilize prey populations and specialist predators tend to destabilize numbers of their primary prey (Krebs 2009).

Population Cycles

Mammalian population cycles are among the most impressive biological phenomena (**Fig. 23-18**). Population cycles, mostly restricted to arvicoline rodents and lagomorphs, are remarkable because they are regular, periodic, high in amplitude, and synchronous across wide geographic areas (King & Schaffer 2001; Korpimäki et al. 2004). Striking changes in density occur primarily in arctic and subarctic areas, with lower amplitude cycles in more temperate regions (Korpimäki et al. 2004). High-latitude areas are characterized by biotic assemblages and food webs that are simpler than those of more temperate and tropical areas. The typical boreal community has a limited biota and supports few species of vertebrates, but some species may, at least periodically, be remarkably abundant (**Table 23-4**). The simplicity of the northern community is seemingly partly responsible for its instability. Where so few kinds of organisms exist, any marked fluctuation in the density of one species seems to disrupt the entire community. Thus, in many northern ecosystems the abundances of several small mammal species fluctuate in synchrony (Korpimäki et al. 2004). Even within limited areas, however, some vole populations may be more stable than others are. In northern Fennoscandia, the bank vole (*Myodes glareolus*) is cyclic, but to the south, it is noncyclic (Bergstedt 1965). As a final complication, some populations of voles have annual cycles over several years and multiannual cycles in other periods (Getz et al. 1987). What mechanisms drive such dramatic population cycles?

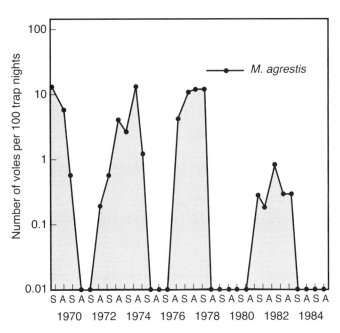

FIGURE 23-18 Cycles of population density in field voles (*Microtus agrestis*) from northern Finland. The abbreviations S and A represent spring and autumn, respectively. (Modified from Boonstra, R., et al., *Ecology* 79 (1998): 1479–1488.)

Characteristics of a Population Cycle

Population cycles in lemmings and voles can be roughly divided into four phases: (1) increase, (2) peak, (3) decline, and (4) low density. The increase phase is a time when densities rise markedly. This phase may continue over several years and may be interrupted annually by short, temporary population declines, or more typically, it may occur within 1 year, with extremely sharp increases over a period of 3 or 4 months (**Fig. 23-19**). In voles at least,

TABLE 23-4 Population Densities of Several Species of Arvicolines (*Microtus* and *Lemmus*)

Density (per Acre)	Species	Region	Reference
1–20	*M. pennsylvanicus*	Northern Minnesota	Beer et al. 1954
6–67	*M. pennsylvanicus*	New York	Townsend 1935
3,000	*M. montanus*	Northwestern United States	Spencer 1958a
200–4,000	*M. montanus*	Oregon	Spencer 1958b
25–81	*M. californicus*	Northern California	Greenwald 1957
425	*M. californicus*	Northern California	Lidicker and Anderson 1962
25–145	*M. ochrogaster*	Kansas	Martin 1956
250–300	*M. agrestis*	England	Chitty and Chitty 1962
1,900	*M. arvalis*	France	Spitz 1963
1,004	*M. guentheri*	Israel	Bodenheimer 1949
2,400	*M.* spp.	Russia	Hamilton 1937
50–100	*L. sibiricus*	Alaska	Rausch 1950
200–300	*L. lemmus*	Sweden	Curry-Lindahl 1962

Data from Aumann (1965).

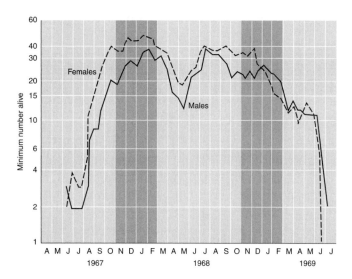

FIGURE 23-19 Changes in the densities of *Microtus pennsylvanicus* during a population cycle in southern Indiana. The shaded period is winter. (Adapted from Gaines, M.S. and Krebs, C.J., *Evolution* 25 (1971): 702–723.)

the increase phase results from changes in the timing of breeding and the rate of sexual maturation. Individuals quickly reach sexual maturity, and breeding begins early in the spring and often continues into the winter. During an increase phase, some Norwegian lemmings (**Fig. 23-20**; *Lemmus lemmus*) were found to be pregnant at 20 days of age (Koshkina & Kholansky 1962). Koshkina (1965) found that the rate of sexual maturation was affected by population density, with early maturation and high survival being typical of an increasing population. There is high dispersal during the increase phase (**presaturation dispersal**), before the habitat is saturated (before reaching its carrying capacity). Such dispersal generally involves vigorous individuals with a good chance of survival (Gaines et al. 1979; Lidicker 1975).

The peak phase is a time of relatively little change in density. The population increase ceases, and the population may remain fairly stable for 1 year or may abruptly swing into a decline. During the peak phase, the breeding season is typically brief. Young animals attain sexual maturity late, and young born at peak times may not mature sufficiently to breed during their first summer. Mortality rates are relatively low during this phase, but dispersal is again high. Dispersal at this time, when the habitat is saturated with voles, is termed **saturation dispersal**. This dispersal typically involves surplus animals (old, young, or social outcasts) that have little chance of survival (Lidicker 1975).

The decline phase varies widely, from precipitous drops in density (population crashes) to uneven declines lasting 1 year or more. As in the peak phase, the decline phase is typified by brief summer breeding and no winter breeding,

FIGURE 23-20 A lemming (*Lemmus lemmus*) from Norway.

and animals reach sexual maturity late. Little dispersal occurs during the decline. Mortality rates are high within the relatively sedentary population.

The low density phase may last from 1 to 3 years, and annual shifts in abundance may occur (Krebs 1966). At this time, the breeding season is short. Animals reach sexual maturity late. Mortality is high, and presaturation dispersal from refugia establishes new colonies.

Regulation in Cyclic Populations

Although population cycles in mammals have been recognized and studied for many decades, considerable controversy remains as to which factors control them, and there are many hypotheses to explain their existence. Factors that regulate population growth cause the mortality rate to increase, or the reproductive rate to decrease, with increasing population density (Krebs 2009). Limiting factors, on the other hand, simply change the average density of the population. Stenseth and Ims (1993) reviewed the competing hypotheses and placed them into three categories: abiotic, intrinsic biotic, and extrinsic biotic. Abiotic factors, such as sunspots and weather conditions, are regarded as unlikely candidates for generating regular population cycles (Korpimäki & Krebs 1996; Ranta et al. 1997).

Intrinsic Biotic Mechanisms The self-regulation of populations by intrinsic biotic processes occurs from within a population. Such intrinsic processes as genotypic, physiological, or behavioral changes have been the focal point of much research. Crowding stress is known to have pronounced effects on some mammals. Pioneering work by Christian (1950) demonstrated increased adrenal activity and other physiological responses in some captive mammals living in dense populations. Christian proposed that these changes were important in controlling population cycles in mammals. Other studies, however, show that the physiological responses to crowding described by Christian are not consistently associated with high population levels under natural conditions (Clough 1965; To & Tamarin 1977). It seems, then, that although the stress syndrome occurs under some conditions, it is not universally important in controlling small-mammal population cycles (Krebs et al. 1992; Lidicker 1988).

Behavioral and morphologic changes during population cycles may also affect population densities. Chitty (1958, 1960) hypothesized that selection for genetically determined behavioral/anatomical features changed with changes in density. For example, both male and female long-tailed voles (*Microtus longicaudus)* in New Mexico were more aggressive at peak population levels than during population declines (Conley 1971). The occurrence of unusually large (or more aggressive) voles at times of peak densities (**Chitty effect**) has been proposed (Krebs & Myers 1974) as part of the pattern of multiannual cycles. Decades later, however, researchers have failed to find substantial support for Chitty's genetic-behavioral hypothesis (Boonstra & Boag 1987; Lidicker & Ostfeld 1991).

Social regulation of reproduction in voles by such responses as disruption of pregnancy may also influence demography and population cycles. The interruption of pregnancy after exposure to strange (unfamiliar) males has been reported for both Old World and New World voles. Heske (1987), who studied voles under simulated natural conditions, found that when a strange male was introduced to an established male and pregnant female pair, the strange male was nearly always defeated and banished by the established male. When the original male (in a pair) was replaced by a strange male, however, the pregnant female usually reacted to the new male's pheromone by terminating pregnancy and conceiving again. This estrus-induction syndrome is perhaps a ubiquitous feature of vole reproduction (Clulow et al. 1982; Richmond & Stehn 1976).

Regulation of populations by immunological dysfunction has been proposed repeatedly (Folstad & Karter 1992; Mihok et al. 1985). The "immunocompetence selection hypothesis" (Lochmiller 1996) proposed that high population densities of herbivores deplete vegetation, leading to protein deficiency and malnutrition. In the decline phase of a cycle, such malnutrition is associated with high risk of infection by opportunistic parasites or disease. Through the various phases of the cycle, selection is thought to favor fecundity at the expense of lowered resistance to opportunistic parasites. Immune-response genes within the **major histocompatability complex** are known to be associated with reproductive performance in a number of vertebrates. For example, genes controlling body size and fertility are within the major histocompatability complex of laboratory rats and mice. During the peak phase, the population might contain an unusually high frequency of animals with high reproductive ability but with low parasite and disease resistance. During the decline phase, then, normally nonpathogenic but opportunistic parasites could cause high mortality, mostly among the newborns. Neonate survival is widely regarded as the prime factor affecting reproductive success and subsequent population densities (Krebs & Myers 1974; Loudan 1985). Individuals stressed by infection may also be highly vulnerable to predation during the decline phase. Lochmiller's model focused on the direct effects of the environment (mainly malnutrition) on survival rates as influenced by host immunity. A key assumption is that selection favoring rapid reproduction will be associated with high frequencies of genotypes conferring reduced immunity to parasites and disease. This association has not been demonstrated for herbivores under natural conditions.

Studies of population dynamics of voles have led to intriguing speculation regarding rodent evolution. Evidence from studies of *Microtus* suggests that rates of evolution differ among populations with different demographic patterns. The frequencies of **allozyme** genotypes change with changes in vole density (Bowen 1982; Gaines 1985). In local populations of California voles, certain rare alleles were often lost during population crashes, indicating that the severity of selection declines during the optimal times that favor population growth. This permits rare genotypes to survive. Relative to stable populations, those given to pronounced cycles may have a greater probability of establishing "evolutionary innovation" in scattered founding populations that survive the crash (Lidicker 1996). Some founding colonies are presumably established by presaturation dispersers with novel genotypes.

Extrinsic Biotic Mechanisms Extrinsic factors are external to the population and occur through the interaction with other species (i.e., predators or pathogens) or by climatic processes that affect food supplies. The quality and abundance of food have been studied in relation to population regulation. Lack (1954) regarded overexploitation of food as a major factor triggering changes in vole and snowshoe hare densities. Much research, however, shows that neither the quantity nor the quality of tundra vegetation limited vole populations. Field experiments that provided excess food during peak densities also failed to slow the population crashes of voles and snowshoe hares (Desy & Batzli 1989; Krebs et al. 1986). Furthermore, Batzli and Pitelka (1970, 1971) found that in one area the favorite food plant of California voles was 10 times more abundant than in another area, yet the two populations of voles underwent similar declines at the same time.

Where rainfall and productivity of vegetation are strongly seasonal, however, plant productivity affects California vole populations. In these California grasslands, vole breeding coincides with the growing season (winter), and floral composition and production influence mean litter size, survival rates, length of residency, and adult sex ratios (Brant 1962; Cockburn & Lidicker 1983; Heske 1987; Lidicker 1976). High vole densities in the growing season reduce future plant productivity (Ford & Pitelka 1984), resulting in low dry-season vole survival and low densities the following wet season. Low winter vole densities, however, lead to good plant productivity, higher survival rates of voles the following summer, and higher vole densities the following winter. Lidicker (1973) recognized the "damaged physiology effect": A population entering the dry season at high density has high mortality rates through this season, and survivors are physiologically damaged. Growth and reproduction are delayed for at least 2 months after the start of the following growing season. At this time, density continues to decline so that not only is the breeding season curtailed but it involves fewer breeders. Such plant–vole interactions may affect short-term population cycles and may influence multiannual cycles. Are other factors controlling population cycles?

Top-down control of small mammal populations by specialist or generalist predators is believed by some researchers to regulate microtine rodent and snowshoe hare cycles, particularly at high latitudes (Korpimäki & Krebs 1996; Korpimäki et al. 2005). The distinct north-to-south gradient in vole cycles, where high-amplitude cycles occur in the north but decline toward southern latitudes, may be associated with the increasing density of generalist predators and the decreasing density of specialist predators in the south (Hanski et al. 1991). Exploitation by generalist predators is thought to synchronize coexisiting prey populations. In western Finland, populations of voles in the genus *Microtus*, bank voles (*Myodes*), Eurasian water voles (*Arvicola*), harvest mice (*Micromys*), and common shrews (*Sorex*) all oscillate in synchrony. Researchers manipulated the densities of the main predators of this small mammal community and demonstrated that predator reduction resulted in increased prey densities and led to the synchronization of the low phase of the population cycle for most of the small mammal species. The authors conclude that "predation mortality and environmental perturbations (e.g., detrimental weather conditions or seasonality) act in concert to induce" synchrony in small mammal populations (Korpimäki et al. 2005).

In contrast, the specialist predator hypothesis, proposes that delayed density-dependent mortality caused by specialist predators drives the dynamics of cyclic vole populations (Hanski et al. 1993). Field experiments in northern Scandinavia in which specialist predators were excluded from certain plots (and compared to control plots with predators) appear to support the specialist predator hypothesis (Klemola et al. 2000; Korpimäki et al. 2004, and references therein). In Finland predator reductions resulted in a fourfold increase in autumn vole densities during the low phase of the cycle, accelerated the increase phase twofold, increased vole densities twofold during the peak phase, and delayed the decline phase of the vole cycle (Korpimäki & Norrdahl 1998; Korpimäki et al. 2002). Similar results were achieved by predator removal experiments in the Canadian arctic. Collared lemming (*Dicrostonyx groenlandicus*) populations were up to ninefold higher in predator exclosure plots relative to control plots during the peak and decline phase of the cycle (Wilson et al. 1999). At another location, collared lemmings exposed to predators declined in the summer, whereas a population protected from predators remained more stable (Reid et al. 1995). Finally, winter food supplementation and predator exclusion experiments in Finland prevented the crash of vole populations, but only in the absence of predators (Huitu et al. 2003). Seemingly predation regulates these cyclic vole and lemming populations.

Such density-dependent regulation by specialist predators may not operate in all cyclic populations (Oli 2003, and references therein). Field voles in northern England typically exhibit cycles every 3 to 4 years with densities ranging from 25 voles per hectare during population lows to more than 400 voles per hectare at the peak phase. Least weasels (a specialist predator; *M. nivalis*; **Fig. 23-21**) were monitored at control sites and removed at manipulated sites (Graham & Lambin 2002). Vole populations in control and predator-removal sites showed similar fluctuations. In addition, predator removal did not prevent vole populations from crashing at any site. Finally, there was no numerical response of weasels to changes in vole density. These results contradicted the specialist predator hypothesis. Graham and Lambin concluded that in northern England "predation by specialist mammalian predators is neither necessary nor sufficient to cause multi-annual fluctuations in the field vole abundance."

In summary, specialist predators appear to play an important role in the population dynamics of small mammals

FIGURE 23-21 A least weasel (*Mustela nivalis*) in its summer pelage in Hackney Marsh, England.

at high latitudes. For the large-scale, high-amplitude cycles of small mammals in boreal regions Korpimäki et al. (2004) suggest that (1) prey population increase is caused largely by increased survival (not increased reproduction); (2) declining food supplies may stop the increase phase; and (3) predator-induced mortality is responsible for the decline phase of the cycle. The regulation of small mammal populations at lower latitudes is still poorly understood.

Multifactor Mechanisms The nature of the human mind is to seek simple, sweeping, universally applicable generalizations or single-factor answers, but the complexity of herbivore cycles, their variability in time and space, and the weight of evidence pointing toward several regulating factors have led many ecologists to adopt a multifactor perspective (Hestbeck 1986; Korpimäki et al. 2005; Lidicker 1988; Stenseth 1985).

In a case involving snowshoe hares (*Lepus americanus*), interactions among three trophic levels (plants, herbivores, predators) seem to regulate herbivore densities (Keith 1981, 1990). The hypothesis that snowshoe hare cycles are controlled by winter food shortages, and predation was tested during a 10-year study in the Yukon (Gilbert & Boutin 1991; Hik 1995; Krebs et al. 1995). In their "predation-risk hypothesis," these authors proposed that in a heterogeneous environment, food is never uniformly in short supply but that hares are forced by risk of predation to avoid the productive open areas in favor of the closed areas with abundant cover but poor food. As predators become more abundant (numerical response) and predation becomes more intense (functional response) during the peak and decline phases of the cycle, the hares become progressively more restricted to safe, food-poor areas. The hares lose weight, and fecundity is low. During the decline phase of the cycle, snowshoe hares are chronically stressed—as indicated by high levels of free cortisol, reduced testosterone response, and overall poor body condition—by elevated predation risk, resulting in a marked deterioration in reproduction (Boonstra et al. 1998). Mortality from predation and the stress induced by predation risk cause the decline phase of the cycle and are regarded as the main driving forces of this system, but because the effects of the nutritional quality of the plants on the condition and reproductive performance of the hares are also vital factors, interactions among trophic levels, rather than predation alone, seem to regulate this cycle.

After decades of research, involving food supplementation and predator removal experiments at a wide variety of locations, there is still no consensus regarding the causes of population cycles in small mammal populations. Advances in our understanding of these complex processes will require long-term, large-scale studies of a broad array of predator–prey interactions. Unfortunately, the time for such studies may be short because climate change may alter boreal food webs and disrupt the dynamics of ecosystems.

Metapopulations

A set of local subpopulations linked together by dispersal is called a metapopulation (Levins 1969). Metapopulation models were originally developed for agricultural pests but are now widely used for any species that inhabits fragmented landscapes (fragmentation may be caused by natural or human agents). Ecologists recognize that populations are frequently composed of subpopulations (or local populations) associated with habitat patches in which the role of dispersal between patches is vital to population maintenance. Instead of focusing on births and deaths within a population, metapopulation theory emphasizes the colonization of empty habitat patches and the extinction of small subpopulations within the larger metapopulation. Consequently, immigration and emigration play a key role in metapopulation theory. Metapopulations may appear stable over long periods even when their subpopulations may be highly ephemeral; subpopulations may go extinct and be recolonized several times within a given time period. Because of its focus on extinction and recolonization of small fragmented populations, metapopulation theory has become an important tool in the practice of conservation biology (Soule 1986).

Despite the popularity of metapopulation theory, few field studies have tested its key assumptions. Large-scale, long-term studies of a field vole (*Microtus agrestis*) metapopulation in the 71-island Tvarminne archipelago in Finland (Crone et al. 2001; Pokki 1981) indicate that this vole metapopulation is controlled by extinctions and colonizations of island subpopulations. Counter to expectations, colonizations from ephemeral small-island subpopulations were most important in maintaining long-term stability of the metapopulation, whereas colonization from stable large-island subpopulations contributed little to the persistence of the vole metapopulation. Overcrowding on large islands led to dispersal from optimal to suboptimal habitat on the same island. On small islands with no suboptimal habitat, voles dispersed to other islands, either by swimming or migrating on ice. Seemingly, within a metapopulation there is a continuum of subpopulation types from those equally prone to extinction and with nearly equal dispersal rates to those with great inequality of these processes.

Small Populations and Extinction Risk

Increasingly today, many medium-sized and larger mammals exist in small populations threatened with extinction (see Chapter 26). Conservation biologists and wildlife managers are interested in the survival of small populations and in reducing population declines. Changes in abundance of large populations, including those in response to environmental changes, are largely deterministic, reflecting average responses. In contrast, changes in small populations tend to be more dependent on chance. Chance (or stochastic) events that threaten persistence of small populations can be divided into environmental, demographic, and genetic categories. Environmental and demographic factors constitute the most pervasive threats to small populations in the wild. In the case of stabilizing and maintaining the smallest populations, such as those in zoos or those restricted to small, highly isolated areas, the focus is on genetic factors, including inbreeding depression and the consequent loss of genetic variability. Genetic variability may be reduced by stochastic processes such as genetic drift (random changes in gene frequency when population size is reduced), non-random mating, and inbreeding. Small populations may

decline because of a train of interactions between genetics and physiology (Krebs 2001). For example, genetic factors that reduce fecundity can prevent a population from increasing, furthering loss of genetic diversity through genetic drift, which further impairs fecundity. Concerns about reduced genetic variability in small populations led Gilpin and Soule (1986) to develop the concept of **minimum viable populations**—the minimum population size needed to have a 90% chance of preventing further loss of genetic variation or suffering chance extinction over an acceptable period time (usually 100, 500, or 1,000 years). A number of stochastic (chance) events can lead to extinction (Shaffer 1981). Natural catastrophes, such as severe weather, floods, droughts, or fires, act disproportionately on small populations, in which the fate of each member is important to the survival of the population. Thus, random declines in birth rate or offspring quality can profoundly affect population survival. This is a case of demographic stochasticity.

Armbruster and Lande (1993) developed a mathematical model predicting requirements for elephant survival in the fluctuating, semiarid environment of Tsavo National Park, Kenya. Their results indicate an equilibrium elephant density of 1.2 elephants per square kilometer at a maximum population growth rate of 3% per year, and they suggest that in semiarid regions a minimum of 2,590 square kilometers is necessary to attain a 99% probability of population persistence for 1,000 years.

Ruggiero and colleagues (1994) introduced a general set of principles regarding population persistence. Populations tend to persist longer (1) in interconnected habitats or those where corridors for dispersal exist, (2) when suitable habitats are close to one another, (3) in larger habitat patches, and (4) when populations have higher reproductive rates. This list is likely incomplete and not universally applicable, but field observations support most of these general principles (see also Chapter 26).

Communities

Biologists have long recognized that animals and plants with similar environmental requirements form identifiable communities. A community is characterized not only by its unique plant and animal assemblage, but by complex interactions between the biota and the physical environment. The term "community" has been used to designate plant–animal assemblages of differing scale and importance (Odum 1971), from the biota of a woodrat nest, for example, to that of the deciduous forests of the eastern United States.

A cautionary note is in order here. The following discussions of communities may give the impression that a community reaches a point of "equilibrium" and that community stability is the rule. The weight of contrary evidence, however, favors the view that communities are in "disequilibrium," continually adjusting to climate and continually falling behind and failing to achieve equilibrium before the onset of a new climatic trend (Davis 1984). Many communities are changing now and will continue to change in response to global climate change, as they always have.

Communities are comprised of multiple species and as such are characterized by several properties not found at the population level, including interspecific interactions, diversity, trophic structure, and succession (Fig. 23-6). Ecologists attempt to describe the biological attributes of a community by such **species diversity** measures as **species richness** (the number of species), **evenness** (relative abundance of individuals), and other indices. Among the many species that make up a community, few are abundant. Most communities contain a few common species and many less abundant ones. In some cases, a single numerically dominant species defines the community. Dominance, however, is only one measure of a species' contribution to a community. A **keystone species** is one that has a critical impact on the community (Krebs 2009).

▪ Species Diversity Patterns

Various diversity measures take into account both the number of species (richness) and how abundant individuals of those species are across the whole community (evenness). Such measures may indicate that two communities have identical species richness but very different evenness. Some diversity indices are used to compare the composition and complexity of different communities. Diversity is determined by a combination of climatic history, geography, latitude, and a host of biotic factors. With a few exceptions, the greatest species richness occurs in tropical areas, with decreasing richness in areas progressively closer to the poles (**Fig. 23-22**). Simpson (1964) recognized several patterns in the North American mammal fauna. First, not all mammal communities show increasing richness toward the equator. The richness of ungulates and shrews, for example, is concentrated in temperate regions, but bats account for much of the species richness in tropical areas. Second, mountainous regions have higher richness than do low-lying areas, which accounts for the higher biotic diversity in western North America. Finally, mammal richness is lower on peninsular areas such as Florida, Nova Scotia, or Baja California. The mammal fauna of South America also shows a strong latitudinal richness gradient. However, this pattern does not hold for all mammal faunas; the highest species richness of small marsupial carnivores occurs in the arid central region of Australia rather than tropical northern Australia (Dickman 1989).

The broad trend toward increased species richness from the poles to the equator—the "latitudinal diversity gradient"—is well-documented, but the explanation for such a pattern remains controversial. One hypothesis is that the higher biological productivity in the tropics results in more energy available to support more species; however,

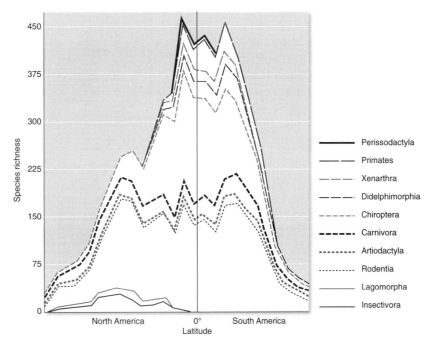

FIGURE 23-22   Latitudinal distribution of species richness of mammalian orders for North and South America. Zero latitude indicates the equator. (Modified from Kaufman, D. M., *J. Mammalogy* 76 (1995): 322–334. American Society of Mammalogists, Allen Press, Inc.)

the positive relationship between primary productivity and species diversity does not hold through the entire latitudinal gradient. In fact, among many animal taxa, the greatest species richness occurs at intermediate levels of productivity (Rosenzweig 1992). Others have suggested that the less seasonal climates of tropical regions allow species to specialize on particular food resources, whereas the highly seasonal climates of the temperate zones cannot sustain specialized diets year round. Alternatively, Kaufman (1995) proposed that in tropical areas, where abiotic conditions are more benign, it is the biotic interactions (e.g., competition, predation) that limit the niche space of each species and result in specialization and increased species packing. In temperate and subarctic areas, in contrast, abiotic conditions (e.g., variation in temperature) are severe and limit the number and types of species that can exist. According to this hypothesis, abiotic conditions set the higher latitude boundaries of species' ranges, whereas biotic interactions limit the distributions of species near the equator. Overall, no single hypothesis to explain the latitudinal species richness gradient has gained universal acceptance.

A similar gradient of species richness occurs with increasing elevation. McCain (2005) conducted a global analysis of elevational diversity trends for **nonvolant** small mammals using 56 data sets from around the world. Her results indicated a clear mid-elevational peak in species richness, with richness peaking at higher elevations on taller mountains. Several factors are important in determining elevational diversity patterns in small mammals, including mountain size, spatial constraints (less area at the top than at the base), and climate.

■ Species–Area Relationships

The 18th and 19th centuries were times of intensive exploration. During this period, naturalists such as Alfred Wallace and Charles Darwin accompanied sailing expeditions to islands around the globe. In the process of cataloging what were new and exotic species to Europeans, certain ecological patterns slowly became apparent. The species–area relationship may be the oldest ecological pattern to be recognized. Johann Rheinhold Forster, the naturalist on Captain Cook's second voyage to the South Pacific (1772), was the first to recognize that "islands only produce a greater or less number of species as their circumference is more or less extensive." With this simple observation, one of the most fundamental ecological relationships was born: As the area of a region increases, so does the number of species encountered.

Ecologists proposed three hypotheses to account for the species–area relationship. The first was the habitat diversity hypothesis (Williams 1964), which states that larger areas have greater habitat diversity and therefore more niches to

be filled. Later, Connor and McCoy (1979) proposed that larger areas simply sampled more individuals and therefore more species; this idea is called the passive sampling hypothesis. Large areas function as "targets" that sample more individuals and hence more species. The dynamic equilibrium hypothesis, proposed originally by Munroe (1948, 1953) and later elegantly dealt with mathematically by MacArthur and Wilson (1963, 1967), has been one of the most influential theories in ecology for more than 4 decades. MacArthur and Wilson proposed that species diversity on islands is determined by island size and two processes: immigration and extinction.

MacArthur and Wilson's (1963) theory proposes that the number of species on an island is determined by a dynamic equilibrium between the rate of immigration of new species to an island and the rate of extinction of resident species. These two rates are affected by factors such as island area, distance from the mainland, population size, number of species, and catastrophic events (**Fig. 23-23**). The model predicts that the species richness of an island will remain relatively constant (i.e., in dynamic equilibrium) but that the list of species may change over time (called the turnover rate).

Habitat "islands" need not be surrounded by water. A patch of rain forest surrounded by farms or mountain tops may serve as an island if it is isolated from other similar habitats. For example, in the Great Basin of the American West the upper elevations of mountain ranges form

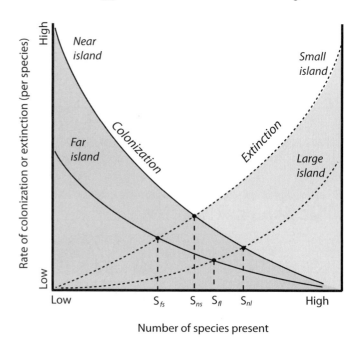

FIGURE 23-23 Graphical representation of the predictions of island biogeography theory. The predicted number of species (*S*) changes with island size and its distance from a "mainland" population. For example, *S* will be greater for large islands that are close to the mainland.

islands of montane habitat isolated from similar "mainland" habitat in the Sierra Nevada and Rocky Mountains by inhospitable desert valleys . Brown (1971, 1978) identified 17 mountain ranges more than 3,000 meters in elevation that rise from a sea of Great Basin desert and are separated from one another and from "mainland" (Rocky Mountains or Sierra Nevada) habitats by desert valleys at least 8 kilometers wide and below 2,300 meters in elevation. The intermountain valleys serve as effective dispersal barriers because suitable habitat for boreal mammals does not exist below 2,300 meters. Boreal mammals were defined as those species that occur at higher elevations in the Rocky Mountains and Sierra Nevada, but not below the piñon-juniper zone (at 2,300 meters) in the Great Basin.

Although mammal species richness was closely correlated with the area of the mountain range, Brown demonstrated that these montane mammal communities do not show a dynamic equilibrium between rates of colonization (= immigration) and extinction. There was no relationship between species saturation on mountaintops and distance to the nearest mainland, suggesting that colonization rates were very low relative to extinction rates. Furthermore, Brown showed that extinction was correlated with both island area and ecological niche. Extinction was most likely in large mammals at higher trophic levels or those with narrow habitat requirements. In the Great Basin ranges, for example, large mammals occur only on the largest islands. Similarly, the cliff chipmunk (*Tamias dorsalis*), a habitat specialist, does not occur on one of the largest ranges (the Ruby Mountains) because piñon-juniper habitat is poorly developed. Finally, Brown noted that 11 species of small mammals that were common in mainland yellow pine forests were completely absent from similar habitat in the Great Basin ranges. Thus, because the extinction rate was far greater than the colonization rate, Brown concluded that mountaintop mammal communities in the Great Basin are not in dynamic equilibrium but are better explained by historical factors, such as the waxing and waning of continental glaciers.

Fossil evidence indicates that cool climates in the late Pleistocene caused boreal conditions to spread over much of the Great Basin, which was then colonized by boreal mammals from "mainland" ranges. As the climate warmed during the Holocene and boreal habitats shrank back to their present distributions, extinction reduced the number of boreal species on isolated mountain ranges to their present levels. Thus, rather than a shifting balance between colonization and extinction rates, as predicted by island biogeography theory, Brown believed that extinction alone shaped these montane communities after the Pleistocene. Recent fossil evidence and the discovery of several "missing" species on ranges in the Great Basin prompted Lawlor (1998) to reevaluate Brown's data (see also Waltari et al.

2007). The reanalysis partly contradicts Brown's hypothesis and shows that modern mountaintop communities are "composed of extinction-resistant woodland species capable of considerable movement among mountain ranges." Brown's work remains important because he recognized that processes other than colonization and extinction have shaped island communities (Brown & Lomolino 2000).

The MacArthur-Wilson equilibrium theory failed to take into account the importance of geologic history and the role of **phylogenesis** (speciation that leads to diversification of a taxonomic group). In the Philippine archipelago, for example, maximal island size was attained during the late Pleistocene when lower sea levels exposed additional shoreline and united some island groups into large "Pleistocene islands." As post-Pleistocene sea levels rose, the modern archipelago, with some 7,000 islands, was formed. To test the importance of historical factors in shaping modern Philippine mammal communities, Heaney et al. (2005) summarized the degree of Pleistocene isolation, colonization ability, and genetic differentiation for six bat species and one muroid rodent. Isolation was considered highest between Pleistocene islands that were continuously separated by deep, wide channels and lowest between islands that were united during the Pleistocene. Colonization ability was based on how widespread each species was across the archipelago and the degree to which it tolerated disturbance. Low gene flow among members of a population results in high genetic differentiation (and vice-versa). As predicted, species with low colonizing ability inhabiting widely separated "Pleistocene islands" had high levels of genetic differentiation. The major pattern of widespread "differentiation between, but not within, Pleistocene islands" led the authors to conclude that speciation is at least as important as colonization and extinction in structuring the mammal communities of the Philippine archipelago. In Southeast Asia, phylogenesis results in as much as 14 times as many species as does colonization (Heaney et al. 2005). Modern views of island biogeography take into account colonization, extinction, historical events, climatic changes, and phylogenesis and recognize that species richness is in a state of dynamic disequilibrium. Constantly changing conditions cause species richness to vary around some point of equilibrium.

Species Interactions

The more diverse a community the more potential interactions there are among its members. Considerations of interactions between one species of mammal and another and between plants and mammals are essential to an understanding of mammalian ecology. A biotic community is a tremendously complex functional unit with an evolutionary history and within which animals live, feed, reproduce, and die. The role of each community member depends on its interactions with other community members and with the physical environment.

The fabric of a community depends on myriad interactions. Consider a typical oak forest community of eastern North America. Oak trees (*Quercus*) produce large numbers of acorns (mast) every few years. These seeds support many animals, including insects, birds, and several mammal species (**Fig. 23-24**). White-footed mice (*Peromyscus leucopus*), chipmunks (*Tamias striatus*), and white-tailed deer (*Odocoileus virginianus*) consume large quantities of acorns during years of heavy acorn production, and rodent density appears to be tightly correlated with acorn production (Ostfeld et al. 1996). The deer and rodents are also the main hosts for parasitic deer ticks (*Ixodes scapularis*) and play an important role as reservoirs of Lyme disease. Tick populations also peak during summers after mast production when host densities are highest. Gypsy moth (*Lymantria dispar*) outbreaks can cause serious defoliation in oak forests, but evidence indicates that small mammal predators, including white-footed mice, play an important role in preventing gypsy moth outbreaks (Ostfeld 1997; Ostfeld et al. 1996). Deer populations also increase in years after

heavy mast production, resulting in overbrowsing of seedlings and saplings, which can eventually alter the species composition of the forest understory. Mast production by oak trees, therefore, has both direct and indirect effects on a number of plant and animals species in the community and on a disease that affects humans.

Interactions involving the effect of one organism on another include interspecific competition, commensalism, mutualism, predation, and parasitism. Interspecific competition (as defined previously) occurs when two or more species use a resource in short supply. **Commensalism** refers to two species that live together; one species gains some advantage, but the other neither gains nor loses. White-footed mice (*Peromyscus*) and shrews (*Notiosorex*) that take shelter in woodrat houses are probably commensals. In **mutualism**, both species benefit: tick birds (*Buphagus*) eat ticks from the bodies of large ungulates, and both birds and ungulates profit. This interaction often shifts to **parasitism** (a parasite feeds on and/or lives in its host, causing it harm), however, when tick birds keep wounds of their hosts open by pecking and eating wound tissue and blood (Maclean, 1993). Thus, species interactions can affect the distribution and abundance of

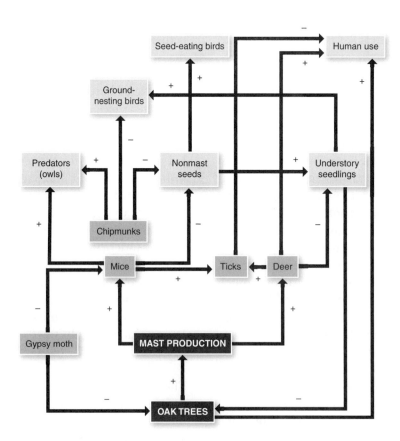

FIGURE 23-24   A model of the relationships between the members of an eastern deciduous forest community in North America. Biomass of a member of an interacting pair will increase if the arrow pointing toward the recipient is positive. For example, acorn (mast) production will increase the biomass of deer and mice, but mice will cause a decrease in biomass of gypsy moths. (Modified from Ostfeld, R. S., et al., *BioScience* 46 (1996): 323–330.)

species within communities. Two of the most common and best studied interactions are interspecific competition and predation.

Interspecific Competition

The intensity of interspecific competition depends on the degree of niche overlap. In practice, no two species ever have identical niches and those with highly similar niches tend to evolve increasingly differentiated niches—a phenomenon known as **niche differentiation**—to alleviate competition. The occurrence of such niche differentiation reflects the strength of competition as a selective force shaping evolution. Nevertheless, when two populations that have evolved similar niches independently suddenly come into contact, complete **competitive exclusion** is also a possibility. In the Spiti Valley in the Indian Trans-Himalaya the wild yak, kiang, Tibetan argali and chiru were probably driven to local extinction by competitive exclusion by domestic livestock (Mishra et al. 2002). Similarly, the more than 50 bat species that occur sympatrically in some tropical regions of South and Central America are segregated into **foraging guilds** according to their preferred habitat, diets, wing morphology, and echolocation calls (Norberg 1974). Siemers and Schnitzler (2004) showed that within bat guilds niches were further divided based on echolocation signals. Each of five species of European *Myotis* that forage by aerial trawling along forest edges had distinctive echolocation calls that contributed to niche separation.

Among 202 local patches of desert in the southwestern United States were 137 different combinations of 29 species of granivorous rodents (Brown 1987; Brown & Kurzius 1987). Despite this complexity, these assemblages seemed to share a series of attributes favoring coexistence.

What resources then are being partitioned? Habitat is often an important factor. Heterogeneous habitats—those with diverse species and sizes of plants and with open spaces of varying sizes—allow for microhabitat segregation among rodents. Quadrupedal pocket mice and harvest mice (*Perognathus*, *Chaetodipus*, and *Reithrodontomys*) forage beneath or near shrubs, whereas bipedal kangaroo rats and kangaroo mice (*Dipodomys* and *Microdipodops*) select open spaces (see Fig. 23-8). Differences in habitat selection (Larsen 1986; Rosenzweig 1973) or partitioning of food resources on the basis of food distribution or foraging efficiency may also favor coexistence (Mares & Williams 1977; Price 1983). The kangaroo rat *D. merriami* learns the locations of seed patches far more quickly than does the pocket mouse *C. intermedius* (Rebar 1995). This may enable the kangaroo rat to forage effectively in open spaces, where seeds occur in scattered clumps. Aggressive interactions involving the use of space are also part of the interaction equation. Although relatively unstudied,

predator-mediated habitat selection must be an additional factor (Kotler 1984; Price et al. 1984). The use of dormancy (torpor, induced by drought or starvation) by some species and different approaches to water conservation may also have an influence.

Experimental field studies in which one species was removed and responses of the other species were monitored have shown that interspecific competition is an important factor in desert rodent communities. In an Arizona community, after exclusion of the large kangaroo rats, four of the five species of small granivorous rodents increased threefold (Brown & Munger 1985; Munger & Brown 1981). With kangaroo rat removal in southern New Mexico, the density of a common pocket mouse (*Chaetodipus penicillatus*) increased two and a half times (Freeman & Lemen 1983). The studies of Brown et al. (1979) and Davidson et al. (1980) suggest the importance of competition between distantly related granivores (seed eaters) in structuring desert communities. Their study area in southern Arizona supported granivorous heteromyid rodents (**Fig. 23-25**), which preferred large seeds, and harvester ants, which preferred small seeds. Replicate experimental plots were established: those with only rodents, only ants, both ants and rodents, or neither species. By comparison with undisturbed control plots, numbers of ant colonies increased sharply (71%) in the plots with no rodents, and rodent biomass increased 29% in the ant-free plots. Seed densities were roughly equal in the ant-free, rodent-free, and undisturbed control plots, whereas they were four times higher in plots with neither ants nor rodents (**Fig. 23-26**). Plant density only doubled on the latter plots, indicating competition between seedlings. These results indicate that rodents and ants compete for seeds and affect seed and plant densities. They also show that plant densities are partly controlled by competition between plants and that the composition of the plant community may be influenced by the seed preferences of rodents and ants.

Competition among rodents may also have selected for differences in morphology or body size (Bowers & Brown 1982; Brown 1973; Hopf & Brown 1986). Body size in coexisting desert rodents is not randomly distributed but tends to be evenly arranged along a size gradient (**Fig. 23-27**). Such size differences occur also among small mustelid and dasyurid carnivores that live sympatrically and depend on small prey (Jones 1997). Size divergence among sympatric mammals with similar ecological requirements has often been regarded as indirect evidence of interspecific competition (Brown & Wilson 1956; Jones 1997; Simberloff & Boecklen 1981).

Predation A predator, in the broadest sense, is any organism that eats all or part of another organism. By this

FIGURE 23-25   A giant kangaroo rat (*Dipodomys ingens*) at the entrance to its burrow in Carrizo Plain National Monument, California.

definition, a pride of lions consuming a zebra, a deer tick sucking the blood of a white-tailed deer, and a koala calmly munching eucalyptus leaves are all acting as predators. Each of these, whether predation, parasitism, or herbivory, benefit one species while harming or reducing the fitness of another species. Unlike predators, which kill and consume

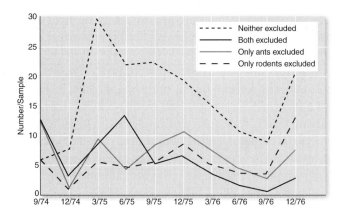

FIGURE 23-26   Comparisons of mean numbers of seeds found in soil samples from plots where neither ants nor rodents were excluded, where both were excluded, where only ants were excluded, and where only rodents were excluded. Note that seed density increased markedly when granivores (seed eaters) were excluded. (Adapted from Brown, J.H., et al., *Am. Zool.* 19 (1979): 1129–1143.)

their prey, herbivores rarely consume the entire plant, and parasites rarely kill their hosts.

Although mammals as a group evolved to exploit many types of food, a single species usually eats a fairly limited array of foods that it is structurally, physiologically, and behaviorally capable of using efficiently. Much of mammalian evolution has been molded by the advantages of achieving the most favorable balance between the energy and time expended in securing and metabolizing food on the one hand (costs) and the energy gained from the food on the other (benefits; Pulliam 1974). The optimal strategy ensures the greatest yield of energy with a minimum of time and energy spent in pursuit, handling, and eating activities (Schoener 1971).

Predatory mammals use a variety of feeding strategies, but two major types represent the extremes. The sit-and-wait predator remains quietly at a vantage point and surveys its surroundings; when prey is detected, the predator makes a brief attack and typically avoids lengthy pursuit. The giant leaf-nosed bat (*Hipposideros gigas*) of Africa exemplifies this predatory style (Vaughan 1977). This bat hangs from an acacia branch and uses echolocation to scan for insects. It is discriminating in its choice of prey and feeds exclusively on large, straight-flying insects. These are captured during brief and precise interception flights, and the bat then returns to its perch to consume the insect. Mountain lions (*Puma concolor*) hunting in the Santa Ana

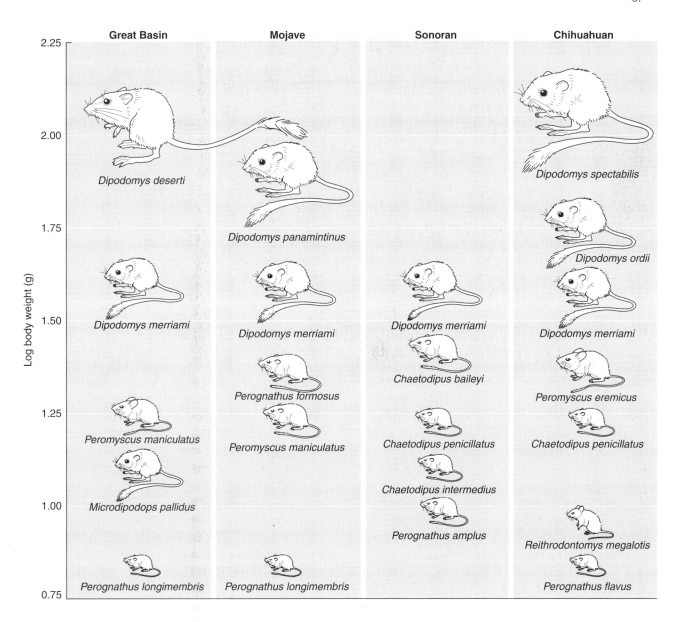

FIGURE 23-27 Species compositions and sizes of the most abundant granivorous rodents that coexist at selected sites in the four subdivisions of the North American Desert: Great Basin Desert, Fish Lake Valley, Nevada; Mojave Desert, near Johannesburg, California; Sonoran Desert, near Silverbell, Arizona; Chihuahuan Desert, near Portal, Arizona. (Modified from Genoways, H. H., and Brown, J. H., *Biology of the Heteromyidae*. American Society of Mammalogists, Allen Press, Inc., 1993.)

Mountains of California waited at vantage points for periods averaging 0.7 hours before moving on (Beier et al. 1995). Such predators are typically solitary and territorial and adjust their watchfulness to periods when prey is most active and consequently most vulnerable. This strategy is profitable when prey is common and mobile. Although the predator spends considerable time waiting for appropriate prey, it is at rest and probably uses little energy. The investment in searching time is great, therefore, but the energy outlay for this activity is low. In contrast, the rate of energy expenditure is high during the capture of prey, but the attack and capture occupy little time.

More common among mammalian predators is the search-and-chase system. Wolves, for example, form social groups capable of searching over wide areas and killing large prey. The pursuit and capture of prey is often a lengthy process that taxes the endurance of the predators and clearly involves the outlay of considerable energy. Although the cost is high, the benefit is great. Packs of wolves probably average only one large kill every several days, but if the prey is an adult moose weighing 400 kilograms, as it often is in the case of the wolves studied by Mech (1966) on Isle Royale, they can gorge on the carcass for at least 2 days. Small carnivores typically hunt solitarily.

Searching may be lengthy, and a series of chases of small prey is usual.

The most abundant and omnipresent foods for terrestrial predators are insects. It is not surprising that the two most important mammalian orders in terms of numbers of species—Rodentia and Chiroptera—depend heavily on this major food source. In addition, members of several other mammalian orders are primarily insect eaters. Insects have probably been a major food source of mammals throughout their evolutionary history. Many mammals have developed special adaptations to exploit common or new insect groups. The early Tertiary adaptive radiation of the highly successful bats probably occurred primarily because of the great abundance of nocturnal moths. The extremely diverse (about 300 species) and nearly cosmopolitan shrews (Soricidae) owe their success to the abundance of insects beneath and in plant litter. In tropical areas, termites and ants are of tremendous importance to mammals today (Redford 1987). At least some members of 10 of the 19 orders of mammals commonly eat termites, and the spiny anteaters (Tachyglossidae), South American anteaters (Myrmecophagidae), pangolins (Manidae), and aardvarks (Orycteropodidae) specialize in termites, as do some members of the orders Erinaceomorpha, Soricomorpha, Afrosoricida, Carnivora, and Macroscelidea.

Many species of mammals are omnivorous and are opportunistic feeders. Such mammals eat a wide variety of plant and animal material. Many marsupials are omnivorous, and omnivores also occur among the orders Erinaceomorpha, Soricomorpha, Afrosoricida, Chiroptera, Primates, Rodentia, Carnivora, and Artiodactyla. Omnivores are typically less specialized in structure than are mammals adapted to narrower diets, and in some orders, the omnivorous mode of life has been highly successful. Among rodents, for example, the nearly ubiquitous North American genus *Peromyscus* includes species seemingly adapted to a wide variety of plant and animal foods. Some terrestrial sciurids are also omnivorous, and in the order Carnivora, such widespread and successful families as Canidae, Ursidae, and Procyonidae have many omnivorous members.

A few mammals have highly specialized diets. The three species of vampire bats have one of the most specialized mammalian feeding techniques: They feed entirely on blood. Some baleen whales eat primarily one kind of planktonic crustacean, at least seasonally. Among the countless filter-feeding animals, the right whales (Balaenidae) are unique: Northern right whales (*Eubalaena glacialis*) feed on tiny organisms to support huge body masses (60,000 to 100,000 kilograms) and a mammalian metabolic rate. In marine ecosystems, trophic efficiency is estimated at 10% (Pauly & Christensen 1995). Accordingly, 1,000 kilograms of phytoplankton can support 100 kilograms of herbivorous zooplankton (e.g., copepod crustaceans), 10 kilograms of zooplankton eaters (such as right whales or herring), and 1 kilogram of piscivores (fish eaters such as halibut or humpback whales). By feeding on zooplankton near the bottom of the food chain, right whales have a relatively plentiful food source. To meet their daily energy needs, these whales must eat an estimated 0.25 to 2.6 billion copepods (*Calanus finmarchicus*) per day (Lockyer 1981). The following discussion of feeding by right whales is based mainly on Baumgartner et al. (2007).

Copepods are among the most abundant zooplankton in the North Atlantic. They occur locally in dense aggregations and are high in energy. Some aggregations are impressive. Baumgartner and Mate (2003) sampled copepod concentrations near feeding right whales and estimated some 15,000 copepods per cubic meter of water. At this density, whales could meet their daily energy needs in 3 hours. Another sample taken under the same conditions had 331,000 *C. finmarchicus* per cubic meter; this would allow a rate of ingestion of 1.4 billion copepods per hour (Beardsley et al. 1996). Later larval stages of this copepod develop oil sacs that provide energy to sustain the copepod. These sacs make this copepod extremely energy-rich relative to other species.

Because of their dependence on *C. finmarchicus*, the feeding patterns of right whales are largely governed by this copepod's life cycle. Through much of winter, the copepods are either newly hatched or are in early larval stages and are so small that they pass through the baleen. At this time, the whales feed on other zooplankton. In late winter, this copepod migrates upward, feeds into the spring on phytoplankton-rich surface waters, and develops into late-stage copepod with oil sacs. In this period, right whales feed at the surface using "skim feeding." During the summer, the entire population of copepods molts into the last larval stage, in which the oil comprises 50% of the body volume, migrates to deep water (deeper than 200 meters), and enters diapause. The right whales respond by shifting their foraging to deep water, where they feed through much of summer and autumn.

At whatever depth they feed, however, right whales depend on concentrations of copepods. If these crustaceans are dispersed, foraging is unprofitable, but aggregations great enough to support feeding whales are patchily distributed in a wide ocean, occurring only where the vagaries of oceanography—such as ocean currents, weather, or ocean-bottom topography—bring huge numbers of planktonic organisms into vertically compressed aggregations (Beardsley et al. 1996).

Many carnivores do considerable scavenging, but none are exclusively scavengers (as are most vultures). Hyenas

(*Crocuta* and *Hyaena*), jackals (*Canis* spp.), and lions (*Panthera leo*) frequently feed on kills made by other carnivores. Animals that steal food secured by other animals are called **kleptoparasites**. On the Serengeti plains, spotted hyenas (*Crocuta crocuta*) often steal prey killed by packs of African wild dogs (**Fig. 23-28**; *Lycaon pictus*). On the open plains, African hunting dogs hunt for an average of 3.5 hours per day, but in order to maintain a positive energy budget, they must increase their hunting effort to nearly 12 hours per day if they lose 25% of their kills to hyenas (Gorman et al. 1998). In wooded areas, visibility is restricted, and hyenas rarely detect kills made by hunting dogs. The decline in hunting dog populations in grassland areas may be the result of high rates of kleptoparasitism in these open habitats.

Herbivory is a special case of predation with plants as prey. Herbivory has many advantages: The biomass of plants vastly exceeds that of animals; plants need not be pursued, and they are equally available throughout the 24-hour cycle. Some members of the Diprotodontia and Primates, as well as the Lagomorpha, Rodentia, Proboscidea, Hyracoidea, Sirenia, Perissodactyla, and most Artiodactyla,

are herbivores. Because the energy yield of plant tissue per unit of weight is low relative to that of animal material and because plants with high levels of secondary compounds must be detected, avoided, or detoxified, herbivores must invest a great deal of time in feeding. The pronghorn (*Antilocapra americana*), an herbivore that feeds primarily on forbs and small shrubs, uses a strategy common to many ungulates. Most of the pronghorn's time is spent eating or processing food. It alternately feeds and beds down through the day and night, and ruminating (chewing the cud) occupies 60% to 80% of the bedding time (Kitchen 1974).

Mammals that eat primarily seeds and grains (**granivores**) allocate time in a markedly different manner. Seeds are the most concentrated source of energy that plants offer; granivores therefore spend relatively little time foraging. In the American Southwest, both pocket mice and kangaroo rats make a series of foraging trips, returning with bulging cheek pouches to deposit seeds in their burrows. This gathering phase requires considerable energy, and during gathering, the rodents are exposed to predation and perhaps to low temperatures, but foraging takes as little

FIGURE 23-28 A pack of African hunting dogs (*Lycaon pictus*) at a water hole in Botswana.

as an hour or two per night. Selectivity in the choice of foraging microhabitat, tactile ability of the forepaws, and the use of high-energy food allow heteromyids to remain in the safety of their burrows for most of the 24-hour cycle. These rodents, however, are at the mercy of the vagaries of desert climates. During years of low rainfall there may be no green annual plants or seed production by annuals. At these times heteromyids do not reproduce and their populations decline (Beatley, 1969).

Heteromyid rodents can markedly influence community structure. Twelve years after three species of kangaroo rats (*Dipodomys*) were removed from experimental plots in the Chihauhuan Desert, the species composition of desert plants was dramatically altered (Brown & Heske 1990). What was once desert shrubland was converted to grassland in the absence of kangaroo rats. The change in plant species composition resulted from reduced seed predation and lack of soil disturbance, which allowed tall grasses to colonize the study plots. Change to a tall-grass community, in turn, allowed colonization by granivorous harvest mice (*Reithrodontomys*). Removal of a single species of kangaroo rat did not cause large changes in plant composition. Brown and Heske (1990) used the term "keystone guild" to describe the three seed-eating species that affected community structure so strongly.

Many tropical plants rely on frugivorous mammals (and birds) to disperse their seeds. These plants produce conspicuous, brightly colored fruits, or fruits with strong scents to attract mammals. Fruit-eating mammals that rely heavily on fleshy fruits for food often have good color vision. Frugivorous mammals tend to process and digest fruits relatively rapidly, resulting in the elimination of the seeds with the feces. Seeds that have been subject to the gentle digestion inside the alimentary tract of a frugivore often germinate at higher rates than seeds that are not eaten.

Fruit specialists have adapted by simplifying the digestive tract to speed the passage of material through the system. Spider monkeys (*Ateles*) are obligate frugivores and have colons with approximately half the surface area of those of leaf-eating howler monkeys (*Alouatta*) (Hladik & Hladik 1969). Although fruit is relatively easy to find and digest, it is usually deficient in protein. To maintain protein balance, large quantities of fruit pulp must move through the digestive tract of a spider monkey each day.

Nectar and pollen are the primary food for some species of pteropodid and phyllostomid bats, and the diet of one phalangerid marsupial (*Tarsipes*) consists partly of nectar. Nectivorous mammals occur in the New World tropics, Old World tropics, and Australasia (Flemming & Muchhala 2008). New World nectar-feeding vertebrates (including birds) are taxonomically and ecologically more diverse than their Old World and Australasian counterparts. Flemming

and Muchhala proposed that the greater flower diversity and resource predictability in the New World favored the radiation of small, hovering nectar feeders with relatively specialized feeding niches. Tropical Africa has similar resources, but nectar-feeding niches tend to be dominated by birds rather than bats. In Australasia, selection favored larger, nonhovering nectar feeders (birds, bats, and marsupials) with more generalized diets.

Given a variety of plants to choose from, most herbivorous mammals are selective foragers (Ward & Keith 1962; Yoakum 1958; Zimmerman 1965). Accordingly, an herbivore may show a great preference for one of the least abundant plants in its habitat. The nutrient content of plants may determine, in part, the choice of food. Interspecific competition may force sympatric species to refine their diets to the point of little dietary overlap. In addition, the choice of specific plants or parts of plants by herbivores is also strongly influenced by the occurrence of defensive chemicals.

Mutualism   One of the most remarkable instances of mutualism concerns a bird, the African honeyguide (*Indicator indicator*), and the African honey badger (*Mellivora capensis*; **Fig. 23-29**). The bird attracts the badger's attention by raucous chattering and then leads the way to a bees' nest. After the nest is torn apart by the mammal (African tribesmen sometimes perform this service), the honeyguide eats bees and their larvae and wax, which it can digest, while the badger eats the honey.

Of particular interest are the close and seemingly mutually beneficial associations between two species of mammals. Herds of impalas (*Aepyceros*) often stay with baboons (*Papio*) through much of the day. DeVore and Hall (1965) judged that the excellent eyesight of the baboons supplemented the acute senses of smell and hearing of the impalas and made the mixed group difficult for a predator to approach undetected. Despite these fascinating examples, truly mutualistic interactions among mammals are rare.

▪ Plant–Mammal Interactions

The development of feeding strategies can be considered as a case of evolutionary gamesmanship between resourceful opponents. The success of every evolutionary move by one species is tested against a position or a countermove by the competing opponent species. A species heading for a desirable square on the evolutionary chessboard may be forced to occupy another square if the first is already occupied by a competing species, or mutual vulnerability to devastating countermoves may make the simultaneous occupancy of adjacent squares by two opponents extremely costly. One result is **coevolution**, the mutual influence of evolution on two different species interacting with each

FIGURE 23-29   An African honey badger (*Mellivora capensis*) in Botswana.

other and reciprocally influencing each other's adaptations. Such coevolutionary patterns are common in many plant–mammal interactions.

Through many millions of years of plant–herbivore interactions, many plants have adapted to mammals by using them as agents of seed dispersal or **pollination**. There is growing evidence that a diversity of mammals (including didelphimorphs, bats, primates, rodents, ungulates, and some carnivorans) improve the reproductive success and dispersal of plants (Fleming & Sosa 1994). This ability of mammals is a result largely of their mobility, the effects of their digestive tracts on seeds, and caching behavior. The Cretaceous ecological revolution that resulted in the domination of terrestrial floras by angiosperms (flowering plants) was perhaps facilitated by the diversification of birds and mammals and their effectiveness as seed dispersers (Regal 1977).

For a particularly interesting example of mutually beneficial interactions between mammals and a plant, we return to Serengeti National Park. Here, and over wide areas of Africa, the umbrella tree (*Acacia tortilis*) is a conspicuous, picturesque, and important savanna tree. The green, leathery pods of the plant are eaten avidly by a variety of herbivores, ranging from the tiny dik-dik (*Madoqua* spp.) to the elephant (*Loxodonta africana*), and elephants often feed heavily on the foliage. Lamprey et al. (1974) found that the germination rates of seeds that had been ingested and eliminated by herbivores were strikingly higher than the germination rates of uningested seeds. In some vertebrate-adapted seeds, the digestive process is known to erode the seed coat and hasten germination, but the high germination rate of ingested acacia seeds is thought by Lamprey et al. not to depend primarily on this effect. Beetles of the family Bruchidae lay their eggs on acacia seed pods, and the larvae feed and grow within the seeds. If the larvae damage the embryo or destroy a large amount of the cotyledon material, the seed will not germinate, but if the seed pods are eaten by herbivores soon after they fall to the ground, as is typically the case in the Serengeti, the digestive process kills the larval bruchids at an early stage of development before they have killed the seeds. Some 500 seed samples were collected from the ground and stored for 1 year; more than 95% of these seeds had bruchid damage, and the germination rate was only 3%. In seeds eaten by impalas (*Aepyceros melampus*), however, the damage rate was 26%, and the germination rate was 28%; in seeds eaten by dik-diks (**Fig. 23-30**), these figures were a 45% damage rate and 11% germination rate. The interactions clearly seem to be mutually advantageous. From the acacia, the herbivores get a seasonally important food, and from the mammals the acacia gains an effective

FIGURE 23-30 A dik dik (*Madoqua kirkii*) in Namibia.

means of escape from a seed predator and wide dissemination of seeds.

In tropical areas, many frugivorous mammals disperse seeds. The Neotropical frugivorous bat *Carollia perspicillata* is an effective seed disperser (Fleming 1988), and this function would be expected of many other frugivorous phyllostomid bats. In a Mexican rainforest, howler monkeys (*Alouatta palliata*) doubled the dispersal rates of seeds of some trees, and some seeds were dispersed 800 meters (Estrada 1991). In Nepal, the one-horned rhinoceros (*Rhinoceros unicornis*) eats quantities of fruit of the successional tree *Trewia nudiflora*, and seeds germinate vigorously from the animal's manure. Because these fruits are too large and hard for bats, birds, or monkeys, the rhinoceros may be the main seed disperser of this tree (Dinerstein & Wemmer 1988).

A remarkable case of plant–mammal coevolution involves a geophyte (a plant with underground storage organs) and mole-rats (Lovegrove & Jarvis 1986). This plant (*Micranthus*; Iridaceae) lives in extremely nutrient-poor soils of the western Cape of South Africa, where there are more than 1,200 species of geophytes. Geophytes increase their nutrient stores (especially nitrogen and phosphorus) over many growing seasons (Dixon 1981). Growing seasons are brief, and the plants are dormant the rest of the year. Geophytes are high in calories, highly digestible (Bennett & Jarvis 1995) and a preferred food of mole-rats. Because of

their great importance to the plant's survival, storage organs of many species are defended against herbivores by toxic or unpalatable compounds. The geophyte–mole-rat association probably began at least 20 million years ago (early Miocene), and mole-rats have developed tolerance to even the most toxic geophytes. *Micranthus*, however, has taken evolutionary gamesmanship in another direction: Although its corms (underground storage organs) are invested by a thick, fibrous, and spiny tunic, the corm segments (up to 30) are highly palatable to mole-rats and are often stored in large numbers (3,043 were found in one *Georychus capensis* cache). Mole-rats must peel the tunic from the entire corm, often dislodging segments that become lost. Because corms are so palatable, they are frequently eaten along the burrows rather than being carried to the cache, thus maximizing chances for dispersal of corm segments. *Micranthus* further favors its survival by producing a cluster of "cormlets" at a depth of 2 to 10 centimeters, well above the larger corms and above the foraging burrows of the mole-rats (at depths of about 95 to 200 centimeters).

Some mutually advantageous plant–animal associations are probably of long standing. Especially noteworthy examples of coevolution are offered by nectar-feeding bats and the plants on which they feed. In the Old World, a number of bats of the family Pteropodidae feed on nectar; in the Neotropics, nectar feeding occurs in the family Phyllostomidae. Most plants are pollinated by many agents, including a number of insects and birds. Such plants are termed **polyphilous**. But some plants depend largely on bats for pollination (**chiropterophily**; discussed by Alcorn et al. 1959; Baker 1961, 1973; Faegri & Van Der Pijl 1966; Fleming 1992; and Ostfeld 1992). Pollination of plants by mammals also evolved in marsupials and lemurs, but only in areas where nectar-feeding bats are uncommon or absent (as in Madagascar and parts of Australia) does pollination by lemurs or marsupials persist.

Most chiropterophilous plants occur in the tropics and subtropics and have the following features: The flowers occur on spreading, often leafless branches. They are whitish and offer visually conspicuous targets at night. They have a strong, often rank smell. Copious amounts of nectar are produced in early evening when the bats forage. The plants produce flowers over long periods, but only a few buds open each night (this favors **traplining**, a nightly patrolling of the same series of plants by the bats).

Two general types of flowers are chiropterophilous. The first type, an example of which is the large, white flower of the African baobab tree (*Adansonia digitata*), accommodates large bats. The feeding bat (often of the genera *Epomophorus* and *Nanonycteris*) clutches the ball of stamens of this flower while lapping nectar from the pillar-like stamen column. The second group of "bat flowers" includes many Neotropical types that have corolla tubes adapted to smaller bats. When a bat pushes its head into the tube, its face is liberally dusted with pollen, which the bat takes to another flower.

Sanborn's long-nosed bat (*Leptonycteris curasoae*) is one of a number of New World phyllostomids that feeds on nectar. Nectar is rich in carbohydrates but usually has no more than trace amounts of protein, a food essential to mammals. *Leptonycteris* ingests large quantities of pollen, probably mostly when grooming the fur, to which pollen adheres readily. Pollen of the plants visited by *Leptonycteris* (saguaro, *Carnegeia gigantea*, and agave, *Agave palmeri*) is much higher in protein than is the pollen of closely related plants that are pollinated mostly by other agents. Of additional interest, "bat pollens" contain at least 18 amino acids and are unusually high in proline, the amino acid that makes up more than 80% of the protein collagen, the connective tissue that braces the wings and tail membranes of bats. *Leptonycteris* has high concentrations of hydrochloric acid-secreting glands in its stomach and often ingests its own urine. Both urea and hydrochloric acid are known to extract protein from pollen.

In the Sonoran Desert, where some bats are effective pollinators of certain columnar cacti, *L. curasoae* has seemingly affected the evolution of **gynodioecy** (separate female and hermaphrodite plants) and **trioecy** (separate males, females, and hermaphrodites) in the columnar cactus *Pachycereus pringlei*. The fitness of males and females of this cactus depends more on pollinators than does fitness in hermaphrodites; thus, the distribution of these reproductive types is not random. Trioecy occurs within the flight distance of bat roosts; away from roosts, this cactus displays gynodioecy (Fleming et al. 1998).

■ Plant Defenses

Mammals began eating plants at least 150 million years ago, and many of the major mammalian radiations have involved the exploitation of plants. Some 63% of the land-dwelling species of mammals are herbivorous. In terms of biomass and ecological impact, herbivores are the most important mammalian members of most terrestrial ecosystems. The order Rodentia alone, nearly all members of which are primarily herbivorous, includes about 44% of the species of land mammals. For some groups of mammals usually regarded as carnivores, such as the Ursidae, Procyonidae, and Canidae, herbivory is important seasonally. Even the grizzly bear is mostly herbivorous in some parts of its range.

Plants have responded to many millions of years of selective pressure from mammals and insects by evolving an arsenal of defenses (**Table 23-5**). Venomous spines, in the case of cacti, formidable thorns and "claws," among many

acacias, and irritating hairs on many plants are familiar deterrents. Less familiar, but of greater importance, is an array of **secondary compounds**, that is, defensive chemicals. These range from toxins that can kill an herbivore (such as those in locoweed, *Oxytropis* sp., and larkspur, *Delphinium* sp.) to a variety of digestion-inhibiting or antimicrobial compounds.

Plant defensive chemicals are of two major types. Ephemeral plant parts (those available briefly), such as flowers, fruits, or new leaves, are typically protected by toxins that probably evolved in response to pressure from dietary generalists. These toxins vary little from plant to plant within a species. Mature leaves, however, are a predictable and abundant food that is available over a relatively long period of time. Probably in response to dietary specialists, mature leaves have evolved secondary compounds with a high degree of individual variation (Rhoades 1979). These chemicals reduce digestibility (Cates & Rhoades 1977; Feeny 1975). Tannins in leaves of a wide array of plants combine chemically with or denature many mammalian digestive and nondigestive enzymes (Pridham 1965). The volatile oils of conifers contain terpenoids, which in some junipers have an antimicrobial action in the rumen of deer (Schwartz et al. 1980a). Because microbial fermentation in the rumen yields the largest share (50% to 70%) of the energy required by ruminants (Annison & Lewis 1959), antimicrobial secondary compounds markedly inhibit ruminant digestion (Nagy et al. 1964). They may also inhibit digestion in the cecum of rodents.

The deactivation of chemical defenses costs herbivores considerable energy. They must balance the intake of nutrients against the intake of defensive compounds. An important cosubstrate used in the breakdown of these compounds is glucuronic acid, a derivative of glucose (Scheline 1991). The usual mechanism is the use of glucuronic acid or other compounds to detoxify defensive chemicals, with the excretion of the byproducts via bile or urine. In the koala (*Phascolarctos cinereus*), a 20% increase in glucose intake may be required to fuel the breakdown of defensive chemicals in the leaves of eucalypts (Cork 1981). The dusky-footed woodrat (*Neotoma fuscipes*), an oak-leaf specialist, also pays a price for its folivory. Seemingly because of high concentrations of tannins and polyphenolics in oak leaves, the assimilation rate of this food by this woodrat is only 53% (Atsatt & Ingram 1983).

Some defensive compounds are altered by microbial action. Oxalates occur in many plants important in the diets of herbivores, and to nonadapted mammals oxalates can be lethal. Some mammals, including rabbits, rodents, pigs, horses, some artiodactyls, and humans, are known to degrade oxalates by microbial action in the large intestine or rumen. Oxalates in the diet favor intestinal or rumen bacteria

TABLE 23-5   Some Plants and Their Defensive Compounds That Either Inhibit Digestion, Are Antimicrobial, or Are Lethal Toxins)

	Genus	Secondary Compounds
Trees		
Mango	*Mangifera*	Steroids, tannins, spirolactones
Cashew	*Anacardium*	Phenolics, flavonoids
Eucalypts	*Eucalyptus*	Phenolics, cyanide precursors, volatile oils
Juniper	*Juniperus*	Tannins, monoterpenes, sesquiterpenes
Oak	*Quercus*	Tannins, polyphenolics
Nux vomica	*Strychnos*	Alkaloids, e.g., strychnine
Shrubs		
Chokecherry	*Prunus*	Tannins, cyanogenic glycosides
Sagebrush	*Artemisia*	Oxygenated monoterpenes, sesquiterpenes, volatile oils
Coffee	*Coffea*	Methylxanthines
Forbs		
Larkspur	*Delphinium*	Delphinine
Locoweed	*Oxytropis*	Toxic alkaloids, oxalates
Milkvetch	*Astragalus*	Cardenolides, alkaloids, organophosphates, glyphosates
Monkshood	*Aconitum*	Aconite
Nightshade	*Solanum*	Toxic alkaloids
Sweetclover	*Melilotus*	Coumarin
Datura	*Datura*	Alkaloids, e.g., atropine
Cacti		
Prickly Pear	*Opuntia*	Oxalates

Data are from many sources.

that use these compounds for growth; high populations of such bacteria result in increased rates of oxalate breakdown (Allison & Cook 1981; Schmidt-Nielsen 1964).

In addition, mammals have evolved morphological and behavioral adaptations to plant secondary compounds. Many herbivorous mammals use an enlarged and subdivided stomach (rumen) or an enlarged diverticulum from the intestine (cecum) as microbial fermentation chambers in which some defensive compounds are detoxified. In general, the length of the gastrointestinal tract is a good predictor of diet; folivores typically have the longest tracts relative to body size (Schwaibold & Pillay 2003).

A widely used strategy of herbivores is selective foraging, the ability to discriminate among individual plants and eat only those with low levels of defensive chemicals. Such finely-tuned foraging depends on intraspecific variation in secondary compounds among plants, and such variation is well-documented. Captive Stephens' woodrats (*Neotoma stephensi*), fed for 2 weeks on the foliage of one juniper, lost weight when changed to a diet of foliage from another juniper (Vaughan & Czaplewski 1985). Glander (1977)

observed that leaf-eating howler monkeys (*Alouatta palliata*) in Costa Rica ate the leaves of some trees but not others of the same species; the petiole was usually eaten instead of the leaf blade, which had higher concentrations of defensive chemicals. The price of carelessness is high: Three of six dead howler monkeys that Glander examined had been eating the leaves of either of two trees with toxic leaves, as had a female that had convulsions and fell from a tree. Plant tissues with high levels of defensive chemicals are not popular foods among mammals. Conifer foliage, for example, is generally either emergency food or one taken in small amounts with an array of other plants. In North America, only three species of small mammals are conifer-leaf specialists.

The life histories of many folivorous mammals strongly reflect the constraints of low-energy diets. Voles (Arvicolinae) are known for high reproductive rates, but two arvicoline depart drastically from this pattern. The Sonoma vole (*Arborimus pomo*) and the red tree vole (*A. longicaudus*) eat the needles of Douglas fir (*Pseudotsuga menziesii*) almost exclusively (Hamilton 1962). The Sonoma vole (the biology of the nearly identical red tree vole is poorly known) strips the resin ducts from each needle and eats prodigious amounts, about 2,400 needles per day. The reproductive rate of this vole is the lowest recorded for any arvicoline (**Table 23-6**) and is probably a response to a diet high in defensive chemicals and low in available energy. Similarly, Stephens' woodrat, the only other North American conifer-leaf specialist, has an unusually low metabolic rate and restricts its locomotor activity (Sorensen et al. 2005a, b). Furthermore, it has the lowest reproductive rate of any species of woodrat; the litter size is usually one. Young are weaned late, at about 40 days, and young do not reach 60% of adult weight until older than 90 days (Vaughan & Czaplewski 1985).

A final example of a lifestyle probably modified by herbivory is that of a primate, the slow loris (*Nycticebus coucang*) of West Malaysia. This arboreal, slow-moving omnivore has a metabolic rate that is less than 60% of the predicted rate (Muller 1979). Its reproductive rate is remarkably low, with a gestation period of 193 days (at least three times that of most comparable-sized eutherians), a litter size of one, and a nursing period of 9 months (Lekagul & McNeely 1977). This mammal's slow pace of life is regarded by Wiens et al. (2006) as a consequence of its diet, which consists mostly of nectar, together with fruit, phloem sap, gum (a wound-closing plant exudate), and insects. The first three foods listed are high in easily digested monosaccharides and disaccharides and are thus a high-energy diet, but seven of the genera that yielded sap or gum eaten by lorises, and flowers and nectars of some other plants, are known to contain toxins or digestion inhibitors. Wiens et al. (2006) hypothesized that slow lorises ingest defensive secondary compounds together with their high-energy diet

and that the slow lifestyle is a result of the energy required to detoxify these compounds. Viewed broadly, mammals "function at an intensity close to maximum potential metabolism set by rate of energy assimilation from food" (McNab 1980).

Halophytic (salt-loving) plants of many desert regions concentrate salts in their above-ground tissues. The halophyte *Atriplex*, the desert saltbush, common in African, Australian, and North and South American deserts, concentrates salt on the outside of its leaves. Several mammals have convergently evolved morphological, behavioral, and physiological mechanisms for dealing with the high salt loads of *Atriplex* (Mares et al. 1997). Chisel-toothed kangaroo rats (*Dipodomys microps*) of the Great Basin Desert use their specialized lower incisors to scrape off the hypersaline outer coating of *Atriplex* leaves (Kenagy 1972, 1973). The North African fat sand rat (*Psammomys obesus*) also eats *Atriplex* and other native halophytes (Degen 1988). This rat is able to reduce salt intake by using its incisors to scrape away external salt from the leaves before eating the inner tissues and by renal adaptations that remove salt from the blood (Degen 1988; Mares et al. 1997). In the deserts of Argentina, the red vizcacha rat (*Tympanoctomys barrerae*) also specializes on *Atriplex* and uses a combination of chisel-shaped lower incisors and a set of specialized hairs just posterior to the upper incisors to strip the salty layer off the leaves before consuming the inner parts (Mares et al. 1997).

Other physical defenses of plants include thorns and seeds with hard or thick coats. In Africa, a great variety of plants including the diverse and widespread acacias bear imposing thorns which probably evolved in response to

TABLE 23-6 Reproductive Data for Some Arvicoline Rodents

Species	Gestation Period (Days)	Litter Size	Age of Weaning (Days)	Female Age at First Reproduction (Days)
Arborimus pomo	27–48	1–2	25–46	60+
Lemmiscus curtatus	25	2–13	21	60
Microtus breweri	21	3.5	14	—
M. californicus	22	4.7	15	21
M. ochrogaster	20–23	3.5	10–14	35
M. oregoni	23–24	3.4	15	24
M. pennsylvanicus	21	4–6	12–14	30
M. pinetorum	20–24	1–4	17	30
M. townsendii	21–24	4–7	15–17	—
Phenacomy intermedius	19	2–9	18	28–42
P. ungava	21	5	21	28–42
Myodes gapperi	17–19	2–8	12–14	60+

Data are from a variety of sources. Note the low reproductive rate of *Arborimus pomo*, an arboreal folivore.

browsing by a diversity of ungulates. The leaves of African acacias have a higher protein content than that of most trees and shrubs (Sauer at al. 1982) and, at least seasonally, are a preferred food of a number of African artiodactyls (Skinner & Smithers 1990).

In summary, mammalian herbivores have been forced to adapt to the nearly ubiquitous defenses of plants. Adaptation of intestinal, cecal, or rumen microbes increases the rate of degradation of some defensive chemicals, and highly selective foraging enables some mammals to feed on those individual plants within a species that have relatively low levels of these chemicals. Some herbivorous mammals avoid ingesting large amounts of any defensive chemical by eating a broad array of plants.

Effects of Mammals on Their Environments

If the importance of a group of animals is equated with its visible effect on the environment, mammals are clearly the most important terrestrial animals. The great impact of mammals on their environment is largely a result of endothermy, with its attending high energy requirements. This impact results from a variety of activities, including feeding, patterns of migration or daily movement, the quest for water, and the construction of shelters or refuges.

Wildebeest (*Connochaetes taurinus*), for example, are the dominant species in the Serengeti Plains in Tanzania and strongly affect the community by virtue of their great numbers. The annual migration of wildebeest from the Serengeti Plains in Tanzania, northward and westward toward the bush country of northern Tanzania and the Masai Mara Game Reserve of Kenya, is a justly famous spectacle (Fig. 23-13). The movement begins at the end of the rainy season, in May or June, and the animals return to the short grass in November. During a 4-day period in May of 1974, grazing by approximately 0.5 million wildebeest reduced the green biomass of this *Themeda-Pennisetum* grassland by 84.9% and the height of the vegetation by 56% (McNaughton 1976; **Table 23-7**). This apparent devastation markedly affected the subsequent growth of grasses and, indirectly, the dry-season distribution of the abundant Thomson's gazelle (*Eudorcas thomsonii*). This antelope, next to the wildebeest, is the most abundant ungulate in the Serengeti Plains. During the 28-day period following the main migration, areas grazed by wildebeest had a net productivity of green vegetation of 2.6 grams per square meter per day, whereas in experimental plots protected from grazing, green biomass declined at a rate of 4.9 grams per square meter per day. A dense mat of new and nutritious vegetation was produced in the grazed area, whereas in the ungrazed plots, the bulk of the biomass was tall, nonnutritious grass stems.

One month after the exodus of the wildebeest, the area is occupied by Thomson's gazelles, which selectively graze the areas of vigorous regrowth, the areas previously grazed heavily by wildebeest. As an indication of the high selectivity, consumption of vegetation by gazelles in these areas averaged 1.05 grams per square meter per day, whereas virtually no grazing by gazelles occurred in the plots where wildebeest had not grazed. McNaughton (1979, 1985) concludes that the wildebeest transformed a senescent grassland into a productive community. This grazing optimization hypothesis has generated much discussion (Belsky 1987; Bergström 1992). The impact of mammals on grasslands and savannas remains an active area of research.

Wildebeest transform African grasslands by virtue of their numbers. In contrast, some species of mammals have a strong impact on their communities regardless of their abundance. The loss of such keystone species profoundly alters the species composition of the remaining community. One striking example occurs along the coastlines of the Aleutian Islands of Alaska, where the sea otter (*Enhydra lutris*) seems to play a critical role in structuring nearshore marine communities (Estes & Duggins 1995; Estes & Palmisano 1974; Steinberg et al. 1995). The Rat Islands of the Aleutian archipelago support high populations of sea otters, and the nearshore community is characterized by abundant beds of macrophytes, consisting mostly of brown algae (kelp), which has leaf-like structures (blades) at the surface and stem-like structures (stipes) that often extend many meters to holdfasts on the sea floor. Filter feeders, such as barnacles and mussels, are scarce, as are motile herbivores, such as sea urchins and chitons in such kelp beds. In the Near Islands, 400 kilometers to the northwest, sea otters are absent from Shemya Island and only recolonized Attu Island in the mid 1960s. Where sea otters are absent, macrophytes were scarce below the lower intertidal zone, and barnacles, mussels, sea urchins, and chitons were many times more abundant than along the Rat Islands. These differences seem related to the activities of otters. Because otters prey heavily on sea urchins (*Strongylocentrotus* spp.), which graze on macrophytes, high populations of otters drastically reduce the abundance of urchins, permitting kelp beds to flourish. Thus, the composition of the nearshore community is conspicuously altered.

Productive kelp beds and a stable nearshore community on the Pacific coast of the United States may be maintained only in the presence of their keystone species, the sea otter; however, a complex web of events has recently altered the structuring of nearshore kelp communities. The typical diet of killer whales (*Orcinus orca*) along the Alaskan coast is Steller sea lions (*Eumetopias jubatus*) and harbor seals (*Phoca vitulina*), populations that have declined in recent years as their food supply dwindled from overfishing by humans. In response to declines in seal populations,

BOX 23-1   A Tale of Two Deserts

The defining feature of deserts is aridity, to which all desert plants and animals must adapt. Thus, strategies for coping with aridity—such as periodic dormancy and the ability to concentrate urine—are shared by many desert mammals (Chapter 21). Because deserts occur on every continent, however, their geological histories and amounts and seasonal distributions of rainfall vary widely, as do the geographic and taxonomic origins of their biotas. Accordingly, mammalian community organization differs markedly among different deserts. With this in mind, let us consider the Sonoran Desert of North America and the Namib Desert of the southwestern African coast.

The Sonoran Desert is not a "typical" desert, in part because it supports a relatively lush and diverse flora. Brief bursts of seed production occur in response to winter and/or summer rains, which average roughly 100 to 300 millimeters per year. Although rainfall is unpredictable, the standing crop of seeds in the soil—especially those of ephemeral forbs—provides a fairly stable resource on which most Sonoran Desert rodents depend (Brown 1973). Windblown seeds collect in large numbers in depressions or on the lee sides of obstacles. Seeds formed from 78% to 94% of the diets of four species of rodents (all heteromyids) that occupied a Sonoran Desert community in Arizona (Reichman 1975). Insects and greenery, not major foods, were eaten only seasonally, and their use fluctuated widely from year to year. For these heteromyids, collecting seeds in the cheek pouches and carrying seeds to scattered, shallow caches (**scatter hoarding**) or to chambers associated with the burrow system (**larder hoarding**) are the primary surface activities. Eating is done underground. Clearly, these species belong to the granivorous guild. All four species are nocturnal. All are independent of free water, and the two smallest species remain underground, presumably in torpor, during the coldest months.

Compared with the Sonoran Desert, Africa's Namib Desert appears overwhelmingly barren (see Fig. 23-5B). Except after the infrequent rains, the stark gypsum flats seem lifeless, with lichen crusts forming the dominant growth; only occasional patches of grass dot the great Namib "dune sea." The extremely low rainfall of this desert (about 10 to 125 millimeters per year) is a result of the South Atlantic anticyclone pressure system and the associated cold (Antarctic) Benguela Current that sweeps up the coast from the south. Upwelling of cold, coastal waters gives rise to persistent winds from the sea, maintaining an inversion layer (a layer of cooler air beneath a layer of warmer air) that reduces the turbulence and cloud formation necessary for rain. The Namib is an ancient desert, having fluctuated from semiarid to arid since about 85 million years ago, when the separation of Africa and South America formed the South Atlantic Ocean. But the Namib's rich biota has a remarkable source of water in addition to the scant rainfall. Fog forms when onshore winds drive warm, moist Atlantic air masses over the cold, coastal Benguela Current. Wind-driven fog sweeps as far inland as 100 kilometers for about 60 nights a year. Many plants and some animals can use condensed fog water (Seely 1981). The frequent winds also transport small fragments of plant and animal material over the dunes and flats. This detritus is the major food source for many Namib animals (Seely 1979).

Seemingly in response to the diversity of arthropods that feed on detritus and the lack of a stable standing crop of seeds in the soil, the gerbils inhabiting the Namib are omnivorous, with arthropods being their most important food (Griffin 1990). About 70% of the diets of two Namib gerbils studied by Perrin et al. (1992) was arthropods; seeds formed up to 33% of the diets in part of the austral winter (May to August), and greenery was of varying importance. Lacking external cheek pouches, gerbils of the Namib rarely cache food, but remains of insects and plants in their burrows indicate that they sometimes eat food there (Downs & Perrin 1989). These gerbils seem to lack the urine-concentrating abilities of those heteromyids fully independent of free water. The gerbil's insect diet yields considerable free water and protein, and they can remove the byproducts of protein metabolism with minimal water loss; however, when insects are scarce, these gerbils must eat succulent vegetation (Downs & Perrin 1991). Seeds, which yield considerable metabolic water, are only sporadically available. That fog affects the water balance of some Namib rodents (including gerbils) is indicated by increased rates of water turnover and greater urine production in these rodents after windborne fogs (Withers et al. 1979). Precipitation (either fog, dew, or rain) occurs at least once each month of the year, a pattern critical to the survival of arthropods, the gerbils' main food. Perhaps this relatively frequent precipitation has also reduced the gerbils' need for physiological adaptations favoring extreme water conservation for long periods.

Viewed broadly then, differences in rodent community organization between the Sonoran Desert and the Namib Desert seem to have resulted from different selective pressures associated with different climates and contrasting food bases. Whereas "in North American deserts seed abundance and distribution serve as the cornerstone of community organization among rodents" (Reichman 1991) and rains are seasonal and unpredictable, "the extensive use of fog water and detritus largely characterize the biology of the Namib" (Seely 1987).

The false assumption is often made that in each major desert of the world specialized seed-eating rodents have evolved. Clearly, this does not hold for the Namib Desert or all other deserts. The Namib gerbils are convergent in some ways with heteromyids (large auditory bullae, long hindlimbs, and the ability to concentrate urine), but the lack of a dependable seed crop precludes granivory and favors omnivory. No one dietary mode is of dominant importance in all deserts.

---

killer whales appear to have shifted their diet from seals to sea otters (Estes et al. 1998), causing otter populations to decline precipitously in the 1990s. Estes et al. studied a sea otter population in a lagoon not accessible to killer whales and another population of otters in an open bay frequented by killer whales. Over a 2-year period, the disappearance rate of sea otters from the open bay was five times that of otters in the inaccessible lagoon. As predicted, in coastal

TABLE 23-7 **Effect on Grassland Vegetation (Largely *Themeda* and *Pennisetum*) of the 4-Day Passage of Wildebeest Herds in the Serengeti Plains**

	Biomass (g/m²)	Height (cm)	Biomass Concentration (mg/10 cm²)
Fenced vegetation, wildebeest excluded:			
Before passage	501.9	64	7.9
After passage	449.2	63	7.1
	NS	NS	NS
Vegetation subject to wildebeest grazing:			
Before passage	457.2	66	6.9
After passage	69.0	29	2.4
	$p = .005$	$p = .005$	$p = .05$

Modified from McNaughton, S.J. 1976. Serengeti migratory wildebeest: facilitation of energy flow by grazing. *Science* 191:92–94. Reprinted with permission from AAAS. p = level of significance; NS = not significant.

areas where sea otter populations have been decimated, sea urchin populations have increased and kelp beds have become depleted. These alterations in the nearshore community further demonstrate the sea otter's role as a keystone species.

In some of the national parks and game preserves of Africa, there are high elephant populations, and because of the encroachment of agriculture, the elephants are no longer free to range widely when pressed by local or seasonal shortages of food or water. Studies of elephants in various parts of East Africa document the impact elephants can have on the landscape under these conditions (Laws 1970). The ideal diet of elephants seems to consist of a mixture of grass and browse from trees and shrubs (Laws & Parker 1968), and the preferred habitat is thus forest edge, woodland, or bush-grass mosaic. During the dry seasons in Tsavo National Park in Kenya, sources of water are not evenly distributed. At these times, elephants need water daily and they concentrate within 20 to 30 kilometers of water. One elephant can eat approximately 140 kilograms of vegetation a day. Thus, when hundreds of elephants congregate in the vicinity of a water hole, the local vegetation is rapidly destroyed. In Tsavo National Park, in the 1960s, 17,000 square miles were occupied by more than 40,000 elephants, and vast areas of bushland were transformed into grassland. Aerial photographic transects studied by Watson (1968) indicated that in a period of 5 years elephants killed from 26% to 28% of the trees above 65 centimeters in crown diameter. Drought and poaching in the 1970s reduced the elephant population to about 10,000, and the trees began regenerating. To the south, in Lake Manyara National Park of Tanzania, Douglas-Hamilton (1973) observed similar destruction: In one area of especially acute damage, elephants killed 8% of the umbrella trees (*Acacia tortilis*) in 1 year. Fortunately this national park has been greatly

enlarged since Douglas-Hamilton's observations and pressure on the acacias has been reduced (Moss 1982). In Tsavo, as elsewhere, elephants alone have not affected the vegetation. Ungulates such as zebras (*Equus burchellii*), Grant's gazelles (*Nanger granti*), and oryx (*Oryx gazella*) were favored by the shift toward grassland (Sheldrick 1972).

In some cases, the effects of mammals on their environment are less obvious (but no less important). Although it is well-known that insectivorous birds significantly reduce insect damage to tropical plants, the role of insect-eating bats has been largely ignored. Research in Panama demonstrate that bats play a significant role in maintaining and reshaping tropical rainforests. Insect-gleaning bats, such as *Micronycteris microtis*, feed extensively on herbivorous insects that damage the leaves of many tropical plants. Using infrared videos of foraging bats, Kalka and Kalko (2006) show that *M. microtis* consumes between 60% and 80% of its body mass in arthropods each night, with herbivorous insects accounting for roughly 70% of the total diet. By comparing data from foraging videos with insect remains in feces collected beneath roosts, the authors also found that fecal analyses missed almost 50% of the herbivorous insects captured by gleaning bats. On Barro Colorado Island in Panama, each *M. microtis* consumes an estimated 700 kilograms (wet weight) of herbivorous insects each year (Kalka & Kalko 2006). With four additional species of insect gleaning bats on the island, an estimated 5,200 kilograms of herbivorous insects are consumed annually, saving approximately 52,000 kilograms of fresh leaf biomass each year.

In a second study, Kalka et al. (2008) used a series of exclosures to separate the impacts of insectivorous birds and bats. Exclosures consisted of covering young plants with nets that allowed insects to pass through, but excluded birds and bats. Bat exclosures were covered during the night (sunset to sunrise). Bird exclosures were covered during the day (sunrise to sunset), and control exclosures were left uncovered during both day and night. Insect abundance was measured repeatedly on each plot over 10 weeks. The results showed that bats were more important than birds in limiting insect abundance and insect herbivory. These results extend to agroforesty systems such as coffee plantations (Williams-Guillen et al. 2008). Here bats also play a more important role than birds do in reducing insect damage to coffee plants during the wet season when many Neotropical migrant birds have left the tropics for their northern breeding grounds.

Reducing the impacts of insect herbivores is not the only means by which bats reshape tropical ecosystems. By dispersing seeds into pastures and other deforested areas, fruit bats aid the regeneration of tropical forest habitats. Kelm et al. (2008) constructed artificial bat roosts near abandoned pastures and sampled seeds in bat feces dropped over open pastures. Pastures near artificial bat roosts had significantly more seed rain than pastures without roosts. Most of the seeds dispersed by bats in this study were early-successional plant species, species most likely to colonize disturbed areas. Thus, fruit bats may speed up the regeneration of tropical forests in disturbed areas.

The story of the wolf in Yellowstone National Park is a striking example of the ecological impact of a large, social carnivore, but the central point here is that predators can play key roles in shaping ecosystems (Smith & Ferguson 2005). Before the reintroduction of wolves in Yellowstone, large herds of elk browsed widely on willows and on cottonwood and aspen saplings, damaging them or destroying them locally, as in the Lamar Valley. Since the return of the wolf and probably because of a variety of causes, elk numbers have declined. Willows and cottonwoods are recovering in places where wolf activity is high and elk are most vulnerable. With willows returning along some Lamar River tributaries, beavers, long absent from these places, are recolonizing. When beavers build dams, the beaver pond and adjacent riparian communities support a rich biota, including insects, fish, amphibians, myriad plants, plus waterfowl and moose. As willow thickets develop, several species of warblers, as well as voles and shrews, are expected to become more abundant. Halfpenny (2003) believes that "the greatest ecology experiment of the 20th century is showing that wolves can modify the entire Yellowstone ecosystem."

Rodents may also have a marked effect on vegetation. Sixty-three years of aerial photographic and soil chemistry data reveal that beavers (*Castor canadensis*) can affect the drainage systems of boreal ecosystems (Naiman et al. 1994). In northern Minnesota, beavers converted more than 13% of the area studied from forest to meadows and ponds. Dams built to retain pond water also prevent the downstream transport of minerals leached from upland soils.

Standing water saturated the soils, resulting in anaerobic conditions, which altered subsequent biogeochemical pathways to the extent that the standing stock of some elements increased by more than 250%. As Naiman et al. note, the influence of beavers on boreal ecosystems has been "spatially extensive and long-lasting, affecting fundamental environmental characteristics of boreal forest drainage networks for decades to centuries."

Plains vizcachas (*Lagostomus maximus*) are key "ecosystem engineers" in the semiarid scrublands of Argentina, Bolivia, and Paraguay (Boogert et al. 2006). These rodents are large (up to 9 kilograms), occur in colonies of 10 to 30 individuals, live in extensive burrow systems, and forage at night on grasses and other understory plants. The burrow systems often underlie areas of at least 600 square meters, and a colony's excavations may displace 80 cubic meters of soil, mostly deposited as surface mounds. In addition, vizcachas void feces in burrows and aboveground, clip small plants they do not eat, and collect quantities of sticks, stones, bones, and so forth that they place on the surface above burrows. The greatest impact of these activities is on shrubs that the vizcachas never eat. The landscape at colony sites studied by Villarreal et al. (2008), compared with landscapes lacking vizcachas, had 14.7% greater biomass, 11.9% more nitrogen, and 6.1% more phosphorus in aboveground vegetation (mostly shrubs). The rodents also altered the spatial occurrence of nutrients, and by redistributing sticks, bones, and such probably changed their rates of weathering and chemical decomposition. Vizcachas graze understory plants and remove sticks from beneath shrubs, thus probably lowering shrub mortality by reducing the frequency and intensity of fires. Because colonies are commonly occupied for many decades and the environmental impact of the colonies persists long after the animals are gone, the effects of vizcachas on landscapes may last for centuries.

Energy Flow and Food Webs

Just as energy transfers from one part of an organism to another are vital to life, complex patterns of energy transfer within a biotic community maintain the system of interdependent and interacting species. The sun provides the ultimate source of energy for life on Earth. Typically, this solar energy is transferred from photosynthetic plants (**primary producers**) to herbivores (**primary consumers**) to primary carnivores (**secondary consumers**) that eat the herbivores, to secondary or perhaps tertiary carnivores in some extended food webs.

The term **food web** has been used to describe the complex pathways of energy transfer that usually occur in nature, often involving predators and primary consumers that figure importantly in more than one level (**Fig. 23-31**). Animals

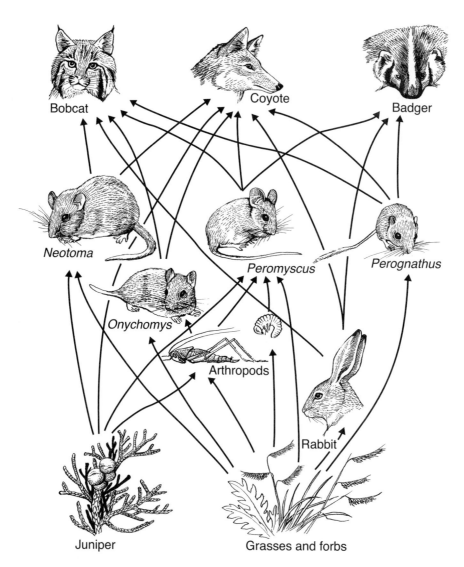

FIGURE 23-31 A simplified food web involving the mammals of a piñon-juniper community in Coconino County, Arizona. The arrows indicate the foods used by the mammals. The plants support arthropods, rabbits, and rodents. The rodents (*Neotoma stephensi, Onychomys leucogaster, Peromyscus truei, Perognathus flavus*) and the rabbits (*Lepus californicus* and *Sylvilagus audubonii*) are preyed upon by bobcats, coyotes, and badger.

that occupy comparable functional positions in the food web—the position of primary consumers, for example—are at the same trophic level. Green plants occupy the first trophic level and are referred to as **autotrophs** (self-feeders). Using mammals as examples, the second trophic level is occupied by herbivorous rodents, rabbits, and ungulates. Small carnivores, such as weasels, occupy the third trophic level, and large carnivores are in the third or fourth level.

The typical relationship of size and abundance of animals in a food web involves small but numerous primary consumers, larger but much less abundant secondary consumers, and still larger but relatively scarce tertiary consumers. Top predators occupy precarious positions because they often depend on animals from high trophic levels, where there is far less biomass and energy than in the lower levels. Because of the great loss of energy accompanying food

transfer between successive trophic levels, the total available energy is largest for consumers at the lower levels. Thus, predators feeding on primary consumers have more energy available to them than do predators feeding on secondary consumers. Considered in this light, the adaptive importance of filter feeding in massive baleen whales becomes clear. Plankton feeders exploit the tremendously larger sources of energy in plankton rather than feeding on larger fish at the secondary or tertiary levels. Only a small fraction of the energy entering a trophic level can be used by the next higher level; this factor limits the length of food chains. Consider the energy flow through an old-field community in Michigan (Golley 1960). Here, roughly 1% of solar energy is converted into plant tissues. Meadow voles (primary consumers, *Microtus pennsylvanicus*) consumed only 1.5% of the plant tissues available to them. Weasels (*Mustela nivalis*), the top vole predator in this community, consumed approximately 30% of the energy available in the voles. Considerable energy is lost at each transfer from one level to another through respiration and through the death of organisms, when their energy goes to decomposer food chains. Herbivores such as voles and lagomorphs, for example, assimilate only about 65% of the energy available from the fibrous plants they eat (Grodzinski & Wunder 1975). Consequently, figures depicting the biomass, caloric content, and energy utilization available at each trophic level typically form pyramids, as shown in **Fig. 23-32** and **Table 23-8**.

Diagrams of trophic structure or food webs, although valuable for purposes of illustration, usually oversimplify what is really an extremely intricate web of interactions (**Fig. 23-33**). A broadly adapted carnivore such as a coyote, or an omnivore such as the opossum (*Didelphis*), may function in all trophic levels above that of the primary producer. For a coyote, the fruit of the prickly pear cactus or juniper berries may form one meal, whereas a jackrabbit or deer fawn may be the next. More frequently, seasonal differences in the position of an animal in the food chain may occur. Johnson (1961, 1964) found that the deer mouse (*Peromyscus maniculatus*) in Colorado and Idaho became strongly insectivorous in the summer and thus functioned during this season as a secondary consumer, whereas it ate largely plant material during the cooler seasons, functioning as a primary consumer at those times.

Some food webs are easily perturbed by the loss of members (local extinctions) or the addition of exotic species. In some cases, a single introduced species can alter an entire ecosystem. Consider the nine Asian mongooses (*Herpestes javanicus*) introduced to Jamaica from India in 1872. Since their introduction, this species spread to other Caribbean islands and was transported to Hawaii (Lever 1994), where they have caused the local extinctions of reptiles, birds, and

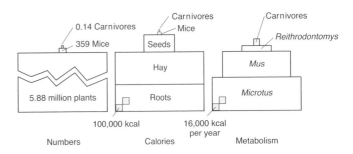

FIGURE 23-32 Pyramids of numbers, calories, and energy utilization for 1 acre of annual grassland near Berkeley, California. The pyramid showing calories is also approximately to scale for biomass. (Modified from Pearson, O. P., *J. Mammalogy*. 45 (1964): 177-188. American Society of Mammalogists, Allen Press, Inc. With kind permission from Anita K. Pearson.)

several native mammals (e.g., endemic inectivores of the West Indies, *Nesophontes*; Hays & Conant 2007).

Within a trophic level, several species often exploit a common set of resources in a similar way, leading to the formation of diverse and conspicuous feeding guilds. Guild members may be from different taxonomic groups. For example, the granivore guild in the deserts of the southwestern United States includes harvester ants, along with several species of birds and rodents (Brown & Davidson 1979). In national parks or reserves in southern Africa the scavenger guild often includes at least 11 vertebrate species (**Table 23-9**). Of these, the vultures (**Fig. 23-34**) are the only strict scavengers; the others are opportunists, scavenging one day and perhaps killing prey the next. In Yellowstone National Park, where wolves were reintroduced in 1995 and are now well-established, scavenger/predators include at least 9 vertebrate species (Table 23-9). Black bears and grizzlies are avid scavengers, and a grizzly can supplant wolves at their kill, thus profiting from a bounty not frequently available otherwise. In 2002, grizzlies trailed a pack of wolves during

TABLE 23-8 Standing Crop of Plants, Prey, and Predators on 1 Acre of California Grassland and Rate of Use of Vegetation by Rodents and of Prey by Carnivores at Peak Population Levels

	Standing Crop		Annual Rate of Use	
	Kg (dry weight)	**Kcal**	**Kcal**	**Percentage of Crop**
Roots	2,131	7,269,000		
Hay	2,097	8,141,000		
Seeds	442.	1,920,000		
Microtus	1.24	6,402	1,368,750	71
Mus	0.88	4,543	876,000	46
Reithrodontomys	0.084	434	81,650	4
Other prey	0.13	671	27,000	
Carnivores	0.126	650	11,700	97

Data from Pearson (1964).

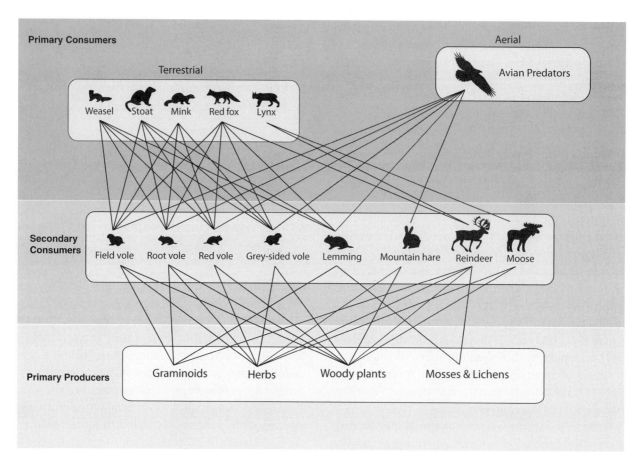

FIGURE 23-33 Complexity of a food web from northeastern Norway. Only three trophic levels are present. (Adapted from Oksanen, L., et al. *Food Webs: Integration of Patterns and Dynamics* (Polis, G.A. and Winemiller, K.O. eds.) Chapman & Hall, 1996.)

much of December and often took over its kills (Halfpenny 2003). In time, this extra food source could increase survival of grizzly cubs and cause changes in the timing of grizzly hibernation.

■ **Landscape Ecology**

The importance of habitat heterogeneity as it affects spatial patterns of distribution of mammals was recognized early in the 20th century by such field biologists as Joseph

TABLE 23-9 **Members of the Scavenger Guild in Etosha National Park, Namibia, and in Yellowstone National Park, United States**

Etosha		Yellowstone	
Lappet-faced vulture	*Torgos tracheliotus*	Common raven	*Corvus corax*
White-headed vulture	*Trigonoceps occipitalis*	Black-billed magpie	*Pica hudsonia*
Cape vulture	*Gyps coprotheres*	Bald eagle	*Haliaeetus leucocephalus*
White-backed vulture	*Gyps africanus*	Golden eagle	*Aquila chrysaetos*
Hooded vulture	*Necrosyrtes monachus*	Coyote	*Canis latrans*
Bateleur	*Terathopius edaudatus*	Red fox	*Vulpes vulpes*
Steppe eagle	*Aquila nipalensis*	Wolverine	*Gulo gulo*
Tawny eagle	*Aquila rapax*	Black bear	*Ursus americanus*
Maribou stork	*Leptoptilos crumeniferus*	Grizzly bear	*Ursus arctos*
Blackbacked jackal	*Canis mesomelas*		
Spotted hyena	*Crocuta crocuta*		

The Yellowstone list is partly from information in Smith and Ferguson (2005)

FIGURE 23-34 Several species of vultures, members of the scavenger guild, feeding on a zebra in Etosha National Park, Namibia.

Grinnell and Aldo Leopold. Important early studies on rodents and disease transmission by the Soviet ecologist Naumov (1936, 1948) demonstrated that rodent population densities were influenced by both habitat and topography, features regarded today as within the realm of landscape ecology.

Landscapes are not defined by size, but include a mosaic of habitats, ecological communities, or ecosystems relevant to the ecological process or pattern being studied. In addition to the spatial scale, there is a temporal component as the ecological processes change over time. Often, landscape ecologists also seek to understand how human modifications to the landscape alter ecological process. A fundamental question is how biological processes vary spatially and how they relate to large-scale, landscape-level dynamics involving metapopulations.

Forman (1995) described the patchwork of habitats viewed from an airplane window as a mosaic of fields, hedgerows, woodlots, housing subdivisions, and forests, broken by rivers, roads, and other barriers. Viewed in this manner, landscapes include patches (of varying scale), corridors between patches, and background matrix. **Patches** are islands of relatively homogeneous habitats surrounded by dissimilar areas. **Corridors** are strips of habitat that connect separate patches. Corridors allow animals to move between patches, increasing gene flow among isolated patches and thereby increasing effective population size. Corridors may also allow the re-colonization of patches where local populations

have been reduced or eliminated. Patches and corridors exist within a background landscape referred to as the matrix, and together these three elements form a mosaic landscape.

The distribution and density of mammals over a heterogeneous landscape depends also on interspecific differences in vagility and habitat requirements. A habitat that is a formidable barrier to one species may be crossed readily by another. In Belgium, for example, the forest-dwelling bank vole (*Myodes glareolus*) seldom occurs in forest fragments surrounded by inhospitable (to the vole) open fields. The wood mouse (*Apodemus sylvaticus*), in contrast, crosses open fields readily and occurs commonly in isolated forest fragments (Geuse et al. 1985). Eastern chipmunks move between forest fragments along fence rows and shelterbelts, but open fields are barriers (Henderson et al. 1985). An excellent example of the reaction of a small mammal to barriers and a patchy landscape is afforded by pikas (*Ochotona princeps*) living in abandoned mine tailings on the eastern slopes of the Sierra Nevada Mountains of California. Pikas are restricted to talus or rock jumbles adjacent to boreal forbs and grasses. Smith (1974) studied an area with 24 "islands" of mine tailings ranging in size from perimeters of a few meters to more than 300 meters and separated by 20 to 300 meters. Two large islands supported permanent populations and provided dispersers that maintained the metapopulation. At a given time, only a fraction of the other islands were occupied, and pika populations were seemingly below carrying capacity. Pikas did not colonize islands separated from others by more than 300 meters. Local extinction was negatively correlated with size of the island, and recolonization was negatively correlated with distance to the nearest island. A fluctuating balance was maintained between extinction and recolonization.

Habitat corridors are especially important to mammals inhabiting highly fragmented landscapes or species that undergo long distance migrations. Among the latter species are pronghorns (*Antilocapra americana*) from the Greater Yellowstone Ecosystem of Wyoming (Berger et al. 2006). Probably for at least the last 6,000 years, one population of pronghorns has moved between their winter ranges in the Green River basin of Wyoming and Grand Teton National Park. Today, this pronghorn population passes through a geographic bottleneck as narrow as 121 meters in the Wind River Mountains. Obstruction of such a narrow corridor (i.e., by roads or mining operations) may result in extirpation of this pronghorn population and the loss of a migration pathway that operated for millennia. In Africa, corridors may be wide, but require crossing through unprotected areas between the relative safety of national parks or reserves. Douglas-Hamilton et al. (2005) show that elephants (*Loxodonta africana*) moved faster in unprotected travel corridors than in protected "home sectors."

Patches are not only separated from one another but are separated from the adjacent ecosystem matrix by a boundary zone or edge. The edge may be abrupt or a zone of varying width along the perimeter of the patch that is influenced by both the patch interior and the surrounding environment. The composition and abundance of species along the edge may be distinct from those in the interior of the patch (**edge effect**). Where two adjacent communities meet, a transitional zone or ecotone is formed. Ecotones may be characterized by sharp physiographic and floral contrasts with adjacent communities. Small mammal communities changed across an ecotone between grassland openings and deciduous forest in North Carolina. Abundance of smoky shrews (*Sorex fumeus*), pine voles (*Microtus pinetorum*), white-footed mice (*Peromyscus leucopus*), and woodland jumping mice (*Napaeozapus insignis*) did not differ across the ecotone, but relative abundances of masked shrews (*S. cinereus*) and red-backed voles (*Myodes*

gapperi) were highest in the ecotone (Menzel et al. 1999). In addition, different small mammal species react differently within the same edge habitat. For example, meadow voles at edges of small patches were larger, stayed in residence longer, and reproduced more often than were voles in continuous habitats (Bowers et al. 1996), but prairie deer mice in the same study area avoided edge habitats where risk of predation was high (Bowers & Dooley 1993).

Humans are causing increasing habitat loss and fragmentation worldwide. This poses an ever greater threat to mammalian biodiversity, especially in tropical regions where deforestation continues to fragment already isolated tropical landscapes (Wright 2005). One consequence is that mammal populations become increasingly subdivided and isolated, resulting in reduced gene flow and genetic diversity. Large mammalian carnivores have large home ranges and small population sizes and are most likely to have negative interactions with humans. Thus, they are particularly vulnerable to habitat fragmentation (Crooks 2002). Black bears (*Ursus americanus*) were once common throughout the southeastern United States, but their numbers were significantly reduced in Florida during the latter half of the 20th century as human populations increased in the state (Dixon et al. 2007 and references therein; Wooding 1993). Today, the Florida black bear exists in nine isolated subpopulations with limited dispersal opportunities. Genetic samples from 339 bears from all nine subpopulations reveal low levels of gene flow due to dispersal barriers (such as high-traffic roads) and habitat fragmentation from large-scale agriculture and housing subdivisions (Dixon et al. 2007).

The detrimental effects of habitat fragmentation and loss are not restricted to large mammalian predators. Tropical bats are diverse, relatively mobile, and fill important roles as pollinators and seed dispersers. Nevertheless, they are at risk from habitat fragmentation. Meyer et al. (2009) used mitochondrial DNA samples from 11 isolated populations of two phyllostomid bats in Panama. They found that bat species with poor dispersal ability (e.g., *Carollia perspicillata*; Fig. 23-14) suffered loss of genetic diversity relative to those with higher vagility (e.g., *Uroderma bilobatum*). In Panama, such "genetic erosion" has occurred in less than 100 years (since the building of the Panama Canal).

SUMMARY

Mammalian ecology, the study of the interactions of mammals with other organisms and their environment, can be subdivided into three broad levels: population, community, and ecosystem ecology. Global climate patterns, including variation in solar radiation and global atmospheric circulation patterns, determine the distribution of terrestrial

biomes. Biomes are characterized by distinctive communities of plants and animals adapted to regional conditions. Mammals inhabit virtually all terrestrial biomes and many marine biomes.

Population ecology seeks to understand the distribution and abundance of species. Mammal populations have specific habitat requirements. Many species attempt to monopolize resources by defending territories or home ranges from conspecifics. Because resources are often limited in a given area, some individuals are forced to disperse or migrate to new areas.

Life history patterns, including reproductive traits and survivorship patterns, form a continuum from those populations with high population growth rates to populations that grow slowly and invest more in each offspring. When resources are plentiful and population densities are low, the per capita growth rate is greater than zero, and populations increase exponentially. Eventually, the population outstrips its resource base (exceeds its carrying capacity), and population growth stops or declines. Density-dependent or density-independent processes may act to regulate populations. Some predator–prey populations cycle periodically between high and low densities. The factors driving such cycles have been studied for several decades, but the causes are not well-understood in all cases.

Populations are not uniformly distributed in a given region. Rather, subpopulations form in appropriate habitats and may exchange genes with other subpopulations by emigration and immigration of members. Such interconnected subpopulations form metapopulations.

Groups of species that interact with one another in a given area form loose ecological communities. Communities are characterized by several properties not found at the population level, including interspecific interactions, diversity, and trophic structure. Ecologists describe the biological attributes of a community by such measures as species richness. Generally, species richness increases from the poles to the equator.

The theory of island biogeography proposes that the number of mammal species on an "island" is determined by a dynamic equilibrium between the rate of immigration of new species to an island and the rate of extinction of resident species. These two rates are affected by factors such as island area, distance from the mainland, population size, number of species, and catastrophic events. The model predicts that the species richness of an island will remain relatively constant but that the list of species may change over time.

Species interactions determine community structure. The more diverse a community the more potential interactions there are among its members. Such interactions include interspecific competition, commensalism, mutualism, predation, and parasitism. Interspecific competition occurs when two or more species use the same limited resource. Two of the most common interactions are interspecific competition and predation. The intensity of interspecific competition depends on the degree of ecological

niche overlap. To alleviate competition, species with similar life styles tend to occupy exclusive niches. Predation, parasitism, or herbivory benefit one species while harming or reducing the fitness of another species. Herbivory is a special case of predation with plants as prey. Herbivory has many advantages: The biomass of plants vastly exceeds that of herbivores, plants need not be pursued, and they are available day and night.

Through millions of years of plant-herbivore interactions, many plants have adapted to mammals by using them as agents of seed dispersal or pollination. Some plant species have also responded by evolving an arsenal of anatomical and chemical defenses. Venomous spines, formidable thorns, and irritating hairs on many plants are familiar deterrents. Less familiar, but of greater importance, are secondary compounds. These range from toxins that can kill an herbivore to digestion-inhibiting or antimicrobial compounds.

As a result of endothermy, with its high energy requirements, mammals have the capacity to radically transform their environment through such activities as feeding, migration or daily movement, and the construction of shelters or refuges. Wildebeests transform African grasslands in part by virtue of their numbers. In contrast, some species of mammals impact their communities regardless of their abundance. The loss of such "keystone" species profoundly alters the species composition of the remaining community. For example, sea otters play a critical role in structuring nearshore marine communities.

Ultimately, virtually all energy on Earth comes from the sun. Solar energy is transferred from photosynthetic plants (primary producers) to herbivores (primary consumers) to primary carnivores (secondary consumers) that eat herbivores to secondary or perhaps tertiary carnivores in some extended food webs. Food webs describe the complex pathways of energy transfer occurring in nature. Animals that occupy comparable functional positions in the food web—primary consumers, for example—are at the same trophic level. Green plants occupy the first trophic level, and herbivores typically occupy the second trophic level. Within a trophic level, several species often exploit a common set of resources in a similar way leading to the formation of diverse and conspicuous feeding guilds.

Landscapes are not defined by size, but include a mosaic of habitats, ecological communities, or ecosystems, and their structure and biotic interactions change over time. Habitat heterogeneity within a landscape affects the spatial patterns and distribution of mammals. Landscapes include patches (of varying scale), corridors between patches, and background matrix. Patches are islands of relatively homogeneous habitats surrounded by dissimilar areas. Corridors are strips of habitat that connect separate patches and allow individuals to move between patches, increasing gene flow among patches that would otherwise be isolated. Corridors may also allow the re-colonization of patches where local populations have been reduced or eliminated. Habitat corridors are especially important to mammals inhabiting highly fragmented landscapes or to species that undergo long-distance migrations. Habitat loss and fragmentation represent an ever-increasing threat to mammalian biodiversity. One consequence is that as mammal populations become increasingly subdivided and isolated, gene flow and genetic diversity are reduced or lost.

Mammalogists have made, and continue to make, significant contributions to our understanding of ecological patterns and processes. We know the most about certain mammal populations, less about more complex community processes, and relatively little about metapopulation dynamics and landscape ecology. Newer technologies, including computers, satellite tags, remote video cameras, genomic techniques, and remote sensing, are opening new avenues for exploration. In summary, mammalian ecology remains an active and important area of research for the curious mammalogy student.

KEY TERMS

Allozyme	Granivore	Pioneer community
Autotroph	Gynodioecy	Pollination
Biome	Herbivory	Polyphilous
Biotic potential	Home range	Population
Carrying capacity	Ideal despotic distribution	Presaturation dispersal
Chiropterophily	Ideal free distribution	Primary consumer
Chitty effect	Inbreeding depression	Primary producer
Climax community	Keystone species	Primary succession
Coevolution	Kleptoparasite	Reproductive potential
Commensalism	Larder hoarding	Saturation dispersal
Competition	Logistic population growth	Scatter hoarding
Competitive exclusion	Major histocompatability complex	Secondary compound
Conspecific	Metapopulation	Secondary consumer
Corridor	Minimum viable population	Secondary succession
Density-dependent	Mutualism	Seral stage
Density-independent	Natal dispersal	Species diversity
Ecological succession	Niche	Species richness
Ecotone	Niche differentiation	Subnivean
Edge effect	Nonvolant	Territory
Evenness	Parasitism	Traplining
Exponential growth	Patch	Trioecy
Food web	Philopatry	
Foraging guild	Phylogenesis	

RECOMMENDED READINGS

Barrett, GW & JD Peles. 1999. *Landscape Ecology of Small Mammals.* Springer, New York.

Gittleman, JL. 1989. *Carnivore Behavior, Ecology, and Evolution.* Cornell University Press, Ithaca, NY.

Halfpenny, JC. 2003. *Yellowstone Wolves in the Wild.* Riverbed Publishing, Helena, MT.

Krebs, CJ, GC Hickman, & SM Hickman. 2002. *Ecology: The Experimental Analysis of Distribution and Abundance.* Benjamin Cummings, Upper Saddle River, NJ.

Kunz, TH & MB Fenton. 2003. *Bat Ecology.* University of Chicago Press, Chicago, IL.

Kunz, TH & PA Racey. 1998. *Bat Biology and Conservation.* Smithsonian Institution Press, Washington, DC.

Lidicker, WZ, Jr. 1995. *Landscape Approaches in Mammalian Ecology and Conservation.* University of Minnesota Press, Minneapolis, MN.

Ricklefs, RE & D Schluter. 1993. *Species Diversity in Ecological Communities, Historical and Geographical Perspectives.* University of Chicago Press, Chicago, IL.

Roff, D. 2001. *Life History Evolution.* Sinauer Associates, Sunderland, MA.

Behavior

The behavior of any mammal is of great interest because natural selection favors behaviors that help the individual survive and reproduce. Consequently, the range of behaviors is wondrously diverse. In the case of the pronghorn (*Antilocapra americana*), for example, great running speed became part of a unified functional system only because of a complex of behaviors that evolved in association with this ability (**Fig. 24-1**). The formation of herds, sexual dimorphism, systems of social behavior, the preference for open habitats, the flashing of the white rump patch as a danger signal to other pronghorn, and the remarkable ability to detect enemies at a distance all allow the pronghorn to use its great speed effectively to escape predators. How a mammal uses its morphologic and physiological equipment is of vital adaptive importance and forms the substance of behavior.

Those who study mammalian behavior seek to understand both the proximate and ultimate mechanisms causing behaviors. Researchers studying the proximate causes ask how stimuli initiate behaviors, what range of responses to those stimuli are possible, or how behaviors are acquired by learning or genetic mechanisms. In contrast, behaviorists addressing the ultimate causes of behavior are interested in understanding why a certain behavior has evolved. How and why mammals make the choices they do about the mates they choose, the places they forage, and the individuals they associate with are within the realm of **behavioral ecology**.

This chapter deals largely with behavioral ecology, the study of the ecological and evolutionary basis for behavior. One might suppose that behavior could be more readily observed and analyzed than could other aspects of biology and that detailed behavioral information on many species would have been assembled relatively early, but this is not the case. Indeed, little is known of the behavior of many mammals that are well known morphologically. Mammalian behaviors are of particular interest because of their flexibility and variability, which make for complex behavioral patterns that often differ widely among species. Remarkably well-developed sense organs, coupled with a brain capable of rapid evaluation of complex sensory information, have enlarged the perceptual sphere of mammals

FIGURE 24-1 A pronghorn (*Antilocapra americana*) running across open grasslands in the Hart Mountain National Antelope Refuge, Oregon.

and have facilitated the evolution of communication and rich social behavior.

Mammalian behaviors are diverse and often exceedingly complex. No single chapter can hope to cover the variety and complexity of behaviors exhibited in nature. Rather, the goal here is to introduce important concepts in behavior and describe typical and unusual behaviors exhibited by mammals. In doing so, hopefully some readers will choose to study mammalian behavior and fill in some of the remaining gaps in our knowledge.

Activity Rhythms

A striking aspect of mammalian behavior is the rhythmic, or cyclic, and predictable pattern of activity. Some species are active at night (nocturnal) and some during the day (diurnal); others are active primarily at dawn and dusk (crepuscular). Mammals also exhibit other kinds of cyclic behavior. Migratory movements are seasonally cyclic. The timing of reproduction is also cyclic, and in some mammals, including some rodents and bats, the physiological processes related to metabolism, such as torpor, exhibit daily or seasonal cycles. Cyclic behaviors generally occur over time scales with either a 24-hour period (**Fig. 24-2**) or a roughly yearly period. **Circannual** cycles are behavior patterns that occur over the course of roughly 1 year and recur year after year. Daily activity rhythms, those based on a 24-hour cycle, are termed **circadian rhythms** and are better understood than are other types of rhythms.

Circadian Rhythms

The Earth revolves slowly on its axis, completing one revolution every 24 hours. Daily rhythms based on the 24-hour light-dark cycles are universal in mammals. Circadian rhythms are adaptations to a particular mode of life and environment and they evolved just as morphologic characteristics have. The question of whether circadian cycles are **endogenous** (internally controlled) or **exogenous** (ultimately regulated by external stimuli) has occupied the attention of many biologists. Clearly, some strong endogenous control is present. Some mammals exhibit relatively constant 24-hour cycles of feeding and sleeping despite

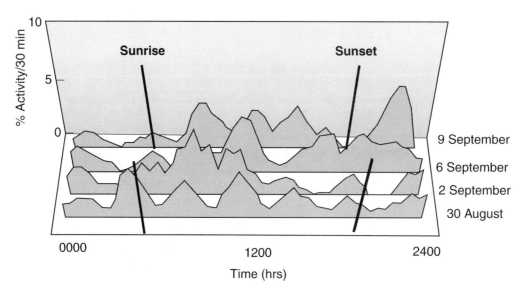

FIGURE 24-2   Summer activity patterns of a population of root voles (*Microtus oeconomus*) during an 8-day period in 1990. Vertical lines indicate the approximate hours of sunset and sunrise. (Modified from Halle, S., *J. Mammalogy* 76 (1995): 88-99. American Society of Mammalogists, Allen Press, Inc.)

being held in the laboratory under constant conditions of photoperiod and temperature. Maintenance of circadian cycles without environmental cues suggests an internal mechanism or "clock." In mammals, the master clock is a cluster of neurons in the superchiasmatic nucleus (SCN) in the brain (Dupre and Loudon 2007). Experimentally destroying these cells results in the loss of circadian cycles. So-called clock genes have been discovered that interact to generate these daily cycles. Remarkably, some mammalian clock genes are homologous with those in fruit flies, suggesting that circadian clocks evolved early in animal history. (A full account of mammalian clock genes can be found in Albrecht 2002 and references therein.)

Mammals held under constant environmental conditions retain daily cycles, but they are typically not exactly 24 hours long. As a result, an individual with a period less than 24 hours long will begin its daily cycle a little earlier each day, as the cycle runs according to its natural period (**free-running cycle; Fig. 24-3**). For example, careful work on the flying squirrel (*Glaucomys volans*) by DeCoursey (1961) showed that, even under constant environmental conditions including continuous darkness, flying squirrels maintained regular activity periods that deviated only 62 minutes from the mean value for activity periods under natural conditions. Similarly, both porcupines and ground squirrels were able to maintain normal daily rest and activity periods over a 24-hour period even during 82 days of constant daylight during the arctic summer (Folk et al. 2006). When a laboratory mammal whose circadian cycle is out of phase with the natural 24-hour light-dark cycle is again exposed to normal day and night conditions, its cycle rapidly shifts and becomes "synchronized," that is, it becomes adjusted and locked (entrained) to the 24-hour cycle (Bruce 1960).

For reindeer and perhaps other herbivorous mammals living in polar regions, the cues necessary for synchronizing the cycle are absent much of the year; in these environments selection for maintaining an internal clock is relaxed. Reindeer (*Rangifer tarandus*) above the Arctic Circle in Norway failed to show normal daily activity rhythms during the constant summer daylight (van Oort et al. 2005). Thus, entrained behaviors for one group may not be adaptive for another.

Daily rhythms differ markedly from one species to another. Even in two nocturnal species circadian rhythms are influenced by interactions between species with similar environmental needs. In some cases, competition between species is reduced or eliminated because their activity cycles are out of phase. Two species of fishing bats (*Noctilio*), both of which feed over water, avoid interfering with one another partly by foraging at different times of the night and also by eating different prey (Hooper and Brown 1968). In general, small mammals such as rodents that are especially vulnerable to predation tend to be nocturnal (most sciurid, arvicoline, and some sigmodontine rodents are exceptions), whereas less vulnerable species such as many ungulates may be active during the day. The activity cycles of carnivorans seem to be geared to the circadian cycles of their prey, that is, to the period when hunting is most rewarding.

As might be expected if circadian cycles are adaptive, the cycles often shift seasonally and depend on an animal's ability to track some environmental variable such as light or temperature. Changing metabolic demands on small mammals accompany the seasonal changes in ambient temperatures, and some shifts in circadian rhythms may allow the animals to avoid or reduce activity during times of most intense temperature stress. For example, many small

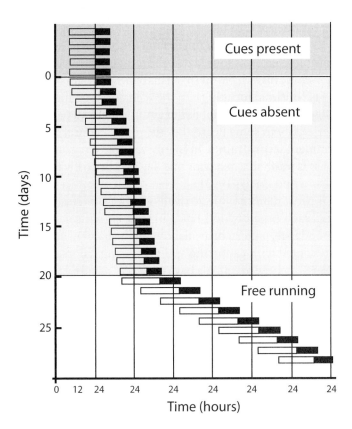

FIGURE 24-3 An example of circadian activity over a 35-day period. Each horizontal bar is one 24-hour period, with the open portion representing daytime activity and the solid part representing nighttime. When photoperiod cues are present, the hypothetical mammal is active on a normal 24-hour cycle (top). When cues are removed (middle), the period is shifted a bit later each day until it eventually becomes a free-running period (bottom).

mammals that are intermittently active day and night in some seasons become more nocturnal during the hottest periods, thereby avoiding heat stress. Likewise, they may become more diurnal in winter, avoiding the coldest hours of night and foraging when temperatures are warmer. Stebbins (1984) reported three types of seasonal changes in circadian rhythms for small mammals in Canada: overall decreased winter activity, changes in percentage of nocturnal and diurnal activity, and changes in peak daily activity. Seasonal shifts in activity probably result in a considerable saving of energy.

Circannual Rhythms

Although less thoroughly studied than circadian rhythms, circannual rhythms (cycles with a period of approximately 1 year) play an equally prominent role in the lives of some mammals. Where environments are highly seasonal, selection favors individuals that can exploit seasons of abundance by rapid growth or development or by timing reproduction so that young have access to more resources. Similarly, selection favors individuals that can avoid or cope with unfavorable seasons through migration or hibernation. To take advantage of seasonal opportunities,

individuals must be able to predict the changing seasons and prepare appropriately. As Bradshaw and Holzapfel (2007) noted, "Fitness in seasonal environments is all about timing: the optimal time to migrate and reproduce, the optimal time to stop reproducing, and the optimal time to migrate again."

Such vital activities as breeding, migration, and hibernation are phased on an annual cycle by exogenous cues. For most mammals, photoperiod (day length) is the environmental cue used for predicting seasonal changes. Other cues include changes in food abundance, rainfall, and temperature. Among mammals, circannual rhythms have been documented for the golden-mantled ground squirrel (*Spermophilus lateralis*; Pengelley and Fisher 1963; Pengelley and Asmundson 1970, 1971), the woodchuck (*Marmota monax*; Davis 1967), and for two species of ground squirrels (*Spermophilus*) and four species of chipmunks (*Tamias*; Heller and Poulson 1970), all of which have periods of hibernation or dormancy in winter.

For hibernators in temperate regions, circannual rhythms make the mammals sensitive to falling temperatures and declining food supplies in the autumn. Thus, the onset of hibernation may be hastened by unfavorable conditions or delayed by favorable temperatures and food supplies. In arctic areas, the extremely harsh environment and the sudden onset of winter, coupled with the brief time available for breeding and for putting on fat in preparation for hibernation, make flexibility nonadaptive. Here the adaptive premium shifts to a precise, inflexible, optimal schedule. Breeding at the optimal time each year, regardless of climatic conditions, probably ensures the greatest reproductive success, and precision in the onset of hibernation ensures maximum overwinter survival. Even in tropical areas, circannual rhythms may be highly adaptive. In Kenya, for example, the African false vampire bat (*Cardioderma cor*) and the giant leaf-nosed bat (*Hipposideros gigas*) become pregnant well before the onset of the late March–April rainy season, seemingly in anticipation of the burst of insect abundance that accompanies the rains.

Photoperiod is probably the most important environmental factor in phasing the underlying circannual rhythm of hibernators (Heller and Poulson 1970). The situation is not entirely simple, however, because temperate-zone hibernators that occupy the same area do not necessarily follow the same circannual rhythms. The golden-mantled ground squirrel (*S. lateralis*), an inhabitant of mountains in the western United States, feeds into autumn and stores relatively nonperishable seeds in its burrow. Entrance into hibernation is relatively tightly scheduled, regardless of environmental cues (Pengelley and Asmundson 1970, 1971). Belding's ground squirrels (**Fig. 24-4**; *S. beldingi*), which often live almost side by side with the golden-mantled ground squirrels, feed on green material that decomposes quickly

FIGURE 24-4   A Belding's ground squirrel (*Spermophilus beldingi*) emerging from its burrow in southern Oregon.

if stored underground. This squirrel feeds as long as possible in autumn, putting on more and more fat, and stores no food. The golden-mantled ground squirrel can perhaps afford greater rigidity in the timing of its hibernation because of the cushion of stored food in the **hibernaculum**, but the Belding's ground squirrel must depend entirely on food stored in the form of body fat and thus feeds as long as such activity is energetically feasible.

Circannual patterns controlled by photoperiod exist for a number of other seasonal processes in mammals. Antler regeneration in male deer is under hormonal control and linked to breeding season (Goss 1982; Price et al. 2005). Seasonal changes in fur thickness, shedding, or changes in pelage color may also be influenced by photoperiod. Mink (*Mustela vison*) exposed to short-day photoperiods that mimicked winter day lengths grew winter pelage 6 weeks earlier than normal. In the same study, mink treated with excess melatonin, a naturally produced hormone that regulates circadian rhythms, also molted their summer pelage earlier and grew thicker winter coats (Rose et al. 1984). Thus, shorter day lengths in autumn alter melatonin levels and affect the timing of winter pelage formation in mink. Finally, embryonic diapause (or delayed implantation, see Chapter 20), which occurs in seven orders of mammals, also appears to be influenced by photoperiod (Tyndale-Biscoe 1980; Thom et al. 2004). In embryonic diapause, the embryo does not immediately implant in the uterus but is maintained in a dormant state that extends the gestation period, delaying birth until environmental conditions are more favorable.

Foraging Behavior

Mammals consume a vast array of food items. Thus, it is not surprising that mammals have evolved a wide range of foraging behaviors. All foraging behaviors involve decisions that influence the costs and benefits of acquiring a meal. Because successful foraging is vital to continued survival, natural selection should favor the foraging behaviors that deliver the highest payoff (in terms of fitness) with minimal costs (**optimal foraging**). In the wild, foraging mammals often approach optimal patterns but usually depart from optimality to some degree because of behavioral or environmental constraints. In highly variable environments, what is profitable behavior one day may be costly the next (Krebs and Davies 1993).

Most mammal species feed on a range of food items, and each type of food has a unique nutritional value, abundance, pattern of availability in the habitat, and risks associated with capturing and processing (Morse 1980). Consequently, mammals have evolved a number of interesting foraging behaviors. Some of these foraging patterns are determined by the location and seasonal abundance of the food, others by the individual's ability to store food or fat for times of future shortage. Predators that hunt have evolved behaviors that help locate and subdue more active prey. In some cases, it is more profitable to hunt in a group. In other cases, solitary hunting yields larger rewards (in fitness). Clearly, a discussion of the full range of mammalian foraging is beyond the scope of this chapter. The following discussion includes examples of some of the diverse foraging behaviors in mammals.

Fossorial Foraging

Some of the most specialized foraging behaviors occur among burrowing rodents. Pocket gophers (Geomyidae), mole-rats (Bathyergidae), and other fossorial rodents dig complex burrow systems, and part or all of their diet consists of underground parts of plants. Because of the tremendous energetic cost of burrowing, fossorial herbivores have evolved behaviors favoring the most efficient system of finding food. The basic geometry of burrow spacing in one species of pocket gopher (*Thomomys bottae*) was found to be remarkably uniform both within one burrow system and from one system to another (Reichman et al. 1982). The basic building unit of the system, as well as the distance between forks and the lengths of the branches, was uniform within and among systems and is analogous to the nodes and internodes of plants. These units can be combined to increase overall burrow length, but the uniform spacing is maintained (**Fig. 24-5**). Such consistent burrow spacing suggests intense selection in pocket gophers for precise, uniform burrowing behavior.

Burrowing behavior is constrained by physiological, morphologic, and climatic factors. In the arid regions of Namibia and the Kalahari, colonies of Damaraland mole-rats (*Fukomys damarensis*, formerly *Cryptomys*) forage underground for tubers and other geophytes. During the dry season, the soils are difficult to excavate and tunnel

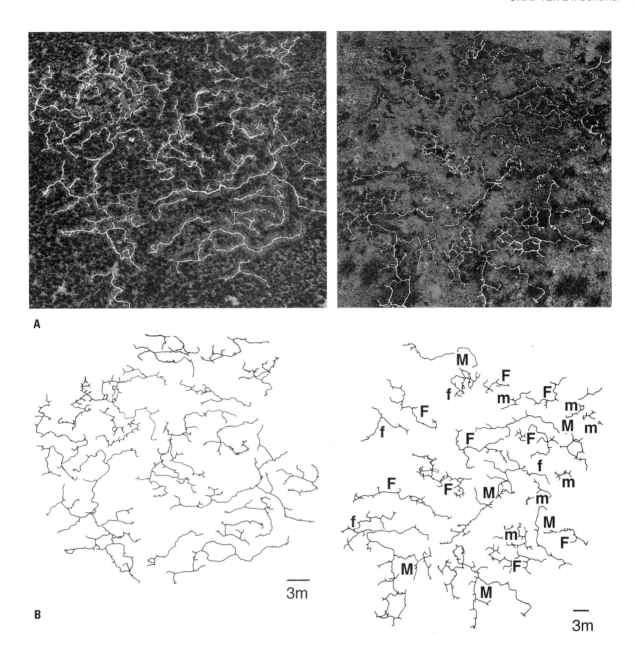

FIGURE 24-5 Aerial photographs and line drawings of pocket gopher (*Thomomys bottae*) burrows excavated at two sites near Cotton-wood, Arizona. Burrow systems were marked with lime and photographed from the air. The diagram on the right designates burrows of adult males (M), adult females (F), nonreproductive males (m), and nonreproductive females (f). (Reproduced from Reichmann, O.J., et al., *Ecology* 63 (1982): 687–695.)

expansion is minimal. After heavy rainfall, soils become moist and burrowing activity increases dramatically (Jarvis et al. 1998). Initially the colony digs long tunnels with few side branches, concentrating their effort on exploratory tunneling with relatively little harvesting of tubers. Jarvis and colleagues (1998) showed that during peak activity a colony of 16 mole-rats was capable of excavating 2.6 tons of soil in only 2 months. As the soil dries and hardens, smaller branch tunnels are created off the main exploratory tunnels and harvesting and storage of geophytes increases. Thus, during the driest periods when continued burrowing is not possible, these mole-rats have already secured sufficient food to last until the rains return.

Caching Behavior

Food storage, or **caching**, is a widely used and seemingly highly adaptive behavior. (Caching, as used here, is the moving of food from one place to another for later consumption.) Caching by mammals takes many forms: A leopard drags its prey into a tree to protect it from hyenas or lions; the short-tailed shrew (*Blarina brevicauda*) buries a mouse immobilized by the shrew's toxic bite (Tomasi 1978); and pikas store "haystacks" for winter consumption beneath rocks. Caching is used by some shrews, moles, and hedgehogs, by many carnivores (canids, ursids, some felids, and some mustelids), and by many rodents. Traveling back and forth from a foraging area to dependent young has probably favored the evolution of caching (Smith and Reichman 1984). Among the advantages of this behavior are protection of food from competitors, protection from predators while eating, and a food supply during lean times in an unpredictable environment.

The food most commonly cached by mammals is seeds, and seed-eating mammals cache food most frequently. Seeds provide a concentrated energy and nutrient reserve and can remain dormant in the soil for long periods. Studies of heteromyid rodents reveal many remarkable behaviors associated with the caching and management of stored seeds. The closely related kangaroo rats and pocket mice (Heteromyidae) travel far from their burrows and gather seeds from the soil by using the long claws of the small forefeet. These rodents rapidly collect seeds in the cheek pouches but tend to gather seeds that are superior energetically to those randomly available in the soil. Typically, many loads are taken to the burrow in an evening. Later, in the safety of the burrow, the animals become even more selective: Of the seeds gathered, the rodents eat those richest in energy (Reichman 1977).

Heteromyids have cached seeds for at least 10 million years (Voorhies 1974), ample time for evolutionary fine-tuning of morphology and behavior associated with caching. In addition to developing anatomic features that further rapid gathering and transport of seeds (dexterous forefeet, external cheek pouches, and elongate hindlimbs), heteromyids have uncoupled food gathering and eating. Seeds are gathered in the cheek pouches, carried underground, and then eaten—soon or even months later—in the safety of the burrow.

These rodents also have the remarkable ability to "manage" the seeds they collect (Reichman et al. 1985). The banner-tailed kangaroo rat (*Dipodomys spectabilis*) typically caches seeds beneath the north or northwest section of their habitation mound at a depth near 30 centimeters or beneath 50 centimeters. Chambers in the soil are humid, and cached seeds rapidly become moldy. More than 100 species of fungi were found in the seed caches of several rodent species (Herrera et al. 1997). In the seed caches of *D. spectabilis*, seeds cached during rainy weather had the greatest abundance and diversity of fungi. In the laboratory, pocket mice (*Chaetodipus intermedius*) ingested slightly moldy seeds in preference to highly moldy or nonmoldy seeds, and wild *D. spectabilis* had similar preferences (Rebar and Reichman 1983; Reichman and Rebar 1985). In the laboratory, seeds were actually managed by the latter species for degree of fungal growth: Sterile seeds were stored in the highest available humidities. Once seeds reached preferred levels of moldiness, they were moved to low humidities that inhibited further fungal growth (Reichman et al. 1986). Other experiments demonstrated that kangaroo rats are sensitive to extremely minor differences in the water

content of seeds, that seeds with the highest water content are preferred, and that some fungal growth is tolerated to get more water (Frank 1988a, 1988b).

The diet of a heteromyid rodent eating a seed cache sometimes is limited because the cache is not replenished daily; therefore, its diet is optimized for the long term rather than the short term. The rodent should not eat all of the preferred or most nutritious items first; it should instead diversify its diet to survive the longest possible time on the cached food items. Just such behavior was demonstrated in the laboratory by Reichman and Fay (1983). A caching pocket mouse (*C. intermedius*) began to diversify its diet by eating less preferred seeds in larger amounts after several days of eating only from its cache. Under the same conditions, a typically noncaching deer mouse (*Peromyscus maniculatus*) consumed the seeds in order of preference, as would be expected for an animal simply foraging for food each day.

Frequent manipulation of cached seeds (in a way yet unknown) by *D. spectabilis* virtually eliminated germination of cached seeds (Reichman et al. 1985). Seeds germinated rapidly in unattended caches, while seeds rarely germinated in caches attended by resident animals. Chipmunks are known to prevent germination and preserve the food value of cached beech seeds by biting off the tips of the embryos (Elliot 1978).

The eastern woodrat (*Neotoma floridana*), which caches mostly vegetation other than seeds, discriminates between foods on the basis of perishability, tending to eat the more perishable foods and to cache the less perishable items (Reichman 1988; Post and Reichman 1991). The short-tailed shrew makes similar decisions with different foods: Seeds are cached before insects, and both are cached before mice, a sequence reflecting degrees of resistance to spoilage (Martin 1984). Probably most caching mammals can recognize items that will not decompose quickly when cached.

The caching behavior of the North American red squirrels (*Tamiasciurus hudsonicus* and *T. douglasii*) is particularly notable (**Fig. 24-6**). These squirrels depend on fir and pine seeds for food. Red squirrels use scatterhoarding and larderhoarding. Scatterhoards consist of a few food items stored at each of many locations throughout the territory, whereas larderhoards are large piles of food items located at a central site within the territory (Hurley and Lourie 1997). Red squirrels dig holes and cache cones in large **middens** formed by the litter of cone fragments that accumulates beneath a squirrel's favorite feeding sites. The larder hoards are frequently 6 to 10 meters in diameter, contain from 2 to 10 bushels of cones, and are in shady situations where the moisture retained in the midden aids in preserving the green cones (Finley 1969). Small numbers of cones are commonly cached in logs or pools of water. The cones are harvested in late summer and autumn and are cut, on occasion, at the rate of 29 per minute (Shaw 1936). The squirrels

FIGURE 24-6 A red squirrel (*Tamiasciurus hudsonicus*) from the northern United States.

are such effective harvesters that one pine in northern California lost 93% of its 926 cones to them (Schubert 1953). Seeds from the cached cones are eaten during the winter. When snow is deep, access burrows are maintained through the snow into the midden.

Scatterhoarding is adaptive only if the cache remains intact and is not lost to conspecifics or other thieves. Hopewell and Leaver (2008) filmed wild gray squirrels (*Sciurus carolinensis*) caching high-quality and low-quality nuts in the presence of conspecifics and alone. When other gray squirrels were present, individuals caching nuts traveled greater distances, were more vigilant, and spent longer camouflaging the cache site (especially for the high-quality nuts). Thus, scatterhoarding squirrels modify their behavior to reduce cache theft when onlookers are present.

Hunting Prey

Predators have evolved behaviors that facilitate the pursuit, capture, and killing of prey, which in most cases have defensive or predator-avoidance strategies. Some behavioral patterns are common to a wide array of carnivores. The neck bite is such a behavior. This killing technique was studied in the house cat by Leyhausen (1956), who presented the predators with normal and headless rats and with rats with the head fastened to the tail end. The cats aimed their bites at any constriction in the body; with normal prey this results in the neck bite. Grabbing prey across the back and shaking it violently is another pattern shared by many carnivores.

Most canids are solitary, cursorial hunters and capture prey by virtue of speed or, occasionally, endurance. Experience and learning are of great importance, and canids are highly adaptable. A coyote with access to a waterfowl marsh may learn to capture molting ducks, while its relative in the desert patrols the perimeters of sand dunes for kangaroo rats. Canids have an extraordinary sense of smell, and prey is often detected initially by windborne scent. Adaptability and diverse tastes, rather than a specialized style of hunting, are the keys to the success of solitary canids.

The cats use more specialized hunting and killing techniques than canids (**Fig. 24-7**). Most cats are not long-distance runners but usually depend on short rushes directed against surprised prey. The sudden rushes of lions seldom cover more than 100 meters, and leopards and smaller felids frequently make only several bounds to reach their prey. The cheetah, an exceptional felid, may chase an antelope several hundred meters at speeds up to 95 kilometers per hour!

To use the typical feline hunting technique effectively, a cat must get close to its prey. The stalking of prey by felids involves a series of beautifully coordinated behaviors, described in detail by Leyhausen (1956). When prey is sighted, the cat crouches low to the ground and approaches, using the "slink-run" and taking advantage of every object

FIGURE 24-7 (A) A pride of lions (*Panthera leo*) attacks an African buffalo (*Syncerus caffer*) on the plains of Duba, South Africa. (B) A group of African wild dogs (*Lycaon pictus*) attacks and kills a wildebeest calf (*Connochaetes taurinus*) in the Serengeti.

offering concealment. At the last available cover, the cat stops, and then "ambushes." The brief rush to the prey ends in a spring; the forefeet clutch the animal, but the hind feet often remain planted and stabilize the cat for the possible struggle. The cat usually makes the kill not by belaboring the prey, as do many canids, but either by a powerful bite at the base of the skull or the neck, which crushes the back of the skull or some of the cervical vertebrae and the spinal cord, or, in the case of large prey, by strangulation (**Fig. 24-8**). It is interesting to note that although pumas (*Puma concolor*) kill large prey by strangulation, when they fight one another (usually in disputes over food or territory) a bite to the head is nearly always the cause of death (Logan and Sweanor 2001). The shortening of the felid jaws is a specialization that contributes to the power of the bite.

Whereas most cats are solitary, the African lion is the only truly social felid. It often hunts in groups in which there is some cooperation among members, with adult females doing most of the killing. The lion typically stalks large prey, often as heavy as or heavier than itself. Although cooperative effort improves success, a prey animal is typically killed by a single lion. Lions attacking prey the size of a zebra or African buffalo attempt to bring the prey to the ground by clutching the rump (Fig. 24-7A), hind legs, or shoulders with the forepaws and throwing the prey off balance (Schaller 1972). When the prey falls, the lion bites the neck or nose and maintains a grip until the prey is suffocated. Schaller pointed out that, by centering the bite on the neck or nose, the lion immobilizes the horns, remains clear of thrashing hooves, and can easily keep the victim on the ground. This specialized killing behavior reduces the risk of serious injury from the powerful prey. Killing large prey is risky nonetheless, as shown by several examples. In one case, the bodies of a mule deer and a puma lay side by side where they had seemingly collided with the trunk of an oak tree (Gashwiler and Robinette 1957). The puma's neck was broken. Similarly, three female New Mexican pumas were killed while attacking mule deer (Logan and Sweanor 2001): Two had severe damage to the rib cage and lungs; the third was stabbed through the braincase by an antler.

Other Foraging Strategies

Unusual behavior patterns enable some carnivorans to break the exoskeletons of invertebrates and the shells of

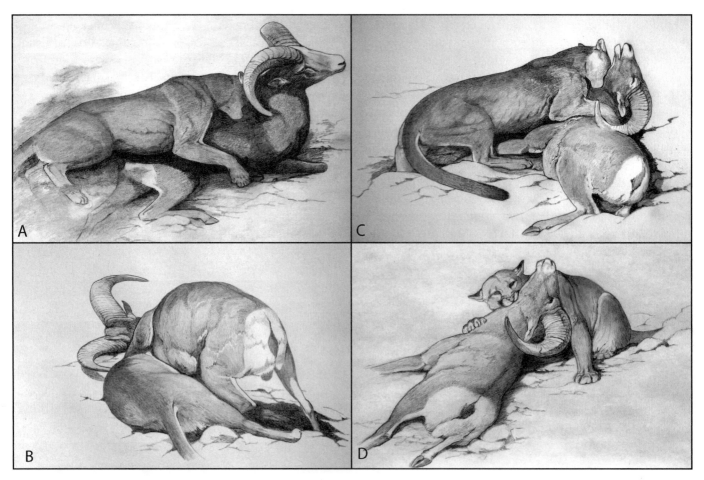

FIGURE 24-8 A series of drawings illustrating the attack of a mountain lion (*Puma concolor*) on a bighorn sheep (*Ovis canadensis*). The mountain lion leapt onto the back of the bighorn (A) and maneuvered the sheep (B–C) until it was able to grasp the throat (D), crushing the windpipe and suffocating the ram.

eggs. Some mongooses use the forefeet to throw objects against hard surfaces (Dücker 1957; Ewer 1968). Similarly, the spotted skunk (*Spilogale putorius*) breaks eggs by kicking them against rocks (Van Gelder 1953). The sea otter (*Enhydra lutris*) smashes the sturdy shells of mollusks by using a tool. The otter floats on its back with a flat stone on its chest, grasps the mollusk with its forepaws, and pounds it against the stone (Fisher 1939). An individual was observed to pound mussels (*Mytilus*) on a stone 2,237 times during a feeding period lasting 86 minutes (Hall and Schaller 1964). These otters are clearly selective in their choice of stones and may use the same one repeatedly.

Mammals that forage underwater also face unique challenges and have evolved some unique methods for capturing prey. Groups of humpback whales (*Megaptera novaeangliae*) encircle schools of prey fish in a "net" of bubbles. Foraging humpbacks dive below a school of fish, emit air through the blowhole as they circle below, creating a wall of bubbles up to 30 meters across. The prey fish apparently perceive the bubbles as a barrier and remain within the bubble "net" (**Fig. 24-9**). As the bubbles rise, entrapping the prey, the whales emit a very loud "feeding" call that seemingly synchronizes the group of whales to lunge upward with mouths agape through the prey school, capturing huge quantities of fish on one pass (Leighton et al. 2004).

An even more unusual use of bubbles for foraging occurs in water shrews (*Sorex palustris*). Water shrews weigh approximately 15 grams and are the smallest mammals to hunt underwater (**Fig. 24-10**). They feed on aquatic invertebrates, including crayfish, and often take small fish. Using high-speed video of captive water shrews foraging in aquaria, Catania and colleagues (2008) showed that these shrews use a variety of cues for locating and capturing prey. Water shrews do not rely heavily on vision underwater;

sonar and electroreception were also ruled out. Instead, the shrews used vibrissae to detect water movements and locate prey. Surprisingly, water shrews also sniffed potential prey underwater by exhaling a bubble onto the prey, presumably picking up the prey's scent, and then inhaling the bubble back into the nostrils. These studies suggest that olfaction is useful underwater is some species (Catania 2006).

Shelter-Building Behavior

Many mammals have evolved elaborate shelter-building behaviors that aid them in maintaining homeostasis, hiding from predators, or protecting offspring. The nests, burrows, or houses of mammals provide insulation that augments the animal's own pelage and saves energy by reducing the rate of thermal conductance from the animal to the external environment or vice versa. The woodrat (*Neotoma*) collects a variety of materials with which it builds houses or improves the shelter provided by rock crevices or vegetation (Fig. 13-27B). Beavers construct the most elaborate shelters of any mammal other than humans, and lodges made of sticks and branches are often more than 3 meters in diameter. Many terrestrial rodents construct nests beneath logs or rocks or in burrows. Arboreal rodents frequently build nests in the branches of trees or in hollows in trees. Some nest-building behaviors are perhaps common to many rodents, but the choice of nesting site seems to be species-specific. For example, red tree voles (*Arborimus longicaudus*) of the humid coastal belt of Oregon and California build their nests only in Douglas firs (*Pseudotsuga menziesii*), the needles of which provide the primary food of the vole (Benson and Borell 1931).

Fossorial rodents follow rather complex patterns of movement when digging. Probably many of the specific

FIGURE 24-9   Bubble net feeding in a group of humpback whales. (A) A spiral-shaped wall of bubbles, which traps a school of fish, is produced by a submerged whale. (B) Humpback whales lunge toward the surface inside the bubble net, and with mouths wide open they scoop up large numbers of trapped fish.

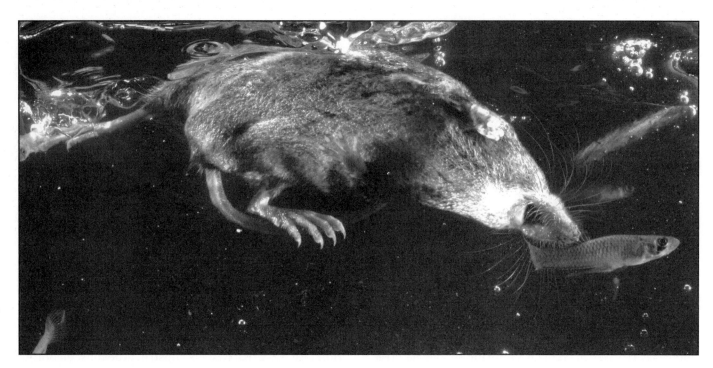

FIGURE 24-10   A water shrew (*Sorex palustris*) attempting to catch a small fish.

components of the total digging sequence are innate behaviors. Pocket gophers (Geomyidae) use the forefeet to loosen the soil by powerful downward sweeps, and the hindlimbs kick the accumulated soil backward from beneath the animal. Pocket gophers periodically eject soil from a burrow entrance by pushing it with the chin and forelimbs. Kennerly (1971) has shown that the long, complex series of behavior patterns associated with mound building are basically innate but may be modified by learning. The pocket gopher characteristically alternates direction in pushing soil from the burrow; it pushes a series of 5 to 20 loads to the right, a similar series to the left, and so on. The frequency distribution of directions of pushing soil indicates that efforts are mainly in three directions: either directly in front of the mouth of the burrow or at an angle of 90° to either side. This results in the fan-shaped mound so typical of pocket gophers. That learning plays a part in burrowing and mound building is suggested by the fact that young animals are less successful in plugging the openings of burrows than are older animals.

The burrowing and mound-building behaviors of some African mole-rats (Bathyergidae) differ markedly from those of pocket gophers. *Heliophobius*, for example, uses its incisors to excavate soil and pushes the dislodged soil in back of its body with its feet. The animal transports a load of soil to the surface by backing up against it with the rump and large hind feet. The forefeet push the animal backward, and the head and upper incisors are braced against the roof of the burrow to gain purchase. A silvery mole-rat (*H. argenteocinereus*) adds up to 1 meter of new tunnels each day and backfills more than 60% of old tunnels

(Skilba et al. 2009). Unlike pocket gophers, which appear briefly at the surface each time they push a load of soil onto the mound, *Heliophobius* pushes a core of soil out onto the mound without appearing on the surface (Jarvis and Sale 1971).

Several kinds of Neotropical bats use leaf shelters, and among these the "leaf-tents" are of special interest (**Fig. 24-11**). White bats (*Ectophylla alba*), for example, roost in groups of one to six individuals beneath the large leaves of several Neotropical plants (Timm 1987). The bats cut the side veins extending from the midrib of the leaf in such a way that the end of the leaf bends down and forms a tent. One colony of bats uses several of these shelters on alternate days (Timm and Mortimer 1976).

Communication

Communication has often been broadly defined to include all interactions between animals that transmit information between them, but if all types of stimulus-reception sequences are regarded as communication, then essentially all behavior of one animal that can be perceived by another must be regarded as communication. For the purposes of discussion here, communication signals are "behavioral, physiological, or morphological characteristics fashioned or maintained by natural selection because they convey information to other organisms" (Otte 1974). Each type of communication—visual, olfactory, auditory, and tactile—is considered separately, but it should be stressed that usually a complex of several kinds of communication signals passes between animals (**Table 24-1**).

FIGURE 24-11   A group of tent-making bats (*Uroderma bilobatum*) clustered under a palm frond in Costa Rica.

▪ Visual Signals

Facial expressions, body postures, and anatomical structures, such as male deer's antlers, are among the many visual signals used by mammals. These visual displays convey information from sender to one or more receivers, helping them coordinate their behaviors. Visual signals may be intraspecific, as in the submissive posture of a lower-ranking member of a wolf pack, who crouches down with tail held between the legs, when encountering an alpha male (the pack's highest-ranking male). Alternatively, visual displays may be interspecific, such as ring-tailed lemurs (*Lemur catta*) responding to alarm calls produced by Verreaux's sifakas (*Propithecus virreauxi*) that have spotted a predator in the neighborhood (Oda and Masataka 1996).

Facial expressions and postures are often fleeting, but flexible, and may convey a wide range of possible signals. In contrast, morphological features are more permanent: Antlers or swollen genitalia are seasonal signals, and the dark mane of an adult male lion is a permanent signal of his maturity. Often visual signals are combined with other types of signals: Male African elephants in musth (breeding condition) secrete chemicals from glands on the face (olfactory signals), trumpet loudly (acoustic signals), and display a number of aggressive postures toward other males (visual signals). The evolution of visual displays is in the direction of reduced ambiguity. However, visual signals are limited in range and typically are less useful underwater or at night, when light levels are low. Nevertheless, visual communication helps sender and receiver coordinate foraging, arbitrate territorial disputes, attract mates, defend themselves, and maintain social bonds (in mammals that live in groups).

Facial expressions are of great importance in communication, and natural selection has favored the development of distinctive facial markings that focus attention on the head. As described by Lorenz (1963), the facial expressions and ear postures of dogs signal degrees of aggressiveness or submissiveness (**Fig. 24-12**). The posture of the head and the facial expression of many ungulates provide visual signals to other members of the herd or to territorial or sexual rivals. An elk (*Cervus elaphus*) ready to run from danger elevates its nose and opens its mouth (McCullough 1969). A Grant's gazelle (*Gazella granti*) holds its head high, elevates its nose, and pricks its ears forward when challenging another male (Estes 1967). The head of both these animals is conspicuous: the elk's because the dark brown head and neck contrast strongly with the pale body, and the gazelle's because of bold black patterns. The intricate facial expressions of primates are frequently made more obvious by

TABLE 24-1   **Properties of Various Modes of Communication**

Property	Visual	Auditory	Chemical	Tactile
Effective distance	Moderate	Long	Long	Very short
Ability to localize	High	Moderate	Variable	High
Ability to go beyond obstacles	Poor	Good[a]	Good	—
Speed to transfer	Fast	Fast	Slow	Fast
Complexity of signal	High	High	Low	Moderate
Persistence of signal	Variable	Low	High	Low

[a] True for moderate- to low-frequency sounds only.

A

C

B

D

FIGURE 24-12 Changes in the facial expressions of a wolf (*Canis lupus*) from passive (A) to aggressive (D).

distinctive and species-specific patterns of pelage color-ation and by brightly colored skin (**Fig. 24-13**).

The effectiveness of visual signaling is heightened in many species of mammals by **weapon automimicry** in the form of striking (and to our eyes handsome) markings (Guthrie and Petocz 1970). The ears are commonly used signaling devices in mammals. Artiodactyls use their ears in signaling, probably because of the proximity of the ears to the horns or antlers. The ears in many species of ungulates are marked or adorned with hair in such a way as to mimic the horns. This probably strengthens the visual signal given by the horns, as well as making the posture of the ears extremely obvious (**Fig. 24-14**). Facial markings may also play a role in automimicry by accentuating the horns. In the sable antelope, oryx, and Grant's gazelle, black markings create a design that extends the contours of the horns (**Fig. 24-15**).

The body is used for signaling in many species. This type of signaling is particularly well-developed in ungulates that inhabit open areas and that gain an advantage from coordinated herd action. Grant's gazelle and Thompson's gazelle (*G. thompsonii*) of Africa, which have two warning displays (Estes 1967), twitch the flank skin (conspicuously marked in *G. thompsonii*) just as they begin to run from a predator that has entered the minimum flight distance (the minimum distance at which an approaching enemy causes the animals to run). The most effective display is a stiff-legged bounding gait, called "stotting," used at times as the gazelles begin to run (Estes 1991). The conspicuousness of this display is enhanced by the erection and flaring of the hair of the white rump patch. In some monkeys and apes, the presentation of the hindquarters as if inviting copulation is a social gesture symbolic of friendship and is accepted by a brief "token" mounting (Heinroth-Berger 1959). Kangaroos threaten one another by standing bipedally at their maximum height, surely an impressive visual signal.

FIGURE 24-13 A mandrill (*Mandrillus sphinx*) from west central Africa showing the conspicuous facial markings on the rostrum.

Olfactory Signals

Eisenberg and Kleiman (1972) defined olfactory communication as "the process whereby a chemical signal is generated by a presumptive sender and transmitted (generally through the air) to a presumptive receiver who by means of adequate receptors can identify, integrate, and respond (either behaviorally or physiologically) to the signal." A chemical signal that elicits a response in a conspecific receiver is known as a pheromone, whereas an **allomone** conveys a message to a receiver of a different species. Olfactory communication is effective because specific chemicals can convey very specific messages, and a scent mark on an object will persist long after it is deposited. Because scent is released into the air and disperses rapidly, however, a receiver must have a sense of smell acute enough to find the source by detecting concentration gradients. Also, olfactory signals broadcasted in the air are as available to a predator as to a conspecific.

Scent marking is used widely as a means of communication and is commonly an expression of dominance. Ralls (1971) indicated that scent marking is used by mammals "in any situation where they are both intolerant of and dominant to other members of their species." Ewer (1973), in discussions of scent marking by carnivores, made a similar point, and the work on European rabbits (*Oryctolagus cuniculus*) by Mykytowycz (1968) and on Mongolian gerbils (*Meriones unguiculatus*) by Thiessen et al. (1971) raised several major points:

FIGURE 24-14 The drooping ear tips of the roan antelope (*Hippotragus equinus*) are an example of automimicry.

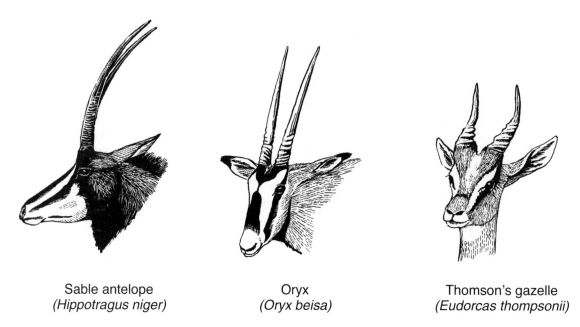

Sable antelope
(*Hippotragus niger*)

Oryx
(*Oryx beisa*)

Thomson's gazelle
(*Eudorcas thompsonii*)

FIGURE 24-15 Facial markings of three antelopes act to extend the line of the horns onto the face.

1. The maturation and use of scent glands are controlled by gonadal hormones produced at sexual maturity.
2. Most scent marking is done by dominant males.
3. Scent marking is often associated with the possession of a territory.

Experimental verification of the relationship between social rank and scent marking has come from studies of house mice (*Mus musculus*). Dominant males avidly marked the entire cage floor, whereas subordinate males voided urine in only a few places in the corners of a cage. Urination by the dominant male was regulated by interactions with another male, previously isolated dominant males immediately increased their urinary scent marking when caged with a subordinate male (Desjardins et al. 1973).

Urine and Fecal Signals

Urine and feces contain metabolic wastes that serve as chemical signals. Many kinds of mammals are highly specific in their choice of urination and defecation sites, and, in some species, a stereotyped routine is associated with urination and defecation. The dik-dik (*Madoqua kirkii*), a small, brush-dwelling African bovid, deposits its feces in conspicuous piles at the borders of its territory, urinates on the piles, and makes scratch marks around them with its hoofs (**Fig. 24-16**). These obvious piles provide both olfactory and visual signals announcing territorial boundaries (Hendrichs and Hendrichs 1971). All defecation and urination by the coyote can be regarded as scent marking. Marking is most common where intrusions into a home range occur (Wells and Bekoff 1981; Gese and Ruff 1997). Peters and

Mech (1975) found that scent marking with urine by wolves was concentrated along the borders of pack territories. Wolves of one pack respect the territorial boundaries of another pack, and buffer zones between neighboring packs serve to reduce interpack conflict. Mech (1994) observed that wolf mortality was concentrated along these territorial boundaries. In aardwolves (*Proteles cristatus*), scent marking serves to intimidate intruders and to synchronize mating (Sliwa and Richardson 1998).

Urine and feces also convey considerable information about an individual's physical condition (Endler 1993). Males of most species of mammals can recognize when a female is in estrus by the smell of her urine, and usually copulation will not be attempted until this time. In coyotes (*Canis latrans*), scent marking establishes reproductive synchrony between dominant (breeding) pack members, serves as an indicator of territorial boundaries, and may provide additional cues about spatial orientation to members of the group (Gese and Ruff 1997).

Scent Glands

Also important as sources of pheromones is a variety of glands. Glands associated with the mouth, eyes, sex organs, anus, and skin are known to produce chemicals used in olfactory communication. Secretions from five locations on the body of the Australian honey glider (*Petaurus breviceps*) serve functions ranging from attracting newborn young, in the case of the pouch gland of the female, to contributing to a community odor within the social group, in the cases of the frontal and sternal glands (Schultze-Westrum 1965). Müller-Schwarze (1971) described a number of pathways of social odors in mule deer (*Odocoileus hemionus*; **Fig. 24-17**).

FIGURE 24-16   A male dik-dik (*Madoqua kirkii*) in Kenya carefully smelling its dung pile (left) and then marking the pile by scratching the soil around it (right).

Ring-tailed lemurs (*Lemur catta*) make wide use of olfactory signals. Both sexes mark tree branches with secretions from the genitals, and, using the palms of the hands, males mark branches with other secretions. Males also have scent glands on the chest and forearms. During aggressive confrontations, the tail is pulled between the forearms, anointed with scent, and then lifted high and waved to disperse the scent. Males indulge in "stink fights," which involve palmar marking, tail marking, and tail waving and often lead to displacement of one animal by the other. The animals face each other when performing the scent marking, and the visual displays by the two animals, each using the conspicuously banded tail, seem to be mirror images of each other. The dominant animal moves forward while the other retreats. Vocalizations are also important during social interactions, and a variety of vocal signals are used. Kappeler (1998) showed that female scents probably function in mate attraction and maintaining female dominance hierarchies, whereas male scents are primarily used in **male–male competition**. Male scents were most often investigated by other males, and male rank and scent marking activity were positively correlated. Recently, Charpentier et al. (2008) showed that the chemical composition of male ring-tailed lemurs' scrotal scents was directly related to genetic quality, thereby providing an honest signal of relatedness and overall health.

One possible mechanism for intrasexual competition among males is illustrated by mouse lemurs (*Microcebus murinus*). Scents from dominant males reduce testosterone levels and sexual activity in subordinate males (Perret 1992). Finally, Kappeler (1998) showed that the recipient of the chemical signal may alter the message being sent. Fresh scent marks are rapidly countermarked by other males, possibly obscuring the original signal.

Vomeronasal Organ

Reproductive behavior in some, and perhaps most, terrestrial mammals is strongly influenced by the sense of smell. The **vomeronasal organ** plays an important sexual role in the male of many species. After smelling the genital area and urine of a female, some mammals, notably perissodactyls, artiodactyls, and some carnivorans, make a characteristic facial expression involving the upward curling of the upper lip and often the lifting of the head (**Fig. 24-18**). This distinctive behavior, called **flehmen** by K. M. Schneider (1930), is thought by some to be important in activating the vomeronasal organ and in perceiving sexual pheromones. This hypothesis has gained support from laboratory studies on the hamster (*Mesocricetus auratus*) by Powers and Winans (1975), who found that destruction of the afferent nerves of the vomeronasal organ produced a disruption of copulatory behavior in one-third of the altered animals. This procedure, coupled with destruction of the afferent nerves of the olfactory bulbs, completely eliminated copulatory behavior in all experimental animals. Powers and Winans suggested that input from both the vomeronasal organ and

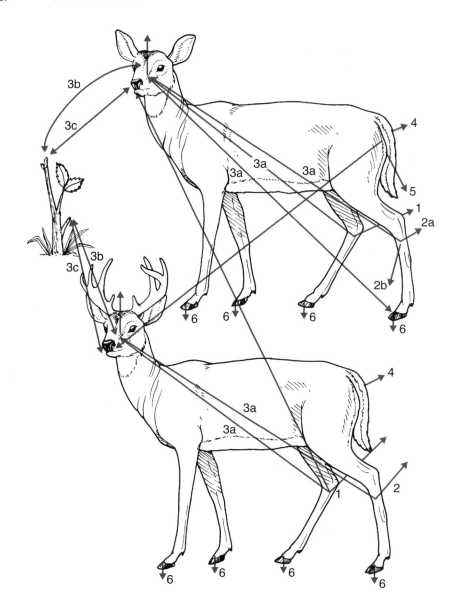

FIGURE 24-17   Sources of scents used in intraspecific communication and pathways of social odors in the mule deer (*Odocoileus hemionus*). The scents of the following are transmitted through the air: tarsal organ (1), metatarsal gland (2a), tail (4), and urine (5). When the animal lies down, the metatarsal gland marks the ground (2b). The hind leg is rubbed against the forehead (3a), and the forehead is rubbed against twigs (3b). Marked objects are sniffed and licked (3c). The interdigital glands (6) deposit scent on the ground. (Reprinted from *Animal Behaviour*, Vol 19, D. Müller-Schwarze, "Pheromones in black-tailed deer (Odocoileus hemionus columbianus)," pp. 141–152, Copyright 1971, with permission from Elsevier.)

the olfactory bulbs is necessary for the arousal of sexual activity. Olfactory communication may also play an important role in the reproductive cycles of some primates. The smell of vaginal secretions of rhesus monkeys (*Macaca mulatta*) in estrus is sexually stimulating to males and promotes copulation (Michael et al., 1971).

Self-Anointing

Self-anointing, an unusual signaling behavior shared by a number of species of hedgehogs (Erinaceidae), involves the smearing of saliva over the quills and results in a pungent smell easily detectable by humans. Brockie (1976) studied this behavior in the European hedgehog (*Erinaceus europaeus*). When removed from their nests, nestlings self-anoint; this seems to aid the mother in recovering them. Self-anointing may also serve as a defensive behavior in hedgehogs. Brodie (1977) found that hedgehogs eat poisonous toads without becoming ill. Indeed, they recycle toxins from the toads' parotid glands by mixing them with their saliva and licking the toxic mixture onto their spines. During laboratory tests, more than half of the Chinese black rats (*Rattus rattus*) that encountered anal gland secretions of weasels (*Mustela sibirica*) anointed their bodies with the scent (Xu et al. 1995). The strong weasel scent may by skin bacteria, these substances produce odor-yielding carboxylic acids. Gorman (1976) suggested that selection has favored the concentration of sebaceous and apocrine glands into discrete organs, where bacteria can produce the carboxylic scents used in olfactory signaling in relatively large amounts.

The capacity to recognize individuals by scent alone is probably important in nocturnal species or those that form large colonies. In particular, the ability of a mother returning from foraging bouts to locate her offspring by scent (and/or sound) in a large colony has obvious fitness benefits. Female degus (*Octodon degus*), a South American rodent, nest communally and mothers are able to recognize their own young from others in the nest by olfactory cues (Jesseau et al. 2008). Ungulates are frequently on the move in search of food. Many species also congregate in large numbers to give birth in synchrony, thereby increasing the odds that their offspring will survive. These ungulates give birth to precocial young capable of following the mother shortly after birth. Mothers have limited supplies of milk and selection favors the ability to recognize their own young from among the many unrelated young in the herd. Mothers recognize their offspring by scent and allow them to suckle, while rebuffing unrelated young. Such exclusive nursing occurs in many artiodactyls and perissodactyls (see Levy et al. 2004 for a review).

Eisenberg and Kleiman (1972) provided a reasonable summary statement of the role of olfactory communication in mammals. They regarded scent "as a means of exchanging information, orienting the movement of individuals, and integrating social and reproductive behavior."

Acoustical Signals

The sense of hearing in mammals is acute, and auditory communication is of great importance. Indeed, the sounds of some rarely seen mammals are commonly heard. Impressive choruses of howling coyotes may be heard nightly in some parts of the western United States, where the animals themselves are only occasionally seen. The importance of nearly constant auditory communication to a herd animal is difficult for one to imagine. Virtually continuous noises made by the members of a herd integrate the group by keeping individuals apprised of each other's location. In caribou (*R. tarandus*), the creaking and snapping of foot bones can be heard for considerable distances and enable scattered members of a herd to remain in auditory contact (Kelsall 1970).

Vocal communication is widely used by mammals. In humans, of course, this type of communication reaches its most complicated development, but some type of vocal communication can be recognized in most mammal species. In some species, short calls are used to warn

FIGURE 24-18 A Uganda Kob displaying flehmen behavior where the upper lip is curled back as it samples the air for pheromones.

mask the rat's own odor and thereby avoid a weasel attack. Young raised in the laboratory with no contact with weasels displayed the same behavior, indicating that it is innate.

Individual Recognition

Some species of mammals are known to discriminate between individuals of their species entirely by scent, and this ability is probably widespread. Gorman and coworkers (Gorman et al. 1974; Gorman 1976) studied the mechanism for this recognition in the small Indian mongoose *Herpestes javanicus*. The anal pockets of this mongoose produce acids used in scent marking objects within the animal's home range. The glands contain a series of six volatile carboxylic acids derived from bacterial decomposition of sebaceous (oil gland) and apocrine (sweat gland) secretions in the pocket. The six acids differ in relative amounts in different individuals, and animals recognize each other by scent on the basis of their unique carboxylic acid profiles. Bacterial production of these acids has been demonstrated in a number of mammals. Sebum (from sebaceous glands) and, to a limited extent, apocrine secretions are waterproofing agents deposited on the pelage of mammals. When metabolized of approaching danger (alarm calls); in other species, the vocal repertoire is large and varied.

Alarm or Contact Calls

Vervet monkeys (*Chlorocebus aethiops*) give different alarm calls to announce different predators, and to each call the monkeys have a different response. They look into the trees in response to the leopard call, at the ground after the snake call, and at the sky after the eagle call (Seyfarth et al. 1980). Gunnison's prairie dog (*Cynomys gunnisoni*) gives alarm calls that are differentiated into local dialects. The complexity of the call is related to the complexity of the habitat: More complex calls are given where shrubs, rocks, and tree stumps are present, whereas less complex calls are given by prairie dogs living in open grassland. The quiet grunt, for example, is used by many primates to maintain contact with each other in dense foliage (Marler 1965), and vocal sounds are used by many mammals to announce their position or to maintain or reestablish contact with one another. This may be one function of howling choruses of canids and the calls of young in a variety of species.

Many mammals have vocalizations that seem to serve primarily as territorial advertisements. Among primates, for example, male howler monkeys (*Alouatta* spp.) of the Neotropics, the woolly lemur (*Indri*) of Madagascar, and the gibbons (*Hylobates* spp.) of southeastern Asia (Marshall and Marshall 1976) make loud, resonant territorial calls. Each species has a different song, the calls of the male and female differ, each individual has a slightly different set of calls, and a subadult will often join in with the female. Territorial songs by a family thus advertise their species and location, the sex and individual identity of each singer, and the presence of a subadult.

Individual Recognition

Vocalizations facilitate individual recognition in many kinds of mammals. To the attentive human ear, the "who-oop" call of the spotted hyena (*Crocuta crocuta*) differs from one individual to another and probably facilitates individual recognition (Molekamp et al. 2007). There are pronounced differences (involving the spacing of pulses and the addition of snorts) among the threat calls of elephant seal (*M. angustirostris*) bulls (**Fig. 24-19**). These differences give each bull a unique vocal signature that allows individual recognition among bulls competing for females on the breeding ground (Shipley et al. 1981). Complex underwater sounds made by dugongs (*Dugong dugon*) in Shark Bay, Western Australia, include not only signals associated with aggression and spacing behavior, but signals that probably facilitate individual recognition (Anderson and Barclay 1995). The calls of the Florida manatee (*Trichechus manatus*) are structurally highly complex, probably provide

for individual recognition by sound, and maintain contact among group members (O'Shea and Poche 2006). The multiple harmonic structure of manatee vocalizations may provide an "adaptive redundancy" that ensures clear signal reception in shallow water, where low-frequency sounds are distorted (attenuated or canceled out).

Matsumura (1981) analyzed in detail mother–infant acoustic communication in a horseshoe bat (*Rhinolophus ferrumequinum*). Of the several kinds of acoustic signals used by this bat in mother–infant communication, some are primarily communicative, whereas others are virtually identical to echolocation sounds. The process of a mother's reuniting with her infant after foraging follows a consistent pattern. In the first stage, the constant-frequency calls made by the flying mother overlap (in frequency) and are temporally out of phase with the broadband sounds of the infant. At a later stage, the mother shortens the duration of the "phrases" of her calls, which the infant's calls tend to follow. After many repetitions of mutual and alternate signaling, both mother and infant shift to single high-pitched, multiharmonic syllables. Eventually, the sounds of the mother and infant overlap precisely and appear to be single sounds. When a mother and her infant are reunited and make body contact, vocalizations gradually cease. A mother and an infant other than her own are unable to synchronize their acoustic phrases and never make body contact. Each mother–infant pair seems to have a fixed and unique range of timing of their sounds during the synchronizing stage. This is perhaps the basis for precise mutual mother–infant recognition within colonies of many, often hundreds of horseshoe bats. Of further importance, the repeated sequences of mutual signaling during an infant's preflight period may be crucial to the infant's development of vocalizations used for echolocation (Matsumura 1981).

The calls of many social mammals within the orders Chiroptera, Primates, Carnivora, Proboscidea, Sirenia, and Rodentia are known to signal individual identity (Palacios et al. 2007), and geographical variation in vocalizations is also known for some species (Janik and Slater 1997). Vocalizations probably impart many kinds of information to conspecifics of social species. Much remains to be learned in this area.

Complex Vocalizations

The Japanese macaque (*Macaca fuscata*) and chimpanzee (*Pan troglodytes*) have complex repertoires of sound signals (Mizuhara 1957; Goodall 1986). The more basic sounds used by the rhesus monkey (*M. mulatta*) may be linked by a series of intermediate sounds, and one basic sound may grade independently into other calls (Rowell 1962). This yields a remarkably complex and rich vocal repertoire. The functional importance of some sounds can be recognized.

FIGURE 24-19 Two male northern elephant seals (*Mirounga angustirostris*) compete for dominance and breeding rites on a beach in California.

The functions of the varied sounds made by cetaceans are not yet well understood, but many are clearly used in communication. Some "vocal" sounds may keep members of a social group aware of one another's position or signal aggression. Some tail slapping and loud splashing by dolphins may provide long-distance communication among scattered members of a foraging group. The complex songs of humpback whales probably represent a form of social communication that has not been deciphered. Differences in the vocal repertoires among neighboring groups of cetaceans have been reported for killer whales (*Orcinus orca*; Ford 1991) and sperm whales (*Physeter catodon*; Weilgart and Whitehead 1997).

Ultrasonic Signals

Ultrasonic signals are best known in the echolocation signals of bats and dolphins (see Chapter 22 for a complete discussion). The vocal repertoire of some rodents also consists basically of ultrasonic signals. Ultrasonics play an important role in the integration of the reproductive behavior of the laboratory rat (*Rattus norvegicus*; Barfield and Geyer 1972): A 50-kilohertz call is associated with aggression and such aspects of sexual behavior as solicitation and mounting, and a 22-kilohertz signal is given by reproductively refractory or unreceptive individuals. Parent–young communication is also based on ultrasonics. Parents respond to the ultrasonic distress vocalizations of helpless young by returning them to the nest. The decrease in the acoustical energy of the calls as the young grow older is associated with the development of homeothermy and the accompanying reduction in vulnerability to cold (Noirot 1969). Some ultrasonic calls by rodents probably serve as territorial announcements (Sewell 1968).

Infrasonic Signals

One of the most exciting discoveries in the realm of mammalian communication comes from Africa. Actually, it comes initially from the Washington Park Zoo in Portland, Oregon, where Katherine Payne sensed that Asian elephants were making vocalizations that could not be heard but could be felt as a throbbing in the air. Recording equipment verified that these elephants were indeed making fairly intense, extremely low-frequency vocalizations (14 to 24 hertz; infrasound), mostly below the threshold of human hearing (Payne et al. 1986). Payne found that similar infrasounds were made by African elephants in Amboseli National Park in Kenya.

Observations of African elephants by scientists and naturalists have revealed an almost science-fiction aspect to elephant communication. Low frequencies are attenuated far more slowly in air than are high frequencies: Theoretically, elephant infrasounds could carry at least 6 miles. African elephants have a varied repertoire of infrasounds that announce locations of individuals or groups and that signal such conditions as alarm, aggressiveness, and reproductive readiness and social identity (Poole et al. 1988; McComb et al. 2003). Radio-tracking studies in Zimbabwe showed that families of different clans would suddenly alter their direction of travel when still several miles apart in an apparent attempt to avoid contact. Also in Zimbabwe, Garth Thompson observed a group of 80 elephants suddenly abandon their habitual home area the same day that many elephants were being shot in a culling operation 90 miles away in Hwange National Park. Several days later, Thompson found the 80 displaced elephants bunched together as far away from Hwange as they could get. Some message of danger and death had seemingly been relayed many miles from group to group of elephants. Long-range communication by infrasound adds yet another dimension to the complex social world of elephants.

Barklow (1997) found that about 80% of hippopotamus (*Hippopotamus amphibius*) vocalizations are given underwater. The physics of sound transmission in water and air is very different because water is more than 800 times denser than air. As a result, no sounds produced in air are heard underwater. Reception of sounds produced underwater is easy, but determining their direction is difficult. Mammals localize airborne sounds by turning the head until the sound waves reach both ears at the exact same time or by comparing in their brain the difference in time of arrival of the sound at each ear. Because of the greater density of water, waterborne sounds are transmitted to the middle ear as vibrations of the skull bones, reaching both ears simultaneously no matter what the position of the head; thus, waterborne sounds appear to come from everywhere. Cetaceans have solved this problem by suspending the middle and inner ear from ligaments, effectively isolating them from direct contact with the skull bones (Chapter 19). In dolphins, waterborne sounds are transmitted via the dentary to a tube of fat tissue on the medial surface of the dentary and via the fat tissue to the middle ear bones. Remarkably, Barklow found that several cranial features important for hearing underwater are shared by hippopotamuses and cetaceans, another example of a characteristic uniting hippos and cetaceans: (1) both have a thin, dishshaped area on the dentary bone; (2) in both, the middle ear is suspended from ligaments; and (3) both share a fatty connection from the middle ear to the dentary (Schwartz 1996). Underwater hippo sounds include squeals, moans, and bursts of staccato clicks reminiscent of the sonar clicks of dolphins. The possibility that "click trains" are used in the murky river water as a form of echolocation was tested by Barklow and his colleagues, but no evidence of echolocation has yet been found.

Underground Communication

Naked mole-rat (*Heterocephalus glaber*) colonies are some of the most complex and highly structured nonhuman vertebrate societies (Jarvis 1981), and vocalizations are seemingly the most important form of communication in mole-rats. Naked mole-rats, which have the richest vocal repertoire of any rodent (18 distinct vocalizations), give calls in a wide variety of behavioral contexts (Pepper et al. 1991). The function of one distinctive "chirp" is to bring a new food source to the attention of fellow colony members (Judd and Sherman 1996). Having found the new food, a scout chirps as it carries a sample back to the nest, where it holds its head high and waves the food for all to smell. Colony members respond to this chirp-and-wave behavior by backtracking the signature scent of the scout to the newly found food. Naked mole-rats dig prodigious distances to find scattered patches of bulbs or tubers. A patch usually provides enough food to support the entire colony, at least for a time. By alerting fellow colony members to a new food patch, the colony is benefited and the inclusive fitness of the scout is increased because the scout is typically closely related to all of its fellow colony members. This behavior is similar to the "waggle dance" of worker honey bees returning to the hive after finding a rich nectar source.

Acoustic communication by fossorial rodents is complicated by the fact that there are no air currents in the sealed burrows of many fossorial species and only a narrow range of frequencies can be transmitted efficiently (Credner et al. 1997; Brückmann and Burda 1997). Under these conditions, low-frequency sounds are optimal, and many species of blind mole-rats are most sensitive to frequencies between 0.6 and 1 kilohertz. Some fossorial mammals that live in closed burrow systems, where visual and olfactory communication is useless, use a combination of low-frequency vocalizations and **seismic signals** (vibrations transmitted through the soil). The solitary blind mole-rat, *Spalax ehrenbergi* (Spalacidae), uses its head to rap out long-distance messages on the ceiling of its burrow (Heth et al. 1987). In addition, a vocal repertoire of some six calls, including a low-frequency (0.5- to 4.5-kilohertz) purr, made mostly by the male and occasionally by the female, reduces agonistic behavior of the female and facilitates copulation (Heth et al. 1988). The solitary Cape mole-rat, *Georychus* *capensis* (Bathyergidae), produces seismic signals by drumming on the burrow floor with its hind feet (Narins et al. 1992). Although territorial drumming rates are similar in both sexes, in the breeding season the drumming rate of the male is nearly twice that of the female (26/s versus 15/s), and the communicating pair (in separate burrow systems) synchronize their "footrolls."

These seismic signals are propagated greater distances through the soil than are airborne signals, which are rapidly attenuated. Laboratory tests show that *Spalax* perceives and reacts to the seismic components of the head-thumping signals but ignores the auditory component transmitted through the air of the burrows (Rado et al. 1987). Banner-tailed kangaroo rats (*Dipodomys spectabilis*) also communicate by footdrumming (Randall 1997; Randall and Matocq 1997). Territorial ownership is typically communicated by airborne footdrumming sounds made by rats drumming on the surface of large seed cache mounds (Randall 1993). Each kangaroo rat has a unique footdrumming signature that allows these rodents to distinguish between neighbors and strangers (Randall 1989, 1995). Kangaroo rats also often footdrum within their sealed burrows (when remaining in the burrows, the animals typically plug the entrance hole with dirt). Randall (1997) found that footdrumming by *D. spectabilis* produces seismic-borne energy nearly 40 decibels greater in peak intensity than the airborne sound that must first pass through the burrow walls. On windless nights, communication between distant territories is via airborne sounds produced by footdrumming on top of the mounds. On windy nights, however, communication among adjacent territories is via seismic signals made within burrows.

Footdrumming also serves as a warning and often deters snakes from pursuing kangaroo rats (Randall and Matocq 1997). The pattern of footdrumming shifts dramatically when the kangaroo rat switches from territorial drumming to footdrumming in the presence of snakes. Snakes can detect the seismic signals of footdrumming through low-frequency vibration detectors in the skin (Hartline 1971).

The low-frequency sensitivity of a number of fossorial or semifossorial mammals (*Georychus*, *Cryptomys*, *Spalax*, *Talpa*, *Dipodomys*, other heteromyids, and gerbils) suggests the widespread importance of seismic communications. Relative to what remains to be learned, we know little about acoustical communication in most mammals under natural conditions. This is a promising area for research.

■ Tactile Signals

The use of tactile communication by mammals is widespread. Information transmitted by touch is especially important in social mammals. Grooming is among the best studied forms of tactile behavior. Another form of bodily contact prior to copulation, serves to stimulate the partner into mating.

Allogrooming

Allogrooming of one individual by another has an especially important social function in primates (Sebeok 1977), and this function far overshadows that of removing parasites. In general, females groom most often and for longer periods than do males, and males groom each other much less often than they groom females. Grooming typically serves to reduce social tension and, in many primates, is important in establishing and maintaining social contact among individuals. Grooming is regularly practiced after aggressive encounters, when it dissipates tensions and re-establishes "amicable" social contact. Often one animal will seek contact with another of higher social rank by grooming it, and mothers will distract young being weaned by grooming the youngster rather than allowing it to nurse.

Sowls (1974) described mutual grooming in collared peccaries (*Pecari tajacu*). Animals rub their heads against each other's flanks and rump, which bears a much-used scent gland, in a ritual that seems to serve as a greeting ceremony but that also has a scent-marking component. The African dik-dik uses its nose to touch various parts of another's body. Often the female rejoining her young or a male rejoining a female will perform this behavior. This seems to be a tactile reassurance and a reassertion of familiarity akin to mutual grooming (Bowker 1977).

Nongrooming Tactile Signals

The sexual behavior of many mammals includes precopulatory activities by the male, such as laying the chin on the female's rump, nuzzling the genital area, or touching various other parts of her body. These behaviors presumably are sexually stimulating to the female or at least cause her to accept mounting by the male. Perhaps tactile stimuli are of greatest importance in connection with sexual activities in most mammals, but in the social mammals they have assumed other roles as well.

Wild Pacific bottlenose dolphins (*Tursiops aduncus*) rub their flippers over the bodies of other dolphins or over each other's flippers (Sakai et al. 2006). Both types of flipper rubbing were most common among individuals of the same sex and age class. Mothers and calves were more likely to engage in flipper-to-body rubbing. The unusual postures and reciprocal rubbing between members of the pair suggest that these dolphins exchange some as yet unknown benefit by taking turns.

Among primates, bonobos (*Pan paniscus*) from central Africa show the widest range of tactile communication. Bonobos form matriarchal groups in which females form and maintain strong social bonds with other females.

Males appear to derive their social status from that of their mother. Bonobo societies are famous for their relative lack of aggression and high degree of sexual contact (DeWaal 2007). A wide variety of sexual interactions (excluding copulation) occur among bonobos, including female–female genital rubbing, male–male genital rubbing, oral-genital contact, and tongue kissing. In addition, these behaviors occur among virtually all combinations of group members, not just between kin. In bonobos, sexual behaviors are integral to the formation of social bonds, help reduce conflicts, and maintain relatively peaceful societies (De Waal 2007 and references therein).

Defensive Behavior

Threat behaviors are among the most familiar activities of mammals. A dog lifts its upper lip to expose the length of its upper canines, a cat opens its mouth and hisses, some rodents grind their teeth, and others perform footdrumming. These actions all signal a readiness to fight or to attack if the antagonist does not retreat or take other appropriate action. A threat is typical of a situation in which conflicting tendencies preclude either an immediate attack or a hasty retreat.

Threats can be simple, as in animals that merely open the mouth wide to display the teeth, or complex, as in some horned artiodactyls in which both distinctive postures and movements are involved that usually seem to advertise the most important offensive weapons. Visual threats may be made more impressive or startling by such audible threats as explosive hisses or growls.

Some mammals have carried this type of defensive behavior a step further by discarding threat displays altogether. Instead they adopt an **appeasement** posture or behavior that serves as a complete surrender and contains no elements likely to trigger an opponent's aggression. Virginia opossums (*Didelphis virginiana*), for example, feign death by becoming nearly catatonic as a defense of last resort. Complete vulnerability is emphasized, and the response on the part of the predator is to cease its hostile activity because the stimulus of a moving prey is absent. Wolves and many other mammals appease their dominant opponents by lying on the back with the vulnerable throat and underside unprotected. In several artiodactyls, lying down serves as appeasement (Burckhardt 1958; Walther 1966), and a subordinate black wildebeest (*Connochaetes gnou*) may roll on its side with its belly toward its superior and the side of its head on the ground (Ewer 1968). In Grant's gazelle, a lowering of the head, the reverse of the high-headed threat posture, is adopted by a submissive animal (Walther 1965). In some primates, the presenting of the rump as the female does prior to copulation is an appeasement gesture. The brightly colored skin on the rump of some Old World

monkeys may serve in part to make the rump conspicuous and thus make "presenting" appeasement gestures more effective. A "grin" serves as an appeasement in some higher primates.

Appeasement behavior clearly helps both participants avoid further conflict. It allows an animal being defeated in a fight to avoid further injury and in many cases allows a subordinate animal to avoid a contest altogether. In highly social species, threat and appeasement behaviors foster the peaceful perpetuation of a dominance hierarchy and allow animals to be close to one another with a minimum of energy wasted on aggressive interactions. Ritualized appeasement behavior may even be important in permitting a subordinate animal to seek social contact without risking attack (Schenkel 1967).

Mating Systems

Because of the high costs of pregnancy and lactation for females, the disparity of costs between the sexes is greater in mammals than almost any animal. Thus, there is a fundamental conflict of interest between males and females with respect to their reproductive success. Males produce large numbers of tiny gametes, and because the cost of gamete production is minimal, males maximize their fitness by fathering as many offspring as possible. Females produce relatively few, much larger gametes than do males (**anisogamy**) and must nourish the developing fetus and the rapidly growing neonate. As a result, females have a greater investment in each offspring and should be selective in their choice of mates. Trivers (1972) proposed that sexual competition occurs because one sex invests much more than the other in the production of the young, a situation that may not be typical of all species. The conflict arises because females are a limiting resource for male reproductive success (Krebs and Davies 1993). Males must compete with each other for the right to mate (male–male competition), while females seek to select the best mating partner from the many courting males (**female choice**). Thus, males and females have competing interests.

Male mammals typically invest less in gametes, mate with multiple females, and contribute little parental care for the young. Males can never be absolutely certain about **paternity** because females may also mate with multiple males (Trivers 1972). For a male, fitness may increase if he mates with as many females as possible. In contrast, females are always certain the offspring are hers. Therefore, her fitness increases if she can ensure that her current offspring reach sexual maturity and reproduce. Maternal care, including prolonged periods of gestation and/or lactation, likely evolved as a way for mothers to ensure that their genes are passed on to future generations through their progeny. In sum, the competing interests of males and females, and differences in their ecology and life history traits, lead to variation in mating systems and the degree of maternal and paternal care.

Clutton-Brock (1989) reviewed the diverse mating systems in mammals and hypothesized that the variation in male mating systems is attributed to (1) the need for males to assist in rearing the young, (2) the ability of males to defend females or their ranges, (3) the size and stability of female groups, and (4) the predictability of locating females (**Fig. 24-20**). When successfully rearing young requires male assistance or when such help yields a higher breeding rate, then monogamy is the predicted strategy. When paternal care is not required, the size of the female's home range and the stability of female groups lead to polygamy.

Mammalian mating systems are diverse and likely form a continuum rather than discrete categories. Even within a species there may be several mating systems depending on the availability of resources or other environmental factors. The following sections describe some of the variation in mammalian mating systems.

Monogamy

Exclusive mating between one male and one female during one or more breeding seasons occurs in a variety of mammals, from some bats and rodents to some artiodactyls. Kleiman (1977) estimated that fewer than 3% of mammalian species are monogamous, although this mating system is widespread among the mammalian orders. In some families of mammals—the marmosets (Callitrichidae) and dogs and foxes (Canidae), for example—monogamy is the major type of mating system. How long monogamous pair bonds last under natural conditions is difficult to determine, but the bonds clearly persist in some species through a number of reproductive seasons or as long as both members of the pair live. However, even in behaviorally monogamous species, recent DNA studies indicate that multiple paternity in the young of supposedly monogamous females is common (Isvaran and Clutton-Brock 2007).

Monogamous mammals typically display little morphologic or behavioral dimorphism, and the members of a pair interact infrequently except during the early stages of pair-bond formation and when rearing young. In **obligate monogamy**, the carrying capacity of a habitat is so low that only a single female can occupy a home range and she cannot raise a litter without help from conspecifics. In this case, an extended period of parent–young association may occur. In the monogamous African yellow-winged bat (*Lavia frons*), unpredictable periods of insect scarcity have seemingly selected for a long period of parent–young

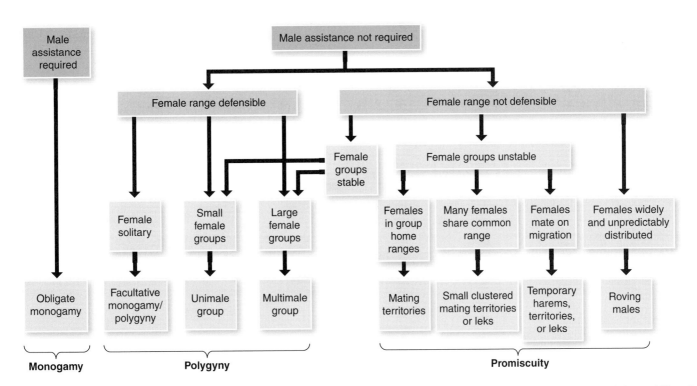

FIGURE 24-20 A flow chart of mating systems based on the requirement for male assistance, female defensibility, and the stability of female groups. (Modified from Clutton-Brock, T.H., *Proc. R. Soc. Lond. B* 236(1989): 339–372.)

association that allows young to become skillful foragers (Vaughan and Vaughan 1987). In some species older juveniles may help raise younger siblings, and the father often aids in the feeding, defense, and socializing of the young.

Obligate monogamy is common in canids, and litter sizes tend to be large. Black-backed jackals (*Canis mesomelas*), for example, are monogamous, but some older offspring may serve as helpers in rearing a new litter (Estes 1991). Survival of the pups is directly associated with the presence of helpers in the Serengeti (Moehlman 1983). Helpers regurgitate food for the pups and can spend as much time at the den guarding the litter as the parents do. Bat-eared foxes (*Otocyon megalotis*) are probably monogamous but can form trios with a single male and two females (**Fig. 24-21**; Van Lawick and Van Lawick-Goodall 1970). Unlike other canids, bat-eared foxes are not territorial; home ranges often overlap those of neighboring family groups. Cubs are not provisioned at the den and mature earlier than many other canids. Lamprecht (1979) suggested that these differences in social organization result from their specialized insectivorous diet. At the other end of the extreme are canids such as African wild dogs (*Lycaon pictus*) and wolves (*Canis lupus*) that form large packs consisting of a breeding pair and several nonbreeding adults that assist in hunting and provisioning the young.

Facultative monogamy occurs when population densities are low and male ranges overlap with the range of a single female. More commonly, male ranges overlap those of several females, resulting in **polygyny**. Where population densities fluctuate widely, males may be more flexible in their mating system. Arvicoline rodents display several types of social and mating systems, from monogamy for the prairie vole (*Microtus ochrogaster*; Thomas and Birney 1979) and pine voles (*M. pinetorum*; McGuire and Novak 1982), to polygyny or **promiscuity** for the taiga vole (*M. xanthognathus*; Wolff 1980) and meadow vole (*M. pennsylvanicus*; Madison 1980). The ability of some arvicolines to be socially flexible is probably associated with their tendency toward population fluctuations. The social biology of the California vole (*M. californicus*), comparatively well-known through the work of Lidicker (1973, 1976, 1979, 1980), provides a model for what seems to be an intermediate system. A family group of California voles maintains and shares a runway system to the exclusion of other individuals and families. The family group has from 2 to 12 (mean 5.6) individuals, including one male and one or more females and their young. Monogamy is the rule at low population densities, but at high densities polygyny is common. A pair bond probably occurs, at least at low

FIGURE 24-21 A pair of bat-eared foxes (*Otocyon megalotis*) resting outside their den in Kenya.

densities, and some paternal care has been observed in this and some other species of *Microtus*. The inflexibility of the male–female pair system was demonstrated by Lidicker (1976, 1979), who introduced several pairs into a small enclosure. The fighting and mortality that ensued usually reduced the population to a single pair, which reproduced successfully. More than 25 years of research by Getz and his colleagues (Getz et al. 1993; Getz and Carter 1996) revealed that the mating system of prairie voles is flexible and probably evolved in response to limited food supplies. Prairie voles can live in monogamous pairs or in large communal groups, depending on food supply and population density (Getz and McGuire 1997).

Polygyny

The vast majority of mammal species are either polygynous or promiscuous (Clutton-Brock 1989; Wolff and MacDonald 2004). Polygyny, in which males mate exclusively with multiple females (females mate with a single male) is common in species where male territories overlap more than one female home range or where males can defend resources needed by females. In contrast, promiscuous mating occurs when both males and females mate with multiple partners during a breeding season.

In species in which females tend to occur in small groups or have small home ranges, males can often monopolize several females, forming defended harems (referred to as **female defense polygyny**). Thanks to the careful work of Bradbury and Emmons (1974), the elaborate social organization of the small insectivorous bat, *Saccopteryx bilineata*, is well known. These bats occur widely in the Neotropics, where they live by day in colonies in the buttress cavities at the bases of large tropical trees. Each colony is organized into a number of harems, and each harem and its territory are maintained as a discrete social unit by a single male, with each unit containing from one to eight adult females. The harem territories are from about 0.10 to 0.36 square meters in area, and females are regularly spaced from 5 to 8 centimeters apart.

The males are first to return to the colony sites at dawn. When females and their young begin to return, the males begin singing and performing various displays. The morning period of displaying and of sorting animals into their territories generally occupies several hours. Females vocalize, are highly aggressive toward one another, and rigidly maintain their individual distances. This tendency is probably a major factor limiting harem size. Territorial defense by males involves a remarkably intricate series of displays. Males bark, confront each other, patrol territorial boundaries, and open the scent gland in the wing and shake the wing as a visual and olfactory signal called "salting." Males

salt both females in their own territory and another male's females across a territorial boundary, and they will also salt another male across a boundary.

In *S. bilineata*, harem size is positively correlated with the frequency of male courtship and defensive flights (Voigt and Helversen 1999). Not surprisingly, both territory maintenance and courtship are energetically expensive for male bats (Voigt et al. 2001) and competition for territories is intense. Using microsatellite DNA to establish paternity for offspring in a large colony consisting of 18 harems, Heckel and von Helversen (2003) showed that 89% of offspring were fathered by males within the colony. However, a harem-holding male fathered only 30% of the young born to females within his harem. Nevertheless, males defending larger harems had higher overall reproductive success. Heckel and von Helversen hypothesized that female mate choice may play an important role in male reproductive success in these colonies; females are behaviorally dominant over males and may seek extra-harem matings during flights outside the colony.

Many bat species form harems. Small harems, consisting of a male and 2 to 7 females, occur in the African vespertilionid *Myotis bocagei*. Harems of the southeast Asian club-footed bats (*Tylonycteris pachypus*) commonly include more than a dozen females. The largest harems are found in tropical phyllostomid bats. Spear-nosed bat (*Phyllostomus hastatus*) harems can have more than 100 females defended by a single male, but harems of 18 to 20 females are the norm (McCracken and Bradbury 1977, 1981; McCracken 1987). The composition of females in these harems is stable from year to year, but the tenure of the male is variable. Harem males father most, but not all, of the offspring born to harem females (McCracken 1987). Juvenile males disperse into groups of bachelor males, and yearling females disperse to form new harems within the same cave system. Juvenile dispersal serves to reduce the potential of inbreeding between father and daughters. In other social phyllostomid species, males do not defend harems. Harem formation depends on the degree to which males can control access to reproductive females, which is, in turn, determined by how long females maintain group tenure (Wilkinson 1987; Fleming 1988).

Polygyny also occurs when males control access to resources, such as food or prime habitats, required by females (**resource defense polygyny**). Many ungulates have breeding cycles that feature harems of females, each maintained by a dominant male. The area occupied by the females is the strongly defended territory of a single male, or in some cases, the male maintains a "breeding territory" that is defended even when females are not present. Breeding is done mostly by mature, vigorous, and aggressive males. In an elk herd, only 12% of the bulls—the largest individuals—did 84% of the observed copulating (Mc-Cullough 1969).

The rutting behavior of the elk (*Cervus elaphus*; called red deer in Europe) is especially well known from the studies of Darling (1937), Graf (1955), McCullough (1969), Struhsaker (1967), and Clutton-Brock et al. (1982). McCullough recognized four main categories of bulls during the breeding season. Primary bulls are powerful, mature individuals that shed the velvet from their antlers early and are the first to establish harems (**Fig. 24-22**). Secondary bulls are large individuals that take over the harems by defeating the primary bulls as the latter become exhausted. Tertiary bulls assume control of the harems after the secondary bulls decline. Opportunist bulls are those whose only contact with cows (females are called cows in North America and hinds in Europe) is by chance. When a bull becomes exhausted through constantly herding cows together, driving rival bulls away, and copulating (all while unable to obtain adequate food and rest), it is beaten in a fight with a fresh bull, who takes over the harem from the deposed master.

In polygynous pinnipeds—the otariids, some phocids, and the walrus—the males are extremely vocal, are much larger than the females, and maintain breeding territories. In the California sea lion (*Z. californianus*), large bulls establish territories adjacent to the water at sites favored as hauling-out places by females, which arrive at the rookery a few days before they give birth (Peterson and Bartholomew 1967). Nonterritorial bulls usually form aggregations apart from the pupping and breeding rookery. Because the same females do not continuously occupy a male's territory and because males make no effective effort to herd females into territories, the term harem does not apply. Roughly 2 weeks after parturition, females enter estrus and copulation follows. Fighting between males occurs during the establishment of territories, and males signal their possession of territories by incessant barking. Little actual fighting occurs once territories are established, but a boundary ceremony between males on adjoining territories periodically reaffirms boundaries. These ceremonies involve an initial charge toward one another, followed by open-mouthed head shaking as the males confront each other at close quarters; in the final standoff the bulls stare obliquely at each other. The ceremony is so precisely ritualized that, should animals get uncomfortably close to one another, they adroitly avoid contact. Females are aggressive toward one another through much of the breeding season; again, however, injury is avoided by ritualized aggressive threats. Although males may be on territories in a rookery from May through August, each male maintains a territory for only 1 or 2 weeks; territories are thus occupied by a succession of males.

In contrast to the pinnipeds mentioned earlier, male elephant seals (*M. angustirostris*) establish a social hierarchy on the breeding ground, but are not territorial (Le Boeuf and Peterson 1969). The highest ranking males stay close to breeding females, and breeding success is closely correlated with social rank. On Año Nuevo Island off the coast of California, four of the highest ranking males, which constituted but 6% of the 71 bulls in the area, copulated with 88% of the 120 females. At another study area on the same island, the alpha bull (the bull at the top of the hierarchy) maintained its rank throughout the breeding season and was involved in 73% of the observed copulations.

Although the advantage of harems to males is obvious, it has often been assumed that females benefit from this arrangement as well (Orians 1969). In 20 years of careful observation and experimentation on yellow-bellied marmots (*Marmota flaviventris*), however, Armitage and

FIGURE 24-22 (A) A bull elk (*Cervus elaphus*) herds his harem of females across a valley in Rocky Mountain National Park. (B) A male elk in Yellowstone National Park samples the air (flehmen) to detect females in estrus.

his coworkers (Downhower and Armitage 1971; Armitage 1986, 1998) were able to show that reproductive success increases with harem size for males but not for females.

Promiscuity

When females are not defensible and their daily movements are not predictable, males cannot rely on territories for access to mates, even in prime locations. Under these conditions, males must resort to following the females or to forming seasonal harems. If females use predictable routes to water holes or foraging areas, males may form small territories along these well-worn paths and attempt to mate with the females as they pass through their small territory. Alternatively, males may congregate on traditional breeding grounds that are used year after year and are known to females. Here, males may form extremely small territories, or arenas, where they display for the gathered females. When topi (*Damaliscus lunatus*; **Fig. 24-23**) densities are high, males may form small territories only 50 meters in diameter (Jewell 1972). Intense competition for the best locations also leads to very small territories in kob (*Kobus kob*), lechwe (*K. leche*), and puku (*K. vardonii*). In extreme cases, an arena, or lek, is shared by 30 to 40 males, with only 15 to 30 meters between adjacent males (Leuthold 1966; Floody and Arnold 1975). Leks are arenas where males congregate (but where no resources are present) and females come to mate with the most aggressive or vigorous males. Male lechwe on leks gain higher mating rates than males that do not aggregate on the lek. When Nefdt and Thirgood (1997) experimentally reduced the number of females visiting the lek, males eventually abandoned their lek territories. In another experiment to test the hypothesis that female fallow

FIGURE 24-23 A group of topi (*Damaliscus lunatus*) survey the plains in the Maasai Mara, Kenya.

deer (*Dama dama*) choose the "best" male and not the resources it defends, Clutton-Brock and coworkers (1989) forced the most successful males to move to another part of the arena. Despite their new locations, females still favored these males.

The amazingly specialized hammer-head bat (*Hypsignathus monstrosus*) forms leks. Males defend no resources and provide no parental care; as with other lek species, after copulation the females raise young alone. Traditional lek sites may be used over and over again for longer than 60 years. Males select sites along streams, and the raucous vocalizations of the male are used not only for attracting females but for establishing dominance ranking and for maintaining a small territory in the lek. Males able to occupy the sites in the lek most favored by females do most of the breeding. Only 6% of the males were responsible for nearly 80% of the copulations in 1974 (Bradbury 1977). Not surprisingly, males may take several years to establish themselves in these favored places.

Parental Care

It is not enough to produce the most offspring, however, because genes can only be passed on to subsequent generations if those offspring survive and reproduce themselves. Parental care helps ensure the survival of offspring at least until weaning. All mammals receive maternal care in the form of gestation and lactation (at a minimum). In mammals, the females must not only carry and nourish the young during the gestation period but must also provide the young with milk for various lengths of time after the birth. It is not surprising, then, that in mammals females perform the bulk of the parental care.

The sex of the care provider and the amount of parental care seem tightly coupled to the type of mating system in mammals (Clutton-Brock 1989; Trivers 1972). The paternity hypothesis, formulated to account for sexual differences in parental care, suggests that a parent's degree of care is correlated with the likelihood that this parent recognizes the offspring as its own (Trivers 1972). Obviously, females can be absolutely certain of maternity, but males rarely can be certain of their paternity. Thus, paternal care (care provided by the presumptive father) is more likely in monogamous mating systems, where males have a higher degree of certainty of paternity.

Male care occurs in some carnivores, rodents, and primates (Woodroffe and Vincent 1991). The care provided by males usually is in the form of food provisioning, defending, grooming, and/or transporting the young. Male cotton-top tamarins (*Saguinus oedipus*) carried older infants more often than did mothers and were more likely to go to the aid of an infant being harassed (Tardif et al. 1990). In fat-tailed dwarf lemurs (*Cheirogaleus medius*), from western Madagascar, both sexes contribute to baby-sitting and guarding of the young (Fietz and Dausmann 2003). In an extreme case of male care, Dyak fruit bats (*Dyacopterus spadiceus*) have functional mammary glands that secrete a liquid, imitating lactation (Francis et al. 1994; Racey et al. 2009).

Care of offspring is also performed by individuals other than the biological parents (called **alloparental care**). In prairie voles, older littermates remain in the nest and groom and brood pups from subsequent litters (Wang and Novak 1992). Alloparental care in mammals may even involve nursing unrelated offspring. Female bighorn sheep (*Ovis canadensis*) that have lost lambs as a result of predation may allow unrelated lambs to nurse. In this case, the females may still be distantly related to the young they suckle because females tend to remain in their natal group (Hass 1990). Alloparents tend to be related to the young they care for. The high risks associated with dispersal, establishing a territory of their own, and finding a mate probably serve to keep nonbreeders with parents (Emlen 1982). The costs and benefits to the helpers are discussed further in the section on altruism.

The amount of parental care given by mammals varies considerably. In the case of Malaysian tree shrews (*Tupaia tana*), mothers provide only minimal parental care (Emmons and Biun 1991). Young are placed in "nursery" nests where the female comes every other day to nurse the young for a few minutes. Absentee parental care is rare in mammals and is taken to extremes in tree shrews: One female spent a total of less than 50 minutes with her young during the first month of their lives.

Social Behavior

Many mammals form social units that exist beyond the mating and offspring rearing periods. Theory predicts that such social groups form when the benefits of group living exceed the costs (Wilson 1975; Silk 2007). The benefits and costs of living in groups are summarized in **Table 24-2**. The exact cost-benefit ratio for each species is unique, and there may be different costs and benefits for males and females of the same species. For many years, biologists assumed that pervasive benefits derived from sociality, that is, some automatic increase in survival and fitness resulted from a social life. In his consideration of the evolution of social behavior, Alexander (1974) stressed that, to the contrary, social living often has important disadvantages. Competition for food, mates, and space is heightened, and the conspicuousness of groups can be disadvantageous to prey and predatory species alike. An additional liability for social animals is the rapid spread of disease or parasites.

Alexander (1974) also discussed three broad advantages of sociality. First, an individual's vulnerability to predation may be reduced by effective group defense or herd behavior. Defense of the group by dominant males is an important antipredator strategy of baboons, and a cohesive, running herd of ungulates presents a problem for predators. "The safety of the herd consists of the cohesive mass of animals running in an organized manner [**Fig. 24-24**]. The animals exposed are only those on the outside, and even these are protected by the number of flying hoofs and the ebbs and surges within the group. The vast array of movement has a disorienting effect on the observer's vision" (McCullough

TABLE 24-2 Potential Costs and Benefits to Group Living in Mammals

Benefits	Costs
Antipredator benefits	**Predator costs**
Increased vigilance	Increased visibility to predator
Prey dilution effect	Increased attack rate
Predator confusion	**Foraging costs**
Selfish herd effects	Increased kleptoparasitism
Group defense	Decreased food availability
Foraging benefits	Increased aggression within group
Group hunting	Increased detection by prey by group
Public sharing of information	predators
Mate choice benefits	Increased visibility of group
Increased choice of mates	predators to prey
Physiological benefits	**Mating costs**
Shared thermoregulation	Increased competition for mates
	Increased chance of cuckoldry

Source: Based on data in Krause and Ruxton (2002).

1969). Under a variety of situations, "hiding" within the herd is an effective means of escape; selection against the straggler or the individual who breaks from the herd is intense (Hamilton 1971). Second, the cooperative effort of a predatory group (such as hyenas or wolves) may be effective in bringing down large prey that could not be killed by a solitary predator. With baboons, scattered but rich sources of food can be found more often by many searchers than by a single animal. Finally, a paucity of safe nocturnal or diurnal retreats may have forced a partly social life on such animals as baboons and many bats.

After groups of mammals form, refinements in social behavior are subject to natural selection. These refinements serve several functions, which may increase the advantages of group living. Such a behavior as the formation of a defensive ring of adults by musk oxen (*Ovibos moschatus*) tends to reduce the vulnerability of the herd to predation by wolves. Further advantages may also be gained by groups of predators. The precise positions and spacing maintained by individuals of some foraging groups of cetaceans may increase the ability of the group to perceive and capture prey.

Most important, evolution of social behavior affects reproductive competition among group members and the reproductive performance of the population at large. The social system of desert-dwelling baboons (*Papio hamadryas*) described by Kummer (1968), for example, is based on the one-male unit. An adult male maintains a group of one to several females, which are threatened or punished when they stray. This is a stable unit, and the male copulates with his females only, but by keeping his females with him constantly and by being aware as they come into estrus, he ensures his fitness. A young male following such a group sometimes has opportunities to copulate with females in estrus, and becomes familiar with the social behaviors that may later be used in gaining and maintaining his own unit. During the evolution of this system, the fitness of the socially integrated individuals was presumably greater than that of the individual who was solitary or did not learn the behavioral tactics associated with the social life.

Many behavioral ecologists agree that the evolution of social systems in mammals is associated with increased inclusive fitness, but a social system in any group of mammals is tested against the constraints imposed by a specific environment. Perhaps as strongly as any factor, the abundance and distribution of food limit evolutionary options. Indeed, the evolution of some social systems may have been influenced primarily by selective pressures imposed by the distribution of food in time and space. Using baboons as examples again, the small, one-male social unit serves well in dry areas, where productivity of the habitat is low and food has a patchy distribution but is nowhere abundant. Savanna-dwelling baboons, however, occupy a

FIGURE 24-24   A group of plains zebra (*Equus burchelli*) fleeing from a waterhole in a "lion panic." The confusing pattern of stripes probably acts to distract predators.

more productive area where food is scattered but a patch may provide abundant food; these animals forage in large social groups.

Social Systems

Patterns of social behavior in mammals are so diverse that broad summary statements are hard to frame. At one extreme are asocial mammals that rarely come into contact, except during the breeding season when temporary pair bonds form prior to copulation. After copulation, the sexes separate and defend separate territories. Contact between individuals outside of the breeding season typically results in aggressive interactions. Cougars (*Puma concolor*) are an example. Other mammal species form seasonal groups centered on clumped resources such as waterholes or roost sites. Hippos congregate in deeper pools as river channels dry up during the dry season, and many vespertilionid bats form large, single-sex roosting colonies. These social groups are persistent and may have relatively stable membership. At the opposite end of the social continuum are complex social systems with dominance hierarchies, stable membership that may include several generations, and complex forms of intragroup communication and social behaviors.

As might be expected, the large, more spectacular mammals and game species have been most thoroughly studied, but some of the most important groups of mammals remain poorly understood. Although rodents and bats together make up more than half of the known species of mammals, we know relatively little of the social behavior in these groups. This section gives a necessarily cursory overview of some of the social systems of mammals. Additional treatments of this subject include works by Eisenberg (1966, 1981), Ewer (1968, 1973), Wilson (1975), Eisenberg and Kleiman (1983), Rubenstein and Wrangham (2006), and Wolff and Sherman (2007).

Metatherians

Among metatherians, social behavior has evolved in two diprotodont families. In the Petauridae, several species are social to some extent, but sociality is perhaps best developed in the honey glider (*Petaurus breviceps*), in which cohesive family units are dominated by males. In the family Macropodidae, sociality is developed to varying degrees, and probably the most highly evolved marsupial social behavior is seen in the whiptail wallaby (*Macropus parryi*). The population studied by Kaufmann (1974) consisted of subunits called mobs. The members of a mob occupy a home range to the near exclusion of members of other mobs, but the area is not defended. The social organization of a mob is loose, but some structure is provided by a rather flexible dominance hierarchy among the males that is maintained by nonviolent, ritualized fighting (Dawson,

1995). (A **dominance hierarchy** is a fairly permanent tiered social system in which each individual recognizes its "position" relative to others above and below its current status.) A female wallaby in estrus is typically accompanied for 1 to 3 days by her dominant-male consort, with exclusive mating rights.

In contrast, wombats (Vombatidae), such as the southern hairy-nosed wombat (*Lasiorhinus latifrons*), exhibit a completely different social structure (Walker et al. 2008). These large (20- to 30-kg) nocturnal marsupials create extensive burrow systems, or warrens. Males are philopatric and share burrows with male kin for several years. Females disperse and are less likely to associate with female relatives. This social system is unusual in mammals.

Chiroptera

Our knowledge of the sociobiology of bats is extremely fragmentary; mating systems are known for only approximately 6% of bat species. However, there is no doubt that this group has a wide array of social systems, and some species have complex social behavior (McCracken and Wilkinson 2000; Burland and Worthington Wilmer 2001). Few male bats provide paternal care and females' ranges are generally not defensible. Thus, most bats should exhibit polygynous or promiscuous mating systems. A few species are completely solitary, except during copulation and when the mother–young bond is maintained briefly. A more common pattern, typical of vespertilionid bats and some species in other families, involves the separation of the sexes during the season when pups are born (**Fig. 24-25**). Females form nursery colonies exclusive of males; the sexes may associate again when the young can fly and forage. Monogamous family groups are formed by a few bats in the families Emballonuridae, Nycteridae, Rhinolophidae, Megadermatidae, and Vespertilionidae.

Bats that construct their own roosts, such as tent-making bats, tend to live in smaller groups. The phyllostomid round-eared bat (*Lophostoma silvicolum*) roosts in active termite nests located on the branches of trees (Dechmann et al. 2005). A single male excavates a roost in the hard termite nest and attracts several female roostmates. Paternity analysis revealed that the male harem holder sired most of the offspring of his roostmates (resource defense polygyny), thereby justifying the high energetic costs of roost construction.

The social system of the Bahamian buffy flower bat (*Erophylla sezekorni*) involves roosting in multimale–multifemale groups. Males congregate on small territories where they display to attract females during a 2-month breeding season (Murray and Fleming 2008). Buffy flower bats exhibit some but not all of the characteristics of a lek breeding system; they are best described as having a promiscuous

FIGURE 24-25 A colony of vespertilionid bats (*Myotis myotis*) roosting communally.

mating system. Of great interest as examples of complicated social behavior are the vampire bats, which show reciprocal altruism (discussed on p. 592).

Rodentia

Widely divergent social systems occur in the Rodentia, from the solitary system of pocket gophers (Geomyidae) to the highly social prairie dogs (Sciuridae) and mole-rats (Bathyergidae). Social evolution in rodents has been influenced by many factors: patterns of daily and seasonal activity, the temporal and spatial distribution of food, environmental constraints, and predation. In solitary rodents, such as pocket gophers, kangaroo rats (Heteromyidae), and New World porcupines (Erethizontidae), each individual is solitary except briefly when the male and female are together during mating.

The black-tail prairie dog (*Cynomys ludovicianus*) is highly social (Hoogland 1995). Prairie dogs formerly occupied many parts of the western United States, where they occurred in large "towns," often including more than 1000 animals and covering many hectares. (Now, lamentably, prairie dog towns are rare in most parts of the West, owing partly to intensive, government-sponsored poisoning campaigns.) The functional social units are **coteries**, which generally consist of an adult male, several adult females, and a group of young. No dominance hierarchy exists in the coterie. The paths, burrows, and food in the area held by a coterie are shared by its members, but hostility with adjacent coteries is the universal pattern. A two-syllable territorial call is used to proclaim ownership of territory. A repetitive, high-pitched yelp is a warning of danger. Members of a coterie become familiar with each other in part by grooming, playing, and "kissing" behaviors. During kissing, the mouth is open and the incisors are bared. This seemingly ritualized method of distinguishing between friend and foe also is used by many species of squirrels (**Fig. 24-26**). Faced with the threatening expression of open mouth and bared teeth, a trespasser retreats, while a fellow coterie or family member meets its "friend" and "kisses." During the spring, when females are pregnant or lactating, the coterie

FIGURE 24-26 Kissing behavior occurs in many sciurid rodents. Shown here is Harris's antelope squirrel (*Ammospermophilus harrisii*) from Arizona.

system partially dissolves and some yearlings and adults establish themselves beyond the territorial limits of their coterie. The personnel of coteries thus may change, but the territory itself is stable.

An individual gains several advantages from this social system. Many eyes are watchful for danger, and many voices are ready to sound a warning. Foraging keeps vegetation low over a wide area and provides terrestrial carnivores with little concealment. Perhaps equally effective in providing for long-term occupancy of an area, the animals are kept spaced so that overuse of food plants is generally avoided.

Recent advances in genetic techniques, notably **DNA fingerprinting**, have allowed researchers to establish the degree of genetic relatedness of colony members. Travis et al. (1996) used such techniques to study mating systems in Gunnison's prairie dogs (*C. gunnisoni*) in Arizona, where animals are colonial and occupy stable territories, which resident males defend aggressively. Based on above-ground behavioral observations, their mating system is usually described as harem polygyny (Rayor 1988; Travis and Slobodchikoff 1993). DNA fingerprint analysis revealed that females within a territory tend to be related to one another, but males sharing territories are not. Furthermore, these analyses showed that females living on a territory often produce litters of mixed male parentage. In fact, female Gunnison's prairie dogs mating with at least three males increase their chances of pregnancy and parturition (Hoogland 1998). An important finding of the DNA fingerprint analysis was that 61% of all offspring were fathered by males from outside the female's territory. This level of

extraterritorial paternity is higher than values reported for any other species (Travis et al. 1996). Resident males aggressively defend their territories from incursion by neighboring males but seem incapable of preventing females from leaving the territory temporarily to seek copulations with other males. Territoriality may provide access to resources other than mates. Reevaluation of the mating system of Gunnison's prairie dogs, in light of the DNA evidence, suggests that these rodents practice "overlap promiscuity," not harem polygyny as previously suggested (Boellstorff et al. 1994; Travis et al. 1996).

The evolution of sciurid social systems has been the subject of a recent review by Blumstein and Armitage (1998). They provide evidence that increasing social complexity among squirrels leads to social systems characterized by fewer breeding females, increased age at first reproduction, decreased litter size, and higher first-year survival rates of offspring (**Table 24-3**). The costs of fewer offspring are offset by the benefits of increased offspring survival. Increased survival in the first year in social sciurids (such as ground squirrels, *Spermophilus*) results from retention of juveniles within the social unit (Armitage 1981; Barash 1989). Increasing social complexity should also foster some form of female reproductive suppression (as in the naked mole-rat, *H. glaber*; Faulkes et al. 1991). Blumstein and Armitage (1998) suggested that reproductively suppressed females remain in the social unit because the costs of dispersal are high. Dispersal increases the risk of predation, and dispersers are subject to increased aggression from residents.

TABLE 24-3 Life History Traits and Social Complexity of Social Sciurids Based on the Work of Blumstein and Armitage (1998)

Species	Social Complexity	Minimum Female Mass (g)	% Females Breeding	Time to First Repro. (yrs)	Gestation Time (days)	Lactation Time (days)	Average Litter Size
Marmota olympus	1.46	1400	45	3	—	—	4.0
Marmota marmota	1.41	2811	48	3	37	45	2.4
Marmota caligata	1.35	3300	45	3	30	28	3.0
Marmota caudata	1.22	1400	14	3	30	—	4.2
Marmota monax	0.27	3314	95	1	34	46	4.0
Cynomys parvidens	1.23	703	80	1	—	—	3.9
Cynomys gunnisoni	1.03	600	66	1	29	39	4.6
Cynomys leucurus	0.84	575	88	1	30	35	5.6
Spermophilus columbianus	0.65	376	75	2	24	28	3.7
Spermophilus beldingi	0.40	211	95	1	27	25	4.2
Spermophilus richardsonii	0.39	269	95	1	22	29	6.8
Spermophilus beecheyi	0.26	486	90	1	28	38	7.5

Source: From data cited in Blumstein and Armitage 1998.

Primates

Even in the most primitive primates, the Lemuridae, sociality is well-developed, but, relative to those of many other primates, lemurid social systems are rather simple. The mouse lemur (*Microcebus murinus*), studied by Petter (1962) and Martin (1973), occurs on Madagascar in dispersed "population nuclei." The proportion of females to males in these nuclei is 4 to 1, and as many as 15 females may occupy the same nest. Surplus males not accompanying groups of females often occupy nests on the periphery of the area. Although mouse lemurs occupy nests together, perhaps because nest sites are at a premium, there is no organized social life, and animals forage alone. This primate, therefore, must be regarded as a basically solitary creature that is flexible enough to be able to occupy nests communally.

The ring-tailed lemur (*Lemur catta*) is far more advanced socially (Jolly 1966, 1972). This lemur lives in troops that range in size from 10 to 20 or more animals. Adult males and females are equally represented in the troop, and their total numbers are usually equaled by the numbers of young. Troops occupy exclusive areas, and there is little intertroop contact. Social organization within a troop is based on dominance patterns. Females are dominant over males, a reversal of the usual primate system. A male dominance hierarchy is established, and dominant males seem to be able to remain for long periods with a troop but (surprisingly) do not always have first access to females in estrus.

Within all of the diverse types of social organization in primates, individuals must learn to be responsive to a complex social milieu. An individual must be constantly aware of the attitudes and displays of most members of the social group and of the social ranks of these animals. Manipulation of the social milieu becomes an important aspect of the behavior of many primates, and even the ranking of an individual may depend in part upon its effectiveness as a manipulator (Lee and Oliver 1979). As a result of enduring social bonds between female baboons, two "friends" can put up an intimidating united front when one is threatened by a third individual. In some primate social systems, the rank of a female depends partly on her close association with a dominant male and on her ability to depend upon his help or protection during aggressive confrontations. Her status may abruptly decline if the male is deposed from his dominant position. In some baboon troops, and in other primate societies in which group structure and cohesiveness are maintained by strong dominance patterns, the dominant male is the focal point of attention. An individual's behavior and the behavior of the entire social group are geared to the responses of this leader.

Baboons that live in savannas and pursue an almost entirely terrestrial life are often vulnerable to attack by predators. Because food frequently is scattered, a troop must forage over wide areas, thus increasing the chance of encounters with predators. A large and tightly organized social group has evolved in these baboons in response to this pressure. These groups include from about a dozen to more than 150 individuals. Each group occupies a largely exclusive and seemingly undefended home range;

neighbors usually respect the boundaries. When a group is moving, males quickly respond to threats from any quarter, and their united action provides the primary defense of the troop.

The pronounced sexual dimorphism in baboons enhances the male's intimidating appearance, as well as his fighting ability. Male baboons are about twice as large as females, are more powerfully built, and have comparatively huge canines. The long fur over the crown, neck, and shoulders of the male accentuates the impression of size.

The mating pattern of baboons seems to be related to dominance ranking. Whereas subadult, juvenile, and less dominant males copulate with females in the early stages of estrus, dominant males have exclusive rights to females during the period of maximal sexual swelling (the time when ovulation occurs). In some groups, only the highest-ranking male copulated with females during the height of the swelling, and, in a group observed by DeVore (1965) in Kenya, not one dominant male attempted copulation until the swelling was at its peak.

Habitat differences between savanna-dwelling baboons and desert-dwelling baboons are associated with certain differences in social behaviors. Group size, for example, differs markedly. Savanna baboons depend on food patches that are typically rich enough to support large social groups: In Amboseli National Park of Kenya, groups average 51 (Altmann and Altmann 1970). Desert-dwelling baboons, on the other hand, have a relatively unpredictable and sparse food supply, and foraging groups are small, from several animals to perhaps 20. In addition, social organization differs between these baboons. Savanna baboons form only one type of social unit, the group, which includes many females, their young, and multiple mature males. The male–female pair bond is brief, lasting only a few hours or days, during the female's estrus period. Desert baboons, in sharp contrast, have four levels of social organization (Kummer 1968, 1984). The smallest and most tightly knit unit is the family group, consisting of a single mature male, one to several adult females, their young, and often a young adult bachelor male "follower." Several such family groups band together to form a less tightly knit foraging unit called the clan, and a number of these clans form a fairly stable traveling unit called a band, which usually includes some 60 baboons. Many bands tolerate each other to sleep in safety on the same cliff at night; this loose aggregation is called a troop and contains several hundred baboons.

The social systems of great apes (Hominidae) do not depart radically from basic primate patterns, but do have some unique features. Groups of the mountain gorilla (*Gorilla gorilla*) include from 2 to 30 animals. Social interplay among members is amiable, and assertions of dominance are low key (Schaller 1963, 1965a, 1965b; Fossey 1972).

Particularly notable is the age-graded male troop, with the nucleus of the group consisting of the dominant silver-backed male (10 years of age or older) and adult females and their young. Additional males, including less dominant silverbacks and black-backed males, attach themselves to the periphery of the group.

The chimpanzee (*P. troglodytes*) has been the subject of considerable field observation by a number of workers, including Izawa and Itani (1966), van Lawick-Goodall (1968, 1973), Izawa (1970), Nishida and Kawanaka (1972), Sugiyama (1973), Goodall (1983, 1986), and Stanford (1995). The basic social unit of chimpanzees is an often-dispersed group of 30 to 80 animals that show considerable fidelity to a large home range. Particularly unusual is the looseness of the organization of the social group, with intricate patterns of establishment and dissolution of small parties. Highly evolved visual, tactile, and vocal communications are used (**Fig. 24-27**). When a party of chimpanzees finds trees bearing fruit, their almost manic vocalizations and actions attract other parties to the bonanza. Male chimpanzees

FIGURE 24-27   A chimpanzee (*Pan troglodytes*) calling to troopmates in Kenya.

are capable of quick and coordinated hunting behavior (Teleki 1973; Stanford 1995; Stanford et al. 1994). Typically chimps hunt opportunistically, such as when they find a young baboon that has become separated from its troop. In this situation, some males will act to block its main path back to the troop while others block off other escape routes. If captured, the prey is killed and consumed by members of the hunting party. Chimpanzee hunting may be more common than was previously thought. Stanford (1995) estimated that Gombe chimps consume approximately 1 ton of meat annually. Lower-ranking males appear to barter meat for sex. Chimpanzees are clearly not always gentle and benign: On several occasions, males of one group have been observed systematically killing males of another group.

The closely related bonobo (*Pan paniscus*) has evolved a highly unusual complex of social behaviors (de Waal 1995). Bonobo social systems are based on sexual contacts between individuals and include such behaviors as mouth-to-mouth "kissing," genitogenital rubbing between adult females, and pseudocopulation between adult males (de Waal 1989, 1995). These behaviors appear to have evolved to minimize conflicts among group members.

Viewed broadly, primates differ from other mammals in the complexity of their social systems and in their highly developed vocal and visual communication. Diurnal feeding and binocular vision are two factors thought to be important in the evolution of their social systems. Much remains to be learned of primate sociobiology. This is especially true for the forest-dwelling species.

Cetacea

Our knowledge of the behavior of cetaceans is as yet extremely incomplete, but current evidence indicates that most species are social (Mann et al. 2000; Pryor and Norris 1998). Not only do some cetaceans travel and forage in social groups in which some consistent spatial organization is evident, but cooperative behavior is known. A review by Connor et al. (1998) compared the convergent evolution of social systems in odontocete whales with those of chimpanzees and elephants. Sperm whales (*Physeter catodon*) and the African elephant both have **matrilineal groups** comprising up to a dozen females, males are largely solitary, and males do not generally have the opportunity to breed until relatively late in life. Chimpanzees (*P. troglodytes*) and bottlenose dolphins (*Tursiops*) share a **fission–fusion** social system in which members of smaller groups freely move between groups.

Carnivora

Although most carnivorans are not social, highly organized social systems have evolved in some, such as lions, wolves, and spotted hyenas. The coyote (*C. latrans*) is socially flexible (Bekoff and Wells 1980), depending on circumstances. When carrion provides large, defensible patches of food, coyotes form small, cohesive packs, but when depending on small, dispersed prey, such as rodents, coyotes are solitary. Some diurnal procyonids are also social. The social structure of the white-nosed coati (*Nasua narica*) is unique within the Carnivora, consisting of groups of up to 30 females and their offspring, called "bands," and solitary males (Gompper et al. 1997). Coatis form female **coalitions** (groups of individuals acting as a unit to access resources or deter opponents) similar to those in primate societies. DNA analyses revealed that not all females within a band are related. Unrelated females experience more aggression and competition for food from subgroups of related females within the band, but they stay with the band because the costs of being solitary are considerably higher. Gompper (1996) showed that solitary females were frequently displaced from fruit patches by larger, more aggressive males and suffered higher parasite loads and increased risk of predation. Coalitions of females were able to displace solitary males and gain greater access to the patchy food resources; simultaneously, they reduced their risk of predation by increased vigilance.

Some diurnal mongooses (Herpestidae) are also highly social. In four genera of African herpestine carnivorans (*Suricata*, *Crossarchus*, *Mungos*, and *Helogale*), highly developed social systems have evolved, perhaps in response to predation. Studies by Rood (1978, 1980; Rasa 1985) have revealed the complex social organization of the dwarf mongoose (*Helogale parvula*). This small diurnal carnivore (about 320 grams) occurs in parts of eastern and southern Africa, usually in open woodland or scrubland with short grasses, where the dens are in termite mounds (**Fig. 24-28**). The packs average about eight members, and include the dominant, breeding (alpha) pair, usually the oldest pack members, plus, typically, an additional male, two adult females, and several juveniles. All of the young born in the pack are those of the alpha pair; adult subordinates are behaviorally and endocrinologically suppressed (Creel et al. 1991, 1993). The mortality rate of young is well over 50% in their first year, mostly as a result of predation. Tight adherence to social rules and cooperation typify the social organization. The pack leader is usually the alpha female, and the pack sleeps together, forages together, and members share in watchfulness for, and defense against, predators. Group cohesion is maintained during foraging by repeated contact calls. An alarm call announces predators. Helpers (other than the alpha pair) do most of the grooming, baby-sitting, and carrying of the young between den sites, and helpers bring all of the young's food after they are weaned. Rood recorded eight cases of subordinate females lactating, perhaps after aborting their litters. In one case, a lactating

FIGURE 24-28   A group of dwarf mongoose (*Helogale parvula*) check for danger from atop a termite mound in Kruger National Park, South Africa.

subordinate saved a litter after the mother died. All of the lactating subordinates were related to the young of their respective packs. DNA fingerprinting demonstrated a high degree of inbreeding among pack members (Keane et al. 1996). Dwarf mongooses are unusual in that subordinates of both sexes often disperse, but this dispersal does not reduce the generally high degree of relatedness among pack members. Seemingly, dwarf mongoose packs do not suffer inbreeding depression, leading Keane et al. (1996) to hypothesize that many generations of mild "inbreeding have purged most of the deleterious recessive alleles from this population."

Among the most highly social canids (the wolf, *C. lupus*; the African hunting dog, *L. pictus*; the dhole of southern Asia, *Cuon alpinus*), a dominance hierarchy gives structure to the pack. By cooperative hunting, these cursorial animals can kill large prey.

Kruuk (1972), who studied spotted hyenas (*Crocuta crocuta*) in the Ngorongoro Crater and the Serengeti Plains of Tanzania, observed a remarkable social system. The basic social unit of the spotted hyena is the clan, which may contain as many as 80 animals. The clan system is periodically disrupted in the Serengeti, where seasonal migrations of wildebeest and zebras result in drastic shifts in food supply. Each hyena clan defends a territory, the boundaries of which are maintained in part by systematic scent marking. Territorial disputes are often violent, and individuals are occasionally killed during border warfare. Females are larger than males and are dominant to them. A rather complex dominance hierarchy exists within a clan, and strong bonds develop between females. At a kill, dominance is often not asserted and competition is based largely on the ability to eat extremely rapidly (aided by profuse

salivation) rather than on fighting prowess or size. Females tend their cubs at a central denning area, and here young receive early training (or practice) in the social rituals of the species.

In South Africa, spotted hyena clans have five social classes: resident females, cubs, and three classes of males (Henschel and Skinner 1987). Some males stay with their natal clans (resident natal males). Alternatively, some males leave their natal clan and attempt to join neighboring clans, but the resulting social status of these males was generally low (peripheral immigrant males) unless they displayed persistent attention to the female clan members, in which case they occasionally gained breeding status within the clan (central immigrant males).

When two hyenas meet, they typically go through a meeting ritual, part of which involves mutual examination of the external genitalia. (The external genitalia of the female hyena mimic the male penis and scrotum: The female's clitoris is very large, resembles a penis, is erectile, and two sacs filled with fibrous tissue form a false scrotum.) A variety of hypotheses have been advanced to explain the male-like genitalia of female spotted hyenas. Kruuk (1972) concluded that the genital mimicry evolved because of its importance during meeting ceremonies. This ritual probably enables individuals to be close to each other briefly and to "identify" each other while attention is attracted to the genitals and to a course of action other than fighting. This cooling-off period perhaps allows aggressive tendencies to subside. The leg-lifting action seems to be an appeasement gesture, hence its initiation typically is by a subordinate animal. The adaptive value of such behavior to a spotted hyena is perhaps heightened by its flexible social life. Although often social, hyenas may be solitary for varying lengths of time, and peaceful meeting behavior and recognition of individuals are frequently important. In predators with such powerful offensive weapons as the teeth of hyenas, control of aggression is of critical importance, and the great array of scents, displays, and vocal signals serves to restrict the use of these weapons.

However, genital masculinization in female hyenas may be a by-product of selection for female aggressiveness (Frank 1997). Selection for aggressiveness in females, and the associated increase in prenatal androgens (male hormones), may have led to enlarged, male-like female genitalia. In other words, the enlarged female genitalia are a byproduct of selection for aggressiveness and larger body size in females; the role of genitals in appeasement behaviors occurred later. Holekamp (2006) pointed out that this hypothesis remains controversial because despite treating pregnant females with "drugs that block the action of androgenic hormones on the fetus, each female offspring of these treated females nevertheless develops a full-sized pseudopenis." There are risks to the female and her offspring of such an elongate structure because during birth the offspring must pass down the long, narrow "phallus" and offspring occasionally suffocate in the process (Holekamp 2006). Such risks make it unlikely that the pseudopenis in females is simply a by-product of selection for other male-like traits. A number of other hypotheses have been advanced in recent years, but none has universal support (Cunha et al. 2005; Holekamp 2005).

In the Serengeti, where prey biomass is low except for short periods of the year, hyena clans are large (median size of seven clans was 47 individuals). During part of the year, clans were forced to leave their territories and commute to migrating herds (**Fig. 24-29**; Hofer and East 1993a, 1993b). Mean commuting distance was approximately 40 kilometers. During the commute, hyenas avoided contact with resident clans. Resident clans tended to ignore commuters in transit through their territories, but they responded aggressively to commuters at kills. The commuting system of Serengeti hyenas in response to a migratory food supply is unique among carnivores. Females with young cubs suffered disproportionate mortality during the commuting season (46% to 62% of the year is spent commuting). Lactating females made briefer, but more frequent, commuting trips than did other members of the clan.

Hyena clans in the Namib Desert are much smaller, consisting of three to five adults. Here the average home range was 570 square kilometers, compared to territories of approximately 50 square kilometers in the Serengeti (Tilson and Henschel 1986; Hofer and East 1993a). In the Namib Desert, home ranges did not overlap and hyenas did not exhibit territorial scent marking. The characteristics exhibited by Namib hyena clans are probably adaptations to the harsh environment and depauperate prey communities (Tilson and Henschel 1986).

Holekamp et al. (2007) regard the societies of spotted hyenas as being similar in complexity to those of cercopithecine primates and find that these groups share the ability to use a variety of sensory cues in the recognition of individuals and of kin, to appreciate the relative value of individuals as social partners, and to use knowledge of relationships among group-mates in guiding social behavior. Dominance in both societies is independent of size or fighting prowess: Dominant individuals are those with the "best network of allies."

Long-term field studies on the lion (*Panthera leo*) in Serengeti National Park of Tanzania by Schaller (1972), Bertram (1973, 1975), Packer (1986), and Heinsohn and Packer (1995) have provided a fascinating picture of the social life of this mammal. The lion pride is a fairly stable social unit, usually with about 3 to 12 adult females, and 1 or more adult males. There is virtually no recruitment of outside females: All females are born and grow up within the pride, and all are closely related. A female reaching 3 years of age is either accepted as a member of the pride or is driven from it. Rejected females and males often become nomadic, follow migrating prey, and make up about 15% of the total lion population in the Serengeti. A pride is usually controlled by several adult males that defend the pride's largely exclusive and fairly stable territory. Although on occasion individuals may hunt alone, or part of a pride may separate from the rest, the members of a pride are familiar with each other, and social contacts are usually peaceful or seemingly affectionate. A member of the central sisterhood of the pride leads a stable, if at times violent, life, and her reproductive life is about 13 years.

FIGURE 24-29   A family of hyenas (*Crocuta crocuta*) move with their prey in Arusha, Tanzania.

A male, on the other hand, does not associate consistently with a single pride throughout life, and his reproductive life may be only 2 or 3 years. Young males stay with their pride until they are about 3 years of age, when they are either forced from the pride or leave it voluntarily. Often several males leave the pride together; these males may be brothers or—because the females of a pride are all grandmothers, mothers, sisters, or daughters—at least closely related. The young outcasts become members of the nomadic population, and, unable to depend on the hunting prowess of the experienced females, often turn to scavenging. After roughly 2 years of nomadic life, these males are approaching the prime of life and are sexually mature and sufficiently formidable to take over a pride.

The pride they take over is almost never their natal pride. A pride lacking males may be taken over peacefully, or several males past the prime of life may be easily displaced, but violent fighting may accompany the displacement. Because a group of males can successfully challenge males holding a pride whereas one or two males cannot, selection favors tight social ties among males. New males disrupt the life of the pride: Pregnant females may abort, the ovarian cycle of females coming into estrus may be interrupted, and the newcomer males may even kill cubs (infanticide). After a few months, however, females again begin coming into estrus, they are bred by the new males, and males often help take care of the cubs. Two or 3 years later, however, when the males are aging, they are driven out by a new group of prime males. The reproductive life of displaced males is over, and, because of their declining physical condition, their life expectancy is not great.

All the females in a pride tend to synchronize the bearing of young, and a cub can nurse from lactating females other than its mother. This communal care of young can perhaps be explained by the close genetic relationships among females of a pride and by the increased survival of cubs with familiar companions (Bertram 1975).

In general, however, lions are inefficient reproductively, and the mortality rate of cubs is high (about 80%). Bertram (1975) detailed the situation as follows: "Assuming that lions mate every 15 minutes for 3 days, that only one in five 3-day mating periods results in cubs, that the mean size of litters is two and a half cubs and that the mortality among cubs is 80 percent, then a male must mate on average some 3,000 times for each of his offspring reared to the next generation." Because each copulation is relatively unimportant and because the males of a pride need each other to maintain control of the pride, pressure on the males to fight for the chance to copulate with a female in estrus is reduced. The lion has few predators, and the size of the pride is perhaps controlled largely by periodic food shortages. The life span of lions is fairly long (12–15 years). Reproductive inefficiency, therefore, does not prejudice the survival of the pride. A critical factor may be the reduction of aggression in a pride to a level permitting the survival of at least some young. Bertram suggests that reproductive inefficiency and reduction of competition in males result in the increased stability of the pride, fewer changes of the male guard, and hence greater chances for the survival of cubs.

Elephants

The social life of African elephants (*Loxodonta africana*) is known through the work of Laws and Parker (1968), Hendrichs and Hendrichs (1971), Douglas-Hamilton (1972, 1973), McKay (1973), Moss (1983), Poole (1987, 1989), and McComb et al. (2003). The elephant social system is structured at several levels. The first level is that of the family group, including an old matriarch and 10 to 20 related females and their offspring. Because of the long life span of elephants, the family unit generally includes grandmothers, mothers, sons, daughters, grandsons, and granddaughters. Lifetime social bonds among females may last 50 years or more. The second level of the social system is the kinship group, consisting of several family groups that remain in the same vicinity and, on occasion, mingle peaceably. Under some conditions, as during migration, many kinship groups may band together to form clans, containing on occasion 100 or more animals. The clan probably has no social cohesion at any level above the kinship group.

Bulls leave the family units when they become sexually mature and assemble in all-male groups (bachelor herds) in which dominance is established by ritualized fighting and sparring. Dominant males are temporarily accepted by family units with a female in estrus. Males in musth (periods of heightened aggression caused by elevated hormone levels) do most of the breeding and are given a wide berth by other bulls.

Elephant societies are maintained by a complex series of tactile, olfactory, vocal, and visual behaviors. Deep rumbling infrasounds (below human hearing thresholds) are used to communicate over long distances. Greeting ceremonies involving inserting the trunk into the other's mouth often occur when animals are reunited. Especially remarkable is the importance of cooperative and apparently **altruistic** behavior within the family unit. A juvenile is allowed to suckle from any lactating female, young females approaching sexual maturity are solicitous of the well-being of small calves, and the safety of a calf seems to be the concern of the entire family unit. When threatened, the family forms a defensive phalanx of adult members.

Ungulates

Ungulate social behavior is of particular interest for several reasons. Many species are large, occupy open habitats, can be observed easily, and have therefore been well-studied. Because these open-country dwellers are probably the most highly social of all ungulates, we have a reasonably good understanding of ungulate social behavior. Further, a growing knowledge of the environments occupied by a variety of ungulates has provided a basis for a theoretical approach to the relationships between ecology and the evolution of social behavior, morphological features, and color patterns (Estes 1974, 1991; Geist 1974; Jarman 1974).

Jarman (1974) and Estes (1991) gave order to our view of bovid behavior by relating the sociobiology of bovids to their ecology (**Table 24-4**). Jarman showed that, in contrast to forests, grasslands (open woodlands, savannas, and plains) support a higher diversity and biomass of bovids. In grasslands, the bulk of the food available to bovids is grass. Whereas a high percentage of each grass plant is edible, much less of a tree or shrub can be eaten. Grasslands therefore produce proportionately more food per growing season for bovids, but growing seasons are short. Although they produce less food for bovids, forests have moisture and plant productivity distributed more evenly throughout the year, better enabling populations to be sustained.

Associated with the differences in density and habitat between large and small bovids are sharp differences in their social organization. The five categories of Jarman (1974), as outlined by Wilson (1975), are shown in **Table 24-5** and indicate the relationships among habitat, feeding style, and social organization in African bovids.

Estes (1974) stressed a major structural and behavioral dichotomy within the family Bovidae. The ancestral bovids were probably small forest dwellers, perhaps resembling today's forest-dwelling duikers. The expansion of grasslands in the Miocene and Pliocene of Eurasia and Africa set the stage for the movement of bovids into open grassland or savanna habitats. This major evolutionary step brought some bovids under the influence of new suites of selective forces and led to their structural and behavioral divergence from the persistently forest-dwelling species. Viewed today, the dichotomy is between forest-dwelling browsers that are generally small, cryptically marked, with simple horns, and which escape from predators by hiding, and the open-country grazers of medium or large size, conspicuously marked, with large and often complex horns, and which use their speed in the open to escape predators. The behavioral dichotomy is also clearly delineated: The forest dwellers either are solitary or live in small family groups and use scent marking as the primary means of communication; the open-country grazers are typically highly gregarious and primarily use visual signals.

As noted and summarized in Table 24-4, the social organization of the classes of bovids recognized by Jarman and Wilson forms a progression from the small social units of the selective feeders of the bush and forest to the very large herds of the less selective feeders of the grassland. To flesh out this outline, comments on the social behavior of several species follow.

Blue duikers (*Cephalophus monticola*) are small, forest-dwelling African antelopes that form monogamous pairs and defend small permanent territories averaging 3 to 4 hectares in size. Territories are maintained in the forest by scent marking. Members of a pair stay close together, and pair bonds are reinforced by behaviors such as social licking and mutual pressing together of preorbital glands (DuBost 1983). Duikers have several alarm calls and respond to danger by seeking cover.

The impala has a somewhat different social organization (Class C in Table 24-5; Jarman 1970; Jarman and Jarman 1974; Hart and Hart 1987, 1988). Dominant male

TABLE 24-4 Behavioral and Ecological Comparison of a Forest Antelope and a Plains Antelope

Forest Duiker *(Cephalophus)*	Plains Oryx *(Oryx)*
Small body size (4–64 kg)	Large body size (150–200 kg)
Horns are short spikes in both sexes	Long horns, straight or curved in both sexes
Preorbital glands are well-developed in both sexes	Preorbital glands are vestigial or absent
Inhabit lowland and montane forests and scrublands	Inhabit arid zone plains and savannas
Concealing coloration (e.g. tan, gray, or reddish-brown pelage)	Conspicuous coloration (e.g. black and white markings on face, sides, and legs)
Water-dependent	Water-independent
Selective browsers	Grazers
Small home ranges	Large territories
Typically live in solitary, monogamous social systems	Live in gregarious, polygynous mating systems
Defend small territory or home range	Nomadic with huge home ranges
Perennial breeding	Seasonal breeding
When threatened, seek cover and hide	When threatened, flee in open

Adapted from Estes (1991).

TABLE 24-5 **Behavioral/Ecological Classification of Some African Bovids**

Social Organization and Feeding Style	Body Size (kg)	Antipredator Behavior	Examples
CLASS A Solitary or in pairs or family groups Small, permanent home range Highly diversified diet, but selective	1–20	Freeze, dash to cover and freeze, or lie down; do not outrun or counterattack predator	Dik-dik (*Madoqua*) Duiker (*Cephalophus*)
CLASS B Several female-offspring units associate Group size 1–12 Permanent home range Males solitary Diversified diet	15–100	Similar to Class A, but with some outrunning of predators for short distances	Reedbucks (*Redunca*) Vaal reedbuck (*Pelea*) Oribi (*Ourebia*) Lesser kudu (*Tragelaphus imberbis*)
CLASS C Larger herds—six to hundreds Males have breeding territories Selective browsers and grazers	20–200	Diverse; hiding used in heavy cover, running used in open areas; communication of alarm behavior important	Kob, Waterbuck, Lechwe (*Kobus*) Gazelles (*Gazella*) Impala (*Aepyceros*) Greater Kudu (*Tragelaphus*)
CLASS D During sedentary times, societies as in Class C species Gigantic herds during migrations Feed on variety of grasses Selective as to plants eaten	100–250	Run from large predators or mount unified counterattack on smaller predators	Wildebeest (*Connochaetes*) Hartebeest (*Alcelaphus*) Topi (*Damaliscus*)
CLASS E Large stable herds of females and young with males organized into dominance hierarchies Herd size up to 2,000 No coalescing of herds during migration Unselective grazers or browsers	200–700	Run from predators or mount unified counterattack even on larger predators; group responds to distress of young	Buffalo (*Syncerus caffer*) Oryx, gemsbok (*Oryx gazella*) Probably Eland (*Taurotragus*)

Data from Tyndale-Biscoe and Renfree (1987) and Renfree (1993).

impalas, constituting about one-third of the population of adult males, maintain territories in the breeding season in the most favorable habitat (**Fig. 24-30**). These territories

FIGURE 24-30 A male impala (*Aepyceros melampus*) from Kenya. Impala typically form large herds where males hold breeding territories (Class C in Table 24-5).

form a mosaic of adjoining areas, and each dominant male defends his area against other males of comparable social status. Females and bachelor males occupy home ranges that typically include a number of territories. A territorial male attempts to round up females that enter this area and keep them within it. In bachelor herds, the hierarchy is based partly on age distinctions, with older, larger-horned animals dominating younger ones. Males at or near the top of the bachelor hierarchy challenge territorial males, and repeated encounters between a territorial male and his challenger might span several weeks. A male holding a prime territory much frequented by impalas is kept busy herding females, checking for females in estrus, and keeping bachelor and competitive adult males at a distance. These males become exhausted and lose their territory more quickly than do males holding less preferable areas. In areas with seasonal precipitation, the territorial system is abandoned during the dry season.

Sinclair (1970, 1974, 1977) studied the advanced social organization of the African buffalo (Class E in Table 24-5) in Tanzania. This nonterritorial animal forms herds of from 50 to 2,000 animals. The size of a given herd is rather constant, at a mean of about 350 animals. For the first 3 years of its life, a young buffalo tends to remain near its mother, and bonds between mothers and daughters seem closer than those between mothers and sons. In the third year of life, males begin to leave their mother, and when they are 4 and 5 years old they form subgroups within the herd. Older adult males that remain with the herd establish a linear dominance hierarchy. The repeated sparring typical of immature males may result in the formation of this hierarchy. The less dominant males are driven from the herd and form small bachelor groups that remain separate from the mixed herds. Old males, older than about 10 years of age, leave the herd and become extremely sedentary; they are either solitary or form small social units. The breeding is done largely by the dominant males of the herd, with the highest-ranking ones having the greatest access to females in estrus.

Especially remarkable is the way in which the herd functions as a tightly knit unit. An entire herd will rally to the defense of a member in distress, and this formidable united front will discourage even the largest predators. A herd also moves and feeds as a closely massed unit. There is little attempt to maintain individual distance, and the bodies of herd members may, on occasion, be touching.

Geist (1974) pointed out that the widespread substitution by bovids of ritualized combat and aggressive displays for damaging physical contact has probably evolved under selection exerted by high densities and high diversities of predators. Bovids that attack and injure others invite damaging counterattack and are likely to be wounded, whereas those that use nondamaging, ritualized fighting are less likely to sustain injury (**Fig. 24-31**). Because predators often concentrate on conspicuously wounded animals, selection by predators strongly favors the adherence by bovids to ritualized intraspecific contests. This great development of ritualized combat in African bovids, which must face many diverse predator populations, contrasts with the more damaging encounters between members of northern species, such as bighorn sheep (*Ovis canadensis*), that are under far less pressure from predators.

Kin Selection

Natural selection favors individuals with life-history strategies that maximize their genetic contribution to future generations. The particular life-history strategy adopted depends on the interaction between the individual and its environment (involving climate, food resources, predators, competitors, and nest sites). The individual's chance of surviving and reproducing depends to a large extent on its behavior (Krebs and Davies 1993). Because behavior has a genetic component, natural selection operates on genetically based behavioral variation in the population, just as it does on genetically based morphologic variation. Often many genes influence the expression of a particular behavioral trait (**polygenic traits**). Those behaviors that increase a mammal's foraging efficiency, predator avoidance,

FIGURE 24-31   A pair of male oryx (*Oryx gazella*) fighting for dominance in Namibia. Fighting is ritualized and seldom leads to severe injury or death.

ability to find mates, and so on, are favored by natural selection. Natural selection results in fitness differences tallied as the lifetime contribution to future generations of one individual's **genotype** relative to the contribution of other genotypes in the population.

Behaviors that appear to be altruistic, those that increase the fitness of the recipient while potentially reducing the fitness of the donor, seem to contradict evolutionary theory. How could the frequencies of genes controlling altruistic behaviors have increased in populations via natural selection? One mechanism proposed by Hamilton (1964) is kin selection. **Kin selection** is a type of natural selection in which "the frequency of a gene in a population will be influenced not only by the effects that gene has on the survival and fertility of individuals carrying it, but also by its effects on the survival and fertility of relatives of that individual" (Maynard Smith 1976). The term "fitness" refers to an individual's success in passing on genes to future generations; inclusive fitness is a subset of fitness in which genes of relatives are promoted. It is relatively easy to see how the fitness of an individual is increased by passing on copies of its genes via its own offspring who share, on average, 50% of their genes in common with each parent. Offspring, however, are only one type of kin. Siblings also share 50% of their genes in common, and cousins share 12.5%. The fitness gained by helping a sibling to survive and reproduce, therefore, is considerably greater than that gained by helping a cousin, but helping a cousin may still result in fitness benefits for the "altruist." Thus, genes for altruistic behavior can increase in frequency in a population if

$$rB - 2C > 0$$

where r is the coefficient of relatedness between the donor and recipient, B is the fitness gained by the recipient, and C is the fitness cost to the donor of a behavior (Hamilton 1964). If an individual helps a sister rear three more offspring ($r = 0.25$ for nieces) than the sister could have raised without the extra help, the fitness gained by the recipient ($B = 3$) is greater than the cost to the helper of raising one less young of her own ($C = 0.5$ for offspring). In this case, the gene for the altruistic behavior will increase in frequency in the population. As the geneticist Haldane (1932) said (anticipating Hamilton's theory), "I would give my life for three brothers or nine cousins!"

The value of kin selection theory is that it allows the quantification of the costs and benefits of a particular behavior in terms of inclusive fitness and explains how apparently altruistic behaviors can evolve via natural selection. Altruistic behaviors in mammals include alarm calling to warn others of the approach of a predator, cooperative breeding (alloparental care) in which helpers assist in the rearing of another's young rather than reproducing themselves, coalition forming to cooperate in acquiring mates, and food sharing.

Alarm Calls

Consider an example of predator alarm calls given by Belding's ground squirrels (*S. beldingi*). These social squirrels inhabit subalpine meadows. Females mate in the spring, establish territories around the nest burrow, and give birth to three to six young; males leave after mating and contribute no parental care. After weaning, juvenile males disperse, leaving the juvenile females with their mothers in the natal area. Females, therefore, are nearly always surrounded by close kin, whereas males seldom interact with close relatives. Sherman (1981a, 1981b) found that closely related females cooperated to defend each other's young against territorial raids by unrelated conspecifics. (Undefended offspring were often killed by young males in search of an easy meal and by unrelated adult females attempting to take over nesting burrows.) Such cooperation in the defense of closely related young is what would be predicted by kin selection theory because an individual can increase its fitness by protecting close relatives.

Belding's ground squirrels also give alarm calls whenever predators are sighted. Individuals give two different types of alarm calls: one for aerial predators and one for terrestrial predators. In the case of aerial predators such as hawks and eagles, significantly more noncallers (28%) were captured than those that gave an alarm whistle (2%), regardless of their relationship to those nearby (Sherman 1985). Here the caller clearly benefits from early detection of the predator, and this behavior is easily explained by individual selection. When terrestrial predators such as coyotes or snakes are considered, however, the outcome is very different. Callers suffered significantly higher predation than did noncallers. A closer look at the genealogies of the ground squirrel population provides an explanation for the apparently altruistic behavior. When the frequency of alarm calls is categorized by sex and kinship, it is clear that Belding's ground squirrels are playing favorites (**Fig. 24-32**). Because daughters tend to stay in the vicinity of the natal territory, the offspring of the female giving the alarm call received the greatest benefit (individual selection), but females without young also benefited to a lesser extent. In the latter case, the evolution of terrestrial alarm calls can be explained by kin selection because adult females in the ground squirrel society are closely related.

Cooperative Breeding

Many other examples of apparent altruism can be explained by kin selection. For example, cooperative breeding, in which nonbreeding helpers assist with the care and defense of the young, would also seem to reduce the fitness of the

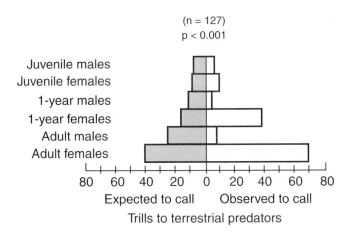

FIGURE 24-32 Observed and expected frequencies of alarm calls given in response to terrestrial predators by Belding's ground squirrels. Females gave more calls than would be expected by chance, and males gave many fewer calls than expected by chance alone. (Modified from Sherman, P., *Behav Ecol Sociobiol.* 17 (1985): 313–323. With kind permission of Springer Science & Business Media and Professor Paul W. Sherman, Cornell University.)

helper. Helpers perform a variety of tasks, including provisioning the young with food, defending the young against predators, or serving as extra baby-sitters, as in the case of the dwarf mongoose. Black-backed jackals (*C. mesomelas*) and African hunting dogs (*L. pictus*) deliver food to the

den via regurgitation. In ring-tailed lemurs (*L. catta*), baby-sitters actually allow infants to nurse (Pereira and Izand 1989). Saddle-backed tamarins (*Saguinus fuscicollis*) give birth to twins, each of which can weigh up to 20% of the mother's weight. Several males other than the father help by carrying these juveniles (Terborgh and Goldizen 1985). In most of these cases, the helpers are genetically related to the offspring they care for and, therefore, are increasing their own inclusive fitness by helping.

In addition to the benefits gained via inclusive fitness, helpers that forgo breeding may receive benefits that are not obvious. African hunting dogs, for example, live in packs containing several adult females that do not breed. These nonreproductive females help rear the young of the dominant female and are often, but not always, closely related to her. Even for distantly related helpers, the short-term fitness cost of not breeding may be more than paid for by the long-term benefits of remaining in the pack. Solitary hunting is not a realistic option for these dogs because they rely on the pack to kill prey much larger than themselves.

Male lions often form long-lasting coalitions that cooperate in taking over prides (**Fig. 24-33**). When coalition partners are closely related (for example, brothers), kin selection might account for the evolution of this cooperative behavior. However, by using DNA fingerprinting, Packer and coworkers (Packer and Pusey 1982; Packer 1986)

FIGURE 24-33 A pair of young male lions (*Panthera leo*) form a coalition to hunt and eventually to take over a pride.

demonstrated that approximately half of the Serengeti male coalitions included one or more unrelated males. Thus, certain types of altruistic or cooperative behavior do not appear to be directed at relatives and cannot be explained by kin selection alone. In lions, such unrelated males are accepted into a coalition because there is a direct benefit to all members of the coalition (individual selection). Larger coalitions have a better chance of taking over a pride and remain in control of the pride for longer periods than smaller coalitions do. Because lion prides are loosely organized, any one of the male partners may be able to find and mate with a female in estrus, and thus all coalition males potentially benefit from cooperation. Coalition-forming males, therefore, probably produce more offspring over the long run than males that do not enter coalitions.

Reciprocal Altruism

Cases in which the short-term cost of providing another with some resource is offset when the recipient returns the favor at a later time are called **reciprocal altruism**. As Trivers (1971) showed, reciprocity can evolve only if the donor and recipient can recognize one another, repayment is likely, and the benefit to the recipient is greater than the cost to the donor. Food sharing by vampire bats (*Desmodus rotundus*; **Fig. 24-34**) is typically considered a case of reciprocal altruism. Vampire bats share blood meals with their unrelated roostmates (Wilkinson 1984, 1990). A bat that has failed to find a meal may beg for food from a roostmate. Begging is stereotypical, beginning with a short period of grooming followed by the recipient licking the donor's lips. The donor then regurgitates a few milliliters of blood, which is usually enough to sustain the recipient until the next night.

There are clear benefits to the recipient because vampire bats can starve to death if they do not find food in two consecutive nights. Wilkinson's (1984) experiments showed that food sharing occurred only between close relatives (explained by kin selection) or between unrelated bats that frequently shared the same roost. Furthermore, starved bats that received a meal were significantly more likely to reciprocate by donating a meal later on. The third condition for reciprocal altruism, that the benefit to the recipient be greater than the cost to the donor, was also demonstrated by Wilkinson's experiments. Weight loss after feeding in vampire bats declines exponentially (**Fig. 24-35**), with starvation occurring at 75% of their prefed weight. The shape of this weight loss curve indicates relatively little cost to a donor that has recently fed. By donating a few milliliters of condensed blood, for example, a recently fed donor may lose 5% of its prefeeding weight, which is equivalent to approximately 6 hours worth of food for the donor. The recipient, on the other hand, not having fed for nearly 48 hours, is down to 80% of its normal prefeeding weight and is in danger of starvation. A blood meal equivalent to 5% of its body weight may result in a 16-hour reprieve from starvation. (A 5% loss to the donor shifts it to the right

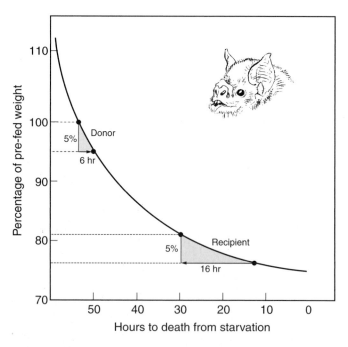

FIGURE 24-35 Model of the blood-sharing behavior of vampire bats (*Desmodus rotundus*). In this model, time to starvation is illustrated by the decreasing exponential curve. Donors who give blood to a roostmate lose 5% of their body weight and decrease their time to starvation by about 6 hours (right arrow). Recipients are already closer to starvation and may be at only 80% of their prefeeding weight. At this point on the curve, a gain of 5% from the donor pushes the time to starvation back by about 16 hours (left arrow). (Data from Wilkinson, G.S., *Nature* 308 (1984): 181–184.)

FIGURE 24-34 A group of vampire bats (*Desmodus rotundus*) roosting together in Calcehtok cave, Yucatan, Mexico.

along the upper part of the curve, but a 5% gain to the recipient shifts it to the left along the lower part of the curve; Fig. 24-35.) Perhaps this system has evolved because blood meals are difficult to obtain, because a bat that donates blood one night may be begging several nights later, and because of the long associations among roostmates (individuals may share the same roost for more than 12 years). Reciprocal food sharing, therefore, can evolve via natural selection acting at the level of the individual.

Kin Recognition

For kin selection or reciprocal altruism to operate, individuals must be able to recognize close relatives or roostmates. Familiarity may be one mechanism allowing kin recognition. Young golden-mantled ground squirrels (*Spermophilus lateralis*) prefer to play with siblings, even when the siblings were reared separately (Holmes 1995). Siblings seemingly use visual or olfactory cues to recognize one another. Hamilton (1964) suggested that an animal might inherit the ability to distinguish kin via so-called recognition alleles. This recognition mechanism would give kin a recognizable label and the ability to perceive others with that label. The alleles for this type of recognition may be associated with a family of genes called the major histocompatibility complex (MHC), a system used by the immune system for self-recognition. The immune system sends out specific types of cells to find and destroy foreign pathogens. The immune system's ability to distinguish self from nonself enables it to selectively destroy these foreign molecules. MHC genes code for molecules embedded in each cell's membrane that act as molecular fingerprints to label the cell as "self." In one experiment, mice (*Mus musculus*) that were inbred so that they differed by a single allele for one of the MHC genes were given a choice of mates. Males typically preferred females that differed in MHC alleles and therefore avoided inbreeding (Yamazaki et al. 1976, 1980; Penn and Potts 1999). Mice in seminatural enclosures also prefer mates with different MHC genes (Potts et al. 1991). In rodents, genetic differences in MHC genes produce subtle differences in urinary odors that can be used to assess relatedness (Boyse et al. 1982). Thus, even a single genetic difference may act as a kin recognition label allowing animals to distinguish kin from nonkin.

Eusocial Mammals

Among the remarkably highly evolved social systems of insects are those in which only a single female colony member (the queen) breeds and the remainder of the colony is divided into **social castes**. (A caste is a group of morphologically distinctive individuals that performs specialized labor in a colony.) These eusocial insects typically have a long life span, overlapping generations, cooperative care of eggs and young, and the reproductive female is the member of the colony least vulnerable to predation or accident. Insects with this system include some bees, some ants, social wasps, and termites. Only two vertebrates, both mammals, are eusocial: the naked mole-rat (*Heterocephalus glaber*) and Damara mole-rats (*Fukomys damarensis*; Jarvis 1978, 1981; Withers and Jarvis 1980; Bennett and Jarvis 1988; Sherman et al. 1991).

Naked mole-rats are small (25 to 50 grams), nearly hairless rodents that occupy the hot, dry parts of Kenya, Ethiopia, and Somalia (**Fig. 24-36**). They live in large colonies

FIGURE 24-36 Two eusocial mammals: (A) the naked mole-rat (*Heterocephalus glaber*) of east Africa; (B) the Damara mole-rat (*Fukomys damarensis*) of southern Africa.

of up to 40 individuals, with each colony occupying an extensive burrow system. The mole-rats eat enlarged roots and tubers of plants (geophytes) that are adapted to long dry seasons, and most burrowing is done when the usually hard soils are made friable by rain.

In many ways, the social system of the naked mole-rat parallels the eusocial systems of insects. The colony is composed of three castes. "Frequent workers" are small (25 to 30 grams) nonbreeders that burrow cooperatively, forage, and build the communal nest. Members of this caste make many trips to the nest with food for the other castes. "Infrequent workers" are slightly larger (about 35 grams) and work at roughly half the rate of frequent workers. "Nonworkers" are the largest colony members (about 46 grams) and rarely work but do care for the young. They are brought food by the frequent workers, and when nonworkers sleep they are often joined by other mole-rats. Naked mole-rats have a low metabolic rate and poor ability to thermoregulate. Their huddling together reduces the energy expended by the colony. The single breeding female performs no colony tasks, breeds with nonworker males, and produces one to four very large litters (up to 24 young per litter) each year. Some individuals in each litter grow more rapidly than their siblings and become larger than the frequent workers; such individuals may replace infrequent workers or nonworkers that die. A laboratory colony studied by Jarvis (1978, 1981) had 16 frequent workers, 9 infrequent workers, and 8 nonworkers.

Suppression of reproduction in all but the single breeding female is probably under pheromonal control. Jarvis showed that physical contact between the breeding female and other females is necessary for such suppression and suggests that the pheromone is carried in the urine and transmitted from the breeding female to others at the communal latrine. All males produce sperm, but the small frequent worker males have difficulty in copulating with the large breeding female. Females other than the single breeding female have ovaries that are seemingly quiescent; they contain many primordial and primary follicles but few mature follicles.

The other eusocial mammal, the Damaraland mole-rat (*F. damarensis*; Fig. 24-36B), is also known through the studies of Jarvis and her colleagues (Bennett and Jarvis 1988; Jacobs et al. 1991; Jarvis and Bennett 1991, 1993). This mole-rat is fully furred and averages 131 grams in weight. It inhabits the red Kalahari sands of some hot and arid deserts of southern Africa, where rain is highly unpredictable. Underground storage organs of various plants are the main food source for *F. damarensis*. Colonies of this mole-rat include up to 41 individuals, all of which share a common, extensive burrow system.

Evidence for eusociality in *F. damarensis* is conclusive. Each colony has but one reproductive female. She breeds three or four times a year, has a mean litter size of three, and remains the sole reproductive female throughout her time in the colony (up to 10 years in a laboratory colony). She initiates courtship. Usually only one male per colony is reproductively active. The reproductive pair are typically the largest colony members and are dominant animals, with nonreproductive females often the lowest members in the hierarchy. Reproduction is suppressed in nonreproductive females; their ovarian follicles do not mature. Although spermatogenesis occurs in both reproductive and nonreproductive males, the latter have smaller testes and different behavior. DNA analysis of entire colonies shows that each colony is composed of the reproductive pair and their offspring.

Nonreproductive members of both sexes are loosely differentiated into two classes, frequent and infrequent workers. The former class does most burrow maintenance work. All colonies studied in the field had overlapping generations. At least one-third of the nonreproductives remain in their natal colonies long enough to help care for four or more litters of siblings.

Eusociality has developed independently in the naked mole-rat and the Damara mole-rat, two evolutionarily divergent species (Faulkes et al. 1997), and in both cases ecological constraints probably played a critical role (Jarvis et al. 1994). Both species dig extensive burrows to find food (geophytes), and both occupy areas with low and erratic rainfall, where droughts of a year or more are common. The soil at the burrow depth of 25 centimeters is only workable when a rain of at least 25 millimeters either softens the hard soil, in the case of *H. glaber*, or makes the sand compact enough to retain a burrow, in the case of *F. damarensis*. At such times, the mole-rats must dig as fast as possible and locate enough food to last until the next heavy rain. During the brief moist period, a single animal could not dig fast enough to find sufficient food to last through a prolonged drought, whereas a group of animals can accomplish this. Jarvis et al. (1998) showed that small colonies fail more frequently than do large ones, and that survival of newly formed colonies depends on the richness of the food supply and the size of the workforce searching for food. Thus, in arid areas, ecological bottlenecks in the form of extended droughts and food scarcity have probably punctuated the evolutionary history of bathyergid mole-rats and may have favored sociality. By contrast, all five species of the three solitary bathyergid genera live in mesic areas, where the soil is friable for extended periods.

Although only the two bathyergid mole-rats described here are known to be eusocial, at least seven additional

FIGURE 24-37 A tuco-tuco (*Ctenomys minutus*) from southern Brazil.

species of *Fukomys/Cryptomys* are colonial, and sociality is known in three other genera of subterranean rodents. In some ways, the behavior of the social tuco-tuco (*Ctenomys sociabilis*; Ctenomyidae) of South America parallels that of the eusocial mole-rats (Lacey et al. 1997). Members of the colonies of *C. sociabilis* share entire burrow systems (**Fig. 23-37**). Colonies typically have multiple adults, from 1 to 18 juveniles, and all colony members shared a single nest. In contrast to the eusocial mole-rats, however, *C. sociabilis* colonies did not have reproductive division of labor, and probably not all colonies had an adult male. Among some

38 species of *Ctenomys*, only *C. sociabilis* is known to be social. Further research may show that this species lives under some of the same ecological constraints as those facing the eusocial bathyergid mole-rats.

Alexander and his colleagues (Alexander et al. 1991) summarized the combination of traits that probably favored the evolution of eusociality: extensive parental care, long-lasting and stable food resources that are easily defensible but that have a patchy distribution, a subterranean lifestyle that reduces predation and allows young to remain in their natal groups, and high risks associated with dispersal and new-colony formation.

SUMMARY

Mammals exhibit a remarkably wide range of behaviors compared to other vertebrates. Some behaviors are initiated by external cues from the environment; these behaviors are often innate (instinctive) and relatively inflexible. Mammal species tend to be nocturnal, diurnal, or crepuscular. Mammals also exhibit various kinds of cyclic behavior, including seasonal migrations, seasonal breeding, and daily or seasonal periods of reduced metabolism (torpor). Cyclic behaviors generally occur over a 24-hour period (circadian) or a roughly yearly period (circannual). Clearly,

some strong endogenous control is present. Some mammals exhibit relatively constant 24-hour cycles of feeding and sleeping despite being held in the laboratory under constant conditions of photoperiod and temperature.

Mammals have evolved a wide range of foraging behaviors that require decisions about the costs and benefits of acquiring a meal. Natural selection favors optimal foraging behaviors—behaviors that deliver the highest payoff (in terms of reproductive fitness) with minimal energetic costs. For example, because fossoriality is energetically costly, burrowing mammals have evolved behaviors favoring the most efficient system of finding food. Thus, basic burrow geometry and spacing tend to be remarkably uniform both within and between burrow systems. Food storage, or caching, is a widely used behavior in many herbivores. Predators have evolved behaviors that facilitate the pursuit, capture, and killing of prey, which in most cases have defensive or predator-avoidance strategies.

Many behaviors result from communication with other individuals. For example, a male elk bugles to signal his domination of harem females (acoustic signal), curls his upper lip in a flehmen response to detect pheromones (olfactory signal), and thrashes his antlers in the underbrush to ward off other males (visual signal). A few mammal species evolved highly specialized forms of communication; ultrasonic signals are used by echolocating bats, infrasonic signals permit elephants to communicate over vast distances, and seismic signals allow footdrumming kangaroo rats to communicate while remaining underground. These and other behaviors involve signals that quickly and accurately communicate complex information about the behavioral state of the sender to one or more recipients. The ability of mammals to interpret varied signals sets the stage for the evolution of complex social systems.

Mammalian social behaviors are often based on learning and experience; individuals learn who is related, who they have formed alliances with, and whom to avoid. Much of this learning takes place early in development. Lactation and the extended maternal care it affords provide young the time to learn from experience and remember the rules of group membership.

Energetic costs of pregnancy and lactation are high. Therefore, females invest more in each offspring than males do and should be selective in their choice of mates. Males must compete with one another for the possibility to mate (male–male competition), while females seek to select the best mating partner from the many courting males (female choice). Thus, males and females have competing interests.

Mammalian mating systems are diverse and likely form a continuum. Monogamy, exclusive mating between one male and one female during one or more breeding seasons, occurs in a variety of mammal groups, but in fewer than 3% of mammalian species. The vast majority of mammal species are either polygynous or promiscuous. Polygyny, in which males mate exclusively with multiple females (females mate with a single male), is common where male territories overlap more than one female's home range or where males can defend resources needed by females. In contrast, promiscuous mating occurs when both male and female mate with multiple partners during a breeding season. When females occur in small groups or have small home ranges, males defend harems (female defense polygyny). Alternatively, males may control access to resources required by females (resource defense polygyny). When females are not defensible, males must resort to following females or forming seasonal harems. In extreme cases, males congregate on an arena, or lek, and females come to mate with the most aggressive or vigorous males.

Parental care behavior helps ensure the survival of offspring at least until weaning. The sex of the care provider and the amount of parental care seem tightly coupled to the type of mating system. All mammals have some degree of maternal care. Females can be certain of maternity, but males rarely can be certain of their paternity. Thus, paternal care is more likely in monogamous mating systems, where males have greater certainty that the offspring are theirs. Offspring care is also performed by nonparents. Such alloparental behavior may even involve nursing unrelated offspring.

Many mammals form social units that exist beyond the mating and offspring-rearing periods when the benefits of group living exceed the costs. Patterns of social behavior in mammals are diverse. At one extreme are asocial mammals that rarely come into contact, except during the breeding season. Other mammal species form seasonal groups centered on clumped resources such as waterholes or roost sites. These social groups are persistent and may have relatively stable membership. At the opposite end of the social continuum are complex social systems with dominance hierarchies, stable membership including several generations, and complex intragroup communication. As an extreme example, naked mole-rats form eusocial colonies where only a single female (the queen) breeds and the remainder of the colony is divided into social castes.

KEY TERMS

Allomone	Caching	DNA fingerprinting
Alloparental care	Circadian rhythm	Dominance hierarchy
Altruistic	Circannual rhythm	Endogenous
Anisogamy	Coalition	Exogenous
Appeasement	Coterie	Facultative monogamy
Behavioral ecology	Diurnal	Female choice

Female defense polygyny
Fission–fusion social system
Flehmen
Free-running cycle
Genotype
Hibernaculum
Kin selection
Male–male competition

Matrilineal group
Midden
Obligate monogamy
Optimal foraging
Paternity
Polygenic trait
Polygyny
Promiscuity

Reciprocal altruism
Resource defense polygyny
Seismic signal
Social castes
Vomeronasal organ
Weapon automimicry

RECOMMENDED READINGS

Clutton-Brock, TH. 1988. *Reproductive Success: Studies of Individual Variation in Contrasting Breeding Systems.* University of Chicago Press, Chicago, IL.

Clutton-Brock, TH. 1991. *The Evolution of Parental Care.* Princeton University Press, Princeton, NJ.

De Waal, F. 2007. *Chimpanzee Politics, Power, and Sex among the Apes.* Johns Hopkins University Press, Baltimore, MD.

Estes, R and EO Wilson. 1992. *The Behavior Guide to African Mammals: Including Hoofed Mammals, Carnivores, Primates.* University of California Press, Berkeley, CA.

Krebs, JR and NB Davies. 1997. *Behavioral Ecology: An Evolutionary Approach.* 4th ed. Blackwell Scientific Publications, Malden, MA.

Mann, J, et al. 2000. *Cetacean Societies: Field Studies of Dolphins and Whales.* University of Chicago Press, Chicago. IL.

McComb, K, et al. 2003. Long-distance communication of acoustic cues to social identity in African elephants. *Animal Behaviour,* 65:317–329.

Rubenstein, DI and RW Wrangham. 2006. *Ecological Aspects of Social Evolution: Birds and Mammals.* Princeton University Press, Princeton, NJ.

Wolff, JO and PW Sherman. 2007. *Rodent Societies: An Ecological and Evolutionary Perspective.* University of Chicago Press, Chicago, IL.

Zoogeography

One of the most familiar kinds of biological information concerns **biogeography**, the study of the geographical distribution of living things in time and space. Children learn that lions and zebras live in Africa and not in North America, and that most rain forests grow in the tropics and not in the arctic. This type of knowledge of the presence or absence of various kinds of animals in different parts of the world is the substance of **zoogeography**, the study of animal distribution and a subdiscipline of biogeography.

Considerations of zoogeography include several major approaches. The first is descriptive and static and seeks to delineate the distributions of living species. The second approach is ecological and historical and attempts to explain the observed distributions of biodiversity. The third and most integrative approach uses the first two approaches merged with a phylogenetic or molecular genetic approach; it attempts to follow evolutionary lineages through time and place, and sometimes goes further to apply the knowledge gained to potential solutions for the conservation of the species. In each case, information is usually gained by fieldwork, laboratory work, and careful observation.

In the first approach, scientists have traditionally produced basic local or regional distribution maps and field guides. Regional guides and field guides are fundamentally important as a basis for education and research, as well as stimulating popular interest so crucial to conservation. High-quality field guides to mammals in many parts of the world have been available for decades (Hickman 1981), and new or improved editions continue to appear regularly (http://www.library.illinois.edu/bix/fieldguides/main.htm). A few countries still remain poorly known even in this regard, and field guides in their native languages made available to local peoples could enhance appreciation for wildlife, especially if combined with their own traditional uses and traditional ecological knowledge, just as they help provide people in developed countries with an appreciation

for nature as the real provider and sustainer of human life. Examining patterns of distributions and endemicity can also be useful for conservation planning on large (country-wide) scales; two excellent mammalian examples of this are studies for Mexico (Escalante et al. 2002) and Peru (Pacheco et al. 2009).

Of course, species' distributions are dynamic, and the maps often found in guides are not intended to be immutable. As the extirpation of mammals from their potential or natural ranges continues, maps frequently show the "historical" or "former range" and the "present range," while some maps indicate the dynamics of species' range boundaries. Studies of "range collapse," the dramatic shrinkage in geographic distribution that accompanies a species' endangerment and disappearance, have implications for the conservation and potential reintroduction of the species (Channell 1998; Lomolino & Channell 1995). Moreover, although climate variability and change have always been a part of earth history, at times, as in the present, change has come quickly. The unpredictable effects of novel new climates and disappearing old climates in the near future (Williams et al. 2007) will further restrict the already shrunken ranges of many species (Parmesan 2006). Novel solutions such as **facilitated dispersal** may help some species reach remnants of suitable habitat from their artificially relict ranges in parks and reserves.

The second, historical and ecological, approach to biogeography often involves syntheses based on diverse lines of evidence. Historical and ecological zoogeography have traditionally been separate. Historical biogeographers attempt to explain how things came to be via originations and extinctions over time. How, when, and from where did various mammals reach the areas they now occupy? Virtually every fauna consists of animals that reached their present ranges at different times, from different regions, and by different means. Our knowledge of the complex history of a fauna basically depends on the completeness of the worldwide fossil record and on our understanding of the geologic history and paleoecology of the major landmasses. Ecological biogeographers investigate present relationships between organism and environment to explain geographic distribution. But the effects of ecology and history can be difficult to distinguish (Endler 1982). Badgely and Fox (2000) convincingly described the interplay of abiotic environmental variables with latitude in determining the **ecological biogeography** of North American land mammals.

Current biogeographic inquiry uses a synthetic perspective and relies on information from a wide array of sources (Riddle & Hafner 2007; Riddle et al. 2008). Increasingly, our knowledge improves as ecological and historical geographic data are integrated with new insights from morphological and molecular phylogenetics. These, in turn, using the perspectives of geochronology, paleontology, and geomorphology (**phylochronology**) can be used to reconstruct and chronicle the geography of a lineage through time. The geographical scale can range from local or regional to continental landforms and even to entire ocean basins, and the lineage scale can likewise vary from local populations of a taxon to continent-wide populations or whole **biotas**. This is an integrative approach sometimes called **phylogeography** that enables biogeographers to reconstruct relatively robust hypotheses regarding the geographic context of the evolution of a taxon or biota. Phylogeography is a dynamic, burgeoning field that is difficult to "pigeonhole" (see **Box 25-1**).

Today mammals occupy all continents, from far beyond the Arctic Circle in the north to the southernmost parts of the continents and large islands in the south. Antarctica presently has no land mammals, although it did in the distant past at times when it was freer of ice; today only seals and whales frequent its icebound margins. In the Western Hemisphere, the northernmost lands—the northern coasts of Greenland and of Ellesmere Island—are today inhabited by the arctic hare (*Lepus arcticus*), collared lemming (*Dicrostonyx torquatus*), wolf (*Canis lupus*), arctic fox (*Alopex lagopus*), polar bear (*Ursus maritimus*), ermine (*Mustela erminea*), caribou (*Rangifer tarandus*), and musk ox (*Ovibos moschatus*). A similar group of mammals, lacking the musk ox, lives on the northern coast of the Taymyr Peninsula in Siberia, which is the northernmost coast of Asia (Berg 1950). The southernmost part of Africa has a rich mammalian fauna. Tasmania, the southernmost part of the Australian region, supports two monotremes, many marsupials, several native rodents, and several bats. On Tierra del Fuego, at the southern tip of South America, occur a bat, several rodents, a fox, otters, and a llama. The chiropteran family Vespertilionidae occurs almost everywhere there is land except in arctic areas. The families

Leporidae, Sciuridae, Canidae, Mustelidae, and Felidae are native to all continents except Antarctica and Australia. (Leporids, canids, and felids were recently introduced to Australia.) All oceans, and all seas connected to the oceans, are inhabited by cetaceans; odontocetes also live in some large rivers and lakes.

One long-recognized phenomenon and popular topic of study in the geographical distribution of life on earth is the "latitudinal species gradient," the increase in the number of species of living things as one moves from the poles to the equator. The arctic regions are relatively poor in species and higher taxa as well as in morphologic variety ("evolutionary novelty"; Jablonski 1993), whereas the tropics are extraordinarily rich, both on land and in the seas. Given the harsher conditions for existence in arctic regions, it is perhaps not surprising that more species should exist in equatorial regions. Yet Earth has not always been so climatically polarized as it is at present. Glacial periods have taken place numerous times throughout Earth history (notably in the Neoproterozoic, Ordovician-Silurian, Pennsylvanian-Permian, and Quaternary). As discussed later in this chapter, during much of the reign of mammals, global climates were uniformly warm and humid from poles to equator in the Cretaceous and early Paleogene, but a long, slow, general climatic cooling has taken place throughout the rest of the Cenozoic (Smith & Uppenbrink 2001; Zachos et al. 2001, 2008).

Explanations of the latitudinal diversity gradient are numerous and hotly debated (Brown & Lomolino 1998). Researchers have suggested that the tropics (1) have higher rates of evolutionary origination (speciation) and act as a "diversity pump" (Darlington 1957; Rohde 1992; Terborgh 1973); (2) have lower rates of extinction and thus tend to accumulate more species (Matthew 1915; Stebbins 1974); or (3) combine the two (Rosenzweig 1992, 1995).

Students of mammalian geography, too, have frequently examined this high- to low-latitude increase in taxonomic and morphologic diversity, especially in North America (**Fig. 25-1**; e.g., Arita et al. 2005; Shepherd 1998). Badgely and Fox (2000) used several statistical treatments to show that most of the latitudinal gradient in North American mammal species diversity can be explained by the variation in a handful of environmental variables such as gradients in the winter temperature, annual moisture, frost-free period, annual potential evapotranspiration, and elevation. The latitudinal diversity gradient is less well examined for land mammals on other continents, and the phenomenon is discussed further in the Ecology chapter (see Chapter 23). For marine mammals (baleen whales and pinniped carnivorans), the pattern is reversed: Greatest diversity occurs at high latitudes, as does the greatest abundance of their planktonic foods (Feldhamer et al. 1999).

BOX 25-1 Phylogeography

The phylogeographic approach is typified by many recent examples, such as those by Cook et al. (2004) on red-backed voles, by Fitzpatrick and Turelli (2006) on range overlap in several clades of mammals, by Hundtermark et al. (2001) on effects of glaciations on moose populations, by Kay et al. (1997a) on Neotropical primate diversity and rainforest primary productivity; by Patton et al. (2000) on Amazonian marsupials and rodents, by Sanchez-Villagra et al. (2006) on talpid moles in Eurasia and North America, and by Velazco and Patterson (2008) on broad-nosed bats in South America. Separations of the major lineages of placental mammals on the continents are reflected in their genomes (Kriegs et al. 2006). Even our own genomic relationships and those of the animals we bring with us are partly determined by human history (Li et al. 2008; Searle et al. 2008a, 2008b).

One example of a phylogeographic study is that by Mustrangi and Patton (1997) of slender mouse opossums (*Marmosops*) in South America. Most species of these tiny, generalized rain-forest-dwelling marsupials inhabit the Amazon Basin north to Central America; but a second group, isolated by dry areas, occupies the Atlantic rain forest in coastal highlands of southeastern Brazil. This area was thought to have just one species of *Marmosops* before Mustrangi and Patton discovered that two similar species occurred there. These differed markedly in the sequence of a mitochondrial gene, indicating a relatively low level of phylogenetic relationship. One species lived mostly in low-lying and inland forests, while the other lived in higher montane forests. A cricetid rodent and a bird had similar distributions. The authors observed great genetic divergence between the Atlantic Forest *Marmosops* and the Amazonian species, indicating a divergence of these species during the Miocene uplift of the coastal highlands. The Atlantic forest biome may have a common biogeographic history dating back to the Miocene.

The Atlantic rain forest is rich in unique species of plants and animals; new species of primates have been discovered in the last decade and other new species of both plants and animals are sure to be found (Costa & Leite 2009; Fonseca 1985). Hopefully, Mustrangi and Patton's study will focus the attention of conservationists and the Brazilian government on a still-salvageable national treasure in imminent danger of destruction by humans.

Dispersal, Vicariance, and Faunal Interchange

Dispersal occurs when an individual or a population moves from its place of origin to a new area. The ability to disperse is as basic as the ability to reproduce and is necessary to the survival of a species. A spacing of members of a population so that each individual can satisfy its environmental needs is critical to all organisms. Territoriality is one familiar means by which this spacing is ensured, and the young of territorial species usually establish home ranges largely separate from those of other individuals, including their parents. The pressures exerted by reproduction and the necessity for the spacing of individuals create a tendency of populations to occupy ever-increasing areas, to colonize unoccupied areas, and to repopulate areas where they were previously extirpated. The more widespread a species, the less likely it is to be forced into extinction by local mortality. As a result, natural selection has usually favored those species that have broad distributions. A high adaptive premium is placed on dispersal ability. Udvardy (1969) stated that "without evolved means of dispersal most animal populations would have succumbed, over a period of time, to the vicissitudes of the environment." A well-documented example of a single species colonizing new areas is the range expansion of the nine-banded armadillo from the Mexico–United States border along the Rio Grande, eastward to Florida and northward as far as Nebraska, between about 1850 and the present day (Taulman & Robbins 1996).

The ability of a population to expand into new areas depends on its innate dispersal ability (which is greater, for example, in fliers than in burrowers), on the breadth of environmental conditions it can tolerate, and on the presence of barriers. Barriers may be ecological; for example, forests are barriers to grassland species. Physical barriers include bodies of water, precipitous cliffs, or barren lava formations. If enough information were available, much of the story of zoogeography could be told by considering the way in which animal dispersal patterns have been modified by the location, effectiveness, and temporal history of barriers.

Vicariance

Barriers may affect the distribution and speciation of animals in a passive sense by **vicariance**. Vicariance biogeography is primarily historical in its approach. Instead of citing active dispersal by an organism, vicariance biogeographers seek to explain the observed distribution patterns of species as the result of the splitting of an area occupied by a species. If a barrier such as a mountain range, shallow sea, or river arises within the distribution of a species that is initially widespread, the species' distribution can become discontinuous or restricted. As a result, gene flow is also restricted or terminated and the separated populations may undergo divergence and speciation. Thus, the phylogeny of

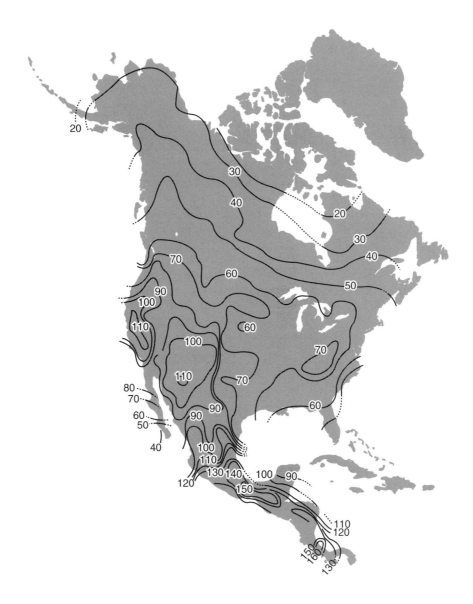

FIGURE 25-1 A representation of the geographic variation in the species density of terrestrial North American mammals. Contour lines enclose areas of equal species richness; numbers indicate the number of mammal species within the contour. Note that diversity increases toward the tropics. But at a given latitude, diversity is also greater in areas of greatest topographic relief, where habitats are the most diverse. (Adapted from Simpson, G. G., *Syst. Zool.* 13 (1964): 57–73 and Brown, J. H. and Lomolino, M. V. *Biogeography, Second Edition.* Sinauer Associates, 1998.)

a taxonomic group is sometimes influenced by vicariance events. Many workers use phylogenetic biogeography to explain the **historical biogeography** of a region, primarily by examining the phylogenetic relationships of its taxa (e.g., Dávalos 2005, 2006; Dubey et al. 2008).

Either vicariance or dispersal might result in the **disjunct** distribution of species. A disjunction exists in the present distribution of tapirs; three modern species *Tapirus pinchaque, T. terrestris,* and *T. bairdii* occur in the New World tropics, whereas the closely related Malay tapir *T. indicus* of tropical Southeast Asia is distantly isolated from the others. Distinguishing between vicariance and dispersal as

the cause of the observed distribution of a given taxonomic group can be complex. Indeed, the difference between the two explanations becomes blurred (and perhaps semantic) when evidence is viewed from a geological time scale, and testability of hypotheses is difficult. Both mechanisms act upon animal (and plant) populations (Cox & Moore 2005; Lomolino et al. 2006; Millien-Parra & Jaeger 1999).

Migration and Faunal Interchange

Certain regions have apparently been major centers of origin of mammalian groups. Many orders and families first appear in the fossil record in Eurasia (Beard 1998), and North America seems also to have been the place of origin for several groups. The present mammalian faunas of regions such as Africa and South America are partly **allochthonous** (some members originated outside the area where they now occur), derived from mammalian migrations from northern continents, but largely derived from **autochthonous** evolution (having originated in those areas in which they now occur). Many families are endemic, or unique to certain continents and islands. (A taxon is endemic to an area if it lives nowhere else.) Despite uncertainty as to the place of origin of many mammalian groups (where a group first appears in the fossil record is generally taken as its place of origin), movements of mammals from place to place are in some cases well documented by the fossil record.

Simpson (1940) recognized several avenues of faunal interchange. The corridor is a pathway that offers relatively little resistance to mammalian migration and along which considerable faunal interchange would be expected to occur. Such a continuous corridor now exists across Eurasia; interchange of animals between Europe and Asia is highly probable and has apparently occurred frequently. A **filter route** allows passage of certain animals, but stops others. Selective filtering has occurred at times along **Beringia**, the land bridge that has periodically connected Siberia and Alaska. When this bridge was present late in the Pleistocene, for example, conditions were such that only animals adapted to cold climates and mammoth-steppe habitats could migrate between these two continents. Mountain ranges, deserts, waterways, tropical areas, or abrupt changes in habitat may also form filter routes. Such features may simultaneously act as corridors, barriers, or filters to different species. The third and most restrictive route is the **sweepstakes route**. This is a pathway that will probably not be crossed by large numbers of any given type of animal but may be followed by an occasional individual. Such a pathway is that between Africa and Madagascar. Dispersal via a sweepstakes route must occur by swimming or flying or by such uncertain means as rafting from one land mass to another on floating vegetation or debris. The probability that an animal will follow a sweepstakes route is extremely low if the route is long, as, for example, from North America to Hawaii, but is increased if an animal is small and can cling to floating material, is aquatic, or can fly. (The only land mammals that reached Hawaii without the help of humans were bats.) Despite the unlikelihood of an animal's dispersal via a sweepstakes route, such dispersal was witnessed during recent (1998) hurricanes in the Caribbean that blew rafts of vegetation and debris carrying iguanas probably 300 kilometers between islands.

Mammals of the Biogeographic Regions

The biogeographical realms or regions shown in **Fig. 25-2** (Olson et al. 2001), which are the basis for the organization of this discussion, are modified from those originally proposed by Wallace (1876) and have been widely used in discussions of zoogeography. Biogeographic realms are based on flowering plants (e.g., Good 1974) as well as animal distributions and tend to be concordant with these zoogeographic regions. The regions also reflect a historical component deeply seated in Earth's geological past with respect to changing continental and sea connections. Global distribution patterns of mammals including relative species richness and species endemism were recently summarized by Olson et al. (2001), with a view toward conservation priorities. Space does not

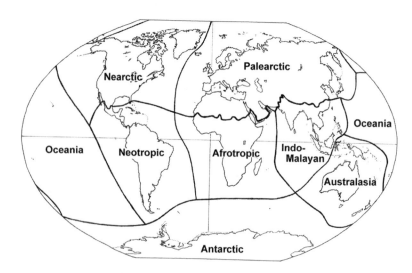

FIGURE 25-2 A map of the world showing the biogeographic regions discussed in this chapter. (Adapted from Olson, DM, et al., *BioScience*, 51 (2001):933–938 and from World Wildlife Fund, 2000. http://www.worldwildlife.org/wildworld).

allow us to discuss the important and complex historical, geologic, and biogeographic events in mammalian evolution that ultimately led to the regions discussed in the following sections. Instead, we attempt to provide references as an entry into the literature pertinent to each region.

Palearctic Region

The Palearctic Region includes much of the northern part of the Old World and is the largest terrestrial zoogeographic region. Included in this vast area are Iceland, Europe, North Africa, the Arabian peninsula (except its southern periphery), Asia (except India, Pakistan, and southeastern Asia), and the Middle East. The climate is largely temperate, but contrasting conditions exist, from the intense heat of North Africa to the arctic cold of northern Siberia. Broad areas of coniferous forests, comparable in many ways to those of northern North America, are typical of much of the northern Palearctic Region, and deserts are widespread in the south. The Palearctic is at present separated from the Afrotropical Region by deserts, from the Indo-Malayan Region by the Himalayas, and from the Nearctic by the Bering Strait.

The Palearctic mammalian fauna includes 36 families and resembles most strongly the Indo-Malayan fauna, with which it shares 76% of its families (**Table 25-1**). Because of repeated faunal interchange across the land bridge that periodically spanned the Bering Strait, the Palearctic shares 46% of its mammalian families with the Nearctic. (Together, the Palearctic and Nearctic Regions or their shared mammal species are sometimes called "Holarctic.") Many genera, and a few species, of the families Soricidae, Vespertilionidae, Canidae, Ursidae, Mustelidae, Felidae, and Cervidae occur in both regions. No family is endemic to the Palearctic Region at present, although a now-extinct lagomorph family, Prolagidae, occurred there until the Holocene. Historical zoogeography of Palearctic Region mammals is treated extensively in Beard and Dawson (1998) and Agustí and Antón (2002).

Nearctic Region

The Nearctic Region includes nearly all of North America north of the tropical sections of Mexico (except for the southernmost part of the Florida peninsula) and Greenland. It contains habitats ranging from semitropical thorn forest to arctic tundra. The mammalian fauna includes 30 families, some of which are mostly tropical in distribution (e.g., Phyllostomidae and Tayassuidae), together with some primarily **boreal** families (typical of north-temperate forests: Dipodidae, Castoridae, and Ursidae). Only two Nearctic families (Aplodontidae and Antilocapridae) are endemic. The mammalian fauna of the Nearctic resembles most closely that of the Neotropical Region (see Table 25-1). The transition between the Nearctic and Neotropical Regions was recently examined with respect to the distribution of bats by Ortega and Arita (1998). The Cenozoic history of mammals in North America is examined in detail in Woodburne (2004). Additional references for the Pleistocene include extensive coverage in Faunmap (Faunmap Working Group 1994) and by Kurtén and Anderson (1980) and Webb and Barnosky (1989).

Neotropical Region

The Neotropical Region features great climatic and biotic diversity and includes all of the Americas from tropical Mexico south, plus the southernmost tip of Florida and the islands of the Caribbean Sea. Much of the area is tropical or subtropical, and broad areas are covered with spectacular evergreen rain forest. Tropical savanna and grasslands occupy parts of the southern half of South America, and some of the driest deserts in the world are in the south and along the western coast. The higher parts of the Andes support montane forests and alpine tundra. The South American part of the Neotropics was isolated from the rest of the world through most of the Cenozoic, but the Isthmus of Panama has provided a connection between South America and Central America since at least the late Pliocene.

TABLE 25-1 Comparison of the Terrestrial Mammals of the Biogeographic Regions

Region	Number of Families	Number of Endemic Families	Percentage of Families Also Found in Region					
			PA	NA	NT	AF	IM	AU
Palearctic (PA)	36	1	—	47	39	89	81	31
Nearctic (NA)	31	2	55	—	77	42	48	13
Neotropic (NT)	56	28	25	43	—	23	27	7
Afrotropic (AF)	58	20	55	22	22	—	55	21
Indo-Malayan (IM)	49	8	59	33	33	65	—	27
Australasian (AU)	35	21	31	14	17	37	37	—

Source: Based on data from Wilson and Reeder (2005). A few of the families became extinct in the Holocene. Percentages do not include species transported by humans in the last two millennia.

This region is second only to the Afrotropical Region in diversity of families of mammals. The Neotropical Region supports 56 families of mammals and has the largest number of endemic families (29). Especially characteristic of the Neotropical Region are marsupials, bats (including three endemic families), primates (four endemic families), xenarthrans (pilosans and cingulatans; four endemic families), and hystricognath rodents (12 endemic or nearly endemic families, one of which became extinct in the Holocene). Two species of the genus *Lama* live in South America and are the only extant Western Hemisphere representatives of the family Camelidae. (Wild Eastern Hemisphere camelids occur only in the Gobi Desert of Mongolia.) Tapirs are restricted to the Neotropical and Indo-Malayan Regions. The Neotropical mammalian fauna most strongly resembles that of the Nearctic, but it also shares one-fourth of its families with the Indo-Malayan Region.

Historical biogeographic treatments of the Neotropics are many. Some of the most comprehensive and recent accounts include Simpson (1980), Pascual (2006), Pascual and Ortiz Jaureguizar (1990), Woodburne and Case (1996), Kay et al. (1997b), MacFadden (2006), and Flynn (2009). The Great American Biotic Interchange has been the focus of much research and is discussed in detail in Carranza-Castañeda and Lindsay (2006), Morgan (2008), Stehli and Webb (1985), Vrba (1992), and Webb and Rancy (1996). A brief synopsis of the Great American Biotic Interchange is given later in this chapter.

Afrotropical Region

The Afrotropical Region includes Madagascar and Africa south of the Sahara, and the southern edge of the Arabian Peninsula. Deserts, tropical savannas, tropical forests, montane forests, and even alpine tundra are all represented, and the most extensive tropical savannas in the world occur in Africa.

The Afrotropical Region has the greatest number of mammalian families (58) of any faunal region and, after the Neotropics, the greatest number of endemic families (20). The impressive array of ungulates that inhabits the savannas of Africa is unmatched elsewhere, and Africa is the last important stronghold of the families Equidae, Rhinocerotidae, Elephantidae, and Hippopotamidae. Although the only endemic artiodactylan family is Giraffidae, nearly all of the African genera of antelopes (Bovidae) are endemic. The primitive lemuroid primates of Madagascar (5 families, all endemic to the island) and the diverse group of cercopithecid primates of Africa are especially typical of the region, and two of the five genera of great apes live only in Africa. Apart from South America, Africa is the only area with a fairly diverse hystricognath rodent fauna. Viverroid carnivorans reach their greatest diversity in the Afrotropical Region, where two families are endemic (Eupleridae and Nandiniidae). The Afrotropic mammalian fauna most closely resembles those of the Palearctic and Indo-Malayan Regions. Accounts of the historical biogeography of the Afrotropical Region include Kingdon (1989), articles in Goldblatt (1993), especially Vrba (1993), Turner and Antón (2004), and a few chapters in Goodman and Benstead (2003).

Indo-Malayan Region

Included in the Indo-Malayan Region are India, Indochina, southern China, the Malay Peninsula, the Philippines Islands, and the islands of Indonesia east to a line (**Wallace's Line**) between Borneo and Sulawesi and between Bali and Lombok. The area is dominated by tropical climates and formerly supported broad areas of tropical forests before extensive land clearing and burning by humans. Deserts occur in the Pakistan area. The Indo-Malayan Region is partly isolated from the Palearctic by deserts in the west and by the Himalayas to the north.

The mammalian fauna of the Indo-Malayan Region includes 49 families (8 endemic) and resembles most strongly that of the Palearctic realm, with which it shares 90% of its families of mammals. Many (63%) of the Indo-Malayan families of mammals also occur in the Afrotropical Region. The most distinctive elements of the Indo-Malayan mammalian fauna are all of tropical affinities. Five families of primates occur in this region. A few Indo-Malayan families occur elsewhere only in the Afrotropical Region (Manidae, Elephantidae, Rhinocerotidae, and Tragulidae). Some aspects of mammalian historical zoogeography in the Indo-Malayan Region are discussed by Corbet and Hill (1992) and Heaney (1986).

Australasian Region

The Australasian Region includes Australia, Tasmania, the islands of Indonesia west of a line (Wallace's Line) between Borneo and Sulawesi and between Bali and Lombok, New Guinea and the islands of Melanesia, and New Zealand. In the area are hundreds of islands of various sizes and degrees of isolation. The island continent of Australia is connected with New Guinea by a broad continental shelf under the Arafura Sea, but these landmasses are presently separated by the Torres Strait, 160 kilometers wide. The northern part of the Australasian realm, including New Guinea and parts of the northeastern coast of Australia, are covered with tropical forest, but much of Australia is tropical savanna and desert. Some of the most arid deserts in the world occur in the interior of Australia.

The Australasian Region is famous for its unusual mammalian fauna, and, to the popular imagination, Australia itself is an area supporting marsupials almost exclusively.

Actually, Australasia has 9 native terrestrial eutherian families (mostly bats) and 18 metatherian families. More than 50% of the extant native families of the Australasian Region are marsupials, and 21 families (the monotremes and marsupials) are endemic. The mammals of the Australasian Region have their closest affinities with those of the Indo-Malayan Region. The historical biogeography of mammals of the Australasian Region is vividly discussed by Archer et al. (1991), Beck et al. (2008), Flannery (1995) Long et al. (2002), and Woodburne and Case (1996).

Antarctic Region

The Antarctic biogeographic realm includes the south polar continent of Antarctica, which has had no land mammals since it became completely covered with ice in the Neogene. After the breakup of the Paleozoic-Mesozoic continental masses (see the section titled "Continental Drift" later in this chapter and **Fig. 25-5**), most of the southern continents occupied temperate and tropical zones, but Antarctica drifted poleward. It then lost its last connections with Australia, New Zealand, and its already tenuous connection with southernmost South America via the Antarctic Peninsula and an island arc now known as the South Sandwich Islands. Movement of the Scotia tectonic plate pushed the Scotia Arc of the South Sandwich Islands eastward, removing this island chain and opening up the Drake Passage. This and the opening of the Tasman seaway between Tasmania–Australia and Antarctica enabled ocean currents to circulate completely around Antarctica. Because of the circum-Antarctic current and attendant climatic changes, as well as the southern arctic position that isolated the continent from faunal exchanges, Antarctica began to be covered by ice about 34 million years ago (Kohn et al. 2004; Liu et al. 2009; Woodburne 2004b). By the late Miocene and Pliocene, the previously existing cool temperate rainforests and resident animals were smothered with ice and Antarctica was left without land mammals. To date, only fossil mammals of Eocene age have been found in Antarctica in the islands along the northern Antarctic Peninsula just outside the Antarctic Circle. The Antarctic fossil mammalian fauna is still small but relatively diverse and includes several metatherians, unique extinct South American–Antarctic ungulates (Litopterna and Astrapotheria) and a possible sloth (Pilosa), showing affinities with the Eocene mammalian faunas of southern South America and Australia (Tejedor et al. 2009; Vizcaíno et al. 1998; Woodburne & Case 1996).

Oceanic Region

The oceans of the world compose the ecoregion known as Oceania. The oceans are much more homogeneous from basin to basin than are the continents because of their broad connections and the obvious potential for global dispersal of drifting and swimming organisms. Only isolated oceanic islands are included in the region with the largest being Fiji and the Hawaiian Islands. In this region live mainly marine mammals: whales and dolphins, some seals, sea lions and walruses. The only land mammals are bats capable of dispersing over great expanses of open ocean, or those brought by humans, such as introduced murid rodents and other mammals associated with humans. Historical zoogeographic accounts of some marine mammals include Arnason et al. (2006), Berta et al. (2006), Deméré et al. (2003), Fordyce and Barnes (1994), and Wynen et al. (2001).

Continental Drift, Mammalian Evolution, and Zoogeography

Continental Drift

The concept of **continental drift** was first proposed and defended by Wegener (1912, 1915, 1966), but until recent decades his views were regarded as heretical. Throughout much of the 20th century, teaching in geology and paleontology in North America was dominated by the view that the positions of the continents and the intervening oceans were fixed, that they had remained immutable back through the vast sweep of geologic time. (Continental drift was taken seriously by some European scientists.) Because they dismissed continental drift, most North American paleontologists were forced to rely on tenuous intercontinental land bridges or sweepstakes dispersal to account for intercontinental movements of terrestrial animals. Within the last 40 years, however, our geologic, paleontologic, and biogeographic perspective has been drastically transformed by convincing evidence in favor of the theory of plate tectonics and associated continental drift.

The discovery by E. H. Colbert in 1969 of a Triassic fossil therapsid (*Lystrosaurus*) in Antarctica put the capstone on the pyramid of evidence supporting continental drift. This nonaquatic dicynodont therapsid had previously been found in Triassic deposits on other southern continents; its distribution could be explained only by assuming that the continents had once been connected (Elliott et al. 1970). Wegener's wild theory was finally vindicated.

During the last half of the 20th century, continental drift and its driving mechanism "plate tectonics" were elaborated and supported by mountains of evidence and form a major unifying theory in geology that explains many other phenomena. The theory of plate tectonics is viewed as the main theme underlying all aspects of the Earth's historical geology just as evolution is the main unifying theory underlying the earth's biology. The paleo-positions of the earth's continents and magnetic poles through much of geological time have been mapped in considerable detail (Smith et al. 1994).

During the early part of the 4.5 billion years of Earth's history, the accreting masses that formed the early planet underwent a density separation. The densest materials sank to the core; the lightest materials where life evolved rose to the outer surface as the lithosphere (crust), hydrosphere (oceans), and atmosphere. The mostly granitic continents are the relatively lightest rocks forming the crust. Denser basalts form most of the ocean floor crust. In the last billion years or so, the crust became divided into a series of about 16 tectonic plates (**Fig. 25-3**). Driven by convective heat undercurrents in the Earth's mantle, these relatively rigid plates move as molten magma wells up from the mantle and new basaltic crust is added at spreading centers called midoceanic ridges (**Fig. 25-4**). Riding atop the plates, the continents are passively carried along at rates of millimeters or centimeters per year and probably have been on the move throughout Earth's history. At the edge of some tectonic plates, opposite the spreading center, one plate may override another. The overriding plate is forced upward and its edge crushed, as evidenced by prominent geological faulting; in the process, ocean bottom sediments containing marine fossils and other rocks from low-lying areas may be pushed up to great elevations as mountains. The plate sucked downward is "subducted" into the hot, dense mantle, where its materials (crustal basalts, ocean floor sediments, and water) are melted and recycled by rising up to

the crust and erupting through it as volcanoes. Landmasses have been rifted, or split apart, at spreading centers, and they have collided and united where adjacent plates converge. As crustal plates move, they buckle, bulge, and warp, causing epicontinental seas to advance onto low-lying areas and retreat from high areas. Entire ocean basins appear and disappear and alter oceanic circulation patterns, which in turn strongly affect climate. Likewise, major mountain chains of the world, formed in part by deformation of the Earth's crust and volcanism, can alter atmospheric circulation and climate.

During the last half billion years, starting in the early Paleozoic, the early continents came together to form a temporary supercontinent called Pangea that was surrounded by a single ocean, sometimes called Panthalassa. As the plates continued to move, the continents began to separate again in the Mesozoic, at first into a mainly northern portion (sometimes called Laurasia) and a southern portion (called Gondwana) that were joined in the west but separated in the east by a sea called Tethys (Fig. 25-5). Further fragmentation in the late Mesozoic resulted in modern continents that were essentially recognizable by the early Cenozoic, but that will continue to change through geological time.

To biologists, the dynamism of the Earth is of tremendous importance. Just as continents have separated, collided, or drifted progressively farther apart, so too have terrestrial biotas been isolated or brought together, entire

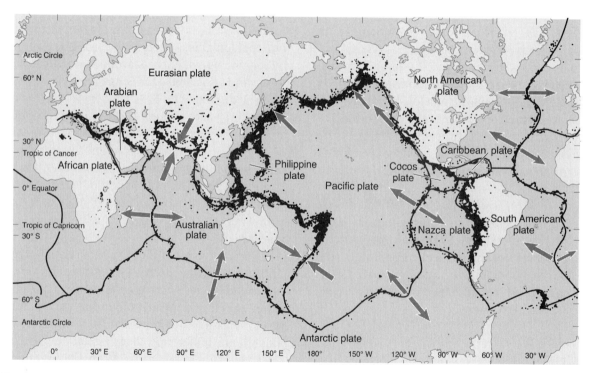

FIGURE 25-3 The present extent of the major tectonic plates and the direction of their movement. Plate movement is influenced by upwelling of molten rock from deep within the Earth along the rift lines between some plates. (Adapted from Colbert, E. H. *Wandering Lands and Animals.* Dutton, 1973.)

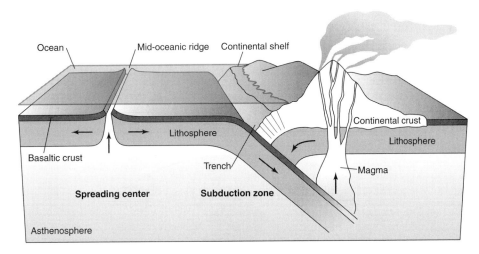

FIGURE 25-4   Simplified dynamics of continental drift as shown in a diagrammatic cross-section through tectonic plates of the earth's crust (not to scale). Rifting between two tectonic plates results from the upwelling of molten rock from the depths, to form a mid-oceanic spreading center (at left). The leftward drift of the plate originating from the right side of the image, carrying a continental block, collides with the rightward-drifting plate originating from the mid-ocean spreading center beneath the ocean. The oceanic plate is forced down (subducted) into the earth's mantle beneath the lithosphere. The subduction zone is marked by the formation of a deep trench and the uplifting and volcanism of the mountains, with resultant frequent earthquakes. Vulcanism results when the subducting plate is melted and recycled as the less dense material penetrates the lithosphere and continental crust, and erupts through the surface. (Adapted from Colbert, E. H. Wandering Lands and Animals. Dutton, 1973.)

distribution patterns of marine biotas profoundly altered, and global ecological diversity shifted. To the evolutionary biologist, the movements of the Earth's crust provide "the stage for all biological activity" (McKenna 1972). It is obvious that, when one considers the biogeography of individual species or of entire biotas, one must take into account plate tectonics and continental drift. The beauty of these concepts is that they provide explanations for a diverse array of biogeographic patterns that long appeared inexplicable. The explosion in fossil discoveries of Mesozoic mammals during the end of the 20th century and beginning of the 21st century has even provided some initial insights into the early biogeography of mammals on continents unfamiliar to many of us (Kielan-Jaworowska et al. 2004; Rich 2008).

A striking feature of mammalian evolution has been duplication of functional and, to some extent, structural types in separate groups. Examples of such convergent evolution are abundant, even in Mesozoic mammals (Luo 2007; see Fig. 3-14). Among extant mammals, numerous unrelated rodents that live in deserts of different continents convergently evolved similar anatomic, physiological, and dietary lifestyles (Mares 1975, 1980, 1993a, 1993b). Perhaps more astounding is convergence among constituents of different higher taxa. Members of several orders specialize in eating ants and termites (**Fig. 25-6**). The orders Diprotodontia, Rodentia, Lagomorpha, Artiodactyla, and Perissodactyla all contain herbivorous, cursorial mammals that pursue basically similar modes of life. Small, terrestrial, insect-eating mammals have developed in at least 11 orders (Didelphimorphia, Paucituberculata, Microbiotheria, Dasyuromorphia, Afrosoricida, Macroscelidea,

Cingulata, Pilosa, Rodentia, Erinaceomorpha, and Soricomorpha). The greatest duplication has occurred in southern landmasses, which have been longer and more completely isolated than have the Nearctic and Palearctic areas. Mammalian diversity, then, may be as much a result of the progressive Mesozoic and Cenozoic separation of the continents as of the structural and functional adaptability of the mammals themselves.

Climate and Mammalian Distribution

The preceding chapters have mentioned the geologic periods of the Mesozoic and Cenozoic in relation to the evolutionary history of various mammalian families and orders. Cenozoic patterns of climatic change had profound effects on the evolution, character, composition, and distribution of plant communities, and the evolutionary patterns and distributions of mammals, in turn, were influenced by these floral changes (e.g., Frakes et al. 1992; Janis 1993; Kröpelin et al. 2008; Woodburne et al. 2009). Although one cannot with assurance account for the myriad patterns of mammalian adaptation by recourse to climatic changes alone, there is no better point of beginning.

In the Cretaceous, climatic conditions were relatively warm, humid, and stable worldwide. There was no pronounced latitudinal gradient in temperature, and the earth was probably ice-free. Sea level was possibly the highest it has been in the last half billion years. Flora and fauna were remarkably uniform over much of the Earth, with subtropical plants and animals existing at latitudes 70° from the equator, although provincialism increased as Gondwana

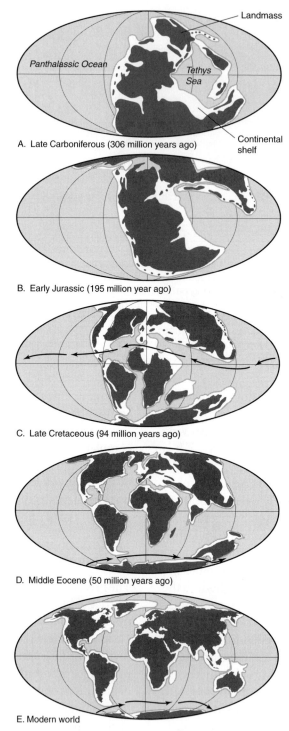

FIGURE 25-5 Paleogeographic maps of the Earth's known land masses in selected geological time periods during the evolution of synapsids, including mammals. (A) The supercontinent Pangea, home to pelycosaurs, formed from the aggregation of earlier Paleozoic continents. Note equatorial position of pre–North American highlands region, Tethys Sea in the east, and southern polar position of much of the land mass. (B) By the time morganucodontids and other early mammals existed, Tethys widened and Pangea had drifted northward. (C) Circumequatorial ocean circulation through shallow seas may have helped to stabilize and equalize global climate. Dinosaurs existed from what is now northern Alaska to Antarctica, Gondwana broke apart, angiosperms radiated, monotremes inhabited South America to Australia, and early metatherian and eutherian mammals began their fundamental diversifications. Note broad epicontinental sea dividing western and eastern portions of North America. (D) Continued separation of former components of Gondwana closed Tethys Seaway and permitted circum-Antarctic ocean circulation instead of circumequatorial. The switch was accompanied by the first Cenozoic appearance in Antarctica of ice, heralding a global general cooling trend lasting to present day. After this time, modern families of mammals rose to dominance—those in the Southern Hemisphere mostly in isolation. Mammals entered the oceanic realm for the first time (Sirenia and Cetacea). (E) The modern world, late in an ice age. Antarctica, former core of Gondwana and forested haven of diverse mammals, has become a deep freeze; amount of northern polar ice fluctuates. All landmasses are dominated by a single, recently evolved species of primate. (Adapted from Dietz, R.S. and Holden, J.C., *Sci Am.* 223 (1970): 30–41.)

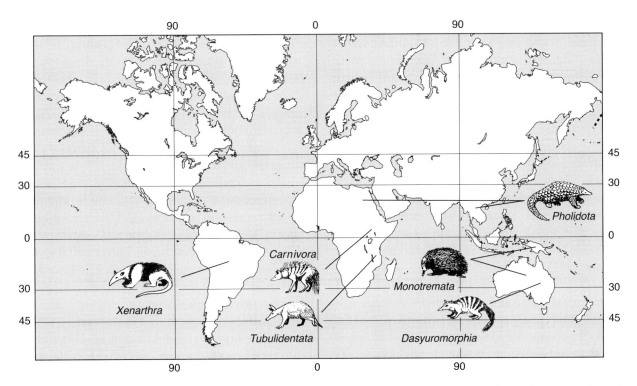

FIGURE 25-6 Members of at least six mammalian groups that occur in southern continents are adapted to eating ants and termites: Xenarthra: Pilosa (all members of the family Myrmecophagidae), Carnivora (the hyaenid *Proteles*), Tubulidentata (the aardvark *Orycteropus*), Monotremata (the echidnas *Tachyglossus* and *Zaglossus*), Dasyuromorphia (the numbat *Myrmecobius*), and Pholidota (the pangolin *Manis*).

fragmented (Hedges et al. 1996; Krause et al. 1997; Parrish 1990; Taylor & Taylor 1990). Although angiosperms (flowering plants) radiated rapidly during the late Cretaceous, they may have been restricted in their distribution to streamside areas, while ferns dominated other habitats, at least in the North American interior (Wing et al. 1993).

The uniformly subtropical climate of the Cretaceous continued through the Paleocene into the early part of the Eocene, to a peak paleoclimatologists refer to as the "Paleocene-Eocene Thermal Maximum." In parts of North America, fossil root traces and soil profiles indicate that humid forests of the Cretaceous and Paleocene gave way to dry woodlands or wooded savannas or scrublands. Loss of the forest canopy of wind-pollinated conifers was followed by paleofloras dominated by wind-pollinated angiosperms (Wolfe 1978). The early Eocene high-latitude forests and warm climates allowed forest-dwelling vertebrates to continue to dwell in arctic zones. For example, an early Eocene vertebrate fauna from Ellesmere and Axel Heiberg Islands (the northernmost part of North America; paleolatitude about 75° N) included catfish, bowfin, several kinds of turtles, crocodilians, and several kinds of mammals, including primitive primates (McKenna 1980; Dawson et al. 1993).

By the middle Eocene (about 40 to 37 million years ago), however, a major rise in sea level, major changes in ocean circulation resulting in part from the breakup of Gondwana and movements of other tectonic plates, and

significant climatic deterioration began. Uplift of the Himalaya Mountains and Tibetan Plateau as the Indian subcontinent collided with and continued to push into Asia strongly affected regional and probably global climate at the Eocene-Oligocene transition (Dupont-Nivet et al. 2007). Year-round ice likely began to accumulate in Antarctica by this time. The northern continents also underwent a cooling and drying after the middle Eocene, resulting in warm subtropical forests changing to dry woodland by the latest Eocene–early Oligocene. A diverse array of warmth-adapted organisms, both marine and terrestrial, underwent concomitant extinction in the late Eocene (Prothero 1994a, 1994b). The development of Antarctic ice and changes in ocean circulation during the middle Eocene to early Oligocene brought about increased seasonality and an overall cooling and drying to the global paleoclimate. Grass pollen and grass fossils are absent from the Eocene record of North America during the early development of savannas, suggesting that Eocene-Oligocene drylands were dominated by low-growing woody scrub vegetation lacking grasses ("rangelands"), rather similar to modern, seasonally dry *Ephedra*–saltbush communities of the Great Basin or bluebush–saltbush communities of central Australia (Wolfe 1994). In Europe, data from paleobotany and fossil vertebrates show no obvious Eocene drying trend, but such a trend is present by the Oligocene there. A drying trend is recorded in the middle Eocene fossil pollen record in Asia, where the Gobi Desert and parts of China were arid or sub-arid. By contrast, in southern South America climate seems to have been stable during the Eocene-Oligocene climate transition (Kohn et al. 2004). No detailed record of Eocene paleoclimate has yet been developed in the rest of South America, Africa, or Antarctica, but Oligocene data indicate that the drying trend had occurred by then. For example, in South America, the beginning of Andean uplift changed vegetation from subtropical woodlands to scrubby, arid savanna woodlands; mammals that were adapted for grazing rather than browsing appeared earlier in South America than in other continents (Pascual & Ortiz Jaureguizar 1990; MacFadden et al. 1996). In Australia, arid-adapted plants greatly increased in importance in the Oligocene (Barker & Greenslade 1982).

As noted earlier, until at least the late Eocene, a land connection or near connection still existed between South America and Antarctica, as evidenced by fossil mammals. But by the Oligocene, the Antarctic circumpolar ocean circulation had isolated Antarctica and led to its more extensive glaciation and an attendant drop in global sea level (Zachos et al. 1992). However, a cool-temperate flora of low species diversity still grew on some parts of Antarctica throughout the Oligocene. Oligocene plant assemblages in Australia also reflect cool-temperate conditions, including plants indicative of greater seasonality; parts of the continent underwent increasing drying that gave an open aspect to the forests. A similar development of semiarid habitats probably continued or occurred in South America too, as indicated there by a fossil record of mammals adapted to grazing on coarse vegetation (Pascual & Ortiz Jaureguizar 1990; Prothero & Berggren 1992).

In the Northern Hemisphere, the Arctic region remained free of ice, but the global cooling trend caused plant zones to shift such that the warm-adapted, mesic-adapted ones narrowed and moved to lower latitudes and were replaced at high latitudes by cool-temperate forests. Increasing aridity in the rain shadow of the Rocky Mountains in central North America resulted in changes from the subtropical forests of the Eocene to dry woodland, wooded shrubland with gallery forest, and even open scrubland, or possibly grassland with gallery woodland, by the middle Oligocene. In Asia, Eocene forests were replaced with Oligocene woody savanna in Kazakhstan; in China, the deserts of the northwest were replaced by woody savanna.

Savanna woodlands continued to dominate North American floras in the Miocene, at least where fossil floras exist in the continental interior, and grasses became more diverse and widespread. The origins of such desert-adapted animals as kangaroo rats (Heteromyidae) can be traced to the Miocene, when semideserts became widespread. After about 5 million to 7 million years ago, grassland-steppe floras became widespread in North America, and grazing hoofed mammals reached their peak in North America and on most other continents at this time (this occurred in the Oligocene in South America). In sections of the Andean cordillera, periods of quiescence were punctuated by periods of rapid uplift in the Miocene and Pliocene; the mountain building affected atmospheric circulation and various aspects of the vegetation and mammal distribution in portions of South America.

Global cooling continued through the Pliocene; glacial ice covered most or all of Antarctica by the middle Pliocene. About 3.5 million years ago, extensive glaciation appeared also in South America and possibly northern Asia and Alaska, although plant fossils indicate that mixed boreal forests still grew in parts of Siberia, northern Alaska, and arctic Canada. For the first time in the Cenozoic, ice began to form in the Arctic during the early to middle Pliocene. This north polar ice produced the Labrador Current, which forced the Gulf Stream southward, altering the climate of northern Europe. By 3 million years ago, glaciers covered parts of Greenland and Iceland. The tectonic closure of the Isthmus of Panama may have affected Atlantic Ocean circulation in such a way as to produce sufficient atmospheric moisture to build the large volumes of ice in Greenland and Iceland, burying the native vegetation. Vegetation in the Arctic included taiga by about 5 million years ago and tundra by 2 million years ago. North American desert scrub communities did not develop until the Pleistocene. In Africa, the northern half of the continent had been relatively mesic and supported lush vegetation in the early Holocene. As late as the middle Holocene, about 6,000 years ago, the region began a drying trend, developing the Sahara Desert by 2,700 years ago (Kröpelin et al. 2008).

Major climate changes can directly affect sea level (as do tectonic events), and sea level can affect the availability of low-lying areas and continental shelves as avenues for intercontinental dispersals. Using data from climatology, sea-level fluctuations, geochronology, and the mammalian fossil record, Woodburne and Swisher (1995) charted overland dispersals of mammals in the Cenozoic between North America and other continents as follows (numbered events correspond to those in **Fig. 25-7**):

1. First immigration of rodents to North America (also several other extinct groups), presumably from Asia and/or Europe via Beringia or the North Atlantic via Greenland.
2. Major dispersal between North America and western Europe via Greenland and Spitsbergen. Last great dispersal event between these continents via the North Atlantic corridor route. This dispersal route was subsequently lost as an ocean barrier was created by the tectonic widening of the North Atlantic Ocean. Generic similarity between North

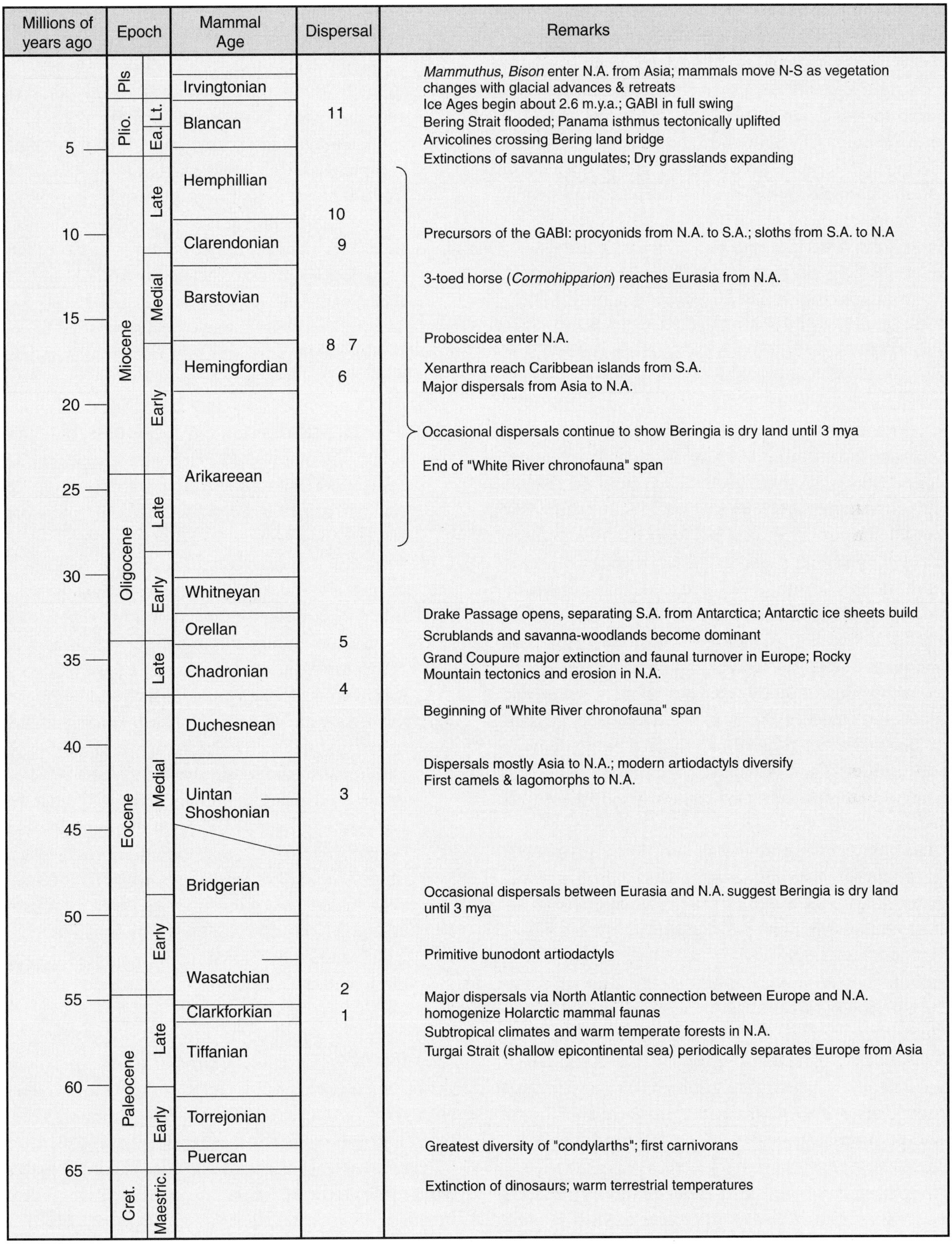

Millions of years ago	Epoch		Mammal Age	Dispersal	Remarks
	Pls		Irvingtonian		*Mammuthus, Bison* enter N.A. from Asia; mammals move N-S as vegetation changes with glacial advances & retreats
5	Plio.	Ea. Lt.	Blancan	11	Ice Ages begin about 2.6 m.y.a.; GABI in full swing Bering Strait flooded; Panama isthmus tectonically uplifted Arvicolines crossing Bering land bridge Extinctions of savanna ungulates; Dry grasslands expanding
10	Miocene	Late	Hemphillian	10	
			Clarendonian	9	Precursors of the GABI: procyonids from N.A. to S.A.; sloths from S.A. to N.A
15		Medial	Barstovian		3-toed horse (*Cormohipparion*) reaches Eurasia from N.A.
		Early	Hemingfordian	8 7 6	Proboscidea enter N.A. Xenarthra reach Caribbean islands from S.A. Major dispersals from Asia to N.A.
20					Occasional dispersals continue to show Beringia is dry land until 3 mya
25	Oligocene	Late	Arikareean		End of "White River chronofauna" span
30		Early	Whitneyan		
			Orellan	5	Drake Passage opens, separating S.A. from Antarctica; Antarctic ice sheets build Scrublands and savanna-woodlands become dominant
35	Eocene	Late	Chadronian	4	Grand Coupure major extinction and faunal turnover in Europe; Rocky Mountain tectonics and erosion in N.A.
40		Medial	Duchesnean		Beginning of "White River chronofauna" span Dispersals mostly Asia to N.A.; modern artiodactyls diversify First camels & lagomorphs to N.A.
45			Uintan Shoshonian	3	
		Early	Bridgerian		Occasional dispersals between Eurasia and N.A. suggest Beringia is dry land until 3 mya
50					Primitive bunodont artiodactyls
55	Paleocene	Late	Wasatchian	2	Major dispersals via North Atlantic connection between Europe and N.A. homogenize Holarctic mammal faunas
			Clarkforkian	1	Subtropical climates and warm temperate forests in N.A.
			Tiffanian		Turgai Strait (shallow epicontinental sea) periodically separates Europe from Asia
60		Early	Torrejonian		
65			Puercan		Greatest diversity of "condylarths"; first carnivorans
	Cret.	Maestric.			Extinction of dinosaurs; warm terrestrial temperatures

FIGURE 25-7 Summary of North American land mammal ages and major intercontinental dispersals of mammals during the Cenozoic Era. See text for explanation of boldface numbered events 1 to 11. Numerous minor dispersal events are not shown. NA = North America; SA = South America. (Adapted from Woodburne, M. O. and Swisher, C. C., III. *Geochronology Time Scales and Global Stratigraphic Correlation.* Society for Sedimentary Geology, 1995.)

America and Europe was greater at this time (late Paleocene) than at any other time in the Cenozoic. Early perissodactyls (*Hyracotherium*), artiodactyls (*Diacodexis*), adapid primates, creodonts, rodents, and many other mammals of primitive aspect were involved.

3. Incursion of Asian mammals of modern aspect via trans-Beringian filter and corridor, especially perissodactyls (rhinocerotoids), but also early camels, rabbits, miacid carnivorans, others.

4. First North American appearance of certain early saber-tooth cats, canids, rabbits, tragulid-like deer, sciurids, tayassuids. Probably via Beringian filter route.

5. The Grand Coupure ("great cut"), a major mammalian faunal turnover event in Europe in which endemic and archaic taxa were abruptly replaced in the fossil record by a host of new forms. Sixty percent of indigenous European mammal genera went extinct, including certain primates, rodents, artiodactyls, perissodactyls, creodonts, condylarths, and others. With the retreat of the shallow Turgai Strait that previously separated Europe and Asia, most of the new arrivals entered Europe from Asia; fewer arrived from North America. The replacements consisted of rhinoceroses, titanotheres, hedgehogs, heterosoricids, muroids, aplodontids, new sciurids, dipodids, castorids, mustelids, viverrids, lagomorphs, ursids, nimravids, felids, and many others. The Grand Coupure is associated with the major climatic cooling (and beginning of continental glaciers in Antarctica) during the late Eocene and early Oligocene. It is also associated with the tectonic uplift of the Alps as the African Plate contacted southwestern Eurasia, bringing with it Gondwanan mammals and other forms of life.

6. Major influx of Palearctic taxa into North America via Beringian corridor. Immigrants included the first North American soricine and other kinds of shrews, pikas, bear-dogs (Amphicyonidae), bears, mustelids, procyonids, a rhinoceros, and others.

7. Another major immigration of Palearctic taxa via Beringian corridor. At this time, North America received its first true felid (*Pseudaelurus*), first petauristine squirrels, an eomyid rodent, and new mustelids, including a lutrine. In an unrelated event, pilosans (megalonychid sloths) dispersed to the Greater Antilles from South America, certainly by the middle Miocene, possibly earlier (MacPhee & Iturralde-Vinent 1994).

8. A lesser but important immigration from Asia. Proboscideans crossed Beringia into North America, but apparently they took another million years to reach the North American interior. At about this time, the first muroid rodent *Copemys* also entered North America from Asia.

9. One of the few known Neogene dispersals from Nearctic to Palearctic, hipparionine horses left via Beringia and thereafter radiated widely in Eurasia, where horses had not previously existed.

10. Limited interchanges with Eurasia and South America over a few million years. North America's first arvicoline rodents arrived via Beringian filter. About 9 million years ago, before a land connection with South America was completely formed, two genera of large megalonychid sloths entered North America (possibly a sweepstakes dispersal), thus beginning a series of intermittent precursory exchanges between the two continents. (Note that sloths from South America reached the Antilles much earlier than North America.) Around 5.8 million to 7.5 million years ago, procyonids strayed across the strait from North America to South America. About 6 million years ago or later, muroid rodents entered South America. These dispersals may have taken place during brief low-stands of sea level (the Messinian low); in any case, they indicate a filter route. The isthmus between South America and Central America was not yet dry land but was developing. The area was (and is) tectonically active and could have provided an island-hopping route.

11. The Great American Biotic Interchange began by at least 2.7 million years ago. The Isthmus of Panama was established as a dry-land connection (filter route), as shown by hordes of immigrants moving from South America to North America and vice versa. North America regained metatherians (Didelphidae), which had become extinct there in the middle Miocene. North America also received its first hystricognath rodents (capybaras and porcupines), additional xenarthrans (glyptodonts, armadillos, more sloths), and toxodont notoungulates (although the huge toxodonts apparently did not move northward beyond central Mexico). Pliocene exchanges via Beringia continued with the bog lemming *Synaptomys* and spectacled bear *Tremarctos* arriving from Asia, and the horse *Equus* exiting to Eurasia. Many other North American exchanges with Eurasia and South America continued intermittently through the Pleistocene.

Beginning about 2.5 million years ago, the late Pliocene and Pleistocene was a time of pronounced climatic shifts, when periods of lowered temperatures alternated with periods of relative warmth. Paleoclimatologists using oxygen isotope ratios and other data from ocean-bottom cores chart at least 11 major episodes of cool climates (4 of which seem to have been accompanied in North America by major continental glacial advances) separated by warm intervals. Accompanying the periods of cooling, which were apparently worldwide, were a number of spectacular environmental changes. Precipitation increased everywhere, and with increased snowfall, continental glaciers developed and pushed southward.

At one time in the Pleistocene, more than 25% of the land surface was covered with glaciers: Eurasia had 3.2 million square miles of ice; the Nearctic ice sheet covered 4.5 million square miles and, during its greatest push southward, reached what is now Kansas. The weight of so much ice actually depressed the covered portions of the continents slightly. Glaciers on Mount Kenya in Africa extended about 1,700 meters below the present vestigial snowfields (at 4500 meters), and New Guinea and Madagascar had montane glaciers. The distributions of floras were changed. Boreal vegetational zones were driven downward on mountainsides, and coniferous forests spread southward over areas that previously supported less boreal floras. Concurrently, tropical floras receded toward the equator, and deserts became far more restricted than they are today. In Africa, the retreat of tropical forests stranded at least one shrew, *Congosorex*, and other organisms in a refugium of ancient forest on the Eastern Arc Mountains of Tanzania, far from the forests of the Guineo-Congolian region (Stanley et al. 2005). In southwestern North America, juniper–piñon woodlands and sagebrush filled many intermountain basins that are occupied today by desert scrub (Betancourt et al. 1990).

The Pleistocene ended 10,000 years ago with the extinction in the Nearctic and Palearctic regions of such common Pleistocene mammals as mammoths, camels, woodland musk ox, shrub ox, ground sloths, horses, saber-toothed cats, and the giant beaver (**Figs. 25-8** and **25-9**; Kurtén & Anderson 1980; Martin & Klein 1984; Webb & Barnosky 1989). Following the retreat of the last continental glaciers at about 10,000 years ago, the present interglacial period, called the Holocene, has been unusually stable compared to the wild fluctuations of at least the last 100,000 years of the Pleistocene. Rapid warming began again in the last century, and is predictably and immediately affecting ecosystems and mammal faunas. For a few examples of the present climate change on mammals, see Adams and Hayes (2008), Moritz et al. (2008), Welbergen et al. (2007), and Williams et al. (2003).

Numerous Pleistocene episodes of faunal interchange between Siberia and North America occurred across the periodically emergent Beringia and had a profound effect on the North American mammalian fauna. At maximal extent, Beringia was a nearly 2,000-kilometer-wide land bridge connecting Eurasia and North America (**Fig. 25-10**).

FIGURE 25-8   Life restorations of a few representative large mammals of the late Pleistocene of North America that became extinct at the end of the Pleistocene. From left to right, the diminutive pronghorn *Capromeryx* (a North American endemic), a horse, *Equus* (from Eurasian immigrant stock), the Columbian mammoth *Mammuthus columbi* (from Eurasian stock), a giant llama, *Camelops* (from Eurasian stock), and a ground sloth (from Neotropical stock).

FIGURE 25-9 A saber-toothed cat, *Smilodon fatalis*, defends its freshly killed horse from dire wolves (*Canis dirus*) in coastal western North America in the late Pleistocene, as interpreted from the extremely rich fossil deposit at Rancho La Brea, California, now occupied by the city of Los Angeles. A large collection of fossil vertebrate, invertebrate, and plant remains recovered from asphalt deposits (tar pits) is preserved in the Los Angeles County Museum and the Page Museum of La Brea Discoveries at the site.

It lasted until some 15,500 years ago, when it was inundated by the rising sea level. The sea off the north (Arctic) coast was frozen most of the year, while the sea off the south (Pacific) coast was also extremely cold. Fossils of marine diatoms from this coast are species typical today of shallow seas that are frozen half of the year (Baldauf 1982). Bitter cold winds from the Cordilleran ice sheet swept westward along the south coast, making the area unsuitable for marine mammals that bear young on land or on ice floes (Pielou 1991). Also, the cold waters precluded the occurrence of the extensive near-shore kelp (giant marine algae) beds typical of the area today. The ranges of the extinct Steller's sea cow (*Hydrodamalis gigas*) and the living sea otter (*Enhydra lutris*), species dependent on kelp, may not then have included the south coast of Beringia.

During at least part of this time, Alaska and the Yukon Territory were cut off from the rest of North America by ice (Fig. 25-10) and were broadly connected by the Bering Isthmus with Eurasia. Biogeographically, Alaska–Yukon was then part of Beringia, faunally more closely associated with Asia than with North America. We have good evidence of what the habitat of the Bering land bridge was actually like only during at least the last glacial maximum 20,500 to 14,500 years ago. Beringia was largely occupied by a "mammoth steppe," a cold, arid, short grassland with areas of bare ground (Cwynar & Ritchie 1980; Guthrie 1990). Some plants, such as blue grama grass (*Bouteloua gracilis*), lived in Beringia several thousand kilometers north of their present range, and fossils of badgers (*Taxidea taxus*) are known from Dawson in the Yukon, far north of their present range (Guthrie 1982). The climate was drier than it is today, and relentless winds buffeted the steppes, as evidenced today by large areas covered with loess (a dust of fine, wind-blown particles of clay, sand, or silt) and widespread sand dunes in Alaska and the Yukon. The Beringian climate at times was apparently cold temperate and permitted the survival of

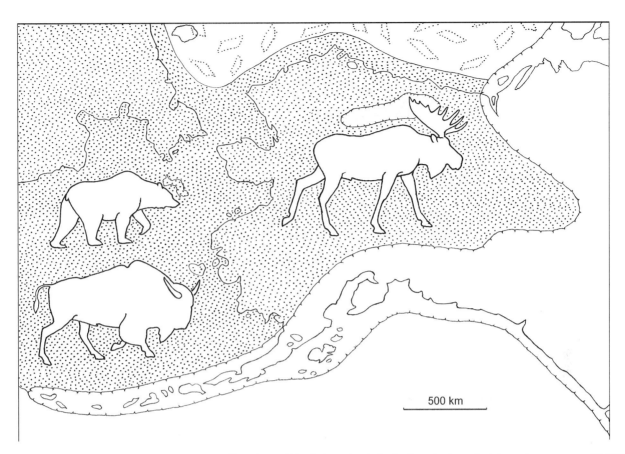

FIGURE 25-10 Beringia, the land corridor between Eurasia and North America (stippled), at the late glacial maximum some 20,000 years ago. Solid lines trace the present-day coastlines of Siberia (left), Alaska (right), and the Aleutian Islands (below). The approximate border of the merged Laurentide and Cordilleran continental glaciers is outlined; the ice covered southern Alaska, much of Canada, and the Aleutian Islands. The sea north of Beringia (the present-day Chukchi Sea and Arctic Ocean) was frozen much of the year; the sea south of Beringia (Pacific Ocean) was probably frozen only in winter. The grizzly bear (*Ursus arctos*), a now-extinct bison (*Bison priscus*), and the moose (*Alces alces*) immigrated to North America across Beringia but were barred from moving southward until ice-free routes developed between the eastern (Laurentide) and western (Cordilleran) ice sheets and along the Pacific Coast of North America.

animals not adapted to arctic conditions; at other times, it was arctic and barred their passage. The mammoth steppe was also inhabited by abundant large mammals, especially bison, mammoths, horses, bonneted musk oxen, saiga antelopes, and many others (Guthrie 1990). Mammals tolerant of arid short grass habitat and cold conditions flourished in the mammoth steppe. Arvicoline rodents have repeatedly used this route, but fewer other mammals have, indicating the importance of this bridge and its function as a filter (Bell 1998). DNA evidence suggests that humans entered the Americas via Beringia in the late Pleistocene, perhaps 20,000 to 30,000 years ago, followed by migration southward along a deglaciated western coastline after 16,500 years ago (Goebel et al. 2008).

Some species of small mammals that live today on islands or occupy small coastal areas are remnants of widespread ancestral Pleistocene populations. The Pribilof Island shrew (*Sorex pribilofensis*) and the St. Lawrence Island shrew (*S. jacksoni*) are island remnants. Each occupies an island in the Bering Sea, and each is closely related to the

cinereus shrew (*S. cinereus*). This shrew occurs today over most of Canada and Alaska and probably lived in Beringia and part of Siberia in the late Pleistocene. When the rising sea level created what is now the Bering Sea, only the Aleutian Islands and a few others were not inundated. Perhaps most of these islands originally had shrews, but they only survived on a few islands. Under complete isolation from mainland *S. cinereus*, the island populations have differentiated and are recognized as distinct species. A coastal remnant is the rock-dwelling Gaspé shrew (*S. gaspensis*), which occupies Cape Breton Island and two other small areas in the Gaspé Peninsula vicinity of eastern Canada. During the peak of the last glacial advance, these areas were probably part of a small, ice-free refugium surrounded by the Laurentide ice sheet. This covered most of Canada's east coast at this time, and glacial ice may have isolated shrews there from the rock-dwelling parent species (the long-tailed shrew *S. dispar*) to the south.

During glacial advances, the ranges of some boreal (northern) mammals extended well south of present limits.

Remains of the musk ox (*Ovibos*), arctic shrew (*Sorex arcticus*), collared lemming (*Dicrostonyx*), and many other species have been found well south of their present northern ranges. Abundant evidence verifies the occurrence of northern assemblages of mammals during the Pleistocene as far south as Kansas and Oklahoma. There were reciprocal northward movements of subtropical or desert mammals during interglacial times, as indicated by the fossil occurrence of such animals as the hog-nosed skunk (*Conepatus*) and jaguar (*Panthera onca*) far north of their present ranges. A fossil record of the jaguar, for example, is from Tennessee, hundreds of kilometers north of the animal's present range. The lion (*Panthera leo atrox*) made its way from the Old World into North America and entered South America via the Panamanian Isthmus, temporarily achieving the title of the world's most widespread mammal species. The environmental fluctuations of the Pleistocene often brought together species that are found today in very different habitats. Like plants, mammal species responded individualistically to environmental change (Graham et al. 1996).

One of the most common and obvious patterns of mammalian distribution—the occurrence of isolated or semi-isolated populations of northern mammals on mountain ranges at fairly low latitudes—is partly the result of Pleistocene southward migrations of boreal faunas. During glacial advances, assemblages of boreal mammals were widespread in lowlands well south of their present ranges. Concurrent with the movements of these mammals northward during the retreat of cool climates were movements of boreal mammals into montane areas. Here, because of the effect of elevation on climate, cool refuges were available. Many of these montane populations have persisted in "boreal islands" far south of the northern stronghold of their closest relatives, and the diversity of mammals on some mountain ranges in the southwestern United States has resulted from a combination of Pleistocene dispersals, vicariant events, differentiations, and extinctions. The North American Southwest has been a favorite area for study among mammalian zoogeographers (Findley 1996; Harris 1990; Patterson 1995; Rickart 2001). Similarly, elevational gradients of diversity for montane mammals are also an active area of research, especially in light of rapidly changing climate (e.g., Colwell et al. 2008; Heaney et al. 2001 and articles therein). A widely publicized mammalian situation involves the pika of western North American mountains: As alpine glaciers and snowfields retreat and subalpine habitats respond by moving upslope, the pikas' habitat disappears. The pikas cannot move any farther upslope and cannot migrate long distances through unsuitable habitats to other mountains. Their future is uncertain.

Humans have obviously reduced the ranges of many mammals and possibly contributed to the Quaternary extinction of others, at least in North America. Climate change probably also contributed to the megafaunal (large mammal) extinctions in South America, North America, and Australia; the culprits are difficult or impossible to sort out. The Quaternary extinctions of large mammals have wide-ranging ecological and evolutionary consequences for surviving plant communities and habitats (Anhuf et al. 2006; Guimarães et al. 2008; Johnson 2009). The size of the largest mammals or other vertebrates in a particular ecosystem is relative, and their impact on their ecosystems equally significant aside from their absolute body size (Hansen & Galetti 2009). Many species are changing their ranges today in response to the warming climatic cycle (Parmesan 2006) but also in response to human alteration of the environment. One potential consequence involves vampire bats, *Desmodus*. In the Pleistocene, vampires were widely distributed in North America much farther north than at present. Presently, common vampire bats are restricted to tropical climates by their intolerance of cool winter temperatures. But under a warming trend in the climate, and given the widespread availability of livestock as prey in northern Mexico and the United States, might they once again invade these areas?

South American Mammals and the Great American Biotic Interchange

The origins of the Neotropical mammalian fauna have long held the interest of distinguished scientists. Wallace (1876) was first to recognize the faunal interchanges that occurred between North and South America late in the Neogene, and intensive paleontological fieldwork in the late 19th century provided a more complete understanding of these events. In 1893, von Zittel wrote that "there was thus accomplished, toward the end of the Pliocene, one of the most remarkable migrations of faunas that geology has been able to record." The classic works of Simpson (especially 1980) did much to clarify our understanding of South American historical zoogeography, and continuing fieldwork and advances in our geologic knowledge have refined previously held views (Flynn et al. 2005b; Lindsey et al. 2009; Morgan 2008).

As discussed earlier in this chapter, North America experienced repeated immigrations from Europe and Asia in the Cenozoic (see Fig. 25-7). In strong contrast, South America was isolated from all other continents throughout most of the Cenozoic. An early connection of South America with North America was lost in the late Paleocene, and the last tenuous connection (possibly via an archipelago) with Antarctica was lost in the late Eocene. No reconnection with North America came about until the late Pliocene, by at least 2.7 million years ago, when the Isthmus of Panama was established. For perhaps 35 million years, the South

American fauna evolved in isolation (with one influx of mammals—primates and hystricognath rodents—in the Oligocene). The late Pliocene emergence of the Panamanian land bridge provided a gateway for an intermingling of North and South American faunas. This classic natural experiment has been called the Great American Biotic Interchange (or GABI).

Before the Great American Biotic Interchange, most of the South American orders, families, and genera of mammals were autochthonous and endemic (Simpson 1980). Metatherians arrived in the late Cretaceous to early Paleocene, before the eutherians (except for notoungulates), and began their fundamental radiation into major lineages (Flynn 2009; Woodburne & Case 1996). The Cretaceous-Cenozoic marsupial radiation in South America—including metatherian carnivores (Sparassodonta) such as the saber-toothed "possums" and doglike borhyaenids, and many others—has been discussed previously (Chapter 6), but the extremely impressive ungulate radiation also deserves mention. Condylarths, ancestors of many ungulates, were the only eutherians to reach South America early in the Paleogene, probably by the middle Paleocene. They radiated rapidly, and by the end of the Paleocene, a diverse series of evolutionary lines was established. In isolation from North American ungulate stocks, the South American ungulates went their unique evolutionary ways. Although they clearly filled many of the same niches occupied by other lineages of ungulates in other parts of the world, many of the South American ungulates were anomalous-looking beasts unlike any ungulates elsewhere.

These South American ungulates spanned a considerable size range. There were rat-sized little ones and tusk-bearing giants (order Astrapotheria) approaching the size of an elephant. Especially successful was the order Notoungulata, which included various herbivorous genera, one of the largest of which was *Toxodon*, a stubby-legged, rhinoceros-like beast some 3 meters in length (**Fig. 25-11**). Another group of notoungulates, the hegetotheres, included a number of small cursorial rabbit-like types. One advanced Miocene genus of the order Litopterna (*Thoatherium*) had one-toed feet that not only were much more specialized than those of the contemporary Miocene North American horses but were even more specialized than the feet of present-day horses (**Figs. 25-12** and **25-13**). Another litoptern had a camel-like body with heavy legs and with the snout lengthened into a proboscis (Fig. 25-11).

The distinctive South American ungulates reached their peak of diversity and numbers in the Oligocene and Miocene, but they declined in the Pliocene, and by the end of the Pleistocene only fossils remained. The decline of the South American ungulates was not because of the invasion of South America by Nearctic ungulates and carnivores.

FIGURE 25-11   A life restoration of the South American notoungulate *Toxodon* (Toxodontidae). *Toxodon* was discovered by Charles Darwin on the Argentine pampas during his service as naturalist on the ship H.M.S. "Beagle."

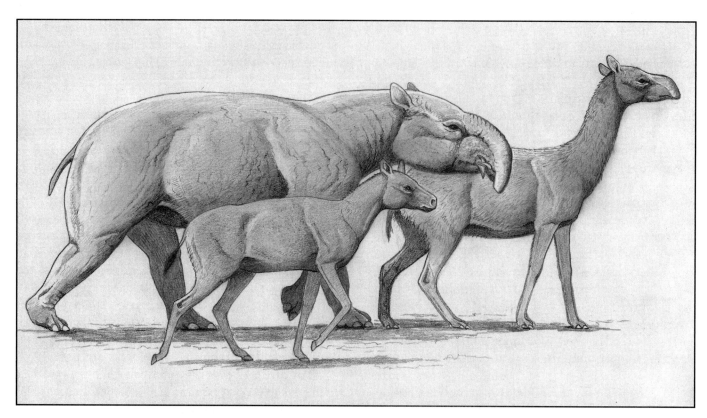

FIGURE 25-12 Life restorations of several Miocene South American ungulates, from left to right: a tusked, trunk-bearing astrapothere *Astrapotherium* (Astrapotheriidae), a highly cursorial litoptern *Thoatherium* (Proterotheriidae), and a litoptern with a small proboscis *Theosodon* (Macraucheniidae).

Major faunal shifts occurred before the emergence of the Panamanian land bridge. By this time, more than half of the mammals occupying the adaptive zone of the large herbivores were not ungulates: Of the 10 families of large Pliocene herbivores, 4 were xenarthrans and 2 were gigantic hystricognath rodents (Vucetich et al. 1999).

The tectonic evolution of Central America and the Caribbean region leading up to the Great American Biotic

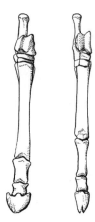

FIGURE 25-13 The hind foot skeleton of a modern horse, *Equus*, left and a Miocene South American litoptern, *Thoatherium*, right. Note that the vestiges of digits 2 and 4 are more strongly reduced in the litoptern than in the horse.

Interchange is complex (Pitman et al. 1993; Marshall & Sempere 1993). After the Cretaceous–Paleocene dry-land connection between South America and North America was lost, island arcs were probably intermittently available between South America–Antilles–Yucatan Peninsula (and possibly Florida Peninsula) from the Eocene to the present (Iturralde-Vinent & MacPhee 1999). The islands were pushed up by the eastward-moving Caribbean tectonic plate. By the early Miocene, most of Central America came into existence as the Chortis tectonic microplate and Middle American Arc combined and moved into place from the west (Pitman et al. 1993). By the late Miocene, a short island arc along Panama was emerging as the final link between South America and North America (Coates 2003; **Fig. 25-14**). It was at this time that the forerunners of the interchange island-hopped between the two continents. By the late Pliocene, the Isthmus of Panama was established and hordes of mammals, other animals, and plants crossed (**Fig. 25-15**).

Not all mammalian taxa participated in the Great American Biotic Interchange, although our knowledge of the interchange may be incomplete because of a very poor fossil record in southern Mexico and Central America. As the fossil record of Mexico, Central America, and northern South America improves, ideas about the interchange are being modified (e.g., Carlini et al. 2008a,b; Flynn et al.

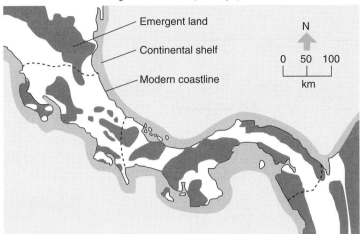

A. Central America during Late Miocene (6-7 mya)

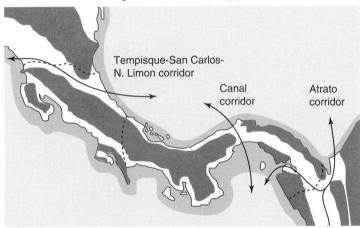

B. Central America during Middle Pliocene (3 mya)

FIGURE 25-14   Paleogeographic map of islands of the developing Central American Isthmus during (A) the late Miocene about 6 to 7 million years ago, when the first forerunners of the Great American Biotic Interchange crossed and (B) the middle Pliocene about 3 million years ago, immediately before the final emergence of the isthmus that allowed overland dispersal in both directions. Emergent land is represented by dark shading; submerged continental shelf areas are indicated by lighter shading. To the northwest and southeast of this region, dry land was already continuous. The modern coastline (solid line) and international boundaries between Nicaragua, Costa Rica, Panama, and Colombia (dashed lines) are shown for reference. Arrows in B show the last remaining corridors for the dispersal of marine organisms between the Caribbean Sea and Pacific Ocean. (Adapted from Jackson, J. B. C., Budd, A. F., and Coates, A. G. *Evolution and Environment in Tropical America.* University Chicago Press, 1996.)

2005b; Lindsey et al. 2009; Rincón & White 2007). As far as is known, none of the endemic South American carnivorous metatherians crossed the land bridge. In fact, the only South American terrestrial predator to reach North America was *Titanis*, a 3-meter-tall, flightless phorusrhacid bird, in the early Pliocene before the land bridge was formed (MacFadden et al. 2007). This huge raptorial bird is known from fossils found along the Gulf of Mexico coast in what are now Texas and Florida. Ceboid monkeys do not seem to have entered Central America from South America until some time after the main interchange had slowed in the middle Pleistocene.

The mammalian participants in the interchange can be divided into two groups on the basis of the means and timing of dispersal. The first group was made up of "waif immigrants" that dispersed along the island arc in the late Miocene and early Pliocene. This group includes the Megalonychidae and Mylodontidae (extinct ground sloths), which dispersed from South America to North America, and the procyonids and cricetids, which moved into South America from North America. The second group includes mammals that dispersed across the Panamanian land bridge at various times after its emergence. A diversity of mammals belongs to this group, including North American taxa that immigrated to South America and South American taxa that entered North America.

This spectacular reciprocal interchange of land mammals (discounting bats and manatees) was roughly

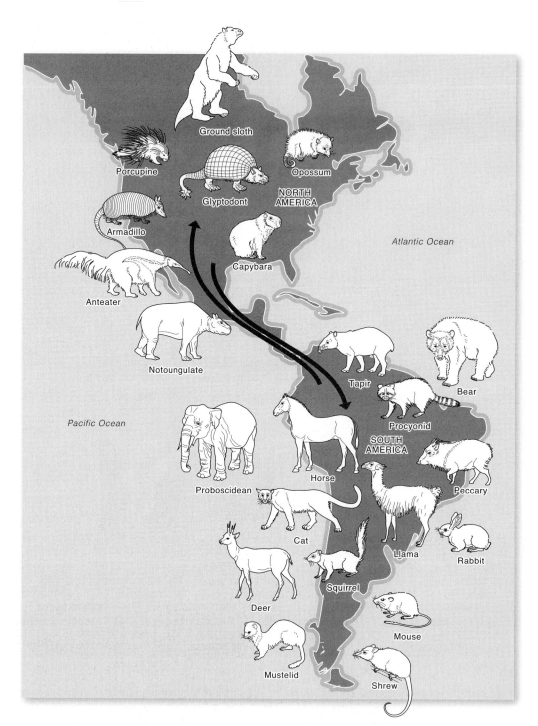

FIGURE 25-15 Representatives of the groups of mammals that were involved in the Great American Biotic Interchange. The interchange took place during the last 9 million years, beginning with a few early dispersals and reaching a climax after the Isthmus of Panama emerged as dry land in the Pliocene about 3 million years ago. Crossing the filter at different times into North America and Central America from South America were xenarthrans such as various ground sloths, glyptodonts, pampatheres (armadillo-like glyptodontoids), armadillos, and anteaters; rodents such as capybaras and porcupines; a notoungulate toxodont; and didelphid opossums. Reaching South America from North America at various times were carnivorans such as procyonids, mustelids, tremarctine bears, cats and saber-tooth cats; rodents such as cricetids and sciurids; proboscideans (gomphotheres); artiodactyls such as deer, camelids, and peccaries; perissodactyls such as tapirs and horses; rabbits; and shrews. (Adapted from Pough, F. H., Janis, C. M., and Heiser, J. B. *Vertebrate Life*. MacMillan Publishers, 1998 and Marshall, L. G., *Am. Sci.*, 76 (1988): 380–388.)

symmetrical at first but decidedly unbalanced later. On both sides, the number of families and genera increased and then declined as a result of extinctions. South American taxa generally diversified little in North America; after a million years, the impact of their intrusion had faded. In contrast, North American taxa radiated explosively in South America (**Fig. 25-16**) throughout the Pleistocene.

Reasons for the imbalance have been debated for decades. An early explanation was that the North American fauna, longer subjected to repeated immigrations, were better competitors and predators. More recently, speculative ecogeographic explanations for the imbalance have predominated. Late Cenozoic mountain-building activity elevated the northern Andes to 4,000 meters, affecting atmospheric circulation and climate in the isthmian region. Glacial phases, too, influenced the tide of dispersal. Webb and Rancy (1996) suggested a three-phase model for the Great American Biotic Interchange after the emergence of the isthmus. The first phase in the late Pliocene coincided with the onset of glacial conditions, during which the greater aridity would have caused forested habitats to shrink and allowed savanna habitats to expand into tropical latitudes (**Fig. 25-17**), establishing a corridor for savanna-adapted mammals along the slopes of the Andes and far into temperate regions on both continents. At such a time, movement was primarily from a larger North American source area southward into high latitudes in South America. Fossil faunas from the late Pliocene in Florida and Argentina show remarkable similarities with one another and are dominated by savanna-adapted mammalian taxa that suggest maximum continuity of savanna habitats between these regions. In the middle Pleistocene, a second phase occurred in which humid interglacial conditions caused closed-canopy rain forests to dominate in tropical

America, thus providing a route for forest-adapted organisms to move primarily from the large reservoir of Amazonia northward. In a third phase in the late Pleistocene, tropical America suffered the extinction of about 56 genera of its mammals (54 of which were of large body size), presumably as a result of a combination of climate change and human overkill. Lindsey et al. (2009) suggested higher initial diversity and taxonomic turnover rates among North American participants in the Interchange than were previously supposed.

It is clear, then, that the South American mammalian fauna has a complex derivation. The unusually large number of endemic Neotropical taxa is a reflection of the degree and duration of separation of South America from other continents and of the evolutionary success of Nearctic invaders there.

The Unusual Mammalian Fauna of Madagascar

Islands long isolated from continents frequently have an unusual mammalian fauna. Such a fauna may be dominated by a group equally important nowhere else, as in the case of the marsupials of Australia, or may be extremely poor in mammals, as in the case of New Zealand, where the only native mammals are bats. Madagascar is an interesting example of a refugium supporting a primitive mammalian fauna with little ordinal diversity, superlative numbers of

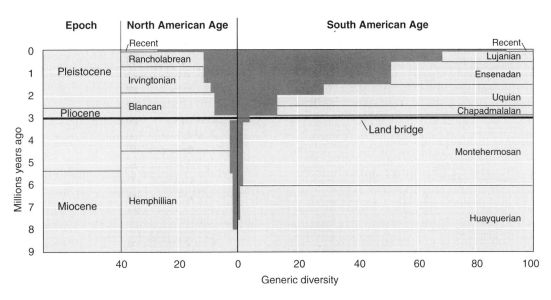

FIGURE 25-16 Graphic comparison of the number of genera of land mammals from North and South America participating in the Great American Biotic Interchange in the late Cenozoic. The initial invasion in each continent is approximately balanced up to and immediately after the appearance of the land bridge. However, the subsequent increase in North America of immigrants from South America faded by the middle Pleistocene; in South America the North American invaders continued to increase exponentially. (Adapted from Marshall, L. G., *Am. Sci.* 76 (1988): 380–388.)

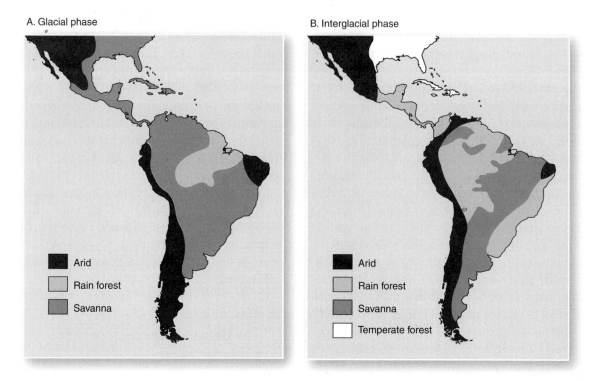

FIGURE 25-17 Approximate distribution of general habitat types in South and southern North America during two phases of the late Cenozoic. In the aridity accompanying a glacial phase (A) dry-land habitats such as savanna woodlands and other open landscapes predominated, creating a barrier to forest-adapted species and an avenue of dispersal for savanna-adapted mammals through the Central American Isthmus from temperate North America to temperate South America and vice versa. During such a phase, the slopes of the Andes Mountains and periphery of Amazonia might have held a mosaic of habitats including savannas. In an interglacial phase (B), Central America and the isthmus may have been clothed in rain forest, much like the present. This situation would have created a filter route across the isthmus, blocking savanna species and allowing passage into Central America of rainforest species from Amazonia that could cross or go around the northern Andes. (Adapted from Webb, S. D., *Paleobiology* 17 (1991): 266–280.)

endemics for its area, and a seemingly incomplete exploitation of habitats (Wilmé et al. 2006).

Madagascar is a large island, 1,600 kilometers in length, with a maximum width of 560 kilometers. It lies 420 kilometers east of the eastern coast of Africa. Madagascar separated from Gondwana as a fragment of the Indian subcontinent by 100 million years ago after that subcontinent had already separated from Africa 160 million years ago (Krause et al. 2006). Madagascar has been isolated from other landmasses throughout the Cenozoic. The island has supported seven orders of land mammals in recent times: Afrosoricida, Chiroptera, Soricomorpha, Primates, Rodentia, Carnivora, and Artiodactyla. Most of the mammals are endemic. The most highly diversified groups are the lemuroid primates (with an astounding 32 species still surviving after humans arrived on the island; Ganzhorn et al. 2006; Goodman & Patterson 1997), which probably arrived in the Eocene, and the tenrecid afrosoricidans, which perhaps dispersed to or from Africa as early as the Paleocene. Many of the Malagasy mammals occupy niches filled elsewhere by mammals of different taxa. There are viverrid carnivorans that resemble cats,

there are lemurs that are variously nocturnal or diurnal, terrestrial or arboreal, and the bizarre aye-aye (*Daubentonia*) has woodpecker-like abilities. The tenrecs include hedgehog-like species, burrowing, mole-like species, and web-footed, amphibious fish-eating species. The tiny, mouse-eared, termite-eating tenrec *Geogale* resembles the North American desert shrew *Notiosorex* and inhabits the arid southwestern part of Madagascar. The only artiodactyl present before the arrival of humans was the now extinct hippopotamus (*Hippopotamus lemelii*), and today the introduced river hog (*Potamochoerus*) is the only wild artiodactyl. The ungulate niche has largely gone unfilled, although a group of strange, large Pleistocene lemurs, now extinct, may have been terrestrial herbivores (Tattersall 2008). There is also a paucity of frugivorous mammals compared to other tropical areas (Goodman & Benstead 2003).

The Island Syndrome

Mammals isolated on islands typically face different selective pressures than do members of parental mainland stocks. On islands competition is usually reduced or may be absent, predators are often absent or few kinds are present, and the flora may be depauperate. In some cases, as on some small desert islands off the east coast of Baja California, one or two species of rodents are the only mammalian inhabitants and just a fraction of the number of species of plants that occur on the mainland is present. Through time, island mammals tend to diverge from parental mainland stocks (**island syndrome**), but the pattern of divergence is not consistent for all species. Mammals (and other land vertebrates) on islands tend to evolve toward a fundamental or optimal body size for a particular ecological strategy or body plan (Lomolino 2005).

Island mammals typically differ in size from mainland relatives, and whereas some mammals become larger on islands, others become smaller. Island rodents and marsupials are generally larger than their mainland relatives; soricomorphs, erinaceomorphs, lagomorphs, carnivores, and artiodactyls, however, are usually smaller. Examples are numerous. The *Peromyscus* (deermice) inhabiting islands off the coast of British Columbia are unusually large, but the caribou that inhabited one of these islands (but that is now extinct) was a dwarfed form. The gray fox (*Urocyon littoralis*) that lives on the Channel Islands off the coast of California is substantially smaller than the mainland sister species *Urocyon cinereoargenteus*, and the mammoth that lived on these islands in the Pleistocene was dwarfed. However, evolutionary patterns on islands are not completely consistent. Counter to the usual trend, not all insectivorous mammals that inhabit islands are small. Unusually large soricomorphs (solenodons) evolved on some of the islands of the West Indies, and the largest erinaceomorph of all time (*Deinogalerix koenigswaldi*) lived on the Mediterranean island of Gargano (now a peninsula). This erinaceid was larger than a fox and probably fed on rodents (Freudenthal 1972).

Sondaar (1977) has discussed the remarkable dwarfed Pleistocene mammals of the Mediterranean islands. In the Pleistocene, elephants (*Elephas*) and deer (*Cervus*) lived on many of these islands, and some islands supported hippopotami (*Hippopotamus*). These mammals must have reached the islands by sweepstakes routes, for the extent to which they diverged morphologically from the mainland stocks and the fact that generally not all types occurred on an island suggests that access to the islands was across water. All of the three types listed are known to be strong swimmers, and a single pregnant female could have founded a population on an island. Of special interest are the similar evolutionary trends exhibited by large mammals on a number of islands between which passage of terrestrial mammals would have been impossible. Elephants on the

Mediterranean islands became strongly dwarfed relative to the parental mainland stock of *Elephas namadicus*. *Elephas falconeri* of Sicily, an example of extreme dwarfism, was roughly 1 meter high, about one-quarter the size of its mainland progenitor, and relative to mainland elephants, *E. falconeri* had short distal segments of the limbs, cheek teeth with fewer enamel ridges, and a much lower skull with a reduction of the elaborate system of air sinuses. Short-leggedness was especially pronounced in the island deer, but the pig-sized island hippopotami also became short-legged.

These patterns of parallel evolution probably resulted from similar selective pressures on the many isolated islands. No large predators were on the islands. Large size is an extremely effective adaptation to avoid predation; without large predators, great size was no longer of advantage. An unreliable food supply for herbivores may have favored smaller size, and, in the absence of predators, overpopulation might have triggered periodic heavy mortality. Beds of deer bones found on the island of Crete are interpreted by some paleontologists as evidence of mass mortality, and abnormalities of the bones suggest starvation as the cause of death. In the dwarf elephant *E. falconeri*, the reduction of the skull crest and the reduced number of enamel ridges on the molars were related to the general dwarfing (Maglio 1973). The marked shortening of the limbs of the deer is thought by Sondaar (1977) to have been the result of two factors: the absence of predators and the consequent lack of need for speed, and the need for sturdy and well-braced limbs with which to negotiate mountainous terrain.

As in other vertebrates, even human populations when isolated on islands may become dwarfed. On the island of Flores in the Lesser Sunda Islands of Indonesia, *Homo floresiensis* was a 1-meter tall, roughly 25-kilogram body weight human that inhabited the island in the late Pleistocene. *Homo floresiensis* shared its island home with giant monitor lizards, the dwarfed elephant *Stegodon*, and giant rodents.

Just as mammals on islands are divergent structurally, some have changed behaviorally. As an example, desert woodrats (*Neotoma lepida* group) on Danzante Island in the Gulf of California have very large home ranges, and males are resource-defense polygynists, whereas this species on the nearby mainland does not have these behaviors (Vaughan & Schwartz 1980). Mammals that live on islands and have no mammalian predators often show little fear of humans. Blake (1887) found island gray foxes on California's Santa Cruz Island to be remarkably fearless, approaching people closely and pulling at their blankets at night. In the face of human invasions and introductions of predators and competitors, island faunas around the world have suffered greatly.

SUMMARY

The study of the distribution of mammals (or other organisms) in space is called biogeography; the biogeography of animals in particular is zoogeography. In historical biogeography, scientists follow the chronicle of changes in the distribution of organisms through time, usually on a long-term or geological time scale. When these time changes are combined with the evolutionary history of diversification of a taxon, lineage of taxa, or regional biota and reconciled with the relevant deep-time geological changes in the surface of the Earth, the result is called phylogeography. On a global scale, Earth's living organisms, including mammals, show continental-scale ranges that allow us to make a map of biogeographic realms or regions with commonalities in their biotas. The biogeographic regions are Afrotropic, Australasian, Indo-Malayan, Nearctic, Neotropic, Palearctic, and Oceanic. The shared families and higher taxa of mammals of these regions reflect the history of diversification of the mammalian groups in those parts of the world. Over long periods of geological time, the biogeographic patterns are affected by tectonic plate movements or continental drift. As continents move, collide, and upwarp or downwarp, there are attendant changes in connections and separations between and among regions, coinciding with changes in the oceans and seas. These changes in turn affect oceanic circulation patterns, atmospheric circulation patterns, and climate. The resulting variability across the surface of the Earth provides changing topography, ecosystems, and habitats, conditions affecting the active dispersal and passive separation (vicariance) of populations and their attendant evolution. Many regions, small islands, or even whole continents have experienced periods of connection by land bridges and periods of isolation from other regions or continents. The changes generate natural experiments in the evolution of the regions' land and sea mammals.

KEY TERMS

Allochthonous

Autochthonous

Beringia

Biogeography

Biota

Boreal

Continental drift

Disjunct

Dispersal

Ecological biogeography

Facilitated dispersal

Filter route

Historical biogeography

Island syndrome

Phylochronology

Phylogeography

Sweepstakes route

Vicariance

Wallace's Line

Zoogeography

RECOMMENDED READINGS

Avise, JC. 2000. *Phylogeography: The History and Formation of Species.* Harvard University Press, Cambridge, MA.

Badgely, C and DL Fox. 2000. Ecological biogeography of North American mammals: species density and ecological structure in relation to environmental gradients. *Journal of Biogeography,* 27:1437–1467.

Cox, CB and RD Moore. 2005. *Biogeography: An Ecological and Evolutionary Approach,* 7th ed. Blackwell Publishing, Malden, MA.

Lomolino, MV, BR Riddle, and JH Brown. 2006. *Biogeography,* 3rd ed. Sinauer Associates, Sunderland, MA.

Olson, DM, et al. 2001. Terrestrial ecoregions of the world: a new map of life on earth. *BioScience,* 51(11):933–938. http://www.worldwildlife.org/wildworld/.

Pascual, R. 2006. Evolution and geography: the biogeographic history of South American land mammals. *Annals of the Missouri Botanical Garden,* 93: 209–230.

Patton, JL, MNF Da Silva, and JR Malcolm. 2000. Mammals of the Rio Juruá and the evolutionary and ecological diversification of Amazonia. *Bulletin of the American Museum of Natural History,* 244:1–306.

Riddle, BR and DJ Hafner. 2007. Phylogeography in historical biogeography: investigating the biogeographic histories of populations, species, and young biotas, Pp. 161–176, in *Biogeography in a Changing World* (MC Ebach and RS Tangney, eds.). CRC Press, Boca Raton, FL.

Riddle, BR, et al. 2008. The role of molecular genetics in sculpting the future of integrative biogeography. *Progress in Physical Geography,* 32(2):173–202.

Scotese, CR. 2002. Paleomap project. Paleogeographic maps of the earth. http://www.scotese.com/.

Mammalian Conservation Ethics

We cannot win the battle to save species and environments without forging an emotional bond between ourselves and nature—for we simply will not fight to save what we do not love.

—Stephen Jay Gould

State of the Earth

As noted occasionally throughout this book, mammals face numerous threats to their continued existence including **habitat degradation** and destruction, overexploitation, loss of genetic diversity, endangerment, and extinction. The main problems confronting not only mammals but all the Earth's biodiversity are clear and obvious everywhere: human population pressure and the way we view and interact with the Earth (**Fig. 26-1**). For thousands of years after we humans evolved, we existed in tenuous, small, isolated groups. During the last 500 years or so, we have gone from a small global population to the dominant ecological force on the planet. By the time of first contact of Europeans with native peoples in North America, **indigenous populations** in the Americas are estimated to have been between 10 million and 100 million, and for thousands of years the natives had been interacting with the Earth in ways that maintained human and environmental health, drinkable streams, and abundant wildlife. Having "turned the corner" a century and a half ago, our population began an exponential increase. Concurrently, Western paradigms of expansionism and for-profit commodification of nature largely replaced the indigenous responsibility connections. Our present globalized economic approach is accelerating global disaster. In a very short span of time (within the average life span of a person), we have converted or destroyed much of nature, except for a few relatively small parks and reserves isolated like islands in a sea of human-dominated ecosystems (Kingdon 1989; Meyer 1996; Morowitz 1991; Turner et al. 1990; Valdez et al. 2006; Vitousek et al. 1997). At present, only about 3% of the land surface worldwide is set aside in protected areas and nature reserves (Soulé 1991), and even less of the oceans is protected. These reserves are increasingly becoming more fragmented and isolated.

One can go nowhere on Earth where the heavy stamp of humanity is not obvious. Even the remotest parts of the planet are no longer particularly remote. Our devotion to mechanized transport has lead to road expansion that negatively affects mammals and other living things across all parts of the United States (Strasburg 2006; Watts et al. 2007). Satellite images of the Earth are particularly poignant reminders of our heavy-handed approach to transforming the environment (e.g., Editors of Collins 2006). Some parts of the environment are so entirely converted over to agricultural, industrial, and urbanized ecosystems that the original ecosystems of those regions are completely unrecognizable and have been gone for decades (e.g., see historical imagery at earth.google.com). The present inhabitants have little idea and little evidence remaining of what was lost. The pressure of our numbers is simply too great, as is the rate of our modifications of the Earth. In the meanwhile, our effects upon global ecosystems are having broader impacts at greater and more critical scales. What were once concerns are now crises (Morrison et al. 2007; O'Shea and O'Dell 2008; Schipper et al. 2008; Stokstad 2009).

When the last edition of this book was published in 2000, the Earth's human population was estimated at 6 billion people. Now, about a decade later, our numbers have grown by another three-quarters of a billion (U.S. Census Bureau World POPClock projection 2009). The demands of our growing population for simple basic resources such as shelter, water, fire, and food threaten to consume the remaining large tracts of undeveloped nature.

Accompanying this global encroachment and overconsumption is the expected mass extinction of one-fourth to one-half of all living species (Soulé 1991; Wilson 1992). Already one-fourth of the world's mammals are in danger of extinction, according to data compiled by the International Union for Conservation of Nature and more than 1,700 collaborators (Schipper et al. 2008). The outlook is grim and the blame is squarely on us: The primary worldwide threats to mammals are habitat loss and fragmentation, habitat degradation, and harvesting (Schipper et al. 2008). Many biologists even contemplate our own extinction (Wilson 1993).

One constant throughout Earth history is change: cosmological, geological, environmental, evolutionary. Based on the oldest records of rocks and fossils, the Earth has changed constantly through its history and will continue to do so. Uncounted thousands of species appeared,

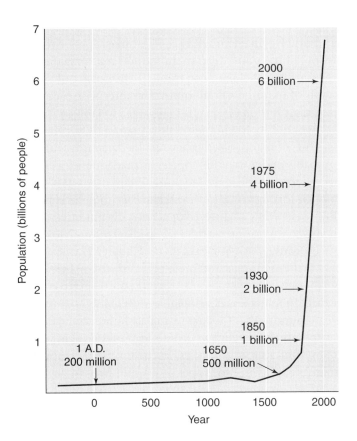

FIGURE 26-1   Growth of the human population during the last 2000 years. In 1999, about 6 billion people inhabited the Earth, and in 2009 the population was increasing by about 162 people every minute (net change of births and deaths). United Nations projections for the next 25 years predict an increase of another 2 billion, with a world population by the year 2100 ranging from a low estimate of 10.4 billion to a high estimate of 17.5 billion people. (Data from United Nations, 1998.)

existed for millions of years, and became extinct. Global climate has changed countless times, even throughout the last 65 million years of the "Age of Mammals" (the Cenozoic era; Zachos et al. 2001, 2008). The normal changes in climate during Earth's history have always caused changes in ecological systems and will continue to reshuffle them in completely new ways in the future (Fox 2007; Moritz et al. 2008).

On a geological time scale, extinction is common. Most extinctions are individual or low-level "background" events, but large-scale extinctions have periodically taken place. As noted by Jablonski (1991), "The most basic observation is simply that mass extinctions have happened: irreversible biotic upheavals have occurred repeatedly in the geological past. Marine and terrestrial biotas are not infinitely resilient, and certain environmental stresses can push them beyond their limits. . . . Survival of species or

lineages during mass extinctions is not strictly random, but it is not necessarily closely tied to success during times of normal background extinction."

Intentionally or unintentionally, we are pushing toward the next mass extinction (Leakey and Lewin 1995), by intensifying global **climate change**. During the last few centuries, as settlers followed explorers into new lands, they decimated or nearly decimated whole faunas and even indigenous peoples. On several continents, colonizers overhunted species with disastrous results to the indigenous wildlife and peoples (Adams and Mulligan 2003). The same befuddling advance was even more calamitous on the world's islands. Many islands, small, fragile, and isolated habitats to begin with, held populations of mammals and other organisms that were wiped out within just a few years, sometimes by overharvesting and sometimes indirectly through the commensal animals (and diseases) the settlers brought with them, such as rats, dogs, cats, goats, and others. As a result, most of the native species of the world's islands are already gone (Prothero 2006).

How can we stop or lessen this outcome? There are no easy answers. The problems and factors involved are immense and complex, but our own survival may depend on our attempts to solve them (Koshland 1991). A major change in our philosophy and our way of interacting with the Earth is necessary if we are to restore the environment to one that will sustain human and other life (Cajete 1999, 2000). Encouraging trends in recent years are the increasing awareness of the problems involved in conserving biodiversity, the proliferation of conservationists, and the progress toward understanding and solving these problems (Groom et al. 2006; Wilson 1992). Many mammalogists and other conservation-minded people are feverishly working to document basic biodiversity while simultaneously developing ideas for its protection, management, and sustainability (for examples, Galliari and Goin 1993; Ginsberg 2001; Medellín 1998; Milner-Gulland and Woodroffe 2001; Patton et al. 1997; Wemmer et al. 1993; Zahler 2001). Another encouraging trend is a decrease in the reproductive rate of humans around the world in recent years (Bongaarts 1998).

Among the conservationists is Michael Soulé, who summarized several major factors of human interference, including the destruction of habitat, fragmentation of habitat, overexploitation, the spread of **exotic species** (introduced and alien) and diseases, air, soil, and water pollution, and climate change. Soulé further attributed these proximate causes of the loss of biodiversity to still more fundamental factors of the human condition. Because human nature is unlikely to change quickly, he proposed a number of tactics to ameliorate the attrition of biotic diversity: "The human condition is dynamic and unpredictable and will remain so for at least a century, if for no other reasons than the momentum of the population explosion and the unsatisfactory economic and social status for billions of people during the 21st century. The 'biotic condition,' therefore, will also be tenuous during this interval. Fortunately, conservationists have an increasing number of tools with which to deal with the crisis" (see Soulé 1991). Limited space does not allow us to discuss them here, but the reader is encouraged to consult the articles (most available online), books, and websites cited in this chapter, and the references therein, for a better understanding of the problems and potential solutions for preserving the Earth's biota.

Many people point up the economic value of nature and natural resources as a rationale for their preservation (Geist 1994; Daily 1997; Reaka-Kudla et al. 1997; Martín-López et al. 2008), and there are different views of "economic value." There are those who believe that if forests and other wildlands are to be saved, they must "pay their own way," that is, provide basic sustenance for some humans (Robinson and Redford 1991) or profits for others. For example, the extraction of DNA from nature produces billions of dollars annually for biotechnology, agriculture, and public health (Myers 1984). Regrettably, these anthropocentric benefits sometimes provide even more reason to exploit the resources, while little effort is reinvested in maintaining them. Even selectively extracting resources alters or erodes the ecology and biodiversity of a habitat. Other people emphasize the ecological "services" of nature, from recreational opportunities to the regulation of carbon dioxide in the atmosphere (Balmford et al. 2002; Costanza et al. 1997; Daily 1997), or the value of the genetic information content of species (Crozier 1997).

Conservationists argue that cultural values are often in conflict with conservation policies and that "a new ethic or a revolutionary change in human consciousness is necessary" before significant progress is possible (Leopold 1948; International Union for the Conservation of Nature [iucn.org] 2009; Devall and Sessions 1985; Soulé 1991; Goin and Goñi 1991; Wilson 2006; terralingua.org; SER 2009). One cultural value that has recently been questioned is the private ownership of land (Steinberg 1995; Mitchell 1998).

Human Impact on Mammals

We humans have long been interested in our fellow mammals and have long exploited them. As many as 4 million years ago, *Australopithecus* was killing and eating baboons and antelope, and the use of mammals for food remains characteristic of most cultures today. Many kinds of mammals have been domesticated, and some are taught to work for their owners. The trained Indian elephant lifts and drags teak logs in the remaining tropical forests of Sri Lanka, where the periodically saturated soil limits the usefulness of

vehicles; dogs help some African hunters capture antelope and other game; trained rhesus monkeys pick coconuts from tall trees and drop them to their masters; and even the unruly camel has been trained. The raising of various kinds of mammals is an important enterprise today. The very distribution of early humans was probably influenced by their ability to kill their fellow mammals, for the skins and furs of mammals may have enabled primitive humans, probably endowed with hopelessly inadequate insulation, to penetrate cool or cold regions. Today, mammals are exploited through ever more intensive technological means, from genetically engineered or **genetically modified organisms** such as transgenic mice for medical research to cloned sheep, cattle, dogs, and others (see Chapter 27 online).

Wild mammals and most other organisms are under greater pressure than ever from humans in all parts of the world. In many less developed areas, the inhabitants hunt year-round; either they are primarily hunters who depend on mammals for much of their food or they hunt to supplement limited food supplies. In the developed countries, many people hunt for sport or trap for furs during regulated seasons. In the United States, where virtually all populations of game mammals are managed, the sale of hunting and trapping licenses contributes to wildlife management and conservation in general. Hunting remains popular in many parts of the United States, and the sale of firearms, ammunition, and other equipment associated with hunting is big business. In places, hunting is a factor forcing the rate and direction of evolution of certain traits in wild populations of mammals and other organisms (Coltman et al. 2003; Darimont et al. 2009).

Our exploitation of the Earth's mammals and other resources is not necessarily something to rebel against or be ashamed of. Like all mammals, as obligate heterotrophic organisms and native inhabitants of Earth, we have no choice but to exploit responsibly the planet's resources for our own survival. The Earth's natural resources are the ultimate origin for everything we need; there is no other source. Unfortunately, we are not presently using them conscientiously but are severely abusing them when we barely fathom their ecological relationships. Even our efforts at conservation sometimes produce unexpected results because of other perturbations and unanticipated linkages between ecosystems (Chapin et al. 1998; for mammalian examples, see Estes et al. 1998; Boccadori et al. 2008). We must reclaim a responsible, conscientious, respectful approach to utilizing other mammals.

In some ways our "consumption" of mammals and other wildlife is pure waste. On just four 0.5-kilometer stretches of a highway passing through rain forest in northeastern Queensland, Australia, during a 38-month period, Miriam Goosem (1997) recorded more than 4000 road-killed vertebrates, including about 500 mammals. **Road kill,** data for roads in the United States are not widely published, but impromptu counts made during mammalogy class field trips are deplorably high in Oklahoma and other parts of the country. When multiplied by the 6.3 million kilometers of roadways in the United States alone, the needless destruction is staggering (Forman et al. 2003). At the same time that the traffic on these roads kills some species, the roads themselves not only fragment the habitat but also present barriers to dispersal and migration for other mammals. For example, Goosem (1997) also learned that musky rat-kangaroos completely avoided the vicinity of the highway; Lumholtz's tree kangaroos, green ringtail possums, and long-tailed pygmy possums avoided the highway or were reluctant to descend to the ground but were able to cross through the forest canopy as long as an "overpass" of contiguous branches was available. In and near national parks in Wyoming, Montana, and Arizona, as well as in other parts of the American West, highways, railroads, pipelines, fencing along transportation corridors and other manmade structures interfere with traditional **migration corridors** used by American pronghorns, elk, bison, caribou, and other ungulates (Berger 2004; Hart et al. 2008).

Wild mammals can be costly to agricultural and ranching businesses. Pocket gophers, rabbits, meadow voles, ground squirrels, and even deer and elk may damage crops or rangeland, and efforts to combat these losses are frequently expensive. In addition, the United States federal government supports considerable research on mammals and, at the same time, finances the local control of virtually all carnivoran species for commercial ranching interests. According to the Predator Defense Institute (predatordefense.org), U.S. Department of Agriculture federal animal damage control programs cost U.S. taxpayers about $36 million per year at the end of the 20th century. Many biologists deplore these losses and regard such programs as a misuse of federal monies.

The long-term exploitation of mammals by humans has had a devastating impact. In the last 500 years, 76 species of mammals have become extinct (Cole et al. 1994; MacPhee and Flemming 1999; Schipper et al. 2008). Of the approximately 5,500 living species of mammals, 188 were classified as "Critically Endangered" in 2008; 29 of these may also already be extinct or have virtually no chance of surviving (Schipper et al. 2008). The conservation status of those species for which data are sufficient are rated according to various "categories of threat" by the IUCN as critically endangered, endangered, vulnerable, and near threatened (see the **IUCN Red List** at iucnredlist.org/mammals). Twenty-five percent of all mammals for which sufficient data are available are in danger of extinction; among those

mammals for which population trends are known, 52% are declining (Schipper et al. 2008). The greatest proportion of threatened species are marine mammals and land mammals in southern and southeastern Asia, and especially members of the orders Primates, Soricomorpha, Artiodactyla, Cetacea, Chiroptera, Carnivora, and Rodentia. Species that are the sole or among the few remaining members of long-distinct evolutionary lineages are considered by some workers to be the most endangered because of their uniqueness (EDGE 2009; see edgeofexistence.org).

Some extinct mammals were disposed of remarkably summarily. Steller's sea cow of the Bering Sea was pushed to extinction only 27 years after its first discovery by whalers. Sea otters, which were hunted along the Pacific Coast of North America at least as early as 1786, were killed for their valuable fur; probably more than 200,000 were killed between 1786 and 1868 (Evermann 1923). By 1900, these animals were rare over much of their range, and they were seemingly lucky to have survived until protected by legislation in the early 1900s. Not so lucky was the grizzly bear in California. In the 1890s, grizzlies still persisted in the San Gabriel Mountains near Los Angeles, but the last known southern California grizzly was killed in 1916, and the last verified occurrence in California was in 1922, in the foothills of the Sierra Nevada in central California (Grinnell et al. 1937). Only about 60 years were required to bring the grizzly in California from fair abundance to total **extirpation**. In Mexico, a population of grizzlies that survived in a small mountain range in central Chihuahua in 1957 was probably wiped out by 1963, the very year when funds were raised by the World Wide Fund for Nature to set aside a refuge for the animals. In Brazil, more than 90% of the Atlantic rain forest has been destroyed by **logging** and other development where 17 of the 23 types of primates and many other mammals are endemic. No one knows how many unnamed species were lost in the process, but Brazilian conservationists are struggling fiercely to protect what remains.

Mammals in the Eastern Hemisphere have fared no better. The quagga, a zebra that inhabited southern Africa, was extirpated in the wild about 1860, and another type of zebra was exterminated by roughly 1910. The Arabian oryx, well on its way to extinction, has been hunted in recent years with machine guns mounted on jeeps. The black rhinoceros of East Africa has been extirpated over broad areas, and its survival in the wild seems unlikely. The rate at which rhinoceros and elephant populations were once shot for sport in Africa is astonishing (Beard 1977). Siberian tigers, which number only a few hundred in the wild, are threatened by logging of old-growth forest in their prime habitat in Russia. The giant panda in China faces habitat destruction as well as a wide array of crucial but nonscientific problems, from political upheavals and inadequate captive conditions to bureaucratic problems and poaching (Schaller 1993). In Australia, introductions of foreign species by humans, whether intentional or inadvertent, have decimated populations of native, endemic species (Short and Smith 1994; Smith and Quin 1996). During the last 170 years, European humans, European rabbits, and red foxes were brought to Australia; the spread of each of these alien species across the continent was similar to the spread of a contagion and was accompanied by dramatic contractions in the ranges of native mammals (Channell 1998) and plants. Similar introductions of alien species (including humans) to many islands around the globe have threatened, wiped out, or replaced their unique and often highly endemic, but easily disrupted, ecosystems, even as mammalogists were conducting initial surveys and describing new species in these island ecosystems (Wiles 1992; Atkinson and Cameron 1993; Heaney 1993; Heaney et al. 1997).

In a few cases, there is reason for hope that some species will be saved. Through persistent hunting by the hundreds of thousands, the blue whale, the largest animal of all times, was reduced by more than 99% to a total population of probably no more than several hundred individuals by the late 1960s. Since the hunting of blue whales was banned in 1968, their population stabilized and is beginning to recover slightly (Baskin 1993; Brower 2009). Similar is the widely known story of the North American bison; from a population of millions these animals were intentionally slaughtered by colonizers to the verge of extinction. Concerted efforts by indigenous Americans and others to bring bison back in parks, preserves, and tribal and private ranches seem to be helping and offer opportunities for restoration of other species sharing their habitat (**Fig. 26-2**; Nabhan and Kindscher 2006; Zontek 2007). Reintroductions of native species to their former ranges from captive breeding populations offer some hope for their future (Stokstad 2009; Short et al. 1992; Lomolino and Channell 1998), but these efforts can be confounded by climate change and attendant habitat change, requiring radical, austere, and triage-like approaches (Hoegh-Guldberg et al. 2008; Marris 2009). Climatic perturbations of species' habitats will affect the species in ways unknown until we witness the changes, and we will need to act accordingly to help species adapt as well as adapting ourselves (Willis and Bhagwat 2009). **Assisted colonization** might help some species (Hoegh-Guldberg et al. 2008).

It is obvious that the fate of wildlife depends on the persistence of appropriate habitat, which is being destroyed over broad areas at an ever-increasing rate by ever-expanding human populations. In many developing countries, where the focal point of the inhabitants' existence is the day-to-day search for food and fuel, the pressure on land and wildlife is intense. This pressure becomes vastly more acute when revolution, struggles for justice and independence, or strife between political factions is associated with reigns of lawlessness. Meanwhile, powerful corporations from the developed nations are a more insidious threat, driven by the addiction of our "advanced civilization" for incredible amounts of energy and materials from developing countries eager to become just like us.

The situation in Amazonia described by Laurance (1998) provides a tragic case in point:

> Today, even the remotest areas of the Amazon are being influenced by human activities. Illegal gold-mining is widespread, with wildcat miners polluting streams with mercury (used to separate gold from sediments) and threatening indigenous Indians through intimidation and introductions of new diseases. A recent government census, for example, tallied more than 3000 illegal miners in the Yanomami Indian Reserve in northern Amazonia. There are also increasing numbers of major mineral, oil, and natural gas developments sanctioned by Amazonian governments. Much of the remote Peruvian Amazon— one of the world's most biologically important areas—has been opened up for oil and gas exploration, with multinational corporations investing hundreds of millions of dollars in the region. Roads created for oil exploration and development in Ecuador have caused a sharp rise in forest colonization, land speculation, and commercial hunting.

Roads also have enabled greater access to forests and markets in tropical parts of the world. As firearms such as shotguns have become readily available to forest people, hunting pressure has grown in tropical forests in the Amazon, southeastern Asia, and western and central Africa (see bushmeat.org). Targeted species include larger primates; tapirs; artiodactyls such as antelopes, deer, pigs, and peccaries; larger rodents; and top carnivores. For many dwellers in and near tropical forests, this **bushmeat** may be an important source of protein. As hunting pressure increases, the structure of the animal communities in the forest can be altered, eliminating species with low reproductive rates, disrupting food webs, and aggravating the effects of **habitat fragmentation** (Laurance et al. 1998; Silvius et al. 2005). Montane forest on the slopes surrounding the Amazon Basin have been cleared at an alarming rate to grow coca for cocaine and opium poppies for heroin to supply the high demand for these drugs in the United States and Europe (Goodman 1993).

While bushmeat hunting, road building, habitat fragmentation, and deforestation continue at an alarming pace in many regions, there are a few encouraging signs for conservationists. When one author (Ryan) began working on

FIGURE 26-2   The American bison or buffalo, *Bison bison*, at the Nature Conservancy's Tallgrass Prairie Preserve in the Osage Plains of northern Oklahoma. As seen here, bison often prefer to graze on the lush green growth that sprouts after prescribed burning on this managed habitat.

mammalian conservation in Madagascar in 1987, there were approximately 760,000 hectares of protected parks and reserves in the entire country. Today, thanks to the foresight and tenacity of conservationists, political leaders, and local Malagasy citizens, more than 1,698,639 hectares are protected in 12 new parks and protected areas. Similarly, in 2005 the African nation of Gabon declared approximately 10% of the country as protected parks to preserve indigenous populations of forest elephants, chimpanzees, gorillas, and other rainforest animals. No country has been more progressive in its efforts to protect its natural heritage than Costa Rica; approximately 27% of the country is protected areas (compared to 3.6% for the United States). Nevertheless, much remains to be done.

The need for the human race to set its own house in order is basic both to our own survival and to the perpetuation of the biological richness of the world (Pimm et al. 2001). Clearly, our ability to solve social problems and the future of wildlife are tightly linked. People not under the pressures and stresses occasioned by high populations and limited resources, and at peace with one another, can work toward saving the biotas of the world—fearful people with empty stomachs make poor conservationists. On the success of the politicians, scientists, and teachers of the world in halting the rise in human populations and stopping strife between peoples hinges the survival of our biotic heritage. We have clearly reached the eleventh hour. If we do not learn from our mistakes, the wildlife of the world will pay

a devastating price and, thus, so will we. We cannot expect that many of the already threatened species will survive, but some species will escape the destruction and will adapt to partially repopulate the remaining habitat (**Fig. 26-3**).

A New Conservation Ethic

Of course, we cannot separate mammals (including ourselves) from all of the other interconnected components of a functioning ecosystem. Ecologists and conservationists have long known that each element of a biota plays an essential part in the ecosystem it occupies and that the loss of even a seemingly insignificant species might tip the delicate biotic balance. There is good evidence that diversity of biota in an ecosystem contributes to the stability and productivity of that ecosystem (Naeem et al. 1994; Reaka-Kudla et al. 1997; Tilman and Downing 1994; Wilson 1992). We do not begin to know the degree of pressure that most ecosystems can tolerate before collapse. Neither do we know the size of the human population that the Earth can support (Cohen 1996). Nevertheless, an open-ended, uncontrolled experiment is probably not the best way to find out.

Biologists have stressed practical problems: The pollution of water should be avoided not only because of the potential for serious public health problems but also because the ecosystem of a stream might be drastically altered and species of fish from which we derive pleasure or some monetary return might disappear. Range managers have

FIGURE 26-3 An exurban ecosystem in the early 21st century, a garbage dump in the North American Southwest. Common species in this environment are feral cats (arrow), dogs, black and barn rats (*Rattus*), house mice (*Mus*), and black vultures (background); native mammals are largely absent. Plant life is a minor component, replaced by dominant ground cover of plastic, cardboard, and scrap metal.

emphasized economic problems: Unwise grazing practices alter a grassland ecosystem to the point where its economic importance is reduced; in concrete terms, the weight that each head of cattle gains per day may be reduced to a point at which ranchers can no longer realize a profit. But can we, with due respect for honesty, justify conservation with only these kinds of arguments?

In his excellent discussion entitled *The Conservation of Non-Resources*, Ehrenfeld (1976) pointed out that attempts to justify the conservation of many species on the basis of their economic importance are unjustifiable scientifically. The conservation doctrines lose force if a species is destroyed without the disruption of its ecosystem. If the ecosystem is destroyed following the loss of the species, however, not only is it too late to save the day, but the cause-and-effect relationship involved can never be proved and may not even be hypothesized.

We must look to a new conservation ethic. Few species can be proved essential to the survival of their ecosystems or to have great economic value; nonetheless, each species forms a part of a biological richness developed over millions of years, and each is worthy of perpetuation at least in part because of what Ehrenfeld termed its "natural art value." Two moving presentations of this view are quoted by Ehrenfeld. In his book, *Ulendo: Travels of a Naturalist in and out of Africa*, Carr (1974) stated,

> It would be cause for world fury if the Egyptians should quarry the pyramids, or the French should loose urchins to throw stones in the Louvre. It would be the same if the Americans dammed the Valley of the Colorado. A reverence for original landscape is one of the humanities. It was the first humanity. Reckoned in terms of human nerves and juices, there is no difference in the value of a work of art and a work of nature. There is this difference though. . . . Any art might somehow, some day be replaced—the full symphony of the savanna landscape, never.

Regarding specific "nonresource" species, in this case, three small primates called lion tamarins, Coimbra-Filho et al. (1975) wrote:

> In purely economic terms, it really doesn't matter if three Brazilian monkeys vanish into extinction. Although they can be (and previously were) used as laboratory animals in biomedical research, other far more abundant species from other parts of South America serve equally well or better in laboratories. Lion tamarins can be effectively exhibited in zoos, but it is doubtful that the majority of zoo-goers would miss them. No, it seems that the main reason for trying to save them and other animals like them is that the disappearance of any species represents a great esthetic loss for the entire world. It can perhaps be compared to the destruction of a great work of art by a famous painter or sculptor, except that, unlike a man-made work of art, the evolution of a single species is a process

that takes many millions of years and can never again be duplicated.

Although effort expended on the perpetuation of some species can be justified on the basis of economic importance, we regard this natural art value of many species as their greatest importance. A more basic argument would insist that a species should be preserved because of a reverence for its vast evolutionary history and that each species has a right to play out its evolutionary role; we have no moral right to set ourselves up as the instrument of their destruction. Certainly, our Western attitudes of intolerance, overuse, and disregard for nature must change if our biotas are to survive, and an incipient rationale or conservation ethic may unite enough of humankind behind the conservation cause to turn the tide.

If the present decimation of the Earth's wildlife is to be curtailed, at least the following steps must be taken:

1. **Human population growth** must be halted and the need for space and resources stabilized.
2. Our credence in never-ending economic expansionism and our greed for profits must be abandoned and replaced with sustainable exploitation.
3. We must view ourselves as an integral, interdependent part of life on Earth and behave accordingly. We must rethink our self-proclaimed superiority over nature and instead treat nature as an extension of ourselves.
4. We must balance convenience with responsibility (Haozous 2009).
5. We must accept the validity of other cultures and other ways of thinking besides Western thought. **Traditional ecological knowledge** and wisdom of indigenous groups can greatly benefit Western science, and vice versa (SER online 2009).
6. Our exploitation of many species must be drastically reduced.
7. Large tracts of land cared for and utilized by humans must be maintained for wildlife, too, as well as connectivity corridors for migrants and between fragmented populations (e.g., Nabhan 2004; White 2009).
8. A broad understanding of ecology must underlie an interest in utilizing and preserving balanced faunas and floras and helping them adapt to new climatic and ecological conditions.
9. Control of animals threatening crops and livestock must be local, with no attempt to exterminate a species over wide areas with little respect to the damage it is doing.
10. The use of **biocides** must be carefully controlled.
11. We must accept some types of economic losses and inconvenience caused by wildlife and must feel that these are more than compensated for by our enjoyment of and belonging within a rich and balanced biota.

Additional worthwhile goals and ideas are discussed in Nabhan and Seibert (2002) and other articles in CSE et al. (2002).

We, as biologists, are among the most acutely aware of the problems facing the Earth's biota. Although the human overpopulation problem has long been recognized, human population control remains controversial and unpopular in many cultures. Despite the recent global decline in human births, higher standards of living and better nutrition and health around the world have led to a decline in global mortality rates and greater longevity, ensuring continued population growth (Bongaarts 1998). If additional diseases such as malaria—which "has killed half of the people who have ever lived" (Palacio 2005) and infects half a billion people per year—are defeated, hundreds of millions more of us will be competing for ever-shrinking basic resources such as shelter, water, food. Biologists and others have dared to speak out to urge continued population control (Meffe et al. 1993; Hardin 1993; Jayaraman 1993a, 1993b) and to plead for educating the world about the fundamental importance of the "balance of nature" and the inescapable fact that it includes us. As stated by Lovejoy (1997), "We must not only recognize that, but also behave as if, we live within ecosystems, rather than perceiving nature as something confined to a few protected areas isolated within a degraded, human-dominated landscape." What we do to the Earth we do to ourselves. The most ethical and exemplary endeavor for present and future mammalogists and all people is in the conservation not just of mammals but of all living things. We must take the lead, as a large group of ecologists recently publicly pledged to do (Bazzaz et al. 1998), in devoting part of our professional (and private) lives to deterring environmental degradation and to teaching the public about the importance of biodiversity. We must boldly and actively bring about slow, orderly changes to restore a balance in an ever-changing world as a way to heal the Earth at ecological speed, and learn directly from the land. We must adopt a holistic, multidisciplinary approach to achieve ecological understanding and sustainable use (Martinez et al. 2008; Salmón 2000). The conservation literature and World Wide Web abound with recent examples of **ecological restoration** and global efforts at local conservation (e.g., see ERI 2009; IUCN/WCPA 2009; PAL-Net 2009; Rodríguez et al. 2007). And on a personal level, we must each ask ourselves, "Will my children's children be happy in the world I left them?" Because the survival of much of the world's life and indeed our very own survival are in our hands, one cannot help but fervently hope that the word *sapiens* (meaning wise) becomes a justly earned part of the name *Homo sapiens*.

In our modern society, we have given up responsibility for convenience, and as a result, we are destroying much of the Earth's biota. Instead of learning to change with the Earth and restore balance to the complex interrelationships within changing ecosystems, we seem to believe that technology will help us through the next major disaster, and we apply it at ever larger and more destructive scales. As a result, we destroy the very source of everything that nurtures us. And yet, we recognize the need to reestablish our connectedness to the Earth. Numerous authors have come to realize the loss: For example, in recent issues of the *Journal of Mammalogy*, Schmidly (2005) and Hafner (2007) and in *Conservation Biology*, Hayes (2009) lament the drop in the number of students becoming naturalists and mammalogists. Pulitzer Prize–winning author and biologist Edward O. Wilson has written extensively about our instinctive connections to the land and living things (e.g., 1984, 1992, 2006) while also lamenting our loss of a deeper, spiritual connection to nature. What happened to this connection? Even the greater society of the United States of America has recognized the problem of "nature-deficit disorder" as it is called by Louv (2005). We "protect" our urbanized children from unstructured, unsupervised time in unmanicured nature, rather than encouraging such activity.

As pointed out by another Pulitzer Prize winner, Diamond (2005) in his book *Collapse: How Societies Choose to Fail or Succeed*, other great human cultures have disappeared. Why should ours be any different? What will be the human cost of our alienation from nature? How can we avoid our own extinction? The key idea in the title of Diamond's book, as well as in a solution to the problem, is choice. We alone can choose between a life of comfort and convenience or one of responsibility and humility. We alone can restore balance to the Earth's ecosystems. Sometimes as scientists we must drop our scientific objectivity and become morally and ethically involved in the biological phenomena we study because we realize it is the only way that we can conserve the Earth and its biota. Impartiality might help us be scientists by the Western way of thinking, but it will not insure that we are responsible stewards of the Earth. Impartiality denies our interdependence with nature and the need to give and take. Numerous research studies show that direct exposure to nature is essential for healthy physical, emotional and spiritual childhood development (Louv 2005), and will promote a common-sense connection with the environment.

A personal commentary seems appropriate here. The senior author (Vaughan) grew up in the 1930s on a hill in Los Angeles where colonies of ground squirrels remained and burrowing owls and roadrunners were common. My neighborhood friends and I, after less than an hour's bicycle ride, could be in "wild country", tracts of chaparral and oak woodland where there were deer, gray foxes, bobcats, the lovely California gray squirrel, and an occasional mountain lion. In an entirely natural way, my young friends and I came to revere and even to love this wild country. Children growing up today in the same neighborhood are completely cut off from what natural areas remain by multi-lane boulevards carrying heavy and dangerous traffic and often, and more importantly, by an indifference toward or even fear of wild places.

We are also finding much help in traditional ecological knowledge and wisdom (TEK and TEKW) where it still remains, from cultures that coevolved with ecosystems around the globe (Berkes et al. 2000; Ford and Martinez 2000; Long et al. 2003; Martinez et al. 2008; Salmón 2000) and are putting their knowledge into practice by fostering participatory research in ecological management and communication with local communities.

Some native cultures "define a sense of personal responsibility to the environment and reinforce the intimate symbiotic relationship between nature and culture" (SER online 2009) that Westerners would do well to emulate. A particularly practical and stimulating model for reconnecting our children and ourselves with the Earth is the approach developed by Young et al. (2008), as well as numerous articles online and courses offered at nature schools in the United States and around the world (e.g., see links at wildernessawareness.org). One of the authors of this textbook (Czaplewski), in addition to teaching university students, is thrilled to be involved in a number of public programs through his employment at a museum. Teaching adult, family, and children's programs is a gratifying and humbling way to try to reconnect people with nature (as well as providing potential jobs for mammalogists!). Young people in particular are naturally prone to give back to the Earth as much as they take from it, an ethic that should be gently and subtly encouraged and guided (Stumpff 2007). Collaborative and community-based environmental restoration activities can accomplish much and stimulate further similar advances.

The Earth will go on, with or without us and the species we take down with us. It is our responsibility to evolve along with the Earth and its climate, not only in our ways of thinking but also in our treatment of the Earth, for the future generations. We have shown our ability to alter the Earth. Are we wise, capable, moral, or compassionate enough to admit our elemental dependence on the Earth and restore balance to it?

There can be no purpose more enspiriting than to begin the age of restoration, reweaving the wondrous diversity of life that still surrounds us.

—E. O. Wilson, *The Diversity of Life*

SUMMARY

Mammals face numerous threats to their continued existence including habitat degradation and destruction, overexploitation, loss of genetic diversity, endangerment, and extinction. The main problems confronting not only mammals but all the Earth's biodiversity are clear and obvious everywhere: human population pressure and the way we humans view and interact with the Earth.

Accompanying the global encroachment and overconsumption is the expected mass extinction of one-fourth to one-half of all living species. Already one-fourth of the world's mammals are in danger of extinction, based on data compiled by the International Union for Conservation of Nature and more than 1700 collaborators. Habitat destruction and fragmentation are the main causes of threats to species. The synergistic effects on the environment of combinations of ongoing climate change and forest fragmentation, fragmentation and fire, highways, logging and hunting, logging roads and spontaneous human colonization, climate change, **invasive species** and **emerging pathogens**, population growth, **economic globalization** and industrialization, and many others aggravate and accelerate pressures on the Earth's biota. As noted by William F. Laurance in a plenary talk at the 10th International Mammalogical Congress in Mendoza, Argentina, in 2009, these **environmental synergisms** become killers, gravely affecting wildlife.

How can we stop or lessen this outcome? There are no easy answers. The problems and factors involved are immense and complex, but our own survival depends on our attempts to solve them before we even understand them. Some authors point up **ecosystem services** and economic reasons for conserving nature, but market-based and profit-oriented conservation often fails to protect species that conflict with human economic interests. Many mammalogists and other conservation-minded people are feverishly working to document basic biodiversity while simultaneously developing ideas for its protection, management, and sustainability. Butler and Laurance recommend public-pressure campaigns for conservation targeting strategic multinational corporations and trade groups that are involved in industrial-scale lumbering, oil and gas development, exotic-tree plantations for biofuels, and large-scale agriculture. Ecological restoration is being implemented in a few places but must be flexible in its goals to account for Earth dynamics and synergistic effects. A change in our philosophy and our way of interacting with the Earth is necessary if we are to restore the environment to one that will sustain human and other life.

Ecologists and conservationists have long known that each element of a biota plays an essential part in the ecosystem it occupies and that the loss of even a seemingly insignificant species can tip the delicate biotic balance. There is good evidence that diversity of biota in an ecosystem contributes to the stability and productivity of that ecosystem. We must save entire ecosystems, not just individual species. Few species can be proved essential to the survival of their ecosystems or to have great economic value; nonetheless, each species forms a part of a biological richness developed over millions of years, and each is worthy of perpetuation at least in part because of its intrinsic "natural art value." We must look to a new conservation ethic, realize the appropriateness and the right of all living things to exist, not just of ourselves, as modeled by the traditional ecological knowledge of many disappearing indigenous peoples who have a long-range view. We must learn from our failures and turn them into positives and successes.

Like the vast majority of species that have ever existed on Earth, we are bound to become extinct eventually. Whether we bring this on ourselves soon or late, it makes little sense to exacerbate our own and many other species' demise through complacency and self-important ignorance. Nevertheless, we can feel solace that the Earth and some of its inhabitants will continue on without us, as has happened for a few billion years.

KEY TERMS

Assisted colonization
Biocide
Bushmeat
Climate change
Ecological restoration
Economic globalization
Ecosystem services
Emerging pathogens

Environmental synergisms
Exotic species
Extirpation
Genetically modified organism
Habitat degradation
Habitat fragmentation
Human population growth
Indigenous populations

Invasive species
IUCN Red List
Logging
Migration corridor
Road kill
Traditional ecological knowledge

RECOMMENDED READINGS

Alliance for Zero Extinction. 2009. www.zeroextinction.org/index.htm.

Altieri, MA. 2009. The ecological impacts of large-scale agrofuel monoculture production systems in the Americas. *Bulletin of Science, Technology, & Society*, 29(3): 2244–2336.

Barnosky, AD. 2009. *Heatstroke: Nature in an Age of Global Warming*. Island Press, Washington, DC.

Berger, J. 2004. The last mile: how to sustain long-distance migration in mammals. *Conservation Biology*, 18:320–331.

Butler, RA and WF Laurance 2008. New strategies for conserving tropical forests. *Trends in Ecology and Evolution*, 15(9):469–472.

Groom, MJ, GK Meffe, & CR Carroll. 2006. *Principles of Conservation Biology*, 3d ed. Sinauer Associates, Sunderland, MA.

Hobbs, RJ. 2009. Looking for the silver lining: making the most of failure. *Restoration Ecology*, 17:1–3.

IUCN Red List of Threatened Species. 2008. www.iucnredlist.org.

Jackson, ST and RJ Hobbs. 2009. Ecological restoration in the light of ecological history. *Science*, 325:567–569. (and other related articles in the same issue)

Laurance, WF and DC Useche. 2009. Environmental synergisms and the extinction of tropical species. *Conservation Biology*, 23:1427–1437.

McCauley, DJ. 2006. Selling out on nature. *Nature*, 443:27–28.

Millennium Ecosystem Assessment. 2009. www.millenniumassessment.org/en/About.aspx.

Moritz, C, et al. 2008. Impact of a century of climate change on small-mammal communities in Yellowstone National Park, USA. *Science*, 322:261–264.

Morrison, JC, et al. 2007. Persistence of large mammal faunas as indicators of global human impacts. *Journal of Mammalogy*, 88:1363–1380.

O'Shea, TJ and DK Odell. 2008. Large-scale marine ecosystem change and the conservation of marine mammals. *Journal of Mammalogy*, 89:529–533.

Schipper, J, et al. 2008. The status of the world's land and marine mammals: diversity, threat, and knowledge. *Science*, 322:225–230.

Society for Ecological Restoration International. www.ser.org/default.asp.

Spellerberg, IF. 1998. Ecological effects of roads and traffic: a literature review. *Global Ecology and Biogeography Letters*, 7(5):317–333.

Sutherland, WJ, et al. 2009. One hundred questions of importance to the conservation of global biological diversity. *Conservation Biology*. http://www.uq.edu.au/spatialecology/docs/Publications/2009_Sutherland_etal_100QuestionsofImportance.pdf.

Vié, J-C, C Hilton-Taylor, and SN Stuart. 2009. *Wildlife in a Changing World: An Analysis of the 2008 IUCN Red List of Threatened Species*. International Union for the Conservation of Nature and Natural Resources. http://www.iucn.org/about/work/programmes/red_list.

Williams, SE, EE Bolitho, and S Fox. 2003. Climate change in Australian tropical rainforests: an impending environmental catastrophe. *Proceedings of the Royal Society*, 270:1887–1892.

Wilson, EO. 1984. *Biophilia*. Harvard University Press, Cambridge, MA.

Zahler, P. 2001. Endangered mammals. Pp. 441–454, in *Encyclopedia of Biodiversity*, vol. 2 (SA Levin, ed.) Academic Press, NY.

Bibliography

Note: This list of references is organized alphabetically by first author's last name. Publications by the same author are listed in chronological order. Following the list of single-authored works by a particular author are the multiauthored publications having the same first author, listed in chronological order. In cases where several authors share the same last name, the references are listed alphabetically by author's first initial. Works with more than three authors are listed alphabetically by the first author's last name followed by et al.

Ables, ED. 1969. Home range studies of red foxes (*Vulpes vulpes*). *Journal of Mammalogy*, 50:108–120.

Ackerman, D. 1991. *A Natural History of the Senses.* Knopf Doubleday Publishing Group, NY.

Adams, R & MA Hayes. 2008. Water availability and successful lactation by bats as related to climate change in arid regions of western North America. *Journal of Animal Ecology*, 77:1115–1121.

Adams, WM & M Mulligan. 2003. *Decolonizing Nature: Strategies for Conservation in a Post-colonial Era.* Earthscan Publications, London.

Adkins, RM & RL Honeycutt. 1993. A molecular examination of archontan and chiropteran monophyly, 227–249, in *Primates and Their Relatives in Phylogenetic Perspective* (RDE MacPhee, ed.). Plenum Press, NY.

Aguilar Soto, N, et al. 2008. Cheetahs of the deep sea: deep foraging sprints in short-finned pilot whales off Tenerife (Canary Islands). *Journal of Animal Ecology*, 77:936–947.

Agustí, J & M Antón. 2002. *Mammoths, Sabertooths, and Hominids: 65 Million Years of Mammalian Evolution in Europe.* Columbia University Press, NY.

Aiello, A. 1985. Sloth hair: unanswered questions, 213–218, in *The Evolution and Ecology of Armadillos, Sloths, and Vermilinguas* (GG Montgomery, ed.). Smithsonian Institution Press, Washington, DC.

Akersten, WA. 1985. Canine function in *Smilodon* (Mammalia; Felidae; Machairodontinae). *Contributions in Science Natural History Museum Los Angeles Co.*, 356:1–22.

Akersten, WA, et al. 2002. How and why do shrews have red teeth? *Journal of Vertebrate Paleontology*, 22(suppl 3):31A.

Akins, JB, et al. 2007. Flight speeds of three species of Neotropical bats: *Glossophaga soricina, Natalus stramineus,* and *Carollia subrufa. Acta Chiropterologica*, 9(2):477–482.

Alberdi, MT, G Leone, & EP Tonni (eds.). 1995. *Evolución Biológica y Climática de la Región Pampeana Durante los Últimos Cinco Millones de Años. Un Ensayo de correlación con el Mediterráneo Occidental.* Monografías Museo Nacional de Ciencias Naturales, Consejo Superior de Investigaciones Científicas, Madrid.

Albrecht, U. 2002. Functional genomics of sleep and circadian rhythm: invited review: regulation of mammalian circadian clock genes. *Journal of Applied Physiology*, 92:1348–1355.

Albuja, VL & BD Patterson. 1996. A new species of northern shrew-opossum (Paucituberculata: Caenolestidae) from the Cordillera del Cóndor, Ecuador. *Journal of Mammalogy*, 77:41–53.

Alcorn, SM, et al. 1959. Pollination requirements of the saguaro (*Carnegiea gigantea*). *Cactus Succulents Journal*, 31:39–41.

Aldridge, HDJN & IL Rautenbach. 1987. Morphology, echolocation and resource partitioning in insectivorous bats. *Journal of Animal Ecology*, 56:763–778.

Alerstam, T, A Hedenstrom, & S Akesson. 2003. Long-distance migration: evolution and determinants. *Oikos*, 103:247–260.

Alexander, RD. 1974. The evolution of social behavior. *Annual Review of Ecology and Systematics*, 5:325–383.

Alexander, RD, KM Noonan, & BJ Crespi. 1991. The evolution of eusociality, 3–44, in *The Biology of the Naked Mole-rat* (PW Sherman, JUM Jarvis, & RD Alexander, eds.). Princeton University Press, Princeton, NJ.

Alexander, RM. 1982. *Locomotion of Animals.* Blackie, Glasgow.

Alexander, RM. 1992. *Exploring Biomechanics: Animals in Motion.* Scientific American Library, NY.

Allard, MW, et al. 1992. DNA systematics and evolution of the artiodactyl family Bovidae. *Proceedings of the National Academy of Sciences*, 89:3972–3976.

Allen, GM. 1940. *The Mammals of China and Mongolia, Part 2.* American Museum Natural History, NY, pp. 621–1350.

Allin, EF. 1975. Evolution of the mammalian middle ear. *Journal of Morphology,* 147:403–437.

Allin, EF & JA Hopson. 1992. Evolution of the auditory system in Synapsida ("mammal-like reptiles" and primitive mammals) as seen in the fossil record, 587–614, in *The Evolutionary Biology of Hearing* (DP Webster, RR Fay, & AN Popper, eds.). Springer-Verlag, NY.

Allison, JJ & HM Cook. 1981. Oxalate degradation by microbes of the large bowel of herbivores: the effect of dietary oxalate. *Science,* 212:675–676.

Altenbach, JS. 1977. Functional morphology of two bats: *Leptonycteris* and *Eptesicus.* Special Publication No. 5, American Society of Mammalogists.

Altenbach, JS. 1979. Locomotor morphology of the vampire bat *Desmodus rotundus.* Special Publication, No. 6, American Society of Mammalogists.

Altenbach, JS & JW Hermanson. 1987. Bat flight muscle function and the scapulo-humeral lock, 100–118, in *Recent Advances in the Study of Bats* (MB Fenton, P Racey, & JMV Rayner, eds.). Cambridge University Press, Cambridge, UK.

Alter, SE, E Rynes, & SR Palumbi. 2007. Evidence for historic population size and past ecosystem impacts of gray whales. *Proceedings of the National Academy of Sciences USA,* 104:15,162–165,167.

Altman, PL & DS Dittmer. 1964. *Biology Data Book.* Federation American Society Experimental Biology, Washington, DC.

Altmann, SA & J Altmann. 1970. *Baboon Ecology.* University of Chicago Press, Chicago, IL.

Amico, G & MA Aizen. 2000. Mistletoe seed dispersal by a marsupial. *Nature,* 408:929–930.

Andersen, DC & JA MacMahon. 1981. Population dynamics and bioenergetics of a fossorial herbivore, *Thomomys talpoides* (Rodentia: Geomyidae), in a spruce-fir sere. *Ecological Monographs,* 51:179–202.

Anderson, HL & PC Lent. 1977. Reproduction and growth of the tundra hare (*Lepus othos*). *Journal of Mammalogy,* 458:53–57.

Anderson, P. 1995. Competition, predation and the evolution and extinction of Steller's sea cow, *Hydrodamalis gigas. Marine Mammal Science,* 11:391–454.

Anderson, PK. 1997. Shark Bay dugongs in summer. I. Lek mating. *Behaviour,* 134:433–462.

Anderson, PK. 2002. Habitat, niche, and evolution of sirenian mating systems. *Journal of Mammalian Evolution,* 44:55–98.

Anderson, PK. 2005. Competition, predation, and the evolution and extinction of Steller's sea cow, *Hydrodamalis gigas. Marine Mammal Science,* 11(3):391–394.

Anderson, PK & RMR Barclay. 1995. Acoustic signals of solitary dugongs: physical characteristics and behavioral correlates. *Journal of Mammalogy,* 76:1226–1237.

Anderson, S. 1984. *Orders and Families of Recent Mammals of the World.* Wiley, NY.

Andersson, A. 1969. Communication in the lesser bushbaby (*Galago senegalensis moholi*). Unpubl. M.S. thesis, University of Witwatersrand.

Andrews, P. 1988. A phylogenetic analysis of the Primates, 143–175, in *The Phylogeny and Classification of the Tetrapods. Vol. 2 Mammals* (MJ Benton, ed.). Systematics Association Spec. Vol. 35B, Clarendon Press, Oxford, UK.

Andrews, P & L Martin. 1987. Cladistic relationships of extant and fossil hominoids, 101–118, in *Primate Phylogeny* (FE Grine, JG Fleagle, & LB Martin, eds.). Academic Press, NY.

Anhuf, D, et al. 2006. Paleo-environmental change in Amazonian and African rainforest during the LGM. *Palaeogeography, Palaeoclimatology, Palaeoecology,* 239:510–527.

Annison, EF & D Lewis. 1959. *Metabolism in the Rumen.* Methuen, London.

Ar, A, R Arieli, & A Shkolnik. 1977. Blood-gas properties and function in the fossorial mole-rat under normal and hypoxic-hypercapnic atmospheric conditions. *Respiratory Physiology,* 30:201–218.

Archer, M. 1984. Origins and early radiations of marsupials, 585–625, in *Vertebrate Zoogeography and Evolution in Australasia* (M Archer & G Clayton, eds.). Hesperian Press, Victoria Park, Australia.

Archer, M & G Clayton. 1984. *Vertebrate Zoogeography and Evolution in Australasia.* Hesperian Press, Victoria Park, Australia.

Archer, M & L Dawson. 1982. Revision of marsupial lions of the genus *Thylacoleo* Gervais (Thylacoleonidae, Marsupialia) and thylacoleonid evolution in the late Cainozoic, 477–494, in *Carnivorous Marsupials* (M Archer, ed.). Royal Zoological Society, New South Wales, Sydney, Australia.

Archer, M, S Hand, & H Godthelp. 1991. *Riversleigh: The Story of Animals in Ancient Rainforests of Inland Australia.* Reed Books, Balgowlah, New South Wales, Sydney.

Archer, M, et al. 1992. Description of the skull and non-vestigial dentition of a Miocene platypus (*Obdurodon dicksoni* n. sp.) from Riversleigh, Australia, and the problem of monotreme origins, 15–27, in *Platypus and Echidnas* (ML Augee, ed.). Royal Zoological Society, New South Wales, Sydney.

Archer, M, et al. 1993. Reconsideration of monotreme relationships based on the skull and dentition of the Miocene *Obdurodon dicksoni,* 75–94, in *Mammal Phylogeny, vol. 1. Mesozoic Differentiation, Multituberculates, Monotremes, Early Therians, and Marsupials* (FS Szalay, MJ Novacek, & MC McKenna, eds.). Springer-Verlag, NY.

Archer, M & J Kirsch. 2006. The evolution and classification of marsupials, 1–21, in *Marsupials* (PJ Armati, CR Dickman, & ID Hume, eds.). Cambridge University Press, Cambridge, UK.

Archibald, JD. 2003. Timing and biogeography of the eutherian radiation: fossils and molecules compared. *Molecular Phylogenetics and Evolution,* 28:350–359.

Archibald, JD, AO Averianov, & EG Ekdale. 2001. Late Cretaceous relatives of rabbits, rodents, and other extant eutherian mammals. *Nature,* 414:62–65.

Archie, EA, et al. 2006. Dominance rank relationships among wild female African elephants (*Loxodonta africana*). *Animal Behaviour,* 71:117–127.

Arey, LB. 1974. *Developmental Anatomy.* 7th ed. W. B. Saunders, Philadelphia, PA.

Argilés, JM, et al. 2007. Antiproteolytic effects of plasma from hibernating bears: a new approach for muscle wasting therapy? *Clinical Nutrition,* 26:658–661.

Arieli, R & A Ar. 1981a. Heart rate responses of the mole rat (*Spalax ehrenbergi*) in hypercapnic, hypoxic, and cold conditions. *Physiological Zoology,* 54:14–21.

Arieli, R & A Ar. 1981b. Blood capillary density in heart and skeletal muscles of the fossorial mole rat. *Physiological Zoology,* 54:22–27.

Arita, HT. 1990. Noseleaf morphology and ecological correlates in phyllostomid bats. *Journal of Mammalogy,* 71:36–47.

Arita, HT & MB Fenton. 1997. Flight and echolocation in the ecology and evolution of bats. *Trends in Ecology and Evolution,* 12:53–58.

Arita, HT, P Rodríguez, & E Vázquez-Domínguez. 2005. Continental and regional ranges of North American mammals: Rapoport's rule in real and null worlds. *Journal of Biogeography,* 32:961–971.

Arlettaz, R, P Christe, & M Desfayes. 2002. 33 years: a new longevity record for a European bat. *Mammalia,* 66(3):441–442.

Armbruster, P & R Lane. 1993. A population viability analysis for African elephant (*Loxodonta africana*): how big should reserves be? *Conservation Biology,* 7:602–610.

Armitage, KB. 1981. Sociality as a life-history tactic of ground squirrels. *Oecologia,* 48:36–49.

Armitage, KB. 1986. Marmot polygyny revisited: determinants of male and female reproductive strategies, 303–331, in *Ecological Aspects of Social Evolution* (DI Rubenstein & RW Wrangham, eds.). Princeton University Press, Princeton, NJ.

Armitage, KB. 1998. Reproductive strategies of yellow-bellied marmots: energy conservation and differences between the sexes. *Journal of Mammalogy,* 79:385–393.

Armitage, KB & OA Schwartz. 2000. Social enhancement of fitness in yellow-bellied marmots. *Proceedings of the National Academy of Sciences,* 106:12149–12152.

Armstrong, E. 1983. Relative brain size and metabolism in mammals. *Science,* 220:1302–1304.

Arnason, U, et al. 2000. The mitochondrial genome of the sperm whale and a new molecular reference for estimating eutherian divergence dates. *Journal of Molecular Evolution,* 50:569–578.

Arnason, U, et al. 2006. Pinniped phylogeny and a new hypothesis for their origin and dispersal. *Molecular Phylogenetics and Evolution,* 41:345–354.

Asdell, SA. 1964. *Patterns of Mammalian Reproduction.* Cornell University Press, Ithaca, NY.

Asher, RJ. 2005. Insectivoran-grade placental mammal, 50–70, in *The Rise of Placental Mammals: Origins and Relationships of the Major Clades* (JD Archibald & K Rose, eds.). Johns Hopkins University Press, Baltimore, MD.

Asher, RJ, et al. 2002. Morphology and relationships of *Apternodus* and other extinct zalambdodonts, placental mammals. *Bulletin American Museum Natural History,* 273:1–117.

Asher, RJ, MJ Novacek, & JH Geisler. 2003. Relationships of endemic African mammals and their fossil relatives based on morphological and molecular evidence. *Journal of Mammalian Evolution,* 10:131–194.

Asher, RJ & M Hofreiter. 2006. Tenrec phylogeny and the noninvasive extraction of nuclear DNA. *Systematic Biology,* 55:181–194.

Asher, RJ, JH Geisler, & MR Sánchez-Villagra. 2008. Morphology, paleontology, and placental mammal phylogeny. *Systematic Biology,* 57:311–317.

Ashton, KG, MC Tracy, & A deQueiroz. 2000. Is Bergman's rule valid for mammals? *American Naturalist,* 156:390–415.

Atkinson, IAE & EK Cameron. 1993. Human influence on the terrestrial biota and biotic communities of New Zealand. *Trends in Ecology and Evolution,* 8:447–451.

Atsatt, PR & T Ingram. 1983. Adaptation to oak and other fibrous, phenolic-rich foliage by a small mammal, *Neotoma fuscipes. Oecologia,* 60:135–142.

Attwell, CAM. 1982. Population ecology of the blue wildebeest *Connochaetes taurinus taurinus* in Zululand, South Africa. *African Journal of Ecology,* 20:147–168.

Au, WWL, et al. 1974. Measurement of echolocation signals of the Atlantic bottlenose dolphin, *Tursiops truncatus* Montagu, in open waters. *Journal of the Acoustical Society America,* 56:1280–1290.

Au, WWL & KJ Benoit-Bird. 2003. Automatic gain control in the echolocation system of dolphins. *Nature,* 423: 861–863.

Audet, D. 1990. Foraging behavior and habitat use by a gleaning bat, *Myotis myotis. Journal of Mammalogy,* 71:420–427.

Audet, D, et al. 1988. Foraging strategies and use of space by the Indian false vampire, *Megaderma lyra* (Megadermatidae). *Bat Research News,* 29:43.

Augee, ML & BA Gooden. 1992. Evidence for electroreception from field studies of the echidna, *Tachyglossus aculeatus,* 211–215, in *Platypus and Echidnas* (ML Augee, ed.). Royal Zoology Society of New South Wales, Sydney, Australia.

Augee, ML, BA Gooden, & AM Musser. 2006. *Echidna—Extraordinary Egg-laying Mammal.* Collingwood, Victoria, Australia.

Aumann, GD. 1965. Microtine abundance and soil sodium levels. *Journal of Mammalogy,* 46:594–612.

Avise, JC. 2000. *Phylogeography: The History and Formation of Species.* Harvard University Press, Cambridge, MA.

Ayala, FJ. 1999. Molecular clock mirages. *BioEssays,* 21:71–75.

Ayala, FJ. 2000. Neutralism and selectionism: the molecular clock. *Gene,* 261:27–33.

Backus, RH & WE Schevill. 1966. *Physeter* clicks, 510–528, in *Whales, Dolphins and Porpoises* (KS Norris, ed.). University of California Press, Berkeley, CA.

Badgely, C & DL Fox. 2000. Ecological biogeography of North American mammals: species density and ecological structure in relation to environmental gradients. *Journal of Biogeography,* 27:1437–1467.

Bailey, WJ, JL Slightom, & M Goodman. 1992. Rejection of the "flying primate" hypothesis by phylogenetic evidence from the e-globin gene. *Science,* 256:86–89.

Baker, HG. 1961. The adaptation of flowering plants to nocturnal and crepuscular pollinators. *Quarterly Review of Biology,* 36:64–73.

Baker, HG. 1973. Evolutionary relationships between flowering plants and animals in American and African tropical forests, 145–159, in *Tropical Forest Ecosystems in Africa and South America: A Comprehensive Review* (BJ Meggers, ES Ayensu, & WD Duckworth, eds.). Smithsonian Institution Press, Washington, DC.

Baker, JR & RM Ranson. 1933. Factors affecting the breeding of the field mouse (*Microtus agrestis*). *Proceedings Royal Society of London,* 113B:486–495.

Baker, RJ & RD Bradley. 2006. Speciation in mammals and the genetic species concept. *Journal of Mammalogy,* 37(4):643–667.

Baker, RJ, et al. 2003. Diversification among New World leaf-nosed bats: an evolutionary hypothesis and classification inferred from digenomic congruence of DNA sequence. *Occasional Papers, Museum of Texas Tech University,* 230:1–32.

Baker, WW, SG Marshall, & VB Baker. 1968. Autumn fat deposition in the evening bat (*Nycticeius humeralis*). *Journal of Mammalogy,* 49:314–317.

Baldauf, J. 1982. Identification of the Holocene-Pleistocene boundary in the Bering Sea by diatoms. *Boreas,* 11:113–118.

Ballard, KA & KM Kovacs. 1995. The acoustic repertoire of hooded seals (*Cystophora cristata*). *Canadian Journal of Zoology*, 73:1362–1374.

Balmford, A, et al. 2002. Economic reasons for conserving wild nature. *Science*, 297:950–953.

Barash, DP. 1989. *Marmots: Social Behavior and Ecology*. Stanford University Press, Stanford, CA.

Barclay, RMR. 1982. Interindividual use of echolocation calls: eavesdropping by bats. *Behavioral Ecology and Sociobiology*, 10:271–275.

Barclay, RMR. 1983. Echolocation calls of emballonurid bats from Panama. *Journal of Comparative Physiology A*, 151:515–520.

Barclay, RMR. 1986. The echolocation calls of hoary (*Lasiurus cinereus*) and silver-haired (*Lasionycteris noctivagans*) bats as adaptations for long- vs. short-range foraging strategies and the consequences of prey selection. *Canadian Journal of Zoology*, 64:2700–2705.

Barclay, RMR. 1989. The effect of reproductive condition on the foraging behavior of female hoary bats, *Lasiurus cinereus*. *Behavioral Ecology and Sociobiology*, 24:31–37.

Barclay, RMR, et al. 1981. Echolocation calls produced by *Trachops cirrhosus* (Chiroptera: Phyllostomatidae) hunting for frogs. *Canadian Journal of Zoology*, 59:750–753.

Barfield, RJ & LA Geyer. 1972. Sexual behavior: ultrasonic postejaculatory song of the male rat. *Science*, 176:1349–1350.

Bargo, MS, N Toledo, & SF Vizcaíno. 2006. Muzzle of South American Pleistocene ground sloths (Xenarthra, Tardigrada). *Journal of Morphology*, 267:248–263.

Barkalow, FS Jr, RB Hamilton, & RF Soots Jr. 1970. The vital statistics of an unexploited gray squirrel population. *Journal of Wildlife Management*, 34:489–500.

Barker, WR & PJM Greenslade. 1982. *Evolution of the Flora and Fauna of Arid Australia*. Peacock Publications, Adelaide, Australia.

Barklow, WE. 1997. Some underwater sounds of the hippopotamus (*Hippopotamus amphibius*). *Marine and Freshwater Behaviour and Physiology*, 29:237–249.

Barklow, WE. 2004. Low-frequency sounds and amphibious communication in *Hippopotamus amphibious* (A). *Journal of the Acoustical Society of America*, 115:2555A.

Barnes, BM. 1989. Freeze avoidance in a mammal: body temperatures below 0° C in an Arctic hibernator. *Science*, 244:1593–1595.

Barnett, R, et al. 2005. Evolution of the extinct sabre-tooths and the American cheetahlike cat. *Current Biology*, 15: R1–R2.

Barnosky, AD. 1982. Locomotion in moles (Insectivora, Proscalopidae) from the middle Tertiary of North America. *Science*, 216:183–185.

Barry, RE & PJ Mundy. 2002. Seasonal variation in the degree of heterosexual association of two syntopic hyraxes (*Heterohyrax brucei* and *Procavia capensis*) exhibiting synchronous parturition. *Behavioral Ecology and Sociobiology*, 52:177–181.

Barsanti, L, et al. 2008. Oddities and curiosities in the algal world. 353–391, in *Algal Toxins: Nature, Occurrence, Effect and Detection*. (V. Evangelista et al., eds.). NATO Science for Peace and Security Series A: Chemistry and Biology. Springer Science + Business Media B. V., The Netherlands.

Bartholomew, GA. 1964. The roles of physiology and behavior in the maintenance of homeostasis in the desert environment. *Symposium Society Experimental Biology*, 18:7–29.

Bartholomew, GA & JW Hudson. 1960. Aestivation in the Mohave ground squirrel, *Citellus mohavensis*. *Bulletin Museum Comparative Zoology*, Harvard, 124:193–208.

Bartholomew, GA & JW Hudson. 1961. Desert ground squirrels. *Scientific American*, 205(5):107–116.

Bartholomew, GA & NE Collias. 1962. The role of vocalization in the social behavior of the northern elephant seal. *Animal Behaviour*, 10:7–14.

Bartholomew, GA, WR Dawson, & RC Lasiewski. 1970. Thermoregulation and heterothermy in some of the smaller flying foxes (Megachiroptera) of New Guinea. *Z. Vergleichende Physiologie*, 70:196–209.

Bartholomew, GA & M Rainy. 1971. Regulation of body temperature in the rock hyrax, *Heterohyrax brucei*. *Journal of Mammalogy*, 52:81–95.

Bartness, TJ & GN Wade. 1984. Photoperiodic control of body weight and energy metabolism in Syrian hamsters (*Mesocricetus auratus*): role of pineal

gland, melatonin, gonads, and diet. *Endocrinology*, 114:492–498.

Baskin, JA. 1978. *Bensonomys, Calomys*, and the origin of the phyllotine group of Neotropical cricetines (Rodentia: Cricetidae). *Journal of Mammalogy*, 59:125–135.

Baskin, JA. 1982. Tertiary Procyoninae (Mammalia: Carnivora) of North America. *Journal of Vertebrate Paleontology*, 2:71–93.

Baskin, JA. 1989. Comments on New World Tertiary Procyonidae (Mammalia, Carnivora). *Journal of Vertebrate Paleontology*, 9:110–117.

Baskin, Y. 1993. Blue whale populations may be increasing off California. *Science*, 260:287.

Bateman, GC & TA Vaughan. 1974. Nightly activities of mormoopid bats. *Journal of Mammalogy*, 55:45–65.

Bateman, JA. 1959. Laboratory studies of the golden mole and the mole rat. *African Wildlife*, 13.

Bates, LA, et al. 2008. African elephants have expectations about the locations of out-of-sight family members. *Biology Letters*, 4:34–36.

Batzli, GO & FA Pitelka. 1970. Influence of meadow mouse populations on California grassland. *Ecology*, 51:1027–1039.

Batzli, GO & FA Pitelka. 1971. Conditions and diet of cycling populations of the California vole, *Microtus californicus*. *Journal of Mammalogy*, 58:141–163.

Batzli, GO, et al. 1980. The herbivore-based trophic system, 378–381, in *An Arctic Ecosystem: The Coastal Tundra at Barrow, Alaska*. (J Brown, PC Miller, LL Tieszen, FL Bunnell, & SF MacLean, eds.). Dowden, Hutchinson, and Ross, Stroudsburg, PA.

Bauchop, T. 1978. Digestion of leaves in vertebrate arboreal folivores, 193–204, in *The Ecology of Arboreal Folivores* (GG Montgomery, ed.). Smithsonian Institution Press, Washington, DC.

Baudinette, RV. 1982. The energetics of locomotion in dasyurid marsupials, 261–265, in *Carnivorous Marsupials*. Vol. 1 (M Archer, ed.). Mosmoan, Royal Zoology Society of New South Wales.

Baudinette, RV & AA Biewener. 1998. Young wallabies get a free ride. *Nature*, 395:653.

Baumgartner, MF & BR Mate. 2003. Summer-time foraging ecology of North Atlantic right whales. *Marine Ecology Progress Series*, 264:123–135.

Baumgartner, MF, CA Mayo, & RD Kenny. 2007. Enormous carnivores, microscopic food, and a restaurant that's hard to find, 138–171, in *The Urban Whale* (SD Kraus and RM Rolland, eds.). Harvard University Press, Cambridge, MA.

Baverstock, PR & B Green. 1975. Water recycling in lactation. *Science*, 187:657–658.

Baverstock, PR, et al. 1979. Water balance of small, lactating rodents: the total water balance picture of the mother–young unit. *Comparative Biochemistry and Physiology*, 63:247–252.

Bazzaz, F, et al. 1998. Ecological science and the human predicament. *Science*, 282:879.

Beard, KC. 1998. East of Eden: Asia as an important center of taxonomic origin in mammalian evolution, 5–39, in *Dawn of the Age of Mammals in Asia* (KC Beard & MR Dawson, eds.). *Bulletin of Carnegie Museum of Natural History* 34, Pittsburgh, PA.

Beard, KC & MR Dawson. 1998. Dawn of the age of mammals in Asia. *Bulletin of Carnegie Museum of Natural History* 34, Pittsburgh, PA.

Beard, PH. 1977. *The End of the Game*. Doubleday, Garden City, NY.

Beardsley, RC, et al. 1996. Spatial variability in zooplankton abundance near feeding right whales in the Great South Channel. *Deep Sea Research II*, 43:1601–1625.

Beatley, JC. 1969. Dependence of desert rodents on winter annuals and precipitation. *Ecology*, 50:721–724.

Beatley, JC. 1976. Rainfall and fluctuating plant populations in relation to distributions and numbers of desert rodents in southern Nevada. *Oecologia*, 24:21–42.

Beck, RMD, et al. 2008. Australia's oldest marsupial fossils and their biogeographical implications. *PLoS ONE* 3(3):e1858.

Bedford, JM. 1977. Evolution of the scrotum: the epididymis as the prime mover? 171–182, in *Reproduction and Evolution* (JH Calaby & CH Tyndale-Biscoe, eds.). Australian Academy of Science, Canberra City.

Bedford, MJ. 2004. Enigmas of mammalian gamete form and function. *Biological Reviews*, 79:429–460.

Beer, JR. 1961. Seasonal reproduction in the meadow vole. *Journal of Mammalogy*, 42:483–489.

Beer, JR, R Lukens, & D Olson. 1954. Small mammal populations on the islands of Basswood Lake, Minn. *Ecology*, 35:437–445.

Beier, P, D Choate, & RH Barrett. 1995. Movement patterns of mountain lions during different behaviors. *Journal of Mammalogy*, 76:1056–1070.

Bekoff, M & MC Wells. 1980. The social ecology of coyotes. *Scientific American*, 242(4):130–148.

Bel'kovich, VM & AV Yablokov. 1963. Marine animals "share experience" with designers. *Nauka Zhizn* 8, 30:61.

Bell, CJ. 1998. North American Quaternary land mammal ages and the biochronology of North American microtine rodents, 2-605 to 2-645, in *Dating and Earthquakes: Review of Quaternary Geochronology and Its Application to Paleoseismology* (J. Sowers, JS Noller, & WR Lettis, eds.). U.S. Nuclear Regulatory Commission NUREG/CR 5562.

Bell, GP. 1982a. Behavioral and ecological aspects of gleaning by the desert insectivorous bat, *Antrozous pallidus* (Chiroptera: Vespertilionidae). *Behavioral Ecology and Sociobiology*, 10:217–223.

Bell, GP. 1982b. Prey location and sensory ecology of two species of gleaning insectivorous bats, *Antrozous pallidus* (Vespertilionidae) and *Macrotus californicus* (Phyllostomatidae). Ph.D. thesis, Carleton University, Ottawa.

Bell, GP & MB Fenton. 1984. The use of Doppler-shifted echoes as a flutter detection and clutter rejection system: the echolocation and feeding behavior of *Hipposideros ruber* (Chiroptera: Hipposideridae). *Behavioral Ecology and Sociobiology*, 15:109–114.

Bell, RHV. 1971. A grazing ecosystem in the Serengeti. *Scientific American*, 225:86–93.

Belsky, AJ. 1987. The effects of grazing: confounding of ecosystem, community, and organism scales. *American Naturalist*, 129:777–783.

Bennett, AF & JA Ruben. 1979. Endothermy and activity in vertebrates. *Science*, 206:649–654.

Bennett, AF & JA Ruben. 1986. The metabolic and thermoregulatory status of therapsids, 207–218, in *The Ecology and Biology of Mammal-like Reptiles* (N Hotton III, PD MacLean, JJ Roth, & EC Roth, eds.). Smithsonian Institution Press, Washington, DC.

Bennett, NC & JUM Jarvis. 1988. The social substructure and reproductive biology of colonies of the mole-rat, *Cryptomys damarrensis* (Rodentia: Bathyergidae). *Journal of Mammalogy*, 69:293–302.

Bennett, NC, JUM Jarvis, & FPD Cotterill. 1993. Poikilothermic traits and thermoregulation in the Afrotropical social subterranean mole-rat (*Cryptomys hottentotus darlingi*) (Rodentia: Bathyergidae). *Journal of Zoology London*, 231:179–186.

Bennett, NC, et al. 1994a. Reproductive repression in eusocial *Cryptomys damarensis* colonies: socially-induced infertility in females. *Journal of Zoology London*, 233:617–630.

Bennett, NC, et al. 1994b. Thermoregulation in three species of Afrotropical subterranean mole-rats (Rodentia; Bathyergidae) from Zambia and Angola and scaling within the genus *Cryptomys*. *Oecologia*, 97:222–227.

Bennett, NC & JUM Jarvis. 1995. Coefficients of digestibility and nutritional values of geophytes and tubers eaten by southern African mole-rats (Rodentia: Bathyergidae). *Journal of Zoology London*, 236:189–198.

Benshemesh, J & K Johnson. 2003. Biology and conservation of marsupial moles (*Notoryctes*), 464–474, in *Predators with Pouches: The Biology of Carnivorous Marsupials* (M Jones, C Dickman, & M Archer, eds.). CSIRO Publishing, Collingwood, Victoria, Australia.

Benson, SB. 1933. Concealing coloration among some desert rodents of the southwestern United States. *University California Publications in Zoology*, 40:1.

Benson, SB & AE Borell. 1931. Notes on the life history of the red tree mouse, *Phenacomys longicaudus*. *Journal of Mammalogy*, 12:226–233.

Benstead, JP, KH Barnes, & CM Pringle. 2001. Diet, activity patterns, foraging movement and responses to deforestation of the aquatic tenrec *Limnogale mergulus* (Lipotyphla: Tenrecidae) in eastern Madagascar. *Journal of Zoology*, 254:119–129.

Benstead, JP & LE Olson. 2003. *Limnogale mergulus*, web-footed tenrec or aquatic tenrec, 1267–1273, in *The Natural History of Madagascar* (SM Goodman & JP Benstead, eds.). University of Chicago Press, Chicago, IL.

Benton, MJ. 1990. *Vertebrate Palaeontology*. Unwin Hyman, London.

Benton, MJ. 1997. *Vertebrate Palaeontology*. 2nd ed. Chapman and Hall, London.

Benton, MJ. 2004. *Vertebrate Palaeontology.* 3rd ed. Wiley-Blackwell, NY.

Berg, LS., 1950. *Natural Regions of the U.S.S.R.* Macmillan, NY.

Berger, J. 2004. The last mile: how to sustain long-distance migration in mammals. *Conservation Biology*, 18:320–331.

Berger, J, SL Cain, & KM Berger. 2006. Connecting the dots: an invariant migration corridor links the Holocene to the present. *Biology Letters*, 2:528–531.

Berger, PJ, et al. 1981. Chemical triggering of reproduction in *Microtus montanus. Science*, 214:69–70.

Berger, PJ, N Negus, & M Day. 1997. Recognition of kin and avoidance of inbreeding in the montane vole (*Microtus montanus*). *Journal of Mammalogy*, 78:1182–1186.

Bergqvist, LP. 2003. The role of teeth in mammal history. *Brazilian Journal of Oral Science*, 2:249–257.

Bergqvist, LP, ÉAL Abrantes, & LdS Avila. 2004. The Xenarthra (Mammalia) of São José de Itaboraí Basin (upper Paleocene, Itaboraian), Rio de Janeiro, Brazil. *Geodiversitas*, 26:323–337.

Bergqvist, LP, A de Lima Moreira, & D Ribeiro Pinto. 2006. Bacia de São José de Itaboraí: 75 anos de história e ciência. *Serviço Geológico do Brasil, Companhia de Pesquisa de Recursos Minerais:* 84.

Bergstedt, B. 1965. Distribution, reproduction, growth and dynamics of the rodent species *Clethrionomys glareolus* (Schreber), *Apodemus flavicollis* (Melchior) and *Apodemus sylvaticus* (Linne) in southern Sweden. *Oikos*, 16:132–160.

Bergström, R. 1992. Browse characteristics and impact of browsing on trees and shrubs in African savannas. *Journal of Vegetation Science*, 3:315–324.

Berkes, F, J Colding, & C Folke. 2000. Rediscovery of traditional ecological knowledge as adaptive management. *Ecological Applications*, 10:1251–1262.

Bernard, HJ & AA Hohn. 1989. Differences in feeding habits between pregnant and lactating spotted dolphins (*Stenella attenuata*). *Journal of Mammalogy*, 70:211–215.

Bernard, RTF & GS Cumming. 1997. African bats: evolution of reproductive patterns and delays. *Quarterly Review of Biology*, 72:253–274.

Berta, A, CE Ray, & AR Wyss. 1989. Skeleton of the oldest known pinniped, *Enaliarctos mealsi. Science*, 244:60–62.

Berta, A, JS Sumich, & KM Kovacs. 2005. *Marine Mammals, an Evolutionary Approach.* 2nd ed. Academic Press, NY.

Berta, A, JL Sumich, & KM Kovacs. 2006. *Marine Mammals: Evolutionary Biology.* 2nd ed. Academic Press, San Diego, CA.

Berta, A, et al. 2006. *Marine Mammals Evolutionary Biology.* Academic Press, NY.

Berthold, P. 2001. *Bird migration: A general survey.* Oxford Ornithology Series, Oxford University Press, NY.

Bertram, BCR. 1973. Lion population regulation. *East African Wildlife Journal*, 11:215–225.

Bertram, BCR. 1975. The social system of lions. *Scientific American*, 232(5):54–65.

Bertram, GCL. 1940. The biology of the Weddell and crabeater seals, with a study of the comparative behavior of the Pinnipedia. *British Museum (Natural History) Science Reports Brit. Graham Land Expedition*, 1934–1937, 1:1–139.

Berzin, AA. 1971. *The Sperm Whale.* Pacific Sci. Res. Inst. Fisheries Oceanogr. Trans. 1972, Israel.

Betancourt, JL, TR Van Devender, & PS Martin. 1990. *Packrat Middens, the Last 40,000 Years of Biotic Change.* University of Arizona Press, Tucson.

Bieber, C. 1998. Population dynamics, sexual activity, and reproductive failure in the fat dormouse (*Myoxus glis*). *Journal of Zoology London*, 244:223–229.

Biewener, AA & RV Baudinette. 1995. In vivo muscle force and elastic energy storage during steady-speed hopping of tammar wallabies (*Macropus eugenii*). *Journal of Experimental Biology*, 198:1829–1841.

Bininda-Emonds, ORP, et al. 2007. The delayed rise of present-day mammals. *Nature*, 446:507–512, and 2008, Corrigendum, 456:274.

Björnhag G, et al. 1994. The gastrointestinal tract of the rock hyrax (*Procavia habessinica*). 1. Morphology and motility patterns of the tract. *Comparative Biochemistry and Physiology A*, 109:649–653.

Blackburn, DG, V Hayssen, & CJ Murphy. 1989. The origins of lactation and the evolution of milk: a review with new hypotheses. *Mammal Review*, 19:1–26.

Blackburn, ST. 2007. *Maternal, Fetal, and Neonatal Physiology, a Clinical Perspective.* Saunders-Elsevier, Philadelphia, PA.

Blair, WF. 1951. Evolutionary significance of geographic variation in population density. *Texas Journal of Science*, 1:53–57.

Blake, EW Jr. 1887. The coast fox. West. *American Scientist*, 3:49.

Blanc, JJ, et al. 2003. *African Elephant Status Report 2002: An Update from the African Elephant Database.* IUCN, Gland and Cambridge.

Blanton, JD, et al. 2008. Rabies surveillance in the United States during 2007. *Journal of the American Veterinary Medical Association*, 233(6):884.

Blatt, CM, CR Taylor, & MB Habal. 1972. Thermal panting in dogs: the lateral nasal gland, a source of water for evaporative cooling. *Science*, 177:804–805.

Blix, AS, HJ Grav, & K Ronald. 1979. Some aspects of temperature regulation in newborn harp seal pups. *American Journal of Physiology*, 236:R188–197.

Bloch, JI & DM Boyer. 2002. Grasping primate origins. *Science*, 298:1606–1610.

Bloch, JI & DM Boyer. 2003. Response to comment on "Grasping Primate Origins." *Science*, 300:741c.

Bloedel, P. 1955. Hunting methods of fish-eating bats, particularly *Noctilio leporinus. Journal of Mammalogy*, 36:390–399.

Blumstein, DT & KB Armitage. 1998. Life history consequences of social complexity: a comparative study of ground-dwelling sciurids. *Behavioral Ecology*, 9:8–19.

Boccadori, SJ, et al. 2008. Yellowstone pronghorn alter resource selection after sagebrush decline. *Journal of Mammalogy*, 89:1031–1040.

Bodenheimer, FS. 1949. *Problems of Vole Populations in the Middle East: Report on the Population Dynamics of the Levant Vole (Microtus guentheri D.).* Azriel Print Works, Jerusalem.

Bodmer, RE & GB Rabb. 1992. *Okapia johnstoni. Mammalian Species*, 422:1–8.

Boellstorff, DE, et al. 1994. Reproductive behavior and multiple paternity of California ground squirrels. *Animal Behaviour*, 47:1057–1064.

Bogdanowicz, W & RD Owen. 1992. Phylogenetic analysis of the bat family Rhinolophidae. *Zeitschrift fur Zoologische Systematik und Evolutionsforschung*, 30:142–160.

Bogdanowicz, W, RD Csada, & MB Fenton. 1997. Structure of noseleaf, echolocation, and foraging behavior in the Phyllostomidae (Chiroptera). *Journal of Mammalogy*, 78:942–953.

Bogdanowicz, W & RD Owen 1998. In the minotaur's labyrinth: phylogeny of the bat family Hipposideridae, 27–42, in *Bat*

Biology and Conservation (TH Kunz & PA Racey, eds.). Smithsonian Institution Press, Washington, DC.

Bogdanowicz, W, MB Fenton, & K Daleszczyk. 1999. The relationships between echolocation calls, morphology, and diet in insectivorous bats. *Journal of Zoology, London*, 247:381–393.

Boisserie, J-R. 2007. Family Hippopotamidae, 106–119, in *The Evolution of Artiodactyls* (DR Prothero & SE Foss, eds.). Johns Hopkins University Press, Baltimore, MD.

Boissere, J-R, F Lihoreau, & M Brunet. 2005. The position of Hippopotamidae within Cetartiodactyla. *Proceedings of the National Academy of Sciences of the United States of America*, 102:1537–1541.

Bonaccorso, FJ & BK McNab. 1997. Plasticity of energetics in blossom bats (Pteropodidae): impact on distribution. *Journal of Mammalogy*, 78:1073–1088.

Bonato, V, K Gomes Facure, & W Uieda. 2004. Food habits of bats of subfamily Vampyrinae in Brazil. *Journal of Mammalogy*, 85:708–713.

Boness, DJ, WD Bowen, & OT Oftedal. 1994. Evidence of a maternal foraging cycle resembling that of otariid seals in a small phocid, the harbor seal. *Behavioral Ecology and Sociobiology*, 34:95–104.

Boness, DJ & WD Bowen. 1996. The evolution of maternal care in pinnipeds. *BioScience*, 46:645–654.

Bongaarts, J. 1998. Demographic consequences of declining fertility. *Science*, 282:419–420.

Boogert, NJ, DM Paterson, & KN Laland. 2006. The implications of niche construction and ecosystem engineering for conservation biology. *BioScience*, 56:570–578.

Boonstra, R & PT Boag. 1987. A test of the Chitty hypothesis: inheritance in life-history traits in meadow vole (*Microtus pennsylvanicus*). *Evolution*, 41:929–947.

Boonstra, R, et al. 1998. The impact of predator-induced stress on the snowshoe hare cycle. *Ecological Monographs*, 79:371–394.

Born, EW, J Teilmann, & F Riget. 2002. Haul-out activity of ringed seals (*Phoca hispida*) determined from satellite telemetry. *Marine Mammal Science*, 18:167–181.

Bouchard, S. 2001. Sex discrimination and roost mate recognition by olfactory cues in the bats *Mops condylura* and *Chaerephon pumilus*. *Journal of Zoology, London*, 254:109–117.

Bouetel, V & C de Muizon. 2006. The anatomy and relationships of *Piscobalaena nana* (Cetacea, Mysticeti), a Cetotheriidae s.s. from the early Pliocene of Peru. *Geodiversitas*, 28:319–395.

Bourliere, F. 1956. *The Natural History of Mammals*. Knopf, NY.

Bourliere, F. 1973. The comparative ecology of rain forest mammals in Africa and tropical America, 279–292, in *Tropical Forest Ecosystems in Africa and South America: A Comparative Review* (BJ Meggers, ES Ayensu, & WD Duckworth, eds.). Smithsonian Institution Press, Washington, DC.

Bowen, BS. 1982. Temporal dynamics of microgeographic structure of genetic variation in *Microtus californicus*. *Journal of Mammalogy*, 63:625–638.

Bowen, WD, OT Oftedal, & DJ Boness. 1992. Mass and energy transfer during lactation in a small phocid, the harbor seal (*Phoca vitulina*). *Physiological Zoology*, 65:844–866.

Bowen, WD, J MacMillan, & R Mohn. 2003. Sustained exponential population growth of grey seals at Sable Island, Nova Scotia. *Journal of Marine Science*, 60:1265–1274.

Bowers, MA & JH Brown. 1982. Body size and coexistence in desert rodents: chance or community structure? *Ecology*, 63:391–400.

Bowers, MA & JL Dooley Jr. 1993. Predation hazard and seed removal by small mammals: microhabitat versus patch scale effects. *Oecologia*, 94:247–254.

Bowers, M, et al. 1996. Use of space and habitats by meadow voles at the home range, patch and landscape scales. *Oecologia*, 105:107–115.

Bowers, MA & SF Matter. 1997. Landscape ecology of mammals: relationships between density and patch size. *Journal of Mammalogy*, 78:999–1013.

Bowker, M. 1977. Behavior of Kirk's dik-dik, *Rhynchotragus kirki*. Unpublished Ph.D. dissertation, Northern Arizona University, Flagstaff.

Bowler, JM. 1982. Aridity in the late Tertiary and Quaternary of Australia, 35–45, in *Evolution of the Flora and Fauna of Arid Australia* (WR Barker & PJM Greenslade, eds.). Peacock Publications, Frewville.

Boyd, IL. 1991. Environmental and physiological factors controlling the reproductive cycles of pinnipeds. *Canadian Journal of Zoology*, 69:1135–1148.

Boyse, EA, et al. 1982. A new aspect of the major histocompatibility complex and other genes in the mouse. *Oncodevelopmental Biology and Medicine*, 4:101–116.

Bradbury, JW. 1977. Social organization and communication, 1–72, in *Biology of Bats*, Vol. 3 (W Wimsatt, ed.). Academic Press, NY.

Bradbury, JW & LH Emmons. 1974. Social organization of some Trinidad bats. I: Emballonuridae. *Zeitschrift fur Tierpsychologie*, 36:137–183.

Bradshaw, GVR. 1962. Reproductive cycle of the California leaf-nosed bat, *Macrotus californicus*. *Science*, 136:645–646.

Bradshaw, WE & CM Holzapfel. 2007. Evolution of animal photoperiodism. *Annual Review of Ecology, Evolution, and Systematics*, 38:1–25.

Brain, C. 1992. Deaths in a desert baboon troop. *International Journal of Primatology*, 13:593–599.

Brain, C & R Bohrmann. 1992. Tick infestation of baboons (*Papio ursinus*) in the Namib Desert. *Journal of Wildlife Diseases*, 28:188–191.

Braithwaite, RW & AK Lee. 1979. A mammalian example of semelparity. *American Naturalist*, 113:151–155.

Bramble, DM. 1989. Cranial specialization and locomotor habit in the Lagomorpha. *American Zoologist*, 29:303–317.

Bramble, DM & FA Jenkins Jr. 1993. Mammalian locomotor-respiratory integration: implications for diaphragmatic and pulmonary design. *Science*, 262:235–240.

Bramble, DM & FA Jenkins Jr. 1998. Locomotor-respiratory integration: implications for mammalian and avian divergence. *Journal of Vertebrate Paleontology*, 18(suppl 3):28A.

Bramble, DM & DE Lieberman. 2004. Endurance running and the evolution of Homo. *Nature*, 432:345–352.

Brant, DH. 1962. Measures of the movements and population densities of small rodents. *University of California Publications in Zoology*, 62:105–184.

Brandt, JH, et al. 2001. Debate on the authenticity of *Pseudonovibos spiralis* as a new species of wild bovid from Vietnam and Cambodia. *Journal of Zoology, London*, 255:437–444.

Braude, S. 2000. Dispersal and new colony formation in wild naked mole-rats: evidence against inbreeding as the system of mating. *Behavioral Ecology*, 11:7–12.

Bravo, A, et al. 2008. Collpas: activity hotspots for frugivorous bats (Phyllostomidae) in the Peruvian Amazon. *Biotropica*, 40:203–210.

Brawand, D, W Wahli, & H Kaessmann. 2008. Loss of egg yolk genes in mammals and the origin of lactation and placentation. *PLoS Biology*, 6:507–517.

Brigham, RM & MB Fenton. 1986. The influence of roost closure on the roosting and foraging behavior of *Eptesicus fuscus* (Chiroptera: Vespertilionidae). *Canadian Journal of Zoology*, 64:1128–1133.

Broadbooks, HE. 1970. Home ranges and territorial behavior of the yellow-pine chipmunk, *Eutamias amoenus*. *Journal of Mammalogy*, 51:310–326.

Brockie, R. 1976. Self-anointing by wild hedgehogs, *Erinaceus europaeus*. *Animal Behaviour*, 24:68–71.

Brodie, ED. 1977. Hedgehogs use toad venom in their own defense. *Nature*, 268:627–628.

Brody, S. 1945. *Bioenergetics and Growth*. Reinhold, NY.

Bromley, PT. 1969. Territoriality in pronghorn bucks on the National Bison Range, Moiese, Montana. *Journal of Mammalogy*, 50:81–89.

Bronner, GN & PD Jenkins. 2005. Order Afrosoricida, 71–81, in *Mammal Species of the World*, Vol. 1 (DE Wilson & DM Reeder, eds.). Johns Hopkins University Press, Baltimore, MD.

Bronson, FH. 1971. Rodent pheromones. *Biology of Reproduction*, 4:344–357.

Bronson, FH. 1979. The reproductive ecology of the house mouse. *Quarterly Review of Biology*, 54:265–299.

Bronson, FH. 1989. *Mammalian Reproductive Biology*. University of Chicago Press, Chicago, IL.

Bronson, FH & JA Maruniak. 1975. Male-induced puberty in female mice: evidence for a synergistic action of social cues. *Biology of Reproduction*, 13:94–98.

Brooke, AP. 1994. Diet of the fishing bat, *Noctilio leporinus* (Chiroptera: Noctilionidae). *Journal of Mammalogy*, 75:212–218.

Brooks, DR & DA McLennan. 1991. *Phylogeny, Ecology, and Behavior: A Research Program in Comparative Biology*. University of Chicago Press, Chicago, IL.

Brower, K. 2009. Still blue. *National Geographic*, 215:134–153.

Brown, J, et al. 1980. *An Arctic Ecosystem: The Coastal Tundra at Barrow, Alaska*.

Dowden, Hutchinson and Ross, Stroudsburg, PA.

Brown, JH. 1968. Adaptation to environmental temperature in two species of woodrats, *Neotoma cinerea* and *N. albigula*. *Miscellaneous Publications of Museum of Zoology, University of Michigan*, 135:1–48.

Brown, JH. 1971. Mammals on mountaintops: non-equilibrium insular biogeography. *American Naturalist*, 105:467–478.

Brown, JH. 1973. Species diversity of seed-eating rodents in sand dune habitats. *Ecology*, 54:775–787.

Brown, JH. 1978. The theory of island biogeography and the distribution of boreal birds and mammals. *Great Basin Naturalist Memoirs*, 2:209–227.

Brown, JH. 1987. Variation in desert rodent guilds: patterns, processes, and scales, 185–203, in *Organization of Communities: Past and Present* (JHR Gee & PS Giller, eds.). Blackwell Scientific Publishers, Oxford, England.

Brown, JH & GA Bartholomew. 1969. Periodicity and energetics of torpor in the kangaroo mouse, *Microdipodops pallidus*. *Ecology*, 50:705–709.

Brown, JH & DW Davidson. 1977. Competition between seed-eating rodents and ants in desert ecosystems. *Science*, 196:880–882.

Brown, JH, DW Davidson, & OJ Reichman. 1979a. An experimental study of competition between seed-eating desert rodents and ants. *American Zoologist*, 19:1129–1143.

Brown, JH, OJ Reichman, & DW Davidson. 1979b. Granivory in desert ecosystems. *Annual Review of Ecology and Systematics*, 10:201–227.

Brown, JH, OJ Reichman, & DW Davidson. 1979c. Granivory in desert ecosystems. *Annual Review of Ecology and Systematics*, 10:201–227.

Brown, JH & JC Munger. 1985. Experimental manipulation of a desert rodent community: food addition and species removal. *Ecology*, 66:1545–1563.

Brown, JH & MA Kurzius. 1987. Composition of desert rodent faunas: combinations of coexisting species. *Annales Zoologici Fennici*, 24:227–237.

Brown, JH & EJ Heske. 1990a. Mediation of a desert-grassland transition by a keystone rodent guild. *Science*, 250:1705–1707.

Brown, JH & EJ Heske. 1990b. Temporal changes in a Chihuahuan rodent community. *Oikos*, 59:290–302.

Brown, JH & BA Harney. 1993. Population and community ecology of heteromyid rodents in temperate habitats, 618–651, in *Biology of the Heteromyidae* (HH Genoways & JH Brown, eds.). Special Publication no. 10, American Society of Mammalogists.

Brown, JH & MV Lomolino. 1998. *Biogeography*. 2nd ed. Sinauer Associates, Sunderland, MA.

Brown, JH & MV Lomolino. 2000. Concluding remarks: historical perspective and the future of island biogeography theory. *Global Ecology & Biogeography*, 9:87–92.

Brown, P. 1976. Vocal communication in the pallid bat, *Antrozous pallidus*. *Zeitschrift fur Tierpsychologie*, 41:34–54.

Brown, WL & EO Wilson. 1956. Character displacement. *Systematic Zoology*, 5:49–64.

Brownell, S. 2003. Blast from the vast: What's the purpose of nature's most powerful sound? *Discover*, Dec:51–57.

Bruce, HM. 1966. Smell as an exteroceptive factor, 83–87, in *Environmental Influences on Reproductive Processes* (W Hansel & RH Dutt, eds.). *Journal of Applied Sciences* (suppl) 25.

Bruce, VG. 1960. Environmental entrainment of circadian rhythms. *Cold Springs Harbor Symposium Quantitative Biology*, 25:29–48.

Brückmann, G & H Burda. 1997. Hearing in blind subterranean Zambian mole-rats (*Cryptomys* sp.): collective behavioural audiogram in a highly social rodent. *Journal of Comparative Physiology A*, 181:83–88.

Bruderer, B & AG Popa-Lisseanu. 2005. Radar data on wing-beat frequencies and flight speeds of two bats. *Acta Chiropterologica*, 7:73–82.

Brunet, M, et al. 2005. New material of the earliest hominid from the Upper Miocene of Chad. *Nature*, 434:752–755.

Bryant, HN. 1991. Phylogenetic relationships and systematics of the Nimravidae (Carnivora). *Journal of Mammalogy*, 72:56–78.

Bubenik, GA & AB Bubenik. 1990. *Horns, Pronghorns, and Antlers: Evolution, Morphology, Physiology, and Social Significance*. Springer-Verlag, NY.

Buchler, ER. 1976. The use of echolocation by the wandering shrew. *Animal Behaviour*, 24:858–873.

Buck, CL & BM Barnes. 1999. Temperatures of hibernacula and changes in body composition of arctic ground squirrels over winter. *Journal of Mammalogy,* 80:1264–1276.

Buck, CL & BM Barnes. 2000. Effects of ambient temperature on metabolic rate, respiratory quotient, and torpor in an arctic hibernator. *American Journal of Physiology Regulatory Integrative Comparative Physiology,* 279:R255–R262.

Buckner, CH. 1969. Some aspects of the population ecology of the common shrew. *Sorex araneus,* near Oxford, England. *Journal of Mammalogy,* 50:326–332.

Buffenstein, R & S Yahav. 1991. Is the naked mole-rat, *Heterocephalus glaber,* an endothermic yet poikilothermic mammal? *Journal of Thermal Biology,* 16:227–232.

Bullen, RD & NL McKenzie. 2008. The pelage of bats (Chiroptera) and the presence of aerodynamic riblets: the effect on aerodynamic cleanliness. *Zoology,* 111:279–286.

Bullock, TH, et al. 1968. Electrophysiological studies of central auditory mechanisms in cetaceans. *Journal of Comparative Physiology A,* 59:117–316.

Burckhardt, D. 1958. Kindliches verhalten als ausdrucksvewegung im fortpflanzungszeremoniell einiger wiederkauer. *Revue Suisse de Zoologie,* 65:311–316.

Burland, TM & JW Wilmer. 2001. Seeing in the dark: molecular approaches to the study of bat populations. *Biological Reviews,* 76:389–409.

Burns, CE, BJ Goodwin, & RS Ostfeld. 2005. A prescription for longer life? Bot fly parasitism in the white-footed mouse. *Ecology,* 86:753–761.

Burns, V, H Burda, & MJ Ryan. 1989. Ear morphology of the frog-eating bat (*Trachops cirrhosus,* Family: Phyllostomidae): apparent specializations for low-frequency hearing. *Journal of Morphology,* 199:103–118.

Burrell, H. 1927. *The Platypus.* Angus and Robertson, Sydney.

Burt, WH. 1943. Territoriality and home range concepts as applied to mammals. *Journal of Mammalogy,* 24:346–352.

Bushmeat Crisis Task Force. Accessed 23 November 2009. http://www.bushmeat. org/.

Butler, PM. 1972. Some functional aspects of molar evolution. *Evolution,* 26:474–483.

Byers, JA. 1997. *American Pronghorn: Social Adaptations and the Ghosts of Predators Past.* University of Chicago Press, Chicago, IL.

Byers, JA. 2003. *Built for Speed: A Year in the Life of a Pronghorn.* Harvard University Press, Cambridge, MA.

Byrnes, G, NT-L Lim, & AJ Spence. 2008. Take-off and landing kinetics of a free-ranging gliding mammal, the Malayan colugo (*Galeopterus variegatus*). *Proceedings of the Royal Society B,* 275:1007–1013.

Byrom, AE & CJ Krebs. 1999. Natal dispersal of juvenile arctic ground squirrels in the boreal forest. *Canadian Journal of Zoology,* 77:1048–1059.

Cajete, G. 1999. *A People's Ecology: Explorations in Sustainable Living.* Clear Light Publishers, Santa Fe, NM.

Cajete, G. 2000. *Native Science: Natural Laws of Interdependence.* Clear Light Publishers, Santa Fe, NM.

Calaby, JH. 1971. The current status of Australian Macropodidae. *Australian Zoology,* 16:17–29.

Calder, WA. 1969. Temperature relations and underwater endurance of the smallest homeothermic diver, the water shrew. *Comparative Biochemistry and Physiology,* 30:1075–1082.

Cameron, RD. 1994. Reproductive pauses by female caribou. *Journal of Mammalogy,* 75:10–13.

Camp, CL & N Smith. 1942. Phylogeny and functions of the digital ligaments of the horse. *Memoirs of the University of California,* 13:69–124.

Canals, M, et al. 2005. Relative size of hearts and lungs of small bats. *Acta Chiropterologica,* 7:65–72.

Canivenc, R & M Bonnin. 1979. Delayed implantation is under environmental control in the badger (*Meles meles* L.). *Nature,* 278:849–850.

Cannon, B & J Nedergaard. 2003. Brown adipose tissue: function and physiological significance. *Physiological Reviews,* 84:277–359.

Cao, J, et al. 2003. The treeshrews: adjuncts and alternatives to primates as models for biomedical research. *Journal of Medical Primatology,* 32:123–130.

Capelli, C, et al. 2006. A nuclear DNA phylogeny of the woolly mammoth (*Mammuthus primigenius*). *Molecular Phylogenetics and Evolution,* 40:620–627.

Capuco, AV & RM Akers. 2009. The origin and evolution of lactation. *Journal of Biology* 8:37.

Cardilio, M, et al. 2006. Latent extinction risk and the future battlegrounds of mammalian conservation. *Proceedings of the National Academy of Sciences USA,*103:4157–4161.

Careau, V, J Morand-Ferron, & D Thomas. 2007. Basal metabolic rate of Canidae from hot deserts to cold arctic climates. *Journal of Mammalogy,* 88:394–400.

Carleton, MD. 1994. Systematic studies of Madagascar's endemic rodents (Muroidea: Nesomyinae): revision of the genus *Eliurus. American Museum Novitates,* 3087:1–55.

Carleton, MD & GG Musser. 1984. Muroid rodents, 289–379, in *Orders and Families of Recent Mammals of the World* (S Anderson & JK Jones, eds.). Wiley, NY.

Carleton, MD & DF Schmidt. 1990. Systematic studies of Madagascar's endemic rodents (Muroidea: Nesomyinae): an annotated gazetteer of collecting localities of known forms. *American Museum Novitates,* 2987:1–36.

Carleton, MD & SM Goodman. 1996. Systematic studies of Madagascar's endemic rodents (Muroidea: Nesomyinae): a new genus and species from the central highlands, 231–256, in *A Floral and Faunal Inventory of the Eastern Slopes of the Reserve Naturelle Integrale d'Andringitra, Madagascar: With Reference to Elevational Variation* (SM Goodman, ed.). *Fieldiana: Zoology,* 85:1–319.

Carleton, MD & SM Goodman. 1998. New taxa of nesomyine rodents (Muroidea: Muridae) from Madagascar's northern highlands, with taxonomic comments on previously described forms, 163–200, in *A Floral and Faunal Inventory of the Reserve Speciale d'Anjanaharibe-Sud, Madagascar: With Reference to Elevational Variation* (SM Goodman, ed.). *Fieldiana: Zoology,* 90:1–246.

Carlini, AA, et al. 1990. The first Paleogene land placental mammal from Antarctica: its paleoclimatic and paleobiogeographical bearings, 325, in *4th International Congress Systematics Evolutionary Biology,* abstracts. University of Maryland, College Park.

Carlini, AA, SF Vizcaíno, & GJ Scillato-Yané. 1997. Armored xenarthrans: a unique taxonomic and ecologic assemblage, 213–226, in *Vertebrate Paleontology in the Neotropics: The Miocene Fauna of La Venta, Colombia* (RF Kay, RH Madden, RL Cifelli, & JJ Flynn, eds.). Smithsonian Institution Press, Washington, DC.

Carlini, AA, AE Zurita, & OA Aguilera. 2008a. North American glyptodontines (Xenarthra, Mammalia) in the upper Pleistocene of northern South America (MR Sánchez-Villagra & OA Aguilera, eds.). *Paläontologische Zeitschrift*, 82(2):125–138.

Carlini, A, A Zurita, & DD Gillette. 2008b. Glyptodontines (Xenarthra, Glyptodontidae) and the Great American Biotic Interchange: a new interpretation. *Journal of Vertebrate Paleontology*, 28(3, suppl):61A.

Caro, T. 2005. The adaptive significance of coloration in mammals. *BioScience*, 55:125–136.

Caroom, D & FH Bronson. 1971. Responsiveness of female mice to preputial attractant: effects of sexual experience and ovarian hormones. *Physiology and Behavior*, 7:659–662.

Carpenter, RE. 1969. Structure and function of the kidney and the water balance of desert bats. *Physiological Zoology*, 42:288–302.

Carr, A. 1974. *Ulendo: Travels of a Naturalist in and out of Africa*. Knopf, NY.

Carranza-Castañeda, Ó & EH Lindsay (eds.). *Advances in Late Tertiary Vertebrate Paleontology in Mexico and the Great American Biotic Interchange*. Universidad Nacional Autónoma de México, Instituto de Geología and Centro de Geociencias, Publicación Especial 4:73–101.

Carraway, LN. 1995. A key to Recent Soricidae of the western United States and Canada based primarily on dentaries. *Occasional Papers Natural History Museum Kansas*, 175:1–49.

Carroll, RL. 1988. *Vertebrate Paleontology and Evolution*. W. H. Freeman, NY.

Carroll, SB. 2005. *Endless Forms Most Beautiful: The New Science of Evo-Devo*. W.W. Norton, NY.

Carter, AM & A Mess. 2008. Evolution of the placenta and associated reproductive characters in bats. *Journal of Experimental Zoology*, 310B:428–449.

Cartmill, M. 1992. New views on primate origins. *Evolutionary Anthropology*, 1:105–111.

Case, TJ. 1978. Endothermy and parental care in the terrestrial vertebrates. *American Naturalist*, 112:861–874.

Catania, KC. 1995. A comparison of the Eimer's organs of three North American moles: the hairy-tailed mole (*Parascalops breweri*), the star-nosed mole (*Condylura cristata*), and the eastern mole (*Scalopus aquaticus*). *Journal of Comparative Neurology*, 354:150–160.

Catania, KC. 2006. Underwater "sniffing" in semi-aquatic mammals. *Nature*, 444:1024–1025.

Catania, KC & JH Kaas. 1996. The unusual nose and brain of the star-nosed mole. *BioScience*, 46:578–586.

Catania, KC & FE Remple. 2005. Asymptotic prey profitability drives star-nosed moles to the foraging speed limit. *Nature*, 433:519–522.

Catania, KC, JF Hare, & KL Campbell. 2008. Water shrews detect movement, shape, and smell to find prey underwater. *Proceedings of the National Academy of Sciences USA*, 105:571–576.

Cates, RG & DF Rhoades. 1977. Patterns in the production of antiherbivore chemical defenses in plant communities. *Biochemical Systematics and Ecology*, 5:185.

Ceballos, G & PR Ehrlich. 2009. Discoveries of new mammal species and their implications for conservation and ecosystem services. *Proceedings of the National Academy of Sciences USA*, 106:3841–3846.

Chaffee, RRJ & JC Roberts. 1971. Temperature acclimation in birds and mammals. *Annual Review of Physiology*, 33:155–202.

Chaimanee, Y, et al. 2003. A middle Miocene hominoid from Thailand and orangutan origins. *Nature*, 422:61–65.

Chang, JL & JE Adams. 2008. Sequencing the armadillo genome, 181–195, in *The Biology of the Xenarthra* (SF Vizcaíno & WJ Loughry, eds.). University Press of Florida, Gainesville, FL.

Channell, RB. 1998. A geography of extinction: patterns in the contraction of geographic ranges. Unpubl. Ph.D. dissertation. University of Oklahoma, Norman.

Chapin, FS, et al. 1998. Ecosystem consequences of changing biodiversity. *Bioscience*, 48:45–52.

Chappell, MA. 1981. Standard operative temperature and cost of thermoregulation in the arctic ground squirrel. *Oecologia*, 49:387–403.

Charles-Dominique, P. 1971. Eco-ethologie des prosimiens du Gabon. *Biologica Gabonica*, 7:121–228.

Charles-Dominique, P & JJ Petter. 1980. Ecology and social life of *Phaner furcifer*, 75–96, in *Nocturnal Malagasy Primates* (P Charles-Dominique et al., eds.). Academic Press, NY.

Charpentier, MJE, M Boulet, & CM Drea. 2008. Smelling right: the scent of male lemurs advertises genetic quality and relatedness. *Molecular Ecology*, 17:3225–3233.

Chew, RM & AE Dammann. 1961. Evaporative water loss of small vertebrates, as measured with an infrared analyzer. *Science*, 133:384–385.

Chiarello, AG. 2008. Sloth ecology: an overview of field studies, 269–280, in *The Biology of the Xenarthra* (SF Vizcaíno & WJ Loughry, eds.). University Press of Florida, Gainesville.

Childs, JE, JN Mills, & GE Glass. 1995. Rodent-borne hemorrhagic fever viruses: a special risk for mammalogists? *Journal of Mammalogy*, 76:664–680.

Chitty, D. 1958. Self-regulation of numbers through changes in viability. *Cold Springs Harbor Symposium Quantitative Biology*, 22:277–280.

Chitty, D. 1960. Population processes in the vole and their relevance to general theory. *Canadian Journal of Zoology*, 38:99–113.

Chitty, D & H Chitty. 1962. Population trends among the voles at Lake Vyrnwy, 1932–1960. *Journal of Animal Ecology*, 35:313–331.

Chivers, DL. 1974. *The Siamang in Malaya: A Field Study of a Primate in a Tropical Rain Forest*. Contributions in Primatology, Vol. 4. S. Karger, Basel.

Choquenot, D & DMJS Bowman. 1998. Marsupial megafauna, aborigines and the overkill hypothesis: application of predator-prey models to the questions of Pleistocene extinction in Australia. *Global Ecology and Biogeography Letters*, 7:167–180.

Chorn, J & RS Hoffmann. 1978. *Ailuropoda melanoleuca*. *Mammalian Species*, 110:1–6.

Chow, M & TH Rich. 1982. *Shuotherium dongi*, gen. et sp. nov., a therian with pseudotribosphenic molars from the Jurassic of Sichuan, China. *Australian Mammalogy*, 5:127–142.

Christian, DP. 1978. Effects of humidity and body size on evaporative water loss in three desert rodents. *Comparative Biochemistry and Physiology*, 60:425–430.

Christian, DP. 1979. Comparative demography of three Namib Desert rodents: responses to the provision of supplementary water. *Journal of Mammalogy*, 60:679–690.

Christian, JJ. 1950. The adreno-pituitary system and population cycles in mammals. *Journal of Mammalogy*, 31:247–259.

Cifelli, RL. 1993a. Theria of metatherian-eutherian grade and the origin of marsupials, 205–215, in *Mammal Phylogeny: Mesozoic Differentiation, Multituberculates, Monotremes, Early Therians, and Marsupials* (FS Szalay, MJ Novacek, & MC McKenna, eds.). Springer-Verlag, NY.

Cifelli, RL. 1993b. Early Cretaceous mammal from North America, and the evolution of marsupial dental characters. *Proceedings National Academy Sciences USA*, 90:9413–9416.

Cifelli, RL, et al. 1996. Origins of marsupial pattern of tooth replacement: fossil evidence revealed by high-resolution X-ray CT. *Nature*, 379:715–718.

Cifelli, RL & C de Muizon. 1997. Dentition and jaw of *Kokopellia juddi*, a primitive marsupial or near-marsupial from the medial Cretaceous of Utah. *Journal of Mammalian Evolution*, 4:241–258.

Cifelli, RL, et al. 1997. High-precision ^{40}Ar/^{39}Ar geochronology and the advent of North America's Late Cretaceous terrestrial fauna. *Proceedings National Academy Sciences USA*, 94:11163–11167.

Cifelli, RL & C de Muizon. 1998. Tooth eruption and replacement pattern in early marsupials. *Comptes Rendus de L' Academie des Sciences Paris*, 326:215–220.

Cifelli, RL & SK Madsen. 1998. Triconodont mammals from the medial Cretaceous of Utah. *Journal of Vertebrate Paleontology*, 18:403–411.

Cifelli, RL & BM Davis. 2003. Marsupial origins. *Science*, 302:1899–1900.

Ciochon, RL & AB Chiarelli. 1980. *Evolutionary Biology of the New World Monkeys and Continental Drift*. Plenum Press, NY.

Clark, CW & JM Clark. 1980. Sound playback experiments with southern right whales (*Eubalaena australis*). *Science*, 207:663–665.

Clark, JD, et al. 2003. Stratigraphic, chronological and behavioural contexts of Pleistocene *Homo sapiens* from Middle Awash, Ethiopia. *Nature*, 423:747–752.

Clark, WEL. 1971. *The Antecedents of Man*. Quadrangle Books, NY.

Clarke, A & P Rothery. 2008. Scaling of body temperature in mammals and birds. *Functional Ecology*, 22:58–67.

Clarke, JR & JP Kennedy. 1967. Effect of light and temperature upon gonad activity in the vole (*Microtus agrestis*). *General and Comparative Endocrinology*, 8:474–488.

Clarke, MR. 1979. The head of the sperm whale. *Scientific American*, 240:128–141.

Clauss, M. 2004. The potential interplay of posture, digestive anatomy, density of ingesta and gravity in mammalian herbivores: why sloths do not rest upside down. *Mammal Reviews*, 34:241–245.

Clemens, WA. 1968. Origin and early evolution of marsupials. *Evolution*, 22:1–18.

Clemens, WA. 1970. Mesozoic mammalian evolution. *Annual Review of Ecology and Systematics*, 1:357–390.

Clementz, MT, et al. 2006. Isotopic records from early whales and sea cows: contrasting patterns of ecological transition. *Journal of Vertebrate Paleontology*, 26:355–370.

Cleveland, CJ, et al. 2006. Economic value of the pest control service provided by Brazilian free-tailed bats in south-central Texas. *Frontiers in Ecology and the Environment*, 4:238–243.

Clough, GC. 1965. Lemmings and population problems. *American Scientist*, 53:199–212.

Clulow, FV, EA Franchetto, & PE Langford. 1982. Pregnancy failure in the red-backed vole *Clethrionomys gapperi*. *Journal of Mammalogy*, 63:499–500.

Clutton-Brock, TH. 1989. Mammalian mating systems. *Proceedings of the Royal Society B*, 236:339–372.

Clutton-Brock, TH, FE Guinness, & SD Albon. 1982. *Red Deer: Behavior and Ecology of Two Sexes*. University of Chicago Press, Chicago, IL.

Clutton-Brock, TH, M Hiraiwa-Hasegawa, & A Robertson. 1989. Mate choice on fallow deer leks. *Nature*, 340:463–465.

Coates, AG (compiler). 2003. *Paseo Pantera: Una Historia de la Naturaleza y Cultura de Centroamérica*. Smithsonian Books, Institution Press, Washington, DC.

Coates, AG & JA Obando. 1996. The geologic evolution of the Central American Isthmus, 21–56, in *Evolution and Environment in Tropical America* (JBC Jackson, AF Budd, & AG Coates, eds.). University of Chicago Press, Chicago, IL.

Cockburn, A. 1992. Habitat heterogeneity and dispersal: environmental and genetic patchiness, 65–95, in *Animal Dispersal* (NC Stenseth & WZ Lidiker Jr, eds.). Chapman and Hall, NY.

Cockburn, A & WZ Lidicker. 1983. Microhabitat heterogeneity and population ecology of an herbivorous rodent, *Microtus californicus*. *Oecologia*, 59:167–177.

Coe, MJ. 1967. Preliminary notes on the spring hare *Pedetes surdaster larvalis* in East Africa. *East African Wildlife Journal*, 5:174–177.

Cohen, JE. 1996. *How Many People Can the Earth Support?* W. W. Norton, NY.

Cohen, MJ. 2007. *Reconnecting with Nature: Finding Wellness through Restoring Your Bond with the Earth*. 3rd ed. Ecopress, Lakeville, MN.

Coimbra-Filho, AF, A Magananini, & RA Mittermeier. 1975. Vanishing gold: last chance for Brazil's lion tamarins. *Animal Kingdom*, Dec. 20.

Colbert, EH. 1973. *Wandering Lands and Animals*. Dutton, NY.

Colbert, EH. 1982. Personal communication.

Cole, FR, DM Reeder, & DE Wilson. 1994. A synopsis of distribution patterns and the conservation of mammal species. *Journal of Mammalogy*, 75:266–276.

Cole, RW. 1970. Pharyngeal and lingual adaptations in the beaver. *Journal of Mammalogy*, 51:424–425.

Collins, LR & JF Eisenberg. 1972. Notes on the behavior and breeding of pacaranas *Dinomys branickii* in captivity. *International Zoo Yearbook*, 12:108–114.

Coltman, DW, et al. 2003. Undesirable evolutionary consequences of trophy hunting. *Nature*, 426:655–658.

Colwell, RK, et al. 2008. Global warming, elevational range shifts, and lowland biotic attrition in the wet tropics. *Science*, 322:258–261.

Comiso, JC & CL Parkinson. 2004. Satellite-observed changes in the Arctic. *Physics Today*, 57:38–44.

Condy, PR. 1978. Annual cycle of the southern elephant seal (*Mirounga leonina*) at Marian Island. *South African Journal of Zoology*, 14:95–102.

Conley, WH. 1971. Behavior, demography, and competition in *Microtus longicaudus* and *M. mexicanus*. Ph.D. thesis, Texas Tech University, Lubbock.

Connor, EF & ED McCoy. 1979. The statistics and biology of the species-area relationship. *American Naturalist*, 113:791–833.

Connor, RC, et al. 1998. Social evolution in toothed whales. *Trends in Ecology and Evolution*, 13:228–232.

Cook, JA, AM Runck, & CJ Conroy. 2004. Historical biogeography at the crossroads of the northern continents: molecular phylogenetics of red-backed voles (Rodentia: Arvicolinae). *Molecular Phylogenetics and Evolution*, 30:767–777.

Cooper, HM, M Herbin, & E Nevo. 1993. Ocular regression conceals adaptive progression of the visual system in a blind subterranean mammal. *Nature*, 361:156–159.

Cooper, KL & CJ Tabin. 2008. Understanding of bat wing evolution takes flight. *Genes & Development*, 22:121–124.

Corbet, GB. 1971. Family Macroscelididae. Part 1.5, 1–6, in *The Mammals of Africa: An Identification Manual* (J Meester & HW Setzer, eds.). Smithsonian Institution Press, Washington, DC.

Corbet, GB. 1988. The family Erinaceidae: a synthesis of its taxonomy, phylogeny, ecology, and zoogeography. *Mammal Review*, 18:117–172.

Corbet, GB & J Hanks. 1968. A revision of the elephant-shrews, Family Macroscelididae. *Bulletin British Museum (National History) Zoology*, 16:1–111.

Corbet, GB & JE Hill. 1992. *The Mammals of the Indomalayan Region*. Oxford, Natural History Museum Publications and Oxford University Press, London.

Cork, SJ. 1981. Digestion and metabolism in the koala (*Phascolarctos cinereus* Goldfuss): an arboreal folivore. Ph.D. thesis, University of New South Wales, Kensington.

Corneli, PS. 2002. Complete mitochondrial genomes and eutherian evolution. *Journal of Mammalian Evolution*, 9:281–305.

Costa, LP & Y Leite. 2009. The Atlantic forest: a hot spot divided in two. Presented at the 10th International Mammalogical Congress, Mendoza, Argentina, 9–14 August. Abstracts pdf, p. 113.

Costanza, R, et al. 1997. The value of the world's ecosystem services and natural capital. *Nature*, 387:253–260.

Cott, HB. 1966. *Adaptive Coloration in Animals*. Methuen, London.

Covey, E. 2005. Neurobiological specializations in echolocating bats. *Anatomical Record Part A: Discoveries in Molecular, Cellular, and Evolutionary Biology*, 287:1103–1116.

Cowan, IMcT. 1936. Nesting habits of the flying squirrel, *Glaucomys sabrinus*. *Journal of Mammalogy*, 17:58–60.

Cox, CB & RD Moore. 2005. *Biogeography: An Ecological and Evolutionary Approach*. 7th ed. Blackwell, Malden, MA.

Crabtree, RL & JW Sheldon. 1999. The ecological role of coyotes on Yellowstone's Northern Range. *Yellowstone Science*, 7:15–23.

Cranford, TW. 1999. The sperm whale's nose: sexual selection on a grand scale? *Marine Mammal Science*, 15:1133–1157.

Cranford, TW. 2000. In search of impulse sound sources in odontocetes, 109–156, in *Hearing by Whales and Dolphins* (WWL Au, AN Popper, & RR Fay, eds.). Springer-Verlag, NY.

Cranford, TW, M Amundin, & KS Norris. 1996. Functional morphology and homology in the odontocete nasal complex: implications for sound generation. *Journal of Morphology*, 228:223–285.

Cranford, TW, et al. 1997. Functional morphology of the dolphins biosonar signal generator studied by high-speed video endoscopy. *Journal of Morphology*, 232:243 (abstract).

Cranford, TW & ME Amundin. 2003. Biosonar pulse production in odontocetes: the state of our knowledge, 27–35, in *Echolocation in Bats and Dolphins* (JA Thomas, CF Moss, & M Vater, eds.). University of Chicago Press, Chicago, IL.

Crasso, PBB & AP Wagner. 2008. Primeiro registro de predação de *Poecilia reticulata* Peters, 1859 e *Phalloceros caudimaculatus* (Hensel, 1868) por *Noctilio leporinus* (Linnaeus, 1758) (Chiroptera, Noctilionidae). *Chiroptera Neotropical*, 14:391–396.

Crawford, EC Jr. 1962. Mechanical aspects of panting in dogs. *Journal of Applied Physiology*, 17:249–251.

Credner, S, H Burda, & F Ludescher. 1997. Acoustic communication underground: vocalization characteristics in subterranean social mole-rats (*Cryptomys* sp., Bathyergidae). *Journal of Comparative Physiology A*, 180:245–255.

Creel, SR, et al. 1991. Behavioral and endocrine mechanisms of reproduction suppression in Serengeti dwarf mongooses. *Animal Behaviour*, 43:231–245.

Creel, SR, SL Monfort, & DE Wildt. 1993. Aggression, reproduction, and androgens in wild dwarf mongooses: a test of the challenge hypothesis. *American Naturalist*, 141:816–825.

Crelin, ES. 1969. Interpubic ligament: elasticity in pregnant free-tailed bat. *Science*, 164:81–82.

Crockett, CM & R Sekulic. 1984. Infanticide in red howler monkeys (*Alouatta seniculus*), 173–191, in *Infanticide: Comparative and Evolutionary Perspectives* (G Hausfater & SB Hrdy, eds.). Aldine, Hawthorne, NY.

Crockett, CM & R Rudran. 1987. Red howler monkey birth data. I. Seasonal variation. *American Journal of Primatology*, 13:347–368.

Crompton, AW. 1971. The origin of the tribosphenic molar, 165–180, in *Early Mammals* (DM Kermack & KA Kermack, eds.). *Journal of the Linnean Society, London, Zoology*, 50.

Crompton, AW. 1974. The dentitions and relationships of the southern African mammals *Erythrotherium parringtoni* and *Megazostrodon rudnerae*. *Bulletin British Museum Natural History (Geology)*, 24:399.

Crompton, AW. 1995. Masticatory function in nonmammalian cynodonts and early mammals, 55–75, in *Functional Morphology in Vertebrate Palaeontology* (JJ Thomason, ed.). Cambridge University Press, Cambridge, UK.

Crompton, AW & FA Jenkins Jr. 1968. Molar occlusion in Late Triassic mammals. *Biological Reviews*, 43:427.

Crompton, AW & K Hiiemae. 1969. How mammalian molar teeth work. *Discovery*, 5:23.

Crompton, AW & K Hiiemae. 1970. Molar occlusion and mandibular movements during occlusion in the American opossum, *Didelphis marsupialis* L. *Journal of the Linnean Society, London, Zoology*, 49:21.

Crompton, AW & FA Jenkins Jr. 1973. Mammals from reptiles: a review of mammalian origins. *Annual Review Earth Planetary Science*, Vol. I. Annual Reviews, Palo Alto, CA.

Crompton, AW & FA Jenkins Jr. 1979. Origin of mammals, 59–73, in *Mesozoic Mammals: The First Two-thirds of Mammalian History* (JA Lillegraven, Z Kielan-Jaworowska, & WA Clemens, eds.). University of California Press, Berkeley, CA.

Crompton, AW & Z Luo. 1993. Relationships of the Liassic mammals *Sinoconodon*, *Morganucodon oehleri*, and *Dinnetherium*, 30–44, in *Mammal Phylogeny: Mesozoic Differentiation, Multituberculates, Monotremes, Early Therians, and*

Marsupials (FS Szalay, MJ Novacek, & MC McKenna, eds.). Springer-Verlag, NY.

Crompton, RH & PM Andau. 1987. Ranging, activity rhythms, and sociality in free-ranging *Tarsius bancanus*: a preliminary report. *International Journal of Primatology*, 8:43–72.

Crone, EE, D Doak, & J Pokki. 2001. Ecological influences on the dynamics of a field vole metapopulation. *Ecology*, 82:831–843.

Crooks, KR. 2002. Relative sensitivities of mammalian carnivores to habitat fragmentation. *Conservation Biology*, 16:488–502.

Crowcroft, P. 1957. *The Life of the Shrew*. Max Reinhart, London.

Crozier, RH. 1997. Preserving the information content of species: genetic diversity, phylogeny, and conservation worth. *Annual Review of Ecology and Systematics*, 28:243–268.

CSE. 2002. *Safeguarding the Uniqueness of the Colorado Plateau.* Center for Sustainable Environments, Northern Arizona University, Grand Canyon Wildlands Council, and Terralingua: Partnerships for Linguistic and Biological Diversity, Flagstaff, AZ.

Cullinane, DM, D Aleper, & JEA Bertram. 1998a. The functional and biomechanical modifications of the spine of *Scutisorex somereni*, the hero shrew: skeletal scaling relationships. *Journal of Zoology, London*, 244:447–452.

Cullinane, DM & D Aleper. 1998b. The functional and biomechanical modifications of the spine of *Scutisorex somereni*, the hero shrew: spinal musculature. *Journal of Zoology, London*, 244:453–458.

Cunha, GR, et al. 2005. The ontogeny of the urogenital system of the spotted hyena (*Crocuta crocuta* Erxleben). *Biology of Reproduction*, 73:554–564.

Cupper, ML & J Duncan. 2006. Last glacial megafaunal death assemblage and early human occupation at Lake Menindee, southeastern Australia. *Quaternary Research*, 66:332–341.

Curry-Lindahl, K. 1962. The irruption of the Norway lemmings in Sweden during 1960. *Journal of Mammalogy*, 43:171–184.

Cutright, WJ & T McKean. 1979. Countercurrent blood vessel arrangement in beaver (*Castor canadensis*). *Journal of Morphology*, 161:169–176.

Cwynar, LC & JC Ritchie. 1980. Arctic steppe-tundra: a Yukon perspective. *Science*, 208:1375–1377.

Czaplewski, NJ. 1987. Deciduous teeth of *Thyroptera tricolor*. *Bat Research News*, 28:23–25.

Czaplewski, NJ. 1997. Chiroptera, 410–431, in *Vertebrate Paleontology in the Neotropics: The Miocene Fauna of La Venta, Colombia* (RF Kay, RH Madden, RL Cifelli, & JJ Flynn, eds.). Smithsonian Institution Press, Washington, DC.

Czaplewski, NJ. 2010. Colhuehuapian bats (Mammalia: Chiroptera) from the Gran Barranca, Chubut province Argentina. In *The Paleontology of Gran Barranca: Evolution and Environmental Change through the Middle Cenozoic of Patagonia* (RH Madden, AA Carlini, MG Vucetich, & RF Kay, eds.). Cambridge University Press, Cambridge, UK.

Dagg, AI. 1962. The role of the neck in the movements of giraffe. *Journal of Mammalogy*, 43:88–97.

Daily, GC. 1997. *Nature's Services: Societal Dependence on Natural Ecosystems.* Island Press, Washington, DC.

Dalquest, WW, JA Baskin, & GE Schultz. 1996. Fossil mammals from a Late Miocene (Clarendonian) site in Beaver County, Oklahoma, 117–137, in *Contributions in Mammalogy.* (HH Genoways & RJ Baker, eds.). Museum Texas Tech University, Lubbock.

Daniel, JC Jr. 1970. Dormant embryos of mammals. *BioScience*, 20:411–415.

Daniel, MJ. 1979. The New Zealand short-tailed bat, *Mystacina tuberculata*: a review of present knowledge. *New Zealand Journal of Zoology*, 6:357.

Daniel, MJ. 1990. Order Chiroptera, 114–137, in *The Handbook of New Zealand Mammals* (CM King, ed.). Oxford University Press, Auckland.

Dapson, RA. 1979. Phenologic influences on cohort-specific reproductive strategies in mice (*Peromyscus polionotus*). *Ecology*, 60:1125–1131.

Darimont, CT, et al. 2009. Human predators outpace other agents of trait change in the wild. *Proceedings of the National Academy of Sciences USA*, 106:952–954.

Darling, FF. 1937. *A Herd of Red Deer.* Oxford University Press, London.

Darlington, PJ. 1957. *Zoogeography: The Geographical Distribution of Animals.* Wiley, NY.

Dashzeveg, D. 1990. New trends in adaptive radiation of early Tertiary rodents (Rodentia, Mammalia). *Acta Zoologica Cracoviensia*, 33:37–44.

Dashzeveg, D, et al. 1995. Extraordinary preservation in a new vertebrate assemblage from the Late Cretaceous of Mongolia. *Nature*, 374:446–449.

Dávalos, LM. 2005. Molecular phylogeny of funnel-eared bats (Chiroptera: Natalidae), with notes on biogeography and conservation. *Molecular Phylogenetics and Evolution*, 37:91–103.

Dávalos, LM. 2006. The geography of diversification in the mormoopids (Chiroptera: Mormoopidae). *Biological Journal of the Linnean Society*, 88:101–118.

David-Gray, ZK, et al. 1998. Light detection in a "blind" mammal. *Nature Neuroscience*, 1:655–656.

Davidson, DW, JH Brown, & RS Inouye. 1980. Competition and the structure of granivore communities. *BioScience*, 30:233–238.

Davies, TJ, et al. 2008. Phylogenetic trees and the future of mammalian biodiversity. *Proceedings of the National Academy of Science USA*, 105:11556–11563.

Davis, DE. 1967. The annual rhythm of fat deposition in woodchucks (*Marmota monax*). *Physiological Zoology*, 40:391–402.

Davis, EB. 2007. Family Antilocapridae, 227–240, in *The Evolution of Artiodactyls* (DR Prothero & SE Foss, eds.). Johns Hopkins University Press, Baltimore, MD.

Davis, MB. 1984. Climatic instability, time labs, and community disequilibrium, 269–284, in *Community Ecology* (J Diamond & TJ Case, eds.). Harper & Row, NY.

Davis, RB, CF Herreid Jr, & HL Short. 1962. Mexican free-tailed bats in Texas. *Ecological Monographs*, 32:311–346.

Davis, RW, et al. 1999. Hunting behavior of a marine mammal beneath the Antarctic fast ice. *Science*, 283:993–996.

Davis, WH & WZ Lidicker Jr. 1956. Winter range of the red bat. *Journal of Mammalogy*, 37:280–281.

Davis, WH & HB Hitchcock. 1965. Biology and migration of the bat, *Myotis lucifugus*, in New England. *Journal of Mammalogy*, 46:296–313.

Dawe, AR & WA Spurrier. 1969. Hibernation induced in ground squirrels by blood transfusion. *Science*, 163:298–299.

Dawson, MR. 1958. Later Tertiary Leporidae of North America. *University of Kansas Paleontological Contributions, Vertebrata,* 6:1–750.

Dawson, MR. 1967. Lagomorph history and stratigraphic record, 287–315, in *Essays in Paleontology and Stratigraphy, Raymond C. Moore Commemorative Volume.* Special Publication No. 2. University of Kansas, Lawrence.

Dawson, MR. 2003. Paleogene rodents of Eurasia. *Deinsea,* 10:97–125.

Dawson, MR & L Krishtalka. 1984. Fossil history of the families of Recent mammals, 11–57, in *Orders and Families of Recent Mammals of the World* (S Anderson & JK Jones, eds.). Wiley, NY.

Dawson, MR, CK Li, & T Qi. 1984. Eocene ctenodactyloid rodents (Mammalia) of eastern and central Asia, 138–150, in *Papers in Vertebrate Paleontology Honoring Robert Warren Wilson* (RM Mengel, ed.). Special Publications 9. Carnegie Museum of Natural History, Pittsburgh, PA.

Dawson, MR, et al. 1993. An early Eocene plagiomenid mammal from Ellesmere and Axel Heiberg Islands, Arctic Canada. *Kaupia, Darmstädter Beiträge zur Naturgeschichte,* 3:179–192.

Dawson, MR & KC Beard. 1996. New Late Paleocene rodents (Mammalia) from Big Multi Quarry, Washakie Basin, Wyoming. *Palaeovertebrata,* 25:301–321.

Dawson, MR, et al. 2006. *Laonastes* and the "Lazarus effect" in Recent mammals. *Science,* 311:1456–1458.

Dawson, TJ. 1995. *Kangaroos: Biology of the Largest Marsupials.* Cornell University Press, Ithaca, NY.

Dawson, TJ & AJ Hulbert. 1970. Standard metabolism, body temperature, and surface areas of Australian marsupials. *American Journal of Physiology,* 218:1233–1238.

Dawson, TJ & CR Taylor. 1973. Energetic cost of locomotion in kangaroos. *Nature,* 246:313–314.

Deavers, DR & JW Hudson. 1979. Water metabolism and estimated field water budgets in rodents (*Clethrionomys gapperi* and *Peromyscus leucopus*) and an insectivore (*Blarina brevicauda*) inhabiting the same mesic environment. *Physiological Zoology,* 52:137–152.

Debruyne, R, V Barriel, & P Tassy. 2003. Mitochondrial cytochrome b of the Lyakhov mammoth (Proboscidea, Mammalia): new data and phylogenetic analyses of Elephantidae. *Molecular Phylogenetics and Evolution,* 26:421–434.

Dechmann, DKN, et al. 2005. Mating system of a Neotropical roost-making bat: the white-throated, round-eared bat, *Lophostoma silvicolum* (Chiroptera: Phyllostomidae). *Behavioral Ecology and Sociobiology,* 58:316–325.

Dechmann, DKN, K Safi, & MJ Vonhof. 2006. Matching morphology and diet in the disk-winged bat *Thyroptera tricolor* (Chiroptera). *Journal of Mammalogy,* 87(5):1013–1019.

DeCoursey, P. 1961. Effect of light on the circadian activity rhythm of the flying squirrel, *Glaucomys volans. Journal of Comparative Physiology A,* 44:331–354.

Deevey, ES. 1947. Life tables for natural populations of animals. *Quarterly Review of Biology,* 22:283–314.

Degen, AA. 1988. Ash and electrolyte intakes of the fat sand rat, *Psammomys obesus,* consuming saltbush, *Atriplex halimus,* containing different water content. *Physiological Zoology,* 61:137–141.

Degen, AA. 1996. *Ecophysiology of Small Desert Mammals.* Springer-Verlag, Berlin.

Dehnhardt, G, B Mauck, & H Bleckmann. 1998. Seal whiskers detect water movements. *Nature,* 394:235–236.

Delany, MJ. 1971. The biology of small rodents in Mayanja Forest, Uganda. *Journal of Zoology London,* 165:85–129.

Dellow, QW & ID Hume. 1982. Studies on the nutrition of macropodine marsupials: IV. digestion in the stomach and intestines of *Macropus giganteus, Thylogale thetis,* and *Macropus eugenii. Australian Journal of Zoology,* 30:767–777.

Delsuc, F, et al. 2001. The evolution of armadillos, anteaters and sloths depicted by nuclear and mitochondrial phylogenies: implications for the status of the enigmatic fossil *Eurotamandua. Proceedings of the Royal Society B,* 268:1605–1615.

Delsuc, F, et al. 2002. Molecular phylogeny of living xenarthrans and the impact of character and taxon sampling on the placental tree rooting. *Molecular Biology and Evolution,* 19:1656–1671.

Delsuc, F & EJP Douzery. 2008. Recent advances and future prospects in xenarthrans molecular phylogenetics, 11–23, in *The Biology of the Xenarthra* (SF Vizcaíno & WJ Loughry, eds.). University Press of Florida, Gainesville.

de Meijere, J. C. 1894. Über die Haare der Säugetiere besonders über ihre Anordnung. *Gegenaurs Morphol. Jahrb.,* 21:312–424.

Deméré, TA, A Berta, & PJ Adams. 2003. Pinnipedimorph evolutionary biogeography, in LJ Flynn (ed.) *Vertebrate Fossils and Their Context: Contributions in Honor of Richard H. Tedford. Bulletin of the American Museum of Natural History* 279:32-76.

Deméré, TA, et al. 2008. Morphological and molecular evidence for a stepwise evolutionary transition from teeth to baleen in Mysticete whales. *Systematic Biology,* 57:15–37.

Derocher, AE, NJ Lunn, & I Sterling. 2004. Polar bears in a warming climate. *Journal of Integrative and Comparative Biology,* 44:163–176.

Desjardins, C, JA Maruniak, & FH Bronson. 1973. Social rank in house mice: differentiation revealed by ultraviolet visualization of urinary marking patterns. *Science,* 182:939–941.

DesRoche, K, MB Fenton, & WC Lancaster. 2007. Echolocation and the thoracic skeletons of bats: a comparative morphological study. *Acta Chiropterologica,* 9(2):483–494.

Desy, EA & GO Batzli. 1989. Effects of food availability and predation on prairie vole demography: a field experiment. *Ecology,* 70:411–421.

Devall, B & G Sessions. 1985. *Deep Ecology.* Gibbs M. Smith, Layton, UT.

Devillard, S, et al. 2004. Does social complexity lead to sex-biased dispersal in polygynous mammals? A test on ground-dwelling sciurids. *Behavioral Ecology,* 15:83–87.

DeVore, I. 1965. *Primate Behavior.* Holt, Rinehart and Winston, NY.

DeVore, I & KRL Hall. 1965. Baboon ecology, 20–52, in *Primate Behavior* (I DeVore, ed.). Holt, Rinehart and Winston, NY.

de Waal, FBM. 1989. *Peacemaking Among Primates.* Harvard University Press, Cambridge, MA.

de Waal, FBM. 1995. Bonobo sex and society. *Scientific American,* March:82–88.

de Waal, F. 2007. *Chimpanzee Politics: Power and Sex Among the Apes.* Johns Hopkins University Press, Baltimore, MD.

Diamond, J. 1993. *The Third Chimpanzee.* Harper Perennial Library, NY.

Diamond, J. 2005. *Collapse: How Societies Choose to Fail or Succeed.* Viking, Penguin Group, NY.

Diamond, J. 2008. The last giant kangaroo. *Nature*, 454:835–836.

Dickman, CR. 1989. Patterns in the structure and diversity of marsupial carnivore communities, 241–251, in *Patterns in the Structure of Mammalian Communities* (DW Morris, Z Abramsky, BJ Fox, & MR Willig, eds.). Texas Tech University Press, Lubbock.

Dieterlen, F. 2005. Family Ctenodactylidae, 1536–1537, in *Mammal Species of the World*, vol. 2 (D Wilson & D Reeder, eds.). Johns Hopkins University Press, Baltimore, MD.

Dijkgraaf, S. 1943. Over een merkwaardige functie wan den gehoorzin bij vleermuizen. *Verslagen Nederlandsche Akademie Wetenschappen Afd. Naturkunde*, 52:622–627.

Dijkgraaf, S. 1946. Die Sinneswelt der Fledermäuse. *Experientia*, 2:438–448.

Dinerstein, E & CM Wemmer. 1988. Fruits *Rhinoceros* eat: dispersal of *Trewia nudiflora* (Euphorbiaceae) in lowland Nepal. *Ecology*, 69:1768–1774.

Dinerstein, E, C Wemmer, & H Mishra. 1988. Adoption in greater one-horned rhinoceros (*Rhinoceros unicornis*). *Journal of Mammalogy*, 69:813–814.

Dixon, JD, et al. 2007. Genetic consequences of habitat fragmentation and loss: the case of the Florida black bear (*Ursus americanus floridanus*). *Conservation Genetics*, 8:455–464.

Dixon, KW. 1981. Western Australian plants with underground fleshy storage organs. Ph.D. thesis, University of Western Australia.

Dixson, AF. 1994. Reproductive biology of the owl monkey, 113–132, in *Aotus: The Owl Monkey* (JF Baer, RE Weller, & I Kakoma eds.). Academic Press, San Diego, CA.

Dobzhansky, T. 1950. Mendelian populations and their evolution. *American Naturalist*, 84:401–418.

Domning, DP. 1972. Steller's sea cow and the origin of North Pacific aboriginal whaling. *Syesis*, 5:187–189.

Domning, DP. 1978. Sirenian evolution in the North Pacific Ocean. *University of California Publications in Geological Science*, 118:1–176.

Domning, DP. 1999. Fossils explained 24: Sirenians (seacows). *Geology Today*, March/April 1999:75–79.

Domning, DP. 2001. The earliest known fully quadrupedal sirenian. *Nature*, 413:625–627.

Domning, DP. 2008. Sirenia, 629–638, in *Evolution of Tertiary Mammals of North America Volume 2: Small Mammals, Xenarthrans, and Marine Mammals* (CM Janis, GF Gunnell, & MD Uhen, eds.). Cambridge University Press, Cambridge, UK.

Domning, DP, J Thomason, & DG Corbett. 2007. Steller's sea cow in the Aleutian Islands. *Marine Mammal Science*, 23:976–983.

Douady, CJ, et al. 2002a. Molecular evidence for the monophyly of Tenrecidae (Mammalia) and the timing of the colonization of Madagascar by Malagasy tenrecs. *Molecular Phylogenetics and Evolution*, 22:357–363.

Douady, CJ, et al. 2002b. Molecular phylogenetic evidence confirming the Eulipotyphla concept and in support of hedgehogs as the sister group to shrews. *Molecular Phylogenetics and Evolution*, 25:200–209.

Douady, CJ & EJP Douzery. 2003. Molecular estimation of eulipotyphlan divergence times and the evolution of "Insectivora." *Molecular Phylogenetics and Evolution*, 28:285–296.

Douady, CJ, et al. 2003. The Sahara as a vicariant agent, and the role of Miocene climatic events, in the diversification of the mammalian order Macroscelidea (elephant shrews). *Proceedings of the National Academy of Sciences USA*, 100: 8325–8330.

Douady, CJ, et al. 2004. "Lipotyphlan" phylogeny based on the growth hormone receptor gene: a reanalysis. *Molecular Phylogenetics and Evolution*, 30:778–788.

Douglas-Hamilton, I. 1972. On the ecology and behavior of the African elephant: the elephants of Lake Manyara. Ph.D. thesis, Oxford University, Oxford.

Douglas-Hamilton, I. 1973. On the ecology and behavior of the Lake Manyara elephants. *East African Wildlife Journal*, 11:401–403.

Douglas-Hamilton, I & O Douglas-Hamilton. 1995. *Among the Elephants*. Book Club Associates, London.

Douglas-Hamilton, I, T Krink, & F Vollrath. 2005. Movements and corridors of African elephants in relation to protected areas. *Naturwissenschaften*, 92:158–163.

Douzery, EJP & D Huchon. 2004. Rabbits, if anything, are likely Glires. *Molecular Phylogenetics and Evolution*. 33:922–935.

Downhower, JF & KB Armitage. 1971. The yellow-bellied marmot and the evolution of polygyny. *American Naturalist*, 105:355–370.

Downs, CT & MR Perrin. 1989. An investigation of the macro- and micro-environments of four *Gerbillurus* species. *Cimbebasia*, 11:41–54.

Downs, CT & MR Perrin. 1991. Urinary concentrating ability of four *Gerbillurus* species of southern Africa arid regions. *Journal of Arid Environments*, 20:71–81.

Dragoo, JW & RL Honeycutt. 1997. Systematics of mustelid-like carnivores. *Journal of Mammalogy*, 78:426–443.

Driscoll, CA, et al. 2007. The near eastern origin of cat domestication. *Science*, 317:519–523.

Duangkhae, S. 1990. Ecology and behavior of Kitti's hog-nosed bat (*Craseonycteris thonglongyai*) in western Thailand. *Natural History Bulletin Siam Society*, 38:135–161.

Dubey, S, et al. 2008. Molecular phylogenetics reveals Messinian, Pliocene, and Pleistocene colonizations of islands by North African shrews. *Molecular Phylogenetics and Evolution*, 47:877–882.

Dubost, G. 1968. Aperçu sur le rythme annuel de reproduction des muridés du nord-est du Gabon. *Biol. Gabonica*, 4:227–239.

Dubost, G. 1983. La comportement de *Cephalophus monticola* Thunberg et *C. dorsalis* Grey, et la place des céphalophes au sein des ruminants, Part I. *Mammalia*, 47:141–177.

Dücker, G. 1957. *Ford und Helligkeits Sehen und Instinkte bei Viverriden und Feliden.* Handbuch fur Zoologie Band, Walter de Gruyter, Germany.

Ducrocq, S, J-J Jaeger, & B Sigé. 1992. Late Eocene southern Asian record of a megabat and its inferences on the megabat phylogeny. *Bat Research News*, 33:41–42.

Ducrocq, S, J-J Jaeger, & B Sigé. 1993. Un megachiroptere dans l'Eocene superieur de Thailand; incidence dans la discussion phylogenique du groupe. *Neues Jahrbuch für Geologie und Paläontologie Mh.*, 1993:561–575.

Ducrocq, S, et al. 1998. The earliest known pig from the upper Eocene of Thailand. *Palaeontology*, 41:147–156.

Dudley, R & P DeVries. 1990. Tropical rain forest structure and the geographical distribution of gliding vertebrates. *Biotropica*, 22:432–434.

Dudley, R, et al. 2007. Gliding and the functional origins of flight: biomechanical novelty or necessity? *Annual Reviews of Ecology, Evolution, and Systematics*, 38:179–201.

Dumont, ER. 2006. The correlated evolution of cranial morphology and feeding behavior in New World fruit bats, 160–177, in *Functional and Evolutionary Ecology of Bats* (A Zubaid, GF McCracken, & TH Kunz, eds.). Oxford, Oxford University Press.

Dunbar, MB & TE Tomasi. 2006. Arousal patterns, metabolic rate, and energy budget of eastern red bats (*Lasiurus borealis*). *Journal of Mammalogy*, 87:1096–1102.

Dung, VV, et al. 1993. A new species of living bovid from Vietnam. *Nature*, 363:443–445.

Dunne, JA, RJ Williams, & ND Martinez. 2004. Network structure and robustness of marine food webs. *Marine Ecology Progress Series*, 273:291–302.

Dunning, DC. 1968. Warning sounds of moths. *Zeitschrift fur Tierpsychologie*, 25:129–138.

Dunning, DC & KD Roeder. 1965. Moth sounds and the insect-catching behavior of bats. *Science*, 147:173–174.

Dupont-Nivet, G, et al. 2007. Tibetan plateau aridification linked to global cooling at the Eocene-Oligocene transition. *Nature*, 445:635–638.

Dupre, SM & ASI Loudon. 2007. Circannual clocks: annual timers unraveled in sheep. *Current Biology*, 17:R216–R217.

Eadie, WR. 1952. Shrew predation and vole populations on a localized area. *Journal of Mammalogy*, 33:185–189.

Eberle, J. 2005. A new tapir from Ellesmere Island, Arctic Canada—implications for northern high latitude palaeobiogeography and tapir palaeobiology. *Palaeogeography, Palaeoclimatology, Palaeoecology*, 227:311–322.

EDGE (Evolutionarily Distinct and Globally Endangered). Website. http://www.edgeofexistence.org/mammals/top_100.php (accessed March 2009).

Editors of Collins. 2006. *Fragile Earth: Views of a Changing World*. Harper Collins Publishers, NY.

Eggert, LS, CA Rasner, & DS Woodruff. 2002. The evolution and phylogeography of the African elephant inferred from mitochondrial DNA sequence and nuclear microsatellite markers. *Proceedings of the Royal Society*, 269:1993–2006.

Ehrenfeld, DW. 1976. The conservation of non-resources. *American Scientist*, 64:648.

Eick, GN, DS Jacobs, & CA Matthee. 2005. A nuclear DNA phylogenetic perspective on the evolution of echolocation and historical biogeography of extant bats (Chiroptera). *Molecular Biology and Evolution*, 22:1869–1886.

Einarsen, AS. 1948. *The Pronghorn Antelope*. Wildlife Management Institute, Washington, DC.

Eisenberg, JF. 1966. The social organization of mammals. *Handbuch der Zoologie*, 10:1–92.

Eisenberg, JF. 1981. *The Mammalian Radiations: An Analysis of Trends in Evolution, Adaptation, and Behavior*. University of Chicago Press, Chicago, IL.

Eisenberg, JF. 1989. *Mammals of the Neotropics: The Northern Neotropics. Vol. 1*, University of Chicago Press, Chicago, IL.

Eisenberg, JF & E Gould. 1970. The tenrecs: a study in mammalian behavior and evolution. *Smithsonian Contributions to Zoology*, 27:1–137.

Eisenberg, JF & DG Kleiman. 1972. Olfactory communication in mammals. *Annual Review of Ecology and Systematics*, 3:1–32.

Eisenberg, JF & DE Wilson. 1981. Relative brain size and demographic strategies in didelphid marsupials. *American Naturalist*, 118:1–15.

Eisenberg, JF & DG Kleiman. 1983. *Advances in the Study of Mammalian Behavior*. Special Publication No. 7, America Society of Mammalogists.

Elbroch, M. 2003. *Mammal Tracks and Sign: A Guide to North American Species*. Stackpole Books, Mechanicsburg, PA.

Elizalde-Arellano, C, et al. 2007. Food sharing behavior in the hairy-legged vampire bat *Diphylla ecaudata*. *Acta Chiropterologica*, 9:314–319.

Ellerman, JR. 1940. *The Families and Genera of Living Rodents, vol I, Rodents Other Than Muridae*. British Museum of Natural History, London.

Elliott, DH, et al. 1970. Triassic tetrapods from Antarctica: evidence for continental drift. *Science*, 169:197–201.

Elliott, L. 1978. Social behavior and foraging ecology of the eastern chipmunk (*Tamias striatus*) in the Adirondack Mountains. *Smithsonian Contributions to Zoology*, 265:1–107.

Elliott, VA. 1976. Circadian rhythms and photoperiodic time measurement in mammals. *Federation Proceedings*, 35:2339–2346.

El-Mogharbel, N, et al. 2007. DMRT gene cluster analysis in the platypus: new insights into genomic organization and regulatory regions. *Genomics*, 89:10–21.

Else, PL & AJ Hulbert. 1981. Comparison of the "mammal machine" and the "reptile machine:" energy production. *American Journal of Physiology*, 240:R3.

Emlen, ST. 1982. The evolution of helping. I. An ecological constraints model. *American Naturalist*, 119:29–39.

Emmons, LH. 1981. Morphological, ecological, and behavioral adaptations for arboreal browsing in *Dactylomys dactylinus* (Rodentia, Echimyidae). *Journal of Mammalogy*, 62:183–189.

Emmons, LH. 1990. *Neotropical Rainforest Mammals: A Field Guide*. University of Chicago Press, Chicago, IL.

Emmons, LH. 1991. Frugivory in treeshrews (*Tupaia*). *American Naturalist*, 138:642–649.

Emmons, LH. 2000. *Tupai: A Field Study of Bornean Treeshrews*. University of California Press, Berkeley, CA.

Emmons, LH & AH Gentry. 1983. Tropical forest structure and the distribution of gliding and prehensile-tailed vertebrates. *American Naturalist*, 121:513–524.

Emmons, LH & A Biun. 1991. Malaysian treeshrews. *National Geographic Research Explorer*, 7:70–81.

Endler, JA. 1982. Problems in distinguishing historical from ecological factors in biogeography. *American Zoology*, 22:441–452.

Endler, J. 1993. Some general comments on the evolution and design of animal communication systems. *Philosophical Transactions Royal Society London B*, 340:215–225.

Engel, SR, et al. 1998. Molecular systematics and paleobiogeography of the South American sigmodontine rodents. *Molecular Biology and Evolution*, 15:35–49.

ERI (Ecological Restoration Institute) and ERI library online. Northern Arizona University, Flagstaff. See especially Restoration Resources and numerous links. http://www.eri.nau.edu (accessed 23 November 2009).

Erickson, CJ. 1991. Percussive foraging in the aye-aye, *Daubentonia madagascariensis*. *Animal Behaviour*, 41:793–801.

Escalante, T, D Espinosa, & JJ Morrone. 2002. Patrones de distribución geográfica de los mamíferos terrestres de México. *Acta Zoologica Mexicana*, 87:47–65.

Espenshade, EB. 1971. *Goode's World Atlas.* 13th ed. Rand-McNally, Chicago, IL.

Essop, MF, EH Harley, & I Baumgarten. 1997. A molecular phylogeny of some Bovidae based on restriction-site mapping of mitochondrial DNA. *Journal of Mammalogy*, 78:377–386.

Estes, JA & JF Palmisano. 1974. Sea otters: their role in structuring nearshore communities. *Science*, 185:1058–1060.

Estes, JA, DO Duggins, & GB Rathbun. 1989. The ecology of extinctions in kelp forest communities. *Conservation Biology*, 3:252–264.

Estes, JA & DO Duggins. 1995. Sea otters and kelp forests in Alaska: generality and variation in a community ecological paradigm. *Ecological Monographs*, 65:75–100.

Estes, JA, et al. 1998. Killer whale predation on sea otters linking oceanic and nearshore ecosystems. *Science*, 282:473–475.

Estes, RD. 1967. The comparative behavior of Grant's and Thompson's gazelles. *Journal of Mammalogy*, 48:189–209.

Estes, RD. 1974. Social organization of the African Bovidae, 166–205, in *The Behavior of Ungulates and Its Relation to Management* (V Geist & FR Walther, eds.). International Union for Conservation of Nature and Natural Resources, Morges, Switzerland.

Estes, RD. 1991. *The Behavior Guide to African Mammals: Including Hoofed Mammals, Carnivores, Primates.* University of California Press, Berkeley, CA.

Estrada, A. 1991. Howler monkeys (*Alouatta palliata*), dung beetles (Scarabaeidae) and seed dispersal: ecological interactions in the tropical rainforest of Los Tuxtlas, Mexico. *Journal of Tropical Ecology*, 7:459–474.

Evans, BK, et al. 1994. Diving ability of the platypus. *Australian Journal of Zoology*, 42:17–27.

Evans, WE. 1973. Echolocation by marine delphinids and one species of fresh-water dolphin. *Journal of the Acoustical Society of America*, 54:191–199.

Evans, WE & BA Powell. 1967. Discrimination of different metallic plates by an echolocating delphinid, 363–382, in *Animal Sonar Systems: Biology and Bionics* (RG Busnel, ed.). Laboratoire de Physiologie Acoustique, Jouy-en-Josas, France.

Evans, WE & J Bastian. 1969. Marine mammal communication: social and ecological factors, 424–475, in *The Biology of Marine Mammals* (HT Anderson, ed.). Academic Press, London.

Evans, WE, JA Thomas, & W Davis. 2004. Vocalizations from Weddell seals (*Leptonychotes weddelli*) during diving and foraging, 541–546, in *Echolocation in Bats and Dolphins* (JA Thomas, CF Moss, & M Vater, eds.). University of Chicago Press, Chicago, IL.

Evans, WE & AV Yablokov. 2004. *Noninvasive Study of Mammalian Populations.* Pensoft Publishers, Sofia, Bulgaria.

Evermann, BW. 1923. The conservation of marine life of the Pacific. *Scientific Monthly*, 16:521.

Ewer, RF. 1968. *Ethology of Mammals.* Plenum Press, NY.

Ewer, RF. 1973. *The Carnivores.* Cornell University Press, Ithaca, NY.

Ewing, WG, EH Studier ,& MJ O'Farrell. 1970. Autumn fat deposition and gross body composition in three species of *Myotis. Comparative Biochemistry and Physiology*, 36:119–129.

Faegri, K & L Van Der Pijl. 1966. *The Principles of Pollination Ecology.* Pergamon Press, NY.

Fall, PL, CA Lindquist, & SE Falconer. 1990. Fossil hyrax middens from the Middle East: a record of paleovegetation and human disturbance, 408–427, in *Packrat Middens: The Last 40,000 Years of Biotic Change* (JL Betancourt, TR Van Devender, & PS Martin, eds.). University of Arizona Press, Tucson.

Faulkes, CG, DH Abbott, & JUM Jarvis. 1991. Social suppression of reproduction in male naked mole-rats. *Journal of Reproduction and Fertility*, 91:593–604.

Faulkes, CG, et al. 1997. Ecological constraints drive social evolution in the African mole-rats. *Proceedings of the Royal Society B*, 264:1619–1627.

Faunmap Working Group. 1994. Faunmap: A database documenting late Quaternary distributions of mammal species in the United States. *Illinois State Museum Scientific Papers* 25 (1 and 2): 690 pp. + 1 diskette.

Faure, PA, JH Fullard, & RMR Barclay. 1990. The response of tympanate moths to the echolocation calls of a substrate-gleaning bat, *Myotis evotis. Journal of Comparative Physiology A*, 166:843–849.

Faure, PA, JH Fullard, & JW Dawson. 1993. The gleaning attacks of the northern long-eared bat, *Myotis septentrionalis*, are relatively inaudible to moths. *Journal of Experimental Biology*, 178:173–189.

Favre, L, et al. 1997. Female-biased dispersal in the monogamous mammal *Crocidura russula*: evidence from field data and microsatellite patterns. *Proceedings of the Royal Society B*, 264:127–132.

Fay, FH. 1981. Walrus—*Odobenus rosmarus*, 1–23, in *Handbook of Marine Mammals* (SH Ridgway & RJ Harrison, eds.). Academic Press, London.

Fay, FH, GC Ray, & AA Kibal'chich. 1984. Time and location of mating and associated behavior of the Pacific walrus, *Odobenus rosmarus divergens* Illiger, in *Soviet-American Cooperative Research on Marine Mammals: vol. 1. Pinnipeds.* NOAA Technical Report NMFS 12.

Fedak, MA, NC Heglund, & CR Taylor. 1982. Energetics and mechanics of terrestrial locomotion. II: kinetic energy changes of the limbs and body as a function of speed and body size in birds and mammals. *Journal of Experimental Biology*, 97:23–40.

Feeny, PP. 1975. Biochemical coevolution between plants and their insect herbivores, 3–19, in *Coevolution of Plants and Animals* (LE Gilbert & PH Raven, eds.). University of Texas Press, Austin.

Feldhamer, GA. 1979. Age, sex ratios, and reproductive potential in black-tailed jackrabbits. *Mammalia*, 43:473–478.

Feldhamer, GA, et al. 1999. *Mammalogy: Adaptation, Diversity, and Ecology.* McGraw-Hill, Boston.

Feldkamp, SD. 1987a. Swimming in the California sea lion: morphometrics, drag, and energetics. *Journal of Experimental Biology*, 131:117–135.

Feldkamp, SD. 1987b. Foreflipper propulsion in the California sea lion, *Zalophus californianus. Journal of Zoology, London*, 212:43–57.

Fell, BF, KA Smith, & RM Campbell. 1963. Hypertrophic and hyperplastic changes in the alimentary canal of the lactating rat. *Journal of Pathology and Bacteriology*, 85:179–188.

Fenton, MB. 1982. Echolocation, insect hearing, and feeding ecology of insectivorous bats, 261–285, in *Ecology of Bats* (T Kunz, ed.). Plenum Press, NY.

Fenton, MB. 1984. Echolocation: implications for ecology and evolution of bats. *Quarterly Review of Biology*, 59:33–53.

Fenton, MB. 1990. The foraging behavior and ecology of animal-eating bats. *Canadian Journal of Zoology*, 68:411–422.

Fenton, MB. 1994. Echolocation: its impact on the behaviour and ecology of bats. *Écoscience*, 1:21–30.

Fenton, MB. 1995. Natural history and biosonar signals, 37–86, in *Hearing in Bats* (AN Popper & RR Fry, eds.). Springer-Verlag, NY.

Fenton, MB & GP Bell. 1979. Echolocation and feeding behavior in four species of *Myotis* (Chiroptera). *Canadian Journal of Zoology*, 57:1271–1277.

Fenton, MB, GP Bell, & DW Thomas. 1980. Echolocation and feeding behavior of *Taphozous mauritianus* (Chiroptera: Emballonuridae). *Canadian Journal of Zoology*, 58:1774–1777.

Fenton, MB & GP Bell. 1981. Recognition of species of insectivorous bats by their echolocation calls. *Journal of Mammalogy*, 62:233–243.

Fenton, MB & JH Fullard. 1981. Moth hearing and feeding strategies of bats. *American Scientist*, 69:266–275.

Fenton, MB, CL Gaudet, & ML Leonard. 1983. Feeding behavior of the bats *Nycteris grandis* and *Nycteris thebaica* (Nycteridae) in captivity. *Journal of Zoology London*, 200:347–354.

Fenton, MB & IL Rautenbach. 1986. A comparison of the roosting and foraging behavior of three species of African insectivorous bats. *Canadian Journal of Zoology*, 64:2860–2867.

Fenton, MB, et al. 1987. Foraging and habitat use by *Nycteris grandis* (Chiroptera: Nycteridae) in Zimbabwe. *Journal of Zoology, London*, 211:709–716.

Fenton, MB, et al. 1990. Foraging behavior and prey selection by large slit-faced bats (*Nycteris grandis*; Chiroptera: Nycteridae). *Biotropica*, 22:2–8.

Fenton, MB, et al. 1993. Variation in foraging behaviour, habitat use, and diet of large slit-faced bats (*Nycteris grandis*). *Zeitschrift Saugetier-kunde*, 58:65–74.

Fenton, MB, et al. 1995. Signal strength, timing, and self-deafening: the evolution of echolocation in bats. *Paleobiology*, 21:229–242.

Fenton, MB & DR Griffin. 1997. High-altitude pursuit of insects by echolocating bats. *Journal of Mammalogy*, 78:247–250.

Ferkin, MH. 1989. Adult-weaning recognition among captive meadow voles (*Microtus pennsylvanicus*). *Behaviour*, 118:114–124.

Ferkin, MH. 1990. Kin recognition and social behavior in microtine rodents, 11–24, in *Social Systems and Population Cycles in Voles* (R Tamarin, RS Ostfeld, SR Pugh, & G Bujalska, eds.). Birkhauser Verlag, Boston, MA.

Fernandez, A, et al. 2005. "Gas and fat embolic syndrome" involving a mass stranding of beaked whales (family Ziphiidae) exposed to anthropogenic sonar signals. *Veterinary Pathology*, 42:446–457.

Fiedler, J. 1979. Prey catching with and without echolocation in the Indian false vampire (*Megaderma lyra*). *Behavioral Ecology and Sociobiology*, 6:155–160.

Fielden, LJ, MR Perrin, & GC Hickman. 1990a. Feeding ecology of the Namib Desert golden mole, *Eremitalpa granti namibensis* (Chrysochloridae). *Journal of Zoology London*, 220:367–389.

Fielden, LJ, MR Perrin, & GC Hickman. 1990b. Water metabolism in the Namib Desert golden mole, *Eremitalpa granti namibensis* (Chrysochloridae). *Comparative Biochemistry and Physiology*, 96A:227–234.

Fielden, LJ, et al. 1990c. Thermoregulation in the Namib Desert golden mole, *Eremitalpa granti namibensis* (Chrysochloridae). *Journal of Arid Environments*, 18:221–237.

Fietz, J. 2003. Primates: *Cheirogaleus*, dwarf lemurs or fat-tailed lemurs, 1307–1309, in *The Natural History of Madagascar* (SM Goodman & JP Benstead, eds.). University of Chicago Press, Chicago, IL.

Fietz, J & JH Dausmann. 2003. Cost and potential benefits of parental care in the nocturnal fat-tailed dwarf lemur (*Cheirogaleus medius*). *Folia Primatologica*, 74:246–258.

Findley, JS. 1996. Mammalian biogeography in the American Southwest, 297–307, in *Contributions in Mammalogy*. (HH Genoways & RJ Baker, eds.). Museum of Texas Tech University, Lubbock.

Findley, JS & DE Wilson. 1974. Observations on the Neotropical disk-winged bat, *Thyroptera tricolor* Spix. *Journal of Mammalogy*, 55:562–571.

Finley, RB Jr. 1969. Cone caches and middens of *Tamiasciurus* in the Rocky Mountain region, 233–273, in *Contributions in Mammalogy*. University of Kansas, Museum of Natural History Miscellaneous Publications No. 51.

Fish, FE. 1979. Thermoregulation in the muskrat (*Ondatra zibethicus*): the use of regional heterothermia. *Comparative Biochemistry and Physiology*, 64:391.

Fish, FE. 1993. Influence of hydrodynamic design and propulsive mode on mammalian swimming energetics. *Australian Journal of Zoology*, 42:79–101.

Fish, FE, et al. 1997. Energetics of swimming by the platypus *Ornithorhynchus anatinus*: metabolic effort associated with rowing. *Journal of Experimental Biology*, 200:2647–2652.

Fisher, EM. 1939. Habits of the southern sea otter. *Journal of Mammalogy*, 20:21–36.

Fitch, HS, R Goodrum, & C Newman. 1952. The armadillo in the southeastern United States. *Journal of Mammalogy*, 33:21–37.

Fitch, WM. 1976. Molecular evolutionary clocks, 160–178, in *Molecular Evolution* (FJ Ayala, ed.). Sinauer Associates, Sunderland, MA.

Fitzgerald, EMG. 2006. A bizarre new toothed mysticete (Cetacea) from Australia and the early evolution of baleen whales. *Proceedings of the Royal Society B*, 273:2955–2963.

FitzGibbon, C. 1995. Comparative ecology of two elephant-shrew species in Kenyan coastal forest. *Mammal Review*, 25:19–30.

FitzGibbon, C. 1997. The adaptive significance of monogamy in the golden-rumped elephant-shrew. *Journal of Zoology, London*, 242:167–177.

Fitzpatrick, BM & M Turelli. 2006. The geography of mammalian speciation: mixed signals from phylogenies and range maps. *Evolution*, 60(3):601–615.

Flannery, T. 1995. *Mammals of New Guinea*. Smithsonian Institution Press, Washington, DC.

Flannery, T. 2002. *The Future Eaters: An Ecological History of the Australasian Lands and People*. Grove/Atlantic Press, NY.

Fleagle, JG. 1976. Locomotion and posture of the Malayan siamang. *Folia Primatologica*, 26:245–269.

Fleagle, JG. 1999. *Primate Adaptation and Evolution*. Academic Press, NY.

Fleagle, JG, RF Kay, & MRL Anthony. 1997. Fossil New World monkeys, 473–495, in *Vertebrate Paleontology in the Neotropics: The Miocene Fauna of La Venta, Colombia* (RF Kay, RH Madden, RL Cifelli, & JJ Flynn, eds.). Smithsonian Institution Press, Washington, DC.

Fleagle, JG, C Janson, & K Reed. 1999. *Primate Communities*. Cambridge University Press, Cambridge, UK.

Fleming, TH. 1970. Notes on the rodent faunas of two Panamanian forests. *Journal of Mammalogy*, 51:473–490.

Fleming, TH. 1971. *Artibeus jamaicensis*: delayed embryonic development in a Neotropical bat. *Science*, 171:402–404.

Fleming, TH. 1988. *The Short-tailed Fruit Bat: A Study in Plant–Animal Interactions*. University of Chicago Press, Chicago, IL.

Fleming, TH. 1992. How do fruit- and nectar-feeding birds and mammals track their food resources? 355–391, in *Effects of Resource Distribution on Animal–Plant Interactions* (MD Hunter, T Ohgushi, & PW Price, eds.). Academic Press, San Diego, CA.

Fleming, TH. 1995. The use of stable isotopes to study the diets of plant-visiting bats. *Symposia Zoological Society London*, 67:99–110.

Fleming, TH, RA Nuñez, & L Sternberg. 1993. Seasonal changes in the diets of migrant and non-migrant nectarivorous bats as revealed by carbon stable isotope analysis. *Oecologia*, 94:72–75.

Fleming, TH & VJ Sosa. 1994. Effects of nectarivorous and frugivorous mammals on reproductive success of plants. *Journal of Mammalogy*, 75:845–851.

Fleming, TH, S Maurice, & JL Hamrick. 1998. Geographic variation in the breeding system and the evolutionary stability of trioecy in *Pachycereus pringlei* (Cactaceae). *Evolutionary Ecology*, 12:279–289.

Fleming, TH & N Muchhala. 2008. Nectar-feeding bird and bat niches in two worlds: pantropical comparisons of vertebrate pollination systems. *Journal of Biogeography*, 35:764–780.

Flexner, LB, et al. 1948. The permeability of the human placenta to sodium in normal and abnormal pregnancies and the supply of sodium to the human fetus as determined with radioactive sodium. *American Journal of Obstetrics and Gynecology*, 55:469–480.

Floody, OR. 1979. Behavioral and physiological analyses of ultrasound produced by female hamsters (*Mesocricetus auratus*). *American Zoologist*, 19:443–455.

Floody, OR & AP Arnold. 1975. Uganda kob (*Adenota kob thomasi*) territoriality and the spatial distribution of sexual and agonistic behaviors at a territorial ground. *Zeitschrift fur Tierpsychologie*, 37:192–212.

Flynn, JJ. 2009. Splendid isolation. *Natural History*, June:26–32.

Flynn, J & H Galiano. 1982. Phylogeny of early Tertiary Carnivora, with a description of a new species of *Protictis* from the Middle Eocene of northwestern Wyoming. *American Museum Novitates*, 2725:1–64.

Flynn, JJ & MA Nedbal. 1998. Phylogeny of the Carnivora (Mammalia): congruence vs incompatibility among multiple data sets. *Molecular Phylogenetics and Evolution*, 9:414–426.

Flynn, JJ, et al. 2000. Whence the red panda? *Molecular Phylogenetics and Evolution*, 17:190–199.

Flynn, JJ, et al. 2003. The Tinguiririca fauna, Chile: biochronology, paleoecology, biogeography, and a new earliest Oligocene South American land mammal age. *Palaeogeography, Palaeoclimatology, Palaeoecology*, 195:229–259.

Flynn, JJ & GD Wesley-Hunt. 2005. Carnivora, 175–198, in *The Rise of Placental Mammals* (KD Rose & JD Archibald, eds.). Johns Hopkins University Press, Baltimore, MD.

Flynn, JJ, et al. 2005a. Molecular phylogeny of the Carnivora (Mammalia): assessing the impact of increased sampling on resolving enigmatic relationships. *Systematic Biology*, 54:317–337.

Flynn, JJ, et al. 2005b. Geochronology of Hemphillian-Blancan aged strata, Guanajuato, Mexico, and implications for timing of the Great American Biotic Interchange. *Journal of Geology*, 113:287–307.

Fogden, M. 1974. A preliminary field study of the western tarsier, *Tarsius bancanus* Horsefield, 151–166, in *Prosimian Biology* (RD Martin, GA Doyle, & AC Walker, eds.). Duckworth, London.

Folk, GE Jr, A. Larson, & MA Folk. 1976. Physiology of hibernating bears, 378–380, in *Bears—Their Biology and Management* (MR Pelton, JW Lentfer, & GE Folk, eds.). Publication No. 40, International Union Conservation Nature.

Folk, GE Jr & HA Semken Jr. 1991. The evolution of sweat glands. *International Journal of Biometeorology*, 35:180–186.

Folk, GE, et al. 2006. Mammalian activity—rest rhythms in Arctic continuous daylight. *Biological Rhythm Research*, 37:455–469.

Folstad, I & AJ Karter. 1992. Parasites, bright males, and the immunocompetence handicap. *American Naturalist*, 139:603–622.

Fons, R, et al. 1997. Rates of rewarming, heart and respiratory rates and their significance for oxygen transport during arousal from torpor in the smallest mammal, the Etruscan shrew *Suncus etruscus*. *Journal of Experimental Biology*, 200:1451–1458.

Fonseca, GAB da. 1985. The vanishing Brazilian Atlantic forest. *Biological Conservation*, 34:17–34.

Ford, J & D Martinez. 2000. Traditional ecological knowledge, ecosystem science, and environmental management. *Ecological Applications*, 10:1249–1250.

Ford, JKB. 1991. Vocal traditions among resident killer whales (*Orcinus orca*) in coastal waters of British Columbia. *Canadian Journal of Zoology*, 69:1454–1483.

Ford, RG & FA Pitelka. 1984. Resource limitation in populations of the California vole *Microtus californicus*. *Ecology*, 65:122–136.

Ford, SM. 1986. Systematics of New World monkeys, 73–135, in *Comparative Primate Biology, Systematics, Evolution, and Anatomy*, vol. 1 (DR Swindler & J Erwin, eds.). Alan R. Liss, NY.

Fordyce, RE. 1992. Cetacean evolution and Eocene/Oligocene environments, 368–381, in *Eocene-Oligocene Climatic and Biotic Evolution* (DR Prothero & WA Berggren, eds.). Princeton University Press, Princeton, NJ.

Fordyce, RE. 2003. Cetacean evolution and Eocene-Oligocene oceans revisited, 154–170, in *From Greenhouse to Icehouse: The Marine Eocene-Oligocene Transition* (DR Prothero, LC Ivany, & ER Nesbitt, eds.). Columbia University Press, NY.

Fordyce, RE & LG Barnes. 1994. The evolutionary history of whales and dolphins. *Annual Review Earth Planetary Science*, 22:419–455.

Fordyce, RE & C de Muizon. 2001. Evolutionary history of cetaceans: a review, 169–233, in *Secondary Adaptation of Tetrapods to Life in Water: Proceedings of the International Meeting Poitiers, 1996* (J-M Mazin & V de Buffrénil, eds.). Verlag Friedrich Pfeil, München.

Forman, RTT. 1995. *Land Mosaics: The Ecology of Landscapes and Regions*. Cambridge University Press, Cambridge, UK.

Forman, RTT, et al. 2003. *Road Ecology: Science and Solutions*. Island Press, Washington, DC.

Formozov, AN. 1946. The covering of snow as an integral factor of the environment and its importance in the ecology of mammals and birds. *Material for Fauna and Flora of the USSR, New Series Zoology,* 5:1–141.

Formozov, AN. 1966. Adaptive modifications of behavior in mammals of the Eurasian steppes. *Journal of Mammalogy,* 47:208–222.

Fossey, D. 1972. Living with mountain gorillas, 208–229, in *The Marvels of Animal Behavior* (PR Marler, ed.). National Geographic Society, Washington, DC.

Fortelius, M & J Kappelman. 1993. The largest land mammal ever imagined. *Zoological Journal of the Linnean Society,* 108:85–101.

Fox, D. 2007. Back to the no-analog future? *Science,* 316:823–825.

Frakes, LA, JE Francis, & JI Syktus. 1992. *Climate Modes of the Phanerozoic: The History of the Earth's Climate Over the Past 600 Million Years.* Cambridge University Press, Cambridge, UK.

Francis, CM. 2008. *A Guide to the Mammals of Southeast Asia.* Princeton University Press, Princeton, NJ.

Francis, CM, et al. 1994. Lactation in male fruit bats. *Nature,* 367:691–692.

Frank, CL. 1988a. The influence of moisture content on seed selection by kangaroo rats. *Journal of Mammalogy,* 69:353–357.

Frank, CL. 1988b. The effects of moldiness level on seed selection by *Dipodomys spectabilis. Journal of Mammalogy,* 69:358–362.

Frank, LG. 1997. Evolution of genital masculinization: why do female hyaenas have such a large "penis?" *Trends in Ecology and Evolution,* 12:58–62.

Frappell, PB & JP Mortola. 2000. Respiratory function in a newborn marsupial with skin gas exchange. *Respiration Physiology,* 120:35–45.

Freeman, PW. 1981. A multivariate study of the family Molossidae (Mammalia: Chiroptera): morphology, ecology, evolution. *Fieldiana Zoology,* 7:1–173.

Freeman, PW. 1984. Functional analysis of large animalivorous bats (Microchiroptera). *Biological Journal Linnean Society,* 21:387–408.

Freeman, PW. 1988. Frugivorous and animalivorous bats (Microchiroptera): dental and cranial adaptations. *Biological Journal of the Linnean Society,* 33:249–272.

Freeman, PW. 1995. Nectarivorous feeding mechanisms in bats. *Biological Journal of the Linnean Society,* 56:439–463.

Freeman, PW. 2000. Macroevolution in Microchiroptera: recoupling morphology and ecology with phylogeny. *Evolutionary Ecology Research,* 2:317–335.

Freeman, PW & C Lemen. 1983. Quantification of competition among coexisting heteromyids in the southwest. *Southwestern Naturalist,* 28:41–46.

French, AR. 1977. Periodicity of recurrent hypothermia during hibernation in the pocket mouse, *Perognathus longimembris. Journal of Comparative Physiology,* 115:87.

French, AR. 1993. Physiological ecology of the Heteromyidae: economics of energy and water utilization, 509–538, in *Biology of the Heteromyidae* (HH Genoways & JH Brown, eds.). Special Publications No. 10, American Society of Mammalogy.

Fretwell, SD. 1972. *Populations in a Seasonal Environment.* Princeton University Press, Princeton, NJ.

Freudenberger, DO, IR Wallis, & ID Hume. 1989. Digestive adaptations of kangaroos, wallabies, and rat-kangaroos, 179–189, in *Kangaroos, Wallabies, and Rat-kangaroos* (G Grigg, P Jarman, & I Hume, eds.). Surrey Beatty and Sons, Chipping Norton, New South Wales, Australia.

Freudenthal, M. 1972. *Deinogalerix koenigswaldi* nov. gen., nov. spec.: a giant insectivore from the Neogene of Italy. *Scripta Geologica, Leiden,* 14:1.

Freyer, C, Q Zeller, & MB Renfree. 2002. Ultrastructure of the placenta of the tammar wallaby, *Macropus eugenii*: comparison with the grey short-tailed opossum, *Monodelphis domestica. Journal of Anatomy,* 201:101–119.

Friend, JA & ND Thomas. 2003. Conservation of the numbat (*Myrmecobius fasciatus*), 452–463, in *Predators with Pouches: The Biology of Carnivorous Marsupials* (M Jones, C Dickman, & M Archer, eds.). CSIRO Publishing, Collingwood, Victoria, Australia.

Friis, EM & WL Crepet. 1987. Time of appearance of floral features, 145–179, in *The Origins of Angiosperms and Their Biological Consequences* (EM Friis, WG Chalconer, & PR Crane, eds.). Cambridge University Press, Cambridge, UK.

Froehlich, DJ. 1999. Phylogenetic systematics of basal perissodactyls. *Journal of Vertebrate Paleontology,* 19:140–159.

Froehlich, DJ. 2002. Quo vadis *eohippus*? The systematics and taxonomy of early Eocene equids (Perissodactyla). *Zoological Journal of the Linnean Society,* 134:141–256.

Fryxell, JM, et al. 1999. Density dependence, prey dependence, and population dynamics of martins in Ontario. *Ecology,* 80:1311–1321.

Fullard, JH. 1982. Echolocation assemblages and their effects on moth auditory systems. *Canadian Journal of Zoology,* 60:2572–2576.

Fullard, JH. 1987. Sensory ecology and neuroethology of moths and bats: interactions in a global perspective, 244–272, in *Recent Advances in the Study of Bats* (MB Fenton, P Racey, & JMV Rayner, eds.). Cambridge University Press, Cambridge, UK.

Fullard, JH, MB Fenton, & JA Simmons. 1979. Jamming bat echolocation: the clicks of arctiid moths. *Canadian Journal of Zoology,* 57:647–649.

Fullard, JH & JJ Belwood. 1988. The echolocation assemblage: acoustic ensembles in a neotropical habitat, 639–643, in *Animal Sonar* (PE Nachtigall & PWB Moore, eds.). Plenum Press, NY.

Fullard, JH & JW Dawson. 1997. The echolocation calls of the spotted bat *Euderma maculatum* are relatively inaudible to moths. *Journal of Experimental Biology,* 200:129–137.

Fullard, JH, et al. 2008. Surviving cave bats: auditory and behavioral defences in the Australian noctuid moth, *Speiredonia spectans. Journal of Experimental Biology,* 211:3808–3815.

Fuller, TK & PW Kat. 1993. Hunting success of African wild dogs in southwestern Kenya. *Journal of Mammalogy,* 74:464–467.

Gaines, MS. 1985. Genetics, 845–883, in *Biology of the New World Microtus* (RH Tamarin, ed.). Special Publication No. 8, American Society of Mammalogists.

Gaines, MS & C.J Krebs. 1971. Genetic changes in fluctuating vole populations. *Evolution,* 25:702–723.

Gaines, MS, AV Vivas, & CL Baker. 1979. An experimental analysis of dispersal in fluctuating vole populations: demographic parameters. *Ecology,* 60:814–828.

Gaisler, J. 1979. Ecology of bats, 281–342, in *Ecology of Small Mammals* (DM Stoddart, ed.). Chapman and Hall, London.

Galliari, CA & FJ Goin. 1993. Conservación de la biodiversidad en la Argentina: el caso de los mamíferos, 367–399, in *Elementos de Politica Ambiental. La Plata* (FJ Goin & RG Goñi, eds.). H. Cámara de Diputados, Provincia de Buenos Aires.

Gallivan, GJ & RC Best. 1986. The influence of feeding and fasting on the metabolic rate and ventilation of the Amazonian manatee (*Trichechus inunguis*). *Physiological Zoology*, 59:552–557.

Gambaryan, PP. 1974. *How Mammals Run: Anatomical Adaptations.* Keter, Jerusalem.

Gambaryan, PP & Z Kielan-Jaworowska. 1997. Sprawling versus parasagittal stance in multituberculate mammals. *Acta Palaeontologica Polonica*, 42:13–44.

Ganzhorn, JU, et al. 2006. Lemur biogeography, in *Primate Biogeography* (SM Lehman & JG Fleagle, eds.). Springer-Verlag, NY.

Gardner, AL. 1993. Order Didelphimorphia, 15–23, in *Mammal Species of the World: A Taxonomic and Geographic Reference* (DE Wilson & DM Reeder, eds.). Smithsonian Institution Press, Washington, DC.

Garstang, M. 2004. Long-distance, low-frequency elephant communication. *Journal of Comparative Physiology A*, 190:791–805.

Gasc, JP, et al. 1986. Morphofunctional study of the digging system of the Namib Desert golden mole (*Eremitalpa granti namibensis*): cinefluorographical and anatomical analysis. *Journal of Zoology, London (A)*, 208:9–35.

Gashwiler, JS & WL Robinette. 1957. Accidental fatalities of the Utah cougar. *Journal of Mammalogy*, 38:123–126.

Gaskin, DE. 1982. *The Ecology of Whales and Dolphins.* Heinemann, Portsmouth, NH.

Gasparini, Z, X Pereda-Suberbiola, & RE Molnar. 1996. New data on the ankylosaurian dinosaur from the Late Cretaceous of the Antarctic peninsula. *Memoirs of the Queensland Museum*, 39:583–594.

Gatesy, J. 1997. More support for a Cetacea / Hippopotamidae clade: the blood clotting protein gene g-fibrinogen. *Molecular Biology and Evolution*, 14:537–543.

Gatesy, J, et al. 1992. Phylogeny of the Bovidae (Artiodactyla, Mammalia), based on mitochondrial ribosomal DNA sequences. *Molecular Biology and Evolution*, 9:433–446.

Gatesy, J, et al. 1996. Evidence from milk casein genes that cetaceans are close relatives of hippopotamid artiodactyls. *Molecular Biology and Evolution*, 13:954–963.

Gatesy, J, et al. 1999. Stability of cladistic relationships between Cetacea and higher-level artiodactyl taxa. *Systematic Biology*, 48:6–20.

Gatesy, J & MA O'Leary. 2001. Deciphering whale origins with molecules and fossils. *Trends in Ecology & Evolution*, 16:562–570.

Gatesy, J, et al. 2002. Resolution of a supertree/supermatrix paradox. *Systematic Biology*, 51:652–664.

Gaudin, TJ. 1999. The morphology of xenarthrous vertebrae (Mammalia, Xenarthra). *Fieldiana (Geology)*, n.s. 41:1–38.

Gaudin, TJ. 2004. Phylogenetic relationships among sloths (Mammalia, Xenarthra, Tardigrada): the craniodental evidence. *Zoological Journal of the Linnean Society*, 140:255–305.

Gaudin, TJ, et al. 1996. Reexamination of the morphological evidence for the cohort Epitheria (Mammalia, Eutheria). *Journal of Mammalian Evolution*, 3:31–79.

Gaudin, TJ & JR Wible. 2006. The phylogeny of living and extinct armadillos (Mammalia, Xenarthra, Cingulata): a craniodental analysis, 153–198, in *Amniote Paleobiology: Perspectives on the Evolution of Mammals, Birds, and Reptiles* (MT Carrano, TJ Gaudin, RW Blob, & JR Wible, eds.). University of Chicago Press, Chicago, IL.

Gaudin, TJ & HG McDonald. 2008. Morphology-based investigations of the phylogenetic relationships among extant and fossil xenarthrans, 24–36, in *The Biology of the Xenarthra* (SF Vizcaíno & WJ Loughry, eds.). University Press of Florida, Gainesville.

Gauthier, J, AG Kluge, & T Rowe. 1988. Amniote phylogeny and the importance of fossils. *Cladistics*, 4:105–209.

Geiser, F & LS Broome. 1993. The effects of temperature on the pattern of torpor in a marsupial hibernator. *Journal of Comparative Physiology B*, 163:133–137.

Geiser, F & CR Pavey. 2007. Basking and torpor in a rock-dwelling desert marsupial: survival strategies in a resource-poor environment. *Journal of Comparative Physiology B*, 177:885–892.

Geisler, JH. 2001. New morphological evidence for the phylogeny of Artiodactyla, Cetacea, and Mesonychidae. *American Museum Novitates*, 3344:1–53.

Geisler, JH & MD Uhen. 2005. Phylogenetic relationships of extinct cetartiodactyls: results of simultaneous analyses of molecular, morphological, and stratigraphic data. *Journal of Mammalian Evolution*, 12:145–160.

Geisler, JH, et al. 2007. Phylogenetic relationships of cetaceans to terrestrial artiodactyls, 19–33, in *The Evolution of Artiodactyls* (DR Prothero & SE Foss, eds.). Johns Hopkins University Press, Baltimore, MD.

Geisler, JH & JM Theodor. 2009. Hippopotamus and whale phylogeny. *Nature*, 458:E1–E4.

Geissmann, T. 2002. Duet-splitting and the evolution of gibbon songs. *Biological Review*, 77:57–76.

Geissmann, T, et al. 2000. *Vietnam Primate Conservation Status Review 2000—Part 1: Gibbons.* Fauna & Flora International, Indochina Programme, Hanoi.

Geist, V. 1974. On the relationship of ecology and behavior in the evolution of ungulates: theoretical considerations, 235–246, in *The Behavior of Ungulates and Its Relation to Management* (V Geist & FR Walther, eds.). International Union for Conservation of Nature and Natural Resources, Morges, Switzerland.

Geist, V. 1994. Wildlife conservation as wealth. *Nature*, 368:491–492.

Geluso, KN. 1975. Urine concentration cycles of insectivorous bats in the laboratory. *Journal of Comparative Physiology*, 99:309–319.

Geluso, KN. 1978. Urine concentrating ability and renal structure of insectivorous bats. *Journal of Mammalogy*, 59:812–822.

Geluso, KN & EH Studier. 1979. Diurnal fluctuation in urine concentration in the little brown bat, *Myotis lucifugus*, in a natural roost. *Comparative Biochemistry and Physiology*, 62:471–473.

Gemmell, RT & RW Rose. 1989. The senses involved in movement of some newborn Macropodoidea and other marsupials from cloaca to pouch, 339–347, in *Kangaroos, Wallabies and Rat-Kangaroos*, (G Grigg, P Jarman, & I Hume, eds.). Surrey, Beatty, Sydney, Australia.

Gengozian, N & CB Merritt. 1970. Effect of unilateral ovariectomy on twinning frequency in the marmoset. *Journal of Reproduction and Fertility*, 23:509–512.

Genoud, M. 1985. Ecological energetics of two European shrews: *Crocidura russula* and *Sorex coronatus* (Soricidae: Mammalia). *Journal of Zoology, London (A)*, 207:63–85.

Genoways, HH & JH Brown. 1993. *Biology of the Heteromyidae.* Special Publication No. 10, American Society of Mammalogy.

Gentry, AW. 1992. The subfamilies and tribes of the family Bovidae. *Mammal Review,* 22:1–32.

Gentry, AW, GE Rössner, & EP Heizmann. 1999. Suborder Ruminantia, 225–258, in *The Miocene Land Mammals of Europe* (GE Rössner & EPJ Heizmann, eds.). Verlag Dr. Friedrich Pfeil, Munich.

George, SB, VR Choate, & HH Genoways. 1986. *Blarina brevicauda. Mammalian Species,* 262:1–9.

George, W & BJ Weir. 1972. The chromosomes of some octodontids with special reference to *Octodontomys* (Rodentia, Hystricomorpha). *Chromosoma,* 37:53–62.

Gerstein, E, L Gerstein, & S Forsythe. 2004. Do manatees utilize infrasonic communication or detection? *Journal of the Acoustical Society of America,* 115:2554–2555.

Gerstell, R. 1938. The Pennsylvania deer problem in 1938. *Pennsylvania Game News,* 9(5):12–13, 31; 9(6):10–11, 27, 32; 9(7):6–7, 29.

Gese, EM & RL Ruff. 1997. Scent-marking by coyotes, *Canis latrans*: the influence of social and ecological factors. *Animal Behaviour,* 54:1155–1166.

Getz, LL, et al. 1987. Fourteen years of population fluctuations of *Microtus ochrogaster* and *M. pennsylvanicus* in east central Illinois. *Canadian Journal of Zoology,* 65:1317–1325.

Getz, LL, CM Larson, & KA Lindstrom. 1992. *Blarina brevicauda* as a predator on nestling voles. *Journal of Mammalogy,* 73:591–596.

Getz, LL, et al. 1993. Social organization of the prairie voles (*Microtus ochrogaster*). *Journal of Mammalogy,* 74:44–58.

Getz, LL & CS Carter. 1996. Prairie-vole partnerships. *American Scientist,* 84:56–62.

Getz, LL & B McGuire. 1997. Communal nesting in prairie voles (*Microtus ochrogaster*): formation, composition, and persistence of communal groups. *Canadian Journal of Zoology,* 75:525–534.

Geuse, P, V Bauchau, & E LeBoulenge. 1985. Distribution and population dynamics of bank voles and wood mice in a patchy woodland habitat in central Belgium. *Acta Zoologica Fennica,* 173:65–68.

Gheerbrant, E, J Sudre, & H Cappetta. 1996. A Palaeocene proboscidean from Morocco. *Nature,* 383:68–70.

Gheerbrant, E, et al. 2002. A new large mammal from the Ypresian of Morocco: evidence of surprising diversity of early proboscideans. *Acta Palaeontologica Polonica,* 47:493–506.

Gheerbrant, E, et al. 2005. Nouvelles données sur *Phosphatherium escuilliei* (Mammalia, Proboscidea) de l'Éocène inférieur du Maroc, apports à la phylogénie des Proboscidea et des ongulés lophodontes. *Geodiversitas,* 27:239–333.

Gheerbrant, E, S Peigne, & H. Thomas. 2007. First description of the skeleton of a Paleogene hyracoid: *Saghatherium* from the early Oligocene of Jebel al Hasawnah, Libya. *Palaeontographica Abt. A,* 279:93–145.

Giannini, NP & NB Simmons. 2003. A phylogeny of megachiropteran bats (Mammalia: Chiroptera: Pteropodidae) based on direct optimization analysis of one nuclear and four mitochondrial genes. *Cladistics,* 19:496–511.

Giannini, NP, JR Wible, & NB Simmons. 2006. On the cranial osteology of Chiroptera. I. *Pteropus* (Megachiroptera: Pteropodidae). *Bulletin of the American Museum of Natural History,* 295:1–134.

Giannini, NP & NB Simmons. 2007. Element homology and the evolution of dental formulae in megachiropteran bats (Mammalia: Chiroptera: Pteropodidae). *American Museum Novitates,* 3559:1–27.

Gibbons, A & E Culotta. 2008. American Association of Physical Anthropologists Meeting. Snapshots from the meeting. *Science,* 320(5786):609.

Gilbert, A. 2008. *What the Nose Knows: The Science of Scent in Everyday Life.* Crown Publishers, NY.

Gilbert, BS & S Boutin. 1991. Effect of moonlight on winter activity of snowshoe hares. *Arctic and Alpine Research,* 23:61–65.

Gilbert, MTP, et al. 2008. Intraspecific phylogenetic analysis of Siberian woolly mammoths using complete mitochondrial genomes. *Proceedings of the National Academy of Sciences USA,* 105(24):8327–8332.

Gillette, DD. 1994. *Seismosaurus: The Earth Shaker.* Columbia University Press, NY.

Gilpin, ME & ME Soule. 2001. Minimum viable populations: processes of species extinction, 19–34, in *Conservation Biology: The Science of Scarcity and Diversity* (ME Soule, ed.). Sinauer, Sunderland, MA.

Gingerich, PD. 1980. Eocene Adapidae: paleobiogeography and the origin of South American Platyrrhini, 123–138, in *Evolutionary Biology of the New World Monkeys and Continental Drift* (RL Ciochon & AB Chiarelli, eds.). Plenum Press, NY.

Gingerich, PD, et al. 2001. Origin of whales from early artiodactyls: hands and feet of Eocene Protocetidae from Pakistan. *Science,* 293:2239–2242.

Gingerich, PD, et al. 2009. New protocetid whale from the middle Eocene of Pakistan: birth on land, precocial development, and sexual dimorphism. *PLoS One,* 4:e4366:1–20.

Ginsberg, JR. 2001. Mammals, biodiversity of, 777–810, in *Encyclopedia of Biodiversity,* vol. 3 (SA Levin, ed.). Academic Press, San Diego, CA.

Gittleman, J. 1993. *Carnivore Behavior, Ecology and Evolution.* Cornell University Press, Ithaca, NY.

Glander, KE. 1977. Poison in a monkey's garden of Eden. *Natural History,* 86:35–41.

Goebel, T, MR Waters, & DH O'Rourke. 2008. The late Pleistocene dispersal of modern humans in the Americas. *Science,* 319:1497–1502.

Goin, FJ. 1997. New clues for understanding Neogene marsupial radiations, 187–206, in *Vertebrate Paleontology in the Neotropics: The Miocene Fauna of La Venta, Colombia* (R. F. Kay, R. H. Madden, R. L. Cifelli, & J. J. Flynn, eds.). Smithsonian Institution Press, Washington, DC.

Goin, FJ. 2003. Early marsupial radiations in South America, 30–42, in *Predators with Pouches: the Biology of Carnivorous Marsupials* (M Jones, C Dickman, & M Archer, eds.). CSIRO Publishing, Collingwood, Victoria, Australia.

Goin, FJ & R Pascual. 1987. News on the biology and taxonomy of the marsupials Thylacosmilidae (late Tertiary of Argentina). *Anales de la Academia Nacional de Ciencias Ex. Fís. Nat., Buenos Aires,* 39:219–246.

Goin, FJ & R Goñi. 1991. Naturaleza, naturalistas, tecnología e innovación, 211–220, in *Ciencia, Tecnología e Innovación: Perspectivas y Estrategias* (RG Goñi & FJ Goin, eds.). La Plata, H. Cámara de senadores, Provincia de Buenos Aires.

Goin, FJ & AA Carlini. 1995. An early Tertiary microbiotheriid marsupial from Antarctic. *Journal of Vertebrate Paleontology*, 15:205–207.

Goldblatt, P. 1993. *Biological Relationships Between Africa and South America*. Yale University Press, New Haven, CT.

Goldman, AS. 2002. Evolution of the mammary gland defense system and the ontogeny of the immune system. *Journal of Mammary Gland Biology and Neoplasia*, 7:277–289.

Goldman, DP, R Giri, & SJ O'Brien. 1989. Molecular genetic-distance estimates among the Ursidae as indicated by one- and two-dimensional protein electrophoresis. *Evolution*, 43:282–295.

Goldman, LJ & OW Henson Jr. 1977. Prey recognition and selection by the constant frequency bat, *Pteronotus parnellii*. *Behavioral Ecology and Sociobiology*, 2:411–420.

Golightly, RT & RD Ohmart. 1983. Metabolism and body temperature of two desert canids: coyotes and kit foxes. *Journal of Mammalogy*, 64:624–635.

Golley, FB. 1960. Energy dynamics of a food chain of an old-field community. *Ecological Monographs*, 30:187–206.

Gompper, ME. 1996. Sociality and asociality in white-nosed coatis (*Nasua narica*): foraging costs and benefits. *Behavioral Ecology*, 7:254–263.

Gompper, ME, JL Gittleman, & RK Wayne. 1997. Genetic relatedness, coalitions and social behavior of white-nosed, coatis, *Nasua narica*. *Animal Behaviour*, 53:781–797.

Good, R. 1974. *The Geography of Flowering Plants*. 3rd ed. Longman, White Plains, NY.

Goodall, J. 1977. Infant killing and cannibalism in free-living chimpanzees. *Folia Primatologica*, 28:259–282.

Goodall, J. 1983. Population dynamics during a 15-year period in one community of free-living chimpanzees in the Gombe National Park, Tanzania. *Primates*, 21:545–549.

Goodall, J. 1986. *Chimpanzees of Gombe*. Harvard University Press, Cambridge, MA.

Goodman, B. 1993. Drugs and people threaten diversity in Andean forests. *Science*, 261:293.

Goodman, SM & BD Patterson (eds.). 1997. *Natural Change and Human Impact in Madagascar*. Smithsonian Institution Press, Washington, DC.

Goodman, SM & JP Benstead. 2003. *The Natural History of Madagascar*. University of Chicago Press, Chicago, IL.

Goodman, SM, F Rakotondraparany, & A Kofoky. 2007. The description of a new species of *Myzopoda* (Myzopodidae: Chiroptera) from western Madagascar. *Mammalian Biology*, 72:65–81.

Goold, JC & SE Jones. 1995. Time and frequency domain characteristics of sperm whale clicks. *Journal of the Acoustical Society of America*, 98:1279–1291.

Goosem, M. 1997. Internal fragmentation: the effects of roads, highways, and powerline clearings on movements and mortality of rainforest vertebrates, 241–255, in *Tropical Forest Remnants: Ecology, Management, and Conservation of Fragmented Communities* (WF Laurance & RO Bierregaard Jr, eds.). University of Chicago Press, Chicago, IL.

Göpfert, MC & LT Wasserthal. 1995. Notes on echolocation calls, food, and roosting behaviour of the Old World sucker-footed bat *Myzopoda aurita* (Chiroptera, Myzopodidae). *Zeitschrft für Säugetier-kunde*, 60:1–8.

Gordon, MS, et al. 1977. *Animal Physiology: Principles and Adaptations*. 3rd ed. Macmillan, NY.

Gorman, ML. 1976. A mechanism for individual recognition by odor in *Herpestes auropunctatus* (Carnivora, Viverridae). *Animal Behaviour*, 24:141–145.

Gorman, ML, DB Nedwell, & RM Smith. 1974. An analysis of the anal scent pockets of *Herpestes auropunctatus* (Carnivora: Viverridae). *Journal of Zoology, London*, 172:388–389.

Gorman, ML & RD Stone. 1990. *The Natural History of Moles*. Cornell University Press, Ithaca, NY.

Gorman, ML, et al. 1998. High hunting costs make African wild dogs vulnerable to kleptoparasitism by hyaenas. *Nature*, 391:479–481.

Gosho, ME, DW Rice, & JM Breiwick. 1984. The sperm whale *Physeter macrocephalus*. Marine Fisheries Review, 46:54–64.

Goss, RJ. 1983. *Deer Antlers—Regeneration, Function and Evolution*. Academic Press, NY.

Gough, WA, AR Cornwell, & LJS Tsuji. 2004. Trends in seasonal sea ice duration in southwestern Hudson Bay. *Arctic*, 57(3):299–305.

Gould, EW. 1937. Occurrence of low growing game foods during the old-field pine-mixed hardwood succession in the Harvard Forest. Master's thesis, Harvard University, Cambridge, MA.

Gould, E. 1965. Evidence for echolocation in the Tenrecidae of Madagascar. *Proceedings of the American Philosophical Society*, 109:352–360.

Gould, E. 1969. Communication in three genera of shrews (Soricidae): *Suncus*, *Blarina*, and *Cryptotis*. *Communications in Behavioral Biology, Part A*, 3:11–31.

Gould, E. 1970. Echolocation and communication of bats, 144–162, in *About Bats* (BH Slaughter & DW Walton, eds.). Southern Methodist University Press, Dallas, TX.

Gould, E. 1971. Studies of maternal-infant communication and development of vocalization in the bats *Myotis* and *Eptesicus*. *Communications in Behavioral Biology, Part A*, 5:263–313.

Gould, E. 1978. The behavior of the moonrat, *Echinosorex gymnurus* (Erinaceidae) and the pentail shrew, *Ptilocercus lowi* (Tupaiidae) with comments on the behavior of other Insectivora. *Zeitschrift fur Tierpsychologie*, 48:1–27.

Gould, E, NC Negus, & A Novick. 1964. Evidence for echolocation in shrews. *Journal of Experimental Zoology*, 156:19–38.

Gould, E & JF Eisenberg. 1966. Notes on the biology of the Tenrecidae. *Journal of Mammalogy*, 47:660–686.

Graf, W. 1955. The Roosevelt elk. *Port Angeles Evening News* (Port Angeles, Washington).

Graham, IM & X Lambin. 2002. The impact of weasel predation on cyclic field-vole survival: the specialist predator hypothesis contradicted. *Journal of Animal Ecology*, 71, 946–956.

Graham, RW, et al. 1996. Spatial response of mammals to late-Quaternary environmental fluctuations. *Science*, 272:1601–1606.

Grand, T, E Gould, & R Montali. 1998. Structure of the proboscis and rays of the star-nosed mole, *Condylura cristata*. *Journal of Mammalogy*, 79:492–501.

Grand, TI & R Lorenz. 1968. Functional analysis of the hip joint in *Tarsius bancanus* (Horsefield, 1821) and *Tarsius syrichta* (Linnaeus, 1758). *Folia Primatologica*, 9:161–181.

Grauer, D & D Higgins. 1994. Molecular evidence for the inclusion of cetaceans within the order Artiodactyla. *Molecular Biology and Evolution*, 11:357–364.

Grayson, DK. 2005. A brief history of Great Basin pikas. *Journal of Biogeography*, 32:2103–2111.

Gregory, JE, et al. 1989. Responses of electroreceptors in the snout of the echidna. *Journal of Physiology*, 414:521–538.

Gregory, WK. 1910. The orders of mammals. *Bulletin American Museum Natural History New York*, 27:1–524.

Green, B. 1997. Field energetics and water fluxes in marsupials, 143–162, in *Marsupial Biology: Recent Research, New Perspectives* (NR Saunders & LA Hinds, eds.). University of New South Wales Press, Sydney, Australia.

Green, B, J Merchant, & K Newgrain. 1997. Lactational energetics of a marsupial carnivore, the eastern quoll (*Dasyurus viverrinus*). *Australian Journal of Zoology*, 45:295–306.

Greenwald, GS. 1956. The reproductive cycle of the field mouse, *Microtus californicus*. *Journal of Mammalogy*, 37:213–222.

Greenwald, GS. 1957. Reproduction in a coastal California population of the field mouse, *Microtus californicus*. *University of California Publications in Zoology*, 54:421–446.

Greenwood, DR, et al. 2005. Chirality in elephant pheromones. *Nature*, 438:1097–1098.

Greenwood, PJ. 1980. Mating systems, philopatry and dispersal in birds and mammals. *Animal Behaviour*, 28:1140–1162.

Griffin, DR. 1958. *Listening in the Dark*. Yale University Press, New Haven, CT.

Griffin, DR & R Galambos. 1940. Obstacle avoidance by flying bats. *Anatomical Record*, 78:95.

Griffin, DR & R Galambos. 1941. The sensory basis of obstacle avoidance by flying bats. *Journal of Experimental Zoology*, 86:481–506.

Griffin, DR & JA Simmons. 1974. Echolocation of insects by horseshoe bats. *Nature*, 250:731–732.

Griffin, M. 1990. A review of taxonomy and ecology of gerbilline rodents of the central Namib Desert, with keys to the species (Rodentia: Muridae), 83–98, in *Namib Ecology: 25 Years of Namib Research* (MK Seely, ed.). Transvaal Museum Monograph No. 7, Transvaal Museum, Pretoria.

Griffiths, M. 1978. *The Biology of the Monotremes*. Academic Press, NY.

Griffiths, M, RT Wells, & DJ Barrie. 1991. Observations on the skulls of fossil and extant echidnas (Monotremata: Tachyglossidae). *Australian Mammalogy*, 14:87–101.

Griffiths, TA. 1997. Phylogenetic position of the bat *Nycteris javanica* (Chiroptera: Nycteridae). *Journal of Mammalogy*, 78:106–116.

Grigg, GC, LA Beard, & ML Augee. 1989. Hibernation in a monotreme, the echidna (*Tachyglossus aculeatus*). *Comparative Biochemistry and Physiology*, 92A:609–612.

Grigg, GC, ML Augee, & LA Beard. 1992. Thermal relations of free-living echidnas during activity and in hibernation in a cold climate, 160–173, in *Platypus and Echidnas* (ML Augee, ed.). Royal Zoology Society of New South Wales, Sydney, Australia.

Grigg, GC, et al. 2003. Body temperature in captive long-beaked echidnas (*Zaglossus bartoni*). *Comparative Biochemistry and Physiology-Part A: Molecular and Integrative Physiology*, 136:911–916.

Grinnell, J. 1922. A geographical study of the kangaroo rats of California. *University of California Publications in Zoology*, 24:1–124.

Grinnell, J & TI Storer. 1924. *Animal Life in the Yosemite*. University of California Press, Berkeley, CA.

Grinnell, J, JS Dixon, & JM Linsdale. 1937. *Fur-bearing Mammals of California*, 2 vols. University of California Press, Berkeley, CA.

Grodzinski, W & BA Wunder. 1975. Ecological energetics of small mammals, 173–204, in *Small Mammals: Their Productivity and Population Dynamics* (FB Golley, K Petrusewicz, & L Ryszkowski, eds.). Cambridge University Press, Cambridge, UK.

Gröger, U & L Wiegrebe. 2006. Classification of human breathing sounds by the common vampire bat, *Desmodus rotundus*. *BMC Biology*, 4:18.

Groom, MJ, GK Meffe & CR Carroll. 2006. *Principles of Conservation Biology*, 3rd ed. Sinauer Associates, Sunderland, MA.

Groves, C. 2001. *Primate Taxonomy*. Smithsonian Institution Press, Washington, DC.

Groves, C. 2005 Order Primates, 111–184, in *Mammal Species of the World: A Taxonomic and Geographic Reference*. 3rd ed. (D Wilson & DM Reeder, eds.). Johns Hopkins University Press, Baltimore, MD.

Groves, C & P Grubb. 1987. Classification of living cervids, 21–59, in *The Biology and Management of the Cervidae* (C Wemmer, ed.). National Zoological Society Symposium Vol. Smithsonian Institution Press, Washington, DC.

Grubb, P. 2005a. Order Perissodactyla, 629–636 in *Mammal Species of the World* (DE Wilson & DM Reeder, eds.). Johns Hopkins University Press, Baltimore, MD.

Grubb, P. 2005b. Order Artiodactyla, 637–722, in *Mammal Species of the World* (D Wilson & D Reeder, eds.). Johns Hopkins University Press, Baltimore, MD.

Grutzner, F, et al. 2004. In the platypus a meiotic chain of ten sex chromosomes shares genes with the bird Z and mammal X chromosomes. *Nature*, 432:913–917.

Grzimek, B. 1990. *Grzimek's Encyclopedia of Mammals*. Vol. 1. McGraw-Hill, NY.

Guimarães, PR, M Galetti, & P Jordano. 2008. Seed dispersal anachronisms: rethinking the fruits extinct megafauna ate. *PLoS ONE*, 3(3):e1745.

Gundersen, G & HP Andreassen. 1998. Causes and consequences of dispersal in root voles, *Microtus oeconomus*. *Animal Behaviour*, 56:1355–1366.

Gunnell, GF & NB Simmons. 2005. Fossil evidence and the origin of bats. *Journal of Mammalian Evolution*, 12:209–246.

Gunnell, GF, EL Simons, & ER Seiffert. 2008. New bats (Mammalia: Chiroptera) from the Late Eocene and Early Oligocene, Fayum Depression, Egypt. *Journal of Vertebrate Paleontology*, 28:1–11.

Gurovich, Y & R Beck. 2009. The phylogenetic affinities of the enigmatic mammalian clade Gondwanatheria. *Journal of Mammalian Evolution*, 16:25-49.

Gursky, S. 2000. Sociality in the spectral tarsier, *Tarsius spectrum*. *American Journal of Primatology*, 51:89–101.

Gursky, S. 2002. Determinants of gregariousness in the spectral tarsier (Prosimian: *Tarsius spectrum*). *Journal of Zoology, London*, 256:401–410.

Gustafson, AW & DA Damassa. 1985. Annual variation in plasma sex steroid-binding protein and testosterone concentrations in the adult male little brown bat: relation to the asynchronous recrudescence of the testis and accessory reproductive organs. *Biology of Reproduction*, 33:1126–1137.

Guthrie, MJ. 1933. The reproductive cycles of some cave bats. *Journal of Mammalogy*, 14:199–216.

Guthrie, RD. 1982. Mammals of the mammoth steppe as paleoenvironmental indicators, 307–326, in *Paleoecology of Beringia* (D Hopkins et al. eds.). Academic Press, NY.

Guthrie, RD. 1990. *Frozen Fauna of the Mammoth-Steppe: The Story of Blue Babe*. University of Chicago Press, Chicago, IL.

Guthrie, RD & RG Petocz. 1970. Weapon automimicry among mammals. *American Naturalist*, 104:585–588.

Habersetzer, J. 1981. Adaptive echolocation sounds in the bat *Rhinopoma hardwickei*: a field study. *Journal of Comparative Physiology A*, 144:559–566.

Habersetzer, J & G Storch. 1987. Klassifikation and functionelle Flügelmorphologie paläogener Fledermäuse (Mammalia, Chiroptera). *Courier Forschungsinstitut Senckenberg*, 91:117–150.

Habersetzer, J & G Storch. 1989. Ecology and echolocation of the Eocene Messel bats, 213–233, in *European Bat Research 1987* (V Hanak, T Horacek, & J Gaisler, eds.). Charles University Press, Prague.

Habersetzer, J, G Richter, & G Storch. 1994. Paleoecology of early Middle Eocene bats from Messel, FRG: Aspects of flight, feeding, and echolocation. *Historical Biology*, 8:235–260.

Hafner, MS. 2007. Field research in mammalogy: an enterprise in peril. *Journal of Mammalogy*, 88:1119–1128.

Haig, D. 1999. What is a marmoset? *American Journal of Primatology*, 49:285–296.

Haldane, JBS. 1932. *The Causes of Evolution*. Longman, London.

Halfpenny, JC. 2003. *Yellowstone Wolves in the Wild*. Riverbed Publishing, Helena, MT.

Hall, BG. 2007. *Phylogenetic Trees Made Easy: A How-to Manual*, 3rd ed. Sinauer Associates, Sunderland, MA.

Hall, ER & KR Kelson. 1959. *The Mammals of North America*. Ronald Press, NY.

Hall, ER & WW Dalquest. 1963. The mammals of Veracruz. *University of Kansas Publications Museum Natural History*, 14:165–362.

Hall, KRL.1968. Behaviour and ecology of the wild patas monkey, *Erythrocebus patas*, in Uganda. 32–119, in *Primates: Studies in Adaptation and Variability* (Jay, PC, ed.). Holt, Rinehart & Winston, NY.

Hall, KRL & GB Schaller. 1964. Tool-using behavior of the California sea otter. *Journal of Mammalogy*, 45:287–298.

Halle, S. 1995. Effect of extrinsic factors on activity of root voles, *Microtus oeconomus*. *Journal of Mammalogy*, 76:88–99.

Hallstrom, BM & A Janke. 2008. Resolution among major placental mammal interordinal relationships with genome data imply that speciation influenced their earliest radiations. *BMC Evolutionary Biology*, 8:162:1–13.

Hamilton, H Jr, et al. 2001. Evolution of river dolphins. *Proceedings of the Royal Society B*, 268:549–556.

Hamilton, WD. 1964. The genetical evolution of social behavior. *Journal of Theoretical Biology*, 7:1–52.

Hamilton, WD. 1971. Geometry for the selfish herd. *Journal of Theoretical Biology*, 31:295–311.

Hamilton, WJ Jr. 1937. The biology of microtine cycles. *Journal of Agricultural Research*, 54:779–790.

Hamilton, WJ III. 1962. Reproductive adaptations of the red tree mouse. *Journal of Mammalogy*, 43:486–504.

Hand, SJ & JAW Kirsch. 1998. A southern origin for the Hipposideridae (Microchiroptera)? Evidence from the Australian fossil record, 72–90, in *Bat Biology and Conservation* (TH Kunz & PA Racey, eds.). Smithsonian Institution Press, Washington, DC.

Hand, SJ, M Archer, & H Godthelp. 2001. New Miocene *Icarops* material (Microchiroptera: Mystacinidae) from Australia, with a revised diagnosis of the genus. *Memoirs of the Association of Australasian Palaeontologists*, 25:139–146.

Hand, S & JAW Kirsch. 2003. *Archerops*, a new annectant hipposiderid genus (Mammalia: Microchiroptera) from the Australian Miocene. *Journal of Paleontology*, 77:1139–1151.

Hanken, J & BK Hall. 1993a. *The Skull: Development*. Vol. 1. University of Chicago Press, Chicago, IL.

Hanken, J & BK Hall. 1993b. *The Skull: Patterns of Structural and Systematic Diversity*. Vol. 2. University of Chicago Press, Chicago, IL.

Hanken, J & BK Hall. 1993c. *The Skull: Functional and Evolutionary Mechanisms*. University of Chicago Press, Chicago, IL.

Hansen, DM & M Galetti. 2009. The forgotten megafauna. *Science*, 324:42–43.

Hansen, RM. 1962. Movements and survival of *Thomomys talpoides* in a mima-mound habitat. *Ecology*, 43:151–154.

Hansen, RM. 1978. Shasta ground sloth food habits, Rampart Cave, Arizona. *Paleobiology*, 4:302–319.

Hansen, RM & AL Ward. 1966. Some relations of pocket gophers to rangelands on Grand Mesa, Colorado. *Colorado Agricultural Experiment Station, Technical Bulletin*, 88:1–20.

Hanski, I, A Peltonen, & L Kaski. 1991. Natal dispersal and social dominance in the common shrew *Sorex araneus*. *Oikos*, 62:48–58.

Hanski, I, et al. 1993. Population oscillations of boreal rodents: regulation by mustelid predators leads to chaos. *Nature*, 364, 232–235.

Haozous, B. 2008. *Unconquered: Allan Houser and the Legacy of One Apache Family*. Quotation from exhibit 2008–2009 at Oklahoma History Center, Oklahoma City, OK.

Happold, DCD. 1987. *The Mammals of Nigeria*. Clarendon Press, Oxford, UK.

Hardin, G. 1993. *Living Within Limits: Ecology, Economics, and Population Taboos*. Oxford University Press, Oxford, UK.

Harris, AH. 1990. Fossil evidence bearing on southwestern mammalian biogeography. *Journal of Mammalogy*, 71:219–229.

Harris, CM. 1996. Absorption of sound in air versus humidity and temperature. *Journal of the Acoustic Society of America*, 40:148–159.

Harris, JM & L Liu. 2007. Superfamily Suoidea, 130–150, in *The Evolution of Artiodactyls* (DR Prothero & SE Foss, eds.). Johns Hopkins University Press, Baltimore, MD.

Harrisingh, MC & MN Nitabach. 2008. Integrating circadian timekeeping with cellular physiology. *Science*, 320:879–880.

Hart, JV, et al. 2008. Effects of fenced transportation corridors on pronghorn movements at Petrified Forest National Park, Arizona, 141–165, in *The Colorado Plateau III: Integrating Research and Resources Management for Effective Conservation* (C van Riper III & MK Sogge, eds.). University of Arizona Press, Tucson, AZ.

Hart, LA & BL Hart. 1987. Species-specific patterns of urine investigation and flehmen in Grant's gazelle (*Gazella granti*), Thomson's gazelle (*Gazella thomsonii*), impala (*Aepyceros melampus*) and eland (*Taurotragus oryx*). *Journal of Comparative Psychology*, 101:229–304.

Hart, LA & BL Hart. 1988. Autogrooming and social grooming in impala. *Annals of the New York Academy of Sciences,* 525:399–402.

Hartenberger, JL. 1985. The order Rodentia: major questions on their evolutionary origin, relationships, and suprafamilial systematics, 1–33, in *Evolutionary Relationships Among Rodents: A Multidisciplinary Analysis* (WP Luckett & JL Hartenberger, eds.). Plenum Press, NY.

Hartley, DJ & RA Suthers. 1987. The sound emission pattern and the acoustical role of the noseleaf in the echolocating bat, *Carollia perspicillata. Journal of the Acoustical Society of America,* 82:1892–1900.

Hartline, PH. 1971. Physiological basis for detection of sound and vibration in snakes. *Journal of Experimental Biology,* 54:349–371.

Hartman, CG. 1933. On the survival of spermatozoa in the female genital tract of the bat. *Quarterly Review of Biology,* 8:185–193.

Harvey, MJ & RW Barbour. 1965. Home ranges of *Microtus ochrogaster* as determined by a modified minimum area method. *Journal of Mammalogy,* 46:398–402.

Hashimoto, K, Y Saikawa, & M Nakata. 2007. Studies on the red sweat of the *Hippopotamus amphibious. Pure and Applied Chemistry,* 79:507–517.

Hasler, JF, AE Buhl, & EM Banks. 1976. The influence of photoperiod on growth and sexual function in male and female collared lemmings (*Dicrostonyx groenlandicus*). *Journal of Reproduction and Fertility,* 46:323–329.

Hass, CC. 1990. Alternative maternal-care patterns in two herds of bighorn sheep. *Journal of Mammalogy,* 71:24–35.

Hatori, M, et al. 2008. Inducible ablation of melanopsin-expressing retinal ganglion cells reveals their central role in non-image forming visual responses. *PLoS ONE,* 3:e2451.

Hatt, RT. 1932. The vertebral column of ricochetal rodents. *Bulletin American Museum Natural History,* 63:599–738.

Hausfater, G & SB Hrdy. 1984. *Infanticide: Comparative and Evolutionary Perspectives.* Aldine, Hawthorne, NY.

Hausman, LA. 1929. The "ovate bodies" of the hair of *Nothrotherium shastense. American Journal of Science,* 18:331–333.

Hawkins, CE, et al. 2006. Emerging disease and population decline of an island endemic, the Tasmanian devil *Sarcophilus harrisii. Biological Conservation,* 131:307–324.

Hayes, MA. 2009. Into the field: naturalistic education and the future of conservation. *Conservation Biology* 23:1075-1079.

Hayman, DL. 1977. Chromosome number—constancy and variation, 27–48, in *The Biology of Marsupials* (B Stonehouse & D Gilmore, eds.). Macmillan, London.

Hays, WTS & S Conant. 2007. Biology and impacts of Pacific island invasive species. 1. A worldwide review of effects of the small Indian mongoose, *Herpestes javanicus* (Carnivora: Herpestidae). *Pacific Science,* 61:3–16.

Hayssen, V, RC Lacy, & PJ Parker. 1985. Metatherian reproduction: transitional or transcending? *American Naturalist,* 126:617–632.

Heaney, LR. 1985. Systematics of oriental pygmy squirrels of the genera *Exilisciurus* and *Nannosciurus* (Mammalia: Sciuridae). *Miscellaneous Publications Museum Zoology University of Michigan,* 170:1–58.

Heaney, LR. 1986. Biogeography of mammals in SE Asia: estimates of rates of colonization, extinction and speciation. *Biological Journal of the Linnean Society,* 28:127–165.

Heaney, LR. 1993. Biodiversity patterns and the conservation of mammals in the Philippines. *Asia Life Science,* 2:261–274.

Heaney, LR, DS Balete, & ATL Dans. 1997. Terrestrial mammals, 141–168, in *Wildlife Conservation of the Philippines.* Philippine Red Data Book. Bookmark, Manila, Philippines.

Heaney, LR, et al. 2001. Frontispiece: Diversity patterns of small mammals along elevational gradients. *Global Ecology & Biogeography: Elevational Gradients in Mammals,* Special Issue 10:1.

Heaney, LR, JS Walsh Jr, & AT Peterson. 2005. The roles of geological history and colonization abilities in genetic differentiation between mammalian populations in the Philippine archipelago. *Journal of Biogeography,* 32:229–247.

Heckel, G & O von Helversen. 2003. Genetic mating system and the significance of harem associations in the bat *Saccopteryx bilineata. Molecular Ecology,* 12:219–227.

Hedges, SB. 2001. Afrotheria: plate tectonics meets genomics. *Proceedings of the National Academy of Sciences USA,* 98:1–2.

Hedges, SB, et al. 1996. Continental breakup and the diversification of birds and mammals. *Nature,* 381:226–229.

Hediger, H. 1950. Gefangenschafts Geburt ein afrikanischen Springhasen. *Zool. Gart. Leipzig,* 17(5).

Heffner, R & H Heffner. 1982. Hearing in the elephant: absolute thresholds, frequency discrimination, and sound localization. *Journal of Comparative Physiology and Psychology,* 96:926–944.

Heglund, NC, CR Taylor, & TA McMahon. 1974. Scaling stride frequency and gait to animal size: mice to horses. *Science,* 186:1112–1113.

Heglund, NC, et al. 1982. Energetics and mechanics of terrestrial locomotion. IV: total mechanical energy changes as a function of speed and body size in birds and mammals. *Journal of Experimental Biology,* 97:57–66.

Heideman, PD. 1989. Delayed development in Fischer's pygmy fruit bat, *Haplonycteris fischeri,* in the Philippines. *Journal of Reproduction and Fertility,* 85:363–382.

Heideman, PD, JA Cummings, & LR Heaney. 1993. Reproductive timing and early embryonic development in an Old World fruit bat, *Otopteropus cartilagonodus* (Megachiroptera). *Journal of Mammalogy,* 74:621–630.

Heideman, PD & KS Powell. 1998. Age-specific reproductive strategies and delayed embryonic development in an old world fruit bat, *Ptenochirus jagori. Journal of Mammalogy,* 79:295–311.

Heim de Balsac, H. 1954. Un genre inedit et inattendu de mammifera (Insectivore Tenrecidae) d'Afrique Occidentale. *Comptes Rendus de L' Academie des Sciences, Paris,* 239.

Heinrich, RE & KD Rose. 1995. Partial skeleton of the primitive carnivoran *Miacis petilus* from the early Eocene of Wyoming. *Journal of Mammalogy,* 76:148–162.

Heinroth-Berger, K. 1959. Beobachtungen an handaufgezogenen Mantelpavianen (*Papio hamadryas* L.). *Zeitschrift fur Tierpsychologie,* 16:706–732.

Heinsohn, R & C Packer. 1995. Complex cooperative strategies in group-territorial African lions. *Science,* 269:1260–1262.

Heldmaier, G, S Ortmann, & R Elvert. 2004. Natural hypometabolism during hibernation and daily torpor in mammals. *Respiratory Physiology and Neurobiology,* 141:317–329.

Helgen, KM. 2005. Order Scandentia, 104–109, in *Mammal Species of the World: A Taxonomic and Geographic Reference,* 3rd ed. (DE Wilson & DM Reeder, eds.). Johns Hopkins University Press, Baltimore, MD.

Heller, HC & TL Poulson. 1970. Circannian rhythms. II: endogenous and exogenous factors controlling reproduction and hibernation in chipmunks (*Eutamias*) and ground squirrels (*Spermophilus*). *Comparative Biochemistry and Physiology*, 33:357–383.

Henderson, MT, G Merriam, & J Wegner. 1985. Patchy environments and species survival: chipmunks in an agricultural mosaic. *Biological Conservation*, 31:95–105.

Hendrichs, H & U Hendrichs. 1971. *Dikdik und Elephanten*. R. Piper, Munich.

Henschel, JR & JD Skinner. 1987. Social relationships and dispersal patterns in a clan of spotted hyaenas *Crocuta crocuta* in the Kruger National Park. *South African Journal of Zoology*, 22:18–24.

Henschel, JR & JD Skinner. 1990. The diet of spotted hyaenas *Crocuta crocuta* in Kruger National Park. *African Journal of Ecology*, 28:69–82.

Henshaw, RE. 1970. Thermoregulation in bats, 188–232, in *About Bats* (BH Slaughter & DW Walton, eds.). Southern Methodist University Press, Dallas, TX.

Heppes, JB. 1958. The white rhinoceros in Uganda. *African Wildlife*, 12:273–280.

Herald, ES, et al. 1969. Blind river dolphin: first side-swimming cetacean. *Science*, 166:1408–1410.

Hermanson, JW. 1981. Functional morphology of the clavicle in the pallid bat, *Antrozous pallidus*. *Journal of Mammalogy*, 62:801–805.

Hermanson, JW & JS Altenbach. 1981. Functional anatomy of the primary downstroke muscles in the pallid bat, *Antrozous pallidus*. *Journal of Mammalogy*, 62:795–800.

Hermanson, JW, et al. 1993. Histochemical and myosin composition of vampire bat (*Desmodus rotundus*) pectoralis muscle targets a unique locomotory niche. *Journal of Morphology*, 217:347–356.

Hermes, R, et al. 2000. Ultrasonography of the estrous cycle in female African elephants (*Loxodonta africana*). *Zoo Biology*, 19:369–382.

Herrera, J, CL Kramer, & OJ Reichman. 1997. Patterns of fungal communities that inhabit rodent food stores: effect of substrate and infection time. *Mycologia*, 89:846–857.

Herrera, M. LG. et al. 2001. Sources of protein in two species of phytophagous bats in a seasonal dry forest: evidence from stable-isotope analysis. *Journal of Mammalogy*. 82(2):352–361.

Hershkovitz, P. 1977. *Living New World Monkeys (Platyrrhini), With an Introduction to the Primates*. Vol. 1. University of Chicago Press, Chicago, IL.

Hershkovitz, P. 1992. Ankle bones: the Chilean opossum *Dromiciops gliroides* Thomas and marsupial phylogeny. *Bonner Zoologische Beitrage*, 43:181–213.

Hershkovitz, P. 1999. *Dromiciops gliroides* Thomas, 1894, last of the Microbiotheria (Marsupialia), with a review of the family Microbiotheriidae. *Fieldiana Zoology* new series no. 93:1–60.

Heske, EJ. 1987a. Spatial structuring and dispersal in a high density population of the California vole, *Microtus californicus*. *Holarctic Ecology*, 10:137–149.

Heske, EJ. 1987b. Pregnancy interruption by strange males in the California vole. *Journal of Mammalogy*, 68:406–410.

Hestbeck, JB. 1986. Multiple regulation states in populations of the California vole, *Microtus californicus*. *Ecological Monographs*, 56:161–181.

Heth, G, et al. 1987. Vibrational communication in subterranean mole rats (*Spalax ehrenbergi*). *Behavioral Ecology and Sociobiology*, 20:31–33.

Heth, G, E Frankenberg, & E Nevo. 1988. "Courtship" calls of the subterranean mole rats (*Spalax ehrenbergi*): physical analysis. *Journal of Mammalogy*, 69:121–125.

Hickey, MBC. 1988. Foraging behavior and use of torpor by the hoary bat (*Lasiurus cinereus*). *Bat Research News*, 29:47.

Hickie, PF. 1936. Isle Royale moose studies. *Transactions of the North American Wildlife Conference*, 1:396–399.

Hickman, GC. 1981. National mammal guides: a review of references to Recent faunas. *Mammal Review*, 11:53–85.

Hieronymous, TL, LM Witmer, & RC Ridgely. 2006. Structure of whole rhinoceros (*Ceratotherium simum*) horn investigated by X-ray computed tomography and histology with implications for growth and external form. *Journal of Morphology*, 267:1172–1176.

Higdon, JW, et al. 2007. Phylogeny and divergence of the pinnipeds (Carnivora: Mammalia) assessed using a multigene dataset. *BMC Evolutionary Biology*, 7:216–225.

Higgins, LV. 1993. The nonannual, nonseasonal breeding cycle of the Australian sea lion, *Neophoca cinerea*. *Journal of Mammalogy*, 74:270–274.

Hik, DS. 1995. Does risk of predation influence population dynamics? Evidence from the cyclic decline of snowshoe hares. *Wildlife Research*, 22:115–129.

Hildebrand, M. 1959. Motions of the running cheetah and horse. *Journal of Mammalogy*, 40:481–495.

Hildebrand, M. 1960. How animals run. *Scientific American*, 202:148–156.

Hildebrand, M. 1965. Symmetrical gaits of horses. *Science*, 150:701–708.

Hildebrand, M. 1974. *Analysis of Vertebrate Structure*. Wiley, NY.

Hildebrand, M. 1985. Walking and running, 38–57, in *Functional Vertebrate Morphology* (M Hildebrand, DM Bramble, KF Liem, & DB Wake, eds.). Belknap Press, Cambridge, MA.

Hildebrand, M. 1987. The mechanics of horse legs. *American Scientist*, 75:594–601.

Hill, AV. 1950. The dimensions of animals and their muscular dynamics. *Scientific Progress*, 38:209.

Hill, CJ. 1941. The development of the Monotremata. Part V. Further observations on the histology and the secretory activities of the oviduct prior to and during gestation. *Transactions of the Zoological Society London*, 25:1–31.

Hill, JE & TD Carter. 1941. The mammals of Angola, Africa. *Bulletin American Museum Natural History*, 78:1–211.

Hill, JE & SE Smith. 1981. *Craseonycteris thonglongyai*. *Mammalian Species*, 160:1–4.

Hill, JE & MJ Daniel. 1985. Systematics of the New Zealand short-tailed bat *Mystacina* Gray, 1843 (Chiroptera: Mystacinidae). *Bulletin British Museum Natural History (Zoology)*, 48:279–300.

Hill, JE & JD Smith. 1992. *Bats: A Natural History*. University of Texas Press, Austin.

Hill, RW 1992. The altricial/precocial contrast in the thermal relations and energetics of small mammals, 122–159, in *Mammalian Energetics: Interdisciplinary Views of Metabolism and Reproduction* (TE Tomasi & TH Horton, eds.). Cornell University Press, Ithaca, NY.

Hinds, DS & RE MacMillen. 1984. Energy scaling in marsupials and eutherians. *Science*, 225:335–337.

Hinds, DS & RE MacMillen. 1985. Scaling of energy metabolism and evaporative water loss in heteromyid rodents. *Physiological Zoology*, 58:282–298.

Hinds, DS, et al. 1993. Maximum metabolism and the aerobic factorial scope of endotherms. *Journal of Experimental Biology*, 182:41–56.

Hladik, A & CM Hladik. 1969. Rapports trophiques entre vegetation et primates dans la foret de Barro Colorado (Panama). *Terre et Vie*,1:25–117.

Hladik, CM, P Charles-Dominique, & JJ Petter. 1980. Feeding strategies of five nocturnal prosimians in the dry forest of the west coast of Madagascar, 41–73, in *Nocturnal Malagasy Primates* (P Charles-Dominique et al., eds.). Academic Press, NY.

Ho, S. 2008. The molecular clock and estimating species divergence. *Nature Education*, 1(1).

Hoagland, JL. 1995. *The Black-tailed Prairie Dog.* University of Chicago Press, Chicago, IL.

Hoagland, JL. 1998. Why do female Gunnison's prairie dogs copulate with more than one male? *Animal Behavior*, 55:351–359.

Hobson, KA & LI Wassenaar (eds). 2008. *Tracking Animal Migration with Stable Isotopes.* Academic Press, NY.

Hoeck, HN. 1982. Population dynamics, dispersal, and genetic isolation in two species of hyrax (*Heterohyrax brucei* and *Procavia johnstoni*) on habitat islands in the Serengeti. *Zeitschrift fur Tierpsychologie*, 59:177–210.

Hoeck, HN. 1989. Demography and competition in Hyrax. *Oecologia*, 79:353–360.

Hoeck, HN, H Klein, & P Hoeck. 1982. Flexible social organization in hyrax. *Zeitschrift fur Tierpsychologie*, 59:265–298.

Hoegh-Guldberg, O, et al. 2008. Assisted colonization and rapid climate change. *Science*, 321:345–346.

Hoekstra, HE & NW Nachman. 2005. Coat color variation in rock pocket mice (*Chaetodipus intermedius*): from genotype to phenotype, 79–100, in *Mammalian Diversification: From Chromosomes to Phylogeography (A Celebration of the Career of James L Patton)* (EA Lacey, & P Myers, eds.). University of California Publications in Zoology, vol. 133.

Hofer, H & ML East. 1993a. The commuting system of Serengeti spotted hyaenas: how a predator copes with migratory prey. I. social organization. *Animal Behavior*, 46:547–557.

Hofer, H & ML East. 1993b. The commuting system of Serengeti spotted hyaenas: how a predator copes with migratory prey. II. intrusion pressure and commuters space use. *Animal Behavior*, 46:559–574.

Holekamp, KE. 1986. Proximal causes of natal dispersal in Belding's ground squirrels (*Spermophilus beldingi*). *Ecological Monographs*, 56:365–391.

Holekamp, KE. 2006. Spotted hyenas. *Current Biology*, 16:R944–R945.

Holekamp, KE, ST Sakeal, & BL Lundrigan. 2007. The spotted hyena (*Crocuta crocuta*) as a model system for study of the evolution of intelligence. *Journal of Mammalogy*, 88:545–554.

Holland, RA, et al. 2006. Bat orientation using the Earth's magnetic field. *Nature*, 444:702.

Hollar, LJ & MS Springer. 1997. Old World fruitbat phylogeny: evidence for convergent evolution and an endemic African clade. *Proceedings of the National Academy of Sciences USA*, 94:5716–5721.

Holling, CS. 1959. The components of predation as revealed by a study of small mammal predation of the European pine sawfly. *Canadian Entomology*, 91:293–320.

Holmes, WG. 1995. The ontogeny of littermate preferences in juvenile golden-mantled ground squirrels: effects of rearing and relatedness. *Animal Behavior*, 50:309–322.

Holmes, WG & PW Sherman. 1983. Kin recognition in animals. *American Scientist*, 71:46–55.

Holroyd, PA & JC Mussell 1995. Macroscelidea and Tubulidentata, 71–83, in *The Rise of Placental Mammals* (KD Rose & JD Archibald, eds.). Johns Hopkins University Press, Baltimore, MD.

Holy, TE, C Dulac, & M Meister. 2000. Responses of vomeronasal neurons to natural stimuli. *Science*, 289:1569–1572.

Honeycutt, RL & RM Adkins. 1993. Higher level systematics of eutherian mammals: an assessment of molecular characters and phylogenetic hypotheses. *Annual Review of Ecology and Systematics*, 24:279–305.

Hoofer, SR & RA Van Den Bussche. 2003. Molecular phylogenetics of the chiropteran family Vespertilionidae. *Acta Chiropterologica*, 5:1–63.

Hooker, JJ. 2005. Perissodactyla, 199–214, in *The Rise of Placental Mammals, Origins, and Relationships of the Major Extant Clades* (KD Rose & JD Archibald, eds.). Johns Hopkins University Press, Baltimore, MD.

Hooker, JJ, et al. 2008. The origin of Afro-Arabian "didelphimorph" marsupials. *Palaeontology*, 51:635–648.

Hooper, ET & JH Brown. 1968. Foraging and breeding in two sympatric species of Neotropical bats, genus *Noctilio*. *Journal of Mammalogy*, 49:310–312.

Hopewell, LJ & LA Leaver. 2008. Evidence of social influences on cache-making by grey squirrels (*Sciurus carolinensis*). *Ethology*, 114:1061–1068.

Hopf, FA & JH Brown. 1986. The bulls-eye method for testing randomness in ecological communities. *Ecology*, 67:1139–1155.

Hopson, JA. 1994. Synapsid evolution and the radiation of non-eutherian mammals, 190–219, in *Major Features in Vertebrate Evolution. Short Courses in Paleontology.* No. 7. (RS Spencer, ed.). Paleontological Society, University of Tennessee, Knoxville.

Hornocker, MG. 1970. An analysis of mountain lion predation upon mule deer and elk in the Idaho Primitive Area. *Wildlife Monographs*, No. 21:1–39.

Horovitz, I & MR Sánchez-Villagra. 2003. A morphological analysis of marsupial mammal higher-level phylogenetic relationships. *Cladistics*, 19:181–212.

Hosley, NW. 1956. Management of the white-tailed deer in its environment, 187–260, in *The Deer of North America* (WP Taylor, ed.). Wildlife Management Institute, Telegraph Press, Harrisburg, PA.

Howell, AB. 1932. The saltatorial rodent *Dipodomys*: functional and comparative anatomy of its muscular and osseous systems. *Proceedings American Academy of Arts and Sciences*, 67:377–536.

Howell, AB. 1944. *Speed in Animals.* University of Chicago Press, Chicago, IL.

Howell, DJ. 1974. Bats and pollen: physiological aspects of the syndrome of chiropterophily. *Comparative Biochemistry and Physiology*, 48A:263–276.

Hrdy, SB. 1976. The care and exploitation of nonhuman primate infants by conspecifics other than the mother. *Advances in the Study of Behavior*, 6:101–158.

Hrdy, SB. 1977. *The Langurs of Abu: Female and Male Strategies of Reproduction.* Harvard University Press, Cambridge, MA.

Hu, Y, et al. 1997. A new symmetrodont mammal from China and its implications for mammalian evolution. *Nature*, 390:137–142.

Huchon, D & EJP Douzery. 2001. From the Old World to the New World: a molecular chronicle of the phylogeny and biogeography of hystricognath rodents. *Molecular Phylogenetics and Evolution*, 20:238–251.

Huchon, D, et al. 2007. Multiple molecular evidences for a living mammalian fossil. *Proceedings of the National Academy of Sciences USA*, 104:7495–7499.

Hudson, JW. 1973. Torpidity in mammals, 97–165, in *Comparative Physiology of Thermoregulation, vol. III*. Academic Press, NY.

Hudson, JW & TJ Dawson. 1975. Role of sweating from the tail in the thermal balance of the rat-kangaroo *Potorous tridactylus. Australian Journal of Zoology*, 23:453–461.

Hugget, AStG & WF Widdas. 1951. The relationship between mammalian foetal weight and conception age. *Journal of Physiology*, 114:306–317.

Hughes, RL. 1984. Structural adaptations of the eggs and the fetal membranes of monotremes and marsupials for respiration and metabolic exchange, 389–421, in *Respiration and Metabolism of Embryonic Vertebrates* (RS Seymour, ed.). Doordrecht, Junk.

Hughes, RL, FN Carrick, & CD Shorey. 1975. Reproduction in the platypus, *Ornithorhynchus anatinus*, with particular reference to the evolution of viviparity. *Journal of Reproduction and Fertility*, 43:374–375.

Hughes, RL, et al. 1990. Observations on placentation and development in *Echymipera kalubu*, 259–270, in *Bandicoots and Bilbies* (JH Seebeck, PR Brown, RL Wallis, & CM Kemper, eds.). Surrey Beatty, & Sons, Sydney, Australia.

Huitu, O, et al. 2003. Winter food supply limits growth of northern vole populations in the absence of predation. *Ecology*, 84:2108–2118.

Hulbert, AJ & TJ Dawson. 1974. Standard metabolism and body temperature of perameloid marsupials from different environments. *Comparative Biochemistry and Physiology*, 47A:583–590.

Hulva, P & I Horacek. 2002. *Craseonycteris thonglongyai* (Chiroptera: Craseonycteridae) is a rhinolophoid: molecular evidence from cytochrome *b. Acta Chiropterologica*, 4:107–120.

Hulva, P, I Horacek, & P Benda. 2007. Molecules, morphometrics, and new fossils provide an integrated view of the evolutionary history of Rhinopomatidae (Mammalia: Chiroptera). *BMC Evolutionary Biology*, 7:165.

Hume, ID. 1982. *Digestive Physiology and Nutrition of Marsupials*. Cambridge University Press, Cambridge, UK.

Hume, I. 2006. Nutrition and digestion, 137–158, in *Marsupials* (P Armati, C Dickman, & I Hume, eds.). Cambridge University Press, Cambridge, UK.

Humphrey, SR. 1974. Zoogeography of the nine-banded armadillo (*Dasypus novemcinctus*) in the United States. *BioScience*, 24:457–462.

Humphrey, SR & JB Cope. 1976. Population ecology of the little brown bat, *Myotis lucifugus*, in Indiana and north-central Kentucky. American Society of Mammalogists, *Special Publication No. 4*:38–53.

Hundtermark, KJ, et al. 2002. Mitochondrial phylogeography of moose (*Alces alces*): late Pleistocene divergence and population expansion. *Molecular Phylogenetics and Evolution*, 22(3):375–387.

Hunt, RM Jr. 1987. Evolution of the aeluroid Carnivora: significance of auditory structure in the nimravid cat *Dinictis. American Museum Novitates*, 2886:1–74.

Hunt, RM Jr & RH Tedford. 1993. Phylogenetic relationships within the aeluroid Carnivora and implications of their temporal and geographic distribution, 53–73, in *Mammal Phylogeny, Placentals* vol. 2. (FS Szalay, MJ Novacek, & MC McKenna, eds.). Springer-Verlag, NY.

Hunter, JP & J Jernvall. 1995. The hypocone as a key innovation in mammalian evolution. *Proceedings of the National Academy of Sciences USA*, 92:10718–10722.

Hurly, TA & SA Lourie. 1997. Scatterhoarding and larderhoarding by red squirrels: size, dispersion, and allocation of hoards. *Journal of Mammalogy*, 78:529–537.

Hurum, JH. 1998. The inner ear of two Late Cretaceous multituberculate mammals and its implications for multituberculate hearing. *Journal of Mammalian Evolution*, 5:65–93.

Hutchinson, JR, et al. 2003. Are fast-moving elephants really running? *Nature*, 422:493–494.

Hutterer, R. 2005. Order Erinaceomorpha, 212–219, in *Mammal Species of the World* (DE Wilson & DM Reeder, eds.). Johns Hopkins University Press, Baltimore, MD.

Hyvärinen, H. 1994. Brown fat and the wintering of shrews, 139–148, in *Advances in the Biology of Shrews* (JF Merritt, GL Kirkland Jr & RK Rose, eds.). *Special Publication No. 18*, Carnegie Museum of Natural History.

Imes, RA, J-A Henden, & ST Killengreen. 2007. Collapsing population cycles. *Trends in Ecology and Evolution*, 23:79–86.

Ingles, LG. 1949. Ground water and snow as factors affecting the seasonal distribution of pocket gophers, Thomomys monticola. *Journal of Mammalogy*, 30:343–350.

International Union for the Conservation of Nature. 1980. *World Conservation Strategy: Living Resources Conserved for Sustainable Development*. Nairobi, Kenya.

IUCN (International Union for the Conservation of Nature). 2008. Red List of Threatened Species. http://www.iucnredlist.org/mammals.

IUCN/WCPA (International Union for the Conservation of Nature). 2010. World Commission on Protected Areas. http://www.iucn.org/about/union/commissions/wcpa/ (accessed March 2009).

Irvine, AB. 1983. Manatee metabolism and its influence on distribution in Florida. *Biological Conservation*, 25:315–334.

Irving, L. 1966. Adaptations to cold. *Scientific American*, 214:94–101.

Irving, L & JS Hart. 1957. The metabolism and insulation of seals as bare-skinned mammals in cold water. *Canadian Journal of Zoology*, 35:497–511.

Isaac, NJB, et al. 2007. Mammals on the EDGE: conservation priorities based on threat and phylogeny. *PLoS ONE*, 2:e296.

Isvaran, K & T Clutton-Brock. 2007. Ecological correlates of extra-group paternity in mammals. *Proceedings of the Royal Society B*, 274:219–224.

Iturralde-Vinent, MA & RDE MacPhee. 1999. Paleogeography of the Caribbean region: implications for Cenozoic biogeography. *Bulletin of the American Museum of Natural History*, 238:1–95.

Izawa, K. 1970. Unit groups of chimpanzees and their nomadism in the Savannah woodland. *Primates*, 11:1–45.

Izawa, K & J Itani. 1966. Chimpanzees in Kasakata Basin, Tanganyika: I. ecological study in the rainy season, 1963–64. *Kyoto University Afr. Studies*, 1:73–156.

Jablonski, D. 1991. Extinctions: a paleontological perspective. *Science*, 253:754–757.

Jablonski, D. 1993. The tropics as a source of evolutionary novelty through geological time. *Nature*, 364:142–144.

Jachmann, H, PSM Berry, & H Imae. 1995. Tusklessness in African elephants: a future trend. *African Journal of Ecology*, 33:230–235.

Jacobs, BF, JD Kingston, & LL Jacobs. 1999. The origin of grass-dominated ecosystems. *Annals of the Missouri Botanical Garden*, 86:590–643.

Jacobs, DS, et al. 1991. The colony structure and dominance hierarchy of the Damaraland mole-rat, *Cryptomys damarensis* (Rodentia: Bathyergidae), from Namibia. *Journal of Zoology, London*, 224:553–576.

Jacobs, GH. 1993. The distribution and nature of colour vision among the mammals. *Biological Reviews*, 68:413–471.

Jacobs, GH, et al. 1993. Photopigments and colour vision in the nocturnal monkey, *Aotus*. *Vision Research*, 33:1773–1783.

Jacobs, GH & JF Deegan. 2001. Photopigments and colour vision in New World monkeys from the family Atelidae. *Proceedings of the Royal Society B*, 268:695–702.

Jameson, EW Jr. 1988. *Vertebrate Reproduction*. Wiley, NY.

Jameson, EW Jr & RA Mead. 1964. Seasonal changes in body fat, water, and basic weight in *Citellus lateralis, Eutamias speciosus* and *E. amoenus*. *Journal of Mammalogy*, 45:359–365.

Janecka, JE, et al. 2007. Molecular and genomic data identify the closest living relative of Primates. *Science*, 318:792–794.

Janik, VM & PJB Slater. 1997. Vocal learning in mammals. *Advances in the Study of Behavior*, 26: 59–99.

Janis, CM. 1993. Tertiary mammal evolution in the context of changing climates, vegetation, and tectonic events. *Annual Reviews of Ecology and Systematics*, 24:467–500.

Janis, CM. 2007. Artiodactyl paleoecology and evolutionary trends, 292–302, in *The Evolution of Artiodactyls* (DR Prothero & SE Foss, eds.). Johns Hopkins University Press, Baltimore, MD.

Janis, C & K Scott. 1987. The origin of the higher ruminant families with special reference to the origin of the Cervoidea and relationships within the Cervoidea. *American Museum Novitates*, 2893:1–85.

Janis, CM & M Fortelius. 1988. On the means whereby mammals achieve increased functional durability of their dentitions, with special reference to limiting factors. *Biological Reviews of the Cambridge Philosophical Society*, 63:197–230.

Janis, CM, RC Hulbert Jr, & M Mihlbacher. 2008. Addendum, 645–693, in *Evolution of Tertiary Mammals of North America Volume 2: Small Mammals, Xenarthrans, and Marine Mammals* (CM Janis, GF Gunnell, & MK Uhen, eds.). Cambridge University Press, Cambridge, UK.

Jansa, SA, SM Goodman, & PK Tucker. 1999. Molecular phylogeny and biogeography of the native rodents of Madagascar (Muridae: Nesomyinae): a test of the single-origin hypothesis. *Cladistics*, 15:253–270.

Jansa, SA & M Weksler. 2004. Phylogeny of muroid rodents: relationships within and among major lineages as determined by IRBP gene sequences. *Molecular Phylogenetics and Evolution*, 31:256–276.

Jansen, T, et al. 2002. Mitochondrial DNA and the origins of the domestic horse. *Proceedings of the National Academy of Sciences*, 99:10,905–910.

Jarman, MV. 1970. Attachment to home area in impala. *East African Wildlife Journal*, 8:198–200.

Jarman, PJ. 1989. On being thick-skinned: dermal shields in large mammalian herbivores. *Biological Journal Linnean Society*, 36:169–191.

Jarman, RJ. 1974. The social organization of antelope in relation to their ecology. *Behaviour*, 48:215–267.

Jarman, RJ & MV Jarman. 1974. Impala behavior and its relevance to management, 871–881, in *The Behavior of Ungulates and its Relation to Management* (V Geist & FR Walther, eds.). International Union for Conservation of Nature and Natural Resources, Morges, Switzerland.

Jaroszewska, M & B Wilczynska. 2006. Dimensions of surface area of alimentary canal of pregnant and lactating female common shrews. *Journal of Mammalogy*, 87:589–597.

Jarvis, JUM. 1978. Energetics of survival in *Heterocephalus glaber* (Rüppell), the naked mole-rat (Rodentia: Bathyergidae). *Bulletin Carnegie Museum Natural History*, 6:81–87.

Jarvis, JUM. 1981. Eusociality in a mammal: cooperative breeding in the naked mole rat. *Science*, 212:571–573.

Jarvis, JUM & JB Sale. 1971. Burrowing and burrow patterns of East African mole-rats *Tachyoryctes, Heliophobius* and *Heterocephalus*. *Journal of Zoology, London*, 163:451–479.

Jarvis, JUM & NC Bennett. 1991. Ecology and behavior of the family Bathyergidae, 66–96, in *The Biology of the Naked Mole-Rat* (PW Sherman, JUM Jarvis, & RD Alexander, eds.). Princeton University Press, Princeton, NJ.

Jarvis, JUM & NC Bennett. 1993. Eusociality has evolved independently in two genera of bathyergid mole-rats but occurs in no other subterranean mammal. *Behavioral Ecology and Sociobiology*, 33:253–260.

Jarvis, JUM, et al. 1994. Mammalian eusociality: a family affair. *Trends in Ecology and Evolution*, 9:47–51.

Jarvis, JUM, NC Bennett, & AC Spinks. 1998. Food availability and foraging by wild colonies of Damaraland mole-rats (*Cryptomys damarensis*): implications for sociality. *Oecologia*, 113:290–298.

Jayaraman, KS. 1993a. Academies urge population control. *Nature*, 365:382.

Jayaraman, KS. 1993b. Science academies call for global goal of zero population growth. *Nature*, 366:3.

Jefferies, RL, DR Klien, & GR Shaver. 1994. Vertebrate herbivores and northern plant communities: reciprocal influences and responses. *Oikos*, 71:193–206.

Jen, PH-S & N Suga. 1976. Coordinated activities of middle-ear and laryngeal muscles in echolocating bats. *Science*, 191:950–952.

Jenkins, FA Jr. 1970a. Limb movement in a monotreme (*Tachyglossus aculeatus*): a cineradiographic analysis. *Science*, 168:1473–1475.

Jenkins, FA Jr. 1970b. Anatomy and function of expanded ribs in certain edentates and primates. *Journal of Mammalogy*, 51:288–301.

Jenkins, FA Jr. 1971. Limb posture and locomotion in the Virginia opossum (*Didelphis marsupialis*) and in other non-cursorial mammals. *Journal of Zoology, London*, 165:303–315.

Jenkins, FA Jr. 1974. *Primate Locomotion*. Academic Press, NY.

Jenkins, FA Jr. 1990. Monotremes and the biology of Mesozoic mammals. *Netherlands Journal of Zoology*, 40:5–31.

Jenkins, FA Jr & FR Parrington. 1976. The postcranial skeletons of the Triassic mammals *Eozostrodon, Megazostrodon*, and *Erythrotherium*. *Philosophical*

Transactions of the Royal Society London, B, 273:387.

Jenkins, FA Jr & DW Krause. 1983. Adaptations for climbing in North American multituberculates (Mammalia). *Science,* 220:712–714.

Jenkins, HO. 1948. A population study of the meadow mice (*Microtus*) in three Sierra Nevada meadows. *Proceedings of the California Academy Sciences,* ser. 4, 26:43–67.

Jenkins, PD. 2003. *Microgale,* shrew tenrecs, 1273–1278, in *The Natural History of Madagascar* (SM Goodman & JP Benstead, eds.). University of Chicago Press, Chicago, IL.

Jenkins, PD, CJ Raxworthy, & RA Nussbaum. 1997. A new species of *Microgale* (Insectivora, Tenrecidae), with comments on the status of four other taxa of shrew tenrecs. *Bulletin of the Natural History Museum London,* 63:1–12.

Jenkins, PD, et al. 2005. Morphological and molecular investigations of a new family, genus and species of rodent (Mammalia: Rodentia: Hystricognatha) from Lao PDR. *Systematic Biodiversity,* 2:419–454.

Jernvall, J & M Fortelius. 2002. Common mammals drive the evolutionary increase of hypsodonty in the Neogene. *Nature,* 417:538–540.

Jesseau, SA, WG Holmes, & TM Lee. 2008. Mother–offspring recognition in communally nesting degus, *Octodon degus. Animal Behaviour,* 75:573–582.

Jewell, PA. 1972. Social organization and movements of topi (*Damaliscus korrigum*) during the rut at Ishasha, Queen Elizabeth Park, Uganda. *South African Journal of Zoology* (formerly *Zool. Afr.*), 7:233–255.

Ji, Q, et al. 2002. The earliest known eutherian mammal. *Nature,* 416:816–822.

Johnsen, S & KJ Lohmann. 2005. The physics and neurobiology of magnetoreception. *Nature Reviews of Neuroscience,* 6:703–712.

Johnson, BJ, et al. 1999. 65,000 years of vegetation change in central Australia and the Australian summer monsoon. *Science,* 284:1150–1152.

Johnson, C. 2006. *Australia's Mammal Extinctions: A 50,000-Year History.* Cambridge University Press, Cambridge, UK.

Johnson, CN. 2009. Ecological consequences of late Quaternary extinctions of megafauna. *Proceedings of the Royal Society B,* 276:2509–2519.

Johnson, DR. 1961. The food habits of rodents on rangelands of southern Idaho. *Ecology,* 42:407–410.

Johnson, DR. 1964. Effects of range treatment with 2,4-D on food habits of rodents. *Ecology,* 45:241–249.

Johnson, WE, et al. 2006 The late Miocene radiation of modern Felidae: a genetic assessment. *Science,* 311:73–77.

Johnson-Murray, JL. 1977. Myology of the gliding membranes of some petauristine rodents (Genera: *Glaucomys, Pteromys, Petinomys,* and *Petaurista*). *Journal of Mammalogy,* 58:374–384.

Johnson-Murray, JL. 1987. The comparative myology of the gliding membranes of *Acrobates, Petauroides,* and *Petaurus* contrasted with the cutaneous myology of *Hemibelideus* and *Pseudocheirus* (Marsupialia: Phalangeridae) and with selected gliding Rodentia (Sciuridae and Anomaluridae). *Australian Journal of Zoology,* 35:101–113.

Johnston, SD, et al. 2007. One-sided ejaculation of echidna sperm bundles. *American Naturalist,* 170:E162–E164.

Jolly, A. 1966. *Lemur Behavior: A Madagascar Field Study.* University of Chicago Press, Chicago, IL.

Jolly, A. 1972. Troop continuity and troop spacing in *Propithecus verreauxi* and *Lemur catta* at Berenty (Madagascar). *Folia Primatologica,* 17:335–362.

Jolly, A. 2003. *Lemur catta,* Ring-tailed lemur, Maky, 1329–1331, in *The Natural History of Madagascar* (SM Goodman & JP Benstead, eds.). University of Chicago Press, Chicago, IL.

Jones, FW. 1924. *The Mammals of South Australia. Part II: The Bandicoots and Herbivorous Marsupials.* Government Printer, Adelaide, Australia.

Jones, G & JMV Rayner. 1988. Flight performance, foraging tactics and echolocation in free-living Daubenton's bats *Myotis daubentoni* (Chiroptera: Vespertilionidae). *Journal of Zoology London,* 215:113–132.

Jones, G & EC Teeling. 2006. The evolution of echolocation in bats. *Trends in Ecology and Evolution,* 21:149–156.

Jones, G & MW Holderied. 2007. Bat echolocation calls: adaptation and convergent evolution. *Proceedings of the Royal Society B,* 274:905–912.

Jones, M. 1997. Character displacement in Australian dasyurid carnivores: size relationships and prey size patterns. *Ecology,* 78:2569–2587.

Jones, ME & DM Stoddart. 1998. Reconstruction of the predatory behaviour of the extinct marsupial thylacine (*Thylacinus cynocephalus*). *Journal of Zoology, London,* 246:239–246.

Jones, RB & NW Nowell 1973a. Aversive effects of the urine of a male mouse upon the investigatory behavior of its defeated opponent. *Animal Behaviour,* 21:707–710.

Jones, RB & NW Nowell. 1973b. The effect of urine on the investigatory behavior of male albino mice. *Physiology and Behaviour,* 11:35–38.

Judd, TM & PW Sherman. 1996. Naked mole-rats recruit colony mates to food sources. *Animal Behaviour,* 52:957–969.

Julliot, C, S Cajani, & A Gautier-Hion. 1998. Anomalures (Rodentia, Anomaluridae) in central Gabon: species composition, population densities, and ecology. *Mammalia,* 62:9–21.

Kaczmarski, F. 1966. Bioenergetics of pregnancy and lactation in the bank vole. *Acta Theriologica,* 11:409–417.

Kadwell M, et al. 2001. Genetic analysis reveals the wild ancestors of the llama and alpaca. *Proceedings of the Royal Society,* 268:2575–2584.

Kahmann, H & K Ostermann. 1951. Perception and production of high tones by small mammals. *Experientia,* 7:268–269.

Kalela, O. 1961. Seasonal change of habitat in the Norwegian lemming *Lemmus lemmus* L. *Annals of the Academy of Science Fennicae,* Ser. A 4(55):1–72.

Kalka, M & EKV Kalko. 2006. Gleaning bats as underestimated predators of herbivorous insects: diet of *Micronycteris microtis* (Phyllostomidae) in Panama. *Journal of Tropical Ecology,* 22:1–10.

Kalka, MB, AR Smith, & EKV Kalko. 2008. Bats limit arthropods and herbivory in a tropical forest. *Science,* 320:71.

Kalko, EKV. 1995. Echolocation signal design, foraging habitats and guild structure in six Neotropical sheath-tailed bats (Emballonuridae). *Symposium Zoological Society London,* 67:259–273.

Kalko, EKV & MA Condon. 1998. Echolocation, olfaction, and fruit display: how bats find fruit of flagellichorous cucurbits. *Functional Ecology,* 12:364–372.

Kalko, EK, et al. 1998. Echolocation and foraging behavior of the lesser bulldog bat, *Noctilio albiventris*: preadaptations for piscivory? *Behavioral Ecology and Sociobiology,* 42:305–319.

Kanwisher, J & G Sundnes. 1966. Thermal regulation in cetaceans, 397–409, in *Whales, Dolphins and Porpoises* (KS Norris, ed.). University of California Press, Berkeley, CA.

Kappeler, PM. 1998. To whom it may concern: the transmission and function of chemical signals in Lemur catta. *Behavioral Ecology and Sociobiology*, 42:411–421.

Kappeler, PM & JU Ganzhorn. 1993. *Lemur Social Systems and Their Ecological Basis*. Plenum Press, NY.

Kappeler, PM & ME Pereira. 2003. *Primate Life Histories and Socioecology*. University of Chicago Press, Chicago, IL.

Kappeler, PM & RM Rasoloarison. 2003. *Microcebus*, mouse lemurs, Tsidy, 1310–1315, in *The Natural History of Madagascar* (SM Goodman & JP Benstead, eds.). University of Chicago Press, Chicago, IL.

Kappelman, J, et al. 2003. Oligocene mammals from Ethiopia and faunal exchange between Afro-Arabia and Eurasia. *Nature*, 426:549–552.

Kardong, KV. 1998. *Vertebrates: Comparative Anatomy, Function, Evolution*. WCB/McGraw-Hill, NY.

Kaufman, DM. 1995. Diversity of New World mammals: universality of the latitudinal gradients of species and bauplans. *Journal of Mammalogy*, 76:322–334.

Kaufmann, JH. 1974. Social ethology of the whiptail wallaby, *Macropus parryi*, in northeastern New South Wales. *Animal Behavior*, 22:281–369.

Kausrud, KL, et al. 2008. Linking climate change to lemming cycles. *Nature*, 456:93–97.

Kay, RF. 1975. The functional adaptations of primate molar teeth. *American Journal of Physical Anthropology*, 43:195–216.

Kay, RF & BA Williams. 1994. The dental evidence for anthropoid origins, 361–446, in *Anthropoid Origins* (J Fleagle & RF Kay, eds.). Plenum Press, NY.

Kay, RF, C Ross & BA Williams. 1997a. Primate species richness is determined by plant productivity: implications for conservation. *Proceedings of the National Academy of Sciences USA*, 94:13023–13027.

Kay, RF, C Ross & BA Williams. 1997b. Anthropoid origins. *Science*, 275:797–804.

Kay, RF, et al. 2004. Anthropoid origins: a phylogenetic analysis, 91–135, in *Anthropoid Origins: New Visions* (CF Ross & RF Kay, eds.). Kluwer/Plenum, NY.

Kays, RW and KM Slauson. 2008 Remote Cameras. 110–140 in *Noninvasive Survey Methods for Carnivores* (RA Long et al., eds.). Island Press, Washington, DC.

Kays, R, et al. 2009a. Estimating animal density from photo rates and animal movement speeds recorded in camera trap videos. Abstract and talk presented at the 10th International Mammalogical Congress, Mendosa, Argentina, 9–14 August 2009.

Kays, R, et al. 2009b. Camera traps as sensor networks for monitoring animal communities. The 34th IEEE Conference on local computer networks:1–8.

Keane, B, SR Creel, & PM Waser. 1996. No evidence of inbreeding avoidance or inbreeding depression in a social carnivore. *Behavioral Ecology*, 7:480–489.

Keast, A. 1972. Australian mammals: zoogeography and evolution, 195–246, in *Evolution, Mammals, and Southern Continents* (A Keast, FC Erk, & B Glass, eds.). State University of New York Press, Albany, NY.

Keith, LB. 1981. Population dynamics of hares, 395–440, in *Proceedings of the World Lagomorph Conference* (K Myers & CD Mac Innes, eds.). University of Guelph, Guelph, Ontario, Canada.

Keith, LB. 1990. Dynamics of snowshoe hare populations. *Current Mammalogy*, 2:119–195.

Kellogg, WN, R Kohler, & HN Morris. 1953. Porpoise sounds as sonar signals. *Science*, 117:239–243.

Kelm, DH, KR Wiesner, & O vonHelversen. 2008. Effects of artificial roosts for frugivorous bats on seed dispersal in a Neotropical forest pasture mosaic. *Conservation Biology*, 22:733–741.

Kelsall, JP. 1970. Migration of the barren-ground caribou. *Natural History*, 79:98–106.

Kemp, GA & LB Keith. 1970. Dynamics and regulation of red squirrel (*Tamiasciurus hudsonicus*) populations. *Ecology*, 51:763–779.

Kemp, TS. 2005. *The Origin and Evolution of Mammals*. Oxford University Press, Oxford, UK.

Kenagy, GJ. 1972. Saltbush leaves: excision of hypersaline tissues by a kangaroo rat. *Science*, 178:1094–1096.

Kenagy, GJ. 1973a. Daily and seasonal patterns of activity and energetics in a heteromyid rodent community. *Ecology*, 54:1201–1219.

Kenagy, GJ. 1973b. Adaptations for leaf eating in the Great Basin kangaroo rat, *Dipodomys microps*. *Oecologia*, 12:383–412.

Kenagy, GJ. 1981. Effect of day length, temperature, and endogenous control on annual rhythms of reproduction and hibernation in chipmunks (*Eutamias* spp.). *Journal of Comparative Physiology*, 141:369–378.

Kenagy, GJ & GA Bartholomew. 1981. Effects of day length, temperature, and green food on testicular development in a desert pocket mouse *Perognathus formosus*. *Physiological Zoology*, 54:62–73.

Kenagy, GJ & D Vleck. 1982. The data on daily rhythms of resting metabolic rate for 18 species of small mammals, 322–338, in *Vertebrate Circadian Systems* (Aschoff, Dann, & Groos, eds.). Springer-Verlag, Berlin.

Kennerly, TR. 1971. Personal communication.

Ketten, DR. 1997. Structure and function in whale ears. *Bioacoustics*, 8:103–135.

Kick, S. 1982. Target-detection by the echolocating bat, *Eptesicus fuscus*. *Journal of Comparative Physiology A*, 145:431–435.

Kielan-Jaworowska, Z. 1997. Characters of multituberculates neglected in phylogenetic analyses of early mammals. *Lethaia*, 29:249–266.

Kielan-Jaworowska, Z, TM Bown, & JA Lillegraven. 1979. Eutheria, 221–258, in *Mesozoic Mammals: The First Two-thirds of Mammalian History* (JA Lillegraven, Z Kielan-Jaworowska, & WA Clemens, eds.). University of California Press, Berkeley, CA.

Kielan-Jaworowska, Z & PP Gambaryan. 1994. Postcranial anatomy and habits of Asian multituberculate mammals. *Fossils and Strata*, 36:1–92.

Kielan-Jaworowska, Z, RL Cifelli, & Z-X Luo. 2004. *Mammals from the Age of Dinosaurs: Origins, Evolution, and Structure*. Columbia University Press, NY.

Kiltie, RA. 1981. The function of interlocking canines in rain forest peccaries (Tayassuidae). *Journal of Mammalogy*, 62:459–469.

King, AA & WM Schaffer. 2001. The geometry of a population cycle: a mechanistic model of snowshoe hare demography. *Ecology*, 82:814–830.

King, JE. 1983. *Seals of the World*. Cornell University Press, Ithaca, NY.

King, SRB & J Gurnell. 2007. Scent-marking behaviour by stallions: an assessment of function in a reintroduced population of Przewalski horses (*Equus ferus przewalskii*). *Journal of Zoology*, 272:30–36.

Kingdon, J. 1971. *East African Mammals: An Atlas of Evolution*. Vol. 1. Academic Press, NY.

Kingdon, J. 1984a. *East African Mammals: An Atlas of Evolution. Vol. 2A, Insectivores and Bats*. University of Chicago Press, Chicago, IL.

Kingdon, J. 1984b. *East African Mammals: An Atlas of Evolution. Vol. 2B, Hares and Rodents*. University of Chicago Press, Chicago, IL.

Kingdon, J. 1989a. *East African Mammals: An Atlas of Evolution. Vol. 3A, Carnivores*. University of Chicago Press, Chicago, IL.

Kingdon, J. 1989b. *East African Mammals: An Atlas of Evolution. Vol. 3B, Large Mammals*. University of Chicago Press, Chicago, IL.

Kingdon, J. 1989c. *Island Africa*. Princeton University Press, Princeton, NJ.

Kingdon, J. 1997. *The Kingdon Field Guide to African Mammals*. Princeton University Press, Princeton, NJ.

Kinnear, JE, et al. 1979. The nutritional biology of the ruminants and ruminant-like mammals: a new approach. *Comparative Biochemistry and Physiology*, 64:357–365.

Kirk, EC, et al. 2003. Comment on "grasping primate origins." *Science*, 300:741.

Kirkland, GL Jr. 1991. Competition and coexistence in shrews (Insectivora: Soricidae), 15–22, in *The Biology of the Soricidae* (JS Findley & TL Yates, eds.). *Special Publications of the Museum Southwestern Biology*, 1:1–91.

Kirsch, JAW. 1977. The six-percent solution: second thoughts on the adaptedness of the Marsupialia. *American Scientist*, 65:276–288.

Kirsch, JAW, et al. 1995. Phylogeny of the Pteropodidae (Mammalia: Chiroptera) based on DNA hybridisation, with evidence for bat monophyly. *Australian Journal of Zoology*, 43:395–428.

Kita, M, et al. 2004. *Blarina* toxin, a mammalian lethal venom from the short-tailed shrew *Blarina brevicauda*: isolation and characterization. *Proceedings of the National Academy of Sciences USA*, 101:7542–7547.

Kitchen, DW. 1974. Social behavior and ecology of the pronghorn. *Wildlife Monographs*, 38:1–96.

Kitching, I, et al. 1998. *Cladistics: Theory and Practice of Parsimony Analysis*. Systematics Association Special, No. 11. Oxford University Press, NY.

Kleiman, DG. 1977. Monogamy in mammals. *Quarterly Review of Biology*, 52:39–69.

Kleiman, DG & TM Davis. 1978. Ontogeny and maternal care, 387–402, in *Biology of Bats of the New World Family Phyllostomatidae. Part III* (RJ Baker, JK Jones Jr, & DC Carter, eds.). Special Publication of the Museum Texas Tech Press, Lubbock.

Klemola, T, et al. 2000. Experimental tests of predation and food hypotheses for population cycles of voles. *Proceedings of the Royal Society B*, 267, 351–356.

Klingener, D. 1964. The comparative myology of four dipodoid rodents (Genera *Zapus, Napeozapus, Sicista*, and *Jaculus*). *Miscellaneous Publications of the Museum Zoology, University of Michigan*, 124:1–100.

Kober, R & HU Schnitzler. 1990. Information in sonar echoes of fluttering insects available for echolocating bats. *Journal of the Acoustical Society of America*, 87:882–896.

Kock, D, et al. 2006. On the nomenclature of Bathyergidae and *Fukomys* n. gen. (Mammalia: Rodentia). *Zootaxa*, 1142:51–55.

Koepfli, K-P, et al. 2007. Phylogeny of the Procyonidae (Mammalia: Carnivora): molecules, morphology and the Great American Interchange. *Molecular Phylogenetics and Evolution*, 43:1076–1095.

Kohn, MJ, et al. 2004. Climate stability across the Eocene-Oligocene transition, southern Argentina. *Geology*, 32(7):621–624.

Kohno, N. 2006. A new Miocene odobenid (Mammalia: Carnivora) from Hokkaido, Japan, and its implications for odobenid phylogeny. *Journal of Vertebrate Paleontology*, 26:411–421.

Kolb, A. 1976. Funktion und wirkungsweise der reichlaute der mausohrfledermaus, *Myotis myotis*. *Zeitschrift fur Saugetierkunde*, 41:226–236.

Komarek, E. V. 1932. Notes on mammals of Menominee Indian Reservation, Wisconsin. *Journal of Mammalogy*, 13:203–209.

Koopman, HN. 1998. Topographic distribution of the blubber of harbor porpoises (*Phocoena phocoena*). *Journal of Mammalogy*, 79:260–270.

Kooyman, GL. 1963. Milk analysis of the kangaroo rat, *Dipodomys merriami*. *Science*, 147:1467–1468.

Kooyman, GL. 1968. An analysis of some behavioral and physiological characteristics related to diving in the Weddel seal, 227–261, in *Biology of the Antarctic Seals III*. Vol. 2 (WL Schmitt & GA Llano, eds.). American Geophysical Union, Washington, DC.

Kooyman, GL. 1975. A comparison between day and night diving in the Weddell seal. *Journal of Mammalogy*, 56:563–574.

Kooyman, GL & HT Andersen. 1969. Deep diving. 65–94 in *The Biology of Marine Mammals* (HT Andersen, ed.). Academic Press, NY.

Kooyman, GL, et al. 1971. Pulmonary function in freely diving Weddell seal, *Leptonychotes weddelli*. *Respiration Physiology*, 17:283–290.

Kooyman, GL, RL Gentry, & DL Urquhart. 1976. Northern fur seal diving behavior: a new approach to its study. *Science*, 193:411–412.

Kooyman, GL, et al. 1980. Aerobic and anaerobic metabolism during voluntary diving in Weddell seals; evidence of preferred pathways from blood chemistry and behavior. *Journal of Comparative Physiology*, 138:335–346.

Korpimäki, E & CJ Krebs. 1996. Predation and population cycles of small mammals, a reassessment of the predation hypothesis. *BioScience*, 46:754–764.

Korpimaki, E & K Norrdahl. 1998. Experimental reduction of predators reverses the crash phase of small-rodent cycles. *Ecology*, 79:2448–2455.

Korpimaki, E, et al. 2002. Dynamic effects of predators on cyclic voles: field experimentation and model extrapolation. *Proceedings of the Royal Society B*, 269:991–997.

Korpimaki, E, et al. 2004. The puzzles of population cycles and outbreaks of small mammals solved? *Bioscience*, 54:1071–1079.

Korpimaki, E, et al. 2005. Predator-induced synchrony in population oscillations of coexisting small mammal species. *Proceedings of the Royal Society B*, 272:193–202.

Korth, WW. 1994. *The Tertiary Record of Rodents in North America*. Plenum Press, NY.

Koshkina, TV. 1965. Population density and its importance in regulating the abundance of the red vole (WA Fuller, trans.). *Bulletin of the Moscow Society of Naturalists*, 70:5–19.

Koshkina, TV & AS Kholansky. 1962. Reproduction of the Norwegian lemming (*Lemmus lemmus* L.) on the Kola Peninsula (WA Fuller, trans.). *Zoological Zhurnal*, 41:604–615.

Koshland, DE Jr. 1991. Editorial: Preserving Biodiversity. *Science*, 253:717.

Kotler, BP. 1984. Predation risk and the structure of desert communities. *Ecology*, 65:689–701.

Krassilov, V. 1973. Mesozoic plants and the problem of angiosperm ancestry. *Lethaia*, 6:163–178.

Krause, DW. 1982. Jaw movement, dental function, and diet in the Paleocene multituberculate *Ptilodus*. *Paleobiology*, 8:265–281.

Krause, DW, et al. 1997. Cosmopolitanism among Gondwanan Late Cretaceous mammals. *Nature*, 390:504–507.

Krause, DW, et al. 2006. Late Cretaceous terrestrial vertebrates from Madagascar: implications for Latin American biogeography. *Annals of the Missouri Botanical Garden*, 93:178–208.

Krause, J, et al. 2006. Multiplex amplification of the mammoth mitochondrial genome and the evolution of Elephantidae. *Nature*, 439:724–727.

Krause, J & GD Ruxton. 2002. *Living in Groups*. Oxford University Press, Oxford, UK.

Krause, WJ & JH Cutts. 1984. Scanning electron microscope observations on the opossum yolk sac chorion immediately prior to uterine attachment. *Journal of Anatomy*, 138:189–191.

Krebs, CJ. 1966. Demographic changes in fluctuating populations of *Microtus californicus*. *Ecological Monographs*, 36:239–273.

Krebs, CJ. 2001. *Ecology: The Experimental Analysis of Distribution and Abundance*. Harper-Collins, NY.

Krebs, CJ. 2009. *Ecology: The Experimental Analysis of Distribution and Abundance*. Benjamin Cummings, San Francisco, CA.

Krebs, CJ & JH Myers. 1974. Population cycles in small mammals. *Advances in Ecological Research*, 8:267–399.

Krebs, CJ, et al. 1986. Population biology of snowshoe hares. I. Demography of food-supplemented populations in the southern Yukon, 1976–1984. *Journal of Animal Ecology*, 55:963–982.

Krebs, CJ, et al. 1992. What drives the snowshoe hare cycle in Canada's Yukon? 886–896, in *Wildlife 2001: Populations*

(DM McCullough & R Barrett, eds.). Elsevier, London.

Krebs, CJ, et al. 1995. Impact of food and predation on the snowshoe hare cycle. *Science*, 269:1112–1115.

Krebs, HA. 1950. Body size and tissue metabolism. *Biochemistry Biophysics Acta*, 4:249–269.

Krebs, JR & NB Davies. 1993. *An Introduction to Behavioral Ecology*. 3rd ed. Blackwell Science Publications, London.

Kriegs, JO, et al. 2006. Retroposed elements as archives for the evolutionary history of placental mammals. *PLoS ONE Biology*, 4(4):e91.

Kronwitter, F. 1988. Population structure, habitat use and activity patterns of the noctule bat, *Nyctalus noctula* Shreb., 1774 (Chiroptera: Vespertilionidae) revealed by radio-tracking. *Myotis*, 26:23–85.

Kropelin, S, et al. 2008. Climate-driven ecosystem succession in the Sahara: the past 6000 years. *Science*, 320:765–768.

Kruger, L. 1966. Specialized features of the cetacean brain, 232–254, in *Whales, Dolphins, and Porpoises* (KS Norris, ed.). University of California Press, Berkeley, CA.

Krumrey, WA & IO Buss. 1968. Age estimation, growth, and relationships between body dimensions of the female African elephant. *Journal of Mammalogy*, 49:22–31.

Kruuk, H. 1972. *The Spotted Hyena: A Study of Predation and Social Behavior*. University of Chicago Press, Chicago, IL.

Kruuk, H. 1976. Feeding and social behaviour of the striped hyaena (*Hyaena vulgaris* Desmarest). *African Journal of Ecology*, 14:91–111.

Kruuk, H & WA Sands. 1972. The aardwolf (*Proteles cristatus* Sparrman) 1783 as predator on termites. *East African Wildlife Journal*, 10:211–227.

Krzanowski, A. 1961. Weight dynamics of bats wintering in the cave at Pulway (Poland). *Acta Theriologica*, 4:249–264.

Kuhn, H-J & U Zeller. 1987. The cavum epitericum in monotremes and therian mammals, 51–70, in *Morphogenesis of the Mammalian Skull, Mammalia Depicta*, vol. 3 (H-J Kuhn & U Zeller, eds.). Verlag Paul Parey, Hamburg.

Kullberg, M, et al. 2008. Phylogenetic analysis of 1.5 Mbp and platypus EST data refute the Marsupionta hypothesis and unequivocally support Monotremata as sister group to Marsupialia/Placentalia. *Zoologica Scripta*, 37:115–127.

Kulzer, E. 1965. Temperaturregulation bein fledermausen (Chiroptera) aus berschiedenen Klimazonen. *Zeitschrift fur Vergleichende Physiologie*, 50:1–34.

Kumar, S. 2005. Molecular clocks: four decades of evolution. *Nature Reviews Genetics*, 6:654–662.

Kummer, H. 1968. *Social Organization of Hamadryas Baboons*. University of Chicago Press, Chicago, IL.

Kummer, H. 1984. From laboratory to desert and back: a social system of hamadryas baboons. *Animal Behaviour*, 32:965–971.

Kunz, TH & KA Nagy. 1988. Methods of energy budget analysis, 277–302, in *Ecological and Behavioral Methods for the Study of Bats* (TH Kunz, ed.). Smithsonian Institution Press, Washington, DC.

Kunz, TH & CA Diaz. 1995. Folivory in fruit-eating bats, with new evidence from *Artibeus jamaicensis* (Chiroptera: Phyllostomidae). *Biotropica*, 27:106–120.

Kurtén, B & E Anderson. 1980. *Pleistocene Mammals of North America*. Columbia University Press, NY.

Kurtén, L & U Schmidt, 1982. Thermoreception in the common vampire bat (*Desmodus rotundus*). *Journal of Comparative Physiology*, 146:223–228.

Kuyper, MA. 1979. A biological study of the golden mole *Amblysomus hottentotus*. M.S. thesis. University of Natal, Pietermaritzburg, South Africa.

Kuyper, MA. 1985. The ecology of the golden mole *Amblysomus hottentotus*. *Mammal Reviews*, 15:3–11.

Lacey, EA, SH Braude, & J Wieczorek. 1997. Burrow sharing by colonial tuco-tucos (*Ctenomys sociabilis*). *Journal of Mammalogy*, 78:556–562.

Lacher, TE. 1981. The comparative social behavior of *Kerodon rupestris* and *Galea spixii* and the evolution of behavior in the Caviidae. *Bulletin of Carnegie Museum of Natural History*, 1:1–71.

Lack, D. 1954. Cyclic mortality. *Journal of Wildlife Management*, 18:25–37.

Lacombat, F. 2005. The evolution of the rhinoceros, 46–49, in *Save the Rhinos: EAZA Rhino Campaign 2005/6* (R Fulconis, ed.). European Association of Zoos and Aquaria, London.

Lacy, RC. 1997. Importance of genetic variation to the viability of mammalian populations. *Journal of Mammalogy*, 78:320–335.

Lambertsen, R, N Ylrich, & J Straley. 1995. Frontomandibular stay of

Balaenopteridae: a mechanism for momentum recapture during feeding. *Journal of Mammalogy,* 76:877–899.

Lamprecht, J. 1979. Field observations on the behavior and social system of the bat-eared fox (*Otocyon megalotis* Desmarest). *Zeitschrift fur Tierpsychologie,* 52:171–200.

Lamprey, HF, G Halevy, & S Makacha. 1974. Interactions between *Acacia,* bruchid seed beetles, and large herbivores. *East African Wildlife Journal,* 12:81–85.

Lancaster, WC, OW Henson, & AW Keating. 1995. Respiratory muscle activity in relation to vocalization in flying bats. *Journal of Experimental Biology,* 198:175–191.

Lancaster, WC & JR Speakman. 2001. Variations in respiratory muscle activity during echolocation when stationary in three species of bat (Microchiroptera: Vespertilionidae). *Journal of Experimental Biology,* 204:4185–4197.

Lang, H & JP Chapin. 1917. The American Museum Congo expedition collection of bats III: field notes. *Bulletin of the American Museum of Natural History,* 37.

Langer, P. 2002. The digestive tract and life history of small mammals. *Mammal Reviews,* 32:107–131.

Langman, VA, et al. 1979. Nasal heat exchange in the giraffe and other large mammals. *Respiration Physiology,* 37:325–333.

Lariviere, S & SH Ferguson. 2003. Evolution of induced ovulation in North American carnivores. *Journal of Mammalogy,* 84:937–947.

Larom, D, et al. 1994. Abiotic controls on elephant communication. *Journal of the Acoustical Society of America,* 96:3297–3298.

Larsen, EC. 1986. Competitive release in microhabitat use among coexisting desert rodents: a natural experiment. *Oecologia,* 69:231–237.

Larsen, G, et al. 2007. Current views on *Sus* phylogeography and pig domestication as seen through modern mtDNA studies, 30–41, in *Pigs and Humans: 10,000 Years of Interaction* (Albarella, Dobney, Ervynck, & Rowley-Cony, eds.). Oxford University Press, Oxford, UK.

Laurance, WF. 1998. A crisis in the making: responses of Amazonian forests to land use and climate change. *Trends in Ecology and Evolution,* 13:411–415.

Laurance, WF & DC Useche. 2009. Environmental synergisms and the extinction of tropical species. *Conservation Biology,* 23:1427–1437.

Laurin, M & RR Reisz. 2007. Synapsida. Mammals and their extinct relatives. Tree of Life Web Project. http://tolweb.org/Synapsida/14845/2007.04.06 (accessed October 29, 2009).

Laursen, L & M Bekoff. 1978. *Loxodonta africana. Mammalian Species,* 92:1–8.

Lavocat, R. 1973. Les rongeurs du Miocene d'Afrique Orientale. I, Miocene inferieur. *Mémoires et Travaux de l'Institut de Montpellier, Ecole Practique des Hautes Etudes,* 1:1–284.

Lavocat, R. 1976. Rongeurs caviomorphes de l'Oligocene de Bolivia. II, Rongeurs du bassin Deseadien de Salla-Luribay. *Palaeovertebrata,* 7:15–90.

Lawlor, TE. 1998. Biogeography of great basin mammals: paradigm lost? *Journal of Mammalogy,* 79:1111–1130.

Lawrence, BD & JA Simmons. 1982. Measurements of atmospheric attenuation at ultrasonic frequencies and the significance for echolocation by bats. *Journal of the Acoustical Society of America,* 71:585–590.

Laws, RM. 1966. Age criteria for the African elephant, *Loxodonta a. africana. East African Wildlife Journal,* 4:1–36.

Laws, RM. 1970. Elephants as agents of habitat and landscape change in East Africa. *Oikos,* 21:1–15.

Laws, RM & ISC Parker. 1968. Recent studies on elephant populations in East Africa. *Symposium of the Zoological Society of London,* 21:319–359.

Laws, RM, ISC Parker, & RCB Johnstone. 1975. *Elephants and Their Habitats: The Ecology of Elephants in North Bunyoro, Uganda.* Clarendon Press, Oxford, UK.

Lay, DM. 1967. *A Study of Mammals of Iran. Fieldiana Zoology,* 54:1–282.

Leakey, R & R Lewin. 1995. *The Sixth Extinction.* Doubleday, NY.

Le Boeuf, BJ & RS Peterson. 1969. Social status and mating activity in elephant seals. *Science,* 163:91–93.

Le Boeuf, BJ & J Reiter. 1988. Lifetime reproductive success in northern elephant seals, 344–362, in *Reproductive Success* (TH Clutton-Brock, ed.). University of Chicago Press, Chicago, IL.

Le Boeuf, BJ, et al. 1993. Sex differences in diving and foraging behavior of northern elephant seals. *Symposium of the Zoological Society London,* 66:149–178.

Lechleitner, RR, JV Tileston, & L Kartman. 1962. Die-off of a Gunnison's prairie dog colony in central Colorado. I. Ecological observations and description of the epizootic. *Zoonoses Research,* 1:185–199.

Lechner, AJ. 1978. The scaling of maximal oxygen consumption and pulmonary dimensions in small mammals. *Respiration Physiology,* 34:29–44.

Lee, AK & RW Martin. 1988. *The Koala: A Natural History.* University of New South Wales Press, Sydney, Australia.

Lee, PC & JI Oliver. 1979. Competition, dominance, and the acquisition of rank in juvenile yellow baboons (*Papio cynocephalus*). *Animal Behaviour,* 27:576–585.

Lee, PC & CJ Moss. 1999. The social context for learning and behavioural development among wild African elephants, 102–125, in *Mammalian Social Learning* (HO Box & KR Gibson eds.). Cambridge University Press, Cambridge, UK.

Lehmann, T, et al. 2005. A new species of Orycteropodidae (Mammalia, Tubulidentata) in the Mio-Pliocene of northern Chad. *Zoological Journal of the Linnean Society,* 143:109–131.

Leighton, DR. 1986. Gibbons: territoriality and monogamy, 135–145, in *Primate Societies* (BB Smuts, DL Cheney, RM Seyfarth, RW Wrangham, & TT Struhsaker, eds.). University of Chicago Press, Chicago, IL.

Leighton, TG, SD Richards, & PR White. 2004. Trapped within a "wall of sound" a possible mechanism for the bubble nets of humpback whales. *Acoustics Bulletin,* 29:24–29.

Leighton, T, et al. 2007. An acoustical hypothesis for the spiral bubble nets of humpback whales, and the implications for whale feeding. *Acoustics Bulletin,* Jan/Feb:17–21.

Leiner, NO, EZF Stez, & WR Sliva. 2008. Semelparity and factors affecting the reproductive activity of the Brazilian slender opossum (*Marmosops paulensis*) in southeastern Brazil. *Journal of Mammalogy,* 89:153–158.

Leitner, P & JE Nelson. 1967. Body temperature, oxygen consumption, and heart rate in the Australian false vampire bat, *Macroderma gigas. Comparative Biochemistry and Physiology,* 21:65–74.

Lekagul, B & J McNeely. 1977. *Mammals of Thailand.* Sahakarnbhat, Bangkok.

Leonard, ML & MB Fenton. 1984. Echolocation calls of *Euderma maculatum* (Vespertilionidae): use in orientation and communication. *Journal of Mammalogy,* 65:122–126.

Leong, KM, et al. 2003. The use of low-frequency vocalizations in African elephant (*Loxodonta africana*) reproductive strategies. *Hormones and Behavior*, 43:433–443.

Leopold, A. 1949. *A Sand County Almanac and Sketches Here and There.* Oxford University Press, London.

Leopold, AS, et al. 1951. The jawbone deer herd. *California Division of Fish and Game, Game Bulletin*, 4:1–139.

Leuthold, W. 1966. Variations in territorial behavior of Uganda kob, *Adenota kob thomasi* (Neumann 1896). *Behaviour*, 27:214–257.

Lever, C. 1985. *Naturalized Mammals of the World.* Longman, London.

Lever, C. 1994. *Naturalised animals: The Ecology of Successfully Introduced Species.* Poyser Natural History, London.

Levins, R. 1969. The effect of random variation of different types on population growth. *Proceedings of the National Academy of Sciences USA*, 62:1061–1065.

Levy, F, M Keller, & P Poindron. 2004. Olfactory regulation of maternal behavior in mammals. *Hormones and Behavior*, 46:284–302.

Lewis-Oritt, N, RA Van Den Bussche, & RJ Baker. 2001. Molecular evidence for evolution of piscivory in *Noctilio* (Chiroptera: Noctilionidae). *Journal of Mammalogy*, 82:748–759.

Leyhausen, P. 1956. Verhaltensstudien an Katzen. *Zeitschrift fur Tierpsychologie*, 2:1–120.

Li, CK & SY Ting. 1985. Possible phylogenetic relationships: eurymylid-rodent and mimotonid-lagomorph, 35–58, in *Evolutionary Relationships Among Rodents* (WP Luckett & JL Hartenberger, eds.). Plenum Press, NY.

Li, CK, et al. 1987. The origins of rodents and lagomorphs, 97–108, in *Current Mammalogy*. Vol 1 (HH Genoways, ed.). Plenum Press, NY.

Li, JZ, et al. 2008. Worldwide human relationships inferred from genome-wide patterns of variation. *Science*, 319:1100–1104.

Lidicker, WZ Jr. 1973. Regulation of numbers in an island population of the California vole: a problem in community dynamics. *Ecological Monographs*, 43:271–302.

Lidicker, WZ Jr. 1975. The role of dispersal in the demography of small mammals, 103–128, in *Small Mammals: Their Productivity and Population Dynamics* (FB Golley, K Petrusewicz, & L Ryzkowski,

eds.). Cambridge University Press, Cambridge, UK.

Lidicker, WZ Jr. 1976. Experimental manipulation of the timing of reproduction in the California vole. *Research Population Ecology*, 18:14–27.

Lidicker, WZ Jr. 1979. Analysis of two freely-growing enclosed populations of the California vole. *Journal of Mammalogy*, 60:447–466.

Lidicker, WZ Jr. 1980. The social biology of the California vole. *Biologist*, 62:46–55.

Lidicker, WZ Jr. 1988. Solving the enigma of microtine "cycles." *Journal of Mammalogy*, 69:225–235.

Lidicker, WZ Jr. 1996. Rodent evolution in evolutionary and ecological time: are there any connections? 203–210, in *Biodiversity and Adaptation* (A Zaime, ed.). Proceedings of the Fifth International Conference of Rodents and Spatium Actes Editions, Rabat, Morocco.

Lidicker, WZ Jr & PK Anderson. 1962. Colonization of an island by *Microtus californicus*, analyzed on the basis of runway transects. *Journal of Animal Ecology*, 31:503–517.

Lidicker, WZ Jr & RS Ostfeld. 1991. Extra-large body size in California voles: causes and fitness consequences. *Oikos*, 61:108–121.

Liebenberg, L. 1990. *The Art of Tracking: The Origin of Science.* David Philip Publishers, Cape Town, South Africa.

Lillegraven, JA. 1974. Biogeographical considerations of the marsupial-placental dichotomy. *Annual Review of Ecology and Systematics*, 5:263–283.

Lillegraven, JA. 1975. Biological considerations of the marsupial-placental dichotomy. *Evolution*, 29:707–722.

Lillegraven, JA. 1979. Reproduction in Mesozoic mammals, 259–276, in *Mesozoic Mammals: The First Two-thirds of Mammalian History* (JA Lillegraven, Z Kielan-Jaworowska, & WA Clemens, eds.). University of California Press, Berkeley, CA.

Lillegraven, JA, Z Kielan-Jaworowska, & WA Clemens. 1979. *Mesozoic Mammals: The First Two-thirds of Mammalian History.* University of California Press, Berkeley, CA.

Lillegraven, JA, MJ Kraus, & TM Bown. 1979. Paleogeography of the world of the Mesozoic, 277–308, in *Mesozoic Mammals: The First Two-thirds of Mammalian History* (JA Lillegraven, Z Kielan-Jaworowska, & WA Clemens,

eds.). University of California Press, Berkeley, CA.

Lillegraven, JA, et al. 1987. The origin of eutherian mammals. *Biological Journal of the Linnean Society*, 32:281–336.

Lillegraven, JA & G Hahn. 1993. Evolutionary analysis of the middle and inner ear of Late Jurassic multituberculates. *Journal of Mammalian Evolution*, 1:47–74.

Lilly, JC. 1962. Vocal behavior of the bottle-nosed dolphin. *Proceedings of the American Philosophical Society*, 106:520–529.

Lilly, JC. 1963. Distress call of the bottle-nosed dolphin: stimuli and evoked behavioral responses. *Science*, 139:116–118.

Lim, BK & JM Dunlop. 2008. Evolutionary patterns of morphology and behavior as inferred from a molecular phylogeny of New World emballonurid bats (tribe Diclidurini). *Journal of Mammalian Evolution*, 15:79–121.

Lim, BK, et al. 2008. Molecular phylogeny of New World sheath-tailed bats (Emballonuridae: Diclidurini) based on loci from the four genetic transmission systems in mammals. *Biological Journal of the Linnean Society*, 93:189–209.

Lim, MLM, NS Sodhi, & JA Endler. 2008. Conservation with sense. *Science*, 319:281.

Lim, N. 2007. *Colugo: The Flying Lemur of South-East Asia.* Draco Publishing and National University of Singapore, Singapore.

Lindberg, DR & ND Pyenson. 2006. Evolutionary patterns in Cetacea: fishing up prey size through deep time. 67-81 in *Whales, Whaling, and Ocean Ecosystems* (JA Estes, DP Demaster, & DF Doak, eds.). University of California Press, Berkeley, CA.

Lindberg, DR & ND Pyenson. 2007. Things that go bump in the night: evolutionary interactions between cephalopods and cetaceans in the Tertiary. *Lethaia*, 40:335–343.

Lindgren, BS. 1992. Attraction of Douglas-fir beetle, spruce beetle and a bark beetle predator (Coleoptera: Scolytidae and Cleridae) to enantiomers of frontalin. *Journal of the Entomological Society of British Columbia*, 89, 13–17.

Lindsey, E, et al. 2009. Reassessing faunal dynamics during the Great American Biotic Interchange using updated data and adjustments for sampling biases. Abstract of presentation at Society of Vertebrate Paleontology annual meeting, September

2009, Bristol, UK. *Journal of Vertebrate Paleontology* 29(suppl 3):135A.

Lindstedt, SL. 1980b. Regulated hypothermia in the desert shrew. *Journal of Comparative Physiology*, 137:173–176.

Lindstedt, SL, et al. 1991. Running energetics in the pronghorn antelope. *Nature*, 353:748–750.

Linzey, DW & AV Linzey, 1967. Maturational and seasonal molts in the golden mouse, *Ochrotomys nuttalli*. *Journal of Mammalogy*, 48:236–241.

Liu, AGSC, ER Seiffert, & EL Simons. 2008. Stable isotope evidence for an amphibious phase in early proboscidean evolution. *Proceedings of the National Academy of Sciences of the USA*, 105:5786–5791.

Liu, L. 2001. Eocene suoids (Artiodactyla, Mammalia) from Bose and Yongle basins, China, and the classification and evolution of the Paleogene suoids. *Vertebrata PalAsiatica*, 39:115–128.

Liu, Z, et al. 2009. Global cooling during the Eocene-Oligocene climate transition. *Science*, 323:1187–1190.

Lochmiller, RL. 1996. Immunocompetence and animal population regulation. *Oikos*, 76:594–602.

Lockyer, C. 1981. Growth and energy budgets of large baleen whales from the Southern Hemisphere, 379–487, in *FAO Advisory Committee on Marine Resource Research, Mammals in the Sea* (Anonymous, ed.).Vol. III, General papers and large cetaceans. FAO, Rome, Italy.

Logan, KA & LL Sweanor. 2001. *Desert Puma, Evolutionary Ecology and Conservation of an Enduring Carnivore*. Island Press, Washington, Covelo, London.

Lomolino, MV. 2005. Body size evolution in insular vertebrates: generality of the island rule. *Journal of Biogeography*, 32:1683–1699.

Lomolino, MV & R Channell. 1995. Splendid isolation: patterns of geographic range collapse in endangered mammals. *Journal of Mammalogy*, 76:335–347.

Lomolino, MV & R Channell. 1998. Range collapse, reintroductions and biogeographic guidelines for conservation: a cautionary note. *Conservation Biology*, 12:481–484.

Lomolino, MV, BR Riddle, & JH Brown. 2006. *Biogeography*. 3rd ed. Sinauer Associates, Sunderland, MA.

Long, J, et al. (eds.). 2002. *Prehistoric Mammals of Australia and New Guinea: One Hundred Million Years of Evolution*.

Johns Hopkins University Press, Baltimore, MD.

Long, JW, A Tecle, & BM Burnette. 2003. Cultural foundations for ecological restoration on the White Mountain Apache Reservation. *Conservation Ecology*, 8(1):4. http://www.consecol.org/vol8/iss1/art4.

Long, RA, TJ Martin, & BM Barnes. 2005. Body temperature and activity patterns in free-living arctic ground squirrels. *Journal of Mammalogy*, 86:314–322.

Long, RA, et al. 2008. *Noninvasive Survey Methods for Carnivores*. Island Press, Washington, DC.

López Antoñanzas, R, S Sen, & P Mein. 2004. Systematics and phylogeny of the cane rats (Rodentia: Thryonomyidae). *Zoological Journal of the Linnean Society*, 142:423–444.

Lorenz, K. 1963. *Das Sogenannte Böse*. G. Borotha-Schoeler, Vienna. (English version, 1966. *On Aggression*. Methuen, London.)

Lorini, M & VG Persson. 1990. Nova especie de *Leontopithecus* Lesson 1840, do sul do Brasil (Primates, Callitrichidae). *Boletím do Museu Nacional, Rio de Janeiro*, 338:1–14.

Loudan, ASI. 1985. Lactation and neonatal survival of mammals. *Symposium of the Zoological Society London*, 54:183–207.

Loudon, A, N Rothwell, & M Stock. 1985. Brown fat, thermogenesis and physiological birth in a marsupial. *Comparative Biochemistry and Physiology*, 81A:815–819.

Louis, EE Jr, et al. 2006. Molecular and morphological analyses of the sportive lemurs (Family Megaladapidae: Genus *Lepilemur*) reveals 11 previously unrecognized species. *Texas Tech University Special Publications*, 49:1–49.

Louv, R. 2005. *Last Child in the Woods: Saving Our Children from Nature-Deficit Disorder*. Algonquin Books, Chapel Hill, NC.

Lovegrove, BC. 2000. Daily heterothermy in mammals: coping with unpredictable environments, 29–40, in *Life in the Cold* (G Heldmaier & M Klingenspor, eds.). 11th International Hibernation Symposium, Springer, NY.

Lovegrove, BG & JUM Jarvis. 1986. Coevolution between mole-rats (Bathyergidae) and a geophyte, *Micranthus* (Iridaceae). *Cimbebasia*, 8:79–85.

Lovegrove, BG, J Raman, & MR Perrin. 2001a. Daily torpor in elephant shrews (Macroscelidea: *Elephantulus* spp.) in response to food deprivation. *Journal of Comparative Physiology Series B*, 171:11–21.

Lovegrove BG, J Raman, & MR Perrin. 2001b. Heterothermy in elephant shrews (*Elephantulus* spp.): hibernation or daily torpor? *Journal of Comparative Physiology B*, 171:1–10.

Lovejoy, TE. 1997. Foreword, ix–x, in *Tropical Forest Remnants: Ecology, Management, and Conservation of Fragmented Communities* (WF Laurance & RO Bierregaard Jr, eds.). University of Chicago Press, Chicago, IL.

Lowe, VPW. 1969. Population dynamic of the red deer (*Cervus elaphus* L.) on Rhum. *Journal of Animal Ecology*, 38:434–444.

Loy, A, E Dupré & E Capanna. 1994. Territorial behavior in *Talpa romana*, a fossorial insectivore from southcentral Italy. *Journal of Mammalogy*, 75:529–535.

Lucas, SG & JC Sobus. 1989. The systematics of Indricotheres, 358–378, in *The Evolution of Perissodactyls* (DR Prothero & RM Schoch, eds.). Oxford University Press, NY.

Luckett, WP. 1980. Monophyletic or diphyletic origins of Anthropoidea and Hystricognathi: evidence from fetal membranes, 347–368, in *Evolutionary Biology of the New World Monkeys and Continental Drift* (RL Ciochon & AB Chiarelli, eds.). Plenum Press, NY.

Luckett, WP. 1993. An ontogenetic assessment of dental homologies in therian mammals, 182–203, in *Mammal Phylogeny: Mesozoic Differentiation, Multituberculates, Monotremes, Early Therians, and Marsupials* (FS Szalay, MJ Novacek, & MC McKenna, eds.). Springer-Verlag, NY.

Luckett, WP. 1994. Suprafamilial relationships within Marsupialia: resolution and discordance from multidisciplinary data. *Journal of Mammalian Evolution*, 2:225–283.

Luckett, WP & JL Hartenberger. 1985. *Evolutionary Relationships Among Rodents: A Multidisciplinary Analysis*, Plenum Press, NY.

Luckett, WP & JL Hartenberger. 1993. Monophyly or polyphyly of the order Rodentia: possible conflict between morphological and molecular interpretations. *Journal of Mammalian Evolution*, 1:127–147.

Luis, C, et al. 2006. Iberian origins of New World horse breeds. *Journal of Heredity*, 97:107–113.

Lund, RD & JS Lund, 1965. The visual system of the mole, *Talpa europaea*. *Experimental Neurology*, 13:302–316.

Lundrigan, B. 1996. Morphology of horns and fighting behavior in the family Bovidae. *Journal of Mammalogy*, 77:462–475.

Luo, Z-X. 2007. Transformation and diversification in early mammal evolution. *Nature*, 450:1011–1019.

Luo, Z-X, RL Cifelli, & Z Kielan-Jaworowska. 2001. Dual evolution of tribosphenic mammals. *Nature*, 409:53–57.

Luo, Z-X, et al. 2003. An Early Cretaceous tribosphenic mammal and metatherian evolution. *Science*, 302:1934–1940.

Lyman, CP & WA Wimsatt. 1966. Temperature regulation in the vampire bat, *Desmodus rotundus*. *Physiological Zoology*, 39:101–109.

Lynch, GR, HW Heath, & CM Johnston. 1981. Effects of geographical origin on the photographic control of reproduction in the white-footed mouse, *Peromyscus leucopus*. *Biology of Reproduction*, 25:475–480.

Lynch, VJ & GP Wagner. 2006. The birth of the uterus. *Natural History*, Dec/Jan:36–41.

MacArthur, RA & LCH Wang. 1973. Physiology of thermoregulation in the pika, *Ochotona princeps*. *Canadian Journal of Zoology*, 51:11–16.

MacArthur, RH & EO Wilson. 1963. An equilibrium theory of insular zoogeography. *Evolution*, 17:373–387.

MacArthur, RH & EO Wilson. 1967. The theory of island biogeography. *Monographs in Population Biology*, No. 1, Princeton University Press, Princeton, NJ.

MacDonald, D. 1984. *The Encyclopedia of Mammals*. Facts on File Publishing, NY.

MacDonald, D. 2001. *The Encyclopedia of Mammals*. Andromedia Oxford Ltm, Oxfordshire, England.

MacFadden, BJ. 1992. *Fossil Horses: Systematics, Paleobiology and Evolution of the Family Equidae*. Cambridge University Press, NY.

MacFadden, BJ. 2000. Cenozoic mammalian herbivores from the Americas: reconstructing ancient diets and terrestrial communities. *Annual Reviews of Ecology and Systematics*, 31:33–59.

MacFadden, BJ. 2005. Fossil horses— evidence for evolution. *Science*, 307:1728–1730.

MacFadden, BJ. 2006. Extinct mammalian biodiversity of the ancient New World tropics. *Trends in Ecology and Evolution*, 21:157–165.

MacFadden, BJ, TE Cerling, & JF Prado. 1996. Cenozoic terrestrial ecosystem evolution in Argentina: evidence from carbon isotopes of fossil mammal teeth. *Palaios*, 11:319–327.

MacFadden, BJ, et al. 2007. Revised age of the lat Neogene terror bird (*Titanis*) in North America during the Great American Interchange. *Geology*, 35(2):123–126.

MacKay, MR. 1970. Lepidoptera in Cretaceous amber. *Science*, 167:379–380.

Maclean, GL. 1993. *Robert's Birds of Southern Africa*. New Holland Publishing, London.

MacMillen, RE. 1964. Population ecology, water relations, and social behavior of a southern California semidesert rodent fauna. *University of California Publications in Zoology*, 71:1–66.

MacMillen, RE. 1972. Water economy of nocturnal desert rodents, 147–174, in *Comparative Physiology of Desert Animals*. Vol. 31 (GMO Maloiy, ed.). Symposium of the Zoological Society London. Academic Press, NY.

MacMillen, RE. 1983a. Water regulation in *Peromyscus*. *Journal of Mammalogy*, 64:38–47.

MacMillen, RE. 1983b. The adaptive physiology of heteromyid rodents. *Great Basin Naturalist*, 7:65–76.

MacMillen, RE & AK Lee. 1967. Australian desert mice: independence of exogenous water. *Science*, 158:383–385.

MacMillen, RE & AK Lee. 1969. Water metabolism of Australian hopping mice. *Comparative Biochemistry and Physiology*, 28:493–514.

MacMillen, RE & JE Nelson. 1969. Bioenergetics and body size in dasyurid marsupials. *American Journal of Physiology*, 217:1246–1251.

MacMillen, RE & AK Lee. 1970. Energy metabolism and pulmocutaneous water loss of Australian hopping mice. *Comparative Biochemistry and Physiology*, 35:355–369.

MacMillen, RE, RV Baudinette, & AK Lee. 1972. Water economy and energy metabolism of the sandy inland mouse, *Leggadina hermannsburgensis*. *Journal of Mammalogy*, 53:529–539.

MacMillen, RE & EA Christopher. 1975. The water relations of two populations of noncaptive desert rodents, 117–137, in *Environmental Physiology of Desert Organisms* (NF Hadley, ed.). Dowden, Hutchinson and Ross, Stroudsburg, PA.

MacMillen, RE & DE Grubbs. 1976. The effects of temperature on water metabolism in rodents, 63–69, in *Progress in Animal Biometeorology*. Vol. 1 (DH Johnson, ed.). Swetz and Zeitlinger, Lisse, The Netherlands.

MacMillen, RE & DS Hinds. 1983a. Water regulatory efficiency in heteromyid rodents: a model and its application. *Ecology*, 64:152–164.

MacMillen, RE & DS Hinds. 1983b. Adaptive significance of water regulatory efficiency in heteromyid rodents. *BioScience*, 33:333–334.

MacMillen, RE & DS Hinds. 1992. Standard, cold-induced, and exercise-induced metabolism of rodents, 16–33, in *Mammalian Energetics: Interdisciplinary Views of Metabolism and Reproduction* (TE Tomasi & TH Horton, eds.). Cornell University Press, Ithaca, NY.

MacMillen, RE & BJ MacMillen. 2007. *Meanderings in the Bush*. Dog Ear Publishing, Indianapolis, IN.

MacPhee, RDE. 1987. The shrew tenrecs of Madagascar: systematic revision and Holocene distribution of *Microgale* (Tenrecidae, Insectivora). *American Museum Novitates*, 2889:1–45.

MacPhee, RDE & MA Iturralde-Vinent. 1994. First Tertiary land mammal from Greater Antilles: an Early Miocene sloth (Xenarthra, Megalonychidae) from Cuba. *American Museum Novitates*, 3094:1–13.

MacPhee, RDE & C Flemming. 1999. Requiem aeternum: the last five hundred years of mammalian species extinctions, 333–371, in *Extinctions in Near Time: Contexts, Causes, and Consequences* (RDE MacPhee, ed.). Plenum Press, NY.

MacPhee, RDE, C Flemming, & DP Lunde. 1999. "Last occurrence" of the Antillean insectivoran *Nesophontes*: new radiometric dates and their interpretation. *American Museum Novitates*, 3261:1–20.

Macri, S, et al. 2002. Risk taking during exploration of a plus-maze is greater in adolescent than in juvenile or adult mice. *Animal Behaviour*, 64:541–546.

Madison, DM. 1980. A review of the social biology of *Microtus pennsylvanicus*. *Biologist*, 62:20–33.

Madsen, O, et al. 2001. Parallel adaptive radiations in two major clades of placental mammals. *Nature*, 409:610–614.

Madsen, PT, M Wahlberg, & B Mohl. 2002. Male sperm whale *(Physeter macrocephalus)* acoustics in a high-latitude habitat: implications for echolocation and communication. *Behavioral Ecology and Sociobiology*, 53:31–41.

Maglio, VJ. 1973. Origin and evolution of the Elephantidae. *Transactions of the American Philosophical Society*, 63:1.

Mahboubi, M, R Ameur, & JY Crochet. 1984. Earliest known proboscidean from Early Eocene of north-west Africa. *Nature*, 308:543–544.

Maier, W & F Schrenk. 1987. The hystricomorphy of the Bathyergidae, as determined from ontogenetic evidence. *Zeitschrift fur Säugetierkunde*, 52:156–164.

Maier, W, P Klingler, & I Ruf. 2002. Ontogeny of the medial masseter muscle, pseudo myomorphy, and the systematic position of the Gliridae (Rodentia, Mammalia). *Journal of Mammalian Evolution*, 9:253–269.

Maina, JN. 2000. What it takes to fly: the structural and functional respiratory refinements in birds and bats. *Journal of Experimental Biology*, 203:3045–3064.

Maley, LE & CR Marshall. 1998. The coming of age of molecular systematics. *Science*, 279:505–506.

Mallory, FF & RJ Brooks. 1978. Infanticide and other reproductive strategies in the collared lemming (*Dicrostonyx groenlandicus*). *Nature*, 273:144–146.

Maloiy, GMO. 1973. The water metabolism of a small East African antelope: the dik-dik. *Proceedings of the Royal Society B*, 184:167–178.

Manger, PR, R Collins, & JD Pettigrew. 1997. Histological observations on presumed electroreceptors and mechanoreceptors in the beak skin of the long-beaked echidna, *Zaglossus bruijnii*. *Proceedings of the Royal Society B*, 264:165–172.

Mann, J, et al. 2000. *Cetacean Societies: Field Studies of Dolphins and Whales.* University of Chicago Press, Chicago, IL.

Marenssi, SA, et al. 1994. Eocene land mammals from Seymour Island, Antarctica: paleobiogeographical implications. *Antarctic Science*, 6:3–15.

Mares, MA. 1975. South American mammal zoogeography: evidence from convergent evolution in desert rodents. *Proceedings of the National Academy of Sciences USA*, 72(5):1702–1706.

Mares, MA. 1977. Water economy and salt balance in a South American desert rodent *Eligmodontia typus*. *Comparative Biochemistry and Physiology*, 56A:325–332.

Mares, MA. 1980. Convergent evolution among desert rodents: a global perspective. *Bulletin of Carnegie Museum of Natural History*, 16:1–51.

Mares, MA. 1993a. Desert rodents, seed consumption, and convergence. *BioScience*, 43:373–379.

Mares, MA. 1993b. Heteromyids and their ecological counterparts: a pandesert view of rodent ecology and evolution, 652–714, in *Biology of the Heteromyidae* (HH Genoways & JH Brown, eds.). Special Publication No. 10, American Society of Mammalogists.

Mares, MA & DF Williams. 1977. Experimental support for food particle size resource allocation in heteromyid rodents. *Ecology*, 58:1186–1190.

Mares, MA, et al. 1997. How desert rodents overcome halophytic plant defenses. *BioScience*, 47:699–704.

Marín, JC, et al. 2008. Mitochondrial DNA variation and systematics of the guanaco (*Lama guanicoe*, Artiodactyla: Camelidae). *Journal of Mammalogy*, 89:269–281.

Markham, RM & CP Groves. 1990. Brief communication: weights of wild orangutans. *American Journal of Physical Anthropology*, 81:1–3.

Marler, PR. 1965. Communication in monkeys and apes, 544–585, in *Primate Behavior* (I DeVore, ed.). Holt, Rinehart and Winston, NY.

Marris, E. 2009. Reflecting the past. *Nature*, 462:30-32.

Marsh, H. 1980. Age determination of the dugong (*Dugong dugon*) (Muller) in northern Australia and its biological implications. *Reports of the International Whaling Commission*, 3:181–201.

Marsh, H, et al. 1982. Analysis of stomach contents of dugongs from Queensland. *Australian Wildlife Research*, 9:55–67.

Marshall, JT & ER Marshall. 1976. Gibbons and their territorial songs. *Science*, 193:235–237.

Marshall, LG. 1972. Evolution of the peramelid tarsus. *Proceedings of the Royal Society Victoria*, 85:51–60.

Marshall, LG. 1977. A new species of *Lycopsis* (Borhyaenidae: Marsupialia) from the La Venta fauna (Miocene) of Colombia, South American. *Journal of Paleontology*, 51:633–642.

Marshall, LG. 1988. Land mammals and the great American interchange. *American Scientist*, 76:380–388.

Marshall, LG, JA Case, & MO Woodburne. 1990. Phylogenetic relationships of the families of marsupials, 433–505, in *Current Mammalogy*. Vol. 2 (HH Genoways, ed.). Plenum Press, NY.

Marshall, LG & T Sempere. 1993. Evolution of the Neotropical Cenozoic land mammal fauna in its geochronologic, stratigraphic, and tectonic context, 329–392, in *Biological Relationships Between Africa and South America* (P Goldblatt, ed.). Yale University Press, New Haven, CT.

Marshall, PT & GM Hughes. 1980. *Physiology of Mammals and Other Vertebrates.* 2nd ed. Cambridge University Press, Cambridge, UK.

Martin, EP. 1956. A population study of the prairie vole (*Microtus ochrogaster*) in northeastern Kansas. *University of Kansas Publications Museum Natural History*, 8:361–416.

Martin, IG. 1984. Factors affecting food hoarding in the short-tailed shrew *Blarina brevicauda*. *Mammalia*, 48:65–71.

Martin, LD. 1980. Functional morphology and the evolution of the cats. *Transactions of the Nebraska Academy of Sciences*, 7:141–154.

Martin, LD. 1989. Fossil history of the terrestrial Carnivora, 536–568, in *Carnivore Behavior, Ecology, and Evolution* (JL Gittleman, ed.). Cornell University Press, Ithaca, NY.

Martin, P. 1971. Movements and activities of the mountain beaver (*Aplodontia rufa*). *Journal of Mammalogy*, 52:717–723.

Martin, PS & RG Klein. 1984. *Quaternary Extinctions: A Prehistoric Revolution.* University of Arizona Press, Tucson.

Martin, RD. 1968. Reproduction and ontogeny in tree shrews (*Tupaia belangeri*), with reference to their general behaviour and taxonomic relationships. *Zeitschrift fur Tierpsychologie*, 25:409–495, 25:505–532.

Martin, RD. 1973. A review of the behavior and ecology of the lesser mouse lemur (*Microcebus murinus*, J. F. Miller 1777), 1–68, in *Comparative Ecology and Behavior of Primates* (RP Michael & JH Crook, eds.). Academic Press, NY.

Martin, RD & SK Bearder. 1979. Radio bushbaby. *Natural History*, 88:77–81.

Martin, T. 1994. African origin of caviomorph rodents is indicated by incisor enamel microstructure. *Paleobiology*, 20:5–13.

Martin, T & M Nowotny. 2000. The docodont *Haldanodon* from the Guimarota Mine, 91–96, in *Guimarota: A Jurassic Ecosystem* (T Martin & B Krebs, eds.). Verlag Dr. Friedrich Pfeil, Munich.

Martinez, D, E Salmón, & MK Nelson. 2008. Restoring indigenous history and culture to nature, 88–115, in *Original Instructions: Indigenous Teachings for a Sustainable Future* (MK Nelson, ed.). Bear and Company, Rochester, VT.

Martín-López, B, C Montes, & J Benayas. 2008. Economic valuation of biodiversity conservation: the meaning of numbers. *Conservation Biology*, 22:624–635.

Martinson, DL. 1968. Temporal patterns in the home ranges of chipmunks. *Journal of Mammalogy*, 49:83–91.

Maser, C & Z Maser. 1988. Interactions among squirrels, mycorrhizal fungi, and coniferous forests in Oregon. *Great Basin Naturalist*, 48:358–369.

Mason, MJ. 2003a. Morphology of the middle ear of golden moles (Chrysochloridae). *Journal of Zoology*, 260:391–403.

Mason, MJ. 2003b. Bone conduction and seismic sensitivity in golden moles (Chrysochloridae). *Journal of Zoology*, 260:405–413.

Matschie, P. 1899. Beitrage zur Kenntnis von *Hypsignathus monstrosus* Allen. *Sitzungsberichte der Gesellschaft naturforschender Freunde*, Berlin.

Matsumura, S. 1981. Mother–infant communication in a horseshoe bat (*Rhinolophus ferrumequinum nippon*): vocal communication in 3-week-old infants. *Journal of Mammalogy*, 62:20–28.

Matthee, CA & TJ Robinson. 1997a. Mitochondrial DNA phylogeography and comparative cytogenetics of the springhare, *Pedetes capensis* (Mammalia: Rodentia). *Journal of Mammalian Evolution*, 4:53–73.

Matthee, CA & TJ Robinson. 1997b. Molecular phylogeny of the springhare, *Pedetes capensis*, based on mitochondrial DNA sequences. *Molecular Biology and Evolution*, 14:20–29.

Matthew, WD. 1910. The phylogeny of the Felidae. *Bulletin of American Museum of Natural History*, 28:289–316.

Matthew, WD. 1915. Climate and evolution. *Annals of the New York Academy of Sciences*, 24:171–318.

Mattingly, DK & PA McClure. 1985. Energy allocation during lactation in cotton rats (*Sigmodon hispidus*) on a restricted diet. *Ecology*, 66:928–937.

Mayer, JJ & PN Brandt. 1982. Identity, distribution, and natural history of the peccaries, Tayassuidae, 433–455, in *Mammalian Biology in South America* (MA Mares & HH Genoways, eds.). Special Publications Series, Pymatuning Laboratory of Ecology Vol. 6. University of Pittsburgh Press, Pittsburgh, PA.

Mayer, WV. 1960. Histological changes during the hibernating cycle in the arctic ground squirrel. *Bulletin of the Museum of Comparative Zoology*, 124:131–154.

Maynard Smith, J. 1976. Evolution and the theory of games. *American Scientist*, 64:41–45.

Mayr, E. 1942. *Systematics and the Origin of Species*. Columbia University Press, NY.

Mayr, E. 1963. *Animal Species and Evolution*. Harvard University Press, Cambridge, MA.

Mayr, E. 1970. *Populations, Species, and Evolution*. Cambridge University Press, Cambridge, UK.

McBrearty, S & NG Jablonski. 2005. First fossil chimpanzee. *Nature*, 437:105–108.

McCarthy, TS, WN Ellery, & A Bloem. 1998. Some observations on the geomorphological impact of hippopotamus (*Hippopotamus amphibious* L.) in the Okavango Delta, Botswana. *African Journal of Ecology*, 36:44–56.

McComb, K, et al. 2003. Long-distance communication of acoustic cues to social identity in African elephants. *Animal Behaviour*, 65:317–329.

McCracken, GF. 1987. Genetic structure of bat social groups, 282–298, in *Recent Advances in the Study of Bats* (MB Fenton, P Racey, & JMV Rayner, eds.). Cambridge University Press, Cambridge, UK.

McCracken, GF & JW Bradbury. 1977. Paternity and genetic heterogeneity in the polygynous bat, *Phyllostomus hastatus*. *Science*, 198:303–306.

McCracken, GF & JW Bradbury. 1981. Social organization and kinship in the polygynous bat, *Phyllostomus hastatus*. *Behavioral Ecology and Sociobiology*, 8:11–34.

McCracken, GF & GS Wilkinson. 2000. Bat mating systems, 321–362, in *Reproductive Biology of Bats* (PH Krutzsch & EG Creighton, eds.). Academic Press, NY.

McCain, CM. 2005. Elevational gradients in diversity of small mammals. *Ecology*, 86:366–372.

McCullough, DR. 1969. The tule elk, its history, behavior, and ecology. *University of California Publications in Zoology*, 88:1–209.

McDonough, CM & WJ Loughry 2008. Behavioral ecology of armadillos, 281–293, in *The Biology of the Xenarthra* (SF Vizcaíno & WJ Loughry, eds.). University Press of Florida, Gainesville.

McDonald, HG & G De Iuliis. 2008. Fossil history of sloths, 39–55, in *The Biology of the Xenarthra* (SF Vizcaíno & WJ Loughry, eds.). University Press of Florida, Gainesville.

McGuire, BA & MA Novak. 1982. A comparison of maternal behavior in three species of voles (*Microtus pennsylvanicus*, *M. pinetorum*, and *M. ochrogaster*) using a laboratory system, 139–145, in *Proceedings of the 6th Eastern Pine and Meadow Vole Symposium* (RE Byers, ed.). Harpers Ferry, WV.

McKay, GM. 1973. The ecology and behavior of the Asiatic elephant in southeastern Ceylon. *Smithsonian Contributions to Zoology*, 125:1–113.

McKean, TA & B Walker. 1974. Comparison of selected cardiopulmonary parameters between the pronghorn and the goat. *Respiration Physiology*, 21:365–370.

McKendrick, JD, et al. 1980. Some effects of mammalian herbivores and fertilization on tundra soils and vegetation. *Arctic and Alpine Research*, 12:565–578.

McKenna, MC. 1972. Possible biological consequences of plate tectonics. *BioScience*, 22:519.

McKenna, MC. 1975. Toward a phylogenetic classification of the Mammalia, 21–46, in *Phylogeny of the Primates: A Multidisciplinary Approach* (WP Luckett & FS Szalay, eds.). Plenum Press, NY.

McKenna, MC. 1980. Eocene paleolatitude, climate, and mammals of Ellesmere Island. *Palaeogeography, Palaeoclimatology, and Palaeoecology*, 30:349–362.

McKenna, MC & SK Bell. 1997. *Classification of Mammals Above the Species Level*. Columbia University Press, NY.

McLaren, A. 1970. The fate of the zona pellucida in mice. *Journal of Embryology and Experimental Morphology*, 23:1–19.

McLean, DC. 1944. The prong-horned antelope in California. Bureau Game Conservation, California Division of Fish Game, San Francisco, 30:221–241.

McNab, BK. 1966. The metabolism of fossorial rodents: a study of convergence. *Ecology*, 47:712–733.

McNab, BK. 1978. The evolution of endothermy in the phylogeny of mammals. *American Naturalist*, 112:1–21.

McNab, BK. 1979. The influence of body size on the energetics and distribution of fossorial and burrowing mammals. *Ecology*, 60:1010–1021.

McNab, BK. 1980a. Energetics and the limits to a temperature distribution in armadillos. *Journal of Mammalogy*, 61:606–627.

McNab, BK. 1980b. Food habits, energetics, and the population biology of mammals. *American Naturalist*, 116:106–124.

McNab, BK. 1982. Evolutionary alternatives in the physiological ecology of bats, 151–200, in *Ecology of Bats* (TH Kunz, ed.). Plenum, NY.

McNab, BK. 1983. Ecological and behavioral consequences of adaptation to various food resources, 664–697, in *Advances in the Study of Mammalian Behavior* (JE Eisenberg & DG Kleiman, eds.). Special Publications No. 7, American Society of Mammalogists.

McNab, BK. 1984. Physiological convergence amongst ant-eating and termite-eating mammals. *Journal of Zoology London*, 203:485–510.

McNab, BK & FJ Bonaccorso. 1995. The energetics of pteropodid bats, 111–122, in *Ecology, Evolution, and Behaviour of Bats*. Symposium of the Zoological Society London, 67:1–421.

McNaughton, SJ. 1976. Serengeti migratory wildebeest: facilitation of energy flow by grazing. *Science*, 191:92–94.

McNaughton, SJ. 1979. Grazing as an optimization process: grass–ungulate relationships in the Serengeti. *American Naturalist*, 113:691–703.

McNaughton, SJ. 1985. Ecology of a grazing ecosystem: the Serengeti. *Ecological Monographs*, 55:259–294.

Mead, JG & RL Brownell Jr. 2005. Order Cetacea, 723–743, in *Mammal Species of the World* (DE Wilson & D Reeder, eds.). Johns Hopkins University Press, Baltimore, MD.

Mead, JG & RE Fordyce. 2009. The therian skull: a lexicon with emphasis on the odontocetes. *Smithsonian Contributions to Zoology*, 627:1-248.

Mead, JI. 1987. Quaternary records of pikas, *Ochotona*, in North America. *Boreas*, 16:165–171.

Mead, RA. 1968a. Reproduction in eastern forms of the spotted skunk (genus *Spilogale*). *Journal of Zoology London*, 156:119–136.

Mead, RA. 1968b. Reproduction in western forms of the spotted skunk (genus *Spilogale*). *Journal of Mammalogy*, 49:373–389.

Mead, RA. 1989. The physiology and evolution of delayed implantation in carnivores, 437–464, in *Carnivores Behavior, Ecology, and Evolution* (JL Gittleman, ed.). Cornell University Press, Ithaca, NY.

Meagher, EM, et al. 2002. The relationship between heat flow and vasculature in the dorsal fin of wild bottlenose dolphins (*Tursiops truncates*). *Journal of Experimental Biology*, 205:3475–3486.

Mech, LD. 1966. *The Wolves of Isle Royale*. U.S. National Park Service, Fauna Series 7.

Mech, LD. 1994. Buffer zones of territories of gray wolves as regions of intraspecific strife. *Journal of Mammalogy*, 75:199–202.

Medellín, RA. 1998. True international collaboration: now or never. *Conservation Biology*, 12:939–940.

Medellín, RA & J Soberón. 1998. Predictions of mammal diversity on four land masses. *Conservation Biology*, 13(1):143–149.

Meehan, TE. 1976. The occurrence, energetic significance and initiation of spontaneous torpor in the Great Basin pocket mouse (*Perognathus parvus*). Ph.D. dissertation, University of California, Irvine.

Meffe, GK, AH Ehrlich, & D Ehrenfeld. 1993. Human population control: the missing agenda. *Conservation Biology*, 7:1–3.

Meier, B & Y Rumpler. 1987. Preliminary survey of *Hapalemur simus* and a new species of *Hapalemur* in eastern Betsileo, Madagascar. *Primate Conservation*, 8:10–43.

Meier, B, et al. 1987. A new species of *Hapalemur* (Primates) from south east Madagascar. *Folia Primatologica*, 48:211–215.

Mein, P, M Pickford, & B Senut. 2000. Late Miocene micromammals form the Harasib karst deposits, Namibia. Part 1—large muroids and non-muroid rodents. *Communications of the Geological Survey of Namibia*, 12:375–390.

Melton, DA. 1976. The biology of aardvarks (Tubulidentata–Orycteropodidae). *Mammal Reviews*, 6:75–88.

Meng, J. 2004. Phylogeny and divergence of basal Glires. *Bulletin of the American Museum of Natural History*, 285:93–109.

Meng, J, et al. 1994. Primitive fossil rodent from Inner Mongolia and its implications for mammalian phylogeny. *Nature*, 370:134–136.

Meng, J & AR Wyss. 1995. Monotreme affinities and low-frequency hearing suggested by multituberculate ear. *Nature*, 377:141–144.

Meng, J & A Wyss. 2001. The morphology of *Tribosphenomys* (Rodentiaformes, Mammalia): phylogenetic implications for basal Glires. *Journal of Mammalian Evolution*, 8:1–71.

Meng, J, Y Hu, & C Li. 2003. The osteology of *Rhombomylus* (Mammalia: Glires): implications for phylogeny and evolution of Glires. *Bulletin of the American Museum of Natural History*, 275:1–247.

Menkhorst, PW. 1995. *Mammals of Victoria: Distribution, Ecology, and Conservation*. Oxford University Press, Oxford, UK.

Menzel, MA, et al. 1999. Forest to wildlife opening: habitat gradient analysis among small mammals in the southern Appalachians. *Forest Ecology and Management*, 114:227–232.

Meredith, RW, et al. 2008. A phylogeny and timescale for marsupial evolution based on sequences for five nuclear genes. *Journal of Mammalian Evolution*, 15:1–36.

Merritt, JF. 1995. Seasonal thermogenesis and changes in body mass of masked shrews, *Sorex cinereus*. *Journal of Mammalogy*, 76:1020–1035.

Merritt, JF & A Adamerovich. 1991. Winter thermoregulatory mechanisms of *Blarina brevicauda* as revealed by radiotelemetry, 47–64, in *The Biology of the Soricidae* (JS Findley & TL Yates, eds.). Special Publications of the Museum of Southwestern Biology, Albuquerque, NM.

Merritt, JF & F Bozinovic. 1994. Thermal biology of free-ranging shrews as revealed by computer-facilitated radiotelemetry: energetic implications, 163–169, in *Advances in the Biology of Shrews* (JF Merritt, GL Kirkland Jr, & RK Rose, eds.). Special Publication No. 18, Carnegie Museum of Natural History, Pittsburgh, PA.

Merritt, JF, GL Kirkland Jr, & RK Rose. 1994. *Advances in the Biology of Shrews*. Special *Publication No. 18*, Carnegie Museum of Natural History, Pittsburgh, PA.

Meserve, PL, et al. 1995. Heterogeneous responses of small mammals to an El Niño southern oscillation event in northcentral semiarid Chile and the importance of ecological scale. *Journal of Mammalogy*, 76:580–595.

Meslow, EC & LB Keith. 1968. Demographic parameters of a snowshoe hare population. *Journal of Wildlife Management*, 32:812–834.

Messer, M, et al. 1998. Evolution of the monotremes: phylogenetic relationship to marsupials and eutherians, and estimation of divergence dates based on a-lactalbumin amino acid sequences. *Journal of Mammalian Evolution*, 5:95–105.

Messier, F. 1994. Ungulate populations models with predation: a case study with the North American moose. *Ecology*, 75:478–488.

Métais, G, et al. 2007. Eocene bunoselenodont Artiodactyla from southern Thailand and the early evolution of Ruminantia in South Asia. *Naturwissenschaften*, 94:493–498.

Métais, G & I Vislobokova. 2007 Basal ruminants, 189–212, in *The Evolution of Artiodactyls* (DR Prothero & SE Foss, eds.). Johns Hopkins University Press, Baltimore, MD.

Meyer, CFJ, EKV Kalko, & G Kerth. 2009. Small-scale fragmentation effects on local genetic diversity in two phyllostomid bats with different dispersal abilities in Panama. *Biotropica*, 41:95–102.

Meyer, WB. 1996. *Human Impact on the Earth*. Cambridge University Press, Cambridge, UK.

Meyer-Rochow, VB & IAN Stringer. 1997. An honorary non-flying mammal pollinator. *Trends in Ecology and Evolution*, 12:277.

Michael, RP, EB Keverne, & RW Bonsall. 1971. Pheromones: isolation of male sex attractants from a female primate. *Science*, 172:964–966.

Michaux, J, A Reyes, & F Catzeflis. 2001. Evolutionary history of the most speciose mammals: molecular phylogeny of muroid rodents. *Molecular Biology and Evolution*, 18:2017–2031.

Michener, GR. 1998. Sexual differences in reproductive effort of Richardson's ground squirrels. *Journal of Mammalogy*, 79:1–19.

Michener, GR. 2004. Hunting techniques and tool use by North American badgers preying on Richardson's ground squirrels. *Journal of Mammalogy*, 85:1019–1027.

Michener, GR & DR Michener. 1977. Population structure and dispersal in Richardson's ground squirrels. *Ecology*, 58:359–368.

Michener, GR & L Locklear. 1990a. Differential costs of reproduction for male and female Richardson's ground squirrels. *Ecology*, 71:855–868.

Michener, GR & L Locklear. 1990b. Over-winter weight loss by Richardson's ground squirrels in relation to sexual differences in mating effort. *Journal of Mammalogy*, 71:489–499.

Migula, P. 1969. Bioenergetics of pregnancy and lactation in the European common vole. *Acta Theriologica*, 14:167–179.

Mihok, SB, N Turner, & SL Iverson. 1985. The characterization of vole population dynamics. *Ecological Monographs*, 55:399–420.

Milinkovitch, MC & JGM Thewissen. 1997. Even-toed fingerprints on whale ancestry. *Nature*, 388:622–624.

Milius, S & M Nweeia. 2006. That's one weird tooth. *Science News*, 169:12.

Millar, JS. 1975. Tactics of energy partitioning in breeding *Peromyscus*. *Canadian Journal of Zoology*, 53:967–976.

Miller, EH. 1985. Airborne acoustic communication in the walrus *Odobenus rosmarus*. *National Geographic Research*, 1:124–145.

Miller, EH & DJ Boness. 1983. Summer behaviour of the Atlantic walrus (*Odobenus rosmarus rosmarus*) on Coats Island, NWT. *Zeitschrift fur Säugetierkunde*, 48:298–313.

Miller, G. 2005. Society for Neuroscience Meeting: bats have a feel for flight. *Science*, 310:1260–1261.

Miller, GJ. 1969. Man and *Smilodon*: a preliminary report on their possible coexistence at Rancho La Brea. *Los Angeles County Museum Contributions to Science*, vol. 163.

Miller, PJO, MP Johnson, & PL Tyack. 2004. Sperm whale behaviour indicates the use of echolocation click buzzes "creaks" in prey capture. *Proceedings of the Royal Society B*, 271:2239–2247.

Miller-Butterworth, CM, et al. 2007. A family matter: conclusive resolution of the taxonomic position of the long-fingered bats, *Miniopterus*. *Molecular Biology and Evolution*, 24:1553–1561.

Millien-Parra, V & J-J Jaeger. 1999. Island biogeography of the Japanese terrestrial mammal assemblages: an example of a relict fauna. *Journal of Biogeography*, 26: 959–972.

Mills, MGL. 1989. The comparative behavioral ecology of hyenas: the importance of diet and food dispersion, 125–142, in *Carnivores Behavior, Ecology, and Evolution* (JL Gittleman, ed.). Cornell University Press, Ithaca, NY.

Milner-Gulland, EJ & R Woodroffe. 2001. Mammals, conservation efforts for, 811–824, in *Encyclopedia of Biodiversity*. Vol. 3 (SA Levin, ed.). Academic Press, NY.

Mishra, C, et al. 2002. A theoretical analysis of competitive exclusion in a Trans-Himalayan large-herbivore assemblage. *Animal Conservation*, 5:251–258.

Mitchell, D, et al. 1997. Activity, blood temperature, and brain temperature of free-ranging springbok. *Journal of Comparative Physiology B*, 167:335–343.

Mitchell, D, et al. 2002. Adaptive heterothermy and selective brain cooling in arid-zone mammals. *Comparative Biochemistry and Physiology B*, 131:571–585.

Mitchell, JH. 1998. *Trespassing: An Inquiry into the Private Ownership of Land*. Addison-Wesley, NY.

Mittermeier, RA. 1987. Rescuing Brazil's muriqui: monkey in peril. *National Geographic*, 171:386–395.

Mittermeier, RA, et al. 1994. *Lemurs of Madagascar*. Conservation International, Washington, DC.

Miyamoto, MM, JL Slightom, & M Goodman. 1987. Phylogenetic relations of humans and African apes from DNA sequences in the psi eta-globin region. *Science*, 238:369–373.

Miyamoto, MM & J Cracraft. 1991. *Phylogenetic Analysis of DNA Sequences*. Oxford University Press, NY.

Mizuhara, H. 1957. *The Japanese Monkey: Its Social Structure*. San-ichi-syobo, Kyota (in Japanese).

Moehlman, PD. 1983. Socioecology of silverbacked and golden jackals (*Canis mesomelas* and *Canis aureus*), 423–453, in *Advances in the Study of Mammalian Behavior* (JF Eisenberg & DG Kleiman, eds.). Special Publication No. 7, American Society of Mammalogy.

Moehlman, PD. 1989. Intraspecific variation in canid social systems, 143–163, in *Carnivores Behavior, Ecology, and*

Evolution (JL Gittleman, ed.). Cornell University Press, Ithaca, NY.

Mohl, B, M Wahlberg, & PT Madsen. 2003. The monopulsed nature of sperm whale clicks. *Journal of the Acoustical Society of America*, 114:1143–1154.

Mohr, E. 1941. Schwanzverlust und Schwanzregeneration bei Nagetieren. *Zoology Anzeiger*, 135:49–65.

Möhres, FP. 1953. Uber die Ultraschallorientierung der Hufeisennasen (Chiroptera— Rhinolophidae). *Zeitschrift fur Vergleichende Physiologie*, 34:547–588.

Möhres, FP. 1967. Communicative characters of sonar signals in bats, 939–945, in *Cours d'Ete' O.T.A.N. sur les Systèmes Sonars Animaux: Biologie et Bionique* (N.A.T.O. Advanced Study Institute). Vol. 2. Laboratoire de Physiologie Acoustique, Paris.

Moll, RJ, et al. 2007. A new 'view' of ecology and conservation through animal-borne videosystems. *Trends in Ecology & Evolution* 22(12):660–668.

Mones, A. 1981. Sinopsis sistematica preliminar de la familia Dinomyidae (Mammalia: Rodentia, Caviomorpha), 605–619, in *Proc. II Congreso Latino-Americano de Paleontologica*, Rio Grande do Sul, Brasil.

Montgelard, C, FM Catzeflis, & E Douzery. 1997. Phylogenetic relationships of artiodactyls and cetaceans as deduced from the comparison of cytochrome b and 12S rRNA mitochondrial sequences. *Molecular Biology and Evolution*, 14:550–559.

Moore, MJ & GA Early. 2004. Cumulative sperm whale bone damage and the bends. *Science*, 306:2215.

Mooser, O & WW Dalquest. 1975. Pleistocene mammals from Aguascalientes, Central Mexico. *Journal of Mammalogy*, 56:781–820.

Mora, EC & S Macías. 2006. Echolocation calls of Poey's flower bat (*Phyllonycteris poeyi*) unlike those of other phyllostomids. *Naturwissenschaften*, 94:380–383.

Morgan, GS. 2008. Vertebrate fauna and geochronology of the Great American Biotic Interchange in North America. *New Mexico Museum of Natural History and Science Bulletin*, 44:93–140.

Morgan, GS & CA Woods. 1986. Extinction and the zoogeography of West Indian land mammals. *Biological Journal of the Linnean Society*, 28:167–203.

Morgan, GS & NJ Czaplewski. 2003. A new bat (Chiroptera: Natalidae) from the early Miocene of Florida, with comments on natalid phylogeny. *Journal of Mammalogy*, 82:729–752.

Moritz, C, et al. 2008. Impact of a century of climate change on small-mammal communities in Yellowstone National Park, USA. *Science*, 322:261–264.

Morlo, M, S Peigne, & D Nagel. 2004. A new species of *Prosansanosmilus*: implications for the systematic relationships of the family Barbourofelidae new rank (Carnivora, Mammalia). *Journal of the Linnean Society*, 140:43–61.

Morowitz, HJ. 1991. Balancing species preservation and economic considerations. *Science*, 253:752–754.

Morrison, JC, et al. 2007. Persistence of large mammal faunas as indicators of global human impacts. *Journal of Mammalogy*, 88:1363–1380.

Morrison, P, FA Ryser, & AR Dawe. 1959. Studies on the physiology of the masked shrew *Sorex cinereus*. *Physiological Zoology*, 32:256–271.

Morrison, P & BK McNab. 1967. Temperature regulation in some Brazilian phyllostomid bats. *Comparative Biochemistry and Physiology*, 21:207–221.

Morse, DH. 1980. *Behavioral Mechanisms in Ecology*. Harvard University Press, Cambridge, MA.

Morton, SR. 1978. Torpor and nest sharing in free-living *Sminthopsis crassicaudata* (Marsupialia) and *Mus musculus* (Rodentia). *Journal of Mammalogy*, 59:569–575.

Morton, SR. 1980. Field and laboratory studies of water metabolism in *Sminthopsis crassicaudata* (Marsupialia, Dasyuridae). *Australian Journal of Zoology*, 28:213–227.

Morton, SR & RE MacMillen. 1982. Seeds as sources of preformed water for desert-dwelling granivores. *Journal of Arid Environments*, 5:61–67.

Moss, CJ. 1982. *Portraits in the Wild*. 2nd. ed. University of Chicago Press, Chicago, IL.

Moss, CJ. 1983. Estrous behavior and female choice in the African elephant. *Behaviour*, 86:167–196.

Moss, CJ. 2001. The demography of an African elephant (*Loxodonta africana*) population in Amboseli, Kenya. *Journal of Zoology*, 255:145–156.

Moss, CJ & JH Poole 1983. Relationships and social structure in African elephants, 315–325, in *Primate Social Relationships:*

An Integrated Approach (RA Hinde, ed.). Blackwell Scientific Publications, Oxford, UK.

Mouchaty, SK, et al. 2000. Phylogenetic position of the tenrecs (Mammalia: Tenrecidae) of Madagascar based on analysis of the complete mitochondrial genome sequence of *Echinops telfairi*. *Zoology Scripta*, 29:307–317.

Mowat, G, BG Slough, & S Boutin. 1996. Lynx recruitment during a snowshoe hare population peak and decline in southwest Yukon. *Journal of Wildlife Management*, 60:441–452.

Moynihan, M. 1964. *Some Behavior Patterns of Platyrrhine Monkeys: I. The Night Monkey (Aotus trivirgatus)*. Smithsonian Institution, Washington, DC.

Muchhala, N. 2006. Nectar bat stows huge tongue in its rib cage. *Nature*, 444:701.

Muijres, FT, et al. 2008. Leading-edge vortex improves lift in slow-flying bats. *Science*, 319:1250–1253.

Muizon, C de. 1982. Phocid phylogeny and dispersal. *Annals of South African Museum*, 89:175–213.

Muizon, C de. 1992. La fauna de mamíferos de Tiupampa (Paleoceno inferior, Formación Santa Lucía), Bolivia. *Revista Técnica de Yacimientos Petrolíferos Fiscales de Bolivia*, 12:575–624.

Muizon, C de. 1994. A new carnivorous marsupial from the Palaeocene of Bolivia and the problem of marsupial monophyly. *Nature*, 370:208–211.

Muizon, C de. 1998. *Mayulestes ferox*, a borhyaenoid (Metatheria, Mammalia) from the early Palaeocene of Bolivia: phylogenetic and palaeobiologic implications. *Geodiversitas*, 20:19–142.

Muizon, C de & IM Brito. 1993. Le bassin calcaire de São José de Itaboraí (Rio de Janeiro, Brésil): ses relations fauniques avec le site de Tiupampa (Cochabamba, Bolivie). *Anais de Paleontology*, 79:233–269.

Muizon, C de & HG McDonald. 1995. An aquatic sloth from the Pliocene of Peru. *Nature*, 375:224–227.

Muizon, C de & B Lange-Badre. 1997. Carnivorous dental adaptations in tribosphenic mammals and phylogenetic reconstruction. *Lethaia*, 30:353–366.

Muizon, C de, RL Cifelli, & R Céspedes Paz. 1997. The origin of the dog-like borhyaenoid marsupials of South America. *Nature*, 389:486–489.

Müller, B, et al. 2009. Bat eyes have ultraviolet-sensitive cone photoreceptors. *PLoS ONE*, 4(7):e6390.

Muller, EF. 1979. Energy metabolism, thermoregulation and water budget in the slow loris (*Nycticebus coucang*, Boddaert 1785). *Comparative Biochemistry and Physiology A*, 64:707–711.

Müller, F. 1969. Verhaltnis von körperentwicklung und cerebralisation in ontogenese und phylogenese der sänger: versuch einer libersicht des problems. *Verhandlungen der Naturforschende Gesellschaft, Basel*, vol. 80.

Muller, LK & DGP Byrnes. 2006. Green or not—bat fur can host symbiotic algae. Abstract from 36th Annual North American Symposium on Bat Research, Wilmington, NC. *Bat Research News*, 47:131.

Müller-Schwarze, D. 1971. Pheromones in the black-tailed deer (*Odocoileus hemionus columbianus*). *Animal Behaviour*, 19:141–152.

Munger, JC & JH Brown. 1981. Competition in desert rodents: an experiment with semipermeable exclosures. *Science*, 211:510–512.

Munroe, EG. 1948. The geographical distribution of butterflies in the West Indies. Ph.D., dissertation, Cornell University, Ithaca, NY.

Munroe, EG. 1953. The size of island faunas. *Proceedings of the 7th Pacific Congress. Vol. IV Zoology*. Whitcome and Tombs, Auckland, New Zealand.

Murata, Y, et al. 2003. Afrotherian phylogeny as inferred from complete mitochondrial genomes. *Molecular Phylogenetics and Evolution*, 28:253–260.

Murie, A. 1944. *The Wolves of Mount McKinley*. U.S. Department of the Interior National Park Service, Fauna Series 5.

Murphy, MJ, et al. 2001. Molecular phylogenetics and the origins of placental mammals. *Nature*, 409:614–618.

Murphy, MR. 1985. History of the capture and domestication of the Syrian golden hamster (*Mesocricetus auratus*, Waterhouse), 3–20, in *The Hamster, Reproduction and Behavior* (HI Siegel, ed.). Plenum Press, NY.

Murphy, WJ, et al. 2001. Resolution of the early placental mammal radiation using Bayesian phylogenetics. *Science*, 294:2348–2351.

Murray, KL & TH Fleming. 2008. Social structure and mating system of the buffy flower bat, *Erophylla sezekorni*

(Chiroptera, Phyllostomidae). *Journal of Mammalogy*, 89:1391–1400.

Musser, AM. 2005. Monotremata. 1-3 in *Encyclopedia of Life Sciences*. Wiley, NY. http://www.els.net and http://mrw.interscience.wiley.com/ emrw/9780470015902/els/article/ a0001553/current/pdf.

Musser, GG & MD Carleton. 1993. Family Muridae, 501–755, in *Mammal Species of the World: A Taxonomic and Geographic Reference*. 2nd ed. (DE Wilson & DM Reeder, eds.). Smithsonian Institution Press, Washington, DC.

Musser, GG & MD Carleton. 2005. Superfamily Muroidea, 894–1531, in *Mammal Species of the World*. Vol. 2 (D Wilson & D Reeder, eds.). Johns Hopkins University Press, Baltimore, MD.

Mustrangi, MA & JL Patton. 1997. Phylogeography and systematics of the slender mouse opossum *Marmosops* (Marsupialia, Didelphidae). *University of California Publications in Zoology*, 130:1–86.

Mutere, FA. 1967. The breeding biology of equatorial vertebrates: reproduction in the fruit bat *Eidolon helvum* at latitude 0° 21' N. *Journal of Zoology London*, 153:153–163.

Mutscher, T & CL Tan. 2003. *Hapalemur*, bamboo or gentle lemurs, 1324–1329, in *The Natural History of Madagascar* (SM Goodman & JP Benstead, eds.). University of Chicago Press, Chicago, IL.

Myers, N. 1984. *The Primary Source: Tropical Forests and Our Future*. Norton, NY.

Mykytowycz, R. 1968. Territorial marking by rabbits. *Scientific American*, 218:116–126.

Myrcha, A, L Ryskowski, & W Walkowa. 1969. Bioenergetics of pregnancy and lactation in white mice. *Acta Theriologica*, 15:161–166.

Mzilikazi, N, BG Lovegrove, & DO Ribble. 2002. Exogenous passive heating during torpor arousal in free-ranging rock elephant shrews, *Elephantulus myurus*. *Oecologia*, 133:307–314.

Mzilikazi, N & BG Lovegrove. 2005. Daily torpor during the active phase in free-ranging rock elephant shrews (*Elephantulus myurus*). *Journal of Zoology, London*, 267:103–111.

Nabhan, GP. 2004. *Conserving Migratory Pollinators and Nectar Corridors in Western North America*. University of Arizona Press and Arizona-Sonora Desert Museum, Tucson.

Nabhan, G & K Kindscher. Bison nation. http://www.slowfoodusa.org/index. php/programs/raft_detail/publications/ (accessed March 2006).

Nabhan, GP & D Seibert. 2002. Living as if biocultural diversity matters: conservation opportunities and recommended actions. 75-80, in *Safeguarding the Uniqueness of the Colorado Plateau*. Center for Sustainable Environments, Northern Arizona University, Grand Canyon Wildlands Council, & Terralingua: Partnerships for Linguistic and Biological Diversity, Flagstaff, AZ.

Naeem, S, et al. 1994. Declining biodiversity can alter the performance of ecosystems. *Nature*, 368:734–737.

Nagy, JG, HW Steinhoff, & GM Ward. 1964. Effects of essential oils of sagebrush on deer rumen microbial function. *Journal of Wildlife Management*, 28:785–790.

Nagy, KA. 1987. Field metabolic rate and food requirement scaling in mammals and birds. *Ecological Monographs*, 57:111–128.

Nagy, KA, VH Shoemaker, & WR Costa. 1976. Water, electrolyte, and nitrogen budgets of jackrabbits (*Lepus californicus*) in the Mojave Desert. *Physiological Zoology*, 49:351–363.

Nagy, KA & CC Peterson. 1980. Scaling of water flux rates in animals. *University of California Publications in Zoology*, 120:1–172.

Nagy, TR, BA Gower, & MH Stetson. 1993. Development of collard lemmings, *Dicrostonyx groenlandicus*, is influenced by pre- and postweaning photoperiods. *Journal of Experimental Zoology*, 267:533–542.

Naiman, RJ, et al. 1994. Beaver influences on the long-term biogeochemical characteristics of boreal forest drainage networks. *Ecology*, 75:905–921.

Naples, VL. 1990. Morphological changes in the facial region and a model of dental growth and wear pattern development in *Nothrotheriops shastensis*. *Journal of Vertebrate Paleontology*, 10:372–389.

Naples, VL. 1999. Morphology, evolution, and function of feeding in the giant anteater (*Myrmecophaga tridactyla*). *Journal of Zoology, London*, 249:19–41.

Narins, PM, et al. 1992. Seismic signal transmission between burrows of the Cape mole-rat, *Georychus capensis*. *Journal of Comparative Physiology A*, 170:13–21.

National Research Council, Food, and Nutrition Board. 1968. National Academy of Science Publications.

Naumov, NP. 1936. On some peculiarities of ecological distribution of mouse-like rodents in southern Ukraine. *Zoology Zhurnal*, 15:675–696.

Naumov, NP. 1948. *Sketches of the Comparative Ecology of Mouse-like Rodents.* Isd-vo-Akademii Nauk, SSSR, Moscow.

Naylor, GJP & DC Adams. 2001. Are the fossil data really at odds with the molecular data? Morphological evidence for Cetartiodactyla phylogeny reexamined. *Systematic Biology*, 50:444–453.

Nefdt, RJC & SJ Thirgood. 1997. Lekking, resource defense, and harassment in two subspecies of lechwe antelope. *Behavioral Ecology*, 8:1–9.

Neff, NA. 1983. The basicranial anatomy of the Nimravidae (Mammalia: Carnivora): character analyses and phylogenetic inferences. Ph.D. dissertation, City University of New York, NY.

Nei, M & S Kumar. 2000. *Molecular Evolution and Phylogenetics.* Oxford University Press, NY.

Nestler, JR. 1990. Relationships between respiratory quotient and metabolic rate during entry to and arousal from daily torpor in deer mice (*Peromyscus maniculatus*). *Physiological Zoology*, 63:504–515.

Nestler, JR, GP Dieter, & BG Klokeid. 1996. Changes in total body fat during daily torpor in deer mice (*Peromyscus maniculatus*). *Journal of Mammalogy*, 77:147–154.

Neubaum, DJ, TJ O'Shea, & KR Wilson. 2006. Autumn migration and selection of rock crevices as hibernacula by big brown bats in Colorado. *Journal of Mammalogy*, 87:470–479.

Nevo, E. 1979. Adaptive convergence and divergence of subterranean mammals. *Annual Review of Ecology and Systematics*, 10:269–308.

New, DAT, M Mizell, & DL Cockroft. 1977. Growth of opossum embryos in vitro during organogenesis. *Journal of Experimental Morphology*, 41:111–123.

Ni, X, et al. 2003. A euprimate skull from the early Eocene of China. *Nature*, 427:65–68.

Nicholas, KR & CH Tyndale-Biscoe. 1985. Prolactin-dependent accumulation of lactalbumin in mammary gland explants from pregnant tammar wallaby (*Macropus eugenii*). *Journal of Endocrinology*, 106:337–342.

Nicol, S, NA Andersen, & U Mesch. 1992. Metabolic rate and ventilatory pattern in the echidna during hibernation and arousal, 150–159, in *Platypus and Echidnas* (ML Augee, ed.). Royal Zoology Society of New South Wales, Sydney, Australia.

Niethammer, G. 1970. Beobachtungen am Pyrenaen-Desman, *Galemys pyrenaica. Bonner Zoologische Beiträge*, 21:157–182.

Nikaido M, AP Rooney, & N Okada. 1999. Phylogenetic relationships among cetartiodactyls based on insertions of short and long interspersed elements: hippopotamuses are the closest extant relatives of whales. *Proceedings of the National Academy Sciences USA*, 96:10261–10266.

Nikaido, M, et al. 2001. Retroposon analysis of major cetacean lineages: the monophyly of toothed whales and the paraphyly of river dolphins. *Proceedings of National Academy of Science USA*, 98:7384–7389.

Nikolai, JC & DM Bramble. 1983. Morphological structure and function in desert heteromyid rodents. *Great Basin Naturalist*, 7:44–64.

Nishida, T & K Kawanaka. 1972. Interunit-group relationships among wild chimpanzees of the Mahali Mountains. *Kyoto University of African Studies*, 7:131–169.

Nishihara, H, et al. 2005. A retroposon analysis of Afrotherian phylogeny. *Molecular Biology and Evolution*, 22:1823–1833.

Noirot, E. 1969. Sound analysis of ultrasonic distress calls of mouse pups as a function of their age. *Animal Behaviour*, 17:340–349.

Noll-Banholzer, U. 1979a. Body temperature, oxygen consumption, evaporative water loss, and heart rate in the fennec. *Comparative Biochemistry and Physiology*, 62:585–592.

Noll-Banholzer, U. 1979b. Water balance and kidney structure of the fennec. *Comparative Biochemistry and Physiology*, 62:593–597.

Norberg, UM. 1969. An arrangement giving a stiff leading edge to the hand wing in bats. *Journal of Mammalogy*, 50:766–770.

Norberg, UM. 1972. Bat wing structures important for aerodynamics and rigidity. *Zoomorphology*, 73:45–61.

Norberg, UM. 1981. Allometry of bat wings and legs and comparison with bird wings. *Philosophical Transactions of the Royal Society of London B*, 292:359–398.

Norberg, UM. 1985a. Evolution of flight in birds: aerodynamic mechanical and ecological aspects, 293–302, in *Major Patterns in Vertebrate Evolution* (MK Hecht, PC Goody, & BM Hecht, eds.). NATO Advanced Study Series No. 14.

Norberg, UM. 1985b. Evolution of vertebrate flight: an aerodynamic model for the transition from gliding to active flight. *American Naturalist*, 126:303–327.

Norberg, UM. 1989. Ecological determinants of bat wing shape and echolocation call structure with implications for some fossil bats, 213–233, in *European Bat Research 1987* (V Hanak, T Horacek, & J Geisler, eds.). Charles University Press, Prague.

Norberg, UM. 1994. Wing design, flight performance, and habitat use in bats, 205–239, in *Ecological Morphology: Integrative Organismal Biology* (PC Wainwright & SM Reilly, eds.). University of Chicago Press, Chicago, IL.

Norberg, UM & JMV Rayner. 1987. Ecological morphology and flight in bats (Mammalia: Chiroptera): wing adaptations, flight performance, foraging strategy, and echolocation. *Philosophical Transactions of the Royal Society B*, 316:335–427.

Norman, JE & MV Ashley. 2000. Phylogenetics of Perissodactyla and tests of the molecular clock. *Journal of Molecular Evolution*, 50:11–21.

Norris, KS. 1964. Some problems of echolocation in cetaceans, 316–336, in *Marine Bioacoustics* (WN Tavolga, ed.). Pergamon Press, NY.

Norris, KS. 1968. The echolocation of marine mammals, 391–423, in *The Biology of Marine Mammals* (HT Andersen, ed.). Academic Press, NY.

Norris, KS, et al. 1961. An experimental demonstration of echolocation behavior in the porpoise, *Tursiops truncatus* (Montagu). *Biological Bulletin*, 120:163–176.

North, GJ & RE Marsh. 1999. Black-tailed jackrabbit, *Lepus californicus*, 699–701, in *The Smithsonian Book of North America Mammals* (DE Wilson & S Ruff, eds.). Smithsonian Institution Press. Washington, DC.

Novacek, MJ. 1985. Cranial evidence for rodent affinities, 59–81, in *Evolutionary Relationships of Rodents* (WP Luckett & JL Hartenberger, eds.). Plenum Press, NY.

Novacek, MJ. 1986. The skull of leptictid insectivorans and the higher-level classification of eutherian mammals. *Bulletin of American Museum of Natural History*, 183:1–111.

Novacek, MJ. 1989. Higher mammal phylogeny: the morphological-molecular synthesis, 421–435, in *The Hierarchy of Life* (B Fernholm, K Bremer, & H Jörnvall, eds.). Elsevier Science Publishers B. V. (Biomedical Division).

Novacek, MJ. 1992a. Fossils, topologies, missing data, and the higher level phylogeny of eutherian mammals. *Systematic Biology*, 41:58–73.

Novacek, MJ. 1992b. Mammalian phylogeny: shaking the tree. *Nature*, 356:121–125.

Novacek, MJ, TM Bown, & D Schankler. 1985. On the classification of the early Tertiary Erinaceomorpha (Insectivora, Mammalia). *American Museum Novitates*, 2813:1–22.

Novacek, MJ & AR Wyss. 1986. Higher-level relationships of the Recent eutherian orders: morphological evidence. *Cladistics*, 2:257–287.

Novacek, MJ, et al. 1997. Epipubic bones in eutherian mammals from the Late Cretaceous of Mongolia. *Nature*, 389:483–486.

Nowacek, SM, et al. 2004. Florida manatees, *Trichechus manatus latirostris*, respond to approaching vessels. *Biological Conservation*, 119:517–523.

Novick, A. 1955. Laryngeal muscles of the bat and production of ultrasonic sounds. *American Journal of Physiology*, 1 83:648.

Novick, A. 1958. Orientation in paleotropical bats. II: Megachiroptera. *Journal of Experimental Zoology*, 137:443–462.

Novick, A. 1970. Echolocation in bats. *Natural History*, 79(3):32–41.

Novick, A & E Gould. 1964. *Comparative Study of Echolocation in the Tenrecidae of Madagascar and Other World Insectivores*. Fort Belvoir Defense Technical Information Center.

Nowak, RM. 1999. *Walker's Mammals of the World*. Vol. 1, 6th ed. Johns Hopkins University Press, Baltimore, MD.

Nowak, RM & JL Paradiso. 1991. *Walker's Mammals of the World*. 5th ed. Vol. 1:1–642; Vol. 2:643–1629. Johns Hopkins University Press, Baltimore, MD.

Nummela, SA, et al. 2004. Eocene evolution of whale hearing. *Nature*, 430:776–778.

Nunes, S, et al. 1997. Why do female Belding's ground squirrels disperse away from food resources? *Behavioral Ecology and Sociobiology*, 40:199–207.

Oates, JF. 1984. The niche of the potto, *Perodicticus potto*. *International Journal of Primatology*, 5:51–61.

O'Brien, SJ, et al. 1985. A molecular solution to the riddle of the giant panda's phylogeny. *Nature*, 317:140–144.

Obrist, MK. 1989. Individuelle Variabilität der Echoortung, Vergleichende Freilanduntersuchungen an vier vespertilioniden Fledermausarten Kanadas. Dissertation, Ludwig-Maximilians-Universität, München.

Obrist, MK. 1995. Flexible bat echolocation: the influence of individual habitat and conspecifics on sonar signal design. *Behavioral Ecology and Sociobiology*, 36:207–219.

Obrist, MK, et al. 1993. What ears do for bats: a comparative study of pinna sound pressure transformation in Chiroptera. *Journal of Experimental Biology*, 180:119–152.

O'Connell, CE, BT Arnason, & LA Hart. 1997. Seismic transmission of elephant vocalizations and movement. *Journal of the Acoustical Society of America*, 102:3124A.

O'Connell, CE, BT Arnason, & LA Hart. 2000. Seismic properties of Asian elephant (*Elephas maximus*) vocalizations and locomotion. *Journal of the Acoustical Society of America*, 108:3066–3072.

Oda, R & N Masataka. 1996. Interspecific responses of ringtailed lemurs to playback of antipredator alarm calls of Verreaux's sifaka. *Ethology*, 102:441–453.

O'Donoghue, M, et al. 1997. Numerical responses of coyotes and lynx to the snowshoe hare cycle. *Oikos*, 80:150–162.

Odum, EP. 1971. *Fundamentals of Ecology*. 3rd ed. W. B. Saunders, Philadelphia, PA.

O'Farrell, MJ & EH Studier. 1970. Fall metabolism in relation to ambient temperatures in three species of *Myotis*. *Comparative Biochemistry and Physiology*, 35:697–703.

O'Farrell, TP. 1965. Home range and ecology of snowshoe hares in interior Alaska. *Journal of Mammalogy*, 46:406–418.

Oftedal, OT. 2002a. The mammary gland and its origin during synapsid evolution. *Journal of Mammary Gland Biology and Neoplasia*, 7:225–252.

Oftedal, OT. 2002b. The origin of lactation as a water source for parchment-shelled eggs. *Journal of Mammary Gland Biology and Neoplasia*, 7:253–266.

Oftedal, OT, WD Bowen, & DJ Boness. 1993. Energy transfer by lactating hooded seals and nutrient deposition in their pups during the 4 days from birth to weaning. *Physiological Zoology*, 66:412–436.

O'Gara, BW. 1978. *Antilocapra americana*. *Mammalian Species*, 90:1–7.

O'Gara, BW, RF Moy, & GD Bear. 1971. The annual testicular cycle and horn casting in the pronghorn (*Antilocapra americana*). *Journal of Mammalogy*, 52:537–544.

Oksanen, L, et al. 1996. Structure and dynamics of Arctic-Subarctic grazing webs in relation to primary productivity, 231–242, in *Food Webs: Integration of Patterns and Dynamics* (GA Polis & KO Winemiller, eds.). Chapman & Hall, NY.

O'Leary, MA & JH Geisler. 1999. The position of Cetacea within Mammalia: phylogenetic analysis of morphological data from extinct and extant taxa. *Systematic Biology*, 48:455–490.

O'Leary, MA & MD Uhen. 1999. The time of origin of whales and the role of behavioral changes in the terrestrial-aquatic transition. *Paleobiology*, 25:534–556.

O'Leary, MA & J Gatesy. 2008. Impact of increased character sampling on the phylogeny of Cetartiodactyla (Mammalia): combined analysis including fossils. *Cladistics*, 23:1–46.

Olesiuk, PF, MA Bigg, & GM Ellis. 1990. Life history and population dynamics of resident killer whales (*Orcinus orca*) in the coastal waters of British Columbia and Washington State. *Report of the International Whaling Commission Special Issue*, 12:209–243.

Oli, MK. 2003. Population cycles of small rodents are caused by specialist predators: or are they? *Trends in Ecology and Evolution*, 18:105–107.

Oliveira, EV & LP Bergqvist. 1998. A new Paleocene armadillo (Mammalia, Dasypodoidea) from the Itaboraí Basin, Brazil. *Asociación Paleontologica Argentina, Publicacion Especial*, 5:35–40.

Olson, DM, et al. 2001. Terrestrial ecoregions of the world: a new map of life on earth. *BioScience*, 51(11):933–938. http://www.nationalgeographic.com/wildworld/terrestrial.html.

Olson, LE & SM Goodman. 2003. Phylogeny and biogeography of tenrecs, 1235–1242, in *The Natural History of Madagascar* (SM Goodman & JP Benstead, eds.). University of Chicago Press, Chicago, IL.

Olson, LE, EJ Sargis, & RD Martin. 2005. Intraordinal phylogenetics of treeshrews (Mammalia: Scandentia) based on evidence from the mitochondrial 12S rRNA gene. *Molecular Phylogenetics and Evolution*, 35:656–673.

Oprea, M, et al. 2006. Bat predation by *Phyllostomus hastatus. Chiroptera Neotropical*, 12:255–258.

O'Riain, MJ, JUM Jarvis, & CG Faulkes. 1996. A dispersive morph in the naked mole-rat. *Nature*, 380:619–621.

O'Riain, MJ, et al. 2000. Morphological castes in a vertebrate. *Proceedings of the National Academy of Sciences USA*, 97:13194–13197.

Orians, GH. 1969. On the evolution of mating systems of birds and mammals. *American Naturalist*, 103:589–603.

Ortega, J & HT Arita. 1998. Neotropical–Nearctic limits in Middle America as determined by distributions of bats. *Journal of Mammalogy*, 79:772–783.

Ortmann, S & G Heldmaier. 2000. Regulation of body temperature and energy requirements of hibernating Alpine marmots (*Marmota marmota*). *American Journal of Physiology, Regulatory Integrative Comparative Physiology*, 278:R698–R704.

O'Shea, TJ. 1994. Manatees. *Scientific American*, 271:66–72.

O'Shea, TJ, et al. 1991. An epizootic of Florida manatees associated with a dinoflagellate bloom. *Marine Mammal Science*, 7:165–179.

O'Shea, TJ, BB Ackerman, & HF Percival. 1995. *Population Biology of the Florida Manatee* (*Trichechus manatus latirostris*). National Biological Service, Information and Technical Report 1.

O'Shea, TJ & LB Poche Jr. 2006. Aspects of underwater sound communication in Florida manatees (*Trichechus manatus latirostris*). *Journal of Mammalogy*, 87:1061–1071.

O'Shea, TJ & DK Odell. 2008. Large-scale marine ecosystem change and the conservation of marine mammals. *Journal of Mammalogy*, 89:529–533.

Ostfeld, RS. 1992. Small-mammal herbivores in a patchy environment: individual strategies and population responses, 43–74, in *Effects of Resource Distribution on Animal–Plant Interactions* (MD Hunter, T Ohgushi, & PW Price, eds.). Academic Press, NY.

Ostfeld, RS. 1997. The ecology of Lyme-disease risk. *American Scientist*, 85:338–346.

Ostfeld, RS & LL Klosterman. 1986. Demographic substructure in a California vole population inhabiting a patchy environment. *Journal of Mammalogy*, 67:693–704.

Ostfeld, RS, CG Jones, & JO Wolff. 1996. Of mice and mast, ecological connections in eastern deciduous forests. *BioScience*, 46:323–330.

Ostrowski, S, JB Williams, & K Ismael. 2003. Heterothermy and the water economy of free-living Arabian oryx (*Oryx leucoryx*). *Journal of Experimental Biology*, 206:1471–1478.

Oswald, C & PA McClure. 1990. Energetics of concurrent pregnancy and lactation in cotton rats and woodrats. *Journal of Mammalogy*, 71:500–509.

Oswald, C, et al. 1993. Lactational water balance and recycling in white-footed mice, red-back voles, and gerbils. *Journal of Mammalogy*, 74:963–970.

Otte, D. 1974. Effects and functions in the evolution of signaling systems. *Annual Review of Ecology and Systematics*, 5:385–417.

Owen, D. 2004. *Tasmanian Tiger: The Tragic Tale of How the World Lost Its Most Mysterious Predator*. Johns Hopkins University Press, Baltimore, MD.

Oxnard, E. 1981. The uniqueness of *Daubentonia. American Journal of Physical Anthropology*, 54:1–21.

Pabst, DA. 1990. Axial muscles and connective tissues of the bottlenose dolphin, 51–67, in *The Bottlenose Dolphin* (S Leatherwood & RR Reeves, eds.). Academic Press, San Diego, CA.

Pabst, DA. 1993. Intramuscular morphology and tendon geometry of the epaxial swimming muscles of dolphins. *Journal of the Zoological Society of London*, 230:159–176.

Pabst, DA, et al. 1995. Thermoregulation of the intra-abdominal testis of the bottlenose dolphin (*Tursiops truncates*) during exercise. *Journal of Experimental Biology*, 198:221–226.

Pacheco, V, et al. 2009. Diversidad y endemismo de los mamíferos del Perú. *Revista Peruana de Biología*, 16(1):5–32.

Pack, AA & LM Herman. 1995. Sensory integration in the bottlenosed dolphin: immediate recognition of complex shapes across the senses of echolocation and vision. *Journal of the Acoustical Society of America*, 98:722–733.

Packer, C. 1986. The ecology of felid sociality, 429–451, in *Ecological Aspects of Social Evolution* (DJ Rubenstein & RW Wrangham, eds.). Princeton University Press, Princeton, NJ.

Packer, C & AE Pusey. 1982. Cooperation and competition within coalitions of male lions: kin selection or game theory? *Nature*, 296:740–742.

Packer, C, et al. 1988. Reproductive success of lions, 363–383, in *Reproductive Success: Studies of Individual Variation in Contrasting Breeding Systems* (TH Clutton-Brock, ed.). University of Chicago Press, Chicago, IL.

Paddle, R. 2000. *The Last Tasmanian Tiger: The History of an Extinction of the Thylacine*. Cambridge University Press, Cambridge, UK.

Paige, KN. 1995. Bats and barometric pressure: conserving limited energy and tracking insects from the roost. *Functional Ecology*, 9:463–467.

Palacio, Z. 2005. Malaria vaccine may be on the way. Voice of America News. http://www.voanews.com/english/archive/2005-04/2005-04-23-voa2.cfm?moddate=2005-04-23 (accessed October 30, 2009).

Palacios, V, E Font, & R Marquez. 2007. Iberian wolf howls: acoustic structure, individual variation, and a comparison with North American populations. *Journal of Mammalogy*, 88:606–613.

Palma, RE, et al. 2002. Phylogenetic and biogeographic relationships of the mouse opossum *Thylamys* (Didelphimorphia, Didelphidae) in southern South America. *Molecular Phylogenetics and Evolution*, 25:245–253.

Palmeirim, JM & RS Hoffmann. 1983. *Galemys pyrenaicus. Mammalian Species*, 207:1–5.

PALNet (Protected Areas Learning Network). http://www.parksnet.org (accessed October 30, 2009).

Palombit, RA. 1994. Dynamic pair bonds in hylobatids: implications regarding monogamous social systems. *Behaviour*, 128:65–101.

Pardiñas, UFJ, G D'Elía, & PE Ortiz. 2002. Sigmodontinos fósiles (Rodentia, Muroidea, Sigmodontinae) de América del Sur: estado actual de su conocimiento y prospectivo. *Mastozoología Neotropical/Journal of Neotropical Mammalogy*, 9:209–252.

Parker, P. 1977. An ecological comparison of marsupial and placental patterns of reproduction, 273–286, in *The Biology of Marsupials* (B Stonehouse & D Gilmore, eds.). MacMillan, London.

Parkinson, CL, et al. 1999. Arctic sea ice extents, areas, and trends, 1978–1996. *Journal of Geophysical Research*, 104:20837–20856.

Parmesan, C. 2006. Ecological and evolutionary responses to recent climate change. *Annual Review of Ecology, Evolution and Systematics*, 27:637–669.

Parrish, JT. 1990. Gondwanan paleogeography and paleoclimatology, 15–26, in *Antarctic Paleobiology: Its Role in the Reconstruction of Gondwana* (TN Taylor & EL Taylor, eds.). Springer-Verlag, NY.

Parry, DA. 1949. The structure of whale blubber and its thermal properties. *Quarterly Journal of Microbiological Science*, 90:13–26.

Pascual, R. 2006. Evolution and geography: the biogeographic history of South American land mammals. *Annals of the Missouri Botanical Garden*, 93:209-230.

Pascual, R & E Ortiz Jaureguizar. 1990. Evolving climates and mammal faunas in Cenozoic South America, 23–60, in *The Platyrrhine Fossil Record* (JG Fleagle & AL Rosenberger, eds.). Academic Press, London.

Pascual, R, et al. 1992a. First discovery of monotremes in South America. *Nature*, 356:67–74.

Pascual, R, et al. 1992b. The first non-Australian monotreme: an early Paleocene South American platypus (Monotremata, Ornithorhynchidae), 2–15, in *Platypus and Echidnas* (ML Augee, ed.). Royal Zoology Society of New South Wales, Sydney, Australia.

Pascual, R, FJ Goin, & AA Carlini. 1994. New data on the Groeberiidae: unique late Eocene–early Oligocene South American marsupials. *Journal of Vertebrate Paleontology*, 14:247–259.

Pascual, R, et al. 2002. New data on the Paleocene monotreme *Monotrematum sudamericanum*, and the convergent evolution of triangulate molars. *Acta Palaeontologica Polonica*, 47:487–492.

Pascual, R & E Ortiz-Jaureguizar. 2007. The Gondwanan and South American episodes: two major and unrelated moments in the history of the South American mammals. *Journal of Mammalian Evolution*, 14:75–137.

Patterson, B. 1965. The fossil elephant shrews (Family Macroscelididae). *Bulletin Museum of Comparative Zoology, Harvard University*, 133:295–335.

Patterson, B. 1975. The fossil aardvarks (Mammalia: Tubulidentata). *Bulletin Museum of Comparative Zoology, Harvard University*, 147:185–237.

Patterson, B & R Pascual. 1972. The fossil mammal fauna of South America, 247–309, in *Evolution, Mammals, and Southern Continents* (A Keast, FC Erk, & B Glass, eds.). SUNY Press, Albany, NY.

Patterson, BD. 1995. Local extinctions and the biogeographic dynamics of boreal mammals in the Southwest, 151–176, in *Storm Over a Mountain Island: Conservation Biology and the Mount Graham Affair* (CA Istock & RS Hoffmann, eds.). University of Arizona Press, Tucson.

Patterson, N, et al. 2006. Genetic evidence for complex speciation of humans and chimpanzees. *Nature*, 441:1103–1108.

Patton, JL, et al. 1997. Diversity, differentiation, and the historical biogeography of nonvolant small mammals of the Neotropical forests, 455–465, in *Tropical Forest Remnants: Ecology, Management, and Conservation of Fragmented Communities* (WF Laurance & RO Bierregaard Jr, eds.). University of Chicago Press, Chicago, IL.

Patton, JL, MNF Da Silva, & JR Malcolm. 2000. Mammals of the Rio Juruá and the evolutionary and ecological diversification of Amazonia. *Bulletin of the American Museum of Natural History*, 244:1–306.

Pauly, D & V Christensen. 1995. Primary production required to sustain global fisheries. *Nature*, 374: 255–257.

Payne, KB, WR Langbauer Jr, & EM Thomas. 1986. Infrasonic calls of the Asian elephant (*Elephas maximus*). *Behavioral Ecology and Sociobiology*, 18:297–301.

Payne, RS. 1970. *Songs of the Humpback Whale.* An LP Record by CRM Records, Del Mar, CA.

Pearson, OP. 1948. Metabolism of small mammals with remarks on the lower limit of mammalian size. *Science*, 108:44.

Pearson, OP. 1959. Biology of the subterranean rodents, *Ctenomys*, in Peru. *Memorias del Museo de Historia Natural "Javier Prado,"* 9:1–56.

Pearson, OP. 1964. Carnivore-mouse predation: an example of its intensity and bioenergetics. *Journal of Mammalogy*, 45:177–188.

Pedersen, SC. 1998. Morphometric analysis of the chiropteran skull with regard to mode of echolocation. *Journal of Mammalogy*, 79:91–103.

Peichl, L. 2005. Diversity of mammalian photoreceptor properties: adaptation to habitat and lifestyle? *Anatomical Record Part A: Discoveries in Molecular, Cellular, and Evolutionary Biology*, 287A:1001–1012.

Pengelley, ET & KC Fisher. 1963. The effect of temperature and photoperiod on the yearly hibernating behavior of captive golden-mantled ground squirrels (*Citellus lateralis tescorum*). *Canadian Journal of Zoology*, 41:1103–1120.

Pengelley, ET & SJ Asmundson. 1970. The effect of light on the free-running circannual rhythm of the golden-mantled ground squirrel, *Citellus lateralis. Comparative Biochemistry and Physiology*, 30:177–183.

Pengelley, ET & SJ Asmundson. 1971. Annual biological clocks. *Scientific American*, 224:72–79.

Penn, DJ & WK Potts. 1999. The evolution of mating preferences and major histocompatibility complex genes. *American Naturalist*, 153:145–164.

Penny, D & MJ Phillips. 2007. Mass survivals. *Nature*, 446:501–502.

Pennycuik, PR. 1969. Reproductive performance and body weights of mice maintained for 12 generations at 34°C. *Australian Journal of Biological Science*, 22:667–675.

Pepper, JW, et al. 1991. Vocalizations of the naked mole-rat, 243–274, in *The Biology of the Naked Mole-rat* (PW Sherman, JUM Jarvis, & RD Alexander, eds.). Princeton University Press, Princeton, NJ.

Pereira, ME & MK Izard. 1989. Lactation and care for unrelated infants in forest-living ringtailed lemurs. *American Journal of Primatology*, 18:101–108.

Perret, M. 1992. Environmental and social determinants of sexual function in the male lesser mouse lemur (*Microcebus murinus*). *Folia Primatologica*, 59:1–25.

Perrin, MR. 1995a. Comparative aspects of the metabolism and thermal biology of elephant-shrews (Macroscelidea). *Mammal Review*, 25:61–78.

Perrin, MR. 1995b. The biology of elephant-shrews—a symposium held during the 6th International Theriological Congress, Sydney, 5 July 1993. *Mammal Review*, 25:1–100.

Perrin, MR, H Boyer, & DC Boyer. 1992. Diets of the hairy-footed gerbils *Gerbillurus paeba* and *G. tytonis* from the dunes of the Namib Desert. *Israel Journal of Zoology,* 38:373–383.

Perrin, MR & LJ Fielden. 1999. *Eremitalpa granti. Mammalian Species,* 629:1–4.

Perrin, N & J Goudet. 2001. Inbreeding, kinship, and the evolution of natal dispersal, 123–142, in *Dispersal* (J Clobert, E Danchin, A Dhondt, & JD Nichols, eds.). Oxford University Press, Oxford, UK.

Perrin, WF, B Würsig, & JGM Thewissen. 2002. *Encyclopedia of Marine Mammals.* Academic Press, San Diego, CA.

Peter, WP & A Feiler. 1994. Hörner von einer unbekannten bovidenart aus Vietnam (Mammalia: Ruminantia). *Faunistische Abhandlungen, Staatliches Museum für Tierkunde Dresden,* 19:247–253.

Peters, RP & LD Mech. 1975. Scent-marking in wolves. *Scientific American,* 63:628–637.

Peterson, RO. 1977. *Wolf Ecology and Prey Relationships on Isle Royale.* National Park Service Monograph Series 11.

Peterson, RO & P Ciucci. 2003. The wolf as a carnivore, 104–130, in *Wolves, Behavior, Ecology, and Conservation* (LD Mech & L Boitani, eds.). University of Chicago Press, Chicago, IL.

Peterson, RS & GA Bartholomew. 1967. *The Natural History and Behavior of the California Sea Lion.* Special Publication No. 1, American Society of Mammalogists.

Petter, JJ. 1962. Ecological and behavioral studies of Madagascar lemurs in the field. *Annals of the N.Y. Academy of Sciences,* 102:267–281.

Philippe, H. 1997. Rodent monophyly: pitfalls of molecular phylogenies. *Journal of Molecular Evolution,* 45:712–715.

Pianka, ER. 1976. Natural selection of optimal reproductive strategies. *American Zoologist,* 16:775–787.

Picard, K, et al. 1994. Bovid horns: an important site of heat loss during winter? *Journal of Mammalogy,* 75:710–713.

Pielou, EC. 1991. *After the Ice Age—The Return of Life to Glaciated North America.* University of Chicago Press, Chicago, IL.

Pimm, SL, JH Lawton, & JE Cohen. 1991. Food web patterns and their consequences. *Nature,* 350:669–674.

Pimm, SL, et al. 2001. Can we defy nature's end? *Science,* 293:2207–2208.

Pine, RH, et al. 1985. Labile pigments and fluorescent pelage in didelphid marsupials. *Mammalia,* 49:249–256.

Pitman, WC, III, et al. 1993. Fragmentation of Gondwana: the separation of Africa from South America, 15–34, in B*iological Relationships Between Africa and South America* (P. Goldblatt, ed.). Yale University Press, New Haven, CT.

Pivorunas, A. 1979. The feeding mechanisms of baleen whales. *American Scientist,* 67:432–440.

Plumptre, AJ, et al. 2003. The current status of gorillas and threats to their existence at the beginning of a new millennium, 414–431, in *Gorilla Biology: A Multi-disciplinary Perspective* (AB Taylor & ML Goldsmith, eds.). Cambridge University Press, Cambridge, UK.

Pochron, ST & PC Wright. 2005. Dance of the sexes. *Natural History,* 6:35–39.

Poinar, HN, et al. 1998. Molecular coproscopy: dung and diet of the extinct ground sloth *Nothrotheriops shastensis. Science,* 281:402–406.

Pokki, J. 1981. Distribution, demography, and dispersal of the field vole, *Microtus agrestis* (L.), in the Tvaerminne Archipelago, Finland. *Acta Zoologica Fennica,* 164:1–48.

Pollak, GD & JH Casseday. 1989. *The Neural Basis of Echolocation in Bats.* Springer-Verlag, Berlin.

Pond, CM. 1977. The significance of lactation in the evolution of mammals. *Evolution,* 31:177–199.

Poole, JH. 1987. Elephants in musth, lust. *Natural History,* 11:46–55.

Poole, JH. 1989. Mate guarding, reproductive success and female choice in African elephants. *Animal Behaviour,* 37:842–849.

Poole, JH & CJ Moss. 1981. Musth in the African elephant (*Loxodonta africana*). *Nature,* 292:830–831.

Poole, JH, et al. 1988. The social contexts of some very low frequency calls of African elephants. *Behavioral Ecology and Sociobiology,* 22:385–392.

Popa-Lisseanu, AG, et al. 2007. Bats' conquest of a formidable foraging niche: the myriads of nocturnally migrating songbirds. *PLoS One,* 2(2):e205.

Porter, CA, et al. 1994. Evidence on primate phylogeny from e-globin gene sequence and flanking regions. *Journal of Molecular Evolution,* 40:30–55.

Post, D & OJ Reichman. 1991. Effects of food perishability, distance, and competitors on caching behavior by eastern woodrats. *Journal of Mammalogy,* 72:513–517.

Potts, WK, CJ Manning, & EK Wakeland. 1991. Mhc genotype influences mating patterns in semi-natural populations of *Mus Nature,* 352:619–621.

Pough, FH, CM Janis, & JB Heiser. 1998. *Vertebrate Life.* MacMillan, NY.

Pournelle, GH. 1968. Classification, biology, and description of the venom apparatus of insectivores of the genera *Solenodon, Neomys,* and *Blarina,* 31–42, in *Venomous Animals and Their Venoms* (W Bucherl, EA Buckley, & V Deulofeu, eds.). Academic Press, NY.

Poux, C & EJP Douzery. 2004. Primate phylogeny, evolutionary rate variations, and divergence times: a contribution from the nuclear gene IRBP. *American Journal of Physical Anthropology,* 124:1–16.

Poux, C, et al. 2005. Asynchronous colonization of Madagascar by the four endemic clades of primates, tenrecs, carnivores, and rodents as inferred from nuclear genes. *Systematic Biology,* 54:719–730.

Powers, JB & SS Winans. 1975. Vomeronasal organ: critical role in mediating sexual behavior of the male hamster. *Science,* 187:961–963.

Prakash, I. 1959. Foods of the Indian false vampire. *Journal of Mammalogy,* 40:545–547.

Price, J & S Allen. 2004. Exploring the mechanisms regulating regeneration of deer antlers. *Philosophical Transactions of the Royal Society, London B,* 359:809–822.

Price, JS, et al. 2005. Deer antlers: a zoological curiosity or the key to understanding organ regeneration in mammals? *Journal of Anatomy,* 207:603–618.

Price, MV. 1978. The role of microhabitat in structuring desert rodent communities. *Ecology,* 59:910–921.

Price, MV. 1983. Laboratory studies of seed size and species selection by heteromyid rodents. *Oecologia,* 60:259–263.

Price, MV, NW Waser, & TA Bass. 1984. Effects of moonlight on microhabitat use by desert rodents. *Journal of Mammalogy,* 65:353–356.

Price, SA, ORP Bininda-Emonds, & JL Gittleman. 2005. A complete phylogeny of the whales, dolphins and even-toed hoofed mammals (Cetartiodactyla). *Biological Reviews,* 80:445–473.

Pridham, J. 1965. *Enzyme Chemistry of Phenolic Compounds.* MacMillan, NY.

Proctor-Grey, E. 1984. Dietary ecology of the coppery brushtail possum, green ringtail possum and Lumholtz's tree-kangaroo in North Queensland, 129–135, in *Possums and Gliders* (AP Smith & ID Hume, eds.). Australian Mammal Society and Surrey Beatty and Sons, Sydney, Australia.

Prothero, DR. 1994a. The Late Eocene–Oligocene extinctions. *Annual Review Earth Planetary Science*, 22:145–165.

Prothero, DR. 1994b. *The Eocene–Oligocene Transition: Paradise Lost.* Columbia University Press, NY.

Prothero, DR. 2005. *The Evolution of North American Rhinoceroses.* Cambridge University Press, NY.

Prothero, DR. 2006. *After the Dinosaurs: The Age of Mammals.* Indiana University Press, Bloomington, IN.

Prothero, DR. 2007. Family Moschidae, 221–226, in *The Evolution of Artiodactyls* (DR Prothero & SE Foss, eds.). Johns Hopkins University Press, Baltimore, MD.

Prothero, DR & WA Berggren. 1992. *Eocene–Oligocene Climatic and Biotic Evolution.* Princeton University Press, Princeton, NJ.

Prothero, DR & SE Foss. 2007 Summary, 303–315, in *The Evolution of Artiodactyls* (DR Prothero & SE Foss, eds.). Johns Hopkins University Press, Baltimore, MD.

Pryor, K & KS Norris. 1998. *Dolphin Societies: Discoveries and Puzzles.* University of California Press, Berkeley, CA.

Pucek, M. 1968. Chemistry and pharmacology of insectivore venoms, 43–50, in *Venomous Animals and Their Venoms* (W Bucherl, EA Buckley, & V Deulofeu, eds.). Academic Press, NY.

Puechmaille, SJ, et al. 2008. Characterization and multiplex genotyping of 16 polymorphic microsatellite loci in the endangered bumble-bee bat, *Craseonycteris thonglongyai* (Chiroptera: Craseonycteridae). *Conservation Genetics*, 10:1073–1076.

Pujos, F, et al. 2007. A peculiar climbing Megalonychidae from the Pleistocene of Peru and its implication for sloth history. *Zoological Journal of the Linnean Society*, 149:179–235.

Pusey, AE & C Packer. 1987. The evolution of sex-biased dispersal in lions. *Behaviour*, 101:275–310.

Pyare, S & WS Longland. 2001. Patterns of ectomycrorrhizal fungi consumption by small mammals in remnant old-growth forests of the Sierra Nevada. *Journal of Mammalogy*, 82:681–689.

Pye, JD. 1988. Noseleaves and bat pulses, 791–796, in *Animal Sonar Systems: Processes and Performances* (PA Nachtgall & PWB Moore, eds.). Plenum Press, NY.

Quilliam, TA. 1966. The mole's sensory apparatus. *Journal of Zoology, London*, 149:76–78.

Quinn, TH & JJ Baumel. 1993. Chiropteran tendon locking mechanism. *Journal of Morphology*, 216:197–208.

Quintana, RD, S Monge, & AI Malvárez. 1998. Feeding patterns of capybara *Hydrochaeris hydrochaeris* (Rodentia, Hydrochaeridae) and cattle in the non-insular area of the lower delta of the Paraná River, Argentina. *Mammalia*, 62:37–52.

Rabinowitz, PD, MF Coffin, & D Falvey. 1983. The separation of Madagascar and Africa. *Science*, 220: 67–69.

Racey, DN, M Peaker, & PA Racey. 2009. Galactorrhoea is not lactation. *Trends in Ecology and Evolution*, 24(7):354.

Racey, PA. 1973. Environmental factors affecting the length of gestation in heterothermic bats. *Journal of Reproduction and Fertility*, 19(suppl):175–189.

Racey, PA. 1982. The ecology of reproduction, 57–104, in *Ecology of Bats* (TH Kunz, ed.). Plenum Press, NY.

Racey, PA & PJ Stephenson. 1996. Reproductive and energetic differentiation of the Tenrecidae of Madagascar. *Biogeographie de Madagascar*, 307–319.

Radespiel, U, et al. 2008. Exceptional diversity of mouse lemurs (*Microcebus* spp.) in the Makira region with the description of one new species. *American Journal of Primatology*, 70:1033–1046.

Radinsky, LB. 1966. The adaptive radiation of the phenacodontid condylarths and the origin of the Perissodactyla. *Evolution*, 20:408–417.

Radinsky, LB. 1984. Ontogeny and phylogeny in horse skull evolution. *Evolution*, 38:1–15.

Rado, R. et al. 1987. Seismic signalling as a means of communication in a subterranean mammal. *Animal Behavior*, 35:1249–1251.

Rageot, R. 1978. Observaciones sobre el monito del monte. *Chile Min. Agric. Corp. Nac. For. Dept. Tecn.* IX. 1–16.

Raghuram, H & G Marimuthu. 2007. Maternal feeding of offspring with vertebrate prey in captive Indian false vampire bat, *Megaderma lyra. Acta Chiropterologica*, 9:437–443.

Rahm, U. 1969. Notes sur la cri du *Dendrohyrax dorsalis* (Hyracoidea). *Mammalia*, 33:68–79.

Rahm, U. 1970. Note sur la reproduction des sciuridés et muridés dans forêt équatoriale au Congo. *Revue Suisse Zoologie*, 77:635–646.

Rajemison, B & SM Goodman. 2007. The diet of *Myzopoda schliemanni*, a recently discovered Malagasy endemic, based on scat analysis. *Acta Chiropterologica*, 9:311–313.

Ralls, K. 1971. Mammalian scent marking. *Science*, 171:443–449.

Ralls, K, PH Harvey, & AM Lyles. 1986. Inbreeding in natural populations of birds and mammals, 35–56, in *Conservation Biology: The Science of Scarcity and Diversity* (M Soule , ed.). Sinauer Associates, Sunderland MA.

Ralls, K, JD Ballou, & A Templeton. 1988. Estimates of lethal equivalents and the cost of inbreeding in mammals. *Conservation Biology*, 2:185–193.

Rambaldini, DA & RM Brigham. 2008. Torpor use by free-ranging pallid bats (*Antrozous pallidus*) at the northern extent of their range. *Journal of Mammalogy*, 89:933–941.

Ramm, SA, GA Parker, & P Stockley. 2005. Sperm competition and the evolution of male reproductive anatomy in rodents. *Proceedings of the Royal Society B*, 272:949–955.

Ramos Pereira, MJ, et al. 2006. Status of the world's smallest mammal, the bumble-bee bat *Craseonycteris thonglongyai*, in Myanmar. *Oryx*, 40:456–464.

Randall, JA. 1989. Individual footdrumming signatures in bannertailed kangaroo rats, *Dipodomys spectabilis. Animal Behaviour*, 38:620–630.

Randall, JA. 1993. Behavioral adaptations of desert rodents (Heteromyidae). *Animal Behaviour*, 45:263–287.

Randall, JA. 1995. Modification of footdrumming signatures by kangaroo rats: changing territories and gaining new neighbors. *Animal Behaviour*, 49:1227–1237.

Randall, JA. 1997. Species-specific footdrumming in kangaroo rats: *Dipodomys ingens, D. deserti, D. spectabilis. Animal Behaviour*, 54:1167–1175.

Randall, JA & MD Matocq. 1997. Why do kangaroo rats (*Dipodomys spectabilis*) footdrum at snakes? *Behavioral Ecology*, 8:404–413.

Randolph, PA, et al. 1977. Energy costs of reproduction in the cotton rat, *Sigmodon hispidus. Ecology*, 58:31–45.

Ranta, E, et al. 1997. Solar activity and hare dynamics: a cross-continental comparison. *American Naturalist,* 149:765–775.

Rasa, OEA. 1985. *Mongoose Watch.* John Murray, London.

Rasmussen, DT. 1998. Evolutionary history of lorisiform primates. *Folia Primatologica,* 69 (suppl 1):250–285.

Rasmussen, K, et al. 2007. Southern Hemisphere humpback whales wintering off Central America: insights from water temperature into the longest mammalian migration. *Biology Letters,* 3:302–305.

Rasmussen, LEL, AJ Hall-Martin, & DL Hess. 1996. Chemical profiles of male African elephants, *Loxodonta africana*: physiological and ecological implications. *Journal of Mammalogy,* 77:422–439.

Rasmussen, LEL, H Riddle, & V Krishnamurthy. 2002. Mellifluous matures to malodorous in musth. *Nature,* 415:975–976.

Rasmussen, LEL & DR Greenwood. 2003. Frontalin: a chemical message of musth in Asian elephants (*Elephas maximus*), *Journal of Chemical Senses,* 28:433–446.

Rasweiler, JJ & NK Badwaik. 1997. Delayed development in the short-tailed fruit bat, *Carollia perspicillata. Journal of Reproduction and Fertility,* 109:7–20.

Rathbun, G. 1978. Evolution of the rump region in the golden-rumped elephant-shrew. *Bulletin of Carnegie Museum of Natural History,* 6:11–19.

Rathbun, G. 1979. The social structure and ecology of the elephant-shrews. *Zeitschrift fur Tierpsychologie,* 20:1–79.

Rathbun, GB & S Kyalo. 2000. Golden-rumped elephant-shrew, 125–129, 340–341, in *Endangered Animals—Conflicting Issues* (RP Reading & BJ Miller, eds.). Greenwood Press, Westport, CT.

Rauhut, OWM, et al. 2002. A Jurassic mammal from South America. *Nature,* 416:165–168.

Rausch, R. 1950. Observations on a cyclic decline of lemmings (*Lemmus*) on the Arctic coast of Alaska during the spring of 1949. *Arctic,* 3:166–177.

Rayner, JMV. 1991a. Complexity and a coupled system: flight, echolocation and evolution in bats, 173–190, in *Constructional Morphology* (N Schmidt-Kittler & K Vogel, eds.). Springer-Verlag, Berlin.

Rayner, JMV. 1991b. Echolocation—the cost of being a bat. *Nature,* 350:383–384.

Rayor, LS. 1988. Social organization and space-use in Gunnison's prairie dog.

Behavioral Ecology and Sociobiology, 22:69–78.

Reaka-Kudla, ML, DE Wilson, & EO Wilson. 1997. *Biodiversity II. Understanding and Protecting Our Biological Resources.* Joseph Henry Press, Washington, DC.

Rebar, CE. 1995. Ability of *Dipodomys merriami* and *Chaetodipus intermedius* to locate resource distributions. *Journal of Mammalogy,* 76:437–447.

Rebar, C & OJ Reichman. 1983. Ingestion of moldy seeds by heteromyid rodents. *Journal of Mammalogy,* 64:713–715.

Redford, KH. 1987. Ants and termites as food: patterns of mammalian myrmecophagy, 349–399, in *Current Mammalogy* (HH Genoways, ed.). Plenum Press, NY.

Redford, KH & JG Dorea. 1984. The nutritional value of invertebrates with emphasis on ants and termites as food for mammals. *Journal of Zoology London,* 203:385–395.

Redman, JP & JA Sealander. 1958. Home ranges of deer mice in southern Arkansas, *Journal of Mammalogy,* 39:390–395.

Reed, CA. 1944. Behavior of a shrew mole in captivity. *Journal of Mammalogy,* 25:196–198.

Reeve, N. 1994. *Hedgehogs.* T & AD Poyser, London.

Regal, PJ. 1977. Ecology and evolution of flowering plant dominance. *Science,* 196:622–629.

Reichman, OJ. 1975. Relation of desert rodent diets to available resources. *Journal of Mammalogy,* 56:731–751.

Reichman, OJ. 1977. Optimization of diet through food preferences by heteromyid rodents. *Ecology,* 58:454–457.

Reichman, OJ. 1988a. Caching behavior by eastern woodrats, *Neotoma floridana,* in relation to food perishability. *Animal Behaviour,* 36:1525–1532.

Reichman, OJ. 1988b. Comparisons of the effects of crowding and pocket gopher disturbance on mortality, growth, and seed production of *Berteroa incana. American Midland Naturalist,* 120:57–69.

Reichman, OJ. 1991. Desert mammal communities, 311–347, in *The Ecology of Desert Communities* (G Polis, ed.). University of Arizona Press, Tucson.

Reichman, OJ & KM Van De Graaff. 1975. Influence of green vegetation on desert rodent reproduction. *Journal of Mammalogy,* 53:503–506.

Reichman, OJ, TG Whitham, & GA Ruffner. 1982. Adaptive geometry of burrow

spacing in two pocket gopher populations. *Ecology,* 63:687–695.

Reichman, OJ & P Fay. 1983. Comparisons of the diets of a caching and a noncaching rodent. *American Naturalist,* 122:576–581.

Reichman, OJ & C Rebar. 1985. Seed preferences by desert rodents based on levels of moldiness. *Animal Behavior,* 33:726–729.

Reichman, OJ, DT Wicklow, & C Rebar. 1985. Ecological and mycological characteristics of caches in the mounds of *Dipodomys spectabilis. Journal of Mammalogy,* 66:643–651.

Reichman, OJ, A Fattaey, & K Fattaey. 1986. Management of sterile and moldy seeds by a desert rodent. *Animal Behaviour,* 34:221–225.

Reid, DG, CJ Krebs, & A Kenney. 1995. Limitation of collared lemming population growth at low densities by predation mortality. *Oikos,* 73: 387–398.

Reid, RT. 1970. The future role of ruminants in animal production, in *Physiology of Digestion and Metabolism in the Ruminant* (AT Phillipson, ed.). Oriel Press, Newcastle-upon-Tyne, England.

Reig, OA. 1970. Ecological notes on the fossorial octodontid rodent *Spalacopus cyanus* (Molina). *Journal of Mammalogy,* 51:592–601.

Reig, OA. 1980. A new fossil genus of South American cricetid rodents allied to *Wiedomys,* with an assessment of the Sigmodontinae. *Journal of Zoology, London,* 192:257–281.

Reijnders, PJH. 1978. Recruitment in the harbour seal (*Phoca vitulina*) population in the Dutch Wadden Sea. *Netherlands Journal of Sea Research,* 12:164–179.

Reiss, KZ. 1997. Myology of the feeding apparatus of myrmecophagid anteaters (Xenarthra: Myrmecophagidae). *Journal of Mammalian Evolution,* 4:87–117.

Reiss, KZ. 2001. Using phylogenies to study convergence: the case of the ant-eating mammals. *American Zoologist,* 41:507–525.

Reiter, RJ. 1983. The role of light and age in determining melatonin production in the pineal gland, 227–241, in *The Pineal Gland and Its Endocrine Role* (J Axelrod, F Fraschini, & GP Velo, eds.). Plenum Press, NY.

Renfree, MB. 1981. Embryonic diapause in marsupials. *Journal of Reproduction and Fertility,* 29(suppl): 67–78.

Renfree, MB. 1993. Ontogeny, genetic control, and phylogeny of female reproduction in monotreme and therian mammals, 4–20, in *Mammal Phylogeny. Vol. 1, Mesozoic Differentiation, Multituberculates, Monotremes, Early Therians, and Marsupials* (FS Szalay, MJ Novacek, & MC McKenna, eds.). Springer-Verlag, NY.

Renfree, MB, et al. 1981. Abolition of seasonal embryonic diapause in a wallaby by pineal denervation. *Nature*, 293:138–139.

Renouf, D & MB Davis. 1982. Evidence that seals may use echolocation. *Nature*, 300:635–637.

Rens, W, et al. 2004. Resolution and evolution of the duck-billed platypus karyotype with an X1Y1X2Y2X3Y3X4Y4X5Y5 male sex chromosome constitution. *Proceedings of the National Academy of Sciences USA*, 101:16257–16261.

Rentz, DC. 1975. Two new katydids of the genus *Melanonotus* from Costa Rica with comments on their life history strategies (Tettigoniidae: Pseudohyllinae). *Entomological News*, 86:129–140.

Repenning, CA. 1976. Adaptive evolution of sea lions and walruses. *Systematic Zoology*, 25:375–390.

Repenning, CA, CE Ray, & D Grigorescuy. 1979. Pinniped biogeography, 357–369, in *Historical Biogeography, Plate Tectonics, and the Changing Environment* (J Gray & AJ Boucot, eds.). Oregon State University Press, Corvallis.

Reppert, SM & DR Weaver. 2002. Coordination of circadian timing in mammals. *Nature*, 418:935–941.

Rhoades, DF. 1979. Evolution of plant chemical defenses against herbivores, 3–54, in *Herbivores: Their Interactions with Plant Secondary Metabolites* (GA Rosenthal & DH Janzen, eds.). Academic Press, NY.

Rich, TH. 2008. The palaeobiogeography of Mesozoic mammals: a review. *Arquivos do Museu Nacional, Rio de Janeiro*, 66(1):231–249.

Rich, TH, et al. 2001. Monotreme nature of the Australian Early Cretaceous mammal *Teinolophus*. *Acta Palaeontologica Polonica*, 46:113–118.

Richard, PB. 1982. La sensibilité tactile de contact chez le desman (*Galemys pyrenaicus*). *Biology of Behavior*, 7:325–336.

Richmond, ME & RA Stehn. 1976. Olfaction and reproductive behavior in microtine rodents, 197–217, in *Mammalian Olfaction, Reproductive Processes, and Behavior* (RL Doty, ed.). Academic Press, NY.

Rickart, EA. 2001. Elevational diversity gradients, biogeography, and the structure of montane mammal communities in the intermountain region of North America. *Global Ecology & Biogeography*, 10:77–100.

Riddle, BR. 2008. What is modern biogeography without phylogeography? *Journal of Biogeography*, 2009. 36:1–2.

Riddle, BR, et al. 2000. Cryptic vicariance in the historical assembly of a Baja California peninsular desert biota. *Proceedings of the National Academy of Sciences USA*, 97(26):14438–14443.

Riddle, BR & DJ Hafner. 2007. Phylogeography in historical biogeography: investigating the biogeographic histories of populations, species, and young biotas, 161–176, in *Biogeography in a Changing World* (MC Ebach & RS Tangney, eds.). CRC Press, Boca Raton, FL.

Riddle, BR, et al. 2008. The role of molecular genetics in sculpting the future of integrative biogeography. *Progress in Physical Geography*, 32(2):173–202.

Rincón, AD & RD White. 2007. Los Xenarthra Cingulata del Pleistoceno tardio (Lujanense) de Cerro Misión, estado Falcón, Venezuela. *Boletín Soc. Venezolana Espeleol.* 41:2–12.

Rinderknecht, A & RE Blanco. 2008. The largest fossil rodent. *Proceedings of the Royal Society B*, 275:923–928.

Rinker, GC. 1954. The comparative myology of the mammalian genera *Sigmodon*, *Oryzomys*, *Neotoma*, and *Peromyscus* (Cricetinae), with remarks on their intergeneric relationships. *Miscellaneous Publications of the Museum of Zoology University of Michigan*, 83:1–124.

Riskin, DK, JEA Bertram, & JW Hermanson. 2005. Testing the hindlimb-strength hypothesis: nonaerial locomotion by Chiroptera is not constrained by the dimensions of the femur or tibia. *Journal of Experimental Biology*, 208:1309–1319.

Riskin, DK, et al. 2006. Terrestrial locomotion of the New Zealand short-tailed bat *Mystacina tuberculata* and the common vampire bat *Desmodus rotundus*. *Journal of Experimental Biology*, 209:1725–1736.

Rissman, EF & RE Johnson. 1985. Female reproductive development is not activated by male California voles exposed to family cues. *Biology of Reproduction*, 32:352–360.

Roberts, MS & JL Gittleman. 1984. *Mammalian Species*. Number 201–306. The American Society of Mammalogists.

Robichaud, WG. 1998. Physical and behavioral description of a captive saola, *Pseudoryx nghetinhensis*. *Journal of Mammalogy*, 79:394–405.

Robinson, JG. 1981. Spatial structure in foraging groups of wedge-capped capuchin monkeys (*Cebus nigrivittatus*). *Animal Behaviour*, 29:1036–1056.

Robinson, JG & KH Redford. 1991. *Neotropical Wildlife Use and Conservation*. University of Chicago Press, Chicago, IL.

Robinson, K & DHK Lee. 1941. Reactions of the dog to hot atmospheres. *Proceedings of the Royal Society of Queensland*, 53:159–170.

Roca, AL, et al. 2004. Mesozoic origin for West Indian insectivores. *Nature*, 429:649–651.

Rodger, JC & JM Bedford. 1982a. Induction of oestrus, recovery of gametes, and the timing of fertilization events in the opossum, *Didelphis virginiana*. *Journal of Reproduction and Fertility*, 64:159–169.

Rodger, JC & JM Bedford. 1982b. Separation of sperm pairs and sperm–egg interaction in the opossum, *Didelphis virginiana*. *Journal of Reproduction and Fertility*, 64:171–179.

Rodrigues, FHG, et al. 2008. Anteater behavior and ecology, 257–268, in *The Biology of the Xenarthra* (SF Vizcaíno & WJ Loughry, eds.). University Press of Florida, Gainesville.

Rodrigues Nogueira, M, L Rabello Monteiro, & AL Peracchi. 2006. New evidence of bat predation by the wooly false vampire bat *Chrotopterus auritus*. *Chiroptera Neotropical*, 12:286–288.

Rodríguez, JP, et al. 2007. Globalization of conservation: a view from the South. *Science*, 317:755–756.

Rodríguez-Durán, A & JA Soto-Centeno. 2003. Temperature selection by tropical bats roosting in caves. *Journal of Thermal Biology*, 28:465–468.

Rodríguez-Herrera, B, RA Medellín, & M Gamba-Ríos. 2008. Roosting requirements of white tent-making bat *Ectophylla alba* (Chiroptera: Phyllostomidae). *Acta Chiropterologica*, 10:89–95.

Roeder, KD. 1965. Moths and ultrasound. *Scientific American*, 212:94–102.

Roeder, KD & AE Treat. 1961. The detection and evasion of bats by moths. *American Scientist*, 49:135–148.

Rogers, L. 1981. A bear in its lair. *Natural History*, 90:64–70.

Rohde, K. 1992. Latitudinal gradients in species diversity: the search for the primary cause. *Oikos*, 65:514–527.

Rojas Bracho, L & A Jaramillo Legorreta. 1999. Familia Phocoenidae, 627–633, in *Mamíferos del Noroeste de México II* (ST Alvarez-Castañeda & JL Patton, eds.). Centro de Investigaciones Biológicas del Noroeste, S.C. La Paz, Baja California Sur.

Romer, AS. 1966. *Vertebrate Paleontology.* University of Chicago Press, Chicago, IL.

Rommel, SA, et al. 2006. Elements of beaked whale anatomy and diving physiology and some hypothetical causes of sonar-related stranding. *Journal of Cetacean Research and Management,* 7:189–209.

Rood, JP. 1970a. Notes on the behavior of the pygmy armadillo. *Journal of Mammalogy,* 51:179.

Rood, JP. 1970b. Ecology and social behavior of the desert cavy (*Microcavia australis*). *American Midland Naturalist,* 83:415–454.

Rood, JP. 1972. Ecological and behavioral comparisons of three genera of Argentine cavies. *Animal Behavior Monographs,* 5:1–83.

Rood, JP. 1978. Dwarf mongoose helpers at the den. *Zeitschrift fur Tierpsychologie,* 48:277–287.

Rood, JP. 1980. Mating relationships and breeding suppression in the dwarf mongoose. *Animal Behaviour,* 28:143–150.

Rose, J, et al. 1984. Induction of winter fur growth in mink (*Mustela vison*) with melatonin. *Journal of Animal Science,* 58:57–61.

Rose, KD. 1982. Skeleton of *Diacodexis*, oldest known artiodactyl. *Science,* 216:621–623.

Rose, KD. 1996. On the origin of the order Artiodactyla. *Proceedings of the National Academy of Science USA,* 93:1705–1709.

Rose, KD. 2006. *The Beginning of the Age of Mammals.* Johns Hopkins University Press, Baltimore, MD.

Rose, KD & EL Simons, 1977. Dental function in the Plagiomenidae: origin and relationships of the mammalian order Dermoptera. *University of Michigan Contributions from the Museum of Paleontology,* 24:221–236.

Rose, KD, A Walker, & LL Jacobs. 1981. Function of the mandibular tooth comb in living and extinct mammals. *Nature,* 289:583–585.

Rose, KD & RJ Emry. 1993. Relationships of Xenarthra, Pholidota, and fossil "edentates": the morphological evidence, 81–102, in *Mammal Phylogeny: Placentals* (FS Szalay, MJ Novacek, & MC McKenna, eds.). Springer-Verlag, NY.

Rose, KD & JD Archibald. 2005. *The Rise of Placental Mammals.* Johns Hopkins University Press, Baltimore, MD.

Rose, KD, et al. 2005. Xenarthra and Pholidota, 106–126, in *The Rise of Placental Mammals: Origins and Relationships of the Major Extant Clades* (KD Rose & JD Archibald, eds.). Johns Hopkins University Press, Baltimore, MD.

Rosenberg, HI & KC Richardson. 1995. Cephalic morphology of the honey possum, *Tarsipes rostratus* (Marsupialia: Tarsipedidae); an obligate nectarivore. *Journal of Morphology,* 223:303–323.

Rosenberger, AL. 1984. Aspects of the systematics and evolution of the marmosets, 159–180, in *Primatologica no Brazil* (MT deMello, ed.). Anais do 1. Congresso Brasileiro de Primatologia, Sociedad de Primatologia.

Rosenthal, MA. 1975. Observations on the water opossum or yapok *Chironectes minimus* in captivity. *International Zoo Yearbook,* 15:4–6.

Rosenzweig, ML. 1973. Habitat selection experiments with a pair of coexisting heteromyid species. *Ecology,* 54:111–117.

Rosenzweig, ML. 1992. Species diversity gradients: we know more and less than we thought. *Journal of Mammalogy,* 73:715–730.

Rosenzweig, ML. 1995. *Species Diversity in Space and Time.* Cambridge University Press, NY.

Rosevear, DR. 1969. *The Rodents of West Africa.* British Museum Natural History, London.

Ross, CF & R Kay. 2004. Anthropoid origins: retrospective and prospective, 699–737, in *Anthropoid Origins: New Visions* (CF Ross & RF Kay, eds.). Kluwer/Plenum Publishing, NY.

Rossie, JB, X Ni, & KC Beard. 2006. Cranial remains of an Eocene tarsier. *Proceedings of the National Academy of Sciences USA,* 103:4381–4385.

Rossiter, SJ, et al. 2005. Mate fidelity and intra-lineage polygyny in greater horseshoe bats. *Nature,* 437:408–411.

Rössner, GE. 2007. Family Tragulidae, 213–220, in *The Evolution of Artiodactyls* (DR Prothero & SE Foss, eds.). Johns Hopkins University Press, Baltimore, MD.

Rovero1, F, et al. 2008. A new species of giant sengi or elephant-shrew (genus *Rhynchocyon*) highlights the exceptional biodiversity of the Udzungwa Mountains of Tanzania. *Journal of Zoology,* 274:126–133.

Rowan, W. 1950. Winter habits and numbers of timber wolves. *Journal of Mammalogy,* 31:167–169.

Rowcliffe, J, et al. 2008. Estimating animal density using camera traps without the need for individual recognition. *Journal of Applied Ecology* 45:1228–1236.

Rowe, MJ & RC Bohringer. 1992. Functional organization of the cerebral cortex in monotremes, 177–193, in *Platypus and Echidnas* (ML Augee, ed.). Royal Zoology Society of New South Wales, Sydney, Australia.

Rowe, T. 1988. Definition, diagnosis and origin of Mammalia. *Journal of Vertebrate Paleontology,* 8:241–264.

Rowe, T. 1993. Phylogenetic systematics and the early history of mammals, 129–145, in *Mammal Phylogeny: Mesozoic Differentiation, Multituberculates, Monotremes, Early Therians, and Marsupials* (FS Szalay, MJ Novacek, & MC McKenna, eds.). Springer-Verlag, NY.

Rowe, T. 1996a. Brain heterochrony and origin of the mammalian middle ear, 71–95, in *New Perspectives on the History of Life* (MT Ghiselin & G Pinna, eds.). Memoirs of the California Academy of Sciences Number 20, California Academy of Sciences, San Francisco, CA.

Rowe, T. 1996b. Coevolution of the mammalian middle ear and neocortex. *Science,* 273:651–654.

Rowe, T & J Gauthier. 1992. Ancestry, Paleontology, and definition of the name Mammalia. *Systematic Biology,* 41:372–378.

Rowe, T, et al. 2008. The oldest platypus and its baring on divergence timing of the platypus and echidna clades. *Proceedings of the National Academy of Sciences USA,* 105:1238–1242.

Rowell, TE. 1962. Agonistic noises of the rhesus monkey (*Macaca mulatta*). *Symposium of the Zoological Society London,* 8:91–96.

Ruberstein, DI. 1989. Life history and social organization in arid adapted ungulates. *Journal of Arid Environments,* 17:145–156.

Rubenstein, DI & M Hack. 2004. Natural and sexual selection and the evolution of multi-level societies: insights from zebras with comparisons to primates, 266–279, in *Sexual Selection in Primates: New and Comparative Perspectives* (PM Kappleler & CP vanSchaik, eds.). Cambridge University Press, Cambridge, UK.

Rubenstein, DI & RW Wrangham. 2006. *Ecological Aspects of Social Evolution: Birds and Mammals.* Princeton University Press, Princeton, NJ.

Rubsamen, K, ID Hume, & W von Engelhardt. 1982. Physiology of the rock hyrax. *Comparative Biochemistry and Physiology A,* 72:271–277.

Rudran, R. 1973. Adult male replacement in one-male troops of purple-faced langurs (*Presbytis senex senex*) and its effect on population structure. *Folia Primatologica,* 19:166–192.

Rudran, R. 1979. Demography and social mobility in a red howler monkey (*A. seniculus*) population, 107–126, in *Vertebrate Ecology in the Northern Neotropics* (JF Eisenberg, ed.). Smithsonian Institution Press, Washington, DC.

Ruff, FJ. 1938. Trapping deer on the Pisgah National Game Preserve, North Carolina. *Journal of Wildlife Management,* 2:151–161.

Ruggiero, LF, GD Hayward, & JR Squires. 1994. Viability analysis in biological evaluations: concepts of population viability analysis, biological population, and ecological scale. *Conservation Biology,* 8:364–372.

Ruxton, G. 2002. The possible fitness benefits of striped coat coloration for zebra. *Mammalian Review,* 32:237–244.

Ryan, JM. 1986. Comparative morphology and evolution of cheek pouches in rodents. *Journal of Morphology,* 190:27–41.

Ryan, JM. 1989. Comparative myology and phylogenetic systematics of the Heteromyidae (Mammalia, Rodentia). *Miscellaneous Publications of the Museum of Zoology University of Michigan,* 176:1–103.

Ryan, JM. 1991a. Morphology of the glans penis in four genera of molossid bats (Chiroptera: Molossidae). *Journal of Mammalogy,* 72:658–668.

Ryan, JM. 1991b. Comparative morphology of the glans penis in *Molossus, Promops, & Eumops* (Chiroptera: Molossidae), 122–137, in *Contributions to Mammalogy in Honor of Karl F. Koopman* (TA Griffiths & DJ Klingener, eds.). Bulletin of the American Museum Natural History, NY.

Rydell, J. 1990. Behavioral variation in echolocation pulses of the northern bat (*Eptesicus nilssoni*). *Ethology,* 85:103–113.

Rydell, J. 1993. Variation in the sonar of an aerial-hawking bat (*Eptesicus nilssoni*). *Ethology,* 93:275–284.

Rydell, J & R Arlettaz. 1994. Low-frequency echolocation enables the bat *Tadarida teniotis* to feed on tympanate insects. *Proceedings of the Royal Society London,* 257B:175–178.

Sadleir, RMFS & CH Tyndale-Biscoe. 1977. Photoperiod and the termination of embryonic diapause in the marsupial *Macropus eugenii. Biology of Reproduction,* 16:605–608.

Sahley, CT & LE Baraybar. 1994. The natural history and population status of the nectar-feeding bat, *Platalina genovensium* in southwestern Peru. Abstract of paper presented at the 24th Annual North American Symposium on Bat Research, Ixtapa, Mexico. *Bat Research News,* 35:113.

Saikawa, Y, et al. 2004. Pigment chemistry: the red sweat of the hippopotamus. *Nature,* 429:363.

Sakai, M, et al. 2006. Flipper rubbing behaviors in wild bottlenose dolphins (*Tursiops aduncus*). *Marine Mammal Science,* 22:966–978.

Sale, JB. 1970. The behavior of the resting rock hyrax in relation to its environment. *Zoologica Africana,* 5:87–99.

Sales, G & D Pye. 1974. *Ultrasonic communication by animals.* Chapman & Hall, London.

Salesa, MJ, et al. 2006. Evidence of a false thumb in a fossil carnivore clarifies the evolution of pandas. *Proceedings of the National Academy of Sciences USA,* 103:379–382.

Salmón, E. 2000. Kincentric ecology: indigenous perceptions of the human–nature relationship. *Ecological Applications,* 10:1327–1332.

Sampson, SD, et al. 1998. Predatory dinosaur remains from Madagascar: implications for the Cretaceous biogeography of Gondwana. *Science,* 280:1048–1051.

Sánchez-Villagra, MR. 2001. The phylogenetic relationships of argyrolagoid marsupials. *Zoological Journal of the Linnean Society,* 131:459–463.

Sánchez-Villagra, MR & RF Kay. 1997. A skull of *Proargyrolagus,* the oldest argyrolagid (late Oligocene Salla Beds, Bolivia), with brief comments concerning its paleobiology. *Journal of Vertebrate Paleontology,* 17:717–724.

Sánchez-Villagra, MR, RF Kay, & F Anaya-Daza. 2000. Cranial anatomy and paleobiology of the Miocene marsupial *Hondalagus altiplanensis* and a

phylogeny of argyrolagids. *Palaeontology,* 43:287–301.

Sánchez-Villagra, MR, O Aguilera, and I Horovitz. 2003. The anatomy of the world's largest rodent. *Science,* 301:1708–1710.

Sánchez-Villagra, MR, I Horovitz, & M Motokawa. 2006. A comprehensive morphological analysis of talpid moles (Mammalia) phylogenetic relationships. *Cladistics,* 22:59–88.

Sánchez-Villagra, MR, Y Narita, & S Kuratani. 2007. Thoracolumbar vertebral number: the first skeletal synapomorphy for afrotherian mammals. *Systematics and Biodiversity,* 5:1–7.

Sánchez-Villagra, MR, et al. 2007a. Exceptionally preserved North American Paleogene metatherians: adaptations and discovery of a major gap in the opossum fossil record. *Biology Letters,* 3:318–322.

Sánchez-Villagra, MR, et al. 2007b. Enigmatic new mammals from the late Eocene of Egypt. *Palaontologische Zeitschrift,* 81:406–415.

Sanders, EH, et al. 1981. 6-Methoxybenzoxazalinone: a plant derivative that stimulates reproduction in *Microtus montanus. Science,* 214:67–69.

Sanyal, S, et al. 1990. The eye of the blind mole rat, *Spalax ehrenbergi.* Rudiment with hidden function? *Investigative Ophthalmology & Visual Science,* 31:1398–1404.

Sapargeldyev, MS. 1984. A contribution to the ecology of the mouse-like hamster *Calomyscus mystax* (Rodentia, Cricetidae) in Turkmenistan. *Zoologicheskij zhurnal,* Moscow, 63:1388–1395.

Saper, CB, TE Scammell, & J Lu. 2005. Hypothalamic regulation of sleep and circadian rhythms. *Nature,* 437:1257–1263.

Sargent, AB & DW Warner. 1972. Movements and denning habits of a badger. *Journal of Mammalogy,* 53:207–210.

Sargis, EJ. 2001. A preliminary qualitative analysis of the axial skeleton of tupaiids (Mammalia, Scandentia): functional morphology and phylogenetic implications. *Journal of Zoology, London,* 253:473–483.

Sargis, EJ. 2002a. Functional morphology of the forelimb of tupaiids (Mammalia, Scandentia) and its phylogenetic implications. *Journal of Morphology,* 253:10–42.

Sargis, EJ. 2002b. Functional morphology of the hindlimb of tupaiids (Mammalia,

Scandentia) and its phylogenetic implications. *Journal of Morphology*, 254:149–185.

Sargis, EJ. 2002c. The postcranial morphology of *Ptilocercus lowii* (Scandentia, Tupaiidae): an analysis of primatomorphan and volitantian characters. *Journal of Mammalian Evolution*, 9:137–160.

Sarich, VM & JE Cronin. 1980. South American mammal molecular systematics, evolutionary clocks, and continental drift, 399–421, in *Evolutionary Biology of the New World Monkeys and Continental Drift* (RL Ciochon & AB Chiarelli, eds.). Plenum Press, NY.

Sasaki, T, et al. 2006. *Balaenoptera omurai* is a newly discovered baleen whale that represents an ancient evolutionary lineage. *Molecular Phylogenetics and Evolution*, 41:40–52.

Sassone-Corsi, P. 1998. Molecular clocks: mastering time by gene regulation. *Nature*, 392:871–874.

Sauer, EGF. 1973. Zum sozialverhalten der kurzohrigen elefantenspitzmaus, *Macroscelides proboscideus*. *Zeitschrift fur Saugetierkunde*, 38:65–97.

Sauer, JJC, JD Skinner, & R Neitz. 1982. Seasonal utilization of leaves by giraffes (*Giraffa camelopardalis*) and the relationship of the seasonal utilization to the chemical composition of leaves. *South African Journal of Zoology*, 17:210–219.

Saunders, JK Jr. 1963. Movements and activities of the lynx in Newfoundland. *Journal of Wildlife Management*, 27:390–400.

Sawyer, H & F Lindzey. 2000. *The Jackson Hole Pronghorn Study*. Wyoming Cooperative Fish and Wildlife Research Unit, University of Wyoming, Laramie.

Scantlebury, M, et al. 2006. Energetics reveals physiologically distinct castes in a eusocial mammal. *Nature*, 440:795–797.

Schaefer, VH & RMFS Sadleir. 1979. Concentrations of carbon dioxide and oxygen in mole tunnels. *Acta Theriologica*, 24:267–276.

Schaller, GB. 1963. *The Mountain Gorilla: Ecology and Behavior*. University of Chicago Press, Chicago, IL.

Schaller, GB. 1965a. The behavior of the mountain gorilla, 324–367, in *Primate Behavior: Field Studies of Monkeys and Apes* (I De Vore, ed.). Holt, Rinehart and Winston, NY.

Schaller, GB. 1965b. *The Year of the Gorilla*. Ballantine Books, NY.

Schaller, GB. 1972. *The Serengeti Lion: A Study of Predator–Prey Relations*. University of Chicago Press, Chicago, IL.

Schaller, GB. 1993. *The Last Panda*. University of Chicago Press, Chicago, IL.

Schaller, G, et al. 1985. *The Giant Pandas of Wolong*. Chicago University Press, Chicago, IL.

Schaller, GB & A Rabinowitz. 1995. The saola or spindlehorn bovid *Pseudoryx nghetinhensis* in Laos. *Oryx*, 29:107–114.

Scheffer, VB. 1958. *Seals, Sea Lions, and Walruses*. Stanford University Press, Stanford, CA.

Scheich, H, et al. 1986. Electroreception and electrolocation in platypus. *Nature*, 319:401–402.

Scheline, RR. 1991. *Handbook of Mammalian Metabolism of Plant Compounds*. CRC Press, Boca Raton, FL.

Schenkel, R. 1967. Submission: its features and functions in the wolf and dog. *American Zoologist*, 7:319–329.

Schevill, WE & B Lawrence. 1949. Underwater listening to the white porpoise, *Delphinapterus leucas*. *Science*, 109:143–144.

Schevill, WE & B Lawrence. 1953. Auditory response of a bottle nosed porpoise, *Tursiops truncatus*, to frequencies above 100 kc. *Journal of Experimental Zoology*, 124:147–165.

Schevill, WE, WA Wadkins, & C Ray. 1963. Underwater sounds of pinnipeds. *Science*, 141:50–53.

Schevill, WE, WA Wadkins, & C Ray. 1966. Analysis of underwater *Odobenus* calls with remarks on the development and function of the pharyngeal pouches. *Zoologica*, 51:103–106.

Schipper, J, et al. 2008. The status of the world's land and marine mammals: diversity, threat, and knowledge. *Science*, 322:225-230.

Schmid, J. 2000. Torpor in the tropics: the case of the gray mouse lemur (*Microcebus murinus*). *Journal of Basic and Applied Ecology*, 2:133–139.

Schmid, J & PM Kappeler. 1994. Sympatric mouse lemurs (*Microcebus* spp.) in western Madagascar. *Folia Primatologica*, 63:162–170.

Schmid, J, T Ruf, & G Heldmaier. 2000. Metabolism and temperature regulation during daily torpor in the smallest primate, the pygmy mouse lemur (*Microcebus myoxinus*) in Madagascar. *Journal of Comparative Physiology B*, 170:59–68.

Schipper, J, et al. 2008. The status of the world's land and marine mammals: diversity, threat, and knowledge. *Science*, 322:225–230.

Schmidly, DJ. 2005. What it means to be a naturalist and the future of natural history at American universities. *Journal of Mammalogy*, 86:449–456.

Schmidt, S. 1988. Evidence for a spectral basis of texture perception in bat sonar. *Nature*, 331:617–619.

Schmidt-Nielsen, K. 1959. The physiology of the camel. *Scientific American*, 201:140–151.

Schmidt-Nielsen, K. 1964. *Desert Animals: Physiological Problems of Heat and Water*. Oxford University Press, NY.

Schmidt-Nielsen, K. 1981. Countercurrent systems in animals. *Scientific American*, 244:118–128.

Schmidt-Nielsen, K, et al. 1956. The question of water storage in the stomach of the camel. *Mammalia*, 20:1–15.

Schmidt-Nielsen, K, et al. 1957. Body temperature of the camel and its relation to water economy. *American Journal of Physiology*, 188:103–112.

Schmidt-Nielsen, K & AE Newsome. 1962. Water balance in the mulgara (*Dasycercus cristicauda*), a carnivorous desert marsupial. *Australian Journal of Biological Science*, 15:683–689.

Schmidt-Nielsen, K & HB Haines. 1964. Water balance in a carnivorous desert rodent, the grasshopper mouse. *Physiological Zoology*, 37:259–263.

Schmidt-Nielsen, K, FR Hainsworth, & DE Murrish. 1970. Counter-current heat exchange in the respiratory passages: effects on water and heat balance. *Respiration Physiology*, 9:263–276.

Schmidt-Nielsen, K, RC Schroter, & A Shkolnik. 1980. Desaturation of exhaled air in camels. *Proceedings of the Royal Society B*, 211:305–319.

Schmitz, J, C Roos, & H Zischler. 2005 Primate phylogeny: molecular evidence from retroposons. *Cytogenetics and Genome Research*, 108:26–37.

Schneider, KM. 1930. Das Flehmen. *Zoology Gart, Leipzig*, 3:183–198; 4:349–364; 5:200–226, 287–297.

Schneider, R, H Jurg Kugn, & G Kelemen, 1967. Die Larynx der *Hypsignathus monstrosus* Allen 1861. Ein Unifum in der Morphologie des Kehlkopfes. *Zeitschrift fur Wissenschaftliche Zoologie*, 175:1–53.

Schnitzler, H-U. 1987. Echoes of fluttering insects: information for echolocating bats, 226–243, in *Recent Advances in the Study of Bats* (MB Fenton, P Racey, & JMV Rayner, eds.). Cambridge University Press, Cambridge, UK.

Schnitzler, H-U & AD Grinnell. 1977. Directional sensitivity of echolocation in the horseshoe bat, *Rhinolophus ferrumequinum*. I. Directionality of sound emission. *Journal of Comparative Physiology*, 116:51–61.

Schnitzler, H-U, et al. 1983. The acoustical image of fluttering insects in echolocating bats, 235–250, in *Neuroethology and Behavioral Physiology: Roots and Growing Pains* (F Huber & H Markl, eds.). Springer, Berlin.

Schoener, TW. 1971. Theory of feeding strategies. *Annual Review of Ecology and Systematics*, 2:370–404.

Scholl, P. 1974. Temperaturregulation beim madegassischen Igeltanrek, *Echinops telfairi* (Martin, 1838). *Journal of Comparative Physiology A*, 89:175–195.

Schröpfer, R, B Klemer-Fringes, & E Naumer. 1985. Locomotion patterns and habitat utilisation of the two jerboas *Jaculus jaculus* and *Jaculus orientalis* (Rodentia, Dipodidae). *Mammalia*, 49:445–454.

Schubert, GH. 1953. Ponderosa pine cone cutting by squirrels. *Journal of Forestry*, 51:202.

Schulte, BA. 1998. Scent marking and responses to male castor fluid by beavers. *Journal of Mammalogy*, 79:191–203.

Schultze-Westrum, T. 1965. Innerartliche Verstandigung durch Diifte beim Gleitbeutler *Petaurus breviceps papuanus* Thomas (Marsupialia, Phalangeridae). *Zeitschrift Verglanen Physiologie*, 50:151–220.

Schusterman, RJ, et al. 2000. Why pinnipeds don't echolocate. *Journal of the Acoustical Society of America*, 107:2256–2264.

Schusterman, RJ, et al. 2004. Pinniped sensory systems and the echolocation issue, 531–535, in *Echolocation in Bats and Dolphins* (JA Thomas, CF Moss, & M Vater, eds.). University of Chicago Press, Chicago, IL.

Schutt, B. 2008. *Dark Banquet: Blood and the Curious Lives of Blood-feeding Creatures*. Harmony Books, NY.

Schutt, WA Jr & JS Altenbach. 1997. A sixth digit in *Diphylla ecaudata*, the hairy legged vampire bat (Chiroptera, Phyllostomidae). *Mammalia*, 61:280–285.

Schutt, WA Jr & NB Simmons. 1998. Morphology and homology of the chiropteran calcar, with comments on the phylogenetic relationships of *Archaeopteropus*. *Journal of Mammalian Evolution*, 5:1–32.

Schutt, WA Jr & NB Simmons. 2006. Quadrupedal bats: form, function, and evolution, 145–159, in *Functional and Evolutionary Ecology of Bats* (A Zubaid, GF McCracken, & TH Kunz, eds.). Oxford University Press, Oxford, UK.

Schwaibold, U & N Pillay. 2003. The gut morphology of the African ice rat, *Otomys sloggetti robertsi*, shows adaptations to cold environments and sex-specific seasonal variation. *Journal of Comparative Physiology B*, 173:653–659.

Schwartz, CC, JG Nagy, & WL Regelin. 1980. Juniper oil yield, terpenoid concentration, and antimicrobial effects on deer. *Journal of Wildlife Management*, 44:107–113.

Schwartz, DM. 1996. Snatching scientific secrets from the hippo's gaping jaws. *Smithsonian*, March:91–105.

Scotese, CR. 2002. Paleomap project. Paleogeographic maps of the earth. http://www.scotese.com.

Searle, JB, et al. 2009. Of mice and (Viking?) men: phylogeography of British and Irish house mice. *Proceedings of the Royal Society B*, 276(1655):201–207. http://rspb.royalsocietypublishing.org/content/276/1655/201.abstract.

Searle, JB, et al. 2009. The diverse origins of New Zealand house mice. *Proceedings of the Royal Society B*, 276(1655):209–217. http://rspb.royalsocietypublishing.org/content/276/1655/209.abstract .

Sears, KE, et al. 2006. Development of bat flight: morphologic and molecular evolution of bat wing digits. *Proceeding of the National Academy of Sciences USA*, 103:6581–6586.

Sebeok, TA. 1977. *How Animals Communicate*. Indiana University Press, Bloomington, IN.

Sedgeley, JA 2006. Roost site section by lesser short-tailed bats (*Mystacina tuberculata*) in mixed podocarp-hardwood forest, Whenua Hou/Codfish Island, New Zealand. *New Zealand Journal of Zoology*, 33:97–111.

Seely, MK. 1979. Ecology of a living desert: 20 years of research in the Namib. *South African Journal of Science*, 75:298–303.

Seely, MK. 1981. Desert plants use fog water. *Scientific Progress*, 14:4.

Seely, MK. 1987. *The Namib*. Shell Namibia, Namibia.

Seiffert, ER. 2007. A new estimate of aftotherian phylogeny based on simultaneous analysis of genomic, morphological, and fossil evidence. *BMC Evolutionary Biology*, 7:224.

Selwood, L & MH Johnson. 2006. Trophoblast and hypoblast in the monotreme, marsupial and eutherian mammal: evolution and origin. *BioEssay*, 28:128–145.

Semaw, S, et al. 2003. 2.6-million-year-old stone tools and associated bones from OGS–6 and OGS–7, Gona, Afar, Ethiopia. *Journal of Human Evolution*, 45:169–177.

Semaw, S, et al. 2005. Early Pliocene hominids from Gona, Ethiopia. *Nature*, 433:301–305.

Sénégas, F. 2004. A new species of *Petromus* (Rodentia, Hystricognatha, Petromuridae) from the early Pliocene of South Africa and its paleoenvironmental implications. *Journal of Vertebrate Paleontology*, 24:757–763.

SER (Society for Ecological Restoration International) online. Indigenous Peoples' Restoration Network. TEK (Traditional Ecological Knowledge) and restoration. http://www.ser.org/iprn/restoration.asp (accessed 2009.)

Sewell, GD. 1968. Ultrasound in rodents. *Nature*, 217:682–683.

Seyfarth, RM, DL Cheney, & P Marler. 1980. Vervet monkey alarm calls: semantic communications in a free-ranging primate. *Animal Behaviour*, 28:1070–1094.

Seymour, RS, PC Withers, & WW Weathers. 1998. Energetics of burrowing, running, and free-living in the Namib Desert golden mole (*Eremitalpa namibensis*). *Journal of Zoology, London*, 244:107–117.

Sharman, GB. 1962. The initiation and maintenance of lactation in the marsupial, *Trichosurus vulpecula*. *Journal of Endocrinology*, 25:375–385.

Sharman, GB. 1970. Reproductive physiology of marsupials. *Science*, 167:1221–1228.

Shaw, G. 1996. The uterine environment in early pregnancy in the tammar wallaby. *Reproduction, Fertility, and Development*, 8:811–818.

Shaw, WT. 1936. Moisture and its relation to the cone-storing habit of the western pine squirrel. *Journal of Mammalogy*, 17:337–349.

Shearman, LP, et al. 2000. Interacting molecular loops in the mammalian circadian clock. *Science*, 288:1013–1019.

Shedlock, AM, MC Milinkovitch, & N Okada. 2000. SINE evolution, missing data, and the origin of whales. *Systematic Biology*, 49:808–816.

Sheftel, BI. 1994. Spatial distribution of nine species of shrews in the central Siberian taiga, 45–56, in *Advances in the Biology of Shrews* (JF Merritt, GL Kirkland Jr, & RK Rose, eds.). Special Publication No. 18, Carnegie Museum of Natural History, Pittsburgh, PA.

Sheldrick, D. 1972. Death of the Tsavo elephant. *Saturday Review*, Sept. 30:29.

Shepherd, UL. 1998. A comparison of species diversity and morphological diversity across the North American latitudinal gradient. *Journal of Biogeography*, 25:19–29.

Sherman, PW. 1981a. Reproductive competition and infanticide in Belding's ground squirrels and other animals, 311–331, in *Natural Selection and Social Behaviour: Recent Research and New Theory* (RD Alexander & QW Tinkle, eds.). Chiron Press, NY.

Sherman, PW. 1981b. Kinship, demography, and Belding's ground squirrel nepotism. *Behavioral Ecology and Sociobiology*, 8:251–259.

Sherman, PW. 1985. Alarm calls of Belding's ground squirrels to aerial predators: nepotism or self-preservation? *Behavioral Ecology and Sociobiology*, 17:313–323.

Sherman, PW & ML Morton. 1984. Demography of Belding's ground squirrels. *Ecology*, 65:1617–1628.

Sherman, PW, JUM Jarvis, & RD Alexander. 1991. *The Biology of the Naked Mole-rat*. Princeton University Press, NJ.

Sherwood, L, H Klandorf, & PH Yancey. 2005. *Animal Physiology, from Genes to Organisms*. Rhomson-Brooks/Cole, Belmont, CA.

Shield, J. 1972. Acclimation and energy metabolism of the dingo *Canis dingo* and the coyote, *Canis latrans*. *Journal of Zoology, London*, 168:483–501.

Shimamura, M, et al. 1997. Molecular evidence from retroposons that whales form a clade within even-toed ungulates. *Nature*, 388:666–670.

Shionhara, A, KL Campbell, & H Suzuki. 2003. Molecular phylogenetic relationships of moles, shrew moles, and desmans from the New and Old Worlds.

Molecular Phylogentics and Evolution, 27:247–258.

Shipley, K, M Hines, & JS Buchwald. 1981. Individual differences in threat calls of northern elephant seal bulls. *Animal Behaviour*, 29:12–19.

Shirer, HW & HS Fitch. 1970. Comparison from radiotracking of movements and denning habits of the raccoon, striped skunk, and opossum in northeastern Kansas. *Journal of Mammalogy*, 51:491–503.

Shkolnik, A & A Borut. 1969. Temperature and water relations in two species of spiny mice (*Acomys*). *Journal of Mammalogy*, 50:245–255.

Short, J, et al. 1992. Reintroduction of macropods (Marsupialia: Macropodoidea) in Australia—a review. *Biological Conservation*, 62:189–204.

Short, J & A Smith. 1994. Mammal decline and recovery in Australia. *Journal of Mammalogy*, 75:288–297.

Shoshani, J. 1998. Understanding proboscidean evolution: a formidable task. *Trends in Ecology and Evolution*, 13:480–487.

Shoshani, J. 2005. Order Sirenia, 92–93, in *Mammal Species of the World* (DE Wilson & DM Reeder eds.). Johns Hopkins University Press, Baltimore, MD.

Shoshani, J & P Tassy. 1996. *The Proboscidea Evolution and Palaeoecology of Elephants and Their Relatives*. Oxford University Press, Oxford, UK.

Sidor, CA. 2003. The naris and palate of *Lycaenodon longiceps* (Therapsida: Biarmosuchia), with comments on their early evolution in the Therapsida. *Journal of Paleontology*, 77:977–984.

Sieffert, ER, EL Simons, & Y Attia. 2003. Fossil evidence for an ancient divergence of lorises and galagos. *Nature*, 422:421–424.

Sieffert, ER, et al. 2005. Additional remains of *Wadilemur elegans*, a primitive stem galagid from the late Eocene of Egypt. *Proceedings of the National Academy of Sciences USA*, 102:11396–11401.

Siemers, BM & H-U Schnitzler. 2004. Echolocation signals reflect niche differentiation in five sympatric congeneric bat species. *Nature*, 429:657–661.

Sigé, B, et al. 1994. Les chiroptères de Taqah (Oligocene inférieur, Sultanat d'Oman). Premier inventaire systématique. Münchner Geowiss. *Abhandlungen*, 26:35–48.

Sigé, B, J-Y Crochet, & A Insole. 1977. Les plus vieilles taupes. *Géobios*, 1:141–157.

Sigogneau-Russell, D. 1995. Two possibly aquatic triconodont mammals from the Early Cretaceous of Morocco. *Acta Palaeontologica Polonica*, 40:149–162.

Silcox, MT, et al. 2005. Euarchonta (Dermoptera, Scandentia, Primates), 127–144, in *The Rise of Placental Mammals* (KD Rose & JD Archibald, eds.). Johns Hopkins University Press, Baltimore, MD.

Silk, JB. 2007. Social components of fitness in primate groups. *Science*, 317:1347–1351.

Silva Taboada, G, W Suárez Duque, & S Díaz Franco. 2007. *Compendio de los mamíferos terrestres autóctonos de Cuba vivientes y extinguidos*. Ediciones Boloña, La Habana, Cuba.

Silvertown, J. 2009. A new dawn for citizen science. *Trends in Ecology & Evolution* 24(9):467–471.

Silvius, KM, RE Bodmer, & JMV Fragoso. 2005. *People in Nature: Wildlife Conservation in South and Central America*. Columbia University Press, NY.

Simberloff, DS & W Boecklen. 1981. Santa Rosalia reconsidered: size ratios and competition. *Evolution*, 35:1206–1228.

Simmons, JA. 1973. The resolution of target range by echolocating bats. *Journal of the Acoustical Society of America*, 54:157–173.

Simmons, JA. 1974. Response of the Doppler echolocation system in the bat, *Rhinolophus ferrumequinum*. *Journal of the Acoustical Society of America*, 56:672–682.

Simmons, JA. 1989. A view of the world through the bat's ear: the formation of acoustic images in echolocation. *Cognition*, 33:155–199.

Simmons, JA, et al. 1974. Target structure and echo spectral discrimination by echolocating bats. *Science*, 186:1130–1132.

Simmons, JA, DJ Howell, & N Suga. 1975. Information content of bat sonar echoes. *American Scientist*, 63:204–215.

Simmons, JA, et al. 1978. Echolocation by free-tailed bats (Tadarida). *Journal of Comparative Physiology*, 125:291–299.

Simmons, JA, MB Fenton, & MJ O'Farrell. 1979. Echolocation and pursuit of prey by bats. *Science*, 203:16–21.

Simmons, JA & RA Stein. 1980. Acoustic imaging in bat sonar: echolocation signals and the evolution of echolocation. *Journal of Comparative Physiology*, 135:61–84.

Simmons, JA, SA Kick, & BD Lawrence. 1984. Echolocation and hearing in the mouse-tailed bat, *Rhinopoma hardwickei*: acoustic evolution of echolocation in bats. *Journal of Comparative Physiology A*, 154:347–356.

Simmons, NB. 1993. Morphology, function, and phylogenetic significance of pubic nipples in bats (Mammalia: Chiroptera). *American Museum Novitates*, 3077:1–37.

Simmons, NB. 1994. The case for chiropteran monophyly. *American Museum Novitates*, 3103:1–54.

Simmons, NB. 2005. Chiroptera, 159–174, in *The Rise of Placental Mammals* (KD Rose & JD Archibald, eds.). Johns Hopkins University Press, Baltimore, MD.

Simmons, NB. 2008. Taking wing. *Scientific American*, December:96–103.

Simmons, NB & RS Voss. 1998. The mammals of Paracou, French Guiana: a Neotropical lowland rainforest fauna, part 1. Bats. *Bulletin of the American Museum of Natural History*, 237:1–219.

Simmons, NB & JH Geisler. 1998. Phylogenetic relationships of *Icaronycteris*, *Archaeonycteris*, *Hassianycteris*, and *Palaeochiropteryx* to extant bat lineages, with comments on the evolution of echolocation and foraging strategies in Microchiroptera. *Bulletin of the American Museum of Natural History*, 235:1–182.

Simmons, NB, et al. 2008. Primitive early Eocene bat from Wyoming and the evolution of flight and echolocation. *Nature*, 451:818–821.

Simons, RS. 1996. Lung morphology of cursorial and non-cursorial mammals: lagomorphs as a case study for a pneumatic stabilization hypothesis. *Journal of Morphology*, 230:299–316.

Simpson, GG. 1940. Mammals and land bridges. *Journal of the Washington Academy of Science*, 30:137.

Simpson, GG. 1945. The principles of classification and a classification of mammals. *Bulletin of the American Museum of Natural History*, 85:1–350.

Simpson, GG. 1951. *Horses.* Oxford University Press, NY.

Simpson, GG. 1964. Species density of North American Recent mammals. *Systematic Zoology*, 13:57–73.

Simpson, GG. 1970. The Argyrolagidae, extinct South American marsupials. *Bulletin Museum of Comparative Zoology*, 139:1–86.

Simpson, GG. 1980. *Splendid Isolation: The Curious History of South American Mammals.* Yale University Press, New Haven, CT.

Sinclair, ARE. 1970. Studies of the ecology of the East African buffalo. Ph.D. thesis, Oxford University, Oxford, UK.

Sinclair, ARE. 1974. The social organization of the East African buffalo, 676–689, in *The Behavior of Ungulates and Its Relation to Management* (V Geist & FR Walther, eds.). International Union for Conservation of Nature and Natural Resources, Morges, Switzerland.

Sinclair, ARE. 1977. *The African Buffalo: A Study in Resource Limitation of Populations.* University of Chicago Press, Chicago, IL.

Skinner, JD, S Davis, & G Ilani. 1980. Bone collecting by striped hyaenas, *Hyaena hyaena*, in Israel. *Palaeontologia Africana*, 23:99–104.

Skinner, JD & RHN Smithers. 1990. *The Mammals of the Southern African Subregion.* University of Pretoria, Pretoria, Republic of South Africa.

Skinner, JD & CH Chimimba. 2005. *The Mammals of the Southern African Subregion.* Cambridge University Press, Cambridge, UK.

Skliba, J, et al. 2009. Home-range dynamics in a solitary subterranean rodent. *Ethology*, 115:217–226.

Skoczen, S. 1958. Tunnel digging by the mole (*T. europaea* Linne). *Acta Theriologica*, 2:235–249.

Slattery, JP & SJ O'Brien. 1995. Molecular phylogeny of the red panda (*Ailurus fulgens*). *Journal of Heredity*, 86:413–422.

Sliwa, A & PRK Richardson. 1998. Responses of aardwolves, *Proteles cristatus*, Sparrman 1783, to translocated scent marks. *Animal Behavior*, 56:137–146.

Slobodchikoff, CN & R Coast. 1980. Dialects in the alarm calls of prairie dogs. *Behavioral Ecology and Sociobiology*, 7:49–53.

Smith, AG, DG Smith, & BM Funnell. 1994. *Atlas of Mesozoic and Cenozoic Coastlines.* Cambridge University Press, NY.

Smith, AP & DG Quin. 1996. Patterns and causes of extinction and decline in Australian conilurine rodents. *Biological Conservation*, 77:243–267.

Smith, AT. 1974. The distribution and dispersal of pikas: consequences of insular population structure. *Ecology*, 55:1112–1119.

Smith, AT. 1978. Comparative demography of pikas *Ochotona*: effect of spatial and temporal age-specific mortality. *Ecology*, 59:133–139.

Smith, AT & Y Xie. 2008. *A Guide to the Mammals of China.* Princeton University Press, Princeton, NJ.

Smith, CC & OJ Reichman. 1984. The evolution of food caching by birds and mammals. *Annual Review of Ecology and Systematics*, 15:329–351.

Smith, DW & G Ferguson. 2005. *Decade of the Wolf.* Lyons Press, Guilford, CT.

Smith, HM. 1960. *Evolution of Chordate Structure.* Holt, Rinehart and Winston, NY.

Smith, J & J Uppenbrink. 2001. Earth's variable climatic past. *Science*, 292:657–659.

Smith, MF & JL Patton. 1999. Phylogenetic relationships and the radiation of sigmodontine rodents in South America: evidence from cytochrome b. *Journal of Mammalian Evolution*, 6:89–128.

Smith, RB. 1971. Seasonal activities and ecology of terrestrial vertebrates in a Neotropical monsoon environment. M.S. thesis, Northern Arizona University, Flagstaff.

Smith, RH. 1975. Nitrogen metabolism in the rumen and the comparative and nutritive value of nitrogen compounds entering the duodenum, 399, in *Digestion and Metabolism in the Ruminant* (IW McDonald & ACI Warner, eds.). University of New England, Armidale, New South Wales, Australia.

Smith, T, et al. 2007. High bat (Chiroptera) diversity in the early Eocene of India. *Naturwissenschaften*, 94:1003–1009.

Smith, WP. 2007. Ecology of *Glaucomys sabrinus*: habitat, demography, and community relations. *Journal of Mammalogy*, 88:862–881.

Smithers, RHN. 1971. The mammals of Botswana. *Museum Memoirs, National Museums and Monuments of Rhodesia*, 4:1–340.

Smuts, GL. 1976. Reproduction in the zebra mare *Equus burchelli antiquorum* from the Kruger National Park. *Koedoe*, 19:89–132.

Smythe, N. 1978. The natural history of the Central American agouti (*Dasyprocta punctata*). *Smithsonian Contributions to Zoology*, 257:1–52.

Society for Ecological Restoration International (SER) Online, Indigenous Peoples' Restoration Network. TEK and ecological restoration. http://www.ser.org/iprn/restoration.asp (accessed October 30, 2009).

Solari, S, et al. 2004. Geographic distribution, ecology, and phylogenetic affinities of *Thyroptera lavali* Pine 1993. *Acta Chiropterologica*, 6:293–302.

Solomon, ME. 1949. The natural control of animal populations. *Journal of Animal Ecology*, 18:1–35.

Solounias, N. 1997. Remarkable new findings regarding the evolution of the giraffe neck. *Journal of Vertebrate Paleontology*, 17(suppl 3):78A.

Solounias, N. 1999. The remarkable anatomy of the giraffe's neck. *Journal of Zoology, London*, 247:257–268.

Solounias, N. 2007a. Family Giraffidae, 257–277, in *The Evolution of Artiodactyls* (DR Prothero & SE Foss, eds.). Johns Hopkins University Press, Baltimore, MD.

Solounias, N. 2007b. Family Bovidae, 278–291, in *The Evolution of Artiodactyls* (DR Prothero & SE Foss, eds.). Johns Hopkins University Press, Baltimore, MD.

Sondaar, PY. 1977. *Insularity and Its Effects on Mammal Evolution*. NATO Advanced Study Institute, Plenum Press, NY.

Sorensen, JS, JD McLister, & MD Dearing. 2005a. Plant secondary metabolites compromise the energy budget of specialist and generalist mammalian herbivores. *Ecology*, 86:125–139.

Sorensen, JS, JD McLister, & MD Dearing. 2005b. Novel plant secondary metabolites impact dietary specialists more than generalists (*Neotoma* spp.). *Ecology*, 86:140–154.

Sorenson, MW & CH Conaway. 1968. The social and reproductive behavior of *Tupaia montana* in captivity. *Journal of Mammalogy*, 49:502–512.

Souders, HJ & AF Morgan. 1957. Weight and composition of organs during the reproductive cycle in the rat. *American Journal of Physiology*, 191:1–7.

Soulé, ME. 1986. *Conservation Biology: The Science of Scarcity and Diversity*. Sinauer Associates, Sunderland, MA.

Soulé, ME. 1991. Conservation: tactics for a constant crisis. *Science*, 253:744–750.

Southern, HN. 1964. *Handbook of British Mammals*. Blackwell, Oxford, UK.

Sowls, LK. 1974. Social behavior of the collared peccary *Dicotyles tajacu*, 144–165, in *The Behavior of Ungulates and Its Relation to Management* (V Geist & FR Walther, eds.). International Union for Conservation of Nature and Natural Resources, Morges, Switzerland.

Speakman, JR. 1993. The evolution of echolocation for predation. *Symposium of the Zoological Society London*, 65:39–63.

Spencer, AW & HW Steinhoff. 1968. An explanation of geographic variation in litter size. *Journal of Mammalogy*, 49:281–286.

Spencer, DA. 1958a. *Preliminary Investigations on the Northwestern Microtus Irruption*. U.S. Fish and Wildlife Service, Denver Wildlife Research Laboratory Special Report.

Spencer, DA. 1958b. Biological and control aspects, in *The Oregon Meadow Mouse Irruption of 1957–1958*. Federal Cooperative Extension Service, Oregon State College, Corvallis, OR.

Spinage, CA. 1972. African ungulate life tables. *Ecology*, 53:645–652.

Spitz, F. 1963. Étude des densities de population de *Microtus arvalis*. Pall. A Saint-Michal-en-L'Hern (Vendu). *Mammalia*, 27:497–531.

Spotorno, AE, et al. 1997. Chromosome divergences among American marsupials and the Australian affinities of the American *Dromiciops*. *Journal of Mammalian Evolution*, 4:259–269.

Sprankel, H. 1965. Untersuchungen an *Tarsius*. I: morphologie des schwanzes nebst ethologischen bemerkungen. *Folia Primatologica*, 3:153–188.

Springer, MS. 1997. Molecular clocks and the timing of the placental and marsupial radiations in relation to the Cretaceous–Tertiary boundary. *Journal of Mammalian Evolution*, 4:285–301.

Springer, MS, M Westerman, & JAW Kirsch. 1994. Relationships among orders and families of marsupials based on 12S ribosomal DNA sequences and the timing of marsupial radiation. *Journal of Mammalian Evolution*, 2:85–115.

Springer, MS, JAW Kirsch, & JA Case. 1997. The chronicle of marsupial evolution, 129–161, in *Molecular Evolution and Adaptive Radiation* (TJ Givnish & KJ Sytsma, eds.). Cambridge University Press, Cambridge, UK.

Springer, MS, et al. 2001. Integrated fossil and molecular data reconstruct bat echolocation. *Proceedings of the National Academy of Sciences USA*, 98:6241–6246.

Springer, MS, et al. 2007. A molecular classification for the living orders of placental mammals and the phylogenetic placements of primates, 1–28, in *Primate Origins: Adaptations and Evolution* (MJ Ravosa & M Dagosto, eds.). Springer, NY.

Stafford, BJ & FS Szalay. 2000. Craniodental functional morphology and taxonomy of dermopterans. *Journal of Mammalogy*, 81:360–385.

Stahl, WR. 1967. Scaling of respiratory variables in mammals. *Journal of Applied Physiology*, 22:453–460.

Stanford, CB. 1995. Chimpanzee hunting behavior and human evolution. *American Scientist*, 83:256–261.

Stanford, CB, et al. 1994. Patterns of predation by chimpanzees on red colobus monkeys in Gombe National Park, Tanzania, 1982–1991. *American Journal of Physical Anthropology*, 94:213–228.

Stanhope, MJ, et al. 1993. A molecular view of primate supraordinal relationships from the analysis of both nucleotide and amino acid sequences, 251–292, in *Primates and Their Relatives in Phylogenetic Perspective* (RDE MacPhee, ed.). Plenum Press, NY.

Stanhope, MJ, et al. 1998a. Highly congruent molecular support for a diverse superordinal clade of endemic African mammals. *Molecular Phylogenetics and Evolution*, 9:501–508.

Stanhope, MJ, et al. 1998b. Molecular evidence for multiple origins of the Insectivora and for a new order of endemic African mammals. *Proceedings of the National Academy of Sciences USA*, 95:9967–9972.

Stanley, WT, MA Rogers, & R Hutterer. 2005. A new species of Congosorex from the Eastern Arc Mountains, Tanzania, with significant biogeographical implications. *Journal of Zoology, London*, 265:269–280.

Stebbins, GL. 1974. *Flowering Plants: Evolution Above the Species Level*. Harvard University Press, Cambridge, MA.

Stebbins, LL. 1984. Overwintering activity of *Peromyscus maniculatus, Clethrionomys gapperi, C. rutilus, Eutamias amoenus*, and *Microtus pennsylvanicus*, 301–314, in *Winter Ecology of Small Mammals* (JF Merritt, ed.). Special Publication No. 10, Carnegie Museum of Natural History Pittsburgh, PA.

Steadman, DW, et al. 2005. Asynchronous extinction of late Quaternary sloths on continents and islands. *Proceedings of the National Academy of Sciences USA*, 102:11763–11768.

Steeman, ME. 2007. Cladistic analysis and a revised classification of fossil and recent mysticetes. *Zoological Journal of the Linnean Society*, 150:875–894.

Stehli, FG & SD Webb. 1985. *The Great American Biotic Interchange.* Plenum Press, NY.

Stehn, RA & FJ Jannett Jr. 1981. Male-induced abortion in various microtine rodents. *Journal of Mammalogy,* 62:369–372.

Stein, BR. 1990. Limb myology and phylogenetic relationships in the superfamily Dipodoidea (birch mice, jumping mice, and jerboas). *Journal of Zoological Systematics and Evolutionary Research,* 28:299–314.

Steinberg, PD, AA Estes, & FC Winter. 1995. Evolutionary consequences of food chain length in kelp forest communities. *Proceedings of the National Academy of Sciences USA,* 92:8145–8148.

Steinberg, T. 1995. *Slide Mountain, or the Folly of Owning Nature.* University of California Press, Berkeley, CA.

Steiper, ME & NM Young. 2006. Primate molecular divergence dates. *Molecular Phylogenetics and Evolution,* 41:384–394.

Stejneger, L. 1887. How the great northern sea-cow (*Rytina*) became exterminated. *American Naturalist,* 21:1047–1054.

Steller, GW. 1751. De bestiis marinis. Novi Commentarii Acad. *Scientiarum Petropolitanae,* 2:289–398.

Stenlund, MH. 1955. *A Field Study of the Timber Wolf (Canis lupus) on the Superior National Forest, Minnesota.* Minnesota Department of Conservation Technical Bulletin 4.

Stenseth, NC. 1985. Mathematical models of microtine cycles: models and the real world. *Acta Zoology Fennica,* 173:7–12.

Stenseth, NC & RA Ims. 1993. *The Biology of Lemmings.* Linnean Society of London, London.

Stephenson, AB. 1969. Temperatures within a beaver lodge in winter. *Journal of Mammalogy,* 50:134–136.

Stephenson, PJ. 2003. Lipotyphla (ex Insectivora): *Geogale aurita,* large-eared tenrec, 1265–1267, in *The Natural History of Madagascar* (SM Goodman & JP Benstead eds.). University of Chicago Press, Chicago, IL.

Stephenson, PJ & PA Racey. 1993. Reproductive energetics of the Tenrecidae (Mammalia: Insectivora). I: the large-eared tenrec, *Geogale aurita. Physiological Zoology,* 66:643–663.

Steppan, S, R Adkins, & J Anderson. 2004. Phylogeny and divergence-date estimates of rapid radiations in Muroid rodents based on multiple nuclear genes. *Systematic Biology,* 53:533–553.

Steppan, SJ, et al. 2005 Multigene phylogeny of the Old World mice Murinae reveals distinct geographic lineages and the declining utility of mitochondrial genes compared to nuclear genes. *Molecular Phylogenetics and Evolution,* 37:370–388.

Sterling, E. 2003. *Daubentonia madagascariensis,* aye-aye, 1348–1351, in *The Natural History of Madagascar* (SM Goodman & J.P Benstead, eds.). University of Chicago Press, Chicago, IL.

Sterling, I. 1969. Ecology of the Weddell seal in McMurdo Sound, Antarctica. *Ecology,* 50:573–586.

Stevenson-Hamilton, J. 1947. *Wildlife in South Africa.* Cassell, London.

Stewart, BS. 1997. Ontogeny of differential migration and sexual segregation in northern elephant seals. *Journal of Mammalogy,* 78:1101–1116.

Stewart, BS & RL DeLong. 1995. Double migrations of the northern elephant seal, *Mirounga angustirostris. Journal of Mammalogy,* 76:196–205.

Stewart, F. 1984. Mammogenesis and changing prolactin receptor concentrations in the mammary glands of the tammar wallaby (*Macropus eugenii*). *Journal of Reproduction and Fertility,* 71:141–148.

Stewart, KM, et al. 2005. Density-dependent effects on physical condition and reproduction in North American elk: an experimental test. *Oecologia,* 143:85–93.

Stirling, I, W Calvert, & C Spencer. 1987. Evidence of stereotyped underwater vocalizations of male Atlantic walruses (*Odobenus rosmarus rosmarus*). *Canadian Journal of Zoology,* 65:2311–2321.

Stirling, I, NJ Lunn, & J Iacozza. 1999. Long-term trends in the population ecology of polar bears in western Hudson Bay in relation to climatic change. *Arctic,* 52:294–306.

Stirling, I & CL Parkinson. 2006. The effects of climate warming on selected populations of polar bears (*Ursus maritimus*) in the Canadian Arctic. *Arctic,* 59:261–275.

Stock, C. 1949. *Rancho La Brea: A Record of Pleistocene Life in California.* Los Angeles County Museum, Science Series, 13:1.

Stokstad, E. 2009. Will many endangered species recover? *Science,* 323:998–999.

Stone, GN & A Purvis. 1992. Warm-up rates during arousal from torpor in heterothermic mammals: physiological correlates and a comparison with heterothermic insects. *Journal of Comparative Physiology B,* 162:284–295.

Storer, TI & RL Usinger, 1965. *General Zoology.* McGraw-Hill, NY.

Storrs, E, HP Burchfield, & RJ W. Rees. 1989. Reproduction delay in the common long-nosed armadillo, *Dasypus novemcinctus* L, 535–548, in *Advances in Neotropical Mammalogy* (KH Redford & JF Eisenberg, eds.). Sandhill Crane Press, Gainesville, FL.

Strahan, R. 1995. *Mammals of Australia.* Smithsonian Institution Press, Washington, DC.

Strait, SG & SC Smith. 2006. Elemental analysis of soricine enamel: pigmentation variation, and distribution in molars of *Blarina brevicauda. Journal of Mammalogy,* 87:700–705.

Strasburg, JL. 2006. Roads and genetic connectivity. *Nature,* 440:875–876.

Strier, KB. 2002. *Primate Behavioral Ecology.* Allyn and Bacon/Longman, Boston, MA.

Struhsaker, TT. 1967. Behavior of elk (*Cervus canadensis*) during the rut. *Zeitschrift fur Tierpsychologie,* 24(1):80–114.

Struhsaker, TT. 1977. Infanticide and social organization in the redtail monkey (*Cercopithecus ascanius schmidti*) in the Kibale Forest, Uganda. *Zeitschrift fur Tierpsychologie,* 4:75.

Strum, SC. 1981. Processes and products of change: baboon predatory behavior at Gilgil, Kenya, 255–302, in *Omnivorous Primates: Gathering and Hunting in Human Evolution* (RSO Harding & G Teleki, eds.). Columbia University Press, NY.

Studier, EH & DJ Howell. 1969. Heart rate of female big brown bats in flight. *Journal of Mammalogy,* 50:842–845.

Stumpff, LM. 2007. Restoring youth: restoring relationships to wildlife and wild places, 397–401, in *Science and Stewardship to Protect and Sustain Wilderness Values* (A Watson, J Sproul, & L Dean, eds.). Eighth World Wilderness Congress Symposium, September 30–October 6, 2005, Anchorage, Alaska. USDA Forest Service Proceedings RMRS-P–49, Fort Collins, Colorado, U.S. Department of Agriculture, Forest Service, Rocky Mountain Research Station.

Sudman, PD, LJ Barkley, & MS Hafner. 1994. Familial affinity of *Tomopeas ravus* (Chiroptera) based on protein electrophoretic and cytochrome B sequence data. *Journal of Mammalogy*, 75:365–377.

Suga, N & T Shimozawa. 1974. Site of neural attenuation of responses to self-vocalized sounds in echolocating bats. *Science*, 183:1211–1213.

Sugiyama, Y. 1973. Social organization of wild chimpanzees, 68–80, in *Behavioral Regulators of Behavior in Primates* (CR Carpenter, ed.). Bucknell University Press, Lewisburg, PA.

Sullivan, J & DL Swofford. 1997. Are guinea pigs rodents? the importance of adequate models in molecular phylogenetics. *Journal of Mammalian Evolution*, 4:77–86.

Sun, K, et al. 2008. Structure, DNA sequence variation and phylogenetic implications of the mitochondrial control region in horseshoe bats. *Mammalian Biology*, 74:130–144.

Surlykke, A & EKV Kalko. 2008. Echolocating bats cry out loud to detect their prey. *PLoS One*, 3(4): e2036.

Sussman, RW & WG Kinzey. 1984. The ecological role of the Callitrichidae: a review. *American Journal of Physical Anthropology*, 64:419–449.

Suthers, RA. 1965. Acoustic orientation by fish-catching bats. *Journal of Experimental Zoology*, 158:319–348.

Suthers, RA. 1967. Comparative echolocation by fishing bats. *Journal of Mammalogy*, 48:79–87.

Suthers, RA & JM Fattu. 1973. Fishing behavior and acoustic orientation by the bat (*Noctilio labialis*). *Animal Behaviour*, 21:61–66.

Suthers, RA & JM Fattu. 1982. Selective laryngeal neurotomy and the control of phonation by the echolocating bat, *Eptesicus*. *Journal of Comparative Physiology*, 145:529–537.

Suwa, G, et al. 2007. A new species of great ape from the late Miocene epoch in Ethiopia. *Nature*, 448:844–845, 921–924.

Swank, WG. 1958. *The Mule Deer in Arizona Chaparral*. Arizona Game and Fish Department, Wildlife Bulletin No. 3.

Swartz, SM, MB Bennett, & DR Carrier. 1992. Wing bone stresses in free flying bats and the evolution of skeletal design for flight. *Nature*, 359:726–729.

Swartz, SM, K Bishop, & M-F I. Aguirre. 2006. Dynamic complexity of wing form in bats: implications for flight performance, 110–130, in *Functional and Evolutionary Ecology of Bats* (A Zubaid, GF McCracken, & TH Kunz, eds.). Oxford University Press, Oxford, UK.

Sweeney, RCH. 1956. Some notes on the feeding habits of the ground pangolin *Smutsria temmincki* (Smuts). *Annals and Magazine of Natural History, London*, 12:893–896.

Swinhoe, R. 1870. On the mammals of Hainan. *Proceedings of the Zoology Society London*, 1870:224–239.

Swofford, DL. 1998. *PAUP 4.0 Phylogenetic Analysis Using Parsimony (and Other Methods)*. Computer software and manual, Sinauer Associates.

Symington, MM. 1988. Food competition and foraging party size in the black spider monkey (*Ateles paniscus* chamek). *Behaviour*, 105:117–134.

Symington, MM. 1990. Fission-fusion social-organization in *Ateles* and *Pan*. *International Journal of Primatology*, 11:47–61.

Szalay, FS. 1977. Phylogenetic relationships and a classification of the eutherian Mammalia, 315–374, in *Major Patterns in Vertebrate Evolution* (MK Hecht, PC Goody, & BM Hecht, eds.). Plenum Press, NY.

Szalay, FS. 1982. A new appraisal of marsupial phylogeny and classification, 612–640, in *Carnivorous Marsupials* (M Archer, ed.). Sydney, Royal Zoology Society of New South Wales, Australia.

Szalay, FS. 1994. *Evolutionary History of the Marsupials and an Analysis of Osteological Characters*. Cambridge University Press, NY.

Szalay, FS & E Delson. 1979. *Evolutionary History of the Primates*. Academic Press, NY.

Szalay, FS, MJ Novacek, & MC McKenna. 1993a. *Mammal Phylogeny. Mesozoic Differentiation, Multituberculates, Monotremes, Early Therians, and Marsupials*. Springer-Verlag, NY.

Szalay, FS, MJ Novacek, & MC McKenna. 1993b. *Mammal Phylogeny. Placentals*. Springer-Verlag, NY.

Taber, AB, et al. 1993. Ranging behavior and population dynamics of the chacoan peccary, *Catagonus wagneri*. *Journal of Mammalogy*, 74:443–454.

Taber, RD & RF Dasmann. 1957. The dynamics of three natural populations of deer *Odocoileus hemionus columbianus*. *Ecology*, 38:233–246.

Taddei, VA & W Uieda. 2001. Distribution and morphometrics of *Natalus stramineus* from South America (Chiroptera, Natalidae). *Iheringia Serie Zoologia*, 91:123–132.

Takai, M & N Shigehara. 2004. The Pondaung primates, enigmatic "possible anthropoids" from the latest Middle Eocene, central Myanmar, 283–321, in *Anthropoid Origins: New Visions* (CF Ross & RF Kay, eds.). Kluwer Academic/ Plenum Publishers, NY.

Talamantes, F. 1975. Comparative study of the occurrence of placental prolactin among mammals. *General and Comparative Endocrinology*, 27:115–121.

Talbot, LM & MH Talbot. 1963. *The Wildebeest in Western Masailand*. Wildlife Monograph, No. 12. The Wildlife Society, Washington, DC.

Tardif, D, RL Carson, & BL Gangaware. 1990. Infant-care behavior of mothers and fathers in a communal-care primate, the cotton-top tamarin (*Saguinus oedipus*). *American Journal of Primatology*, 22:73–85.

Tate, GHH. 1933. A systematic revision of the marsupial genus *Marmosa*. *Bulletin of the American Museum of Natural History*, vol. 66.

Tate, GHH & R Archbold. 1937. Results of the Archbold expeditions. 16: some marsupials of New Guinea and Celebes. *Bulletin of the American Museum of Natural History*, 73:331–476.

Tattersall, I. 1982. *The Primates of Madagascar*. Columbia University Press, NY.

Tattersall, I. 2008. Reconstruction of an extraordinary extinct primate from Madagascar. *Proceedings of the National Academy of Sciences USA*, 105(31):10639–10640.

Taulman, JF & LW Robbins. 1996. Recent range expansion and distributional limits of the nine-banded armadillo (*Dasypus novemcinctus*) in the United States. *Journal of Biogeography*, 23:635–648.

Taylor, CR. 1968a. Hygroscopic food: a source of water for desert antelopes? *Nature*, 219:181–182.

Taylor, CR. 1968b. The minimum water requirements of some East African bovids. *Symposium of the Zoological Society London*, 21:195.

Taylor, CR. 1969. The eland and the oryx. *Scientific American*, 220:89–95.

Taylor, CR. 1972. The desert gazelle: a paradox resolved, 215–227, in *Comparative Physiology of Desert Animals* (GMO Maloiy, ed.). Symposium of the Zoological Society London, 31. Academic Press, NY.

Taylor, CR. 1974. Exercise and thermoregulation, 163–184, in *Environmental Physiology* (D Robertshaw, ed.). International Review of Science, Environment and Physiology, Butterworth, London.

Taylor, CR, K Schmidt-Nielsen, & JL Raab. 1970. Scaling of energetic cost of running to body size in mammals. *American Journal of Physiology*, 219:1104–1107.

Taylor, CR, et al. 1971. Effect of hyperthermia on heat balance during running in the African hunting dog. *American Journal of Physiology*, 220:823–827.

Taylor, CR & CP Lyman. 1972. Heat storage in running antelopes: independence of brain and body temperatures. *American Journal of Physiology*, 222:114–117.

Taylor, CR, NC Heglund, & GMO Maloiy. 1982. Energetics and mechanics of terrestrial locomotion. I: metabolic energy consumption as a function of speed and body size in birds and mammals. *Journal of Experimental Biology*, 97:1–21.

Taylor, E, JR Potter, & M Chitre. 1997. Ambient noise imaging potential of marine mammals. *Proceedings of the Underwater Bio-Sonar and Bioacoustics Symposium*, Loughborough, UK.

Taylor, TN & EL Taylor. 1990. *Antarctic Paleobiology: Its Role in the Reconstruction of Gondwana*. Springer-Verlag, NY.

Taylor, WA & JD Skinner. 2001. Associative feeding between anteating chats, *Myrmecocichla formicivora*, and aardvarks, *Orycteropus afer*. *Ostrich*, 72:199–200.

Taylor, WA & JD Skinner. 2003. Activity patterns, home ranges and burro use of aardvarks (*Orycteropus afer*) in the Karoo. *Journal of Zoology, London*, 261:291–297.

Tedford, RH. 1967. The fossil Macropodidae from Lake Menindee, New South Wales. *University of California Publications in Geological Science*, 64:1–165.

Teeling, EC, et al. 2003. Nuclear genes confirm an ancient link between New Zealand's short-tailed bat and South American noctilionoid bats. *Molecular Phylogenetics and Evolution*, 28:308–319.

Teeling, EC, et al. 2005. A molecular phylogeny for bats illuminates biogeography and the fossil record. *Science*, 307:580–584.

Teerink, BJ. 1991. *Hair of West-European Mammals: Atlas and Identification Key*. Cambridge University Press, Cambridge, UK.

Tejedor, A. 2005. A new species of funnel-eared bat (Natalidae: *Natalus*) from Mexico. *Journal of Mammalogy*, 86:1109–1120.

Tejedor, A. 2006. The type locality of *Natalus stramineus* (Chiroptera: Natalidae): implications for the taxonomy and biogeography of the genus *Natalus*. *Acta Chiropterologica*, 8:361–380.

Tejedor, MF. 1998. The evolutionary history of platyrrhines: old controversies and new interpretations. *Neotropical Primates*, 6:77–82.

Tejedor, MF, et al. 2009. New early Eocene mammalian fauna from western Patagonia, Argentina. *American Museum Novitates*, 3638:1–43.

Teleki, G. 1973. *The Predatory Behavior of Wild Chimpanzees*. Bucknell University Press, Lewisburg, PA.

Terborgh, J. 1973. On the notion of favorableness in plant ecology. *American Naturalist*, 107:481–501.

Terborgh, J. 1983. *Five New World Primates*. Princeton University Press, Princeton, NJ.

Terborgh, J & AW Goldizen. 1985. On the mating system of the cooperatively breeding saddle-backed tamarin (*Saguinus fuscicollis*). *Behavioral Ecology and Sociobiology*, 16:293–299.

Terman, CR. 1992. Reproductive inhibition in female white-footed mice from Virginia. *Journal of Mammalogy*, 73:443–448.

Thabah, A, et al. 2007. Diet, echolocation calls, and phylogenetic affinities of the great evening bat (*Ia io*; Vespertilionidae): another carnivorous bat. *Journal of Mammalogy*, 88:728–735.

Theodor, JM. 2002. Crowning glories. *Nature*, 417:498–499.

Theodor, JM & SE Foss. 2005. Deciduous dentitions of Eocene cebochoerid artiodactyls and cetartiodactyl relationships. *Journal of Mammalian Evolution*, 12:161–181.

Theodor, JM, KD Rose, & J Erfurt. 2005. Artiodactyla, 215–233, in *The Rise of Placental Mammals* (KD Rose & JD Archibald, eds.). Johns Hopkins University Press, Baltimore, MD.

Theodor, JM, J Erfurt, & G Métais. 2007. The earliest artiodactyls, 32–58, in *The Evolution of Artiodactyls* (DR Prothero & SE Foss, eds.). Johns Hopkins University Press, Baltimore, MD.

Thewissen, JGM. 1985. Cephalic evidence for the affinities of Tubulidentata. *Mammalia*, 49:257–284.

Thewissen, JGM & SA Etnier. 1995. Adhesive devices on the thumb of vespertilionoid bats (Chiroptera). *Journal of Mammalogy*, 76:925–936.

Thewissen, JGM & S Bajpai. 2001. Whale origins as a poster child for macroevolution. *BioScience*, 51:1037–1049.

Thewissen, JG, et al. 2001. Skeletons of terrestrial cetaceans and the relationship of whales to artiodactyls. *Nature*, 413:277–281.

Thewissen, JGM & EM Williams. 2002. The early radiations of Cetacea (Mammalia): evolutionary pattern and developmental correlations. *Annual Review of Ecology and Systematics*, 33:73–90.

Thewissen, JGM, et al. 2006. Developmental basis for hind-limb loss in dolphins and origin of the cetacean bodyplan. *Proceedings of the National Academy Sciences USA*, 103:8414–8418.

Thewissen, JGM, et al. 2007. Whales originated from aquatic artiodactyls in the Eocene epoch of India. *Nature*, 450:1190–1194.

Thewissen, JGM & S Nummela. 2008. *Sensory Evolution on the Threshold: Adaptations in Secondarily Aquatic Vertebrates*. University of California Press, Berkeley, CA.

Thies, W, EKV Kalko, & HU Schnitzler. 1998. The roles of echolocation and olfaction in two Neotropical fruit-eating bats, *Carollia perspicillata* and *C. castanea*, feeding on Piper. *Behavioral Ecology and Sociobiology*, 42:397–409.

Thiessen, DD, K Owen, & G Lindzey. 1971. Mechanisms of territorial marking in the male and female Mongolian gerbils (*Meriones unguiculatus*). *Journal of Comparative and Physiological Psychology*, 77:38–47.

Thom, MD, DDP Johnson, & DW MacDonald. 2004. The evolution and maintenance of delayed implantation in the mustelidae (Mammalia: Carnivora). *Evolution*, 58:175–183.

Thomas, DW, GP Bell, & MB Fenton. 1987. Variation in echolocation call frequencies recorded from North American vespertilionid bats: a cautionary note. *Journal of Mammalogy*, 68:842–847.

Thomas, JA & EC Birney. 1979. Parental care and mating system of the prairie vole, *Microtus ochrogaster*. *Behavioral Ecology and Sociobiology*, 5:171–186.

Thomas, JA & MS Jalili. 2004. Echolocation in insectivores and rodents, 27–35, in *Echolocation in Bats and Dolphins*

(JA Thomas, CF Moss, & M Vater, eds.). University of Chicago Press, Chicago, IL.

Thomas, JA, CF Moss, & M Vater. 2004. *Echolocation in Bats and Dolphins*. University of Chicago Press, Chicago, IL.

Thomas, SP & RA Suthers. 1972. The physiology and energetics of bat flight. *Journal of Experimental Biology*, 57:317–335.

Thompson, SD. 1982. Structure and species composition of desert heteromyid rodent species assemblages: effects of a simple habitat manipulation. *Ecology*, 63:1313–1321.

Thompson, SD, et al. 1980. The energetic cost of bipedal hopping in small mammals. *Nature*, 287:223–224.

Thorington, RW Jr, & S Anderson. 1984. Primates, 187–217, in *Orders and Families of Recent Mammals of the World* (S Anderson & JK Jones, eds.). Wiley, NY.

Thorington, RW Jr, K Darrow, & CG Anderson. 1998. Wing tip anatomy and aerodynamics in flying squirrels. *Journal of Mammalogy*, 79:245–250.

Tian, X, et al. 2006. Direct measurements of the kinematics and dynamics of bat flight. *Bioinspiration and Biomimetics*, 1:S10–S18.

Tilman, D & JA Downing. 1994. Biodiversity and stability in grasslands. *Nature*, 367:363–365.

Tilson, RL & JR Henschel. 1986. Spatial arrangement of spotted hyaena groups in a desert environment, Namibia. *African Journal of Ecology*, 24:173–180.

Timm, RM. 1987. Tent construction by bats of the genera *Artibeus* and *Uroderma*, 187–212, in *Studies in Neotropical Mammalogy: Essays in Honor of Philip Hershkovitz* (BD Patterson & RM Timm, eds.). Fieldiana Zoology No. 39, Field Museum Natural History, Chicago, IL.

Timm, RM & J Mortimer. 1976. Selection of roost sites by Honduran white bats, *Ectophylla alba* (Chiroptera: Phyllostomatidae). *Ecology*, 57:385–389.

Timm, RM & SE Lewis. 1991. Tent construction and use by *Uroderma bilobatum* in coconut palms (*Cocos nucifera*) in Costa Rica. *Bulletin of the American Museum of Natural History*, 206:251–260.

To, LP & RH Tamarin. 1977. The relation of population density and adrenal gland weight in cycling and non-cycling voles (*Microtus*). *Ecology*, 58:928–934.

Tobias, JA, et al. 2008. Comment on "The latitudinal gradient in recent speciation and extinction rates of birds and mammals." *Science*, 319:901c.

Tomasi, TE. 1978. Function of venom in the short-tailed shrew, *Blarina brevicauda*. *Journal of Mammalogy*, 59:852–854.

Torres, H. 1992. *South American Camelids: An Action Plan for Their Conservation*. IUCN/ SSC South American Camelid Specialist Group, IUCN, Gland, Switzerland.

Tosini, G & M Menaker. 1996. Circadian rhythms in cultured mammalian retina. *Science*, 272:419–421.

Townsend, MT. 1935. Studies on some small mammals of central New York. *Roosevelt Wildlife Annual*, 4:1–120.

Tracy, CR, JS Turner, & RB Huey. 1986. A biophysical analysis of possible thermoregulatory adaptations in sailed pelycosaurs, 195–206, in *The Ecology and Biology of Mammal-like Reptiles* (N Hotton III, PD MacLean, JJ Roth, & EC Roth, eds.). Smithsonian Institution Press, Washington, DC.

Tracy, RL & GE Walsberg. 2002. Kangaroo rats revisited: re-evaluating a classic case of desert survival. *Oecologia*, 133:449–457.

Travis, SE & CN Slobodchikoff. 1993. Effects of food resource distribution on the social system of Gunnison's prairie dog (*Cynomys gunnisoni*). *Canadian Journal of Zoology*, 71:1186–1192.

Travis, SE, CN Slobodchikoff, & P Keim. 1996. Social assemblages and mating relationships in prairie dogs: a DNA fingerprint analysis. *Behavioral Ecology*, 7:95–100.

Trivers, RL. 1971. The evolution of reciprocal altruism. *Quarterly Review of Biology*, 46:35–57.

Trivers, RL. 1972. Parental investment and sexual selection, 136–179, in *Sexual Selection and the Descent of Man, 1871–1971* (B Campbell, ed.). Aldine, Chicago, IL.

Troughton, E. 1947. *Furred Animals of Australia*. Scribner's, NY.

Troyer, JL, et al. 2004. Patterns of feline immunodeficiency virus multiple infection and genome divergence in a free-ranging population of African lions. *Journal of Virology*, 78:3777–3791.

Tschapka, M, AP Brooke, & LT Wasserthal. 2000. *Thyroptera discifera* (Chiroptera: Thyropteridae): observations on echolocation and new record in Costa Rica. *International Journal of Mammalian Biology*, 65:193–198.

Tschapka, M, et al. 2008. Diet and cranial morphology of *Musonycteris harrisoni*, a highly specialized nectar-feeding bat in western Mexico. *Journal of Mammalogy*, 89:924–932.

Turner, A. 1997. *The Big Cats and Their Fossil Relatives: An Illustrated Guide to Their Evolution and Natural History*. Columbia University Press, NY.

Turner, A & M Antón. 2004. *Evolving Eden: An Illustrated Guide to the Evolution of the African Large-Mammal Fauna*. Columbia University Press, NY.

Turner, BL, et al. 1990. *The Earth as Transformed by Human Action: Global and Regional Changes in the Biosphere Over the Past 300 Years*. Cambridge University Press, Cambridge, NY.

Turner, JS & CR Tracy. 1986. Body size, homeothermy, and the control of heat exchange in mammal-like reptiles, 185–194, in *The Ecology and Biology of Mammal-like Reptiles* (N Hotten III, PD MacLean, JJ Roth, & EC Roth, eds.). Smithsonian Institution Press, Washington, DC.

Turney, CSM, et al. 2008. Late-surviving megafauna in Tasmania, Australia, implicate human involvement in their extinction. *Proceedings of the National Academy of Sciences USA*, 34:12150–12153.

Turvey, ST, et al. 2007. First human-caused extinction of a cetacean species? *Biology Letters*, 3:537–540.

Tuttle, MD & MJ Ryan. 1981. Bat predation and the evolution of frog vocalizations in the Neotropics. *Science*, 214:677–678.

Tyack, PL, et al. 2006. Extreme diving of beaked whales. *Journal of Experimental Biology*, 209:4238–4253.

Tyndale-Biscoe, CH. 1980. Photoperiod and the control of seasonal reproduction in marsupials, 277–282, in *Endocrinology 1980. Proceedings of the VI International Congress of Endocrinology, Melbourne, Australia* (IA Cummin, JW Funder, & FAO Mendelsohn, eds.). Elsevier, Amsterdam, The Netherlands.

Tyndale-Biscoe, CH. 1984. Mammals—marsupials, 386–454, in *Marshall's Physiology of Reproduction* (GE Lamming, ed.). Churchill Livingstone, Edinburgh.

Tyndale-Biscoe, H. 2005. *Life of Marsupials*. CSIRO Publishing, Collingwood, Victoria, Australia.

Tyndale-Biscoe, CH & M Renfree. 1987. *Reproductive Physiology of Marsupials*. Cambridge University Press, Cambridge, UK.

Tyrrell, K. 1988. The use of prey-generated sounds in flycatcher-style foraging by *Megaderma spasma*. *Bat Research News*, 29:51.

Uchida, TA & T Mori. 1987. Prolonged storage of spermatozoa in hibernating bats, 351–365, in *Recent Advances in the Study of Bats* (MB Fenton, P Racey, & JMV Rayner, eds.). Cambridge University Press, Cambridge, UK.

Udvardy, M. D. F. 1969. *Dynamic Zoogeography*. Van Nostrand Reinhold, New York.

Ulanovsky, N & CF Moss. 2008. What the bat's voice tells the bat's brain. *Proceedings of the National Academy of Sciences USA*, 105:8491–8498.

U.S. Census Bureau. POPClock. http://www.census.gov/ipc/www/popclockworld.html (accessed Feb. 2010).

Vachrameev, VA & MA Akhmet'yev. 1972. The development of floras on the boundary of the Late Cretaceous and the Paleocene (on data from the study of leaf remains), in *The Development and Replacement of the Organic World on the Boundary of the Mesozoic and Cenozoic* (VN Shimansiy, & AN Solov'yev, eds.). (Conference April, 1972.) Abstract, Akademii Nauk SSSR, Moscow Soc. Natur. (In Russian.)

Valderrama, X, et al. 2000. Seasonal anointment with millipedes in a wild primate: a chemical defense against insects? *Journal of Chemical Ecology*, 12:2781–2790.

Valdez, R, et al. 2006. Wildlife conservation and management in Mexico. *Wildlife Society Bulletin*, 34:270–282.

Valentine, DE & CF Moss. 1998. Sensorimotor integration in bat sonar, 220–230, in *Bat Biology and Conservation* (TH Kunz & PA Racey, eds.). Smithsonian Institution Press, Washington, DC.

Van De Graaff, K & RP Balda. 1973. Importance of green vegetation for reproduction in the kangaroo rat, *Dipodomys merriami merriami*. *Journal of Mammalogy*, 54:509–512.

Van Den Bussche, R & SR Hoofer. 2004. Phylogenetic relationships among recent chiropteran families and the importance of choosing appropriate out-group taxa. *Journal of Mammalogy*, 85:321–330.

van der Klaauw, C. 1931. The auditory bulla in some fossil mammals. *Bulletin of the American Museum of Natural History*, 62:1–341.

Van Deusen, HM. 1967. Marsupials, 61–86, in *Recent Mammals of the World* (S Anderson, ed.). Ronald Press, NY.

Van Gelder, RG. 1953. The egg-opening technique of a spotted skunk. *Journal of Mammalogy*, 34:255–256.

Van Horn, RC, TL McElhinny, & KE Holekamp. 2003. Age estimation and dispersal in the spotted hyena (*Crocuta crocuta*). *Journal of Mammalogy*, 84:1019–1030.

Van Lawick, H & J Van Lawick-Goodall. 1970. *Innocent Killers*. Collins, London.

Van Lawick-Goodall, J. 1968. The behavior of free-living chimpanzees in the Gombe Stream Reserve. *Animal Behavior Monographs*, 1:1–311.

Van Lawick-Goodall, J. 1973. The behavior of chimpanzees in their natural habitat. *American Journal of Psychiatry*, 130:1–12.

van Oort, BE, et al. 2005. Circadian organization in reindeer. *Nature*, 438:1095–1096.

Van Roosmalen, MGM, et al. 1998. A new and distinctive species of marmoset (Callitrichidae, Primates) from the lower Rio Aripuanã, state of Amazonas, central Brazilian Amazonia. *Goeldiana Zoologia*, 22:1–27.

Van Valen, L. 1967. New Paleocene insectivores and insectivore classification. *Bulletin of the American Museum of Natural History*, 135:217–284.

Van Valen, L & RE Sloan, 1966. The extinction of the multituberculates. *Systematic Zoology*, 15:261–278.

Vartanyan, SL, et al. 2008. Collection of radiocarbon dates on the mammoths (*Mammuthus primigenius*) and other genera of Wrangel Island, northeast Siberia, Russia. *Quaternary Research*, 70:51–59.

Vater, M. 1987. Narrow-band frequency analysis in bats, 200–225, in *Recent Advances in the Study of Bats* (MB Fenton, P Racey, & JMV Rayner, eds.). Cambridge University Press, Cambridge, UK.

Vater, M. 1998. Adaptation of the auditory periphery of bats for echolocation, 231–245, in *Bat Biology and Conservation* (TH Kunz & PA Racey, eds.). Smithsonian Institution Press, Washington, DC.

Vaughan, TA. 1954. Mammals of the San Gabriel Mountains of California. *University of Kansas Publications, Museum Natural History*, 7:513–582.

Vaughan, TA. 1959. Functional morphology of three bats: *Eumops, Myotis, Macrotus*. *University of Kansas Publications, Museum Natural History*, 12:1–153.

Vaughan, TA. 1967. Food habits of the northern pocket gopher on shortgrass prairie. *American Midland Naturalist*, 77:176–189.

Vaughan, TA. 1969. Reproduction and population densities in a montane small mammal fauna, 51–74, in *Contributions in Mammalogy*. Miscellaneous Publications of the Museum of Natural History, University of Kansas, No. 51.

Vaughan, TA. 1970a. Adaptations for flight in bats, 127–143, in *About Bats* (BH Slaughter & DW Walton, eds.). Southern Methodist University Press, Dallas, TX.

Vaughan, TA. 1970b. The skeletal system. The muscular system. Flight patterns and aerodynamics, 97–138, in *Biology of Bats* (WA Wimsatt, ed.). Academic Press, NY.

Vaughan, TA. 1976. Nocturnal behavior of the African false vampire bat (*Cardioderma cor*). *Journal of Mammalogy*, 57:227–248.

Vaughan, TA. 1977. Foraging behavior of the giant leaf-nosed bat (*Hipposideros commersoni*). *East African Wildlife Journal*, 15:237–249.

Vaughan, TA. 1980. Opportunistic feeding by two species of *Myotis*. *Journal of Mammalogy*, 61:118–119.

Vaughan, TA & GC Bateman. 1970. Functional morphology of the forelimbs of mormoopid bats. *Journal of Mammalogy*, 51:217–235.

Vaughan, TA & MM Bateman. 1980. The molossid wing: some adaptations for rapid flight, 69–78, in *Proceedings of the Fifth International Bat Research Conference* (D Wilson & A Gardner, eds.). Texas Tech Press, Lubbock.

Vaughan, TA & ST Schwartz. 1980. Behavioral ecology of an insular woodrat. *Journal of Mammalogy*, 61:205.

Vaughan, TA & WP Weil. 1980. The importance of arthropods in the diet of *Zapus princeps* in a subalpine habitat. *Journal of Mammalogy*, 61:122–124.

Vaughan, TA & NJ Czaplewski. 1985. Reproduction in Stephens' woodrat: the wages of folivory. *Journal of Mammalogy*, 66:429–443.

Vaughan, TA & RP Vaughan. 1986. Seasonality and the behavior of the African yellow-winged bat. *Journal of Mammalogy*, 67:91–102.

Vaughan, TA & RP Vaughan. 1987. Parental behavior in the African yellow-winged bat (*Lavia frons*). *Journal of Mammalogy*, 68:217–223.

Vehrencamp, SL, FG Stiles, & JW Bradbury. 1977. Observations on the foraging behavior and avian prey of the Neotropical carnivorous bat, *Vampyrum spectrum*. *Journal of Mammalogy*, 58:469–478.

Velazco, PM & BD Patterson. 2008. Phylogenetics and biogeography of the broad-nosed bats, genus *Platyrrhinus* (Chiroptera: Phyllostomidae). *Molecular Phylogenetics and Evolution*, 49:749–759.

Vermeulen, HC & JAJ Nel. 1988. The bush Karoo rat, *Otomys unisulcatus*, on the Cape West coast South Africa. *Journal of Zoology*, 23:103–111.

Vieira, EM & D Astúa de Moraes. 2003. Carnivory and insectivory in Neotropical marsupials, 271–284, in *Predators with Pouches: The Biology of Carnivorous Marsupials* (M Jones, C Dickman, & M Archer, eds.). CSIRO Publishing, Collingwood, Victoria, Australia.

Vigne, J-D, et al. 2004. Early taming of the cat in Cyprus. *Science*, 304:259.

Vila, C, et al. 2001. Widespread origins of domestic horse lineages. *Science*, 291:474–477.

Villa-Ramirez, B. 1966. *Los Murciélagos de Mexico*. Inst. Biol., UNAM.

Villa-Ramirez, B & EL Cockrum, 1962. Migration in the guano bat, *Tadarida brasiliensis mexicana*. *Journal of Mammalogy*, 43:43–64.

Villarreal, D, et al. 2008. Alterations of ecosystem structure by a burrowing herbivore, the plains vizcacha (*Lagostomus maximus*). *Journal of Mammalogy*, 89:700–711.

Visser, IN, et al. 2007. Antarctic peninsula killer whales (*Orcinus orca*) hunt seals and a penguin on floating ice. *Marine Mammal Science*, 24:225–234.

Vitousek, PM, et al. 1997. Human domination of earth's ecosystems. *Science*, 277:494–499.

Vizcaíno, SF. 1994. Sistematica y anatomia de los Astegotheriini Ameghino, 1906 (Nuevo rango) (Xenarthra, Dasypodidae, Dasypodinae). *Ameghiniana*, 31:3–13.

Vizcaíno, SF, et al. 1998. Antarctica as background for mammalian evolution. In *Paléogeno de América del Sur y de la Península Antártica*. Asociación Paleontológica Argentina Publicación Especial 5:199–209.

Vizcaíno, SF & WJ Loughry. 2008. *The Biology of the Xenarthra*. University Press of Florida, Gainesville.

Vleck, D. 1979. The energy cost of burrowing by the pocket gopher *Thomomys bottae*. *Physiological Zoology*, 52:122–136.

Vogel, P. 1983. Contribution a l'écologie et a la zoogéographie de *Micropotamogale lamottei* (Mammalia, Tenrecidae). *Rev. Ecol. (Terre Vie)*, 38:37–49.

Vogel, S. 1988. *Life's Devices: The Physical World of Animals and Plants*. Princeton University Press, Princeton, NJ.

Vogl, RJ. 1973. Ecology of the knobcone pine in the Santa Ana Mountains, California. *Ecological Monographs*, 43:125–143.

Voigt, CC & O von Helversen. 1999. Storage and display of odour by male *Saccopteryx bilineata* (Chiroptera, Emballonuridae). *Behavioral Ecology and Sociobiology*, 47:29–40.

Voigt, CC, et al. 2001. The economics of harem maintenance in the sac-winged bat, *Saccopteryx bilineata* (Emballonuridae). *Behavioral Ecology and Sociobiology*, 50:31–36.

Voigt, CC & JR Speakman. 2007. Nectar-feeding bats fuel their high metabolism directly with exogenous carbohydrates. *Functional Ecology*, 21:913–921.

Voigt, CC, et al. 2008. Songs, scents, and senses: sexual selection in the greater sac-winged bat, *Saccopteryx bilineata*. *Journal of Mammalogy*, 89:1401–1410.

Von Muggenthaler, E, et al. 2003. Songlike vocalizations from the Sumatran rhinoceros (*Dicerorhinus sumatrensis*). *Acoustics Research Letters*, 4:83–88.

von Zittel, KA. 1893. *Handbuch der Palaeontologie. I Abteilung. Palaeozoologie. IV Band. Vertebrata (Mammalia)*. R Oldenbourg, Munich.

Vorhies, MR. 1974. Fossil pocket mice burrows in Nebraska. *American Midland Naturalist*, 91:492–498.

Vrba, ES. 1992. Mammals as a key to evolutionary theory. *Journal of Mammalogy*, 73:1–28.

Vrba, ES. 1993. Mammal evolution in the African Neogene and a new look at the Great American Interchange, 393–432, in *Biological Relationships between Africa and South America* (P. Goldblatt, ed.). Yale University Press, New Haven, CT.

Vucetich, MG, DH Verzi, & J-L Hartenberger. 1999. Review and analysis of the radiation of the South American Hystricognathi (Mammalia, Rodentia). *C.R. Acad. Sci Paris*, 329:763–769.

Wada, S, M Oishi, & TK Yamada. 2003. A newly discovered species of living baleen whales. *Nature*, 426: 278–281.

Wahlert, JH. 1985. Skull morphology and relationships of geomyoid rodents. *American Museum Novitates*, 2819:1–20.

Wai-Ping, V & MB Fenton. 1989. Ecology of spotted bats (*Euderma maculatum*): roosting and foraging behavior. *Journal of Mammalogy*, 70:617–622.

Walker, A. 1969. The locomotion of the lorises, with special reference to the potto. *East African Wildlife Journal*, 8:1–5.

Walker, FM, AC Taylor, & P Sunnucks. 2008. Female dispersal and male kinship-based association in southern hairy-nosed wombats (*Lasiorhinus latifrons*). *Molecular Ecology*, 17:1361–1374.

Wall, CE & DW Krause. 1992. A biomechanical analysis of the masticatory apparatus of *Ptilodus* (Multituberculata). *Journal of Vertebrate Paleontology*, 12:172–187.

Wallace, AR. 1876. *The Geographical Distribution of Animals*. 2 vols. Harper, New York. Reprinted by Hafner, NY.

Wallace, SC & X Wang. 2004. Two new carnivores from an unusual late Tertiary forest biota in eastern North America. *Nature*, 431:556–559.

Waltari, E, et al. 2007. Locating Pleistocene refugia: comparing phylogeographic and ecological niche model predictions. *PLoS One*, 7(e563):1–11.

Walther, F. 1965. Verhaltensstudien an der Grantgazell (*Gazella granti* Brooke, 1872) im Ngorongoro-Krater. *Zeitschrift fur Tierpsychologie*, 22:167–208.

Walther, F. 1966. Zum liegeverhalten des weisschwanzgnus (*Connochaetes gnou* Zimmerman, 1780). *Zeitschrift fur Säugetierkunde*, 31:1–16.

Walton, LR, et al. 2001. Movement patterns of barren-ground wolves in the central Canadian arctic. *Journal of Mammalogy*, 82:867–876.

Wang, X & M Novak. 1992. Influence of the social environment on parental behavior and pup development of meadow voles (*Microtus pennsylvanicus*) and prairie voles (*M. ochrogaster*). *Journal of Comparative Psychology*, 106:163–171.

Wang, X & RH Tedford. 2008. *Dogs, Their Fossil Relatives and Evolutionary History*. Columbia University Press, NY.

Wang, Y, et al. 2007. Bats respond to polarity of a magnetic field. *Proceedings of the Royal Society B*, 274:2901–2905.

Ward, AL & JO Keith. 1962. Feeding habits of pocket gophers on mountain grasslands. *Ecology*, 43:744–749.

Ward, S & MB Renfree. 1986. Some aspects of reproduction in *Acrobates pygmaeus*. *Australian Mammal Society Bulletin*, 9(suppl): Abstract 20.

Warnecke, L, JM Turner, & F Geiser. 2008. Torpor and basking in a small arid zone marsupial. *Naturwissenschaften*, 95:73–78.

Warren, WC, et al. 2008. Genome analysis of the platypus reveals unique signatures of evolution. *Nature*, 453:175–184.

Watkins, WA. 1977. Acoustic behaviors of sperm whales. *Oceanus*, 20:50–58.

Watkins, WA, et al. 1993. Sperm whales tagged with transponders and tracked underwater by sonar. *Marine Mammal Science*, 9:55–67.

Watson, RM. 1968. Report on aerial photographic studies of vegetation carried out in the Tsavo area of Kenya. Typescript. (Cited by Laws, 1970).

Watts, RD, et al. 2007. Roadless space of the conterminous United States. *Science*, 316:736–738.

Wayne, RK, et al. 1989. Molecular and biochemical evolution of the Carnivora, 465–494, in *Carnivore Behavior, Ecology, and Evolution* (JL Gittleman, ed.). Cornell University Press, Ithaca, NY.

Webb, SD. 1985. The interrelationships of tree sloths and ground sloths, 105–112, in *The Evolution and Ecology of Armadillos, Sloths, and Vermilinguas* (GG Montgomery, ed.). Smithsonian Institution Press, Washington, DC.

Webb, SD. 1991. Ecogeography and the Great American Interchange. *Paleobiology*, 17:266–280.

Webb, SD. 2006. The Great American Biotic Interchange: patterns and processes. *Annals of the Missouri Botanical Garden*, 93:245-257.

Webb, SD & BE Taylor. 1980. The phylogeny of hornless ruminants and a description of the cranium of *Archaeomeryx*. *Bulletin of the American Museum of Natural History*, 167:117–158.

Webb, SD & AD Barnosky. 1989. Faunal dynamics of Pleistocene mammals. *Annual Review of Earth Planetary Science*, 17:413–438.

Webb, SD & A Rancy. 1996. Late Cenozoic evolution of the Neotropical mammal fauna, 335–358, in *Evolution and Environment in Tropical America* (JBC Jackson, AF Budd, & AG Coates, eds.). University of Chicago Press, Chicago, IL.

Weber, NS & JS Findley. 1970. Warm-season changes in fat content of *Eptesicus fuscus*. *Journal of Mammalogy*, 51:160–162.

Webster, FA & DR Griffin. 1962. The role of flight membranes in insect capture by bats. *Animal Behaviour*, 10:332–340.

Wecker, SC. 1963. The role of early experience in habitat selection by the prairie deer mouse, *Peromyscus maniculatus bairdi*. *Ecological Monographs*, 33:307–325.

Wegener, A. 1912. Die entstehung der Kontinente. *Geologische Rundschau*, 3:276–292.

Wegener, A. 1915. *Die Entstehung der Kontinente und Ozeane*. Braunschweig, Vieweg.

Wegener, A. 1966. *The Origin of Continents and Oceans* (Translation of 1929 edition by J. Biram.) Dover Publications, NY.

Weilgart, L & H Whitehead. 1997. Group-specific dialects and geographical variation in coda repertoire in South Pacific sperm whales. *Behavioral Ecology and Sociobiology*, 40:277–285.

Weinbeer, M & EKV Kalko. 2007. Ecological niche and phylogeny: the highly complex echolocation behavior of the trawling long-legged bat, *Macrophyllum macrophyllum*. *Behavioral Ecology and Sociobiology*, 61:1337–1348.

Weir, BJ. 1974. The tuco-tuco and plains viscacha, 113–130, in *The Biology of the Hystricomorph Rodents* (IW Rowlands & BJ Weir, eds.). Academic Press, NY.

Weir, JT & D Schluter. 2007. The latitudinal gradient in recent speciation and extinction rates of birds and mammals. *Science*, 315:1574–1576.

Weir, JT & D Schluter. 2008. Response to comment on "The latitudinal gradient in recent speciation and extinction rates of birds and mammals." *Science*, 319:901d.

Welbergen, JA, et al. 2008. Climate change and the effects of temperature extremes on Australian flying-foxes. *Proceedings of the Royal Society B*, 275(1633):419–425. http://rspb.royalsocietypublishing.org/content/275/1633/419.full.pdf+html?sid=55daa2c9-2d5b-4ea6-866e-9222bdd0ba19.

Wells, MC & M Bekoff. 1981. An observational study of scent marking in coyotes, *Canis latrans*. *Animal Behaviour*, 29:332–350.

Wells, PV & CD Jorgensen. 1964. Pleistocene wood rat middens and climatic change in the Mojave Desert: a record of juniper woodlands. *Science*, 143:1171–1174.

Welsey-Hunt, GD & JJ Flynn. 2005. Phylogeny of the Carnivora: basal relationships among the carnivoramorphans, and assessment of the position of "Miacoidea" relative to Carnivora. *Journal of Systematic Paleontology*, 3:1–28.

Wemmer, C, et al. 1993. Training developing-country nationals is the critical ingredient to conserving global biodiversity. *BioScience*, 43:762–767.

Werdelin, L & A Nilsonne. 1999. The evolution of the scrotum and testicular descent: a phylogenetic view. *Journal of Theoretical Biology*, 196:61–72.

West, SD. 1977. Midwinter aggregation in the northern red-backed vole (*Clethrionomys rutilus*). *Canadian Journal of Zoology*, 55:1404–1409.

Westerman, M & D Edwards. 1992. DNA hybridization and the phylogeny of monotremes, 28–34, in *Platypus and Echidnas* (ML Augee, ed.). Royal Zoology Society of New South Wales, Sydney, Australia.

Wetterer, AL, MV Rockman, & NB Simmons. 2000. Phylogeny of phyllostomid bats (Mammalia: Chiroptera): data from diverse morphological systems, sex chromosomes, and restriction sites. *Bulletin of the American Museum of Natural History*, 248:1–200.

Wetzel, RM. 1977. The Chacoan peccary *Catagonus wagneri* (Rusconi). *Bulletin of Carnegie Museum of Natural History*, 3:1–36.

Wetzel, RM, et al. 1975. *Catagonus*, an "extinct" peccary, alive in Paraguay. *Science*, 189:379–381.

Wharton, CH. 1950. Notes on the life history of the flying lemur, *Cynocephalus volans*. Journal of Mammalogy, 31:269–273.

Whitaker, JO Jr. 1963. Food, habitat and parasites of the woodland jumping mouse in central New York. *Journal of Mammalogy*, 44:316–321.

Whitaker, JO Jr. 1968. Parasites, 254–311, in *The Biology of Peromyscus* (JA King, ed.). Special Publication No. 2., American Society of Mammalogists.

White F. 1996. Comparative socio-ecology of *Pan paniscus*, 29–41, in *Great Ape Societies* (WC McGrew, LF Marchant, & T Nishida, eds.) Cambridge University Press, Cambridge, UK.

White, JL. 1997. Locomotor adaptations in Miocene xenarthrans, 246–264, in *Vertebrate Paleontology in the Neotropics: The Miocene Fauna of La Venta, Colombia* (RF Kay, RH Madden, RL Cifelli, & JJ Flynn, eds.). Smithsonian Institution Press, Washington, DC.

White, M. 2009. Path of the jaguar. *National Geographic*, 215:122–133.

White, TD, et al. 2003. Pleistocene *Homo sapiens* from Middle Awash, Ethiopia. *Nature*, 423:742–747.

White, TD, et al. 2006. Asa Issie, Aramis and the origin of *Australopithecus*. *Nature*, 440:883–889.

Whittington, CM, et al. 2008a. Defensins and the convergent evolution of platypus and reptile venom genes. *Genome Research*, 18:986–994.

Whittington, CM, et al. 2008b. Expression patterns of platypus defensin and related venom genes across a range of tissue types reveal the possibility of broader functions for OvDLPs than previously suspected. *Toxicon*, 52:559–565.

Whittow, GC. 1974. Sun, sand, and sea lions. *Natural History*, 83:56–63.

Wible, JR. 1991. Origin of Mammalia: the craniodental evidence re-examined. *Journal of Vertebrate Paleontology*, 11:1–28.

Wible, JR. 2003. On the cranial osteology of the short-tailed opossum *Monodelphis brevicaudata* (Didelphidae, Marsupialia). *Annals of Carnegie Museum*, 72:137–202.

Wible, JR. 2008. On the cranial osteology of the Hispaniolan solenodon, *Solenodon paradoxus* Brandt, 1833 (Mammalia, Lipotyphla, Solenodontidae). *Annals of Carnegie Museum*, 77:321–402.

Wible, JR & HH Covert. 1987. Primates: cladistic diagnosis and relationships. *Journal of Human Evolution*, 16:1–22.

Wible, JR, D Miao, & JA Hopson. 1990. The septomaxilla of fossil and Recent synapsids and the problem of the septomaxilla of monotremes and armadillos. *Zoology Journal of the Linnean Society*, 98:203–228.

Wible, JR, et al. 1995. A mammalian petrosal from the Early Cretaceous of Mongolia: implications for the evolution of the ear region and mammaliamorph interrelationships. *American Museum Novitates*, 3149:1–19.

Wible, JR & TJ Gaudin. 2004. On the cranial osteology of the yellow armadillo *Euphractus sexcinctus* (Dasypodidae, Xenarthra, Placentalia). *Annals of Carnegie Museum*, 73:117–196.

Wible, JR, et al. 2005. Cranial anatomy and relationships of the new Ctenodactyloid (Mammalia, Rodentia) from the early Eocene of Hubei Province, China. *Annals of Carnegie Museum*, 74:91–150.

Wible, JR, et al. 2009. The eutherian mammal *Maelestes gobiensis* from the Late Cretaceous of Mongolia and the phylogeny of Cretaceous Eutheria. *Bulletin of the American Museum of Natural History*, 327:1-123.

Wickler, W Von & D Uhrig. 1969. Verhalten und okologische Nische der Gelbflugelfledermaus, *Lavia frons* (Geoffroy) (Chiroptera, Megadermatidae). *Zeitschrift fur Tierpsychologie*, 26:726–736.

Wildman, DE, et al. 2006. Evolution of the mammalian placenta revealed by phylogenetic analysis. *Proceedings of the National Academy of Sciences USA*, 103:3203–3208.

Wiens, F, A Zitzmann, & NA Hussein. 2006. Fast food for slow lorises: is low metabolism related to secondary compounds in high-energy plant diet? *Journal of Mammalogy*, 87:790–798.

Wiens, F, et al. 2008. Chronic intake of fermented floral nectar by wild treeshrews. *Proceedings of the National Academy of Sciences USA*, 105:10426–10431.

Wilderness Awareness School. http://www.wildernessawareness.org/links_wilderness_programs.html (accessed March 2009).

Wiles, GJ. 1992. Recent trends in the fruit bat trade on Guam, 53–60, in *Pacific Flying Foxes: Proceedings of an International Conservation Conference* (DE Wilson & GL Graham, eds.). U.S. Fish and Wildlife Service, Biological Report 90(23).

Wiley, EO. 1981. *Phylogenetics: The Theory and Practice of Phylogenetic Systematics*. Wiley, NY.

Wilkinson, GS. 1984. Reciprocal food sharing in the vampire bat. *Nature*, 308:181–184.

Wilkinson, GS. 1987. Altruism and co-operation in bats, 299–323, in *Recent Advances in the Study of Bats* (MB Fenton, P Racey, & JMV Rayner, eds.). Cambridge University Press, NY.

Wilkinson, GS. 1990. Food sharing in vampire bats. *Scientific American*, 262:76–82.

Willerslev, E, et al. 2007. Ancient biomolecules from deep ice cores reveal a forested southern Greenland. *Science*, 317:111–113.

Williams, CB. 1964. *Patterns in the Balance of Nature*. Academic Press, NY.

Williams, DF. 1991. Habitat of shrews (genus *Sorex*) in forest communities of the western Sierra Nevada, California, 1–14, in *The Biology of the Soricidae* (JS Findley & TL Yates, eds.). Special Publication No. 1, Museum Southwestern Biology, Albuquerque, NM.

Williams, GC. 1967. Natural selection, the costs of reproduction, and a refinement of Lack's principle. *American Naturalist*, 100:687–690.

Williams, JB, et al. 2002. Energy expenditure and water flux of Ruppell's foxes in Saudi Arabia. *Physiological and Biochemical Zoology*, 75:479–488.

Williams, JW, ST Jackson, & JE Kutzbach. 2007. Projected distributions of novel and disappearing climates by 2100 AD. *Proceedings of the National Academy of Sciences USA*, 104(14):5738–5742.

Williams, SE, EE Bolitho, & S Fox. 2003. Climate change in Australian tropical rainforests: an impending environmental catastrophe. *Proceedings of the Royal Society B*, 270:1887–1892.

Williams, TC, LC Ireland, & JM Williams. 1973. High altitude flights of the free-tailed bat, Tadarida brasiliensis, observed with radar. *Journal of Mammalogy*, 54:807–821.

Williams, TM. 1989. Swimming by sea otters: adaptations for low energetic cost locomotion. *Journal of Comparative Physiology A*, 164:815–824.

Williams-Guillen, K, I Perfecto, & J Vandermeer. 2008. Bats limit insects in a Neotropical agroforestry system. *Science*, 320:70.

Willis, KJ & SA Bhagwat. 2009. Biodiversity and climate change. *Science*, 326:806-807.

Wilmé, L, SM Goodman, & JU Ganzhorn. 2006. Biogeographic evolution of Madagascar's microendemic biota. *Science*, 312:1063–1065.

Wilson, DE. 1979. Reproductive patterns, 317–378, in *Biology of Bats of the New World Family Phyllostomatidae. Part III* (RJ Baker, J Jones Jr, & DC Carter, eds.). Special Publication of the Museum, No. 16. Texas Tech Press, Lubbock.

Wilson, DJ, CJ Krebs, & T Sinclair. 1999. Limitation of collared lemming populations during a population cycle. *Oikos*, 87:382–398.

Wilson, DE & DM Reeder. 2005. *Mammal Species of the World: A Taxonomic and Geographic Reference.* 3rd ed. Johns Hopkins University Press, Baltimore, MD.

Wilson, EO. 1975. *Sociobiology: The New Synthesis.* University of Chicago Press, Chicago, IL.

Wilson, EO. 1984. *Biophilia.* Harvard University Press, Cambridge, MA.

Wilson, EO. 1992. *The Diversity of Life.* Belknap Press of Harvard University Press, Cambridge, MA.

Wilson, EO. 1993. Is humanity suicidal? *New York Times Magazine,* May 30:24–29.

Wilson, EO. 2006. *The Creation: An Appeal to Save Life on Earth.* W. W. Norton, NY.

Wilson, JA & AC Runkel. 1991. Prolapsus, a large sciuravid rodent and new eomyids from the Late Eocene of Trans-Pecos Texas. *Pearce-Sellards Series,* No. 48:1–30.

Wilson, M, et al. 2007. Intense ultrasonic clicks from echolocating toothed whales do not elicit anti-predator responses or debilitate the squid *Loligo pealeii. Biology Letters,* 3:225–227.

Wilson, RW. 1949. Early Tertiary rodents of North America. *Carnegie Institute of Washington,* 584:67–164.

Wilson, RW. 1960. *Early Miocene Rodents and Insectivores from Northeastern Colorado.* University of Kansas Paleontological Contributions Vertebrata, 7.

Wilson, RW. 1972. Evolution and extinction in early Tertiary rodents. *Proceedings of the International Geology Congress,* 24:217–224.

Wimsatt, WA. 1944. Further studies on the survival of spermatozoa in the female reproductive tract of the bat. *Anatomical Record,* 88:193–204.

Wimsatt, WA. 1945. Notes on breeding behavior, pregnancy, and parturition in some vespertilionid bats of eastern United States. *Journal of Mammalogy,* 26:23–33.

Wimsatt, WA. 1969a. Transient behavior, nocturnal activity patterns, and feeding efficiency of vampire bats (*Desmodus rotundus*) under natural conditions. *Journal of Mammalogy,* 50:233–244.

Wimsatt, WA & B Villa-Ramirez, 1970. Locomotor adaptations in the disc-winged bat. *American Journal of Anatomy,* 129:89–119.

Wing, SL, LJ Hickey, & CC Swisher. 1993. Implications of an exceptional fossil flora for late Cretaceous vegetation. *Nature,* 363:342–344.

Winge, H. 1941. *The Interrelationships of the Mammalian Genera.* Vol. 1. C.A. Reitzels Forlag, Copenhagen. (Danish translated by E Deichmann, & GM Allen.)

Winter, Y, J Lopez, & O von Helversen. 2003. Ultraviolet vision in a bat. *Nature,* 425:612–614.

Wislocki, GB. 1942. Studies on the growth of deer antlers. I: On the structure and histogenesis of the antlers of the Virginia deer (*Odocoileus virginianus borealis*). *American Journal of Anatomy,* 71:371–415.

Wislocki, GB, JC Aub, & CM Waldo. 1947. The effects of gonadectomy and the administration of testosterone proprionate on the growth of antlers in male and female deer. *Endocrinology,* 40:202–224.

Withers, PC. 1978. Bioenergetics of a "primitive" mammal, the Cape golden mole. *South African Journal of Science,* 74:347–348.

Withers, PC. 1992. *Comparative Animal Physiology.* Saunders College, Philadelphia, PA.

Withers, PC, GN Louw, & J Henschel. 1979. Energetics and water relations of Namib Desert rodents. *South African Journal of Zoology,* 15:131–137.

Withers, PC & JUM Jarvis. 1980. The effect of huddling on the thermoregulation and oxygen consumption for the naked mole-rat. *Comparative Biochemistry and Physiology,* 66:215–219.

Witmer, LM, SD Sampson, & N Solounias. 1999. The proboscis of tapirs (Mammalia: Perissodactyla): a case study in novel narial anatomy. *Journal of Zoology,* 249:249–267.

Wojcik, JM & M Wolsan. 1998. *Evolution of Shrews.* Mammal Research Institute, Polish Academy of Sciences, Bialowieza.

Wolfe, JA. 1978. A paleobotanical interpretation of Tertiary climates in the Northern Hemisphere. *American Scientist,* 66:694–703.

Wolfe, JA. 1994. Tertiary climatic changes at middle latitudes of western North America. *Palaeogeography, Palaeoclimatology, and Palaeoecology,* 108:195–205.

Wolff, JO. 1980. Social organization of the taiga vole (*Microtus xanthognathus*). *Biologist,* 62:34–45.

Wolff, JO. 1997. Population regulation in mammals: an evolutionary perspective. *Journal of Animal Ecology,* 66:1–13.

Wolff, JO & WZ Lidicker. 1981. Communal winter nesting and food sharing in taiga voles. *Behavioral Ecology and Sociobiology,* 9:237–240.

Wolff, JO, JL Lundy, & R Baccus. 1988. Dispersal, inbreeding avoidance and reproductive success in white-footed mice. *Animal Behaviour,* 36:456–465.

Wolff, JO & DW MacDonald. 2004. Promiscuous females protect their offspring. *Trends in Ecology and Evolution,* 19:127–134.

Wolff, JO & PW Sherman. 2007. *Rodent Societies: An Ecological and Evolutionary Perspective.* University of Chicago Press, Chicago, IL.

Wood, AE. 1935. Evolution and relationships of the heteromyid rodents with new forms from the Tertiary of western North America. *Annals of Carnegie Museum,* 24:73–262.

Wood, AE. 1957. What, if anything, is a rabbit? *Evolution,* 11:417–425.

Wood, AE. 1965. Grades and clades among rodents. *Evolution,* 19:115–130.

Wood, AE. 1980. The origin of caviomorph rodents from a source in Middle America: a clue to the area of origin of the platyrrhine primates, 79–91, in *Evolutionary Biology of the New World Monkeys and Continental Drift* (RI Ciochon & AB Chiarelli, eds.). Plenum Press, NY.

Wood, DH. 1970. An ecological study of *Antechinus stuartii* (Marsupialia) in a south-east Queensland rain forest. *Australian Journal of Zoology,* 19:347–353.

Wood, FG Jr. 1959. Underwater sound production and concurrent behavior of captive porpoises, *Tursiops truncatus* and *Stenella plagiodon. Bulletin of Marine Science of the Gulf and Caribbean,* 3:120–133.

Woodall, PF. 1995. The male reproductive system and the phylogeny of elephant-shrews (Macroscelidea). *Mammal Review,* 25:87–93.

Woodburne, ME & RH Tedford, 1975. The first Tertiary monotreme from Australia. *American Museum Novitates,* 2588:1–11.

Woodburne, MO (ed.). 2004a. *Late Cretaceous and Cenozoic Mammals of North America: Biostratigraphy and Geochronology.* Columbia University Press, NY.

Woodburne, MO (ed.). 2004b. Global events and the North American mammalian biochronology, 315–343, in *Late Cretaceous and Cenozoic Mammals of North America: Biostratigraphy and*

Geochronology. Columbia University Press, NY.

Woodburne, MO & CC Swisher III. 1995. Land mammal high-resolution geochronology, intercontinental overland dispersals, sea level, climate, and vicariance, in *Geochronology Time Scales and Global Stratigraphic Correlation.* SEPM Special Publication No. 54. Society for Sedimentary Geology, Tulsa, OK.

Woodburne, MO & WJ Zinsmeister. 1982. Fossil land mammal from Antarctica. *Science,* 218:284–286.

Woodburne, MO & JA Case. 1996. Dispersal, vicariance, and the Late Cretaceous to early Tertiary land mammal biogeography from South America to Australia. *Journal of Mammalian Evolution,* 3:121–161.

Woodburne, MO, TH Rich, & MS Springer. 2003. The evolution of tribosphery and the antiquity of mammalian clades. *Molecular Phylogenetics and Evolution,* 28:360–385.

Woodburne, MO, et al. 2009. Climate directly influences Eocene mammal faunal dynamics in North America. *Proceedings of the National Academy of Sciences USA,* 106(32):13399–13403.

Wooding, JB. 1993. *Management of the Black Bear in Florida: A Staff Report to the Commissioners.* Florida Game and Fresh Water Fish Commission, Tallahassee, FL.

Woodroffe, R & A Vincent. 1991. Mother's little helpers: patterns of male care in mammals. *Trends in Ecology and Evolution,* 9:294–297.

Woods, CA. 1982. The history and classification of South American hystricognath rodents—reflections on the far away and long ago, 377–392, in *Mammalian Biology in South America* (MA Mares & HH Genoways, eds.). Special Publication Pymatuning Laboratory of Ecology, University of Pittsburgh Press, Pittsburgh, PA.

Woods, CA. 1984. Hystricognath rodents, 389–446, in *Orders and Families of Recent Mammals of the World* (S Anderson & J Jones, eds.). Wiley, NY.

Woods, CA. 1993. Suborder Hystricognathi, 771–806, in *Mammal Species of the World* (DE Wilson & DM Reeder, eds.). Smithsonian Institution Press, Washington, DC.

Woods, CA, JA Ottenwalder, & WLR Oliver. 1985. Lost mammals of the Greater Antilles: summarized findings of a ten week field survey in the Dominican Republic, Haiti, and Puerto Rico. *Dodo* (Jersey Wildlife Preservation Trust), 22:23–42.

Woodsworth, GC, GP Bell, & MB Fenton. 1981. Observations on the echolocation, feeding behavior and habitat use of *Euderma maculatum* in southcentral British Columbia. *Canadian Journal of Zoology,* 59:1099–1102.

Woolley, P. 1966. Reproduction in *Antechinus* spp., and other dasyurid marsupials. *Symposium of the Zoological Society of London,* 15:281–294.

Worthy, TH, MJ Daniel, & JE Hill. 1996. An analysis of skeletal size variation in *Mystacina robusta* Dwyer, 1962 (Chiroptera: Mystacinidae). *New Zealand Journal of Zoology,* 23:99–110.

Wozencraft, WC. 1989. The phylogeny of the Recent Carnivora, 495–535, in *Carnivores; Behavior, Ecology, and Evolution* (JL Gittleman, ed.). Cornell University Press, Ithaca, NY.

Wozencraft, WC. 2005. Order Carnivora, 532–628, in *Mammal Species of the World* (DE Wilson & DM Reeder eds.). Johns Hopkins University Press, Baltimore, MD.

Wright, PC. 1994. The behavior and ecology of the owl monkey, 97–112, in *Aotus: The Owl Monkey* (JF Baer, RE Weller, & I Kakoma, eds.). Academic Press, San Diego, CA.

Wright, PC. 1996. The neotropical primate adaptation to nocturnality: feeding in the night (*Aotus nigriceps* and *A. azarae*), 369–381, in *Adaptive Radiations of Neotropical Primates* (MA Norconk, AL Rosenberger, & PA Garber, eds.). Plenum Press, NY.

Wright, S. 1977. *Evolution and the Genetics of Populations: Experimental Results and Evolutionary Deductions.* University of Chicago Press, Chicago, IL.

Wright, SJ. 2005. Tropical forests in a changing environment. *Trends in Ecology and Evolution,* 20:553–560.

Wyatt, KB, et al. 2008. Historical mammal extinction on Christmas Island (Indian Ocean) correlates with introduced infectious disease. *PLoS One,* 3:e3602:1–9.

Wynen, L, et al. 2001. Phylogenetic relationships within the eared seals (Otariidae: Carnivora): implications for the historical biogeography of the family. *Molecular Phylogenetics and Evolution,* 21(2):270–284.

Wyss, AR & JJ Flynn. 1993. A phylogenetic analysis and definition of the Carnivora, 32–53, in *Mammal Phylogeny, Placentals. Vol. 2* (FS Szalay, MJ Novacek, & MC McKenna, eds.). Springer-Verlag, NY.

Xu, Z, et al. 1995. Self-anointing behavior in the rice-field rat, *Rattus rattoides. Journal of Mammalogy,* 76:1238–1241.

Yaakobi, D & A Shkolnik. 1974. Structure and concentrating capacity in kidneys of hedgehogs. *American Journal of Physiology,* 226:948–952.

Yalden, DW. 1966. The anatomy of mole locomotion. *Journal of Zoology London,* 149:5–64.

Yamazaki, K, et al. 1976. Control of mating preferences in mice by genes in the major histocompatibility complex. *Journal of Experimental Medicine,* 144:1324–1335.

Yamazaki, K, et al. 1980. The major histocompatibility complex as a source of odors imparting individuality among mice, 267–273, in *Chemical Signals* (D Müchwartze & RM Silverstein, eds.). Plenum Press, NY.

Yoakum, J. 1958. Seasonal food habits of the Oregon pronghorn antelope (*Antilocapra americana oregona* Bailey). *International Antelope Conference Transactions,* 9:47–59.

Ylonen, H, et al. 2002. Predation risk and habitat selection of Australian house mice *Mus domesticus* during an incipient plague: desperate behavior due to food depletion. *Oikos,* 99:284–289.

Yoder, AD. 2003a. The phylogenetic position of genus *Tarsius*: whose side are you on? 161–175, in *Tarsiers: Past, Present, and Future* (PC Wright, EL Simons, & S Gursky, eds.). Rutgers University Press, New Brunswick, NJ.

Yoder, AD. 2003b. Phylogeny of the lemurs, 1242–1247, in *The Natural History of Madagascar* (SM Goodman, & JP Benstead, eds.). University of Chicago Press, Chicago, IL.

Yoder, AD & JJ Flynn. 2003. Origin of Malagasy Carnivora, 1253–1256, in *The Natural History of Madagascar* (SM Goodman & JP Benstead, eds.). University of Chicago Press, Chicago, IL.

Yoder, AD, et al. 2003. Single origin of Malagasy Carnivora from an African ancestor. *Nature,* 421:734–737.

Yoder, AD & Z Yang. 2004. Divergence dates for Malagasy lemurs estimated from multiple gene loci: geological and evolutionary context. *Molecular Ecology,* 13:757–773.

York, AE & VB Scheffer. 1997. Timing of implantation in the northern fur seal, *Callorhinus ursinus. Journal of Mammalogy,* 78:675–683.

Young, J, E Haas, & E McGown. 2008. *Coyote's Guide to Connecting with Nature for Kids of All Ages and Their Mentors.* OWLink Media, Shelton, WA.

Young, RA. 1976. Fat, energy, and mammalian survival. *American Zoology,* 16:699–710.

Yu, L & Y Zhang. 2006. Phylogeny of the caniform carnivora: evidence from multiple genes. *Genetica,* 127:65–79.

Zachos, JC, JR Breza, & SW Wise. 1992. Early Oligocene ice-sheet expansion on Antarctica. *Geology,* 20:569–573.

Zachos, J, et al. 2001. Trends, rhythms, and aberrations in global climate 65 Ma to present. *Science,* 292:686–693.

Zachos, JC, GR Dickens, & RE Zeebe. 2008. An early Cenozoic perspective on greenhouse warming and carbon-cycle dynamics. *Nature,* 451:279–283.

Zack, SP, et al. 2005. Affinities of "hyopsodontids" to elephant shrews and a Holarctic origin of Afrotheria. *Nature,* 434:497–501.

Zahler, P. 2001. Endangered mammals, 441–454, in *Encyclopedia of Biodiversity* (SA Levin, ed.). Academic Press, NY.

Zbinden, K. 1989. Field observations on the flexibility of the acoustic behavior of the European bat *Nyctalus noctula* (Schreber, 1774). *Revue Suisse Zoology,* 96:335–343.

Zhang, J, et al. 2006. Constant darkness is a circadian metabolic signal in mammals. *Nature,* 439:340–343.

Zhang, X & S Firestein. 2002. The olfactory receptor gene superfamily of the mouse. *Nature Neuroscience,* 5:124–133.

Zhanxiang, Q. 2003. Dispersals of Neogene carnivorans between Asia and North America. *Bulletin of the American Museum of Natural History,* 279:18–31.

Zhao, H, et al. 2009. The evolution of color vision in nocturnal mammals. *Proceedings of the National Academy of Sciences USA,* 106(22):8980–8985.

Zhuang, Q & R Muller. 2006. Noseleaf furrows in a horseshoe bat act as resonance cavities shaping the biosonar beam. *Physical Review Letters,* 97:218701-1 to 218701-4.

Zimmer, C. 1998. *At the Water's Edge: Fish Without Fingers, Whales with Legs, and How Life Came Ashore But Then Went Back to Sea.* Touchstone Books, NY.

Zimmerman, EG. 1965. A comparison of habitat and food of two species of *Microtus. Journal of Mammalogy,* 46:605–612.

Zontek, K. 2007. *Buffalo Nation: American Indian Efforts to Restore the Bison.* University of Nebraska Press, Lincoln, NE.

Zook, JM. 2005. The neuroethology of touch in bats: cutaneous receptors of the wing. *Neuroscience Abstracts,* 78:21.

Zortéa, M & S Lucena Mendes. 1993. Folivory in the big fruit-eating bat, *Artibeus lituratus* (Chiroptera: Phyllostomidae) in eastern Brazil. *Journal of Tropical Ecology,* 9:117–120.

Zuckerkandl, E & L Pauling. 1965. Evolutionary divergence and convergence in proteins, in *Evolving Genes and Proteins* (V Bryson & HJ Vogel, eds.). Academic Press, NY.

Zurita, A, A Carlini, & D Gillette. 2008. *Glyptotherium-Glyptodon* (Xenarthra, Glyptodontidae, Glyptodontinae): anatomy and palaeobiogeography. *Journal of Vertebrate Paleontology,* 28(3, suppl):165A.

Glossary

A

Abomasum The fourth and most posterior chamber of the ruminant stomach. It is homologous with the stomach of other mammals.

Acetabulum The lateral socket on the pelvis that accommodates the ball of the femur.

Acromion A continuation of the scapular spine that projects above the shoulder capsule.

Adaptive radiation The evolutionary diversification of numerous species from a common ancestor that colonized a new environment. A rapid cladogenetic event.

Adductor Muscles that act to move a skeletal element toward the midline of the body.

Age cohort Individuals in a population of a particular age group.

Alleles One of several alternative forms of a gene.

Allochthonous Having originated outside the area in which it now occurs.

Allomone A chemical released by one species that serves as a communication signal to another species.

Alloparental care The assistance in the care of offspring by individuals other than the parents.

Allotonic frequencies Echolocation frequencies of bats that are above or below the range of frequencies used by moths to detect predators.

Allozyme A particular amino acid sequence of an enzyme produced by a given allele at a gene locus when there are different possible forms of the enzyme.

Altricial Neonates that are born relatively helpless and require extended periods of parental care.

Altruistic behavior A behavior that enhances the evolutionary fitness of an unrelated individual while simultaneously decreasing the fitness of the individual performing the behavior.

Amniote A vertebrate whose embryo is enclosed in a fluid-filled membrane called the "amnion."

Analogous Structures in two or more organisms that perform a similar function but the similarity is not the result of descent from a common ancestor.

Ancestral Of, or pertaining to, or inherited from, a common ancestor.

Ancient DNA DNA recovered from archaeological and historical specimens such as mummified tissues or tissues from archival collections.

Angle of attack The angle that the leading edge of the wing or airfoil makes with the oncoming airstream.

Anisogamy The condition of having gametes of different sizes. Typically, the male gamete or sperm cell is much smaller that the female gamete or ovum.

Anticoagulant A substance that prevents or delays the clotting or coagulation of the blood.

Apnea The temporary cessation of breathing.

Apomorphy In cladistics, a derived character state, modified from the ancestral state.

Aponeurosis A broad, flat sheet of tendon that may serve as the origin or insertion of a muscle.

Appeasement A behavior that serves to reduce the potential for conflict between conspecifics.

Arboreal Living mainly within the crowns of trees.

Aspect ratio The ratio of the wing span to the average wing chord in an airfoil.

Assisted colonization *See* Facilitated dispersal.

Asymptomatic Neither exhibiting nor causing symptoms of disease.

Attenuated To be reduced or diminished in force or intensity.

Autochthonous Having originated within the area or habitat in which it now occurs.

Autotroph An organism that is capable of assimilating energy from either sunlight or inorganic compounds (i.e., green plant).

B

Baculum A bone within the penis found in certain orders of mammals; os penis.

Baleen plates The keratinized straining plates that form from the integument of the upper jaw in mysticete whales. They are used to filter small invertebrates from the seawater.

Bandwidth Of or pertaining to the breadth or range of frequencies produced.

Basal metabolic rate The minimum metabolic rate of an animal measured when the animal is resting, postabsorptive (finished digesting food), and within its thermal neutral zone.

Beach cast Something washed up on a beach.

Behavioral ecology The field of study concerned with the evolution and ecology of animal behaviors.

Behavioral thermoregulation The ability to change behaviors to maintain an appropriate thermal balance.

Beringia The geographic region now comprising eastern Siberia, Alaska, the Yukon Territory, and the continental shelf between them (sea floor of the Bering Strait) that was periodically dry land during vast intervals of the Cenozoic era, forming a land bridge between Asia and North America at times when sea level was lowered.

Bilophodont Cheek teeth having an occlusal pattern with paired transverse ridges or lophs.

Biocides Chemicals capable of selectively killing living organisms, plant or animal.

Biodiversity The variety of living organisms considered at all levels in all habitats and ecosystems.

Biogeography The study of the geographical distributions of organisms, their habitats, and the historical and ecological factors that produced the distributions.

Biome A large community of plants and animals that occupies a distinct region.

Biota The complete set of organisms of a geographic area or time period. Subsets such as the fauna and flora indicate the animals and plants of the area or time period, respectively.

Biotic potential The exponential growth of a population growing in an ideal and unlimited environment.

Bipedal locomotion Walking, hopping, or running using only the two hindlimbs.

Blastocyst An early stage in the developing mammalian embryo consisting of an outer trophoblast layer and the inner cell mass.

Blastodisc A late stage in the development of the inner cell mass in a mammalian embryo in which these cells differentiate into two layers of cells that resemble the flattened disk of cells sitting atop the yolk in avian embryos.

Boreal Refers to one of the major phytogeographical areas characterized by cold temperate regions of the Northern Hemisphere with a flora consisting of coniferous forests or taiga.

Brachiation Arboreal locomotion by an alternating series of arm swings and grasps in which the body is suspended below the branches.

Brachydont Pertaining to cheek teeth with low crowns. Typical of mammals with omnivorous diets.

Broadband In echolocation, refers to signals that span a wide range of frequencies.

Bruce effect An effect demonstrated in mice by which the presence of a strange male or his odor causes a female to abort her pregnancy and become receptive.

Bulla A typically rounded series of bones that partly or completely cover the middle and inner ear region in mammals.

Bunodont Pertaining to low-crowned teeth with rounded or blunt cusps, used for crushing.

Bushmeat Meat of wild animals, killed for subsistence or commercial purposes.

C

Caching The storage of particular food items for future use when food is less abundant.

Caecum (or Cecum) A blind diverticulum extending from the junction of the small and large intestines. Often contains symbiotic microorganisms in herbivores that aid in digestion of cellulose.

Calcar A cartilaginous or bony process that projects medially from the ankle in many species of microchiropteran bats and helps support the uropatagium. Analogous to the uropatagial spur of megachiropteran bats.

Camber The front to back curvature of an airfoil.

Camera trap A noninvasive capture technique that uses a camera placed in a field location to remotely photograph wild animals.

Canine A tooth posterior to the incisors and anterior to the premolars that is usually elongated, single-rooted, and single-cusped, and which is rooted in the maxilla or dentary.

Caniniform Canine-shaped.

Cannon bone A bone in the lower limbs resulting from the fusion of metacarpals III and IV or metatarsals III and IV. Typically found in cursorial artiodactyls.

Carnassials The cofunctioning pair of bladelike shearing teeth of carnivorans, including the last upper premolar and the first lower molar.

Carotid rete An interwoven network of blood vessels formed from the carotid artery.

Carrion Dead animal matter used as a food source by scavengers.

Carrying capacity The maximum population size that can be sustained by the resources available in the environment.

Castoreum The oily fluid secreted from the preputial glands of beavers.

Catarrhine A member of a group of primates that includes the Old World monkeys (Cercopithecidae), gibbons (Hylobatidae), apes, and humans (Hominidae).

Caudal Pertaining to the tail; toward the tail.

Cementum The relatively soft bony material on parts of a tooth in some mammals, with a structure different from enamel or dentine.

Cervical Of or pertaining to the neck; of or pertaining to the cervix of the uterus.

Character A feature of an organism that can be described, measured, or effectively communicated between scientists. For example, pigmentation of incisor enamel.

Character state One of two or more alternative forms of a character. For example, incisor enamel (a character) with no pigmentation, slight orange pigmentation, or heavy reddish brown pigmentation (character states).

Cheek teeth The premolar and molar teeth in mammals.

Chimera An organism composed of two or more genetically distinct tissues.

Chiropatagium The portion of the wing membrane of a bat that extends between the digits.

Chiropterophily Phenomenon describing plants that rely wholly or in part on bats for pollination.

Chitty effect The pattern in cyclic vole and lemming populations where adults reach higher body masses when populations are at higher densities.

Chorioallantoic placenta A type of placenta, found in eutherians and to a lesser extent in peramelemorph metatherians, composed of an outer chorionic layer and an inner vascularized allantois.

Chorionic villi Finger-like projections (villi) from the chorion that invade the maternal tissues and form the placenta.

Choriovitelline placenta A type of placenta, often called a "yolk sac placenta," found in metatherians (except bandicoots) in which there are no villi and there is only a weak connection to the uterus.

Circadian rhythm A biological rhythm or cycle with a periodicity of approximately 24 hours.

Circannual rhythm A cycle or rhythm with a periodicity of approximately 1 year.

Citizen science The collection of scientific data by interested members of the public (as volunteers) or students for the public good. Such projects when mammal-related often pertain to issues of ecological restoration, conservation, natural resource management, or wildlife monitoring.

Clade A group of species or higher taxa consisting of a single common ancestor and all its descendants.

Cladist A person who practices the cladistic approach to phylogenetic reconstruction.

Cladistics A method of reconstructing a phylogenetic hypothesis that is based on grouping taxa solely by their shared derived character states.

Cladogram A branching diagram that illustrates hypothetical relationships between taxa and shows the evolution of lineages of organisms that have diverged from a common ancestor.

Climatic change An alteration in the global or regional pattern of prevailing long-term (annual or multiyear) weather conditions. In recent years, the term has been used to refer to changes in global climate patterns worsened by increased levels of atmospheric carbon dioxide and other materials produced by the burning of fossil fuels during the last 6 decades.

Climax community The end point in a successional sequence of ecosystems that has reached a steady state under a particular set of environmental conditions.

Cloaca A common chamber into which the gut, urinary tubes, and reproductive tubes empty their contents prior to leaving the body.

Coalition An alliance among individuals, during which the individuals cooperate to enhance their own self-interests.

Cochlea The spiral portion of the bony labyrinth of the inner ear that contains the organ of Corti, with sound-sensitive hairs connected to auditory-nerve endings.

Coevolution The mutual influence on the evolution of two different species interacting with each other and reciprocally influencing each other's adaptations.

Commensals Organisms having a symbiotic relationship in which one organism benefits from the relationship and the other organism is neither helped nor harmed by the relationship.

Competition A contest between individuals or groups for access to limited resources, such as territories, mates, or food.

Competitive exclusion The theoretical principle that no two species can coexist if they require the exact same suite of resources.

Conservation medicine An interdisciplinary field that studies the relationship between human and animal health, and environmental conditions.

Conspecific An animal belonging to the same species as another.

Constant frequency signal Sound signals that typically span less than 10 kilohertz of bandwidth.

Continental drift A geological theory that describes the movement of continents and crustal plates. It is one part of the larger theory of plate tectonics.

Convergence The evolution of similar characteristics for similar functions in unrelated animals as a result of adaptation to similar environmental conditions or natural selection pressures.

Convergent evolution *See* Convergence.

Coprophagy The eating of feces.

Copulation The sexual coupling of two individuals.

Coracoid A bone in the pectoral girdle of monotremes (and reptiles and birds) between the scapula and the sternum. In mammals other than monotremes, a process anterior to the glenoid fossa on the scapula.

Corpus callosum A broad band of nerve fibers that interlinks the right and left cerebral hemispheres.

Corpus luteum The progesterone-secreting mass of follicle cells that develops in the ovary after the egg has been released at ovulation.

Corridor A broad and more or less continuous connection between adjacent land masses or habitat types that allows for dispersal of organisms between the adjacent areas.

Cortex The outer layer or portion of an organ, bone, or hair.

Coterie In the society of prairie dogs, the basic, small group of individuals that occupies communal burrows.

Cotyledonary placenta A type of chorioallantoic placenta in which the villi are grouped into tufts or balls separated by regions of smooth chorion.

Countercurrent heat exchange The exchange of thermal energy between two fluid streams that are traveling in opposite directions in adjacent conduits (or blood vessels).

Crepuscular Active mostly near dawn and dusk.

Crescentic Crescent shaped.

Crown groups Term used in phylogeny reconstruction to designate the smallest monophyletic group of organisms that includes the last common ancestor of all extant members of the group, and all of that ancestor's descendants. An example within

mammals is Rodentia, which includes all living rodents and fossil rodents that are descended from the hypothetical ancestor of Rodentia.

Cryptic species Biological species that are morphologically indistinguishable, but are genetically and reproductively distinct.

Cursorial Adapted for running.

Cutaneous evaporation Water loss across the skin's surface.

Cuticular scale One of a series of keratinous plates covering the shaft of a hair.

D

Deciduous teeth Teeth that are replaced usually early in a mammal's life.

Deciduous placenta A type of placenta in which a portion of the uterine wall is lost at birth.

Delayed development A type of embryonic development in which the growth rate of the embryo slows following implantation in the uterine lining.

Delayed fertilization A situation in which mating occurs and sperm are deposited in the uterine tract of the female, but ovulation and subsequent fertilization does not occur for several months. The sperm remain viable in the female's reproductive tract during this time.

Delayed implantation A delay in the embedding of the blastocyst into the uterine lining that may last several days or up to several months.

Density-dependent Any characteristic of a population that varies with the density of the population.

Density-independent Any characteristic that affects the size of a population but is not influenced by the density of the population.

Dental formula A shorthand expression of the characteristic number of each type of teeth on one side of the skull in mammals. For example, the dental formula for a soricid shrew is 3/1, 1/1, 3/1, 3/3 = 32, which is shorthand for three upper incisors over one lower incisor, one upper and one lower canine, three upper over one lower premolar, and three upper and lower molars on each side of the jaw, for a total of 32 teeth on both sides.

Dentary The single bone making up one-half of the mandible, or lower jaw, of mammals.

Dentine A bonelike material that forms the body of a tooth. It is hard and made of hydroxyapatite crystals and collagen and differs from bone in that it lacks osteocytes and osteons. The dentine of a tooth is often covered by harder enamel.

Derived Refers to a character state that is a modified version of, and differs from, that in the ancestral stock.

Diaphysis The shaft portion of a long bone.

Diastema A space between the teeth, usually the incisors and cheek teeth. Typical of rodents and lagomorphs but also found in artiodactyls, perissodactyls, and other mammals.

Dichromatic Color vision based on the ability to detect two colors.

Didactyl Refers to the condition in there are two digits.

Diestrus The last stage in the estrus cycle in which progesterone levels increase and then decline. The duration of this period can vary.

Diffuse placenta A type of chorioallantoic placenta in which the villi are spread over the entire surface of the chorion.

Digitigrade A foot posture in which the balls of the feet (metapodial pads) support the weight and the "heel" of the hand or foot is off the ground, as in cats and dogs.

Diphyodont A pattern of tooth replacement involving only two sets of teeth, typically a set of deciduous ("milk") teeth and a set of permanent teeth.

Diploid number The total number of chromosomes in the cell nucleus of a somatic cell.

Discoidal placenta A type of chorioallantoic placenta in which the villi are restricted to a disk-shaped region of the chorion.

Disjunct Distinctly separate or discontinuous ranges in which one or more populations are separated from other populations by sufficient distance to prevent gene flow between them.

Dispersal The one-way movement or spreading of organisms from the natal area to new areas.

Diurnal Active primarily during daylight hours.

DNA fingerprinting A technique that uses the unique pattern of DNA base pairs to identify a specific individual. Typically used to establish paternity.

Domestication The process of taming an animal (or plant)—by generations of selective breeding—to live in close association with human beings, usually creating a dependency, so that the animal loses its ability to live in the wild.

Dominance hierarchy A type of social structure in which each animal in a group holds a rank, with some individuals dominant over others in the group, and some individuals submissive to those above them in the hierarchy.

Doppler shift The change in frequency of a sound produced by a moving object such that a pulse of sound hitting a target that is moving toward the source is compressed, resulting in a higher frequency than the original sound.

Drag Resistance to the motion of a body through water or air.

E

Echolocation The process of emitting sounds and using the information from the returning echoes to sense the surrounding environment.

Ecological biogeography The study of the geographic distribution of species in relation to their life histories, dispersal abilities, abiotic factors, and interspecific relationships.

Ecological restoration The process of intentionally renewing the health, integrity, and sustainability of a degraded, damaged, or destroyed area or ecosystem.

Ecological succession The process by which species in a habitat are gradually replaced through a regular progression culminating in a stable climax community.

Economic globalization Process whereby a capitalistic philosophy and ideology has become adopted by cultures worldwide and which promotes trade among nations leading to a single world market.

Ecosystem services Natural processes of living systems as viewed through the perspective of their monetary value and cultural or economic benefit as consumable resources or commodities.

Ecotone A habitat created by the juxtaposition of distinctly different habitats, such as a zone of transition between woodland and grassland.

Ectotherm An animal whose body temperature is determined primarily by passive heat exchange with its environment.

Edge effect The occurrence of greater species diversity and density in an ecotone than in the adjacent ecological communities.

Egg tooth A temporary toothlike structure at the tip of the bill (in birds), used to crack and open the egg during hatching.

Eimer's organ A specialized touch receptor located on the snouts of moles and desmans.

Electroreceptor A sensory receptor that responds to changes in electric field intensity.

Embryonic diapause A period of arrested development of the blastocyst typical of macropods (kangaroos and wallabies).

Embryotroph Any nutritive material supplied to a mammalian embryo during development.

Emerging infectious disease An infectious disease that has appeared, or whose incidence has increased in the past 20 years and threatens to increase in the near future.

Emerging pathogens *See* Emerging infectious disease.

Emigration Movement of individuals or groups out of one area or population to settle in another region.

Enamel Hard crystalline material in the teeth of vertebrates, similar in composition to bone, but without bone-forming osteocytes.

Endemic Pertaining to a species that is native to a given geographic region and is not found in any other regions.

Endogenous Originating within the organism.

Endometrial gland Secretory gland of the mucous membrane lining of the uterus.

Endometrium The mucous membrane lining the inner surface of the uterus into which the blastocyst implants during gestation.

Endotheliochorial placenta An arrangement of the chorioallantoic placenta in which the chorion of the embryo is in direct contact with the maternal capillaries.

Endotherm An animal whose body temperature is elevated substantially above the ambient temperature by internal, metabolic heat production.

Entoconid A major cusp found in the lingual portion of the talonid of the lower molars.

Environmental synergisms An expression of the profound complexity and interconnectedness of processes of the Earth's lithosphere (rocks and soils), hydrosphere (waters), atmosphere, and biota that have evolved together and continue to evolve through Earth history. A change in any of these, such as climate change or habitat conversion, will have inevitable interacting or mutually stimulating effects on the others.

Epidemic A disease affecting an atypically large number of individuals within a population, community, or region at the same time.

Epiphysis The head(s) of a bone, usually bearing an articular surface. A secondary center of ossification of a bone during growth.

Epipubic bones A pair of bones that extend anteriorly from the pubic bones of the pelvis in reptiles, monotremes, and most metatherians.

Epitheliochorial placenta A type of chorioallantoic placenta in which the villi rest in pockets in the endometrium, but the fetal blood supply is separated from the maternal blood supply by six tissue layers.

Epizootic Pertaining to disease that spreads rapidly through an animal population.

Estrus The period during which a female mammal will permit copulation. Also called "heat" in domesticated mammals.

Eupnea Normal, quiet breathing.

Eusocial Pertaining to a social system in which individuals live in groups of overlapping generations and in which one or a few individuals produce all the offspring and the rest of the colony members serve as functionally sterile helpers in rearing young, procuring food, and defending the colony.

Evaporative cooling Loss of heat through the evaporation of sweat or saliva from the skin, or of water vapor from the nasal mucosa or lungs.

Evenness A measure of the relative abundance of the different species making up the species richness of an area or community. A community with four species each having 25 individuals has a higher evenness than a community with four species, where one species has 97 individuals and the other three species are represented by a single individual each.

Ever-growing teeth Teeth that continue to grow throughout the life of the animal. Sometimes called "rootless" or "hypselodont."

Exogenous Originating from outside the organism.

Exotic species A species living outside its native range, and typically brought there either accidentally or intentionally by human activity.

Exponential growth The accelerating increase in the size of a population occurring when conditions for growth are optimal and per capita growth rate is constant and positive.

Extension Movement of a jointed appendage resulting in an increase in the angle of the joint.

External auditory meatus A passageway that leads from the base of the pinna or surface of the head to the tympanic membrane.

Extirpation The extermination of a population or taxon from a given area.

F

Facilitated dispersal Also known as assisted colonization. A controversial process whereby humans translocate organisms whose survival is increasingly threatened by habitat fragmentation and degradation to other areas of putatively suitable habitat in which the organism might continue to survive. The process is highly controversial because of numerous problems with ecological uncertainties such as moving species to new areas that may not be truly suitable; climate changes that may perturb ecological communities and different areas in different and unpredictable ways; or introducing a species into an area in which it might act like an invasive species by disrupting or causing significant damage to the native species in the area into which it is moved. The process is also considered a desperate, triage-like effort to save species for which dispersal cannot be accomplished on the species' own efforts because of the need to cross increasingly large expanses of unsuitable habitat to find diminishing areas of suitable habitat.

Facultative monogamy Monogamous depending on the circumstances or conditions.

Fecundity The number of offspring that an individual produces during a given amount of time.

Feeding guilds See foraging guilds

Female choice The situation in which a female selects a mate based on his superior quality relative to those she does not select to mate with.

Female defense polygyny A mating system in which males control access to females by directly interfering with other males.

Fenestrated Perforated by one or many openings (referring to a bone surface). "Windowed."

Feral Existing in a wild or natural state, or having reverted to the wild state from domestication.

Fertilization Process by which the nucleus of the sperm fuses with the nucleus of the oocyte to produce a zygote.

Filter route A narrow corridor of suitable habitat that acts as a selective filter blocking the passage of some but not all animals that attempt to disperse across it.

Fission-fusion social system A social system found in some primate species in which group members leave temporarily and then may rejoin the group later.

Flehmen A behavior done by male mammals in which the upper lip is retracted after smelling the urine (or pheromones) of a reproductively receptive female.

Flexion A movement that reduces the angle between two articulating bones.

Fluke The horizontal tail "fin" of a cetacean or sirenian.

Folivore An animal whose primary diet consists of leaves.

Food web A representation of the many paths of energy flow from one organism to another in a community by eating or being eaten.

Foraging guild A group of species that feed on common resources in a similar manner.

Foramen An opening or passage through a bone. Typically, the opening transmits nerves or blood vessels.

Fossorial Pertaining to an animal that digs burrows for shelter and forages underground.

Fovea centralis The portion of the retina providing the sharpest vision, with the highest concentration of cones in those species with color vision.

Free-running cycle A cycle that is not subject to periodic resetting by environmental factors such as photoperiod.

Frequency In acoustics, the number of wavelengths per second (expressed in hertz or kilohertz).

Frequency modulated signal Acoustic signals that vary in frequency over time, spanning more than 10 kilohertz.

Frugivorous Having a diet consisting primarily of fruit.

Fundamental frequency In acoustics, the lowest or root tone of a chord.

Fusiform A streamlined shape in which both ends are tapered.

G

Gametogenesis The formation of the gametes (male and female reproductive cells).

Gastric fermentation The breakdown of cellulose by bacteria and other microbes that live in chambers modified from the stomach or esophagus.

Genetically modified organisms Organisms whose genetic material has been altered using recombinant DNA technology.

Genome A complete genetic sequence on one set of chromosomes, including genes and noncoding sequences.

Genotype The genetic constitution of an individual or one of its cells, often referring to alleles of one or more genes.

Geochronology The measurement of geological time as recorded in the Earth's rock layers.

Geophyte A plant that grows from and produces underground storage organs such as tubers, corms, or bulbs.

Gestation The period of time from fertilization to birth in mammals that have embryos that develop within the body of the mother.

Granivore An animal that feeds primarily on seeds or grains.

Graviportal A mode of locomotion in large terrestrial mammals in which the limbs are straight and column-like to support the great mass of the animal.

Gregarious Living in groups or herds.

Guano Feces, especially that of bats or birds, that is harvestable for its nitrates and phosphates.

Gynodioecy A condition in some plant species in which there are separate females and hermaphrodites.

H

Habitat degradation The reduction in habitat quality and its ability to support a diverse biological community.

Habitat fragmentation A process by which an area in a natural state is disintegrated and separated into small detached parts or segments.

Hallux The most medial (first) digit of the hind foot. "Big toe."

Halophytic Refers to plants that can live in salty soils.

Harem A group of females guarded by a male who prevents other males from mating with the females in his group.

Harmonics In acoustics, frequencies that are integral multiples of the fundamental frequency.

Hemochorial placenta A type of chorioallantoic placenta in which the villi are in direct contact with the maternal blood supply.

Herbivory The process of eating plants.

Heterodont Pertaining to teeth that vary in structure in different parts of the tooth row; for example, the teeth of a mammal are usually differentiated into incisors, canines, premolars, and molars.

Heterothermic Referring to animals that normally have high body temperatures regulated by metabolic heat production, and which can also allow their body temperatures to drop close to ambient temperatures.

Heterozygous A genotype or individual that possesses different alleles on each of the two chromosomes at a particular gene locus.

Hibernaculum The place in which an animal hibernates.

Hibernation A period of inactivity normally induced by cold ambient temperatures characterized by lowered body temperatures and a depressed metabolic rate.

High duty cycle Refers to bat echolocation signals that are emitted continuously but in such a way that outgoing pulses are separated in frequency from the incoming echoes. In high duty cycle signals there is overlap in time between the strong outgoing pulses and the weaker returning echoes, but the pulse frequencies are outside the range of the maximum hearing sensitivity. Such signals tend to be long constant frequency signals that end with a brief frequency modulated sweep.

Historical biogeography The study of the distributions of species or higher taxa in terms of their past distributions and the physical history of the Earth.

Holarctic A zoogeographical region comprising the Palearctic and Nearctic regions (i.e., temperate and arctic Eurasia and North America).

Home range The area in which an individual or group spends the bulk of its time.

Homeotherm An animal that can maintain a nearly constant body temperature by physiological means, regardless of the ambient temperature.

Homodont Pertaining to teeth that do not vary in structure in different parts of the tooth row, such as the numerous and identical conical teeth of many dolphins.

Homologous Pertaining to structures or properties having a similar phylogenetic origin but not necessarily retaining a similar function, behavior, or identical structure.

Homoplasy A similarity in a character in two different species that arises from evolutionary convergence or parallelism, not from common ancestry.

Host An animal on or in which a parasitic organism lives.

Human population growth Process by which the total number of *Homo sapiens* living on Earth increases through reproduction and survival. Represents the balance between birth rate and death rate. (See www. poodwaddle.com/worldclock.swf).

Hygroscopic Able to extract moisture from the atmosphere.

Hyoid apparatus A series of bones derived from the gill arch supports that support the muscles of the tongue.

Hypercarnivorous Having characteristics adapted for a diet solely of meat.

Hyperphalangy The presence of extra bones (phalanges) in the digits.

Hyperthermia Having an unusually high body temperature.

Hyphae The threadlike filaments forming the mycelium of a fungus.

Hypocone A main cusp found on the distal lingual side of the upper molars.

Hypoconid The main cusp on the labial side of the talonid of the lower molars.

Hypoconulid A prominent cusp on the posterior portion of the talonid of the lower molars.

Hypothermia A condition of lower than normal body temperature in mammals.

Hypsodont Pertaining to cheek teeth with high crowns, adapted to a diet of relatively abrasive plant material.

Hystricognathous A condition in certain rodents in which the angular process of the mandible is usually lateral to the plane of the alveolus of the lower incisor.

Hystricomorphous A condition in certain rodents in which the infraorbital foramen is greatly enlarged and transmits a portion of the medial masseter muscle.

I

Ideal despotic distribution A theoretical model that predicts the distribution of wild animals, where dominant individuals occupy the best sites, forcing other individuals to use lower quality sites.

Ideal free distribution A theoretical model that predicts the distribution of wild animals, where animals aggregate in patches proportionately to the amount of resources available in each patch.

Ilium The largest and most dorsal of the three bones of the pelvis. Its fusion with the sacrum solidly braces the hind limbs with the vertebral column.

Immune response A response to a foreign substance or antigen by the body's immune system in which antibodies and immune system cells are mobilized to fight off the foreign invader.

Immunoglobin One of a class of proteins comprising the antibodies.

Inbreeding Mating among related individuals.

Inbreeding depression The decrease in the relative fitness of an individual as a result of breeding with close relatives (inbreeding).

Incisor A tooth in mammals located anterior to the canines and rooted in the premaxillae (upper) or dentary (lower).

Incubation period The time elapsed between exposure to a pathogen or disease and when symptoms first appear.

Indigenous populations Populations originating and living or occurring naturally in an area or environment, and thus coevolved with the biota of the area and environment.

Induced ovulation Release of an ovum triggered by the act of copulation.

Infanticide The killing of offspring.

Infectious disease A disease resulting from the presence of a pathogen such as a virus, bacterium, fungus, protozoan, multicellular parasite, or prion.

Infrasound Sound of very low frequency (less than 20 Hz; below the range of human hearing), such as that produced by elephants or giraffes for long-distance communication.

Infundibulum A funnel-shaped opening of the oviduct near the ovary that receives the oocytes at ovulation.

Ingroup A group of organisms that is under cladistic analysis; a set of taxa that are presumed to be more closely related to one another than any is to an outgroup.

Interspecific Between or among individuals of different species.

Interstitial fluid The fluid in the tissues that fills the spaces between cells.

Intestinal fermentation Anaerobic digestion of food by microbes in the cecum.

Intraspecific Between individuals of the same species; within one species.

Invasive species A nonnative or nonindigenous species that colonizes a particular habitat.

Ischial callosities Pads of tough, horny skin attached internally to the flattened parts of the ischium.

Ischium The posteriormost of the three bones of the pelvis.

Island syndrome A phenomenon whereby isolation on islands causes populations of organisms living there to experience changes in aspects of their morphology, behavior, reproduction, and demography relative to mainland populations of the same species. Some examples in mammals include increased or decreased body size, reduced dispersal ability, reduced reproductive output, increased tameness, and higher population density.

IUCN Red List The world's most comprehensive inventory of the global conservation status of plant and animal species, complied by the International Union for the Conservation of Nature and Natural Resources (IUCN).

J

Joey A young kangaroo that is still nursing but is no longer restricted to the pouch.

K

Karyotype The characteristic chromosome complement of a cell, individual, or species.

Keystone guild A group of species that, acting together, have a disproportionately large effect on other species in a community.

Keystone species A species that has a disproportionately large effect on other species in a community.

Kin Related individuals.

Kin selection Natural selection that affects the survival and reproductive success of genetically related individuals.

Kleptoparasite Individual that takes prey away from another individual that has caught, killed, or otherwise stored the prey or food item.

L

Lactation A unique feature of mammals in which milk is formed and secreted by the female's mammary glands for nourishing the developing young after birth.

Lambdoidal crest A bony crest or ridge between the parietal and occipital bones of the skull at the posterior end of the sagittal crest. Also called the "nuchal crest."

Land bridge A dry land connection between land masses that forms a potential migration or dispersal route.

Larderhoarding Storing large quantities of food at a central location within an animal's territory.

Lek A communal courtship area that is regularly used by several males to attract and mate with females.

Lingual Of, or pertaining to, the tongue. Sometimes used to refer to a structure that is on the tongue side of the toothrow.

Locus The location of a gene on a chromosome; the location of a tooth in the toothrow.

Logging The legal or illegal cutting and removal of trees.

Logistic population growth A model of population growth described by an S-shaped curve with an upper asymptote.

Lophodont Pertaining to cheek teeth in which there are a series of transverse ridges or lophs, on the occlusal (chewing) surface.

Low duty cycle Refers to bat echolocation signals that are emitted so that outgoing pulses are separated in time from incoming echoes by an interpulse interval (typically this requires short pulses and relatively long interpulse intervals).

Lower critical temperature Ambient temperature below which the basal metabolic rate becomes insufficient to balance heat loss.

Lumbar Pertaining to the lower back region and the vertebrae between the rib cage and the pelvis.

M

Major histocompatibility complex A series of surface antigens that plays an important role in the coordination and activation of the immune response to foreign substances.

Male–male competition Refers to competition between males for access to females during the breeding season.

Mammary gland Milk-producing gland found in female mammals. The growth and activity of mammary glands are governed by several reproductive hormones.

Manus The hand or forefoot.

Marsupium An external pouch formed by folds of skin in the abdominal wall that encloses the mammary glands and serves as a protective incubation chamber for the young in many metatherians and some monotremes.

Mastication The process of chewing.

Matriarchal Pertaining to a social system in which the bulk of activities and behaviors are centered around a dominant female.

Matrilineal group Social group in which rank in the dominance hierarchy is passed down from a mother to her offspring.

Medulla The inner layer or core of an organ or hair.

Menstruation A special type of estrus cycle found in some female primates in which the endometrial lining of the uterus regresses after ovulation and breaks down, resulting in the passage of blood and cellular debris through the vaginal opening.

Meroblastic A type of embryonic cleavage in which the cleavage plane does not pass completely through the embryo before the next cleavage event begins, usually because of the large amount of yolk.

Mesaxonic Pertaining to a type of foot in which the plane of symmetry passes through the third digit.

Mesic Refers to habitats with moderate rainfall and humidity.

Mesocarnivorous An animal that consumes between 50% and 70% meat for its diet.

Metabolic water Water produced by the aerobic breakdown of food.

Metacone A main cusp on the posterior, labial side of the upper molars.

Metaconid A main cusp on the posterior, lingual side of the trigonid in the lower cheek teeth.

Metaphysis The region of active growth between the epiphysis and diaphysis where the epiphyseal plate is located in developing long bones.

Metapopulation A network of semi-isolated populations with some level of regular gene flow and in which individual populations may go extinct but can be recolonized from one of the adjoining populations.

Metestrus The third stage in the estrus cycle, in which the corpora lutea are fully formed and progesterone levels are high.

Microbial fermentation The anaerobic breakdown of nondigestible food, especially cellulose, by symbiotic protozoans and bacteria.

Midden An accumulation of refuse and stored food materials in or near the nest of a wood rat or other mammal.

Migration corridor An area of suitable habitat that allows the movement of individuals between seasonal ranges.

Minimum viable population A theoretical model that predicts the smallest size at which a population can exist without facing extinction for a specified period of time (customarily 100 years).

Molar A cheek tooth located posterior to the premolars and having no deciduous precursor.

Molariform Pertaining to teeth that have the shape and appearance of molars but which are not molars.

Molecular clock A hypothetical means of measuring evolutionary time that is based in the assumption that the rate at which mutational changes accumulate is relatively constant over time and therefore the changes are useful for dating the divergence of lineages. Also, a self-sustaining, autonomous, genetically regulated timekeeper within the hypothalamus of the brain that can be adjusted by daylight as sensed by the retina and perhaps by other environmental cues.

Molecular phylogeny A hypothetical representation of the evolutionary history of a group of organisms based on characters defined at the molecular level.

Monestrus Having a single estrus period per breeding period.

Monodactyl Having a single functional digit in the manus or pes.

Monogamous Pertaining to a mating system in which one male and one female remain together to rear at least one litter.

Monophyletic Refers to a group of organisms whose members are all descended from (and including) a common ancestor.

Morula A cluster of cells (blastomeres) produced by the early mitotic division of the zygote.

Musk gland A specialized gland, found in various bodily locations in different taxa, that secretes a scent.

Musth A period of heightened reproductive activity in male elephants.

Mutualism A relationship between two species in which both benefit from the association.

Myometrium The thick layer of smooth muscle in the wall of the uterus.

Myomorphous A pattern of jaw musculature in rodents in which a slip of the medial masseter passes through an oval or V-shaped infraorbital foramen.

Mystacial pad The region on the snout from which most of the vibrissae originate.

Mysticeti One of two suborders of cetaceans whose living members are characterized by the presence of baleen plates used for filtering food from water. Baleen whales.

N

Narrowband In echolocation, refers to signals that span a narrow range of frequencies.

Natal Of or pertaining to birth.

Natal dispersal A permanent movement away from an individual's birth site to a new place where the individual reproduces or would have if it had survived the relocation.

Neonate Newborn animal.

Neopallium The expanded nonolfactory portion of the cerebral cortex in mammals.

Niche The ecological position or function of an organism in a community of plants and animals.

Niche differentiation The process by which natural selection forces competing species into different ecological niches (also called niche segregation or niche partitioning).

Nictitating membrane A transparent membrane beneath the eyelid of some vertebrates that can cover and protect the eye. "Third eyelid."

Nocturnal Active mainly during the night.

Node Refers to the place on a cladogram where two lineages diverge.

Nondeciduous placenta A type of placenta that separates easily from the uterine wall, resulting in little or no damage to the uterus.

Nonshivering thermogenesis A process in which fats are oxidized by enzymes to produce heat.

Nonvolant Lacking the ability to fly.

Normothermia A state in which the body temperature is within the normal or preferred range of temperatures.

O

Obligate monogamy A form of monogamy in which a male and female must pair in order to rear offspring.

Odontoceti One of two suborders of cetaceans whose living members are characterized by having teeth instead of plates of baleen. Dolphins and toothed whales.

Omasum The third chamber of the stomach of a ruminant artiodactyl.

Omnivorous Having a diet consisting of both animal and plant material.

Ontogenetic Refers to the course of development from a zygote to an adult.

Oocyst A thick-walled spore phase of certain protists that can survive for lengthy periods outside a host.

Oocyte A developing ovum or egg cell.

Optimal foraging A theory in ecology that predicts that organisms will behave in such a way as to maximize their energy intake per unit time.

Organogenesis The process during embryonic development in which the major organs are formed from endoderm, mesoderm, ectoderm, and neural crest.

Os penis *See* Baculum.

Ossicle Any small bone, but usually refers to one of the three middle ear bones.

Osteoderm A thin, bony scalelike plate in the skin serving as protective armor.

Outbreak A term used in epidemiology to describe an occurrence of disease greater than would otherwise be expected in a particular time and place.

Outgroup In phylogeny reconstruction, a group used for comparison that is related to but not part of the group under study.

Oviparous A reproductive pattern involving the laying of eggs.

Ovoviviparous A reproductive pattern in which the young hatch from eggs that are retained within the mother's uterus.

Ovulation The process by which a mature ovum is released from the ovarian follicle.

P

Palmar Of, or pertaining to, the ventral surface of the hand or foot.

Pandemic An epidemic of infectious disease spreading among humans across a large region, a continent, or on a worldwide scale.

Panting A method of cooling by rapid, shallow breathing that increases the rate of evaporation of water from the respiratory surfaces.

Paracone A main cusp anterior to the protocone and on the labial side of the upper molars.

Paraconid A main cusp of the lower molars on the anterior and lingual part of the trigonid.

Parallelism Evolutionary change in two or more related lineages such that the corresponding features undergo equivalent alterations.

Paraphyletic A taxonomic group in which all members are descendants of a single common ancestor, but the group does not include all of the descendants of the most recent common ancestor.

Parasagittal A dorsoventrally oriented plane parallel to the long axis of the body and parallel to the midsagittal plane, or plane of bilateral symmetry.

Parasitism A type of relationship between two species where one species benefits (parasite) at the expense of the other species (host).

Paraxonic A type of foot anatomy in which the plane of symmetry passes between a pair of similar-sized digits.

Parturition The process of giving birth.

Patch In ecology, refers to a discrete area of habitat of any size.

Paternity The state of being a father to an offspring.

Pathogen A biological agent that causes disease or illness to its host.

Pedicel A bony supporting structure for an antler.

Pelage All of the hairs on an individual mammal. The fur.

Peristalsis A traveling wave of contraction within a tubular structure such as the walls of the digestive tract.

Pes The hind foot.

Pheromone A volatile chemical signal used to communicate a physiological or behavioral state to another member of the same species.

Philopatric The behavior of remaining, or returning to, an individual's birthplace (site fidelity).

Philopatry *See* Philopatric.

Photoperiod A physiological response to day length.

Phylochronology The use of phylogeny reconstruction and population genetics to study the spatial history and timing of phylogenetic events in an evolutionary lineage.

Phylogenesis The sequence of events involved in the evolutionary development of a species or group of organisms.

Phylogenetic relationship *See* Phylogeny.

Phylogeny The evolutionary history of an organism or group of organisms with respect to ancestor-descendant relationships; also, a hypothesis graphically describing such relationships, usually in a treelike or bush-like diagram.

Phylogeography An approach to biogeography that integrates morphologic, molecular genetic, historical geologic, paleontological, ecological, and especially phylogenetic data in interpreting the distribution of common units of taxa or biotas.

Phytoestrogen A plant secondary compound that mimics the effects of estrogen hormones.

Pinna The external ear (an anatomical structure of cartilage, skin, and hair) that surrounds the external auditory canal and serves to funnel sound waves to the tympanic membrane.

Pinna gain The ability of the pinna to focus and amplify sound pressures at the tympanic membrane.

Pioneer community The first assemblage of species to colonize an area and initiate succession.

Plagiopatagium The portion of a bat wing membrane that extends between the body and hindlimbs to the arm and fifth digit.

Plantigrade A style of locomotion in which the entire sole of the foot (including the heel or proximal ends of the metapodials) touches the ground.

Plesiomorphy A character state that is ancestral for the organism under study.

Polarity The direction of evolutionary transformation in a phylogenetic character.

Pollex The most medial digit of the manus (forefoot). "Thumb."

Pollination The process by which pollen is placed on the stigma of the carpal of a flower, resulting in fertilization.

Polyestrous Having several estrous cycles annually or during a breeding season.

Polyandrous (polyandry) A mating system in which females acquire more than one male mate.

Polygamous A mating system in which both males and females mate with multiple members of the opposite sex.

Polygenic trait Trait in which two or more genes have additive effects on a single phenotypic characteristic.

Polygynous (polygyny) A mating system in which males mate with more than one female and females typically provide most of the parental care.

Polyphilous The condition of a plant species having more than one pollinator.

Population Groups of individuals of the same species inhabiting the same place and time.

Postpartum Refers to the period after birth.

Precocious Refers to young that are born relatively well developed and require minimal parental care.

Prehensile Capable of grasping by wrapping around an object. Usually refers to tails that are capable of grasping and holding onto objects.

Premolar Cheek tooth located anterior to the molars and posterior to the canines.

Preorbital vacuity An opening in the skull bones anterior to the orbit. Found in bovid artiodactyls.

Prepuce A fold of skin covering the glans penis.

Presaturation dispersal Movement away from the natal area that occurs before the population reaches peak density.

Primary consumer Animal that feeds on plants or algae (e.g., herbivore).

Primary producer Organism that is capable of synthesizing complex organic molecules from simple inorganic materials (e.g., autotrophs).

Primary succession The ecological process whereby plant life colonizes a substrate having minimal soil and lacking plant life.

Prion Infectious agent composed of a misfolded protein.

Proboscis A long, more or less flexible snout such as those found in tapirs and elephants.

Procumbent Pertaining to teeth that project forward.

Proestrus The initial stage in the estrus cycle when estrogen, progesterone, and luteinizing hormone are elevated.

Promiscuity A mating system in which multiple matings by both sexes occur and there is no prolonged association between the mating pair.

Propatagium The small portion of a bat wing membrane that extends between the arm and the occipitopollicalis muscle along the leading edge of the wing.

Prophylaxis Any preventative treatment designed to preserve health and prevent the spread of disease.

Protocone The major cusp on the lingual side of upper cheek teeth.

Protoconid A main cusp on the labial side of lower cheek teeth, located at the apex of the trigonid in the molars.

Protozoan A member of a group of single-celled, usually microscopic, eukaryotic organisms, including amoebas, ciliates, flagellates, and sporozoans.

Protrogomorphous The ancestral condition in some rodents in which the masseter muscles arise solely from the zygomatic arch and do not penetrate the infraorbital foramen.

Protrusible Capable of being thrust forward.

Puberty The period during which an animal becomes sexually mature and capable of reproducing.

Pubis One of the three bones of the pelvis.

Pulmocutaneous evaporation Water loss that occurs across the pulmonary and skin surfaces.

R

Reciprocal altruism A situation in which the short-term costs of providing some other individual with some resource or beneficial behavior is offset when the recipient returns the favor at a later time.

Recombinant DNA technology A group of molecular techniques for combining DNA sequences that do not normally occur in nature.

Refugium Geographic area that provides temporary shelter or protection.

Regional heterothermy The ability of some mammals to allow the temperatures of the skin or extremities to drop well below the core body temperature.

Regurgitation A reverse movement of food from the stomach back to the oral cavity, produced by reverse peristalsis.

Reproductive potential *See* Biotic potential.

Reservoir species Species capable of maintaining a disease within its populations and also capable of passing on the disease to other host species.

Resource defense polygyny A type of mating system in which males defend clumped resources needed by females in order to gain access to multiple fertile females during the breeding season.

Reticulum The second of four chambers in a ruminant artiodactyl's stomach.

Rhinarium An area of moist, hairless skin and cartilage surrounding the nostrils.

Ricochetal A style of saltatorial locomotion that involves quick changes in the direction of travel.

Riparian Habitat along the banks and floodplain of a waterway.

Road kill Organism killed by a vehicle as it attempted to cross a road.

Rumen The first of four chambers in the ruminant stomach, actually formed from an enlargement of the lower part of the esophagus.

Ruminant An artiodactyl that has a multichambered stomach and is able to rechew regurgitated plant material (cud) to improve its digestibility by symbiotic microbes.

S

Sacral Of or pertaining to the vertebrae that are fused into a single sacrum to which the pelvic girdle attaches.

Sagittal crest A vertical flange of bone on the dorsal midline of the braincase of a mammalian skull that increases the area of origin for the temporalis muscle.

Saltatorial A style of locomotion involving repeated jumping or leaping.

Saturation dispersal Movement out of the natal area that occurs when the population is at its peak density.

Scatterhoarding The storage of a few food items at many scattered locations within an animal's territory.

Scent gland Sweat or sebaceous gland modified for the production of odoriferous secretions.

Sciurognathous A type of mandible in which the angular process is medial to or approximately in line with the plane of the incisor alveolus.

Sciuromorphous A condition in which the masseter muscles originate on the zygomatic arch and from an expanded zygomatic plate up onto the rostrum, but no portion of the masseter passes through the infraorbital foramen.

Sclerotic cartilage A ring of cartilage associated with the tough outer layer of the eye that forms the white area.

Scrotum The pouch that encloses the testes in those species with descended testes.

Sebaceous gland A skin gland that secretes the waxy or oily substance called sebum.

Secondary compounds Chemicals produced by plants as offshoots of basic metabolism and which often deter herbivores.

Secondary consumers Organisms that eat herbivores (e.g., carnivores).

Secondary palate The bony partition comprising parts of the premaxilla, maxilla, and palatine bones that separates the dorsal airway from the nostrils from the oral cavity.

Secondary succession The gradual colonization of a habitat following an ecological disturbance whereby some species are replaced by others over time.

Sectorial Pertaining to a tooth adapted for cutting.

Seismic signal Communication signal made by kangaroo rats and other burrowing rodents that involves striking the ground to produce a series of low-frequency vibrations that travel through the ground.

Selective breeding The process of breeding animals (or plants) for particular genetic traits (e.g., artificial selection).

Selenodont An occlusal pattern of a tooth in which longitudinally arranged, crescent-shaped ridges or lophs are formed on the tooth surface. Typically occurs in artiodactyls.

Semifossorial Refers to mammals that are partially, but not completely, adapted for life underground or for digging.

Seral stage Successional stage on the way to formation of a climax community.

Serological data Any data derived from blood tests.

Serrate Condition in which the edge of a structure is toothed or notched as in a saw blade.

Sexual dimorphism The condition that exists when there are externally apparent differences between the males and females of a species.

Sister group In a phylogeny, the monophyletic group most closely related to a monophyletic group under study. One of two clades that resulted from the splitting of a single lineage.

Social castes A biologically or functionally distinct group of individuals within a species of eusocial animal.

Spatulate Broad and rather flattened with a narrowed base.

Species A named kind of organism; the basic unit of biological classification. A group of potentially interbreeding natural populations that are capable of producing viable offspring, but not capable of reproducing with other such groups.

Species diversity A mathematical index that combines the number of species in an area (richness) with their relative abundance (evenness).

Species richness The number of different species in a defined region.

Spermaceti organ An organ found in the head of certain odontocete whales that contains a waxy liquid.

Spermatogenesis The series of cell divisions that leads to the production of mature sperm.

Spontaneous ovulation Ovulation that occurs without a triggering event such as copulation.

Stable isotope One of two or more nondecaying (i.e., nonradioactive) atomic versions of a chemical element, each of which has a different atomic mass. In natural Earth systems, stable isotopes can provide clues to past atmospheric temperatures, plant photosynthetic pathways, animal diets and migration, and other phenomena. Some common stable isotopes in natural systems include isotopes of carbon, oxygen, nitrogen, and hydrogen.

Stem group Term used in phylogeny reconstruction to designate an extinct member or members of a group known only by fossils, and which is probably close to but not part of a monophyletic crown group. An example within mammals is the extinct gliriform *Tribosphenomys*, which is a stem taxon that is related to rodents, rabbits, and extinct relatives, but that is outside of the crown groups Rodentia and Lagomorpha on a separate side branch.

Streamlined *See* Fusiform.

Stridulating organ An organ capable of producing sound by the rubbing together of modified hairs (in tenrecs) or parts of the exoskeleton (in insects).

Subnivean Refers to the zone under a snow layer.

Subspecies A relatively uniform and genetically distinct population of a species, often in a specific geographic region.

Suckling The process whereby young mammals nurse and take nourishment from milk produced in the mother's mammary glands (breasts or teats).

Suture In osteology, a contact line or tight joint between two bones such as the bones of the skull.

Sweat gland Long tubular gland that extends from the dermis to the skin surface and secretes perspiration or scent.

Sweepstakes route A chance, or accidental, crossing of a water barrier or other geographic barrier.

Symbiosis An interaction between two species in which one benefits while the other either benefits, is harmed, or is unaffected.

Sympatric Refers to two or more populations that occupy the same or overlapping geographic areas.

Symplesiomorphy In phylogenetics, an ancestral character shared by two or more groups.

Synapomorphy In phylogenetics, a derived, homologous character shared by two or more groups.

Syndactylous A condition in which certain of the digits in metatherians (usually the second and third digits) are fused and share a common sheath of skin. *See* Didactylous.

Systematics The study of patterns and processes of evolution that are used to construct phylogenies and classify organisms.

T

Talonid The "heel" or back half of a tribosphenic lower molar that occludes with the protocone of an upper molar.

Tapetum lucidum A reflective choroid layer in the eyes of nocturnal animals that aids in night vision.

Target ranging The ability of an echolocating mammal to estimate the distance to a target by comparing the time it takes a sound pulse to reach the target with the time it takes the echo to return.

Tarsus (or tarsals) The ankle bones.

Taxon A named group of organisms of a given category (such as of a species, genus, family, etc.) of classification; e.g., dogs, Artiodactyla, *Myotis lucifugus*. Plural: taxa.

Taxonomy The practice of naming and classifying organisms.

Telolecithal Refers to eggs that have large amounts of yolk at the vegetal pole relative to the amount of active cytoplasm at the animal pole.

Temporal fenestra A hole or opening in the temporal region of the skull.

Territory An area occupied and defended by an animal or group of animals.

Thermal conductance Heat loss from the skin surface to the environment.

Thermal inertia A phenomenon by which animals of large body size are able to retain body heat longer than smaller animals are.

Thermal neutral zone A range of ambient temperatures within which an animal maintains its basal metabolic rate at a relatively constant and minimal level.

Thermogenesis The generation of heat.

Thoracic Pertaining to the region of the thorax.

Titer A measure of the concentration of a solution determined by titration.

Torpor A type of adaptive hypothermia (dormancy) in which heart rate, body temperature, and respiration are reduced.

Traditional ecological knowledge A knowledge system through which people are strongly tied to a place and its environment over a long period of time and learn to interact with, and adapt to changes in, that place through careful observation, participation, and experience as an integral part of the ecosystem. The knowledge is passed down from one generation to another.

Tragus The projection from the lower medial border of the pinna found in many microchiropteran bats.

Transgenic organisms *See* Genetically modified organisms.

Traplining A foraging behavior in which the animal daily follows a particular path from one food resource to the next, stopping to feed at each.

Trenchant Sharp; able to cut.

Tribosphenic Pertaining to a type of molar in which the protocone of a three-cusped upper tooth fits into a basin in the talonid of a lower tooth (for crushing food), and in which crests between other cusps shear past one another (for cutting food).

Trichromatic vision Color vision based on the presence of three main photopigments.

Trigonid The front half of a tribosphenic lower molar, which includes a triangle formed by three cusps.

Trioecy A reproductive pattern in plants in which individual plants can be either males, females, or hermaphrodites.

Trophoblast The superficial layer of the blastocyst that is involved with implantation, placenta formation, and hormone production.

Tuberculosectorial A kind of tooth that has cusps (tubercles) for crushing and crests or ridges for shearing or sectioning food.

Turbinal bones Convoluted or scroll-shaped bones within the nasal passage of the skull that provide increased surface area for moisturizing and warming inhaled air.

U

Ultrasonic Referring to sounds at frequencies beyond human hearing capability (i.e., above 20 kilohertz).

Ungulate Mammal with hoofs.

Unguligrade A type of locomotion in which only the hoofs (nails of the terminal phalanges) touch the ground.

Unicuspid Referring to a tooth with a single cusp, such as a canine tooth.

Upper critical temperature The temperature at which an animal must dissipate heat to maintain thermal homeostasis.

Uropatagial spur A cartilaginous spur or rod that projects from the gastrocnemius tendon, found in megachiropteran bats. Analogous to the calcar of microchiropteran bats.

Uropatagium A flight membrane in some bats that extends between the hindlimbs and the tail.

Uterine mucosa The inner lining of the uterus.

V

Valvular Capable of being temporarily closed off, such as in the nostrils of sirenians.

Vasoconstriction The narrowing of a blood vessel.

Vectors Any organism that transmits a pathogen.

Venter The belly surface.

Vermiform Having a long, thin, and flexible or wormlike form.

Vibrissae The long, stiff hairs on the snout that serve as tactile receptors.

Vicariance In biogeography and evolution, the divergence of two populations from one original population that was split by the formation of a natural barrier.

Virulent Extremely infectious or poisonous.

Viviparity (viviparous) The ability to give birth to live young rather than laying eggs.

Vomeronasal organ A branch of the nasal epithelium that forms a pocket that may open to the mouth cavity. Important in sensing chemical behavioral cues.

W

Waif dispersal A chance movement to a new area. May be achieved, for example, by drifting on the ocean currents.

Wallace's line An imaginary line that separates the Oriental and Australian biogeographical regions. The line passes between islands of the Philippines and Maluku (Moluccas) in the north and between Sulawesi and Borneo and between Lombok and Bali in the south.

Warren A communal series of burrows used by a group of rabbits.

Water balance A suite of behavioral and physiological responses to prevent dehydration.

Weapon automimicry A feature in the appearance of an animal (such as facial stripes or ear shape) that mimics the possession of a "weapon" such as a horn.

Wing loading The body mass of a flyer divided by the area of its airfoils.

X

Xenarthrous A condition in which extra articular surfaces similar to zygapophyses are found on the posterior trunk vertebrae of cingulatan and pilosan mammals (sloths, armadillos, and anteaters).

Z

Zalambdodont A type of upper molar characterized by a protocone and a V-shaped crest (an ectoloph) with the largest cusp at the apex of the V.

Zona pellucida A thick, elastic covering of the egg cell.

Zonary placenta A type of chorioallantoic placenta in which the villi surround the fetus in a band.

Zoogeography The study of the geographical distribution of animals and their ecological communities.

Zoonoses (zoonotic disease) Any animal disease capable of being transmitted to human beings.

Zygapophysis One of the articular surfaces between two adjacent vertebrae.

Zygomatic arch An arch of bone on the side of a mammal skull that is formed by the jugal bone and a process of the squamosal bone. "Cheekbone."

Zygote A fertilized egg.

Index

Locators starting with W (for example, W-4) refer to pages in the book's Web-based chapters.

I

Photo Credits

Part Opener and Chapter Opener Art

Courtesy of Terry Vaughan

Chapter 1

1-1 Courtesy of Roland Kays

Chapter 2

2-5 Courtesy of Terry Vaughan.

Chapter 5

5-5 Courtesy of Michael L. Augee; 5-7A © Robert Gubiani/Dreamstime.com; 5-7B © WildLife/Peter Arnold, Inc.

Chapter 6

6-11 Courtesy of Phil Myers, Curator of Mammals, Museum of Zoology, University of Michigan; 6-12A © ARCO/W. Layer/age fotostock; 6-12B Courtesy of R. Rotsaert; 6-12C Courtesy of Darren Bos; 6-12D Courtesy of Anthony Robinson; 6-15A © Patsy A. Jacks/ShutterStock, Inc.; 6-15B © FLPA/Hugh Clark/age fotostock; 6-16 © FLPA/Tom and Pam Gard/age fotostock; 6-17 Courtesy of Anthony Robinson; 6-19 © Timothy Craig Lubcke/ShutterStock, Inc.; 6-20 © Liv Falvey/ShutterStock, Inc.; 6-22 Courtesy of Anthony Robinson; 6-23A © Stuart Wilson/Photo Researchers, Inc.; 6-23B © Gary Unwin/ShutterStock, Inc.; 6-24 Courtesy of Pavel German; 6-25 © Alan & Sandy Carey/Photo Researchers, Inc.; 6-27A © Gary Unwin/ShutterStock, Inc.; 6-27B © FLPA/David Hosking/age fotostock; 6-27C © John Cancalosi/age fotostock

Chapter 8

8-2A © John F. Eisenberg and the Mammal Images Library of the American Society of Mammalogists; 8-2B Courtesy of M. Andera; 8-2C Courtesy of A. Kitching; 8-3 Courtesy of James Ryan; 8-4 Reproduced from Martin E. Nicoll and Galen B. Rathbun. *African Insectivora and Elephant-Shrews: An Action Plan for their Conservation.* IUCN, 1990. Photo courtesy of Peter Vogel; 8-5A Courtesy of J. Jarvis; 8-5B Courtesy of G. Rathbun; 8-7 Courtesy of Terry Vaughan; 8-8A Courtesy of G. Rathbun; 8-8B Courtesy of G. Rathbun; 8-10 Courtesy of Terry Vaughan; 8-11 Courtesy of A. Taylor

Chapter 9

9-5A Courtesy of Terry Vaughan; 9-5B Courtesy of Rosie Lamb; 9-8 Courtesy of A. Stewart; 9-9 Courtesy of Terry Vaughan; 9-10A © Wayne Johnson/ShutterStock, Inc.; 9-10B Courtesy of D. Kendzia; 9-12 Courtesy of Terry Vaughan; 9-14A Courtesy of Dr. Hendrik Hoeck; 9-14B Courtesy of Dr. Hendrik Hoeck; 9-15A Courtesy of Dr. Hendrik Hoeck; 9-15B Courtesy of Dr. Hendrik Hoeck

Chapter 10

10-11A © Laura Hart/Dreamstime.com; 10-11B © Andrea Schneider/Dreamstime.com; 10-13A © Wayne Lynch/age fotostock; 10-13B © Alexey Stiop/ShutterStock, Inc.; 10-15 © Alvaro Pantoja/Dreamstime.com; 10-17 © Biosphoto/Compost Alain/Peter Arnold, Inc.

Chapter 11

11-1A Courtesy of Norman Lim; 11-1B Courtesy of Norman Lim; 11-3 Courtesy of M. W. Sorenson

Chapter 12

12-6 Courtesy of James Ryan; **12-7** © Simone van den Berg/ShutterStock, Inc.; **12-10** Courtesy of James Ryan; **12-12** © Smellme/Dreamstime.com; **12-13** © Photodisc; **12-14** © Nico Smit/Dreamstime.com; **12-18** © Holger Mette/Dreamstime.com; **12-20A** © John Arnold/Shutterstock, Inc.; **12-20B** © Mikhail Levit/ShutterStock, Inc.; **12-22** © jaana piira/ShutterStock, Inc.; **12-23A** © Jean-marc Strydom/Dreamstime.com; **12-23B** © Martina Berg/Dreamstime.com; **12-27** © Benjamin Schalkwijk/Dreamstime.com; **12-28** Courtesy of Andrew Nicholson; **12-29** © Kate Duffell/Dreamstime.com

Chapter 13

13-6 Courtesy of A. Fryer; **13-9A** Courtesy of Terry Vaughan; **13-9B** Courtesy of W. Robinette; **13-11** Courtesy of M. Andera; **13-12** © 3483253554/ShutterStock, Inc.; **13-16A** Courtesy of Terry Vaughan; **13-16B** Courtesy of M. Andera; **13-19** Courtesy of Rosemary Vaughan; **13-20A** Courtesy of M. Andera; **13-20B** Courtesy of M. Andera; **13-23** Courtesy of J. Jarvis; **13-24** Courtesy of M. Andera; **13-25** Courtesy of M. Andera; **13-27A** Courtesy of Terry Vaughan; **13-27B** Courtesy of Rosemary Vaughan; **13-28A** Courtesy of M. Andera; **13-28B** Courtesy of M. Andera; **13-31** Courtesy of M. Andera; **13-34** Courtesy of M. Korinek; **13-38A** Courtesy of J. Jarvis; **13-38B** Courtesy of J. Jarvis; **13-38C** Courtesy of Rosemary Vaughan; **13-40A** © Dusan Zidar/Shutterstock, Inc.; **13-40B** © ARCO/R Müller/age fotostock; **13-41** Courtesy of Rosemary Vaughan; **13-44** © Micha Fleuren/ShutterStock, Inc.; **13-46** © Lee Torrens/ShutterStock, Inc.; **13-47A** Courtesy of W. Quatman; **13-47B** © Juniors Bildarchiv/age fotostock; **13-49** Courtesy of M. Andera; **13-52** Courtesy of V. Motycka; **13-56** © Twildlife/Dreamstime.com; **13-58A-B** Courtesy of Terry Vaughan

Chapter 14

14-4 Courtesy of M. Andera; **14-5** Courtesy of Konstance Wells; **14-6** Courtesy of C. Wemmer; **14-7A** Courtesy of M. Andera; **14-7B** Courtesy of M. Andera; **14-10** Reproduced from David Stone. *Eurasian Insectivores and Tree Shrews.* IUCN, 1995. Photo courtesy of Peter Vogel; **14-11** Courtesy of M. Andera; **14-14** Courtesy of M. Andera; **14-15A** Reproduced from Catania, K. C., and Kass, J. H., "The unusual nose and brain of the star-nosed mole," *BioScience* 46 (1996): 578-586. © 1996 American Institute of Biological Sciences; **14-15B** Reproduced from Catania, K. C., and Kass, J. H., "The unusual nose and brain of the star-nosed mole," *BioScience* 46 (1996): 578-586. © 1996 American Institute of Biological Sciences; **14-16A** Reproduced from Catania, K. C., and Kass, J. H., "The unusual nose and brain of the star-nosed mole," *BioScience* 46 (1996): 578-586. © 1996 American Institute of Biological Sciences

Chapter 15

15-2 Reproduced from Wimsatt, W. (ed). *Biology of Bats.* Academic Press, 1977. Photo courtesy of Terry Vaughan; **15-7** Courtesy of J. Scott Altenbach; **15-11A** Courtesy of J. Scott Altenbach; **15-11B** Courtesy of J. Scott Altenbach; **15-11C** Courtesy of J. Scott Altenbach; **15-11D** Courtesy of J. Scott Altenbach; **15-12** Courtesy of Dr. Nancy B. Simmons, AMNH Department of Mammalogy; **15-13A** Courtesy of Dr. Brock Fenton, University of Western Ontario; **15-13B** Courtesy of Charles M. Francis; **15-17** Courtesy of Charles M. Francis; **15-19A** Courtesy of Bruce Thomson; **15-19B** Courtesy of Bruce Thomson; **15-19C** Courtesy of Bruce Thomson; **15-19D** Courtesy of Rosemary Vaughan; **15-19E** Courtesy of Bruce Thomson; **15-19F** Courtesy of Bruce Thomson; **15-22A** Courtesy of Bruce Thomson; **15-22B** Courtesy of Charles M. Francis; **15-23** Courtesy of Charles M. Francis; **15-28** Courtesy of Lutz T. Wasserthal, from Göpfert, M. and L.T. Wasserthal, 1995: Notes on echolocation calls and roosting behaviour of the old world sucker-footed bat Myzopoda aurita (Chiroptera, Myzopodidae). *Mamm. Biol.* 60: 1-8. Copyright Elsevier; **15-29** Courtesy of Bruce Thomson; **15-30** Courtesy of Lutz T. Wasserthal, from Tschapka, M., Brooke, A.P., Wasserthal, L.T (2000): Thyroptera discifera (Chiroptera: Thyropteridae): Observations on echolocation and new record in Costa Rica. *Mamm. Biol.* 65: 193-198. Copyright Elsevier; **15-31** Courtesy of Frank J. Bonaccorso; **15-34A** Courtesy of J. Scott Altenbach; **15-34B** Courtesy of J. Scott Altenbach; **15-34C** Courtesy of J. Scott Altenbach; **15-35** Courtesy of Bruce Thomson; **15-38** Courtesy of Dr. Brock Fenton, University of Western Ontario; **15-39** Courtesy of Bruce Thomson

Chapter 16

16-3A: Courtesy of Terry Vaughan; **16-3B** © Purestock/age fotostock; **16-3C** Courtesy of Terry Vaughan; **16-3D** © Philip Lane/Dreamstime.com; **16-7** © Photodisc; **16-8A** Courtesy of K. Wells; **16-8B** Courtesy of T. R. Huels; **16-11A** Courtesy of Terry Vaughan; **16-11B** © EcoPrint/ShutterStock, Inc.; **16-12A** Courtesy of M. Korinek; **16-12B-C** Courtesy of James Ryan; **16-13** Courtesy of Terry Vaughan; **16-15A** © Johan Reineke/Dreamstime.com; **16-15B-D** Courtesy of Terry Vaughan; **16-18A** © Karel Bro /ShutterStock, Inc.; **16-18B** © Mike Flippo/ShutterStock, Inc.; **16-22** © Morten Hilmer/ShutterStock, Inc.; **16-24A** © Galina Barskaya/Dreamstime.com; **16-24B** © Pichugin Dmitry/ShutterStock, Inc.; **16-25** Courtesy of Terry Vaughan; **16-27A** © Greg Payan/Dreamstime.com; **16-27B** Courtesy of Terry Vaughan; **16-28A** © L.L. Masseth/ShutterStock, Inc.; **16-28B** © Musat Christian/Dreamstime.com; **16-30A** © Photodisc; **16-30B** © Serg Zastavkin/ShutterStock, Inc.

Chapter 17

17-12A Courtesy of Terry Vaughan; **17-12B** © ARCO/I. Schulz/age fotostock; **17-13** © Jan Gottwald/Dreamstime.com; **17-14** © Ferenc Cegledi/ShutterStock, Inc.; **17-16A** Courtesy of Terry Vaughan; **17-16B** © Carlos Arguelles/ShutterStock, Inc.

Chapter 18

18-7A © Paul Cowan/ShutterStock, Inc.; **18-7B** © Eduardo Rivero/ShutterStock, Inc.; **18-10** © Teo Boon Keng Alvin/ShutterStock, Inc.; **18-12** © EcoPrint/ShutterStock, Inc.; **18-13** Courtesy of Terry Vaughan; **18-16** © Tom McHugh/Photo Researchers, Inc.; **18-17** Courtesy of Terry Vaughan; **18-20** Courtesy of Terry Vaughan; **18-21A** © Ferenc Cegledi/ShutterStock, Inc.; **18-21B** Courtesy of Pam Wood; **18-21C** © Joe Ferrer/ShutterStock, Inc.; **18-21D** Courtesy of Terry Vaughan; **18-23A** Courtesy of Terry Vaughan; **18-23B** © Bambi L. Dingman/Dreamstime.com; **18-24A** © Photos.com; **18-24B** Courtesy of Terry Vaughan; **18-25A-D** Courtesy of Terry Vaughan; **18-26** Courtesy of Terry Vaughan; **18-28** © Roland Seitre/Peter Arnold, Inc.

Chapter 19

19-14 Courtesy of S. Mizroch, NOAA, NMFS, National Marine Mammal Laboratory; **19-15** © Kelvin Aitken/Peter Arnold, Inc.; **19-16** © Melissaf84/Dreamstime.com; **19-18A** © Ferderic/Dreamstime.com; **19-18B** Courtesy of S. Mizroch, NOAA, NMFS, National Marine Mammal Laboratory; **19-19** © David Fleetham/Visuals Unlimited

Chapter 20

20-2 Courtesy of James Ryan; **20-4** Courtesy of P. Rismiller and M. McKelvey; **20-9** © Kitchner Bain/Dreamstime.com; **20-13** Courtesy of Professor Geoff Shaw, University of Melbourne, Australia; **20-19** Courtesy of Ashley Herrod

Chapter 21

21-5 © Photodisc; **21-9** Courtesy of M. Andera; **21-6** Courtesy of Terry Vaughan; **21-17** © T. Allofs/Peter Arnold, Inc.; **21-19** Courtesy of Terry Vaughan; **21-21** © Nico Smit/Dreamstime.com; **21-23** Courtesy of Dr. Hendrik Hoeck; **21-25** Courtesy of Guy Haimovitch; **21-30** Courtesy of Terry Vaughan; **21-32** Courtesy of J. Roser; **21-34** Courtesy of Drew Gardner

Chapter 22

22-4 Courtesy of O. W. Henson, Jr.; **22-10A** Courtesy of Dr. Brock Fenton, University of Western Ontario; **22-10B** Courtesy of M. Andera

Chapter 23

23-5A-C Courtesy of Terry Vaughan; **23-10A** Courtesy of Eric Rosenberg; **23-10B** Courtesy of Victor Scheffer, University of Washington Libraries, Special Collections, UW28226z; **23-13** © Paul Banton/Shutterstock, Inc.; **23-14** Courtesy of W. Quatman; **23-20** Courtesy of K. Gleditsch; **23-21** Courtesy of E. McDaid; **23-25** Courtesy of J. Roser; **23-28** Courtesy of R. Lamb; **23-29** © Corbis/age fotostock; **23-30** Courtesy of Terry Vaughan; **23-34** Courtesy of Rosemary Vaughan

Chapter 24

24-1 Courtesy of J. Irvine; **24-4** Courtesy of Ben Amstutz; **24-5** Modified from Reichmann, O.J., et al., *Ecology* 63 (1982): 687–695; **24-6** © David P. Lewis/ShutterStock, Inc.; **24-7A** Courtesy of Jose Cortes; **24-7B** Courtesy of Jose Cortes; **24-9A** Courtesy of Kirsten Graham; **24-9B** Courtesy of Kirsten Graham; **24-10** Courtesy of K. Catania; **24-11** Courtesy of A. M. Mento; **24-13** © Photodisc; **24-14** Courtesy of R. Bowker; **24-16A** Courtesy of M. Bowker; **24-16B** Courtesy of M. Bowker; **24-18** Courtesy of Lukas Vermeer; **24-19** Courtesy of Michael L. Baird; **24-21** Courtesy of A. Stewart; **24-22A** © Tony Callahan/Dreamstime.com; **24-22B** Courtesy of L. Gerbrandt; **24-23** Courtesy of A. Stewart; **24-24** Courtesy of Terry Vaughan; **24-25** Courtesy of M. Andera; **24-26** Courtesy of Terry Vaughan; **24-27** Courtesy of G. Chandler; **24-28** Courtesy of C. Morgan; **24-29** Courtesy of Helena Stephenson; **24-30** Courtesy of James Ryan; **24-31** Courtesy of Terry Vaughan; **24-33** Courtesy of Donna Perdue/Aradise; **24-34** Courtesy of B. Thomson; **24-36A** Courtesy of D. Higgins; **24-36B** Courtesy of Rosemary Vaughan; **24-37** Courtesy of T. Noviski Fornel

Chapter 26

26-2 Courtesy of Nicholas Czaplewski; **26-3** Courtesy of Nicholas Czaplewski

Chapter 27 (on-line)

Opener © Robert Scoverski/Dreamstime.com; **27-1A** Courtesy of Dave Pape, University at Buffalo; **27-1B** Courtesy of Scott Bauer/USDA ARS; **27-1C** Courtesy of Alexandre Dulaunoy; **27-1D** Courtesy of Keith Weller/USDA ARS; **27-3A** Courtesy of Wally Slowik; **27-3B** Courtesy of Vicky Hugheston; **27-3C** Courtesy of John Leslie; **27-3D** Courtesy of Christine Olson; **27-5** © siloto/ShutterStock, Inc.; **27-6** Courtesy of Emilio Labrador; **27-7** Courtesy of Walter Frisch <http://www.aueroxen.de/>; **27-8A** © Jens Klingebiel/Dreamstime.com; **27-8B** © John Carnemolla/Shutterstock, Inc.; **27-9A** © Dmytro Korolov/Shutterstock, Inc.; **27-9B** © Pichugin Dmitry/Shutterstock, Inc.; **27-10A** © Elifranssens/Dreamstime.com; **27-10B** © Pablo H. Caridad/Shutterstock, Inc.; **27-11A** © Efremova Irina/Shutterstock, Inc.; **27-11B** © Konstantin Karchevskiy/Shutterstock, Inc.; **27-12** Courtesy of Christopher Soghoian; **27-13** © Serge Vero/Alamy Images

Chapter 28 (on-line)

Opener Courtesy of Frederick Murphy/CDC; **28-7** Courtesy of Janice Haney Carr/CDC; **28-9** Courtesy of Scott Bauer/USDA ARS; **28-11** Courtesy of Nancy Heaslip, New York Department of Environmental Conservation/U.S. Fish and Wildlife Service; **28-12** Courtesy of Dr. Menna Jones, University of Tasmania

Unless otherwise indicated, all photographs and illustrations are under copyright of Jones and Bartlett Publishers, LLC.